Physical Constants

The values presented in this table are those used in computations in the text. Generally, the physical constants are known to much better precision.

Quantity	*Symbol*	*Value*	*SI Unit*
Speed of light in vacuum	c	3.00×10^{8}	m/s
Permittivity of vacuum	ϵ_0	8.85×10^{-12}	F/m
Coulomb constant, $1/4\pi\epsilon_0$	k	9.0×10^{9}	$N \cdot m^2/C^2$
Permeability of vacuum	μ_0	1.26×10^{-6} ($4\pi \times 10^{-7}$ exactly)	H/m
Elementary charge	e	1.60×10^{-19}	C
Planck constant	h	6.63×10^{-34}	J·s
	$\hbar = h/2\pi$	1.05×10^{-34}	J·s
Electron rest mass	m_0	9.11×10^{-31}	kg
		5.49×10^{-4}	u
Proton rest mass	m_p	1.67265×10^{-27}	kg
		1.007276	u
Neutron rest mass	m_n	1.67495×10^{-27}	kg
		1.008665	u
Electron charge-to-mass ratio	e/m_0	1.76×10^{11}	C/kg
Avogadro constant	N_A	6.02×10^{23}	mol^{-1}
Molar gas constant	R	8.31	J/mol·K
Boltzmann constant	k	1.38×10^{-23}	J/K
Stefan-Boltzmann constant	σ	5.67×10^{-8}	$W/m^2 \cdot K^4$
Faraday constant	F	9.65×10^{4}	C/mol
Molar volume of ideal gas at STP	V_m	22.4	L/mol
		2.24×10^{-2}	m^3/mol
Rydberg constant	R	1.10×10^{7}	m^{-1}
Bohr radius	a_0	5.29×10^{-11}	m
Electron Compton wavelength	λ_c	2.43×10^{-12}	m
Gravitational constant	G	6.67×10^{-11}	$m^3/kg \cdot s^2$

· PRINCIPLES OF PHYSICS ·

Frank J. Blatt

Michigan State University

Allyn and Bacon, Inc.

Boston London Sydney Toronto

Library of Congress Cataloging in Publication Data

Blatt, Frank J.
Principles of physics.

Includes index.
1. Physics. I. Title.
QC23.B675 1983 530 82-13820
ISBN 0-205-07588-6

Printed in the United States of America

10 9 8 7 6 5 4 3 87 86 85 84 83

COVER PHOTO Molten gold being poured and cast

Credits

CHAPTER 1 FIG. 1.1, p. 4—Courtesy of the National Bureau of Standards. FIG. 1.2, p. 4—Courtesy of the National Bureau of Standards. FIG. 1.3, p. 5—Courtesy of the New York Academy of Science.

CHAPTER 2 QUOTES, p. 28—Leon N. Cooper, *An Introduction to the Meaning and Structure of Physics*, Harper & Row, New York, 1968, pp. 2, 10 (Galileo, Dialogue Concerning Two New Sciences, Henry Crew and Alfonso de Salvio, trans., Macmillan, New York, 1914, p. 155). QUOTES, p. 29—Cooper, p. 10. FIG. 2.10, p. 29—F. W. Sears and M. W. Zemansky, *University Physics*, Addison-Wesley, Reading, Mass. FIG. 2.12, p. 31—Physical Science Study Committee, *Physics*, D. C. Heath, Boston, 1967. FIG. 2.13, p. 31—Courtesy of Dr. Harold E. Edgerton, M.I.T., Cambridge, Mass.

CHAPTER 3 FIG. 3.1(a), p. 43—Philadelphia Museum of Art. Photo by A. J. Wyatt, Staff Photographer. FIG. 3.1(b), p. 43—F. J. Blatt. QUOTE, p. 44—J. Bronowski, *The Ascent of Man*, Little Brown, Boston, 1973. FIG. 3.9, p. 54—Courtesy of NASA.

CHAPTER 4 FIG. 4.1, p. 71—Wayne L. Westcott, *Strength Fitness*, Allyn and Bacon, Boston, 1982. FIG. 4.8, p. 77—Courtesy of Boeing Aircraft Co. FIG. 4.10, p. 78—Courtesy of Dr. Harold E. Edgerton, M.I.T., Cambridge, Mass. FIG. 4.16, p. 85—Miriam Rothschild et al., "The Flying Leap of the Flea," *Scientific American*, November 1973, Copyright © 1973 by Scientific American, Inc.

CHAPTER 5 FIG. 5.1, p. 94—Courtesy of Dr. Harold E. Edgerton, M.I.T., Cambridge, Mass. FIG. 5.2, p. 94—Courtesy of Dr. Harold E. Edgerton, M.I.T., Cambridge, Mass. QUOTES, p. 95—W. F. Magie, *A Source Book in Physics*, Harvard University Press, Cambridge, Mass., 1963, pp. 32, 37. FIG. 5.5, p. 98—Courtesy of NASA. FIG. 5.6, p . 98—Courtesy of Harold Wes Pratt.

CHAPTER 6 FIG. 6.6, p. 117—Susan Dodington. FIG. 6.11(b), p. 121—Courtesy of Cedar Point Amusement Park, Sandusky, Ohio. FIG. 6.13(a), p. 124—Stephen J. Potter/Stock, Boston. FIG. 6.14, p. 124—Culver Pictures. FIG. 6.15, p. 125—Courtesy of Cessna Aircraft Company.

CHAPTER 7 QUOTE, p. 130—Leon N. Cooper, *An Introduction to the Meaning and Structure of Physics*, Harper & Row, New York, 1968, p. 56 (Sir *Isaac Newton's Mathematical Principles of Natural Philosophy and His System of the World*, Andrew Motte, trans., University of California Press, 1962). QUOTE, p. 131 (top)—Cooper, p. 57 (Charles C. Gillispie, *The Edge of Objectivity*, Princeton University Press, 1960). QUOTES, p. 131 (middle and bottom)—J. Bronowski, *The Ascent of Man*, Little, Brown, Boston, 1973, pp. 223, 226, 223. FIG. 7.1, p. 132—Culver Pictures. FIG. 7.2, p. 132—The Granger Collection. FIG. 7.4(a) p. 135—Copyright by Deutsches Museum, Munich. FIG. 7.6, p. 136—Courtesy of NASA. FIG. 7.8, p. 138—Courtesy of NASA.

CHAPTER 8 FIG. 8.19, p. 155—Courtesy of U.S. Geological Survey. FIG. 8.21, p. 151—Courtesy of Sperry Corporation.

CHAPTER 9 FIG. 9.8(a), p. 178—Photo Researchers, Inc. FIG. 9.8(b), p. 178—Courtesy of Dr. Harold E. Edgerton, M.I.T. Cambridge, Mass. FIG. 9.9(b), p. 179—Photo Researchers, Inc.

continued on page 814

• CONTENTS •

• 5 •

Impulse and Momentum 93

• 6 •

Circular Motion 113

• 7 •

Gravitation 130

• 8 •

Rotational Equilibrium; Dynamics of Extended Bodies 142

• 9 •

Mechanical Properties of Matter 169

• 10 •

Hydrostatics and Hydrodynamics 188

• 11 •

Thermal Properties, Calorimetry, and the Mechanical Equivalent of Heat 226

• 12 •

The Ideal Gas Law and Kinetic Theory 255

• 13 •

Thermodynamics 277

• 14 •

Oscillatory Motion 296

• 15 •

Mechanical Waves 316

• 16 •

Sound 339

• 21 •

Direct-Current Circuits 449

• 22 •

Magnetism 465

• 23 •

Electromagnetic Induction 495

• 24 •

Time-Dependent Currents and Voltages: AC Circuits 518

• 25 •

Electromagnetic Waves and the Nature of Light 547

• 26 •

Geometrical Optics: Optical Instruments 576

• 27 •

Physical Optics 607

• 28 •

* Relativity 647

• 29 •

Origins of the Quantum Theory 673

• 30 •

Atomic Structure and Atomic Spectra 700

• 31 •

Aggregates of Atoms: Molecules and Solids 723

• 32 •

Nuclear Physics and Elementary Particles 743

• INDEX OF TABLES •

• PREFACE •

I was motivated to write this book because I believe there is a need for a text that not only presents a clear and complete account of the basic concepts of physics but stresses understanding of these concepts through the use of sound physical arguments, with minimal reliance on mathematical crutches. The aim of a beginning physics course should be not merely to acquaint the students with the subject matter but to help them develop physical insight and intuition, and to guide them in the use of general principles such as dimensional analysis, symmetry, and conservation laws. Finally, especially in a text for non-physics majors, the inclusion of examples that demonstrate the role of physics in other disciplines is of paramount importance.

This book is intended for students in the biological, environmental, earth, and social sciences, and is suitable as a text for a one-year introductory course. Mathematical techniques are limited to algebra and elementary trigonometry; calculus is not used. SI units are employed throughout.

This text has several distinctive features. First, it is characterized by a consistent emphasis on physical insight. Important derivations are preceded by qualitative discussions and arguments based on physical reasoning. This allows students to anticipate the final result and prepares them for the subsequent, more rigorous derivation. In particular, the all-too-common "It can be shown . . ." is studiously avoided. The reader is never presented with *ad hoc* statements that must be accepted at face value. Important results are carefully analyzed in light of their relation to earlier material, and qualitative predictions deduced from these results are stressed. For example, the intimate relation between average kinetic energy and temperature is anticipated using dimensional arguments before its derivation from kinetic theory. Following that derivation, the conclusions that may be drawn concerning molecular speeds and heat capacities of gases are discussed. In the few instances when satisfactory derivations cannot be given without the use of more sophisticated mathematics, for example, Poisseuille's law or the exponential decay of the current in an RC circuit, lucid physical and dimensional arguments are employed that allow the student to understand these relations in light of previously established physical concepts.

The same philosophy is also evident in many of the examples, which not only illuminate new concepts but develop and demonstrate their relation to material presented at an earlier stage. Students are warned against blindly plugging numbers into equations. Instead, through the use of examples, they are shown how symmetry and conservation principles often greatly simplify calculations, and are encouraged to apply similar procedures. This approach should also help students improve their facility in analyzing and solving problems outside the physics classroom.

Many examples are taken from biology, medicine, astronomy, archaeology, and the earth sciences. They have been carefully selected with a view toward general interest, timeliness, and basic simplicity, and are analyzed in sufficient detail to permit the student to comprehend the technique or process that is involved in each case. These examples do not intrude or detract, but are woven into the fabric of the text.

While problem solving is unquestionably an essential part of the learning process, the ability to carry a problem to its correct solution is not the only, sometimes not even the best way to test understanding of physical concepts. In problem solving, mathematical manipulation all too often demands a disproportionate effort and assumes unwarranted importance. The second novel feature of this text is the inclusion of many multiple choice questions at the end of each chapter. I know that many instructors think poorly of multiple choice problems; I am among that group. The multiple choice questions included are not problems but qualitative questions that probe the student's comprehension of concepts, principles, and the physical significance of derived relations. They could have been phrased in the form of discussion questions, and can serve that purpose as well. The multiple choice format, however, has the further benefit that it permits students to test the level and depth of their understanding directly, and can thus function as an effective device for self-evaluation. It is recommended that the multiple choice questions be studied and answered before the end-of-chapter problems are attempted.

The problems at the end of each chapter are numerous and span a wide range of difficulty, from simple exercises to some that will prove challenging even to the best student. In all but a few brief chapters, problems are grouped according to the principal chapter sections to assist the instructor in selecting a representative group for class assignment. The difficulty of each problem is indicated by a simple code (simple exercise—no mark; average difficulty—one dot; difficult—two dots). Answers to odd-numbered problems and to all multiple choice questions are given following the Appendixes.

Another noteworthy feature of this text is the historical material, including direct quotations from original work, which is carefully interlaced in the text and contributes positively to topic development. Historical commentary not only enlivens the discussion but also shows dramatically that science does not always progress logically in step-by-step advances along a well-defined path. All too often, students come away from a physics course with the impression that the laws of physics are immutable rules rather than the formal expressions of our current perception of the world in which we live. Many physicists at the turn of this century were totally unprepared for the revolutionary work of Einstein and Planck because they had been conditioned to think of Newtonian physics as the ultimate revelation in science. Consequently, they rejected relativity and the quantum theory, and were unable to contribute effectively to twentieth-century advances in science. A proper historical perspective may prepare the student to view future drastic departures from accepted dogma with an open mind.

As Alice said, "What good is a book without pictures and conversation?" Though the historical introductions may be a poor substitute for "conversation," this book certainly does not lack pictures. It is profusely illustrated with over 1,100 drawings and air-brush renditions of superior quality, and with numerous photographs.

This book contains undoubtedly more than can be covered in most one-year courses. That is as it should be. An elementary text that must be augmented by auxiliary material obviously fails to meet the needs of the instructor. The topics included encompass all those traditionally taught (and listed in the MCAT manual). There are also many sections that are optional, and these are indicated by an asterisk (*). Some are of special interest to a particular group of students (such as those in biology, or archaeology) but can be omitted without loss of continuity. No material needed in subsequent chapters is contained in the optional sections.

The entire Chapter 28, *Relativity*, has been marked optional. Many students express a desire to learn about this subject, and this chapter attempts to meet that need. It gives a careful, concise account of the problems that led to the theory, and, without mathematical complexity, shows how the central results of the theory can be deduced. The subject matter is, however, far more difficult than any other topic covered here and requires a level of sophistication few beginning students have achieved. Hence, the optional designation.

I have attempted to use a style of writing that is fluent, vivid, and informal, yet clear and concise. This is a textbook; it is also, I hope, a readable and occasionally entertaining account of the development of physics from Archimedes to the present.

Many people have made important contributions to the finished text. Professors John Paul Barach (Vanderbilt University), Paul Bender (Washington State University), Keith Brown (California State Polytechnic University), Michael Browne (University of Idaho), Roger Clapp, Jr. (University of South Florida), James Conley (City College of San Francisco), J.P. Davidson (University of Kansas), A.L. Ford (Texas A & M University), James Gerhart (University of Washington), John Heil (University of Wisconsin Center–Fond du Lac), David Markowitz (University of Connecticut), Edward Nelson (University of Iowa), M.E. Oakes (University of Texas–Austin), William Savage (University of Iowa), Michael Schick (University of Washington), Peter Trower (Virginia Polytechnic Institute and State University), Gilbert Ward (The Pennsylvania State University), Norman Wessells (Stanford University), and Stanley Williams (Iowa State University) have read all or portions of the manuscript and have made numerous valuable suggestions. Professor Clapp also worked every problem and answered every question, and made many thoughtful recommendations that helped clarify problem statements. I appreciate too the assistance of Professor Jerry Peterson (University of Colorado–Boulder) in preparing solutions for a sizable portion of the problem sets.

The excellent illustrations were prepared by Scientific Illustrators, under the direction of Mr. George Morris.

I wish to thank the staff at Allyn and Bacon, in particular, Mary Beth Finch, Judith Fiske, Gary Folven, Nancy Murphy, and James Smith who gave me much encouragement and worked diligently on this project. Through the years of manuscript preparation, class testing, and production, developmental editor Jane Dahl maintained a cheerful and calm disposition despite my occasional ill-tempered outbursts, and labored unceasingly to make this a book of superior quality. To her I owe a very special debt of gratitude.

Last, but not least, I wish to thank the many unnamed, though not forgotten students who, during the year of class testing, discovered errors in a pre-publication version of the book and by their thoughtful criticisms helped me to sharpen arguments and clarify explanations.

This project could never have been completed without the support and sympathetic understanding of my wife who accepted the fate of temporary neglect with equanimity.

1

Units, Dimensions, Vectors, and Other Preliminaries

There shall be standard measures of wine, beer, and corn—the London quarter—throughout the whole of our kingdom, and a standard width of dyed, russet and halberject cloth—two ells within the selvedges; and there shall be standard weights also.

MAGNA CHARTA (1215)

1.1 Units

The observations and experiences that interest scientists span an immense range. Distances extend from the incredibly small dimensions of subnuclear particles to the thousands of light years that separate galaxies of the universe; times encompass those of stellar evolution and the almost infinitesimally short lifetimes of some "elementary particles." Similarly, enormous ranges of masses, electric charges, magnetic fields, pressures, densities, or other variables engage the attention of physicists in their daily work. The ranges of distance, mass, and time of phenomena studied today are listed in Table 1.1.

To describe and characterize these phenomena, scientists must agree on a consistent set of *units* with which measurements are to be compared. Masses, lengths, times, currents, velocities become meaningful only in comparison with ones familiar to us. The unit is simply the standard yardstick with which a particular event is contrasted.

Table 1.1 Ranges of length, time, and mass in the universe

Length (in meters)	
10^{-17}	Present experimental limit in determining nuclear structure
10^{-15}	Diameter of proton
10^{-10}	Diameter of atom
10^{-8}	Length of ribosome
10^{-6}	Wavelength of visible light; length of bacterium
1	Height of man
10^{7}	Radius of earth (6371 km)
10^{11}	Radius of earth's orbit (149×10^6 km)
10^{16}	One light year
10^{22}	Distance to nearest galaxy (M31 in Andromeda)
10^{26}	Radius of universe
	Range of $10^{26}/10^{-17} = 10^{43}$
Time (in seconds)	
10^{-23}	Time for light to cross a proton
10^{-15}	Period of light wave
10^{-8}	Time for emission of photon from excited atom
10^{-2} } 10^{9}	Human time scale: range between reaction time to visual or other stimulus and average life span
10^{7}	One year (3.16×10^7 s)
10^{16}	Solar system completes one turn about galactic center
10^{17}	Age of earth
10^{18}	Age of the universe
	Range of $10^{18}/10^{-23} = 10^{41}$
Mass (in kilograms)	
10^{-30}	Mass of electron
10^{-27}	Mass of proton
10^{-21}	Mass of ribosome
10^{-15}	Mass of bacterium
10^{2}	Mass of man
10^{25}	Mass of earth (5.98×10^{24} kg)
10^{30}	Mass of sun (1.99×10^{30} kg)
10^{41}	Mass of our galaxy
10^{52}	Mass of universe
	Range of $10^{52}/10^{-30} = 10^{82}$

Unfortunately, in the historical development of science different systems of units were used in different parts of the world, and in the same country by different professions. To confuse matters further, the fact that certain physical quantities like electric current and magnetic field are not independent but fundamentally related was not recognized when they were first studied. As a result, several diverse units have been in common use. These are now being replaced, under international agreement, by units of the *Système International,* or SI units. In this system, the *meter, kilogram,* and *second* are the fundamental units of *length, mass,* and *time,* respectively.

Students in the United States are more conversant with the pound, foot, and quart of the British system than with the newton, meter, and liter. However, the SI units are now widely used throughout the world, and will soon be in use in this country as well. The study of physics is a good initiation to the "metric" system; moreover, use of that system makes the physics itself more lucid and the computations more tractable. We shall therefore rarely mention the British units, and then only to compare them with their metric equivalents.

Although we shall stay with the SI, it is important to know how to convert from one unit to another. Conversion factors are listed in Table A inside the front cover. Two examples will illustrate the method.

Example 1.1 A car is driving at a speed of 50 miles per hour (mph). What is the speed of the car in kilometers per hour and in meters per second?

Table A gives the conversion between miles and kilometers, namely, 1 mile = 1.61 km. Denoting the speed of the car by v, we have

$$v = \left(\frac{50 \text{ miles}}{1 \text{ h}}\right)\left(\frac{1.61 \text{ km}}{1 \text{ mile}}\right) = 80.5 \text{ km/h}$$

Note that the unit, miles, cancels out in the conversion.

To convert to meters per second, we note that there is one hour per 60 minutes and one minute per 60 seconds, and that there are 1000 meters per kilometer. Thus

$$80.5 \text{ km/h} = \left(\frac{80.5 \text{ km}}{1 \text{ h}}\right)\left(\frac{1 \text{ h}}{60 \text{ min}}\right)\left(\frac{1 \text{ min}}{60 \text{ s}}\right)\left(\frac{1000 \text{ m}}{1 \text{ km}}\right)$$
$$= 22.4 \text{ m/s}$$

Example 1.2 What is the conversion factor between cubic feet and liters?

One liter (L) is defined as 1000 cm³. To get the answer, we must therefore first determine the number of cubic centimeters contained in one cubic foot. Since 1 ft = 30.48 cm, it follows that

$$(1 \text{ ft})^3 = (30.48 \text{ cm})^3 = 28{,}320 \text{ cm}^3 = 28.32 \text{ L}$$

One of the attractive and convenient features of the SI is that it is a *decimal* system. Kilometers, micrograms, nanoseconds, megawatts are all derived from basic units by multiplication by integral powers of *ten.* This makes computation much simpler than in the British system, in which the inch, foot, yard, rod, chain, and mile bear no such simple relation to each other. With SI units we can then use "scientific notation" to advantage. Another convenience of SI units is the existence of a standard prefix and symbol for each important power of ten. These are listed in Table 1.2.

Table 1.2 *Prefixes and their symbols used to designate decimal multiples and submultiples*

Factor	*Prefix*	*Symbol*	*Factor*	*Prefix*	*Symbol*
10^{18}	exa	E	10^{-1}	deci	d
10^{15}	peta	P	10^{-2}	centi	c
10^{12}	tera	T	10^{-3}	milli	m
10^{9}	giga	G	10^{-6}	micro	μ
10^{6}	mega	M	10^{-9}	nano	n
10^{3}	kilo	k	10^{-12}	pico	p
10^{2}	hecto	h	10^{-15}	femto	f
10^{1}	deca	da	10^{-18}	atto	a

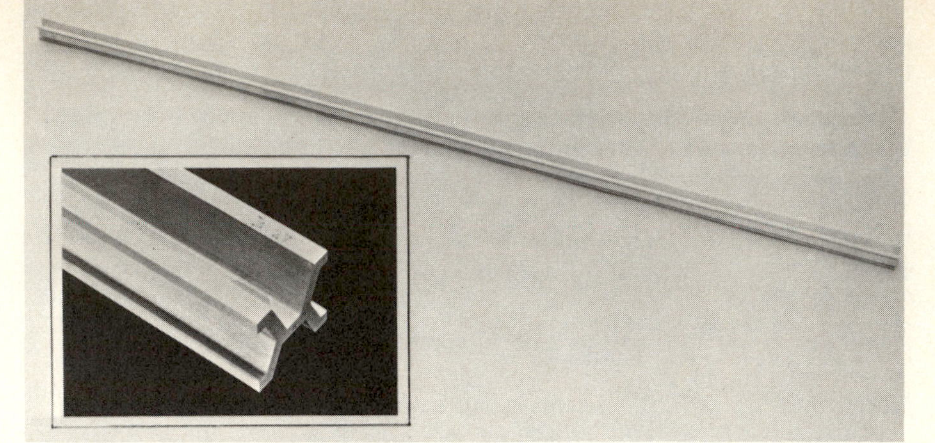

Figure 1.1 The standard meter bar, made of platinum-iridium alloy and housed at the U.S. Bureau of Standards. It is a national standard of the international prototype at the National Bureau of Weights and Measures, Sèvres, France.

1.2 Fundamental Units

Forces, velocities, pressures, energies—indeed all mechanical properties—can be expressed in terms of three basic quantities: mass, length, and time. In the SI, the corresponding units are

Kilogram—Mass

Meter—Length

Second—Time

These are known as *fundamental units.*

For many years the *standard meter* was defined as the distance between two hairline scratches on a bar of platinum-iridium alloy maintained at constant temperature in a vault of the Bureau of Weights and Measures in Sèvres near Paris. Similarly, the *standard kilogram* is a solid platinum-iridium alloy cylinder, also carefully preserved at Sèvres. Since it is not practical for scientists to make regular pilgrimages to Paris, secondary standards of the meter and kilogram, carefully prepared to replicate the primary ones, are kept at the U.S. Bureau of Standards and similar establishments throughout the world.

Figure 1.2 The standard kilogram housed at the International Bureau of Weights and Measures, Sèvres, France. This standard is also made of platinum-iridium alloy.

The *second* was originally defined as 1/86,400 of the mean solar day. Subsequently, when it became apparent that the rate of rotation of the earth was very slowly decreasing, lengthening the mean solar day, the second was redefined in 1960 as 1/31,556,925.9747 of the year 1900.

The frequencies of light emitted by properly stimulated atoms are often very sharply defined, as are the corresponding wavelengths of these light waves in vacuum. Atoms have, as far as we know, remained unchanged since creation and will, we confidently believe, remain unchanged in future centuries. Moreover, atoms are not subject to the hazards of destruction by fire, earthquake, or war and other forms of vandalism. Consequently, these atomic radiations are excellent, truly permanent time and length standards. Today the meter is officially defined as 1,650,763.73 times the wavelength of a certain emission line from krypton, and the second is defined as 9,192,631,770 times the period (duration) of an oscillation associated with a particular atomic transition of the cesium atom. Cesium clocks are now generally used in all experiments requiring the most precise determination of long time intervals.

Thus, though chronometers and secondary standards of the meter are still very useful for many purposes, it is only the kilogram standard that is really needed today.

1.3 Derived Units and Dimensional Analysis

Quantities that concern scientists are not limited to mass, length, and time. We often describe the behavior of objects in terms of their *velocities;* we need to identify the *forces* that act on bodies; we pay for the *energy* consumed by appliances and are curious about the *power* a motor can deliver; atmospheric *pressure* is a useful indicator of weather conditions. All these apparently disparate properties, measured in the units meters per second (velocity), newton (force), joule (energy), watt (power), and pascal (pressure), are ultimately expressible as products of powers of mass, length, and time. These units are therefore known as *derived units,* to distinguish them from the three fundamental units.

This fact has some useful consequences. First, it would clearly be senseless, indeed wrong, to devise standard units of pressure or energy similar to the standard meter or kilogram. Second, since in any equation relating various physical quantities *the units on both sides of the equation must be the same,* we can often use this requirement to check and even derive relations. This is *dimensional analysis,* which we shall use often.

Suppose you seek an equation that relates a velocity, say v, to an energy E, a time T, and a force F. As we shall see, the *dimensions* of velocity, energy, and force, expressed in the fundamental units of mass $[M]$, length $[L]$, and time $[T]$,* are

$$\text{velocity} = \frac{[L]}{[T]}$$

$$\text{energy} = \frac{[M][L]^2}{[T]^2}$$

$$\text{force} = \frac{[M][L]}{[T]^2}$$

The only way E, T, and F can be combined to yield v is in an expression of the form

$$v = \frac{E}{F \times T} = \frac{[M][L]^2}{[T]^2} \times \frac{[T]^2}{[M][L]} \times \frac{1}{[T]} = \frac{[L]}{[T]}$$

because only then does the right-hand side have the dimension $[L]/[T]$.

1.4 Scaling

Related to dimensional analysis is a procedure known as *scaling.* Although the basic idea of scaling is simple, it can be a powerful analytical and predictive tool if used properly. It can also generate much misinformation and nonsense if applied blindly.

Figure 1.3 shows the skeletons of two prehistoric animals drawn so both are of the same size. Yet we immediately gain the impression that the animal belonging to the skeleton at the top was substantially larger

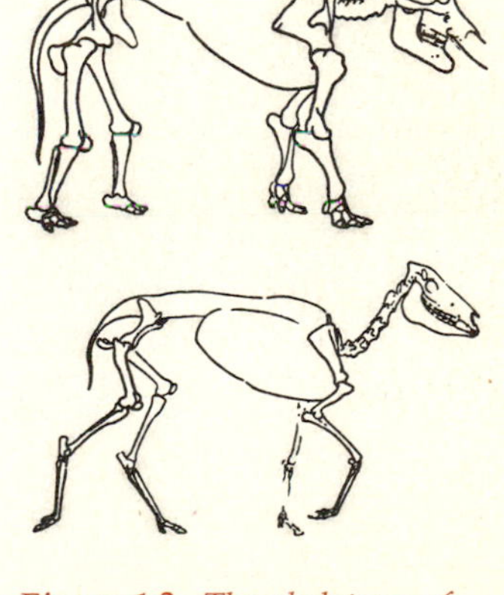

Figure 1.3 *The skeletons of two prehistoric animals, scaled so both appear of equal size. It is apparent that the skeleton at the top belonged to a much larger animal than the one below.*

* We shall always use square brackets, [ ], to indicate the dimensions of a variable.

than the other; and indeed it was. The massive skeleton is that of a mastodon; the other belonged to a *Neohippanus,* a small ancestor of the horse. The perception of great size for the first is associated with the heaviness, the greater diameter of the supporting bone structure. Intuitively you sense this; yet if you were asked to draw the skeleton of King Kong, you would most likely take the skeleton of a gorilla and simply "scale it up" as though you were making a photographic enlargement. Such an animal could not possibly exist! Let's see why.

The function of the skeleton is to provide the mechanical, structural support for the body. The resistance of a bone to buckling under the weight it supports is roughly proportional to the cross-sectional area of the bone; that seems certainly reasonable. Thus we expect that since the cross section of a cylinder is proportional to the square of its diameter d, then d^2 should be proportional to W, the animal's weight. But an animal's weight is crudely proportional to the volume it occupies; and volume, in turn, has the dimension of a length cubed, $[L]^3$. Hence we should expect that for animals of roughly similar shape, for example, terrestrial quadrupeds,

$$d^2 \propto L^3 \qquad \text{or} \qquad d \propto L^{3/2} \tag{1.1}$$

where d is the diameter of a principal structural bone and L is a characteristic length, for instance the height of the beast. Note that the diameter of the bone is *not* proportional to the size of the animal but increases rather more rapidly. Thus, if the skeletal structure of King Kong were a linearly scaled-up version of a gorilla's, the sci-fi beast's bones would be unable to support the mammoth hulk.

1.5 Order-of-Magnitude Estimates

Commonplace caricatures show the scientist as a little bald man in an oversized lab coat searching for the ultimate in precision. Yet, precise measurement is perhaps the least important and surely the least exciting scientific endeavor. Scientists are far more concerned with qualitative than quantitative characterizations of nature. Numbers alone are barren. Whether a stone takes 1.01 s or 1.06 s to fall through a given height matters far less than that two stones of different masses fall a given distance in the same time. Precision is sometimes essential but assumes great importance only when the difference between two nearly identical numbers implies a choice between two competing theories.

Before spending a lot of time and money on a project, a scientist must first estimate the magnitude of the effect he hopes to observe and measure. Unless he can convince himself, and perhaps a granting agency, that present-day technology can provide the necessary tools, and more important, that the effect is theoretically large enough to penetrate the noise due to statistical fluctuations, he had better not even start the project. The ability to make sound order-of-magnitude estimates, educated guesses, is the trademark of the mature, experienced scientist. This knack, this physical insight, is not an inherent talent but requires a good grasp of basic concepts, and where and how to apply them. It does not come quickly, but it can be learned.

Throughout this book we shall, whenever appropriate, make order-of-magnitude estimates. They will prove especially valuable and revealing in describing the properties of atoms and nuclei.

1.6 Significant Figures

At some stage, results more precise than order-of-magnitude estimates are demanded. Even then, it makes little sense to calculate and give an answer to greater accuracy than justified by the original data.

Suppose you are asked to determine the area of a rectangle. You measure the lengths of the sides with a meter stick, but realize that your measurement is reliable only to within the millimeter divisions on the stick. Your measurement gives the lengths of the sides as 43.9 cm and 3.4 cm. You conclude that the area $A = (43.9\text{ cm})(3.4\text{ cm}) = 149.26\text{ cm}^2$. To give this answer without qualification would be wrong, since it asserts that A is not 150.85 cm^2. But since the short side might be 3.44 cm and the long side 43.93 cm (both within the limits of accuracy of reading the meter stick), the area could be greater than 151 cm^2. That is, the five-figure accuracy is not justified by the precision of the initial measurements.

The following is a rough guide, a good rule of thumb.

The number of significant figures of the product of several quantities should be the same as the number of significant figures of the least precise factor.

In the example just cited, the length of the shorter side is known to two significant figures; the answer for the area should therefore also be given to only two significant figures.

So you give the answer as 150 cm^2, and though the casual reader might well interpret this as an answer accurate to two significant figures, it does still imply that the last figure, the zero, is important, that is, that the area is not 151 cm^2. How can we convey the correct impression? We use scientific notation and state the result as

$$A = 1.5 \times 10^2\text{ cm}^2$$

and there can now be no doubt about the number of significant figures of the result.

How should you proceed if the computation involves not the product but the sum or the difference of two or more numbers? Suppose you are asked to give the circumference, $2L + 2W$, of the same rectangle. Should the result contain only two significant figures since W is known to only two-figure accuracy? Evidently not, since you do know each length to within 1 mm. The correct answer is 94.6 cm; three-figure accuracy is justified now.

In addition(or subtraction), the number of decimal places in a sum of two or more numbers should equal the smallest number of decimal places of any term of that sum.

In this book, we shall generally limit answers to three significant figures.

1.7 Scalars and Vectors

In the following chapters we shall use some mathematics, though we shall always try to place the emphasis where it properly belongs, on physical concepts. Still, in physics, as in chemistry, biology, or demography, we are ultimately forced to resort to mathematical formulation simply because it is such a convenient shorthand notation and computational tool. There *is* a difference between the mass of a VW bug and of a Rolls-Royce, between the pull exerted on your hand by a kite string and by the leash at which your dog strains. Using the language of mathematics, we can express these differences precisely and with economy of effort.

We assume familiarity with algebra and elementary trigonometry. The only trigonometry we shall need are the definitions of sine, cosine, and tangent, at least for the first ten chapters. These definitions and a few trigonometric identities that will be useful later are summarized in Appendix A. For convenient reference, the definitions of the three most common trigonometric functions are given below. The defining relations are

$$\sin \theta = \frac{a}{c} \tag{1.2a}$$

$$\cos \theta = \frac{b}{c} \tag{1.2b}$$

$$\tan \theta = \frac{a}{b} \tag{1.2c}$$

where a, b, and c refer to the lengths of the sides and hypotenuse of the right triangle of Figure 1.4.

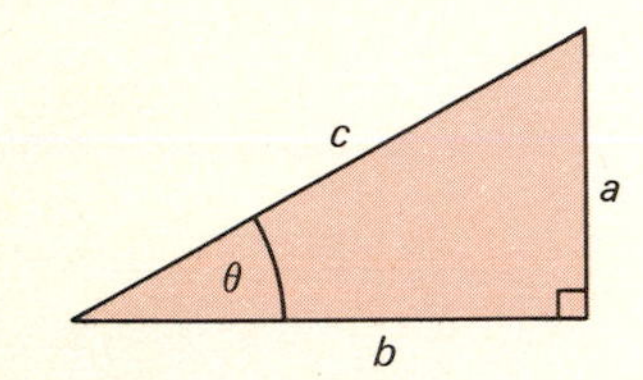

Figure 1.4 A right triangle is used to define trigonometric functions, as follows:

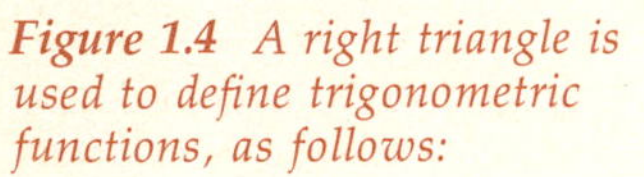

$$\sin \theta = \frac{a}{c}$$

$$\cos \theta = \frac{b}{c}$$

$$\tan \theta = \frac{a}{b}$$

We shall use two mathematical constructs, scalars and vectors. *Scalars* are the familiar "numbers," rational and irrational, that prove so useful in daily affairs—when we buy five pounds of potatoes, cash a check for $25.63 at the bank, or have the tank filled with ten gallons of gasoline. In each instance the single number multiplying the proper unit (pound, dollar, gallon) suffices to specify unequivocally the quantity involved. In physics, many variables—among them mass, volume, energy, temperature, and time—can be completely identified by a single number. We call these *scalar properties.*

Suppose a stranger asks you "Where is the library?" and you reply "It is half a kilometer from here," and then walk away. He will feel miffed,

Figure 1.5 The forces exerted by a kite string and by a chained dog differ in magnitude as well as direction unless the dog is really quite spectacular.

and for good reason. To get to the library he must know in which direction he should proceed. Similarly, the forces exerted on your hand by the kite string and by the leashed dog differ not only in magnitude but also in direction.

A *vector* is a mathematical construct used to characterize those properties with which we associate not only *magnitude* but also *direction*. An arrow is used to represent a vector. The arrow's direction is that of the vector and its length represents the vector's magnitude. Although vectors are only mathematical abstractions, they are very convenient for characterizing a large class of physical quantities such as displacement, force, velocity, acceleration, and momentum. We shall use **bold-face letters (A, a)** to designate vectors and *italic letters* (A, a) to designate their magnitudes, which are scalars.

What mathematical rules do these constructs obey? In other words, how do we add, subtract, or multiply vectors? You already know how to manipulate scalars, and we shall now develop the corresponding formalism for adding and subtracting vectors. There are also rules for vector multiplication, but we need not concern ourselves with them here, except for the following.

DEFINITION: The product of a vector **A** and a scalar b is a vector whose magnitude is bA and whose direction is that of **A**.

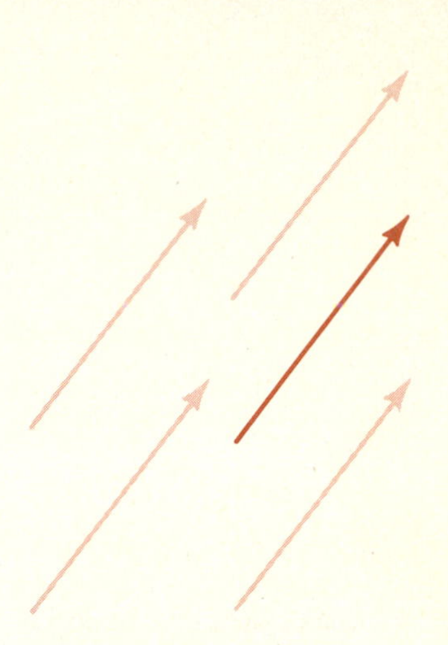

Figure 1.6 *A vector is represented by an arrow. It is not changed by translation parallel to itself. The arrows shown all represent the same vector.*

1.8 Vector Addition and Subtraction

Suppose you walk in an open field 8 paces east, stop and turn north and go another 6 paces. Though you have traveled a total distance of 14 paces, your *displacement* from the point of origin is less than that, and is neither due east nor due north. Applying Pythagoras' theorem, you determine that the magnitude of the displacement is

$$d = \sqrt{8^2 + 6^2} = \sqrt{100} = 10 \text{ paces}$$

To specify the direction of the displacement, we can turn to one of the trigonometric functions, for example, the tangent. From (1.2c) the angle θ of Figure 1.7 is the angle whose tangent is $(6/8) = 0.75$; that is,

$$\theta = \tan^{-1}(0.75)$$

and a glance at a table of trigonometric functions (or using a hand calculator with trigonometric functions) shows that $\theta = 36.9° \simeq 37°$.

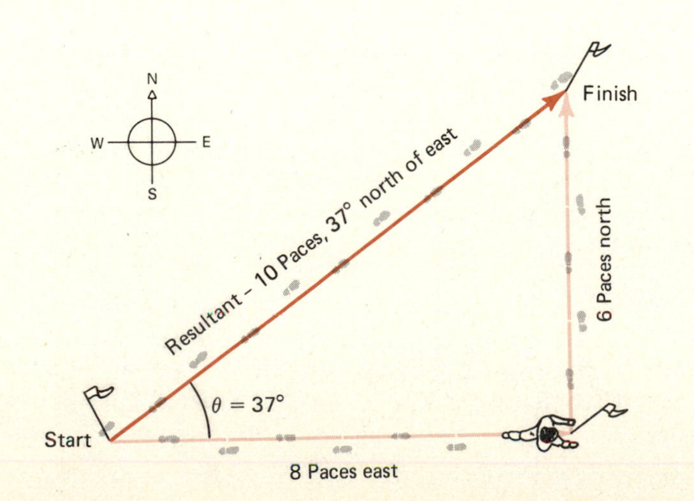

Figure 1.7 *The result of walking 8 paces east and then 6 paces north is a displacement of 10 paces at 37° north of east.*

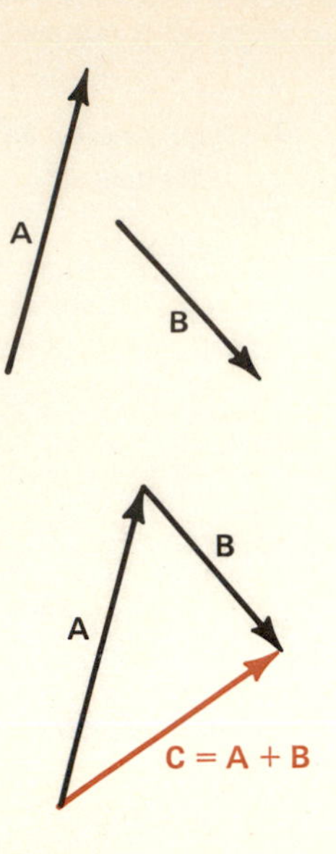

Figure 1.8 *Two vectors are added by placing them head to tail. Here* **C** = **A** + **B**. *Note that the magnitude of* **C** *is less than that of* **A** *in this instance. The magnitude of the vector sum of two vectors depends on their magnitude and direction.*

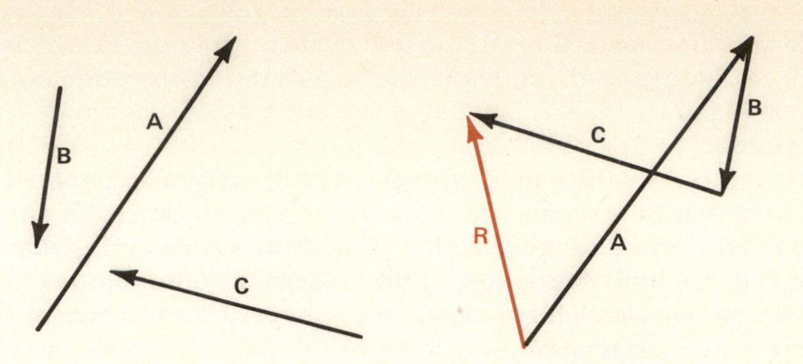

Figure 1.9 *Addition of three vectors:* **R** = **A** + **B** + **C**.

Thus the resultant of the two individual displacements of 8 and 6 paces east and north is a displacement of 10 paces, 37° north of east. This simple example suggests a useful definition for the addition of two vectors.

DEFINITION: If **A** and **B** are two vectors, the vector **C** = **A** + **B** is obtained by placing the vector **B** so that its origin coincides with the end of vector **A**. The vector **C** is then the vector drawn from the origin of **A** to the terminus of **B**.

This geometrical definition for adding two vectors is completely general. In the earlier example, the two displacements of 8 and 6 paces were at right angles; but two vectors may make any angle and we can still determine their sum by the above procedure, as shown in Figure 1.8.

This process can be extended to encompass the sum of any number of vectors. To obtain the vector **R** = **A** + **B** + **C** (Figure 1.9), we shift **B** parallel to itself until its origin coincides with the terminus of **A**; we then shift **C** parallel to itself until its origin falls at the terminus of **B**, placed as indicated. The resultant **R** is then the vector from the origin of **A** to the terminus of **C**.

Note that the associative and commutative laws apply to vector addition; that is,

$$\mathbf{A} + \mathbf{B} + \mathbf{C} = (\mathbf{A} + \mathbf{B}) + \mathbf{C} = \mathbf{A} + (\mathbf{B} + \mathbf{C}) \tag{1.3}$$

$$\mathbf{A} + \mathbf{B} = \mathbf{B} + \mathbf{A} \tag{1.4}$$

In ordinary number theory, one defines zero so that for all numbers N

$$0 + N = N$$

In the realm of vectors, we similarly define the *null vector*, **0**, so that

$$\mathbf{0} + \mathbf{A} = \mathbf{A} \tag{1.5}$$

for any vector **A**. The null vector has zero magnitude; i.e., it is a point.

We can now enlarge the rule for adding vectors to include subtraction. We postulate that for every vector **A** there exists a vector (−**A**), its *negative*, so that

$$\mathbf{A} + (-\mathbf{A}) = \mathbf{0} \tag{1.6}$$

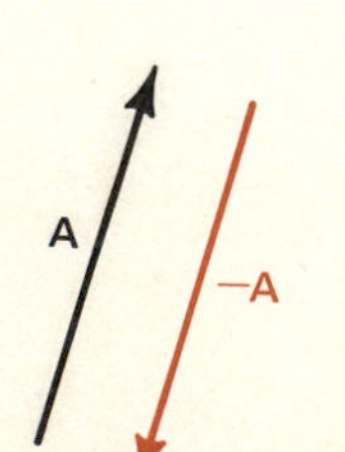

Figure 1.10 *The negative of a vector is another vector of the same magnitude but opposite direction.*

This is exactly analogous to the definition of negative scalars. From the procedure for adding two vectors and from the definition of the null vector, it follows that −**A** is the vector that extends from the terminus of **A** to its origin. That is, −**A** is the vector that has a magnitude equal to that of **A** and a direction exactly opposite to that of **A** (Figure 1.10).

To find the vector $\mathbf{C} = \mathbf{A} - \mathbf{B}$, we write

$$\mathbf{C} = \mathbf{A} - \mathbf{B} = \mathbf{A} + (-\mathbf{B}) \tag{1.7}$$

and proceed by first constructing the vector $-\mathbf{B}$, which we then add to $\mathbf{A}$ according to the earlier prescription. Note that the magnitude of $\mathbf{A} - \mathbf{B}$ may be greater than, equal to, or smaller than the magnitude of $\mathbf{A} + \mathbf{B}$.

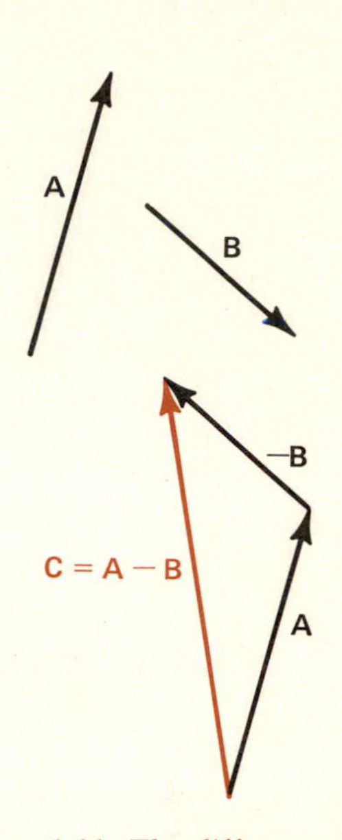

Figure 1.11 *The difference between two vectors is obtained by adding to the first vector the negative of the second. Here* $\mathbf{C} = \mathbf{A} - \mathbf{B} = \mathbf{A} + (-\mathbf{B})$ *is a vector whose magnitude is greater than that of either* **A** *or* **B**.

Adding vectors by geometrical construction is often not convenient and never the most precise procedure. There is, however, a simple analytic process by which the sum $\mathbf{A} + \mathbf{B}$ can be obtained. It involves decomposing a vector into *orthogonal components.* (*Orthogonal* means mutually perpendicular.)

Any vector can be viewed as the sum of two or more vectors. In Figure 1.12, the vector $\mathbf{A}$ is shown as the sum of $\mathbf{B}_1 + \mathbf{B}_2$ and of $\mathbf{C}_1 + \mathbf{C}_2$. Evidently an infinite number of pairs of vectors exist whose sum equals $\mathbf{A}$. One class of pairs, however, is particularly useful for our purposes, namely two vectors that are orthogonal and whose sum equals $\mathbf{A}$, for example, the pair $\mathbf{A}_x$ and $\mathbf{A}_y$ of Figure 1.13. These two vectors are called the *x and y component vectors of* $\mathbf{A}$; their lengths along the positive x and y directions are called the *x and y components of* $\mathbf{A}$. Their values can readily be determined if we know the magnitude and direction of $\mathbf{A}$. If the angle between $\mathbf{A}$ and the x axis is θ_A, then from the definition of the sine and cosine, Equations (1.2a) and (1.2b),

$$A_x = A \cos \theta_A \tag{1.8}$$

$$A_y = A \sin \theta_A \tag{1.9}$$

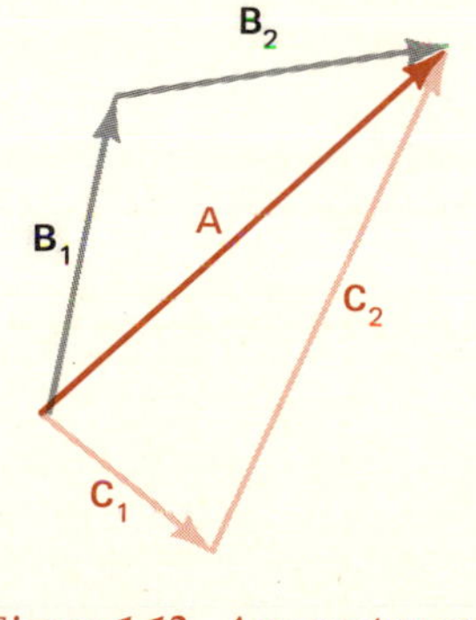

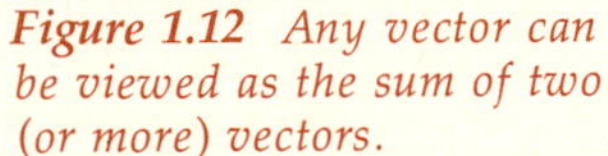

Figure 1.12 *Any vector can be viewed as the sum of two (or more) vectors.*

Well, what's so magical about this decomposition? Nothing, really, except that it is convenient. Suppose we have a second vector $\mathbf{B}$, as in Figure 1.14, and want to form the sum $\mathbf{C} = \mathbf{A} + \mathbf{B}$. Each vector is itself the sum of two component vectors, $\mathbf{A}_x + \mathbf{A}_y$ and $\mathbf{B}_x + \mathbf{B}_y$. Thus,

$$\begin{aligned} \mathbf{C} = \mathbf{A} + \mathbf{B} &= \mathbf{A}_x + \mathbf{A}_y + \mathbf{B}_x + \mathbf{B}_y \\ &= (\mathbf{A}_x + \mathbf{B}_x) + (\mathbf{A}_y + \mathbf{B}_y) = \mathbf{C}_x + \mathbf{C}_y \end{aligned} \tag{1.10}$$

where we have used (1.3) and (1.4). But since $\mathbf{A}_x$ and $\mathbf{B}_x$ are collinear, C_x, the component of $\mathbf{C}$ along the x direction, is just the *algebraic* sum of the x components of $\mathbf{A}$ and $\mathbf{B}$; that is, $C_x = A_x + B_x$. Similarly, $C_y = A_y + B_y$. The vector $\mathbf{C}$ is the vector sum $\mathbf{C}_x + \mathbf{C}_y$. These component vectors are orthogonal, and so we can write

$$C = \sqrt{C_x^2 + C_y^2} \tag{1.11}$$

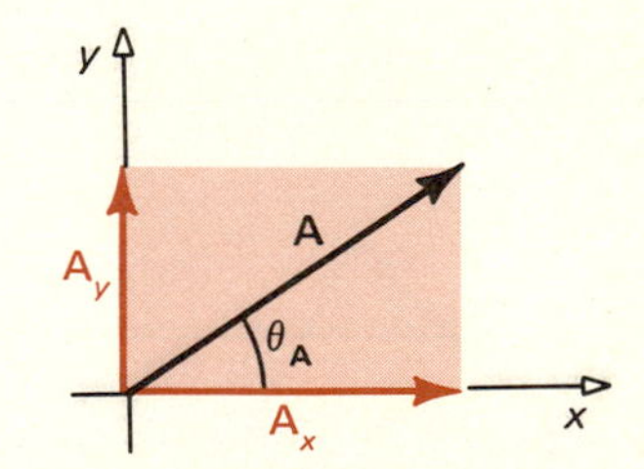

Figure 1.13 *Any vector can be decomposed into orthogonal components. The vector* **A** *is the vector sum of the component vectors* $\mathbf{A}_x$ *and* $\mathbf{A}_y$. *The algebraic values of* $\mathbf{A}_x$ *and* $\mathbf{A}_y$ *are given by*

$$A_x = A \cos \theta$$

$$A_y = A \sin \theta$$

and are called the x and y components of vector **A**.

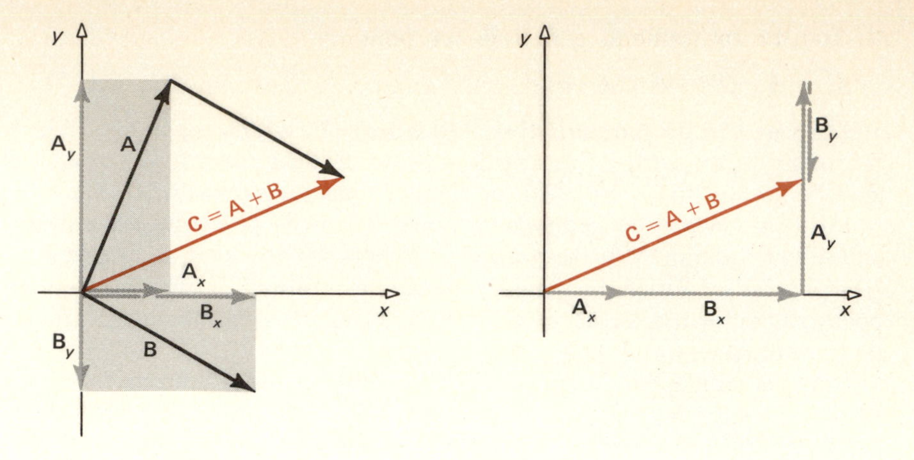

Figure 1.14 Two vectors may be added by taking components and then adding the respective components algebraically. Here $C_x = A_x + B_x$ and $C_y = A_y + B_y$ (B_y is negative).

The angle θ_C that **C** makes with the positive x direction is given by

$$\theta_C = \tan^{-1}\frac{C_y}{C_x} \qquad \textbf{(1.12)}$$

If you are using a hand calculator with trigonometric functions, it is quicker first to find θ_C, using (1.12), and then to evaluate C by

$$C = \frac{C_x}{\cos\theta_C} \quad \text{or} \quad C = \frac{C_y}{\sin\theta_C} \qquad \textbf{(1.13)}$$

than it is to square the components and take the square root of their sum.

This technique can be used in adding or subtracting any number of vectors. Moreover, the method is easily generalized to three dimensions, so that even if three vectors are not coplanar, one can determine their sum by taking components, now along three orthogonal axes, x, y, and z.

● ***Example 1.3*** Find the sum of **A** and **B**, where $A = 8$ units and $\theta_A = 45°$, $B = 12$ units and $\theta_B = -60°$; the angles are measured relative to the positive x direction.

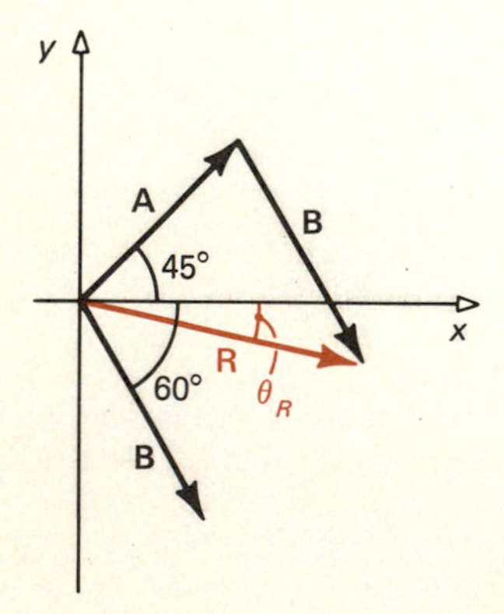

Figure 1.15 A rough sketch can help eliminate mistakes in vector addition. The resultant **R** is about equal in magnitude to **B** and θ_R is approximately −20°.

Before starting to solve any problem involving vector addition, it is wise to make a rough sketch, showing the vectors approximately to scale, and to perform the addition geometrically; you then have a fair idea of what the result should be. Such a sketch, Figure 1.15, suggests that **A** + **B** will be a vector of about the same magnitude as **B**, making an angle of about −20° with the x direction.

We next compute the x and y components of **A** and **B**, using (1.8) and (1.9):

$$\begin{aligned} A_x &= A\cos\theta_A = (8\text{ units})\cos 45° \\ &= (8\text{ units})0.707 = 5.66\text{ units} \end{aligned}$$

$$\begin{aligned} A_y &= A\sin\theta_A = (8\text{ units})\sin 45° \\ &= (8\text{ units})0.707 = 5.66\text{ units} \end{aligned}$$

$$\begin{aligned} B_x &= B\cos\theta_B = (12\text{ units})\cos(-60°) \\ &= (12\text{ units})0.5 = 6.0\text{ units} \end{aligned}$$

$$\begin{aligned} B_y &= B\sin\theta_B = (12\text{ units})\sin(-60°) \\ &= (12\text{ units})(-0.866) = -10.4\text{ units} \end{aligned}$$

Adding the components gives the components of the resultant **R**.

$$R_x = A_x + B_x = 5.66 \text{ units} + 6.0 \text{ units} = 11.66 \text{ units}$$

$$R_y = A_y + B_y = 5.66 \text{ units} + (-10.4 \text{ units}) = -4.74 \text{ units}$$

Last, using (1.12) and (1.13), we find

$$\theta_R = \tan^{-1}\frac{R_y}{R_x} = \tan^{-1}\left(\frac{-4.74}{11.66}\right) = \tan^{-1}(-0.407) = -22.1°$$

and

$$R = \frac{11.66 \text{ units}}{\cos(-22.1°)} = \frac{11.66 \text{ units}}{0.927} = 12.6 \text{ units}$$

Comparison with estimates based on our sketch confirms that we have made no glaring mistake. But suppose you had erroneously written $\cos(-60°) = -0.5$, a common error made by students not familiar with trigonometric functions. In that event you would find $R_x = -0.34$ unit. The magnitude of the resultant would have been about 5 units, and arriving at a result less than half your geometrical estimate should have alerted you to a likely error. ●

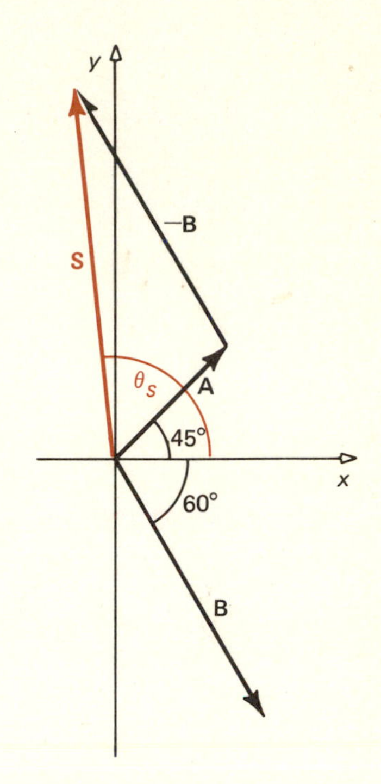

Figure 1.16 A sketch of the difference between vectors **A** *and* **B**. *The angle θ_S is slightly greater than 90°.*

Subtraction of vectors by decomposition into orthogonal components poses no problem. From (1.6),

$$\mathbf{C} + (-\mathbf{C}) = \mathbf{0} = \mathbf{C}_x + \mathbf{C}_y + (-\mathbf{C})_x + (-\mathbf{C})_y$$

We see that the x and y components of $-\mathbf{C}$ are the negatives of the corresponding components of **C**. Thus if $\mathbf{S} = \mathbf{A} - \mathbf{B}$, we obtain **S** by finding $S_x = A_x - B_x$ and $S_y = A_y - B_y$, where B_x and B_y are the x and y components of **B**.

● ***Example 1.4*** Find the vector $\mathbf{S} = \mathbf{A} - \mathbf{B}$, where **A** and **B** are the same vectors as in the preceding example.

From our sketch, Figure 1.16, we see that **S** has a magnitude greater than A or B, perhaps about 15 units, and points nearly along the positive y direction.

The components of **A** and **B** are as given before. Thus

$$S_x = A_x - B_x = 5.66 \text{ units} - 6.0 \text{ units} = -0.34 \text{ unit}$$

$$S_y = A_y - B_y = 5.66 \text{ units} - (-10.4 \text{ units}) = 16.06 \text{ units}$$

A sketch is especially useful here, because we see now that the angle $\theta_S = 90° + \phi$, where ϕ is the small angle whose tangent is the ratio $0.34/16.06 = 0.021$. Hence $\phi = 1.2°$, and $\theta_S = 91.2°$. The magnitude of **S** is

$$S = \frac{16.06 \text{ units}}{\cos 1.2°} = 16.06 \text{ units}$$

(to four-significant-figure accuracy).

The numerical result and the sketch agree. ●

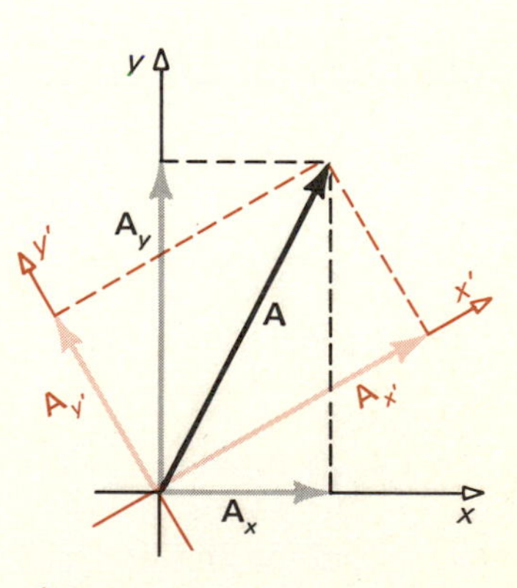

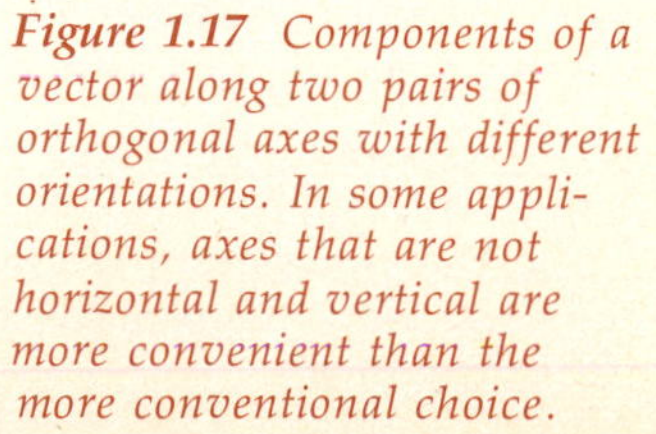

Figure 1.17 Components of a vector along two pairs of orthogonal axes with different orientations. In some applications, axes that are not horizontal and vertical are more convenient than the more conventional choice.

One final word: In the decomposition of vectors, it is natural to select the x and y axes as the horizontal and vertical directions. Often, these are the most convenient. However, any set of mutually perpendicular directions can be used, and sometimes axes other than horizontal and vertical prove more advantageous. The decomposition of a vector along two different sets of orthogonal axes is illustrated in Figure 1.17.

Summary

Several systems of units are in use today, among them the "British" system, based on the foot, pound, and second, and the SI, with the meter, kilogram, and second as the *fundamental units.* The SI has been accepted internationally and will be used in this book. A table of conversion factors for different units is given in Table A.

All mechanical properties can be expressed in units that are products of powers of the meter, kilogram, and second. Units such as those for force (newton), energy (joule), and power (watt) are called *derived units.*

The *dimension* of any mechanical quantity is the product of powers of mass, length, and time. In any equation relating various physical properties, the dimensions on both sides must be the same. Dimensional analysis is useful for verifying relations and can also be employed to derive them, except for multiplicative constants.

Frequently it is of interest to determine the magnitude of quantities without performing detailed, numerical calculations. Order-of-magnitude estimates are very valuable in physics and other sciences.

The result of detailed numerical computations should be expressed only to as many *significant figures* as are justified by the precision of the original data. Rough guidelines for determining the proper number of significant figures are given on page 7.

Vectors are mathematical constructs used to represent physical properties that require both magnitude and direction for their specification. Examples of vector quantities are displacement, velocity, and force.

The vector $-\mathbf{A}$ is a vector that has the same magnitude as **A** but opposite direction.

Two vectors **A** and **B** can be added by the parallel displacement of vector **B** so that its beginning coincides with the terminus of **A**. The vector $\mathbf{A} + \mathbf{B}$ is then the vector that extends from the origin of **A** to the terminus of **B**.

Subtracting vector **B** from vector **A** is equivalent to adding **A** and $-\mathbf{B}$.

Vector addition and subtraction are conveniently done by taking components of the constituent vectors along orthogonal axes. The components of the resultant vector are given by the algebraic sum of the corresponding components of the constituent vectors. Equations (1.8) and (1.9) and (1.11)–(1.13) give the relations between a vector and its components.

Problems

(Section 1.1)

1.1 A car is traveling at 35 mph. What is its speed in (a) kilometers per hour? (b) meters per second?

1.2 Find the area of a circle of 50-cm radius. Give your answer in (a) square meters and (b) inches.

1.3 A cube has sides of length 2 ft. Find its volume in (a) cubic feet; (b) cubic meters; (c) cubic centimeters; (d) liters; and (e) gallons.

1.4 The wavelength of green light is 0.5 micrometer (μm). How many such wavelengths fit into (a) 1 cm? (b) 1 inch?

1.5 A small electric lamp draws a current of 25 milliamperes (mA). Express this current in (a) amperes; (b) microamperes; (c) kiloamperes.

1.6 A motor can develop 4 horsepower (hp). How much power is that when expressed in (a) kilowatts? (b) milliwatts?

1.7 A basketball coach insists that his players must be at least 207 cm tall. Would a student 6 ft 1.5 in. tall qualify for the team?

1.8 The distance between the earth and the moon is 384,000,000 m. Give this result in kilometers. Also express the result in meters and in kilometers in scientific notation.

1.9 The wavelength of light from a laser is 632.8 nm. Express this length in meters, using scientific notation.

1.10 Find the answers to the following numerical problems.

(a) $(9.0 \times 10^7)(3.0 \times 10^{-6})$

(b) $\dfrac{(6.0 \times 10^{12})(4.0 \times 10^{-8})(0.0040)}{1.5 \times 10^5}$

(c) $\dfrac{(3.6 \times 10^5)^{1/2}(4 \times 10^{-2})^{-3}}{(3 \times 10^3)^{-2}}$

1.11 The distance between London and New York is 3480 miles. What is this distance in kilometers?

1.12 The radius of the earth is 6380 km. What is the radius in miles? in feet?

1.13 Convert to meters per second: 40 mph; 20 ft/s; 50 km/h.

1.14 What is the weight in newtons of a person whose weight is 145 lb?

1.15 The peak of Mt. Blanc is 15,772 ft above sea level. What is the height of Mt. Blanc in meters?

(Section 1.3)

1.16–1.23 The following are the dimensions of various quantities that will be discussed in later portions of the text:

velocity (v)	$[L]/[T]$
force (F)	$[M][L]/[T]^2$
energy (E)	$[M][L]^2/[T]^2$
power (P)	$[E]/[T]$
acceleration (a)	$[L]/[T]^2$
pressure (p)	$[F]/[A]$
viscosity (η)	$[p][T]$
area (A)	$[L]^2$

where $[M]$, $[L]$, and $[T]$ denote the dimensions of mass, length, and time.

1.16 The distance an object falls from rest under gravity is given by $s = \frac{1}{2}gt^2$, where g is the acceleration of gravity of 9.8 m/s². Show that this expression is dimensionally correct.

• **1.17** What combination of mass, velocity, and acceleration can appear in expressions for energy and power?

• **1.18** Power may be expressed as the product of force and another variable. What is that variable?

• **1.19** The ideal gas law is $PV = nRT$, where P is the pressure, V the volume, n a number that tells the amount of gas present, R the gas constant, and T the absolute temperature. Show that the quantity on the left-hand side has the dimension of energy.

• • **1.20** The amount of fluid that flows out of the open end of a pipe of a certain length ℓ in 1 s depends on the pressure applied to the fluid at the other end of the pipe, on the length of the pipe, on the diameter of the pipe, and on the viscosity of the fluid. On the assumption that the flow rate, expressed in liters per second, is given by an expression of the form

$$Q = \text{constant} \times \frac{pD^n}{\ell\eta^m}$$

where D is the pipe diameter and m and n are exponents, what must the values of these exponents be?

• • **1.21** A small smooth steel sphere from a ball bearing, when dropped into a bottle of clear syrup, is observed to descend at constant speed after a very brief initial interval. This so-called sedimentation velocity is inversely proportional to the viscosity of the syrup and depends on the mass of the sphere, its radius, and the acceleration of gravity g; the value g denotes the acceleration experienced by an object in free fall (with air friction neglected). Deduce an expression, correct to within a multiplicative constant, for the sedimentation velocity. Is this the only possible combination of the variables?

• • **1.22** Give an expression for the sedimentation velocity that uses the mass density ρ (mass per unit volume) instead of the total mass of the sphere.

• • **1.23** When you maneuver your car around a curve, you seem to experience a force pushing you to the outside of the curve. This centrifugal force evidently depends on the speed of the car and on the radius of the curve. What other variable must also be involved in determining this force? Give a simple expression, correct to within a multiplicative constant, that relates this centrifugal force to the car's speed, the turning radius, and this other variable.

(Section 1.4–1.5)

1.24 Estimate the number of piano tuners in a big city of the United States, for example, Chicago. (Compare your estimate by looking in the yellow pages of the telephone directory.)

1.25 Estimate the annual budget of your college. (Compare with a recent annual report.)

• **1.26** At a given airspeed, the lift provided by an airplane's wing is roughly proportional to its surface area. Suppose a new airplane is to be designed whose fuselage is to be twice that of a Boeing 737, both in length and diameter. How should this larger plane compare (approximately) in wing area with a 737?

• **1.27** Estimate the number of popcorn kernels in a 1-qt container.

• **1.28** Estimate the number of liters of water in a full bathtub of standard dimensions.

• **1.29** Estimate the total number of persons in the age group 6 to 17 in the United States. Then estimate the total number of public school teachers in the United States.

(Section 1.6)

1.30 A right circular cylinder has a height of 10.04 cm and a diameter of 5.2 cm. What is the volume of this cylinder?

1.31 The radius of a circle is 0.04 m. What are the circumference and area of this circle?

• **1.32** A pharmacist mixes 2.670 grams (g) of A, 16.42 grams of B, and 0.025 grams of C. What is the mass of the total mixture and what are the percentages of A, B, and C contained therein?

• **1.33** The diameter of a sphere is 0.250 m. What are the surface area and volume?

• **1.34** A steel rod is 4.5062 m long at 20 °C. As a result of thermal expansion, its length increases by 0.0025% as the temperature is raised to 60 °C. What is its length at 60 °C?

• **1.35** What is the circumference, in meters, of a circle whose radius is 1 ft 2.5 inches?

• **1.36** What is the surface area and the volume of a sphere of radius 4.7 m?

• **1.37** Calculate the average density of the earth, using the data of Appendix B. (The density is defined as the mass per unit volume.)

• • **1.38** There are 6×10^{28} aluminum atoms in 1 $m^3$of aluminum. What is the volume occupied by one aluminum atom? What is the radius of a sphere of the same volume? Give your answers in scientific notation.

(Section 1.8)

1.39 A car travels 5 km east and then 12 km north. Find the displacement of the car.

1.40 Give the magnitude and direction of the following vectors, whose cartesian components are

(a) $A_x = 14$ m, $A_y = -9$ m.

(b) $B_x = -9$ m, $B_y = -12$ m.

(c) $C_x = -10$ m, $C_y = 26$ m.

The following problems should first be done graphically. Following the graphical solution, analytic solution is recommended.

1.41–1.50 The vectors **A**, **B** *and* **C** *are described as follows.* **A** *has x and y components of 4 and 3 units, respectively ($A_x = 4$, $A_y = 3$).* **B** *is of length 8 units and points in the negative x direction.* **C** *is of length 6 units and makes an angle of 30° with the x axis.*

• **1.41** Find the magnitude of **A** and the x and y components of **B** and **C**.

• **1.42** Find the vector **D** = **A** + **C**.

• **1.43** Find the vector **F** = **A** + **B** + **C**.

• **1.44** Find the vector **G** = **B** − **A**.

• **1.45** Find the vector **P** so that **P** + **B** + **C** = 0.

• **1.46** Find a vector of magnitude 8 units so that when this vector is added to **A** the resultant points along the positive y direction.

• **1.47** Find the vector **Q** = −(**B** − **C**).

• **1.48** Find the angle the vector **A** makes with the positive x direction.

• • **1.49** Sketch a new set of coordinate axes x'' and y'' so that the vector **A** has a positive x'' component and no y'' component. What is the angle between the x'' and the x axis? What are the components of the vector **B** in this new coordinate system?

• • **1.50** Sketch the vectors **A**, **B**, **C**, using the usual x and y axes. Then draw a new set of axes, x' and y', where x' and y' are mutually perpendicular, and where the angle between x' and x is 60°, with x' pointing into the first quadrant of the x-y coordinate system.

What are the x' and y' components of these three vectors? What are their lengths in this new coordinate system?

• **1.51** A car travels 7 km north. It then changes direction of travel and ultimately stops when it is 17 km southwest of its starting point. Find the displacement of the car on the second leg of the trip.

• **1.52** A mail truck drives 2 km north, then 3 km east, and then 1.8 km southwest. Find the displacement of the truck, using a graphical construction and using the method of components.

• **1.53** A man walks 50 m west, then 40 m southwest. In what direction and how far should he walk on the third leg so that he will be 25 m due west of his starting point?

• **1.54** Forces are measured in newtons (N) and added vectorially. Find the magnitude of the resultant of two forces of 30 N each that make an angle of (a) 37°, (b) 90°, (c) 150° with each other.

• **1.55** Two forces, $\mathbf{F}_1$ and $\mathbf{F}_2$, act on an object. The resultant of the two forces is $\mathbf{F}_R = 45$ N and is directed along the positive y axis. If $\mathbf{F}_1 = 35$ N and is directed along the positive x axis, what is the magnitude and direction of $\mathbf{F}_2$?

• • **1.56** Suppose that in Problem 1.55 the direction of $\mathbf{F}_2$ is known to be 60° from the positive x axis but the direction of $\mathbf{F}_1$ is not given. Find the magnitude of $\mathbf{F}_2$ and the direction of $\mathbf{F}_1$ so that their resultant is the specified force $\mathbf{F}_R$. If there is more than one possible answer, give as many as you can.

• • **1.57** A boat heads due north for 3 km, then 30° south of west for 4 km. In what direction and how far should it go so that it will then be at a point 8 km northwest of its starting point?

• **1.58** Vectors **A** and **B** are as shown in Figure 1.18. Find the vectors **C** = **A** + **B** and **D** = **B** − **A**.

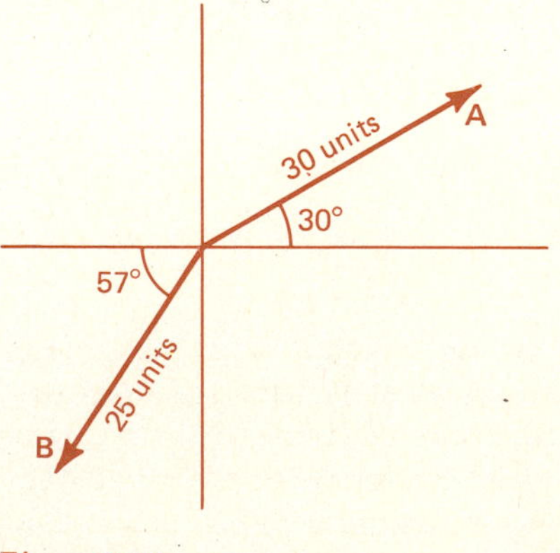

Figure 1.18

• • **1.59** A boat steers due north and travels for 40 km over the water. The currents carry the boat 20 km southeast during this time. How far is the boat from its point of departure?

•2•

Kinematics

To understand motion is to understand nature.

LEONARDO DA VINCI

2.1 Introduction

Mechanics, the study of the motion of bodies, is conveniently divided into two parts. One is purely descriptive; the other is concerned with causality, what causes and changes the course of objects in motion. *Kinematics,* the name for the descriptive part, is restricted to answering the question: Given certain initial conditions and the acceleration of an object at $t = 0$ and at all subsequent times, what are its position and velocity as functions of time? Kinematics does not inquire into the reasons why objects accelerate; it only describes their behavior.

When you throw a ball, not only does the ball follow a certain trajectory in space but it may also be spinning about an axis through its center while flying through the air. Since we do not now want to complicate the description of motion by including this rotation, we shall treat all objects as though they were "point objects," or particles of negligible dimensions. (We shall consider rotational motion in Chapter 8.)

2.2 Velocity

The location of a particle is given by its *position vector,* which is the vector from the origin of the coordinate system to the particle.

When a particle moves, its location changes with time. If at time $t = 0$ s it was located at point A, and two seconds later, when $t = t_f = 2$ s, the particle is at B, its displacement $\Delta\mathbf{s}$ during that time is the vector that points from A to B. Its *average velocity* is*

$$\langle \mathbf{v} \rangle = \frac{\mathbf{s}_f - \mathbf{s}_0}{t_f - t_0} = \frac{\Delta\mathbf{s}}{\Delta t} \qquad \textbf{(2.1)}$$

where $\mathbf{s}_f$ and $\mathbf{s}_0$ are the position vectors of the particle at the final and initial times, t_f and t_0, and $\Delta\mathbf{s}$ denotes the displacement during the time interval Δt.

DEFINITION: The average velocity is the displacement vector divided by the time interval during which this displacement occurred.

From the definition it follows that the dimension of velocity is $[L]/[T]$; the SI unit for velocity is meter per second (m/s). Since displacement is a vector and time a scalar, the average velocity is also a vector quantity; it has a certain magnitude and its direction is that of the displacement.

Note that though we know the average velocity and the initial and final positions of the particle, we can say nothing about the particle's path during that interval of 2 s. It might have traveled at constant velocity in a straight line from A to B; it might have gone from A to B in one second and come to rest at B; it might also have followed some tortuous route, as indicated in Figure 2.1.

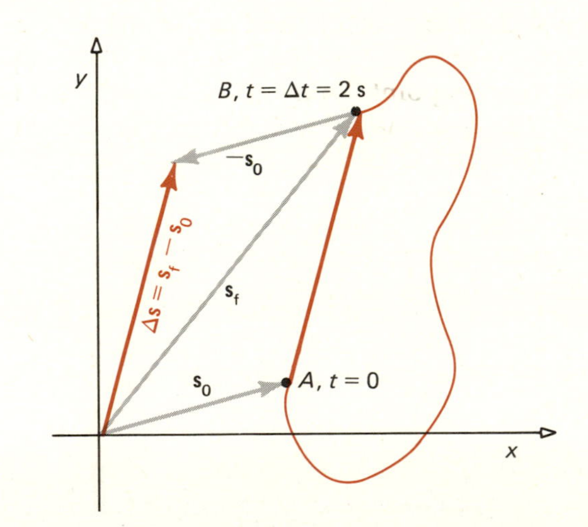

Figure 2.1 *A particle is at point A at t = 0 and is observed at point B at time Δt. The displacement during Δt is **Δs**. The average velocity is*

$$\langle \mathbf{v} \rangle = \frac{\Delta\mathbf{s}}{\Delta t}$$

Knowing the initial and final positions of a particle and its average velocity provides no information on its location at intermediate times.

We can try to remove this uncertainty by observing the position of the particle after only 0.5 s has elapsed. Perhaps we shall then find the particle at point C, as shown in Figure 2.2. Though we still cannot precisely specify the velocity at $t = 0$, we do feel that we now have a more reliable estimate of its velocity near $t = 0$ than if we wait 2 s before locating the particle. Note that as we reduced the time interval, the displacement also diminished.

We see now how we might, in principle, determine the *instantaneous*

* We use the symbol ⟨ ⟩ to indicate averages.

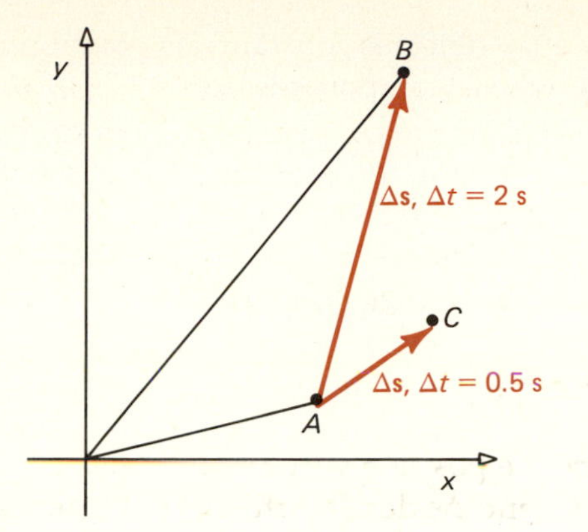

Figure 2.2 The same particle is observed to be at point C half a second after it was at point A.

velocity at $t = 0$. We reduce the time interval between successive position observations until it is infinitesimally small, and though $\Delta\mathbf{s}$ also approaches zero as we do this, the ratio $\Delta\mathbf{s}/\Delta t$ will remain finite. In formal mathematical symbolism, we express this process by

$$\mathbf{v}(0) = \lim_{\Delta t \to 0} \left(\frac{\Delta\mathbf{s}}{\Delta t}\right)_{t=0} \tag{2.2}$$

In words, this equation says:

The instantaneous velocity at the time $t = 0$ is obtained by forming the ratio $\Delta\mathbf{s}/\Delta t$ near the time $t = 0$ and allowing the time interval Δt to approach zero.

We can repeat this process at other times, for example, at $t = 1$ s, $t = 5$ s, and so on, and thus determine the instantaneous velocity at any moment.

The process just described is identical to the one by which one finds the slope of a curve. For the curve, one computes the slope of a chord between two neighboring points on the curve and then allows the second point to approach the first; that is, one reduces the length of the chord until it has diminished to a point, the point of tangency. This is illustrated in Figure 2.3.

Equation (2.2) is really a shorthand notation for *three equations*,

$$v_x = \lim_{\Delta t \to 0} \left(\frac{\Delta x}{\Delta t}\right), \qquad v_y = \lim_{\Delta t \to 0} \left(\frac{\Delta y}{\Delta t}\right), \qquad v_z = \lim_{\Delta t \to 0} \left(\frac{\Delta z}{\Delta t}\right) \tag{2.2a}$$

where Δx, Δy, Δz are the x, y, and z components of the displacement $\Delta\mathbf{s}$ during the time Δt.

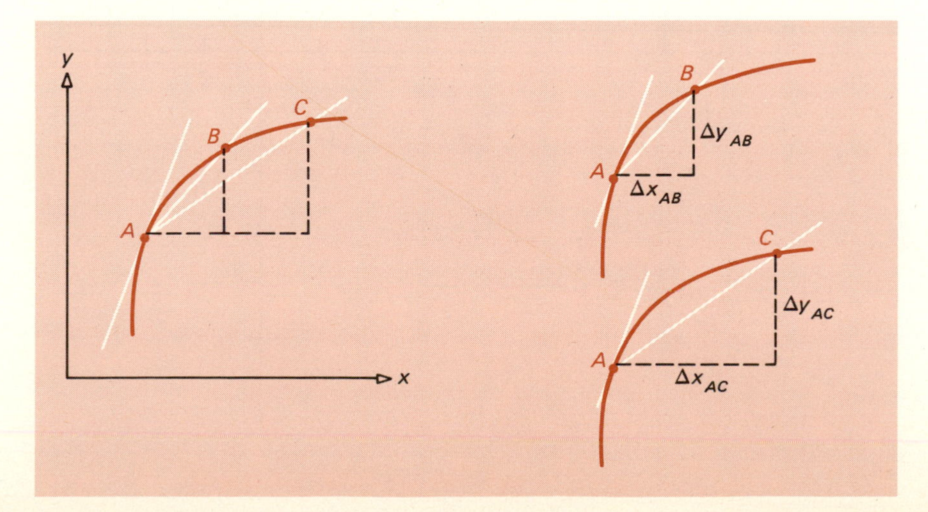

Figure 2.3 The average slope between points A and C is given by the ratio $\Delta y_{AC}/\Delta x_{AC}$. To obtain the slope at the point A, one allows point C to approach point A, reducing the length of the chord until it is infinitesimally small. The chord is then the tangent of the curve at point A.

If the velocity of an object is constant, its position is then given by the position vector **s**, whose components are

$$x = x_0 + v_x t, \qquad y = y_0 + v_y t, \qquad z = z_0 + v_z t \tag{2.3}$$

Equation (2.3) is correct only if the velocity is a constant.

● ***Example 2.1*** A train is traveling due east at a speed of 120 km/h. How far does this train travel in six seconds?

It is best to convert first to the units meters per second or kilometers per second, since the time interval is given in seconds.

$$120 \text{ km/h} = \left(\frac{120 \text{ km}}{1 \text{ h}}\right)\left(\frac{1 \text{ h}}{3600 \text{ s}}\right)\left(\frac{1000 \text{ m}}{1 \text{ km}}\right) = 33.3 \text{ m/s}$$

In 6 s, the train covers a distance of

$$s = vt = (33.3 \text{ m/s})(6 \text{ s}) = 200 \text{ m}$$ ●

● ***Example 2.2*** A car is 5 km east and 3 km south of a given point at $t = 0$. Thirty minutes later it is observed at the point 20 km west and 15 km south of the reference point. What was its average velocity during those 30 min? If it continues to travel with that velocity, where will it be at $t = 40$ min?

The displacement vector between $t = 0$ and $t = 30$ min $= 0.5$ h is a vector whose component along the west-to-east direction is -25 km and along the south-to-north direction is -12 km. The displacement vector $\Delta\mathbf{s}$ is therefore directed at an angle

$$\theta = \tan^{-1}\frac{12 \text{ km}}{25 \text{ km}} = \tan^{-1} 0.48 = 25.6°$$

south of west and has a magnitude of

$$\Delta s = \frac{25 \text{ km}}{\cos 25.6°} = 27.7 \text{ km}$$

Hence the average velocity is

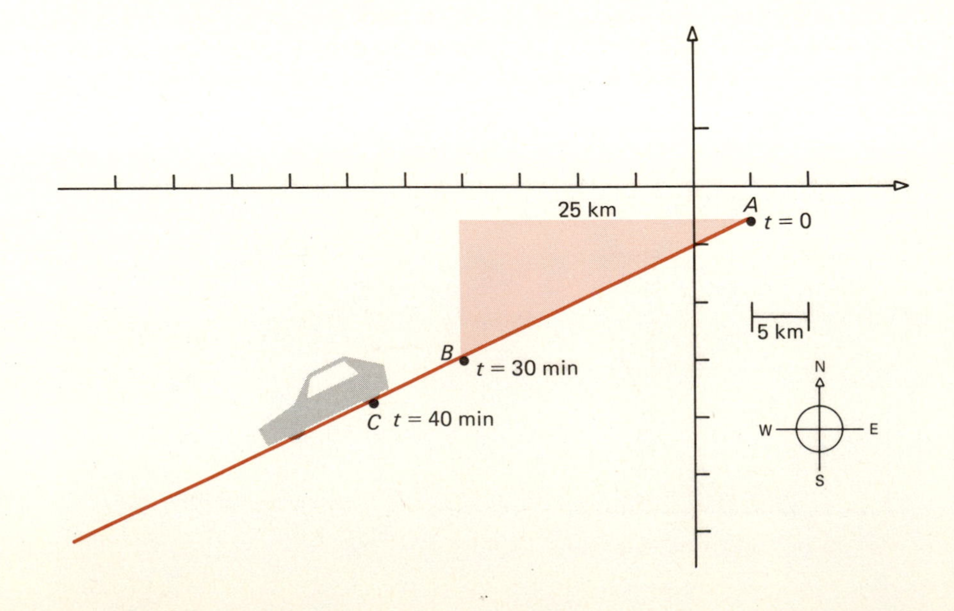

Figure 2.4 *A car traveling at constant velocity is at A at time $t = 0$, at B at time $t = 30$ min. Its velocity is the displacement* **AB** *divided by the elapsed time, 0.5 h. At time $t = 40$ min, it will be at C.*

$$\langle \mathbf{v} \rangle = \frac{27.7 \text{ km}}{0.5 \text{ h}} = 55.4 \text{ km/h, } 25.6° \text{ south of west}$$

If the car continues to travel at this velocity, it will cover a distance of $(55.4 \text{ km/h})(10 \text{ min})(\frac{1}{60} \text{ h/min}) = 9.2$ km during the next 10 min. Therefore, it will then be at a point 36.9 km, 25.6° south of west from its starting point at $t = 0$, or at the coordinates

$$(36.9 \text{ km}) \cos 25.6° - 5 \text{ km} = 28.3 \text{ km west}$$

and

$$(36.9 \text{ km}) \sin 25.6° + 3 \text{ km} = 18.9 \text{ km south}$$

of the reference point. ●

In ordinary conversation one frequently uses the words speed and velocity almost interchangeably. For example, we often say that a bullet is traveling at some high velocity, without stating its direction of motion. In physics we use the word *speed* to designate the *magnitude of the velocity; velocity* always characterizes the *vector* **v**, whose magnitude *and* direction must be given. Similarly, *distance* describes the *length of the path* traversed by an object; *displacement* specifies the *vector difference* between the positions of the object at the final and initial times. Students should develop the habit of using these and other terms with precision.

Thus, although a satellite orbiting the earth may be traveling at constant speed of about 11.1×10^3 km/h, its *velocity* is not constant since the direction of its motion is continually changing. It will complete one orbit in 24 h, so that its total *displacement* during that time is zero even though it has traveled a *distance* of about 266×10^3 km. Consequently, its *average velocity* during that time interval is also zero, even though its speed has remained 11.1×10^3 km/h throughout.

2.3 Addition of Velocities; Reference Frames

Suppose you paddle a canoe at 6 km/h on a heading due north across a stream that flows due east at 3 km/h. If the stream is 1 km wide, you will take just $(1 \text{ km})/(6 \text{ km/h}) = \frac{1}{6} \text{ h} = 10$ min to reach the opposite shore. During that time, the water will have carried the canoe downstream a distance of $(3 \text{ km/h})(\frac{1}{6} \text{ h}) = 0.5$ km. Thus the displacement of the canoe

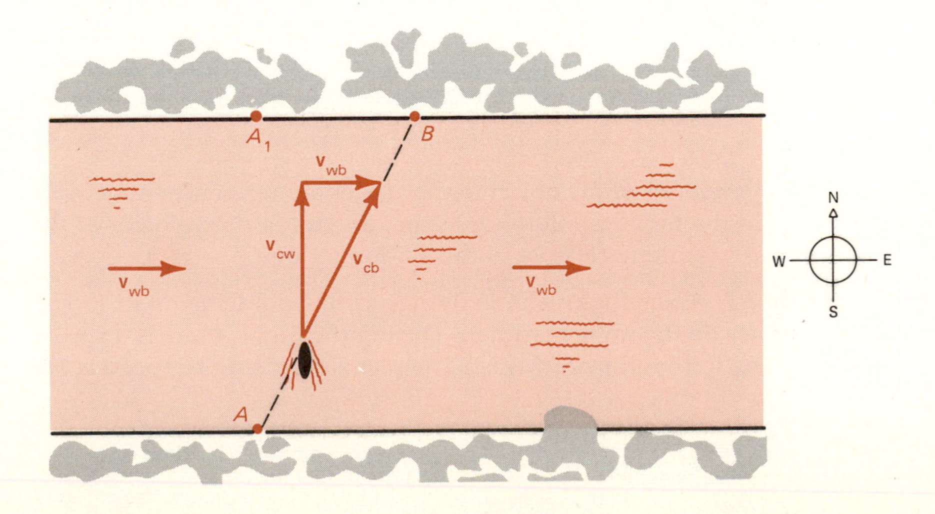

Figure 2.5 A canoe is heading due north with a velocity $\mathbf{v}_{cw}$ relative to the water across a river, which flows with a velocity $\mathbf{v}_{wb}$ relative to the bank. The velocity of the canoe relative to the bank is $\mathbf{v}_{cb}$. The canoe starts at point A and heads toward A_1; it will reach the opposite shore at point B. To an observer floating on a raft, the canoe appears headed due north.

relative to the initial starting point during that 10 min is in a direction given by

$$\theta = \tan^{-1} 0.5 = 26.6^\circ \text{ east of north}$$

and has the magnitude

$$\Delta s = \frac{1 \text{ km}}{\cos 26.6^\circ} = 1.12 \text{ km}$$

Since the actual displacement of the canoe is given by these values, the average velocity of the canoe during those 10 min is

$$\langle \mathbf{v} \rangle = \frac{1.12 \text{ km}}{\frac{1}{6} \text{ h}} = 6.72 \text{ km/h, } 26.6^\circ \text{ east of north}$$

We see that the velocity of the canoe with respect to the bank is the sum of two contributions, the velocity relative to the water on which it floats and the velocity of the water relative to the bank.

Although the canoe is traveling in a direction 26.6° east of north, the person paddling it is heading due north, perhaps using a compass on his lap. Moreover, to a second observer floating on a raft in the river, the canoe appears to travel due north. It is only relative to the bank that the canoe has the velocity east of north, and the raft a velocity due east.

The coordinate system used to locate an object's position is called its *frame of reference*. The canoe traversing a moving stream is an example of an object moving with constant velocity in one reference frame, the stream, which is itself moving at a different constant velocity relative to another reference frame, the surface of the earth. Reference frames that move relative to each other at constant velocity are called *inertial reference frames*, and it is these with which we shall be primarily concerned. Other examples of two different inertial reference frames are a train traveling at constant velocity (frame 1) relative to the ground (frame 2), and an elevator rising at constant velocity (frame 1) relative to the building in which it is located (frame 2).

The velocity of an object relative to a moving reference frame, the velocity of the moving reference frame, and the velocity of the object relative to the stationary reference frame are simply related.* Suppose in the example the canoe were at rest relative to the water. Its velocity relative to the bank would then be the velocity of the stream, which we denote by $\mathbf{v}_R$, the velocity of the reference frame. If the canoe now has a velocity $\mathbf{v}_r$ relative to the moving reference frame, this velocity must be added vectorially to $\mathbf{v}_R$ to obtain the velocity relative to the stationary reference frame. Hence, the relation between the three velocities is

$$\mathbf{v} = \mathbf{v}_r + \mathbf{v}_R \tag{2.4}$$

where $\mathbf{v}$ is the velocity of the object in the stationary reference frame, $\mathbf{v}_r$ is its velocity in the moving reference frame, and $\mathbf{v}_R$ is the velocity of the moving reference frame relative to the stationary one.

Example 2.3 A small airplane is flying north from Cincinnati to Detroit, two cities on the same meridian. During the flight, there is a steady wind of 80 km/h from the northwest. The plane's cruising airspeed is 175

* There is only one caveat to what follows; all speeds must be small compared with the speed of light, $c = 3.0 \times 10^8$ m/s. If relative speeds approach c, relativistic corrections must be applied. These will be detailed in Chapter 28.

km/h. What should the heading of the plane be? With that heading, what is its ground speed?

We take the ground as our stationary reference frame and the air in which the plane is flying as the moving reference frame. Since the direction of the plane's velocity in the stationary frame should be due north, we have the situation shown in Figure 2.6.

The vector $\mathbf{v}_R$ has magnitude 80 km/h and points 45° south of east. To this vector we must add a second vector $\mathbf{v}_r$, whose magnitude is 175 km/h, so that the sum is a vector $\mathbf{v}$ that points due north. This is best visualized using the geometrical construction shown in Figure 2.6(*a*). To solve the problem analytically, we proceed as follows.

Since $\mathbf{v}$ has no east-west, or x, component, the sum of the x components of $\mathbf{v}_R$ and $\mathbf{v}_r$ must vanish. From Figure 2.6, it follows that since the x component of $\mathbf{v}_R$ is

$$v_{Rx} = (80 \text{ km/h}) \cos 45° = 56.6 \text{ km/h}$$

the x component of $\mathbf{v}_r$ must be -56.6 km/h. The magnitude of $\mathbf{v}_r$ is 175 km/h. Consequently, the heading of the plane should be at an angle

$$\theta = \sin^{-1} \frac{56.6}{175} = \sin^{-1} 0.323 = 18.8° \text{ west of north}$$

The ground speed v is now given by the sum of the north, i.e. y, components of $\mathbf{v}_R$ and $\mathbf{v}_r$:

$$\begin{aligned} v &= (175 \text{ km/h}) \cos 18.8° - (80 \text{ km/h}) \cos 45° \\ &= 165.6 \text{ km/h} - 56.6 \text{ km/h} \\ &= 109 \text{ km/h} \end{aligned}$$

We have examined two instances in which an object moves with a certain velocity in one frame of reference while that frame of reference is itself in motion relative to some other reference frame. The chair on

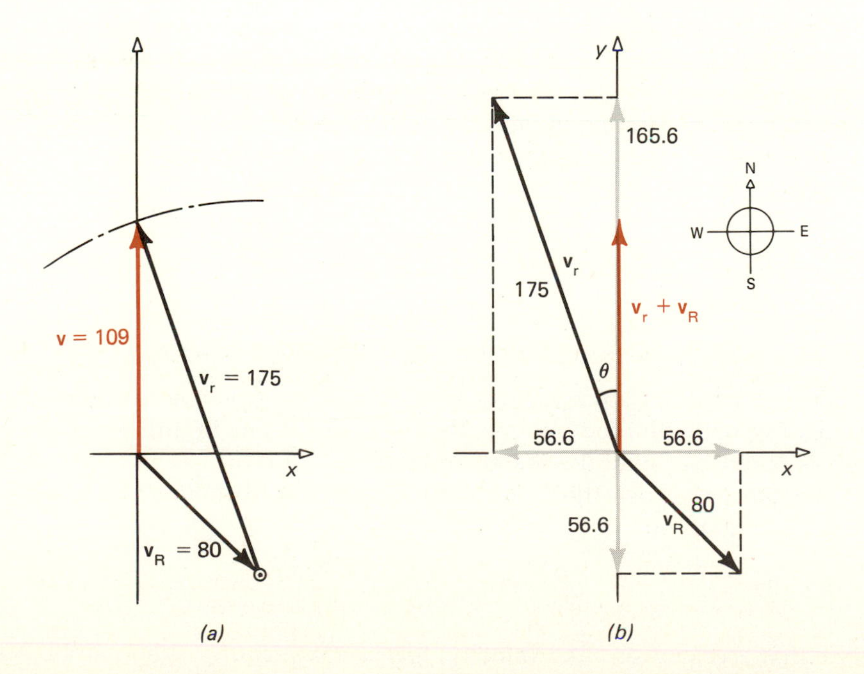

Figure 2.6 *An airplane with a cruising airspeed of $\mathbf{v}_r$ in a wind whose velocity is $\mathbf{v}_R$ must maintain a bearing given by the angle θ to fly due north relative to the ground. The velocity relative to the ground is the vector sum of $\mathbf{v}_r$ and $\mathbf{v}_R$. (a) The correct heading is determined by a geometric construction. (b) The vector $\mathbf{v}_R$ is decomposed into east-west and north-south components. Since $\mathbf{v}_r + \mathbf{v}_R$ must have zero east-west component, the east-west component of $\mathbf{v}_r$ is the negative of the corresponding component of $\mathbf{v}_R$.*

which you sit may appear to be at rest, and so it is—in relation to the room and the surface of the earth. But the earth is rotating on its axis and is also revolving about the sun, which in turn is moving relative to the center of our galaxy. So when we say that an object is at rest or moving with a specified velocity, we should really identify the reference frame which we use to characterize this state of motion. Generally, that reference frame is the surface of the earth, and we then simply omit mention of the reference frame.

One reason why we often do not need to specify the reference frame is that the laws of physics are unaltered as we change from one reference frame to another, provided that the two are inertial frames. Physicists say that the laws of nature are *invariant* with respect to a change of inertial reference frames. This is not a particularly difficult or strange concept. Indeed we accept it as a common fact though we may not express the idea in quite those terms. Once an airplane is airborne and traveling at constant velocity, we are unaware of its motion unless we glance out the window. The behavior of objects within the plane is no different from what it is on the ground; spilled coffee lands in your lap just as it would in your living room at home.

Invariance of physical laws with respect to inertial frames conforms with common experience; it is also a fundamental and powerful principle. First, it tells us that no matter how hard we try, we shall never succeed in identifying the state of "absolute rest"—the absence of motion. Rest, or zero velocity, has meaning only in relation to a particular reference frame. Second, the principle also tells us that our description of nature by physical laws must be such that these laws will be equally valid in any inertial reference frame. Invariance with respect to inertial frames was the firm foundation upon which Einstein constructed his theory of special relativity.

2.4 Accelerated Motion

In most situations of interest, objects do not maintain uniform motion but suffer various changes in velocity; that is, they are accelerated.

The definition of acceleration is analogous to that of velocity. As velocity characterizes the change of position with time, so acceleration characterizes the change of velocity with time.

DEFINITION: The *average acceleration* during a time interval Δt is given by

$$\langle \mathbf{a} \rangle = \frac{\mathbf{v}_f - \mathbf{v}_0}{\Delta t} = \frac{\Delta \mathbf{v}}{\Delta t} \tag{2.5}$$

where $\mathbf{v}_f$ and $\mathbf{v}_0$ are the final and initial velocity vectors and Δt is the time interval during which the velocity changes from $\mathbf{v}_0$ to $\mathbf{v}_f$.

From Equation (2.5) it follows that the dimensions of acceleration are those of velocity divided by time, that is, $[L]/[T]^2$. The SI unit of acceleration is meter per second squared (m/s^2).

We can again use the same limiting process to define an instantaneous acceleration:

$$\mathbf{a} = \lim_{\Delta t \to 0} \frac{\Delta \mathbf{v}}{\Delta t} \tag{2.6}$$

Note that acceleration, like velocity, is a *vector quantity*. When an object experiences an acceleration, its speed or its direction of motion or both

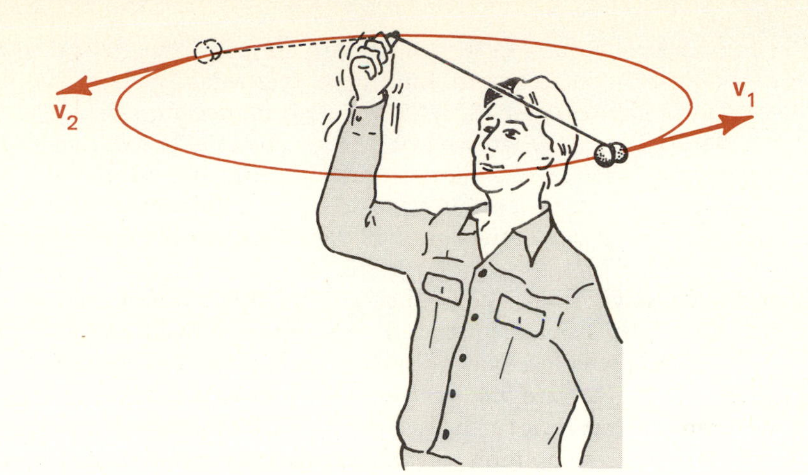

Figure 2.7 A stone moving at constant speed at the end of a string does not move at constant velocity. The direction of its velocity is continually changing.

may change. A stone rotating in a circular path at the end of a string may be moving at constant speed; its *velocity* is, however, continually changing direction. The stone is therefore accelerating.

Equation (2.6) is also a shorthand notation for the *three equations*

$$a_x = \lim_{\Delta t \to 0} \left(\frac{\Delta v_x}{\Delta t} \right), \qquad a_y = \lim_{\Delta t \to 0} \left(\frac{\Delta v_y}{\Delta t} \right), \qquad a_z = \lim_{\Delta t \to 0} \left(\frac{\Delta v_z}{\Delta t} \right) \qquad \textbf{(2.6a)}$$

2.5 Motion in One Dimension

Natural phenomena take place in three dimensions. We shall, however, focus first on the simplest situation, linear motion, before tackling the more complicated, two- and three-dimensional problems.

Before we consider analytic solutions, let us first make certain we understand the meanings of *velocity* and *acceleration* by studying a graphical presentation of the position of a particle as time progresses. Such a curve for the x coordinate as a function of time is shown in Figure 2.8(*a*). What can we say about the motion of this particle, its velocity and acceleration, by studying this curve?

At $t = 0$, the slope of the curve is zero; therefore, $v(0) = 0$; the particle starts from rest. Between $t = 0$ and $t = t_1$, the slope is gradually increasing. During this interval, the velocity is increasing from zero to some positive value. Therefore the acceleration is positive between $t = 0$ and $t = t_1$. Between $t = t_1$ and $t = t_2$, the slope of the curve is constant; the velocity is constant and the acceleration therefore zero. Between $t = t_2$ and $t = t_3$, the acceleration is negative and the velocity diminishes to zero. Between t_3 and t_4, both velocity and acceleration are zero; the particle is again at rest. Between t_4 and t_5, the acceleration is negative and the particle acquires a negative velocity; this negative velocity has a constant magnitude between t_5 and t_7 and is substantially larger than the positive velocity between t_1 and t_2. At the instant t_6, the particle is at its starting position; the average velocity during the interval $\Delta t = t_6 - t_0$ is zero. Between t_7 and t_9, the acceleration is again positive and the velocity changes from a negative to a positive value; it is momentarily zero at $t = t_8$. Thereafter, for $t > t_9$, the acceleration is zero and the particle continues to move with a constant positive velocity in the x direction.

Curves of velocity and acceleration corresponding to Figure 2.8(*a*) are shown in Figures 2.8(*b*) and 2.8(*c*). (We have drawn these curves assum-

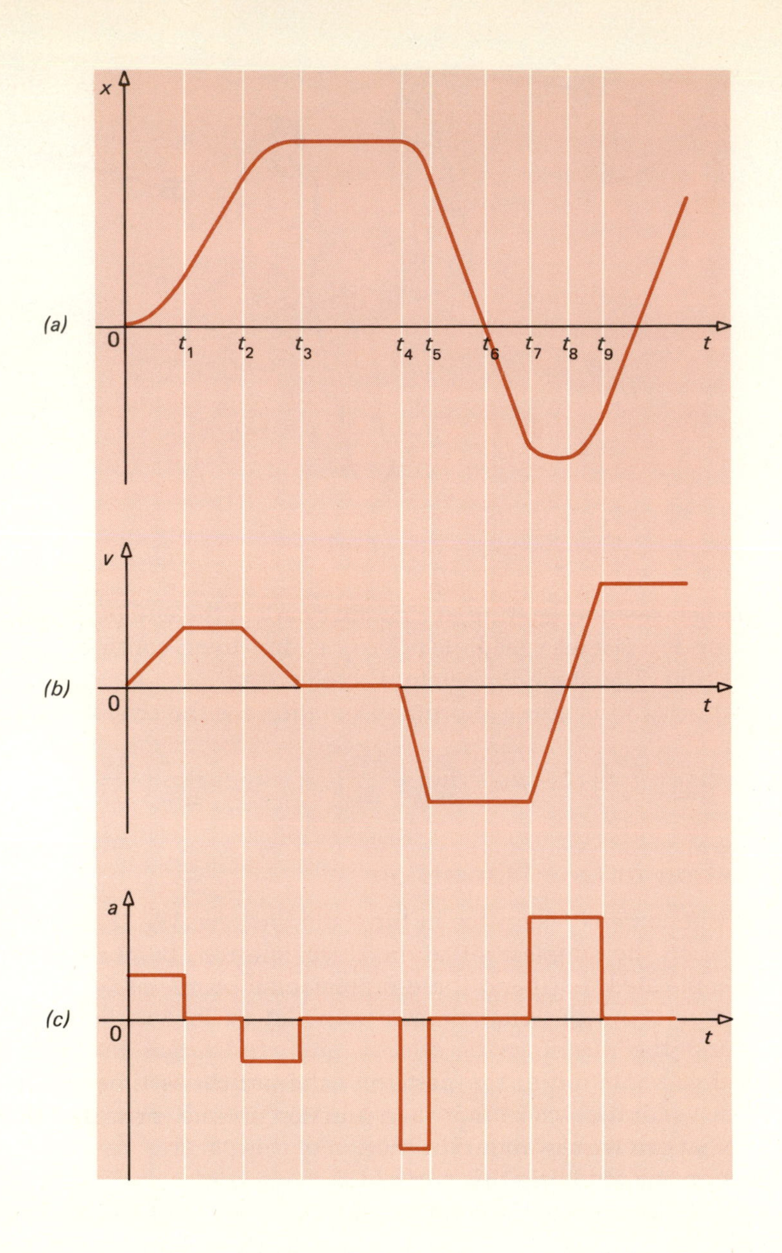

Figure 2.8 (*a*) *Curve showing displacement of an object moving along a straight line as a function of time. The corresponding curves of velocity and acceleration are* (*b*) *and* (*c*).

ing that the acceleration during the given time intervals is constant.) You should study these curves carefully until you really understand the information conveyed. Note especially the following:

1. The fact that an object is at the position designated by $x = 0$ does not imply that either its velocity or its acceleration is zero.

2. A particle may have a positive velocity and at the same time a positive, zero, or negative acceleration.

3. At some instant a particle may have zero velocity and yet have either positive, zero, or negative acceleration.

You should locate the points or regions of Figure 2.8 where these various conditions prevail.

2.5 (a) Uniform Acceleration

The aim of kinematics is to generate relations that specify the position and motion of objects under given initial conditions and acceleration. Although the general problem can be complex, it becomes relatively simple when the acceleration is independent of time. This is a common situation, and we shall here confine our attention to constant acceleration. Since we are still considering only motion in one dimension, we can dispense with vector notation.

From the definition of average velocity, Equation (2.1), we have

$$x(t) = x_0 + \langle v \rangle t \tag{2.7}$$

where $\langle v \rangle$ is the average velocity in the x direction during the interval between time zero and time t.

If the acceleration is zero, the velocity is constant; then, and *only* then, is $\langle v \rangle = v$. If $a \neq 0$, the velocity v changes with time. In the special case we are considering, uniform acceleration, $a = \langle a \rangle$. It then follows from Equation (2.5) that

$$v(t) = v_0 + \langle a \rangle t = v_0 + at \tag{2.8}$$

The velocity changes linearly with time, as in Figure 2.8(b), increasing or decreasing according to the sign of a. When the velocity is a linear function of time, the average velocity is just the algebraic mean of the initial and final velocities. That is,

$$\langle v \rangle = \frac{v(t) + v_0}{2} \tag{2.9}$$

We now substitute (2.8) into (2.9),

$$\langle v \rangle = \frac{v_0 + at + v_0}{2} = v_0 + \frac{at}{2} \tag{2.10}$$

and using this result in Equation (2.7), obtain

$$x(t) = x_0 + \left(v_0 + \frac{at}{2}\right) t$$

$$x(t) = x_0 + v_0 t + \tfrac{1}{2}at^2 \tag{2.11}$$

Equation (2.11) gives the position of the particle whose initial position is x_0 and initial velocity is v_0 at all later times t, provided that the acceleration is independent of time.

Another relation that is also useful is derived by rewriting (2.8),

$$t = \frac{v(t) - v_0}{a}$$

and substituting this expression for t into (2.11). The result

$$x(t) = x_0 + \frac{1}{2a}[v^2(t) - v_0^2]$$

or

$$2as = v^2(t) - v_0^2 \tag{2.12}$$

relates the initial and final velocities to the acceleration and the displacement $s = x(t) - x_0$.

The several useful relations that are valid for linear, uniformly accelerated motion are summarized below.

$$x(t) = x_0 + \langle v \rangle t; \qquad s = \langle v \rangle t \tag{2.13a}$$

$$x(t) = x_0 + v_0 t + \tfrac{1}{2}at^2; \qquad s = v_0 t + \tfrac{1}{2}at^2 \tag{2.13b}$$

$$\langle v \rangle = \tfrac{1}{2}[v(t) + v_0] \tag{2.13c}$$

$$v(t) = v_0 + at \tag{2.13d}$$

$$v^2(t) = v_0^2 + 2a[x(t) - x_0]; \qquad 2as = v^2(t) - v_0^2 \tag{2.13e}$$

Here the symbol $s = x(t) - x_0$ denotes the displacement of the particle during the time interval t.

One of the most common and practically important examples of linear, uniformly accelerated motion is the vertical movement of an object in the gravitational field of the earth. The effect of gravitational attraction is to impart to all unsupported bodies an equal downward acceleration of 9.8 m/s^2 (with the retarding effect of air friction neglected). It is conventional to assign the special symbol g to this acceleration.

It is sometimes stated that Galileo, dropping two stones of dissimilar mass from the leaning tower of Pisa, was the first to demonstrate the fallacy of the Aristotelian dogma that

> the downward movement of a mass of gold or lead, or of any other body endowed with weight, is quicker in proportion to its size.

But already in the sixth century John Philoponus had written:

> Here is something absolutely false, and something we can better test by observed fact than by any demonstration through logic. If you take two masses differing in weight and release them from the same elevation, you will see that the ratio of the times in their movements does not follow the ratio of the weights, but the difference in time is extremely small; so that if the weights do not greatly differ, but one, say, is double the other, the difference in times will be either none at all or imperceptible.

What is so remarkable about Galileo is that he progressed well beyond the qualitative and semiquantitative observations of his predecessors and was able to describe the motion of bodies in considerable mathematical detail. In the *Dialogue on the Great World Systems,* the book that brought him before the Inquisition, we read:

> Some superficial observations have been made, as, for instance, that the free motion of a heavy body is continuously accelerated; but just to what extent this acceleration occurs

Figure 2.9 Galileo Galilei (1564–1642).

> has not yet been announced; so far as I know, no one has yet pointed out that *the distance[s] traversed, during equal intervals of time, by a body falling from rest, stand to one another in the same ratio as the odd numbers beginning with unity.* [Our italics.]

Moreover, he continues,

> It has been observed that missiles and projectiles describe a curved path of some sort; however, no one has pointed out that this path is a parabola.

And with the prophetic vision of a Greek oracle, he anticipates Newton's genius:

> This and other facts, not few in number or less worth knowing, I have succeeded in proving; and what I consider more important, there have been opened up to this vast and most excellent science, of which my work is merely the beginning, ways and means by which other minds more acute than mine will explore its remotest corners.

In the following pages, we shall consider the behavior of freely falling bodies and shall describe the trajectories of "missiles and projectiles." We begin by showing that Galileo's statement quoted above is consistent with the equations for uniformly accelerated motion that we have just derived.

That statement applies to bodies falling *from rest.* Hence $v_0 = 0$ and Equation (2.13b) becomes

$$s = \tfrac{1}{2}gt^2$$

where we have substituted the acceleration of gravity g for a in that equation. At the end of one second, $s = g/2$; two seconds after release, $s = (g/2) \times 4$; at the end of three seconds, $s = (g/2) \times 9$; and so on. During the first second of fall, the distance traversed is $g/2$; during the next second, the *additional* distance the body drops is $[(g/2) \times 4] - (g/2) = 3 \times (g/2)$; during the third second, the distance fallen is $(g/2) \times 9 - (g/2) \times 4 = 5 \times (g/2)$. We see that indeed, "the distance[s] traversed, during equal intervals of time, stand to one another in the ratio of the odd numbers beginning with unity."

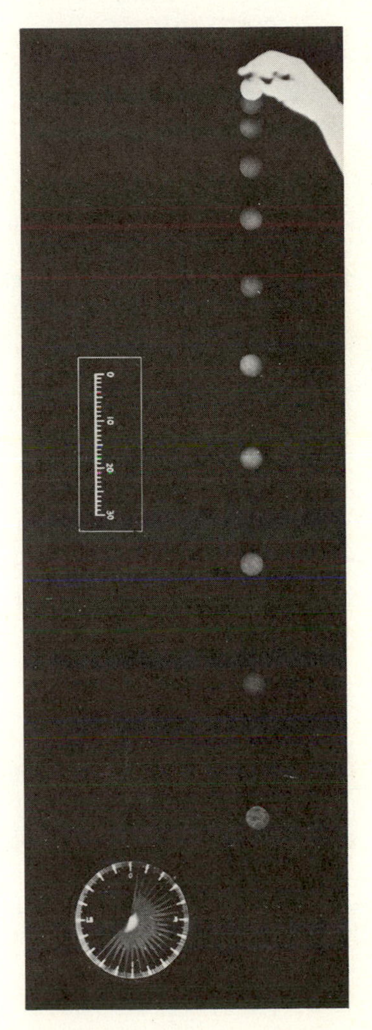

Figure 2.10 *". . . the distance[s] traversed, during equal intervals of time, by a body falling from rest, stand to one another in the same ratio as the odd numbers beginning with unity." Stroboscopic multiple-exposure photograph of a golf ball falling from rest.*

● ***Example 2.4*** An automobile manufacturer claims that his car can accelerate from rest to 60 km/h in 8 s. If the acceleration is constant, what is its value and what is the distance traveled during those 8 s?

From (2.13d), we obtain

$$a = \frac{v(t) - v_0}{t} = \frac{60 \text{ km/h}}{8 \text{ s}} = 7.5 \text{ km/(h·s)}$$

While this is a correct statement, it is confusing to have a result in mixed units, in this case hours and seconds. We can express it in m/s^2.

$$a = \frac{(7.5 \text{ km/[h·s]})(1000 \text{ m/km})}{3600 \text{ s/h}} = 2.08 \text{ m/s}^2$$

or just a bit over one-fifth of g.

Since the acceleration is constant, the distance traveled is given by Equation (2.13b):

$$s = v_0t + \tfrac{1}{2}at^2 = \tfrac{1}{2}(2.08 \text{ m/s}^2)(8 \text{ s})^2 = 66.6 \text{ m}$$ ●

Let us now do a slightly more difficult problem.

● ***Example 2.5*** A stone is thrown vertically from the roof of a building 50 m above the ground. The stone strikes the ground 5 s after it is thrown. With what speed was it thrown, what was the maximum height it attained, and with what speed did it strike the ground?

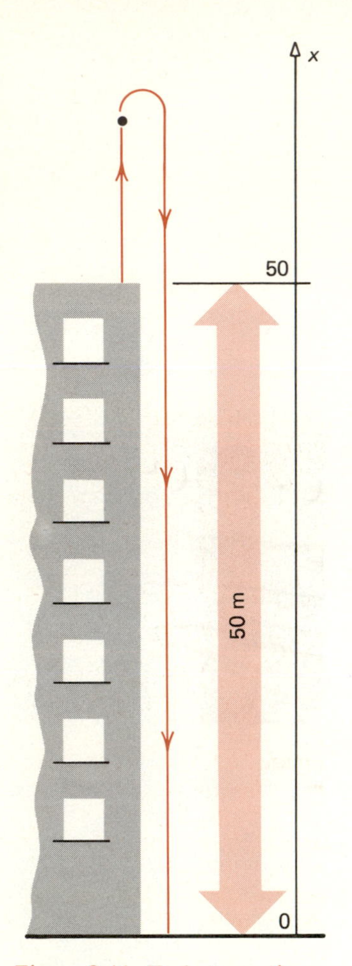

Figure 2.11 Trajectory of a stone thrown straight up from the edge of a building.

If the stone were thrown upward, its path would be as shown in Figure 2.11. However, we do not assume this direction of the initial velocity v_0; its magnitude and direction will be given by the solution. Whereas in Example 2.4 the definition of the positive direction was so natural that we made no mention of it, in this case we must be careful to specify our origin and select one direction as positive; we shall then assign positive or negative values to displacements, velocities, and accelerations accordingly. *This is extremely important!* Failure to establish a convention for the appropriate coordinate system and to *stay with that convention throughout the solution of a problem* is one of the most common errors committed by students.

The choice of the positive direction and of the coordinate origin is completely arbitrary, and the final result does not depend on this whim. We mention two possibilities and shall solve the problem using one of them. You are urged to repeat the solution with another choice.

One possible origin of coordinates is the top of the roof, a possible choice of the positive direction is downward. Another is to locate the origin at the ground level and use upward as the positive direction. There are still other options, equally good. We shall now use the second. Accordingly, $x_0 = 50$ m, $a = g = -9.8\ \text{m/s}^2$. The acceleration of gravity is *negative* because we have chosen *up* as our *positive direction.* We know that at the end of 5 seconds, $x = 0$ m; i.e., $x(5) = 0$ m.

We can now solve for the initial velocity by substituting into Equation (2.13b):

$$x(5) = (50\ \text{m}) + (5\ \text{s})v_0 + \tfrac{1}{2}(-9.8\ \text{m/s}^2)(5\ \text{s})^2 = 0\ \text{m}$$

which yields $v_0 = 14.5$ m/s. The initial velocity was 14.5 m/s and positive, that is, directed upward.

To find the maximum height the stone reached, we follow the trajectory in our mind. The stone starts upward with an initial speed of 14.5 m/s, but since the acceleration is negative, the velocity decreases as soon as the stone leaves the hand and at the top of the trajectory is momentarily zero. Thereafter, the velocity is negative and increases in magnitude. The top of the path is therefore that point at which $v = 0$.

We can find the instant when $v(t) = 0$ from Equation (2.13d):

$$v(t) = v_0 + at = 0 = (14.5\ \text{m/s}) + (-9.8\ \text{m/s}^2)t$$

$$t = \frac{14.5\ \text{m/s}}{9.8\ \text{m/s}^2} = 1.48\ \text{s}$$

Note that we had to use the proper sign for a.

The coordinate of the stone at that instant is given by (2.13b) as

$$\begin{aligned} x(1.48) &= (50\ \text{m}) + (14.5\ \text{m/s})(1.48\ \text{s}) + \tfrac{1}{2}(-9.8\ \text{m/s}^2)(1.48\ \text{s})^2 \\ &= 60.7\ \text{m} \end{aligned}$$

Since the stone starts at a height of 50 m above ground, the maximum height reached is 10.7 m above the roof.

Note that we could also have obtained this last result using Equation (2.13a). The average velocity between $t = 0$ and $t = 1.48$ s is

$$\langle v \rangle = \frac{v(t) + v_0}{2} = \frac{0\ \text{m/s} + 14.5\ \text{m/s}}{2} = 7.25\ \text{m/s}$$

Hence,

$$s = \langle v \rangle t = (7.25\ \text{m/s})(1.48\ \text{s}) = 10.7\ \text{m}$$

To find the speed of the stone just before it strikes the ground, we can use (2.13e). The distance s could be either 50 m or 60.7 m. If we use the first, we must take $v_0 = 14.5$ m/s; and if we use the second, we can take $v_0 = 0$ m/s. In either case, we find

$$v(5) = -34.5\ \text{m/s}$$

which corresponds to a downward velocity of about 77 mph. A stone thrown from a tall building can be a lethal projectile. ●

2.6 Uniformly Accelerated Motion in Two Dimensions

A ball thrown horizontally, a bullet fired at a distant target, a bomb dropped from a plane, all follow a path described by Galileo as a parabola. Since each of these objects experiences a downward acceleration, only the vertical component of velocity changes as time progresses. *The horizontal velocity component remains constant.* If it was zero initially, as in the preceding examples, it remains zero. If it has some finite value, that is also the value at all subsequent times during the trajectory. These simple facts suffice to characterize the projectile's motion completely.

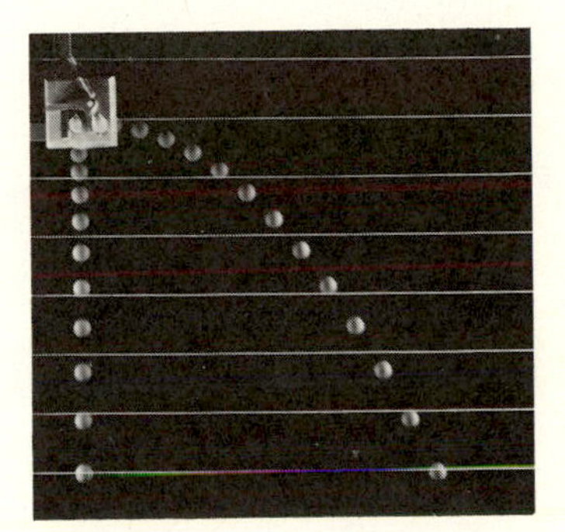

Figure 2.12 *Stroboscopic multiple-exposure photograph of two balls. One ball is dropped from rest; the other is projected horizontally. The vertical component of the motion is the same for both balls.*

To determine the trajectory of a particle that is subject to a uniform acceleration, one proceeds as follows. The motion is decomposed into orthogonal components, with one velocity component in the direction of the acceleration. *This is the only component that changes with time.* The velocity components in the plane perpendicular to the acceleration remain constant.

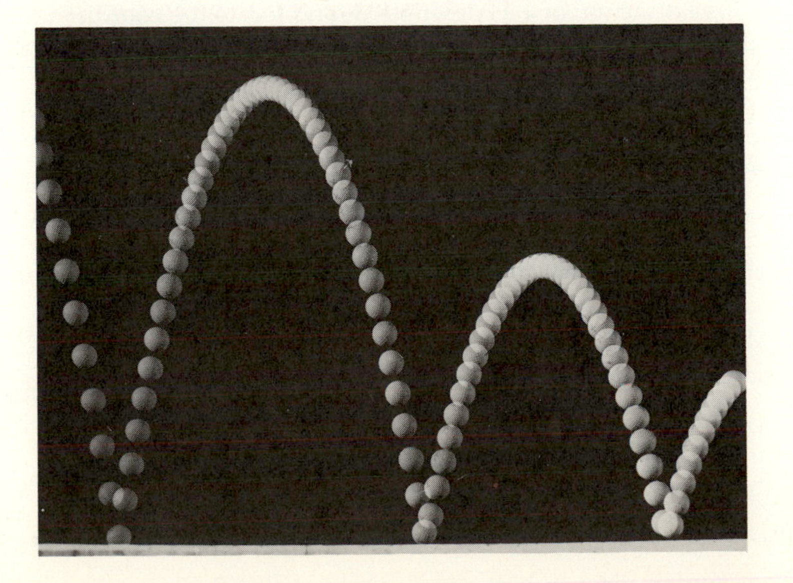

Figure 2.13 *Stroboscopic multiple-exposure photograph of a bouncing ball. The time intervals between individual frames are identical. Note that the horizontal displacement between adjacent positions is the same although the vector displacement changes as time progresses.*

As these arguments suggest, we proceed by decomposing the initial velocity into its horizontal and vertical components. The horizontal component remains constant. The vertical component does change, but its value can be found with one of Equations (2.13). To find the velocity at some time t, we calculate the horizontal and vertical components of $\mathbf{v}$ separately and then combine them according to the now familiar rules, Equations (1.11)–(1.13).

The procedure is best explained by a few examples.

● ***Example 2.6*** Show that the trajectory of a projectile is a parabola. The equation of a parabola is

$$y = ax + bx^2 \tag{2.14}$$

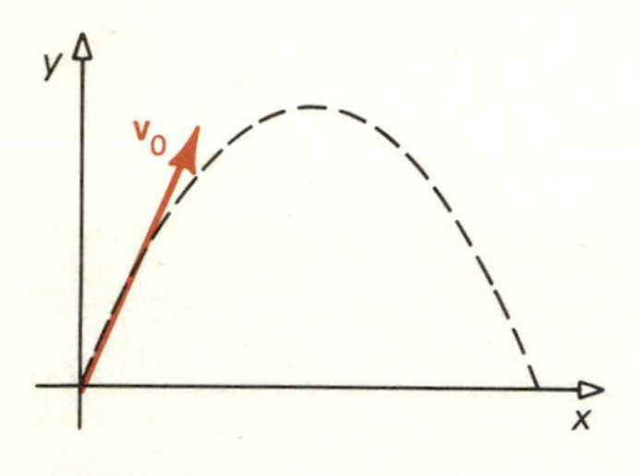

Figure 2.14 Trajectory of a projectile whose velocity at the origin is $\mathbf{v}_0$.

We decompose the initial velocity of the projectile into its horizontal and vertical components. The horizontal component remains constant; the vertical component varies according to Equation (2.13). The x and y coordinates of the projectile are

$$x = v_{0x}t, \qquad y = v_{0y}t - \tfrac{1}{2}gt^2$$

where we have assumed that the projectile was fired from the origin, and have taken the positive y direction to be up. The projectile's initial velocity components are represented by v_{0x} and v_{0y}. If we now substitute $t = x/v_{0x}$ into the expression for y, we obtain

$$y = v_{0y}\left(\frac{x}{v_{0x}}\right) - \tfrac{1}{2}\,g\left(\frac{x}{v_{0x}}\right)^2 = \left(\frac{v_{0y}}{v_{0x}}\right)x - \left(\frac{g}{2v_{0x}^2}\right)x^2$$

which is of the form (2.14) with $a = (v_{0y}/v_{0x})$ and $b = -(g/2v_{0x}^2)$. The trajectory is therefore a parabola, as Galileo stated. ●

● ***Example 2.7*** An object is projected horizontally from a bridge 20 m above a river. The initial speed of the object is 30 m/s. What is the horizontal distance from the bridge to the point where the object strikes the water, what is the velocity at impact, and what is the time elapsed between release and impact?

We designate the horizontal and vertical directions by x and y, take the positive direction of y as downward, and locate the coordinate origin at the point from which the object is released. From the stated initial conditions, we then have

$$x_0 = 0; \qquad y_0 = 0; \qquad v_{0x} = 30\ \text{m/s}; \qquad v_{0y} = 0$$

Following release of the object

$$a_x = 0; \qquad a_y = g = 9.8\ \text{m/s}^2$$

(Note the positive sign; we have taken downward as the positive direction.)

At the moment of impact, $y(t) = 20$ m. From (2.13b),

$$20\ \text{m} = \tfrac{1}{2}(9.8\ \text{m/s}^2)t^2; \qquad t = \sqrt{(40\ \text{m})/(9.8\ \text{m/s}^2)} = 2.02\ \text{s}$$

Just over 2 s after the object leaves the bridge, it strikes the water. During that time it has traveled at constant velocity in the x direction. It has therefore covered a distance of

$$v_{0x}t = (30\ \text{m/s})(2.02\ \text{s}) = 60.6\ \text{m}$$

The horizontal distance between the bridge and the point of impact is therefore 60.6 m.

The velocity at the instant of impact has two components, v_x and v_y. Here, $v_x = v_{0x} = 30$ m/s; and v_y is given by (2.13d),

$$v_y(2.02) = v_{0y} + at = (0 \text{ m/s}) + (9.8 \text{ m/s}^2)(2.02 \text{ s}) = 19.8 \text{ m/s}$$

The velocity vector is therefore directed at an angle

$$\theta = \tan^{-1}\frac{19.8}{30} = \tan^{-1} 0.66 = 33.4°$$

below the horizontal, and its magnitude is

$$v = \frac{30 \text{ m/s}}{\cos 33.4°} = 35.9 \text{ m/s}$$

Note that the time required for the stone to reach the water is independent of the horizontal velocity; it will hit the water at the very same instant as a stone dropped simultaneously from the bridge. ●

● ***Example 2.8*** Suppose that the object is thrown from the bridge with the same initial speed of 30 m/s not horizontally but upward at an angle of 37° with the horizontal. Again, we want to know the time of flight, the horizontal range, and the final velocity at impact.

Figure 2.15 shows the expected trajectory. The initial conditions are now

$$x_0 = 0; \qquad y_0 = 0$$

$$v_{0x} = (30 \text{ m/s}) \cos 37° = 24 \text{ m/s}$$

$$v_{0y} = -(30 \text{ m/s}) \sin 37° = -18 \text{ m/s}$$

The component v_{0y} is negative because we have selected downward as the positive direction.

As before, we could determine first the time of flight. Instead, let us use a different approach just to show that here, as in most problems, there is no unique "correct" method. Several avenues can usually be followed. Some may be more direct or more apparent to the student, but all are equally valid provided that they conform to the conditions of the problem.

From the initial conditions and Equation (2.13e), we can determine the vertical component of the velocity at impact:

$$v_y^2(t) = (18 \text{ m/s})^2 + 2(9.8 \text{ m/s}^2)(20 \text{ m}) = 716 \text{ m}^2/\text{s}^2$$

$$v_y(t) = 26.76 \text{ m/s}$$

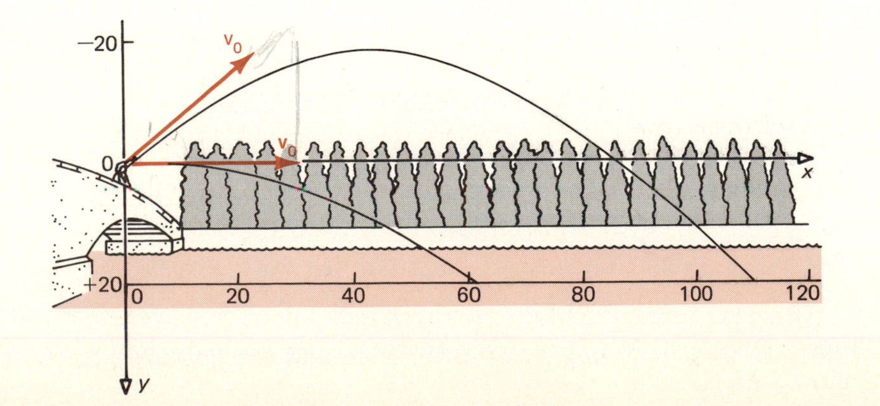

Figure 2.15 *Trajectories of objects thrown from a bridge with the same initial speed but in different directions.*

Here t is the time at which the object strikes the water. It is then traveling with a positive y component of velocity, i.e., downward. Thus the average value of v_y during the time of flight is

$$\langle v_y \rangle = \frac{-18 \text{ m/s} + 26.76 \text{ m/s}}{2} = 4.38 \text{ m/s}$$

and consequently, the time of flight is

$$t = \frac{s_y}{\langle v_y \rangle} = \frac{20 \text{ m}}{4.38 \text{ m/s}} = 4.57 \text{ s}$$

The horizontal range is

$$x = v_{0x}t = (24 \text{ m/s})(4.57 \text{ s}) = 110 \text{ m}$$

The direction of the velocity vector at impact is at an angle

$$\theta = \tan^{-1}\frac{26.76}{24} = 48.1^\circ \text{ below the horizontal}$$

and the magnitude of the velocity on impact is

$$v = \frac{(24 \text{ m/s})}{\cos 48.1^\circ} = 35.9 \text{ m/s}$$

The object remained in the air much longer than before, as we should expect. It also traveled farther in the horizontal direction even though the x component of the velocity was less than before, 24 m/s instead of 30 m/s; again, that result is not unexpected. We know that to achieve maximum range we should throw a ball up as well as forward. And lastly we note that the speed with which the object strikes the water is *the same as before,* again 35.9 m/s, though the velocity vector makes a different angle with the horizontal. That last result may be surprising. Had you been asked to predict whether the speed on impact would be greater than, equal to, or less than that in the preceding example, odds are heavily in favor of the answer "greater than." We shall prove (Chapter 4) that this is not some accident, valid only for our choice of angle, but a consequence of a basic conservation law.

You may have heard or read that, in the absence of air friction, the maximum range of a projectile fired from a gun is attained if the gun is aimed at an angle of 45° above the horizontal. This is true only if the gun and the point of impact are at the same elevation. If the gun is at a higher elevation, the greatest range obtains when the angle is less than 45°. You can verify this statement by calculating the range of our object, taking as the initial velocity direction 45° above the horizontal (Problem 2.65). ●

● ***Example 2.9*** An arrow is shot toward a wall that is 50 m distant with an initial velocity that makes an angle of 45° with the horizontal. The arrow strikes the wall 35 m above ground. Assuming that the arrow was released from ground level and neglecting air friction, find the initial speed of the arrow.

In this instance we do not know the initial velocity. Let us designate it by $\mathbf{v}_0$. We only know that (taking upward positive)

$$v_{0x} = v_0 \cos 45^\circ = 0.707v_0; \qquad v_{0y} = v_0 \sin 45^\circ = 0.707v_0$$

We do not know the time of flight. We do know, however, that at the instant when $x = 50$ m, $y = 35$ m. Hence we can write the following two equations:

$$x(t) = v_{0x}t = 0.707v_0t = 50 \text{ m}$$

$$y(t) = v_{0y}t + \tfrac{1}{2}at^2 = 0.707v_0t - \tfrac{1}{2}gt^2 = 35 \text{ m}$$

(The negative sign is used here because the acceleration is directed downward.)

These two simultaneous equations are readily solved by substituting the first into the second. We then have

$$\tfrac{1}{2}(9.8 \text{ m/s}^2)t^2 = 50 \text{ m} - 35 \text{ m} = 15 \text{ m}$$

$$t^2 = 3.06 \text{ s}^2$$

$$t = 1.75 \text{ s}$$

Finally, substitution of this result into either one of the two equations yields

$$v_0 = 40.4 \text{ m/s}$$

We conclude this section with another example, one that demonstrates important physical reasoning and approximation procedures based on order-of-magnitude arguments.

Example 2.10 An archer tries to hit a shoulder-high target 15 m away. The speed of the arrow as it leaves the bow is 30 m/s. Neglecting air friction, estimate the angle at which the archer should aim to compensate for the fall of the arrow due to the acceleration of gravity.

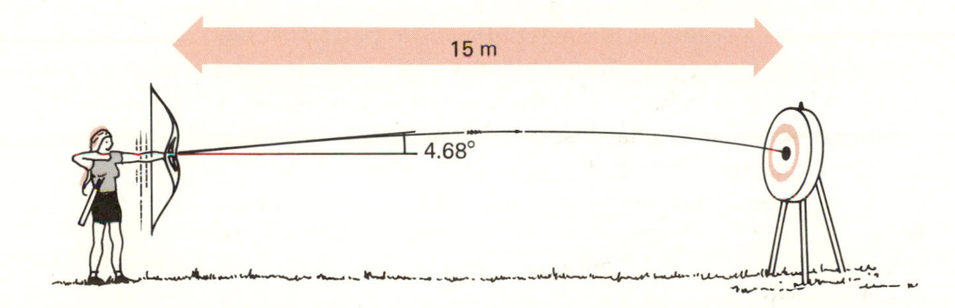

Figure 2.16 *Trajectory of an arrow shot with an initial speed of 30 m/s toward a shoulder-high target. Drawing is approximately to scale. At the top of the trajectory, the arrow is about 0.6 m above its starting point. Under the stated conditions the horizontal component of the velocity very nearly equals the magnitude of the initial velocity.*

This is a nasty problem if we try to solve it exactly. The reason is that as the aim is raised, both the horizontal and vertical components of the initial velocity change, and so does the time of flight. Let us start by asking what would happen if the arrow were fired horizontally. In that case it will not hit the target's center because it will drop slightly during flight. We can, however, get some idea of what to expect by determining the time required for the arrow to travel the 15 m and then calculating how much it has fallen in that time.

Traveling with a horizontal velocity component of 30 m/s, the arrow will strike the target 0.5 s after it leaves the bow. During that time it will fall a distance of

$$y(t) = \tfrac{1}{2}gt^2 = \tfrac{1}{2}(9.8 \text{ m/s}^2)(0.5 \text{ s})^2 = 1.23 \text{ m}$$

Since this distance is relatively small compared with 15 m, we expect that aiming just slightly above the horizontal should suffice to counterbalance the effect of gravity on the arrow.

If the angle θ is small, the horizontal component of the initial velocity, $v_{0x} = v_0 \cos\theta$, is very nearly equal to v_0 because for small angles $\cos\theta \simeq 1$. Hence the actual time of flight will also be almost exactly 0.5 s.

To hit the shoulder-high target, the arrow should rise during the first half of the trajectory, reach its maximum height at the midpoint, then drop back to shoulder height. Since the arrow must have zero vertical component of velocity at the midpoint, that is, at $t = 0.25$ s, we can now calculate v_{0y} from Equation (2.13d):

$$0 = v_{0y} - (9.8\ \text{m/s}^2)(0.25\ \text{s}); \qquad v_{0y} = 2.45\ \text{m/s}$$

Thus the angle θ should be

$$\theta = \sin^{-1}\frac{2.45}{30} = \sin^{-1} 0.0817 = 4.68°$$

We can confirm that v_{0x} is indeed very nearly 30 m/s. Its precise value is

$$v_{0x} = \frac{30\ \text{m/s}}{\cos 4.68°} = 29.9\ \text{m/s}$$

This is close enough to the assumed value of 30 m/s for us to be confident that the angle we have calculated is correct to within 1 percent. ●

Summary

Displacement is a vector quantity. The displacement $\Delta\mathbf{s}$ of an object during the time interval Δt is the vector difference $\mathbf{s}_f - \mathbf{s}_0$, where $\mathbf{s}_f$ and $\mathbf{s}_0$ are the final and initial position vectors.

Distance is a scalar; it is the length of the path traversed by an object.

The dimension of displacement and distance is length $[L]$. The unit of displacement and of distance is the meter.

Velocity is a vector quantity. The average velocity $\langle\mathbf{v}\rangle$ during the time interval Δt is

$$\langle\mathbf{v}\rangle = \frac{\mathbf{s}_f - \mathbf{s}_0}{\Delta t} = \frac{\Delta\mathbf{s}}{\Delta t}$$

The instantaneous velocity is defined as the limit of the average velocity as the time interval Δt approaches zero:

$$\mathbf{v} = \lim_{\Delta t \to 0}\left(\frac{\Delta\mathbf{s}}{\Delta t}\right)$$

The *speed* of an object is the magnitude of its velocity. Speed is a scalar quantity.

The dimension of velocity and of speed is $[L]/[T]$. The SI unit of velocity and of speed is meter per second.

Acceleration is a vector quantity. The average acceleration $\langle\mathbf{a}\rangle$ during the time interval Δt is

$$\langle\mathbf{a}\rangle = \frac{\mathbf{v}_f - \mathbf{v}_0}{\Delta t} = \frac{\Delta\mathbf{v}}{\Delta t}$$

where $\mathbf{v}_f$ and $\mathbf{v}_0$ are the final and initial velocities and $\Delta\mathbf{v}$ is the change in velocity during the interval Δt. The instantaneous acceleration is defined as the limit of the average acceleration as the time interval Δt approaches zero. Thus,

$$\mathbf{a} = \lim_{\Delta t \to 0}\left(\frac{\Delta\mathbf{v}}{\Delta t}\right)$$

The dimension of acceleration is $[L]/[T]^2$. The SI unit of acceleration is meter per second squared, m/s^2.

An object is accelerating if its velocity changes with time even though its speed may remain constant.

For *uniformly accelerated* motion in one dimension, the equations relating displacement, velocity, acceleration, and time are summarized in Equations (2.13).

The effect of gravitational attraction near the surface of the earth is to impart to all objects a uniform downward acceleration of $9.8\ m/s^2$; this acceleration is given the symbol g.

In the absence of air friction, the trajectory of a projectile is a parabola. During flight, the horizontal component of the projectile's velocity is constant; the vertical component of its velocity changes linearly with time due to the constant downward acceleration of gravity. Motion in the horizontal plane is unaffected by the vertically directed acceleration or the vertical component of the velocity, and vice versa.

Multiple Choice Questions

Neglect air friction for all questions and problems.

2.1 Two balls are projected horizontally from a tall building at the same instant, one with speed v_0 and the other with speed $v_0/2$.

(a) Both balls hit the ground at the same time.
(b) The ball with initial speed $\frac{1}{2}v_0$ reaches the ground first.
(c) The ball with initial speed v_0 reaches the ground first.
(d) One cannot say without knowing the height of the building.

2.2 A car traveling with an initial speed v comes to a stop in a time interval t. If the deceleration during this time t is constant, which of the following statements is correct for that time interval?

(a) The average speed is v/t.
(b) The distance covered is $(vt)/2$.
(c) The acceleration is $-v/2$.
(d) The distance covered is $(vt^2)/2$.

2.3 A ball is thrown vertically upward, reaches its highest point and falls back down. Which of the following statements is true?

(a) The acceleration is always in the direction of motion.
(b) The acceleration is always opposite to the velocity.
(c) The acceleration is always directed up.
(d) The acceleration is always directed down.

2.4 An object is dropped from rest. If it falls a distance s_1 during the first second and an additional distance s_2 in the next second, the ratio s_2/s_1 is

(a) 1
(b) 2
(c) 3
(d) 5

2.5 One stone is projected horizontally from a 20-m-high cliff with an initial speed of 10 m/s. A second stone is simultaneously dropped from that cliff. Which of the following is true?

(a) Both strike the ground below with the same velocity.
(b) Both strike the ground below with the same speed.
(c) During the flight, the change in speed of both stones is the same.
(d) During the flight, the change in velocity of both stones is the same.

2.6 A baseball, after being struck by a batter, is traveling to the outfield. The acceleration of the baseball during flight

(a) depends on how it was hit.
(b) depends on whether the ball is going up or coming down.
(c) is greatest at the top of the trajectory.
(d) is the same during the entire flight.

2.7 A ball is thrown upward. After it has left the hand, its acceleration

(a) is zero.
(b) increases.
(c) remains constant.
(d) decreases.

2.8 Figure 2.17 shows the trajectory of a ball. At the highest point A,

(a) the velocity is zero, but the acceleration is not zero.
(b) the velocity is not zero, but the acceleration is zero.
(c) the speed is less than at B, but the acceleration is greater than at B.
(d) velocity and acceleration are perpendicular to each other.

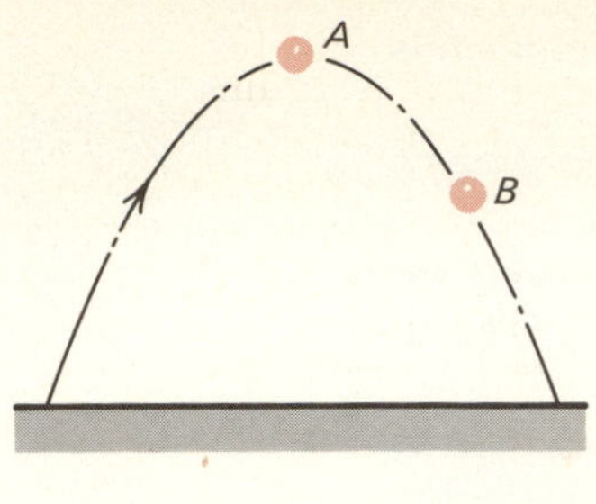

Figure 2.17

2.9 A flowerpot falls off the ledge of a fifth-floor window. Just as it passes the third-floor window someone accidentally drops a glass of water from that window. Which of the following statements is true?

(a) The flowerpot hits the ground first and with a higher speed than the glass.

(b) The flowerpot hits the ground at the same time as the glass, but the speed of the flowerpot is greater.

(c) The flowerpot and the glass hit the ground at the same instant and with the same speed.

(d) The glass hits the ground before the flowerpot.

Problems

(Section 2.2)

2.1 A runner completes two circuits around a track of 200-m circumference in 75 s. What was his average speed and average velocity?

2.2 A girl is traveling from home to her school, a distance of 240 km. She travels at nearly constant speed except for a 30-min coffee break after she has been on the road for 1.5 h. The trip, including the break, takes her 2 h 45 min. What was her average speed while she was on the road and for the trip as a whole?

2.3 An athlete runs along a straight track and covers 1 mile in exactly 4 min. What is his average speed in miles per hour and in meters per second?

2.4 An athlete runs a mile around a track whose circumference is one-quarter mile in 4 minutes. What is his average speed in m/s? What is his average velocity in m/s?

• **2.5** At time $t = 0$, a plane is observed to be 20 km west and 30 km south of Springfield, Ill. Twenty minutes later it is seen 130 km east and 90 km south of Springfield. What was the average velocity of the plane during this time? If it maintained this velocity, where would it be after 10 more minutes?

• **2.6** A boat is heading across a calm lake at a bearing of 60° east of north with a speed of 12 km/h. If the boat started at the shoreline and the shoreline is straight, running due east-west, how far from shore is the boat after 10 min of travel?

• **2.7** An object moves along the x direction. Its position as a function of time is given by $x = 4t - 2t^2$, where x is in meters and t in seconds. Find the average velocity of the object between $t = 0$ and $t = 1$ s, between $t = 1$ and $t = 2$ s, and between $t = 2$ and $t = 4$ s.

• • **2.8** Two students are distance runners; one is able to maintain a speed of 5 m/s, the other only 4 m/s. The slower one starts out first, and 10 min later the faster one follows. How long after the second one starts will they pass each other, and how far will they then be from the starting point?

• • **2.9** The two students of Problem 2.8 compete in the 800-m race, with the faster one under a handicap; he starts only after the slow runner has passed a certain marker on the track. How far from the starting line should this marker be so that both runners finish at the same time?

(Section 2.3)

2.10 Two cars are approaching each other along a straight road. Car A is traveling at a speed of 100 km/h, car B at 60 km/h. At $t = 0$, the two cars are 680 m apart. What is the relative speed with which the two cars approach each other? At what instant will they pass on the road? (We presume they will not crash.)

• **2.11** An airplane cruises at an airspeed of 400 km/h. If its destination is due east of its starting point and there is a 40-km/h wind blowing from the northwest, how should the plane head, and what is its ground speed toward the east?

• **2.12** If you have not solved Problems 2.8 and 2.9 using the concept of relative velocity, do so now.

• **2.13** At several airport buildings, "moving sidewalks" have been installed. If the moving belt travels at a speed of 1.3 m/s and covers 85 m, how much time will a person spend on the belt if he walks with a relative speed of 1.8 m/s in the direction of the belt? How long will it take that person to move from one end of the belt to the other if he walks opposite to the motion of the belt?

• **2.14** Two boat landings are exactly 1 km apart on the same bank of a stream that flows at 1.4 km/h. A motorboat that maintains a constant speed relative to the water makes the round trip in 30 min. Determine the speed of the motorboat relative to the water.

• **2.15** How should a boat, whose speed relative to the water is 18 km/h, head across a stream that is flowing at a rate of 5 km/h due south so that it reaches the opposite shore at a point exactly west of its point of departure? How long does the boat take to cross if the stream is 3 km wide at that point?

• • **2.16** A pilot sets his course due east and flies at an airspeed of 400 km/h for 1.5 h. He is then 560 km east and 60 km north of his starting point. What was the wind velocity during this time? What is the compass

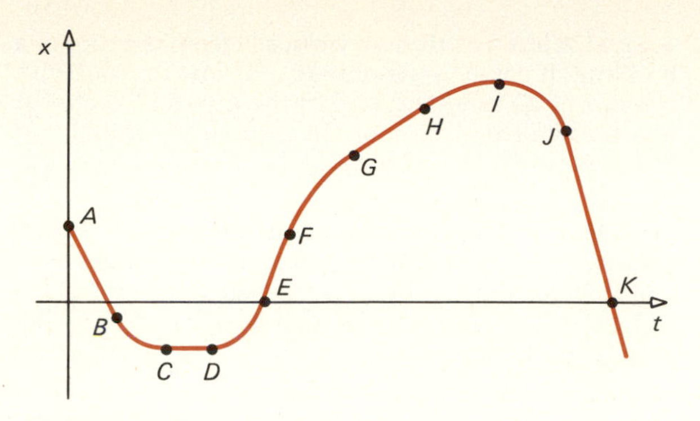

Figure 2.18

direction in which the pilot should have headed the plane to keep on a true easterly course?

• • **2.17** A sailboat, caught in a fog just 5 km south of its destination, heads due north and travels 5.3 km (according to its log) relative to the water before the fog lifts. The skipper then sees the harbor just 1 km northwest of his position. What was the displacement of the boat due to the currents during the time of fog?

• **2.18** The compass on an airplane shows that it is heading due south. Its airspeed indicator registers 180 km/h. Local weather stations report a wind of 45 km/h from the west. What is the velocity of the plane relative to the ground?

• • **2.19** Boat *A* starts out from port *P* at 8 A.M., bearing due north and maintaining a speed of 15 km/h. At the same moment, boat *B* leaves port *Q*, which is due northeast of *P* and 60 km away as the crow flies. Boat *B* cruises at 18 km/h. How should boat *B* head if it wishes to rendezvous with boat *A*? At what time will the two boats meet?

• • **2.20** A plane takes off from airport *A*, destined for airport *B*, which is 250 km away and due north of *A*. The plane's airspeed is 180 km/h, and there is a wind of 50 km/h from the southeast that day. What should the compass heading of the plane be? How long will the flight take?

• • **2.21** A sailor is headed for a harbor 20 km northwest of his position when he is suddenly engulfed in thick fog. Using his compass, he maintains a heading of northwest; his knotmeter shows that he is traveling over the water at 8 km/h. At the end of 3 h, he comes on land but at a point exactly 5 km south of his destination.

(a) What was the average velocity of the current during those 3 hours?

(b) What heading should he have maintained to reach his destination? How long would the trip have lasted?

(*Hint:* To do part (b) it is most convenient to use a reference frame that points SW-NE and SE-NW.)

• • **2.22** A swimmer sets out across a 300-m-wide river. The swimmer maintains a speed of 2 m/s relative to the water. If he starts from the west bank of the river, which flows south at 1 m/s, in what direction should he head to reach the other bank at a point just opposite to his starting point? How long will it take him to make this trip? Can he reach the opposite bank faster if he heads in some other direction? Explain.

(Sections 2.4, 2.5)

2.23 Figure 2.18 shows the displacement of an object from $x = 0$ as a function of time t. Identify the regions and points where $v = 0$, $v > 0$, $v < 0$, $a = 0$, $a > 0$, $a < 0$.

2.24 A block moves at constant acceleration, passing in 2 s two points separated by 50 m. What is its average speed?

2.25 A block moves at constant acceleration, passing in 2 s two points separated by 50 m. If it passes the first point with an instantaneous velocity of 10 m/s, what is its velocity when it passes the second point? What is its acceleration?

• **2.26** A car starts from rest heading due east. It first accelerates at 2 m/s² for 10 s and then continues without further acceleration for 15 s. It then brakes for 3 s, reducing its speed by 50 percent, and travels at the reduced speed for another 5 s. Make a sketch showing the position of the car as a function of time.

• **2.27** A ball is thrown straight up with an initial velocity of 12 m/s. What maximum height does it reach, and how long does it remain in the air?

• **2.28** An object starts moving north at 120 m/s. It experiences a constant acceleration of 8 m/s to the south. How much time elapses before it returns to its starting point?

• **2.29** An object is dropped from rest on a planet on which the acceleration of gravity is unknown. If it drops 0.4 m during the first second, how far will it have dropped at the end of 4 seconds?

• **2.30** A stone is released from the roof of a 40-m-high building. It strikes the ground 2 s later. What was its initial velocity?

• **2.31** Repeat Problem 2.30 for a time of flight of 4 s.

• **2.32** A lady drops a coin from her hand while she is in an elevator that is descending at a constant speed of 1.5 m/s. If her hand is 1.2 m above the floor of the elevator when the coin is released, how long will it be before the coin lands on the floor of the elevator? Suppose the elevator were ascending at 1.5 m/s? Describe the path of the coin as seen by a stationary observer in each instance.

• **2.33** The displacement of an object is given by $x = b + ct^2$, where $b = 8$ m, $c = 2.5$ m/s², and t is measured in seconds.

(a) Find the instantaneous velocity and acceleration of the object at $t = 0$ and at $t = 2$ s.

(b) What is the average velocity of the object between $t = 2$ s and $t = 4$ s?

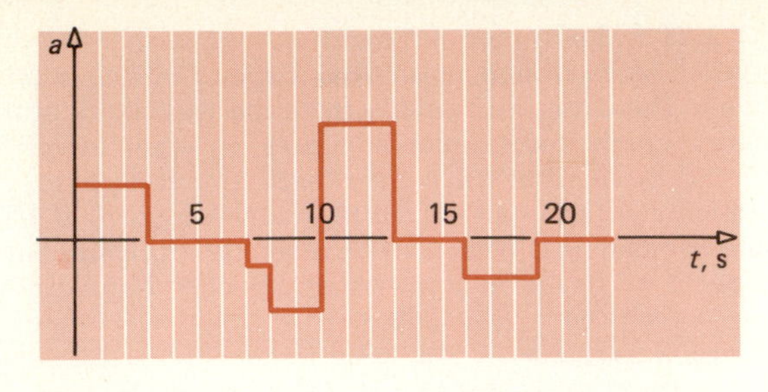

Figure 2.19

• **2.34** Figure 2.19 shows the acceleration of a body as a function of time. Assuming that $x_0 = v_0 = 0$, sketch its position and velocity as a function of time.

• **2.35** Write an expression for the position of a particle that is subject to a constant acceleration of 5 m/s^2 in the negative x direction; the position of this particle at $t = 0$ is $x_0 = 3$ m, $y_0 = 4$ m, and its velocity at $t = 0$ has identical x and y components of 10 m/s. Give the position of this particle at $t = 2$ s and at $t = 3$ s.

• **2.36** A car accelerates uniformly from 20 km/h to 80 km/h in 6 s. What is the acceleration in that interval, and how far does the car travel in those 6 seconds?

• **2.37** A car traveling at 75 km/h can come to rest on a dry pavement in 3 s. What is the average deceleration in m/s^2? How far did the car travel during those 3 s of deceleration?

• **2.38** With what initial speed must a ball be thrown upward to reach a height of 24 m? How long will this ball stay in the air?

• **2.39** A block slides down an inclined plane, starting from rest. It is 2 m from its starting position at the end of 4 s. Assuming that it accelerated uniformly, find its acceleration, average velocity, and velocity at the end of the 4 s.

• **2.40** A car travels east at a speed of 24 m/s. If it cuts its speed in half in 3 s, what is its acceleration vector, and what distance did the car cover during those 3 seconds?

• **2.41** A car travels east at 25 m/s. The driver applies brakes and slows the car to 9 m/s in a distance of 256 m. What are the magnitude and direction of the acceleration, and the time that elapsed? If the brakes are applied until the car is at rest, and the acceleration is constant, what is the total distance covered?

• • **2.42** A car and a motorcycle start from rest at the same moment, with the motorcycle 16.0 m behind the car. The car accelerates at 3.6 m/s^2, the motorcycle at 4.8 m/s^2. How much time will have elapsed when the motorcycle passes the car? How far will each vehicle have traveled by then?

• **2.43** A stone is dropped from the roof of a building. It strikes the ground with a speed of 24.5 m/s. How much time elapsed between release of the stone and its impact with the ground, and how tall is the building?

• • **2.44** A stone is thrown vertically from the roof of a building. It passes a window 10 m below the roof with a speed of 20 m/s, and strikes the ground 2 s after it was thrown from the roof. Determine the initial and final velocities of the stone and the height of the building.

• **2.45** A ball is thrown downward with an initial velocity of 8 m/s from the top of a building that is 45 m tall. How far above ground is the ball at the end of 1 s, and what is its velocity at that instant? With what speed will it strike the ground?

• **2.46** You direct a stream of water from a garden hose nearly vertically up. The water splashes on the cement nearby. At $t = 0$, you suddenly redirect the hose but hear the water splashing down near you for another 1.5 s. What is the maximum height reached by the stream of water?

• **2.47** A stone is thrown upward from the roof of a 20-m-high building with an initial speed of 8 m/s. How much time elapses before it hits the ground, and what is the speed on impact? What is the maximum height attained by the stone?

• • **2.48** A car stopped at a traffic light accelerates at 2 m/s^2 just as soon as the light turns green. One second after the light has turned green a truck passes the car; the truck is traveling at a constant speed of 50 km/h. When and where will the car overtake the truck again, and what will be the speed of the car at that instant?

• • **2.49** A stone is dropped from rest from a high cliff and, 1.5 s later, a second stone is thrown straight down with a speed of 24 m/s. Will the second stone ever overtake the first, and if so, when and where? If the second stone is thrown down with an initial speed of 5 m/s, can it still catch up with the first? Is there a minimum initial downward speed that must be imparted to the second stone to permit it to overtake the first?

• • **2.50** A balloonist who is ascending at a steady rate of 4 m/s releases some ballast (a sandbag) when the balloon is 30 m above ground. What is the velocity of the sandbag 0.5 s, 1 s, and 2 s following release? How long after its release will it strike the ground, and what will its impact speed be?

• • **2.51** A ball that is thrown straight up passes a point that is a third of its maximum height just 1 s after its release. What maximum height does it attain, and what was the initial velocity of the ball?

• **2.52** Figure 2.20 shows the velocity of a body as a function of time. Assuming that $x_0 = 0$, sketch its position and its acceleration as a function of time. Will this body ever return to its initial position, and if so, when?

• **2.53** A bus starts from rest, accelerates at 2.4 m/s^2 for 6 s, then travels at constant velocity for 24 s; it then accelerates at $-4 m/s^2$ until it comes to rest. Determine the distance covered.

• • **2.54** The reaction time of the average driver is 0.8 s; that is, the average driver, on seeing an obstacle in the road, will apply the brakes 0.8 s later. On dry pavement a car can attain a maximum deceleration of 6 m/s^2 and on wet pavement, 3 m/s^2. What is the distance needed to bring a car to rest after seeing a red light if the car is traveling at 60 mph? Do the calculation for wet and for dry pavement.

• • **2.55** A passenger train is traveling at 28 m/s when its engineer sights a freight train on the track ahead. At that moment, the two trains are 350 m apart; that is, the distance separating the engine of the passenger train and the caboose of the freight train is 350 m. If the freight train is traveling at 6 m/s in the direction of the passenger train and if the maximum acceleration of the passenger train is $-0.7\ m/s^2$, will the two trains collide? Assume the reaction time of the engineer to be (a) 0 s and (b) 0.9 s. If a collision takes place, where in relation to the initial point of observation does this collision occur? What is the relative speed of the two trains at the moment of collision?

• • **2.56** A car and motorcycle start from rest at the same moment, with the motorcycle 40 m behind the car. The car has a constant acceleration of 3 m/s^2. The motorcycle passes the car after the car has traveled 96 m. How much time elapsed before the two vehicles passed, and what was the acceleration of the motorcycle? What are the speeds of the car and of the motorcycle when they are abreast?

• • **2.57** A stone is thrown vertically from the roof of a 24.5-m-tall building. The stone passes a window 10 m below the roof with a speed of 20 m/s. How much time elapses between release of the stone and its impact with the ground? Was the stone thrown vertically upward or downward?

(Section 2.6)

• **2.58** A gun fires a projectile due east with an initial horizontal speed of 250 m/s. Four seconds later the projectile strikes the target. Locate the position of the gun relative to the target. What was the speed of the projectile when it struck?

• **2.59** Dive-bombing was a common practice during World War II. Suppose a bomber is on a course toward the target, diving at an angle of 37° below the horizontal with a speed of 250 m/s. It releases a bomb when it is at an altitude of 600 m, and the bomb hits the intended target. Where was the target in relation to the plane at the moment of the bomb's release? How much time elapsed between the release and the impact of the bomb? What was the bomb's speed at the moment of impact?

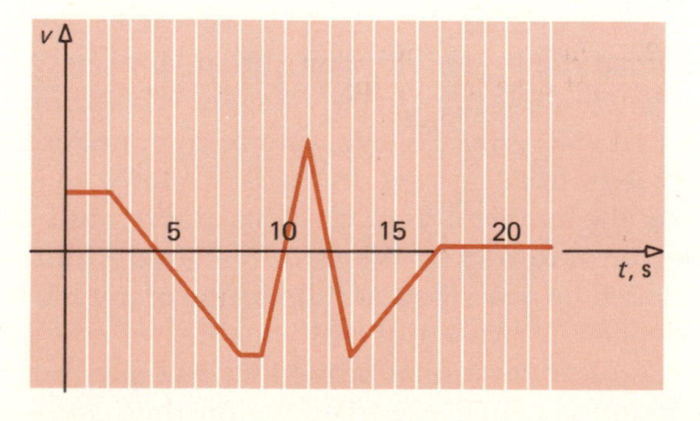

Figure 2.20

• **2.60** A train robber throws a bag of loot off the moving train as the train travels at a speed of 90 km/h on an embankment 15 m above ground. The robber throws the bag horizontally with an initial speed of 15 m/s perpendicular to the train track. Where, in relation to its point of release, will the bag land?

• **2.61** When a spring gun is held at an angle of 45° above the horizontal, the range of its projectile is 25 m. What is the muzzle velocity of the projectile?

• • **2.62** At what angle should the spring gun of Problem 2.61 be held so that the range is reduced to 18 m?

• **2.63** An airplane that is diving at an angle of 37° relative to the horizontal is moving at a speed of 800 km/h. Just before it pulls out of the dive, it releases a bomb that strikes the ground 6 s later. At what altitude was the plane when it released the bomb? Where, in relation to the point of release, did the bomb hit the ground?

• **2.64** A projectile is fired with an initial velocity whose horizontal and vertical components are $v_x = 160$ m/s and $v_y = 75$ m/s.

(a) Determine the magnitude of $\mathbf{v}_0$ and the angle of $\mathbf{v}_0$ with the horizontal.

(b) What is the speed of the projectile at maximum elevation?

(c) What is the maximum elevation of the projectile?

(d) What is the range of the projectile, if the target is at the same level as the gun?

• • **2.65** An object is projected from a bridge 20 m above the water. The initial velocity of the object is 30 m/s, and 45° above the horizontal. How far from the bridge will the object strike the surface of the water, and what is its velocity at that moment? Compare your answer with that for the corresponding result of Example 2.8.

• • **2.66** An archer shoots at a squirrel perched atop a 15-m-tall telephone pole that is 20 m distant. If the squirrel sees the archer fire and drops to the ground just as the arrow leaves the bow, in which direction would the archer have to have aimed to hit the squirrel? If the initial speed of the arrow is 25 m/s, will the arrow hit the squirrel before it drops all the way to the ground? How far above ground is the squirrel when it is struck, if it is hit at all while falling?

• • **2.67** A Frisbee is caught in a tree at a height of 15 m above shoulder level. If you try to dislodge the Frisbee by throwing a stone from a horizontal distance of 20 m, what should the initial velocity of the stone be if it is to strike the Frisbee while its instantaneous velocity is in the horizontal direction? With what velocity will it strike the Frisbee, and what will have been the time of flight?

• 3 •

Dynamics

In this philosophy particular propositions are inferred from the phenomena and afterwards rendered general by induction.

ISAAC NEWTON

3.1 Introduction

In 1642, the same year Galileo died—an old man, blind, broken, and under house arrest—Isaac Newton was born in the village of Woolsthorpe, Lincolnshire. Newton studied at Trinity College, Cambridge, under the famous mathematician Professor Isaac Barrow. In the fall of 1665, an outbreak of the plague forced the university to close its doors, and the twenty-three-year-old Newton returned to Woolsthorpe. There, during the following 18 months, which he described as "the prime of my life for invention," Newton developed the branch of mathematics we now call calculus, discovered the law of gravitational attraction, and performed and interpreted a series of fundamental observations on the nature of light. Shortly after Newton's return to Cambridge, Barrow resigned his chair (some think with Newton in mind), and the twenty-seven-year-old Newton assumed it.

(a)

(b)

Figure 3.1 (a) Portrait of Isaac Newton. (b) Manor house at Woolsthorpe, Lincolnshire, the birthplace of Isaac Newton.

It would be difficult to exaggerate the far-reaching impact of Newton's work not only on science of the eighteenth and nineteenth centuries, but on Western thought in every sphere of intellectual activity. With the comprehensive synthesis of the monumental *Philosophiae Naturalis Principia Mathematicae* (Mathematical Principles of Natural Philosophy) as a point of departure, succeeding generations of scientists explained and illuminated apparently all natural phenomena, from the motion of the heavenly bodies to that of atoms and molecules, in terms of a completely mechanistic theory. As a result of these successes, practically every scientist and philosopher subscribed to the prevalent belief that Newton had explained the fundamental workings of the universe and that any physical event could, in principle, be understood in every detail if one could only carry the necessary complicated and lengthy calculations to completion. Echoing this view, Michelson, at the dedication of the Ryerson Physical Laboratory at the University of Chicago in 1894, expressed the conviction that further discoveries and "future truths of Physical Science are to be looked for in the sixth place of decimals."* This deterministic philosophy was most pointedly expressed by Laplace, who when asked by Napoleon where God might fit into his scheme of the universe, replied, "Je n'ai pas besoin de cette hypothèse." (I do not need that hypothesis.)†

The three laws of motion that bear Newton's name are really very simple statements. Their profound value derives precisely from this elemental simplicity and consequent generality. To propound a theory to explain a particular observation is rarely difficult; but if each new observation needs a new theory, we have surely made little progress toward

* A.A. Michelson, whose area of specialization was optics, was one of the most renowned American physicists of the nineteenth century and the recipient of the 1907 Nobel Prize in physics, the first American so honored. Perhaps his most famous experiment was the measurement of the "ether drift" (described in Chapter 28), which proved a great disappointment to Michelson because of its null result.

† Laplace may have consciously paraphrased Jean Baptiste Coffinthal, who, on ordering the execution of the chemist Lavoisier in 1794, proclaimed, "La Republique n'a pas besoin de savants."

understanding nature's ways. Bronowski* put it succinctly: "All science is the search for unity in hidden likenesses." It is this comprehensive unity that lends Newton's work its majestic beauty. Encompassed therein are Galileo's rule for freely falling bodies, his observation that trajectories of projectiles are parabolas, Kepler's three laws of planetary motion, Bernoulli's and Torricelli's theorems and equations concerning the behavior of liquids, Boyle's and Gay-Lussac's laws on the behavior of gases, and numerous other equations and laws of physics. No wonder many nineteenth-century scientists felt depressed, even cheated, in the pursuit of their work, convinced that nothing of really fundamental significance remained to be discovered.

3.2 Newton's Laws of Motion

3.2 (a) The First Law

Every body continues in its state of rest, or of uniform motion in a straight line, unless it is compelled to change that state by forces impressed on it.

Newton's first law appears to contain two separate statements. The first, that a body at rest will remain at rest unless it is pushed, pulled, or otherwise impelled by forces acting on it, is certainly consistent with common sense and everyday experience. It is even consonant with Aristotelian philosophy, which maintained that rest is the "natural" state of objects.

The second part, however, is less obvious and apparently contradicts experience. Any object we set in motion, a ball rolling along the ground or a hockey puck sliding across the ice, will come to rest after some time. Though each may have started with the same initial speed, one will generally travel longer and farther than the other, which suggests that whatever the retarding action, it is not identical in every situation. Newton recognized the retarding force of friction and abstracted to the ideal case in which friction vanishes. He concluded that in this instance an object would move in a straight line at constant speed.†

An object that does not change its state of motion is said to be in *equilibrium*. Contrary to common parlance, *equilibrium* as defined and used in physics does not demand that the object be at rest. Though this may not seem the "natural" way to define equilibrium, this broader definition conforms to our earlier discussion (Chapter 2), in which we pointed out that the state of rest and of uniform motion differ only as viewed through "the eye of the beholder." They appear different only because the same state is seen from two different inertial reference frames. This concept is already implicit in the first law.

* J. Bronowski, mathematician, philosopher, humanist, and historian, was research professor and fellow of the Salk Institute in La Jolla, California, and its director of the Council for Biology in Human Affairs at the time of his death in 1974. He was the author of about a dozen books, but is best remembered for the superb BBC television series "The Ascent of Man," which was aired on PBS in the United States, and which he wrote and narrated.

† In all fairness, it must be said that the first law was not original with Newton. Decades earlier, Galileo wrote in his *Two New Sciences*, "Any velocity once imparted to a moving body will be rigidly maintained as long as the external causes of retardation are removed."

3.2 (b) The Second Law

If a net force acts on a body, it will cause an acceleration of that body. That acceleration is in the direction of the net force, and its magnitude is proportional to the magnitude of the net force and inversely proportional to the mass of the body.

Before we examine this law, which is the foundation of all dynamical theories, we must first point out that we have here paraphrased Newton. As he stated it, the second law is more general but not so easily explained. We shall return to the more general, and correct, formulation of the second law in a later chapter.

Whereas the first law characterizes the motion of bodies in the absence of forces, the second law addresses itself to the cause-and-effect relation between forces and motion. It relates acceleration, the departure from equilibrium, to the net force impressed on the body and to its mass. It serves, in fact, as the vehicle for defining both force and the unit of force.

According to the second law, if a particular body of given mass is observed to accelerate at, say, 2 m/s^2 as a result of our applying a force **F**, it will accelerate at 4 m/s^2 if we apply a force 2**F**. That is,

$$\mathbf{a} \propto \mathbf{F} \tag{3.1}$$

where the symbol $\propto$ means "is proportional to."

Also, if a force **F** applied to an object of 1-kg mass produces an acceleration of 2 m/s^2, the same force applied to a mass of 2 kg will cause it to accelerate at 1 m/s^2 only. That is,

$$\mathbf{a} \propto \frac{1}{m} \tag{3.2}$$

These two statements can be combined:

$$\mathbf{a} \propto \frac{\mathbf{F}}{m} \tag{3.3}$$

Although Newton's second law did not state this explicitly, it clearly implied that the acceleration depends *only* on force and mass. Acceleration does not depend on the kind of force, whether of gravitational, electric, mechanical, magnetic, or other origin. Nor does the acceleration depend on the shape of the body or its constitution, whether it is lead or wood, or for that matter, its state, whether solid, liquid, or gas. Consequently, the second law can be applied, for example, to calculating fluid flow as well as the motion of solid bodies due to specified forces.

The proportionality symbol of Equation (3.3) can be replaced by an equality sign provided we agree on an appropriate *unit of force*. The SI unit of acceleration is the meter per second squared, that of mass is the kilogram. If, then, we write the second law in the more familiar form,

$$\mathbf{F} = m\mathbf{a} \tag{3.4}$$

the unit of force must be the force that will, when applied to a mass of 1 kg, impart to it an acceleration of 1 m/s^2. Thus the proper unit is the kilogram meter per second square ($kg{\cdot}m/s^2$). This SI unit has been assigned the name *newton,* abbreviated N. One newton acting on a mass of 1 kg will give it an acceleration of 1 m/s^2. The dimension of force is

$$[F] = \frac{[M][L]}{[T]^2}$$

Equation (3.4) is a relation between two vector quantities, force **F** and acceleration **a**. But if two vectors are equal, their respective components along a set of mutually orthogonal axes must also be equal. Thus the vector equation (3.4) is equivalent to the three algebraic equations

$$F_x = ma_x \tag{3.4a}$$

$$F_y = ma_y \tag{3.4b}$$

$$F_z = ma_z \tag{3.4c}$$

Many problems are most conveniently solved by applying Newton's laws to the orthogonal components separately.

3.2 (c) Mass and Weight

On the surface of the earth and also in space, every object of finite mass experiences a gravitational pull toward the center of the earth. Although gravitational forces are almost negligibly small when compared with other forces in nature such as electromagnetic forces and nuclear forces, the force of gravity acting on ordinary objects on earth is substantial, since we are normally concerned with masses containing 10^{20} or more individual atoms interacting with a very massive sphere, the earth. *This force that the earth exerts on an object of given mass we call the object's weight on earth.*

It cannot be emphasized too strongly that *mass and weight are separate, distinct attributes.* Mass is the property that lends a body its inertia, its reluctance to change its state of motion. To cause a change of motion we must exert a force, and that force must be greater the greater the mass. This is true in equal measure on the surface of our planet and in deep space. *Mass is an inherent property of a particular body.*

Since weight is a force, the SI unit of weight is the newton, not the kilogram. On the surface of the earth, all objects not otherwise restrained accelerate downward at $g = 9.8\ \text{m/s}^2$. Consequently, the weight W, that is, the force of gravity acting on an object of mass m is, according to Equation (3.4),

$$W = mg \tag{3.5}$$

The weight of an object of mass m kg *is mg* N.

In the more familiar British system, the unit of force is the pound (lb). When we say that an object weighs one pound, we mean that the force of attraction between it and the earth is one pound. What, then, is its mass? From the definition of weight and the second law, we conclude that the mass is

$$m = \frac{W}{g} \tag{3.6}$$

where g is the acceleration of gravity, which we must now express in British units, $32\ \text{ft/s}^2$. The unit of mass, one pound-second²/foot, is called one *slug*. Accordingly, an object that weighs 64 lb on earth has a mass of 2 slugs. If we transport that object to the moon, its mass remains 2 slugs but its weight will be substantially less.

One pound is equivalent to 4.45 N, and 1 N equals 0.225 lb. To help you get a feeling for the SI unit, you might bear in mind that a "quarter-pounder" may some day be known as a "Newtonburger," especially as inflation progresses and portions slowly shrink.

3.2 (d) The Third Law

To every action there is always opposed an equal reaction; or, the mutual actions of two bodies upon each other are always equal, and directed to contrary parts.

This law states that when one body exerts a force on another, the second exerts an equal, oppositely directed force on the first. Like the first law, the third formalizes a commonplace observation: if you push a car, the car pushes back against your hand; if you support a weight by a rope, the rope pulls down on your hand; a book resting on a table pushes down on the table, and the table, in turn, pushes up against the book; the earth pulls on the moon, holding her in a nearly circular orbit, and the moon pulls on the earth, causing tides.

The third law differs from the other two in one important respect. Whereas the first and second laws are concerned with the behavior of a single body, *the third involves two separate, distinct bodies.* The action force exerted by body A on body B is equal in magnitude but opposite in direction to the reaction force exerted by body B on body A. The inherent symmetry of the action-reaction couple precludes identifying one as action and the other as reaction. Whichever one we designate the action force is arbitrary and a matter of personal preference.

Consider, for example, the simple system of a mass M on a horizontal surface and pulled to the right by a rope. This system is shown in Figure 3.2(a), and the action-reaction couples of this system are illustrated in Figure 3.2(b). Note that in each case the two forces of the pair are of *equal magnitude* and *oppositely directed,* and *they act on different bodies.*

In Figure 3.2(b), the subscripts indicate on which part of the system the force acts and what part of the system exerts the force. Thus, $\mathbf{F}_{SM}$ is the force exerted on the *s*urface by the *m*ass, in this instance, the weight of the mass, mg.

The two forces $\mathbf{F}_{RM}$ and $\mathbf{F}_{RH}$, which act on the two ends of the rope, *do not* constitute an action-reaction couple. They act on the same body, the rope, and moreover, are not necessarily of equal magnitude. If the system is accelerating to the right and the rope has a finite mass M_R, then according to Newton's second law, the product $M_R\mathbf{a}$ must equal the net force on the length of rope; i.e., $M_R\mathbf{a} = \mathbf{F}_{RH} - \mathbf{F}_{RM} \neq \mathbf{0}$.

In Figure 3.2(a), the rope transmits the force exerted by the hand at one point in space to another location, the point at which the rope is attached to M. The rope is under tension, and the tension at any point is the magnitude of the two oppositely directed forces that would be needed to hold the two parts of the rope together if the string were cut at that point. A rope or string, in contrast to a solid rod, can transmit only tensile forces, and only along the direction of the string. By passing an ideal (massless) string over an ideal (frictionless and massless) pulley, the direction of this tensile force may be changed without affecting its magnitude.

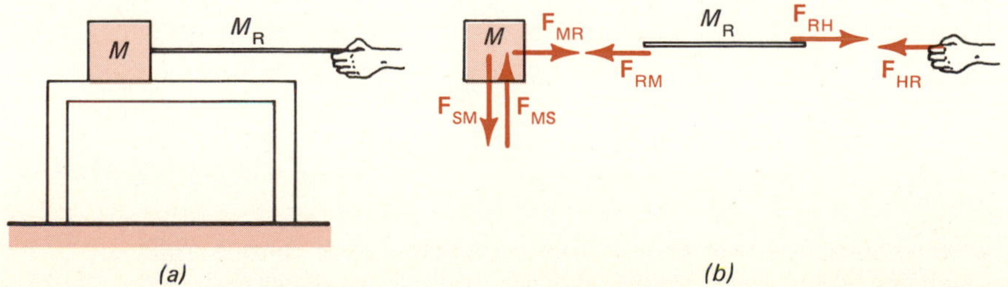

Figure 3.2 Newton's third law. (a) The mass M, supported on a friction-free horizontal surface, is pulled to the right by means of the rope of mass M_R. (b) The various action and reaction forces acting on parts of the system.

Note that the reaction force which the surface exerts on the mass is perpendicular to the plane of the surface. A line that is perpendicular to a surface is said to be *normal,* and this reaction force is often referred to as the *normal reaction force.* A friction-free surface can exert only a normal force.

3.3 Application of Newton's Laws

The remainder of this chapter is devoted to applications of the laws of motion in particle dynamics. We begin with a class of examples that may seem to have no relation to dynamics, namely examples of equilibrium.

An object is in equilibrium if it is not accelerating; therefore, no *net* force must act on it. That does not mean that no forces may be applied to the body. If several forces act simultaneously, equilibrium demands only that the *net* force, that is, the vector sum of the various forces, vanish. Thus, *one of the conditions of equilibrium is**

$$\sum_i \mathbf{F}_i = 0 \tag{3.7}$$

In mathematics, the capital Greek letter *sigma,* Σ, is conventionally used to indicate *summation.* The left-hand side of Equation (3.7) is therefore an instruction to take all the force vectors—$\mathbf{F}_1$, $\mathbf{F}_2$, $\mathbf{F}_3$, . . .—and add them, following the well-established rules for vector addition (described in Section 1.7). Equation (3.7) is equivalent to the three scalar equations

$$\sum_i F_{ix} = 0 \tag{3.7a}$$

$$\sum_i F_{iy} = 0 \tag{3.7b}$$

$$\sum_i F_{iz} = 0 \tag{3.7c}$$

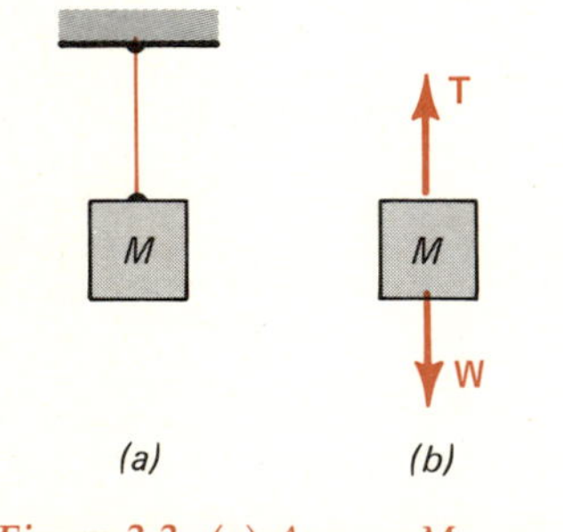

Figure 3.3 (a) A mass M suspended by a string. (b) The corresponding "free-body" diagram.

where x, y, and z denote any three orthogonal axes and F_{ix}, F_{iy}, and F_{iz} are the x, y, and z components of the ith force.

Let us now look at a few simple examples. The first, concerning an object suspended by a string, is almost trivially simple. We want to know the tension in the string.

To solve this or any other equilibrium or dynamics problem we need to know what forces act on the object.

To place these forces clearly in evidence, we *isolate the body;* that is, we display it without ropes or other attachments. If the attachments transmit forces to the body, we then show these forces by drawing the representative force vectors.

In our example, two forces act on the body, as shown in Figure 3.3. One is the gravitational force, the weight **W**. The other force is the tension **T** in the supporting string. These are the only forces acting on the isolated body. Hence Equation (3.7) reads

$$\mathbf{W} + \mathbf{T} = 0$$

and therefore,

$$\mathbf{T} = -\mathbf{W} \tag{3.8}$$

* There is another condition for equilibrium that must also be satisfied. This concerns rotational motion, and we shall consider it in Chapter 8.

That is, the tension in the rope acting on the body is equal in magnitude to the weight of the body but oppositely directed, pointing upward.

All this is simple enough. Still, we sometimes forget that when an object rests on a table, there are forces acting on it, namely its weight and the upward reaction force of the table on the object. Without the latter, the object could not remain at rest but would drop under the pull of gravity, as it does if we remove the supporting tabletop.

● ***Example 3.1*** A slightly more complicated problem is illustrated in Figure 3.4(*a*), which shows an object Q of unknown weight W suspended from a string attached to two other strings at A. One string is fastened to the wall at B, the other passes over a friction-free pulley P to another weight of 15 N. The system is in equilibrium when the strings PA and BA make angles of 37° and 53°, respectively, with the horizontal. The problem is to determine the unknown weight W.

We again begin by isolating the body but are immediately faced with a dilemma. If we isolate the body Q as in Figure 3.4(*b*), we see that there are two forces acting on it, $\mathbf{W}$ pointing down and $\mathbf{T}_A$, the tension in the supporting string, pointing up. We can conclude that $\mathbf{T}_A = -\mathbf{W}$; but since neither is known, we have made little progress.

In a situation of this kind, the "body" that should be viewed in isolation is the point A, the junction of the strings that transmit the forces. This *free-body diagram* is shown in Figure 3.4(*c*).

We already know from the preceding example and from the third law that the force acting on A by the string supporting Q equals $W = m_Q g$. The force this string exerts on Q is upward, but the force the string exerts

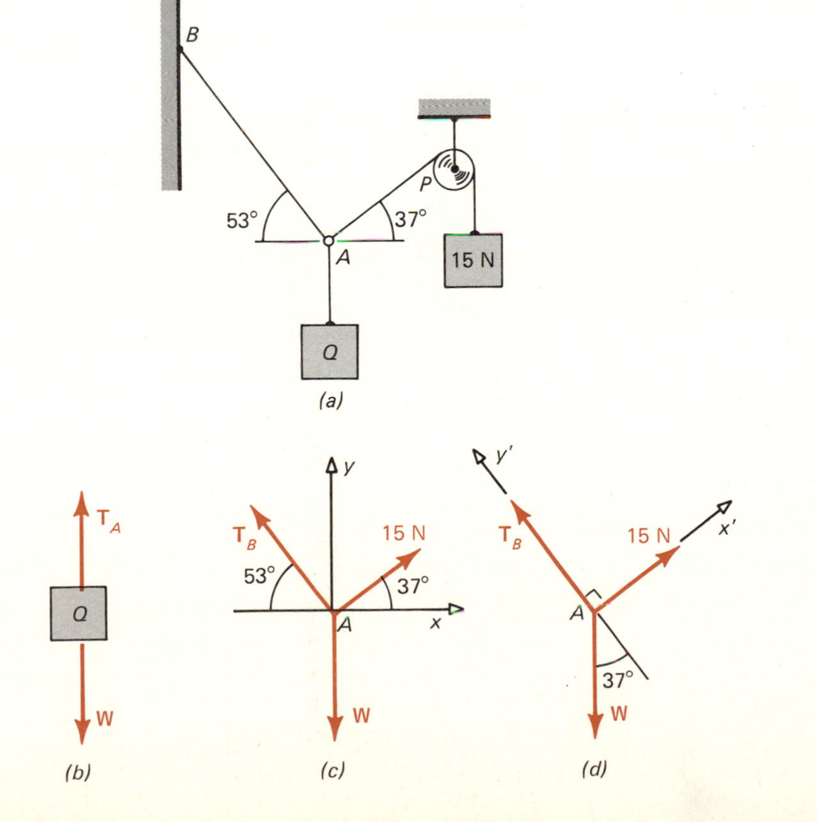

Figure 3.4 (*a*) A body Q supported by two strings making angles of 53° and 37° with the horizontal. The system is in equilibrium. (*b*) The free-body diagram for the mass Q. (*c*) The free-body diagram for the junction point of the strings supporting the mass Q. The coordinate axes are along the horizontal and vertical. (*d*) The same free-body diagram as in (*c*); here, however, the coordinate axes point along the directions of the two strings, which make a right angle with each other.

on the point of support A is oppositely directed, or downward. Since the 15-N weight attached to the string that passes over the pulley P is in equilibrium, the tension in that string must be 15 N, as shown in Figure 3.4(*c*). Finally, there is the tension $\mathbf{T}_B$ in the string AB; its magnitude is unknown.

We now write Equations (3.7a) and (3.7b) for this situation:

$$x \text{ components: } (15 \text{ N}) \cos 37° - T_B \cos 53° = 0$$

$$y \text{ components: } (15 \text{ N}) \sin 37° + T_B \sin 53° - W = 0$$

The first of these can be solved for T_B and gives $T_B = 20$ N. Substituting this value into the second equation and solving for W, we obtain $W = 25$ N. The solution of the problem is that Q weighs 25 N.

We shall now repeat this problem, solving it with a different set of orthogonal axes, x' and y', shown in Figure 3.4(*d*). Since $37° + 53° = 90°$, the two supporting strings lie along the x' and y' axes. The equilibrium equations now read

$$x' \text{ components: } (15 \text{ N}) - W \sin 37° = 0$$

$$y' \text{ components: } T_B - W \cos 37° = 0$$

where we have decomposed $\mathbf{W}$ into its x' and y' components.

Since we are not interested in the tension $\mathbf{T}_B$, we need only solve the first equation. We find immediately that

$$W = \frac{15 \text{ N}}{\sin 37°} = \frac{15 \text{ N}}{0.6} = 25 \text{ N}$$

The result is the same as before (it had better be so), but was deduced by solving one equation instead of two. Sometimes a slightly unorthodox choice of coordinate axes can be advantageous, though you can solve any problem no matter what axes you select if you exercise proper care.●

● ***Example 3.2*** This last example on equilibrium has important practical implications. A long rope is stretched between points A and B, Figure 3.5(*a*). At each end the rope is tied to a spring scale that measures the force the rope exerts on the supports. Suppose the rope is pulled sideways at its midpoint with a force of 400 N (approximately 90 lb), producing a deflection such that the two segments make angles of 5° with the line AB. What is the reading of the spring scales?

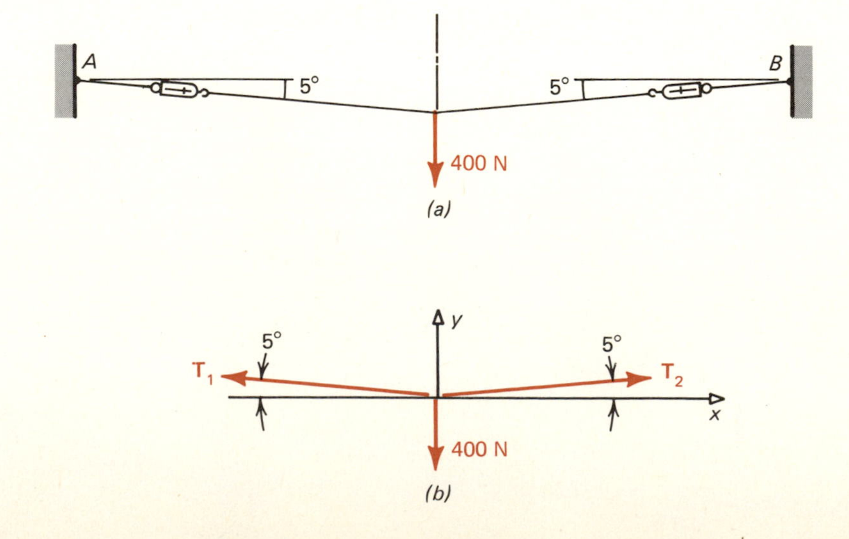

Figure 3.5 *(a) A long rope attached to points A and B by two spring scales, which measure the tension in the rope. A sidewise pull of 400 newtons gives rise to a slight displacement of the center of the rope, so that each segment makes an angle of 5° with the line AB. (b) The free-body diagram corresponding to (a).*

We again isolate the point of contact of the forces; the force diagram is shown in Figure 3.5(*b*). The equilibrium equations are

$$x \text{ components: } T_2 \cos 5° - T_1 \cos 5° = 0$$

$$y \text{ components: } T_1 \sin 5° + T_2 \sin 5° - (400 \text{ N}) = 0$$

The first equation tells us that $T_1 = T_2$, a result apparent from symmetry. Consequently, the second equation becomes

$$2T_1 \sin 5° = 400 \text{ N}; \qquad T_1 = \frac{200 \text{ N}}{\sin 5°} = 2295 \text{ N}$$

The tension in the rope and therefore the force registered on the spring scales is 2295 N (approximately 500 lb). A force of 400 N applied perpendicular to the line *AB* caused a tension of nearly 2300 N, more than five times the applied force in magnitude!

There is a practical lesson to be learned here. Carry a long stout rope in your car at all times. Then, if you are stuck in snow or mud, tie one end to the bumper and the other to a tree or other firm post, leaving as little slack as possible. Ask a passerby or friend to pull the rope *sideways*—not away from the car—near the midpoint of the rope. He will probably be able to extricate you without suffering a hernia, and you will save the cost of a wrecker. Physics is not altogether without merit! ●

We now turn to the second law and illustrate applications with several examples, beginning with one that is very simple.

● ***Example 3.3*** A block of mass $m = 2$ kg rests on an ideal, friction-free horizontal surface. What is the acceleration of the block if a horizontal force of 10 N is applied to it?

We begin by isolating the body, drawing a free-body diagram that exhibits all the forces acting on it (Figure 3.6). There are three forces: the weight $\mathbf{W} = m\mathbf{g}$; the reaction force of the supporting surface $\mathbf{R}$; and the applied force of 10 N. Since the block moves neither up nor down, $\mathbf{R} = -\mathbf{W}$. The net force acting on the body is the horizontal force of 10 N. Hence,

$$\mathbf{a} = \frac{\mathbf{F}}{m} = \frac{10 \text{ N}}{2 \text{ kg}} = 5 \text{ m/s}^2$$

in the direction of $\mathbf{F}$. ●

● ***Example 3.4*** Suppose that instead of applying a horizontal force, we now pull on the block with a force *F* of 10 N by a rope that makes an angle of 37° with the horizontal. What is the acceleration of the block? What is the magnitude of the normal reaction force of the supporting surface on the block?

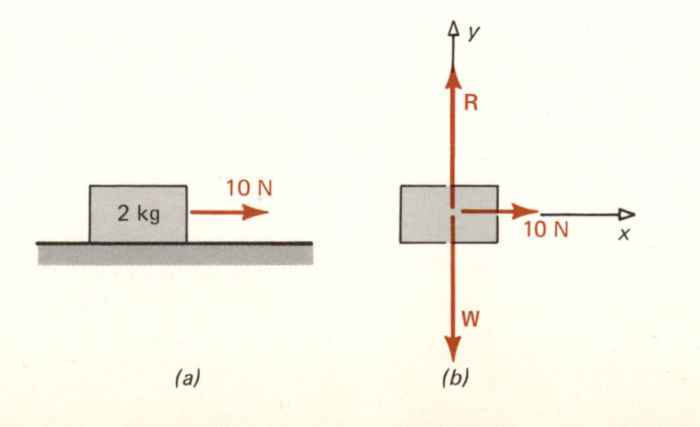

Figure 3.6 (*a*) A 2-kg mass on a friction-free horizontal surface is subjected to a horizontal pull of 10 N. (*b*) The free-body diagram.

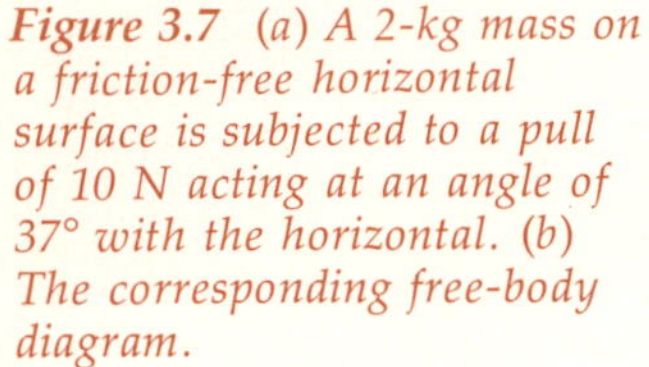

Figure 3.7 (a) A 2-kg mass on a friction-free horizontal surface is subjected to a pull of 10 N acting at an angle of 37° with the horizontal. (b) The corresponding free-body diagram.

The free-body diagram is shown in Figure 3.7. We decompose the forces into their horizontal (x) and vertical (y) components and add the components to find those of the net force.

x components: $F \cos 37° = (10 \text{ N}) \cos 37° = 8 \text{ N}$

$$y \text{ components: } F \sin 37° + R - W = (10 \text{ N})0.6 + R - (2 \text{ kg})(9.8 \text{ m/s}^2) = R - 13.6 \text{ N}$$

Since the block remains on the horizontal surface, $F_y = 0$ and $R = 13.6$ N. Note that this reaction force is less than in Example 3.3, in which $R = W = 19.6$ N. In this instance, a portion of the weight of the block is supported by the tension in the rope.

The acceleration in the x direction is now given by

$$a_x = \frac{F_x}{m} = \frac{8 \text{ N}}{2 \text{ kg}} = 4 \text{ m/s}^2$$

We are familiar with the fact that though it may be possible to lift a fairly heavy object with a relatively thin string if we do this slowly, the string usually breaks if we jerk the object up suddenly. Example 3.5 shows why this happens.

Example 3.5 A thin string that breaks under a tension exceeding 25 N is attached to the roof of an elevator. What is the largest mass that this string can support if the acceleration of the elevator at the beginning of its ascent is 3 m/s²?

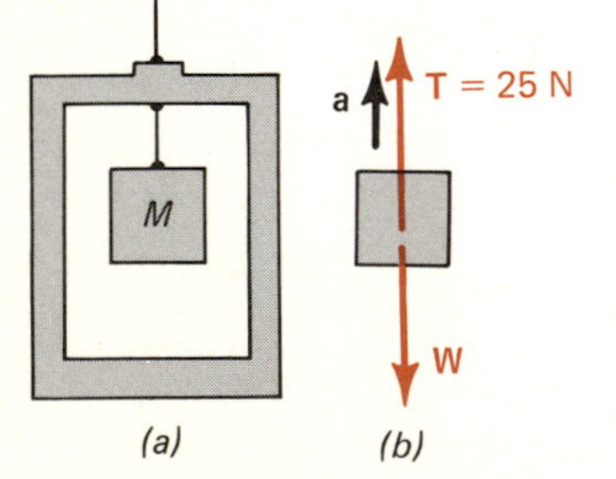

Figure 3.8 (a) A mass M is suspended from the ceiling of an elevator. As the elevator accelerates upward, the mass must also accelerate upward; consequently, the tension in the supporting string must then be greater than the weight of the mass. (b) The free-body diagram for the mass M.

While the elevator is at rest or moving at constant velocity, the system is in equilibrium. Consequently the string can then support a weight of 25 N, corresponding to a mass of $(25 \text{ N})/(9.8 \text{ m/s}^2) = 2.55$ kg.

However, to accelerate the suspended mass upward at 3 m/s² the string must provide a force greater than the weight. From (3.7) we have

$$\mathbf{F} = \mathbf{T} - \mathbf{W} = M\mathbf{a}$$

and substituting the maximum tension, 25 N, for T and (9.8 m/s²) for W, we can now solve for M when $a = 3$ m/s²:

$$25 \text{ N} - (9.8 \text{ m/s}^2)M = (3 \text{ m/s}^2)M$$

$$M = \frac{25 \text{ N}}{12.8 \text{ m/s}^2} = 1.95 \text{ kg}$$

Though the string can support a mass greater than 2.5 kg at rest, it will break during the upward acceleration of the elevator if the mass exceeds 1.95 kg.

In Example 3.5, the upward acceleration of the elevator appears to have the same effect on the string as a sudden increase in the gravitational attraction between that mass M and the earth. This brings to mind the more basic question: How are masses and weights determined?

One normally measures weights of objects. Weighing involves a comparison of forces using a suitable balance. Can one, however, determine *mass* directly, even in the absence of gravitational attraction?

The answer is yes, and Newton's second law tells us how. We simply apply a known force to the object and measure the resulting acceleration. With this procedure, one can then compare this *inertial mass* with the *gravitational mass* obtained by the usual weighing. Whenever such comparison has been made, the two masses have proved identical. This result, which may hold no surprise for you, has nevertheless very deep and important implications. It suggests that it is impossible to tell the difference between the force that must be applied to an object to maintain its position fixed in an *accelerating frame of reference* and the force due to a gravitational attraction on that same mass. In othe words, if you were confined to a spaceship and that spaceship were accelerating through space at 9.8 m/s², you could perform no experiment within the ship that would permit you to decide if you are indeed accelerating through space or merely at rest on the surface of the earth. This is but one example of the *principle of equivalence*, which stands as the cornerstone of the theory of general relativity. It leads to numerous seemingly bizarre conclusions, such as the possible existence of *black holes* (see Chapter 28).

On a more mundane level, we could apply the principle of equivalence in Example 3.5. Since the effect of gravitational attraction is equivalent to a reference frame accelerating upward at 9.8 m/s², the effective gravitational force during the acceleration of the elevator is $(9.8 + a)M = 12.8M$. Hence the string must be able to withstand a tension of $12.8M$.

● ***Example 3.6*** Shortly after blastoff, the upward acceleration of the Saturn V rocket is 80 m/s². What is then the apparent weight of an astronaut of 75-kg mass (approximately 165-lb weight)?

The effective gravitational acceleration is $g + 80\ \text{m/s}^2 = 89.8\ \text{m/s}^2$. The astronaut's apparent weight is therefore $(89.8\ \text{m/s}^2)(75\ \text{kg}) = 6735\ \text{N}$ (about 1514 lb). This is the force the seat exerts on him to keep him at rest in the accelerating reference frame. It is this sensation of increased weight that tells the astronaut, and to a much lesser extent the airplane passenger on takeoff, that he is in an accelerating reference frame.

This large upward force must act on all parts of the astronaut's body if the body is to accelerate as a unit; the force acts on his skeleton and internal organs and on the blood in his arteries and veins. If he were standing upright during blastoff, his skeletal structure would not suffer—bones and joints can withstand great stress, but the connecting tissues and membranes that hold kidneys, pancreas, bowels, and so on in place might well be strained excessively, causing severe internal damage. The increased weight of the blood would force it into the blood vessels of the legs, distending veins and rupturing capillaries. The heart could not provide enough pressure to force the blood through the circulatory system, and blood flow to the eyes and brain would fall precipitously, causing temporary blindness and loss of consciousness. Astronauts are therefore in a reclining position during blastoff and reentry to minimize physiological stresses. During World War II, fighter and

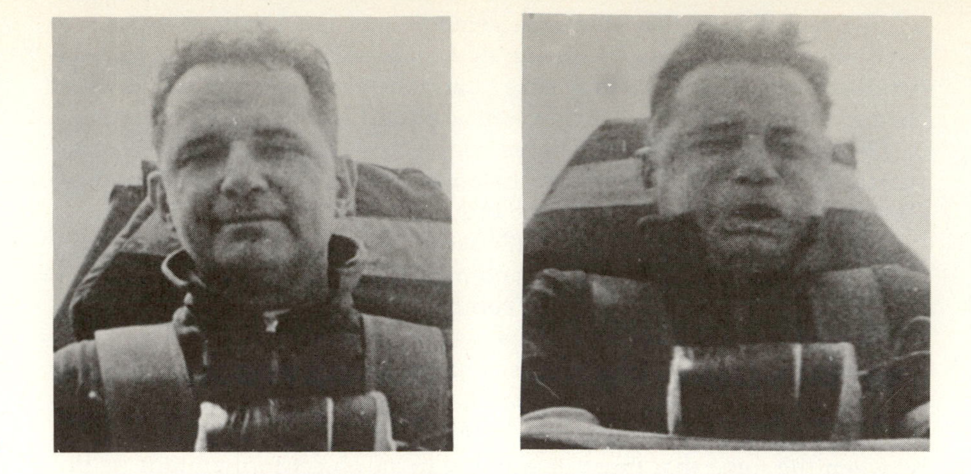

Figure 3.9 Astronaut at takeoff. (NASA photograph.)

dive-bomber pilots frequently experienced brief periods of blackout as they pulled their craft out of steep dives. ●

The preceding examples were limited to the motion of a single body. The next few deal with a more general situation, that of two masses connected by a string. We shall assume that the mass of the string is negligibly small and shall sometimes refer to such connecting members as "massless" strings. The technique for solving these problems that we now demonstrate can be extended to more complicated systems involving three or more objects.

● ***Example 3.7*** Two masses of 5 kg and 8 kg are connected by a massless string. They are supported on a friction-free horizontal surface. A force of 20 N is applied to the 8-kg mass, as shown in Figure 3.10(*a*). We want to know the tension in the string between the two masses.

We proceed by drawing two free-body diagrams, one for each mass, as shown in Figures 3.10(*b*) and (*c*). In diagrams (*b*) and (*c*) the normal reaction force of the supporting surface is equal in magnitude to the

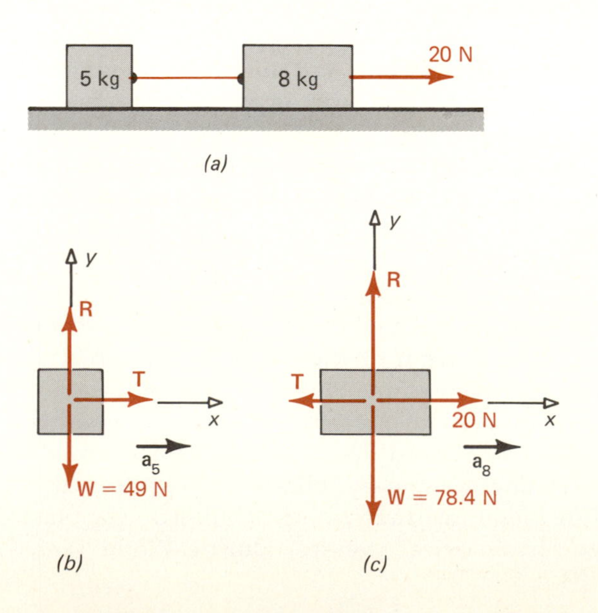

Figure 3.10 (*a*) Two masses, 5 kg and 8 kg, are connected by a string and are pulled along a horizontal friction-free surface by another string to which a tension of 20 N is applied. (*b*) The free-body diagram for the 5-kg mass of (*a*). (*c*) The free-body diagram for the 8-kg mass of (*a*).

weight, (5 kg)(9.8 m/s²) = 49 N and (8 kg)(9.8 m/s²) = 78.4 N. The motion in x direction is given by

$$\text{5-kg mass: } T = (5\text{ kg})a_5$$

$$\text{8-kg mass: } (20\text{ N}) - T = (8\text{ kg})a_8$$

where a_5 and a_8 are the accelerations of the 5-kg and 8-kg masses, and T is the tension in the string that connects the two masses. But the masses are tied together so that $a_5 = a_8 = a$. If we now write, from the first equation, $a = T/5$ kg and substitute this into the second, we obtain

$$T = \frac{(20\text{ N})(5\text{ kg})}{13\text{ kg}} = 7.69\text{ N}$$

An alternative approach is to determine the acceleration of the total mass of the system subject to the applied force of 20 N. We then find

$$a = \frac{F}{m} = \frac{20\text{ N}}{13\text{ kg}} = 1.54\text{ m/s}^2$$

Since both masses must accelerate equally, the tension in the string that pulls the 5-kg mass must be

$$T = (5\text{ kg})(1.54\text{ m/s}^2) = 7.69\text{ N}$$

The second method is apparently simpler and more direct. Nevertheless, it is probably better to follow the somewhat more cumbersome procedure of treating each part of a many-body system separately, taking care to show all the forces acting on each part. For example, if there is friction between the surface and the 5-kg mass and none between the surface of the table and the 8-kg mass, treating each mass separately is the most reliable procedure for arriving at the correct result (Problem 3.39). ●

● ***Example 3.8*** Two masses are connected by a massless string that passes over a massless, friction-free pulley [Figure 3.11(*a*)]. We want to know the acceleration of the masses and the tension in the string.

The two free-body diagrams are shown in Figure 3.11(*b*). The tension in the string is denoted by T, and in this idealized case, it is the same on both sides of the pulley and independent of the position of the masses.*

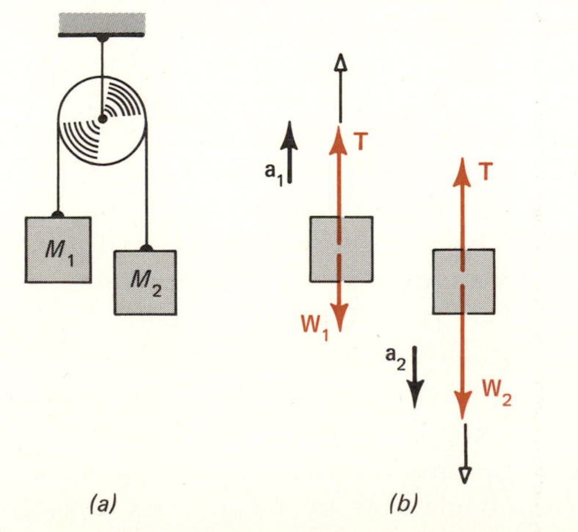

Figure 3.11 (*a*) Two masses, M_1 and M_2, are connected by a thin (massless) string that passes over a light (massless) pulley on friction-free bearings. Such an arrangement is sometimes called an Atwood machine and can be used to measure the acceleration of gravity, at least approximately, since the motion is relatively slow if the masses start from rest, and are large and nearly equal. (*b*) The free-body diagrams for the two masses, M_1 and M_2. Note that the positive direction is taken up for one mass (M_1) and down for the other (M_2); if M_1 moves up M_2 must move down.

* If the string and pulley were not massless or the bearings not friction-free, the tensions would differ on the two sides. We shall later repeat this example, using then a pulley of finite mass.

Having drawn the free-body diagrams, we must now select the positive directions of our coordinate systems. These need not be the same for both diagrams, and it is sensible in this instance to use the upward direction for one and the downward for the other. The reason is that since the masses must move with the same acceleration (magnitude) and speed, one will rise as the other falls. The assumed positive directions are shown in Figure 3.11(*b*).

We now write Equation (3.7b) for each body:

$$T - W_1 = T - M_1g = M_1a_1 = M_1a \tag{3.9a}$$

$$W_2 - T = M_2g - T = M_2a_2 = M_2a \tag{3.9b}$$

where we have used the fact that $a_1 = a_2 = a$.

We have here two linear algebraic equations with two unknowns, T and a. We can eliminate the tension T by adding the two equations; we then get

$$M_2g - M_1g = M_1a + M_2a$$

$$(M_2 - M_1)g = (M_1 + M_2)a \tag{3.10}$$

with the result

$$a = \frac{M_2 - M_1}{M_1 + M_2}g \tag{3.11}$$

We determine the tension by substituting the above expression for a into either of the initial equations. For example,

$$T - M_1g = M_1a = M_1\left[\frac{M_2 - M_1}{M_1 + M_2}\right]g$$

$$T = \frac{M_1M_2 - M_1^2}{M_1 + M_2}g + M_1g = \frac{2M_1M_2}{M_1 + M_2}g \tag{3.12}$$

So much for the general solution of this problem. Let us see if the results are reasonable, if they conform to what we might expect.

First, we have arbitrarily selected downward as the positive direction for the free-body diagram of mass M_2. From (3.11) we see that a is positive (M_2 accelerates down, M_1 accelerates up) if M_2 is greater than M_1; a is negative if $M_2 < M_1$. That is certainly what we should predict. Second, if for instance $M_2 > M_1$, then according to (3.12), the tension T is less than M_2g but greater than M_1g. That, too, makes sense. If M_1 is to accelerate upward, as it does under these conditions, the tension must be greater than the weight M_1g. Similarly, if M_2 accelerates downward but with an acceleration less than g (free fall), T must be less than M_2g but greater than zero. If, however, $M_1 = M_2$, Equations (3.11) and (3.12) give $a = 0$, $T = M_1g = M_2g$; the system is now in equilibrium and the tension is such as to exactly balance the weight of each mass. Finally, Equation (3.10) can be given a simple physical interpretation. The left-hand side is the net gravitational force, $M_2g - M_1g$, acting on a system of total mass $M_1 + M_2$; Equation (3.10) is then a statement of Newton's second law as it applies to this system. ●

● ***Example 3.9*** A similar problem is shown in Figure 3.12(*a*). Here a 10-kg mass is supported on a friction-free inclined plane and is connected to a second mass M by means of the ideal string-and-pulley arrangement. The problem is to find the mass M, the reaction force of the inclined plane on the 10-kg mass, and the tension in the string, given that the acceleration of the mass M is 3 m/s² upward.

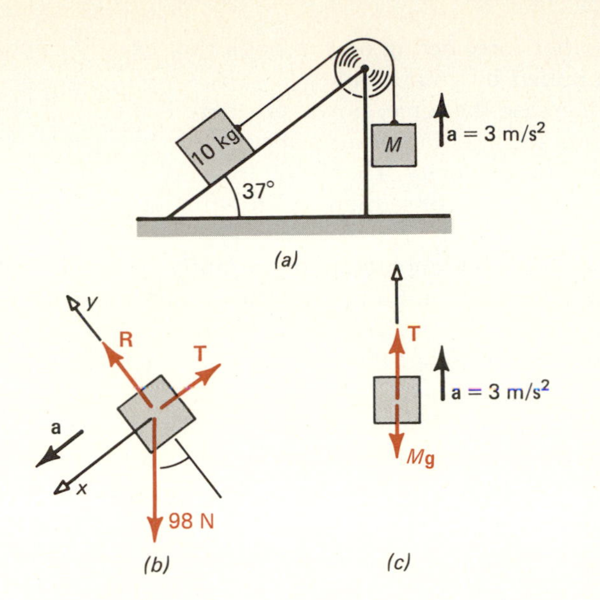

Figure 3.12 (*a*) *A 10-kg mass is supported on an inclined plane and is connected by a thin string to another mass M.* (*b*) *The free-body diagram for the 10-kg mass. The coordinate axes are chosen so that one (the x axis) is in the direction of the known acceleration and the other is perpendicular to the plane. Since the plane is friction-free, it can exert only a force perpendicular to itself, the normal reaction force* **R**. *The weight of the 10-kg mass, 98 newtons, has components along both the x and y axes.* (*c*) *The free-body diagram for the unknown mass M.*

We again draw two separate free-body diagrams [Figures 3.12(*b*), (*c*)]. Since the motion of the 10-kg mass is along the plane, the most convenient choice of orthogonal axes for that body is with axes parallel and perpendicular to the plane, not horizontal and vertical. The positive direction selected is the known direction of motion.

The free-body diagram for the 10-kg mass deserves a few comments. The forces acting on the mass are $m\mathbf{g} = (10\text{ kg})(9.8\text{ m/s}^2) = 98\text{ N}$, the tension in the string, and the reaction force of the plane on the mass. That reaction force *must* be perpendicular to the plane; a perfectly smooth, friction-free plane can exert no tangential force.

The equations of motion now read

$$y \text{ direction: } R - (10\text{ kg})(9.8\text{ m/s}^2)\cos 37° = R - 78.4\text{ N} = 0$$

$$x \text{ direction: } (10\text{ kg})(9.8\text{ m/s}^2)\sin 37° - T = 58.8\text{ N} - T$$

$$= (10\text{ kg})a = (10\text{ kg})(3\text{ m/s}^2) = 30\text{ N}$$

$$T - Mg = Ma = (3\text{ m/s}^2)M$$

where the first two equations are obtained from Figure 3.12(*b*) and the third from Figure 3.12(*c*).

The first equation reflects the fact that the 10-kg mass remains on the inclined plane, that is, has zero acceleration in the y direction of Figure 3.12(*b*). The reaction force is therefore

$$R = 78.4\text{ N}$$

From the second equation, we obtain

$$T = 28.8\text{ N}$$

and substituting this value in the last equation, we find

$$M = 2.25\text{ kg}$$

3.4 Friction

Place a book on the table and push against it horizontally, gradually increasing the force you exert. At first, the book remains stationary even though you apply a small force. From the first law you conclude that there must be another force acting on the book that exactly opposes and balances that applied by your hand.

As you increase the force, however, there is a critical point at which the book starts to slide, but you must continue to push it to maintain its motion at constant velocity. Therefore, to keep the book in equilibrium, you exert a force, and consequently, an equal opposing force must also act on the book.

If you do this simple experiment carefully, you will notice that the force needed to initiate motion is slightly greater than the force to maintain it.

The forces that opposed the motion are *friction forces and are always parallel to the surface of contact and directed opposite to the direction of motion or intended motion.* The friction force that prevents initiation of motion is called the force of *static friction,* denoted f_s. The friction force that opposes motion once initiated is called the force of *kinetic friction,* denoted f_k.

Next, place another book on top of the first and repeat the experiment, pushing on the lower book. Both the force required to initiate motion and the force to maintain it will have increased. Hence, f_s as well as f_k appear to depend on the force with which the object presses on the supporting surface, that is, on the normal reaction force of the surface against the object.

The physical origin of friction forces is the irregularity of the contacting surfaces. In some instances these irregularities are evident to the naked eye, for example, the rough surface of an upholstered chair or the cement of a sidewalk. But irregularities also exist on apparently smooth surfaces, although they become evident only on microscopic examination. When one object rests on another, contact between small protrusions and depressions calls forth some effort to shift the surfaces relative to each other. This effort will be greater the greater the force with which the surfaces are pressed together and the more pronounced the surface irregularities.

Over a moderate range of relative velocities, the frictional force between two contacting surfaces is well approximated by

$$f_k = \mu_k R \tag{3.13}$$

where R is the normal (perpendicular) reaction force and μ_k is a constant of proportionality whose value depends on the properties of the contacting surfaces. This proportionality constant μ_k is called the *coefficient of kinetic friction.*

Similarly, the *maximum* possible force of static friction can be written

$$f_s^{\max} = \mu_s R \tag{3.14}$$

where μ_s is the *coefficient of static friction,* whose value depends on the nature of the two contacting surfaces. Note that f_s *may be less than* $\mu_s R$! Under static conditions, f_s is just equal but opposite to the applied force

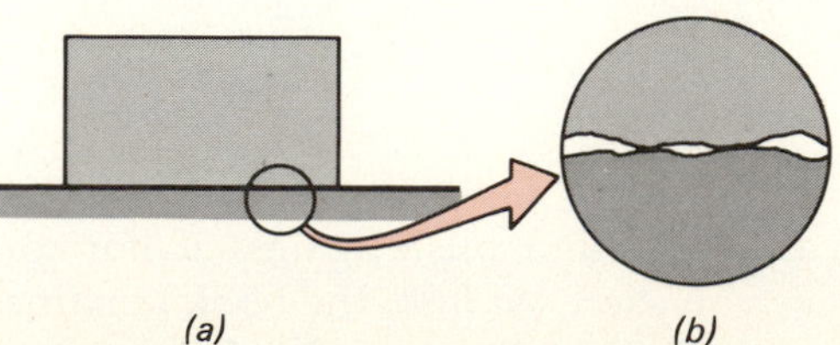

Figure 3.13 *(a) Two surfaces in contact. (b) Enlarged inset, showing surface roughness that results in static and kinetic friction.*

that tends to move the object parallel to the contacting surface but fails to do so. If the applied force is less than $\mu_s R$, so is f_s.

Although we often view friction as an undesirable evil and go to great pains to reduce it by lubrication, we would really be in deep trouble without it. The simplest actions, riding in a car or even walking, would be quite impossible if friction were eliminated. An oil slick or a sheet of ice on the pavement, greatly reducing friction, often spells disaster.

● ***Example 3.10*** An upholstered block of wood is placed on a plank that is covered with a woolen cloth. One end of the plank is raised slowly, and the block starts to slide just when the angle the plank makes with the horizontal is 53°. What is the coefficient of static friction between the upholstery and the wool?

At first glance the problem may seem insoluble; the weight of the block is not given and so we do not know the normal reaction force R. We shall see, however, that the mass of the block cancels out; that is, any block, no matter what its mass, that is upholstered with the same material, will start to slide down this plank when the angle of elevation exceeds 53°.

The free-body diagram of the system is shown in Figure 3.14(b). Since the block is just barely at rest when the elevation angle is 53°, the sum of the forces perpendicular and parallel to the plane must vanish:

x components: $Mg \sin 53° - f_s = 0; \qquad f_s = Mg \sin 53°$

y components: $Mg \cos 53° - R = 0; \qquad R = Mg \cos 53°$

Since f_s at this angle has attained its maximum value, we have, by the defining equation (3.14),

$$\mu_s = \frac{f_s^{\max}}{R} = \frac{Mg \sin 53°}{Mg \cos 53°} = \tan 53° = 1.33$$

In many instances the coefficient of friction is less than one; in this case it exceeds one. ●

Let us now find the force of static friction that acts on the same block when the angle of elevation is less than 53°, say 37° (see Figure 3.15 on the next page). In that case $R = Mg \cos 37° = 0.8Mg$. We might then thoughtlessly conclude that

$$f_s = \mu_s R = 1.33 \times 0.8Mg = 1.07Mg$$

But this *cannot* be correct.

The force pulling the block downward along the plane is the component of Mg parallel to the plank, $Mg\sin 37° = 0.6Mg$. Thus if $f_s = 1.07Mg$, pointing upward along the plank, the net force on the block would be $(1.07 - 0.6)Mg = 0.47Mg$ *up the plank*. The block would then

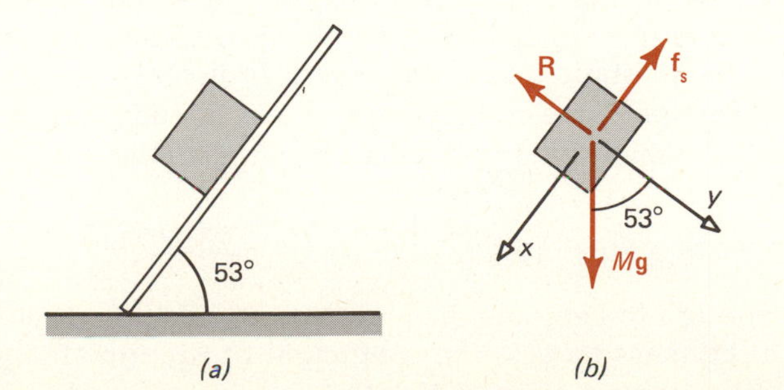

Figure 3.14 *(a) An upholstered block of wood rests on a steep inclined plank that is covered with a woolen cloth. Since the angle shown, 53°, is the critical angle for equilibrium, the force of static friction has attained its maximum value for this system. (b) The free-body diagram corresponding to (a). The force of static friction must equal, in magnitude, the x component of* **Mg**.

have to accelerate up the inclined plane, in defiance of the law of gravity! The answer is that $f_s = 0.6Mg$; *the force of static friction is not equal to its maximum possible value* $f_s^{max} = \mu_s R$ but is just enough to maintain the block at rest.

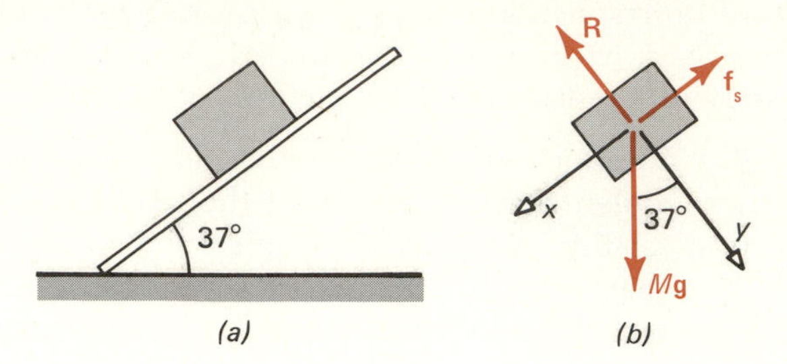

Figure 3.15 *(a) The same block of wood and inclined plank as in Figure 3.14. The elevation angle of the plank is now 37°. (b) The free-body diagram corresponding to (a). The component of* **Mg** *along the plane is now less than when the elevation angle was 53°; hence* f_s *is also smaller.*

● ***Example 3.11*** We return to an earlier example, that of the 10-kg mass on an inclined plane (Example 3.9). We found that if we neglect friction, this mass accelerates down the plane at 3 m/s² when a mass of 2.25 kg is attached to the end of the string that passes over the ideal pulley at the top of the incline. Let us now introduce friction. We assume that the system is initially at rest and that the coefficients of static and kinetic friction between the 10-kg mass and the plane are $\mu_s = 0.25$ and $\mu_k = 0.15$. Will the system remain at rest? If not, what are the acceleration and the tension in the string?

Since, as we already know from the earlier example, the 10-kg mass tends to slide down the plane, the direction of the friction forces must be upward along the plane. Let us first determine the magnitude of the static friction force needed to keep the mass at rest. *If* the system is at rest, the forces acting on the 10-kg mass parallel to the plane are

(a) the component of its weight along the plane
(b) the tension in the string
(c) the force of static friction f_s

For the system to remain at rest, these forces must add to zero.
The first force is

$$(10 \text{ kg})(9.8 \text{ m/s}^2) \sin 37° = 58.8 \text{ N}$$

The tension in the string just equals the weight of the 2.25-kg mass, since this mass is now supported against gravity by the string. It is therefore $T = 2.25g = 22.1$ N. Note that this tension is not the same calculated earlier, 28.8 N, but less. We are *assuming* that the system is in equilibrium, whereas before, the 2.25 kg had an upward acceleration.

Applying now the condition for equilibrium, we have

$$58.8 \text{ N} - 22.1 \text{ N} - f_s = 0$$
$$f_s = 36.7 \text{ N}$$

This is the force of static friction *necessary to prevent motion.* The maximum static friction force is $0.25R$; we have already calculated R and found its value, namely 78.4 N. Hence the largest possible force of static friction is

$$f_s^{max} = 0.25(78.4 \text{ N}) = 19.6 \text{ N}$$

This is not enough to keep the mass at rest; it will slide down the plane.

As the 10-kg mass moves, the coefficient of friction changes to $\mu_k = 0.15$. Hence the forces acting on the 10-kg mass are as shown in Figure

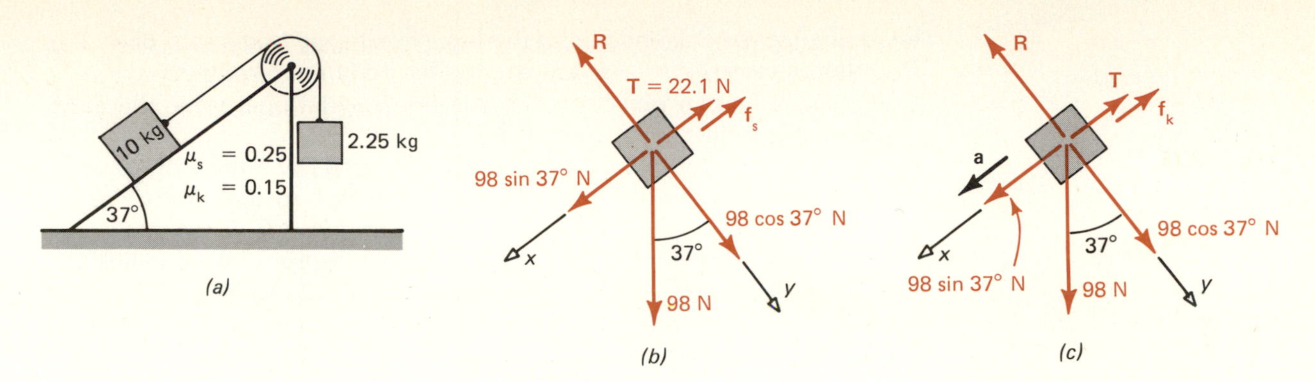

Figure 3.16 (*a*) The same condition as in Figure 3.12 except that now the inclined plane is not friction-free. The coefficient of static friction between the 10-kg mass and the plane is 0.25, and the coefficient of kinetic friction between the 10-kg mass and the plane is 0.15. (*b*) The free-body diagram for the 10-kg mass of (*a*). It is assumed that the system is in equilibrium as a result of the combined effects of the tension in the string that supports the 2.25-kg mass and the force of static friction. In that case, the tension in the string would just equal the weight of the 2.25-kg mass. (*c*) The free-body diagram for the 10-kg mass of (*a*) under the actual conditions. The system accelerates so that the 10-kg mass slides down the plane. The forces acting upward along the plane are the tension in the string and the force of kinetic friction. Because the 10-kg mass is accelerating down the plane and the 2.25-kg mass therefore accelerating upward, the tension in the string will be greater than if the system were in equilibrium.

3.16(*c*). The friction force acting in the direction opposite the motion is

$$f_k = \mu_k R = 0.15(78.4\ \text{N}) = 11.8\ \text{N}$$

We now apply Newton's second law to the two isolated bodies:

$$58.8\ \text{N} - 11.8\ \text{N} - T = (10\ \text{kg})a$$

$$T - (2.25\ \text{kg})(9.8\ \text{m/s}^2) = T - 22.1\ \text{N} = (2.25\ \text{kg})a$$

We again solve by adding the two equations, and obtain

$$58.8\ \text{N} - 11.8\ \text{N} - 22.1\ \text{N} = 24.9\ \text{N} = (12.25\ \text{kg})a$$

$$a = 2.03\ \text{m/s}^2$$

$$T = (2.25\ \text{kg})(2.03\ \text{m/s}^2) + 22.1\ \text{N} = 26.7\ \text{N}$$

As expected, friction has reduced the acceleration of the system, from 3 m/s² to nearly 2 m/s². Moreover, the tension in the string is also reduced, with a value between 22.1 N, the tension when the system is kept at rest, and 28.8 N, the tension when there is no friction and the acceleration is greatest. ●

Summary

Newton's first law states that any body at rest will remain at rest and a body in motion will continue in straight-line motion at constant velocity unless a net force acts on it.

A body that does not change its state of motion is said to be in *equilibrium.* A body need not be at rest to be in equilibrium.

One of the conditions for equilibrium of a body is that the vector sum of all the forces acting on it vanish.

Newton's second law, somewhat rephrased, states that if a force acts on a body of mass m, that body will suffer an acceleration in the direction of the applied force whose magnitude is proportional to the magnitude of the force and inversely proportional to the mass. Newton's second law is commonly written

$$\mathbf{F} = m\mathbf{a}$$

It serves as a defining equation for force. The SI unit of force is the newton, N; a force of one newton will cause a mass of 1 kg to accelerate at 1 m/s².

Newton's third law states that to every action force there is an equal but oppositely directed reaction force. This law refers to the mutual interaction between *two* bodies; if the force exerted *by* body A *on* body B is $\mathbf{F}_{AB}$, then a force $\mathbf{F}_{BA} = -\mathbf{F}_{AB}$ is exerted *by* body B *on* body A.

The *weight* of an object on earth is the gravitational force the earth exerts on the object.

The *principle of equivalence* holds that the effect of a uniform gravitational field is indistinguishable from the effect of an accelerating reference frame.

Surface roughness produces friction forces. One distinguishes between *static friction* and *kinetic friction*. The maximum value of the force of static friction is given by $f_s = \mu_s R$; the force of kinetic friction is given by $f_k = \mu_k R$. Here μ_s is the coefficient of static friction, μ_k the coefficient of kinetic friction, and R the normal reaction force between the contacting surfaces.

In solving dynamics problems involving one or more bodies, one proceeds as follows. Each body is isolated and a free-body diagram is drawn on which all the forces acting on that body are shown. The equation of motion, that is, Newton's second law, is written for each body, and these equations are then solved.

Multiple Choice Questions

3.1 An object is thrown straight up. At the top of its trajectory the object is

(a) instantaneously at rest.
(b) instantaneously in equilibrium.
(c) (a) and (b) are both correct.
(d) Neither (a) nor (b) are correct statements.

3.2 An object is moving at constant velocity. The total force F acting on that object is given by

(a) $F = v^2/2m$
(b) $F = mv$
(c) $F = 0$
(d) $F = mg$

3.3 Which of the following statements describing an object in equilibrium is *not* true?

(a) The vector sum of all the forces acting on the body is zero.
(b) The body is moving at constant speed.
(c) The body must be at rest.
(d) The body is moving at constant velocity.

3.4 A block of mass M is sliding along a friction-free inclined plane, as shown in Figure 3.17. The reaction force exerted by the plane on the block is

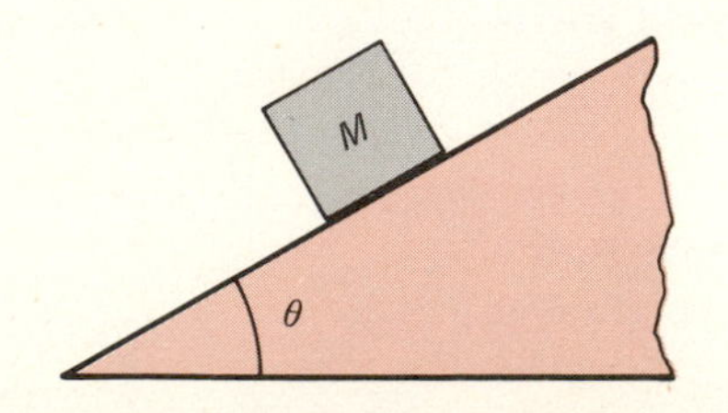

Figure 3.17

(a) $g \sin\theta$
(b) $Mg \sin\theta$
(c) $Mg \cos\theta$
(d) zero because the plane is friction-free.

3.5 A mass is suspended from a string and accelerates downward with an acceleration of $0.7g$. It follows that the tension in the string is

(a) zero.
(b) not zero but less than the weight of the mass.
(c) greater than the weight of the mass.
(d) equal to the weight of the mass.

3.6 A block of mass m rests on an inclined plane that makes an angle of 30° with the horizontal, as shown in Figure 3.18. Which of the following statements about the force of static friction is true?

(a) $f_s \geqq mg$
(b) $f_s \geqq mg \cos 30°$
(c) $f_s = mg \cos 30°$
(d) $f_s = mg \sin 30°$

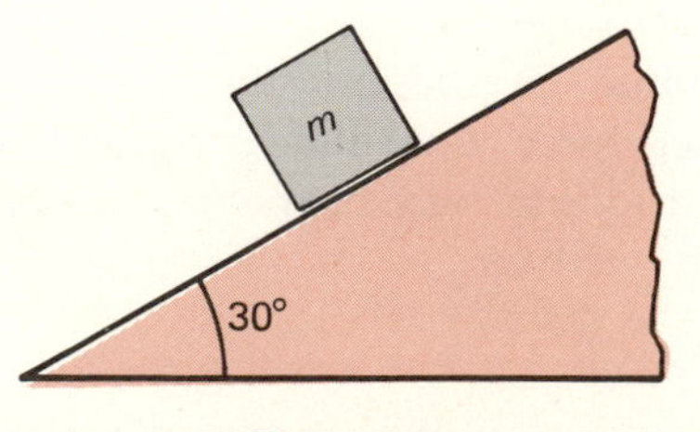

Figure 3.18

3.7 Figure 3.19 shows a system that is in equilibrium. There is no friction between the block of mass M_1 and the inclined plane, and the pulley is friction-

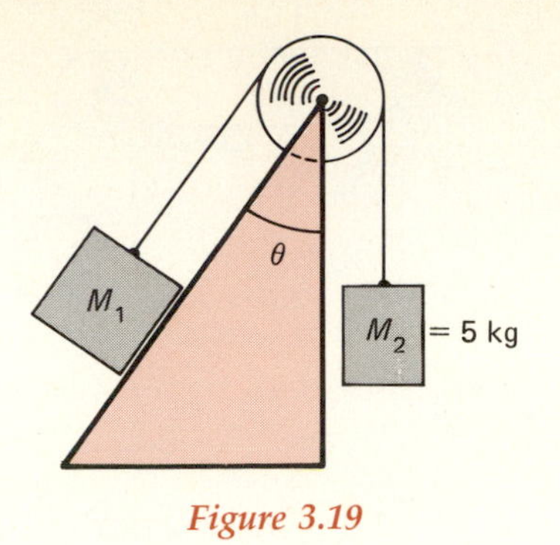

Figure 3.19

free. Mass M_2 = 5 kg; mass M_1 is not known. The tension in the string

(a) is 5g N.

(b) is 5g cos θ N.

(c) is 5g sin θ N.

(d) cannot be determined because M_1 is not given.

3.8 A block of mass M is pulled along a surface, as shown in Figure 3.20. The velocity of the block is constant. The coefficient of kinetic friction between the block and the surface is μ, and the tension in the rope is T. The tension T is then given by

(a) $T = Mg/\mu$

(b) $T = \mu g$

(c) $T = \mu Mg$

(d) none of the above.

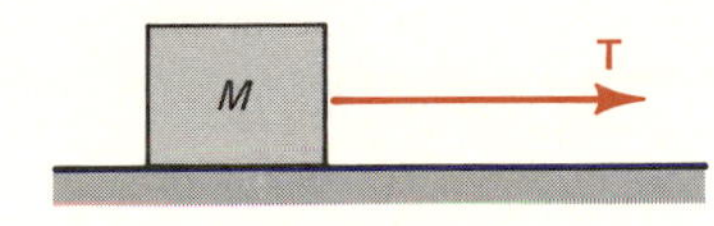

Figure 3.20

3.9 A block of mass m is pulled along a surface as shown. The coefficient of kinetic friction between the block and the surface is μ, and the tension in the rope is T. The acceleration of the block is then

(a) $a = T \cos \theta/\mu mg$

(b) $a = T \cos \theta/\mu g$

(c) $a = T \cos \theta/m + \mu mg$

(d) none of the above.

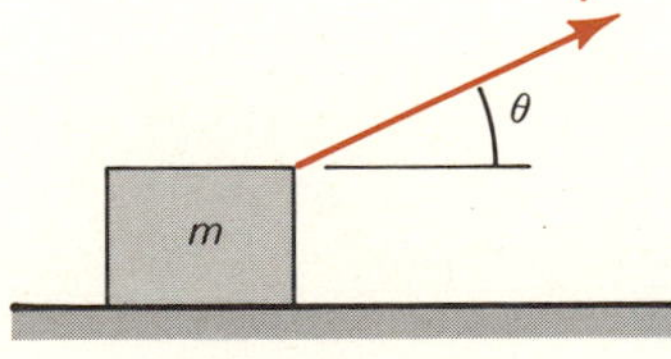

Figure 3.21

3.10 An object is sliding on a horizontal surface, following a push that imparted an initial velocity v in the positive x direction. If the coefficient of kinetic friction between the object and the surface is μ, the acceleration of the object is

(a) $a_x = -\mu m$

(b) $a_x = -\mu g$

(c) $a_x = -\mu mg$

(d) $a_x = -g/\mu$

3.11 Two masses M and m, with $M > m$, are hung over a massless, friction-free pulley, as shown in Figure 3.22. The downward acceleration of the mass M is

(a) g

(b) $\frac{M}{m} g$

(c) $\frac{M - m}{Mm} g$

(d) $\frac{M - m}{M + m} g$

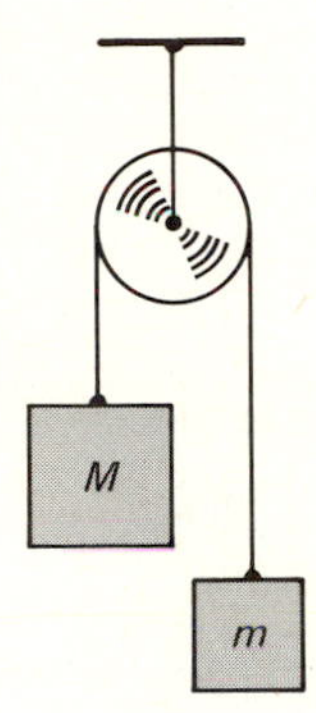

Figure 3.22

3.12 Two masses m_1 and m_2 are accelerated uniformly on a frictionless surface as shown. The ratio of the tensions T_1/T_2 is given by

(a) m_1/m_2

(b) m_2/m_1

(c) $(m_1 + m_2)/m_2$

(d) $m_1/(m_1 + m_2)$

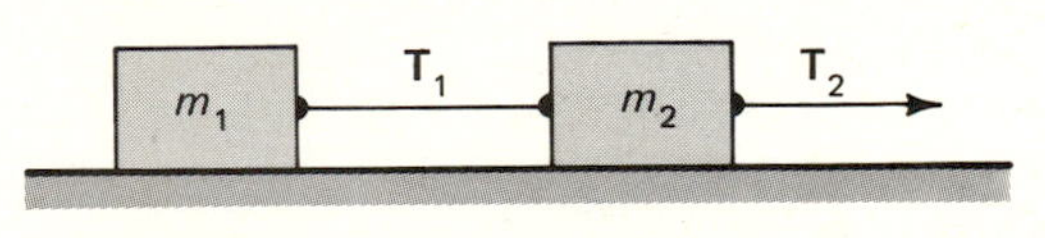

Figure 3.23

3.13 At the surface of a certain planet, acceleration due to gravity is only one-quarter of that on Earth. A 4-kg brass ball is transported to this planet. Which of the following statements is *not* true?

(a) The mass of the brass ball on this planet is only a quarter of its mass as measured on Earth.

(b) The weight of the brass ball on this planet is only a quarter of the weight as measured on Earth.

(c) The brass ball has the same mass on the other planet as on Earth.

Problems

(Sections 3.2, 3.3)

3.1 A ball rolls on a flat table and then falls to the floor. What are the action-reaction forces (a) while it is rolling on the table? (b) as it is falling to the floor? (c) after it has come to rest on the floor?

3.2 A 4-kg bag of potatoes is held by a string. If the tension in the string is 39.2 N, what is the state of motion of the bag? What should the tension in the string be so that the bag accelerates upward at 1.4 m/s^2?

3.3 What is the mass of an object that weighs 24.5 N on earth?

3.4 What is the weight of an object whose mass is 0.4 kg?

3.5 You hold an 0.8-kg mass in your hand. (a) If your hand is at rest, what force does the mass exert on your hand, and what is the reaction force? (b) Suppose you exert an upward force of 10 N on this mass; what is then the motion of the object, and what is then the force exerted by the earth on the mass? Is that latter force equal and opposite to the force exerted by the mass on the earth?

3.6 A 16,000-kg airplane is in level flight at constant velocity. Make a sketch showing the forces that act on this plane. Is the plane in equilibrium?

3.7 A 6-kg objects rests on a frictionless horizontal surface. What horizontal force is required to impart to it an acceleration of 3 m/s^2?

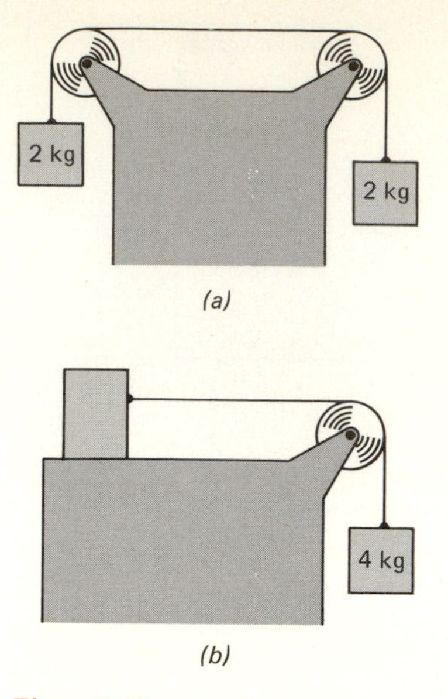

Figure 3.24

3.8 Two equal masses of 3 kg each are connected to the ends of a massless string that passes over a pulley attached to the ceiling by a stout rope. Find the tension in the string and in the rope.

• **3.9** Find the tension in the strings of Figures 3.24(*a*) and (*b*).

• **3.10** Find the tension in the strings of Figures 3.25(*a*)–(*d*).

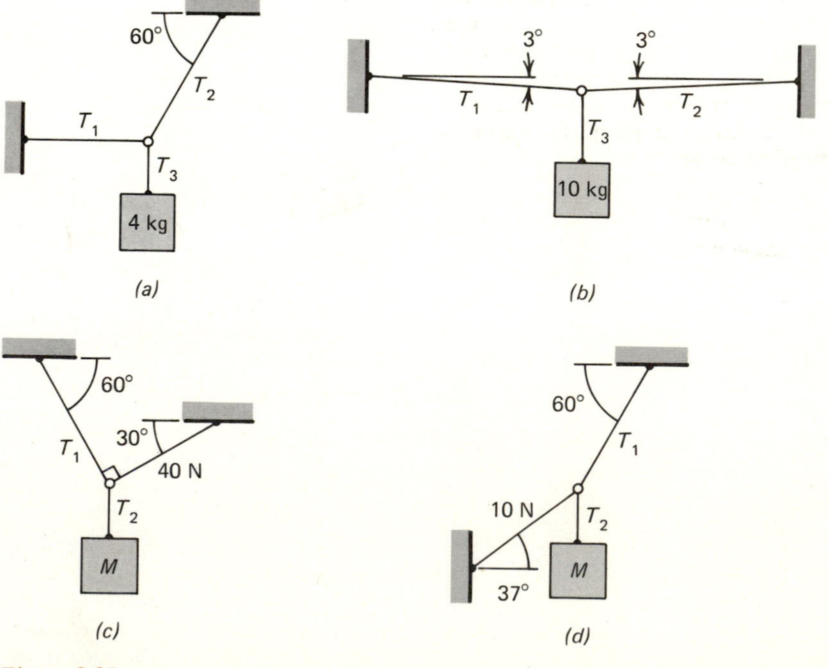

Figure 3.25

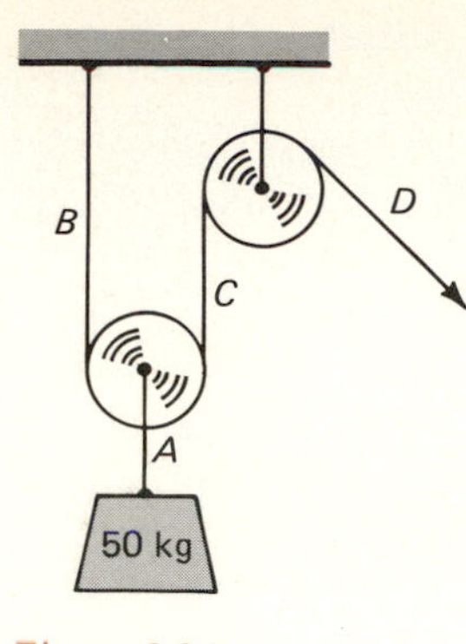

Figure 3.26

• **3.11** The system of two pulleys shown in Figure 3.26 is used to lift a mass of 50 kg. At equilibrium, what are the tensions at the points A, B, C, and D?

• **3.12** A tightrope acrobat, whose mass is 50 kg, crosses a 30-m-long high wire. When the acrobat is at the midpoint, the wire has sagged 0.8 m below the level of the supports. Find the tension in the wire at that instant.

• **3.13** A block, initially at rest on a friction-free horizontal plane, is acted on by a horizontal force of 25 N. If during 4 s the block travels 50 m, what is its mass?

• **3.14** A 1500-kg elevator accelerates upward at 1.6 m/s^2. Find the tension in the cable that lifts the elevator.

• **3.15** Suppose you find yourself stuck in the snow and follow the suggestion on page 51. You tie a 40-m-long rope to the front bumper of the car and the other end to a lamppost further down the street. The rope is now in line with the direction the car is headed. You now push sideways at the midpoint of the rope, and exerting a force of 360 N, you displace the midpoint by 3 m. Calculate the force on the car in the direction it is heading as well as the tension in the rope.

• **3.16** The system shown in Figure 3.27 is in equilibrium. What is the angle θ and what is the tension T in the string?

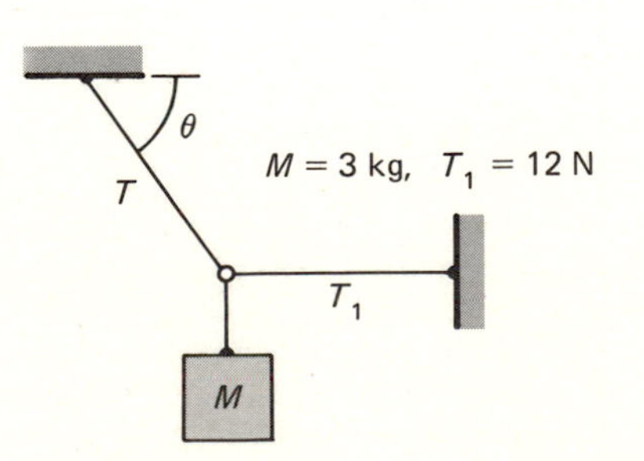

Figure 3.27

• **3.17** Figure 3.28 shows a device often used to provide support and traction for an injured leg. Determine the tension in the rope and the traction force (tension) on the leg.

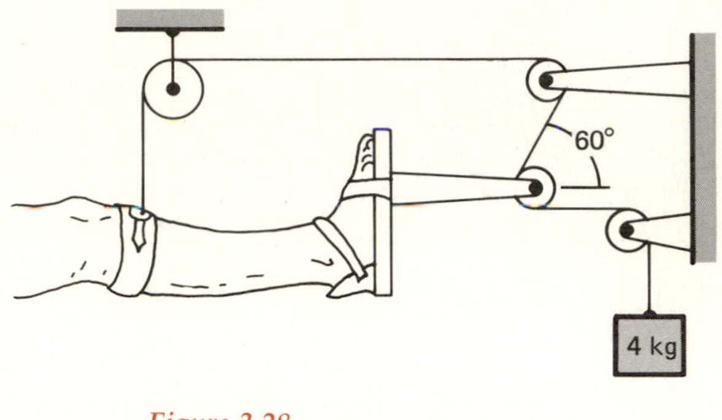

Figure 3.28

• **3.18** A climber rappels from a vertical cliff, using the technique indicated in Figure 3.29(a). If it is assumed that her weight is 500 N and that the forces exerted by her feet are perpendicular to the face of the cliff, what is the tension in the rope?

Figure 3.29

• • **3.19** Suppose that in Problem 3.18 the angle between cliff face and rope remains 12°, but the cliff face makes an angle of 15° with the vertical, as shown in Figure 3.29(b). If the force exerted by the climber's feet is again perpendicular to the cliff face, what is the tension in the rope now?

• **3.20** A string can withstand a tension of 70 N before breaking. If this string is used to support a 5-kg mass, what is the maximum upward acceleration that can be given to that mass?

• •**3.21** In Figure 3.33 the mass M_1 is 5 kg and is initially at rest. At $t = 4$ s, it is just 0.8 m below its starting point. Find the value of M_2 and the tension in the string.

• •**3.22** Mass M_1 rests on a friction-free horizontal surface and is connected to mass M_2 by a frictionless and massless pulley and massless string, as shown in Figure 3.30. If $M_1 = 3$ kg and $M_2 = 2$ kg and the system is initially at rest, what will the velocity of M_1 be after 2 s, and what is the tension in the string?

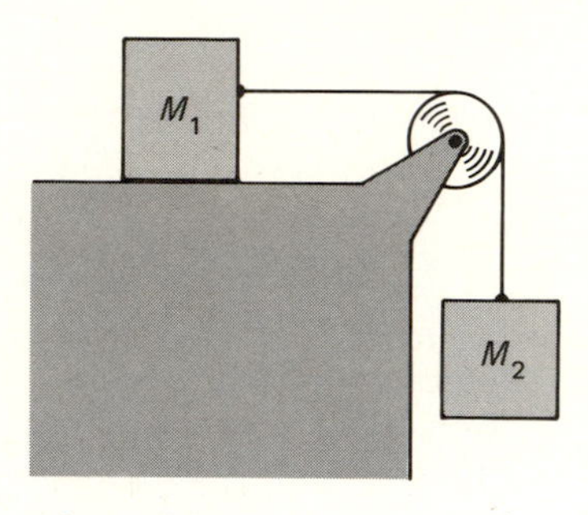

Figure 3.30

• **3.23** A 1200-kg elevator carries a maximum additional load of 500 kg. When the elevator is ascending, the initial upward acceleration is 2.2*g*. What tension can the cable supporting the elevator withstand? Assume the elevator has been designed with a safety factor of 2.

• **3.24** In Figure 3.31, the system starts from rest. What should the mass of M_2 be so that the 8-kg mass drops 0.98 m in exactly 1 s?

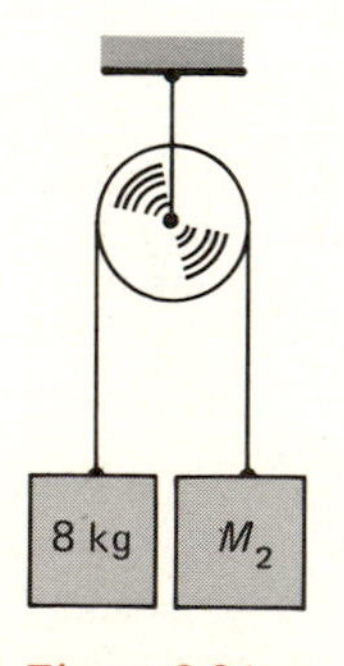

Figure 3.31

• **3.25** Find the value of the unknown weight in each of the situations depicted in Figure 3.32.

• •**3.26** A 5-kg mass slides on a frictionless inclined plane. Its initial velocity is 0.6 m/s, directed upward along the plane. If the angle between the inclined plane and the horizontal is 37°, how far up the plane does the mass move before it descends the plane? How much time elapses before the mass reaches its starting point?

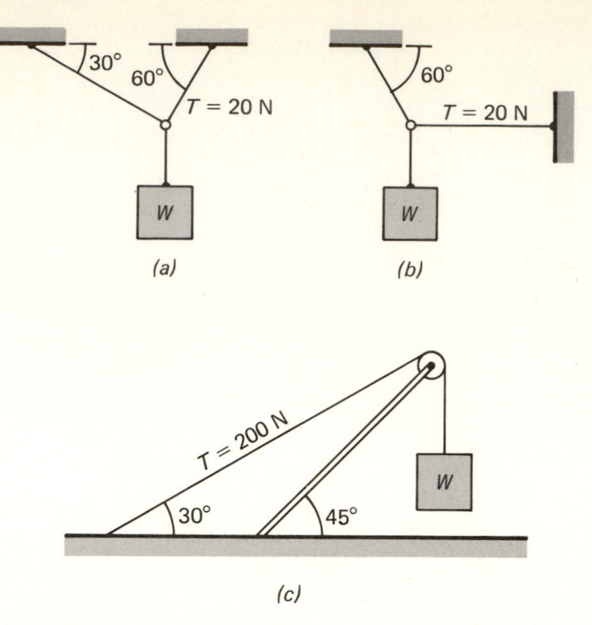

Figure 3.32

• •**3.27** The mass M_2 is 4 kg and the tension in the string in Figure 3.33 is 50 N. If at $t = 0$, the mass M_2 is descending with a velocity of 3 m/s, where will it be in relation to its starting point at $t = 2$ s, and at $t = 6$ s? What is the value of M_1?

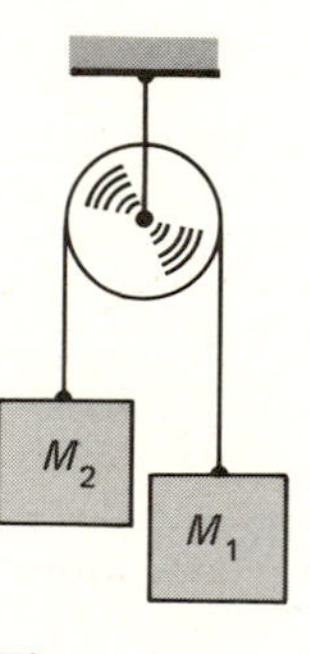

Figure 3.33

• •**3.28** What is the mass M_2 in Figure 3.34 if the tension in the string is 35 N? In which direction will the system move under these conditions, and what will be the acceleration? There is no friction between the blocks and inclined planes.

• **3.29** A woman whose mass is 60 kg is standing in an elevator that moves up at a constant velocity of 5 m/s. If she were standing on a scale while in the elevator, what would be the reading of the scale? As the elevator approaches the proper floor, it decelerates at 3 m/s^2. What would the scale reading be while the elevator is decelerating?

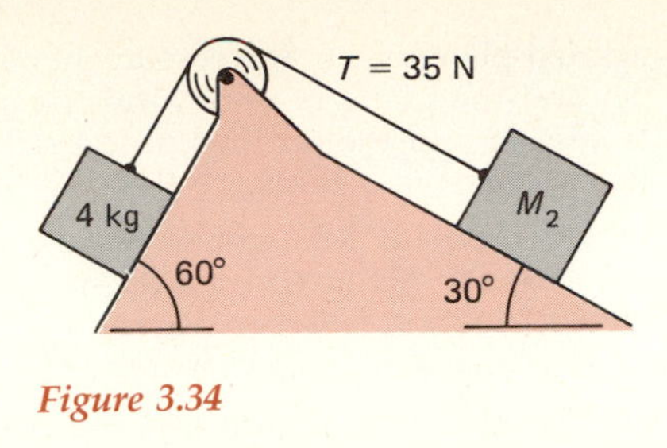

Figure 3.34

• **3.30** The three blocks *A*, *B*, and *C* of Figure 3.35 have masses of 5 kg, 8 kg, and 12 kg, respectively. If the horizontal surface on which block *B* rests is frictionless, and the system is assumed to start from rest, what is

(a) the acceleration of the system?

(b) the tension in the strings attached to block *B*?

(c) the time that elapses before block *C* has descended 4.9 m?

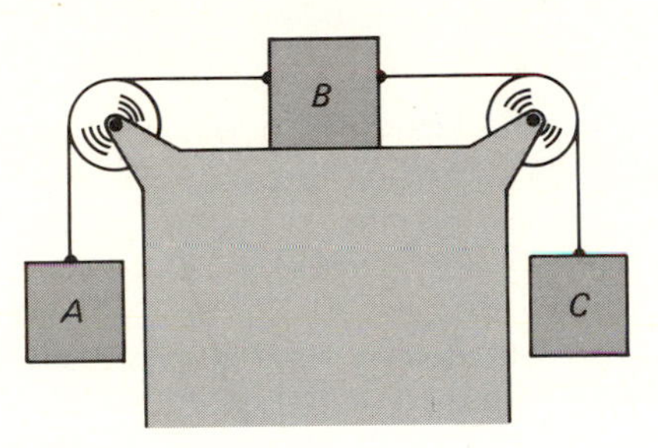

Figure 3.35

(Section 3.4)

• **3.31** Repeat Problem 3.30 for a coefficient of kinetic friction between block *B* and the supporting surface of 0.3.

3.32 A block of 10-kg mass is resting on an inclined plane that makes an angle of 30° with the horizontal. The coefficients of static and kinetic friction between block and plane are 0.6 and 0.2, respectively. Compute all forces acting on the block.

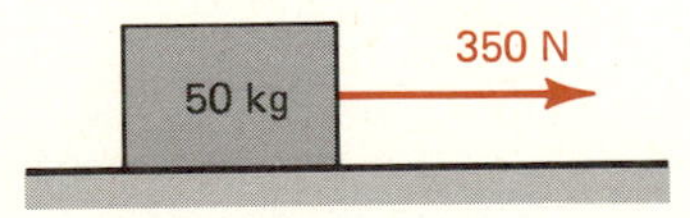

Figure 3.36

3.33 A 50-kg block rests on a horizontal surface. If this block is pulled by a string, as shown in Figure 3.36, it will not move unless the tension exceeds 350 N. What is the coefficient of static friction between the block and the surface?

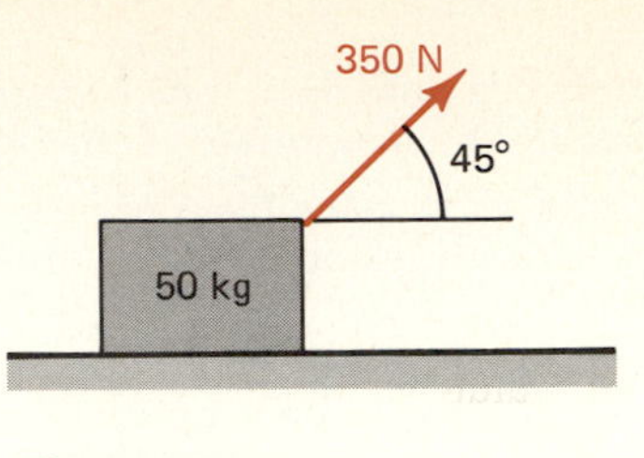

Figure 3.37

• **3.34** In Problem 3.33, the angle the string makes with the horizontal is changed from 0° to 45°, with the tension kept at 350 N. Will this block move under these conditions?

• **3.35** A mass of 4 kg slides down an inclined plane that makes an angle of 30° with the horizontal. The velocity of the mass is observed to be a constant. Is it possible from this information to determine the coefficient of kinetic friction? If so, what is its value? If a 10-kg block of the same material were placed on this inclined plane, would it move at constant velocity or not?

• **3.36** A block of mass *M* is given a push on a horizontal surface so that its initial speed is 1.5 m/s. It then travels 4.5 m before coming to rest. What is the coefficient of friction between the mass and the surface?

• **3.37** A 10-kg object slides on a horizontal surface. The coefficient of kinetic friction between that object and the surface is 0.2. If the object's initial speed was 8 m/s, how far will it slide before coming to rest, and how long will it have been in motion before coming to rest?

• **3.38** A man brings a refrigerator up on his porch by laying it on its side and then pushing it up a wooden ramp. The mass of the refrigerator is 115 kg, the angle the plank makes with the horizontal is 23°, and the coefficients of static and kinetic friction between the refrigerator and ramp are $\mu_s = 0.4$ and $\mu_k = 0.3$. (a) How much force, parallel to the ramp, must be applied to the refrigerator to start it moving up the ramp? (b) If the same force is applied after the refrigerator begins to move, what is its acceleration? (c) If it is to move up the ramp at constant speed, how much pushing force is needed once it has begun to move? (d) If part way up the ramp the man wants to take a breather, can he safely walk away from the refrigerator? (e) If not, how much force must he continue to exert to prevent its crashing down on him?

• **3.39** Suppose that in Example 3.7 the coefficient of friction between the surface and the 8 kg mass is zero, but that $\mu_k = 0.3$ between the surface and the 5 kg mass. What is then the tension in the string between the two masses?

• **3.40** A 50-kg stone falls into a mud pot from a height of 6 m. It sinks 0.5 m below the mud surface before coming to rest. Assuming that a constant frictional force between mud and stone brought the stone to rest, determine the magnitude of that force.

• • **3.41** The smooth wooden block of mass M of triangular cross section slides on a friction-free horizontal table (Figure 3.38).

(a) Describe qualitatively the motion of the block if the block and the mass m on its inclined surface are left free to move under the influence of gravity. (There is also no friction between the mass and the inclined surface of the block.)

(b) How large a horizontal force F must be applied to the triangular block to prevent its motion?

(c) What should the acceleration of the block be to maintain the mass at constant height?

(d) How large a force is needed to provide the acceleration calculated in (c)?

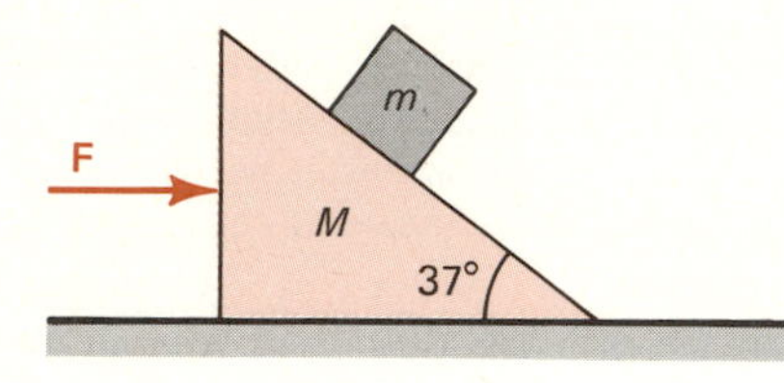

Figure 3.38

• **3.42** A mass of 20 kg is pulled across the floor with a force of 100 N directed as shown in Figure 3.39. Under these conditions the mass accelerates at 1.5 m/s^2. What is the coefficient of friction between the mass and the floor?

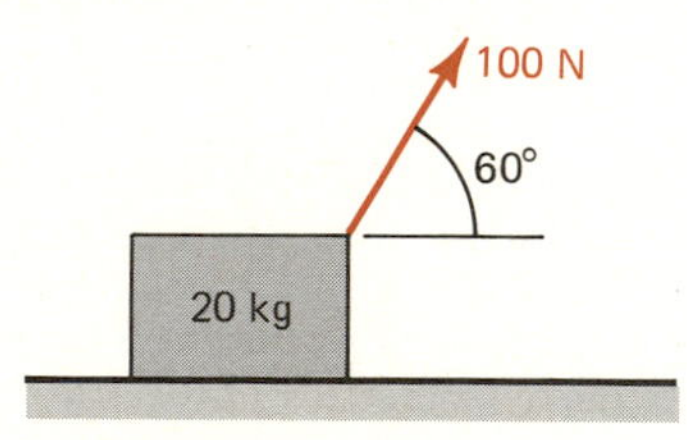

Figure 3.39

• • **3.43** Two masses of 4 kg each are connected by a thin string and are pulled up a 37° incline, as indicated in Figure 3.40. The coefficient of friction between M_1 and the inclined plane is 0.6; the coefficient of friction between M_2 and the plane is zero. Find the acceleration of the system and the tension in the string connecting the two masses if the force F is 120 N.

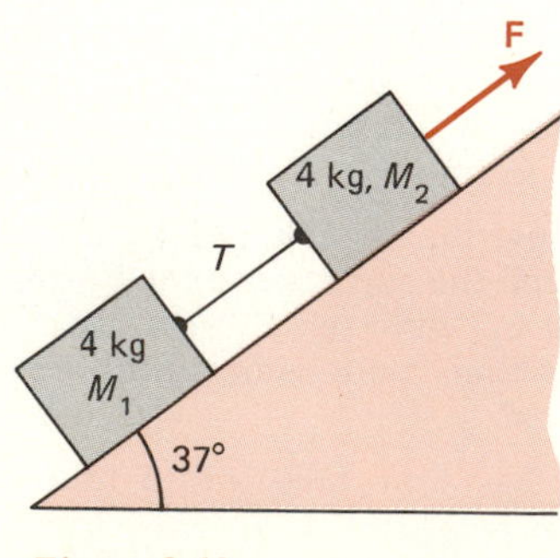

Figure 3.40

• • **3.44** Repeat Problem 3.43 under the assumption that M_1 slides on the plane without friction and that the coefficient of friction between M_2 and the plane is 0.6.

• • **3.45** The two masses in Figure 3.41 slide down the plane at constant speed. The coefficient of friction between the 5-kg mass and the plane is 0.6. Find the coefficient of friction between the 4-kg mass and the plane, and find the tension in the string between the two masses.

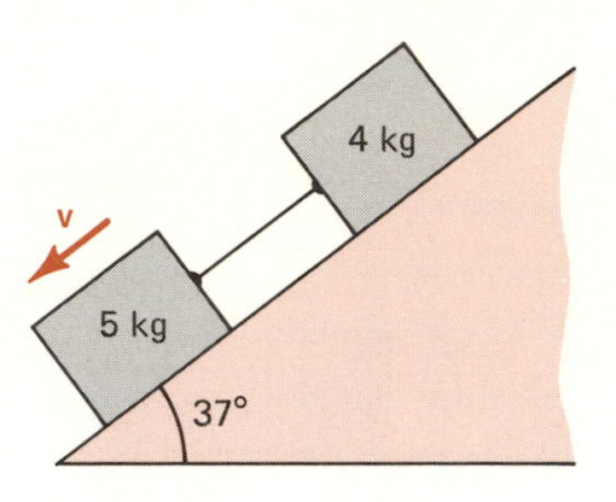

Figure 3.41

• • **3.46** The coefficients of static and of kinetic friction between the 10-kg block and the inclined plane of Figure 3.42 are 0.4 and 0.2, respectively. (a) How large a mass may block A have for the system to remain at rest? (b) If the mass of block A exceeds this limiting value by 2 kg, what are then the acceleration of the system and the tension in the string?

• • **3.47** A 20-kg box sits on the bed of a truck, 6 m from the open tailgate. The coefficients of static and kinetic friction between the box and the truck bed are 0.4 and 0.25, respectively.

(a) What is the maximum acceleration the truck can have without shifting the position of the box?

(b) Suppose the truck accelerates forward at 5 m/s^2 for 1 s, and then continues to move at a constant speed of 5 m/s. Describe the behavior of the box quantitatively.

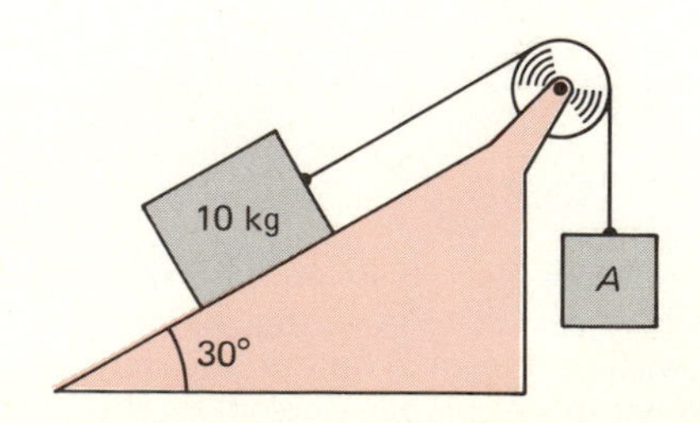

Figure 3.42

(c) Suppose the truck accelerates forward at 5 m/s^2 for 4 s. Will the box slip off the tailgate? If so, how many seconds following the initial acceleration of the truck will it do so?

• • **3.48** Blocks M_1 and M_2 of Figure 3.43 have masses of 3 kg and 5 kg, respectively. Assuming that the coefficients of kinetic friction between the two masses and between mass M_2 and the supporting surface are the same and equal to 0.3, find the tension in the string pulling the block M_2 if this block has an acceleration of 3 m/s^2, in cases (*a*), (*b*), and (*c*). In case (*a*), M_1 remains at rest relative to M_2; i.e., it moves with M_2.

• • **3.49** In each case shown in Figure 3.43, a tension T of 80 N produces an acceleration of 5 m/s^2 of block M_2. Assuming that the coefficients of kinetic friction between blocks M_1 and M_2 and between block M_2 and the supporting surface are the same, determine this frictional coefficient in each case. Also calculate the tensions in the string attached to block M_1 in cases (*b*) and (*c*). In case (*a*), M_1 remains at rest relative to M_2.

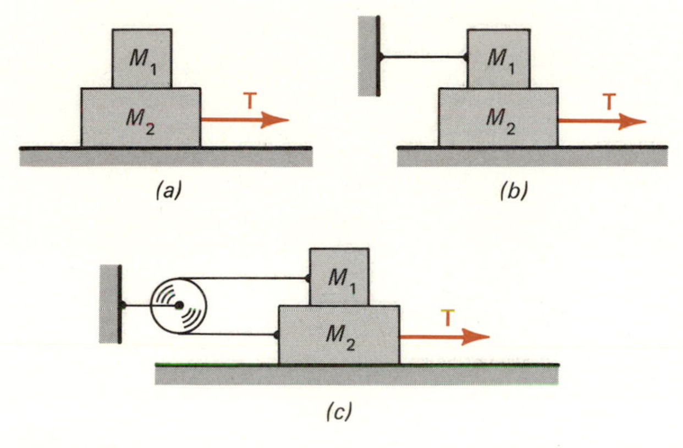

Figure 3.43

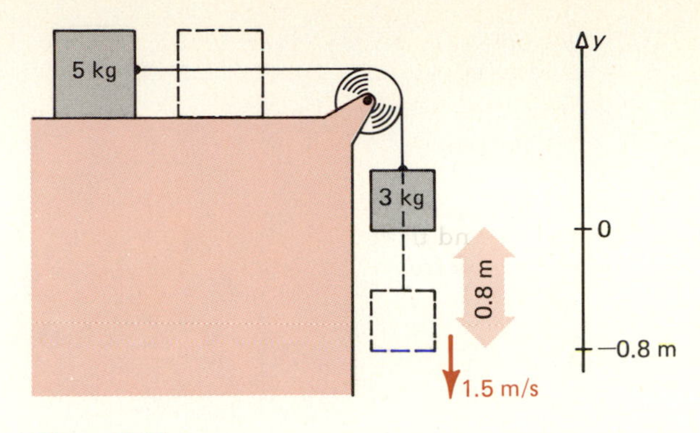

Figure 3.44

• • **3.50** A truck, whose mass is 2000 kg is used to pull a stalled car to a garage. The mass of the car is 800 kg, and the rope used to pull the car breaks if it is subjected to a tension in excess of 1700 N. The coefficient of static friction between the truck tires and the road is 0.4. What limits the acceleration that can be imparted to the car, and what is this maximum acceleration?

• • **3.51** A 5-kg mass supported on a horizontal table is connected by a thin string to a 3-kg mass that hangs freely, as shown in Figure 3.44. The system starts from rest. When the 3-kg mass has dropped through 0.8 m, its speed is 1.5 m/s. Is the horizontal table frictionfree, and if not, what is the frictional force between the 5-kg mass and the table?

Work, Energy, and Power

Plus ça change, plus c'est la même chose.

MONTAIGNE

(The more things change, the more they remain the same.)

4.1 Work

The dictionary definition of *work* is "that on which exertion or labor is expended," where it is understood that mental as well as physical effort is included. Thus, as you sit and study these pages, you are working, in the colloquial sense of the word, though not by the physicist's definition. Work, in physics, is defined much more precisely and narrowly.

DEFINITION: The amount of work performed by a force is the product of the force and the component, in the direction of the force, of the displacement of the point at which this force is applied.

Let us examine this definition carefully. First, work requires the action of a force. Without a force, no work is done. Second, the application of a force is a necessary but not sufficient condition for work. Work is done only if there is displacement of the point of application of the force, and then only if this displacement has a component along the line of action of the force.

According to this definition, if you simply hold a book at arm's length, applying an upward force equal to the weight of the book, you are doing no work though your muscles may soon tire from the effort. Moreover, if you carry this book some horizontal distance you have still performed no work on it, because the displacement (horizontal) has no component in the direction of the force (vertical). Only if the point of application of the force moves such that the displacement has a nonvanishing component in the direction of the force does this force perform work.

Figure 4.1 This man did much work in raising the weight from the floor to its present position. He is doing no work holding it steady above his head.

The definition of work can be expressed by

$$\Delta W = F\,\Delta s\cos\theta \qquad (4.1)$$

where ΔW is the amount of work done *by* the force of magnitude F during a small displacement of magnitude Δs. The angle θ in Equation (4.1) is the angle between the vector $\mathbf{F}$ and the vector $\Delta\mathbf{s}$. (See Figure 4.2(*a*).)

Equation (4.1) suggests two alternative ways of interpreting ΔW. The first is that work is the product of F and the component of $\Delta\mathbf{s}$ in the direction of $\mathbf{F}$. Another is that ΔW is the product of Δs and the component of $\mathbf{F}$ in the direction of $\Delta\mathbf{s}$. That is, we could equally well write

$$\Delta W = F_s\,\Delta s \qquad \text{(Figure 4.2(b))};$$

$$\text{or} \quad \Delta W = F\,\Delta s_F \qquad \text{(Figure 4.2(c))} \qquad (4.2)$$

where the subscripts on s and F indicate the components in the designated directions. If the force $\mathbf{F}$ is constant while acting over a displacement $\mathbf{s}$, the total work done by that force is

$$W = Fs\cos\theta = F_s s = Fs_F \qquad (4.3)$$

Although work is the product of two vector quantities, it is a scalar. The SI unit of work, the newton-meter, or kg·m²/s², has been given the name *joule* (J) in honor of James Prescott Joule (1818–1889), who performed the first direct measurement of the mechanical equivalent of heat energy. One joule is the amount of work done by a force of one newton acting over a distance of one meter in the direction of the displacement.

From Equation (4.3), it follows that the dimension of work is

$$[W] = [F][L] = \left(\frac{[M][L]}{[T]^2}\right)[L] = \frac{[M][L]^2}{[T]^2}$$

that is, the dimension of mass times velocity squared.

● ***Example 4.1*** How much work is done in lifting a 3-kg mass a height of 2 m and in lowering it to its initial position?

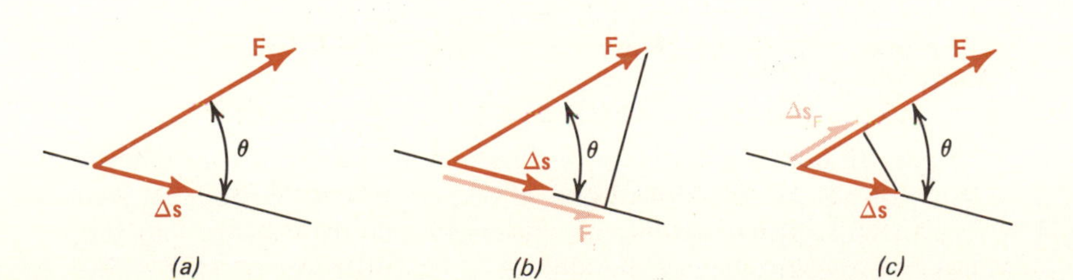

Figure 4.2 (*a*) The work done by the force $\mathbf{F}$ as its point of application moves through the displacement $\Delta\mathbf{s}$ is given by

$$\Delta W = F\,\Delta s\cos\theta$$

(*b*) The same amount of work can also be written

$$\Delta W = F_s\,\Delta s$$

where $F_S = F\cos\theta$ is the component of $\mathbf{F}$ in the direction of $\Delta\mathbf{s}$; or (*c*) as

$$\Delta W = F\,\Delta s_F$$

where $\Delta s_F = \Delta s\cos\theta$ is the component of $\Delta\mathbf{s}$ in the direction of $\mathbf{F}$.

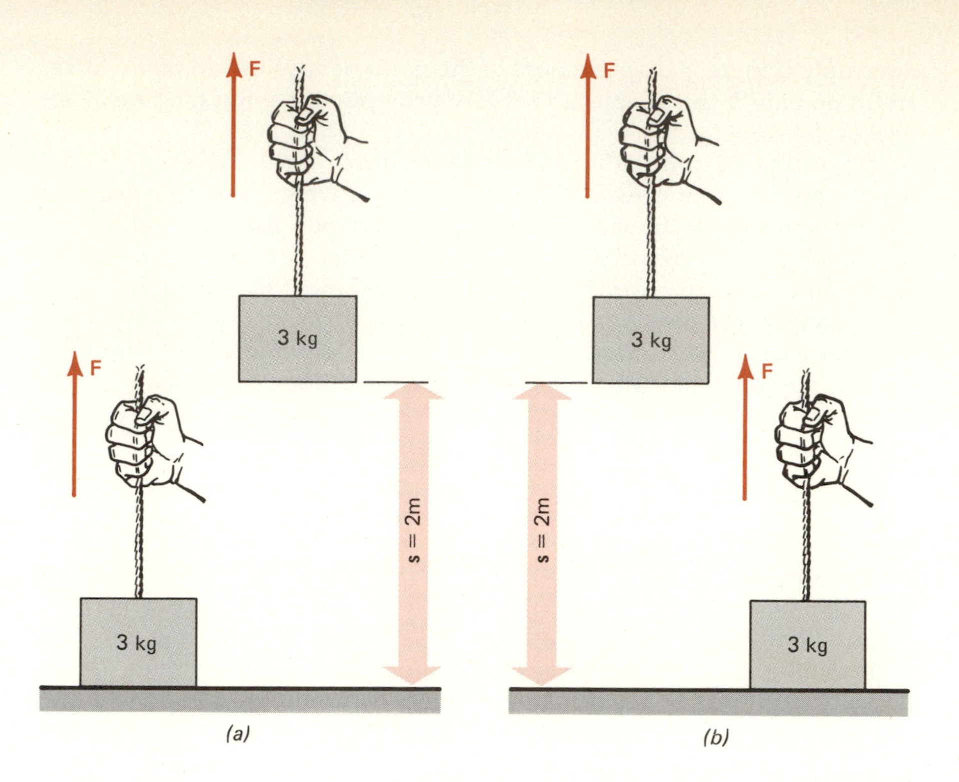

Figure 4.3 (*a*) *To raise the 3 kg mass a force* $F = 3 \times 9.8$ *N must be applied; the work done in raising the mass a height* $s = 2$m *is*

$$W = Fs = 58.8 \text{ J}$$

(*b*) *If the 3 kg mass is slowly lowered by 2 m an* upward *force of* $3 \times 9.8 = 29.4$ *N must be exerted. Since the displacement is now in a direction opposite to that of the applied force, the work done by that force is*

$$W = -58.8 \text{ J}$$

The force that must be applied is an upward force equal in magnitude to the weight mg. (We neglect the infinitesimal initial incremental force needed to give the mass a small upward velocity.) Since the force is directed up and the displacement is in the same direction, $\theta = 0$. Hence,

$$W = mgs = (3 \text{ kg})(9.8 \text{ m/s}^2)(2 \text{ m}) = 58.8 \text{ J}$$

Suppose we now slowly lower this mass to its original position. Again we must apply an upward force of mg to prevent it from dropping. How much work is done by this force?

Now the angle between that force and the displacement is $\theta = 180°$, and since $\cos 180° = -1$, we have

$$W = -58.8 \text{ J}$$

The negative sign tells us that some other agent, in this case gravity, has done work on the body. In this simple example there are two forces that act on the 3-kg mass: the force of gravity, which points downward, and the tension in the string, which pulls upward. If we had asked for the work done by the force of gravity, it would have been negative during the lifting of weight and positive as the weight was lowered. ●

● ***Example 4.2*** A man pulls his child on a sled along a horizontal snow-covered street at constant velocity. The child's mass is 40 kg and that of the sled 5 kg. The coefficient of kinetic friction between the sled runners and the snow is 0.1, and the angle that the pulling rope makes with the horizontal is 30°. How much work does the man do in pulling the child 100 m?

The solution of this problem requires some material from Chapter 3. The study of physics is a cumulative effort; what was learned last week or last month will appear again, sometimes in a slightly disguised form.

First, we recognize that the sled is in equilibrium; it is traveling at constant velocity. Therefore, the net force acting on it must vanish. The

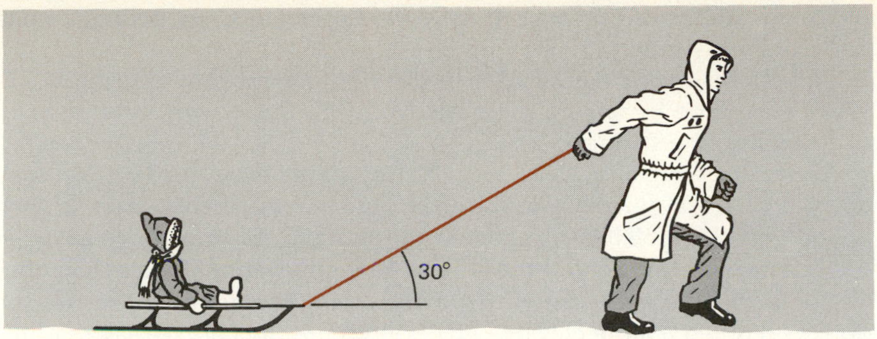

Figure 4.4 The man pulls his child on a sled at constant velocity. The net force acting on the sled is therefore zero.

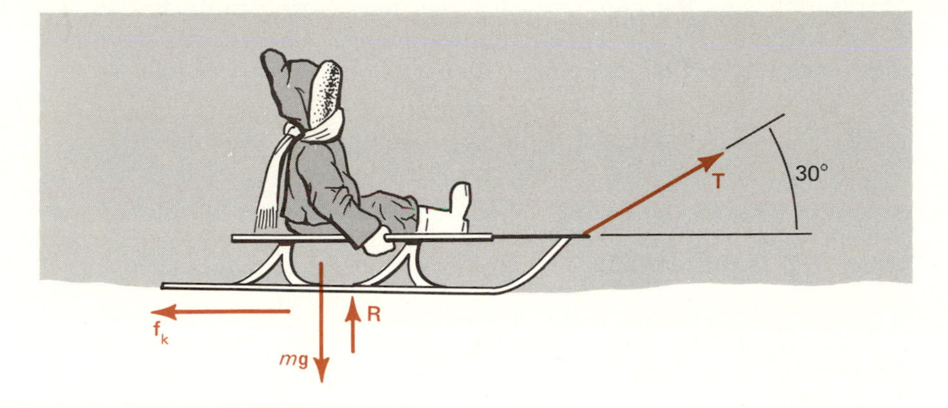

Figure 4.5 The free-body diagram for the sled, showing all the forces acting on it. Note that the reaction force R is less than the weight mg (why?), and the friction force is directed opposite to the velocity of the sled.

free-body diagram for the sled is shown in Figure 4.5. The forces acting are the weight of the sled, including the weight of the child, the normal reaction force of the snow against the sled runners, a force of kinetic friction directed opposite to the velocity, and the tension in the rope. We now write the equations for equilibrium (see p. 48).

x direction: $T \cos 30° - f_k = 0; \qquad 0.866T - 0.1R = 0$

y direction: $T \sin 30° + R - mg = 0; \qquad 0.5T + R - 441\text{ N} = 0$

Again we have two equations with two unknowns, R and T. Since we are not interested in R, we should solve by eliminating this unknown. For example, we can multiply the first equation by 10 and then add it to the second. The result is

$$9.16T - 441\text{ N} = 0; \qquad T = 48.1\text{ N}$$

The work done by this force is

$$W = Ts \cos \theta = (48.1\text{ N})(100\text{ m}) \cos 30° = 4165\text{ J}$$

4.2 Potential Energy and Conservative Forces

Energy is defined as the capacity to do work. A system may have mechanical energy by virtue of its position, its internal structure, or its motion. There are also other forms of energy besides mechanical, namely chemical, electrical, nuclear, and thermal energy. In this chapter we focus on mechanical potential and kinetic energies.

Potential energy (PE) refers to the ability of a system to do work by virtue of its position or internal structure. A mass some height above the floor has PE; if dropped, it can crush another object or drive a nail into a plank. A compressed spring can do work, for instance, propelling a small ball from a toy gun.

In both instances, work had to be done to give those systems their PE. Work was done in raising the mass from the floor; work was done in compressing the spring.

If we lift a weight of 100 N to a tabletop 0.6 m above the floor, we shall do 60 J of work against gravity. (See Figure 4.6.) If this mass is then pushed off the tabletop and falls to the floor, a constant force of mg = 100 N acts on it throughout the fall and that gravitational force then does 60 J of work. The work done by gravity was made possible because the object occupied a more elevated position before the fall. We say that initially the object, when on the tabletop, had greater PE than when at rest on the floor. As the gravitational force does work during the object's descent, the object loses gravitational potential energy. We define the difference in gravitational potential energy of a mass m as follows.

DEFINITION: The difference in gravitational PE of an object between points A and B is the negative of the work done by gravity in moving the object from A to B.

$$\Delta PE_{AB} = -W^g_{AB} = PE(B) - PE(A) \qquad \textbf{(4.4)}$$

where W^g_{AB} is the work done *by* gravity as the object moves from A to B.

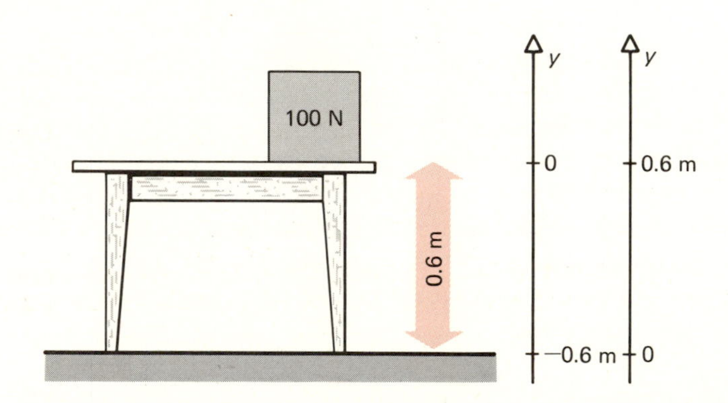

Figure 4.6 *The* PE *depends on where we place the zero of our coordinate system. If the zero is at the height of the tabletop, the* PE *of the* 100-N *weight is zero when it rests on the table and* −60 J *when it is on the floor; if we place the zero of coordinates at the level of the floor the* PE *of the weight is* +60 J *when it rests on the table and zero when it rests on the floor.*

Note that the above defines only *the change in PE*, the difference between the PE at B and at A. That is all we can do because zero PE depends on where we place the zero of our coordinate system. We might put it at the level of the floor, or at the tabletop, or at sea level, and each of these equally good choices corresponds to a different zero of gravitational PE.

In a completely analogous manner one defines the PE difference of a spring ΔPE_{AB} as the negative of the work done by the elastic forces of the spring as its length changes from L_A to L_B. (We shall be concerned with the PE of springs in Chapter 14. However, in this and the next five chapters our attention is centered on the behavior of objects in the earth's gravitational field, and we shall mean gravitational PE whenever we use the term PE without a qualifying adjective.)

Whether a body falls vertically or slides down an inclined plane, the work done on it by gravity depends only on its mass and on the difference in height between the initial and final positions. This follows directly from the defining equation for work, Equation (4.2). The direction of $m\mathbf{g}$ is downward. Hence, in calculations of ΔW^g, only the vertical component of $\Delta \mathbf{s}$ plays a role. In Figure 4.7, the work done by gravity as

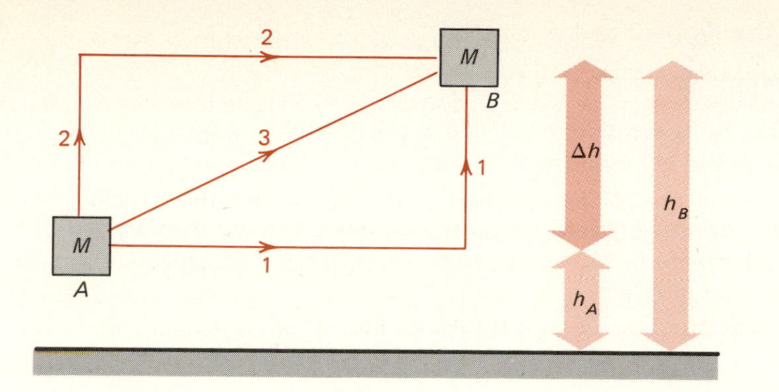

Figure 4.7 The mass M is moved from location A at a height h_A above some reference level to position B at a height h_B above the same reference. The work done is the same regardless of the path followed.

the mass is moved from A to B is

$$W^g_{AB} = -m\mathbf{g}(h_B - h_A) = -m\mathbf{g}\,\Delta h \tag{4.5}$$

regardless of the path followed. The PE difference is therefore

$$\Delta \mathrm{PE}_{AB} = m\mathbf{g}\,\Delta h \tag{4.6}$$

If we measure height from some arbitrary origin, e.g. sea level, and use that origin also for our zero of PE, the PE of a body of mass m at an elevation h is

$$\mathrm{PE}(h) = mgh \tag{4.7}$$

Note that the PE of an object depends only on its location, not on the route by which it arrived at that point. It follows that if a body is transported around a closed path, the change in PE vanishes. For example, in Figure 4.7, $\Delta\mathrm{PE} = 0$ if the mass moves from A to B via path 1 and is brought back to A via path 3. The PE is independent of the previous history because the gravitational force is *conservative.*

DEFINITION: A force is said to be conservative if the work W_{AB} done by the force in moving a body from A to B depends only on the position vectors $\mathbf{r}_A$ and $\mathbf{r}_B$.

In particular, a conservative force must not depend on time, or on the velocity or acceleration of the body. Kinetic friction is an example of a nonconservative force. It acts only if the object is moving and is always directed opposite to the velocity. This force is evidently a velocity-dependent force. For this nonconservative force, a retracing of path does not lead to zero net work, as it does for the gravitational force. We shall do work against friction in pulling the sled of Example 4.2 100 m east, and shall do more work in pulling it back to its starting point.

4.3 Kinetic Energy

We saw in Chapter 3 that when a body initially at rest is subjected to a constant net force, it gathers speed. Moreover, in this special case of a constant force, and consequently constant acceleration, we found a simple expression, Equation (2.13e), that relates the initial and final speeds to the acceleration and the displacement:

$$v_f^2 - v_0^2 = 2as \tag{2.13e}$$

If we now multiply both sides of (2.13e) by the mass m and make use of the second law, the right-hand side becomes $2mas = 2Fs$, and this is just twice the work done by the force F that gave rise to the acceleration a. We now divide by 2 and obtain

$$\tfrac{1}{2}mv_f^2 - \tfrac{1}{2}mv_0^2 = Fs = W \qquad \textbf{(4.8)}$$

The right-hand side is the work done by the net force **F**; consequently, the left-hand side must have the dimension of work (as it does) and must also correspond to some energy. This energy depends not on the position of the body but on its speed. We are thereby led to another form of mechanical energy, one associated with speed of motion, the *kinetic energy* (KE).

We define the KE of a body of mass m by

$$\text{KE} = \tfrac{1}{2}mv^2 \qquad \textbf{(4.9)}$$

where v is its speed. Equation (4.8) can be given a simple interpretation:

The change of a body's KE equals the work done on that body by the net force acting on it.

Although we derived Equation (4.8) using results valid only for uniformly accelerated motion, it can be shown that the increase in KE equals the work done even in the more general case of an arbitrary net force.

Before going on, let's try to answer one question that may be nagging you at this point. In the previous examples, various forces were acting on objects, doing work, and yet the KE was unchanged. The point is that in those examples we focused attention on the work done by *one* specific force, for example the force needed to lift some weight slowly against gravity. The *net* force acting on that weight was zero, and hence the KE was not changed.

● ***Example 4.3*** A 2000-kg car starts from rest with constant acceleration of 3 m/s^2. Find the change in KE of the car during the first second and during the following second.

At the end of the first second, the car is traveling at a speed of $v(1) = v_0 + at = (0\ \text{m/s}) + (3\ \text{m/s}^2)(1\ \text{s}) = 3\ \text{m/s}$. Its KE is then

$$\text{KE}(1) = \tfrac{1}{2}mv^2(1) = 9000\ \text{J}$$

It is tempting now to conclude that since the acceleration is constant, the additional KE gained during the next one second time interval will be another 9000 J; but that is incorrect! Though the velocity of the car increases linearly with time, the KE does not, because it depends on the square of the velocity. Thus, at the end of two seconds, $v(2) = 6$ m/s and

$$\text{KE}(2) = \tfrac{1}{2}(2000\ \text{kg})(6\ \text{m/s})^2 = 36{,}000\ \text{J}$$

Hence the change in KE during this second interval is

$$\Delta\text{KE} = \text{KE}(2) - \text{KE}(1) = (36{,}000\ \text{J}) - (9000\ \text{J}) = 27{,}000\ \text{J}$$ ●

Why should the change in KE be three times greater during the second time interval than during the first? We can understand this result using some of the knowledge gained earlier.

You will recall Galileo's statement that the distances traveled during equal intervals of time by an object accelerating uniformly from rest stand to each other as the odd integers, 1, 3, 5, ... Thus during the second second, the car travels three times as far as during the first. Since the acceleration is constant, the same force acts throughout, and the work done by this force, $W = Fs$, must therefore also increase with each succeeding time interval according to the series of odd numbers beginning with unity.

Figure 4.8 *What is the* KE *of passengers seated in an airplane in flight?*

From our definition of KE, Equation (4.9), it seems that in contrast to the PE whose zero depended on an arbitrary choice of coordinate origin, the zero of KE can be unequivocally identified as that state of a body in which its velocity vanishes. "Seems" so, but is not really true. Whereas the zero of PE depends on the choice of coordinate origin, the zero of KE is determined by the choice of inertial reference frame. What is the KE of your body when you sit relaxed in an airplane that is traveling at 800 km/h? It is only because we are so accustomed to the earth as our reference frame that we tend to associate zero KE with the state of rest *relative to the earth's surface*. It is an arbitrary choice, but the most natural and convenient in most instances; and unless specifically stated otherwise, we shall be using this convention.

These dependences of PE and KE on reference frames are specific examples of a more general experience. Practically all measurements performed in science involve the observation of differences—in energy, position, velocity, etc. Only rarely can one even define absolute quantities, let alone measure them.

As the term *kinetic energy* implies, a body can do work by virtue of its motion. Very substantial frictional forces between brake shoe and drum are required to bring a fast-moving car to rest, and in the process much heat is generated. A stone thrown upward gains PE as it rises, i.e., does work against gravitational attraction; but as it does so its KE correspondingly diminishes. Having reached its peak, the stone gains speed and KE as it falls, and its PE decreases. We can express these qualitative statements in precise mathematical terms, as follows.

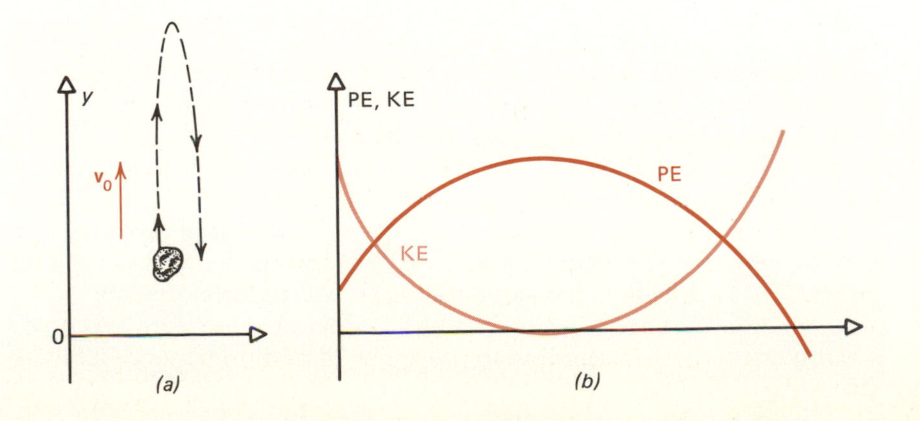

Figure 4.9 *(a) A stone is projected upward. As it rises, its speed decreases, and so does its* KE. *Simultaneously, its* PE *increases. At the top of its trajectory, the* KE *is zero and its* PE *is a maximum. As the stone falls from this point, the* KE *increases and its* PE *decreases. (b) Curves showing the* KE *and* PE *of the stone as functions of time. Note that the* PE *is not zero at the start because the stone starts from some height above our arbitrary zero. As the stone falls, its* PE *goes through zero and becomes negative while its* KE *increases above its initial value.*

For a body subject only to the constant force of gravity, we can replace F in Equation (4.8) by its weight mg, and we can replace the distance over which this constant force acts by the difference in height of the object between its initial and final positions. Thus we obtain

$$\tfrac{1}{2}mv_f^2 - \tfrac{1}{2}mv_0^2 = mg(h_0 - h_f) \quad \textbf{(4.10)}$$

From the definitions of KE and PE, this can be rewritten

$$\text{KE}_f - \text{KE}_i = \text{PE}_i - \text{PE}_f$$

and rearranging terms gives

$$\text{KE}_f + \text{PE}_f = \text{KE}_i + \text{PE}_i \quad \textbf{(4.11)}$$

where the subscript "i" stands for *initial* and "f" for *final*.

Equation (4.11) says that the total mechanical energy of a body, the sum of the potential and kinetic energies, remains constant provided that the only force acting on the body is the conservative force of gravity.

4.4 Conservation of Energy

Equation (4.11) expresses a very general principle, the conservation of energy, in a special case. In the preceding section we restricted ourselves to the motion of a body under the influence of gravity. The real world is, however, more complicated. Forces other than gravity, some conservative and others nonconservative, act on objects. Moreover, energy appears in forms other than mechanical. In general, then, the principle states: The total amount of energy within an isolated system is constant. Another way of phrasing the principle is:

Energy in an isolated system cannot be created or destroyed. All one can do is change energy from one form to another.

For example, we can change gravitational PE to KE and vice versa. We can burn gasoline, changing chemical energy to thermal energy, convert part of that energy to mechanical energy, and by means of an alternator, to electrical energy. But whatever we do, the total energy of an isolated system remains constant.

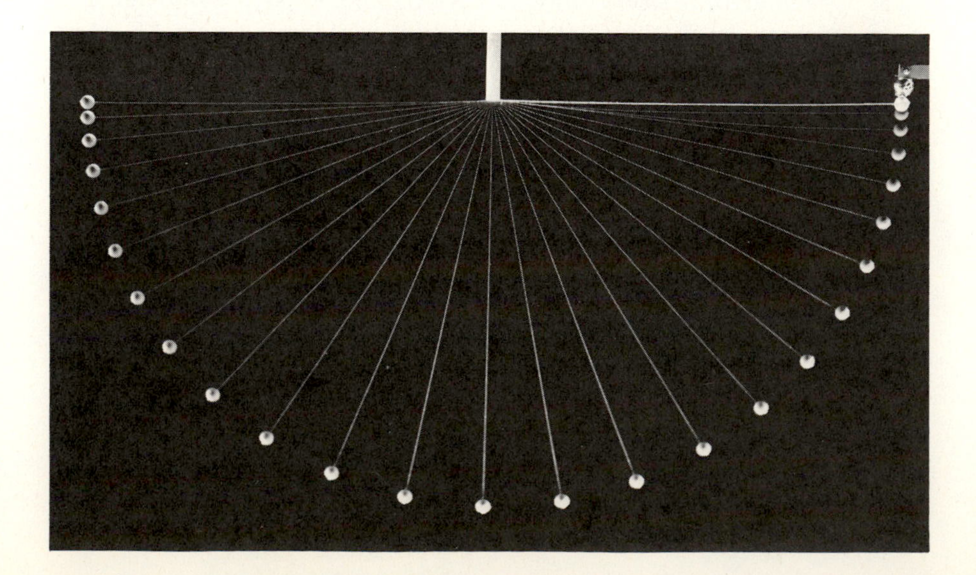

Figure 4.10 *Multiple-exposure stroboscopic photograph of a swinging pendulum. The speed of the bob is greatest at the bottom of the swing (as indicated by the larger separation between the bob's positions) and zero at the top of its swing. As the pendulum swings, PE is converted into KE and then back to PE, and so forth.*

Figure 4.11 *Hoover Dam. The hydroelectric station is located at the bottom of the dam; here the PE of the water at the top of the dam is converted to electrical energy.*

Figure 4.12 *A pole vaulter in action. Chemical energy metabolically generated and stored in his body is first converted to KE as he runs toward the jump. That energy is then changed to PE of the vaulter, and initially also to elastic PE of the deformed pole. This elastic PE is gradually changed to the PE of the athlete, who also expends additional chemical energy as he pushes himself further up along the pole. That PE is later changed to KE as he drops off on the opposite side; and it is ultimately converted to heat as he lands in the sand pit.*

Transformation of energy among its diverse forms is a routine event in nature. Indeed, it plays the cardinal role in every act, and without this constant metamorphosis man, our planet, and the universe itself could not exist.

In general, conservative as well as nonconservative forces act on a system. Whereas the effect of the former on energy can always be represented in terms of a PE, the same is not possible for the latter. Work done by nonconservative forces, for example friction, which ultimately manifests itself as heat, must be included in calculations employing energy conservation. If we denote by W_n the work done *by* the system against *nonconservative* forces, the energy conservation equation reads

$$E_i = KE_i + PE_i' = KE_f + PE_f' + W_n = E_f \quad \textbf{(4.12)}$$

where E_i and E_f are the initial and final total energies, and PE_i' and PE_f' include all PEs, gravitational and elastic, as well as electric and chemical. Since it is often not apparent how these nongravitational PEs are to be determined, an alternative equation

$$KE_i + PE_i = KE_f + PE_f + W \quad \textbf{(4.13)}$$

is more useful. Here PE_i and PE_f refer to the *gravitational* PE only, and W is the work done *by* the system against any *nongravitational* forces, such as driving an electric generator or compressing a gas into a cylinder, as well as work done in overcoming friction.

● ***Example 4.4*** We return to Example 2.7, the trajectory of a stone thrown from a bridge. The stone is projected horizontally with a speed of 30 m/s from a point 20 m above the surface of the water. Suppose that we want to know only the speed of the stone at the moment of impact with the water.

We can solve this problem most readily by using conservation of energy. As the stone drops, it loses PE and gains KE. (We neglect air friction.)

We choose our zero of PE to be the energy at the water's level. The total initial energy is then

$$\begin{aligned} KE_i + PE_i &= \tfrac{1}{2}mv_0^2 + mgh_0 \\ &= \tfrac{1}{2}m(30\text{ m/s})^2 + m(9.8\text{ m/s}^2)(20\text{ m}) \\ &= 646m\text{ J} \end{aligned}$$

At the moment of impact with the water, the PE is zero. Consequently,

$$\begin{aligned} KE_f + PE_f &= \tfrac{1}{2}mv_f^2 + 0 = 646m\text{ J} \\ v_f^2 &= 1292\text{ m}^2/\text{s}^2 \\ v_f &= 35.9\text{ m/s} \end{aligned}$$ ●

The answer is identical to the one we obtained previously, but we arrived at the result with much less effort. Moreover, you will have noted that the *direction* of the initial velocity never appeared in the computation; though included in the statement of the problem, this information was never used. It follows that the speed on impact with the water must be independent of the direction of the initial velocity, and the apparently surprising result obtained earlier is surprising no longer.

But a small effort also gains small dividends. Though we know the speed of the stone on impact, we can say nothing about the direction the stone is traveling as it strikes the water, about the intervening trajectory, or about the time of flight. To answer these questions, we must do more; we must examine the solution of the equations of motion, as we did in Chapter 2.

● ***Example 4.5*** A 5-kg mass supported on a horizontal table is connected by a thin string to a 3-kg mass that hangs freely, as shown in Figure 4.13. The system starts from rest. When the 3-kg mass has dropped through 0.8 m, its speed is 1.5 m/s. Is the horizontal table friction-free, and if not, what is the frictional force between the 5-kg mass and the table?

We could proceed as you perhaps did in solving this problem (Problem 3.51), finding the acceleration of the 5-kg mass and the net force acting on it, and equating this force to the difference between the tension in the string and the frictional force. Instead, let us apply the principle of energy conservation.

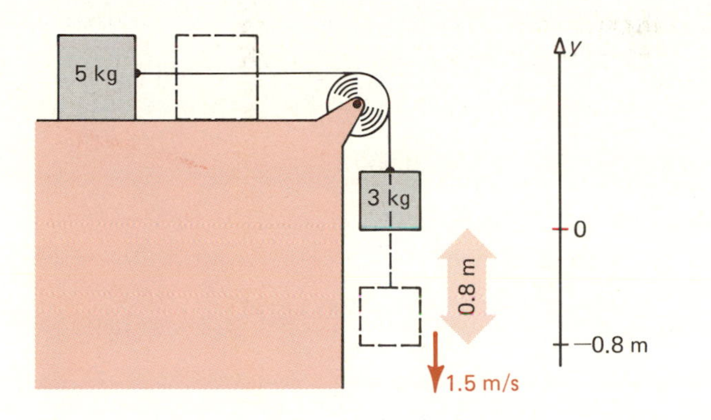

Figure 4.13 *Two masses are connected by a string; the 5-kg mass rests on a horizontal surface, and the 3-kg mass hangs freely. They are shown in their initial position, at rest, and shortly after release; they both have moved a distance of 0.8 m and have acquired a speed of 1.5 m/s.*

We place the vertical origin of our coordinate system at the initial position of the 3-kg mass and take upward for the positive direction. The initial energy of the system is then zero. Thus

$$KE_f + PE_f + W = 0$$

$$\tfrac{1}{2}(5\text{ kg} + 3\text{ kg})(1.5\text{ m/s})^2 - (3\text{ kg})(9.8\text{ m/s}^2)(0.8\text{ m}) + W = 0$$

$$W = 14.52\text{ J}$$

That W is not zero shows that the system has done 14.52 J of work against external forces, in this case friction. Since the force of friction acted over 0.8 m, we have

$$(0.8\text{ m})f_k = 14.52\text{ J}$$

$$f_k = 18.15\text{ N}$$

●

● ***Example 4.6*** A 500-kg roller coaster starts from rest at point A of Figure 4.14. If throughout its travel the average retarding force due to air friction, friction in the wheel bearings, etc., is 200 N, what is its speed at points B and C, and how far will it travel at the elevation of C before coming to rest?

As in the previous example, W does not vanish since the system does work against friction in its travel. We select the elevation of point B as our zero of PE. Then

$$KE_i + PE_i = 0 + (500\text{ kg})(9.8\text{ m/s}^2)(20\text{ m}) = 98{,}000\text{ J}$$

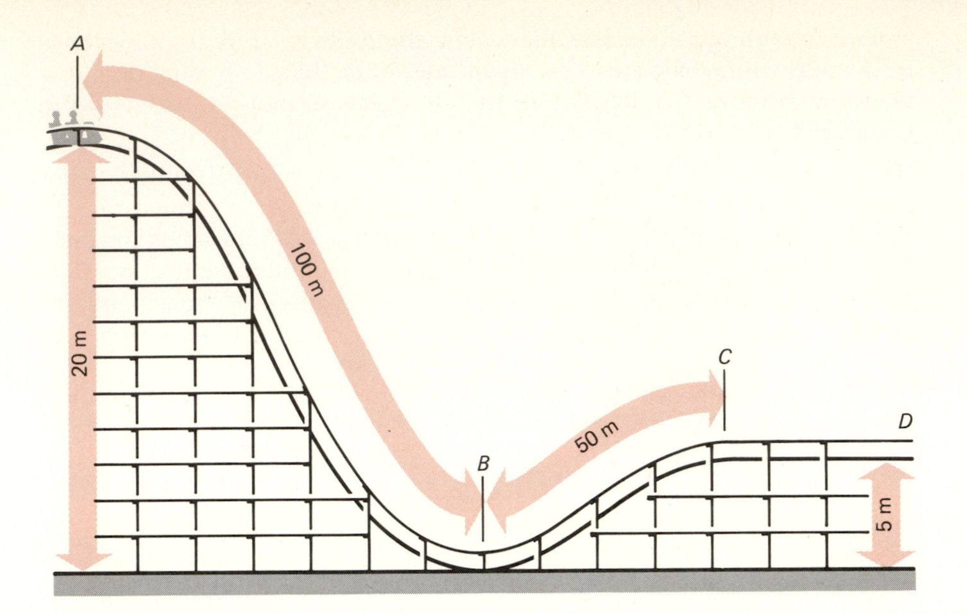

Figure 4.14 Sketch of a simple roller coaster track. The car starts from rest at point A, where its PE is greatest.

The work done against friction is $W = F_f s$, where s is the distance measured along the track. Hence in going from A to B, this work equals

$$W_{AB} = (200 \text{ N})(100 \text{ m}) = 20{,}000 \text{ J}$$

Since the PE at B is zero,

$$\text{KE}_B + 20{,}000 \text{ J} = 98{,}000 \text{ J}$$

$$\text{KE}_B = 78{,}000 \text{ J} = \tfrac{1}{2}(500 \text{ kg})v_B^2$$

$$v_B = 17.7 \text{ m/s}$$

By the time the car has reached point C, a total of (200 N)(150 m) = 30,000 J of energy have been expended against friction. Hence,

$$98{,}000 \text{ J} = \tfrac{1}{2}(500 \text{ kg})v_C^2 + (500 \text{ kg})(9.8 \text{ m/s}^2)(5 \text{ m}) + 30{,}000 \text{ J}$$

$$v_C^2 = \frac{98{,}000 \text{ J} - 30{,}000 \text{ J} - 24{,}500 \text{ J}}{250 \text{ kg}} = 174 \text{ m}^2/\text{s}^2$$

$$v_C = 13.2 \text{ m/s}$$

Thereafter there is no further change in PE. Hence all the available KE at point C, 43,500 J, must be converted to heat as a result of friction as the car comes to rest. The additional distance of travel is therefore

$$s_{CD} = \frac{43{,}500 \text{ J}}{200 \text{ N}} = 217.5 \text{ m}$$

Applying energy conservation has permitted the determination of certain dynamic features of the car's behavior but has also left many questions unanswered. Though we know the speed of the car at various points along its path, we do not know, for example, how long the car will take to reach points B or C.

4.5 Power

Frequently, we are as much concerned with the speed with which a task can be accomplished as with the total energy that will be consumed in its performance. Faced with a flooded basement, we may use a bucket or a motor-driven pump to raise the PE of the water to that at ground level, that is, to get it out of the basement. The energy expended, friction losses neglected, will be the same whatever method is employed; but given a choice, one would not hesitate to use the faster, more powerful method.

So far we have been concerned only with the energy of an object and with the work done by or against external forces. We have not asked how quickly the energy is attained, how rapidly work is done. The rate at which work is performed is called *power.* That is, power, denoted by the letter P, is the work done per unit time:

$$\langle P \rangle = \frac{\Delta W}{\Delta t} \tag{4.14}$$

is the average power developed during the time interval Δt in which an amount of work ΔW was done. The instantaneous power is

$$P = \lim_{\Delta t \to 0} \frac{\Delta W}{\Delta t} \tag{4.14a}$$

The dimension of power is

$$[P] = \frac{[\text{Energy}]}{[\text{Time}]} = \frac{[M][L]^2}{[T]^3}$$

The SI unit of power, the joule per second, is called a watt (W), commemorating James Watt (1736–1819), who perhaps more than any other man, showed the world the benefits to be derived by harnessing the power of steam engines.*

Another unit of power is the horsepower (hp), which equals 746 W.

● ***Example 4.7*** A manually operated winch is used to lift a 200-kg mass to the roof of a 10-m-tall building. Assuming that you can work at a steady rate of 200 W, how long will it take you to lift the object to the roof? Neglect frictional losses.

We assume that as the mass starts to move upward, it is kept moving at constant speed, and that this speed is relatively low. We shall therefore neglect the KE of the mass during its ascent. The small energy expended initially to give it this KE will be recovered at the end as the mass is

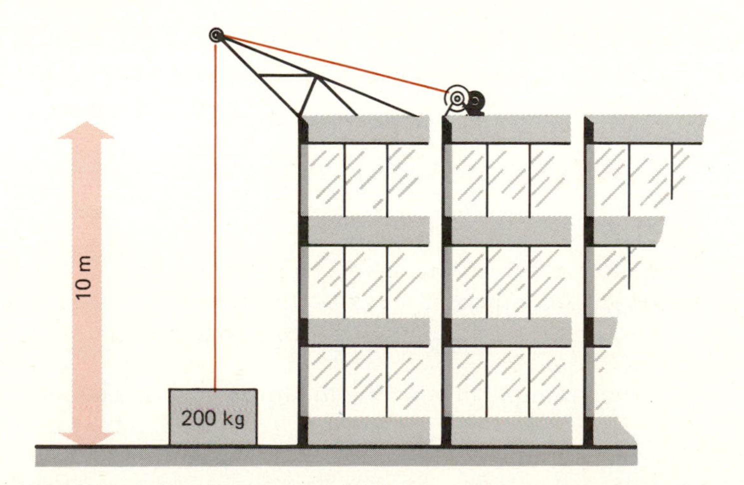

Figure 4.15 *The 200-kg mass is raised to the level of the roof by a winch at a rate of energy expenditure of 200 watts.*

* In 1776, James Watt joined Matthew Boulton in partnership to manufacture the steam engines he had invented. To James Boswell, who asked what they were selling, Boulton proudly replied, "I sell here, Sir, what all the world desires—power."

brought to rest on the roof. At the conclusion of the computation, we shall estimate the error that neglect of KE might introduce in this instance.

The work done in this case equals the increase in PE of the 200-kg mass, namely,

$$W = mgh = (200 \text{ kg})(9.8 \text{ m/s}^2)(10 \text{ m}) = 19{,}600 \text{ J}$$

Since this work is done at a constant rate of 200 W,

$$200 \text{ W} = \frac{19{,}600 \text{ J}}{\Delta t}$$

where Δt is the time required to lift the mass. Thus

$$\Delta t = \frac{19{,}600 \text{ J}}{200 \text{ W}} = 98 \text{ s}$$

It will take about 1.5 min to raise this mass to the roof if you work at a rate of 200 W. A person in excellent physical condition can for a brief time develop 200 W with his arms, but most people would find this task too demanding.

Before we leave this problem, let us just see how large an error may have been made by neglecting the KE of the mass during the ascent. The average speed of the object is

$$v = \frac{10 \text{ m}}{98 \text{ s}} = 0.102 \text{ m/s}$$

The KE during ascent is therefore

$$\tfrac{1}{2}(200 \text{ kg})(0.102 \text{ m/s})^2 = 1.04 \text{ J}$$

an amount negligibly small compared with the change of 19,600 J in PE. We can therefore safely neglect this small amount of KE in the problem's solution

Example 4.8 A car has an engine that can deliver 100 hp at the drive shaft. If the mass of the car plus driver is 1800 kg and the engine maintains constant maximum power output, what is the speed of the car at the end of 2 s and at the end of 4 s? (Assume that the car starts from rest and travels on level ground, and neglect air friction and other frictional losses.)

During 2 s, the work done by the drive shaft is

$$W = Pt = (100 \text{ hp})(746 \text{ W/hp})(2 \text{ s}) = 1.49 \times 10^5 \text{ J}$$

All that energy appears as KE of the car:

$$\tfrac{1}{2}mv^2(2) = 1.492 \times 10^5 \text{ J}$$
$$v^2(2) = 1.66 \times 10^2 \text{ m}^2/\text{s}^2$$
$$v(2) = 12.9 \text{ m/s} = 46.4 \text{ km/h}$$

As in an earlier example, it is wrong to jump to the conclusion that the speed of the car at the end of 4 s is $2v(2) = 92.8$ km/h. At the end of 4 s, the KE of the car will be $2\text{KE}(2) = 2.984 \times 10^5$ J, double its value at the end of 2 s, and therefore, $v^2(4) = 2v^2(2)$. The speed of the car at the end of 4 s is only $\sqrt{2} = 1.41$ times the speed at the end of 2 s; i.e.

$$v(4) = 65.6 \text{ km/h}$$

At constant power, the acceleration of the car is *not* constant but continually diminishing. Hence the force driving the car forward is also not constant but gradually decreasing, even though the power, the rate of energy production by the engine, is constant.

A simple relation between power and speed that we can easily derive helps in resolving this apparent paradox. Since work is given by $\Delta W = F_s \, \Delta s$, we can write the power as

$$P = \frac{F_s \, \Delta s}{\Delta t} = F_s v = F_v v \qquad \textbf{(4.15)}$$

where we have replaced $\Delta s/\Delta t$ by v, the speed of the object, and F_v is the component of the force in the direction of the velocity.

Note that if the power is used to accelerate an object from rest along a horizontal straight track, the product of the accelerating force and the speed v will be constant if P is constant. As the object gathers speed and v increases, F and, therefore, a must decrease if P remains constant. This is a result that is unquestionably familiar to you though you may not have thought about it in these terms.

When an airplane takes off, you feel the acceleration most, are pushed back into the seat most forcefully, at the start of the takeoff run, even though the engines actually develop greater power as the plane's speed increases. Also, you get a sense of much greater acceleration when you floor the accelerator in a car that is moving slowly than when you try to pass a car traveling at 100 km/h and you are already moving at this speed. Your car's engine is not failing you at this higher speed; the increase in KE of the car as it accelerates from 100 to 120 km/h is very much greater than if it accelerates from 10 km/h to 30 km/h.

● ***Example 4.9*** Finally, let us examine the energy balance in a biological system. The common flea is well known for its astounding athletic achievement, the ability to jump more than one hundred times its height. Some years ago, several researchers used high-speed cinephotography to record and study the takeoff pattern of the jump, and obtained the following results.

Starting from rest, the flea attains its peak vertical velocity in one millisecond. At that instant, its speed is about 1 m/s, and this carries the little beast to an elevation of about 3.5 cm.

We shall also need the following information. The flea's mass is about 450 micrograms; less than one-fifth of that mass comprises the leg muscles. Finally, from other studies it is known that the maximum power that insect muscles can develop is about 60 W/kg.

We now want to answer the following questions:

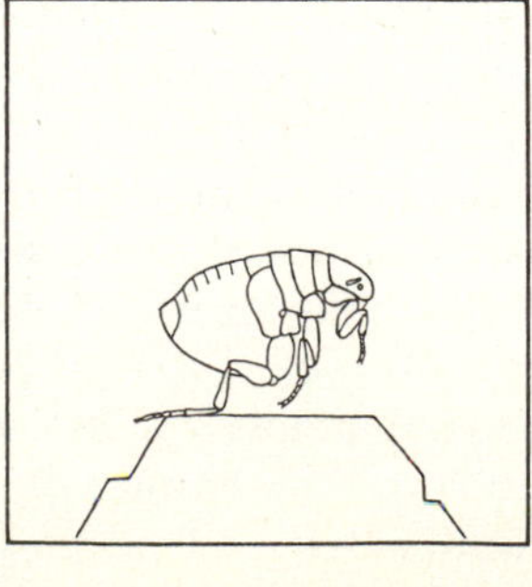
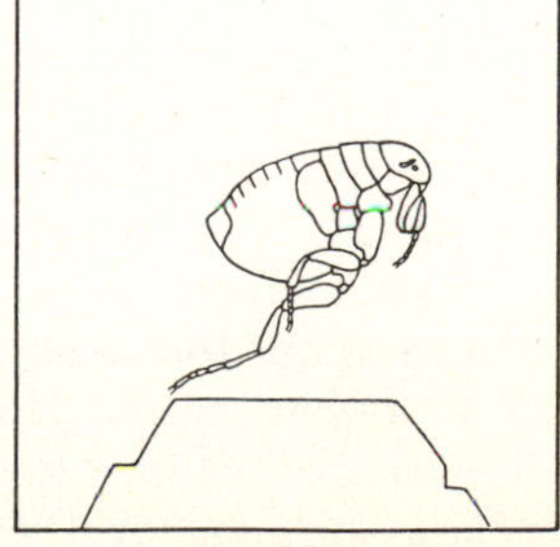
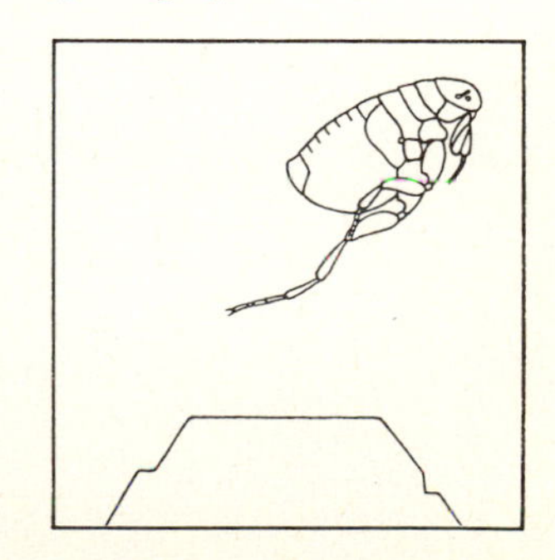

Figure 4.16 *Drawings of a flea on take off, based on a series of high-speed cinephotographs.*

1. What is the average acceleration on takeoff?

2. What is the average energy loss due to air friction in a vertical jump?

3. Can the leg muscles produce enough energy in 1 ms to power the jump?

The average acceleration is quickly determined. In a time interval of 10^{-3} s the flea changes its velocity from zero to 1 m/s. Hence $\langle a \rangle = 1/10^{-3} = 1000\ \text{m/s}^2$, roughly $100g$. No vertebrate could possibly survive that acceleration; at liftoff, the acceleration of the Apollo rocket was about $5g$.

If there were no air friction, an initial upward speed of 1 m/s would take the animal to a height given by

$$mgh = \tfrac{1}{2}mv_0^2$$

$$h = \frac{v_0^2}{2g} = \frac{(1\ \text{m/s})^2}{2(9.8\ \text{m/s}^2)} = 0.051\ \text{m} = 5.1\ \text{cm}$$

Since the maximum height attained is only 3.5 cm, the energy loss due to air friction during ascent is

$$(4.5 \times 10^{-7}\ \text{kg})(9.8\ \text{m/s}^2)(0.051\ \text{m} - 0.035\ \text{m}) = 7.06 \times 10^{-8}\ \text{J}$$

If it is assumed that the same amount of energy is lost during descent, the total energy loss is about 14×10^{-8} J. Not much energy, to be sure, but then neither is the initial energy large; the energy loss is about 60 percent of the initial energy. In fact, it is somewhat less than our estimate because during descent the average speed is substantially less than during ascent and the force due to air friction depends on the speed of the object in air (see p. 215).

The rate of energy release during takeoff is given by

$$P = \frac{\Delta W}{\Delta t} = \frac{\tfrac{1}{2}(4.5 \times 10^{-7}\ \text{kg})(1\ \text{m/s})^2}{10^{-3}\ \text{s}} = 2.25 \times 10^{-4}\ \text{W}$$

A power of 225 microwatts does not seem enormous, but then neither is the power plant. A fair estimate of the muscle power that the animal can produce is

$$(60\ \text{W/kg})(4.5 \times 10^{-7}\ \text{kg})(\tfrac{1}{5}) = 5.4 \times 10^{-6}\ \text{W}$$

an amount only about one-fiftieth that required for takeoff. So how does the little beast accomplish this acrobatic feat?

The answer to this puzzle was found in the small pad of elastic material called *resilin,* located at the base of the large hind leg. Following each jump, the flea slowly bends the hind legs, compressing the resilin pads and storing energy in the form of compressional elastic potential energy. The cuticles of the legs are fitted with a ratchetlike structure that catches into a latch and maintains the resilin pad in its compressed state without further muscular effort. When the flea jumps again, a different muscle contracts, releasing the catch, and the resilin pad quickly expands, liberating this stored energy. Since the process of resilin compression takes a few tenths of a second, the leg muscles can generate quite enough power for that task.

Other jumping insects apparently use this same technique. Next time you are out in a field during the summer or fall, you might observe a grasshopper. You will notice that on landing from a jump, its hind legs are fairly well extended and that it then slowly bends these legs in preparation for the next leap. Resilin pads are also found at the base of insect wings.

What is perhaps most remarkable about this little story is that though the amazing jumping power of fleas was known for centuries, it was as recently as 1967 that the energetics of the process was investigated and it was discovered that simple muscular contraction could not possibly power the takeoff.

Summary

The *work* ΔW done by a force **F** is defined by

$$\Delta W = F\,\Delta s \cos\theta = F_s\,\Delta s = F\,\Delta s_F$$

where $\Delta\mathbf{s}$ is the displacement of the point of application of the force **F**, and θ is the angle between the vectors **F** and $\Delta\mathbf{s}$. This is also equal to the product of the displacement Δs and the component of **F** in the direction of $\Delta\mathbf{s}$, or alternatively, to the product of F and the component of $\Delta\mathbf{s}$ in the direction of **F**.

Work is a scalar quantity. The SI unit of work is the *joule.* One joule equals one newton meter. The dimension of work is $[M][L]^2/[T]^2$.

To move a body of given mass from its initial position to some final position it may be necessary to perform work against gravity or have work done by gravity. This work depends only on the initial and final locations and not on the path followed in shifting the position of the mass. The gravitational force is an example of a *conservative force.*

Energy is defined as the capacity of a body (or a system of bodies) to do work.

Mechanical energy is conventionally categorized as *potential energy* (PE) and *kinetic energy* (KE).

The PE of a body is its capacity to do work by virtue of its position or configuration. The gravitational PE of a body is given by

$$\text{PE} = mgh$$

where h is the height of the body of mass m relative to some arbitrary zero level. It is therefore impossible to specify a unique PE; in all natural events, it is *changes* in PE that matter.

Kinetic energy is the capacity of a body to do work by virtue of its motion. The KE is given by

$$\text{KE} = \tfrac{1}{2}mv^2$$

where v is the speed of the body of mass m.

The KE of an object depends on the choice of inertial reference frame; generally, the surface of the earth is used as the "rest" frame of reference.

The units and dimensions of PE and KE are the same as those of work. PE and KE are scalar quantities.

Practically every natural event involves the transformation of energy—between different forms of mechanical energy, and between mechanical, chemical, electrical, thermal or nuclear energies. In every instance the system conforms with the *principle of conservation of energy:*

The total energy of an isolated system is constant.

We can write, therefore,

$$\text{KE}_\text{i} + \text{PE}_\text{i} = \text{KE}_\text{f} + \text{PE}_\text{f} + W$$

where the subscripts "f" and "i" denote the final and initial conditions and W is the work done *by* the system against other forces besides gravity.

Power (P) is the rate at which work is done, that is,

$$P = \frac{\Delta W}{\Delta t}$$

The SI unit of power is the *watt;* one watt equals one joule per second. The dimension of power is $[M][L]^2/[T]^3$. Power is a scalar quantity.

An alternative expression for power that is often useful is

$$P = Fv \cos \theta = F_v v = F v_F$$

where **F** is the net force acting on the body, **v** is its velocity, and θ is the angle between the vectors **F** and **v**; F_v is the component of the force in the direction of **v**, and v_F is the component of **v** in the direction of the force **F**.

Multiple Choice Questions

4.1 Which of the following is *not* a vector quantity?
(a) velocity
(b) power
(c) acceleration
(d) displacement

4.2 Which of the following is *not* a unit of energy?
(a) W·s
(b) kg·m/s
(c) N·m
(d) J

4.3 The dimension of power is
(a) $[M][L]/[T]$
(b) $[M][L]^2/[T]^2$
(c) $[M][L]^2/[T]^3$
(d) None of the above.

4.4 Abe and Ben move identical boxes equal distances in a horizontal direction. Abe slides the box on a surface that is frictionless. Ben lifts his box, carries it that distance and sets it down again.
(a) Abe does more work than Ben.
(b) Abe does less work than Ben.
(c) Neither Ben nor Abe do any work.
(d) The amount of work done by each depends on the time taken.

4.5 Assume that a pole vaulter obtains all his height by the complete conversion of his KE into PE. If his speed just before placing his pole down is v, the height reached is given by
(a) $v/2g$
(b) $\sqrt{2vg}$
(c) $v^2/2g$
(d) $2g/v^2$

4.6 The work expended to accelerate a car from 0 to 30 m/s is
(a) less than that required to accelerate it from 30 m/s to 60 m/s.
(b) equal to that required to accelerate it from 30 m/s to 60 m/s.
(c) more than that required to accelerate it from 30 m/s to 60 m/s.
(d) may be any of the above depending on the time taken to achieve this change in speed.

4.7 A car starts from rest and travels forward with constant acceleration. Which of the following is correct?
(a) The power delivered by the drive shaft to the wheels is constant.
(b) The power delivered by the drive shaft to the wheels increases as the car gains speed.
(c) The kinetic energy of the car is proportional to the time.
(d) None of the above is correct.

4.8 A pendulum bob of mass M is suspended by a string of length L. The bob is pulled to one side so that it is a height $L/4$ above its freely hanging level. If the bob is now released from rest, the speed of the bob at its lowest point is given by
(a) $v = MgL/8$
(b) $v = \sqrt{MgL/2}$
(c) $v = \sqrt{gL/2}$
(d) $v = \sqrt{gL/8}$

4.9 Assume that when brakes are applied, a constant frictional force is exerted on the wheels of a car. If that is so, it follows that
(a) the car loses KE at a constant rate.
(b) the distance the car travels before coming to rest is proportional to the square of the speed just before the brakes are applied.
(c) the distance the car travels before coming to rest is proportional to the speed of the car just before the brakes are applied.

(d) the KE of the car is inversely proportional to the time, with $t = 0$ being the instant the brakes are applied.

4.10 Two masses are released from a height H above ground: M_1 slides down a frictionless inclined plane that makes an angle of 30° with the horizontal; M_2 slides down a similar plane that makes an angle of 45° with the horizontal. Which of the following is true?

(a) M_1 reaches the bottom after M_2 but both arrive with the same speed at that point.

(b) M_1 and M_2 reach the bottom at the same time and with the same speed.

(c) M_1 arrives at the bottom after M_2, and the speed of M_1 at that point is less than that of M_2.

(d) None of the above is correct.

4.11 Jane, Mike, and Ed load identical concrete blocks from the ground onto the back of a truck. Jane lifts her blocks almost straight up from the ground to the bed of the truck. Mike slides his blocks up a rough plank. Ed slides his blocks up an inclined plane with frictionless rollers.

Mike's plank has the same length as Ed's frictionless-roller inclined plane. All three load the same number of identical blocks. Which of the following statements is true?

(a) Jane does more work than Mike, and Mike does more work than Ed.

(b) Jane and Ed do the same work, but Mike does more.

(c) Mike does more work than Ed, and Ed more than Jane.

(d) Jane, Mike, and Ed all do the same amount of work.

Problems

(Sections 4.1–4.3)

4.1 How much work must be done to raise a 3-kg mass from the floor to the top of a 1.5-m-high table?

4.2 A tractor is towing a barge through a canal. The towrope makes an angle of 26° with the direction of motion of the barge. The tension in the rope is 8000 N. How much work is done in moving the barge 400 m?

4.3 A helium nucleus has a mass of 6.7×10^{-27} kg. If this nucleus has a KE of 8×10^{-13} J, what is its speed?

4.4 An oxygen molecule at room temperature moves with an average speed of 380 m/s. The oxygen molecule has a mass 32 times that of a proton. What is the KE of such a molecule?

4.5 A 1500-kg car is moving at a speed of 100 km/h. What is its KE? Compare this KE with the energy dissipated in 10 min by a 1-kW space heater.

4.6 Calculate the KE of a 1400-kg car traveling at 60 km/h.

• **4.7** The force required to compress a spring a distance x from its undeformed length is proportional to x. That is, this force is $F = kx$, where k is known as the spring constant. Plot F as a function of the compression x, and use this to show that the work done in compressing the spring is given by $W = \frac{1}{2}kx^2$.

(Section 4.4)

4.8 A monkey drops 3 m, grabs hold of a vine, and swings down another 2 m, at which point the monkey is moving horizontally. What is its speed at that moment?

4.9 A child is swinging on a garden swing. At the lowest point the speed of the swing is 3 m/s. The mass of the child and swing seat is 40 kg. How high above this low point will the child and swing rise?

4.10 The average person in reasonably good physical condition can develop about 1 hp for a brief time (about $\frac{1}{2}$ min or so). Estimate the time required to run up three flights of stairs (a total height of about 12 m). Time yourself to verify this estimate.

• **4.11** A stone is thrown from the roof of a 20-m-tall building with a speed of 10 m/s. What is the speed of the stone when it hits the ground?

• **4.12** A rope is suspended from a branch of a tree. The rope is 15 m long. Suppose you wish to raise yourself above ground by running along and grabbing hold of the rope and swinging up. How fast must you run to raise yourself by 2 m?

• **4.13** A 3-kg mass starts from rest at the top of an inclined plane that is 5 m long and makes an angle of 30° with the horizontal. Find the speed of the mass at the bottom if (a) the plank is frictionless; (b) the coefficient of friction between plank and mass is 0.15.

• **4.14** A 20-kg girl slides down a playground slide that is 2.5 m high; she comes to the bottom with a speed of 2.2 m/s. How much energy was dissipated in heat?

• **4.15** A 4-kg block, initially at rest, is pushed by a constant force on a level surface. After the block has moved 6 m, the force is removed. If the block is then traveling at a speed of 10 m/s, how large a force was applied to the block? How does this force compare with the weight of the block? Neglect friction between the block and supporting surface.

• **4.16** Suppose that in Problem 4.15 the coefficient of friction between block and surface is 0.3. What force would then have to be applied to impart to this block a speed of 10 m/s as it moves 6 m, starting from rest? After the force is removed, how far will this block slide before coming to rest again?

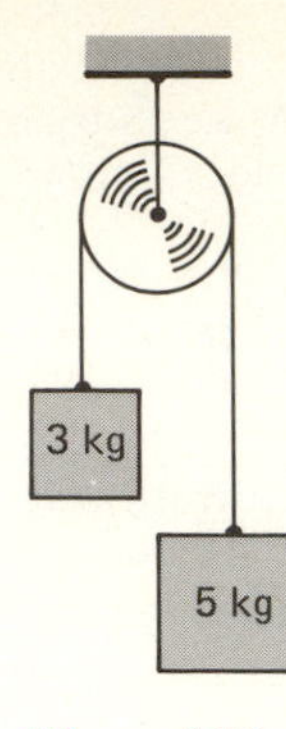

Figure 4.17

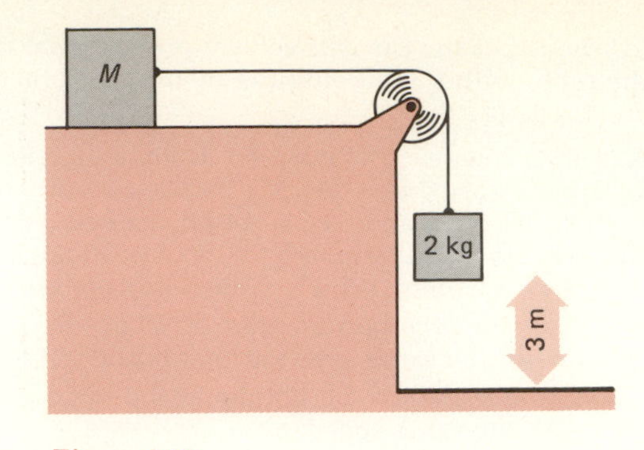

Figure 4.18

• **4.17** At $t = 0$, the 5-kg mass of Figure 4.17 is moving upward at a speed of 2 m/s. How far will this mass rise before it descends again? (Neglect friction and assume that the masses of the pulley and string are negligible.)

• **4.18** A projectile is fired from a gun pointed at an angle of 60° with the horizontal. What is the muzzle velocity if the maximum height the projectile attains is 3000 m?

• **4.19** A 5-kg bob is attached to a 4-m-long string to form a simple pendulum. What is the angle of swing of the pendulum if the maximum KE of the bob is 49 J?

• **4.20** The PE stored in the deformation of a spring is given by PE $= \frac{1}{2}kx^2$, where k is the spring constant. Calculate the energy stored in compressing a spring having $k = 2000$ N/m by 10 cm. If this spring is used in a toy gun to propel a projectile of 20-g mass, what will the muzzle velocity of the projectile be if this spring is compressed (a) 10 cm? (b) 15 cm?

• **4.21** A skier starts at the top of a hill that is 50 m high and has an average elevation angle of 37°. He pushes off with his poles so that his initial speed is 1.5 m/s. At the bottom of the run, his speed is 23 m/s. Find the average value of the coefficient of friction between his skis and the snow, neglecting air friction.

• **4.22** A car is going up a hill with a 7 percent grade (a rise of 7 m for every 100 m along the road) at 60 km/h. There is a stop sign at the top of the hill. The driver is keen on saving both gasoline and his brakes and decides to coast to a stop. If friction losses are negligible, how far from the stop sign should the driver put the car into neutral?

• **4.23** The system shown in Figure 4.18 is released from rest with the 2-kg mass 3 m above the ground. When it is just 1 m above ground, its speed is 3 m/s. With what speed will it strike the ground, and what is the magnitude of the mass M? (Neglect friction.)

• • **4.24** A gun fires a 2.5-g steel bullet with a muzzle velocity of 400 m/s into a block of wood. The bullet penetrates the wood a distance of 4.5 cm. Determine the average force of friction between the wood and the bullet. Assuming that a gun of larger caliber fires a heavier bullet with the same muzzle velocity, estimate the depth of penetration of a 10-g bullet fired into the same block of wood. (*Hint:* Assume that the force of friction is proportional to the area of contact between bullet and wood and that all bullets have the same shape.)

• **4.25** An 85-kg crate is to be lifted onto the bed of a truck. What is the force needed to lift it straight up? How much work is done if the bed of the truck is 0.8 m above ground? Suppose the crate is pushed up a fairly smooth inclined plane that makes an angle of 30° with the horizontal. If the coefficient of kinetic friction between plank and crate is 0.15, how large a force must be applied parallel to the plank to move the crate up the plank at constant speed, and how much work is done in moving the crate into the truck?

• • **4.26** A sailing cruiser comes into harbor under power. Its engine is providing 1 hp at the propeller as the boat maintains a speed of 9 km/h. The boat has a total mass of 8000 kg. Determine the KE of the boat and the retarding force due to drag of the water against the hull. If the skipper stops the engine when the boat is 60 m from the dock, what will the speed of the boat be as it moves alongside the dock, and what is then its kinetic energy? (Assume—contrary to fact—that the retarding force due to drag of the water is independent of the boat's speed.) Would you expect the boat, after drifting 60 m, to be moving faster or more slowly than what has been calculated? Explain.

• • **4.27** A playground slide starts at a height of 3 m. The angle of the slide with the horizontal is 30° (Figure 4.19). Suppose that the frictional force between child and slide is $Mg/4$, where M is the mass of the child, and that this force is independent of the angle of the slide with the horizontal. What is the speed at the bottom of the slide of a child who starts from rest at the top? How long should the horizontal end portion of the slide be so that the child's speed is no more than 1 m/s at the point where the slide ends?

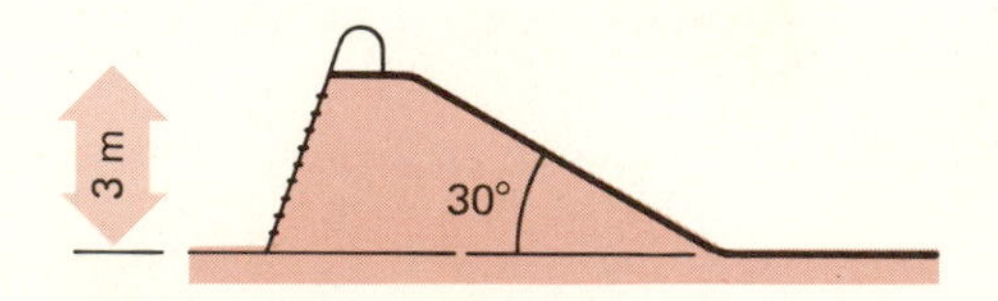

Figure 4.19

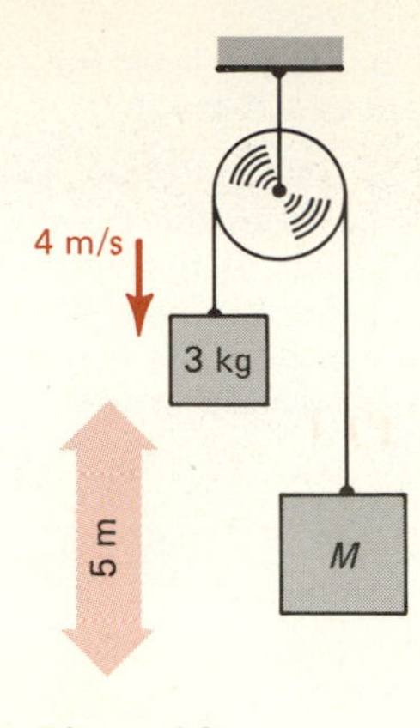

Figure 4.20

• • **4.28** A 200-kg crate hangs from a crane by a 15-m-long cable. A rope is attached to the crate and is used to pull it sideways. If the tension in the rope is 400 N, how far from the vertical has the crate been pulled, and how much work was done in pulling the crate to this position? (*Note:* As the crate is pulled sideways, the tension in the rope is not constant.)

• • **4.29** At the instant shown in Figure 4.20, the 3-kg mass is moving downward with a speed of 4 m/s. When it reaches a point 5 m below the one it occupied at $t = 0$, it is instantaneously at rest. Determine the mass M and the amount of time that elapses before the 3-kg mass passes the starting point on its way back up. Neglect friction.

• **4.30** A mass of 2 kg is released from rest at the top of the inclined plane (Figure 4.21). At the bottom it has a speed of 5 m/s. Determine the coefficient of kinetic friction between plane and mass.

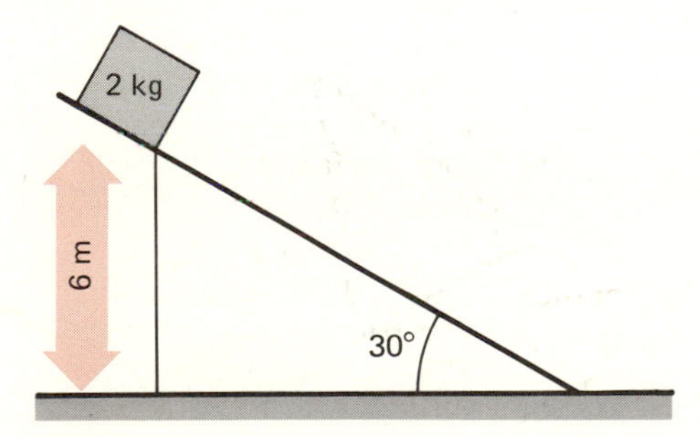

Figure 4.21

• • **4.31** Figure 4.22 shows a sketch of a simple roller coaster. The loaded coaster has a mass of 2000 kg. The starting point of the run is 22 m above point A; the run from B to C is 120 m long, the run from C to D is 100 m. At D, brakes are applied, bringing the coaster to a halt 15 m further along, at E. The average frictional force between B and D is 500 N. Find (a) the work done in raising the roller coaster to its starting point; (b) the speed of the roller coaster at C and at D; and (c) the average braking force applied between D and E.

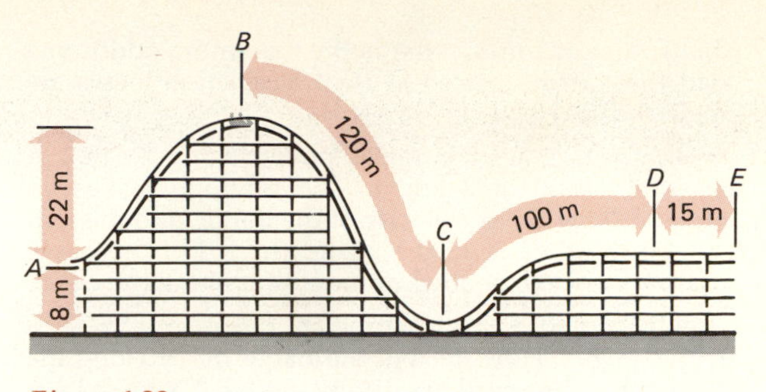

Figure 4.22

• **4.32** The 7-kg mass of Figure 4.23 starts from rest 1.5 m above ground. Just before it strikes the ground, its speed is 2 m/s. What is the value of the mass M if the coefficient of friction between M and the inclined plane is (a) zero? and (b) 0.4?

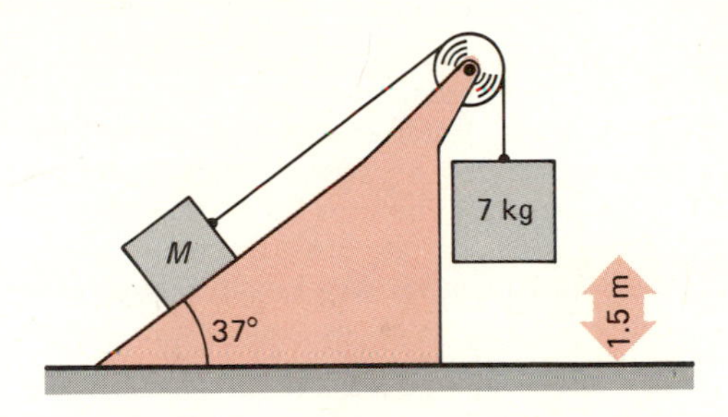

Figure 4.23

(Section 4.5)

4.33 In a car traveling on a level road at 60 km/h, the engine generates 18 hp. Determine the average retarding force due to wind resistance and friction.

• **4.34** An 1800-kg car is equipped with a 35-hp engine. Neglecting all friction losses, determine the maximum speed at which this car can climb a 15° grade.

• **4.35** A sump pump can lift 10 gallons of water per minute to a height of 3 meters. Suppose the efficiency of the pump is 75 percent; what should the minimum power rating of the pump motor be?

• **4.36** Two cars are identical except for the power of their engines. The mass of each car is 1500 kg. Car A's engine generates 45 hp, car B's engine generates 60 hp at the drive shaft. Calculate the time required for car A to reach a speed of 50 km/h under constant, full power. If both cars started at the same instant, what would be the speed of car B at that time if it accelerated under full power? (Neglect losses due to friction.)

• **4.37** A useful rule of thumb for rate of climb is 1000 ft/h for an average climber who is carrying a moderately heavy pack. What is the rate at which a climber of 80-kg mass carrying a 20-kg backpack works in climbing?

• **4.38** A 20-hp motor is used to lift an elevator whose mass, when empty, is 1200 kg. If this elevator should rise at 1 m/s, what is the maximum additional load that can be placed in it, if all friction losses are neglected?

• **4.39** A 60,000-kg barge is pulled along a canal at constant speed of 3 km/h. The towrope makes an angle of 24° with the velocity of the barge, and the tension in the rope is 1200 N. Find the KE of the barge and the power expended in moving it along the canal.

• **4.40** Suppose the towrope in Problem 4.39 suddenly breaks. How far will the barge move along the canal before its speed has diminished to 0.5 km/h? Assume that the frictional drag of the water is independent of speed.

• • **4.41** A loaded jet liner has a total mass of 30,000 kg. Within 15 min after takeoff, it has attained a cruising altitude of 8000 m and a cruising speed of 800 km/h. What was the average power developed by the plane's engines during this time, with all friction losses neglected? Why is this a very unrealistic estimate?

• • **4.42** An enterprising couple decide to use the PE of their bathwater to power some home appliances. The bathtub, located on the second floor, holds 350 liters. (One liter of water has a mass of 1 kg.) They install a water-wheel–driven generator in the basement, 6.5 m below the tub. At what rate must the water drain from the bathtub to power a 1-kW toaster? How long can they run the toaster on a full bathtub? (Assume that the turbine is 100 percent efficient.)

4.43 A 1500-kg car (including the mass of the driver) has an engine that provides 60 kW at the drive shaft. If the car starts from rest, what is its velocity after 3 s? after 6 s? What is the average acceleration during the first 3-s interval and during the second equal interval?

• • **4.44** Suppose the maximum power output of a car's engine is 55 hp. The car, whose mass is 1400 kg, is going at a speed of 80 km/h on a level road, and its engine is generating 22 hp.

(a) Determine the retarding force due to wind resistance and friction.

(b) If these retarding forces do not change substantially with speed (*not* a valid assumption in practice), what is the least time it takes this car to accelerate to 100 km/h?

• • **4.45** Figure 4.24 shows a block and tackle used to lift a mass of 300 kg from the floor to the top of an 0.8-m-high work bench. If the worker operating the block and tackle works at a rate of 120 W, how long will it take to lift the mass to the benchtop? What is the average force with which the worker pulls on the rope?

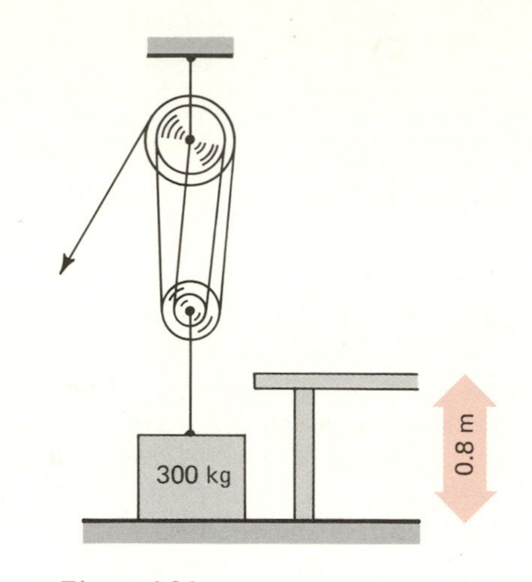

Figure 4.24

• • **4.46** A chairlift at a ski area is to have the following characteristics. It should carry as many as 24 skiers per minute to the top of a 250-m-high hill, in a double-chair arrangement (2 skiers per chair). The length of the lift is 600 m, and the overall efficiency of the machinery is 80 percent when the lift is operating at full capacity. Design such a lift; that is, specify the speed at which the cable carrying the chairs should run, the distance between chairs and the power requirements of the electric motor driving the lift. How long will a skier sit in the chair in going up to the top of the hill? (Note: There is no single "correct" answer to this problem; you should use good judgment. For example, you should allow 5 to 8 seconds between loading of chairs.)

•5•

Impulse and Momentum

I have concluded that this question of impulsive forces is very obscure, and I think that, up to the present, none of those who have treated this subject have been able to clear up its dark corners which lie almost beyond the reach of human imagination.

GALILEO GALILEI

5.1 Introduction

When you practice your tennis serve against a backboard, you hit the ball toward the board at a certain velocity; it rebounds toward you at a speed only slightly diminished. If you play golf, you strike a small elastic ball with a heavy club; immediately thereafter the ball leaves the tee at high speed, traveling through the air a hundred meters or more, a distance far greater than you could throw it. Yet the initial KE of the ball derives from energy expended by your body. If you fire a rifle, it recoils against your shoulder as the bullet travels the length of the barrel and emerges at the muzzle.

These diverse examples have certain common features. First, in each instance an object—tennis ball, golf ball, or bullet—experiences a drastic change in velocity and suffers a very large acceleration. Second, the time interval during which this acceleration takes place is relatively short. Consequently, the average force acting on the object must be fairly

Figure 5.1 *A golf club striking a golf ball off the tee.*

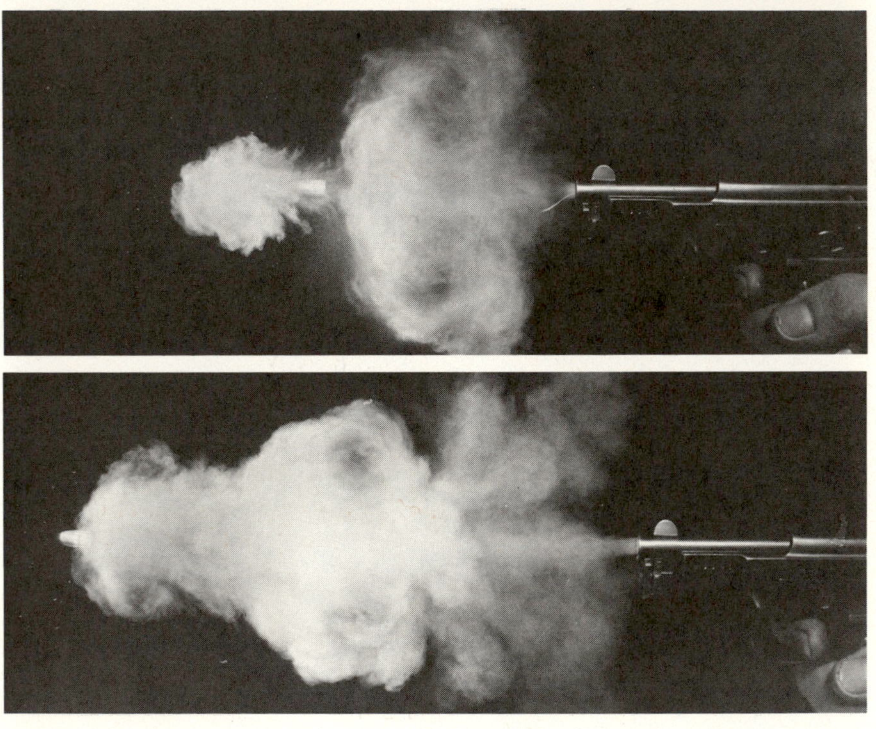

Figure 5.2 *A bullet being fired from a gun.*

large, but in each instance it would be incorrect to assume that this force is constant during that short time interval. Third, in each case a second object, which by Newton's third law must have experienced a reaction force of equal magnitude but opposite direction, manifests a substantially smaller change in velocity—the recoil velocity of the rifle, the change in velocity of the head of the golf club, and the apparently zero velocity of the backboard.

Our aim is to develop a technique for analyzing these and similar events. In the process we shall come across another basic conservation law of physics.

5.2 External and Internal Forces

We shall be concerned here with the dynamics not only of a single body but also of a system of two or more bodies such as a golf club and golf ball, rifle and bullet, or two billiard balls. In analyzing the behavior of a multicomponent or many-body system, it is useful to distinguish between internal and external forces. *Internal forces* are those by which the various parts of the system act on each other. *External forces* are those that agencies outside the system exert on one or more bodies of the system or on the system as a whole.

Suppose that the system under study consists of two masses M_1 and M_2 attached to each other by a massless spring, as shown in Figure 5.3. The action of the spring is to push the two masses apart if the spring is compressed to a length shorter than its equilibrium length L_0, and to exert an attractive force on each of the masses if the spring's extension is greater than L_0. For this system, the force the spring exerts on each of the masses is an internal force. In addition to this internal force, there is also the very weak gravitational attraction between the two masses, which also depends on their separation. The gravitational attraction between each of the masses and the earth, however, is an external force that acts on the system, the two masses and spring.

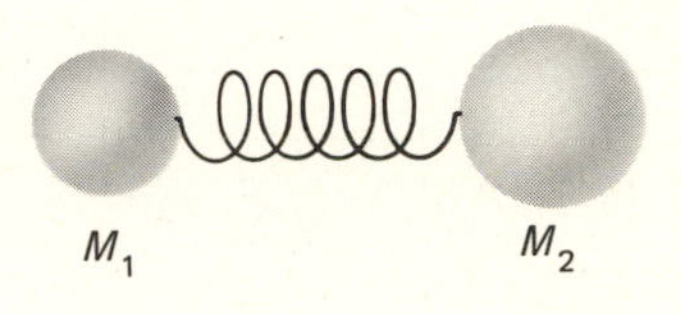

Figure 5.3 *Two masses M_1 and M_2 attached to each other by a massless spring.*

5.3 Newton's Second Law Revisited; Linear Momentum

> The alteration of motion is ever proportional to the motive force impressed; and is made in the direction of the right [straight] line in which that force is impressed.

This is the famous second law as phrased by Newton. To understand its meaning, we must first ascertain what Newton meant by "motion." He defined that word as follows:

> The quantity of motion is the measure of the same, arising from the velocity and quantity of matter conjunctly.

That is, *motion* as Newton used the word means the product of mass and velocity, a quantity we call today the *linear momentum* of the mass; it is defined as follows.

DEFINITION: The linear momentum of an object of mass m and velocity $\mathbf{v}$ equals the product of the mass and velocity;

$$\mathbf{p} = m\mathbf{v} \tag{5.1}$$

The symbol $\mathbf{p}$ is conventionally used for linear momentum. Note that $\mathbf{p}$ is a vector that points in the same direction as $\mathbf{v}$. The SI unit of $\mathbf{p}$ is kg·m/s; the dimension of $\mathbf{p}$ is $[M][L]/[T]$.

The *precise* translation of Newton's second law into mathematical language is therefore not $\mathbf{F} = m\mathbf{a}$ but

$$\mathbf{F} = \lim_{\Delta t \to 0} \left(\frac{\Delta \mathbf{p}}{\Delta t}\right) \tag{5.2}$$

that is, the force equals the rate of change of the momentum.* If the change in momentum is entirely due to a change in velocity, the mass remaining constant, then we can write

$$\mathbf{F} = \lim_{\Delta t \to 0} \left(\frac{\Delta \mathbf{p}}{\Delta t}\right) = \lim_{\Delta t \to 0} \frac{\Delta(m\mathbf{v})}{\Delta t} = m \lim_{\Delta t \to 0} \left(\frac{\Delta \mathbf{v}}{\Delta t}\right) = m\mathbf{a} \tag{5.2a}$$

arriving at the now familiar form of Newton's law.

* The adjective *linear* was included earlier to distinguish this quantity from another, the angular momentum, which we shall encounter later. It is common practice to omit the adjective *linear;* whenever *momentum* is used by itself, "linear momentum" is understood.

Equation (5.2), like Equation (3.4), is shorthand for three equations, one for each of the components of the vectors **F** and $\Delta\mathbf{p}$.

Frequently, as in firing a bullet, we do not know the details of the force acting on the object. All we know is that in a brief time Δt, the momentum of the object changes by $\Delta\mathbf{p}$. Consequently, the average force during the time Δt is then

$$\langle\mathbf{F}\rangle = \frac{\Delta\mathbf{p}}{\Delta t} \tag{5.3}$$

and if the mass remains unchanged during that time Δt,

$$\langle\mathbf{F}\rangle = m\frac{\Delta\mathbf{v}}{\Delta t} = m\langle\mathbf{a}\rangle \tag{5.3a}$$

We rewrite Equation (5.3) in the form

$$\Delta\mathbf{p} = \langle\mathbf{F}\rangle\ \Delta t \tag{5.4}$$

to emphasize that the change in momentum is the product of the time-average force $\langle\mathbf{F}\rangle$ and the time interval Δt over which this force acts. The product $\langle\mathbf{F}\rangle\ \Delta t$ is called *impulse.*

Impulse, for which no conventional symbol is in use, is a vector quantity, directed along the average force $\langle\mathbf{F}\rangle$. It has the same units and dimension as momentum, although it is common practice in referring to impulse to use the unit newton second rather than kilogram meter per second.

In the preceding we have used the phrase "time average" on several occasions. What is meant thereby? To illustrate, let us consider Figure 5.4, which shows the force that might act on a tennis ball as it rebounds from a hard surface. What is then the magnitude of the impulse?

The total impulse that acts can be viewed as the sum of impulses acting successively during the much smaller time intervals δt_i. For example, during the interval δt_4, the average force is F_4 and the corresponding impulse $F_4\ \delta t_4$. In the next interval, the average force is F_5 and

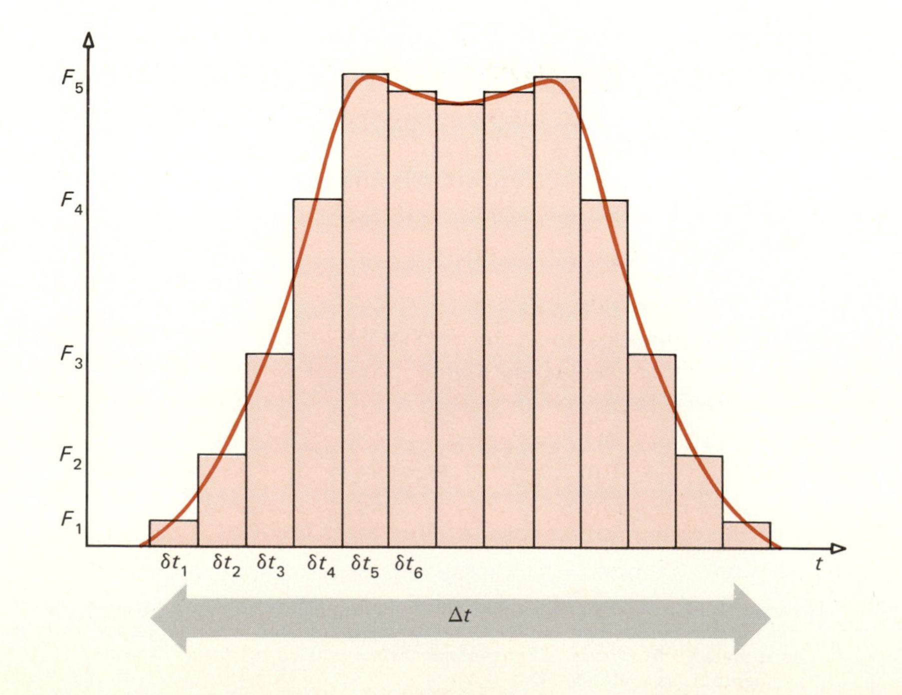

Figure 5.4 *The force acting on a tennis ball as it rebounds from a wall as a function of time (schematic). The short time interval Δt during which this force acts can be subdivided into intervals δt_i. The total impulse $F_{av}\ \Delta t$ is obtained by summing the smaller impulses $F_i\ \delta t_i$.*

the impulse $F_5\,\delta t_5$, and so on. The total impulse during $\Delta t = \Sigma_i\,\delta t_i$ is the sum of the individual small impulses; i.e.,

$$\text{Total impulse} = \langle \mathbf{F} \rangle\, \Delta t = \sum_i \mathbf{F}_i\, \delta t_i \tag{5.5}$$

where $\langle \mathbf{F} \rangle$ is the force averaged over the time interval $\Delta t = \Sigma_i\,\delta t_i$. Equation (5.5) really defines what is meant by the time average, namely,

$$\langle \mathbf{F} \rangle = \frac{\sum_i \mathbf{F}_i\, \delta t}{\sum_i \delta t_i} \tag{5.5a}$$

It is apparent from Figure 5.4 that there is a simple geometric interpretation of impulse; it is the area under the curve of force versus time. The procedure indicated here will give a close approximation of the true area under the $\mathbf{F}(t)$ curve. The precision can be improved without limit by reducing the time interval δt and simultaneously increasing the number of small rectangular elemental areas, the $\mathbf{F}_i\,\delta t_i$. In the limit as δt approaches zero and the number of terms in the summation of Equation (5.5) approaches infinity, this sum will equal the area under the curve exactly.*

5.4 Conservation of Linear Momentum

The value of these new concepts becomes apparent when we focus attention not on a single body but on a system of many bodies that interact with each other but on which no net external forces are acting. For clarity, we begin by limiting discussion to the simplest such system, that consisting of two bodies of masses m_1 and m_2.

If these two objects collide, their individual momenta will change. But from Newton's third law, the force $\mathbf{F}_{12}$, which m_1 exerts on m_2, must be equal in magnitude but opposite in direction to $\mathbf{F}_{21}$, the force m_2 exerts on m_1. That is,

$$\mathbf{F}_{12} = -\mathbf{F}_{21}$$

Since no external force acts on the system, the total force is

$$\mathbf{F} = \mathbf{F}_{12} + \mathbf{F}_{21} = 0 = \frac{\Delta \mathbf{p}}{\Delta t} = \frac{\Delta(\mathbf{p}_1 + \mathbf{p}_2)}{\Delta t} \tag{5.6}$$

Consequently,

$$\Delta \mathbf{p} = 0 \qquad \text{or} \qquad \mathbf{p}_1 + \mathbf{p}_2 = \text{constant}$$

By induction, the conclusion that the total momentum remains constant in the absence of an external force, applies to a system with any number of interacting particles, such as a container filled with gas molecules. The system obeys the *principle of conservation of linear momentum:*

In the absence of a net external force, the total linear momentum of a system remains constant.

Conservation of momentum is as fundamental as conservation of energy, although the latter is perhaps better known by the general public. As the examples we have cited suggest, momentum conservation is the

* This is the definition of the integral in calculus.

cornerstone on which one constructs the solution to a variety of problems involving two or more interacting bodies. Moreover, it also illuminates rocket and jet propulsion, the pressure a gas exerts on the walls of a container, and numerous other phenomena.

Momentum conservation arguments can often be conveniently applied in situations where **F** does not vanish but one or two components of it do. In such cases, the corresponding components of **p** are conserved. The problem of a projectile exploding in flight is one example in which this approach is useful (Example 5.5). The force on the system is not zero, since it is subjected to gravity. That force, however, has no horizontal component, and consequently, the horizontal components of **p** are conserved.

Figure 5.5 *Saturn V rocket at blastoff. In this process, the mass of the rocket is continually diminishing as the fuel is burned. Total momentum is conserved, since the momentum of the hot exhaust gases ejected to the rear equals the forward momentum gained by the rocket.*

5.5 Elastic and Inelastic Collisions

Let us now consider two colliding bodies more carefully. Suppose two objects, of masses 1 kg and 2 kg, are moving along a line toward each other, with speeds of 4 m/s and 2 m/s, respectively (Figure 5.7). The total momentum before the collision is

$$\mathbf{p} = m_1\mathbf{v}_1 + m_2\mathbf{v}_2$$

$$p = (1\text{ kg})(4\text{ m/s}) + (2\text{ kg})(-2\text{ m/s}) = 0\text{ kg}\cdot\text{m/s}$$

where we have taken the direction to the right as positive.

Following the collision, the total momentum must still be zero. For example, m_1 might have a velocity of -4 m/s, and m_2 a velocity of 2 m/s. But this is by no means the only pair of velocities that satisfies momentum conservation. For instance, $v_{1f} = -2$ m/s, $v_{2f} = 1$ m/s, or $v_{1f} = v_{2f} = 0$, are possible (here the subscript "f" denotes the final value, following the collision). And though the solution $v_{1f} = -10$ m/s, $v_{2f} = +5$ m/s, might appear improbable, it is perfectly consistent with momentum conservation. Which of these results is "correct"?

Any one could be. The true solution depends on an aspect of the event we have not yet mentioned, the character of the collision, whether *elastic* or *inelastic*. *An elastic collision is one in which the total KE as well as the total momentum of the system is conserved. Any collision in which the total KE is not conserved is called an inelastic collision.* A collision in which the bodies stick together on colliding and proceed thereafter with a common velocity is said to be *perfectly inelastic*.

Figure 5.6 *Squid in an aquarium. These marine animals move by jet propulsion, ejecting water by violent contraction of the body wall (the mantle).*

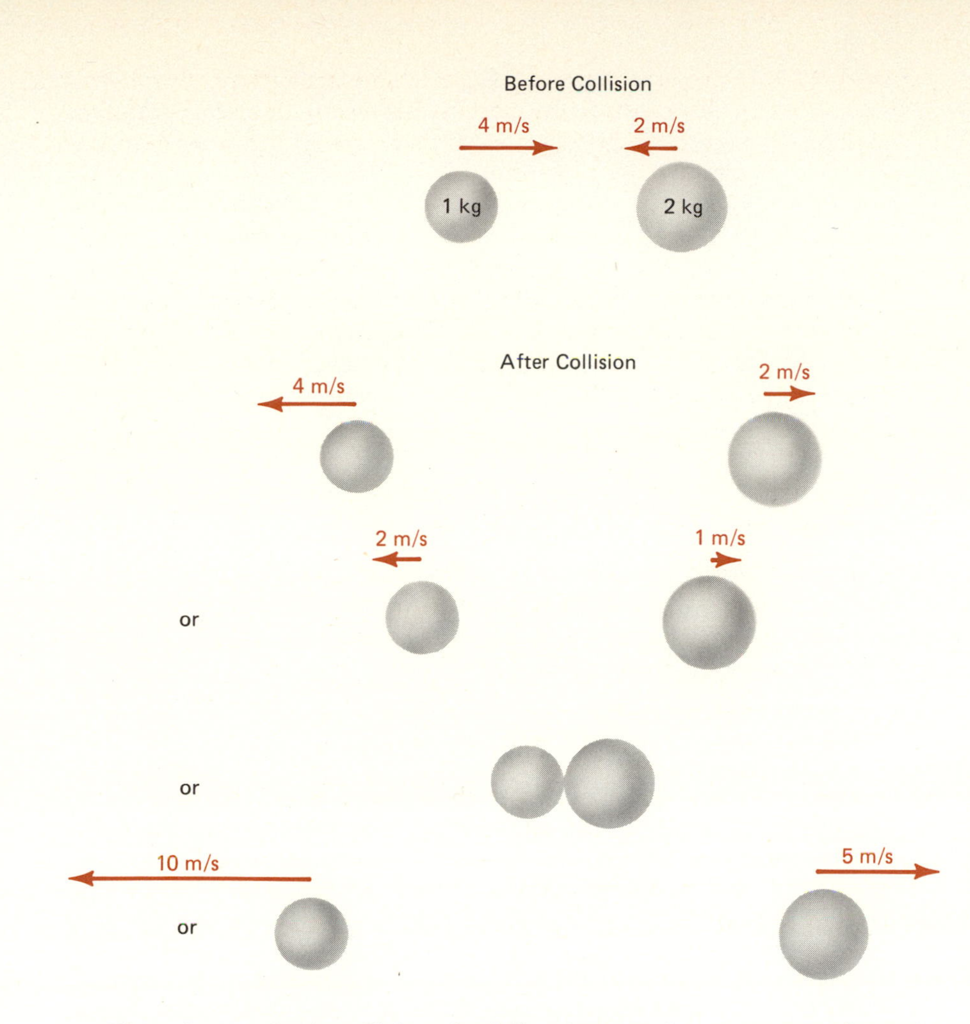

Figure 5.7 A head-on collision of a 1-kg mass and a 2-kg mass. The initial momentum of the system is zero. Following the collision, the momentum of the system must again be zero; the four final configurations all satisfy momentum conservation.

Thus, for an *elastic collision*, the following requirements must be satisfied by the system of interacting bodies:

$$\begin{aligned} \mathbf{p}_i &= \mathbf{p}_f \qquad \text{MOMENTUM CONSERVATION} \\ KE_i &= KE_f \qquad \text{CONSERVATION OF KINETIC ENERGY} \end{aligned} \tag{5.7}$$

In an *inelastic collision*

$$\begin{aligned} \mathbf{p}_i &= \mathbf{p}_f \qquad \text{MOMENTUM CONSERVATION} \\ KE_i &\neq KE_f \qquad \text{KINETIC ENERGY IS NOT CONSERVED} \end{aligned} \tag{5.8}$$

In the case just considered, if the collision were elastic, $v_{1f} = -4$ m/s, $v_{2f} = 2$ m/s, would be the correct solution; if it were perfectly inelastic, $v_{1f} = v_{2f} = 0$ would be the correct result.

Let us now examine a few situations that demonstrate conservation of momentum.

● ***Example 5.1*** A 0.1-kg ball moving at a speed of 5 m/s in the $+x$ direction collides head-on with a 0.3-kg ball that is at rest. Assuming that the collision is elastic, determine the final velocities of the two balls.

Since the collision is elastic, we must satisfy the two conditions of Equation (5.7).

The total initial momentum of the system is

$$p_i = (0.1 \text{ kg})(5 \text{ m/s}) + (0.3 \text{ kg})(0 \text{ m/s}) = 0.5 \text{ kg·m/s}$$

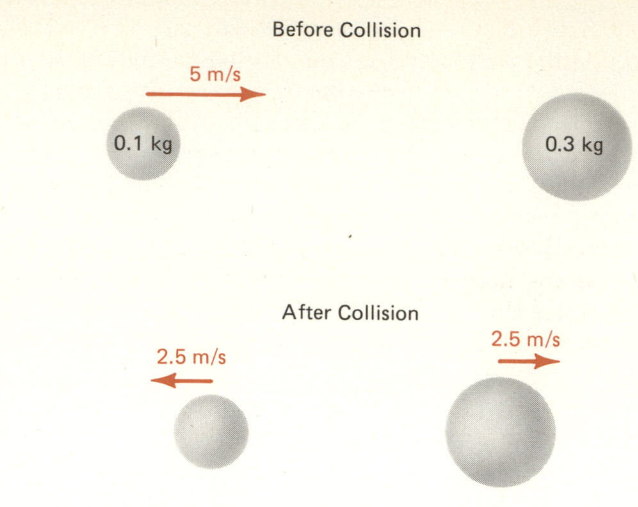

Figure 5.8 *A 0.1-kg ball approaches a 0.3-kg ball initially at rest. They make a head-on elastic collision. After the collision, the balls move with the velocities shown.*

The total initial kinetic energy is

$$\mathrm{KE_i} = \tfrac{1}{2}(0.1\ \mathrm{kg})(5\ \mathrm{m/s})^2 = 1.25\ \mathrm{J}$$

We now express the final velocities by v_{1f} and v_{2f}, where "1" refers to the 0.1-kg ball and "2" to the 0.3-kg ball. Thus,

$$p_f = (0.1\ \mathrm{kg})v_{1f} + (0.3\ \mathrm{kg})v_{2f} = p_i = 0.5\ \mathrm{kg{\cdot}m/s}$$

$$\mathrm{KE_f} = \tfrac{1}{2}(0.1\ \mathrm{kg})v_{1f}^2 + \tfrac{1}{2}(0.3\ \mathrm{kg})v_{2f}^2 = \mathrm{KE_i} = 1.25\ \mathrm{J}$$

We have here two equations with two unknowns, v_{1f} and v_{2f}. In contrast to previous examples involving two equations with two unknowns, one of the equations here is quadratic. This complicates matters a bit, but the solution is not very difficult. We proceed as follows.

From the first of the two equations, expressing momentum conservation, we obtain

$$v_{1f} = 5\ \mathrm{m/s} - 3v_{2f} \tag{5.9}$$

We now substitute this expression into the quadratic equation wherever v_{1f} appears, i.e., in the first term; on multiplying by 2, we have

$$(0.1\ \mathrm{kg})(5\ \mathrm{m/s} - 3v_{2f})^2 + (0.3\ \mathrm{kg})v_{2f}^2 = 2.5\ \mathrm{J}$$

Expanding the parenthesis and multiplying both sides by 10 gives

$$(25 - 30v_{2f} + 9v_{2f}^2 + 3v_{2f}^2)\ \mathrm{J} = 25\ \mathrm{J}$$

$$2v_{2f}^2 - 5v_{2f} = 0$$

This equation is readily factored to $v_{2f}(2v_{2f} - 5) = 0$, yielding two solutions:

$$v_{2f} = 0 \qquad \text{and} \qquad v_{2f} = 2.5\ \mathrm{m/s}$$

The first of these corresponds to the initial condition, when the 0.3-kg ball is at rest. The second solution gives the value of v_2 after the collision. It tells us that the 0.3-kg ball will move off to the right at 2.5 m/s.

The velocity of the 0.1-kg ball after the collision is now obtained by substituting the value of v_{2f} just determined into Equation (5.9). We then find that

$$v_{1f} = -2.5\ \mathrm{m/s}$$

The negative sign indicates that the velocity is directed to the left; the incident ball rebounds with a speed half as great as its incident speed.

The appearance of the initial condition as one of the two solutions of the quadratic equation is related to the *reversibility* of this event. When we say that an event is reversible, we mean that if following this event, we were to reverse the time sequence, as though we were running a film backwards through a projector, the process would reverse also. In this case, such a reversal means the following. After the collision has occurred, we reverse the motion of each of the balls at the same instant. They will then retrace their paths and collide head-on, after which the 0.3-kg ball will be at rest and the 0.1-kg ball will move off to the left at 5 m/s.

Before we leave this example, let us evaluate the impulse received by each ball. For the 0.1-kg ball the change in momentum is

$$\begin{aligned}\Delta p_1 &= p_{1f} - p_{1i} \\ &= (0.1\text{ kg})(-2.5\text{ m/s}) - (0.1\text{ kg})(5\text{ m/s}) \\ &= -0.75\text{ kg}\cdot\text{m/s}\end{aligned}$$

and for the 0.3-kg ball

$$\begin{aligned}\Delta p_2 &= p_{2f} - p_{2i} = (0.3\text{ kg})(2.5\text{ m/s}) - 0 \\ &= 0.75\text{ kg}\cdot\text{m/s}\end{aligned}$$

The changes in momenta are of equal magnitude but opposite direction, resulting, as required, in a total change of momentum for the system of zero. The magnitude of the impulse received by each ball equals the momentum change of each, namely 0.75 N·s. ●

● ***Example 5.2*** This example concerns an inelastic collision, one that occurs all too frequently. A 20,000-kg truck and a 1500-kg car collide head-on. Just before the impact, the truck's speed was 5 m/s (18 km/h), and the car's speed was 8 m/s (28.8 km/h). The two vehicles mesh during the collision and stick together. The time between contact and the

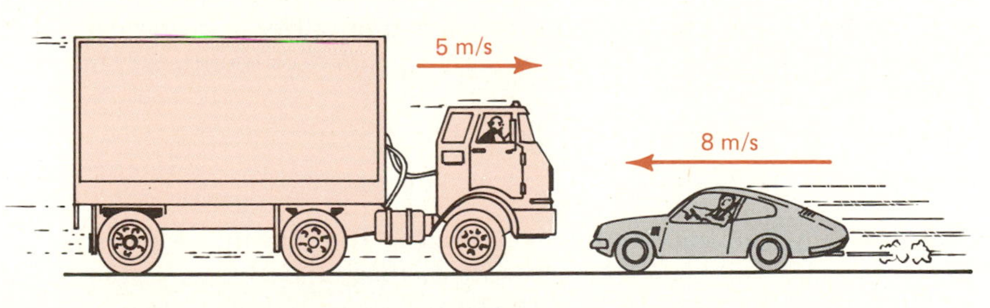

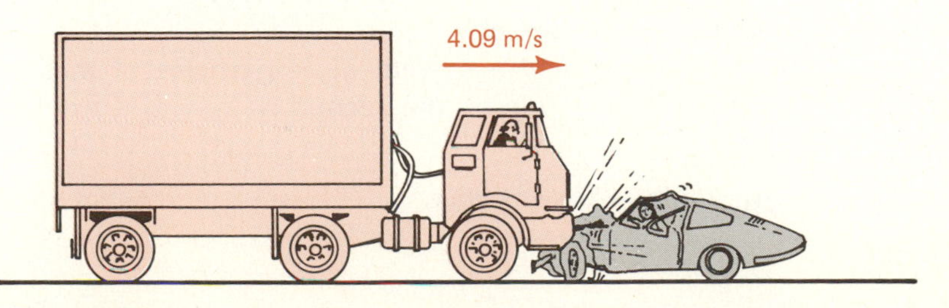

Figure 5.9 *An unpleasant inelastic collision. After the collision, the two vehicles mesh and move together with the same velocity, in this case 4.09 m/s to the right.*

fully meshed configuration is 0.2 s. The mass of each driver is 100 kg, and this mass is included in the masses given above.

We want to know the velocity of the wreck immediately after the collision and the force acting on each driver during the collision, assuming that both were restrained by safety harnesses.

The collision is perfectly inelastic. Hence, the only conservation law that applies is momentum conservation. Taking as our positive direction that of the truck before impact, we have

$$p_i = (20{,}000 \text{ kg})(5 \text{ m/s}) + (1500 \text{ kg})(-8 \text{ m/s}) = 88{,}000 \text{ kg·m/s}.$$

Following the collision, the two vehicles move with the same velocity. Consequently,

$$p_f = (20{,}000 \text{ kg} + 1500 \text{ kg})v_f = 88{,}000 \text{ kg·m/s}$$

from which we find

$$v_f = 4.09 \text{ m/s}$$

The wreck moves in the same direction as the initial velocity of the truck with a speed only slightly less than that of the truck before impact.

The impulses sustained by the two vehicles are equal in magnitude and oppositely directed. But what about the impulses suffered by the two drivers?

The change in momentum of the truck driver is

$$\Delta p_T = (100 \text{ kg})\Delta v_T = (100 \text{ kg})(4.09 \text{ m/s} - 5 \text{ m/s}) = -91 \text{ N·s}$$

whereas for the driver of the car

$$\Delta p_C = (100 \text{ kg})\Delta v_C = (100 \text{ kg})[(4.09 \text{ m/s}) - (-8 \text{ m/s})] = 1209 \text{ N·s}$$

During the 0.2 s of the collision, the truck driver experienced an average force of 455 N, about half his weight, but the driver of the car was subjected to an average force of 6045 N, more than six times his weight. Put another way, the truck driver suffers about 0.5*g* acceleration, but the unfortunate driver of the car more than 6*g*! Clearly, drivers of light sports cars should be more conservative, at least in their driving habits, than owners of Cadillacs and Lincoln Continentals. Unfortunately, they rarely are. ●

5.6 Center of Mass

An especially helpful concept in analyzing the motion of a system of many particles, or of an extended body, is the *center of mass*, abbreviated CM hereafter. Although the CM is particularly useful in treating rotation (Chapter 8), it also greatly simplifies the analysis of collisions, and we therefore introduce the concept now.

The position of the CM of a system of N particles of masses m_1, m_2, . . . , m_N at locations given by the vectors $\mathbf{r}_1$, $\mathbf{r}_2$, . . . , $\mathbf{r}_N$ is given by

$$M\mathbf{r}_{CM} = m_1\mathbf{r}_1 + m_2\mathbf{r}_2 + \cdots + m_N\mathbf{r}_N \qquad \textbf{(5.10)}$$

where

$$M = m_1 + m_2 + \cdots + m_N \qquad \textbf{(5.11)}$$

is the total mass of the system.

As these particles move under the influence of external and internal forces, their positions change with time. If during the brief interval Δt, the position vectors change by $\Delta \mathbf{r}_1, \Delta \mathbf{r}_2, \ldots, \Delta \mathbf{r}_N$, the corresponding change in the position of the CM, $\Delta \mathbf{r}_{CM}$, is then given by

$$M\,\Delta \mathbf{r}_{CM} = m_1\,\Delta \mathbf{r}_1 + m_2\,\Delta \mathbf{r}_2 + \cdots + m_N\,\Delta \mathbf{r}_N \tag{5.12}$$

Dividing by Δt and recalling the definition of velocity, we obtain

$$M\mathbf{v}_{CM} = m_1\mathbf{v}_1 + m_2\mathbf{v}_2 + \cdots + m_N\mathbf{v}_N \tag{5.13}$$

which can be written

$$\mathbf{p}_{CM} = \mathbf{p}_1 + \mathbf{p}_2 + \cdots + \mathbf{p}_N \tag{5.14}$$

In Section 5.4 we showed that in the absence of external forces, the total momentum of a system remains constant. Since $\mathbf{p}_{CM}$ is, in fact, equal to the total momentum of the system, we conclude that in the absence of external forces, the CM of a system at rest remains at rest, and if the CM is in motion it will maintain that motion. Moreover, if a net external force does act, the CM will move in accordance with Newton's second law. In particular, if the total mass does not change with time, the acceleration of the CM is given by

$$\mathbf{a}_{CM} = \frac{\mathbf{F}_{ext}}{M} \tag{5.15}$$

where $\mathbf{F}_{ext}$ is the net external force acting on the system.

● ***Example 5.3*** A mass of 4 kg is located at $x = 0.2$ m, $y = z = 0$ m, and a second mass of 6 kg is at $x = 0.8$ m, $y = z = 0$ m. Locate the CM.

Equation (5.10) is, like all vector equations, really a set of three equations, one for each coordinate. Since the y and z coordinates of the two masses are zero, $y_{CM} = z_{CM} = 0$. The equation for x_{CM} is

$$(10\text{ kg})x_{CM} = (4\text{ kg})(0.2\text{ m}) + (6\text{ kg})(0.8\text{ m})$$

giving $x_{CM} = 0.56$ m. ●

Generally, the individual parts of a system interact with each other through internal forces, thereby changing their individual velocities and momenta as time goes on. However, those interactions will not influence the CM motion. Whenever one can separate the forces acting on a system of two or more particles into internal and external forces, one can simplify the dynamic problem by asking, and answering, two separate questions: What is the motion of the CM? and, What is the motion of the separate parts relative to the CM?

● ***Example 5.4*** A stationary 2-kg mass containing a small explosive charge of negligible mass disintegrates into three fragments. Two have identical masses of 0.5 kg each; the third has a mass of 1 kg. The velocities of the 0.5-kg fragments make an angle of 60° with each other, and their magnitudes are 100 m/s. What is the velocity of the 1-kg fragment?

The situation is shown schematically in Figure 5.10. The y axis is the line bisecting the angle between the velocities of the 0.5-kg fragments.

Since $\mathbf{v}_{CM} = \mathbf{0}$ before the explosion, it must also be zero thereafter. From Equation (5.13), we have

$$(0.5\text{ kg})\mathbf{v}_1 + (0.5\text{ kg})\mathbf{v}_2 + (1\text{ kg})\mathbf{v}_3 = \mathbf{0}\text{ kg·m/s}$$

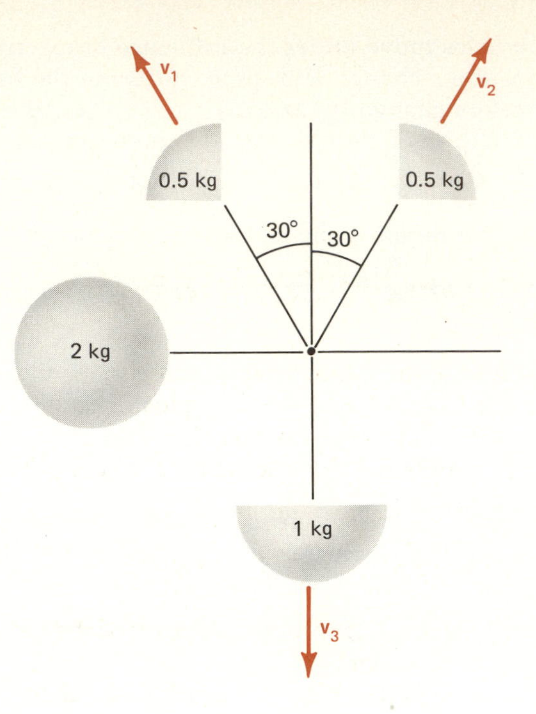

Figure 5.10 *A 2-kg mass, initially at rest, disintegrates by an internal explosion into three segments. The initial momentum is zero; the vector sum of the momenta of the separate parts must add to zero. The CM remains at rest.*

or in component form,

$$(0.5\text{ kg})(v_{1x} + v_{2x}) + (1\text{ kg})v_{3x} = 0\text{ kg·m/s}$$

$$(0.5\text{ kg})(v_{1y} + v_{2y}) + (1\text{ kg})v_{3y} = 0\text{ kg·m/s}$$

Since $v_{1x} = -v_{2x}$ (see Figure 5.10), $v_{3x} = 0$ m/s. Also,

$$v_{1y} = v_{2y} = (100\text{ m/s})\cos 30° = 86.6\text{ m/s}.$$

Hence, $v_{3y} = -86.6$ m/s. ●

● ***Example 5.5*** A projectile is fired from a gun with an initial velocity of 300 m/s at an angle of 60° above the horizontal. When the projectile is at the top of its trajectory, an internal explosion causes a separation into two parts of equal masses. One part falls to the ground as though it had been released from rest at that point. Find the velocity of the second part immediately after the explosion and the distance between the gun and the point of impact. (Neglect air friction.)

This is a slightly more difficult problem. Still it is not complicated if we just think about it a bit and do not blindly plug numbers into equations.

In projectile motion, only the vertical component of velocity changes; the horizontal component remains constant. Moreover, at the peak of the trajectory, the vertical component is instantaneously zero. Consequently, at the moment of the explosion, the velocity of the projectile is

$$v_{\mathrm{i}} = v_{\mathrm{i}x} = (300\text{ m/s})\cos 60° = 150\text{ m/s}$$

First we shall complete this problem using momentum conservation; then we shall repeat the solution using CM motion. The momentum just before the explosion is

$$\mathbf{p}_{\mathrm{i}} = 150M\text{ kg·m/s}$$

directed along x, where M is the mass of the projectile. Immediately following the explosion, there are two parts, each with mass $M/2$, whose

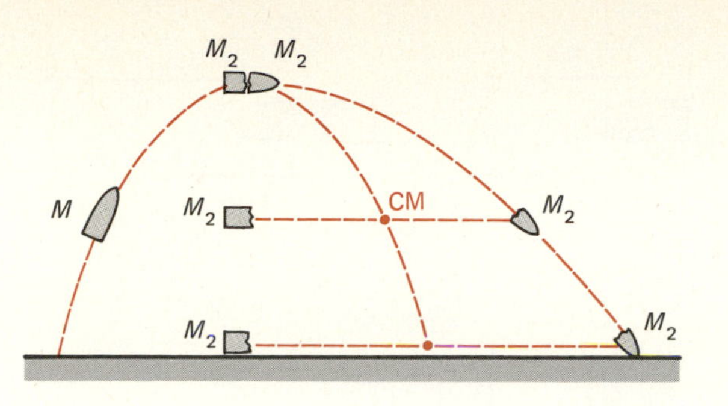

Figure 5.11 *A projectile explodes at the top of its trajectory into two portions of equal mass. One portion falls vertically to the ground as though released from rest. The CM continues its motion as though the explosion had not occurred.*

momenta must add to $\mathbf{p}_i$. Since one part falls as though released from rest, its velocity immediately after the explosion must be zero; its contribution to the total momentum is therefore also zero. Momentum conservation then demands that the momentum of the other part equal $\mathbf{p}_i$. Thus

$$\left(\frac{M}{2}\right)\mathbf{v}_{f2} = 150M\ \text{kg·m/s} \qquad \text{and} \qquad \mathbf{v}_{f2} = 300\ \text{m/s}$$

directed along x, where $\mathbf{v}_{f2}$ denotes the velocity of the second part the instant following the explosion.

The distance a projectile travels is quickly calculated once we know the time of flight. At the top of the projectile motion, $v_y = 0$, whereas initially we had $v_{0y} = 300 \sin 60° = 260$ m/s. From Equation (2.13d), $v(t) = v_0 + at$, the time required to reach the point of explosion is

$$t = \frac{260\ \text{m/s}}{9.8\ \text{m/s}^2} = 26.5\ \text{s}$$

During that time, the projectile has traveled a horizontal distance of

$$s_x = (150\ \text{m/s})(26.5\ \text{s}) = 3975\ \text{m}$$

At that instant, the explosion occurs. Since the two parts have zero vertical component of velocity immediately after the explosion, the time before each reaches the ground is another 26.5 s. The part that has a horizontal velocity of 300 m/s just after the explosion will therefore travel another $300 \times 26.5 = 7950$ m horizontally before striking the ground. Hence the total distance from the gun to the point of impact is 3975 m + 7950 m = 11,925 m, or about 12 km.

Using the CM method, we would proceed as follows. Having established that immediately before the explosion the CM had a horizontal velocity of 150 m/s in the positive x direction, we transform to a reference frame that is moving with that velocity. In this so-called *CM reference frame,* the CM is at rest.

In the CM frame, the portion of the projectile that falls as though from rest in the earth frame now has an initial velocity of −150 m/s. Consequently, to maintain the condition $\mathbf{p}_{CM} = \mathbf{0}$ in the CM frame, the other segment of mass $M/2$ must have an initial velocity of +150 m/s in this CM frame.

To determine the initial velocity of this segment in the earth frame, often called the *rest frame,* we now add to its velocity in the CM frame the velocity of the CM itself. The result is that this segment has an initial horizontal velocity of 300 m/s in the positive x direction. ●

Examples 5.2, 5.4, and 5.5 were all instances of inelastic collisions. In one, two objects, the truck and car, stuck together after the collision; in

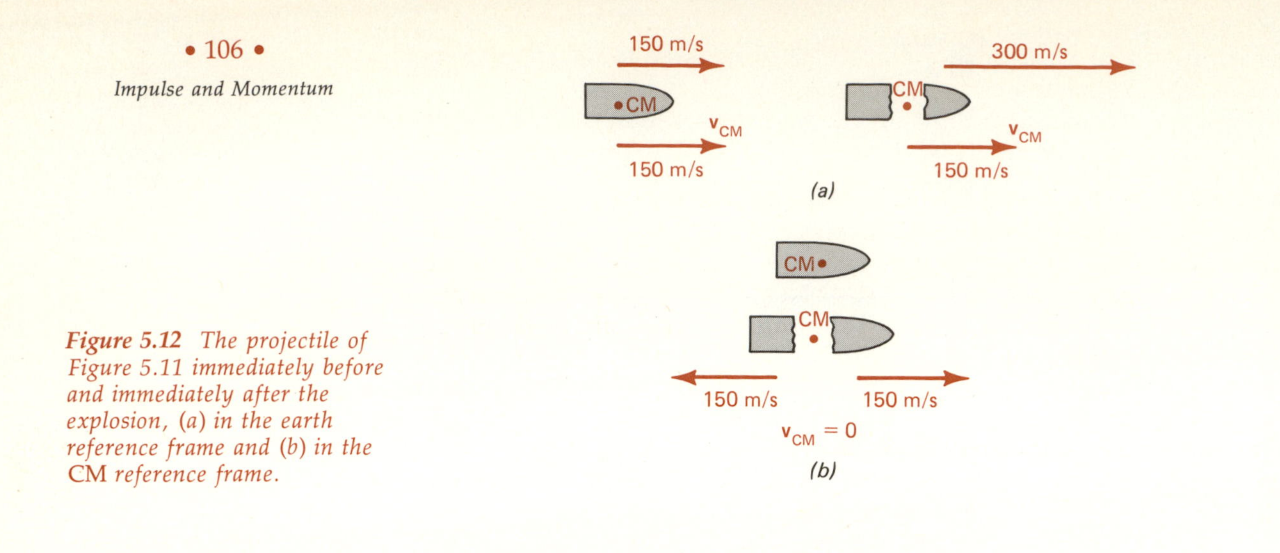

Figure 5.12 *The projectile of Figure 5.11 immediately before and immediately after the explosion, (a) in the earth reference frame and (b) in the CM reference frame.*

the others, the objects moved initially as a unit and then separated into several portions. In each case, momentum was conserved but the KE presumably was not. Let us now calculate the KE changes in two of these cases.

In the truck-car collision, the initial KE of the system is

$$KE_i = \tfrac{1}{2}(20{,}000\ \text{kg})(5\ \text{m/s})^2 + \tfrac{1}{2}(1500\ \text{kg})(8\ \text{m/s})^2 = 298{,}000\ \text{J}$$

and the final KE is

$$KE_f = \tfrac{1}{2}(21{,}500\ \text{kg})(4.09\ \text{m/s})^2 = 177{,}200\ \text{J}$$

Thus, 120,800 J of KE were dissipated in the collision—in work to deform the metal, break glass, scrape rubber from the tires, and make the sound of the crash, as well as in heat generated by friction.

When a projectile explodes, chemical energy stored in the TNT or gunpowder is converted, at least in large part, into KE of the fragments. Consequently, the total KE of the fragments should be greater than the KE of the projectile before the explosion. In the first of these examples, this is obviously true since the initial KE is zero, the 2-kg mass being at rest. In the other case, the initial KE of the projectile, immediately before the explosion, is

$$KE_i = \tfrac{1}{2}M(150\ \text{m/s})^2 = 11{,}250M\ \text{J}$$

After the explosion, one part has zero KE since it drops as though from rest; that is, its velocity the instant after the explosion is zero. Does the other part really have more KE than the unexploded projectile? Let's see. Its KE is

$$KE_f = \tfrac{1}{2}(M/2)(300\ \text{m/s})^2 = 22{,}500M\ \text{J}$$

Indeed, the KE of the system after the explosion is greater than the initial KE; in fact, it is twice as great. In this "inelastic collision," the change in KE was 11,250*M* J.

Viewed in the CM frame, the initial KE of the system is zero since the CM is now at rest. Following the explosion, the two parts move off with identical speeds of 150 m/s. Hence the total KE in the CM frame after the explosion is

$$KE_f = 2[\tfrac{1}{2}(M/2)(150\ \text{m/s})^2] = 11{,}250M\ \text{J}$$

that is, just equal to the increase in KE calculated above.

This is a special case of a general result.

The total KE of any system of particles equals the KE of its parts in the CM frame of reference plus the KE of the CM.

● ***Example 5.6*** The *ballistic pendulum* is a simple device used to measure the velocity of a bullet. A block of wood, suspended from a group of light strings, is initially at rest when a bullet is fired horizontally into the wood. The bullet embeds itself in the block, which then swings in the direction of the projectile's velocity, attaining an easily determined height relative to its initial position.

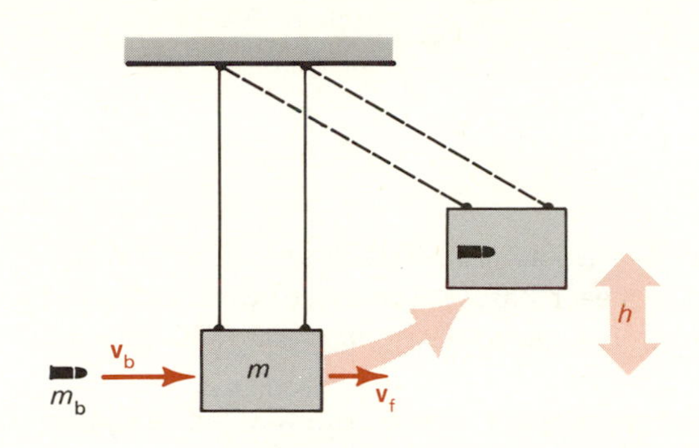

Figure 5.13 The ballistic pendulum. The light bullet approaching with high speed from the left embeds itself in the heavy block of wood. The block with the bullet in it swings to the right, reaching a maximum height h. Measurement of h and knowledge of the masses of bullet and wood allow calculation of the bullet's speed prior to impact.

Suppose the mass of the block is 1.5 kg, that of the bullet is 0.01 kg, and the height of the swing is 0.35 m. What was the speed of the bullet before it buried itself in the wood?

Again, this is a perfectly inelastic collision. As before, $\mathbf{p}_i = \mathbf{p}_f$, where now $\mathbf{p}_i$ is the momentum of the bullet before striking the block (the block has zero momentum), and $\mathbf{p}_f$ is the momentum of the block, with the bullet in it, immediately thereafter. To determine the speed of the bullet, we must therefore determine $\mathbf{p}_f$. Once this is known, the bullet's velocity is given by

$$\mathbf{v}_b = \frac{\mathbf{p}_i}{m} = \frac{\mathbf{p}_i}{0.01 \text{ kg}} = \frac{\mathbf{p}_f}{0.01 \text{ kg}}$$

From the height reached by the block, we know its maximum PE relative to the initial position, namely,

$$Mgh = [(1.5 + 0.01) \text{ kg}](9.8 \text{ m/s}^2)(0.35 \text{ m}) = 5.18 \text{ J}$$

where M is the total mass, block plus bullet. This is the KE that the wood-plus-bullet must have had at the bottom of the swing, that is, just after the bullet had come to rest within the block. Consequently,

$$\frac{1}{2}Mv_f^2 = \frac{\frac{1}{2}p_f^2}{M} = 5.18 \text{ J}$$

where v_f is the speed of this system at the bottom of the pendulum's swing. We can now solve for p_f:

$$p_f = \sqrt{2M(\text{KE})} = \sqrt{2(1.51 \text{ kg})(5.18 \text{ J})} = 3.95 \text{ kg}\cdot\text{m/s}$$

and therefore,

$$v_b = \frac{3.95 \text{ kg}\cdot\text{m/s}}{0.01 \text{ kg}} = 395 \text{ m/s}$$

●

Summary

The *linear momentum* $\mathbf{p}$ of a body is defined by

$$\mathbf{p} = m\mathbf{v}$$

where m is its mass and $\mathbf{v}$ its velocity. The linear momentum is a *vector* whose direction is that of $\mathbf{v}$. It is conventional to omit the adjective *linear;* when referring to momentum, "linear momentum" is understood.

The SI unit of momentum is the kilogram meter per second; no special name has been given this unit. The dimension of momentum is $[M][L]/[T]$.

The proper mathematical statement of Newton's second law is

$$\mathbf{F} = \frac{\Delta \mathbf{p}}{\Delta t}$$

or

$$F_x = \frac{\Delta p_x}{\Delta t}, \qquad F_y = \frac{\Delta p_y}{\Delta t}, \qquad F_z = \frac{\Delta p_z}{\Delta t}$$

This reduces to $\mathbf{F} = m\mathbf{a}$ only if m is constant.

The product $\langle \mathbf{F} \rangle \, \Delta t$, the average force multiplied by the interval of time during which it acts, is called the *impulse*. From Newton's second law, it follows that

$$\Delta \mathbf{p} = \langle \mathbf{F} \rangle \, \Delta t$$

that is, the change in momentum equals the impulse.

Impulse is a vector quantity, directed along the average force. No special symbol is used for impulse. The unit of impulse is the newton second; no special name has been assigned to this unit. Impulse has the same dimension as momentum, $[M][L]/[T]$.

An *isolated system* is defined as a system of bodies on which no net external force is acting.

From Newton's second and third laws, one obtains the *principle of conservation of momentum:*

The total momentum of an isolated system of interacting bodies is constant.

If a net force does act on the system but has a vanishing component in some direction, the corresponding component of the total momentum is conserved.

Momentum conservation is the fundamental concept used in solving collision problems. One distinguishes between *elastic* and *inelastic collisions.* In both types of collision, total momentum is conserved. An *elastic collision* is defined as one in which the total KE is also conserved. An *inelastic collision* is any collision in which the total KE of the system changes. A *perfectly* inelastic collision is one in which the two colliding bodies stick together and consequently move with a common velocity after the collision.

The position of the *center of mass* (CM) of a system of masses is given by

$$M\mathbf{r}_{\mathrm{CM}} = \sum_i m_i \mathbf{r}_i, \qquad M = \sum_i m_i$$

where $\mathbf{r}_{\mathrm{CM}}$ is the position vector of the CM, and M is the total mass of the system consisting of several masses m_i located at positions specified by the vectors $\mathbf{r}_i$.

The total momentum of a system equals the velocity of the CM, $\mathbf{v}_{CM}$, multiplied by the total mass M. Consequently, $\mathbf{v}_{CM}$ of an isolated system of fixed total mass is constant.

It is generally advantageous to solve collision problems by transforming to a coordinate system in which the CM is at rest, the "CM reference frame."

In a perfectly inelastic collision, the final velocities of the colliding bodies are zero in the CM reference frame.

In all cases, the total KE of a system is the sum of the KEs of the individual components in the CM reference frame plus the KE of the CM. Since in a collision $\mathbf{v}_{CM}$ is constant, the change in KE in an inelastic collision is often most simply determined by performing the calculation in the CM reference frame.

Multiple Choice Questions

5.1 A car of mass M traveling with velocity v strikes a car of mass M that is at rest. The two cars' bodies mesh in the collision. The loss of KE in the collision is

(a) a quarter of the initial KE.
(b) half of the initial KE.
(c) all the initial KE.
(d) zero.

5.2 An object initially at rest explodes, disintegrating into three parts of equal mass. Parts 1 and 2 have the same initial speed v, and the velocity vectors $\mathbf{v}_1$ and $\mathbf{v}_2$ are perpendicular to each other. Part 3 will have an initial speed of

(a) $\sqrt{2}v$
(b) $v/2$
(c) $v/\sqrt{2}$
(d) $\sqrt{2v}$

5.3 A 0.3-kg ball drops to the floor and bounces back without loss of KE. Just before it strikes the floor, its speed is 10 m/s. The impulse the ball imparts to the floor is

(a) 0 kg·m/s.
(b) 3 kg·m/s, directed up.
(c) 6 kg·m/s, directed down.
(d) 6 kg·m/s, directed up.

5.4 $[M][L]/[T]$ is the dimension of

(a) force
(b) potential energy
(c) power
(d) impulse

5.5 A block of mass 1 kg moving at a speed of 2 m/s to the right on a frictionless plane collides with and sticks to a block of mass 2 kg, which was initially at rest. Following the collision,

(a) the KE of the system is 2 J.
(b) the momentum of the system is 6 kg·m/s.
(c) the momentum of the system is less than 2 kg·m/s.
(d) the KE of the system is less than 2 J.

5.6–5.11 *A stone is projected vertically upward. The graphs shown in Figure 5.14 relate to the following six questions.*

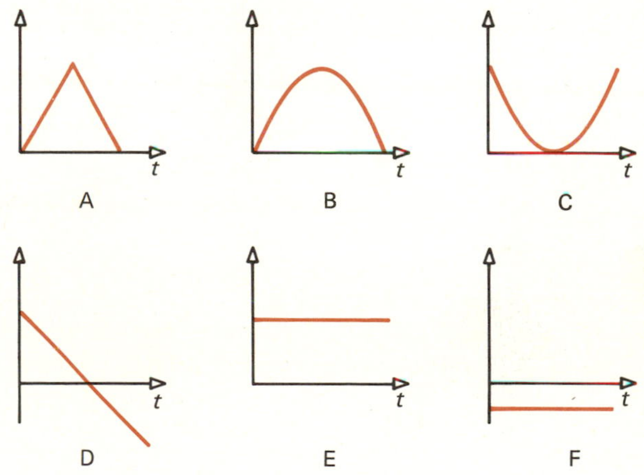

Figure 5.14

5.6 The graph that best displays the position of the stone as a function of time is

(a) A
(b) B
(c) C
(d) D

5.7 The graph that best displays the acceleration of the stone as a function of time is

(a) F
(b) D
(c) B
(d) A

5.8 The graph that best displays the PE of the stone as a function of time is

(a) B
(b) F
(c) D
(d) E

5.9 The graph that best displays the momentum of the stone as a function of time is

(a) D
(b) E
(c) A
(d) F

5.10 The graph that best displays the KE of the stone as a function of time is

(a) C
(b) E
(c) D
(d) F

5.11 The graph that best displays the total energy of the stone as a function of time is

(a) B
(b) A
(c) E
(d) D

5.12 A 4-kg ball makes an elastic head-on collision with a 1-kg ball. Before the collision, the speed of the 4-kg ball is 10 m/s and the speed of the 1-kg ball is zero. The speed of the 1-kg ball after the collision is

(a) zero.
(b) less than 10 m/s.
(c) equal to 10 m/s.
(d) greater than 10 m/s.

5.13 A 2-kg mass which is at rest receives an impulse of 10 N·s. Following this impulse, the

(a) speed of the mass is 20 m/s.
(b) speed of the mass is 10 m/s.
(c) momentum of the mass is 20 kg·m/s.
(d) momentum of the mass is 10 kg·m/s.

5.14 A 0.1-kg mass travels along a horizontal air track at a speed of 1 m/s. It makes an elastic collision with another, identical mass that is initially at rest on the track. Following the impact

(a) the total momentum and KE are the same as before the impact.
(b) the total momentum is the same as before the impact, but the total KE is less.
(c) the KE is conserved, but the momentum after the impact is less than before.
(d) the momentum is shared equally between the two masses after the impact.

Problems

(Section 5.3)

5.1 What is the momentum of a 1500-kg car that is traveling at a speed of 60 km/h?

5.2 A 1500-kg car and a 6000-kg truck have the same momentum. The car's KE is 3×10^5 J. Find the KE of the truck.

5.3 A 0.05-kg ball is dropped and strikes the floor with a speed of 10 m/s. It rebounds vertically with an initial speed of 8 m/s after being in contact with the floor for 0.01 s. Find the average force exerted by the ball on the floor.

5.4 A 0.4-kg object is moving on a frictionless surface with a speed of 25 m/s. A force of 2 N is applied continually until the velocity of the object has been reversed. How long was this force applied?

5.5 On a serve, a tennis ball leaves the racket of a very good player with a speed of 60 m/s. If the time of contact between ball and racket is 25 ms and the mass of the ball is 55 g, what is the average force exerted by the racket?

• **5.6** A body initially at rest on a level surface is pushed by a constant force of 30 N for 10 s. At the end of that time, its speed is 15 m/s. What is the mass of the body (a) if the surface is friction-free? and (b) if the coefficient of kinetic friction is 0.2?

• **5.7** A pitcher throws a baseball (mass of 0.22 kg) so that it passes the plate in a direction 15° below the horizontal, and has a speed of 33 m/s. The batter hits the ball so that it exactly reverses direction and leaves the bat with a speed of 54 m/s. What was the change in momentum of the ball? If the time of contact between ball and bat was 1.5 ms, what was the average force the bat exerted on the ball?

• **5.8** The force on a bullet in a gun barrel diminishes as the bullet travels forward. In rough approximation this force is given by $F = 540 - 9 \times 10^4 t$, where F is in newtons and t in seconds. At the end of the barrel the force can be presumed to vanish. (a) Plot the force as a function of time and calculate the total impulse on the bullet during its path in the barrel. (b) If the mass of the bullet is 5 g, what are the muzzle velocity and the length of the gun barrel?

• • **5.9** A 40-g bullet traveling with a speed of 380 m/s strikes an 0.8-kg wooden block that rests on a flat surface.

(a) If the block is firmly fixed to the surface and the bullet penetrates the block a distance of 8 cm before coming to rest, what is the acceleration of the bullet, assuming that it is constant? Also determine the force acting on the bullet, and the time elapsed before it comes to rest in the wood.

Compare the product of force and time with the change in momentum of the bullet.

(b) If the block is able to slide on the flat surface and the coefficient of kinetic friction between block and surface is 0.25, what is (i) the initial velocity of the block just after the bullet has come to rest in the block? (ii) the distance the block travels before coming to rest? (iii) the energy dissipated in heat within the block as the bullet comes to rest, and that dissipated in heat as a result of friction between block and flat surface?

(Section 5.4)

• **5.10** A 60-kg boy and a 50-kg girl face each other on roller skates. The girl pushes the boy, who moves away at a speed of 2 m/s. What is the girl's speed and how much work did she do?

• **5.11** A 10-g bullet is fired into a 0.4-kg wooden block which is initially at rest on a horizontal surface. The bullet passes through the block, leaving with a speed of 250 m/s. The block acquires a speed of 8 m/s in the event. Calculate the impact speed of the bullet and the energy dissipated as heat within the block of wood.

• **5.12** An empty freight car rolls on a flat, friction-free track beneath a coal chute. The initial speed of the car is 0.5 m/s. While the car is under the chute, 8000 kg of coal are dumped into the car, which, when empty, has a mass of 10,000 kg. What is the speed with which the freight car leaves this loading area?

• **5.13** A machine gun sits on a light wagon on a railroad track. The wagon, gunner, and gun have a total mass of 200 kg. If the mass of each bullet is 60 g and its muzzle velocity is 800 m/s, what is the recoil force if the gun can fire 3 rounds per second? If the gun fires for 15 s in the direction of the track, what is the speed of the wagon following that firing, if the wagon was initially at rest?

• •**5.14** A 34-g bullet is fired up into a block of wood of mass 1.25 kg. The bullet lodges in the wood, which rises to a height of 6 m. Find the speed of the bullet on impact with the wood.

• •**5.15** A light wagon carrying a passenger is moving along a level, friction-free track at a speed of 5 m/s. The total mass of the wagon and load is 200 kg. What are the final speed and KE of the wagon if the passenger throws a mass of 10 kg

(a) forward (in the direction of travel) with a speed of 10 m/s relative to track?

(b) backward with a speed of 10 m/s relative to the track?

(c) sideways, so that it moves perpendicular to the direction of the track, with an initial speed of 10 m/s.

(Sections 5.5–5.6)

5.16 A rifle fires a bullet of mass 0.01 kg with a muzzle velocity of 420 m/s. The mass of the rifle is 4 kg. What is the rifle's recoil velocity?

5.17 A bullet of mass 0.008 kg is fired into a ballistic pendulum of mass 0.60 kg. If the speed of the bullet at impact is 400 m/s, what is the maximum height of the pendulum's swing?

5.18 Three equal masses of 0.2 kg each are placed at the coordinates (0,0), (2,0), and (0,2). Locate the center of mass.

5.19 Repeat Problem 5.18 for a mass of 0.4 kg at the origin, and masses of 0.2 kg at (2,0) and (0,2).

• **5.20** One mode of radioactivity is so-called α decay. In this process an α particle, that is, a helium nucleus of atomic mass 4, is emitted by the radioactive nucleus. The resulting "daughter" nucleus has an atomic mass number that is 4 units less than that of the parent. An example of such a decay is

$$^{212}_{83}\text{Bi} \rightarrow {}^{208}_{81}\text{Tl} + {}^{4}_{2}\alpha$$

In that reaction equation, the superscripts give the atomic mass numbers, the subscripts the atomic numbers (see Chapter 32). It is a fair approximation to assume that the mass of the thallium, Tl, nucleus is $208/4 = 52$ times that of the helium nucleus, and that the mass of the bismuth, Bi, nucleus is 53 times that of a helium nucleus. The speed of the emitted helium nucleus is 1.7×10^7 m/s. Assuming that the Bi nucleus was initially at rest, determine the recoil speed of the Tl nucleus. What is the KE of the α particle and of the recoiling Tl nucleus?

• **5.21** A mass of 2 kg moving to the right at a speed of 4 m/s collides with and sticks to a mass of 4 kg that is initially at rest. What are the momentum and KE of the system after the collision?

• **5.22** A loaded freight car whose mass is 18,000 kg rolls along a level track at a speed of 1.2 m/s toward an empty stationary freight car. The two couple and proceed at a speed of 0.8 m/s. Determine the mass of the empty freight car. Calculate the initial and final kinetic energies.

• **5.23** A 1200-kg car, traveling due east at a speed of 20 km/h, collides at an intersection with a 1800-kg car traveling due north at 15 km/h. In the collision, the two cars mesh. What is the velocity of the wreck immediately after the collision? If the coefficient of friction between tires and pavement is 0.5, how far will the cars slide before coming to rest?

• **5.24** A bullet is fired horizontally into the 2.5-kg block of a ballistic pendulum. The mass of the bullet is 20 g, and the pendulum swings to a height of 1.2 m. What was the speed of the bullet just before it entered the block of wood?

• **5.25** A wooden puck slides due east on an air table with a speed of 1.5 m/s. At the center of the table it collides with an identical puck, which is at rest. The collision is elastic, and following the impact, the two pucks move with equal speeds. Determine the velocity of each puck.

• **5.26** A 75-kg man stands on a rectangular raft of mass 45 kg. The man then walks from one end of the 4-m-long raft to the other. How far has the raft moved in the process?

• **5.27** A mass of 0.6 kg makes a perfectly elastic head-on collision with a mass of 0.4 kg that is initially at rest. Following the collision, the 0.6-kg mass moves at a speed of 2 m/s. Find the initial speed of the 0.6-kg mass and the final speed of the 0.4-kg mass.

• **5.28** A mass of 0.6 kg, moving at a speed of 10 m/s, collides head-on with a mass of 0.4 kg, which is at rest. What are the velocities of the two masses following the collision if the collision is (a) elastic? (b) inelastic?

• • **5.29** A Fourth of July rocket of mass M is fired vertically. When its upward velocity is 10 m/s, it explodes into two fragments of masses $M/3$ and $2M/3$. Immediately after the explosion, the velocity of the lighter fragment is 20 m/s, directed horizontally due east. (a) What is the velocity of the other fragment immediately after the explosion? (b) What is the change in the KE of the system as a result of the explosion? (c) Will the two fragments reach the ground at the same time? If not, which will hit the ground first?

• • **5.30** A 50-g block collides with a 20-g block on a friction-free air table. The initial velocity of the 20-g block is 5 m/s. Following the collision, the 50-g block is at rest. Find the initial velocity of the 50-g block for (a) a perfectly elastic collision; (b) a perfectly inelastic collision.

• • **5.31** Show that when a particle of mass M, moving with velocity $\mathbf{v}$, makes a head-on elastic collision with a particle of mass m that is initially at rest, the final velocity of the particle of mass m is nearly $2\mathbf{v}$ if $M \gg m$.

• • **5.32** Two masses of 1 kg and 2 kg, respectively, approach each other on a head-on collision course on a frictionless surface. The collision is elastic, and following the collision the 1-kg mass is at rest and the 2-kg mass travels with a speed of 5 m/s. What were the initial speeds of the two masses and what was the KE of the CM?

• • **5.33** A projectile is fired at a speed of 250 m/s and at an angle of 53° with the horizontal. At the top of the trajectory it explodes into two pieces of equal mass. One of these lands 10 s after the explosion at a point immediately below the point of the explosion. Where will the other piece land? (Neglect air friction.)

·6·

Circular Motion

If they be two, they are two so
As stiffe twin compasses are two,
Thy soule the fixt foot, makes no show
To move, but doth, if the other doe.

And though it in the center sit,
Yet when the other far doth rome,
It leans, and hearkens after it,
And growes erect, as it comes home.

Such wilt thou be to mee, who must
Like th'other foot, obliquely runne;
Thy firmness makes my circle just,
And makes me end, where I begunne.

A Valediction Forbidding Mourning
JOHN DONNE

6.1 Kinematics of Circular Motion

Suppose a small massive stone is constrained to a circular path by a massless (i.e., extremely light) rod pivoted about a fixed center at O (Figure 6.1). To specify the position of the stone at any instant, we could give its x and y coordinates. However, a much more natural and convenient description is in terms of the angle θ between the rod and some arbitrary direction, such as the line OA. If θ and the length of the rod r are given, we then know precisely where we must look for the stone.

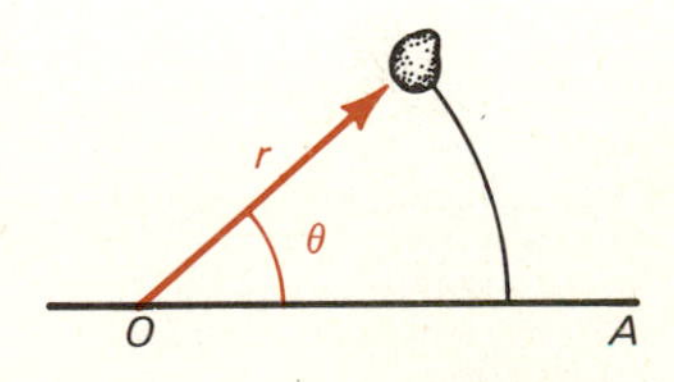

Figure 6.1 The location of the stone relative to the coordinate origin O is uniquely specified by the length r and the angle θ.

The reason the polar coordinates, r and θ, are preferred is that in circular motion only one of these changes with time, and it does so in a simple manner if this motion is uniform. By contrast, as the stone traverses its circular path, the x and y coordinates are both changing; they are given by

$$x = r\cos\theta$$
$$y = r\sin\theta \qquad \textbf{(6.1)}$$

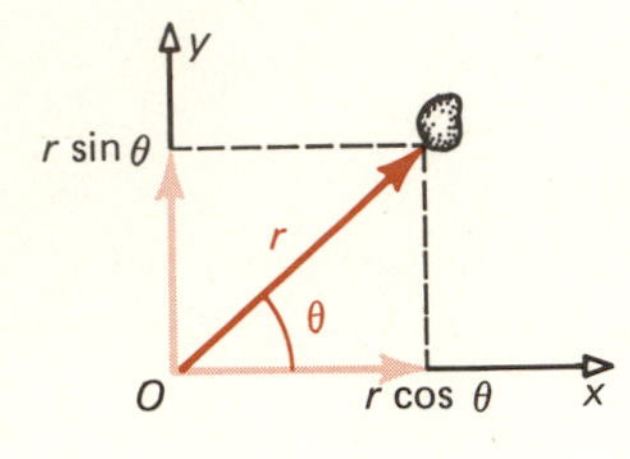

Figure 6.2 The Cartesian coordinates x and y can also be used to locate the stone. As the stone rotates about O, its x and y coordinates change continuously as indicated.

The radius of the circular path is measured in the usual unit of length, the meter. Two measures are common for specifying *angle*. One, probably the more familiar, is the *degree, defined as* $\frac{1}{360}$ *of a complete circle, or full revolution*. That is, as the stone completes one full circular transit, the angle θ changes by 360°.

The other measure of angle is the *radian*.

DEFINITION: One radian is the angle subtended by the arc whose length equals the radius of the circle.

That is, the angle θ, measured in radians, is given by the ratio of arc length to radius

$$\theta = \frac{\text{arc length}}{\text{radius}} = \frac{s}{r} \qquad \textbf{(6.2)}$$

Since the arc length of a full circle of radius r is $2\pi r$, the conversion between radians and degrees is given by the condition

$$2\pi \text{ radians} = 360°$$

or

$$1 \text{ radian} = \frac{360}{2\pi} = 57.3°$$

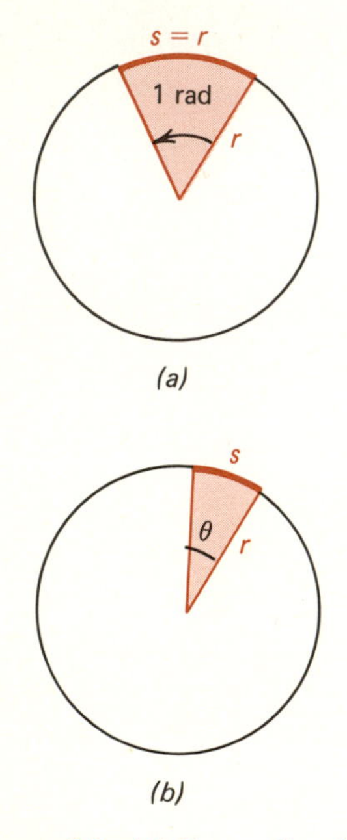

Figure 6.3 (a) One radian is the angle subtended by the arc s whose length equals the radius of the circle. (b) The angle θ, in radians, equals the ratio s/r.

The following table may be helpful for keeping the radian measure in perspective.

$$360° = 2\pi \text{ radians} \qquad 60° = \frac{\pi}{3} \text{ radians}$$

$$180° = \pi \text{ radians} \qquad 45° = \frac{\pi}{4} \text{ radians}$$

$$90° = \frac{\pi}{2} \text{ radians} \qquad 30° = \frac{\pi}{6} \text{ radians}$$

We shall express all angles in radians throughout our discussion of circular motion. The reasons for and advantages of using the radian measure will become apparent shortly.

Since the angle θ is defined by the ratio of two lengths, arc length and radius, *angle is a dimensionless variable*.

As with linear motion, we can define an *angular velocity*, designated by ω (lowercase Greek letter *omega*). The average angular velocity is given by

$$\langle\omega\rangle = \frac{\theta_f - \theta_0}{t_f - t_0} = \frac{\Delta\theta}{\Delta t} \qquad \textbf{(6.3)}$$

and

$$\theta(t) = \theta_0 + \langle\omega\rangle t \qquad \textbf{(6.4)}$$

where, as in Chapter 2, the subscripts "f" and "0" denote final and initial values of the variables and Δ the interval between them.

The instantaneous angular velocity is obtained by the same limiting process that was used to arrive at the instantaneous velocity:

$$\omega = \lim_{\Delta t \to 0} \frac{\Delta \theta}{\Delta t} \tag{6.5}$$

The unit of angular velocity is the radian per second. Its dimension is $[T]^{-1}$.

The angular velocity is generally not constant, although there are many important situations when it is, for example, a satellite in a circular orbit about the earth. On the other hand, the angular velocity of a car's wheels as the car accelerates, or of a circular saw blade as the motor is turned on or off, are not constant. The *angular acceleration* expresses the change of angular velocity with time. The average angular acceleration $\langle \alpha \rangle$ is defined by

$$\langle \alpha \rangle = \frac{\omega_f - \omega_0}{t_f - t_0} = \frac{\Delta \omega}{\Delta t} \tag{6.6}$$

and the instantaneous angular acceleration by

$$\alpha = \lim_{\Delta t \to 0} \frac{\Delta \omega}{\Delta t} \tag{6.7}$$

The unit of angular acceleration is the radian per second squared. Its dimension is $[T]^{-2}$.

If α is constant, we obtain from these definitions

$$\omega(t) = \omega_0 + \alpha t \tag{6.8}$$

and

$$\langle \omega \rangle = \frac{\omega(t) + \omega_0}{2} \tag{6.9}$$

Comparison of the above definitions and the relations among the angular variables θ, ω, and α with the variables for linear motion, s, v, and a, shows a one-to-one correspondence. The algebraic substitutions that led to the kinematic equations for uniformly accelerated translation must therefore also hold for uniformly accelerated angular motion, provided we replace the linear variables with their angular analogs. Thus, for circular motion, we arrive at the results given in Equations (6.10), which are shown alongside their translational counterparts.

$$\theta(t) = \theta_0 + \langle \omega \rangle t; \quad \Delta\theta = \langle \omega \rangle t \qquad x(t) = x_0 + \langle v \rangle t; \; s = \langle v \rangle t \tag{6.10a}$$

$$\theta(t) = \theta_0 + \omega_0 t + \tfrac{1}{2}\alpha t^2 \qquad x(t) = x_0 + v_0 t + \tfrac{1}{2}at^2 \tag{6.10b}$$

$$\langle \omega \rangle = \tfrac{1}{2}[\omega(t) + \omega_0] \qquad \langle v \rangle = \tfrac{1}{2}[v(t) + v_0] \tag{6.10c}$$

$$\omega(t) = \omega_0 + \alpha t \qquad v(t) = v_0 + at \tag{6.10d}$$

$$\begin{aligned} \omega^2(t) &= \omega_0^2 + 2\alpha[\theta(t) - \theta_0] \qquad & v^2(t) &= v_0^2 + 2a[x(t) - x_0] \\ 2\alpha\,\Delta\theta &= \omega^2(t) - \omega_0^2 \qquad & 2as &= v^2(t) - v_0^2 \end{aligned} \tag{6.10e}$$

Like their translational counterparts, the angular kinematic variables are really vector quantities. The directions of $\boldsymbol{\theta}$, $\boldsymbol{\omega}$, and $\boldsymbol{\alpha}$ are defined by the following convention.

The direction of $\boldsymbol{\theta}$ is the direction along which a right-handed screw would advance if turned through the angle $\boldsymbol{\theta}$.

The direction of $\boldsymbol{\omega}$ is the direction of $\Delta\boldsymbol{\theta}$. That is, $\boldsymbol{\omega}$ points along $\boldsymbol{\theta}$ if the angle is increasing as defined above.

The direction of $\boldsymbol{\alpha}$ is the direction of $\Delta\boldsymbol{\omega}$.

Note that just as a change in direction of velocity, without a change in speed, required an acceleration, so a change in the direction of rotation, without a change in the rotational speed, also calls for a finite angular acceleration.

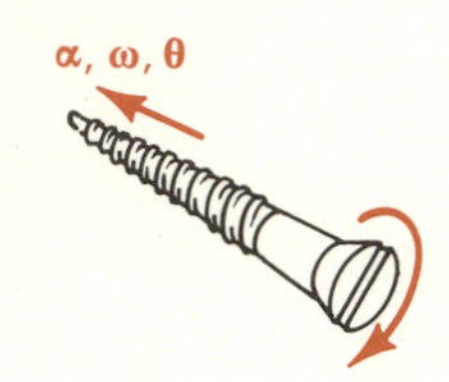

Figure 6.4 *The direction of* $\boldsymbol{\theta}$, $\boldsymbol{\omega}$, *and* $\boldsymbol{\alpha}$ *are the directions in which a right-handed screw advances if turned clockwise for positive,* $\boldsymbol{\theta}$, $\boldsymbol{\omega}$, and $\boldsymbol{\alpha}$.

6.2 Connection between Rotational and Translational Motion

In many practical, commonplace situations, circular motion is somehow related to linear motion. For example, winding the reel of a fishing rod causes the hook (with or without fish) to move in a straight line toward the rod; the rotation of a bicycle's wheels results in its translation. In each instance, the faster the rotation, the faster the translation. Obviously there is some relation between these two motions, but what is this relation?

Figure 6.5 shows a drum with one point on its surface marked. As the drum turns through one complete revolution, the point P traverses a circular path of radius r and moves a distance $2\pi r$. The path length traversed by a point on the drum is just the arc length s, and from the definition of the radian, this length is related to the angular displacement θ by

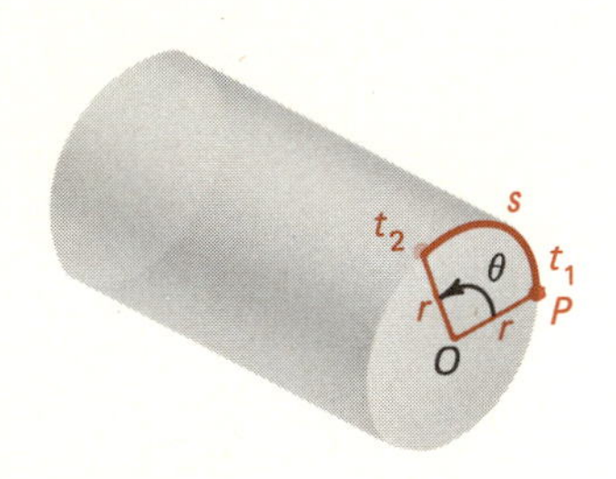

Figure 6.5 *During the time* $t_2 - t_1$ *the point* P *travels the distance* $s = r\theta$ *as the drum turns through the angle* θ.

$$s = r\theta \qquad \textbf{(6.11)}$$

This simple relation between arc length and angle holds only if the angle is measured in radians. Equation (6.11) is the reason that the radian measure is so convenient here.

What about the linear distance covered by the center of a wheel that rolls, without slipping, along the ground? That distance is also given by (6.11). To see this, imagine a thread wound about the circumference of the wheel and fixed to the ground at the initial point of contact between wheel and ground. As the wheel rolls forward, the length of thread that stretches along the ground is the arc length $s = r\theta$, so that after one full turn, the wheel has moved laterally a distance of $2\pi r$ meters.

● ***Example 6.1*** The great pyramids of Egypt have fascinated archaeologists for centuries. One of the curious aspects of these structures is their uniformity of design. Although they differ vastly in size, all pyramids except two are of identical shape, so that the ratio of the circumference of the square base to the height is exactly 2π. The angle of elevation of each side is 52° (51° 52′, to be precise). That's an odd angle; one might more readily understand choosing angles of 45° or 53° (the Pythagorean 3-4-5 right triangle, known to the Egyptians), but why 52°? Had the ancient architects calculated π so precisely that early in history and decided to impress future generations with their mastery of mathematics by erecting great monuments to celebrate this achievement?

A more likely explanation was suggested by T. E. Connely. Suppose that the height of the pyramid was set at $4nL$, where n is an integer and L is a standard length. If the Egyptian builders measured the distance from the center of the planned structure to its side by rolling a drum of diameter L exactly n complete revolutions along the ground, the length of each side would come to $2 \times (n\pi L)$. Consequently, the ratio of circumference to

(a)

(b)

Figure 6.6 The pyramids at Giza seen from the air and from the ground. These pyramids all have identical shapes but greatly different sizes.

height would be exactly $4 \times 2n\pi L/4nL = 2\pi$, and the elevation angle inevitably $\tan^{-1}(4/\pi) = 51° 52'$! This explanation derives further support from the fact that the one pyramid, the so-called Red Pyramid, which deviates from this pattern, has an elevation angle of 43.5°, and not 45°; this angle is precisely $\tan^{-1}(3/\pi)$ and would be obtained if the height were chosen to be $3nL$ instead of $4nL$.

We have seen that the *tangential distance* along a circle of radius r that corresponds to a rotation about an angle of θ radians is given by $s = r\theta$.

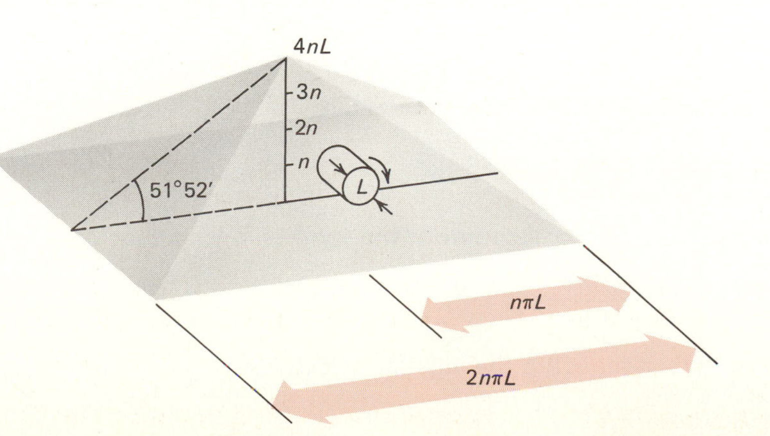

Figure 6.7 If the lateral dimension of the pyramid is fixed by rolling a drum of standard diameter an integral number of turns along the ground and the height equals an integral multiple of the drum diameter, then the ratio of base circumference to height will always equal π times an integer.

Similar simple relations connect the *tangential velocity* of a point on a wheel's perimeter with its angular velocity ω. From the definitions of v and ω, we have

$$v = \lim_{\Delta t \to 0} \frac{\Delta s}{\Delta t} = \lim_{\Delta t \to 0} \frac{\Delta(r\theta)}{\Delta t} = r \lim_{\Delta t \to 0} \frac{\Delta \theta}{\Delta t} = r\omega \tag{6.12}$$

where the third step is justified since the radius of the wheel does not change as the wheel rotates.

Similarly, the *tangential acceleration* a_t is given by

$$a_t = \lim_{\Delta t \to 0} \frac{\Delta v}{\Delta t} = \lim_{\Delta t \to 0} \frac{\Delta(r\omega)}{\Delta t} = r \lim_{\Delta t \to 0} \frac{\Delta \omega}{\Delta t} = r\alpha \tag{6.13}$$

Thus, angular displacement, angular velocity, and angular acceleration of an object showing circular motion are simply related to the corresponding tangential displacement, velocity, and acceleration. Again, these simple relations are valid only if the angular variables are given in radian units.

● ***Example 6.2*** A drum of radius 0.4 m starts from rest at the top of an inclined plane and rolls down without slipping (Figure 6.8). The time between its release and arrival at point B, 8 m further down the plane, is 10 s. Find the angular acceleration, the angular velocity at B, and the number of revolutions the drum has made in traveling from A to B, assuming that the drum proceeds at constant acceleration down the plane.

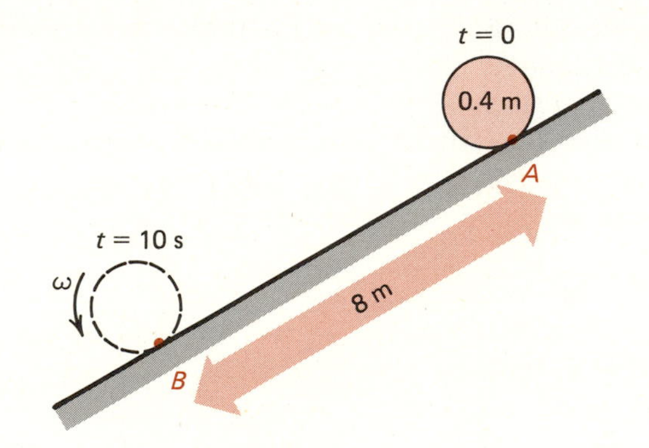

Figure 6.8 *The drum starts from rest at point A; ten seconds later it passes point B.*

Since the elapsed time is 10 s, the average translational velocity is $\langle v_t \rangle$ = (8 m)/(10 s) = 0.8 m/s. Since

$$\langle v_t \rangle = \frac{v_t(10) + v_t(0)}{2} = \frac{v_t(10)}{2}$$

$$v_t(10) = 2\langle v_t \rangle = 2(0.8 \text{ m/s}) = 1.6 \text{ m/s}$$

The tangential acceleration is therefore

$$a_t = \frac{\Delta v_t}{\Delta t} = \frac{1.6 \text{ m/s}}{10 \text{ s}} = 0.16 \text{ m/s}^2$$

and the angular acceleration, given by $\alpha = a_t/r$, is then

$$\alpha = \frac{0.16 \text{ m/s}^2}{0.4 \text{ m}} = 0.4 \text{ rad/s}^2$$

From Equation (6.12), the angular velocity at point B is

$$\omega(10) = \frac{v_t(10)}{r} = \frac{1.6 \text{ m/s}}{0.4 \text{ m}} = 4 \text{ rad/s}$$

Lastly,

$$\theta = s/r = \frac{8 \text{ m}}{0.4 \text{ m}} = 20 \text{ rad}$$
$$= \frac{20 \text{ rad}}{2\pi \text{ rad/rev}}$$
$$= 3.18 \text{ rev}$$

Note that we could have obtained $\theta(10)$ and $\omega(10)$ from Equations (6.10b) and (6.10d). Setting $\omega_0 = \theta_0 = 0$, we have

$$\theta(10) = \tfrac{1}{2}\alpha t^2 = \tfrac{1}{2}(0.4 \text{ rad/s}^2)(10\text{s})^2 = 20 \text{ rad}$$

$$\omega(10) = \alpha t = (0.4 \text{ rad/s}^2)(10\text{s}) = 4 \text{ rad/s}$$

6.3 Dynamics of Uniform Circular Motion

Now that we have the appropriate kinematic equations, we turn our attention to the dynamics of uniform circular motion. That is, we want to find the answer to the question: What force is needed to maintain a body moving about a fixed center at constant speed? This is the central problem that occupied astronomers since ancient times, and its solution was one of the monumental achievements of Isaac Newton. To Aristotle and most of his successors this question posed no difficulty; they disposed of it simply by asserting that since the circle is the most perfect geometrical figure, circular paths are "natural" for celestial bodies, requiring no force whatever.

But some external force is required. We know this, because the velocity of an object moving around a circular path is continually changing though its speed may be fixed (Figure 6.9(*a*)). A change in velocity implies an acceleration, and to accelerate a body must experience a net force.

To determine that force, we must first know the acceleration. Figure 6.9(*b*) shows the difference between the two velocity vectors, $\mathbf{v}_f$ and $\mathbf{v}_0$, of Figure 6.9(*a*), determined by the geometrical construction described in Chapter 1. The average acceleration in the time interval Δt during which this change in velocity occurred is then $\langle \mathbf{a} \rangle = \Delta \mathbf{v}/\Delta t$ and must be in the direction of $\Delta \mathbf{v}$, that is, along the base of the isosceles triangle whose

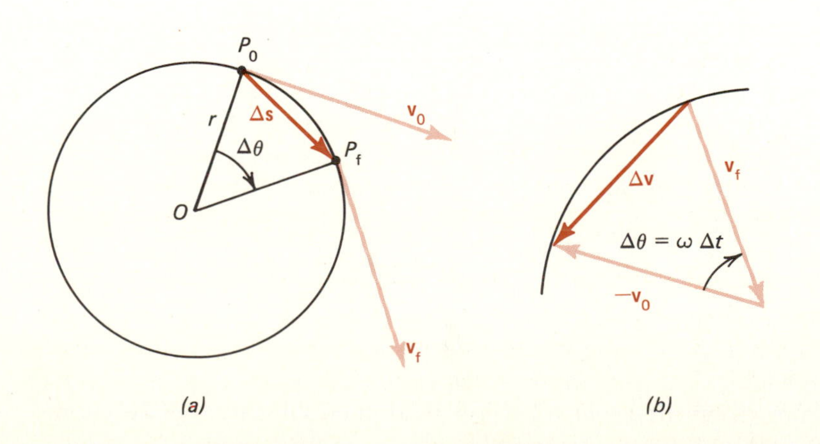

Figure 6.9 *(a) In the time interval Δt, the distance traversed is $r\Delta\theta = r\omega\Delta t$. The velocity vector changes direction, from $\mathbf{v}_0$ to $\mathbf{v}_f$. (b) The triangle formed by the velocity vectors $\mathbf{v}_f$, $-\mathbf{v}_0$, and $\Delta\mathbf{v}$ is similar to the triangle OP_0P_f of (a). In the limit $\Delta t \to 0$ the arc length $v\,\Delta\theta \to \Delta v$. The centripetal acceleration is*

$$a_c = \lim_{\Delta t \to 0}\left(\frac{\Delta v}{\Delta t}\right)$$
$$= v \lim_{\Delta t \to 0}\left(\frac{\Delta \theta}{\Delta t}\right) = v\omega$$

sides are $\mathbf{v}_f$ and $-\mathbf{v}_0$. It is apparent from this construction that $\Delta\mathbf{v}$ is perpendicular to $\Delta\mathbf{s}$. Hence, the average acceleration between P_0 and P_f is directed toward the center of the circular path.

The force that results in this motion must then also point toward the center. That this is the proper direction for the force should be clear if we just imagine whirling an object attached to a string of fixed length. The string is under constant tension, and this tension is what "forces" the object into the circular motion. We know from everyday experience that the moving object pulls outward on the hand holding the string. From Newton's third law, it follows that the force the hand exerts on the object via the string must be an equal inward pull. We call this inwardly directed force that acts on the revolving object the *centripetal force,* and the inwardly directed acceleration of the object its *centripetal acceleration* $\mathbf{a}_c$.

So far so good. But how large is the centripetal force and how does it depend on the angular velocity and the radius of the circle, if at all? Everyday experience gives some qualitative hints. If you take a heavy bolt and attach it to a piece of twine and whirl the bolt about your head, you will find that the greater the angular velocity, the greater the pull on your hand. Also, if the length of twine is increased and you maintain the same angular velocity, the pull on your hand increases. Thus the centripetal force increases with increasing radius and angular velocity.

To derive an expression for the centripetal force $\mathbf{F}_c$, we return to the centripetal acceleration, $\mathbf{a}_c$. In the time Δt, the object travels a distance equal to the arc length $s = r\,\Delta\theta = r\omega\,\Delta t$ (see Figure 6.9(a)). In this same time, the velocity vector has changed direction through the same angle $\Delta\theta$; that is, the angle between $\mathbf{v}_f$ and $\mathbf{v}_0$ in Figure 6.9(a) is $\Delta\theta$. Now as we allow Δt to become ever smaller to obtain the instantaneous acceleration, the chord length, Δv of Figure 6.9(b) approaches ever more closely the arc length $v\,\Delta\theta$. Thus in the limit we have

$$a_c = \lim_{\Delta t \to 0} \frac{\Delta v}{\Delta t} = v \lim_{\Delta t \to 0} \frac{\Delta\theta}{\Delta t} = v\omega \qquad \textbf{(6.14)}$$

We can restate this result in two convenient forms, using the relation $v = r\omega$, Equation (6.12). We obtain

$$a_c = \frac{v^2}{r} \qquad \textbf{(6.15)}$$

$$a_c = r\omega^2 \qquad \textbf{(6.16)}$$

The vector $\mathbf{a}_c$ is directed along the radius toward the center of the circle.

Note that in deriving Equations (6.14)–(6.16), we used (6.11) and (6.12), *which are valid only if angles are expressed in radians.* Consequently, to apply Equations (6.14)–(6.16) *one must express the angular velocity in radians per second,* not degrees or revolutions per second.

The centripetal force is now given from Newton's second law by

$$F_c = ma_c = \frac{mv^2}{r} = mr\omega^2 \qquad \textbf{(6.17)}$$

● ***Example 6.3*** Two masses of 1 kg and 0.5 kg are attached to each other by a massless string that passes through a hole of a horizontal friction-free table. The 1-kg mass is suspended below the table and is in equilibrium when the other mass moves in a circular path of 20-cm radius on the table. What is the angular velocity of the 0.5-kg mass about the hole in the table?

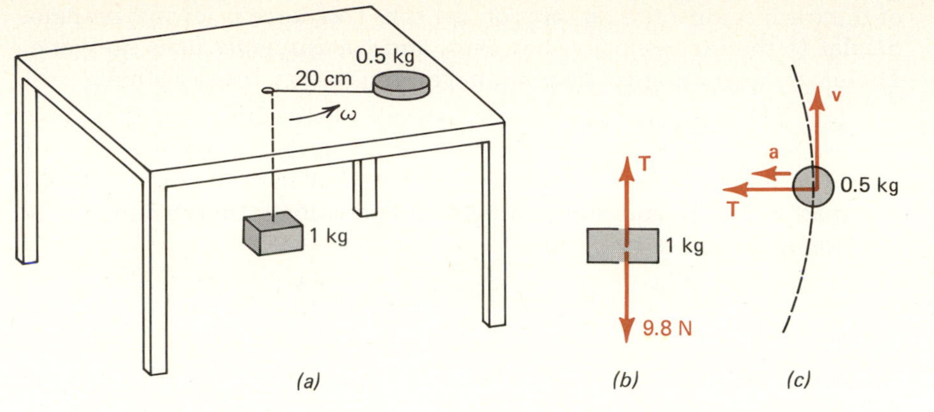

Figure 6.10 (a) The 0.5-kg mass rotates about the hole in the center of the friction-free table at constant angular velocity. The requisite centripetal force is supplied by the weight of the 1-kg mass supported by the string that connects the two masses. (b) The free-body diagram for the 1-kg mass. (c) The free-body diagram for the 0.5-kg mass.

Since the lower mass is in equilibrium, its weight, (1 kg)(9.8 m/s²) = 9.8 N, must equal the tension in the string. This is, then, also the centripetal force the string exerts on the rotating mass of 0.5 kg. Therefore, from Equation (6.17),

$$9.8 \text{ N} = (0.5 \text{ kg})(0.2 \text{ m})\omega^2$$

$$\omega^2 = 98 \text{ rad}^2/\text{s}^2$$

Thus, $\omega = 9.9$ rad/s = 1.58 rev/s. ●

● ***Example 6.4*** A toy car is on a track that forms a circular loop in the vertical plane; the diameter of the loop is 50 cm. If the car is to maintain contact with the track at the top of the loop, what must be its minimum speed at the bottom of the loop? Neglect friction losses.

For the car to remain on the track at the top of the loop, the centripetal acceleration must be at least as great as the acceleration of gravity. Or put another way, if the force of gravity acting on the toy car is greater than the centripetal force needed to keep it in its circular path, the car will fall below the track. Hence the critical condition at the top of the loop is

$$a_c = g$$

From (6.15), we have

$$v_{\text{top}}^2 = ra_c = (0.25 \text{ m})(9.8 \text{ m/s}^2) = 2.45 \text{ m}^2/\text{s}^2$$

At the bottom of the loop, the speed of the car will be greater than at the top because the car, in climbing the 0.5 m, gains potential energy and loses kinetic energy. Throughout its motion a net force, the reaction force

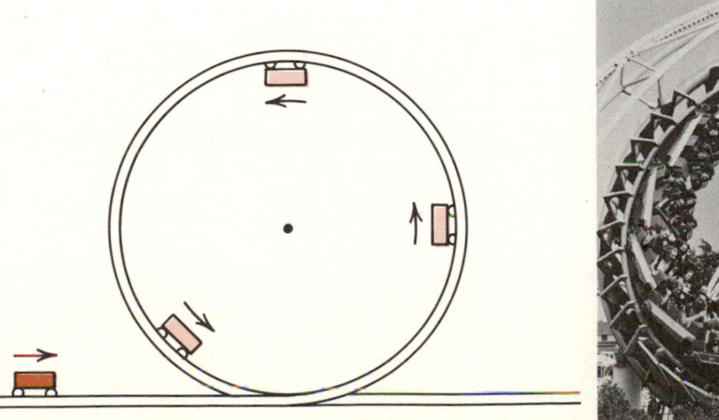

Figure 6.11 (a) A toy car on a loop-the-loop track. (b) The "Corkscrew" roller coaster at Cedar Point Amusement Park near Toledo, Ohio.

of the track against the car, acts on the car. This force is always perpendicular to the car's velocity, however, and consequently does no work. Therefore, neglecting friction, we have from energy conservation

$$\mathrm{KE}_{\mathrm{bottom}} = \mathrm{KE}_{\mathrm{top}} + mgh$$
$$\tfrac{1}{2}mv^2_{\mathrm{bottom}} = \tfrac{1}{2}mv^2_{\mathrm{top}} + mgh$$

The masses cancel, and multiplying by 2 and substituting the appropriate numerical values we have

$$v^2_{\mathrm{bot}} = (2.45\ \mathrm{m^2/s^2}) + 2(9.8\ \mathrm{m/s^2})(0.5\ \mathrm{m}) = 12.25\ \mathrm{m^2/s^2}$$
$$v_{\mathrm{bot}} = 3.5\ \mathrm{m/s}$$

To maintain contact with the track at the top, the car must have a speed of at least 3.5 m/s at the bottom of the loop. ●

*6.4 Banking of Curves

When a car negotiates a curve, a centripetal force must act on the vehicle to force it into that circular course. On a flat pavement, this force is provided by static friction between the tires and pavement. Under normal conditions, the portion of the tire that contacts the pavement is momentarily at rest. Hence, the critical friction coefficient is μ_s, not μ_k. Only during an undesirable skid are the surfaces of contact between tire and pavement in relative motion.

If μ_s is small or the speed of the car excessive, the friction force may be inadequate to pull the car around the curve. The tires will then slip, and the driver may lose control of the car. Skidding on snow and ice is a dangerous and all too common event.

So that cars can go around a curve at relatively high speed, the pavement is banked; that is, the surface of the pavement is tilted so that the outer part is higher than the inner. The normal reaction force between car and banked pavement then has a horizontal as well as a vertical component. If the horizontal component equals the required centripetal force, no friction force need come into play.

Suppose a mass m slides at constant speed at a fixed height h on the inside of a frictionless spherical surface, as shown in Figure 6.12. The only forces acting on the mass are its weight $m\mathbf{g}$, and the normal reaction

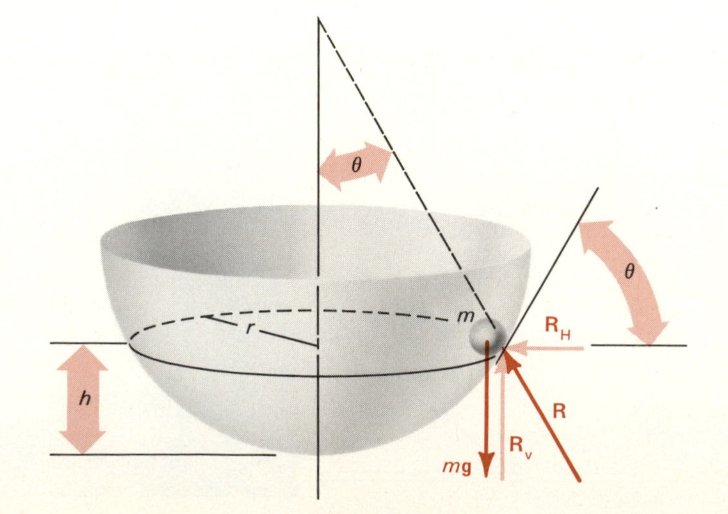

Figure 6.12 *A mass m slides inside a smooth spherical surface so that its trajectory is a circle of radius r in the horizontal plane. The banking angle is the same as the angle between the vertical and the normal to the supporting surface. That angle is given by $\theta = \tan^{-1}(v^2/rg)$, where v is the speed of the mass in its circular path.*

(a)

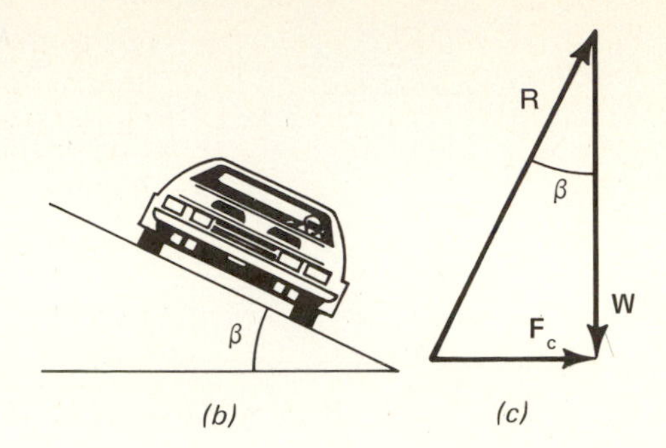

Figure 6.13 (a), (b) Banking of curves. (c) As indicated, the horizontal component of the reaction force **R** equals the centripetal force.

force of the surface. If, as assumed, the mass travels in the horizontal plane, the net vertical force must be zero. The vertical component of **R** must equal the weight mg. For the mass to maintain contact with the supporting surface, the horizontal component of **R** must equal the centripetal force mv^2/r, where r is the radius of the circular path in the horizontal plane. Thus, we have

$$R_V = R \cos \theta = mg; \quad R_H = R \sin \theta = \frac{mv^2}{r}$$

Dividing the second equation by the first yields

$$\tan \theta = \frac{v^2}{rg}$$

$$\theta = \tan^{-1} \frac{v^2}{rg} \qquad \textbf{(6.18)}$$

which gives the correct banking angle for a curve of radius r and vehicle speed v. When the pavement is banked in that manner, the normal force on the tires keeps the car on its curved course, and no force tangent to the pavement is needed.

Note that Equation (6.18) does not contain the mass of the car! The proper banking angle is the same for all vehicles traveling at the same speed.

● ***Example 6.5*** What is the minimum coefficient of static friction between tires and horizontal pavement to permit a car to negotiate a curve of 30-m radius (about 100 ft) at 50 km/h (about 30 mph)? Compare the result with the coefficients of friction between rubber and dry and wet pavement. Determine the correct banking angle so that the normal reaction force of the pavement, not friction, supplies the requisite centripetal force.

To begin, we first express all variables in consistent units; 50 km/h = (50 km/h)(1000 m/km)/(3600 s/h) = 13.9 m/s. If the car has a mass M, the necessary centripetal force is

$$F_c = \frac{Mv^2}{r} = \frac{M(13.9 \text{ m/s})^2}{30 \text{ m}} = 6.44\, M \text{ N}$$

Since this force is due to friction between tires and pavement, we must have*

$$F_c = f_s \leq \mu_s R = \mu_s Mg = 9.8\, \mu_s M \text{ N}$$

Consequently, the coefficient of static friction must be at least

$$\mu_s = \frac{6.44}{9.8} = 0.66$$

The coefficient of friction between rubber and dry cement is in the range 0.7–0.9, and the car can therefore negotiate the curve at 50 km/h without skidding, though perhaps just barely. However, if the pavement is wet, the coefficient of friction drops to a value between 0.4 and 0.8, depending on the type of pavement and the condition of the tires, and chances are the car will go into a skid; it surely will skid if there is snow or ice on the road, when μ_s may drop below 0.1.

The proper banking angle is given by Equation (6.18). The argument of the inverse tangent is just the ratio of the centripetal acceleration to the acceleration of gravity, and for this case is 6.44/9.8 = 0.66. Hence the correct banking angle is

$$\beta = \tan^{-1} 0.66 = 33.5°$$

In practice, highway curves are never banked so steeply as prescribed by Equation (6.18) for normal highway speed of 90 km/h. If this were done, a driver proceeding at that speed would be almost unaware of going into the curve, and such steep banking would only encourage speeding instead of safe driving habits. Moreover, a car moving around the curve under hazardous road conditions such as ice and snow would tend to slip to the inside of the curve if its speed were significantly below 90 km/h. Steep banking is, however, used on Olympic bobsled runs and indoor bicycle tracks; and airplanes must bank correctly to perform a normal turn.

Figure 6.14 Bobsled negotiating a curve on a bobsled run.

* Recall that $\mu_s R$ is the *maximum* value of the force of static friction.

Figure 6.15 *Airplane banking in a turn.*

Summary

The location of an object traveling in a circular path is most conveniently specified by the circular coordinates, r and θ. The angular coordinate may be given in units of degrees or radians. The angle θ in radians is given by

$$\theta = \frac{s}{r}$$

where s is the arc length subtended by θ on a circle of radius r.

$$1 \text{ radian} = 57.3^\circ$$

$$2\pi \text{ radians} = 360^\circ$$

The average and instantaneous angular velocities are defined by

$$\langle \omega \rangle = \frac{\theta_f - \theta_0}{t_f - t_0} = \frac{\Delta\theta}{\Delta t}$$

$$\omega = \lim_{\Delta t \to 0} \frac{\Delta\theta}{\Delta t}$$

The average and instantaneous angular accelerations are defined by

$$\langle \alpha \rangle = \frac{\omega_f - \omega_0}{t_f - t_0} = \frac{\Delta\omega}{\Delta t}$$

$$\alpha = \lim_{\Delta t \to 0} \frac{\Delta\omega}{\Delta t}$$

The kinematic equations relating θ, ω, and α are analogs of the corresponding equations for x, v, and a (Chapter 2) and are summarized in Equation (6.10).

Tangential displacement, velocity, and acceleration of a wheel are related to its angular displacement, angular velocity, and angular acceleration by

$$s = r\theta$$

$$v = r\omega$$

$$a = r\alpha$$

where r is the radius of the wheel and the angular variables are expressed in radian measure.

A body undergoing uniform circular motion about some central pivot experiences a centripetal (centrally directed) acceleration

$$a_c = \frac{v^2}{r} = r\omega^2$$

where v is its translational speed. Consequently, a centripetal force

$$F_c = ma_c = \frac{mv^2}{r} = mr\omega^2$$

must act on a body of mass m to maintain uniform circular motion.

On a horizontal pavement, the centripetal force needed to bring a car around a curve is provided by friction between tires and pavement. The horizontal component of the normal reaction force provides this force if the curve is properly banked. The ideal banking angle is given by

$$\theta = \tan^{-1}\frac{v^2}{rg}$$

where r is the radius of the curve.

Multiple Choice Questions

6.1 A small mass is placed on a turntable that is rotating at 45 rpm. The acceleration of the mass is

(a) greater the closer the mass is to the center of the table.

(b) greater the farther the mass is from the center of the table.

(c) independent of the location of the mass.

(d) zero.

6.2 A wheel has an angular velocity of 2 rad/s. At the end of 5 s it will have made

(a) 10π

(b) $10/\pi$

(c) $5/\pi$

(d) 20π

revolutions.

6.3 A wheel is subjected to uniform angular acceleration about its axis. Initially its angular velocity is zero. In the first 2 s, it rotates through an angle θ_1; in the next 2 s, it rotates through an additional angle θ_2. The ratio θ_2/θ_1 is

(a) 1

(b) 2

(c) 3

(d) 5

6.4 A mass m on a string is released from rest at point A (Figure 6.16). As it passes the lowest point B, the tension in the string is

(a) mg

(b) $2mg$

(c) $3mg$

(d) Can't tell; the answer depends on the length R.

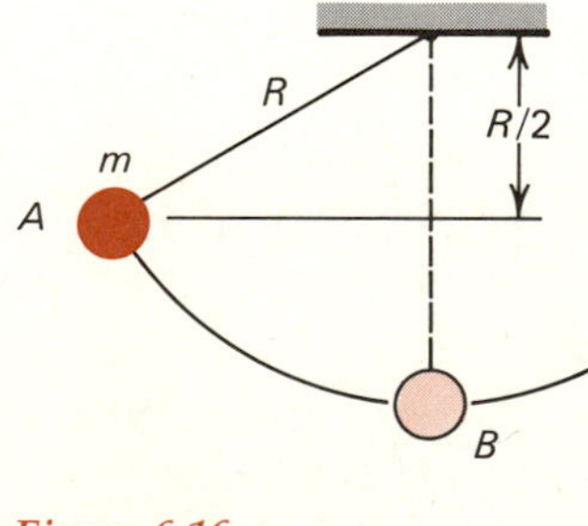

Figure 6.16

6.5 The angular velocity of the earth's rotation about its axis is

(a) $12/\pi$ rad/h

(b) $\pi/12$ rad/h

(c) 48π rad/h

(d) 0.5 deg/min

6.6 A mass is traveling in a circular path at constant tangential speed. It follows that

(a) the acceleration of the mass is zero.

(b) the acceleration is finite and directed to the center of the circular path.

(c) the acceleration is finite and directed outward from the center of the circular path.

(d) the velocity of the mass is constant.

6.7 Figure 6.17 shows a pendulum of length L suspended from the top of a flat beam of height $L/2$. The bob is pulled away from the beam so it makes an angle θ with the vertical. It is then released from rest. If ϕ is the maximum angular deflection to the right, then

(a) $\phi = \theta$

(b) $\phi < \theta$

(c) $\theta < \phi < 2\theta$

(d) $\phi \geqq 2\theta$

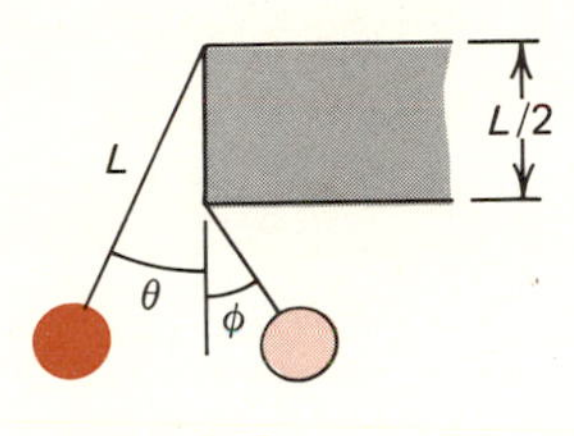

Figure 6.17

6.8 A flywheel that is rotating slows down at a constant rate due to friction in its bearings. After one minute its angular velocity has diminished to 0.70 of its initial value ω_0. At the end of the second minute, the angular velocity is

(a) $0.49\omega_0$

(b) $0.40\omega_0$

(c) $0.35\omega_0$

(d) $0.10\omega_0$

6.9 A mass is supported on a frictionless horizontal surface. It is attached to a string and rotates about a fixed center at an angular velocity ω_0. If the length of the string and the angular velocity are doubled, the tension in the string, which was initially T_0, is now

(a) again T_0

(b) $T_0/2$

(c) $4T_0$

(d) $8T_0$

6.10 A driver finds that he can just negotiate an unbanked curve at 40 km/h without skidding. If he now increases the mass of the car by loading sandbags in the trunk and body of the car,

(a) he will be able to travel safely around the curve at a speed in excess of 40 km/h.

(b) he will skid if he now tries to negotiate the curve at 40 km/h.

(c) he will find that having placed the sandbags in the car makes no perceptible change on the speed with which he can round the curve.

(d) (a), (b), or (c) could be correct depending on the radius of the curve.

Problems

(Sections 6.1–6.2)

6.1 A cyclist starts from rest and pedals so that the wheels have an angular acceleration of 3 rad/s^2. Find the number of revolutions that the wheels will have made at the end of 10 s.

6.2 A helicopter blade rotates at 90 rpm. What is the value of ω in radians per second? If the diameter of the blade is 5 m, what is the tangential speed at the tip of the blade?

6.3 Find the angular velocity in radians per second of a 45-rpm record.

6.4 A router operates at 3600 rpm. Through what angle does the shaft turn in 1 ms?

6.5 The tangential velocity at the rim of a wheel is 8 m/s when it turns at 3 rev/s. Find the diameter of the wheel.

6.6 Change the following to radians: 30°, 45°, 153°, 1.7 revolutions, and 2.0 revolutions.

6.7 A bicycle wheel has a radius of 36 cm. What is its angular velocity if the tangential velocity of a point on its rim is 10 m/s?

6.8 The tires of a car have an outer radius of 28 cm. What is the angular velocity of each wheel about its axis when the car is traveling at a constant speed of 60 km/h?

6.9 What is the tangential velocity of an LP phonograph record at its perimeter? The diameter of the record is 12 in and the angular velocity is 33.3 rpm.

6.10 The turntable of a record player reaches its rated speed of 33.3 rpm in 2.2 s, starting from rest. What is the average angular acceleration during this time, expressed in radian measure?

6.11 A bicycle has 68-cm-diameter wheels. If the bicycle is traveling at a speed of 20 km/h, what is the angular velocity of the wheels about their axes?

• **6.12** The bicycle of Problem 6.11 is approaching an intersection. If the rider applies brakes 40 m from the intersection and decelerates uniformly, reaching 3 km/h at the intersection, what is the value of α? If the bicycle was traveling to the north, what are the directions of $\boldsymbol{\omega}$ and of $\boldsymbol{\alpha}$?

• **6.13** A drum of 1.5-m diameter that is turning at 25 rpm is decelerating at constant rate to 10 rpm. If during this time, rope is winding up on the drum, and the drum takes up 120 m of rope, what was the value of α?

• **6.14** An electric grinder uses a grinding wheel of 8-cm radius. It takes the electric motor 2.5 s to reach its rated speed of 1200 rpm starting from rest. Assuming that the angular acceleration is constant during that 2.5 s, find the angular acceleration, the total angular displacement, and the final angular velocity in radian measures. What is the tangential velocity at the rim of the grinding wheel at rated speed?

(Section 6.3)

6.15 A 0.2-kg mass is attached to a string of length 20 cm whose other end is fixed at the center of a friction-free horizontal table. If the tension in the string is 3 N, what is the angular velocity of the mass and string?

6.16 A particle 15 cm from the center of a centrifuge experiences an acceleration of 50,000g. What is the angular velocity of the centrifuge in revolutions per minute?

6.17 A plane diving at 1400 km/h pulls out of the dive by following a circular trajectory in the vertical plane. What should the radius of the trajectory be so that the acceleration does not exceed 4g?

• **6.18** One simple demonstration of the effect of centripetal acceleration involves swinging a pail of water in a vertical plane. Suppose that the length of your arm is 85 cm, and the level of the water is 25 cm below the handle of the pail. How rapidly must you swing the pail so as to avoid getting drenched during this demonstration? Give your answer in radians per second, and assume that the pail is swung at constant angular speed.

• **6.19** A small coin is placed on a record that is rotating at 33.3 rpm. If the coefficient of static friction between record and coin is 0.15, how far from the center of the record can one place the coin without having it slip off?

• **6.20** A mass of 0.5 kg is attached to a 4-m-long thin rod whose other end is attached to a long vertical shaft by a small hinge. The shaft is then rotated until the angle that the rod makes with the vertical is 30°. Determine the angular velocity of the shaft and the tension in the rod under those conditions. (Neglect the mass of the rod.)

• • **6.21** The radius of curvature of a loop-the-loop roller coaster is 6 m. If at the top of the loop the passengers are held in their seats by an effective radial acceleration of 0.5g, what is then the speed of the roller coaster? From what initial elevation should the roller coaster start (assuming that it starts from rest) to achieve this speed? Neglect friction losses.

• **6.22** A mass of 0.6 kg, supported on a frictionless horizontal surface, is attached to a light string of 0.4-m length that is fixed to a post on the surface. The mass revolves about the post at 40 rpm. Find the magnitude of ω in radians per second and the tension in the string. If the string can support a mass of no more than 1 kg without breaking, at what rotational velocity will this string fail?

• **6.23** A 0.5-kg mass, supported on a frictionless horizontal table, is connected to a second mass of 0.8 kg by a 1-m-long light string that runs through a hole in the table, as shown in Figure 6.18. The 0.5-kg mass revolves about the hole at constant angular velocity so that the 0.8-kg mass is supported by the string just 20 cm below the surface of the table. Find the angular velocity of the 0.5-kg mass.

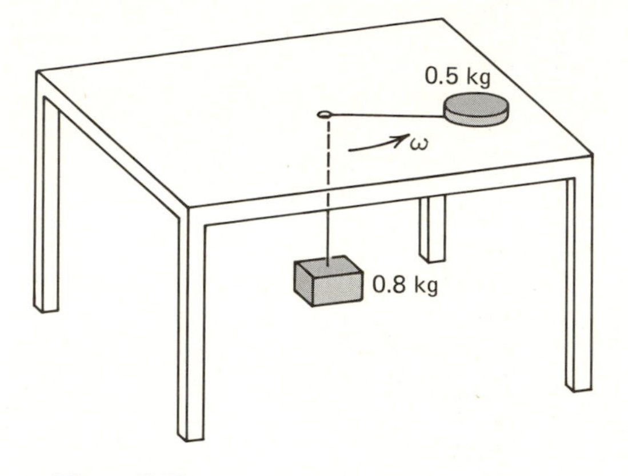

Figure 6.18

• **6.24** A small bug sits on a turntable, 12 cm from the center. What is the centripetal acceleration of the bug when the turntable rotates at 33.3 rpm?

• • **6.25** A stunt pilot is diving toward the earth at an angle of 45°, traveling at a speed of 1000 km/h. At an altitude of 800 m, he pulls back on the stick and puts the plane into a circular turn in the vertical plane. What is the maximum acceleration to which his body will be subjected if he is to avoid crashing the plane?

• **6.26** A space station is constructed in the shape of a doughnut whose outer diameter is 240 m and inner diameter 200 m. At what rate should this space station rotate so that its occupants will experience an acceleration of g when they are at the outer diameter of the station? By what percentage will the weight of an individual change as he moves from the outer to the inner part of the doughnut?

• **6.27** An 0.8-kg pendulum bob is supported by a 1.5-m-long string from a fixed point. The bob is pulled to one side, so that the string makes an angle of 37° with the vertical, and then released from rest. Find the speed of the bob at the lowest point of its swing, the instantaneous angular velocity of the pendulum at that point of its swing, and the tension in the string at that moment.

• • **6.28** A mass of 0.4 kg is rotating in a vertical plane at the end of a 1-m-long string. What must the speed of the mass at the top of the loop be so that the tension in the string is then 2.5 N? If the only force acting on the system is gravity, what is the tension in the string when it makes an angle of 180° relative to the previous direction?

• • **6.29** A mass m slides on a frictionless track with a circular loop, as shown in Figure 6.19. If the mass is released from rest from a height h, what is the largest value of the diameter d for which m remains in contact with the track at all times?

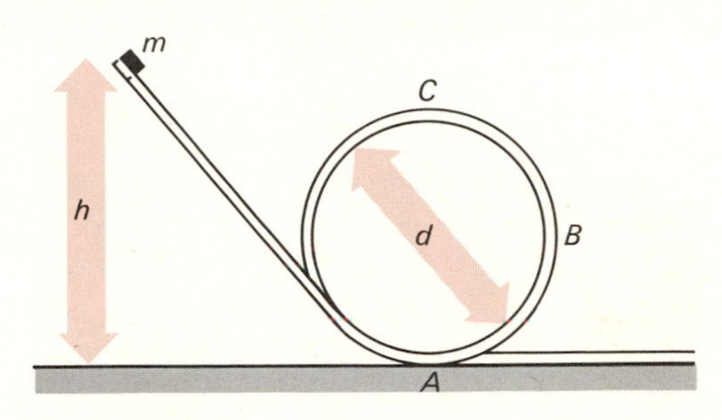

Figure 6.19

• • **6.30** If in Figure 6.19 $m = 0.2$ kg, $h = 1$ m, and $d = 0.25$ m, what is the force that the track exerts on m at points A, B, and C?

• • **6.31** A small mass is precariously balanced at the top of a large spherical smooth globe (Figure 6.20). It is now given a minute push so that it starts to slide down the frictionless surface of the globe. The mass is observed to part contact with the globe at an elevation that is greater than R, where R is the radius of the globe. Explain why that should be. Then calculate the elevation at which mass and globe lose contact.

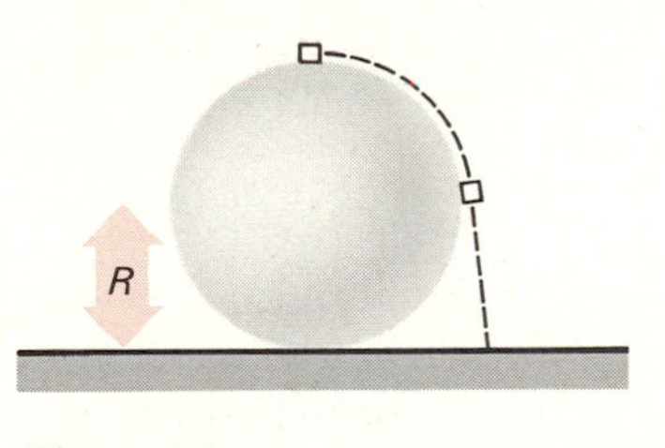

Figure 6.20

(Section 6.4)

• **6.32** An unbanked curve has a radius of curvature of 130 m. If the coefficient of static friction between a car's tires and the pavement is 0.55, what is the maximum speed at which the car can negotiate this curve without skidding?

• **6.33** A car enters an unbanked curve of 80-m radius at a speed of 30 km/h. If the car does not skid, what is the minimum coefficient of static friction between tires and pavement?

• **6.34** At the two ends of the oval of an indoor bicycle track, the radius of curvature of the track is 9 m. What should the banking angle be if the normal racing speed averages 45 km/h?

• • **6.35** A massive bob tied to a massless string whose other end is fixed at the ceiling is called a *conical pendulum* when the bob is made to rotate in a horizontal plane. The time for one complete rotation of the bob is called the *period* of the pendulum. Derive an expression which relates the period T, the length of the string L, and the angle θ between the string and the vertical. (*Note:* Will such a pendulum work in the absence of gravity? The answer is no. Hence, the acceleration of gravity g should appear in the expression for T. Knowing this, combine T, g, L, and θ so that the equation is dimensionally correct. Then derive it from first principles.)

• **6.36** An airplane flying at a speed of 700 km/h makes a turn that changes its direction by 180°. The plane takes 1 min to complete the turn. Determine the radius of the turn and the banking angle of the plane. What is the per cent change in effective weight which the passengers experience during the turn?

• **6.37** A curve whose radius is 250 m is banked at an angle of 15°. What is the optimum speed for negotiating this curve?

• • **6.38** What should the angular velocity of a conical pendulum of 20-cm length be so that the tension in the string is ten times the weight of the bob?

• **6.39** What is the correct banking angle for a curve of 300-m radius and a speed of 60 km/h?

• • **6.40** A curve of radius 120 m is banked at an angle of 12°. If a car of mass 1200 kg travels around this curve at 40 km/h without slipping what is the frictional force between car tires and pavement? In what direction does this force point?

• • **6.41** Repeat Problem 6.40 for a car of the same mass, traveling around the curve at 90 km/h. What is the minimum value of the coefficient of static friction that will prevent skidding?

• **6.42** A curve on a highway is banked at an angle of 15°. The radius of the curve is 80 m. At what speed should a car travel so that the force between tires and pavement is exactly perpendicular to the pavement?

• • **6.43** If a 1200-kg car travels around the curve of Problem 6.42 at 30 km/h, what are the components of the force, parallel and perpendicular to the pavement, that the pavement exerts on the tires?

7

Gravitation

One had to be a Newton to notice that the Moon is falling, when everyone sees that it doesn't fall.

PAUL VALERY

7.1 Introduction

By the time of Newton's birth the Ptolemaic universe with its equants, deferents, and epicycles within epicycles, though tenaciously championed by the Church in Rome, was increasingly questioned by scientists of northern Europe in favor of the Copernican model and its refinement by Kepler. The critical problem confronting these men was, How are the planets maintained in their nearly circular orbits about the sun?

It was this problem, among others, that occupied Newton during his seclusion at Woolsthorpe in 1666. His starting point for understanding planetary motion was the moon, that large satellite circling our globe once every $27\frac{1}{4}$ days. Newton knew that a force must act on the moon, but

> if this force was too small, it would not sufficiently turn the Moon out of rectilinear course; if it was too great it would turn it too much and draw down the Moon from its orbit toward the Earth.

According to legend, Newton conceived the idea of gravity on seeing an apple fall, and indeed, the story is essentially correct. However, the concept of a gravitational pull on objects by the earth was not original with Newton. What was so profoundly original was his brilliant speculation that

> as this power [of gravity] is not found sensibly diminished at the remotest distance from the center of the Earth to which we can rise, neither at the tops of the loftiest buildings, not even the summits of the highest mountains; it appeared reasonable to conclude that this power must extend much farther than is usually thought; why not as high as the Moon; and if so, her motion must be influenced by it; perhaps she is retained in her orbit thereby.

Newton promptly calculated the period of the moon's orbit, using the known distance between earth and moon and assuming that the gravitational attraction between two objects is inversely proportional to the square of their separation. In fact, he had already convinced himself that the inverse square law is consistent with Kepler's laws of planetary motion:

> I *deduced* [our emphasis] that the forces which keep the planets in their orbs must be reciprocally as the squares of their distances from the centers about which they revolve; and thereby compared the force requisite to keep the Moon in her orb with the force of gravity at the surface of the Earth; and found them answer pretty nearly.

Although Newton now knew that his conjecture that gravitational attraction kept the moon and the planets in their stable orbits was indeed correct, he refrained from publishing these revolutionary ideas. His first publications dealt instead with optics, particularly the origin of rainbow colors produced when white light passes through a glass prism (see Chapter 25). Here, too, he proposed a radical new concept, dispersion, for which he was widely criticized by his contemporaries.

> I was so persecuted with discussions arising from the publication of my theory of light that I blamed my own imprudence for parting with so substantial a blessing as my quiet to run after a shadow.

As a result, his most profound contributions remained secreted in his mind for two decades. Then, in 1684, an argument arose between Christopher Wren, Robert Hooke, and Edmund Halley concerning the motion of planets assuming an inverse square law of attraction to the sun.* To resolve it, Halley traveled to Cambridge to put the matter before Newton. As John Conduitt recounted:

> After they had been some time together, the doctor [Halley] asked him what he thought the curve would be that would be described by the planets, supposing the force of attraction toward the Sun to be reciprocal to the square of their distance

* Christopher Wren, best remembered as the architect of St. Paul's Cathedral, the Greenwich Observatory, and other marvelous buildings, was also professor of astronomy at London and later at Oxford University. Robert Hooke was a contemporary of Newton and one of his severest critics; we shall encounter Hooke's law in a later chapter. Halley was the best known astronomer of his time; one of the largest and most spectacular comets bears his name.

Figure 7.1 Edmund Halley (1656–1742). Among his many contributions to science, perhaps the greatest was financing the publication of Newton's Principia.

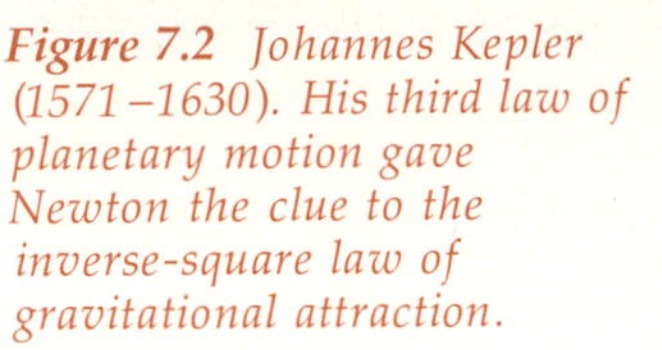

Figure 7.2 Johannes Kepler (1571–1630). His third law of planetary motion gave Newton the clue to the inverse-square law of gravitational attraction.

> from it. Sir Isaac replied immediately that it would be an ellipsis. The doctor, struck with joy and amazement, asked him how he knew it. "Why," saith he, "I have calculated it." Whereupon Dr. Halley asked him for his calculation without any further delay. Sir Isaac looked among his papers but could not find it, but he promised him to renew it, and then send it to him.

This took three more years, but what Halley finally wheedled from Newton was the monumental *Principia,* wherein the law of gravitational attraction is but one of numerous immensely significant discoveries and concepts.

7.2 Law of Universal Gravitational Attraction

The law of gravitational attraction states that any two objects exert an attractive force on each other, and that this force is proportional to the product of their masses and inversely proportional to the square of the distance between them. In mathematical language, this statement reads

$$F = G\frac{m_1 m_2}{r^2} \tag{7.1}$$

where r is the distance that separates masses m_1 and m_2, and G is a constant of proportionality, the universal gravitational constant. The force F is always attractive; that is, the force on m_1 points toward m_2, the force on m_2 points toward m_1.

As discussed in Chapter 3, an object's weight is a consequence of this gravitational attraction between two bodies, in this case the object and the earth. The weight, which is the net attractive force exerted by the earth on the object just above its surface, is, however, not so readily calculated. The difficulty arises from the fact that different parts of the earth exert forces on that object whose magnitudes and directions depend on the relative positions of the object and the part of the earth under consideration. To arrive at the total force experienced by a body above the earth's surface one must evaluate the vector sum of all these diverse forces, as indicated in Figure 7.3.

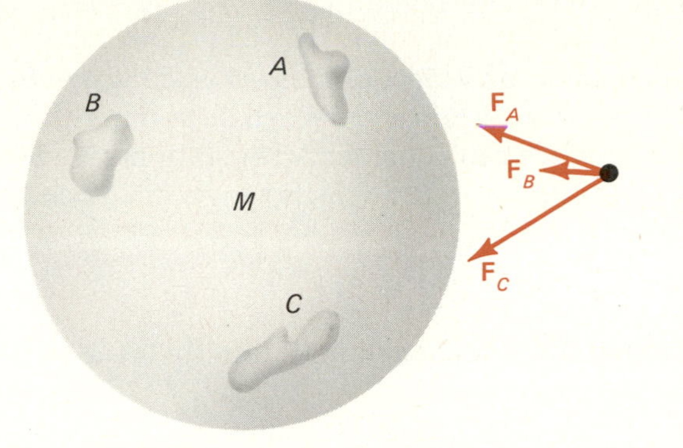

Figure 7.3 A small mass, attracted to a large, massive sphere M. The total force of attraction is the vector sum of the forces due to various portions of the large sphere. These vary in magnitude and direction according to Equation (7.1).

The calculation requires the mathematical technique of integral calculus. This, too, Newton developed in those miraculous months at Woolsthorpe. After much effort, he was able to demonstrate that *the force of gravity that a spherically symmetric body of total mass M exerts on an object of mass m, located outside the sphere, is given by Equation (7.1), where r is the distance between the mass m and the center of the sphere of mass M.*

Note that the large mass need be only spherically symmetric, not homogeneous. That is, the constitution may depend on distance from the center, so long as it does not depend on angular coordinates. This has great practical importance, since the density of the earth and of other planets as well as of the sun is not uniform but increases substantially toward their respective centers.

The weight of an object at the surface of the earth, $W = mg$, is the force of attraction between the earth and the mass m. We therefore have

$$W = mg = G\frac{mM_E}{R_E^2} \tag{7.2}$$

where M_E and R_E are the mass and radius of the earth. It follows that the relation between the acceleration of gravity g and the gravitational constant G is

$$g_P = \frac{GM_P}{R_P^2} \tag{7.3}$$

where the subscript "P" is used to indicate that the relation is valid not only for Earth but for any planet, or for that matter, any star or other spherically symmetric body.

To determine the universal gravitational constant G, one must measure the force of attraction between two objects of *known* masses. Since this attractive force is extremely small for masses normally handled in the laboratory, the experimental measurement of G is a very difficult and challenging experimental project. We shall describe the earliest measurement shortly. First, however, it is instructive to see how Equation (7.1) leads to Kepler's third law, at least for circular orbits.

Kepler's third law states that for a circular orbit, the square of the period of a planet is proportional to the cube of its distance from the sun. That is,

$$\frac{r^3}{T^2} = \text{constant} \tag{7.4}$$

where T denotes the period. The *period* of a planet's motion is the time required for one complete revolution about the sun; Earth's period is $365\frac{1}{4}$ days.

We consider, then, an object of mass m circling another, much more massive object M at a distance r. We shall assume that M is so much greater than m that we can consider M as stationary.* Since the mass m completes one revolution in time T, its tangential speed is

$$v = \frac{2\pi r}{T} \tag{7.5}$$

The centripetal force needed to maintain the mass m in orbit about M is

$$F = \frac{mv^2}{r} = \frac{m(2\pi r/T)^2}{r} = \frac{4\pi^2 rm}{T^2}$$

Now, if $F = k/r^2$, where k is a constant,

$$\frac{4\pi^2 mr}{T^2} = \frac{k}{r^2}$$

or

$$T^2 = \frac{4\pi^2 m}{k} r^3$$

That is, T^2 is proportional to r^3, as Kepler had concluded. In particular, if $F = GmM/r^2$, then

$$T^2 = \frac{4\pi^2}{GM} r^3 \tag{7.6}$$

We can now understand how Newton "deduced that the forces which keep the planets in their orbs must be reciprocally as the squares of their distances from the centers about which they revolve."

7.3 Gravitational Field

At the surface of the earth, the force acting on a 1-kg mass due to the gravitational attraction between it and the earth is 9.8 N. Actually, the magnitude of this force depends somewhat on location, and especially on altitude; it is slightly less atop a high tower than at ground level. At every point of space one can, in principle, measure this force, or calculate it with Equation (7.1), by setting $m_1 = M_E$ and $m_2 = 1$ kg. One obtains then a map of the force due to M_E on the "test mass" of 1 kg for all space. Such a map is called a *force field,* in this case, a gravitational force field.

The magnitude of the field due to a mass M is proportional to the mass M and inversely to the square of the distance from M. The force field is directed toward the source of the gravitational attraction, i.e.,

* In fact, m and M revolve about their common CM, but if $M \ggg m$, the CM of the system will nearly coincide with the position of M.

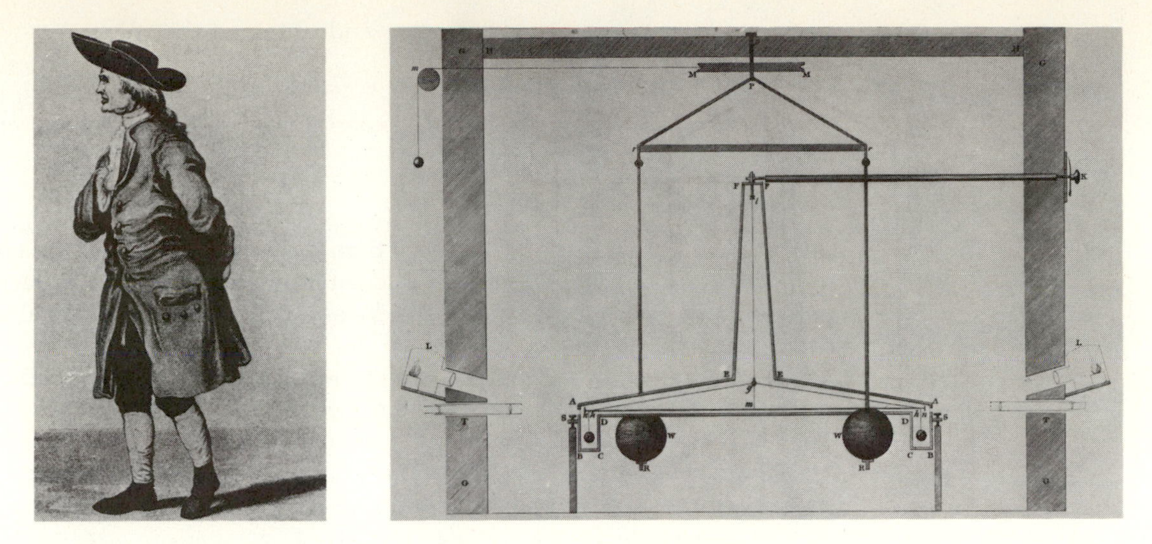

Figure 7.4 (a) Henry Cavendish (1731–1810). (b) A drawing of the apparatus he used to measure the universal gravitational constant G.

toward M. The gravitational field, which is the force per unit mass, is therefore a *vector field*. The field is a convenient concept, because once it is specified, the force acting on an object at some location is just the product of its mass and the field at that point.

A force field that points radially toward or away from the source is called a *central force field,* and the force is referred to as a *central force*. The gravitational force is the prime, though by no means the only example of a central force.

7.4 Determination of G: "Weighing the Earth"

More than a century elapsed following the publication of the *Principia* before Henry Cavendish performed the first successful measurement of the universal gravitational constant. His experimental arrangement is illustrated in Figure 7.5. It contains a sensitive torsion balance with two equal masses m_1 and m_2 attached to the ends of a thin rod supported at its center by a fine fiber. Two large masses M_1 and M_2 are mounted so that they can be brought near m_1 and m_2, as indicated in the figure. Gravitational attraction between the large and small masses deflects the torsion balance first one way, then in the opposite sense when the large masses are repositioned, as indicated in Figure 7.5. Cavendish painstakingly

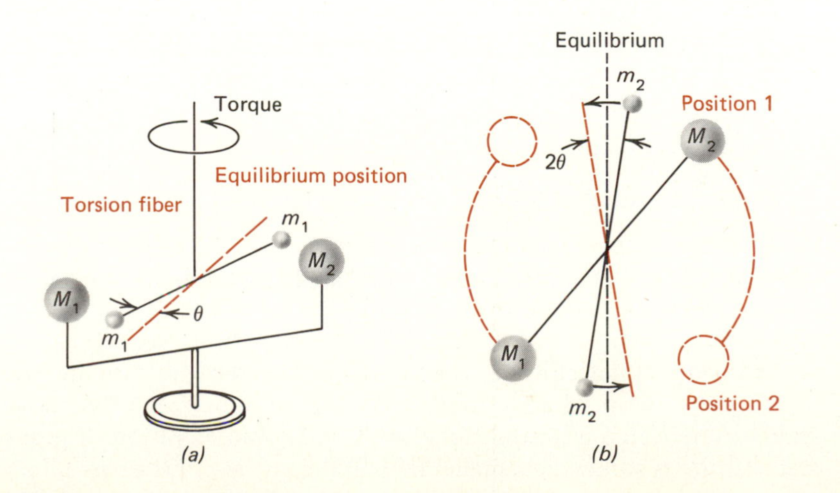

Figure 7.5 (a) Schematic perspective and (b) top view of the Cavendish balance. When the large masses are in the position indicated by 1, the deflection of the balance is clockwise. The large masses are then shifted to position 2, causing a counterclockwise deflection of the torsion balance.

performed the tedious measurements and obtained the value

$$G = 6.71 \times 10^{-11}\ \text{N·m}^2/\text{kg}^2$$

It is a tribute to his remarkable experimental skill that the currently accepted value of this constant,

$$G = 6.672 \times 10^{-11}\ \text{N·m}^2/\text{kg}^2$$

differs from Cavendish's by no more than 0.6%. To appreciate the enormous experimental task, let us calculate the gravitational attraction between two lead spheres of mass 10 kg and 80 kg when they are almost touching.

For this calculation, we must know the radii of the two spheres; the separation between their centers when they touch is the sum of the radii. The radii can be readily determined from the known density of lead, 11,340 kg/m^3 (Table 9.2, p. 173). The total mass of a sphere equals the product of the density and the volume of the sphere. The latter is given by

$$V = \tfrac{4}{3}\pi r^3 \qquad \textbf{(7.7)}$$

Thus, for the 10-kg sphere we have

$$10\ \text{kg} = \tfrac{4}{3}\pi r^3(1.134 \times 10^4\ \text{kg/m}^3)$$

and solving for r, we find

$$r_{10} = 5.95 \times 10^{-2}\ \text{m} = 5.95\ \text{cm}$$

The radius of the 80-kg sphere is twice as great, 11.9 cm, since the volume is proportional to r^3.

The gravitational force between these spheres when they almost touch is therefore

$$F = \frac{(6.67 \times 10^{-11}\ \text{N·m}^2/\text{kg}^2)(10\ \text{kg})(80\ \text{kg})}{(0.179\ \text{m})^2} = 1.67 \times 10^{-6}\ \text{N}$$

This force is to be compared with the weight of the 10-kg sphere of nearly 100 N. That is, the attractive force between the 10-kg and 80-kg spheres is just a bit more than 10^{-8} (one hundred millionths) of the weight of the smaller sphere. Little wonder physicists refer to gravitation as a "weak" interaction. It is only when the larger sphere is replaced by one as massive as the earth that this weak interaction produces a substantial force, 98 N for the 10-kg sphere.

Once G has been determined, one can compute the mass of the earth. Using Equation (7.3) and the known radius of the earth, we can solve for M_E and obtain

$$M_E = \frac{(9.8\ \text{m/s}^2)(6.37 \times 10^6\ \text{m})^2}{6.67 \times 10^{-11}\ \text{N·m}^2/\text{kg}^2} = 5.96 \times 10^{24}\ \text{kg}$$

When Cavendish published his results he titled the paper "Weighing the Earth."

Figure 7.6 An artificial earth satellite. Numerous such satellites circle our globe today. Most have periods of the order of a few hours.

● ***Example 7.1*** What is the period of a satellite that circles the earth at an altitude of 400 km above sea level?

At the specified altitude, the distance between the satellite and the center of the earth is $6.37 \times 10^6\ \text{m} + 4 \times 10^5\ \text{m} = 6.77 \times 10^6\ \text{m}$. Since the gravitational force is proportional to $1/r^2$, the acceleration of gravity at that altitude is somewhat smaller than at the surface of the earth. Its value is

$$g(400 \text{ km}) = g\left(\frac{6.37}{6.77}\right)^2 = (9.8 \text{ m/s}^2)0.885 = 8.68 \text{ m/s}^2$$

This is just the centripetal acceleration of the satellite in its orbit. Hence, from Equation (6.16) we have

$$(6.77 \times 10^6 \text{ m})\omega^2 = 8.68 \text{ m/s}^2$$
$$\omega = 1.13 \times 10^{-3} \text{ rad/s}$$

The period, that is, the time required for one complete revolution or 2π radians, is

$$T = \frac{2\pi}{\omega} = 5.55 \times 10^3 \text{ s} = 1.54 \text{ h}$$

The satellite will circle the earth once every hour and 32.4 minutes. ●

● ***Example 7.2*** Syncom, a synchronous communication satellite, is placed in orbit directly above the equator and its orbit radius adjusted so that it circles our globe once every 24 h. Consequently, its position relative to the earth remains fixed; it is permanently above some particular meridian. What must the radius of its orbit be?

We could proceed as we did in the preceding example; knowing the period, 24 h, we could calculate r. Instead, let us use Kepler's law. We know that the moon, whose orbit radius is 3.84×10^8 m, has a period of 27.25 days; the satellite, whose orbit radius we do not yet know, should have a period of one day. Both satellites are maintained in their stable orbits by the gravitational attraction to Earth. We can apply Kepler's law, Equation (7.4), and write

$$\frac{R_M^3}{T_M^2} = \frac{R_s^3}{T_s^2}, \qquad \text{or} \qquad \frac{(3.84 \times 10^8 \text{ m})^3}{(27.25 \text{ days})^2} = \frac{R_s^3}{(1 \text{ day})^2}$$

where R_s is the radius of the orbit of Syncom. We find

$$R_s = 4.24 \times 10^7 \text{ m}$$

The speed of the satellite is

$$v = \frac{2\pi(4.24 \times 10^7 \text{ m})}{24 \text{ h}} = 1.11 \times 10^7 \text{ m/h} = 11{,}100 \text{ km/h}$$ ●

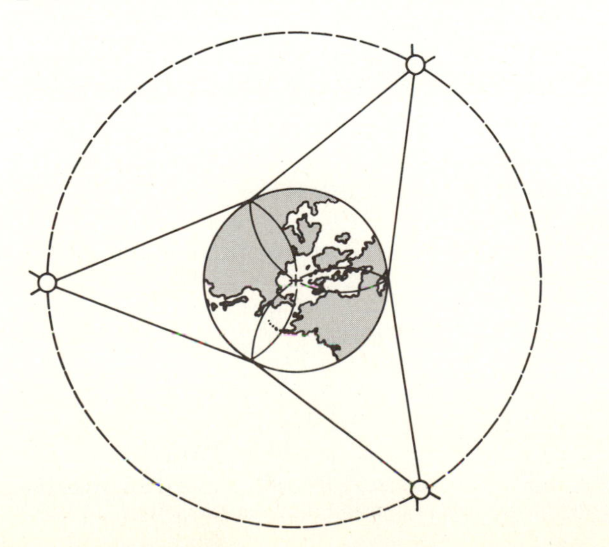

Figure 7.7 Earth and the orbits of Syncom satellites (not to scale). Three such satellites suffice to permit communication by microwave radio between all parts of our globe.

● ***Example 7.3*** What is the acceleration of gravity on the surface of a planet that has a radius half that of Earth and the same average density as Earth?

At the surface of this planet, the acceleration of gravity is

$$g_P = \frac{GM_P}{R_P^2}$$

On the surface of Earth, it is

$$g_E = \frac{GM_E}{R_E^2}$$

Since both bodies have the same density, their masses are proportional to their volumes, that is, proportional to the cube of their radii. Hence, we obtain

$$\frac{g_P}{g_E} = \frac{R_P^3/R_P^2}{R_E^3/R_E^2} = \frac{R_P}{R_E} = \frac{1}{2}$$

$$g_P = \tfrac{1}{2}g_E = 4.9 \text{ m/s}^2$$ ●

● ***Example 7.4*** The planet Mars, whose average radius is 3400 km, has two satellites, Deimos and Phobos.* The orbit radius of Deimos is 23,500 km and its period is 30.3 h. Find the mass of Mars, its average density, and the acceleration of gravity on the surface of Mars.

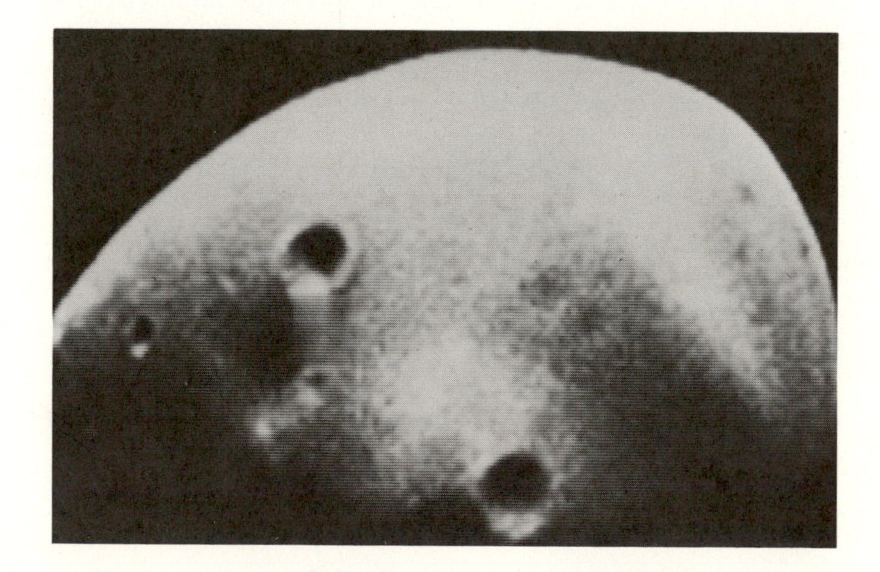

Figure 7.8 *The Martian satellite Deimos, photographed by Viking Orbiter I, showing considerable detail on the surface. This very small satellite has an average diameter of only about 15 km; the illuminated portion is about 12 × 8 km. Measurement of its orbit radius and period permits determination of the mass of Mars.*

Equation (7.6) holds for stable orbits about any center of attraction of mass M. The mass of Mars is therefore

$$M = \frac{4\pi^2 r^3}{GT^2}$$

$$= \frac{4\pi^2(2.35 \times 10^7 \text{ m})^3}{(6.67 \times 10^{-11} \text{ N}\cdot\text{m}^2/\text{kg}^2)[(30.3 \text{ h})(3600 \text{ s/h})]^2}$$

$$= 6.46 \times 10^{23} \text{ kg}$$

* In Greek mythology, Deimos (Panic) and Phobos (Fear) were companions of Ares, the god of war, whose Roman counterpart was Mars.

The average density of Mars is obtained by dividing its mass by its volume:

$$V = \frac{4}{3}\pi r_M^3 = \frac{4}{3}\pi(3.4 \times 10^6 \text{ m})^3 = 1.65 \times 10^{20} \text{ m}^3$$

We obtain

$$\text{Density of Mars} = \frac{6.46 \times 10^{23}}{1.65 \times 10^{20}} = 3.92 \times 10^3 \text{ kg/m}^3$$

The average density of Earth is 5.5×10^3 kg/m³. Mars therefore has a constitution substantially different from that of our planet.

Finally, using Equation (7.3), we find that the acceleration of gravity on the surface of Mars is

$$g_M = \frac{(6.67 \times 10^{-11} \text{ N}\cdot\text{m}^2/\text{kg}^2)(6.46 \times 10^{23} \text{ kg})}{(3.4 \times 10^6 \text{ m})^2} = 3.73 \text{ m/s}^2$$

about 38% of that on Earth. ●

In the centuries following the publication of the *Principia*, Newton's law of universal gravitation was fully confirmed by precise astronomical observations. Newton had already explained the small deviations from Kepler's elliptical planetary orbits that are occasioned by the gravitational interactions between the planets. These are relatively small, since planetary masses are much smaller than the mass of the sun. Nevertheless, they do produce measurable perturbations, which depend on the instantaneous locations of these celestial bodies.

On March 13, 1781, Sir William Herschel discovered a seventh planet, Uranus, during routine observations. Subsequent careful measurements of its position showed that its orbit departed ever so slightly from that calculated even when the influence of all other planets was included in the painstaking calculations. Two theoretical astronomers, Adam in England and Leverrier in France, working independently, concluded that the solar system must contain another, still undiscovered planet, and they were able to predict its location. On September 23, 1846, Newtonian mechanics scored another triumph, when the German observational astronomer Galle found the planet Neptune, exactly where Leverrier had told him to look!

Summary

The *law of universal gravitational attraction* states that any two bodies attract each other with a force given by

$$F = G\frac{m_1 m_2}{r^2}$$

where m_1 and m_2 are the masses of the two bodies and r is the distance separating them. The *universal gravitational constant* G has the value

$$G = 6.67 \times 10^{-11} \text{ N}\cdot\text{m}^2/\text{kg}^2$$

The gravitational attraction exerted by a spherically symmetric object of total mass M on a body of mass m located outside M and a distance r from the center of M is the same as that calculated under the assumption that all the mass M is concentrated at the center of the spherically symmetric object.

The acceleration of gravity at the surface of a planet of mass M_P and radius R_P is

$$g_P = \frac{GM_P}{R_P^2}$$

Kepler's third law of planetary motion states that

$$\frac{r^3}{T^2} = \text{constant}$$

where r is the radius of the (circular) planetary orbit and T is the period of its revolution about the sun. This relation follows directly from the inverse square law of attraction.

Once Cavendish had measured G, the mass of Earth could be calculated, and the masses of other planets determined by measuring the periods and orbit radii of their satellites.

The *gravitational force field,* or *gravitational field,* of a mass M is the three-dimensional map of the force on a unit mass due to the presence of the mass M. That field is a *vector field,* which points toward the source M. A force field that is directed radially along the line from the source is called a *central force field.*

Multiple Choice Questions

7.1 Suppose a planet exists that has half the mass of Earth and half its radius. On the surface of that planet, the acceleration due to gravity is

(a) twice that on Earth.
(b) the same as that on Earth.
(c) half that on Earth.
(d) one-fourth that on Earth.

7.2 Astronomical measurements give precise values for the orbital diameter and period of the small Jovian satellite Ganymede, one of the Galilean moons. With those data and our knowledge of the value of G,

(a) it is possible to calculate the mass of Jupiter and of the moon Ganymede.
(b) it is possible to calculate the mass of Jupiter but not that of the moon Ganymede.
(c) it is possible to calculate the mass of the moon Ganymede but not that of Jupiter.
(d) it is not possible to calculate the masses of these objects.

7.3 Some comets have highly eccentric elliptical orbits. If we take the PE to be zero at infinite distance from the sun, then when these comets are very far from the sun, their

(a) KE and PE are large and positive.
(b) KE is small and their PE is small and negative.
(c) KE is small but their PE is large and negative.
(d) KE is small but their PE is large and positive.

7.4 Two planets are made from the same material. Consequently, their masses are proportional to the cube of their radii, designated r_1 and r_2. It follows that g_1/g_2, the ratio of the acceleration of gravity at the surfaces of the two planets, is

(a) r_1/r_2
(b) r_2/r_1
(c) $(r_1/r_2)^2$
(d) $(r_2/r_1)^2$

7.5 Two satellites orbit well outside the atmosphere at distances r_1 and r_2 from the center of the earth. The ratio v_1/v_2 of the tangential speeds of the two satellites is

(a) r_1/r_2
(b) r_2/r_1
(c) $(r_2/r_1)^2$
(d) none of the above.

7.6 An astronaut on a strange planet with no atmosphere measures the acceleration of gravity at its surface and finds that it is 6 m/s^2. What explanation could account for this observation?

(a) The mass of the planet is smaller than that of Earth and its radius is the same as that of Earth.
(b) The mass of the planet is the same as that of Earth, but its radius is smaller than that of Earth.
(c) The astronaut's watch is running more slowly than it should.
(d) Either (a) or (c) could account for the observation.

7.7 Two stars of masses M and m, with $M > m$, are separated a distance D. At some point between the two stars their gravitational fields are equal in magnitude but oppositely directed. At that point, a test object will feel no net force. This point is

(a) at the CM of the two-star system.

(b) between the CM and the midpoint between the two stars.

(c) between the CM and the star of mass M.

(d) between the midpoint and the star of mass m.

Problems

(Sections 7.2–7.4)

7.1 An object that weighs 50 N on the earth is transported to the moon. There its weight is 8.3 N. What is the acceleration of gravity at the surface of the moon?

7.2 The period of Neptune is 164.8 years. Determine the radius of Neptune's orbit.

7.3 A 2000-kg satellite is in a circular orbit of radius 60,000 km about the earth. Determine the gravitational force on the satellite and its period of revolution.

7.4 From the data given in Appendix B calculate the acceleration of gravity on the surface of the moon and on the surface of Jupiter.

7.5 What is the force of gravitational attraction between two uniform spheres, each having a mass of 9 kg and a diameter of 0.4 m, when they just touch?

7.6 What is the acceleration of gravity 8400 km above the surface of the earth?

• **7.7** Once G is known, not only can the mass of the earth be calculated, but so can the mass of the sun. Show that the sun's mass is given by $M_S = 4\pi^2R^3/GT^2$, where T is the period of a planet in circular orbit of radius R about the sun. Calculate the mass of the sun, using the average orbit radius of the earth, 1.49×10^{11} m.

• **7.8** The average orbit radius of Venus is 1.08×10^{11} m. What is the length of the Venusian year?

• **7.9** How strong is the gravitational attraction between moon and sun, compared with the attraction between moon and earth? (See Appendix B for data on masses and distances; the moon-to-sun and earth-to-sun distances are practically identical.)

• **7.10** During the lunar mission of Apollo II, Armstrong and Aldrin descended to the moon's surface in the lunar module while Collins remained in orbit in the command module. If the command module's period was 2 hours and 20 minutes, how high above the moon's surface did it orbit?

• **7.11** Europa, one of the satellites of Jupiter, has an orbit diameter of 1.34×10^9 m and a period of 3.55 days. The radius of Jupiter is 7.14×10^7 m. What is the mass of Jupiter, its average density and the acceleration of gravity at its surface?

• **7.12** There are many binary star systems in our galaxy. These are two stars held in a stable configuration by their mutual gravitational attraction so that they revolve about their common center of mass. Suppose that such a system consists of two stars of which one has the mass of the sun and the other has a mass of 0.6 solar mass, and their separation is 10^{12} m. Locate the CM of this system and calculate the gravitational field at that point in space (neglecting all other astronomical objects).

• **7.13** What is the centripetal acceleration, due to the planet's rotation, of a point on the equator of Earth? What fraction of g is this?

• **7.14** What is the orbit radius of a 300 kg satellite that completes one circular orbit about Earth once each 2.4 hrs? What is the gravitational force that acts on this satellite?

• • **7.15** At what point between Earth and Moon is the net gravitational attraction on an object due to these two bodies exactly zero?

·8·

Rotational Equilibrium; Dynamics of Extended Bodies

Give me where to stand, and I will move the earth.

ARCHIMEDES

8.1 Torques and Rotational Equilibrium

Suppose you push with one finger against a book that is lying on the table. You will notice that the resulting motion depends not only on the direction and magnitude of the force you apply but also on the point of application of that force. Only if the line of action of the force passes through the CM of the book does the book move forward without rotation. Generally, the push you exert causes rotation as well as translation of the book.

Consider a solid uniform bar of length L that is acted on by two forces, as shown in Figure 8.2. The sum of the forces is zero, yet we know that the bar, if initially at rest, does not remain so but rotates clockwise. We are therefore forced to conclude that the condition for equilibrium, Equation (3.7), which is valid for point objects, must be augmented for extended objects to include a statement about rotational equilibrium.

Experience shows that the action causing rotation depends not only on the magnitude and direction of the applied force but also on the point of application. When you enter a revolving door, for example, you almost instinctively push with your hand near the periphery of the door. You know that a much larger force is needed to turn the door if you push near

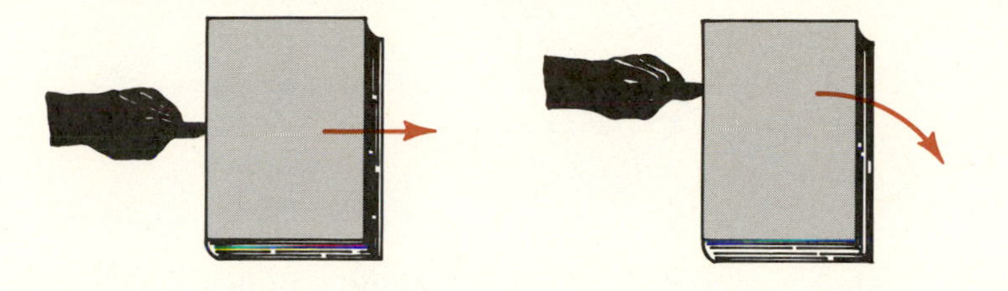

Figure 8.1 *A book resting on the table and pushed with a finger will move in the direction of the force only if that force acts through the center of the book. A push near the edge of the book results in a rotation as well as a forward motion.*

its axis of rotation. Two factors determine the rotational impetus, the force that is applied and the perpendicular distance between the line along which the force acts and the axis about which the body rotates. This perpendicular distance is called the *lever arm*, or *moment arm*, and is denoted by ℓ. The efficacy of a force in producing rotation is found to be proportional to the product of the force F and the moment arm ℓ. This product is called the *torque* about the specified axis of rotation. Torque is denoted by the symbol τ.

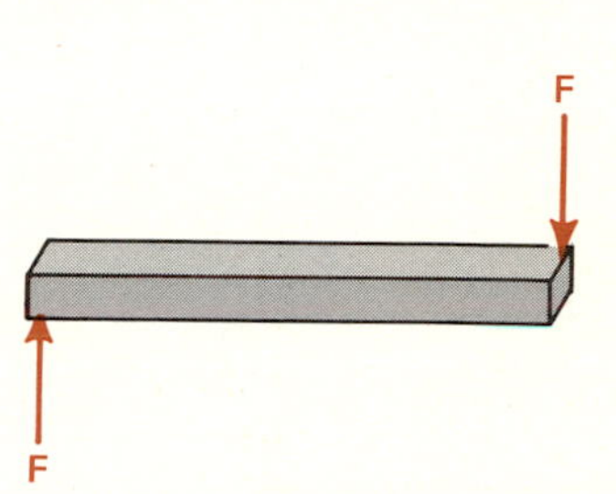

Figure 8.2 *The vector sum of the two forces is zero. Nevertheless, the bar is not in equilibrium, but tends to rotate clockwise.*

Torque = force × moment arm

$$\tau = F\ell \tag{8.1}$$

The dimension of torque is $[M][L]^2/[T]^2$. The unit of torque is the newton meter.

Torque is a vector quantity. The torque vector acts along the axis of rotation (not along the force) *and points in the direction a right-handed screw*

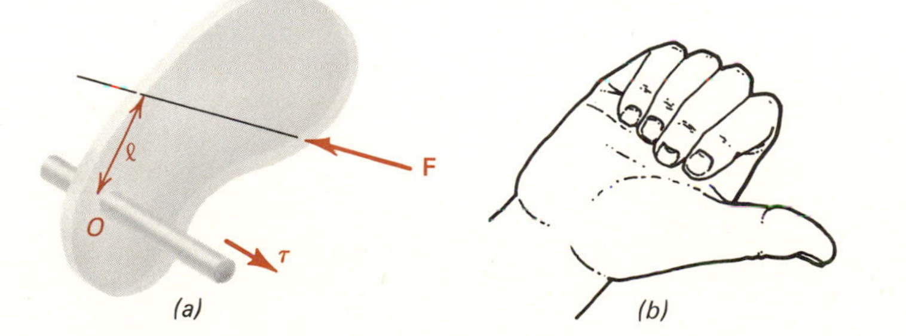

Figure 8.3 *(a) The force* **F** *gives rise to a torque* $\boldsymbol{\tau}$ *about the pivot O. The magnitude of the torque is the product of the force F and the perpendicular distance between the line of action of* **F** *and the axis of rotation. In this case,* $\tau = F\ell$*. The distance* ℓ *is called the moment arm. The direction of the torque is the direction along which a right-handed screw would advance if it were turned by that torque. In this instance,* $\boldsymbol{\tau}$ *points as shown. (b) The direction of the torque can be determined by applying the right-hand rule: The fingers of the right hand point in the direction of the rotation produced by the torque, with the thumb along the axis of rotation. The torque vector then points in the direction of the thumb.*

would advance if turned by that torque. The right-hand rule is a good way to determine the direction of $\boldsymbol{\tau}$. If the fingers of the right hand point in the direction of the rotation that the torque induces, the thumb will point in the direction of the torque vector, as illustrated in Figure 8.3(*b*).

● ***Example 8.1*** A bar is pivoted as shown in Figure 8.4. A force of 5 N acts along the horizontal direction at a point 2 m from the pivot as measured along the bar. What are the magnitude and direction of the torque?

The torque is given by (8.1). The moment arm ℓ of the 5-N force is, however, not 2 m but (2 m) (sin 60°). Hence

$$\tau = (5\text{ N})(2\text{ m})(\sin 60°) = 8.66\text{ N}\cdot\text{m}$$

The direction of the torque is that in which a right-handed screw would move if turned counterclockwise, i.e., out of the page. ●

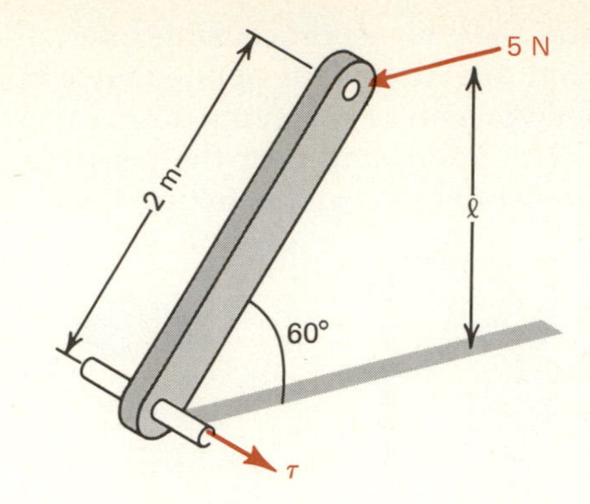

Figure 8.4 The torque acting on the rod is equal to (5 N)(2 m)(sin 60°) = 8.66 N·m and points out of the paper.

● *Example 8.2* A string is wound about a pulley of 1.5-m radius. A mass of 2 kg is supported by the string. What torque does the string exert on the pulley? (Assume that the pulley is clamped to a fixed axle and thereby prevented from turning.)

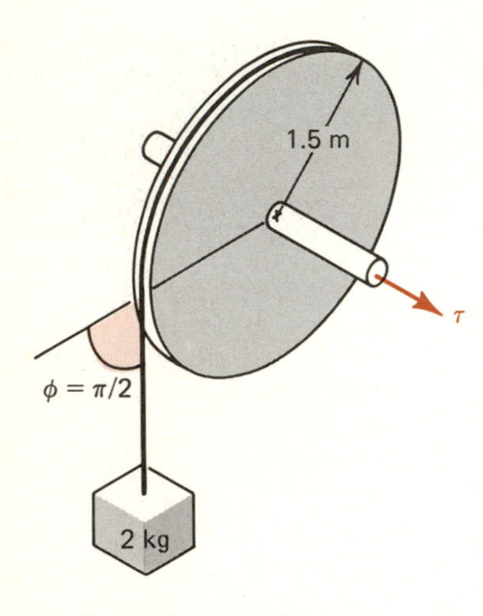

Figure 8.5 A 2-kg mass supported by a string that is wound about a pulley of 1.5 m radius.

The tension in the string is $T = (2 \text{ kg})(9.8 \text{ m/s}^2) = 19.6$ N. Here the moment arm is the radius of the drum, 1.5 m. The torque is therefore (19.6 N)(1.5 m) = 29.4 N·m, directed as shown. ●

We can now answer the question posed at the beginning of this chapter: What additional conditions assure equilibrium of an extended body? The answer is that in addition to the condition that the sum of all *forces* acting on the body vanish, the sum of all *torques* acting on the body must also vanish. Thus for equilibrium we have

$$\sum_i \mathbf{F}_i = 0 \tag{8.2}$$

$$\sum_i \tau_i = 0 \tag{8.3}$$

As with the force equation, the torque equation can also be expressed in component form, that is,

$$\sum_i \tau_{ix} = 0 \tag{8.4a}$$

$$\sum_i \tau_{iy} = 0 \tag{8.4b}$$

$$\sum_i \tau_{iz} = 0 \tag{8.4c}$$

Again, we must remember that the direction of the torque vector is along the rotation axis, not along the force that produces the torque.

● *Example 8.3* A very light (massless) plank of 5-m length is supported at its ends by two ropes. A man weighing 800 N stands on the plank 2 m from the right-hand end. Find the tension in the two ropes.

Figure 8.6(b) shows the free-body diagram for the plank with all the forces acting on it. We now write the equations for equilibrium.

The force equation (8.2) gives for the vertical component of $\Sigma \mathbf{F}_i$

$$\Sigma F_{iv} = T_A + T_B - W = 0$$

$$T_A + T_B = 800 \text{ N}$$

Clearly, this equation alone is insufficient to determine the unknown tensions T_A and T_B. The second equation we need is provided by (8.3).

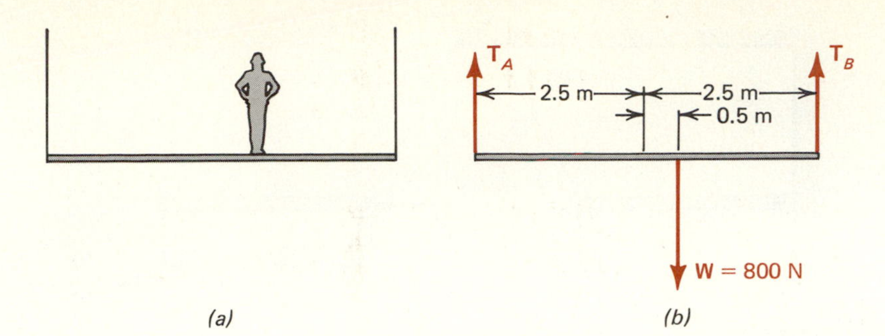

Figure 8.6 (a) A man standing on a very light (massless) plank. (b) The free-body diagram of the plank of (a).

Let us take as our axis of rotation the one that passes through the center of the plank and choose the direction into the page (i.e., a clockwise torque) as positive. We then have

$$\Sigma\tau = (2.5 \text{ m})T_A + (0.5 \text{ m})W - (2.5 \text{ m})T_B = 0$$

$$(2.5 \text{ m})(T_A - T_B) = -400 \text{ N·m}$$

Solving the force and torque equations, we obtain $T_A = 320$ N, $T_B = 480$ N. ●

In solving Example 8.3, we chose to calculate torques about an axis through the center of the plank. Could we have selected some other axis? Indeed we could have. A system in equilibrium is not experiencing angular acceleration about *any* axis. Consequently, any axis can be used in writing the torque equation. For example, we could have considered rotation about an axis through the left-hand end of the plank. In that case, the torque equation would have read

$$(3 \text{ m})(800 \text{ N}) - (5 \text{ m})T_B = 0$$

$$T_B = \frac{2400 \text{ N·m}}{5 \text{ m}} = 480 \text{ N}$$

leading to the same result.

In equilibrium problems, the selection of the axis about which torques are evaluated is completely arbitrary.

Note, however, that the second choice of axis in Example 8.3 simplified the solution considerably. One of the unknown forces was eliminated from the torque equation because with that choice of axis its moment arm vanished. The choice of axis is optional; however, a particular axis is often advantageous because it eliminates one or more unknown forces from the torque equation.

● ***Example 8.4*** A very light (massless) rod of length L is hinged to a wall by a pivot at one end. At its other end hangs a mass M, and the system is maintained in equilibrium by a string that is attached to the ceiling and makes an angle of 60° with the horizontal. What are the tension in the string and the vertical and horizontal components of the reaction force of the hinge on the rod?

We solve this problem by again writing the two equilibrium equations, (8.2) and (8.3), for the isolated body. Before we do so, we examine Figure 8.7(*b*) and note that if we take our axis of rotation not through the hinge but through the right-hand end of the plank, all forces except the vertical component of the reaction force R_V pass through the axis of rotation. Thus the torque equation reduces to

$$R_V \times L = 0$$

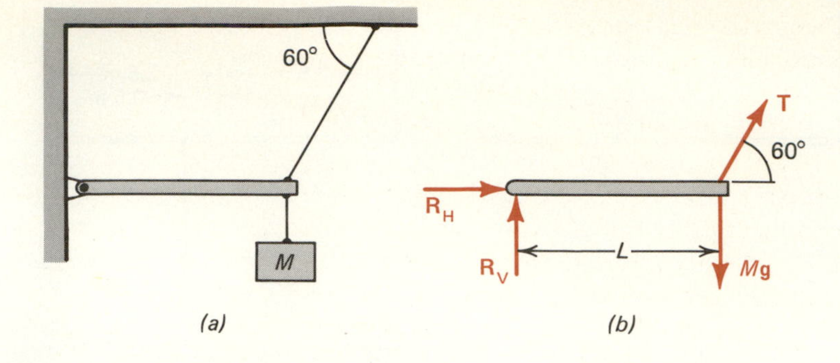

Figure 8.7 (*a*) A mass *M* is hung from the end of a light rod that is hinged at the wall and supported from the ceiling. (*b*) The free-body diagram of the rod of (*a*).

and since $L \neq 0$, $R_V = 0$.

The tension in the string is now readily determined. From (8.3) we have, for the vertical components of the forces,

$$T_y - Mg = 0$$
$$T \sin 60° = Mg$$
$$T = \frac{Mg}{\sin 60°}$$

Lastly, the horizontal components of the forces must vanish:

$$R_H + T_x = 0$$
$$R_H = -T \cos 60°$$
$$= -\frac{Mg \cos 60°}{\sin 60°} = -\frac{Mg}{\tan 60°}$$

The negative sign shows that the force is directed to the left, not to the right as drawn in Figure 8.7(*b*).

8.2 Center of Mass and Center of Gravity

In Chapter 5 we found that for calculating the motion of a system of particles due to a net external force, the total mass can be thought as concentrated at the center of mass (CM). An extended body is just such a system of particles.

The center of gravity (CG) of a body is the point at which the body can be suspended without experiencing any torque due to gravitational forces. If the gravitational field is uniform, the CG and the CM coincide.

The location of the CG is illustrated in Figure 8.8. The object is suspended from different points, and the intersection of the plumb lines drawn from the points of support under equilibrium conditions locates the CG.

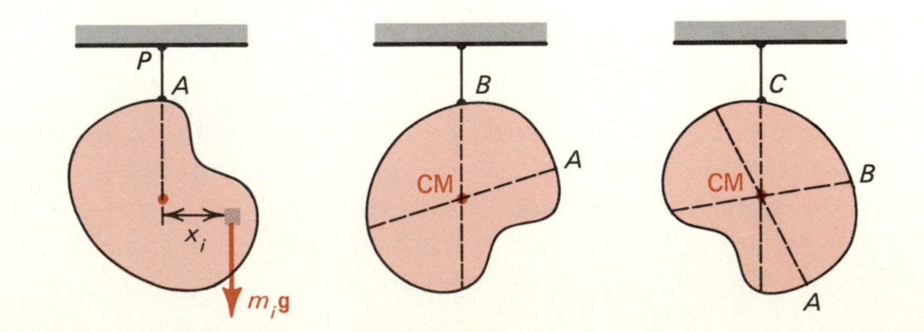

Figure 8.8 When a flat object is suspended from any point, the vertical line passing through the point of support must also pass through the center of gravity. This method of locating the center of gravity, which, in a uniform gravitational field, is at the center of mass, can also be used for three-dimensional objects.

To see why that procedure works, consider a two-dimensional object of arbitrary shape (Figure 8.8) suspended from some point such as P. The object can be thought of as made up of many small masses m_i, as indicated. In the *uniform* gravitational field of the earth, a vertical force $m_i g$ acts on each of these small masses. Each such force exerts a torque of $m_i g x_i$ about the point P; here x_i, the moment arm about P, is the distance between m_i and the vertical line passing through P. The torque is positive if m_i is to the right and negative if m_i is to the left of that line, where the direction into the page is taken as positive.

But the object is at rest, in equilibrium under the influence of all forces and torques acting on it. Consequently, the total torque about P must vanish. That is,

$$\sum_i m_i g x_i = 0, \qquad \text{or} \qquad \sum_i m_i x_i = 0 \tag{8.5}$$

Comparison with Equation (5.10), which defines the location of the CM, shows that (8.5) will hold only if the vertical line through P passes through the CM. Since this argument remains valid for any other randomly selected point of support, the plumb lines from these points must intersect at the CM when the object is at rest.

All the mass of an extended body can be considered concentrated at its CM for purposes of calculating its motion or gravitational energy in a uniform gravitational field. Thus when such an extended body travels through the air, twisting or turning as it does so, its CM describes the parabolic trajectory of a point object.

Figure 8.9 Sketch of a diver performing a one-and-a-half-turn somersault. Throughout this maneuver, the CM follows a parabolic trajectory. Note, however, that the CM does not stay fixed relative to the diver's body, but changes as the diver alters her configuration in space.

● ***Example 8.5*** We shall now apply the conditions of equilibrium for rigid bodies to calculate the forces exerted by certain muscles and skeletal joints in the human body.

Figure 8.10 shows the pelvis and upper part of the thigh bone, the femur. The head of the femur fits into a socket in the pelvis, the acetabulum. Approximately 7 cm from the center of the femoral head is a protrusion, the greater trochanter, to which the tendons of the principal hip

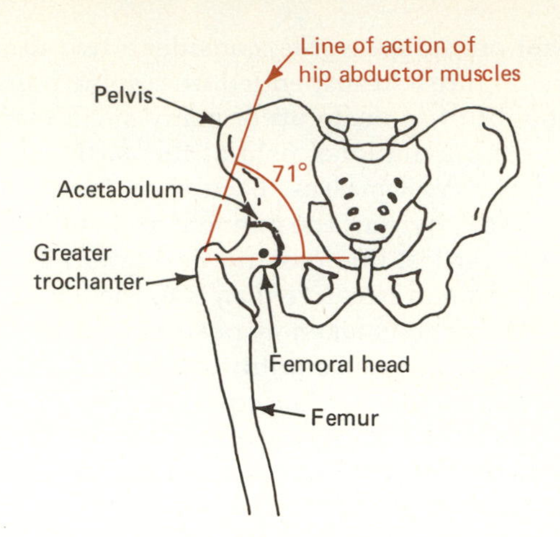

Figure 8.10 *Diagram showing part of the pelvis and femur. The femoral head fits into a hemispherical depression in the pelvis, the acetabulum. The principal hip abductor muscles are attached to the femur at the protrusion called the greater trochanter.*

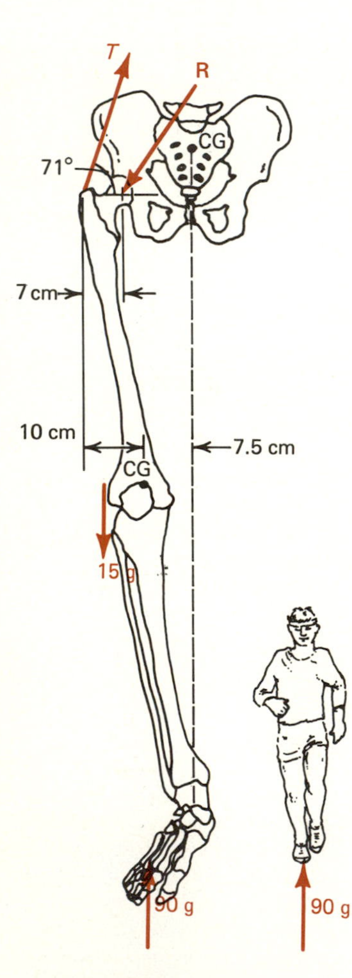

Figure 8.11 *A person resting his weight on the right foot. It is assumed that the torso is vertical. The right foot must rest directly below the CM of the body. (Why?) Typical dimensions needed to evaluate the tension in the hip abductor muscle and the reaction force between femoral head and acetabulum are shown.*

abductor muscles are attached.* The angle of the resultant force of these three muscles with the horizontal line from the pivot point of the femoral head to the greater trochanter is 71° when a person is standing.

We first calculate the force exerted by the hip abductor muscles and the reaction force between femoral head and acetabulum for the case of a man of normal weight who stands on one foot. We assume a total body mass M of 90 kg (weight of 882 N). For an average person, the mass of each leg is about $\frac{1}{6}M$, or 15 kg in our example, and the CM of the leg is located just above the knee joint, as shown in Figure 8.11.

As indicated in the figure, for the unilateral stance assumed here, the person can be in equilibrium only if the right foot is located directly below the CM of the body.

The schematic free-body diagram for the right leg is shown in Figure 8.12. The forces acting on the leg are the normal reaction force of the floor, $N = Mg = 882$ N, the weight of the leg, $W_L = (15)g = 147$ N, acting at its CM, the tension **T** of the hip abductor muscles acting at the greater trochanter in the direction shown, and **R**, the reaction force of the acetabulum on the femoral head. The equations for static equilibrium are now

$$\sum_i F_{ix} = 0 \quad T \cos 71° - R_x = 0 \tag{8.6}$$

$$\sum_i F_{iy} = 0 \quad T \sin 71° + 882 \text{ N} - 147 \text{ N} - R_y = 0 \tag{8.7}$$

$$\sum_i \tau_i = 0 \quad (7 \text{ cm})T \sin 71° + (3 \text{ cm})(147 \text{ N}) - (10.5 \text{ cm})(882 \text{ N}) = 0 \tag{8.8}$$

where we have taken torques about the pivot point in the femoral head so as to eliminate the reaction force **R** from this equation.

Solving for T from (8.8), we find

$$T = 1333 \text{ N} = 1.51Mg$$

and from (8.6) and (8.7),

$$R_x = 434 \text{ N}, \qquad R_y = 1995 \text{ N}$$

and

* An *abductor muscle* pulls a bone away from the body's midline.

$$R = \sqrt{R_x^2 + R_y^2} = 2042 \text{ N} = 2.31Mg$$

The tension in the hip abductor muscles is about 1.5 times the body weight, and the reaction force between acetabulum and femoral head more than twice the body weight!

If the man were to support even a relatively small fraction of his total weight on a cane in his left hand, permitting the right foot to rest more nearly below the femoral head, the strain on his hip muscles would be greatly reduced (see Problem 8.16).

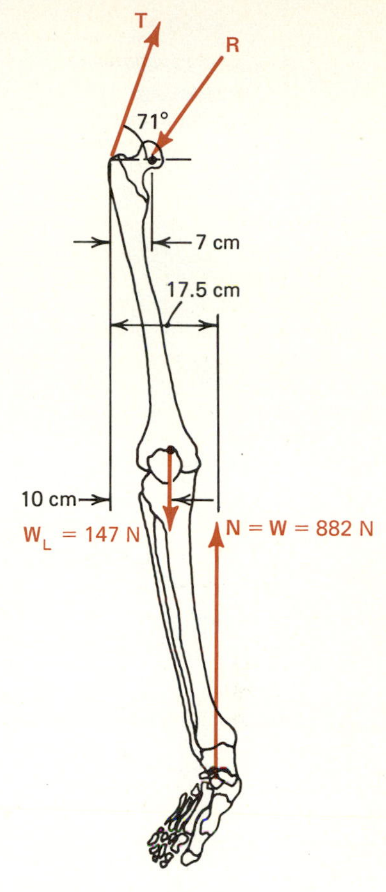

Figure 8.12 *Free-body diagram of the right leg, used in calculating the tension* **T** *in the hip abductor muscle and the reaction force* **R** *of the acetabulum on the femoral head.*

8.3 Rotational Dynamics

In Chapter 6 we exhibited the close parallel between the kinematic equations for uniformly accelerated motion and for rotational motion at constant angular acceleration. It is natural to look for a similar parallelism in the dynamical description, that is, to look for an equation relating the *angular* acceleration $\boldsymbol{\alpha}$ to the driving action, the torque $\boldsymbol{\tau}$, analogous to Newton's second law—which relates the *linear* acceleration $\mathbf{a}$ to the driving force $\mathbf{F}$.

To do this we consider again a very simple system, consisting of a mass m, attached to a massless rod of length r pivoted about its other end (Figure 8.13), to which we apply a tangential force $\mathbf{F}$ of constant magnitude. Under the influence of this force, the mass m experiences a tangential acceleration related to $\mathbf{F}$ by

$$\mathbf{F} = m\mathbf{a}_t \tag{8.9}$$

The magnitude of the tangential acceleration and of the angular acceleration about the pivot are related by

$$a_t = r\alpha \tag{8.10}$$

We also note that the force $\mathbf{F}$ produces a torque of magnitude $\tau = Fr$ about the pivot.

If we now multiply Equation (8.9) by r, we obtain

$$Fr = \tau = mra_t = mr^2\alpha$$

which we can rewrite as

$$\boldsymbol{\tau} = I\boldsymbol{\alpha} \tag{8.11}$$

where

$$I = mr^2 \tag{8.12}$$

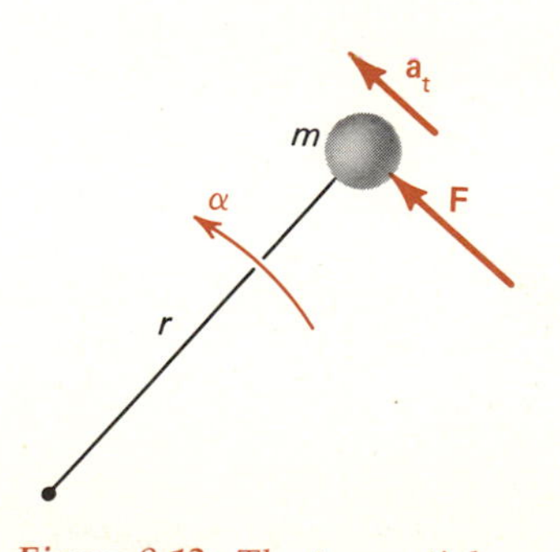

Figure 8.13 *The tangential force* **F** *acting on the mass m produces a tangential acceleration* $\mathbf{a}_t$. *The dynamics of this system can also be expressed in terms of the torque about the pivot point due to the force* **F** *and the angular acceleration* $\boldsymbol{\alpha}$ *of this system.*

Equation (8.11) is of the desired form, relating the torque to the angular acceleration. The proportionality factor is now not the mass but the product of the mass and the square of the moment arm from the axis of rotation. This proportionality factor is called the *moment of inertia* and is designated by the symbol I.

For the point particle considered here, Equation (8.11) is identical in every respect to Equation (8.9) from which it was derived. There is no particular advantage in using one over the other as the dynamical equation of motion. The value of Equation (8.11) and of the concept of moment of inertia becomes apparent when one considers rotation of extended rigid bodies, such as disks, rods, spheres, parallelepipeds, or objects of irregular shape.

For example, suppose we want to know how much torque is needed to impart an angular acceleration α to the object of Figure 8.14, rotating about the axis AA'. Direct application of Newton's second law is now difficult because different parts of the object travel at different speeds, and their accelerations differ in magnitude as well as direction. The speed of some small element of mass δm_i depends on $\mathbf{r}_i$; the speed of a point near the periphery is much greater than the speed of one near the axis. Similarly, the magnitude as well as the direction of the acceleration of δm_i depends on $\mathbf{r}_i$ as well as on α. Thus we cannot write a meaningful expression of the form $F = ma$ that will help us find an answer to the problem.

Provided the system is rigid, however, the *angular* acceleration and *angular* velocity about the axis AA' are the same for every part of this object. Therefore, we *can* write an expression of the form $\boldsymbol{\tau} = I\boldsymbol{\alpha}$ that is valid and meaningful for this body. That is, Equation (8.11) is not limited to the simple case of a point object; it is generally valid provided that one uses the moment of inertia calculated according to the prescription we shall now give.

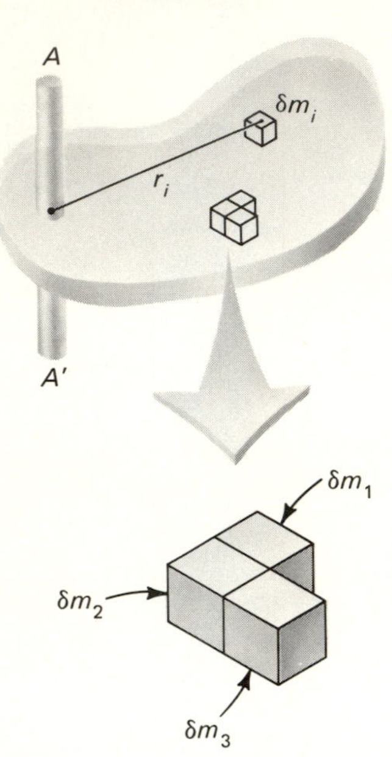

Figure 8.14 The moment of inertia of an irregularly shaped flat object about the axis AA' is the sum of the individual moments of inertia of the small elements of mass, δm_i, located a distance r_i from the axis of rotation.

8.4 Moment of Inertia

We have just examined a particularly simple case, a point mass constrained to move about some specified axis. The extension of the concept of moment of inertia to solid objects of arbitrary shape is straightforward in principle, although in practice, calculating moments of inertia may well be complicated, requiring integral calculus. The process essentially is one of dividing the total mass into infinitesimally small elements, calculating the contribution of each small element to the moment of inertia, and then adding all these contributions.

For example, to determine the moment of inertia of the irregularly shaped flat object shown in Figure 8.14 about the axis AA', one imagines the object divided into very small, contiguous masses δm_1, δm_2, δm_3, . . . , δm_n. The moment of inertia δI_i due to the small mass δm_i, located a distance r_i from the axis AA', is

$$\delta I_i = \delta m_i r_i^2$$

and the total moment of inertia of the object about AA' is

$$I = \sum_i \delta I_i = \sum_i \delta m_i r_i^2 \qquad \textbf{(8.13)}$$

It is this summation that may pose difficulties. We shall not carry out such calculation here, but give the moments of inertia of some simple geometrical objects in Table 8.1.

The fact that the elemental moments of inertia δI_i depend on the distances r_i of the masses δm_i from the rotation axis means that *the moment of inertia of an object is not a fixed quantity but depends on the location of the axis about which the moment of inertia is calculated.* A particular object does have a unique mass; it does *not* have a unique moment of inertia. Its moment of inertia depends not only on its total mass and the distribution of that mass in space—that is, the shape of the object—but also on the orientation and position of the axis of rotation.

● ***Example 8.6*** Four identical masses m are fixed at the corners of a square of sides b. Calculate the moment of inertia of this system for rotation about the axes AA', BB', and CC' shown in Figure 8.15.

Axis AA'. In this case, each of the four masses is at a distance $b/\sqrt{2}$ from the axis of rotation. Each mass therefore contributes an amount

Table 8.1

Cylindrical shell about axis	$I = MR^2$
Solid cylinder about axis	$I = \frac{1}{2}MR^2$
Thin rod about perpendicular line through center	$I = \frac{1}{12}ML^2$
Thin rod about perpendicular line through one end	$I = \frac{1}{3}ML^2$
Thin spherical shell about diameter	$I = \frac{2}{3}MR^2$
Solid sphere about diameter	$I = \frac{2}{5}MR^2$
Solid rectangular parallelepiped about axis through center perpendicular to face	$I = \frac{1}{12}M(a^2 + b^2)$

$mb^2/2$ to the total moment of inertia. Hence

$$I_{AA'} = 4\left(\frac{mb^2}{2}\right) = 2mb^2$$

Axis BB'. Now each mass is exactly $b/2$ from the axis of rotation and consequently contributes $mb^2/4$ to the moment of inertia about this axis. Hence,

$$I_{BB'} = 4\left(\frac{mb^2}{4}\right) = mb^2$$

Axis CC'. Now two masses lie on the axis of rotation, and their moment arms vanish. Each of the other two masses contributes mb^2 to the moment of inertia. Consequently,

$$I_{CC'} = 2(mb^2) = 2mb^2$$

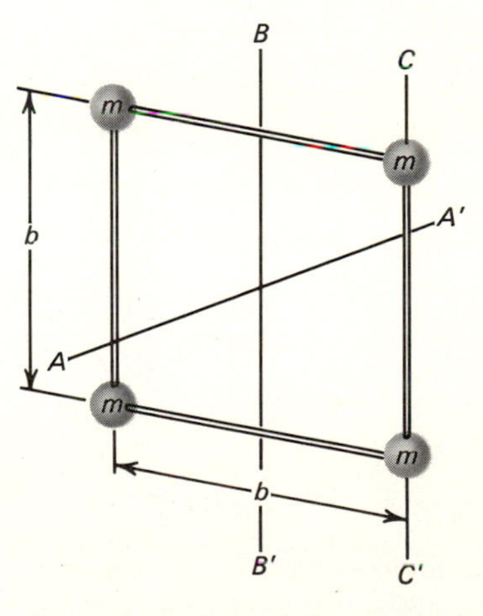

Figure 8.15 Four masses arranged in a square configuration. The moment of inertia of this system depends on the choice of axis. The values $I_{AA'}$, $I_{BB'}$, and $I_{CC'}$ are calculated.

8.5 Angular Momentum

To set a body initially at rest into rotation requires applying a torque; and correspondingly, a torque must be applied to bring a body that is rotating to rest. For example, to stop the wheels of a moving car from turning, friction forces between brake shoe and brake drum must be applied; these are tangential forces whose lever arm is the radius of the brake drum.

The tendency of a rotating object to maintain its rotational motion is reminiscent of the tendency of a body in translational motion to continue on its rectilinear course. For linear motion, we found it convenient to define a quantity, the linear momentum $\mathbf{p} = m\mathbf{v}$. Newton's second law, in its general form, then led to the law of conservation of linear momentum, which proved extremely useful in many instances.

It is natural, then, to ask if a similar conservation law might apply also to rotational motion. To derive this conservation law, we examine again the simple system of a single point mass at a distance r from a fixed axis. Again, we assume that a tangential force $\mathbf{F}$ acts on this mass. Then, Newton's law states,

$$\langle \mathbf{F} \rangle = \frac{\Delta(m\mathbf{v})}{\Delta t} = \frac{\Delta \mathbf{p}}{\Delta t} \tag{8.14}$$

Multiplying Equation (8.14) by r and using the relations $v = r\omega$ and $\tau = Fr$, we obtain

$$\langle \boldsymbol{\tau} \rangle = \frac{\Delta(mr^2\boldsymbol{\omega})}{\Delta t} = \frac{\Delta(I\boldsymbol{\omega})}{\Delta t}$$

Therefore, if, in analogy to the linear momentum $\mathbf{p} = m\mathbf{v}$, we define the *angular momentum* by

$$\mathbf{L} = I\boldsymbol{\omega} \tag{8.15}$$

we arrive at the rotational counterpart of (8.14):

$$\langle \boldsymbol{\tau} \rangle = \frac{\Delta \mathbf{L}}{\Delta t} \tag{8.16}$$

and as $\Delta t \to 0$,

$$\boldsymbol{\tau} = \lim_{\Delta t \to 0} \left(\frac{\Delta \mathbf{L}}{\Delta t} \right) \tag{8.16a}$$

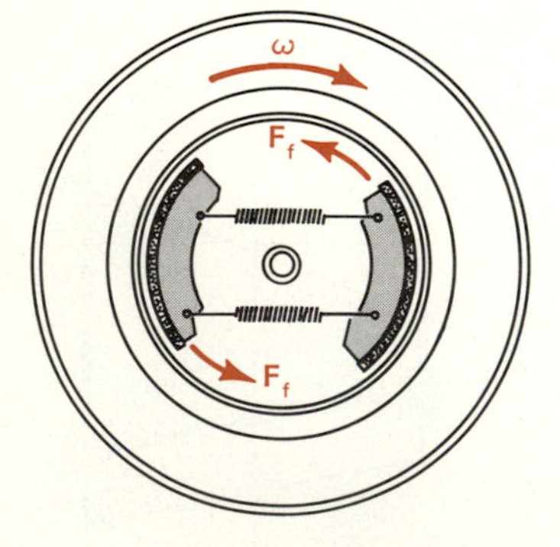

Figure 8.16 A rotating wheel has angular momentum about its axis. A torque must be applied to stop this rotation. In this case, the torque derives from the frictional force between brake shoes and brake drum.

Angular momentum, like angular velocity, is a vector quantity. For symmetrical systems such as we shall be concerned with here, $\mathbf{L}$ has the same direction as $\boldsymbol{\omega}$. Equation (8.16) states that applying a torque to a system with angular momentum $\mathbf{L}$ will cause a change in the angular momentum. As with the linear case, that change in the vector $\mathbf{L}$ may be in any direction, depending on the direction of the torque vector. Thus, a torque is required not only to change the magnitude of the angular momentum of a rotating body but also to change the direction of the axis of rotation even though the magnitude of the angular momentum remains constant.

Equation (8.16) also shows that in the absence of a torque the angular momentum of a system remains constant. That is, we arrive at the *law of conservation of angular momentum:*

The angular momentum of an isolated system is conserved.

It is very important to keep in mind that it is $I\boldsymbol{\omega}$, the *product* of the moment of inertia and the angular velocity that is conserved, and not the angular velocity $\boldsymbol{\omega}$. In many interesting situations, internal rearrangement of the masses of a system may change its moment of inertia; *when that happens the angular velocity changes even though no external torque is applied to the system.*

Figure 8.17 *Conservation of angular momentum. The torque due to friction between the ice skate and the ice is small. Consequently, the angular momentum of the figure skater remains nearly constant. She starts the pirouette in a crouch, arms and one leg extended, rotating relatively slowly because her moment of inertia is large. As she stands up and draws in the leg and arms, her moment of inertia decreases, and the angular velocity must increase to conserve angular momentum.*

For example, a figure skater usually starts a pirouette in a crouch, rotating on one skate with the other leg and both arms extended. She then slowly rises, pulling the extended leg and arms to her body, thus reducing her moment of inertia about the axis of rotation. As she does so, her angular velocity increases substantially. You can easily perform a similar experiment by sitting on a good swivel chair or piano stool while holding some fairly heavy objects in your outstretched hands. If you then start yourself spinning, you will notice that your rotational speed increases perceptibly if you bring your arms and the heavy objects close to your body.

By applying the equations of rotational kinematics and Equation (8.16), the rotational analog of Newton's second law, we can solve various dynamical problems, of which the following are a few examples.

● ***Example 8.7*** A uniform disk of 30-kg mass and 0.5-m radius is driven by a belt connected to a motor. If the disk, starting from rest and accelerating uniformly, attains an angular speed of 20 rev/s in 15 s, what is the tension in the belt?

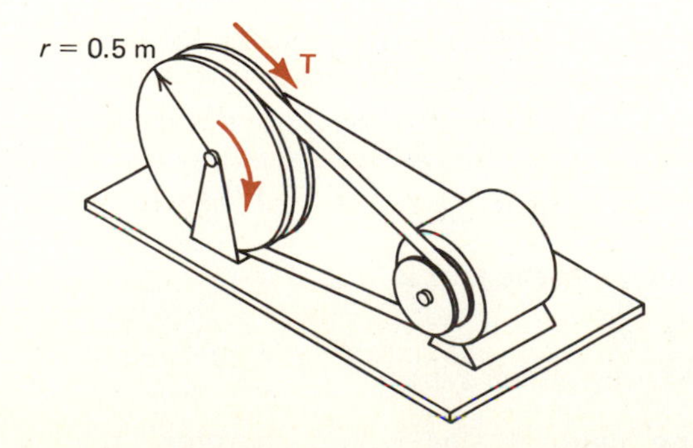

Figure 8.18 *The disk of radius 0.5 m is accelerated by the tension in the belt.*

The angular acceleration of the disk is

$$\alpha = \frac{\Delta\omega}{\Delta t} = \frac{2\pi(20 \text{ rev/s})}{15 \text{ s}} = 8.38 \text{ rad/s}^2$$

The moment of inertia of the disk is

$$I = \tfrac{1}{2}Mr^2 = \tfrac{1}{2}(30 \text{ kg})(0.5 \text{ m})^2 = 3.75 \text{ kg·m}^2$$

Hence,

$$\tau = I\alpha = (3.75 \text{ kg·m}^2)(8.38 \text{ rad/s}^2) = 31.4 \text{ N·m}$$

That torque is provided by the tension in the belt, which has a moment arm of 0.5 m about the axis of rotation. Therefore, the tension in the belt must be

$$T = \frac{\tau}{r} = \frac{31.4 \text{ N·m}}{0.5 \text{ m}} = 62.8 \text{ N}$$

● ***Example 8.8*** At one stage in stellar evolution, a star exhausts all its nuclear fuel. (Most stars, including the sun, derive their energy by a process that converts hydrogen gas to helium.) When that happens, the temperature within the star drops drastically and so does the internal pressure. Under gravitational attraction, the gaseous mass of the star then collapses toward the center, and under proper conditions, the star's density becomes enormous and its diameter extremely small. Such objects are known as neutron stars.

If the sun were to collapse to a neutron star its radius would be barely 20 km. What would be the angular velocity of that neutron star?

The sun is an inhomogeneous sphere of 6.96×10^8 m radius, which rotates about its axis with a period of 27 days. Its density increases dramatically toward the center so that 98% of its mass is confined to a sphere of about one-fifth the solar radius. As a rough approximation we can neglect the mass outside the central region and write (See Table 8.1).

$$I_{\text{sun}} = \frac{2}{5} M_{\text{sun}} \left(\frac{1}{5} R_{\text{sun}}\right)^2$$

Once collapsed to a neutron star, the density of that star is constant and so

$$I_{\text{ns}} = \frac{2}{5} M_{\text{sun}} R_{\text{ns}}^2$$

where we have assumed that no mass is lost during the collapse.

Since no external torque acts on the system during the gravitational collapse, angular momentum is conserved, i.e.,

$$I_{\text{sun}}\omega_{\text{sun}} = I_{\text{ns}}\omega_{\text{ns}}$$

Consequently,

$$\omega_{\text{ns}} = \omega_{\text{sun}} \left[\frac{(6.96 \times 10^8/5)^2}{(2 \times 10^4)^2}\right] = 4.84 \times 10^7 \, \omega_{\text{sun}}$$

The angular velocity of the sun is

$$\frac{1 \text{ rev}}{(27 \text{ days})(24 \text{ h/day})(3600 \text{ s/h})} = 4.29 \times 10^{-7} \text{ rev/s}$$

The resulting neutron star would therefore revolve at a rate of

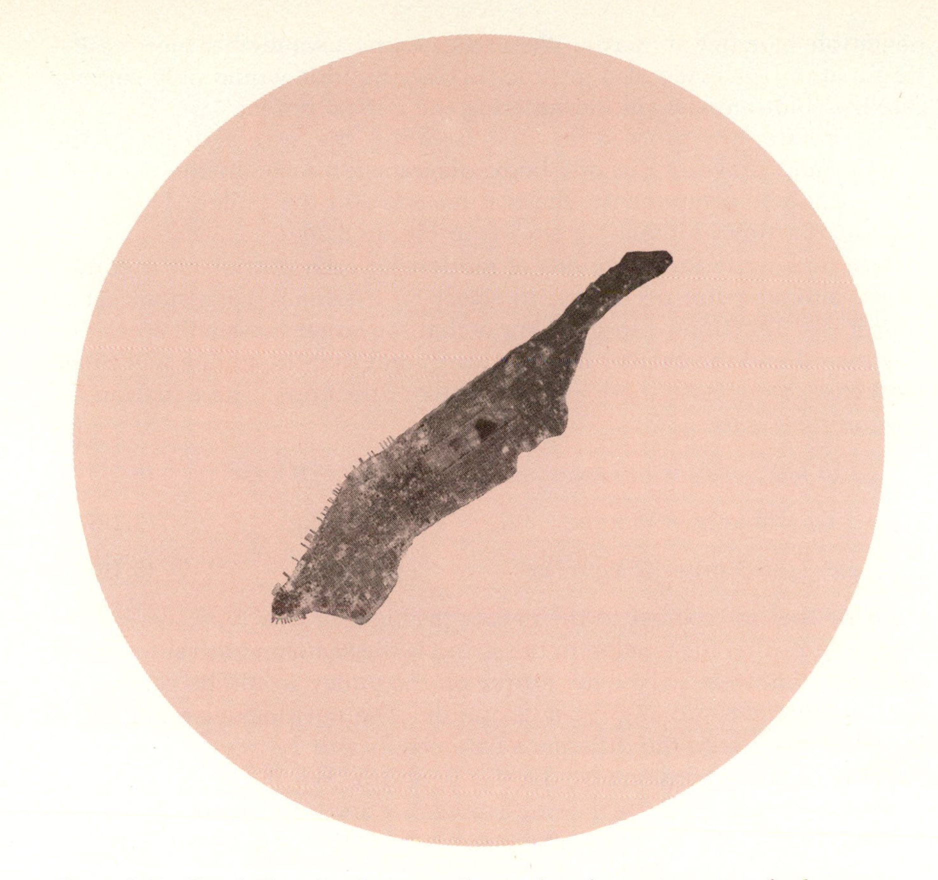

Figure 8.19 Sketch showing the approximate size of a neutron star of solar mass. Its diameter of about 40 km is here contrasted to the length of Manhattan Island.

4.84×10^7 (4.29×10^{-7} rev/s) = 21 rev/s! It is difficult to imagine an object of that size, with a mass equal to that of the sun (2×10^{30} kg) rotating at about 20 rps. One might well ask if such objects are but figments of our imagination.

In 1967, Jocelyn Bell Burnell, then a doctoral student at Cambridge, discovered a stellar radio source that sent signals in regularly spaced pulses once every 1.34 seconds. Though variable astronomical objects had been known for decades, their periods were generally of the order of days or months. Within a few years, many more such *pulsars* were found, one in the region of the Crab Nebula with a period of only 0.033 s. The explanation of these observations posed an exciting, challenging, and difficult problem for theoretical astronomers and astrophysicists, but it is now generally agreed that these pulsating sources are rapidly rotating neutron stars. The neutron star in the Crab Nebula is the remnant of a supernova that occurred in the year 1060 and was recorded by early astronomers in China, America, and other parts of the world. ●

● ***Example 8.9*** Two masses of 5 kg and 8 kg are connected by a thin (massless) string that passes over a pulley of 0.3-m radius and moment of inertia of 2 kg·m². Determine the tension in the string on each side of the pulley and the acceleration of each mass, neglecting friction of the pulley bearings and assuming that the string does not slip on the pulley.

This problem is similar to one we considered earlier, except that previously we neglected the mass of the pulley; that is, we assumed it had a negligible moment of inertia. We now examine a somewhat more realistic situation; we could also include friction, but this would only complicate the solution without illuminating the critical points.

As before, we proceed by first isolating the several parts of the system, and drawing the free-body diagram for each. In this instance there are three components, the two masses and the pulley. These and the forces acting on them are shown in Figure 8.20.

We now write the equations of motion for each portion. It is convenient, although not necessary, to select an internally consistent set of coordinate directions. Since we know that the larger mass will accelerate downward and the smaller mass upward, and that the pulley will rotate clockwise, we choose these as our positive directions. The equations of motion now read

$$(8\text{ kg})g - T_8 = (8\text{ kg})a; \qquad 78.4\text{ N} - T_8 = (8\text{ kg})a \tag{8.17}$$

$$T_5 - (5\text{ kg})g = (5\text{ kg})a; \qquad T_5 - 49\text{ N} = (5\text{ kg})a \tag{8.18}$$

$$(T_8 - T_5)(0.3\text{ m}) = (2\text{ kg}\cdot\text{m}^2)\alpha \tag{8.19}$$

Note that in contrast to the earlier problem of this type, we cannot now assume that the tension in the string is the same on both sides of the pulley. If that were so, the net torque on the pulley would be zero and it would experience no angular acceleration. The net clockwise torque on the pulley is due to the difference between T_8 and T_5.

There are four unknowns, T_8, T_5, a, and α. We therefore need one more relation, and that is the one between a and α:

$$a = r\alpha, \qquad \alpha = \frac{a}{0.3\text{ m}} \tag{8.20}$$

Equation (8.19) can then be rewritten,

$$(T_8 - T_5)(0.3\text{ m}) = \frac{(2\text{ kg}\cdot\text{m}^2)a}{0.3\text{ m}} \tag{8.19a}$$

There are several ways for solving these three equations. For example, adding the first two, one obtains

$$T_5 - T_8 + 29.4\text{ N} = (13\text{ kg})a$$

and from (8.19a),

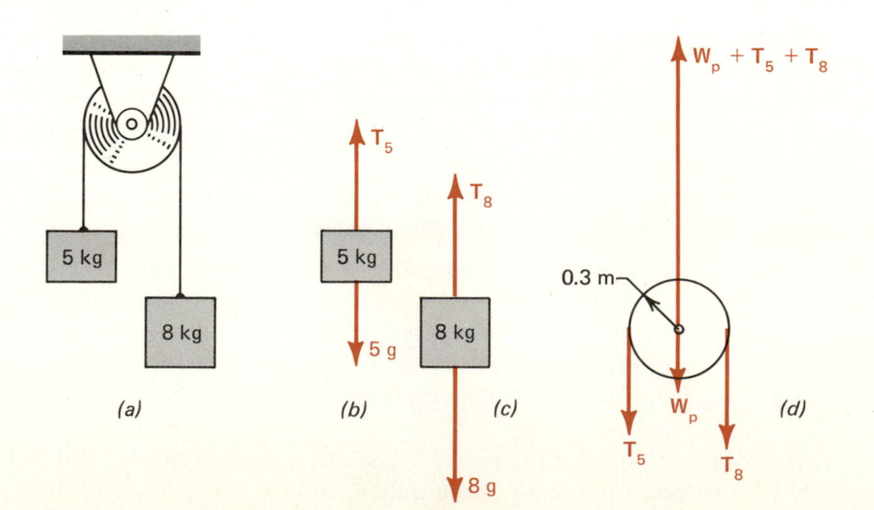

Figure 8.20 *(a) Two unequal masses, 5 kg and 8 kg, are attached to a thin string that passes over a pulley of finite mass and therefore finite moment of inertia. The free-body diagrams for (b), the 5-kg mass, (c), the 8-kg mass, and (d), the pulley. Since a torque is needed to give the pulley an angular acceleration, the tensions in the strings supporting the two masses cannot be the same.*

$$T_8 - T_5 = (22.2\ \text{kg})a$$

Adding these two equations, we have

$$29.4\ \text{N} = (35.2\ \text{kg})a$$
$$a = 0.835\ \text{m/s}^2$$

Substituting this value into Equations (8.17) and (8.18) gives

$$T_8 = 71.72\ \text{N}, \qquad T_5 = 53.17\ \text{N}$$

To check our solution, we can quickly calculate the angular acceleration of the pulley and confirm that it equals a/r:

$$(71.72\ \text{N} - 53.17\ \text{N})(0.3\ \text{m}) = (2\ \text{kg}\cdot\text{m}^2)\alpha$$

hence

$$\alpha = 2.783\ \text{rad/s}^2 = \frac{0.835\ \text{m/s}^2}{0.3\ \text{m}} = \frac{a}{r}$$

This has been a lengthy problem, requiring substantial algebraic manipulation. Let us therefore review the solution and stress the salient points. First, the system consists of three parts; two show translational motion, the third rotates about a fixed axis. The rotation of the pulley is intimately related to the translational motion of the masses through Equation (8.20). Because the pulley must have an angular acceleration, a torque must act on it. Consequently, the tension in the string on the two sides of the pulley cannot be the same; the difference in the tensions multiplied by the radius of the pulley is the torque acting on the pulley. ●

Angular momentum imparts stability to rotating systems since to change the direction of the angular momentum vector a torque must be applied. The bicycle rider can maintain himself on his cycle with relative ease while it is moving but must be an acrobat to keep his balance when it is at rest.

The gyroscope, which serves as the central part of an airplane's automatic pilot and a missile's inertial guidance system, also relies on the vector property of angular momentum.

A gyroscope consists of a well-balanced disk that is mounted in a system of nearly friction-free gimbals so that the axis of rotation of the disk is free to point in any direction independent of the orientation of the mounting. If the disk is set spinning with its axis pointing in a particular direction in space, it will thereafter continue to point in that direction even though the mount may alter its orientation. A gyroscope is a stable pointer, unaffected by changes in magnetic field that influence a magnetic compass.

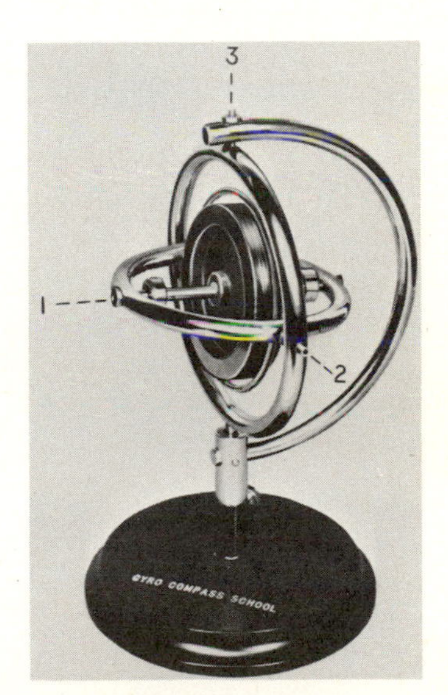

Figure 8.21 A gimbal-mounted gyroscope. When the gyroscope is set into rotation, the direction of its axis of rotation is independent of the orientation of the base. Modern gyroscopes are frequently supported by magnetic means to minimize frictional losses.

8.6 Rotational Kinetic Energy

From the analogy between rotational and translational kinematic variables and from the expressions relating angular acceleration, angular momentum, and torque with their translational counterparts, we can make an educated guess for the expression for rotational KE. Equation (8.11) $\boldsymbol{\tau} = I\boldsymbol{\alpha}$, looks like $\mathbf{F} = m\mathbf{a}$, Equation (3.4), and $\mathbf{L} = I\boldsymbol{\omega}$, Equation (8.15), looks just like $\mathbf{p} = m\mathbf{v}$, Equation (5.1); one need only replace the linear variables $\mathbf{F}$, m, $\mathbf{a}$, and $\mathbf{v}$ with their angular analogs, $\boldsymbol{\tau}$, I, $\boldsymbol{\alpha}$, and $\boldsymbol{\omega}$. We therefore anticipate that the rotational KE will be given by

$$\mathrm{KE}_{\mathrm{rot}} = \tfrac{1}{2} I\omega^2 \qquad \textbf{(8.21)}$$

in analogy with

$$\mathrm{KE}_{\mathrm{tran}} = \tfrac{1}{2} mv^2$$

Dimensionally, Equation (8.21) is correct. The dimensions of moment of inertia and of angular velocity are

$$[I] = [M][L]^2, \qquad [\omega] = \frac{1}{[T]}$$

so that

$$[I\omega^2] = \frac{[M][L]^2}{[T]^2}$$

which is the dimension of energy.

It is, however, not difficult to derive Equation (8.21), assuming a constant torque, and we do so now. We return to an earlier example, the wheel driven by a belt. Suppose that the moment of inertia of the wheel is I, its radius r, and the tension in the belt T. The torque acting on the wheel is then

$$\tau = Tr$$

and the resulting angular acceleration is

$$\alpha = \frac{\tau}{I} = \frac{Tr}{I}$$

If the disk was initially at rest ($\omega_0 = 0$), then at time t its angular velocity will be

$$\omega = \alpha t$$

The work done by the tension T during this time is the product of T and s, where s is the distance a point on the belt has moved in that time. But

$$s = r\theta = r\langle\omega\rangle t = r\left(\frac{\omega + \omega_0}{2}\right) t = \frac{1}{2} r\omega t$$

Consequently,

$$W = Ts = \frac{Tr\omega t}{2} = \frac{\tau\omega t}{2} = \frac{I\alpha\omega t}{2} = \frac{1}{2} I\omega^2$$

Since the work done by the belt was converted to rotational KE of the wheel, we can conclude from conservation of energy that

$$\mathrm{KE}_{\mathrm{rot}} = \tfrac{1}{2} I\omega^2 \qquad \textbf{(8.21)}$$

In many practical situations, the total KE of a system is shared between rotational and translational motion. For example, when a wheel rolls on a surface, its center of mass is moving with some translational velocity and the wheel has translational KE, $\mathrm{KE}_{\mathrm{tran}}^{\mathrm{CM}}$; the wheel is also rotating about its CM, and it must, therefore, also have rotational KE, $\mathrm{KE}_{\mathrm{rot}}^{\mathrm{CM}}$. The total KE is simply the sum of the two contributions:

$$\mathrm{KE} = \mathrm{KE}_{\mathrm{tran}}^{\mathrm{CM}} + \mathrm{KE}_{\mathrm{rot}}^{\mathrm{CM}} \qquad \textbf{(8.22)}$$

Frequently these two contributions are simply related because, as for the wheel just described, the speed v of the CM equals $r\omega$. In such a case, the fraction of the total KE due to motion of the CM and due to rotation about the CM depends on the mass distribution, that is, on the expression for the moment of inertia.

● ***Example 8.10*** A hollow cylinder of mass M_h and radius R_h, a uniform solid cylinder of mass M_c and radius R_c, and a uniform sphere of mass M_s and radius R_s start from rest at the same point on an inclined plane. All three roll down without slipping. Will they arrive at the bottom of the incline at the same instant? If not, in what order will they arrive?

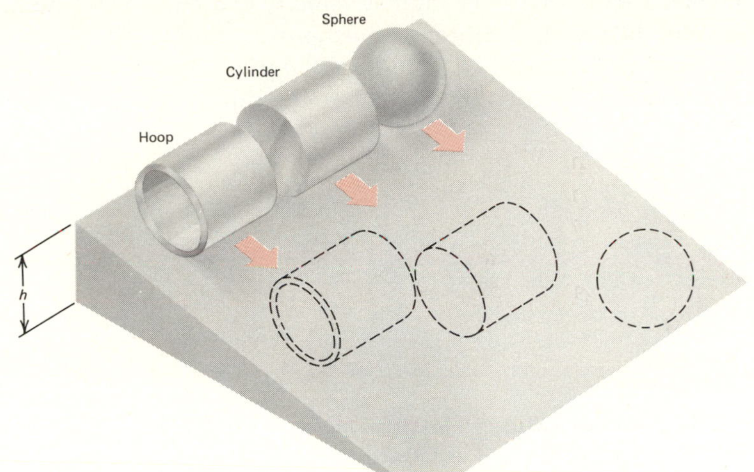

Figure 8.22 *A hollow cylinder, a solid cylinder, and a solid sphere are simultaneously released from rest at the top of the inclined plane. When they reach the bottom of the incline, their linear velocities differ, because the distribution of the total KE between translational and rotational KE's depends on the shape of the objects. The sphere will arrive at the bottom first with the greatest linear velocity, followed by the solid cylinder, and last, the hoop. Note that this conclusion is independent of the mass and radius of the objects.*

All three objects start from rest and descend the same height h. From conservation of energy, we have

PE(at top of incline) = total KE (at bottom of incline)

$$M_h gh = \tfrac{1}{2}M_h v_h^2 + \tfrac{1}{2}I_h\omega_h^2 \quad \textbf{(8.23a)}$$

$$M_c gh = \tfrac{1}{2}M_c v_c^2 + \tfrac{1}{2}I_c\omega_c^2 \quad \textbf{(8.23b)}$$

$$M_s gh = \tfrac{1}{2}M_s v_s^2 + \tfrac{1}{2}I_s\omega_s^2 \quad \textbf{(8.23c)}$$

where the v's and ω's are the translational and angular speeds at the bottom of the incline.

To obtain the translational speeds, we shall replace each ω by the appropriate value of v/R; we also substitute the proper expressions for the moments of inertia according to Table 8.1 (p. 151). Thus,

$$M_h gh = \frac{1}{2} M_h v_h^2 + \frac{1}{2}(M_h R_h^2)\left(\frac{v_h}{R_h}\right)^2 = M_h v_h^2 \quad \textbf{(8.24a)}$$

$$M_c gh = \frac{1}{2} M_c v_c^2 + \frac{1}{2}\left(\frac{1}{2} M_c R_c^2\right)\left(\frac{v_c}{R_c}\right)^2 = \frac{3}{4} M_c v_c^2 \quad \textbf{(8.24b)}$$

$$M_s gh = \frac{1}{2} M_s v_s^2 + \frac{1}{2}\left(\frac{2}{5} M_s R_s^2\right)\left(\frac{v_s}{R_s}\right)^2 = \frac{7}{10} M_s v_s^2 \quad \textbf{(8.24c)}$$

Solving for these speeds, we obtain

$$v_h = \sqrt{gh} \quad \textbf{(8.25a)}$$

$$v_c = 1.155\sqrt{gh} \quad \textbf{(8.25b)}$$

$$v_s = 1.195\sqrt{gh} \quad \textbf{(8.25c)}$$

From Equations (8.25) it follows that the sphere will have the greatest speed, the hollow cylinder the lowest speed at the bottom, and consequently, also the greatest and smallest average speeds, respectively. Therefore, the sphere will arrive first, followed by the solid cylinder, and last by the hollow cylinder. Note that this result is independent of the relative masses and also of the relative radii. ●

● ***Example 8.11*** Two equal masses of 0.1 kg each are located at the ends of a very light rod of 1-m length, which is pivoted on a friction-free bearing at its center as shown in Figure 8.23. A mechanism internal to the rod is capable of moving the masses toward the center along the rod. The system is rotating with an angular velocity of 8 rad/s with the masses at the ends of the rod. If the internal mechanism is now activated and the masses moved so that each is 0.25 m from the pivot, what will the angular speed of the system be then, and what are the initial and final KE's of the system?

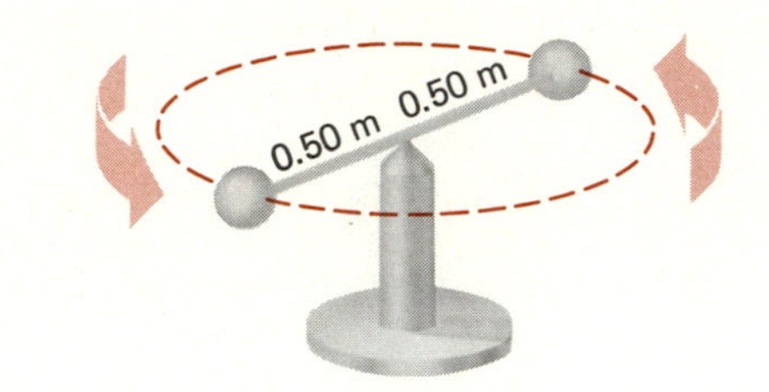

Figure 8.23 *Two masses are attached to the end of a 1-m-long rod that rotates about its midpoint. The masses can be pulled in toward the center by means of an internal mechanism. If they are pulled in, the angular velocity increases, and so does the rotational KE!*

We note that no external torques act on the system. Therefore its angular momentum must be constant:

$$I_1\omega_1 = I_2\omega_2$$

The initial moment of inertia is

$$I_1 = 2(0.1\text{ kg})(0.5\text{ m})^2 = 0.05\text{ kg}\cdot\text{m}^2$$

and the final one

$$I_2 = 2(0.1\text{ kg})(0.25\text{ m})^2 = 0.0125\text{ kg}\cdot\text{m}^2 = \frac{I_1}{4}$$

Since the moment of inertia of the system has been diminished by a factor of 4, the angular velocity must increase by a factor of 4 to conserve angular momentum. Hence, the final angular velocity will be $\omega_2 = 4(8\text{ rad/s}) = 32\text{ rad/s}$.

The initial and final kinetic energies are given by (8.21):

$$\text{KE}_1 = \tfrac{1}{2}(0.05\text{ kg}\cdot\text{m}^2)(8\text{ rad/s})^2 = 1.6\text{ J}$$

$$\text{KE}_2 = \tfrac{1}{2}(0.0125\text{ kg}\cdot\text{m}^2)(32\text{ rad/s})^2 = 6.4\text{ J}$$

Kinetic energy has not been conserved! The final KE is four times as much as the initial. How did this come about? Have we made some subtle error in the calculation, or is energy not conserved? We have applied no torques, the angular momentum has been conserved, and yet if we believe energy conservation, some work appears to have been done on the system. As soon as we think in these terms, the answer to the puzzle is not hard to find. To move the masses closer to the pivot, work had to be done by the inwardly directed centripetal force. This work done *on* the system appeared as an increase in the rotational KE *of* the system. The forces that pull the masses toward the center of the rod pass through the pivot and consequently produce no torque; yet work is done by them.

In passing, we might also note that here is another instance in which energy conservation can give an answer that would be hard to get by a more direct approach. At first one might think that one could also calculate the work done by the centripetal forces by directly evaluating $W = Fs$. The problem with that scheme is that as the masses move toward the pivot and the angular speed increases, so does the centripetal force. In other words, one cannot evaluate W so readily, because the force F changes continuously as the masses are pulled inward. Energy conservation circumvents this difficulty and gives the correct result without any need for knowing the forces. ●

Summary

A stationary body is set in rotation by the application of a torque $\boldsymbol{\tau}$. The magnitude of the torque about an axis produced by a force **F** is given by the product of F and its moment arm relative to the axis of rotation. The moment arm ℓ is the perpendicular distance between the axis and the line of action of the force **F**.

$$\tau = F\ell$$

Torque is a vector quantity. The direction of $\boldsymbol{\tau}$ is that along which a right-handed screw would advance if turned by the torque.

An *extended* body is in equilibrium if the vector sum of the forces *and* the vector sum of the torques that act on it vanish.

In solving equilibrium problems, torques can be calculated about any axis. Generally, the most desirable choice of axis is one through which one or more of the forces is acting.

The equations characterizing rotational dynamics are analogous to equations for translational dynamics; they are summarized in Table 8.2.

Table 8.2 Equations for rotational dynamics

Torque	$\tau = F\ell = Fr \sin\theta$ $\boldsymbol{\tau} = I\boldsymbol{\alpha}$
Moment of inertia	$I = \Sigma m_i r_i^2$
Angular momentum	$\mathbf{L} = I\boldsymbol{\omega}$
Angular dynamics	$\boldsymbol{\tau} = \frac{\Delta \mathbf{L}}{\Delta t}$
Conservation of angular momentum	If $\boldsymbol{\tau} = 0$, $\mathbf{L}$ is constant.
Conditions for equilibrium	$\Sigma \mathbf{F}_i = \mathbf{0}$ and $\Sigma \boldsymbol{\tau}_i = \mathbf{0}$
Rotational KE	$KE_{rot} = \frac{1}{2}I\omega^2$

The rotational analog of $\mathbf{F} = m\mathbf{a}$ is $\boldsymbol{\tau} = I\boldsymbol{\alpha}$, where I is the *moment of inertia* of the body about the axis of rotation. For a point mass rotating about a fixed axis at a radius r, $I = mr^2$. The moment of inertia of an extended object depends not only on the mass distribution within the object but also on the orientation and position of the axis of rotation.

The *angular momentum* **L** of an object rotating with angular velocity $\boldsymbol{\omega}$ is defined by

$$\mathbf{L} = I\boldsymbol{\omega}$$

The change of angular momentum is related to the torque by

$$\langle \boldsymbol{\tau} \rangle = \frac{\Delta \mathbf{L}}{\Delta t}$$

The angular momentum is a vector pointing in the direction of $\boldsymbol{\omega}$. A torque must act on a body to change its angular momentum in magnitude or direction. If $\boldsymbol{\tau} = 0$, the angular momentum remains constant. Hence, by the principle of conservation of angular momentum, *the angular momentum of an isolated system is conserved.*

A rotating body has KE due to its motion even though its CM may be at rest. The rotational KE is given by

$$KE_{rot} = \tfrac{1}{2} I \omega^2$$

The total KE of a body whose motion is a combination of translation and rotation is

$$KE = KE_{tran} + KE_{rot}$$

Multiple Choice Questions

8.1 A uniform pole of mass M is supported by two guy wires, as shown in Figure 8.24. The reaction force of the ground on the pole

(a) depends on the tension in the guy wires but has no horizontal component.

(b) depends on the tension and has a horizontal component that also depends on the coefficient of friction between ground and pole.

(c) has a horizontal component that does not depend on wire tension.

(d) cannot be described by any of the above statements.

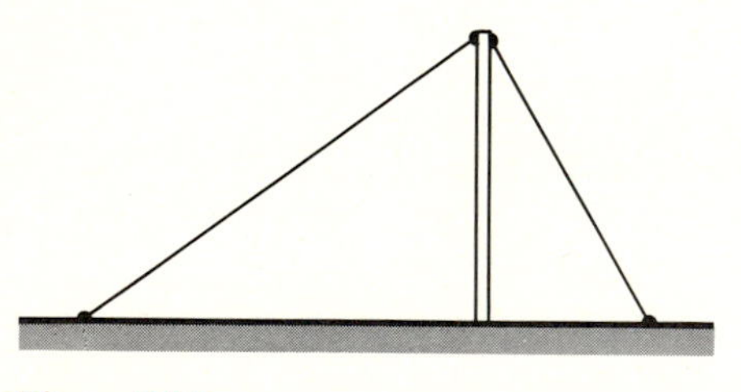

Figure 8.24

8.2 Two identical masses are connected to the horizontal thin (massless) rod of Figure 8.25. When their distance from the pivot is D, a torque τ produces an angular acceleration of α_1. If the masses are now repositioned so that they are $2D$ from the pivot, the same torque will produce an angular acceleration α_2 given by

(a) $\alpha_2 = 4\alpha_1$

(b) $\alpha_2 = \alpha_1$

(c) $\alpha_2 = \frac{1}{2}\alpha_1$

(d) $\alpha_2 = \frac{1}{4}\alpha_1$

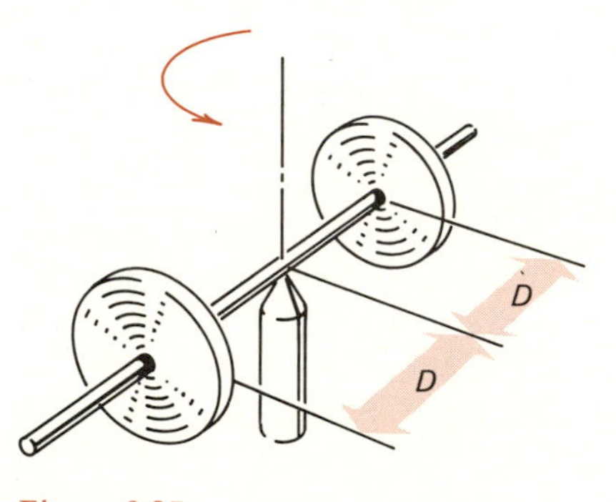

Figure 8.25

8.3 Suppose you sit on a rotating piano stool and hold a 2-kg mass in each outstretched hand. If without moving your arms relative to your body you now drop these masses,

(a) your angular velocity increases.

(b) your angular velocity remains unchanged.

(c) your angular velocity decreases, but your KE increases.

(d) your KE and angular velocity increase.

8.4 A stone is tied to a string and swung in a horizontal circle at constant angular velocity. During the motion,

(a) linear and angular momentum are constant.

(b) linear momentum is constant but angular momentum is changing.

(c) angular momentum is constant but linear momentum is changing.

(d) both linear and angular momentum are changing.

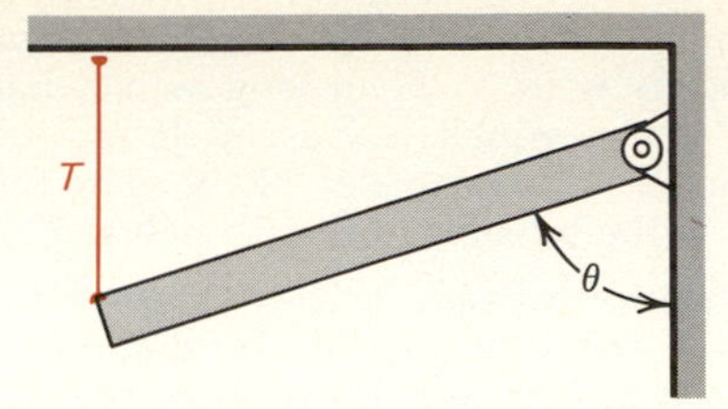

Figure 8.26

8.5 A uniform beam of mass M is hinged at the wall at one end and supported from the ceiling by a vertical string, as shown in Figure 8.26. The reaction force exerted by the hinge on the beam

(a) has a vertical component equal to the tension T and a horizontal component that depends on θ.

(b) has horizontal and vertical components that depend on the angle θ.

(c) has a vertical component that depends on θ and no horizontal component.

(d) has a vertical component equal to T and no horizontal component.

8.6 A uniform cylinder of mass M and radius R and a hollow cylinder of mass M' and the same radius R are released from rest at the same elevation on an inclined plane and roll without slipping down the plane.

(a) The hollow cylinder will reach the bottom first.

(b) The uniform cylinder will reach the bottom first.

(c) The two cylinders will arrive at the same time.

(d) The result of this race will depend on the ratio M/M'.

8.7 A skater is spinning with her arms outstretched. When she brings her arms close to her body,

(a) her angular momentum and angular velocity remain constant.

(b) her angular momentum remains constant.

(c) her angular velocity remains constant.

(d) her angular momentum increases.

(Neglect friction when answering this question.)

8.8 A particle of mass m moves in a circular path of radius r at constant speed v. Its KE is

(a) $(mv^2)/r$

(b) $\frac{1}{2}mv^2$

(c) $\frac{1}{2}I\omega$

(d) $\frac{1}{2}mr\omega^2$

8.9 The dimension of moment of inertia is

(a) $[M][L][T]^2$

(b) $[M][L]^2$

(c) $[M]/[T]^2$

(d) none of the above.

8.10 A block of mass M slides down a frictionless inclined plane while a solid cylinder rolls, without slipping, down a similarly inclined plane. Both start simultaneously from rest at height h.

(a) Both will reach the bottom at the same time.

(b) The block will reach the bottom first.

(c) The block will reach the bottom first only if it has a mass at least as great as that of the cylinder.

(d) The cylinder will reach the bottom first.

8.11 A sphere rolls without slipping down an inclined plane, starting from rest. The ratio of the distance it travels during the second of two equal time intervals to the distance it travels during the first is

(a) determined by the ratio of translational to rotational KE at any given instant.

(b) the same as for a sliding block, namely 3/1.

(c) less than 3/1.

(d) greater the greater the radius of the sphere.

8.12 A uniform solid cylinder, a hollow cylinder, and a uniform solid sphere are rolling on a horizontal table without slipping. They all have the same mass, and the velocities of their CM's are the same. The angular momentum about the CM is greatest for

(a) the solid cylinder.

(b) the hollow cylinder.

(c) the sphere

(d) The answer depends on their relative radii.

8.13 For the three objects of Question 8.12, the object with the greatest KE is

(a) the sphere.

(b) the hollow cylinder.

(c) They all have the same KE.

(d) The answer depends on their relative radii.

8.14 A hoop and a solid cylinder have the same mass. Both roll, without slipping, on a horizontal surface. If their KE's are equal,

(a) the hoop has a greater translational velocity than the cylinder.

(b) the solid cylinder has a greater translational velocity than the hoop.

(c) the translational velocities of the objects are the same.

(d) (a), (b), or (c) could be correct depending on the relative values of the radii.

8.15 A mass M is supported by a massless string that is wound on a uniform cylinder of equal mass M and radius R (Figure 8.27). The system is released from rest. The acceleration of the mass M is

(a) g

(b) $\frac{1}{2}g$

(c) $\frac{2}{3}g$

(d) dependent on the radius of the pulley.

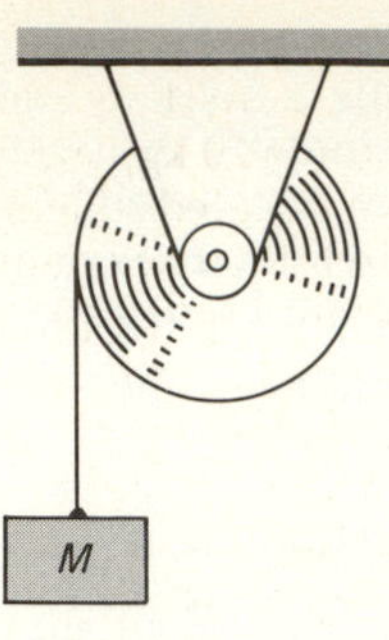

Figure 8.27

Problems

(Unless specifically stated otherwise, neglect friction when solving problems.)

(Section 8.1)

8.1 Two masses, $M_1 = 0.3$ kg and $M_2 = 0.9$ kg, are suspended from the ends of a very thin (effectively massless) aluminum rod whose length is 1.2 m (Figure 8.28). The rod itself is supported by a string from the ceiling. Find the position on the rod at which the string should be attached so that the rod will be horizontal, and find the tension in the supporting string.

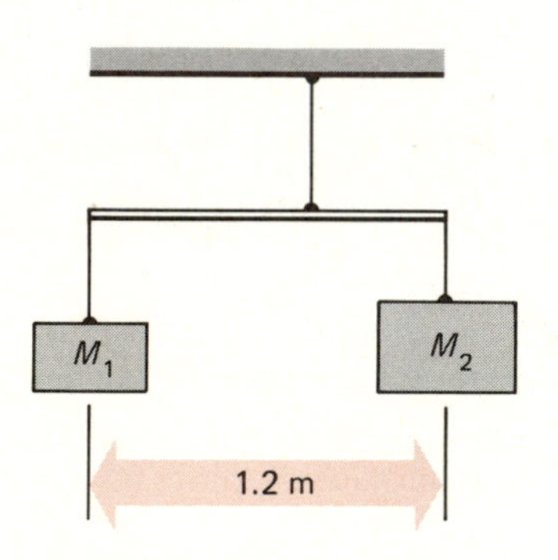

Figure 8.28

8.2 Find the tension in the supporting wires in Figure 8.29.

8.3 A sailor raises the jib using a halyard winch that he turns with a 25-cm-long winch handle. The diameter of the winch is 7.5 cm. If he pulls on the winch handle with a force of 200 N as he tightens the jib, what is the tension in the jib halyard?

• **8.4** A uniform beam of 20-kg mass is hinged to a vertical wall and supported at its other end by a wire, as indicated in Figure 8.30. What is the value of the mass supported from the beam if the tension in the wire is 500 N? What is the reaction force at the hinge that acts on the beam?

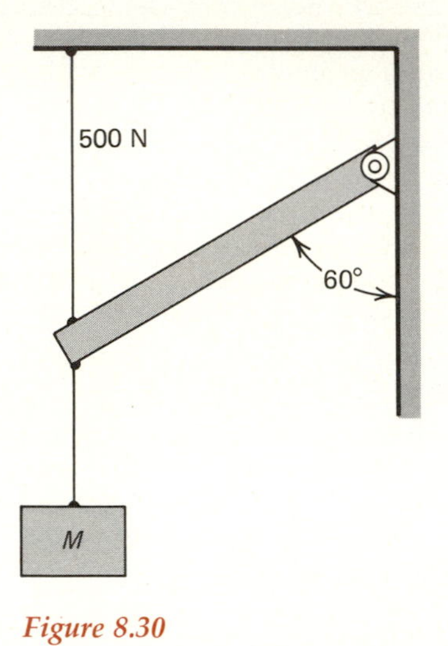

Figure 8.30

• **8.5** An 80-kg mast is maintained vertical by means of guy wires, as shown in Figure 8.31. The tension in the shorter guy wire is 800 N. Find the tension in the other wire and the other forces acting on the mast.

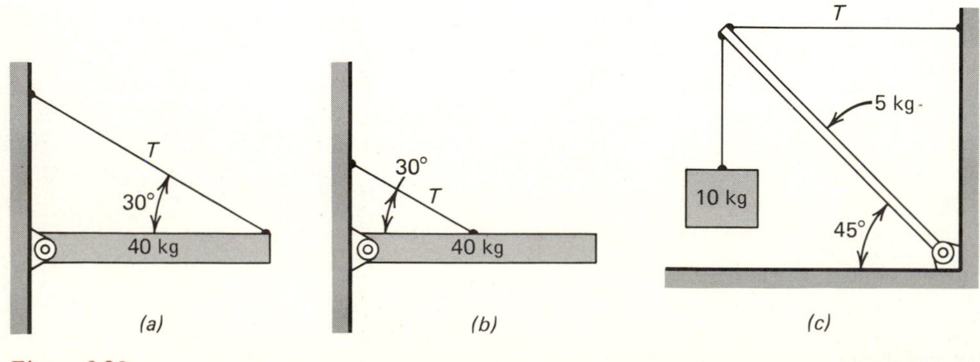

Figure 8.29

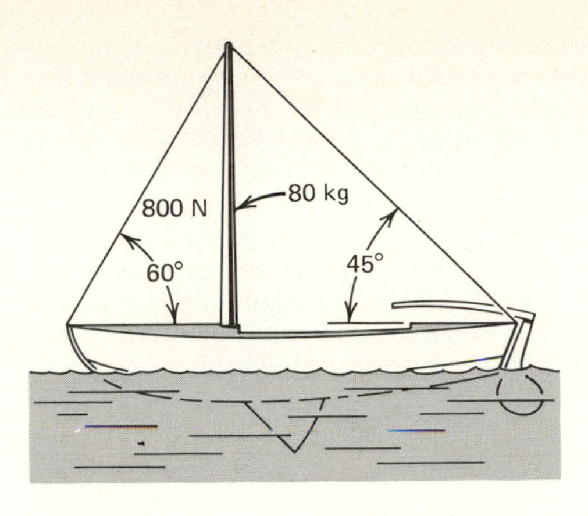

Figure 8.31

• **8.6** A uniform diving board whose mass is 40 kg is held fixed at two points, as shown in Figure 8.32. If a 50-kg diver stands at the edge of the board, what are the forces acting at the points of support?

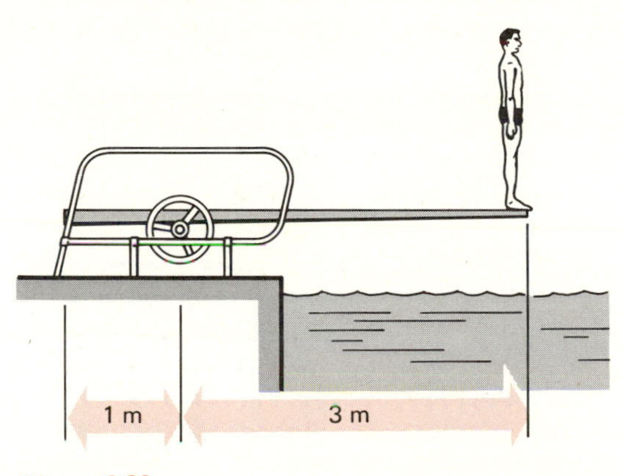

Figure 8.32

• **8.7** A uniform beam of mass 20 kg is hinged to a wall at one end and supported from the ceiling at the other end as shown in Figure 8.33. A 15-kg mass is hung from the beam. Find the tension in the rope supporting the beam.

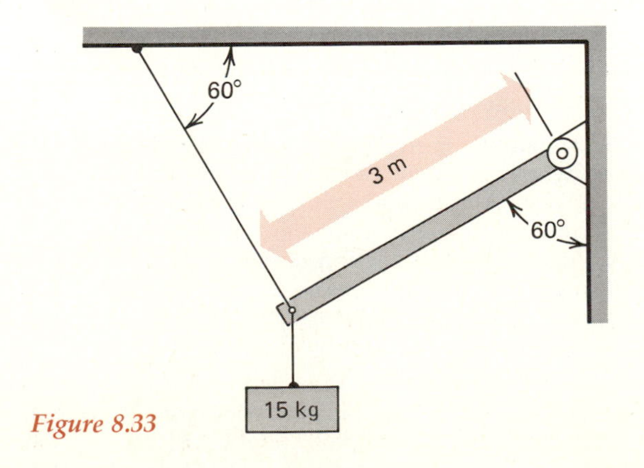

Figure 8.33

• **8.8** A refrigerator is pushed along the floor by applying a horizontal force 1 m above ground. The mass of the refrigerator is 70 kg, and it can be thought of as a uniform parallelepiped of length 60 cm, width 60 cm, and height 2 m. If the coefficient of kinetic friction between refrigerator and floor is 0.25, locate the point at which the normal reaction force of the floor acts on the refrigerator? What force must be applied to keep the refrigerator moving at constant speed?

8.9 A refrigerator like that of Problem 8.8 is moved along the floor and comes up against a small ridge in the floor. The person pushing the unit is unaware of the ridge and increases the force applied 1 m above ground and directed horizontally. How large will the force be when the unit begins to tip?

• **8.10** A refrigerator of the kind described in Problem 8.8 rests on the floor. The coefficients of static and kinetic friction between the refrigerator and the floor are 0.75 and 0.5, respectively. Can this refrigerator be moved by applying a horizontal force 1 m above ground? If not, where must the point of application be so that the unit does not tip over?

• **8.11** For the system shown in Figure 8.34 find the tension T and the reaction force **R** at the wall.

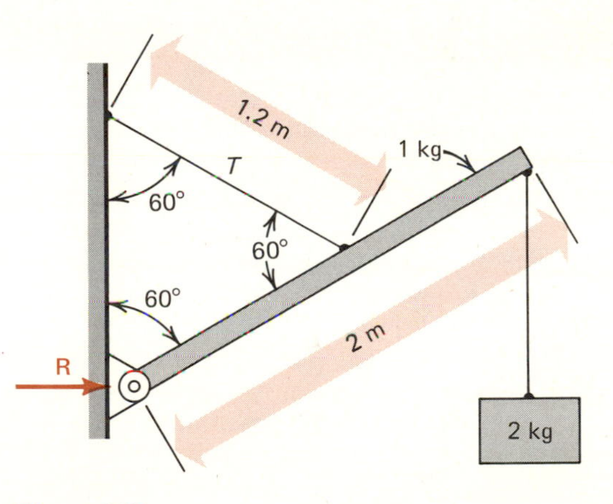

Figure 8.34

• • **8.12** A volleyball net is strung between two poles that are supported by two guy wires each, as shown in Figure 8.35. What should the tension in each of the guy wires be so that the tension in the net is 150 N?

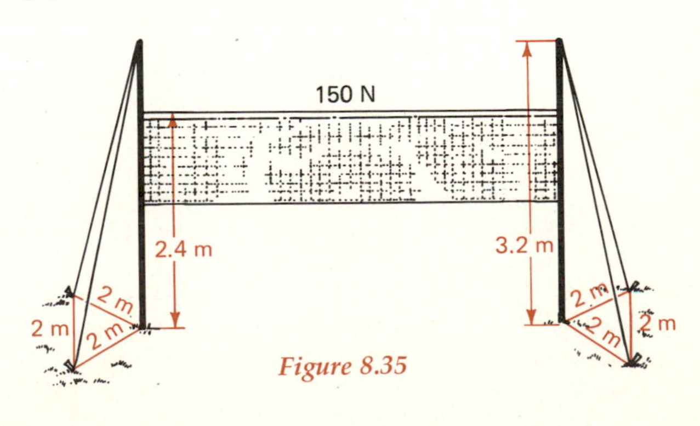

Figure 8.35

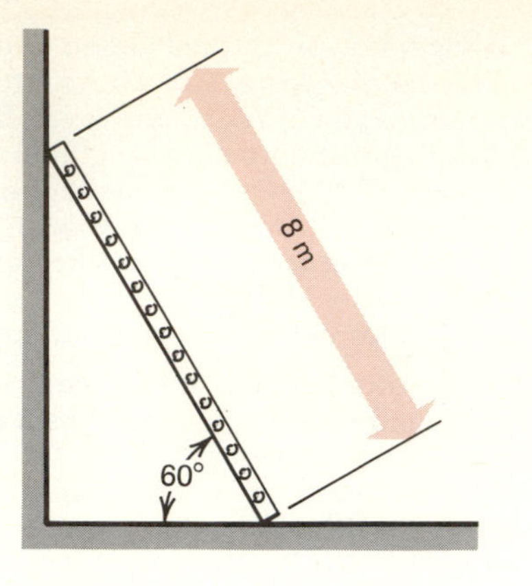

Figure 8.36

• • **8.13** A 10-kg uniform ladder of length 8 m leans against a smooth wall (Figure 8.36). The coefficient of friction between ladder and ground is only 0.3, and the angle the ladder makes with the horizontal is 60°. A 30-kg child climbs this ladder. Can he climb safely to the top? If not, at what height should he stop to avoid serious injury?

• • **8.14** A 10-m-long ladder has a mass of 20 kg. It leans against a smooth vertical wall, with its bottom resting on the ground 3 m from the wall. What must the minimum coefficient of static friction between ladder and ground be so that a 60-kg person can climb safely 80 percent of the way up the ladder?

• • **8.15** If the coefficient of static friction between ladder and ground is 0.27, how far up the ladder can a 50-kg person climb safely? How far up the ladder could an 80-kg person climb safely? Assume that the ladder is the same as in Problem 8.14 and placed in the same way.

• • **8.16** Repeat the example of the man standing on one foot (page 148), under the assumption that he supports one-fifth of his weight on a cane held in his left hand and placed 25 cm to the left of the right foot.

(Sections 8.3, 8.4)

8.17 Four masses of 1.2 kg each are placed at the corners of a square whose sides are 20 cm long. Determine the moment of inertia of this system for rotation about (a) an axis along the diagonal of the square; (b) an axis parallel to a side and passing through the center of the square; (c) an axis passing through the center of the square and perpendicular to its plane.

8.18 Two masses M_1 and M_2 are attached to a massless rod of length L. Locate the position of the CM and then find an expression for the moment of inertia of the system for rotation about the CM.

8.19 Find an expression for the moment of inertia of the system of Problem 8.18 for rotation about an axis perpendicular to the rod L and passing through the mass M_2.

8.20 Two masses, $M_1 = 0.4$ kg and $M_2 = 0.6$ kg, are attached to the ends of a massless rod 0.8 m long. Determine the moment of inertia of the system for rotation about an axis perpendicular to the rod and (a) passing through the CM; (b) passing through the 0.6-kg mass.

• **8.21** A mass of 1.2 kg is attached to a steel rod of 0.6-m length and 0.4-kg mass. Find the moment of inertia of this system for rotation about an axis perpendicular to the rod and passing through the end opposite the 1.2-kg mass.

(Sections 8.5, 8.6)

8.22 Assuming that the effective frictional braking torque acting on the rotor of a centrifuge is 0.6 N·m, and that the rotor has a mass of 1.2 kg and a moment of inertia of 4×10^{-3} kg·m², determine the time required for the rotor to come to rest if it is spinning at 8000 rpm.

8.23 If the centrifuge of Problem 8.22 is to be brought from rest to its operating speed of 8000 rpm in 20 s under conditions of constant angular acceleration, what torque must be applied?

8.24 Determine the torque that must be applied to the turbine wheel of a jet engine, whose moment of inertia is 25 kg·m², so that the wheel will reach an angular velocity of 125 rad/s in 16 s, starting from rest and under conditions of constant angular acceleration.

8.25 A merry-go-round consists of a circular piece of wood about 5 cm thick and 4 m in diameter. If the mass of this disk is 100 kg, what is its moment of inertia?

• **8.26** Four children run along the circumference of the merry-go-round of Problem 8.25. Initially the merry-go-round was rotating at 0.3 rad/s. How hard must each child push in order to cause the merry-go-round to come up to a speed of 1 rad/s within 20 s?

• **8.27** A mass of 2 kg is attached to a string that is wound about a homogeneous pulley of 20-cm radius that is free to rotate about its axis. If the system starts from rest and the pulley has turned through 4 complete revolutions after 2.5 s, what is

(a) the tension in the string?
(b) the angular acceleration of the pulley?
(c) the moment of inertia of the pulley?
(d) the mass of the pulley?

• **8.28** A simple merry-go-round whose diameter is 4 m and whose moment of inertia is 180 kg·m² is initially at rest. A young girl pushes the platform along its circumference with a constant tangential force of 30 N. Assuming that friction is negligible, determine how long it will take her before the merry-go-round is rotating at 0.5 rev/s. How much work will she have done, and how far will she have walked or run alongside the merry-go-round?

• **8.29** Masses of 0.4 kg each are placed at the corners of a regular hexagon (Figure 8.37) whose sides are each

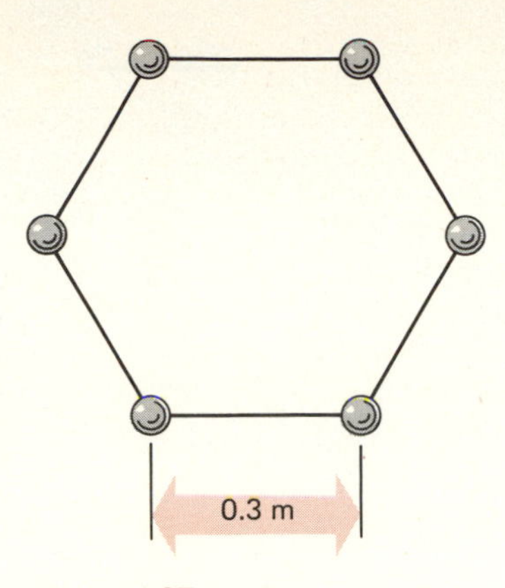

Figure 8.37

0.3 m long. What are the moments of inertia for rotation about an axis through the center of the hexagon and perpendicular to its plane, and about an axis that passes through two opposite vertices?

• **8.30** Figure 8.38 shows a wheel that consists of an 0.8-m-diameter rim of mass 6 kg supported by six spokes made of uniform steel rods, each of mass 1.5 kg. Find the moment of inertia of the wheel about its axis.

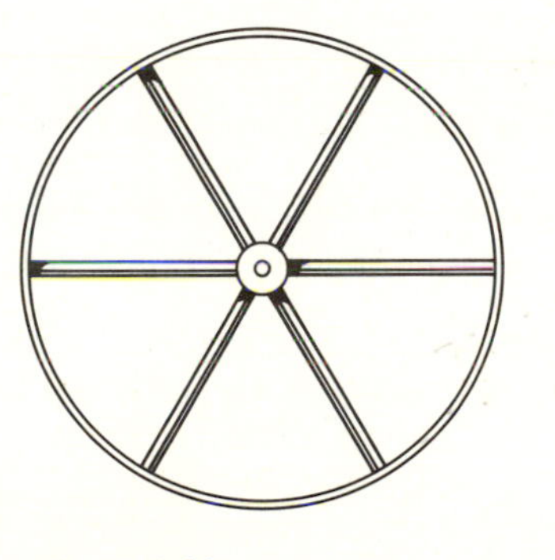

Figure 8.38

• **8.31** A mass of 2 kg is attached to a long string that is wound about a uniform cylinder; this cylinder rotates about a fixed horizontal axis on frictionless bearings. The radius of the cylinder is 0.5 m. When the 2-kg mass is left free to move under gravity, it accelerates at 4 m/s². Find the tension in the string and the mass of the cylinder.

• **8.32** Each of the four children of Problem 8.26 has a mass of 25 kg. When the disk is rotating at 1 rad/s, they all jump on the perimeter of the merry-go-round. If they then make their way to a point 0.75 m from the center of the merry-go-round, what will be the angular velocity of the system? What is the energy of the system when they are at the periphery and when they are nearer to the center of the disk?

• **8.33** A mass of 3 kg is attached to a massless long string that is wound on a drum, as in Figure 8.39. The

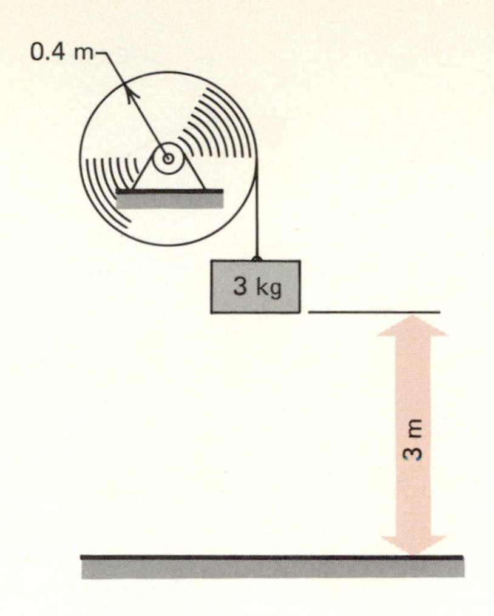

Figure 8.39

drum turns on a fixed horizontal axis on frictionless bearings. The mass starts from rest 3 m above the floor. It strikes the floor with a speed of 4 m/s. Find the moment of inertia of the drum.

• **8.34** If for the system of Figure 8.39 the tension in the string is 10 N, what is the moment of inertia of the drum? What fraction of the total KE of the system is due to the rotation of the drum?

• • **8.35** A boy of 40-kg mass decides to jump on the moving platform of a merry-go-round. The merry-go-round is rotating at 0.6 rev/s and its moment of inertia about its axis of rotation is 150 kg·m². The boy stands alongside the platform and jumps radially inward, landing 1.7 m from the center. Find the angular velocity of the merry-go-round after the boy has jumped aboard, and the kinetic energy of the system before and after the boy has jumped on the platform.

• **8.36** A small bug climbs onto a lazy susan (a circular dish mounted on a vertical axis by frictionless bearings). The dish, 40 cm in diameter, has a mass of 0.3 kg and a moment of inertia of 3×10^{-3} kg·m². Once the bug gets to the rim of the dish, it starts to crawl at a speed of 1 cm/s clockwise (as viewed from above the dish). If the mass of the bug is 10 g, what is the angular velocity of the dish if it was initially at rest?

• **8.37** Suppose the young girl of Problem 8.28 has a mass of 35 kg. Just as the merry-go-round reaches an angular speed of 0.5 rev/s she swings herself onto its edge, having run alongside it. She then walks radially from the edge to a point 0.5 m from the center of the disk. What are the angular velocity and the KE of the system after the girl has swung onto the edge and after she has moved toward the center of the merry-go-round?

• • **8.38** According to Kepler's second law, the orbits of planets about the sun are such that the radius vector sweeps out equal areas in equal intervals of time. Figure 8.40 shows an eccentric elliptical orbit, and the planet near *perihelion* (point nearest the sun) and *aphelion* (point farthest from the sun). By comparing the

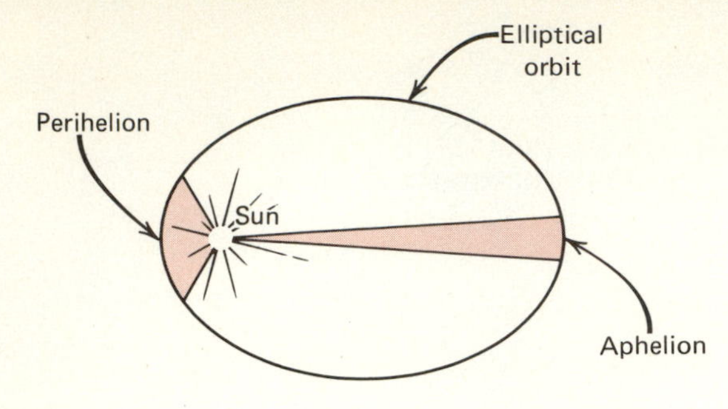

Figure 8.40

areas of the two shaded isosceles triangles, show that conservation of angular momentum about the sun demands that the areas of the two triangles be equal.

• • **8.39** A young girl whose mass is 35 kg stands at the edge of a merry-go-round that is rotating at 1 rad/s. The merry-go-round has a mass of 100 kg, a diameter of 4 m. The girl jumps off the merry-go-round, landing on the ground with zero horizontal velocity relative to the ground. After the girl has jumped off, the merry-go-round rotates at a speed of 1.75 rad/s. Find the moment of inertia of the merry-go-round, and the kinetic energy of the system (including the girl), before and after the girl has jumped from the platform.

• **8.40** Normally when brakes are applied to a car, the vehicle's kinetic energy is converted to heat as a result of friction between the brake lining and brake drum. A German manufacturer has developed a bus that converts the kinetic energy of translation of the bus into kinetic energy of rotation of two flywheels as the bus is made to slow down. This stored energy can then be used to accelerate the vehicle after it has discharged and taken passengers aboard. If the mass of the bus, including passengers, is 4000 kg and its cruising speed is 30 km/h, what would the maximum speed of rotation of the two flywheels be if each has a moment of inertia of 300 kg·m^2? Why are two flywheels used instead of one?

• • **8.41** A flywheel has a mass of 80 kg, a moment of inertia of 50 kg · m,^{2} and a diameter of 1.8 m. It can be stopped by pressing two brake shoes against its periphery. The maximum normal force applied to each brake shoe is 250 N, and the coefficient of friction between flywheel and brake shoe is 0.6. If the wheel is rotating at 12 rev/s when the brake is applied, how long will it take before the wheel comes to rest, and how much energy must be dissipated in heat during the process?

• • **8.42** A marble of uniform density rolls down a loop-the-loop track like that of Problem 6.29. If the diameter of the loop is 20 cm and the marble rolls without slipping and starts from rest, from what altitude should it be released so that it never loses contact with the track? Compare the result with that obtained for Problem 6.29 and account for any difference.

• • **8.43** A long rod of uniform construction is pivoted and supported at one end. Its lower end is now pulled to one side, so that the rod makes an angle of 60° with the vertical, and released from rest. If the length of the rod is 1.8 m, what is its angular velocity at the bottom of its swing? Contrast this result with the one for an ordinary pendulum having a massive bob supported on a string 1.8 m long.

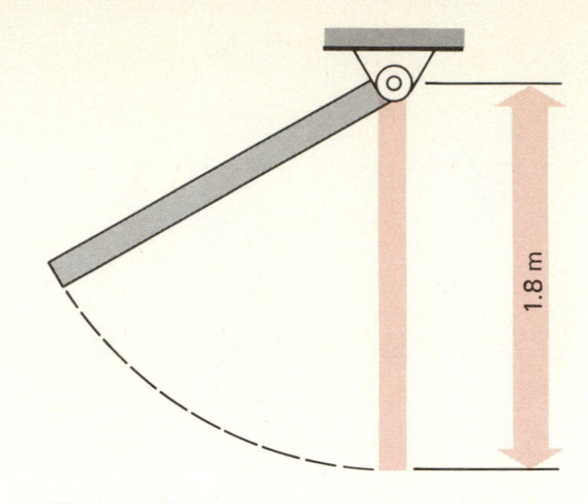

Figure 8.41

• • **8.44** A mass of 6 kg and a mass of 4 kg are tied together by a light string that is looped over a pulley of 0.4-m diameter. At $t = 0$, the 6-kg mass has a speed of 2 m/s and is moving upward. It reaches a maximum height 2 m above its starting point. Find the moment of inertia of the pulley, the tension in the string above the 4-kg mass, and the instant when the system is momentarily at rest.

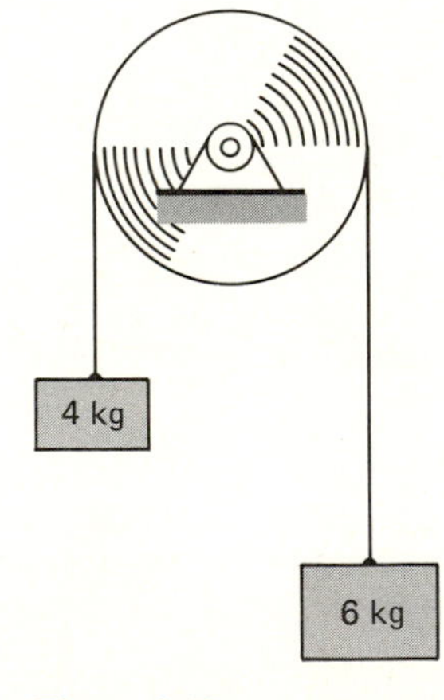

Figure 8.42

• • **8.45** A mass of 3 kg is supported by a massless string wound on a pulley whose diameter is 0.8 m and whose moment of inertia is 0.4 kg·m^2. If initially the system is at rest, what are the tension in the string, the velocity of the 3-kg mass, and the total kinetic energy of the system when the 3-kg mass has reached a point just 2 m below its starting point? How much time has elapsed between the start and the instant the 3-kg mass passes this point?

·9·

Mechanical Properties of Matter

It often matters vastly with what others,
In what arrangements the primordial germs
Are bound together, and what motions, too,
They exchange among themselves,
for these same atoms
Do put together sky, and sea, and lands,
Rivers and suns, grains, trees and
breathing things.
But yet they are commixed in different ways
With different things, with motions each its own.

LUCRETIUS (55 B.C.)

9.1 Introduction

Heretofore we focused on the *motion* of particles and extended rigid bodies. We did not inquire into the internal structure of objects nor into the deformations of bodies due to forces acting on them.

Yet, here lies one of the oldest challenges to man's understanding of nature. Why is water a liquid at atmospheric pressure between 0 °C and 100 °C? Why is it transparent to visible but relatively opaque to infrared radiation? Why and how does water change from liquid to solid at 0 °C and to vapor at high temperature? Why are metals like copper and aluminum good conductors, and glass, quartz, and sapphire electrical insulators?

There are literally hundreds of questions of this sort that one can, and should, ask. You probably know the answers to some already. Others will be answered in later chapters. Many questions we shall not pose because finding answers would take us too far afield. Lastly, there are the *really* important questions, the ones we should ask but don't because we don't know what they are! Here is the real challenge. To pose the right question is the first, essential step toward a new discovery or theory.

In this and the next few chapters we examine some properties, such as elasticity, surface tension, and thermal expansion, that depend critically on internal structure. Atoms are the basic building blocks of all matter. They can combine to form molecules, whose properties are generally very different from those of the constituent atoms. Table salt, for example, a simple chemical compound formed from chlorine and sodium, resembles neither the poisonous gas nor the highly reactive metal. A molecule is the smallest entity of a substance that displays the same *chemical* properties as the substance itself. The emphasis here is on chemical, not optical, mechanical, or electrical. The latter properties usually change quite drastically with change of state.

All substances can exist in the *solid, liquid,* and *gaseous state.* Often the term *phase* is used instead of state, and evaporation and melting are called *phase changes,* or *phase transitions.* But there are other phase transitions also. For example, at exactly 770 °C iron loses its unusual magnetic properties, and above that temperature, it responds magnetically like most other metals, such as zinc, tungsten, and aluminum. Another interesting phase transition is that of lead at quite low temperatures; its electrical resistance abruptly vanishes completely; it becomes a *superconductor.*

9.2 Solids, Liquids, and Gases

A solid is any substance that responds elastically to a shear stress.

A deformation is said to be *elastic* if (1) the deformation is proportional to the force causing it; and (2) on removal of that force, the deformation vanishes.

We shall define shear stress shortly. For the moment, you may think of a shear as a twist. *Fluids*—the generic name that includes liquids and gases—*flow when subjected to a shear stress.* They do not deform elastically.

The trouble with these definitions is that some materials such as glass and plastics flow at elevated temperatures and become gradually more viscous as the temperature is lowered; their mechanical response eventually becomes indistinguishable from that of more conventional solids like iron or table salt. The transition from liquid to solid is not sharply defined for glass.

An alternative definition of a solid, based on its microscopic structure, is:

A solid is any substance wherein the average position of the constituent atoms forms a specific lattice structure.

This definition is sometimes preferred because it avoids the ambiguity just mentioned. It too is imperfect, however, because there are materials, so-called liquid crystals, that do have well-defined crystal structures and also flow like liquids if sheared along certain crystallographic planes.

Models of a few lattices are shown in Figure 9.1. It is important to remember that atoms in a crystal are not immobile but vibrate about their equilibrium positions with amplitudes that depend on temperature. It is only the *average* position that is uniquely characterized by the crystal structure.

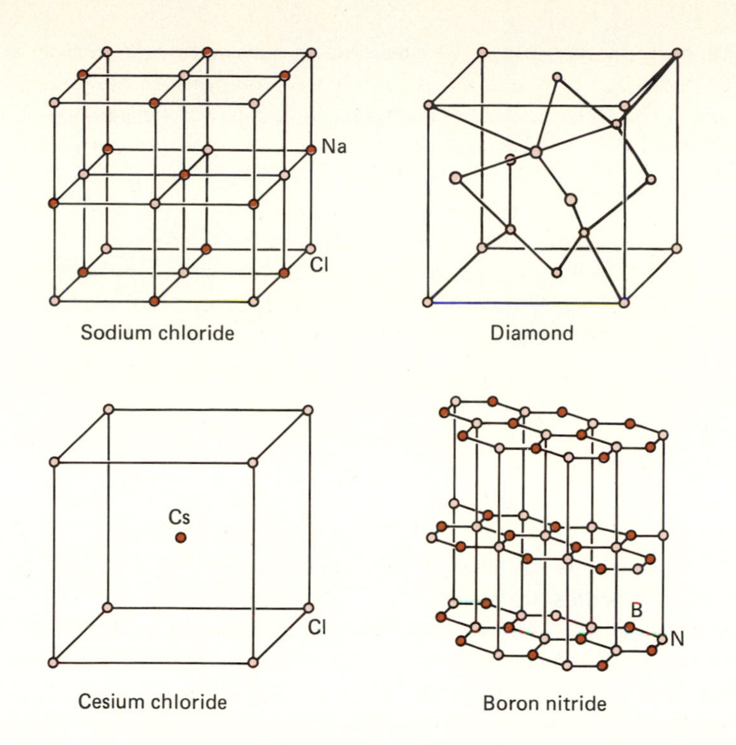

Figure 9.1 Crystal structures of some common crystalline solids. The structures of NaCl (table salt) and CsCl are such that each negatively charged chloride ion is immediately surrounded by positively charged alkali metal ions, and vice versa. The crystal structure of diamond, which is also that of germanium and silicon, is consistent with the valence of four of these atoms (C, Ge, Si). Each carbon atom has four nearest neighbors with which it forms chemical bonds, as indicated.

In fluids—liquids or gases—the positions of molecules are randomly arranged. Only the average number density, that is, the average number of molecules per unit volume, is well defined. As in solids, the atoms or molecules are in continual motion.

We normally regard liquids and gases as quite distinct states of matter. Generally, we have no difficulty deciding whether a substance is one or the other. The change of phase takes place at some specific temperature and pressure. However, if a gas is highly compressed and its temperature properly adjusted, one reaches a condition, the critical point, at which the distinction between the two phases disappears.

From this observation and also from the generic term *fluid,* it is apparent that in some respects at least, the liquid and gaseous phases are more closely allied to each other than to the solid phase. What are these similarities?

First, some mechanical properties are qualitatively similar although there are obvious and important quantitative differences. Liquids and gases are fluids; they flow. Solids do not.

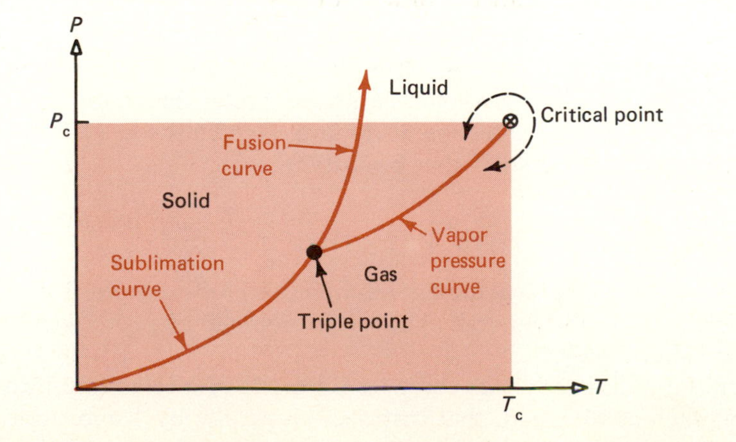

Figure 9.2 A typical phase diagram. The lines are the demarcations between the solid, liquid, and gaseous phases. Note that above the critical point, at temperature T_c and pressure P_c, it is impossible to make a clear distinction between these two phases. Note also the existence of the triple point, at which the solid, liquid, and gaseous phases coexist. Triple points are used in precise calibrations of thermometers.

Table 9.1 Mass densities and number densities of a few substances at 0 °C and 1-atm pressure

Substance	*State*	*Mass density ρ, kg/m³*	*Number density, molecules/m³*
Air	Gas	1.3	2.7×10^{25}
Helium	Gas	0.18	2.7×10^{25}
Hydrogen	Gas	0.09	2.7×10^{25}
Oxygen	Gas	1.43	2.7×10^{25}
Carbon dioxide	Gas	2.0	2.7×10^{25}
Mercury	Liquid	1.4×10^{4}	4.2×10^{28}
Water	Liquid	1.0×10^{3}	3.3×10^{28}
Ice	Solid	9.2×10^{2}	3.1×10^{28}
Aluminum	Solid	2.7×10^{3}	6.0×10^{28}

Second, there are similarities also on the molecular scale. In liquids and gases, molecules are arranged at random and can move about within the substance with relative ease. This is evidenced by the fact that a drop of red food dye placed in a glass of water after a while colors the entire liquid uniformly even without stirring; a bit of hydrogen sulfide released in a corner quickly asserts its noxious odor everywhere in a room. This mixing process, called *diffusion,* is virtually nonexistent in solids at temperatures well below the melting point.

So much for some of the similarities. Wherein do liquids and gases differ? The principal difference is in the particle density. As indicated in Table 9.1, the mass and number densities of a gas at atmospheric pressure and normal temperature are typically a thousand times smaller than in liquids or solids. In liquids and solids the atoms or molecules are relatively close together; in gases they are, on the average, very far apart. This great difference in the number density has some immediate consequences. Though both resist compression, gases are *far* more compressible than liquids. Also, the density change on melting is generally only a few percent; most substances contract as they solidify, though a few, notably water, expand, but the amount of contraction or expansion is relatively small. By contrast, the transition from the liquid to the gaseous phase involves usually a very great change in density.

The second important difference between gases and liquids is closely related to the comparatively high density of liquids. Whereas gases expand to fill the volume in which they are contained, the mutual attractive forces acting on the molecules of a liquid prevent a corresponding space-filling expansion. In the gravitational field of the earth, liquid poured into a container fills it from the bottom up. Without gravitation, however, liquids assume a spherical shape. That spherical shape is a manifestation of the *surface tension* at the interface between a liquid and the gas, other liquid, or solid that bounds it.

9.3 Density and Elastic Properties of Solids

When selecting an object for a given function, for example, a thin wire to support a mass, one generally knows beforehand what properties the object should display.

One might specify that the wire be made of a relatively impervious metal, such as tungsten, that it stretch elastically by 1 mm under a tension of 50 N, and that it be capable of withstanding a tension of 300 N. One could then proceed to test a wide selection of tungsten wires until one is found that meets the stated specifications. In fact, there is an infinite number of tungsten wires that meet the requirements.

The empirical trial-and-error procedure is not necessary; to decide on a wire of proper dimensions, one needs to know only one critical intrinsic parameter characteristic of any and all samples of tungsten, whatever their shape and dimensions. It is therefore convenient to identify and define this and other *intrinsic material properties* that can be used to predict the mechanical response of any sample made of a given material.

9.3 (a) Density

Density is one of these intrinsic material properties—the one most widely used in everyday life and most easily understood. We can often guess whether a piece of automobile trim is made of metallized plastic or steel by just holding it, estimating its weight, and contrasting this with its apparent size. We are then, in effect, estimating its density—to be precise, its weight density.

The *mass density* ρ of a substance is defined as the *mass per unit volume*. That is,

$$\rho = \frac{m}{V} \tag{9.1}$$

The mass of a given object is therefore given by

$$m = \rho V \tag{9.2}$$

and its weight is

$$W = mg = \rho g V$$

where V is the volume of the object. The SI unit of mass density is kilogram per cubic meter; its dimension is $[M]/[L]^3$.

The mass densities of several common solids and liquids are given in Table 9.2. Note that the values of ρ apply at a specific temperature. Most

Table 9.2 *Densities of some common substances at 0 °C and 1-atm pressure*

Solids	Density, kg/m^3	Liquids	Density, kg/m^3	Gases	Density kg/m^3
Copper	8.93×10^3	Ethanol	0.79×10^3	Air	1.3
Glass	$(2.4–6.0) \times 10^3$	Ether	0.74×10^3	Ammonia	0.77
Gold	19.3×10^3	Glycerol	1.26×10^3	Carbon dioxide	2.0
Lead	11.3×10^3	Oil	$(0.8–0.95) \times 10^3$	Helium	0.18
Nickel	8.8×10^3	Water	1.0×10^3	Hydrogen	0.09
Platinum	21.4×10^3	Mercury	1.4×10^4	Oxygen	1.43
Silver	10.5×10^3				
Tungsten	19.3×10^3				
Uranium	18.7×10^3				
Wood	$(0.25–1.0) \times 10^3$				
Zinc	6.9×10^3				
Iron (steel)	7.8×10^3				
Aluminum	2.7×10^3				
Ice	0.92×10^3				

substances expand with increasing temperature, and consequently their densities decrease as the temperature is raised. There are a few exceptions to this rule; the most notable is water, which contracts with increasing temperature in the range from 0 °C to 4 °C.

One cubic meter is a rather large volume, about 264 gallons. A more convenient unit of density than the kilogram per cubic meter is the gram per cubic centimeter. From the definition of the gram and centimeter it follows that

$$1\ \mathrm{g/cm^3} = 1000\ \mathrm{kg/m^3} \tag{9.3}$$

In this larger unit, the density of water at 4 °C is 1 $\mathrm{g/cm^3}$.

The *specific gravity* of a substance is defined as the ratio of its density to the density of water at 4 °C. Specific gravity is therefore a dimensionless number; its value equals the mass density expressed in grams per cubic centimeter.

Example 9.1 What is the length of a cube of aluminum whose weight equals that of a cubic centimeter of gold?

We require that

$$\rho_{\mathrm{Al}}(L_{\mathrm{Al}})^3 = \rho_{\mathrm{Au}}(1)^3$$

Hence,

$$L_{\mathrm{Al}} = \sqrt[3]{\frac{\rho_{\mathrm{Au}}}{\rho_{\mathrm{Al}}}} = \sqrt[3]{\frac{19.3}{2.7}} = 1.93\ \mathrm{cm}$$

9.3 (b) Elastic Deformation and Elastic Moduli

If a rod or wire of length L_0 is subjected to tension, its length increases by an amount ΔL. This length change depends not only on the substance of which the sample is made and on the tensile force, but also on the sample length L_0 and its cross section A. It is convenient, in discussing elastic deformation, to introduce two new terms, *stress* and *strain*. We define the tensile stress by

$$\text{Tensile stress} = \frac{\text{tensile force}}{\text{area}} = \frac{F}{A} \tag{9.4}$$

and the tensile strain by

$$\text{Tensile strain} = \frac{\text{elongation}}{\text{unstressed length}} = \frac{\Delta L}{L_0} \tag{9.5}$$

The unit of tensile stress is newton per square meter and the dimension is $[M]/[L][T]^2$. Tensile strain is a dimensionless number, the ratio of two lengths.

For any given material, whatever the sample shape, one finds that if the deformation is elastic,

$$\text{Stress} = \text{constant} \times \text{Strain} \tag{9.6}$$

The constant in Equation (9.6) is called an *elastic modulus*, and the relation is referred to as *Hooke's law*. The usefulness of Equation (9.6) arises from the fact that the elastic modulus depends on the type of deformation and on the substance, not on the geometry of the sample. That is, elastic moduli are intrinsic material properties.

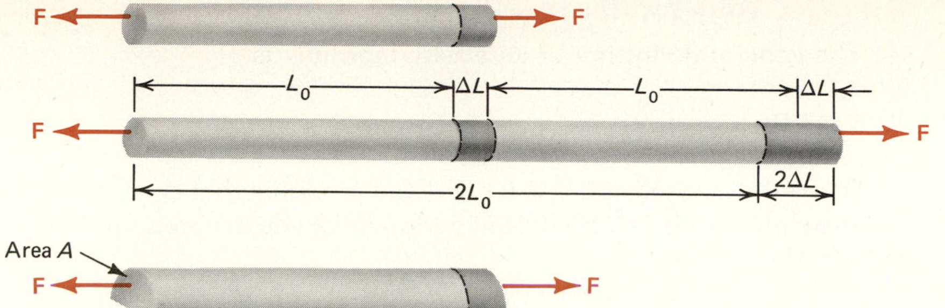

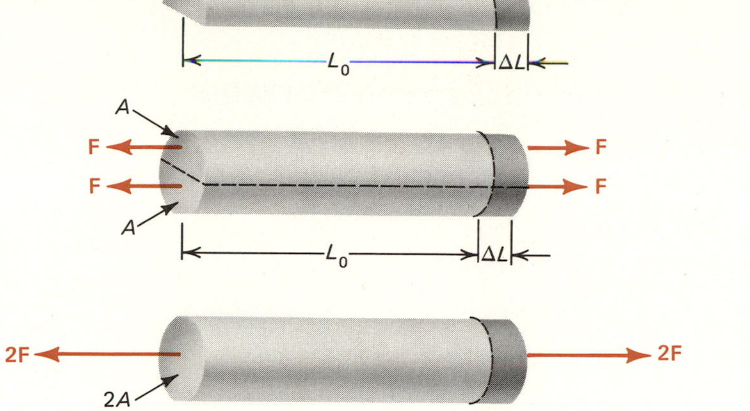

Figure 9.3 The extension of a rod depends on the length of the rod and on the tensile force. For a rod of a given material, however, the fractional change in length is the same if the tensile stress, the tension per unit area, is the same.

Table 9.3 Elastic moduli of some substances. These are typical values; there is much variation among different samples

Substance	*Young's modulus,* N/m^2	*Bulk modulus,* N/m^2	*Shear modulus,* N/m^2
Aluminum	7.0×10^{10}	7.5×10^{10}	2.7×10^{10}
Gold	8.0×10^{10}	16.5×10^{10}	2.8×10^{10}
Copper	12.0×10^{10}	14.0×10^{10}	4.0×10^{10}
Steel	20×10^{10}	17×10^{10}	8×10^{10}
Glass	7×10^{10}	5×10^{10}	3×10^{10}

Figure 9.4 Typical stress-strain curve for a ductile material. The elastic limit is usually reached when the strain is of order 10^{-3}. Some materials can sustain considerable plastic deformation before fracture; 10% strain before fracture is not uncommon for ductile metals. For brittle materials, the fracture point and elastic limit are very close.

The linear relation between stress and strain holds over a limited range only. If the stress exceeds the elastic limit, the material deforms *plastically*, and if the stress is increased further, *fracture* occurs. Plastic deformation means that the sample does not assume its initial shape when the applied stress is removed.

9.3 (c) Young's, Shear, and Bulk Modulus

The general definition of an elastic modulus is

$$\text{Modulus} = \frac{\text{stress}}{\text{strain}} \tag{9.7}$$

When the applied stress is tensile, the modulus that characterizes the strain (elongation) is called *Young's modulus*, designated by the symbol Y.* Thus

$$Y = \frac{F/A}{\Delta L/L_0} \tag{9.8}$$

where F/A is the tensile stress. Young's moduli for several materials are listed in Table 9.3.

● ***Example 9.2*** An 8-kg mass is supported by a steel wire whose diameter is 0.8 mm and whose length is 1.5 m. How much will this wire stretch under the load?

From Equation (9.8), we have

$$\Delta L = \frac{L_0(F/A)}{Y}$$

Young's modulus for steel is 2×10^{11} N/m². The cross-sectional area of the wire is $\pi r^2 = \pi(4 \times 10^{-4}\ \text{m})^2 = 50.3 \times 10^{-8}\ \text{m}^2$. The force of tension is $mg = (8\ \text{kg})(9.8\ \text{m/s}^2) = 78.4$ N. Inserting these values, and $L_0 =$ 1.5 m, into the above expression gives

$$\Delta L = \frac{(1.5\ \text{m})(78.4\ \text{N})}{(50.3 \times 10^{-8}\ \text{m}^2)(2 \times 10^9\ \text{N/m}^2)} = 1.17\ \text{mm}$$ ●

Suppose we take an object of, say, cubic shape and apply equal but oppositely directed forces **F** that are perpendicular to each pair of opposite sides. The stress is then a compression

$$P = \frac{F}{A} \tag{9.9}$$

commonly called *pressure*. Here A is the area over which the force **F** is applied. The SI unit of pressure is the *pascal* (Pa), which is equal to 1 N/m². Under this uniform pressure, the object will contract, and its fractional change in volume, $\Delta V/V_0$, is the *compressional strain*. The corresponding elastic modulus is called the *bulk modulus* B and is defined by

$$B = -\frac{P}{\Delta V/V_0} \tag{9.10}$$

The negative sign is used in the defining relation so that B will be a positive quantity; an increase in pressure results in a decrease in volume, i.e., ΔV is negative when P is positive.

If you place a book on a rough table and, with your hand flat on the top of the book, push as indicated in Figure 9.6, the book will deform as indicated, static friction between the bottom of the book and the table providing the reaction force that prevents lateral motion. The pair of opposing forces, $\mathbf{F}_s$ and $\mathbf{f}_s$, produce a *shear stress* whose value is

* Thomas Young (1773–1829) was a British scientist who made many important contributions to mechanics, acoustics, and optics.

Figure 9.5 *Uniform pressure applied to an object causes a decrease of its volume.*

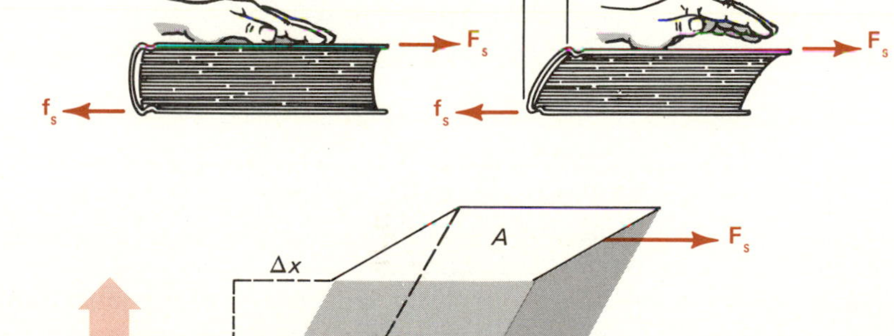

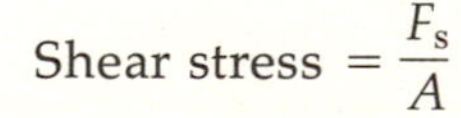

Figure 9.6 *An example of a shear stress. Note that a shear stress involves the application of a force parallel to the surface over which it acts. The resulting shear strain is defined as the ratio of the displacement Δx to the separation L between the two planes over which the shear forces act.*

$$\text{Shear stress} = \frac{F_s}{A} \qquad \textbf{(9.11)}$$

where A is the area over which the shear force $\mathbf{F}_s$ acts. Note that in contrast to tensile and compressional stresses, a shear is produced by a force acting *in the plane of the area A*. The *shear strain* is defined by the ratio $\Delta x/L$, where now L is the separation between the two areas over which the two opposing tangential forces are acting. The *shear modulus S* is defined by

$$S = \frac{F_s/A}{\Delta x/L} = \frac{F_s/A}{\tan \phi} \qquad \textbf{(9.12)}$$

● ***Example 9.3*** A cylindrical column of 0.05-m² cross section is made of disks stacked as shown in Figure 9.7. The planes of the disks are at a bias, making an angle of 60° with the axis of the cylinder. What is the stress

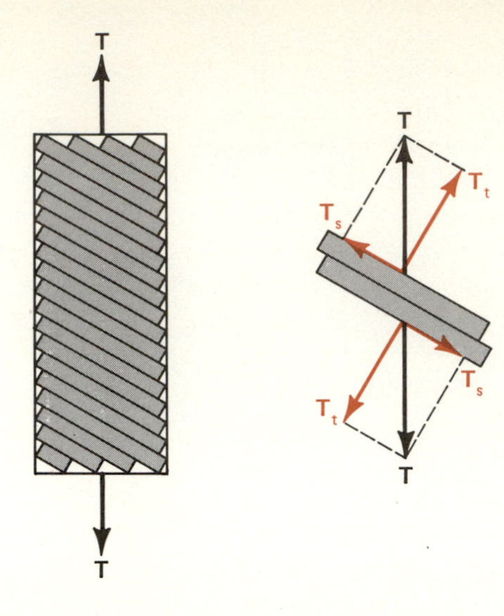

Figure 9.7 A tensile force T acting on a column composed of disks stacked on a bias. This tension can be resolved into two components, perpendicular and parallel to the plane of the disks; they correspond to tension and shear forces acting on the disks.

that acts between adjacent planes of disks if the cylinder is subjected to a tension along its axis of 60 N?

The tension of 60 N can be decomposed into two components, perpendicular and tangential to the planes of the disks. The perpendicular component provides a tension tending to separate the disks, and its value is (60 N) sin 60° = 52 N. The shear component of the applied tension is (60 N) cos 60° = 30 N. Both the tensile and the shear components act over an area of $(0.05\ \text{m}^2)/\sin 60° = 0.0577\ \text{m}^2$. Hence the tensile and shear stresses acting on the stacked disks are 900 N/m^2 and 520 N/m^2, respectively. ●

9.4 Surface Tension

Examine a slowly dripping faucet carefully. You will note that as each drop forms, it first grows into a roughly hemispherical shape, balloons down as it gathers more liquid, but clings tenaciously to the faucet until it finally breaks away under the pull of gravity. As it does so, the top surface pulls taut, reshaping the elongated drop so that it becomes nearly spherical.

You can take a razor blade or sewing needle and float these on a dish of water even though steel has a much greater density than water. Look at any pond on a summer day, and you will probably see numerous in-

(*a*)

(*b*)

Figure 9.8 (*a*) A spider walks on the surface of a pond, depressing the surface of the water as you might depress the surface of a trampoline. (*b*) A drop of milk forms.

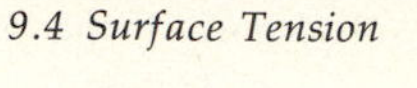

sects walking across its surface as you might walk across a huge trampoline.

These and many more phenomena result from *surface tension.* Before discussing the technique for measuring surface tension, let us first see how we can understand and explain it qualitatively.

Gases condense to liquids because individual molecules exert attractive, *cohesive* forces on each other. A molecule in the interior of a drop of liquid is surrounded on all sides by other molecules and experiences attractive forces from all sides, which cancel vectorially. If we tried to remove this single molecule, we would have to overcome these cohesive forces, break all these bonds, a process requiring a significant energy expenditure.

Removing a molecule from the surface layer, however, requires much less energy, because this molecule has only about half as many neighbors as one in the interior, and only about half as many cohesive bonds need be broken. Put another way, a molecule in the interior sits in a fairly deep potential energy trough; the potential energy trough of a molecule at the surface is more shallow. Consequently, a given mass of liquid, in the absence of external forces, will attain its lowest energy by maximizing the volume-to-surface ratio. Hence, in free fall in vacuum or in a gravity-free environment, liquids assume spherical shape.

(a)

(b)

Figure 9.9 *(a) Cohesive forces acting on a molecule in the interior of a drop of liquid cancel vectorially. Cohesive forces acting on a molecule at the surface have a resultant that points to the interior. (b) Small drops of water on blades of grass. Even in the gravitational field of the earth, a small mass of water will tend to assume a spherical shape.*

On the earth, the gravitational force is so strong that large amounts of liquid seek the lowest gravitational potential energy consistent with the shape of the container, presenting a flat horizontal surface. Small amounts of liquid, however, often maintain almost spherical shape even on earth.

One can quantify these qualitative arguments in terms of the difference in cohesive potential energy between molecules on the surface and in the interior. To enlarge the surface area of a given volume of liquid, for example, to change its shape from a sphere to an oblong ellipsoid, one must bring molecules from the interior to the surface layer. In the interior no *net* force acts on the molecule, but as it approaches the surface the attractive forces pulling it back dominate; therefore, work must be done on the molecule to bring it to the surface. This work is just the difference between the potential energy of cohesion at the surface and in the interior.

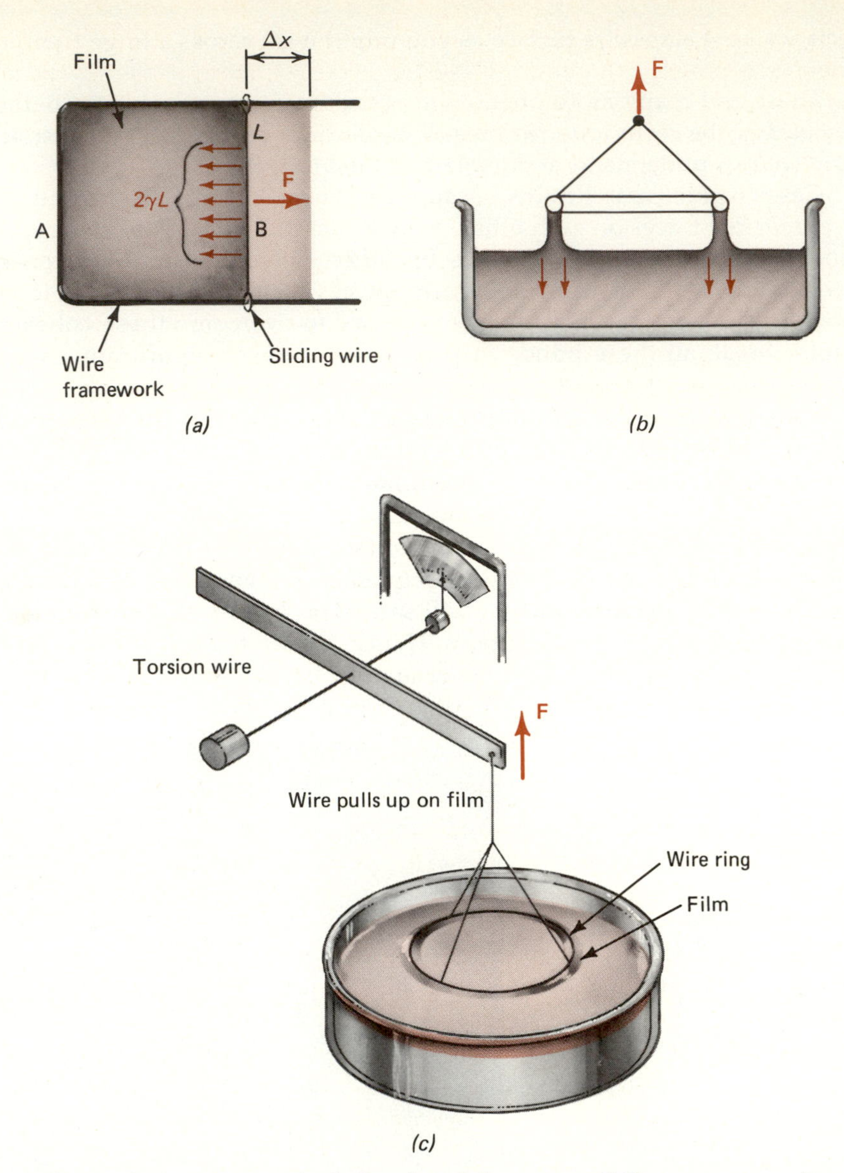

Figure 9.10 *A force* **F** *is required to balance the surface tension of the two sides of a soap film. (a) If the slide wire is pulled to the right by an amount* Δx*, the work done is* $F\ \Delta x$*. The energy of the film is increased by this same amount. This technique, with suitable refinements, can be used to measure the surface tension* γ*. (b) and (c) An alternative method for measuring surface tension. The force acting on the wire ring is measured just before it pulls free of the liquid.*

The surface tension γ is defined as this energy difference per unit surface area:

$$\gamma = \frac{\Delta E}{\Delta A} \tag{9.13}$$

where ΔE is the increase in energy of the liquid due to an increase of its surface area by the amount ΔA.

Still, the term *surface tension* conjures a vision of a membrane under tension, and the preceding discussion had to do with energy, not tension. To bring these concepts into a consistent framework, let us look at surface tension from another viewpoint. In so doing we shall also sketch a conventional method for measuring γ.

Suppose one constructs a rectangular wire frame like that in Figure 9.10, with one arm free to slide along the sides of the rectangle. If a drop or two of soap solution is deposited between the wires A and B when they are close together, the liquid will quickly spread and fill the narrow rectangular region. If a small force is now applied to the movable wire, it can be drawn away, and a soap film then forms in the rectangular area.

Suppose the force required to maintain the movable wire in a given position is F. Since the wire is in equilibrium, there must be an equal but oppositely directed force that also acts on it. That force is due to the surface tension of the soap film. If we define the surface tension constant γ as the force acting along unit length of the film, we have

$$F = 2\gamma L$$

where the factor of 2 appears here because the soap film has two surfaces, top and bottom, each exerting a force γL on the wire of length L. This technique, with some refinements, is a common one for measuring surface tension.

If we increase the applied force F by an infinitesimal amount, the wire will then move to the right. As it moves a distance Δx, the amount of work done by the force F is

$$\Delta W = F\,\Delta x = 2\gamma L\,\Delta x = \gamma\,\Delta A \tag{9.14}$$

where ΔA is the total increase in film surface, top and bottom. From energy conservation, it follows that this must equal the increase in energy of the film, ΔE, and so we arrive at the earlier definition of surface tension, Equation (9.13).

Surface tension may be given in newtons per meter, or equivalently, joules per square meter. The dimension of γ is $[M]/[T]^2$. The surface tensions of some liquids in contact with air are given in Table 9.4.

Table 9.4 *Surface tension of some liquids in contact with air*

Liquid	*Interface*	*Temperature, °C*	*Surface tension γ, N/m*
Benzene	Air	20	29×10^{-3}
Ethanol	Air	20	22.3×10^{-3}
Glycerin	Air	20	63×10^{-3}
Mercury	Air	20	465×10^{-3}
Water	Air	0	75.6×10^{-3}
Water	Air	20	72.8×10^{-3}
Water	Air	100	58.9×10^{-3}

*9.5 Cohesion, Adhesion, Contact Angle, and Capillarity

The attractive forces between like molecules responsible for condensation are known as cohesive forces, and the energy required to remove a molecule from the interior of a liquid or solid is called the *cohesive energy per molecule*. Attraction is, however, not limited to like molecules but also acts between molecules of different kind. Although the fundamental interaction responsible for attraction is the same in both cases, the term *adhesion* is employed when the molecules are different.

If the adhesive force between liquid molecules and molecules of the substrate is greater than the cohesive force, the liquid is said to *wet the surface*. In that case, in the earth's gravitational field, the liquid surface near the container wall curves upward. If, on the other hand, the adhesive force is less than the cohesive, the liquid *does not wet the surface*, and the liquid surface near the container wall curves downward.

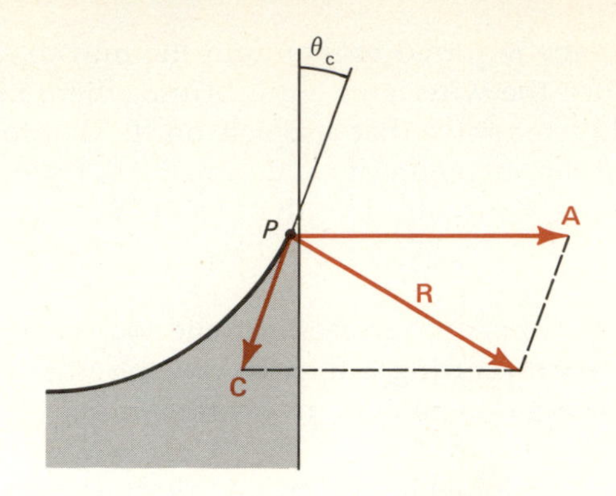

Figure 9.11 A liquid in contact with a solid surface, to which it adheres. The resultant force on a molecule at the surface of the liquid is composed of the net cohesive force **C** and the adhesive force **A**. The surface of the liquid is perpendicular to the resultant **R**. The contact angle between liquid and solid surface is also shown.

As indicated in Figure 9.11, the contact angle between substrate and liquid surface is a measure of the ratio of the adhesive force A to the cohesive force C. If $A \gg C$, the contact angle θ_c is almost 0°; if $A = C$, $\theta_c = 90°$; and if $A < C$, $\theta_c > 90°$, as it is for mercury-glass and water-paraffin interfaces.

Table 9.5 Contact angles

Liquid	Interface	Angle of contact
Water	Clean glass	0°
Ethanol	Clean glass	0°
Mercury	Clean glass	140°
Water	Silver	90°
Water	Paraffin	107°
Kerosene	Clean glass	26°

Capillary action, the rise of liquid in a narrow tube, is a direct consequence of surface tension and adhesion. Figure 9.12 shows a capillary tube in a container of liquid. We shall assume that the liquid wets the inner surface of the tube, so that the contact angle $\theta_c < 90°$. The forces acting on the liquid-air interface are then the surface tension and the weight of the column of liquid pulling the surface down as a result of the cohesive forces. (There is also the pressure of the atmosphere acting on the liquid in the tube, but this is balanced by the same pressure on the liquid in the container.)

At equilibrium, the net force on the surface must vanish. Hence,

$$2\pi\gamma R \cos \theta_c - \rho\pi R^2 hg = 0 \tag{9.15}$$

Here $2\pi R\gamma$ is the total surface tension force acting over a line of length $2\pi R$, the circumference of the inside of the tube; multiplying by $\cos \theta_c$ gives the vertical component of this force (the horizontal components cancel over the circumference of the circle). The second term—the product of mass density ρ, the volume of the column of liquid $\pi R^2 h$, and g—is just the weight of the liquid column.

Solving this equation for the height of the column, we obtain

$$h = \frac{2\gamma \cos \theta_c}{\rho R g} \tag{9.16}$$

Common examples of capillary action are the ability of a blotter and towel to absorb water by pulling it into the narrow channels between

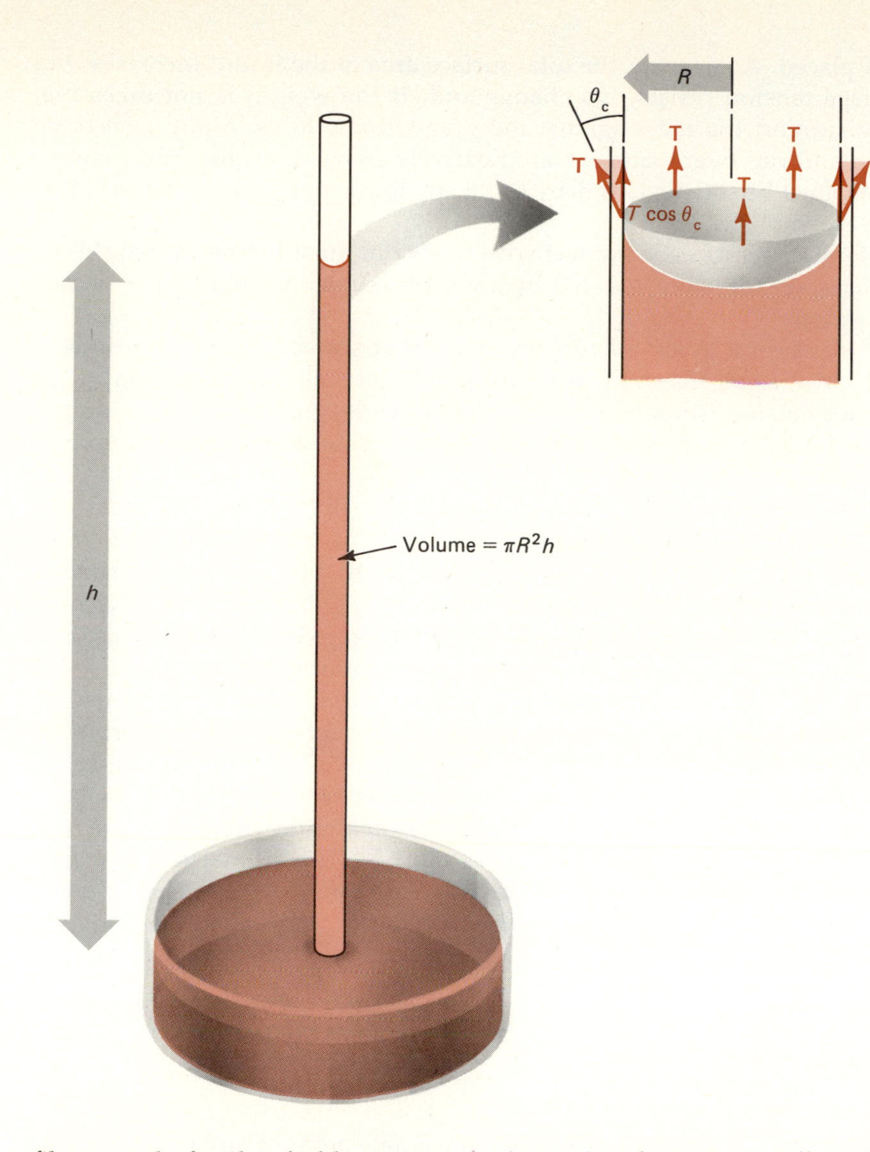

Figure 9.12 The rise of liquid in a narrow tube, which it wets, is known as capillarity. The height to which the liquid rises depends on the contact angle, the surface tension, the radius of the tube, and the density of the liquid.

fibers, and of soil to hold water in the interstices between small, tightly packed particles.

● ***Example 9.4*** A 0.1-mm-diameter glass tube is dipped into a dish of water. To what height above the liquid level in the dish does the water rise in the capillary tube?

The surface tension of water in contact with air is about 70×10^{-3} N/m. The contact angle between clean glass and water is close to 0°. Substituting the appropriate numbers into Equation (9.16), we obtain

$$h = \frac{2(7 \times 10^{-2} \text{ N/m}^2)}{(10^3 \text{ kg/m}^3)(5 \times 10^{-5} \text{ m})(9.8 \text{ m/s}^2)} = 0.286 \text{ m} = 28.6 \text{ cm}$$

In this narrow tube, water will rise nearly a foot due to capillary action. ●

9.5 (a) Walking on Water

We can now understand how an object heavier than water can be supported on the surface of a pond by surface tension. A small mass that is not wet by the liquid tends to depress the surface of the liquid on which it is placed. As a result, the total surface area of the liquid increases, but surface tension resists this change and, if the weight is not excessive, may support the mass against the gravitational force. Many insects do walk and run over water; all are relatively small and light. What conditions must be satisfied, and are there any limits to the size of insects that can accomplish this feat?

First, the surface in contact with the water must not be wetted. Insect cuticles are therefore covered by a waxlike substance that is not wetted by water.

Let us do a quick calculation to see that surface tension can indeed support a small insect on water. Though generally the direction of the surface tension is not vertical, it would have its greatest upward component if it were so directed; we make this assumption to simplify the computation.

The surface tension of water is about 70×10^{-3} N/m. The mass of a good-sized pond skater is about 50 mg. The supporting force must therefore be $(50 \times 10^{-6}\text{ kg})(9.8\text{ m/s}^2) \simeq 5 \times 10^{-4}$ N. The total perimeter of the supporting legs must be at least $(5 \times 10^{-4}\text{ N})/(7 \times 10^{-2}\text{ N/m}) \simeq 7 \times 10^{-3}\text{ m} = 7$ mm. Six legs, each 1 mm long, with a total perimeter of 12 mm, will support the little beast quite comfortably.

Suppose the animal grows to twice its normal size. If the growth is isometric,* the volume and hence the mass increases in proportion to $[L]^3$, i.e., eightfold. The supporting surface tension force must then be $8 \times 5 \times 10^{-4} = 4 \times 10^{-3}$ N. But the perimeter of the legs is only twice as great as before, 24 mm now, so that the maximum supporting force is $(24 \times 10^{-3}\text{ m})(7 \times 10^{-2}\text{ N/m}) \simeq 1.7 \times 10^{-3}$ N. The animal will sink and drown; nature sometimes deals harshly with gluttony! The laws of physics apply not only to inanimate objects but also to the evolution of living systems.

Summary

A molecule, which may consist of a single atom or any finite number of atoms arranged in a unique pattern, is the smallest unit that displays the same chemical properties as the substance itself.

All substances can exist in the solid, liquid, and gaseous states, or phases. Changes of state, called phase transitions, occur at well-defined conditions of pressure and temperature.

Solids are materials that deform elastically under a shear stress.

Liquids and gases, called fluids, do not deform elastically but flow when subjected to a shear stress. The average position of molecules in fluids is random.

Liquids have densities comparable to those of solids. Liquids are also relatively difficult to compress.

Densities of gases are generally several orders of magnitude less than those of liquids. Gases are also easily compressed, compared to liquids, and fill the volume of a container uniformly.

Density is defined as the mass per unit volume:

* An *isometric* growth is one that preserves shape; that is, all dimensions increase in the same proportion.

$$\rho = \frac{m}{V}$$

Specific gravity is the ratio of density to the density of water at 4 °C. At 4 °C, the density of water is 1000 kg/m^3 = 1 g/cm^3. Consequently, the specific gravity is numerically equal to the density expressed in grams per cubic centimeter.

Stress is defined as a force divided by the area over which the force acts:

$$\text{Stress} = \frac{F}{A}$$

Strain is the response of a material to an applied stress; strain is defined as the fractional change of a critical dimension, for example, length or volume.

$$\text{Tensile strain} = \frac{\Delta L}{L_0}$$

$$\text{Compressional strain} = \frac{\Delta V}{V_0}$$

$$\text{Shear strain} = \frac{\Delta x}{L}$$

An elastic deformation is one for which the strain is proportional to the stress. For elastic deformations, *Hooke's law,*

$$\text{Stress} = \text{constant} \times \text{Strain}$$

is obeyed.

An *elastic modulus* is the ratio of stress to strain:

$$\text{Modulus} = \frac{\text{stress}}{\text{strain}}$$

Young's modulus, the bulk modulus, and the shear modulus characterize the response of a material when subjected to a tensile, compressional, and shear stress, respectively.

Cohesive forces between molecules give rise to *surface tension* of liquids. The surface tension γ of a liquid surface is defined as the increase in energy of the liquid per unit increase in surface area:

$$\gamma = \frac{\Delta E}{\Delta A}$$

Adhesion is the tendency of a liquid to adhere to a solid surface or another liquid interface as a result of intermolecular forces. If the adhesive force is greater than the cohesive force, the liquid wets the surface it contacts. Adhesion and surface tension result in *capillary action,* the rise of a liquid in a narrow tube. The height of the liquid column in a capillary tube is given by

$$h = \frac{2\gamma \cos \theta_c}{\rho R g}$$

where γ is the surface tension of the liquid, θ_c the contact angle between the liquid and the capillary tube of radius R, and ρ the density of the liquid.

Multiple Choice Questions

9.1 Two wires of equal length are made of the same material. Wire A has a diameter that is twice as great as that of wire B. If identical weights are suspended from the ends of these wires, the increase in length is

(a) four times as great for wire A as for wire B.

(b) twice as great for wire A as for wire B.

(c) half as great for wire A as for wire B.

(d) one-fourth as great for wire A as for wire B.

9.2 An elastic modulus is the constant of proportionality in a relation of the form

(a) stress = constant × strain.

(b) strain = constant × stress.

(c) stress × strain = constant.

(d) The form of the relation depends on whether the constant is the bulk, shear, or Young's modulus.

9.3 Two wires are made of metals A and B. Their lengths and diameters are related by $L_A = 2L_B$ and $D_A = 2D_B$. When the wires are subjected to the same tensile force, the ratio of the elongations is $\Delta L_A/\Delta L_B = 1/2$. The ratio of Young's moduli Y_A/Y_B is

(a) 1

(b) 2

(c) 1/2

(d) $1/\sqrt{2}$

9.4 When alcohol, whose molecules contain several atoms, vaporizes at room temperature and atmospheric pressure,

(a) most of the molecules dissociate into their atomic constituents.

(b) the average distance between molecules is then roughly equal to the molecular diameter.

(c) the molecules of the vapor concentrate near the bottom of the container as a result of gravitational forces.

(d) None of the above is a correct statement.

9.5 It is observed that when a 60-cm-long capillary tube is inserted into a beaker of water so that 10 cm of the tube is below the level of the water in the beaker, the water level in the tube rises 40 cm above the level of water in the beaker. If a similar glass tube, whose total length is only 40 cm, is placed into this beaker, again with 10-cm submerged below the water level

(a) water in this tube will rise and then flow over the top back down on the outside of the tube.

(b) water will rise in this tube just to the top of the tube.

(c) water will rise in this tube to just 10 cm below the top of the tube, as in the longer tube.

(d) There is not enough information given to come to any unequivocal conclusion.

Problems

9.1 Lead is commonly used for radiation shielding. In some large installations, lead bricks, each with a mass of 10 kg, are sometimes used to build an enclosure around the source of radiation. If the ratio length/depth/height of each brick is 4/2/1, what is the dimension of each brick?

9.2 A sailboat has a hollow keel whose internal cross section is rectangular—120 cm long and 15 cm wide. The keel should have a total mass of 1500 kg. Find the depth to which the keel must be filled if (a) lead and (b) iron is used for ballast.

9.3 What mass of glycerol completely fills a container whose length is 0.8 m and whose cross section is a semicircle of radius 0.4 m?

9.4 An aluminum wire of 1.5-m length is observed to stretch by 0.3 cm when a mass of 4 kg is hung from its end. Find the tensile stress in the wire and determine its diameter.

9.5 A steel rod of 1-mm diameter and 2-m length is subjected to a tension of 400 N. What is the elongation due to this tension?

9.6 A steel wire stretches by 0.02 percent when a mass of 50 kg is suspended from it. Determine the diameter of the wire.

9.7 Calculate the force required to stretch a copper wire of 0.1-mm diameter by 0.04 percent.

9.8 A steel rod is used to support a mass of 300 kg. If this rod has a length of 2 m and should change its length under load by no more than 0.4 mm, what must the minimum diameter of this rod be?

9.9 A block of copper is subjected to a uniform pressure of such magnitude that its density increases by 0.005 percent. Determine the pressure and compare it with that of one atmosphere, which is about 10^5 Pa.

9.10 A sphere of gold is subjected to a pressure of 10^{10} Pa. What is the resulting fractional change in volume of the sphere? What is the fractional change in the sphere's radius?

9.11 A sphere of aluminum is placed in a closed container filled with oil, and the pressure is then increased by means of a piston. What pressure must be applied to reduce the diameter of the sphere by 0.001

percent? Compare your result with normal atmospheric pressure, $10^5\ N/m^2$.

• **9.12** The cylindrical drum of a winch is attached to the spokes of the wheel by six steel rivets, each 0.8 cm in diameter. If the winch is used to raise a 2000-kg mass, what sort of stress acts on each rivet and what is the magnitude of this stress? (Assume that the rivets share the total stress equally.)

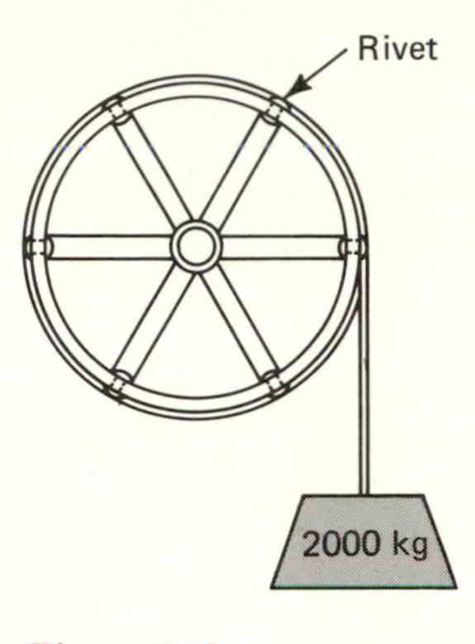

Figure 9.13

• **9.13** A steel wire can withstand a tensile stress of $5 \times 10^8\ N/m^2$. If such a wire is to be used to support a mass of 1000 kg, what is the minimum diameter wire that can be used?

• **9.14** To what height will water rise in a clean glass capillary tube of 0.4-mm diameter?

• **9.15** A 0.1-mm-diameter glass tube is dipped into a pool of ethanol. To what height above the level of ethanol in the pool will the liquid in the capillary tube rise?

• **9.16** In vascular plants, water is brought to the top of the plant through very fine *xylem* tubes, whose diameter may be as small as 4×10^{-6} m. To what height will water rise in such a xylem tube under capillary action? Is it reasonable to presume that capillary action is the principal mechanism by which water is transported to the top of trees?

• **9.17** If a glass tube is lowered into a dish of mercury, one finds that inside the tube the level of mercury is depressed below that in the dish. What is the level difference if the diameter of the glass tube is 2 mm?

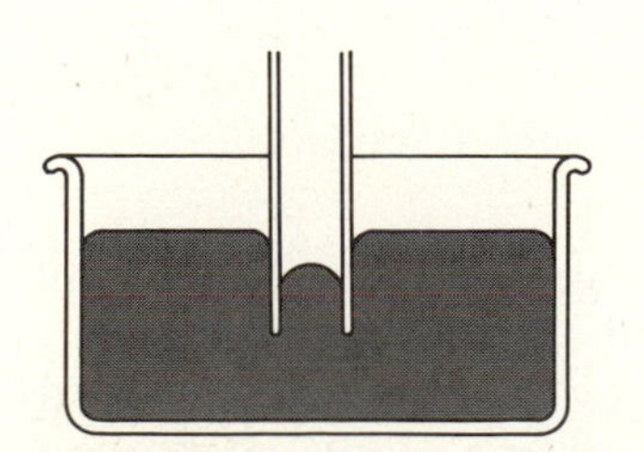

Figure 9.14

• **9.18** Two strips of aluminum are joined near their ends by four rivets of steel. The diameter of the rivets is 0.8 cm. If the shear stress on each rivet is to be less than 1 percent of the fracture strength of $3 \times 10^8\ N/m^2$ and each rivet can be assumed to carry one-fourth of the total load, what is the maximum tension that can be applied to this system?

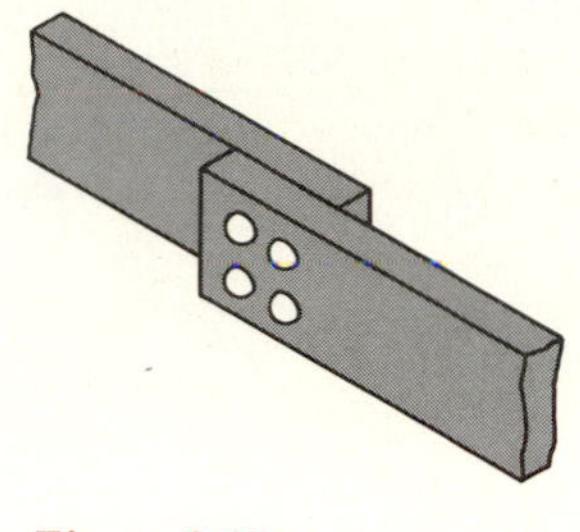

Figure 9.15

• **9.19** An aluminum wire of 1-m length and 1-mm diameter is joined to a steel wire of identical dimensions, one end to the other, so that the length of the composite is 2 m. If a 6-kg mass is suspended from this 2-m-long wire, what is the total elongation of that supporting wire?

• **9.20** A capillary glass tube is dipped into a bowl of water. If the level of water in the capillary rises to 3 cm above the level in the bowl, what is the diameter of the capillary?

• •**9.21** A thin cylindrical steel rod (e.g. a sewing needle) is covered with a very thin layer of paraffin. Determine the largest-diameter rod that will be supported by surface tension when it is placed on the surface of water.

• •**9.22** We have seen that if a pond skater grows isometrically, it may sink and drown. Suppose such a bug does grow to twice the normal body size, what should be the fractional change in its legs to provide the same assurance of support as for the smaller insect?

• •**9.23** A mass is supported by two wires, as shown in Figure 9.16. The two wires have exactly the same unstretched and stretched lengths and the same diameter. One is made of copper, the other of aluminum. Determine the fraction of the total mass supported by each wire.

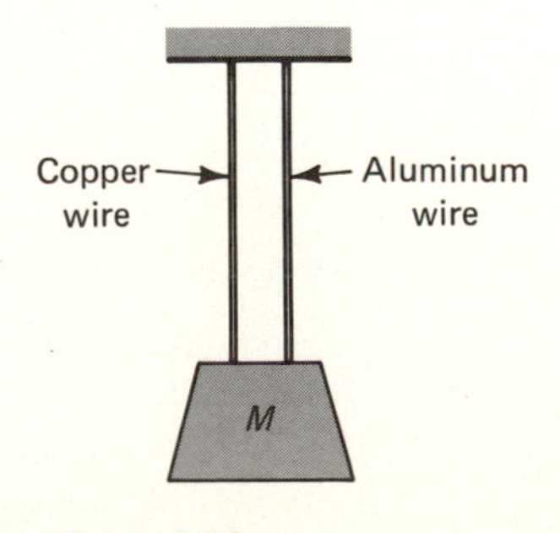

Figure 9.16

·10·

Hydrostatics and Hydrodynamics

If a vessel full of water, otherwise completely closed, has two openings, one of which is one hundred times as large as the other; by putting in each of these a piston which fits it exactly, a man pushing on the small piston will exert a force equal to that of one hundred men pushing on the piston which is one hundred times as large, and will overcome the force of ninety nine men.

BLAISE PASCAL

Taken literally, the title suggests that this chapter addresses only the properties of stationary and moving water. *Hydrostatics* and *hydrodynamics* signify, however, the study of equilibrium and dynamic properties of fluids in general, including gases as well as liquids. This chapter begins with a treatment of atmospheric pressure, an equilibrium property of a gas, but then focuses largely, though not exclusively, on liquids. Discussion of gases is deferred to subsequent chapters in which the relations between the pressure, volume, and temperature of an ideal gas are considered in some detail.

10.1 Atmospheric Pressure

On September 22, 1648, the townspeople of the French village of Clermont witnessed a remarkable, and also historic event. At eight o'clock that morning a group of men, led by M. Perier, arrived in the garden of the local monastery of Minimes, carrying several glass vessels and long glass tubes and a bottle filled with mercury. Under the watchful eyes of several monks, Perier filled the tubes with mercury, and, careful to avoid the entry of air, inverted them into a dish also filled with mercury. As he later wrote to his brother-in-law,

> When I brought the two tubes near each other without lifting them out of the vessel, it was found that the quicksilver which remained in each of them was at the same level, and that it stood above the quicksilver in the vessel twenty-six inches and three lines and a half [26.3″]. . . . I marked on the glass the height of the quicksilver and begged the Rev. Father Chastin, a man as pious as he is capable, to observe it from time to time during the day, so as to see if any change occurred.

Then, with the full solemnity of a religious procession, Perier and his colleagues, carrying the remaining glassware and mercury, ascended the Puy-de-Dôme, a mountain whose peak stands about 1200 m above Clermont. There the experiment was repeated

> five times more, with great accuracy, at different places on the top of the mountain, once under cover of the little chapel which is there, once exposed, once in a shelter, once in the wind, once in good weather, and once during the rain and mists which came over us some times; in all these trials there was found the same height of the quicksilver, twenty-three inches two lines [23.17″], which makes a difference of three inches one line and a half from the twenty-six inches three lines and a half which were found at the Minimes; this result fully satisfied us.

The experiment described in such vivid detail was instigated by Blaise Pascal (1623–1662), to whom Perier was related by marriage. Pascal had learned of the work of Evangelista Torricelli, a pupil of Galileo, who was reputed to have created a vacuum by raising a tube of glass out of a long dish of mercury; he had also been told of Torricelli's revolutionary suggestion that the mercury was supported in the tube not by the force of

Figure 10.1 Blaise Pascal (1623–1662).

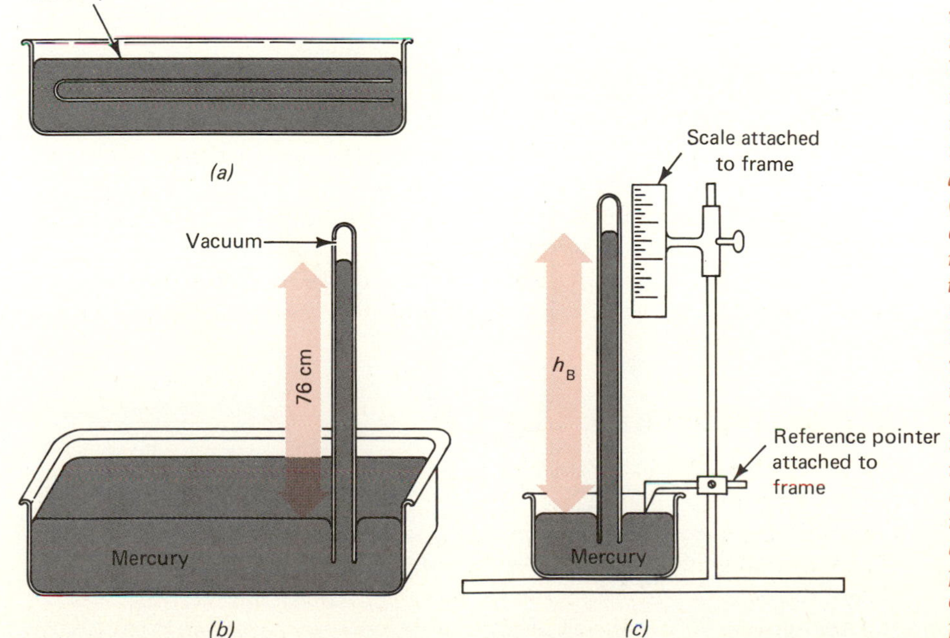

Figure 10.2 Torricelli's experiment and the barometer of Torricelli. (a) A long tube was filled with mercury by submerging it in a trough. (b) When the closed end of the tube was raised out of the trough, the mercury in the tube remained at a level 76 cm above that in the trough. (c) The Torricelli barometer consists of a reservoir of mercury, a long glass tube that was first completely filled with mercury and then inverted into the reservoir without permitting air to enter the tube. By means of a reference pointer that locates the level of Hg in the reservoir and visual observation of the level of Hg in the glass tube at height h_B above the reservoir, the pressure difference corresponding to atmospheric pressure is determined.

the vacuum—supposedly "abhorred by nature"—but by the pressure of the "sea of elemental air" pressing on the surface of the mercury in the open dish. Pascal reasoned that at the top of a mountain the pressure of the atmosphere should be less than in the valley and therefore capable of supporting only a shorter column of mercury. An invalid himself, he wrote to his brother-in-law, who performed the critical experiment with "results that fully satisfied us."

Today the mercury barometer and manometer are familiar laboratory instruments. Altimeters, whose operation relies on the dependence of atmospheric pressure on height above sea level, are also widely used, especially in air travel.

Under normal weather conditions, atmospheric pressure at sea level, $P_A(0)$, is 1.01×10^5 Pa.* This is the force exerted on 1 m^2 of surface by the mass of air in the column above this surface. Since the amount of air above a given height h decreases as h increases, atmospheric pressure diminishes with altitude as deduced by Pascal and demonstrated by Perier. This dependence is, however, not linear, because the air above a given point, pressing down on the gas below, compresses this gas and therefore increases its density. As a result, a given volume of air weighs more at lower altitude. Under certain simplifying assumptions, one obtains

$$P_A(h) = P_A(0)e^{-h/h_0} \qquad \textbf{(10.1)}$$

where $P_A(h)$ is the atmospheric pressure at an altitude h above sea level; the constant h_0 is called the scale height and is about 8.6 km, a rather considerable distance, roughly equal to the height of Mt. Everest. For altitude changes of 2000 m or less one may, to fairly good approximation, replace the exponential behavior by a straight line;† that is,

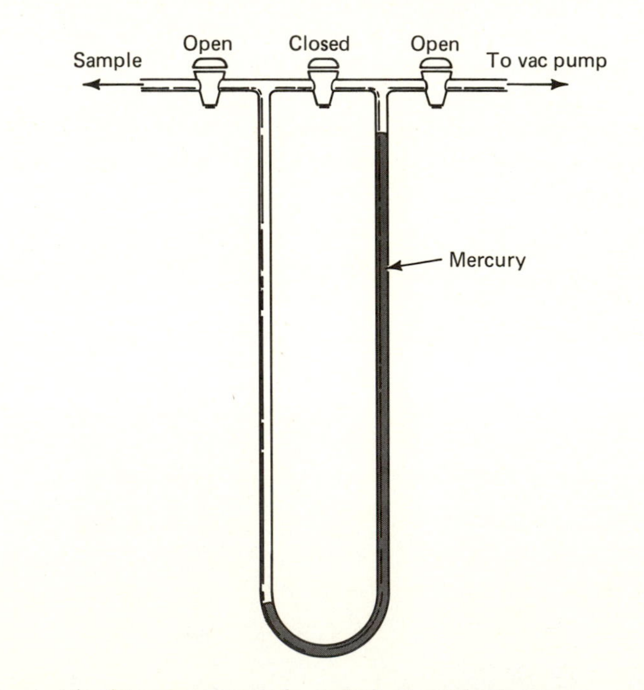

Figure 10.3 *Another simple device for measuring gas pressure is the manometer. The manometer is a glass tube in the shape of a long U and filled with a liquid of known density. One side of the manometer is connected to a good vacuum pump, the other side to the volume of gas whose pressure is to be measured. The difference in the liquid level on the two sides of the manometer is proportional to the pressure exerted by the gas (the pressure on the side connected to the vacuum pump is presumed to be negligibly small).*

* In the past (and to some extent also today), atmospheric pressure was expressed in millimeters of mercury; that is, the pressure was characterized by the height of the column of mercury that this pressure would support. One atmosphere of pressure was defined as equivalent to 760 mmHg, and this is also equal to 1.01×10^5 Pa = 101 kPa.

† See Appendix A.

$$P_A(h + \Delta h) \simeq P_A(h)\left[1 - \frac{\Delta h}{h_0}\right] \quad \textbf{(10.2)}$$

Accordingly, near sea level, atmospheric pressure decreases by about 0.09 mm of mercury per meter, or roughly 1 mm for every 10 m of altitude change.

The exponential dependence of $P_A(h)$ rests on several simplifying assumptions. First, the change in atmospheric temperature with altitude is neglected. Second, the gradual decrease of gravitational attraction with distance from the center of the earth has been neglected. Third, the relative composition of the atmosphere (fraction of O_2, N_2, and other gases) has been assumed independent of altitude. If suitable corrections are made, the curve for $P_A(h)$ deviates slightly from the exponential, as shown in Figure 10.4.

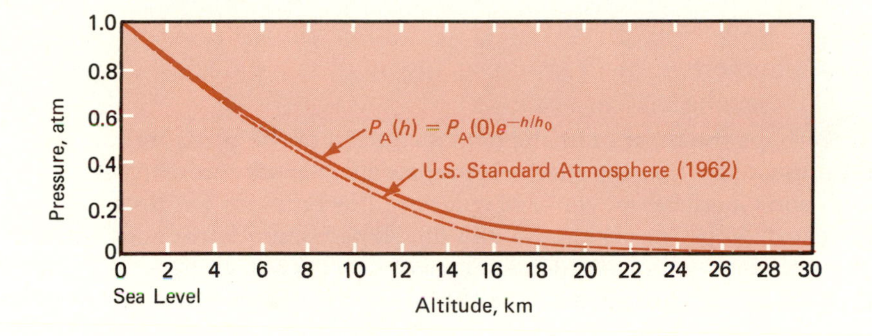

Figure 10.4 *Dependence of air pressure on altitude. The solid line is the exponential dependence obtained if the variation of atmospheric temperature and of g with altitude are neglected. The dashed line is the result of a more reliable calculation, which does not make these assumptions.*

● ***Example 10.1*** From the data provided by M. Perier, calculate the difference in elevation between the peak of Puy-de-Dôme and the monastery of Minimes in Clermont.

We shall first use the approximate expression, Equation (10.2) and then the more precise formula, Equation (10.1). We can rewrite Equation (10.2) in dimensionless form:

$$\frac{P_A(h + \Delta h)}{P_A(h)} = 1 - \frac{\Delta h}{h_0}$$

We see that there is no need to convert the pressures to metric units. We obtain

$$\frac{23.17}{26.3} = 0.881 = 1 - \frac{\Delta h}{8600\text{ m}}$$

$$\Delta h = (8600\text{ m})(0.119) = 1023\text{ m}$$

This elevation difference is substantially less than the 2000-m altitude change for which the linear relation should be a good approximation to Equation (10.1).

Equation (10.1) is readily generalized to

$$P_A(h + \Delta h) = P_A(h)e^{-\Delta h/h_0}$$

(Problem 10.8). Dividing by $P_A(h)$ and taking logarithms of both sides, we have

$$-\frac{\Delta h}{h_0} = \ln\frac{P_A(h + \Delta h)}{P_A(h)}$$

or

$$\frac{\Delta h}{h_0} = \ln \frac{P_A(h)}{P_A(h + \Delta h)}$$

Performing the indicated operations gives

$$\Delta h = 1090 \text{ m}$$

The approximate result, 1023 m, is well within 10 percent of the more precise value.

Although we often think of the atmosphere as "pressing down," the pressure acts uniformly in all directions. To see this, consider a large container of fluid, as in Figure 10.5, and a small imaginary cubic volume. If the pressure on one face were not balanced by pressure on the opposite face, the fluid within this cubic volume would experience a net force and would therefore flow in the direction of that net force. From such argument one concludes that

At equilibrium, the pressure at any point in a fluid is the same in all directions.

Perhaps the most dramatic demonstration of the pressure exerted by the atmosphere was the spectacle prepared in the seventeenth century by Otto von Guericke, mayor of Magdeburg, for Emperor Ferdinand III and his court. Von Guericke, who had developed some fairly good vacuum pumps, cast two hollow hemispheres of bronze, which he sealed with a mixture of wax, grease, and turpentine. When the spherical volume was

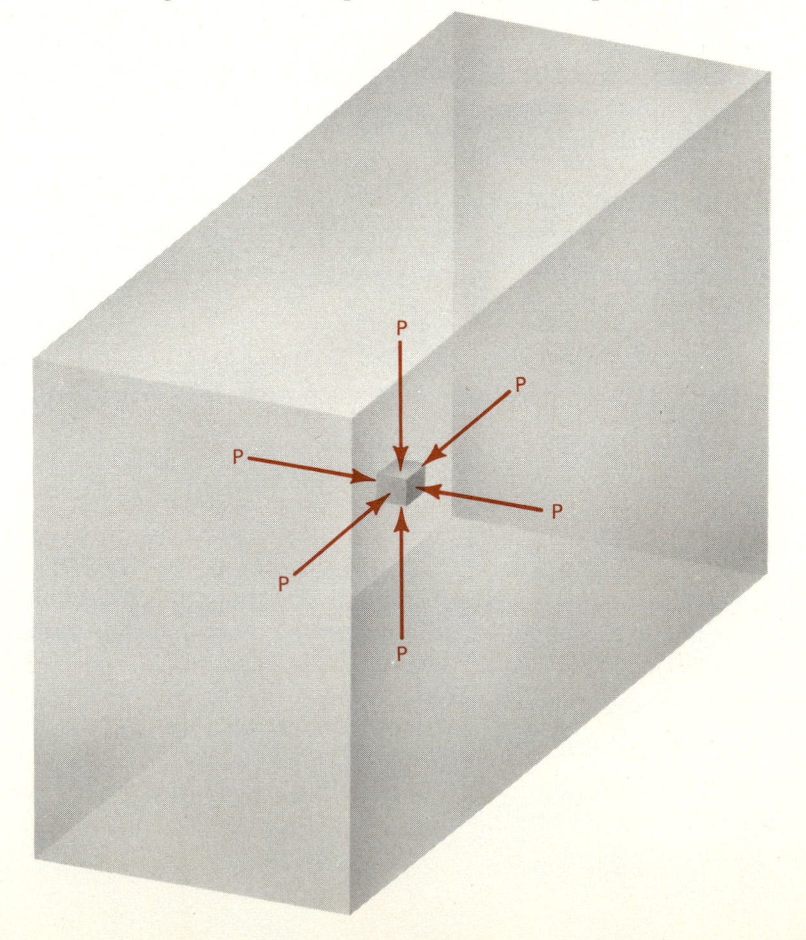

Figure 10.5 The pressures on a very small cubic volume of fluid inside a large container of this fluid.

(a)

(b)

Figure 10.6 (a) Woodcut of the dramatic demonstration of atmospheric pressure prepared by Otto von Guericke in 1642. Two teams of eight horses were unable to separate the hemispheres when the spherical volume within them had been evacuated. (Did von Guericke really need to employ two teams of horses?) (b) The "Magdeburg hemispheres" used by von Guericke.

evacuated, two teams of horses, eight on each side, could not separate the hemispheres; when a valve was opened, admitting air, they readily fell apart.

10.2 Pressure in a Liquid

As anyone who has dived into a swimming pool can attest, the pressure increases perceptibly as one descends even a few feet below the surface.

Figure 10.7 shows a cylindrical container of liquid open to the air. We wish to determine the pressure a distance h below the surface under equilibrium conditions. To do so, we exhibit in Figure 10.8 the liquid as a free body, replacing the container by the forces its walls exert on the fluid.

The force acting over the top surface of the liquid is $P_A A$, where P_A is the atmospheric pressure and A is the surface area of the cylinder. The force supporting the liquid cylinder is $P'(h)A$, where $P'(h)$ is the pressure exerted on the lower surface of the free body at a depth h below the surface. By Newton's third law, this equals, in magnitude, the pressure this

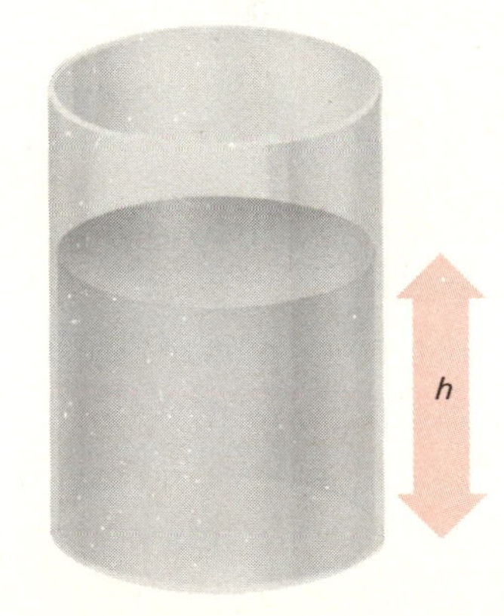

Figure 10.7 A cylindrical container filled to a height h with liquid of density ρ.

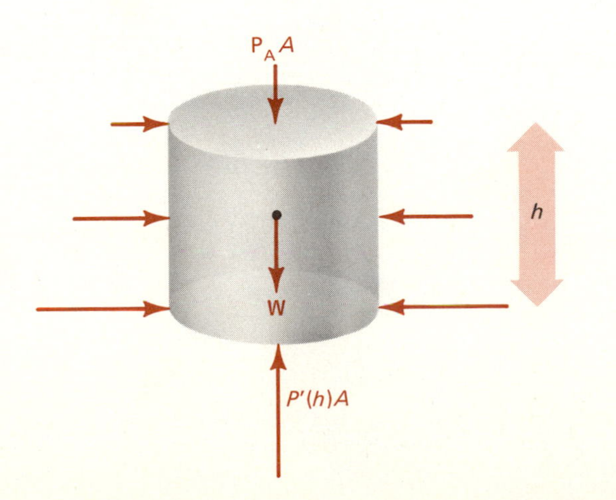

Figure 10.8 Free-body diagram of the liquid in the container of Figure 10.7 showing the forces that must act on the liquid.

liquid exerts on the supporting surface. In addition to the force $P_A A$, the lower surface must also support the weight of the cylindrical column of liquid, namely $W = mg = \rho V g = \rho A h g$. Hence, at equilibrium,

$$P'(h)A = P_A A + \rho g h A = P_A A + P(h)A$$

where

$$P(h) = \rho g h \tag{10.3}$$

Equation (10.3) shows that the *additional* pressure in excess of atmospheric pressure at a depth h below the surface of a liquid is proportional to g, the density ρ of the liquid, and the depth h.* The reason $P(h)$ is, in this instance, a linear and not an exponential function of h is that liquids resist compression far more than gases. Liquids are, to good approximation, incompressible, and their density is constant, independent of h in contrast to the atmosphere, whose density diminishes significantly with increasing altitude.

It follows from Equation (10.3) that the difference in pressure between two points whose depths differ by Δh is

$$\Delta P = \rho g \, \Delta h \tag{10.4}$$

Thus the pressure at A and B in Figure 10.9 is the same even though the amount of liquid is less *directly* above B than above A. The equality of pressure must hold because if the pressure at B were, say, less than at A then the liquid within the imaginary horizontal tube of Figure 10.9 would flow from A to B. Such reasoning leads to the following simple rule first enunciated by Pascal.

PASCAL'S PRINCIPLE: Under equilibrium conditions, a change in pressure at any point in an incompressible fluid is transmitted uniformly to all parts of the fluid.

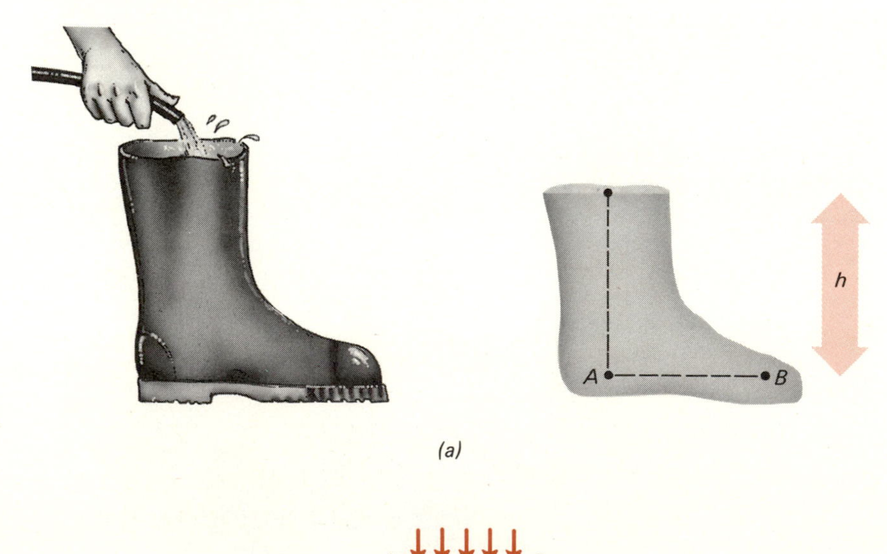

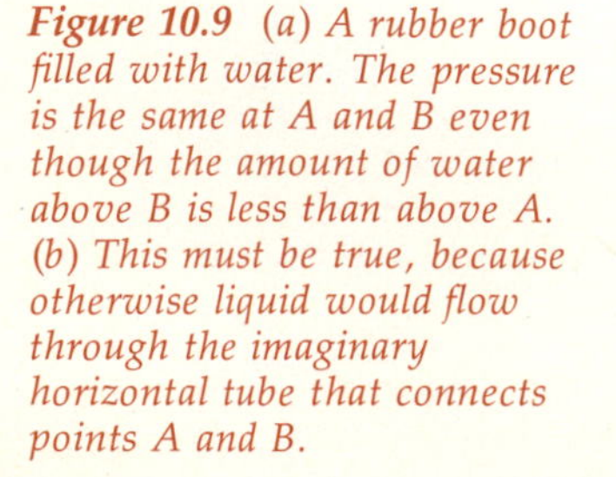

Figure 10.9 *(a) A rubber boot filled with water. The pressure is the same at A and B even though the amount of water above B is less than above A. (b) This must be true, because otherwise liquid would flow through the imaginary horizontal tube that connects points A and B.*

* The difference between total pressure and atmospheric pressure is often referred to as the *gauge pressure,* since it is the pressure that is read by standard gauges, such as tire gauges.

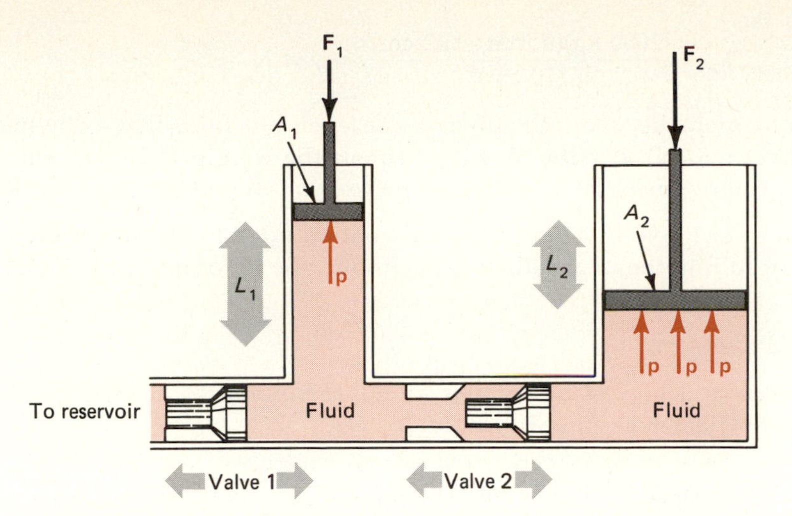

Figure 10.10 Hydraulic lift. Hydraulic fluid is pulled into the small cylinder from a larger reservoir through valve 1 when the piston of area A_1 is raised; valve 2 is then closed. When the piston in the small cylinder is pushed down, valve 1 closes and valve 2 opens, and liquid is forced into the larger cylinder, raising its piston.

The best known application of Pascal's principle is the hydraulic lift, shown schematically in Figure 10.10. If a force F_1 is applied to the piston of area A_1, the pressure in the liquid is increased throughout by F_1/A_1. To maintain the liquid in equilibrium, the upward pressure of the liquid on the piston of area A_2 must be balanced by an equal downward pressure. The downward force on that piston must therefore satisfy the condition

$$\frac{F_2}{A_2} = \frac{F_1}{A_1}, \qquad \text{or} \qquad F_2 = \frac{A_2}{A_1} F_1 \tag{10.5}$$

Hence, if $A_2 >> A_1$ then $F_2 >> F_1$. Thus a relatively small force F_1 can support (or lift) a very large weight resting on the piston of area A_2. The walls of the container must, of course, also be strong enough to withstand the large pressure without rupturing. The same hydraulic principle is also employed in the brake system of a car.

● ***Example 10.2*** A 2000-kg automobile is lifted by a manually operated hydraulic jack (Figure 10.11). The diameter of the shaft of the lift is 10 cm; the diameter of the piston connected to the pump that forces liquid into the system is 2 cm. The pump is operated by the lever arrangement shown in Figure 10.11. What force must be exerted on the end of the lever to raise the car?

By Pascal's principle, the force that must be exerted against the 2-cm-diameter piston is

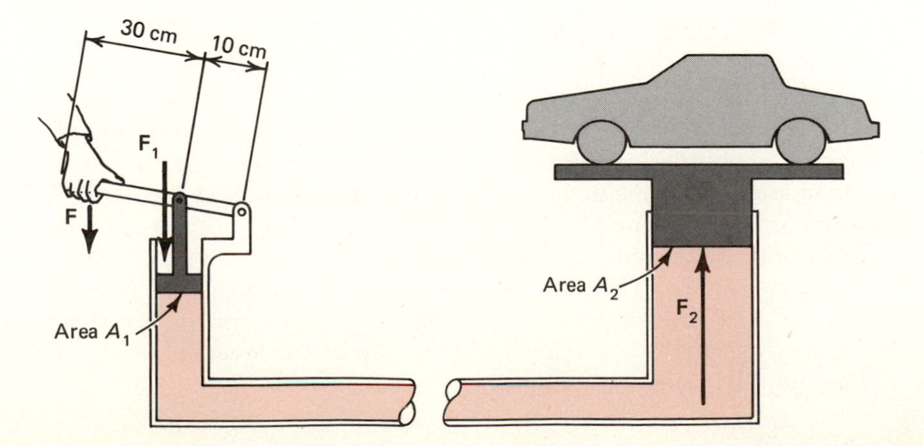

Figure 10.11 Manual hydraulic jack.

$$F_1 = \frac{(2000\ \text{kg})(9.8\ \text{m/s}^2)(2\ \text{cm})^2}{(10\ \text{cm})^2} = 784\ \text{N}$$

Taking moments about the pivot of the lever, one finds that a downward force of (784 N)/4 = 196 N, which equals the weight of 20 kg, will produce the requisite force of 784 N on the 2-cm-diameter piston. ●

If in Example 10.2, an additional volume A_2h_2 of incompressible fluid is forced into the larger-diameter cylinder, the PE of the car is increased by

$$\Delta\text{PE} = mgh_2$$

This is achieved by forcing an equal volume of liquid out of the smaller-diameter cylinder. Hence,

$$A_1h_1 = A_2h_2 \qquad \text{and} \qquad h_1 = \frac{A_2}{A_1}h_2$$

and the work done by the force F_1 is

$$\Delta W = F_1h_1 = \left[\frac{A_1}{A_2}(mg)\right]\left[\frac{A_2}{A_1}(h_2)\right] = mgh_2 = \Delta\text{PE}$$

where we have made use of Equation (10.5). Energy is conserved. The work done by the force F_1 equals the increased PE of the car. Though the force F_1 may be smaller than the weight mg that is raised, the distance over which F_1 must act is correspondingly greater.

10.3 Buoyancy

In 214 B.C., during the Second Punic War between Rome and Carthage, the Romans dispatched an expeditionary force under Claudius Marcellus to capture Syracuse, whose King Hieron had recently renewed his allegiance with Carthage. Marcellus and his Roman legions expected only token opposition from the small kingdom, but were surprised by a rain of catapult balls and darts that greeted the soldiers attacking by land. The few that succeeded in reaching the walls of the city saw their scaling ladders pounded to kindling by heavy stones suspended from cranes that projected over the walls. Those approaching the city by sea fared even worse. Syracusans lowered grapnels from cranes mounted on the cliffs above the sea and caught the bows of the landing ships, lifting them right out of the water by means of multiple pulleys until the would-be attackers were dumped unceremoniously into the sea. Before long the proud Roman legions retreated as soon as they spied so much as a rope projecting over the city's wall.

In the end, after a siege of two years, Syracuse finally succumbed to hunger. The plundering Roman soldiers, entering the palace, came on an old man studying some geometrical figures in the sand on the floor. The soldiers ordered the man to follow them to Marcellus, but the old fellow would not be distracted. "Do not touch my circles," he cried, whereupon the soldiers, annoyed by such insubordination, slew the old man. His name was Archimedes, and he had been responsible for the Roman debacle, having taught the Syracusans the use of the lever, pulleys, catapults, and other mechanical devices.

Figure 10.12 Archimedes (287–212 B.C.).

Archimedes, born in Syracuse in 287 B.C., a friend and relation of King Hieron II, was perhaps the greatest scientist of antiquity. Among his many practical inventions, including the pulley and catapult, is the Archimedes screw, a device for raising water that was widely employed to pump water out of mines as recently as 200 years ago, and is still used in many parts of the world by nomadic tribes. He also devised a procedure for calculating π (he obtained the bounds $3\frac{10}{71} < \pi < 3\frac{1}{7}$, or $3.1408451 \ldots < \pi < 3.1428571 \ldots$), which was not surpassed until Newton applied his genius to the problem nearly eighteen hundred years later.

According to legend, King Hieron, suspecting that his crown might not be of pure gold, asked Archimedes to confirm or allay his suspicions, but do so without damaging the crown. The solution, so the story goes, came to Archimedes as he was taking a bath, whereupon shouting

Figure 10.13 The helical pump invented by Archimedes, also known as Archimedes' screw.

"Heureka," he ran naked through the streets of Syracuse to tell King Hieron. The solution is based on Archimedes' principle.

ARCHIMEDES' PRINCIPLE: A body completely or partially submerged in a fluid is buoyed up by a force equal to the weight of the fluid it displaces.

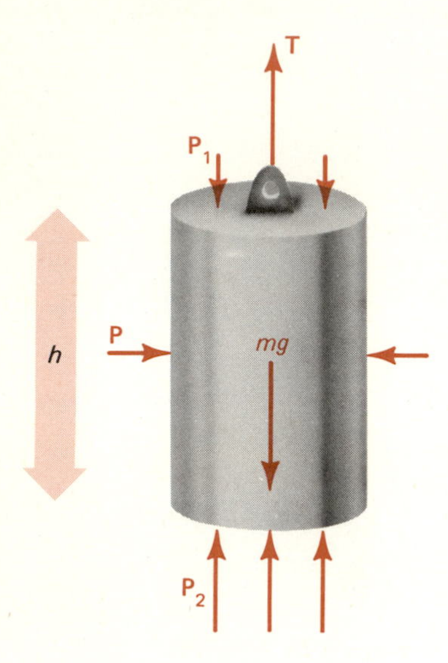

Figure 10.14 Free-body diagram of a solid cylinder suspended by a string (under tension T) below the surface of a liquid of density ρ. The pressure P_2 acting on the lower surface of the cylinder equals $P_1 + \rho gh$, where P_1 is the pressure acting on the top surface and h is the height of the cylinder.

Buoyancy is a familiar phenomenon. Our bodies float in water, as does wood, and even a canoe made of aluminum (unless it capsizes and fills with water). An anchor can be raised to the surface relatively easily, but bringing it aboard requires much greater exertion. The upward force that lifts a helium-filled toy balloon is the buoyant force of the air displaced by the balloon.

This upward force is simply explained. Since pressure in a fluid increases with depth, the lower surface of a submerged object experiences an upward pressure greater than the downward pressure that acts on the upper surface; the resultant force is then directed up.

One can see this most clearly for a simple geometrical shape such as the cylinder, shown in Figure 10.14. If the pressure at the top of the cylinder is P_1, the pressure at the bottom of the cylinder is $P_1 + \rho gh$, where ρ is the density of the fluid.

The difference between the upward force acting on the lower surface and the downward force acting on the upper surface is the *buoyant force* F_B exerted by the fluid. Its magnitude is given by

$$F_B = \rho ghA = \rho gV = m_L g = W_L$$

where $V = hA$ is the volume of the cylinder, $m_L = \rho V$ is the mass, and W_L the weight of the fluid that would occupy a volume V.

Thus, the effect of submersion in the fluid is to provide an upward force F_B equal in magnitude to the weight of the displaced fluid.

To demonstrate the validity of Archimedes' principle in general, consider a body of irregular shape, such as shown in Figure 10.15, submerged in a liquid. Imagine now replacing this body with another of identical shape but composed of the liquid itself. Since this mass of liquid is in equilibrium with its surroundings, the resultant force due to the pressure of the surrounding liquid must be a net upward force just equal to the weight of the liquid contained within the irregular boundary. The same buoyant force will also act on any other object of the same shape placed in this position because the pressure exerted by the surrounding fluid does not depend on the composition of the material within the boundary. We are thus led to Archimedes' principle: The buoyant force equals the weight of the displaced fluid.

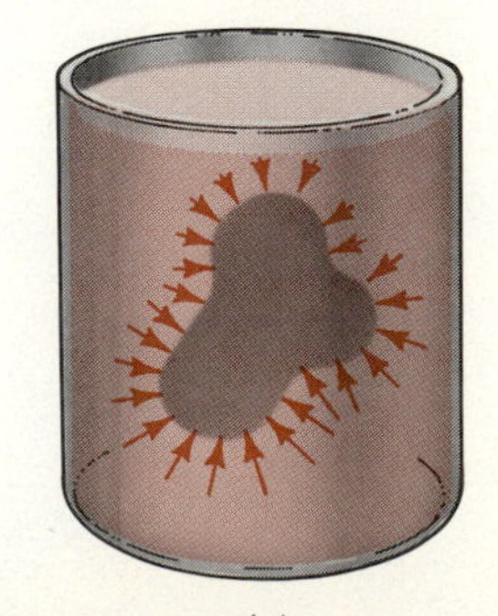

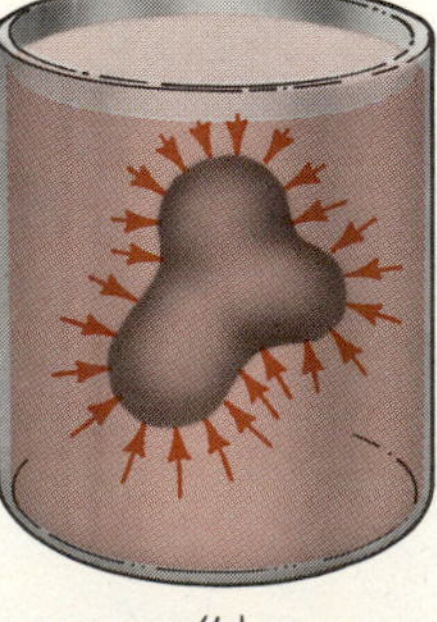

Figure 10.15 Proof of Archimedes' principle for a submerged body of arbitrary shape. (a) An irregularly shaped surface encloses a volume of liquid that is in equilibrium with the surrounding liquid. Since the liquid within this surface is neither falling nor rising in the container, it must experience a buoyant force equal to its own weight. That buoyant force arises from the pressure exerted by the surrounding liquid on the imaginary surface. (b) If the fluid within that surface is now replaced by a solid, or other object of the same shape, the pressure exerted by the surrounding liquid is unchanged, and so is, therefore, the buoyant force. Q.E.D.

10.3 (a) Density Determination by Archimedes' Principle

What aroused Archimedes' joy was not the recognition of buoyancy per se—that fact had been known since men first embarked in ships to cross the seas—but the quantitative insight he had achieved, which allowed him to determine the density of the king's crown without melting it into a simple shape. How is that accomplished?

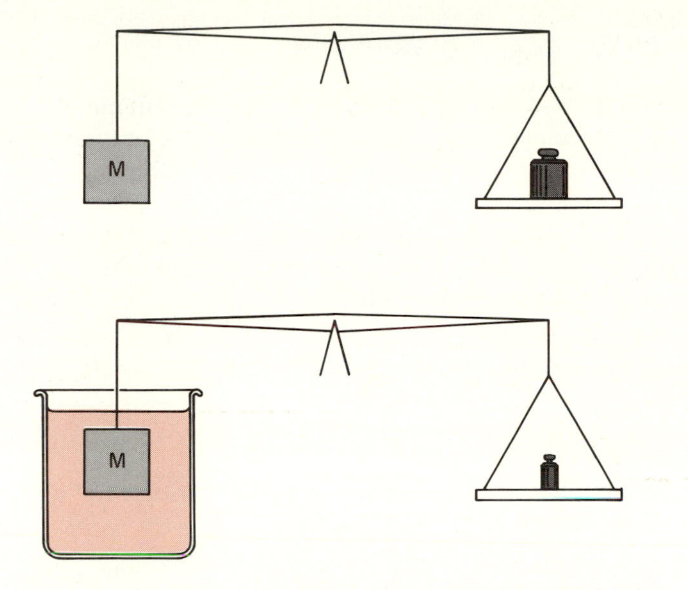

Figure 10.16 Density determination using Archimedes' principle. The object M is weighed (a) in air and (b) when submerged in a liquid of density ρ_L.

Suppose we weigh an object in the usual manner on a balance and find its weight is W_0. If we now support this object from the balance while it is fully submerged in liquid, as in Figure 10.16(*b*), its apparent weight will be diminished by the buoyant force. This buoyant force is $V\rho_L g$, where V is the volume of the object and $V\rho_L$ the mass of the displaced liquid of density ρ_L. Thus the *apparent* weight when submerged in the liquid is

$$W_a = W_0 - V\rho_L g$$

and the volume of the object is therefore

$$V = \frac{W_0 - W_a}{\rho_L g} \tag{10.6}$$

The mass density of the object is $M/V = W_0/Vg$ and is thus given by

$$\rho = \frac{W_0}{W_0 - W_a}\rho_L \tag{10.7}$$

If the object is submerged in water, the ratio ρ/ρ_L is by definition the same as the specific gravity of the object; that is,

$$\text{Spec. grav.} = \frac{W_0}{W_0 - W_a(\text{w})} \quad \text{where } W_a(\text{w}) = \text{apparent weight in water} \tag{10.8}$$

With a slight modification of this technique, involving measurement of the apparent weight of an object in water and in a liquid of unknown density, the density of that liquid can be deduced. The next few examples illustrate several applications of Archimedes' principle.

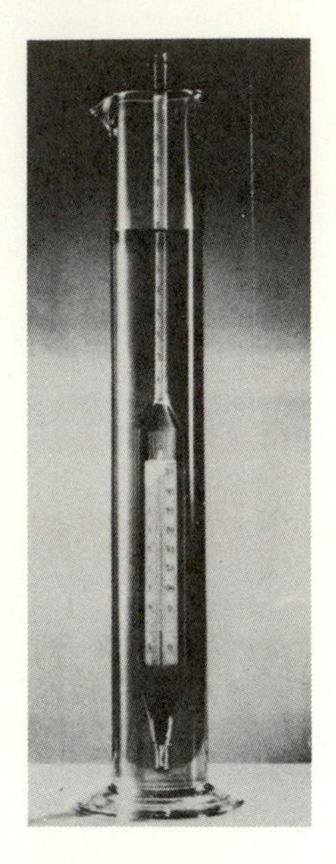

Figure 10.17 *An application of Archimedes' principle—the hydrometer. The level of the liquid surface when the hydrometer floats in the liquid is a measure of the density of the liquid.*

Example 10.3 A piece of metal weighs 45 N in air and 28.3 N when fully submerged in water. Determine the specific gravity of the metal; and assuming that the object is of uniform density, determine what this metal is.

From Equation (10.8), we have

$$\text{Spec. grav.} = \frac{45\text{ N}}{45\text{ N} - 28.3\text{ N}} = 2.69$$

Comparing this value with the values for common metals, we see that aluminum is the most likely candidate. (It is not the only candidate, since a suitable alloy of various other metals could have the same specific gravity.)

Example 10.4 An object that weighs 40 N in air "weighs" 20 N when submerged in water and 30 N when submerged in a liquid of unknown density. What is the density of that liquid?

The apparent weight of the object in water tells us that its volume is

$$V = \frac{W_0 - W_a(\text{w})}{\rho(\text{w})g}$$

If we now solve Equation (10.6) for ρ_L and substitute the above expression for V, we obtain

$$\rho_L = \frac{W_0 - W_a}{Vg} = \frac{W_0 - W_a}{W_0 - W_a(\text{w})} \times \frac{\rho(\text{w})g}{g}$$

$$= \left(\frac{W_0 - W_a}{W_0 - W_a(\text{w})}\right)(1000\text{ kg/m}^3)$$

For this particular example, then,

$$\rho_L = \left(\frac{40\text{ N} - 30\text{ N}}{40\text{ N} - 20\text{ N}}\right)(1000\text{ kg/m}^3) = 500\text{ kg/m}^3$$

and the specific gravity of the unknown liquid is 0.5.

Example 10.5 An object floats on water with 20 percent of its volume above the waterline. What is the average density of the object?

In this case we cannot apply any of the formulas (10.6)–(10.8); these are based on the assumption that the object is fully submerged. Since this object *floats* when 80 percent of its volume is submerged, we know that $F_B = Mg = V\rho g$.

This force equals the weight of the displaced fluid, i.e.

$$F_B = 0.8V(1000\text{ kg/m}^3)g$$

Hence

$$0.8V(1000\text{ kg/m}^3) = V\rho$$

$$\rho = 800\text{ kg/m}^3$$

10.3 (b) Center of Buoyancy and the Stability of Boats

The point through which the buoyant force acts is called the *center of buoyancy* (CB). *The* CB *is at the* CM *of the displaced liquid* and coincides with the CM of the object only if that object is homogeneous and completely submerged in the liquid. The relative locations of the CM and CB are critical in the design of boats and determine their stability against pitch and roll. Whereas the CM of the boat does not change position, the CB can shift quite significantly as the boat rolls.

Figure 10.18 compares the hull shapes of the lighter, flat-bottom sailboats such as the Sunfish with the wineglass shape of a larger cruising sailboat with lead keel. Whereas the relatively broad, flat-bottom hulls provide better "heel resistance," or initial stability against small roll, they have poor ultimate stability. The deep-keel boat has poor initial and good ultimate stability; it will heel readily for the first few degrees but it will rarely capsize. If by some freak wave it is capsized, or "knocked

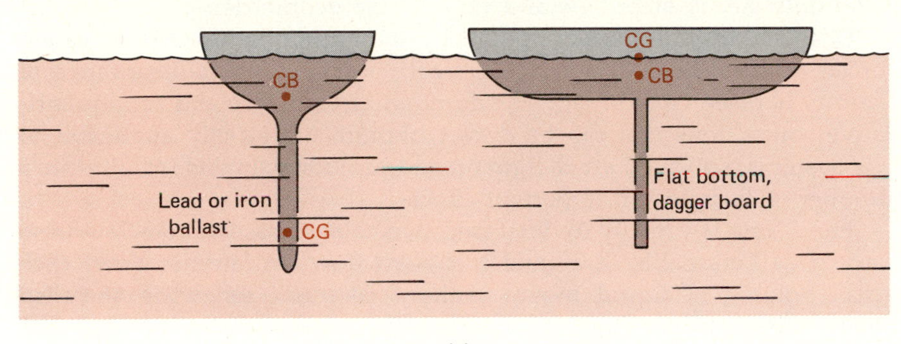

(a)

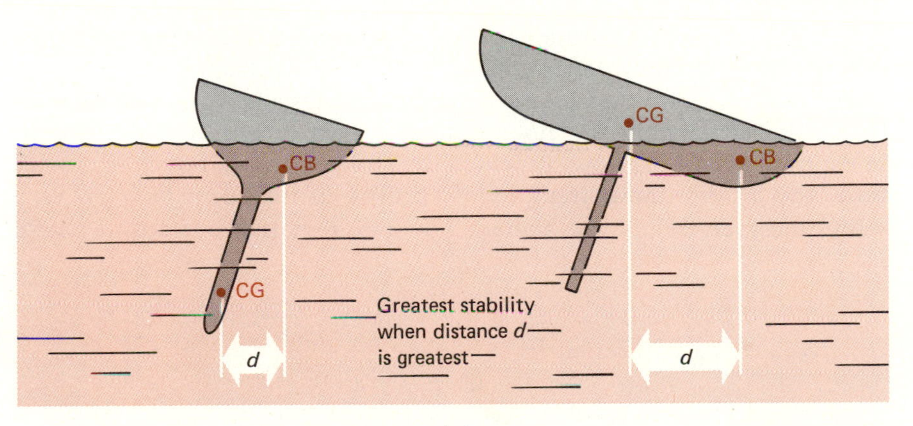

(b)

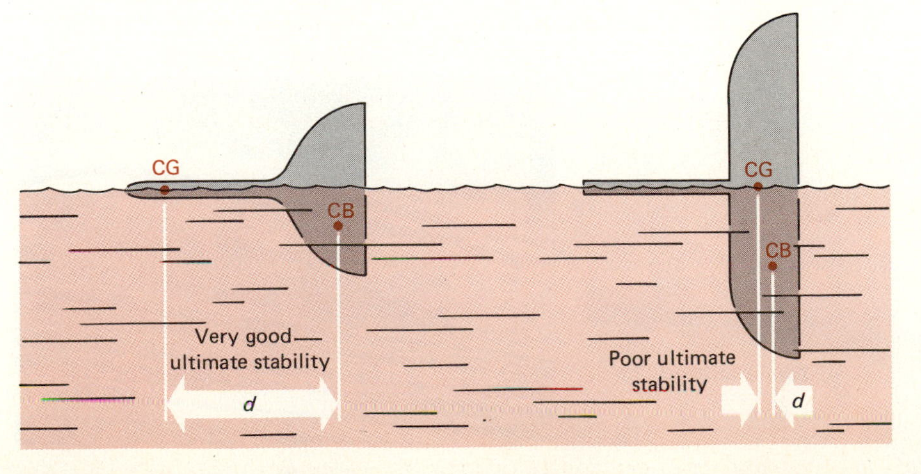

(c)

Figure 10.18 (a) Comparison of the stability of two hull designs. (b) The keel boat has a poorer initial stability than the flat-bottom boat even though in (a) the center of gravity (CG) of the flat-bottom boat is above its center of buoyancy (CB). (c) On the other hand, the keel boat has much better ultimate stability.

down," it will come quickly back up, provided that the cabin is not flooded.

10.4 Hydrodynamics; Laminar and Turbulent Flow

The formal description of fluid flow is expressed by relations between pressure, density, and velocity that hold at every point in space and at any instant. These equations of motion, derived from Newton's laws, are known as the Navier-Stokes equations; their solution poses formidable mathematical difficulties. We shall restrict attention to a few special cases and rely on various simplifying approximations. For example, we consider only steady-state flow and very simple geometries.

Despite the limitations imposed by the approximations that we are forced to use, we shall nevertheless arrive at relations whose range of validity is reasonably broad. The dominant qualitative and semiquantitative conclusions that can be drawn, though not strictly applicable to real systems, still shed much light on and provide valuable insight into a number of curious and important effects.

Fluid flow is broadly divided into two categories, laminar flow (also known as Poisseuille flow) and turbulent flow. In laminar flow, each small volume of liquid moves without rotation, following so-called *streamlines*. Laminar flow is therefore also called *streamline flow*. The streamlines describe the flow pattern in a given situation. Laminar flow obtains when the flow velocity is small.

If the flow velocity through a pipe or past some object is greater than a certain value, the flow pattern changes dramatically from laminar to turbulent, when, as implied by the adjective, eddies and vortices appear in the fluid. Turbulent flow is not a state of dynamical equilibrium; that is, in contrast to laminar flow, in which the streamlines do not change with time, the location and strengths of vortices in turbulent flow do vary with time even though the average flow rate remains constant. A useful

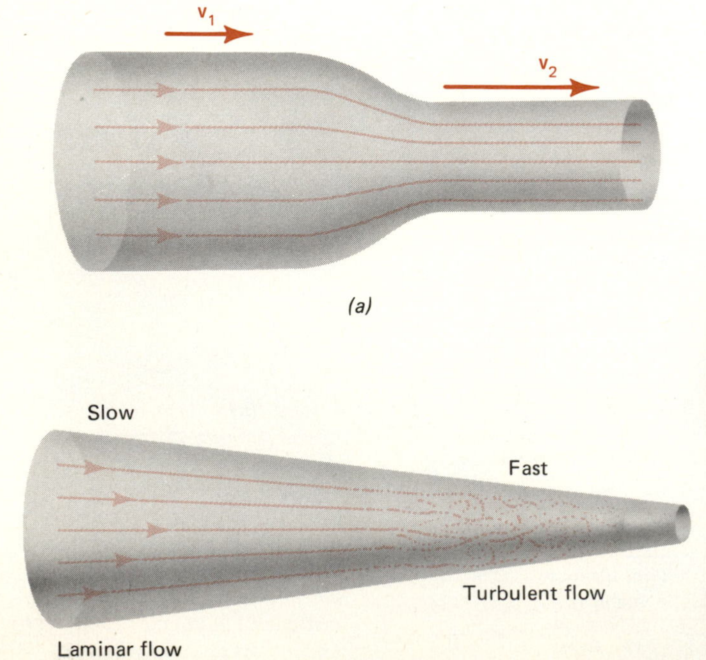

Figure 10.19 *(a) Streamlines of laminar flow of a liquid in a tube with a constriction. (b) Flow of a liquid in a tapered tube. Fluid flow is laminar where the cross section is large and flow velocity small; as the cross section decreases and the flow velocity increases, turbulence sets in. (c) Flow lines around a wing section in a wind tunnel. In the upper photograph the flow is laminar. Turbulence is evident in the lower picture.*

dimensionless parameter that serves as a measure of the character of fluid flow is the Reynolds number (see page 213); when this number exceeds about 2000, the flow pattern changes from laminar to turbulent. In the following pages we limit the discussion to laminar flow.

10.5 Bernoulli's Effect

> Beginning around 1848, when railway trains began to attain speeds of fifty or even seventy miles an hour, a bizarre and inexplicable phenomenon was noted. Whenever a fast-moving train passed a train standing at a station, the carriages of both trains had a tendency to be drawn together in what was called "railway sway." In some cases the carriages heeled over in such a pronounced fashion that passengers were alarmed, and indeed there was sometimes minor damage to coaches. Railway engineers, after a period of technical chatter, finally admitted their perplexity outright. No one had the slightest idea why "railway sway" occurred, or what to do to correct it. . . .
>
> However, by 1851 most engineers had decided correctly that railway sway was an example of Bernoulli's Law, a formulation of a Swiss mathematician of the previous century which stated, in effect, that the pressure within a moving stream of air is less than the pressure of the air surrounding it. [Michael Crichton, *The Great Train Robbery* (New York: Alfred A. Knopf, 1975).]

What is this mysterious Bernoulli effect, and what is its explanation? As we shall see shortly, the effect manifests itself in diverse situations. It accounts for part of the lift on airplane wings as well as on Frisbees and boomerangs. Atomizers, carburetors, and numerous other gadgets all pay practical tribute to the genius of Daniel Bernoulli (1700–1782), as does every pitcher who throws a mean curve.*

Bernoulli's equation, which we shall now derive, is not a fundamental relation but is, like all equations of hydrodynamics, a logical consequence of Newton's laws of motion. In this instance, the derivation is most easily performed by application of the principle of energy conservation.

Consider the steady laminar flow of an incompressible, *inviscid* liquid of density ρ through the pipe shown in Figure 10.19. (The adjective *inviscid* means that the liquid has vanishing viscosity and slides along the walls of the pipe without any viscous drag opposing its motion. Viscosity will be discussed in the next section.) During a small time interval Δt, the mass of liquid shown shaded in Figure 10.20(*a*) shifts to the position shown in Figure 10.20(*b*). The result is the same as though the mass $\Delta M = A_1 v_1 \Delta t \rho$, moving with velocity v_1 at the height h_1 through the pipe of cross section A_1, were removed and an equal mass $\Delta M = A_2 v_2 \Delta t \rho$, now moving with velocity v_2 at the height h_2, placed in the pipe of cross section A_2.

The work done on that mass of liquid during this time interval is

$$\Delta W = P_1 A_1 v_1 \, \Delta t - P_2 A_2 v_2 \, \Delta t \tag{10.9}$$

* Daniel Bernoulli was a member of history's most distinguished family of mathematicians, which included his grandfather Nicolaus, his father Jean, who had posed the problem of the brachistochrone, which Newton solved in one evening, and three brothers, all brilliant mathematicians. Even today, members of the family are carrying the family tradition forward in their native Switzerland.

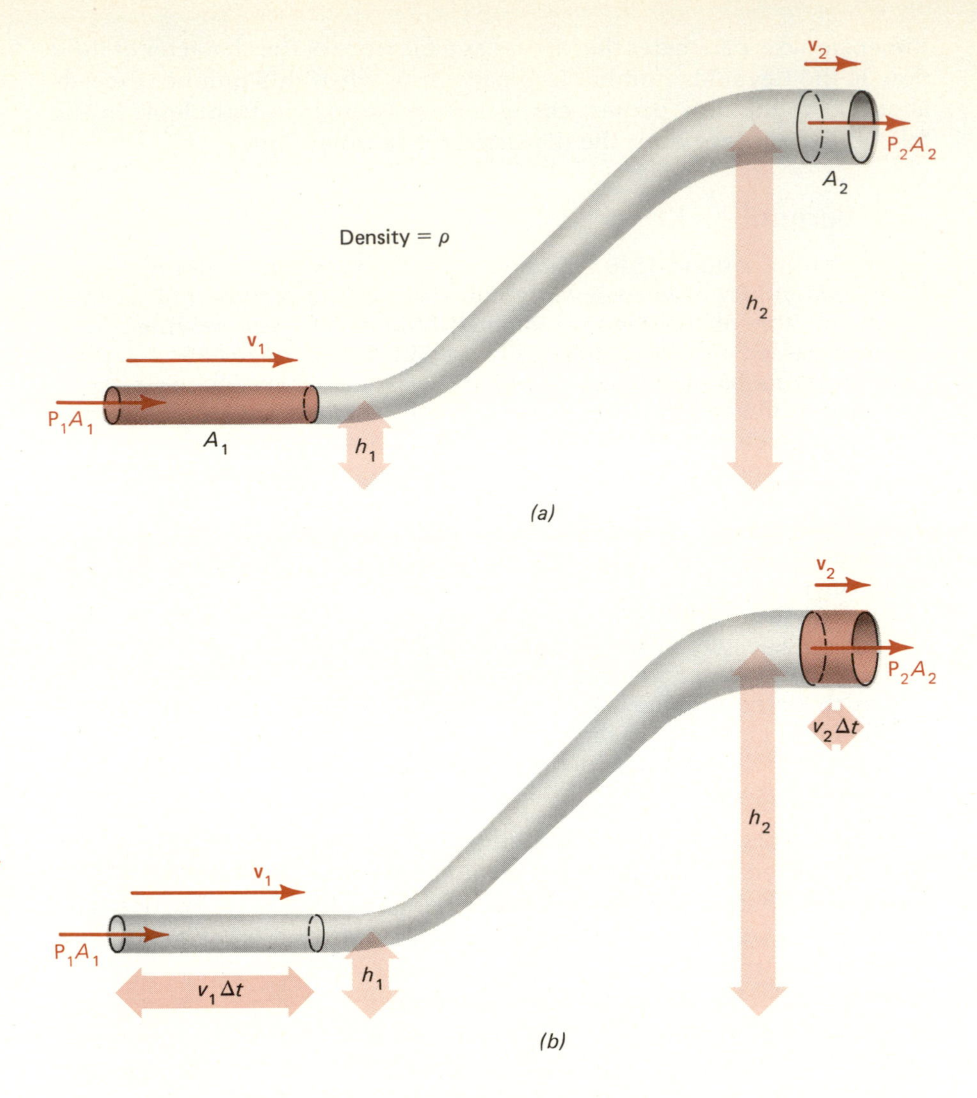

Figure 10.20 *Diagram for calculating the Bernoulli effect. An inviscid, incompressible fluid flows in the pipe. During the time Δt, the liquid shown in (a) has moved to the position shown in (b). The effect of this flow is the same as though the liquid in the volume $A_1v_1\Delta t$ had been transferred to the volume $A_2v_2\,\Delta t$.*

where we have made use of the definition of work, $\Delta W = F\,\Delta s$, and of pressure, $P = F/A$. This work done on the system must equal the gain in energy of the liquid.

There are two contributions to the energy change in the liquid, a change in PE, equal to $\Delta Mg(h_2 - h_1)$, and a change in KE, equal to $\frac{1}{2}\Delta M(v_2^2 - v_1^2)$. Energy conservation therefore leads to

$$P_1A_1v_1\,\Delta t - P_2A_2v_2\,\Delta t =$$
$$A_2v_2\,\Delta t\rho g(h_2 - h_1) + \tfrac{1}{2}A_1v_1\,\Delta t\rho(v_2^2 - v_1^2)$$

Since the liquid is assumed incompressible, the volume $A_2v_2\,\Delta t$ must equal the volume $A_1v_1\,\Delta t$. We can then cancel these common factors, and arrive at Bernoulli's equation by a simple rearrangement of terms:

$$P_1 + \rho g h_1 + \tfrac{1}{2}\rho v_1^2 = P_2 + \rho g h_2 + \tfrac{1}{2}\rho v_2^2 \qquad \textbf{(10.10)}$$

or equivalently,

$$P + \rho g h + \tfrac{1}{2}\rho v^2 = \text{constant} \qquad \textbf{(10.11)}$$

Equation (10.11) is Bernoulli's equation. The first two terms repeat the now familiar result of hydrostatics, Equation (10.3); the last term shows that the pressure in a moving fluid depends on its speed. In particular,

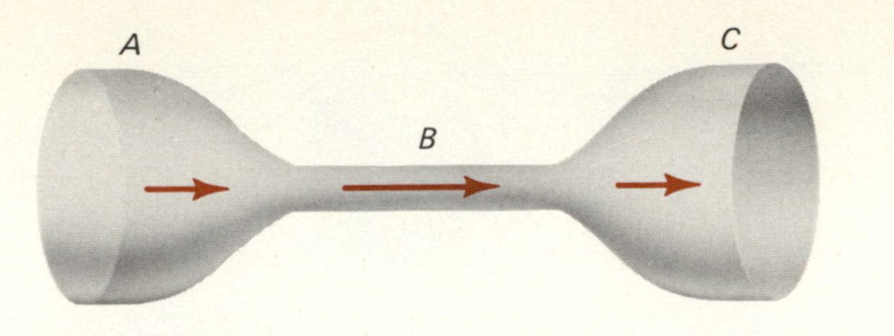

Figure 10.21 *Fluid flowing through a constricted pipe. If the flow is laminar throughout, Bernoulli's equation applies; it predicts that the pressure in the constricted region B will be less than in sections A or C.*

since an incompressible liquid's flow rate is greater in a constricted region than in a region of large cross section, the pressure is *less* within a constricted region, such as that of Figure 10.21, than in the wider portion. Though this conclusion may at first seem to contradict one's intuitive feeling, it is readily demonstrated with the simple apparatus shown in Figure 10.22.

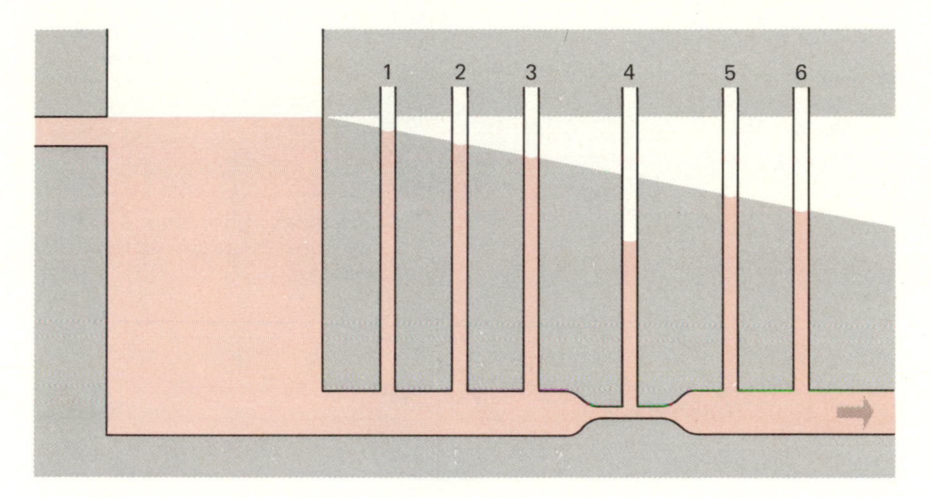

Figure 10.22 *Simple apparatus for demonstrating Bernoulli's effect. The main flow pipe is first closed off at the right, and the system is filled to maximum height h. The liquid will then stand at the same level in all the glass tubes. The stopper is then removed, and as the liquid flows out through the horizontal pipe, the liquid will stand at the indicated levels in the vertical tubes. The gradual decrease in pressure is related to viscous resistance (discussed in Section 10.6). The large drop in pressure in pipe 4 is the result of the pressure reduction in the constricted region of the flow pipe.*

We can now understand why the train standing at the station would tilt toward one rushing past it on a neighboring track. In the space between the two trains, the air is dragged along with the moving train and thus moves at a high average velocity; the air on the other side of the stationary train is at rest. According to Bernoulli's equation, the pressure exerted by the moving air is less than that exerted by the stationary, and consequently, the stationary carriages will be pushed toward the moving train. This is but one of numerous examples of Bernoulli's effect some of which are discussed in the following section.

*10.5 (a) Venturi Meter and Atomizer

The venturi meter, commonly employed in measuring fluid flow, is a U-shaped glass manometer tube partially filled with mercury or some other liquid and connected to a pipe containing a constriction (see Figure 10.24). Provided the constriction is not so severe as to cause turbulence, we can derive the relation between flow velocity of the fluid of density ρ_L in region 1 and the level difference in the manometer tube holding the manometer liquid of density ρ_M.

Since the average level of the flowing liquid is the same in the wide and constricted regions, Bernoulli's equation reads

$$P_1 + \tfrac{1}{2}\rho_L v_1^2 = P_2 + \tfrac{1}{2}\rho_L v_2^2$$

If the difference in the manometer liquid levels is h, there must be a pressure difference $P_1 - P_2$ so that a liquid of density $\rho_M > \rho_L$ may re-

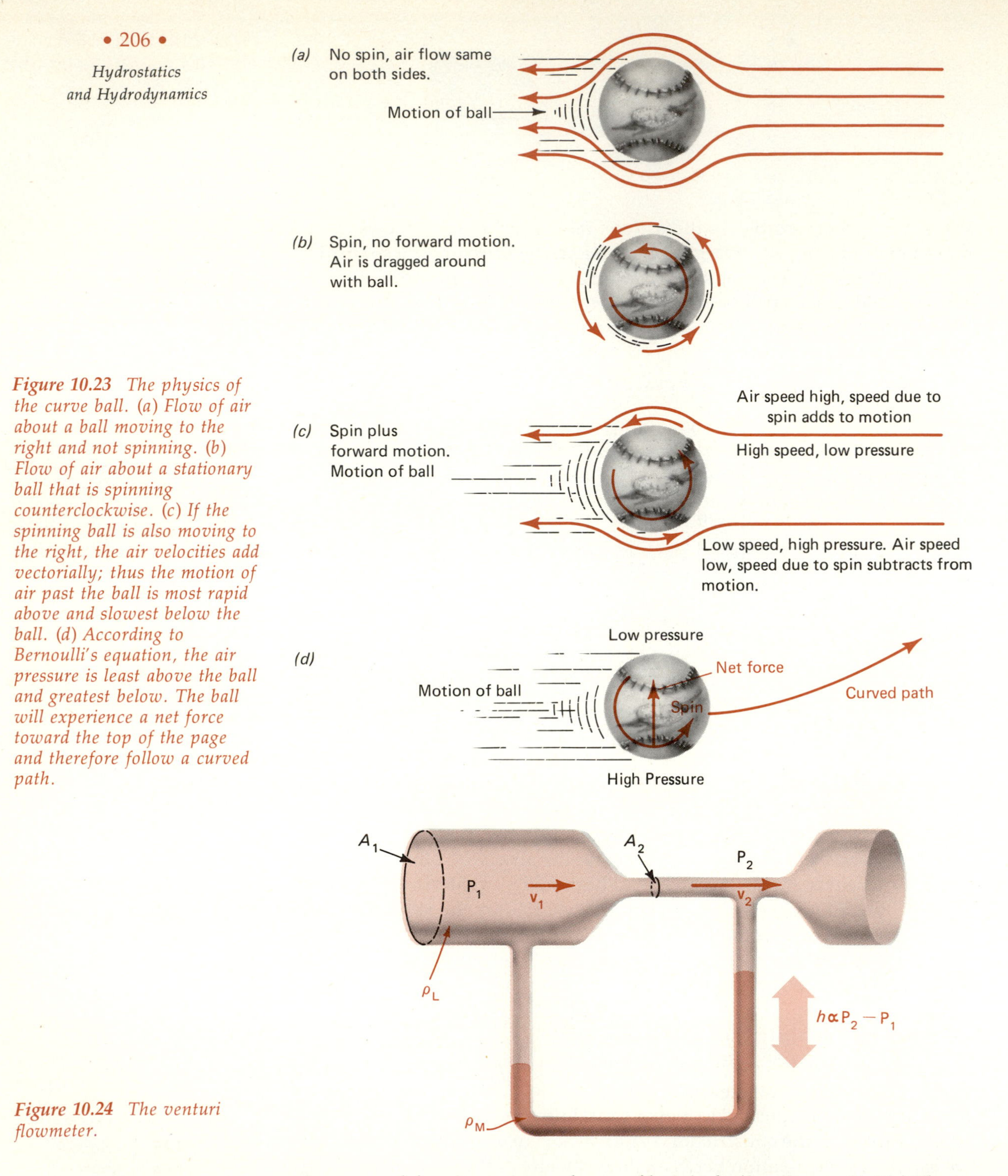

Figure 10.23 The physics of the curve ball. (a) Flow of air about a ball moving to the right and not spinning. (b) Flow of air about a stationary ball that is spinning counterclockwise. (c) If the spinning ball is also moving to the right, the air velocities add vectorially; thus the motion of air past the ball is most rapid above and slowest below the ball. (d) According to Bernoulli's equation, the air pressure is least above the ball and greatest below. The ball will experience a net force toward the top of the page and therefore follow a curved path.

Figure 10.24 The venturi flowmeter.

place one of density ρ_L in a column of height h. (See Figure 10.24.) Therefore,

$$P_1 - P_2 = (\rho_M - \rho_L)gh$$

From Bernoulli's equation, we can then write

$$(\rho_M - \rho_L)gh = \tfrac{1}{2}\rho_L(v_2^2 - v_1^2)$$

or

$$\left(\frac{\rho_M}{\rho_L} - 1\right) gh = \frac{1}{2} v_1^2 \left(\frac{v_2^2}{v_1^2} - 1\right)$$

Since $A_1 v_1 = A_2 v_2$, the preceding equation can be rewritten

$$2(r - 1)gh = v_1^2(b^2 - 1)$$

where $b = (A_1/A_2)$ and $r = (\rho_M/\rho_L)$.

Solving for v_1, we obtain

$$v_1 = \sqrt{\frac{2(r - 1)gh}{(b^2 - 1)}} \qquad \textbf{(10.12)}$$

● ***Example 10.6*** A venturi meter with a pipe diameter of 8 cm and a throat constriction of 5.8-cm diameter is used to measure the flow rate of water. The liquid in the manometer tube is mercury, and the difference in the mercury level is 2.5 mm. Determine the flow velocity of the water in the 8-cm tube.

Since the liquid in the venturi is water, the ratio r in Equation (10.12) is just the specific gravity of mercury, namely, 13.5. The ratio b equals the square of the ratio of the pipe diameters, i.e.,

$$b = \left(\frac{8}{5.8}\right)^2 = 1.90$$

Hence,

$$v_1 = \sqrt{\frac{2(13.5 - 1)(9.8 \text{ m/s}^2)(2.5 \times 10^{-3} \text{ m})}{(1.90)^2 - 1}} = 0.484 \text{ m/s}$$ ●

In an atomizer a rapidly flowing gas is forced through a nozzle, reducing the pressure in the nozzle. This drop in pressure entrains another

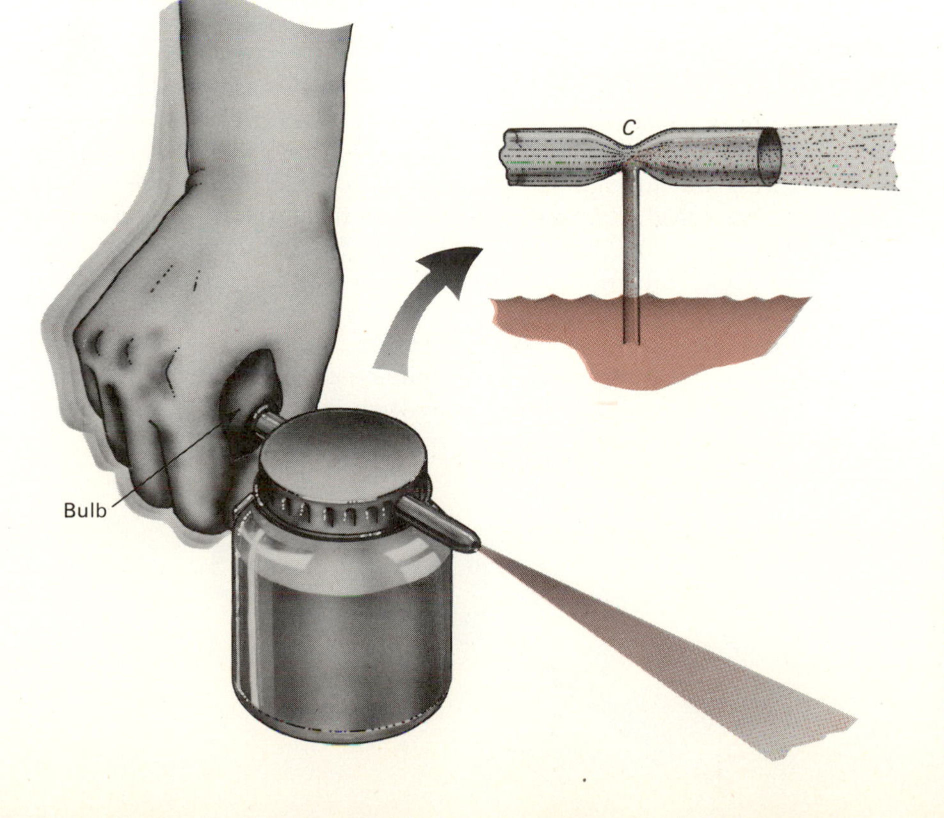

Figure 10.25 Schematic diagram of an atomizer. The air forced to the right by squeezing the bulb is at a pressure well below atmospheric in the constricted region C. As a result, liquid from the container below is pulled into the airstream, and a spray of air and fine droplets is ejected.

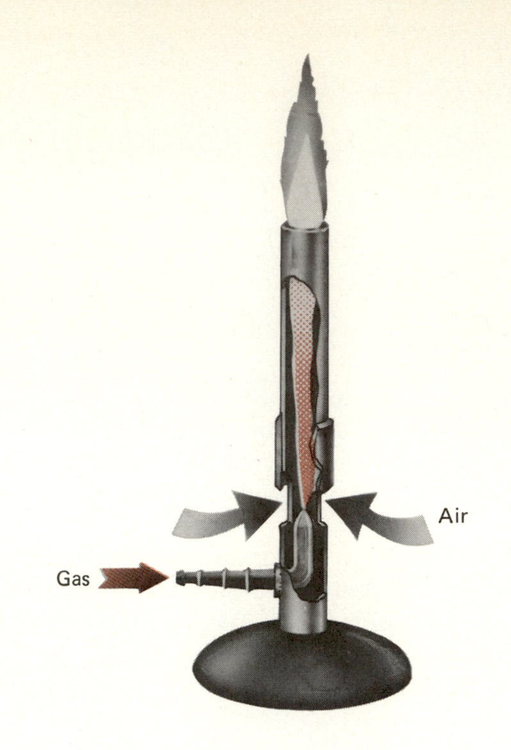

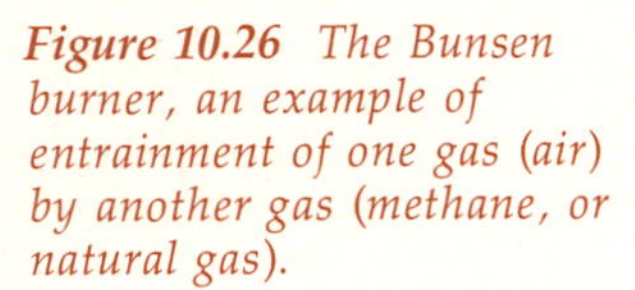

Figure 10.26 The Bunsen burner, an example of entrainment of one gas (air) by another gas (methane, or natural gas).

fluid, in this case a liquid, which emerges from the atomizer as a fine spray. Two other common examples of entrainment are the carburetor (entrainment of gasoline into a flow of air) and the Bunsen burner (entrainment of air into a stream of natural gas).

Bernoulli's effect also plays a critical role in some physiological and pathological conditions. For example, the rapid opening and closing of the air passage between the vocal cords in the larynx is due to Bernoulli's effect. This generates the basic vibrations that are then modulated by the tongue and the nasal and oral cavities into recognizable speech.

The Bernoulli effect is also responsible for a symptom of advanced arteriosclerosis, known as vascular flutter. In this disease, plaque collects within an artery, constricting the vessel with an immediate deleterious effect on the heart. It is intuitively reasonable that a constant flow rate can be maintained through such a constricted vessel only if the driving pressure is increased. Because this pressure increases in proportion to $1/R^4$ (see Section 10.6), where R is the radius of the cylindrical tube, even a relatively slight reduction in arterial radius makes a great demand on the heart muscle.

In some cases, local occlusion is so severe that blood velocity is very high at that point during contraction of the heart. According to Equation (10.11), the pressure in that region may then fall below that of the surrounding body fluids and tissues, compressing and briefly closing the

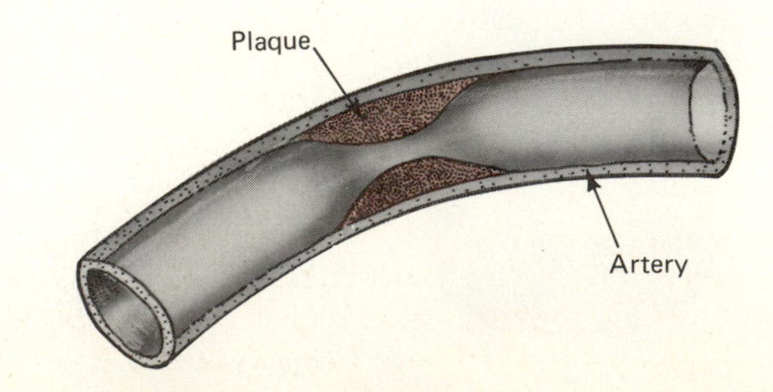

Figure 10.27 Blood flow through an artery constricted through the internal deposition of plaque. If the flow rate in the constricted region becomes sufficiently great, the artery may collapse under the external pressure and the flow of blood is momentarily interrupted. The artery then reopens, and the process starts anew, an oscillatory behavior known as vascular flutter.

artery completely. As soon as the flow stops, the Bernoulli effect disappears and arterial pressure forces a reopening; but as blood spurts through the constriction, the internal pressure again drops drastically and the vessel closes once more. Such vibration is readily discernible through a stethoscope.

10.6 Viscous Flow

In Chapter 9 we defined a fluid as a substance that cannot sustain a shear stress in equilibrium. The crucial qualifier here is the phrase "in equilibrium." In fact, a shear stress is normally required to maintain steady flow of one layer of fluid past another, and the magnitude of that shear stress is a measure of the fluid's *viscosity*. The *coefficient of viscosity* (commonly referred to as the viscosity) is defined by

$$\eta = \frac{F/A}{v/\ell} \tag{10.13}$$

where F/A is the shear stress needed to maintain laminar flow such that two planes of fluid separated by a distance ℓ have a relative velocity of v (see Figure 10.28).

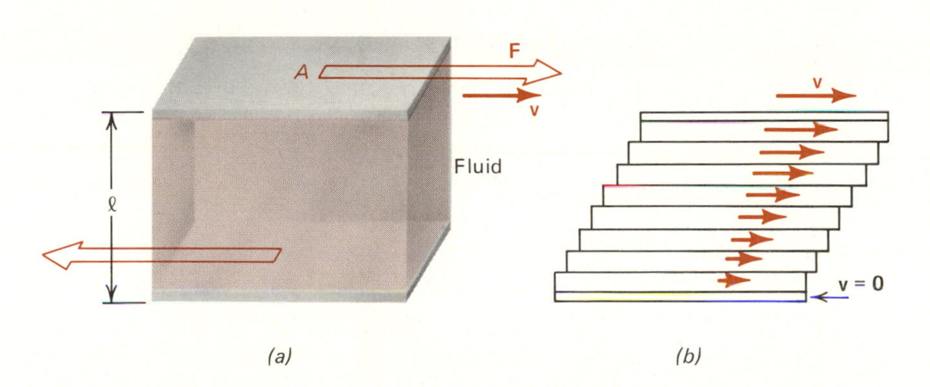

Figure 10.28 *The definition and measurement of viscosity. (a) To maintain the upper plate in motion while the lower is at rest, one must apply a horizontal shear force* **F** *per unit area. (b) In laminar flow, the motion of the fluid between the plates is shown schematically here. The layers immediately adjacent to the bounding surfaces have the same velocity as these surfaces. The velocity of the intermediate layers varies linearly from one surface to the other.*

The SI unit of viscosity is the pascal second, Pa·s, which has the dimension $[M]/([L][T])$. An older, and still widely used unit of viscosity is the *poise*, 1 poise = 0.1 Pa·s; liquid viscosities are often given in centipoises. (One centipose = 10^{-2} poise = 1 millipascal second.) Viscosities of water and of some other liquids are listed in Table 10.1. The viscosity of liquids is a fairly sensitive function of temperature, a fact well known

Table 10.1 *Viscosity coefficients of air and some liquids at 30 °C*

Substance	*Viscosity, Pa·s*
Air	1.9×10^{-5}
Acetone	2.95×10^{-4}
Methanol	5.1×10^{-4}
Water	8.0×10^{-4}
Ethanol	1.0×10^{-3}
SAE No. 10 oil	2.0×10^{-1}
Glycerin	6.29×10^{-1}
Glucose	6.6×10^{10}

Table 10.2 Viscosity coefficients of air and some liquids as functions of temperature (in Pa·s)

Temperature, °C	Castor oil	Water	Blood	Air
0	5.3	1.79×10^{-3}	—	1.7×10^{-5}
20	0.986	1.0×10^{-3}	3.015×10^{-3}	1.8×10^{-5}
37	—	0.695×10^{-3}	2.084×10^{-3}	—
40	0.231	0.656×10^{-3}	—	1.9×10^{-5}
60	0.080	0.469×10^{-3}	—	2.0×10^{-5}
80	0.030	0.357×10^{-3}	—	2.1×10^{-5}
100	0.017	0.284×10^{-3}	—	2.2×10^{-5}

to automobile owners in the northern sections of the country, where during the winter months crankcase oil sometimes becomes so viscous that rotating the crank shaft is very difficult. Not only oil but also water changes viscosity substantially with temperature, as indicated in Table 10.2.

When a viscous liquid flows through a pipe, the monomolecular layer of liquid immediately adjacent to the stationary wall is also at rest. The flow velocity is greatest in the center of the pipe, and the overall velocity profile is roughly as indicated in Figure 10.29. The simple fact that excess pressure must be maintained at one end of the pipe to sustain the flow means that there are frictional forces acting on the liquid that oppose the motion and must be overcome. This is the meaning of viscous drag.

These arguments suggest that for a given pressure differential along a pipe, the average flow speed will increase with increasing pipe diameter. We can, however, make a more precise prediction, using dimensional analysis.

The *flow rate* Q is the volume of liquid that passes a given point per second. The driving force that pushes the liquid through the pipe of length L is the pressure difference between $x = 0$ and $x = L$, $\Delta P = P(0) - P(L)$. We expect that Q is proportional to some power of ΔP, i.e., to $(\Delta P)^m$. It is also to be expected that the longer the pipe, the lower the value of Q, because the increased length increases the frictional loss due to viscous drag. Moreover, viscous drag is proportional to the viscosity coefficient. Hence, we expect Q to be proportional to $(\Delta P)^m$ and inversely proportional to η and L.

But that cannot be the whole story; Q must surely depend also on the radius of the cylindrical pipe. Suppose $\langle v \rangle$ is the average velocity of fluid flow in the pipe—some suitable average between $v = 0$ at the wall and $v = v_{\max}$ at the axis of the pipe. The volume of liquid that moves past a given point in a time Δt is then equal to the volume of the cylinder, whose radius is R and whose length is $\langle v \rangle\, \Delta t$. Since this volume is proportional to R^2 as well as $\langle v \rangle$, it follows that Q must also depend on the product of these two quantities.

Now, as mentioned earlier, the layer of liquid immediately adjacent to the wall of the pipe is stationary. Hence, for a given value of $(\Delta P)^m/\eta L$, the velocity profile will reach a higher $v_{\max}$, and a higher average velocity will be attained if the radius of the pipe is increased. Consequently, Q should depend not merely on the cross-sectional area, that is, on R^2, but on some higher power of R to account for the increase in $\langle v \rangle$ with increasing R.

We are thus led to look for a relation of the form

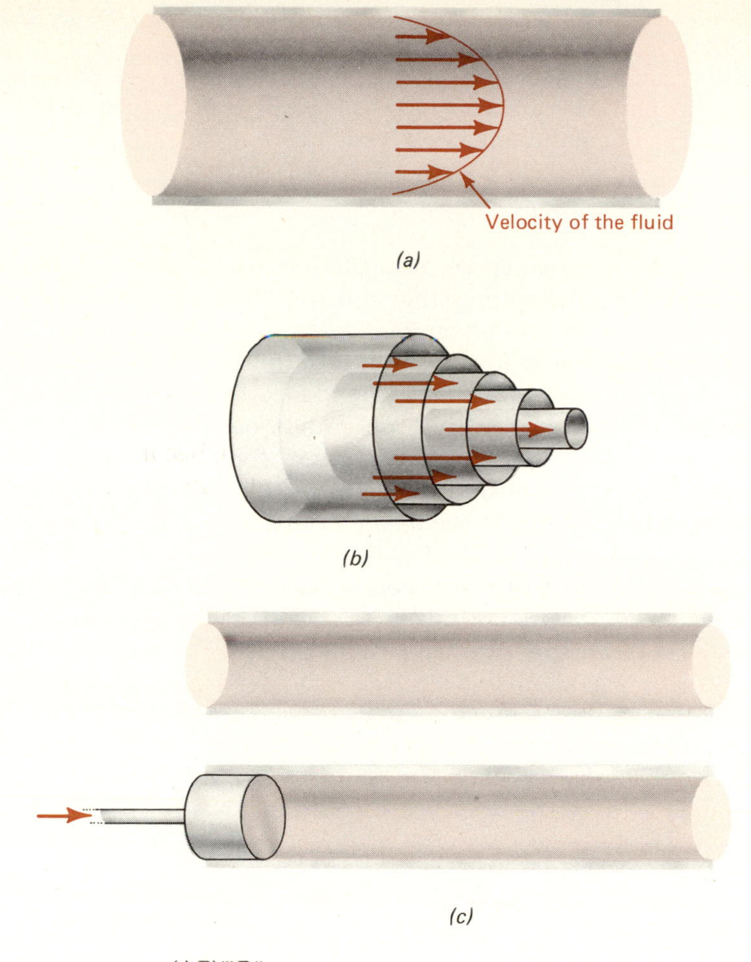

Figure 10.29 *(a) Velocity profile for laminar flow in a pipe of circular cross section. (b) Note that the liquid does not move through the pipe as a simple plug, but that the portion along the axis moves most rapidly and therefore farthest. (c) A simple demonstration of this fact is quickly prepared. A glass tube is partly filled with clear oil, and a layer of oil that is colored with food coloring is then poured on very carefully. If the glass tube is now set down and the oil forced to flow by a plunger inserted in one end, the velocity profile will become apparent. (d) Glacier, showing laminar flow.*

$$Q = \text{const} \times \frac{(\Delta P)^m R^n}{L\eta} \tag{10.14}$$

But what are the exponents m and n? We can deduce the answer by dimensional analysis, assuming, of course, that the logical sequence that led to Equation (10.14) is valid. The dimensions of the relevant variables are as follows:

$$Q = \frac{\text{volume}}{\text{time}} = \frac{[L]^3}{[T]}$$

$$\Delta P = \frac{\text{force}}{\text{area}} = \frac{[M][L]}{[T]^2} \times \frac{1}{[L]^2} = \frac{[M]}{[T]^2[L]}$$

$$L = \text{length} = [L]$$

$$R = \text{length} = [L]$$

$$\eta = \text{viscosity} = \frac{[M]}{[T][L]}$$

Dimensionally, Equation (10.14) reads

$$\frac{[L]^3}{[T]} = \left(\frac{[M]}{[T]^2[L]}\right)^m \times \frac{1}{[L]} \times \frac{[T][L]}{[M]} \times [L]^n = \frac{[M]^{m-1}[L]^{n-m}}{[T]^{2m-1}}$$

The left-hand side is independent of $[M]$; therefore, $m - 1 = 0$, or $m = 1$. With $m = 1$, we obtain $[T]$ in the denominator. The $[L]^3$ in the numerator now requires that $n - m = n - 1 = 3$, or $n = 4$! That is, Q depends on R^4.

What dimensional analysis cannot do is to tell us the value of the multiplicative constant of Equation (10.14). For a pipe of circular cross section, detailed mathematical analysis leads to

$$Q = \frac{\pi R^4 \, \Delta P}{8L\eta} \tag{10.15}$$

Equation (10.15) is known as *Poisseuille's law* or equation. It is valid provided the fluid is incompressible and the flow is laminar. Laminar flow is often referred to as Poisseuille flow.

The most significant and also the most striking prediction of Poisseuille's law is the very dramatic dependence of flow rate on cross-sectional area of the pipe. Offhand, one might have anticipated that at constant pressure difference per unit length, the volume of liquid transported would be proportional to the cross-sectional area of the pipe. In fact, the flow rate increases as the *square* of this area, that is, very much more rapidly.

One immediate consequence of Poisseuille's law is that it is generally more economical to transport fluids through a single large-diameter pipe than through several pipes of smaller diameter. Even though the smaller pipes may have a total cross-sectional area identical with that of the larger pipe, they require a much greater pressure gradient to provide the same flow rate. Correspondingly, relatively short constricted sections can introduce a substantial resistance to fluid flow. We have already commented on this in reference to cardiac strain resulting from arteriosclerosis.

● ***Example 10.7*** Glycerin is poured into a liter bottle from a container through a funnel, as shown in Figure 10.30. The neck of the funnel is 1 cm long and 1 cm in diameter. If the funnel is kept filled, that is, if the level of glycerin is always 10 cm above the neck, how long will it take to fill the liter bottle? The specific gravity of glycerin is 1.26.

In this case the major limitation to the flow rate is the somewhat narrow neck of the funnel; we shall not make a grievous error by neglecting the resistance to flow in the rest of the funnel.

Except for ΔP, all the quantities on the right-hand side of (10.15) are known; $R = 0.5$ cm, $L = 1$ cm, $\eta = 0.629$ Pa·s (Table 10.1). The pressure acting on the liquid as it leaves the bottom of the funnel is just atmospheric pressure. The pressure on the liquid that forces it through the neck of the funnel is atmospheric pressure augmented by the pressure of 10 cm of glycerin (we neglect here the small correction due to the 1 cm of liquid in the neck itself). The pressure difference that forces the liquid through the neck is

$$\Delta P = \rho g h = (1.26 \times 10^3 \text{ kg/m}^3)(9.8 \text{ m/s}^2)(0.1 \text{ m}) = 1.235 \times 10^3 \text{ Pa}$$

Substituting into Equation (10.15), we have

$$Q = \frac{\pi(5 \times 10^{-3} \text{ m})^4(1.235 \times 10^3 \text{ Pa})}{8(1 \times 10^{-2} \text{ m})(0.629 \text{ Pa}\cdot\text{s})} = 4.8 \times 10^{-5} \text{ m}^3/\text{s} = 48 \text{ mL/s}$$

At a flow rate of 48 mL/s, the time needed to fill a 1-L (1000 mL) bottle is

$$t = \frac{1000 \text{ mL}}{48 \text{ mL/s}} = 21 \text{ s}$$

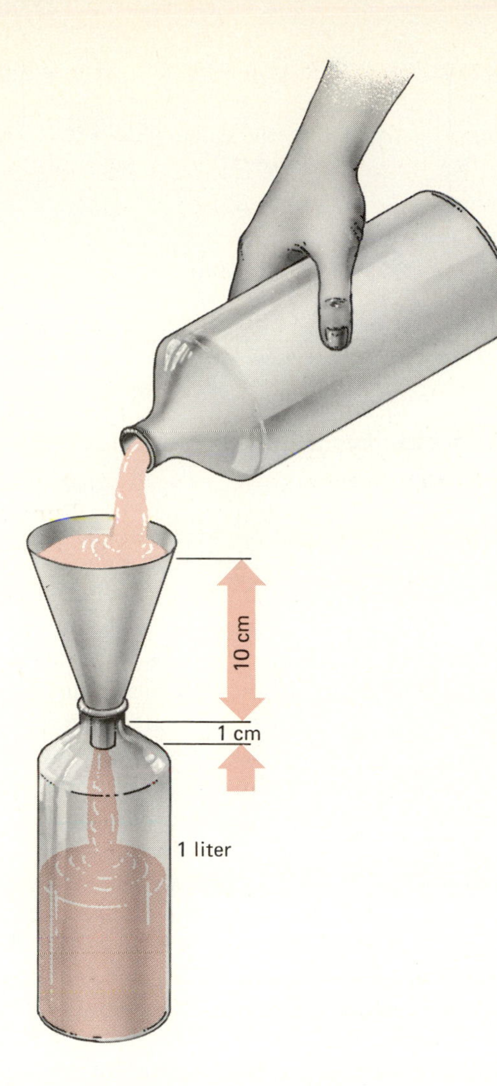

Figure 10.30 A one-liter flask is to be filled with glycerin poured into the top of the funnel. The principal resistance to the flow of glycerin is the narrow neck of the funnel.

Since glycerin is a rather viscous liquid, it is not unreasonable that nearly half a minute may be needed to pour a liter through this funnel.

10.7 Turbulence

When the flow rate exceeds a certain critical velocity, the flow ceases to be laminar and becomes turbulent. The parameter that is useful in characterizing fluid flow is Reynolds number, defined by

$$R_{\mathrm{n}} = \frac{\rho v \ell}{\eta} \qquad \textbf{(10.16)}$$

where ρ is the density of the fluid, η its viscosity, v its velocity, and ℓ a characteristic length of the vessel carrying the flow; for flow through a cylindrical pipe of circular cross section, that length is the radius of the pipe.

It is an experimental observation that in a pipe with smooth walls turbulence generally sets in when R_{n} is of order 2000. If the pipe is not smooth or has bends, turbulence appears at a lower Reynolds number.

The resistance to flow is greater if the flow pattern is turbulent than if it is laminar. Thus, if Poisseuille's equation is applied in situations where the velocity exceeds the critical velocity for turbulence, the results will be in error.

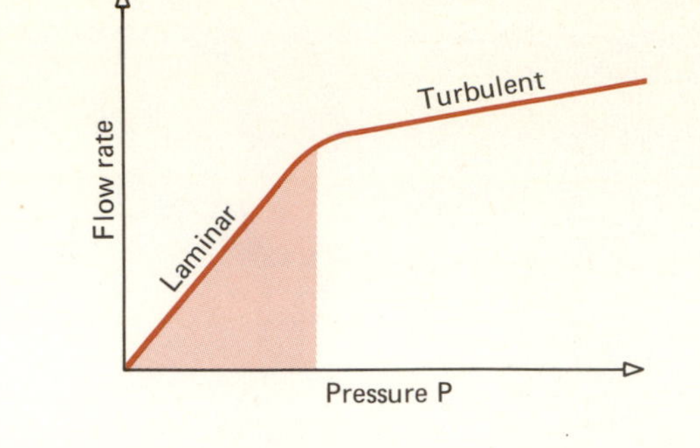

Figure 10.31 *Flow rate as a function of pressure per unit length. Notice that once turbulence sets in, the pressure gradient increases much more rapidly for a given change in flow rate than during the portion characterized by laminar flow.*

*10.8 Terminal Velocity, Sedimentation, and Centrifugation

If you simultaneously drop a Ping-Pong ball and a stone of roughly the same size from a modest height, you will find that the two do not reach the ground at the same instant; the Ping-Pong ball arrives a bit later. You can also try watching raindrops fall during a summer rain, and if you do so carefully, you will note that the larger ones fall more rapidly than the small ones. In the first instance, the two objects are of equal size but unequal density, and the object of lower density falls more slowly. In the second, the density is the same but the sizes of the two objects, the raindrops, differ. In both cases the more massive object falls at a greater rate, and one might be tempted to conclude that mass is all that matters; but that would be incorrect as we shall soon see.

These observations, which seem to support Aristotle's conclusion, are the result of air friction, of the fact that these objects are moving through a medium, although a rarefied one. Air friction is not, however, a frictional force of the sort considered earlier (Chapter 3) and characterized by a coefficient of friction. We can quickly convince ourselves that this could not be the correct description.

If air friction could be written as the product of a frictional coefficient and weight, in a relation like Equation (3.13), the resulting motion would necessarily fall into one of two categories: (1) If the frictional coefficient is equal to or greater than one, the object will experience no net downward force and will therefore simply remain at rest. (2) If the frictional coefficient is less than one, there is a net downward force $mg - \mu mg$ acting on the object; hence it will accelerate downward at constant acceleration $a = (1 - \mu)g$ and never reach a terminal velocity. Yet, measurements of velocities of bodies falling through air or other fluids show that such bodies do approach a limiting, terminal velocity.

Consequently, the resisting force of air friction cannot be independent of the motion. It is reasonable to assume that *this force is velocity-dependent*, becoming greater as the velocity increases. For example, if F_a, the force due to air friction, were proportional to $|v|$ and directed opposite to $\mathbf{v}$, then as the speed increased initially under the action of gravity, the opposing force of friction would also gradually increase, reducing the magnitude of the downward acceleration. Ultimately, the speed would attain a value for which the downward force mg and the upward frictional force would be equal in magnitude; the object is then in equilibrium and according to Newton's first law, continues its motion at constant velocity; that is, it will have attained its *terminal velocity*. So now that we understand qualitatively how a terminal velocity arises, let us determine the relation between F_a and v.

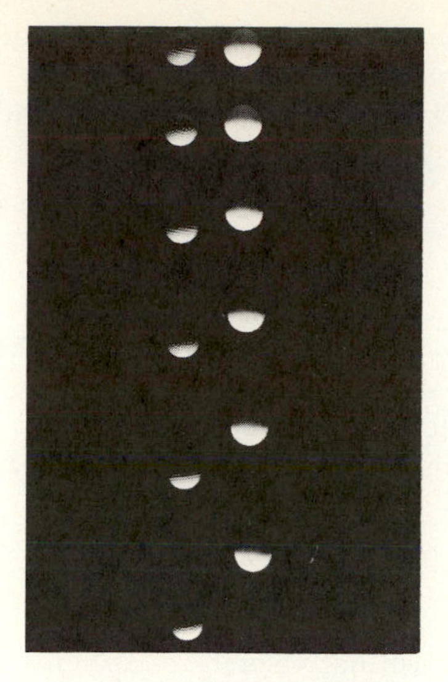

Figure 10.32 A golf ball and a polystyrene sphere are released simultaneously and fall through air. The multiple-exposure stroboscopic photograph shows that the golf ball accelerates throughout the sequence (the distance between successive positions always increases), whereas the Styrofoam sphere appears to fall at about the same speed during the last two or three intervals. The polystyrene sphere has attained terminal velocity.

Instead of examining an object, such as a solid sphere, moving with speed v through stationary air, let's examine the same physical event from a reference frame in which the sphere is at rest and the air is streaming past it with a speed v. The average momentum of each air molecule is $m_a v$, where m_a is the average mass of an air molecule. Let us assume that whenever an air molecule strikes the sphere, it sticks to it momentarily and loses its forward momentum; to be precise, it imparts its forward momentum to the combined system of sphere-plus-molecule in this inelastic collision. Thus, the momentum gained by the sphere with *each* such inelastic event is proportional to the velocity of the air rushing past the sphere.

The number of these collisions that occur each second depends on the cross-sectional area of the sphere and on the velocity of the air flow; the greater the wind speed, the greater the number of air molecules that strike the sphere each second.

Now Newton's second law, in its more general form, states that the force acting on an object equals the rate of change of its momentum (see Equation (5.2)). Accordingly, the force due to air friction equals the momentum imparted to the object per second by the air molecules that strike it. From the foregoing, it follows that

$$F_a = CAv^2 \qquad \textbf{(10.17)}$$

where A is the cross-sectional area of the sphere (or for a less regularly shaped object, the area projected on a plane perpendicular to the air flow), and v is the speed of the object relative to the air. Notice that F_a is a function of v, but depends on v^2, not v. One factor of v appears because the average momentum of *each* air molecule is proportional to v, the other because the *number* of molecules striking the object per second is also proportional to v. The constant C depends on the density of the gas. For an object falling through air at atmospheric pressure, $C \simeq 1\ \text{kg/m}^3$.

Suppose several small aluminum spheres are released simultaneously from the same great height. The spheres have different radii. Will all spheres arrive at the ground at the same time? If not, which will take the longest to fall a given distance?

The object with the greater terminal velocity will reach the ground first. Therefore, we must determine how the terminal velocity of a sphere depends on its radius. Of course, we already know the answer to the

Figure 10.33 In a reference frame moving at the same velocity as a sphere, the sphere is stationary and the air moves past it with the speed that the sphere has in the rest frame of the air. As the air molecules strike this sphere, they impart a small amount of momentum to the sphere with each impact. The amount of momentum per impact is proportional to the velocity of the airflow (or in the reference frame in which the air is at rest, to the velocity of the sphere). The number of molecules that strike the sphere per second is proportional to the cross-sectional area of the sphere and also to the velocity of the airflow.

qualitative question from our earlier observation that small raindrops have a lower terminal speed than larger ones.

The terminal velocity, v_t is reached when F_a exactly balances the force of gravity mg, i.e., when

$$F_a = mg, \qquad CAv_t^2 = mg \tag{10.18}$$

Solving for v_t, we have

$$v_t = \sqrt{\frac{mg}{CA}} \tag{10.19}$$

The mass of a sphere equals its density ρ times its volume $\frac{4}{3}\pi R^3$; its cross-sectional area A is πR^2. Inserting these expressions into (10.19), we obtain

$$v_t = \sqrt{\frac{4R\rho g}{3C}} \tag{10.20}$$

Equation (10.20) shows that the terminal velocity of spheres falling through air is proportional to the square root of the product $R\rho$. Since ρ for all the spheres of our example is the same, the smallest sphere will have the lowest terminal velocity and consequently will be the last to reach the ground. This is a particular case of a general rule, namely, *for objects of the same shape and density the smallest ones will have the lowest terminal velocity.*

● ***Example 10.8*** Estimate the terminal velocity of a raindrop. Compare the terminal velocity of this raindrop with that of a hailstone of equal size.

To answer this question, we must first estimate the size of a raindrop; a reasonable estimate is a sphere of 2-mm diameter. The density of water is 10^3 kg/m^3. Putting these values into (10.20), we obtain

$$v_t(\text{drop}) = \sqrt{\frac{4(10^{-3}\text{ m})(10^3\text{ kg/m}^3)(9.8\text{ m/s}^2)}{3(1\text{ kg/m}^3)}} = 3.6\text{ m/s}$$

which is a fair estimate.

A hailstone, a sphere of ice, if it is of the same size will have a slightly lower terminal velocity because it is about 10 percent less dense than water. ●

10.8 (a) Sedimentation and Centrifugation

When the relative velocity between an object and a fluid is so low that the flow is laminar, the frictional force derives from viscosity; it is then called viscous drag. From the definition of viscosity coefficient (see Section 10.6) it follows that in this limit the retarding frictional force is proportional to v and not v^2.

In this case, the viscous drag force is of the form

$$F_v = \text{const} \times Rv\eta \tag{10.21}$$

where R is some characteristic length. For a smooth sphere of radius R,

$$F_v = 6\pi Rv\eta \qquad \text{(spherical object)} \tag{10.22}$$

Suppose we have a suspension of very small spherical particles of density ρ in a liquid of density ρ_L. If $\rho > \rho_L$, these small particles will, under the influence of gravity, slowly descend to the bottom of the con-

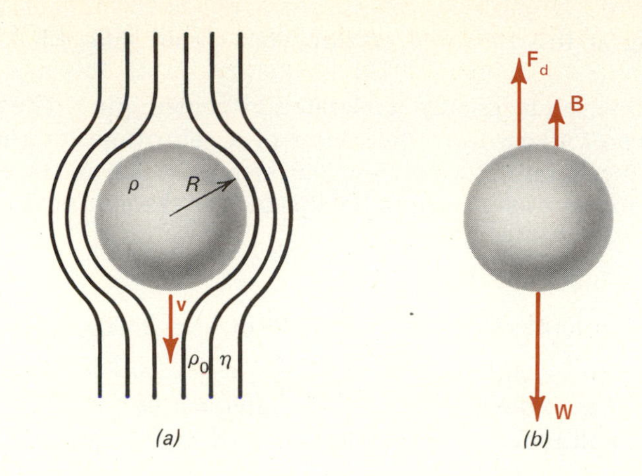

Figure 10.34 (*a*) *A sphere falling through a viscous medium. At sufficiently low speeds, the dominant retarding force is due to viscous drag.* (*b*) *The forces acting on the sphere. They are the weight mg, the buoyant force, and the viscous drag force. The sphere attains terminal velocity when the vector sum of these three forces is zero.*

tainer, forming a sediment. The sedimentation rate is determined by the *sedimentation velocity* v_s, which is just another name for terminal velocity.

This sedimentation velocity is attained when the net force on the particle vanishes. There are three forces we must consider, namely, gravity, buoyancy, and viscous drag. (We really should have included buoyancy in our discussion of objects falling through air, but that force is usually, though not always, negligibly small.) The total downward force, which must vanish when $v = v_s$, is then the sum of the following three terms:

$$\tfrac{4}{3}\pi R^3 \rho g - \tfrac{4}{3}\pi R^3 \rho_L g - 6\pi R \eta v_s = 0$$

Here the first term is the weight of the sphere, the second the buoyant force exerted on the sphere by the displaced liquid, and the last the viscous drag when the sphere is moving through the liquid with speed v_s. Solving for v_s gives

$$v_s = \frac{2R^2 g(\rho - \rho_L)}{9\eta} \qquad (10.23)$$

Equation (10.23) has several noteworthy features:

1. The sedimentation velocity v_s is proportional to g. If we can increase the *effective* gravitational force, for example, by centrifuging, we can enhance the sedimentation rate.

2. If ρ, ρ_L, and η are known, measuring v_s can provide information on the size of the spherical particle. R. A. Millikan employed this technique in his historic measurement of the electronic charge.

● ***Example 10.9*** An erythrocyte (red blood cell) can be approximated by a sphere of 4-μm diameter of density 1.3×10^3 kg/m^3. The density of blood is about 1.06×10^3 kg/m^3. What is the sedimentation velocity of red blood cells in blood (a) under gravity? (b) in a centrifuge vial that is 10 cm from the center of the centrifuge and rotating at 5000 rpm?

We shall calculate the values at 20 °C rather than at body temperature, since we are presumably concerned with results in a laboratory environment. The viscosity of blood at 20 °C is 3×10^{-3} Pa·s, and substituting the numerical values into Equation (10.23), we find that

$$v_s = \frac{2(2 \times 10^{-6}\ \text{m})^2(9.8\ \text{m/s}^2)[(1.3 - 1.06) \times 10^3\ \text{kg/m}^3]}{9(3 \times 10^{-3}\ \text{Pa}\cdot\text{s})}$$
$$= 7 \times 10^{-7}\ \text{m/s}$$

Sedimenting at this rate, the erythrocytes would take 4 h to descend 1 cm.

The process can be greatly accelerated by centrifuging. The effective g at a distance of 10 cm from the center of a centrifuge rotating at 5000 rpm = (5000 rev/min)(2π rad/rev)/(60 s/min) = 524 rad/s equals the centripetal acceleration at that radial distance in the spinning centrifuge, namely,

$$\begin{aligned} g_{\text{eff}} &= a_c = r\omega^2 \\ &= (0.1 \text{ m})(524 \text{ rad/s})^2 = 2.75 \times 10^4 \text{ m/s}^2 = 2800g \end{aligned}$$

Accordingly, the sedimentation velocity in the centrifuge is also 2800 times as rapid as under gravity, i.e., $v_s(\text{centr}) = 1.96 \times 10^{-3}$ m/s, and the erythrocytes will move 1 cm radially outward in only about 5 s. ●

There is, however, one important fact that we neglected in the preceding discussion of sedimentation. In any mixture there is always a natural tendency toward a random distribution of the constituents. If there is a concentration gradient, that is, if the distribution of the particles is not uniform, there will be a flow of particles such as to homogenize the distribution. The process is called diffusion and is characterized by a diffusion coefficient. This randomizing mechanism will clearly counteract sedimentation, which tends to establish a nonuniform distribution.

In a liquid, the diffusion coefficient is inversely proportional to the radius of the particles which we assume are spherical. Thus, as R becomes very small, not only does v_s diminish according to (10.23), but worse yet, the countervailing effect of diffusion prevents any significant sedimentation altogether. It is especially in these cases, for example, in concentrating proteins from an aqueous solution, that an increase in the driving force of sedimentation by centrifuging is absolutely essential. Without it we would never observe sedimentation, no matter how patiently we waited.* In many research applications the ordinary laboratory centrifuge cannot generate sufficiently high effective gravitational forces; ultracentrifuges, whose rotational speeds range between about 50,000 and 100,000 rpm, routinely achieve accelerations of $3 \times 10^5 g$.

Summary

To good approximation, the dependence of atmospheric pressure on height is given by

$$P_A(h) = P_A(0)e^{-h/h_0}$$

where $P_A(h)$ is the atmospheric pressure at an elevation h, and h_0 is a constant equal to 8.6 km. For small changes in h, the exponential may be approximated by a linear function, that is,

$$P_A(h + \Delta h) \simeq P_A(h)\left[1 - \frac{\Delta h}{h_0}\right], \qquad \Delta h \ll h_0$$

* Since oxygen molecules are somewhat heavier than nitrogen molecules, sedimentation should ultimately result in an atmosphere consisting of O_2 near ground and N_2 at high altitudes. Yet, despite the billions of years that have elapsed since vegetation appeared, generating O_2, the O_2/N_2 ratio in the atmosphere is practically independent of altitude, a result of gaseous diffusion.

The pressure at a depth h below the surface of a liquid is

$$P(h) = P_A + \rho g h$$

where P_A is the atmospheric pressure at the surface of the liquid and ρ is the mass density of the liquid. The pressure difference between two points in a liquid that differ in depth by Δh is given by

$$\Delta P = \rho g \, \Delta h$$

Two important principles of hydrostatics are:

Isotropy of pressure. *At equilibrium, the pressure at any point in a fluid is the same in all directions.*

Pascal's principle. *Under equilibrium conditions, a change in pressure at any point in an incompressible fluid is transmitted uniformly to all parts of the fluid.*

In a gravitational field, a body partly or completely submerged in a fluid experiences a buoyant force. Archimedes' principle states that this *buoyant force equals the weight of the displaced fluid.* The buoyant force acts through the *center of buoyancy,* which is at the center of mass of the displaced fluid. Buoyancy can be used to determine the average density of bodies.

Laminar flow (also called *streamline,* or *Poisseuille* flow) is characteristic of small flow velocities. At high velocities, fluid flow becomes turbulent.

Bernoulli's equation

$$P + \rho g h + \tfrac{1}{2}\rho v^2 = \text{constant}$$

shows that the pressure in a fluid is related to its location in the gravitational field and to its flow velocity. Atomizers, carburetors, airplane wings, venturi flowmeters, and many other devices depend on the effect elucidated by Bernoulli. Bernoulli's equation is valid only if the flow is laminar.

Viscosity opposes fluid flow. The *viscosity coefficient,* defined by

$$\eta = \frac{F/A}{v/\ell}$$

is the shear stress F/A per unit velocity gradient v/ℓ.

For laminar flow of a viscous fluid through a cylindrical pipe, the flow rate Q is given by Poisseuille's equation:

$$Q = \frac{\pi R^4 \, \Delta P}{8 L \eta}$$

where R is the radius of the pipe, ΔP the pressure difference over the length L of the pipe, and η the viscosity of the fluid. Like Bernoulli's equation, Poisseuille's equation is also valid only if the fluid flow is laminar.

Generally, *turbulence* will appear in fluid flow through a smooth pipe when *Reynolds number*

$$R_n = \frac{\rho v \ell}{\eta}$$

exceeds a value of about 2000.

An object falling in a fluid under gravity experiences a velocity-dependent retarding force. If the velocity is small, this force is due to viscous drag and is proportional to the relative velocity between fluid and object. At high relative velocity, the retarding force is proportional to the square of the velocity. In either case, a freely falling body attains a *terminal velocity* that depends on its density and dimensions as well as on the properties of the fluid through which it is moving.

For small particles, such as finely divided powder, descending through a fluid, the process is called *sedimentation,* and the terminal velocity is known as the *sedimentation velocity.* Since the sedimentation velocity is proportional to the acceleration of gravity g, sedimentation can be much enhanced by increasing the effective g through centrifuging.

Multiple Choice Questions

10.1 Water flows through the pipe shown in Figure 10.35. The flow is laminar. The pressure

(a) is greater at A than at B.
(b) at A equals that at B.
(c) is less at A than at B.
(d) at A is unrelated to that at B.

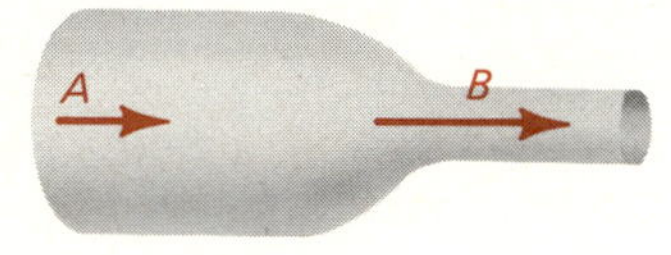

Figure 10.35

10.2 A small sphere of mass M is dropped from a great height. After it has fallen 100 m, it has attained its terminal velocity and continues to fall at that speed. The work done by air friction against the sphere during the first 100 m of fall is

(a) greater than the work done by air friction in the second 100 m.
(b) less than the work done by air friction in the second 100 m.
(c) equal to $100Mg$.
(d) greater than $100Mg$.

10.3 An object of mass M is suspended from a spring balance that reads 25 N. When the mass is completely submerged in water, the spring balance reads 5 N. The specific gravity of the object

(a) is 2.0.
(b) is 1.5.
(c) is 1.25.
(d) cannot be determined from the available data.

10.4 A cube is immersed in water. The pressure is greatest against

(a) the sides of the cube.
(b) the top of the cube.
(c) the bottom of the cube.
(d) None of the above; the pressure is the same on all six sides.

10.5 An object of uniform density floats on water with three-fourths of its volume submerged. Its specific gravity is

(a) $\frac{1}{4}$.
(b) $\frac{3}{4}$.
(c) 1.
(d) $\frac{4}{3}$.

10.6 A piece of wood of density 0.8 g/cm^3 (specific gravity = 0.8) is observed to float on a liquid whose specific gravity is 1.2. The fraction of the wood that is submerged below the surface of the liquid

(a) is 33%.
(b) is 67%.
(c) is 80%.
(d) cannot be stated unless the volume of the wood is known.

10.7 Suppose you release two steel marbles in a highly viscous liquid such as glycerin. One marble has a diameter twice as great as the other. The sedimentation velocity of the larger marble will be

(a) half
(b) twice
(c) four times
(d) one-fourth

the sedimentation velocity of the smaller one.

10.8 Two pistons exert forces F_1 and F_2 on a fluid at points 1 and 2, producing pressures P_1 and P_2 over areas A_1 and A_2 as indicated in Figure 10.36. If the

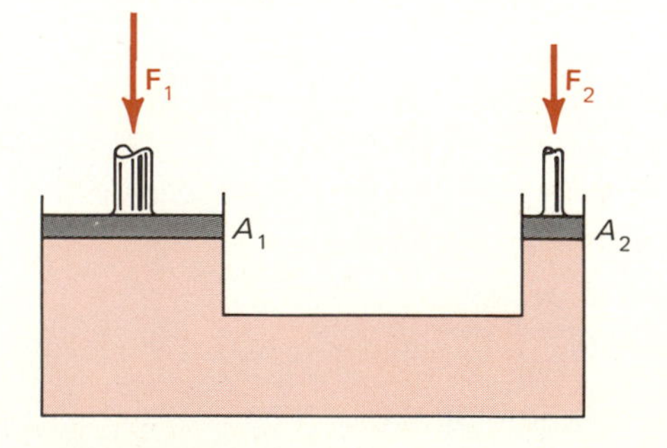

Figure 10.36

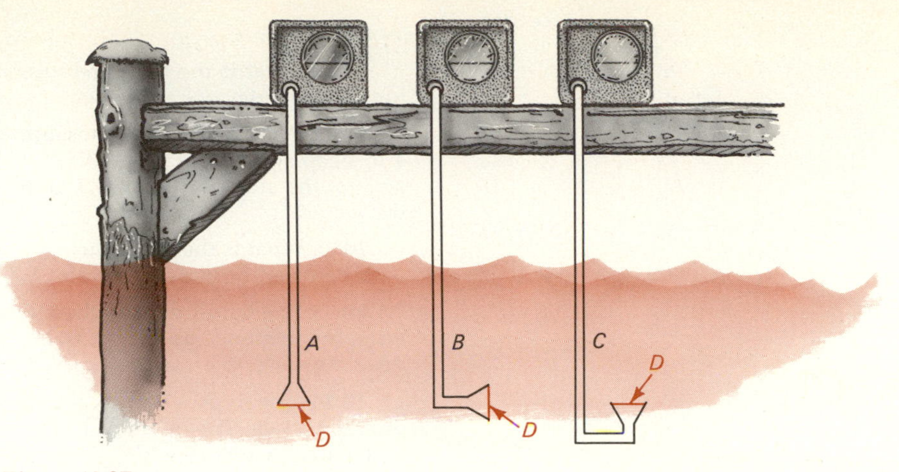

Figure 10.37

pistons do not move under the influence of these forces it then follows that

(a) $F_1 = F_2$

(b) $F_1 = (A_2/A_1)F_2$

(c) $P_1 = (A_1/A_2)P_2$

(d) $P_1 = P_2$

10.9 A block of wood and a 1-kg mass of lead are placed in a bowl that is filled to the brim with water. The 1-kg mass is lifted from the water by a thin wire, and as this is done, the level of the water drops a bit. The lead mass is now placed on the block of wood, which remains afloat while supporting the mass. As the lead is placed on the wood and floated,

(a) the level of the water rises again but does not reach the brim.

(b) the level of the water rises exactly to the brim, as before.

(c) some water spills over the edge of the bowl.

(d) There is not enough information provided to decide what would happen.

10.10 Pressure-sensing devices are placed below the surface of a lake, as shown in Figure 10.37. Each device registers the pressure exerted on the small diaphragm *D*. The following will be observed:

(a) $P_A = P_B$; $P_A > P_C$

(b) $P_A = P_C$; $P_A > P_B$

(c) $P_B = P_C$; $P_A < P_B$

(d) $P_A = P_B = P_C$

10.11 Two Styrofoam spheres of radii R_1 and R_2, $R_1 > R_2$, are released simultaneously from the roof of a 30-m-tall building.

(a) Both spheres will reach the ground at the same moment.

(b) Sphere 1 will arrive at the ground first.

(c) Sphere 2 will arrive at the ground first.

(d) The answer depends on the local atmospheric pressure.

10.12 A large, completely enclosed plastic box of 1-m^3 volume is placed on one pan of a balance. The balance is in equilibrium when a 3-g brass mass is placed on the other pan.

(a) The mass of the box is exactly 3 g.

(b) The weight of the box is exactly 0.0294 N.

(c) The mass of the box is greater than 3 g.

(d) The weight of the box is less than 0.0294 N.

10.13 If a mixture of water and very minute oil droplets (an emulsion of oil in water) is placed in a centrifuge and spun at high angular speed,

(a) the oil will move rapidly toward the periphery.

(b) the oil will move rapidly toward the center.

(c) the oil will move toward the periphery, but more slowly than it would without centrifuging.

(d) the oil will move toward the center, but more slowly than it would without centrifuging.

10.14 A Ping-Pong ball is attached to the bottom of a pail by a rubber band. The pail is then filled with water until the ball is at rest below the surface of the water, as shown in Figure 10.38. The pail, with water, ball, and rubber band, is then carried to the top of a tall building and released from rest at the edge of the roof. The motion of the ball relative to the pail is then best described as follows.

(a) The ball remains at rest below the surface of the water.

(b) The ball initially moves toward the bottom of the pail and then oscillates about the old equilibrium position.

(c) The ball moves initially toward the surface of the water and then oscillates about a new equilibrium position.

(d) None of the above correctly describes the motion of the ball during the fall off the roof.

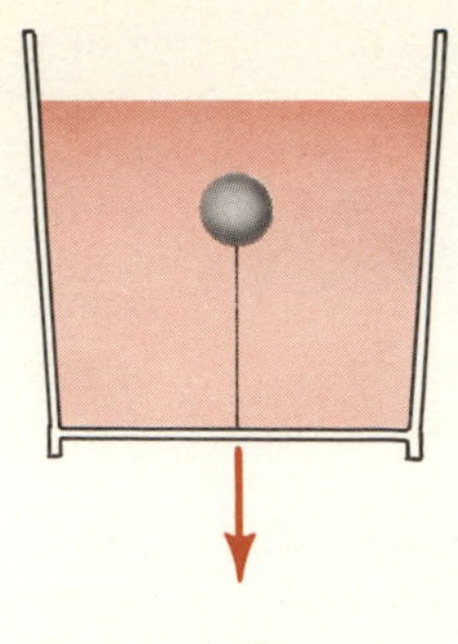

Figure 10.38

Problems

Unless stated otherwise, assume T = 20 °C.

Section 10.1

10.1 If atmospheric pressure at sea level is 10^5 N/m^2, what is the atmospheric pressure at an altitude of 200 m (Chicago), 1600 m (Denver), and 2400 m (Aspen)?

10.2 At a given temperature, the density of N_2 and O_2 in the atmosphere is proportional to pressure. Determine the density of O_2 in the atmosphere at elevations of 2400 m and 4000 m (near the peak of Mt. Rainier) as fractions of the oxygen density in the atmosphere at sea level. What effect might this lower oxygen concentration have on the heart?

10.3 The height of the Sears Tower in Chicago is 443 m. What is the pressure at the top of the tower when the pressure at ground level is 760 mmHg?

10.4 A U-tube manometer is filled with water. If the height of the U tube is 1.2 m, what is the maximum pressure difference that can be measured with this device?

10.5 The gauge pressure in the tires of a car is 26 lb/in.^{2}. What is this pressure in pascals and in atmospheres?

10.6 Arterial blood pressure (gauge pressure) varies between about 120 mmHg and 70 mmHg in the normal adult. The higher pressure is the systolic (on contraction of the heart); the lower, the diastolic (on expansion of the heart). What are these pressures in pascals and atmospheres?

• **10.7** Determine the force that would be required to separate the two halves of the Magdeburg sphere. Assume that the sphere's diameter is 1.2 m and that the pressure inside the sphere is negligible.

• **10.8** Show that if $P_A(h) = P_A(0)e^{-h/h_0}$, then the pressure at an altitude $h + \Delta h$ is related to the pressure at the altitude h by $P_A(h + \Delta h) = P_A(h)e^{-\Delta h/h_0}$.

• **10.9** The gauge pressure in the tires of a 2400-kg car is 26 lb/in^2. How much area of each tire is in contact with the pavement?

Section 10.2

10.10 What is the total pressure at a depth of 80 m below the surface of a freshwater lake?

10.11 What is the total pressure at the bottom of a 4-m-deep swimming pool? What is the total pressure against the side of this pool 3 m below the surface of the water?

10.12 What is the difference in average blood pressure between head and foot of a 1.7-m-tall person?

• **10.13** The tallest dam in the world today is at Dixence, Switzerland. Its height is 285 m. What pressure does the water exert at the base of the dam?

• **10.14** One type of micromanipulator uses a hydraulic system. The operator has her hand on a handle that moves a piston while she watches the microtool under the microscope. That tool is attached to another piston of the hydraulic system. The diameter of the piston to which the tool is attached is 6 mm. What should the diameter of the piston attached to the driving handle be so that a motion of 1 cm of the handle causes a displacement of 0.05 mm of the tool?

• • **10.15** A beam balance is designed to exhibit small changes in weight. Its construction is shown in Figure 10.39. If the system is initially balanced, and a mass of 0.2 g is then added to the right-hand pan, what is the resulting change in the level of the water in the cylinder as the piston moves to its new equilibrium position? Show that this balance arrangement reaches a stable equilibrium, but that the equilibrium is unstable if the diameter of the upper portion is greater than that of the lower.

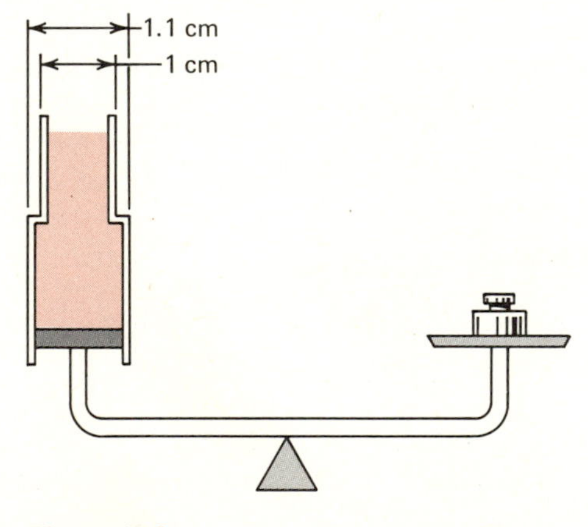

Figure 10.39

• **10.16** A hydraulic lift is operated by means of a pump capable of delivering hydraulic fluid at a gauge pressure of 8 × 10^5 Pa. The output pipe from the pump has a diameter of 3 cm. What is the smallest diameter of a piston that can raise the platform while supporting a car if the combined mass of platform and car is 2500 kg?

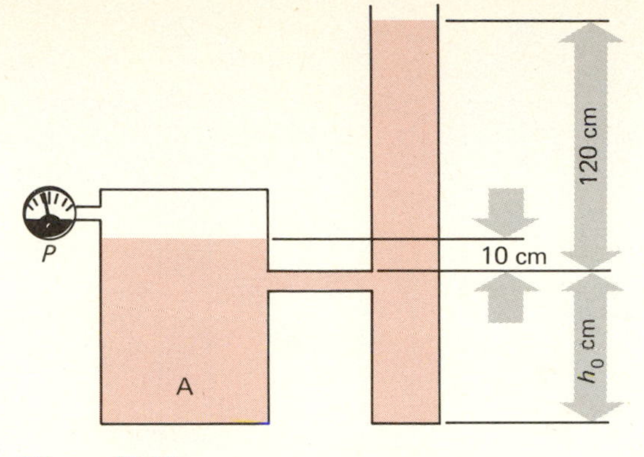

Figure 10.40

• • **10.17** When water is poured into the vessel of Figure 10.40, air is trapped in the container *A* once the liquid level is above the height h_0. What is the reading of the pressure gauge when the liquid level is 10 cm above h_0 in vessel A and 120 cm above h_0 in B? Pressure gauges read pressures relative to atmospheric pressure, not absolute pressure; that is, when connected to the atmosphere, the gauge will read zero pressure.

Section 10.3

10.18 An object weighs 86 N in air. When the object is submerged below the surface of water, its apparent weight is reduced to 57 N. What is the average density of this object?

10.19 A block of Styrofoam floats on water with only 8% of its volume submerged. What is the average density of Styrofoam?

10.20 A 1-kg block of copper is suspended from a spring balance. When the block is fully submerged in a liquid, the spring balance reads 6.5 N. What is the specific gravity of the liquid?

• **10.21** A piece of wood in the shape of a cube has a density of 800 kg/m³. When it is floated on water in a cylindrical container of 15-cm radius, the level of the water is seen to rise by 2 cm. What is the length of the cube's edge?

• **10.22** An object suspended from a spring balance weighs 22 N. When the object is fully submerged in water, the weight registered by the spring balance is only 17 N; when it is fully submerged in some other liquid, the spring balance reads 14 N. Find the average density of the object and the density of the other liquid.

• **10.23** An object floats on water with 68% of its volume submerged. When placed on the surface of some other liquid it again floats, but now has 93% of its volume below the surface of that liquid. Find the density of the object and the specific gravity of the liquid.

• **10.24** A beaker with water is placed on a beam balance, and balance is attained when a mass of 0.8 kg is put on the other pan. A piece of metal that weighs 40 N is now lowered into the beaker so that it is submerged but does not touch the bottom of the beaker. If the tension in the string supporting the metal is 25 N, how much more mass should be placed on the pan opposite the beaker to maintain balance?

• **10.25** The specific gravity of seawater is 1.03 and of ice is 0.92. What fraction of an iceberg is above the surface of water?

• **10.26** A piece of wood floats on the surface of water with 70% of its volume submerged. If the mass of this piece is 1.2 kg, what is its volume?

• **10.27** A block of plastic floats on water with 60% of its volume submerged. If this block is placed in another liquid, it is observed to float with 75% of its volume submerged. What is the specific gravity of the other liquid?

• • **10.28** A large, closed plastic container, of cubic shape and measuring 25 cm on a side, is weighed on an ordinary beam balance. The balance is at equilibrium when a 20-g brass mass is placed in the opposite pan of the balance. What is the true mass of the container, including its contents? (The density of air at normal temperature and atmospheric pressure is 1.3 kg/m³.)

• **10.29** A block of wood whose mass is 1.5 kg floats on the surface of water with 60% of its volume submerged. What mass of lead should be placed on top of the block so that the wood is fully submerged?

• • **10.30** In Problem 10.29, if a mass of lead is suspended from the wood rather than supported by it, what mass of lead is needed so that the top of the wooden block is just below the surface of the water?

Section 10.5

10.31 Show that Equation (10.12) is dimensionally correct.

10.32 Calculate the lift in N/m² on an airplane's wing if the average flow speed is 250 m/s over the top surface and 220 m/s over the bottom. (The density of air is 1.3 kg/m³.)

• **10.33** A venturi flowmeter has a cross section of 50 cm² at its entry and exit ports and a cross section of 30 cm² in its constricted region. What is the maximum flow rate of water for which the flow is laminar in this instrument? At that flow rate, what is the pressure difference between the wide and the constricted regions?

• **10.34** Water is flowing through a 3-mm-diameter pipe that has a constriction of 2.5-mm-diameter. If the level of the water in the vertical tubes is 1.21 m and 1.20 m, as shown in Figure 10.41, what is the velocity of the water in the 3-mm-diameter pipe?

• • **10.35** A venturi flowmeter is to be used to measure the flow velocity of oil whose density is 800 kg/m³. The manometer fluid is water; the diameters of the large and constricted regions of the meter are 3 cm and

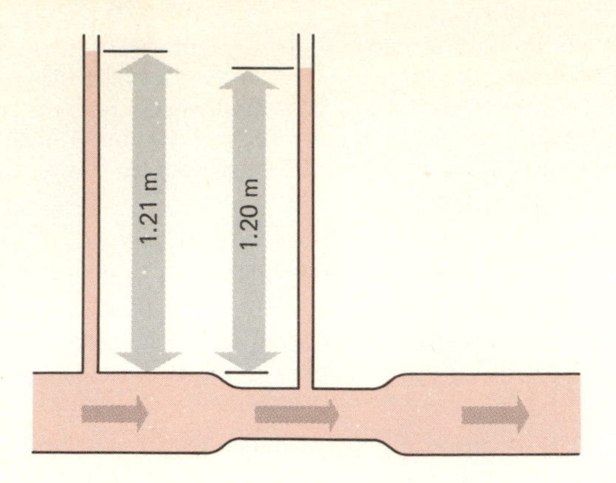

Figure 10.41

2.5 cm respectively. The height of the manometer U tube is 20 cm. (*a*) What is the flow velocity in the 3-cm diameter pipe when the manometer reads a level difference of 8 cm? (*b*) What is the maximum flow rate that can be measured with this instrument?

• **10.36** Figure 10.42 shows a large tank filled with liquid to a height h_1. If a spigot at height h_2 is opened, the liquid will spurt out of the tank with some speed v. A century before Bernoulli, Torricelli concluded that the speed of efflux is the same as that of a body that is dropped from rest through the height $h_1 - h_2$. That is, $v = \sqrt{2g(h_1 - h_2)}$.

Use Bernoulli's equation to prove Torricelli's theorem.

Figure 10.42

• •**10.37** Water flows in a 6-cm diameter horizontal pipe which has a local constriction of 5-cm diameter. If the pressure difference between the two regions of the pipe is 200 Pa, what is the flow rate in the pipe? How much time will elapse in filling a 50-L vat from this pipe?

• •**10.38** Water flows in a pipe that dips 1 m below the horizontal, as shown in Figure 10.43. The horizontal section of the pipe has a diameter of 10 cm, and the gauge pressure in this pipe is 0.2 atm. If the flow rate of the water is 0.3 L/s, what should the diameter be at the bottom of the U section so that the gauge pressure at that point is also 0.2 atm?

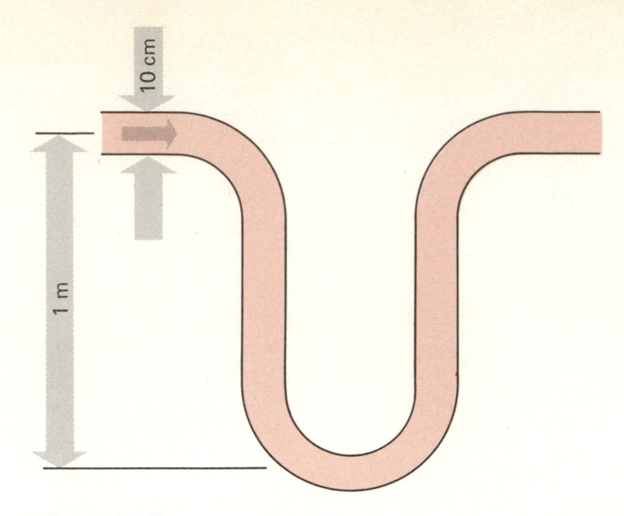

Figure 10.43

Sections 10.6, 10.7

• **10.39** A patient is given sucrose intravenously. If her venous pressure is 18 mmHg, and if the elevation difference between the intravenous needle and sucrose bottle is 0.8 m and the rate of sucrose flow is to be 2 mL/min, what should the diameter of the 4-cm-long needle be? Assume the sucrose has the same viscosity and density as blood.

• **10.40** What is the maximum flow rate of water in a smooth pipe of 8-cm diameter if the flow is to be laminar?

• •**10.41** A patient is given a blood transfusion through a hypodermic needle inserted in a vein. The inner diameter of the 3-cm-long needle is 0.65 mm, and the venous pressure of the patient is 20 mmHg. What should the elevation difference be between the needle and the bottle containing the blood so that the transfusion proceeds at the rate of 20 mL/min? Assume that the plastic tube carrying the blood from the bottle to the needle introduces negligible resistance to flow.

• •**10.42** If in Problem 10.41 the plastic tube between bottle and needle has an inner diameter of 7 mm and a length of 1.5 m, what fractional contribution does this tube make to the resistance to flow in the transfusion?

• •**10.43** A pipeline is to handle a flow rate of 500 gal/min of oil of density 0.8×10^3 kg/m^3 and viscosity 0.2 Pa·s. Determine the minimum pipe diameter needed if the Reynolds number is to be no greater than 1000. What pressure difference must then be maintained over a length of 1 km of the pipe? What pressure difference would be needed to maintain the same flow rate if the pipe diameter were increased by a factor of 1.2?

• •**10.44** Water is brought to the top of a 20-m-tall building through four 5-cm-diameter pipes. What must the velocity of the water be in the pipes so that each pipe will provide a flow rate of 10 L/min? Neglecting viscous losses, determine what pressure is needed at the bottom of the building to assure this flow rate. Could the same total flow rate be achieved using a single pipe of 10-cm diameter and a lower pressure?

Section 10.8

• **10.45** What is the terminal velocity of an air bubble of 1-mm diameter in a jar of vegetable oil whose specific gravity is 0.87 and viscosity is 0.15 Pa·s? Compare your result with a rough measurement using commercial vegetable oil. (Air bubbles can be introduced by vigorous stirring or with an eye dropper.)

• **10.46** A 1-cm-diameter sphere is dropped into a glass cylinder filled with a viscous liquid. The mass of the sphere is 4 g, and the density of the liquid is 1.25×10^3 kg/m^3. If this sphere descends with a terminal velocity of 2 cm/s, what is the viscosity of the liquid?

• **10.47** In his measurement of electronic charge, R. A. Millikan sprayed very small droplets of oil into a small chamber and studied their behavior when subjected to electric forces. He determined the mass of the droplets by observing their rate of descent in air under gravity. If the density of the oil is 0.9×10^3 kg/m^3 and the oil droplets have diameters of 5 μm, what is their rate of descent? The density of the air is 1.3 kg/m^3.

• **10.48** Estimate the terminal velocity of a raindrop, of a 7-mm-diameter hailstone, assumed to be a solid sphere of ice, and of a 1-cm-diameter steel sphere in air.

• •**10.49** Finely divided particles of copper are suspended in water. The particles are spherical and have diameters of 7 μm. Find the sedimentation velocity under normal conditions. Find it also for a suspension placed in a centrifuge cuvette, if the average distance of the liquid from the center of the centrifuge is 8 cm and the centrifuge is spinning at 2000 rev/min.

• •**10.50** In a classic experiment, Millikan measured the charge on an electron. In this experiment, he had to determine the mass of tiny oil droplets. Millikan did this by measuring their terminal velocity. What is the mass of an oil drop that falls at a speed of 0.5 mm/s when it has attained its terminal velocity? Assume that the oil has a specific gravity of 0.9.

• •**10.51** A centrifuge is used to separate 8-μm-diameter spherical particles suspended in water. The density of the particle substance is 1.15×10^3 kg/m^3. How fast should the centrifuge rotate so that the sedimentation velocity of particles 12 cm from the center of the centrifuge is 6 cm/min?

·11·

Thermal Properties, Calorimetry, and the Mechanical Equivalent of Heat

I have long held an opinion, almost amounting to conviction, in common, I believe, with many other lovers of natural knowledge, that the various forms under which the forces of matter are made manifest have one common origin; or, in other words, are so directly related and mutually dependent that they are convertible, as it were, one into another, and possess equivalents of power in their actions.

MICHAEL FARADAY, 1846

11.1 Introduction

It is no accident that the first half of the nineteenth century witnessed many advances and profound insights into the nature of heat. By the late 1700s, the Industrial Revolution had spread from England to the Continent and across the Atlantic. Successful transition from production based on human and animal energies to techniques employing steam engines depended on developing better and more efficient machines. Little progress could be expected until scientists understood the fundamental processes involved in converting heat into mechanical energy.

Before 1830, heat and thermal properties of substances were believed to be unrelated to mechanical, electrical, and magnetic phenomena. According to the contemporary *caloric theory,* a body was endowed with heat in proportion to the amount of caloric fluid it contained. The greater its temperature, the more caloric fluid. Thermal expansion, a well-known phenomenon, was explained as expansion needed to accommodate additional caloric fluid. That heating resulted in no measurable change of mass did pose a troublesome dilemma, but proponents of the caloric theory disposed of it with the assertion that caloric was an "imponderable," or "igneous," fluid—i.e., massless.

Although the caloric theory was put to rest before the middle of the nineteenth century, its legacy—the unit of heat, the calorie—is still widely used today.

Definition: One calorie is the amount of heat required to raise the temperature of one gram of water by 1 °C.*

A larger unit, the kilocalorie = 10^3 calories, is generally used by dietitians, who refer to it as *Calorie,* with a capital C.

The fact that a steam engine, using heat from burning wood or coal, performs mechanical work suggests an intimate relation between work and heat. Though the idea that heat is some form of motion was advanced by Francis Bacon early in the seventeenth century, the two men generally credited with dealing the caloric theory its coup de grâce are Count Rumford, born Benjamin Thompson, and James Prescott Joule.

Thompson, a poor but ambitious and energetic New England farm boy, decided early in life that wealth is the secret to success. At 19, he married the 33-year-old widow of a colonel of the Royal Militia, and thus became overnight one of the wealthiest landowners in what is now New Hampshire. In 1775, he found it expedient to go to England, and after some time in the service of King George, he accepted employment as aide to Karl Theodor, Elector of Bavaria. There, in 1793, he was awarded the formidable title Count of the Holy Roman Empire, and took the name Rumford, after his birthplace, now known as Concord, New Hampshire. Count Rumford was, by all accounts, a thoroughly unpleasant, conceited, abrasive, and overbearing man—but all the same, a superb scientist and imaginative inventor.

> It was by accident that I was led to make the experiments of which I am about to give an account. . . . Being engaged lately in superintending the boring of cannon, . . . I was struck with the very considerable degree of Heat which a brass gun acquires in a short time in being bored, and with the still more intense Heat . . . of the metallic chips separated from it by the borer. . . . From *whence comes* the Heat actually produced in the mechanical operation . . . mentioned?

Figure 11.1 *Caricature of Benjamin Thompson, Count Rumford, published in 1800, showing the count warming himself at a stove of his own design.*

To answer this question, Rumford carried out numerous careful and ingenious experiments. Summarizing the results of these investigations, he concluded:

> By meditating on the results of all these experiments, we are naturally brought to that great question which has often been the subject of speculation among philosophers; namely, What is Heat? Is there such a thing as an *igneous fluid?* Is there anything that can with propriety be called *caloric?* . . . We must not forget . . . that the source of the Heat generated by friction . . . appeared evidently to be

* The precise definition states that one calorie is the amount of heat needed to raise the temperature of one gram of water from 14.5 °C to 15.5 °C at atmospheric pressure. The heat needed to raise 1 g of water by 1 °C depends slightly on the average temperature of the water, but that dependence is sufficiently weak that we shall neglect it.

Figure 11.2 James Prescott Joule (1818–1889).

inexhaustible. It is hardly necessary to add, that anything which an *insulated* body, or system of bodies, can continue to furnish *without limitation*, cannot possibly be *a material substance;* and it appears to me to be extremely difficult, if not quite impossible, to form any distinct idea of anything capable of being excited and communicated in the manner the Heat was excited and communicated in these experiments, except it be *MOTION*.

James Prescott Joule was a man truly possessed by a scientific mission, namely to determine the mechanical equivalent of heat. Joule performed many experiments. He compared the heat generated by the passage of an electric current (now called the Joule heat) with the mechanical energy necessary to drive the electric generator; he compared the heating of water forced through pipes with the work to sustain the flow; he compared the heating of water by turning a paddle wheel with the energy consumed in turning the shaft (see Fig. 11.3); and he made many more similar studies.*

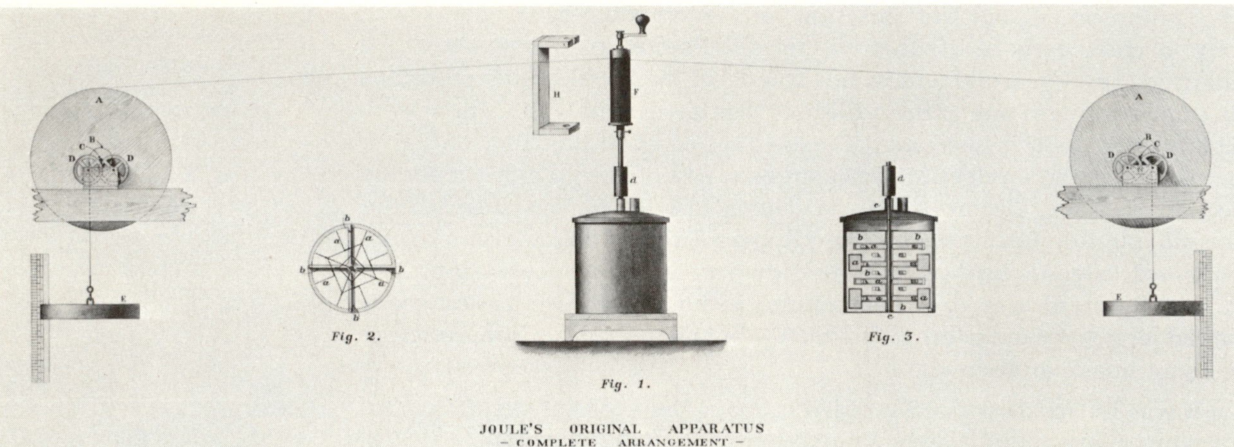

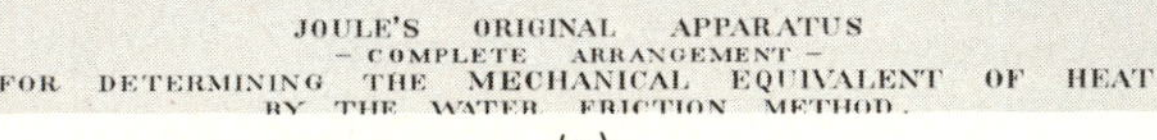

(*a*)

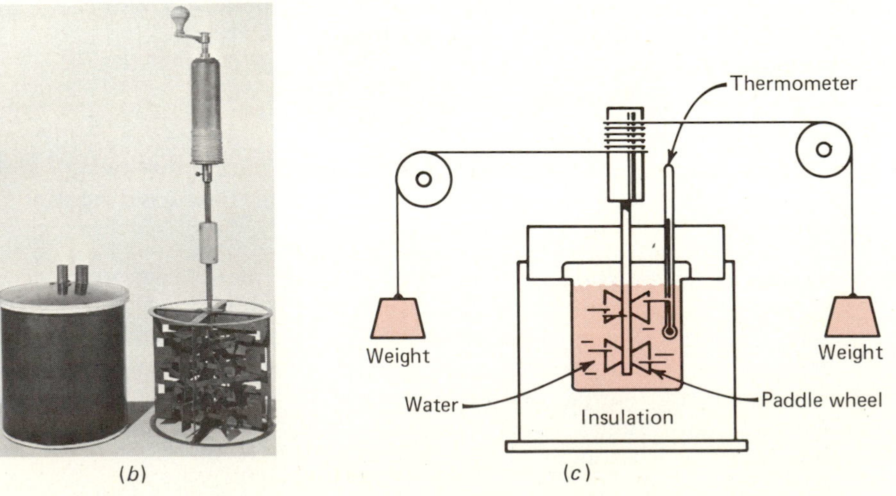

(*b*) (*c*)

Figure 11.3 (*a*) Arrangement of the apparatus used by Joule to measure the mechanical equivalent of heat. (*b*) Close-up view of the calorimeter and of the paddle wheel by which the water inside the can was churned as the weights descended a specific distance. (*c*) Schematic sketch of the calorimeter.

* Joule was so preoccupied by this work that on his honeymoon at Chamonix, armed with a huge thermometer, he measured the temperature of the water above and below the 110-m waterfall at Sollanches. Joule had calculated that when water fell that distance, the conversion of its potential energy to heat should result in a temperature increase of about half a degree Fahrenheit.

Joule's experiments yielded the result that

> the heat required to raise 1 lb of water 1° [F] is . . . 772 foot-pounds.

The currently accepted value is nearly the same, 778 foot pounds. Converted to SI units, the equivalence is

$$1 \text{ calorie} = 4.18 \text{ joules} \quad \textbf{(11.1)}$$

That is, the thermal energy required to raise the temperature of one gram of water by one degree Celsius equals 4.18 joules, more energy than needed to lift 100 g a height of 4 m.

11.2 Temperature

Temperature is a familiar concept. It is simply a measure of how hot or cold an object is. The instrument used to measure temperature, the thermometer, is also familiar. We must, however, extend our ideas of temperature beyond these broad generalities if we want to examine possible relations between physical and thermal properties of a system.

That the body temperature of a sick person is perceptibly higher than normal was known in antiquity; it was common practice then, as it is today, for the attending physician to place his hand on the forehead of the patient to judge the seriousness of the illness. Eventually physicians developed various devices for measuring temperature, all of them relying on the experimental fact that liquids and gases expand with increasing temperature. But these early thermometers lacked uniformity so that comparison of temperatures was difficult.

Figure 11.4 Woodcut showing a medieval physician taking a patient's temperature.

The first reliable mercury thermometer was constructed by Daniel Fahrenheit (1686–1736), who chose as his two fixed temperature points the "most intense cold obtained artificially by a mixture of water, ice and of sal-ammoniac or even sea-salt, and the limit of the heat which is found in the blood of a healthy man."

Reliance on the physical properties of matter for fixed temperature points was initiated by Anders Celsius (1701–1744), who constructed the scale named in his honor. Celsius divided the temperature interval between melting ice, the zero on the Celsius scale (0 °C), and boiling water at atmospheric pressure (100 °C) into one hundred equal parts. Since on the Fahrenheit scale the melting point of ice is 32 °F and water boils at 212 °F, 180 Fahrenheit degrees correspond to 100 Celsius degrees.

(a) (b)

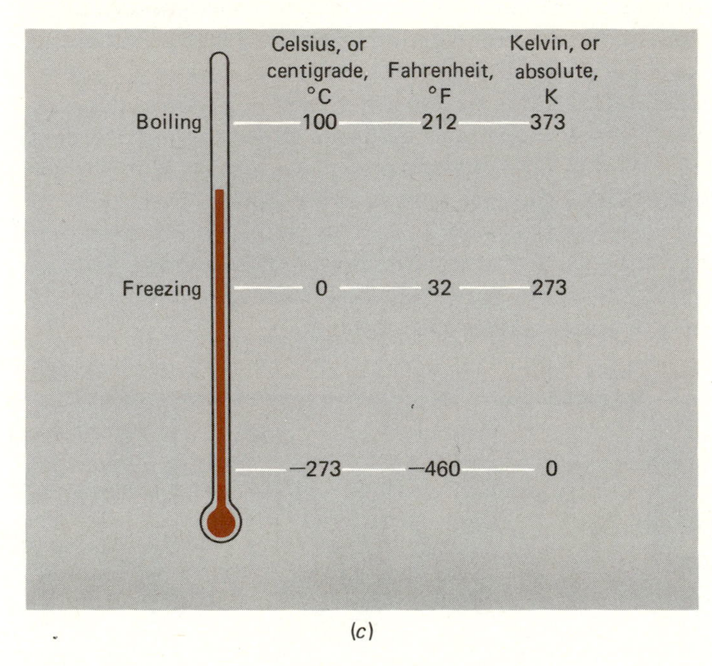

(c)

Figure 11.5 (a) Anders Celsius. (b) Lord Kelvin. (c) Comparison of the Celsius, Fahrenheit, and Kelvin temperature scales.

The conversion between the two temperature scales is therefore expressed by

$$T_F = 32 + \tfrac{180}{100}T_C = 32 + \tfrac{9}{5}T_C \qquad T_C = \tfrac{5}{9}(T_F - 32) \qquad \textbf{(11.2a)}$$

One other thermometric scale widely used in scientific work is the absolute, or Kelvin, scale. The Kelvin scale differs from the Celsius scale only in a shift of the zero; that is, 0 °C = 273.2 K, or 0 K = −273.2 °C. The Kelvin intervals are the same as the Celsius intervals. Thus the conversion between Celsius and Kelvin temperatures is simply

$$T_K = T_C + 273.2 \qquad T_C = T_K - 273.2 \qquad \textbf{(11.2b)}$$

(The significance of the Kelvin scale will become manifest in the following chapter.)

● ***Example 11.1*** Aluminum melts at 933 K and boils at 2720 K at atmospheric pressure. Express these temperatures in degrees Celsius and degrees Fahrenheit.

To convert to Celsius degrees, we only need to subtract 273 from the values in Kelvin. Hence, the melting and boiling points of aluminum are

$$T_m = 933 - 273 = 660\ °C \qquad T_b = 2720 - 273 = 2447\ °C$$

The temperatures in Fahrenheit degrees are obtained by applying Equation (11.2):

$$T_m = 32 + \tfrac{9}{5}(660) = 1220\ °F \qquad T_b = 32 + \tfrac{9}{5}(2447) = 4437\ °F$$

11.3 Thermometric Methods

Thermometry rests on the basic concept of thermal equilibrium. It is a common observation that a cup of hot coffee will cool and a glass of ice water will warm until both have attained the temperature of the surroundings.

DEFINITION: A system is said to be in thermal equilibrium when it has achieved that steady-state condition in which no net energy exchange takes place between any parts of the system. When two or more systems are in thermal equilibrium with each other, their temperatures are identical.

It follows that two systems can attain thermal equilibrium with each other only if there exists a mechanism for energy transfer. If such a mechanism does exist, the two systems are said to be in *thermal contact*. For example, when the oral temperature of a person is being measured, the mercury in the thermometer bulb, though not in direct contact with the mucous membranes of the mouth, is in thermal contact with these surfaces through the glass encapsulation, which can transport energy in the form of heat. After a brief time, the mercury and the mouth come to

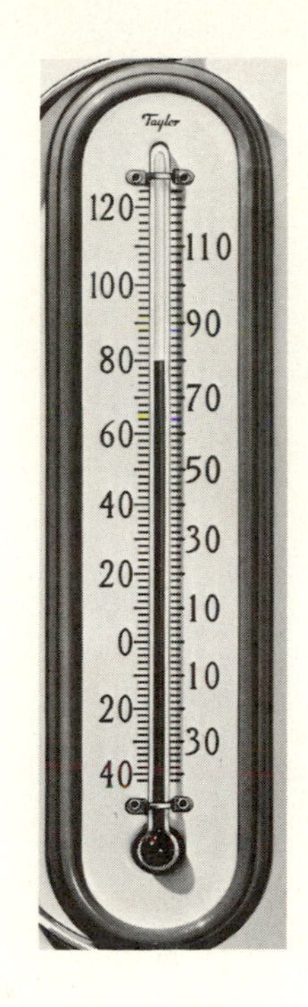

Figure 11.6 A mercury-in-glass thermometer widely used to measure temperatures in the vicinity of room temperature.

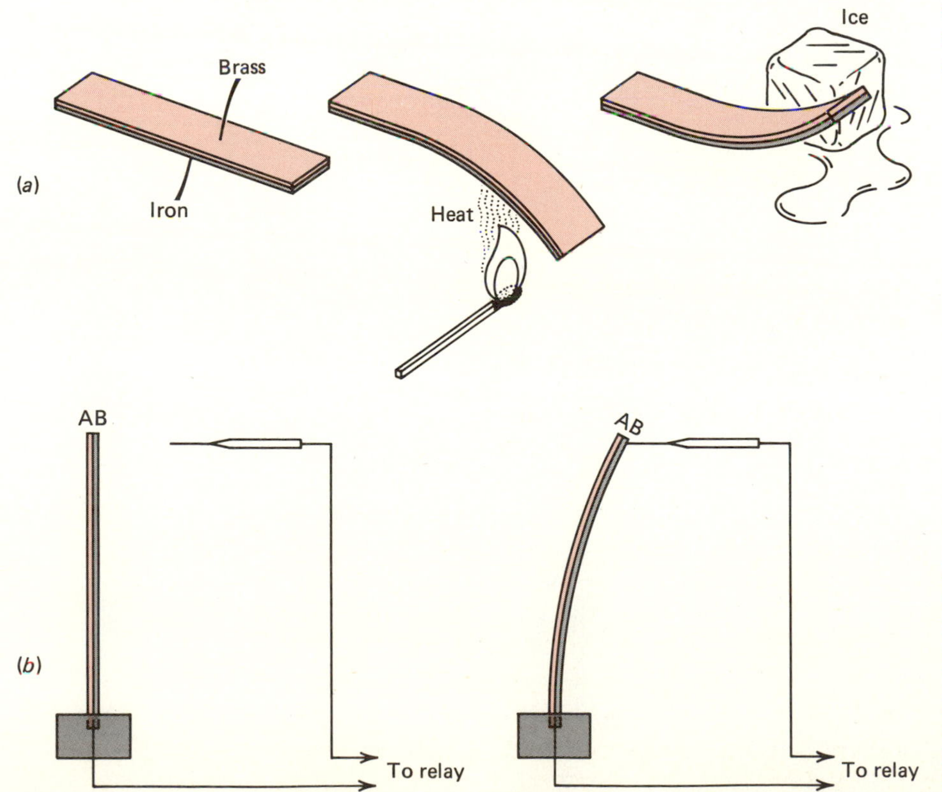

Figure 11.7 (a) Brass expands (or contracts) more with increasing (or decreasing) temperature than iron. A bimetallic strip of brass and iron will bend as indicated when heated or cooled. (b) A bimetallic strip can be used to control an electrical circuit.

thermal equilibrium and the temperature registered by the instrument is then the same as that of the mouth.

The best-known thermometer is the *mercury thermometer,* which relies on the expansion of mercury with temperature. It consists of a bulb that contains nearly all the mercury in the instrument and which connects to a fine capillary tube. As the mercury expands, the column rises in the capillary, and its height is a direct measure of the temperature of the instrument.

Some temperature-sensing devices use the thermal expansion of metallic strips. Two strips of dissimilar metals with different thermal expansion coefficients (see next section) are welded together. If this *bimetallic strip* is clamped firmly at one end and the entire length is heated, the difference in expansion causes the strip to bend. Such units are frequently used in thermal switches, for example, in thermostats that control home furnaces.

The *thermocouple* relies on the thermoelectric effect for its signaling ability. Whenever a temperature difference is maintained along the length of a conductor, a small voltage appears between its two ends. If two wires of different metals are joined at one end while the other terminals are connected to a voltmeter, the voltmeter reading will be proportional to the difference in temperature between the junction and the volt-

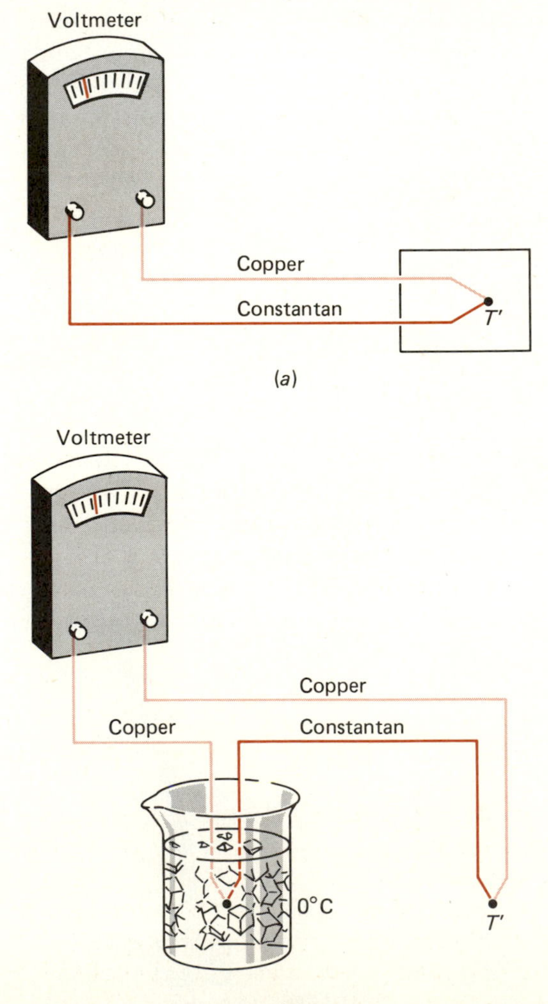

Figure 11.8 *Measuring temperature with a thermocouple. (a) For a rough measurement of the temperature of some region—for instance, a furnace at temperature* T′*—the copper and constantan arms of a commercial thermocouple may be connected directly to the measuring instrument. (b) Temperature can be more reliably and precisely determined if both terminals of the measuring instrument are connected to copper wires and one junction of the copper-constantan thermocouple is immersed in an ice bath.*

meter (room temperature). Thermocouples are generally used to control industrial furnaces, and they have many laboratory applications.

The electrical resistance of metallic wires and other conductors is temperature-dependent. (See Section 20.3.) *Resistance thermometers* made of platinum are frequently used as temperature standards, especially at low temperatures. Other devices, so-called *thermistors,* are made of small, glass-encapsulated pellets of a semiconductor whose resistance changes substantially with temperature. (See Section 31.5.) Thermistors find numerous industrial, research, and medical applications.

Thermistors and also thermocouples have several advantages over conventional mercury thermometers. First, because they are small, they can be placed in regions not readily accessible to the larger, less flexible units. Second, they come to thermal equilibrium very quickly and can therefore follow fairly rapid temperature changes. Finally, their signal output is electrical instead of visual (mercury thermometer) or mechanical (bimetallic strip); these units are especially well suited when temperature is to be controlled by electronic means.

11.4 Thermal Expansion

It is an experimental fact that nearly all substances expand with increasing temperature. If at temperature T_0 a steel rod has a length L_0, then at a temperature $T_0 + \Delta T$ its length will have increased to $L_0 + \Delta L$. The

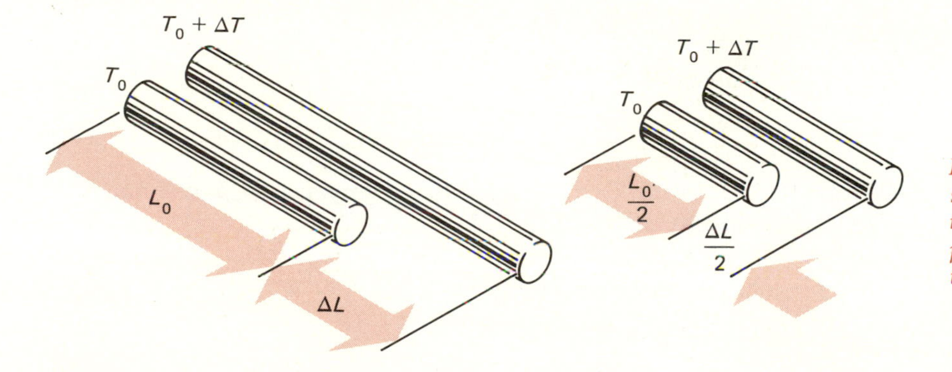

Figure 11.9 Expansion of a steel rod due to an increase of temperature by ΔT is proportional to the length of the rod. Hence,

$$\frac{\Delta L}{L_0} = \frac{\Delta L/2}{L_0/2} = \alpha\,\Delta T$$

length change ΔL depends, however, not only on the temperature change ΔT, but also on L_0, the initial length of the rod. If a 1-m rod increases in length by 1 mm for a given temperature change, a 2-m-long rod will increase by 2 mm under the same conditions. It is the *relative* change in length per unit temperature change that is the intrinsic material property; the linear thermal expansion coefficient α is therefore defined by

$$\alpha = \frac{1}{L_0}\frac{\Delta L}{\Delta T} \tag{11.3}$$

The unit of α is degree^{-1}. Values of α for some materials are listed in Table 11.1.

Thermal expansion, or contraction, is generally so small that we often neglect this effect; sometimes, however, such oversight can have disastrous consequences, as shown in Figure 11.10.

An object that is heated expands not merely in one direction but in all. Thus, it is not only the length but also the diameter of the steel rod of Figure 11.9 that increases with increasing temperature, and consequently, the volume of the rod increases as well. The volume thermal ex-

Figure 11.10 Failure to allow for thermal expansion can have disastrous consequences.

Table 11.1 Thermal expansion coefficients of some materials

Material	α (10^{-6}/K)	β (10^{-6}/K)
Aluminum	24	—
Brass	19	—
Copper	17	—
Glass (ordinary)	9	—
Glass (Pyrex)	3	—
Iron (steel)	12	—
Lead	30	—
Silver	20	—
Ice	51	—
Ethanol	—	1100
Mercury	—	180
Water	—	210

NOTE: These are average values for the range 0 °C to 100 °C, except that the range for ice is −10 °C to 0 °C.

pansion coefficient is defined by

$$\beta = \frac{1}{V_0}\frac{\Delta V}{\Delta T} \qquad (11.4)$$

Provided the material is isotropic (expands uniformly in all directions) and $\alpha \ll 1$,

$$\beta = 3\alpha \qquad (11.5)$$

● ***Example 11.2*** At 20 °C, the volume of a copper can is exactly 1 L. What is its volume at 100 °C?

The volume expansion coefficient of copper is

$3(17 \times 10^{-6}\ °C^{-1}) = 51 \times 10^{-6}\ °C^{-1}$

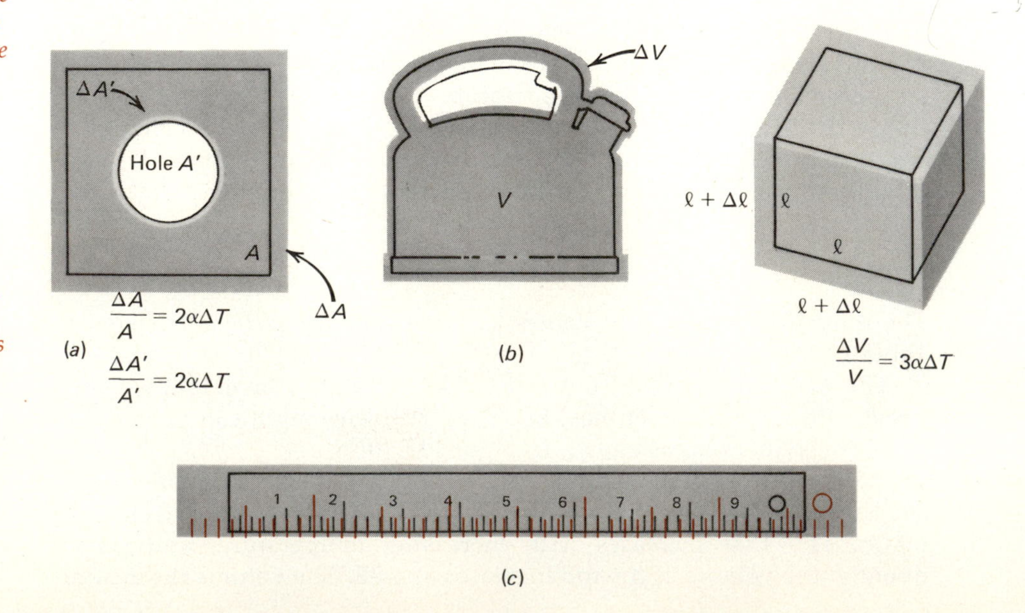

Figure 11.11 Thermal expansion increases every dimension of an object proportionately. (a) The surface area of the square as well as that of the hole in the square is increased in the same proportional amount, equal to $2\alpha\,\Delta T$. (b) The volumes of the cube and of the coffee pot increase acording to $\Delta V/V = 3\alpha\,\Delta T$. (c) The length of the steel rule, the separation between individual divisions, the width of the rule, the diameter of the hole in one end, and the thickness all increase by the same fraction. This is true for isotropic materials, and many materials encountered in everyday life fall in this category—glass, copper wire, aluminum pots, steel tools. Some materials expand anisotropically; the most common is wood, which expands less along the grain than perpendicular to the grain.

Consequently, an 80 °C change in temperature will produce a volume increase of

$$\Delta V = \beta V_0 \, \Delta T = (51 \times 10^{-6} \, ^\circ\text{C}^{-1})(1 \text{ L})(80 \, ^\circ\text{C}) = 4.08 \times 10^{-3} \text{ L}$$

and the volume at the higher temperature will be 1.00408 L. ●

11.5 Heat Capacity, Latent Heat, and Calorimetry

The heat capacity of an object is defined by

$$C = \frac{\Delta Q}{\Delta T} \tag{11.6}$$

where ΔQ is the amount of heat that must be supplied to effect a temperature change ΔT. The unit of C is calories per degree.

Evidently, the larger the object to be heated, the greater the amount of heat that must be supplied. It is therefore sensible to refer everything to unit mass. The *specific heat capacity,* or simply *specific heat,* is this intrinsic material property. It is defined by

$$c = \frac{C}{m} = \frac{1}{m}\frac{\Delta Q}{\Delta T} \tag{11.7}$$

and its unit is calorie per gram degree or kilocalorie per kilogram degree; the value of c is the same in either unit. It follows from this definition that the amount of heat needed to raise the temperature of m grams of a substance by an amount ΔT is

$$\Delta Q = cm \, \Delta T \tag{11.8}$$

Suppose we place a block of aluminum at temperature T_2 in a copper container partly filled with liquid at temperature T_1. We shall assume that the container is thermally insulated from its surroundings. We know from experience that after a while the aluminum block, copper can, and liquid will all attain the same temperature T_f. Thermal equilibrium is established between these substances in thermal contact by the transfer of energy, in this case heat. To calculate the equilibrium temperature, we turn to conservation of energy. Since no mechanical work is performed, the total thermal energy must be conserved. Consequently we can write

Heat gain (by one part of system) = heat loss (by other part)

We can rewrite this conservation equation in terms of the masses, specific heats, and temperature changes of the various components, using Equation (11.8). For the case considered here, the hot aluminum block cools from T_2 to T_f as the liquid and copper can warm from T_1 to T_f. We can write

$$M_{\text{Cu}}c_{\text{Cu}}(T_f - T_1) + M_L c_L(T_f - T_1) = M_{\text{Al}}c_{\text{Al}}(T_2 - T_f) \tag{11.9}$$

Here, the first term on the left-hand side is the heat gained by the copper can of mass M_{Cu} and specific heat c_{Cu}, and the second, the heat gained by the liquid of mass M_L and specific heat c_L, as both increase their temperature from T_1 to T_f. The term on the right-hand side is the heat lost by the aluminum block of mass M_{Al} and specific heat c_{Al} as it cools from T_2 to T_f.

Table 11.2 Specific heat capacities of some substances (Values are near 0° C)

Material	*Heat capacity, cal/g·K*
Aluminum	0.21
Asbestos	0.19
Glass	0.1–0.2
Copper	0.094
Silver	0.056
Mercury	0.033
Lead	0.031
Water	1.00
Ice	0.50
Ethanol	0.58

● ***Example 11.3*** If the aluminum block in the preceding discussion had a mass of 0.2 kg and the copper can a mass of 0.05 kg, and if the can contained 0.2 L of ethyl alcohol at 20 °C, what equilibrium temperature is reached if the initial temperature of the aluminum block was 80 °C?

We shall use Equation (11.9) and solve for T_f, with $T_1 = 20$ °C and $T_2 = 80$ °C. First, however, we must determine the mass of ethyl alcohol in the can. The density of ethyl alcohol is 810 kg/m³ = 0.81 g/cm³. Hence, the mass of liquid is (0.81 g/cm³) (200 cm³) = 162 g. The specific heats of ethyl alcohol, copper, and aluminum are given in Table 11.2. Substituting the appropriate numerical values into Equation (11.9) gives

$$(50\ \text{g})(0.094\ \text{cal/g}\cdot\text{K})(T_f - 20°\text{C}) + (162\ \text{g})(0.6\ \text{cal/g}\cdot\text{K})(T_f - 20°\text{C}) = (200\ \text{g})(0.22\ \text{cal/g}\cdot\text{K})(80\ °\text{C} - T_f)$$

and solving for T_f, one obtains the result

$$T_f = 38.1\ °\text{C}$$

Under certain conditions adding substantial heat to a system produces no observable change in its temperature. For example, the temperature of a well-stirred mixture of ice and water placed over a flame remains at 0 °C until all the ice has melted. The appearance of infinite heat capacity, that is, the ability of a system to absorb heat with no change in temperature is always associated with a phase change, as when ice melts, or solid lead changes to liquid, or water boils and evaporates.

The amount of heat absorbed (or evolved) depends on the mass undergoing the phase change, the type of material, and the kind of phase change. For example, to melt 1 g of ice at 0 °C and convert it to water at 0 °C, one must supply 80 cal; conversely, when 1 g of water freezes and solidifies, 80 cal of heat will be released by the water.

Heat absorbed or released in a phase transition at fixed temperature is called *latent heat*. The latent heats of melting and vaporization of a few materials are listed in Table 11.3.

Table 11.3 *Latent heats of melting and vaporization at atmospheric pressure*

Material	*Melting point, °C*	*Latent heat of melting, cal/g*	*Boiling point, °C*	*Latent heat of vaporization, cal/g*
Helium	—	—	−269	5
Nitrogen	−210	6.1	−196	48
Oxygen	−219	3.3	−183	51
Water	0	80	100	540
Mercury	−39	2.8	357	65
Lead	327	5.9	1620	218
Ethanol	−114	25	78	204
Silver	961	21	2193	558
Gold	1063	15.4	2660	377

What happens to the heat supplied during the phase change? Why should a substance absorb heat when melting and yet the temperature remain constant? The answer is that the greater mobility of atoms or molecules of liquids compared with solids arises from disruption of some bonds between molecules of the solid. Further breaking of cohesive

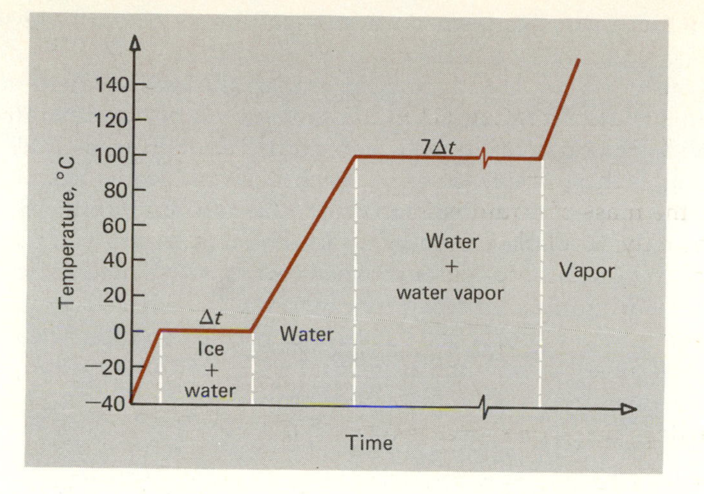

Figure 11.12 The temperature of a fixed mass of water as a function of time when heat is supplied at a constant rate the pressure is maintained at one atmosphere. The first plateau occurs when the ice melts, the second when the water boils. Note the break in the time scale; the second plateau should be nearly seven times as long as the first. Note also the steeper slope when $T < 0$ °C and when $T > 100$ °C.

bonds between molecules is required for vaporization. Such breaking of bonds consumes energy, and this energy may be supplied as heat.

● ***Example 11.4*** 250 g of molten lead at its melting point is poured on a 50-g block of ice in a copper container of 10-g mass. The temperature of the ice and copper container is initially −30 °C, and the copper container is thermally insulated from the surroundings. What is the final constitution and temperature of the system?

As the hot lead strikes the ice, it first releases the latent heat of melting and solidifies, then releases additional heat as the now solid but hot lead cools to its final temperature T_f. At the same time, the copper can and ice absorb that heat and increase their temperature to 0 °C; the ice then melts, absorbing more heat in the process, and finally, *provided that the lead has not already cooled to 0 °C,* the water and copper can will warm and the lead cool to an equilibrium temperature $T_f > 0$ °C.

We balance the heat lost by the lead and gained by the ice and copper can, writing the following equation:

$$\begin{aligned}(250\text{ g})(5.9\text{ cal/g}) &+ (250\text{ g})(0.031\text{ cal/g·K})(327\text{ °C} - T_f)\\ &= (50\text{ g})(0.5\text{ cal/g·K})(30\text{K}) + (10\text{ g})(0.094\text{ cal/g·K})(30\text{ °C})\\ &+ (50\text{ g})(80\text{ cal/g}) + (50\text{ g})(1\text{ cal/g·K})(T_f - 0\text{ °C})\\ &+ (10\text{ g})(0.094\text{ cal/g·K})(T_f - 0\text{ °C})\end{aligned}$$

The left-hand side represents the heat liberated by the molten lead; the first term is the latent heat of melting given up as the liquid solidifies, the second term the heat released as the solid lead cools to the temperature T_f from the melting temperature of 327 °C. The right-hand side represents the heat gained by the ice and copper can; the first two terms correspond to the heat absorbed by the ice and copper can on warming from −30 °C to 0 °C, the third term represents the heat absorbed by the 50 g of ice on melting to 50 g of water, and the last two terms give the heat absorbed by the water and copper can on warming from 0 °C to T_f. Performing the indicated arithmetic, we obtain

$$\begin{aligned}58.7T_f &= -768.7\text{ °C}\\ T_f &= -13.1\text{ °C}\end{aligned}$$

That answer must be wrong! If the final temperature is less than 0 °C, none of the ice could have changed to water. Yet checking the arithmetic, we find that there has been no numerical error. Where have we gone astray?

The answer is not hard to find. In writing the earlier equation, we *assumed* that all the ice would melt. But the amount of heat liberated as the lead solidifies and cools to 0 °C may not be enough to melt all the ice, even though there is five times as much mass of lead as ice.

We must repeat the calculation under the new assumption that only a fraction of the available ice liquefies. In that case we know the final temperature. It must be 0 °C, since we shall be left with a block of lead, some water, and a bit of ice in a copper can, all at the same temperature; ice and water can coexist only at 0 °C at atmospheric pressure. The unknown quantity is now the amount of ice that melts in this process. The new heat conservation equation takes the form

$$\begin{aligned}(250\text{ g})(5.9\text{ cal/g}) &+ (250\text{ g})(0.031\text{ cal/g}\cdot\text{K})(327\text{ °C}) \\ &= (50\text{ g})(0.5\text{ cal/g}\cdot\text{K})(30\text{ °C}) + (10\text{ g})(0.094\text{ cal/g}\cdot\text{K})(30\text{ °C}) \\ &\quad + M_W(80\text{ cal/g})\end{aligned}$$

where M_W is the amount of ice that liquefies. Solving for M_W, we find

$$M_W = 40.4\text{ g}$$

Thus, we shall be left with the lead at the bottom of the copper can submerged in a bath of ice water, 40.4 g of water and 9.6 g of ice! A remarkable and perhaps unexpected result, considering that we started out with 250 g of molten lead at 327 °C and only 50 g of ice and 10 g of copper at −30 °C. It is apparent, on reviewing the solution, that by far the largest heat sink, the term on the right-hand side that corresponds to the dominant heat absorption, is the latent heat of melting of the ice. ●

A glance at the latent heats of melting and of evaporation, and also the specific heats, listed in Tables 11.2 and 11.3, point to the interesting fact that water is unique among the prevalent natural liquids and solids; it has a large specific heat as well as large latent heats of melting and vaporization.

This special property of water has tremendously important consequences on the ecology and meteorology of our planet. Large bodies of water, such as the Great Lakes and the oceans, are vast heat reservoirs, capable of absorbing or releasing enormous quantities of heat with only a small change in temperature. Thus, regions near the coast experience relatively small changes in temperature through the seasons, although locations only a few hundred kilometers further inland may experience extreme temperature fluctuations between winter and summer. The Pacific Northwest of the United States is a perfect example; during the summer the temperature in Seattle rarely reaches even 30 °C (86 °F), and snowfall in the city is equally rare, although plenty of snow does fall on the nearby Cascade and Olympic mountain ranges. Only 100 km to the east, along the banks of the Columbia River, summer temperatures well above 40 °C (104 °F) are common, and during cold snaps in February the temperature frequently drops to −30 °C (−22 °F).

The unusually large latent heats of water also play a crucial role in ecology. A large latent heat of melting means that substantial heat must be liberated from water already cooled to 0 °C to cause solidification. That fact, and the relatively low heat conductivity of ice, result in a rather slow increase in the depth of the ice layer on a lake during the winter. Similarly, the melting of winter snows in spring is relatively slow as each gram of ice crystals must absorb 80 cal as the crystals change to water. If the latent heat of melting of ice were significantly less, spring floods would be far more sudden and severe.

Finally, one important physiological application of vaporization of water concerns temperature control in mammals and birds. Body temperature must be maintained constant to within close tolerance, whatever the surrounding temperature. If our brain and vital organs are to remain at 37 ± 0.5 °C despite wide variations in ambient temperature and in rate of metabolic energy production, there must be various mechanisms that the body can call into play to control its temperature. When the core temperature (temperature of the central portion of the body) rises above 37 °C and the surrounding air is at or above 37 °C, the only effective method of cooling is evaporation of water. As sweat evaporates from the skin, each gram of water vapor carries with it about 580 cal, the latent heat of vaporization at 40 °C.

The rate of evaporation from a surface depends on the amount of water vapor already in the air. If there is much water vapor in the air just above the liquid surface, the evaporation rate is low. It is this dependence on humidity that is responsible for the cooling sensation due to a breeze even on a very warm day. The blowing wind in no way changes the amount of energy carried away by evaporation of one gram of water, but it does increase the evaporation rate by removing air saturated with water vapor from the region just above the skin, replacing it with drier air capable of accepting more vapor.

● *Example 11.5* The rate of metabolic heat production varies with activity, ranging, in the human body, from a low of about 65 kcal/h during sleep to as much as 1400 kcal/h during strenuous exercise. If a person performs normal activity, corresponding to metabolic heat production of 275 kcal/h on a hot summer day when the temperature is 37 °C, how much water must that person drink after three hours of work to replenish loss due to evaporation?

Since the air temperature is the normal body temperature, the only mechanism for heat loss is evaporation of sweat. During three hours, the body will generate a total of

$$(3\ \text{h})(275 \times 10^3\ \text{cal/h}) = 825 \times 10^3\ \text{cal}$$

Consequently,

$$\frac{825 \times 10^3\ \text{cal}}{580\ \text{cal/g}} = 1422\ \text{g}$$

of water at 40 °C must be evaporated from the skin to remove this amount of heat. 1422 g of water equals about 3 pints, or roughly 6 full tumblers. ●

11.6 Heat Transport

We now know the meaning of thermal equilibrium and have seen how one can predict the equilibrium temperature attained by several objects, initially at different temperatures, that are brought into thermal contact. We now consider the mechanisms by which this equilibrium is achieved. There are three distinct physical processes of heat transport, convection, conduction, and radiation. Generally, one of these will be dominant, but occasionally two or all three mechanisms must be included in calculations of heat transport.

11.6 (a) Convection

In convection, heat is transferred by the actual displacement of part of the system that is in thermal equilibrium at a high temperature to a region of lower temperature. Since convection requires motility of the medium, it can take place only in a fluid. It arises either because a change in density of the fluid with temperature causes a mass flow in the gravitational field, a process called *natural convection,* or because some external stirring device mixes hot and cold portions, in which case one speaks of *forced convection.*

A simple demonstration of natural convection is indicated in Figure 11.13. The rectangular glass tube is filled with water or other fluid and heat is supplied at point *A* by a bunsen burner or a small torch. Since above 4 °C, water expands with increasing temperature, the liquid near *A* has a lower density than elsewhere and consequently rises in the tube. It is replaced by cool liquid from the region near *B,* which on being heated expands and rises. The process thus leads to a gentle counterclockwise circulation in the tube. That current can be made visible by introducing a bit of food coloring with an eyedropper at the opening *C.*

Figure 11.13 *Demonstration of convection. Water, heated at A by the bunsen burner, expands and rises in the Pyrex tube, only to be replaced by cooler water from the region near B. The counterclockwise circulation can be made visible by introducing a bit of food coloring at the opening C.*

Similarly, on a hot sunny day, the ground and rocks that are heated by the sun's rays transfer the heat to the surrounding air; that air then rises and is replaced by cooler air, which is then also warmed. The resulting updrafts can be quite strong, especially in mountainous regions, where the slopes tend to establish well-defined patterns for airflow. These updrafts are sought out by soaring enthusiasts, who can reach great altitudes by holding their sailplanes in a circular pattern inside such air currents. In flat terrain, glider pilots generally look for updrafts above large freshly plowed fields; the dark, rough ground absorbs the sun's rays most effectively and the air above such a field is warmer than the surrounding air, and updrafts result.

Natural convection requires not only a fluid medium but also a gravitational field and moreover, a suitable variation of density with temperature. It is the buoyant force exerted by the surrounding fluid that forces the more rarefied, lighter portion to rise. Without gravity, there is no buoyant force and hence no natural convection. However, even in a gravitational field, natural convection is by no means assured.

(a)

(b)

(c)

Figure 11.14 Other examples of convective currents. (a) Convective circulation in a beaker of liquid heated near its center. (b) Direction of surface wind at the shore of a large body of water. (i) On a hot, sunny day, the air above the ground warms more readily than the air above the surface of the water, and the cooler air from the water moves toward the shore. (ii) After sunset, as the ground cools more rapidly than the water, the direction of airflow reverses. (c) Sailplane pilots search for warm air updrafts that keep their gliders aloft.

Water, at atmospheric pressure, attains its greatest density not at the freezing point but at 4 °C; in the range $0° < T < 4$ °C, $\alpha < 0$. This anomalous negative thermal expansion coefficient of water makes possible the wide variety of aquatic life in the less temperate regions of the globe. As the air above a lake cools, so does the water near the surface; that water then sinks and is replaced by warm water rising from greater depths. These convection currents persist until the entire lake reaches 4 °C. Thereafter, further cooling of the surface layer, say to 2 °C, only *reduces* the density of this layer, which now floats above the heavier water at 4 °C. Continued cooling leads to the formation of ice, which since it is lighter than water at any temperature, floats and forms a rigid cover over the lake. This ice sheet serves two important functions. First, it prevents stirring of the lake, that is, forced convection by the strong winds that often accompany winter storms. Second, it provides a surface to support accumulation of snow, an excellent thermal insulator. As a result, even during prolonged periods of subzero temperatures, moderately large bodies of water freeze over but do not freeze solid.

During spring, the sun's rays and the warmer air melt the snow and ice, and the surface layer of the water gradually warms to temperatures well above 4 °C. Again, natural convection does not take place, and the

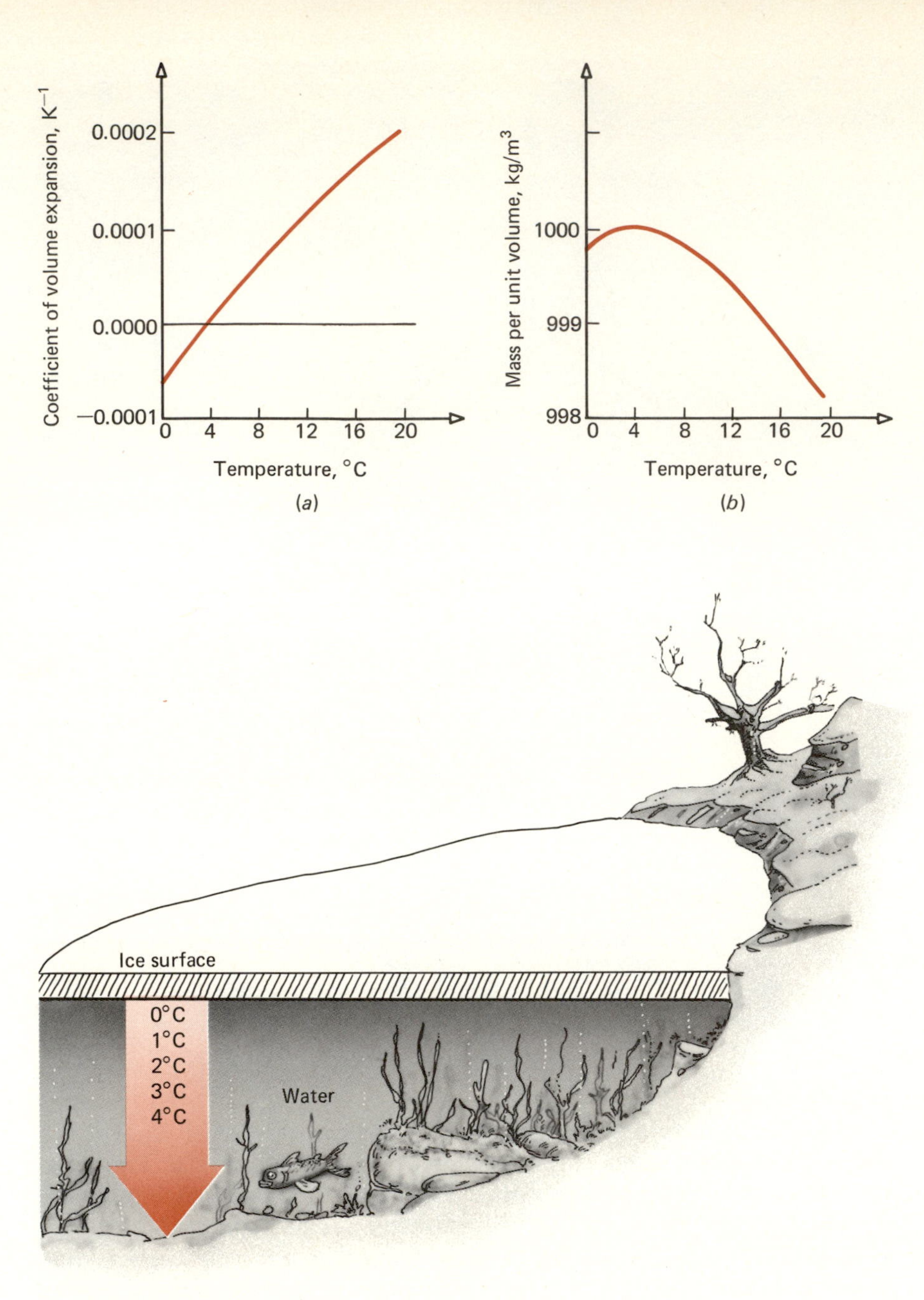

Figure 11.15 *(a) Coefficient of volume expansion of water as a function of temperature. Note that β changes sign near 4 °C and is negative between 0 °C and 4 °C. Consequently, as shown in (b), the density of water reaches a maximum at 4 °C.*

Figure 11.16 *The negative expansion coefficient of water below 4 °C results during the winter months, in the temperature profile shown here.*

temperature of the lower region of the lake increases much more slowly than if convective currents were set up by the thermal gradient. The presence of such thermal gradients is well known to anyone who has swum in a lake during the summer. Relatively cool regions even during hot summer months are also necessary for aquatic life, since many fish can survive only in relatively cold water which has a high oxygen content.

A much less beneficial effect of inhibition of natural convection is familiar to residents of Los Angeles and other large cities. Under certain meteorological conditions, a so-called inversion layer is formed; this consists of a body of cool air beneath a layer of warmer and usually polluted air. The pollutants of the upper layer absorb much of the sun's thermal energy, heating this layer further and preventing heating of the lower, stagnant air. The arrest of normal convection currents prevents

Figure 11.17 Air pollution in central St. Louis.

dispersion of automobile exhaust fumes and other noxious gases, with sometimes lethal consequences to persons suffering from respiratory ailments.

Convective heat transfer can be accelerated by stirring or other forced agitation. Such action is essential when natural convection is altogether absent, as for example, aboard an orbiting satellite.

11.6 (b) Conduction

If you stir hot coffee with a metal spoon, you may notice that the handle of the spoon becomes warm; if the spoon is sterling silver, it may become uncomfortably hot. This is but one of many examples of heat transport by conduction, a mechanism in which thermal energy is transferred without macroscopic movement of the medium that carries the heat flow. On a microscopic scale, heat is transported by conduction as follows.

As will be discussed in greater detail in a later chapter, the atoms of any substance engage in random motion, which is increasingly violent as the temperature increases. In a crystalline solid, the atoms, though their *average* location is fixed, vibrate about these positions with amplitudes that increase with increasing temperature. Since neighboring atoms interact with each other as though connected by tiny springs, the more violent motion of atoms at the hot end is gradually transmitted down the rod of Figure 11.19, just as jiggling a bedspring at one end soon generates vibrations at the opposite. In metals, much of the heat is carried by electrons that are relatively free to move about within the material and are also responsible for the high electrical conductivities of these substances. In fact, in metals good electrical and thermal conductivities go hand in hand. However, nonmetallic solids such as sapphire, quartz, and other pure crystals are also very good heat conductors, especially at very low temperatures.

The total heat conducted per second depends on several variables. There is first the intrinsic material property, its thermal conductivity, denoted by κ. Next, as one might expect, the greater the cross section of the conductor, the greater the amount of heat transported. Last, the heat

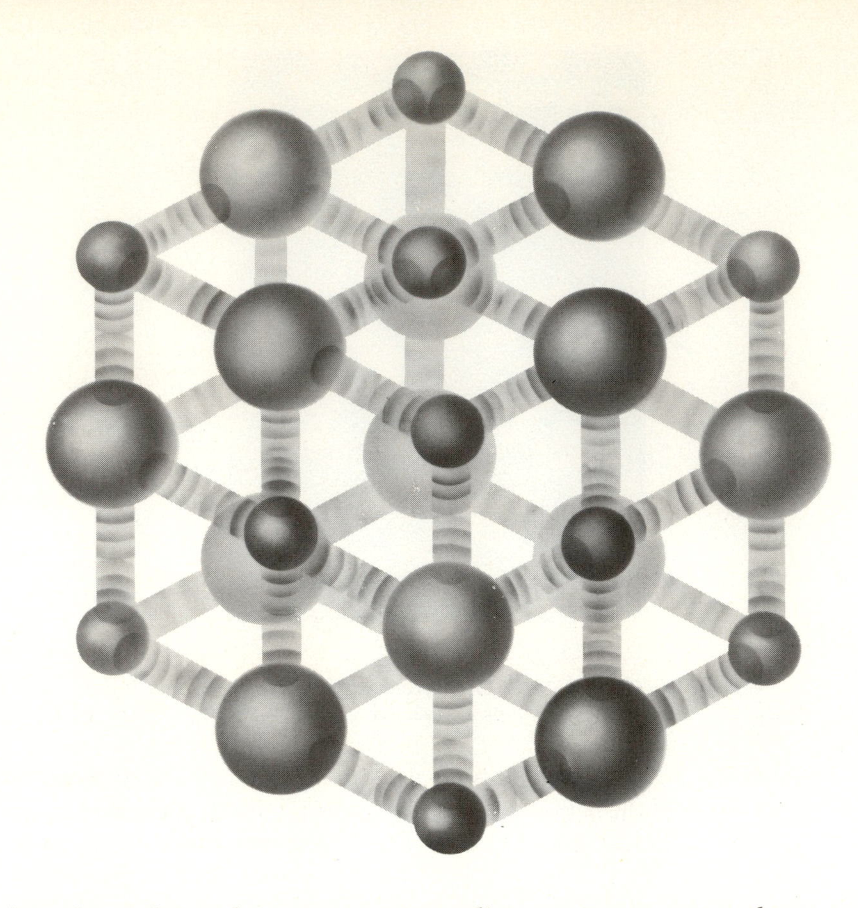

__Figure 11.18__ Model of a crystalline solid. The atoms maintain their average positions but can vibrate about these locations. The amplitude of the vibrational motion increases with increasing temperature; this vibrational energy is then transmitted along the solid as a result of the coupling between neighboring atoms.

flow depends on the temperature gradient, or temperature change per unit length, which is the driving force. That is,

$$\frac{\Delta Q}{\Delta t} = -\kappa A \frac{\Delta T}{\Delta x} \tag{11.10}$$

Here ΔQ is the amount of heat transported across the area A of Figure 11.19 in time Δt as a result of the temperature gradient $\Delta T/\Delta x$, and κ is the thermal conductivity of the material that carries the heat current. The negative sign in Equation (11.10) reminds us that the direction of heat flow is opposite to the temperature gradient; that is, heat flows "down-

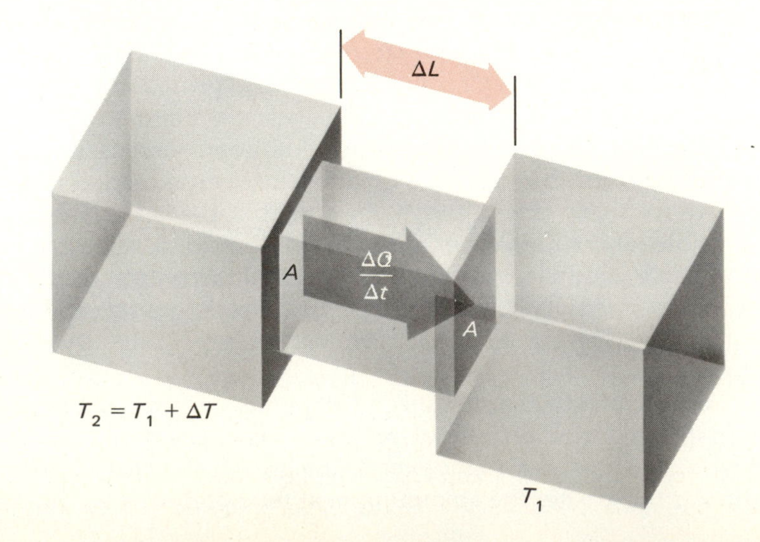

__Figure 11.19__ Conduction of heat between two heat reservoirs. The amount of heat transported by the rectangular rod depends on its cross section A, the temperature difference ΔT divided by the length of the rod ΔL, and the thermal conductivity κ of the rod material.

Table 11.4 Thermal conductivities of some substances at 0 °C

Material	Thermal conductivity, W/m·K
Aluminum	238
Brass	120
Copper	400
Silver	418
Lead	35
Iron	82
Titanium	20
Glass (Pyrex)	1
Asbestos	0.09
Brick	0.04
Cork	0.03
Ice	2.2
Air	0.024

hill,'' from the hot to the cold end and not the reverse. Units of thermal conductivity are cal/s·m·deg, or W/m·deg.

Thermal conductivities for various substances at 0 °C are listed in Table 11.4. The data reflect some general patterns. First, pure metals are the best thermal conductors, followed, in descending order, by metallic alloys, nonmetallic solids, liquids, and lastly, gases. It is important to remember that for fluids, the values given indicate the amount of heat carried by conduction only; often convection is the primary thermal transport mechanism. Thermal conductivities are temperature-dependent, and that dependence differs from one class of material to another. Thus an insulating crystal of quartz may have a low thermal conductivity compared with brass at room temperature and yet be a much better thermal conductor at very low temperatures, 4 K or below.

● ***Example 11.6*** A medium-size classroom in a building constructed during the 1950s has single-paned glass windows covering half of one 16-ft × 30-ft wall. If the thickness of the glass is $\frac{1}{4}$ in., how much heat is lost per hour through the window glass when the temperature drops to −10 °C outside and the inside is maintained at 20 °C. How many watts must be dissipated in the classroom to maintain its temperature at 20 °C under these conditions?

From (11.10), we have

$$\Delta Q = -\kappa A \frac{\Delta T}{\Delta x} \Delta t$$

In this problem, $\Delta T/\Delta x = (30/0.25)$ °C/in. $= 4.72 \times 10^3$ °C/m. The area through which the heat is conducted is 16 ft × 15 ft = 22.3 m², $\Delta t =$ 3600 s, and $\kappa \simeq 0.2$ cal/m·s·deg. Substituting these numerical values into the equation, we find that $\Delta Q = 7.58 \times 10^7$ cal! That is, the amount of heat lost through the window panes in one hour is enough to bring more than 120 L (over 30 gal) of water to a boil, starting at room temperature, *and evaporate this water completely!* The heat loss is at a rate of 2.1×10^4 cal/s $= 8.8 \times 10^4$ J/s = 88 kW. If the room were heated electrically,

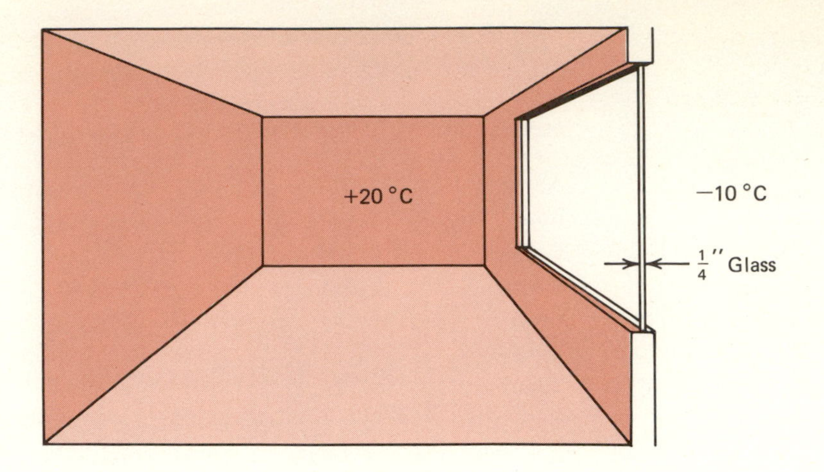

Figure 11.20 Heat flows across a $\frac{1}{4}$-in. pane of glass from the room, at 20 °C, to the outside, at −10 °C.

then at a rate of 4 ¢/kW·h, it would cost 88 × 0.04 × 24 ≃ $84 every day just to maintain this one classroom at a comfortable temperature during a moderately severe cold spell in one of the northern states.

That seems an unreasonably high cost for heating, and so it is. The presumption here is that the outer surface of the glass is at −10 °C and the inner surface at 20 °C, that is, at the same temperatures as the air outside and inside. That would be true only if on both sides of the glass there were a strong wind blowing. In fact, near the inner glass surface, there is a layer of air that is much colder than most of the air in the room; you can readily confirm this by placing your hand near a window on a cold day without actually touching the pane. This layer of cold air provides considerable thermal insulation; unfortunately, the air is not stagnant but circulates slowly by convection, so that warm air from the interior of the room is continually brought toward the cold window pane. Similarly, on the outside there is a thin layer of air that is warmer than −10 °C, and the rate at which this layer dissipates depends on the wind speed. Thus, although the actual conditions are not as outrageously bad as the above calculation might suggest, this example does demonstrate dramatically the benefit of double-paned windows. Since air has a thermal conductivity only about one-fortieth that of glass, two panes of

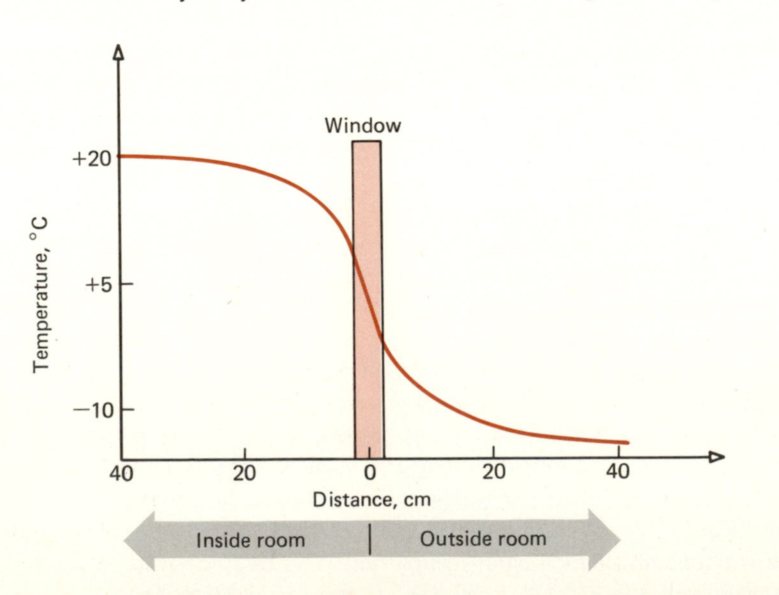

Figure 11.21 An approximate temperature profile when one takes into account the layers of air adjacent to the window on either side. It is assumed that there is no forced convection. (The thickness of the glass pane has been greatly exaggerated in the drawing.)

$\frac{1}{8}$-in. thickness with a $\frac{1}{4}$-in. layer of air trapped between them reduce the conductive heat loss by about a factor of 40. Though the gain is not quite so great, since the air between the two glass sheets transports heat by convection as well as conduction, double-paned glass or storm windows are nevertheless well worth the initial investment in all but the most temperate climates. ●

11.6 (c) Radiation

Though the space between our planet and the sun is a vast expanse of almost perfect vacuum, we can still feel the sun's warming rays. Indeed, all the energy we now use (except for the small contribution of nuclear reactors) is directly or indirectly derived from that supplied by the sun now or at some earlier time. Neither conduction nor convection, both requiring the presence of ponderable material, can account for this constant transport of energy; the mechanism responsible is *radiation*.

Thermal radiation, like light, consists of *electromagnetic waves* akin to radio, TV, and radar waves. The only difference between these radiations is quantitative; their frequencies and wavelengths belong to different parts of the spectrum. Qualitatively, they are identical. It is logical to delay detailed discussion until we have had an opportunity to study electric and magnetic phenomena and the relations between electricity and magnetism. Here we confine ourselves to a few remarks.

First, energy transport by radiation *needs no material medium*, although radiation *can* propagate through many media.

Second, the rate of an object's energy loss by radiation depends on the temperature of that object and increases with increasing temperature. This is a well-established finding, which is readily verified simply by holding one's hand near an electric space heater whose temperature can be adjusted by controlling the current through the heating element. What is not immediately evident is that this temperature dependence is really very rapid. The relation between power radiated and temperature is Stefan's law:

$$P_r = \sigma \epsilon A T^4 \qquad (11.11)$$

Here P_r is the rate at which energy is radiated, A is the surface area of the object whose surface temperature is T K, and $\sigma = 5.67 \times 10^{-8}\ \mathrm{W/m^2 \cdot K^4}$ is Stefan's constant. Note that in Equation (11.11), the temperature must be given in kelvin. The symbol ϵ denotes the *emissivity* of the surface, a parameter, characteristic of the surface, that can take on any value between 0 and 1. The emissivity is a measure of the efficacy of the surface as an emitter of radiation. One can show that surfaces that absorb radiation well are also good emitters and surfaces that reflect most incident radiation are poor emitters. The requirement that two objects in an isolated evacuated enclosure shall attain thermal equilibrium by exchange of radiant energy leads to the condition that the emissivity equal the absorptivity. Absorptivity is defined as the ratio of the absorbed to the incident energy. A highly polished mirror has an absorptivity near zero, and an equally low emissivity; good absorbers, such as a piece of black velvet or a lump of coal, whose black color indicates low reflectance, at least of visible radiation, have emissivities close to 1. This is why the ideal, perfect radiator with $\epsilon = 1$ is called a *black body*.

A good example of heat transport and the attainment of thermal equilibrium by radiation is our solar system. The temperature at the sun's surface, the photosphere, is about 5800 K. Assuming that the sun radiates as a spherical black body of 700,000-km radius, one can calculate the total power radiated by that star and the power absorbed by Earth. Since Earth has by now achieved thermal equilibrium, we can also calculate the average surface temperature of our planet.

The surface area of a sphere is $A = 4\pi R^2$. Hence from (11.11), the power radiated by the sun is

$$P_r = (5.67 \times 10^{-8}\ \text{W/m}^2\cdot\text{K}^4)4\pi(7 \times 10^8\ \text{m})^2(5.8 \times 10^3\ \text{K})^4$$
$$= 3.95 \times 10^{26}\ \text{W}$$

an almost unimaginably huge rate of energy loss; and yet, it has been going on for billions of years.

The distance between Sun and Earth is about 1.5×10^{11} m and Earth's radius is about 6.37×10^6 m. The surface area of a disk corresponding to the projection of Earth is $\pi(6.37 \times 10^6\ \text{m})^2 = 1.27 \times 10^{14}\ \text{m}^2$. The total surface area of a sphere whose radius is the Sun-Earth distance is $4\pi(1.5 \times 10^{11}\ \text{m})^2 = 2.83 \times 10^{23}\ \text{m}^2$. Thus the fraction of the sun's radiation that is intercepted by Earth is $1.27 \times 10^{14}/2.83 \times 10^{23} = 4.5 \times 10^{-10}$. If the absorptivity of Earth were 1, the planet would receive energy at the rate of

$$4.5 \times 10^{-10}(3.95 \times 10^{26}\ \text{W}) = 1.78 \times 10^{17}\ \text{W}$$

In fact, about one-fourth that energy is reflected into space.

Astronomers refer to the ratio of reflected to incident radiant energy as the *albedo*. Jupiter has a high albedo, of 51 percent, one reason it ap-

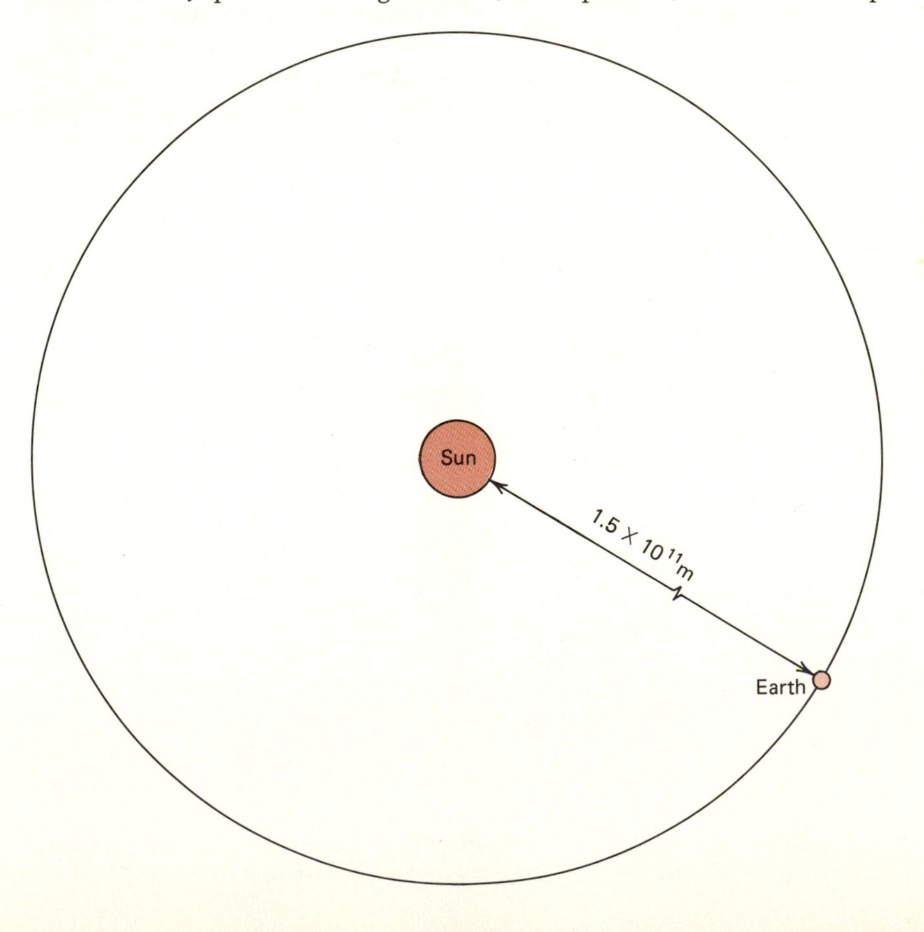

Figure 11.22 *All our energy, except for the comparatively minute nuclear energy resource now in use, derives directly or indirectly from the energy radiated by the sun now or in past ages. Earth intercepts only a tiny fraction of the power radiated by the sun.*

pears so bright; Mercury and the moon have an albedo of about 6 percent. (If the moon had an albedo comparable to that of Jupiter or even Earth, reading by moonlight would be no problem.) The albedo of Earth varies substantially, depending on weather conditions; as any air passenger knows, clouds reflect sunlight much more than the ground beneath. If we take as a reasonable average albedo a value of 25 percent, the *average* energy incident on each square meter of Earth's surface per second is

$$\frac{(1.78 \times 10^{17}\ \text{W})(1 - 0.25)}{4\pi(6.37 \times 10^{6}\ \text{m})^{2}} = 262\ \text{W/m}^{2}$$

If Earth simply continued to absorb this energy, it would gradually become hotter and hotter. Since it appears to have achieved an equilibrium temperature, the energy gain must equal the energy loss. Again, although on the planet heat transport by convection and conduction as well as radiation is possible, Earth as a whole can release energy into space only via radiation. We complete this exercise by estimating the average surface temperature of our globe. To do this we write Equation (11.11) as

$$T^{4} = \frac{(P_{r}/A)}{\sigma\epsilon}$$

If Earth is to maintain constant average temperature, the power radiated per square meter P_r/A must equal 262 W. If we assume that the emissivity for radiation characteristic of Earth's surface temperature is the same as the absorptivity for the sun's radiation, we obtain

$$T^{4} = \frac{262\ \text{W/m}^{2}}{(5.67 \times 10^{-8}\ \text{W/m}^{2}\cdot\text{K}^{4})(0.75)} = 6.157 \times 10^{9}\ \text{K}^{4}$$

and

$$T = 280\ \text{K} = 7\ ^{\circ}\text{C}$$

a reasonable result.

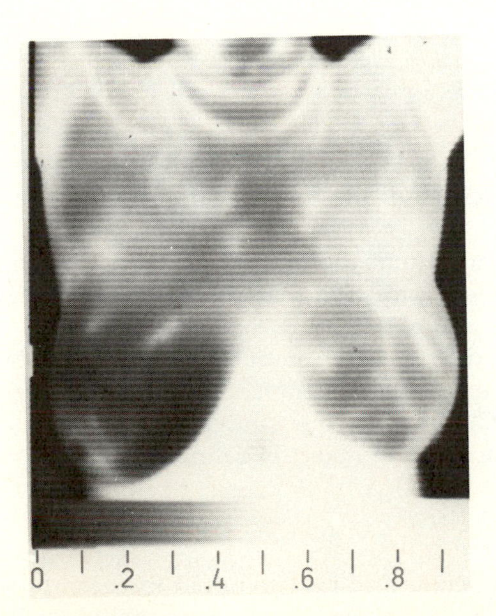

Figure 11.23 The amount of energy radiated depends on the temperature of the radiating surface. Using infrared-sensitive detectors, one can obtain a pictorial representation of temperature distribution. On such a thermogram, warm regions appear bright, cool regions dark. The thermogram of a woman's chest reveals an elevated temperature of the left breast (possibly indicative of a tumor).

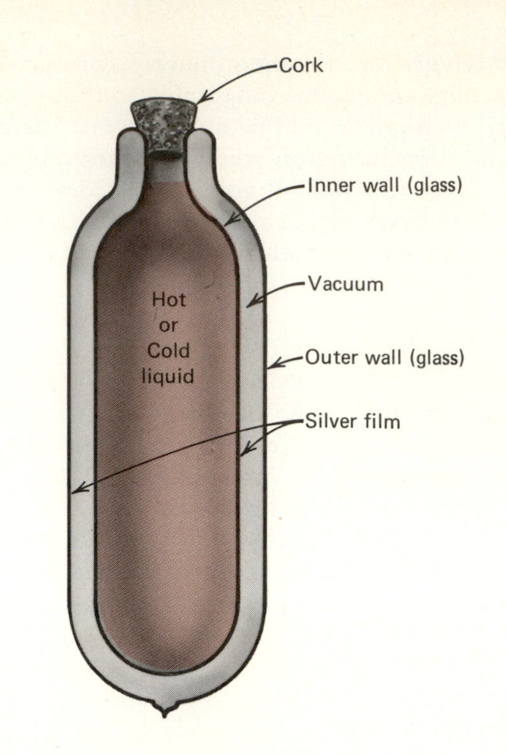

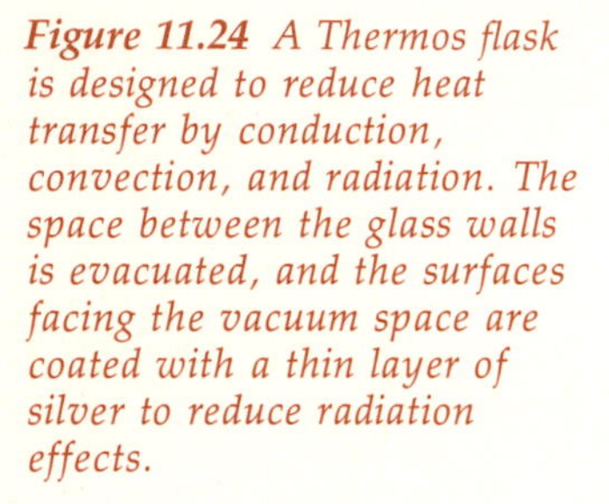

Figure 11.24 *A Thermos flask is designed to reduce heat transfer by conduction, convection, and radiation. The space between the glass walls is evacuated, and the surfaces facing the vacuum space are coated with a thin layer of silver to reduce radiation effects.*

Summary

The relation between temperature as measured on the Celsius, Kelvin, and Fahrenheit scales is given by

$$T_F = 32 + \tfrac{9}{5}T_C \qquad T_C = \tfrac{5}{9}(T_F - 32) \qquad T_K = T_C + 273.2$$

where T_C, T_K, and T_F are the Celsius, Kelvin, and Fahrenheit temperatures.

Besides the common mercury-in-glass thermometers, bimetallic strips, thermocouples, and thermistors are among the numerous devices for measuring temperature.

Many thermometers depend on *thermal expansion.* The *linear thermal expansion coefficient* α is defined by

$$\alpha = \frac{1}{L_0}\frac{\Delta L}{\Delta T}$$

where L_0 is the length of the object at the reference temperature T_0 for which α is given.

The surface and volume thermal expansion coefficients are equal to 2α and 3α, respectively, provided that the linear thermal expansion coefficient is isotropic and $\alpha \ll 1$.

One *calorie* is defined as the amount of heat (thermal energy) required to raise the temperature of one gram of water by one degree Celsius, from 14.5 °C to 15.5 °C. One calorie is equivalent to 4.18 joules.

The *heat capacity* of an object is defined by

$$C = \frac{\Delta Q}{\Delta T}$$

where ΔQ is the amount of heat required to accomplish a temperature change of ΔT. The *specific heat capacity,* or simply *specific heat,* is the heat capacity per gram. The SI unit of heat is the kilocalorie, which is 1000 calories; the specific heat in SI units is given in kilocalories per kilogram degree. The specific heat is the same whether measured in calories per gram degree or kilocalories per kilogram degree. The symbol for specific heat is c.

If no work is done on or by a system, thermal energy is conserved. Under these conditions, if two bodies, not in thermal equilibrium, are placed in thermal contact, the heat gained by one must equal the heat lost by the other.

During a phase change, for example, melting or vaporization, there may be absorption or release of heat without a change in temperature. The heat associated with that phase change is called the *latent heat* of the transformation.

The three mechanisms for heat transport are *convection, conduction,* and *radiation.* In convection, heat is transported by the actual translocation of matter. If this movement is the result of buoyancy effects due to thermal expansion, it is called *natural convection;* if it is brought about by stirring, it is called *forced convection.*

In conduction, thermal energy is transported by the interaction of atoms or molecules as they vibrate about their equilibrium positions in a solid, or collide or otherwise interact in a fluid. Conduction of heat does not require bulk motion. The *thermal conductivity* of a substance is defined by

$$\frac{\Delta Q}{\Delta t} = -\kappa A \frac{\Delta T}{\Delta x}$$

where $\Delta Q/\Delta t$ is the heat transport per second across the area A as a result of a temperature gradient $\Delta T/\Delta x$ in a direction perpendicular to the area A, and κ is the thermal conductivity.

Radiative heat transfer involves the energy transport by propagation of electromagnetic waves. The amount of heat radiated per unit area per unit time is given by *Stefan's law,*

$$P_r = \sigma \epsilon A T^4$$

where P_r is the power emitted, A is the area of the emitting surface, $\sigma = 5.67 \times 10^{-8}$ W/m²·K⁴ is *Stefan's constant,* T is the surface temperature in Kelvin, and ϵ is the *emissivity* of the surface, a parameter that depends on the surface characteristics and can take on values between 0 and 1.

Multiple Choice Questions

11.1 Heat transfer by convection

(a) requires no significant displacement of molecules.

(b) cannot be an effective mechanism in solids.

(c) is proportional to $T_2^4 - T_1^4$, where T_1 and T_2 are the lower and higher temperatures between which the heat flow occurs.

(d) is the only possible heat-transfer mechanism in an evacuated region.

11.2 Wires A and B have identical lengths and have circular cross sections. The radius of A, R_A, is twice the radius of B; i.e., $R_A = 2R_B$. For a given temperature difference between the two ends, both wires conduct heat at the same rate. The relation between the thermal conductivities is given by

(a) $\kappa_A = 4\kappa_B$

(b) $\kappa_A = 2\kappa_B$

(c) $\kappa_A = \kappa_B/2$

(d) $\kappa_A = \kappa_B/4$

11.3 A thermally insulated container holds 50 g of ice at 0 °C. If 50 g of water at 100 °C is poured into the container, and the heat capacity of the container is negligible, the final temperature of the system will be

(a) 0 °C.

(b) greater than 0 °C but less than 20 °C.

(c) 20 °C.

(d) greater than 20 °C.

11.4 If the surface temperature of the sun were to drop by a factor of 2, the radiant energy impinging on Earth per second would be reduced by a factor of

(a) 2
(b) 4
(c) 8
(d) 16

11.5 Conduction is a process of heat transport that

(a) can proceed in and through vacuum.
(b) involves translocation of mass.
(c) is prevalent in solids.
(d) depends on the fourth power of the absolute temperature.

11.6 The linear thermal expansion coefficient of an isotropic material is α at 0 °C. The volume thermal expansion coefficient β of this material at 0 °C is then, assuming $\alpha \ll 1$,

(a) α^3
(b) 3α
(c) α
(d) $(\alpha)^{1/3}$

11.7 Aluminum has a specific heat more than twice that of copper. Identical masses of aluminum and copper, both at 0 °C, are dropped together into a can of hot water. When the system has come to equilibrium,

(a) the aluminum is at a higher temperature than the copper.
(b) the copper is at a higher temperature than the aluminum.
(c) the aluminum and copper are at the same temperature.
(d) The difference in temperature between aluminum and copper depends on the amount of water in the can.

11.8 Two cylindrical rods of the same substance have diameters d_1 and d_2. The amounts of heat conducted by these two rods, for the same temperature difference between the two ends, will be equal if their lengths are related by

(a) $(L_1/L_2) = (d_1/d_2)$
(b) $(L_1/L_2) = (d_1/d_2)^2$
(c) $(L_1/L_2) = (d_2/d_1)$
(d) $(L_1/L_2) = (d_2/d_1)^2$

11.9 The planet Earth loses heat mainly by

(a) conduction.
(b) convection.
(c) radiation.
(d) All three processes contribute significantly.

11.10 The temperature intervals of one Celsius degree (C°), one Kelvin degree (K°), and one Fahrenheit degree (F°) are so related that

(a) C° = K° = F°
(b) C° = K° < F°
(c) C° < K° > F°
(d) C° = K° > F°

11.11 Aluminum has a specific heat more than twice that of copper. Two blocks of copper and aluminum, both of the same mass and both at 0 °C, are dropped into two different calorimeters. Each calorimeter is filled with 100 g of water at 60 °C, and the calorimeter cans have negligible heat capacities. After equilibrium has been attained,

(a) the copper has a higher temperature than the aluminum.
(b) the copper has a lower temperature than the aluminum.
(c) the temperatures of the two calorimeters are the same.
(d) The answer depends on the masses of the metal blocks.

Problems

Sections 11.1, 11.2

11.1 Normal human body temperature is 98.6 °F. Express this temperature in Celsius degrees and in Kelvin.

11.2 The boiling point of hydrogen at atmospheric pressure is 20.3 K. Express this temperature in Celsius and Fahrenheit degrees.

11.3 The metal gallium melts at 303 K. What is the melting point of gallium on the Celsius and Fahrenheit temperature scales?

11.4 A temperature scale that was used for some time, especially in France, is that due to Réaumur. On the Réaumur scale, the ice point is 0 °R (as on the Celsius scale), but the boiling point of water at atmospheric pressure is assigned the value of 80 °R. Derive expressions similar to Equations (11.1) and (11.2) for converting between the Réaumur scale and the Celsius, Fahrenheit, and Kelvin scales. Find the normal human body temperature in Réaumur degrees.

11.5 Victoria Falls, on the Zambezi River, is 108 m high. If all the potential energy of the water were converted to thermal energy, what would the temperature difference be between the top of the waterfall and the bottom? Why is the temperature difference probably much less than that calculated with the above assumption?

11.6 The British thermal unit (Btu) is still widely used to specify the heating ability of fuels. One Btu is defined as the thermal energy needed to raise the temperature of 1 lb of water by 1 °F. Determine the number of calories in one Btu.

• **11.7** A normal adult can work at a rate of about 300 W for a moderate time. If all this work goes into heating 6 L of water to be poured into a wash basin, how long must that person work to heat this water from 10 °C to 45 °C?

11.8 A sick person has a temperature of 103 °F. What would this temperature be on the Celsius scale?

11.9 A patient's temperature reads 42 °C. Is this person seriously ill or does that temperature represent only a minor infection?

• **11.10** A 75-gal water heater draws 1800 W. Neglecting heat losses through the walls and the heat capacity of the tank itself, determine how long it takes to raise the temperature of the water in this heater from 15 °C to 70 °C?

11.11 What is 0 K on the Fahrenheit scale?

Section 11.4

11.12 A brass rule is exactly 1.25 m long at 20 °C. What is its length at 80 °C?

11.13 Copper has a density of 8.93×10^3 kg/m^3 at 0 °C. What is its density at 60 °C?

• **11.14** An aluminum rule that is 1 m long at 20 °C is used to measure the length of a piece of plastic. At 20 °C the plastic is exactly 83 cm long as measured by this ruler. When the system is heated to 140 °C, the plastic appears to be 83.14 cm long. What is the linear expansion coefficient of the plastic?

• **11.15** A 1-L cylindrical container made of aluminum is filled to the brim with mercury. The container and mercury are at 20 °C. The container and mercury are then heated to 320 °C. Describe what will happen quantitatively.

• **11.16** An aluminum can is filled to the brim with water at 20 °C. If can and water are heated to 60 °C, what fraction of the water, if any, will overflow?

• **11.17** An aluminum scale has been carefully calibrated at 20 °C. If this scale is used to measure length at 10 °C, will the reading be too high or too low, and by what percentage?

• **11.18** A brass collar is to be tightly fitted about a steel shaft. The diameter of the shaft is 5.0000 cm at 20 °C. The inner diameter of the brass collar is 4.9985 cm at 20 °C. To what temperature must the brass collar be raised so that it will just slip over the shaft? (Assume that the shaft remains at 20 °C.)

• • **11.19** If the brass collar and steel shaft of Problem 11.18 are heated simultaneously, to what temperature must they be raised so that the collar will just slide over the shaft?

• • **11.20** Prove that if $\alpha \ll 1$, then $\beta = 3\alpha$.

• • **11.21** Suppose that an iron rod of 1-cm^2 cross section and 0.5-m length is held rigidly between two concrete posts. Suppose that the temperature of the rod is raised by 15 °C. What compressive force must be applied to the rod by the concrete pillars to maintain the separation of 0.5 m?

• • **11.22** To avoid the accident depicted in Figure 11.10 space is left between rails to allow room for thermal expansion. If each steel rail is 25 m long and is placed into position on a day when the temperature is 20 °C, how much space should be left between the rails to permit their temperature to rise to 45 °C without damage to the rail bed? What will then be the space between the rails if the temperature drops to −30 °C?

Section 11.5

11.23 How much heat must be supplied to raise the temperature of 6 L of water from 20 °C to 60 °C?

11.24 Heat in the amount of 35 kcal is supplied to 5 L of ethanol. If the temperature of the liquid is 45 °C after equilibrium has been established, what was the initial temperature?

11.25 How much heat must be supplied to change a 5-kg block of ice at −5 °C to water at +5 °C?

• **11.26** A 20-g lead bullet is fired into the wooden block of a ballistic pendulum and comes to rest within that block. If the initial velocity of the bullet is 500 m/s and half its initial kinetic energy is converted to thermal energy in the bullet, what is the temperature of the bullet immediately after coming to rest?

• **11.27** An 80-g metal block is submerged in boiling water until thermal equilibrium has been attained. The metal is then quickly transferred to a calorimeter that contains 40 g of water. Before the hot metal is placed in the calorimeter, the 15-g copper calorimeter can and the 40 g of water are at 20 °C. After thermal equilibrium has been reestablished, the temperature of the water is 42 °C. Determine the specific heat of the metal.

• **11.28** A well-insulated container holds 0.5 L of water at 20 °C. How much ice must be dropped into the container to produce water at 5 °C after equilibrium has been achieved? Assume that the temperature of the ice is 0 °C and neglect the heat capacity of the container.

• **11.29** A 1-L glass flask has a mass of 0.2 kg when empty. If this flask contains 0.5 L of water at 20 °C, how much heat must be supplied to raise the temperature of the flask and water to 30 °C? Assume that the specific heat of glass is 0.16 cal/g·K.

• **11.30** How many calories must be supplied to 15 g of ice at −20 °C if the water is to be evaporated at 100 °C?

• **11.31** One gram of water at 0 °C is poured into 10 g of water at 50 °C. What is the final temperature of the 11 g of water?

• **11.32** 300 g of mercury at 20 °C has been placed in a thermos flask. Liquid nitrogen at atmospheric pressure is slowly poured into the flask and on the mercury. Estimate the amount of liquid nitrogen needed to solidify the 300 g of mercury. Why is it difficult to calculate this precisely, even assuming that the heat capacity of the thermos bottle may be neglected?

• •**11.33** A 40-g aluminum calorimeter contains water at a temperature of 30 °C. When 100 g of copper at a temperature of 80 °C is dropped into the calorimeter, the final equilibrium temperature that is attained is 42 °C. How much water was in the calorimeter can?

• •**11.34** A calorimeter can is made of copper and has a mass of 30 g. A block of 80 g of some material is placed in this can together with 40 g of water. The system is initially at 30 °C. At this point 100 g of water at 70 °C is poured into the calorimeter can. After thermal equilibrium has been achieved, the temperature is 57 °C. What is the specific heat of the unknown material?

Section 11.6

• **11.35** A 50-g cube of ice is placed in a small cup atop an aluminum rod of 5-cm diameter and 15-cm length whose lower end is in contact with boiling water. Determine the time that elapses before the ice has completely melted.

• **11.36** A light bulb has a straight tungsten filament whose length is 20 cm and whose diameter is 1.5 mm. The filament is heated to 2700 K by passing an electric current through it. If this filament is in an evacuated enclosure, how much power must be supplied to this lamp to maintain its operation?

• **11.37** It is claimed that turning down the thermostat at night saves energy. If the thermostat is set to 20 °C for 16 h and 14 °C for the remaining time, estimate the saving as a percentage of the cost of maintaining the setting at 20 °C for 24 h. Assume that the external temperature averages 0 °C during the day (16 h) and −4 °C during the night (8 h).

• **11.38** You wish to determine the thermal conductivity of some insulating material that comes in sheets 1.5 cm thick. To do this you construct a cubical box whose exterior lengths are 25 cm, and place a thermometer and a 100-W heater inside the box. After thermal equilibrium has been attained, the temperature inside the box is 80 °C when the external temperature is 20 °C. Determine the thermal conductivity of this insulating material. Is it a good thermal insulator?

• •**11.39** A freshwater lake has a layer of ice that is 10 cm thick over its surface. Suddenly a severe storm comes up. A strong wind blows air at a temperature of −30 °C over the surface of the ice. Calculate the rate at which the thickness of the ice layer increases under these conditions.

• **11.40** To measure the thermal conductivity of a metal rod, an experimenter takes the rod, which has a cross section of 4 cm^2 and a length of 0.8 m, and places one end in contact with boiling water. He then makes a small cup-shaped depression at the other end and places a 2-g ice cube in this depression. He finds that all the ice has melted at the end of 6 min. Assuming that extraneous heat losses are negligible, determine the thermal conductivity of the metal of which the rod is made.

• **11.41** An adult person sits in a doctor's examining room completely undressed. Suppose that the temperature of the surroundings is 20 °C, that of the adult is 37 °C. Estimate the rate of heat loss from the body of the person due to radiation.

• •**11.42** A hot-water tank is cylindrical and has the following dimensions on the inside: 0.36-m diameter and 1.35-m height. The tank is insulated with a layer of 4-cm-thick glass wool, whose thermal conductivity is 0.038 W/m·K. The metal interior and exterior walls of the tank have thermal conductivities that are several orders of magnitude greater than the value for glass wool. If the inside temperature of the tank is kept at 75 °C and the surrounding temperature is 20 °C, at what rate must energy be supplied to the tank?

• **11.43** The orbit radius of the planet Jupiter is 7.78×10^8 km. Estimate the surface temperature of Jupiter.

• •**11.44** The period of Neptune is 165 yr. Estimate its surface temperature.

• •**11.45** A double-glazed window has two panes of glass, each of thickness t_1. The air space between the two panes is of thickness t_2. Show that the rate of energy loss through the window is given by

$$\frac{\Delta Q}{\Delta t} = \frac{A}{[(2t_1/\kappa_1 + (t_2/\kappa_2)]}(T_i - T_o)$$

where A is the area of the window, κ_1 and κ_2 are the thermal conductivities of glass and air, and T_i and T_o are the temperatures inside and outside.

·12·

The Ideal Gas Law and Kinetic Theory

There was no "One, two, three, and away," but they began running when they liked, and left off when they liked, so that it was not easy to know when the race was over.

LEWIS CARROLL

12.1 Introduction

Although recognition of heat as another form of energy derived from experiments on liquids and solids, it is the study of the thermal properties of gases that most clearly illuminates the relations between thermal energy, molecular motion, and mechanical work. In this chapter we focus largely on an idealized model of a real gas. The *ideal gas* consists of very many identical, infinitesimally small (point) particles whose mutual interactions are limited to occasional elastic "billiard ball" collisions. We resort to this idealization so that we can direct our attention to the central issues and fundamental concepts and not become diverted by difficult but less essential details.

The chapter is divided into three principal portions. First, we review the experimental evidence that bears on the relation between pressure, volume, and temperature of an ideal gas, and use these data to obtain the equation of state of the ideal gas. Next, we derive this same equation of state by means of a few plausible physical assumptions about the microscopic nature of the gas; consideration of the dynamical properties of the molecules and application of Newton's laws yield a relation between pressure, volume, and the average KE of translation of a molecule.

Lastly, we merge the two approaches into one coherent theory.

12.2 Molar Quantities

In earlier chapters it often proved most convenient to reduce equations relating various physical variables to unit mass, length, time, or volume. In Chapter 11, for example, heat capacities and latent heats were given for *unit mass.* In Bernoulli's equation it is the density, or the *mass per unit volume,* that appears. The gravitational constant G is defined in terms of the force acting between *unit masses* separated by *unit distance.*

Figure 12.1 *Postage stamp commemorating the centenary of the death of Amadeo Avogadro. "Equal volumes of all gases at the same temperature and pressure contain equal numbers of molecules."*

For our purposes now, the most convenient measure is not a unit of mass but rather, a unit based on a *fixed number of particles,* the number of molecules per gram mole. One gram mole, or simply *mole,* (abbr. mol) is defined as the amount of material whose mass in grams equals the molecular weight of the constituent molecules.

Molecular weight is really a misnomer, since it is not really a weight at all.

DEFINITION: The molecular weight is the ratio of the mass of the molecule to one-twelfth the mass of a carbon atom whose nucleus is the most abundant isotope of carbon, ^{12}C.

For present purposes it is sufficiently accurate to think of the molecular weight as the ratio of the molecular mass to the mass of a hydrogen atom. The molecular mass of a monatomic molecule is just the atomic mass; for a polyatomic molecule, the molecular mass is the sum of the atomic masses of the atoms of the molecule.

The number of molecules per mole is *Avogadro's number,*

$$N_A = 6.02 \times 10^{23} \text{ molecules/mol} \tag{12.1}$$

This is the number of molecules of H_2 in 2 g of hydrogen gas; or the number of H_2O molecules in 18 g of water, ice, or water vapor. The state of the substance is immaterial. One mole of mercury contains 6.02×10^{23} Hg atoms, whether in the solid, liquid, or gaseous phase.

● ***Example 12.1*** How many water molecules are there in 1 g of water?

The molecular weight of water is that of two hydrogen atoms plus one oxygen atom, that is, $2 \times 1.0 + 16.0 = 18.0$. Hence, 18 g of water contains N_A molecules, and consequently, 1 g of water contains $(\frac{1}{18}) \times 6.02 \times 10^{23} = 3.34 \times 10^{22}$ molecules. ●

12.3 Equation of State of the Ideal Gas

Even in the relatively small volume of 1 cm³, there are 2.7×10^{19} gas molecules at 0 °C and 1 atm pressure (see Table 9.1). It is clearly impractical (even if it were possible) to calculate precisely the dynamical behavior of every gas particle. Fortunately, the bulk properties of a large collection of particles can be deduced without detailed knowledge of the behavior of each individual particle. These properties, for example, pressure, volume, and temperature, are known as *state functions* of the system, and the relationship between them is called an *equation of state*.

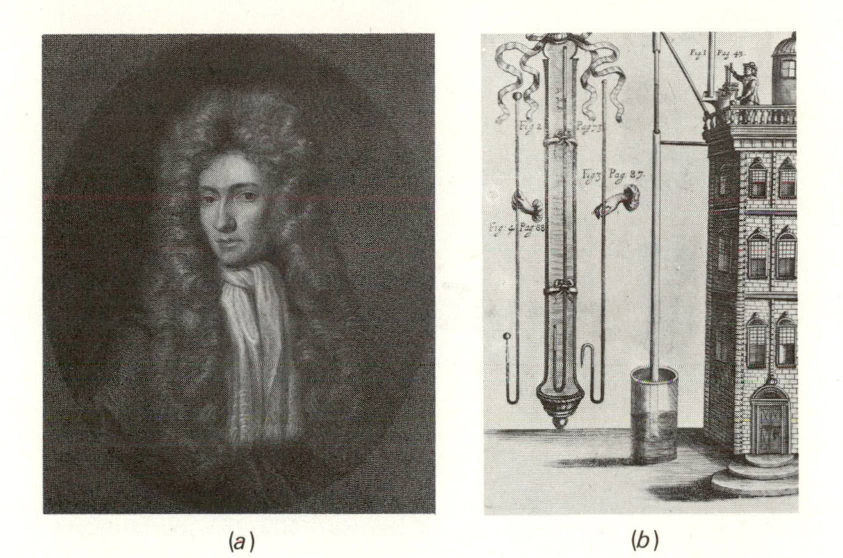

(*a*) (*b*)

Figure 12.2 (*a*) The Honorable Robert Boyle (1627–1691). (*b*) A plate from Boyle's A Continuation of New Experiments Physico-Mechanical Touching the Spring and Weight of the Air.

Among the earliest careful experiments on the behavior of gases were those of the Hon. Robert Boyle (1627–1691), youngest son of the Earl of Cork, a man of independent means and thus able to pursue scientific inquiry as an avocation. Boyle performed a series of measurements on "The Spring and Weight of Air" and found that a simple mathematical formula could express the relation between the pressure and volume of a fixed amount of air maintained at constant temperature. That relation, now known as *Boyle's law*, is

$$Pv = \text{constant} \quad \textbf{(12.2)}$$

or

$$P \times (nv) = PV = n \times \text{constant} \quad \textbf{(12.2a)}$$

where v is the *molar volume*, the volume occupied by 1 mol of air, and P the pressure exerted by the gas against the walls of the container. Equation (12.2a) states that at the same temperature and pressure, 2 mol of air occupy twice as much volume as 1 mol.

In the years following Boyle's work, it was found that Equation (12.2) was valid for any reasonably dilute gas at a temperature well above the

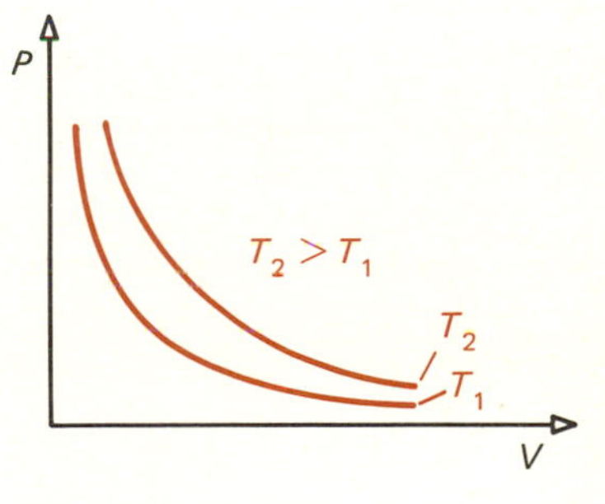

Figure 12.3 The pressure-volume relation according to Boyle's law. The relation PV = constant is the curve of a hyperbola on a P-V diagram. The location of the hyperbola depends on the value of the constant, and this, in fact, depends on the temperature of the gas. We show here two such curves corresponding to two different temperatures.

condensation temperature and moreover, that the constant in Equation (12.2) *was the same for all gases.*

There the matter rested for about a century. Then, near the beginning of the nineteenth century, Jacques Charles (1746–1832) and Joseph Louis Gay-Lussac (1778–1850) independently discovered that the volume of a gas maintained at constant pressure varied linearly with temperature, as indicated in Figure 12.4. To quote directly from Gay-Lussac:

> All gases, whatever may be their density . . . expand equally between the same degrees of heat [temperature] . . . For the permanent gases the increase of volume received by each of them between the temperature of melting ice and that of boiling water is equal to 100/266.66 of the original volume for the centigrade thermometer.

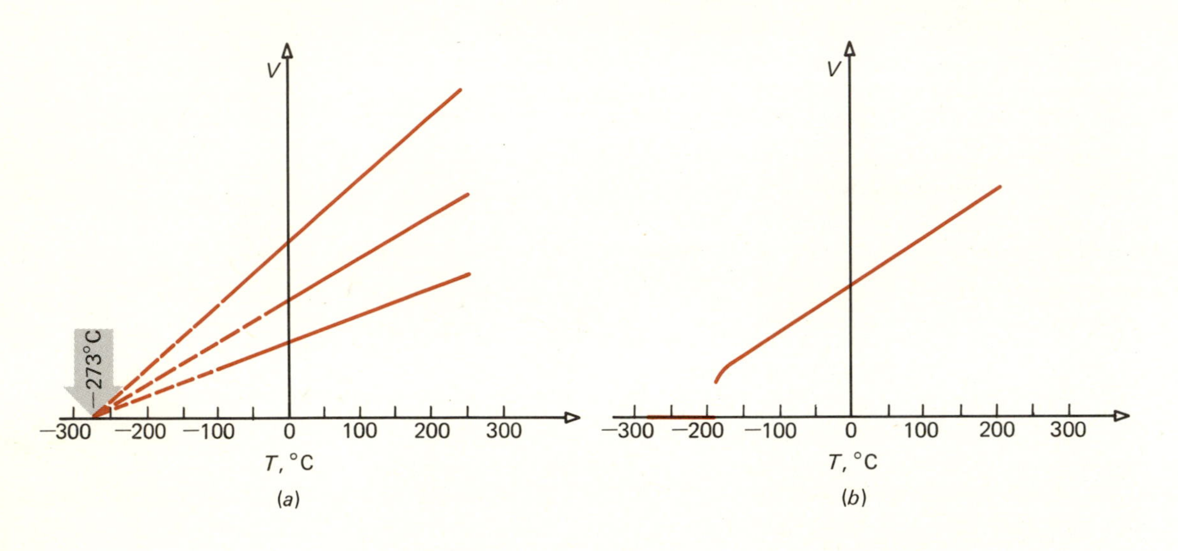

Figure 12.4 *(a) The dependence of the volumes of three ideal gases on temperature under constant pressure. In each case, the volume-vs-temperature curve is a straight line that, if extrapolated to zero volume, intersects the temperature axis at $T = -273.16$ °C. (b) A schematic curve showing the volume-temperature curve of a real gas, nitrogen, at atmospheric pressure. At −195 °C (78 K) and atmospheric pressure, nitrogen liquefies, and thereafter the volume does not change substantially as the temperature is lowered further.*

More precise measurements give the ratio 100/273.16. Thus, though we may use *different* gases and start with *different* initial volumes, a plot of volume versus temperature would yield a straight line for each, and for each one the fractional change in volume between the ice point and the boiling point of water would be the same, 100/273.16. Consequently, if for these gases, maintained at fixed pressures, we plotted volume versus temperature on the Celcius scale, the individual straight lines would intersect the ordinate at different values of volume at 0 °C, but *each line would intersect the abscissa at the same point, namely at −273.16 °C ≃ −273 °C.*

This result implies that if an *ideal gas* were cooled to −273 °C, its volume would shrink to zero, and it suggests that −273 °C might be the lowest attainable temperature. This temperature, which is the zero of a temperature scale based on the behavior of an ideal gas, is also the absolute zero of the thermodynamic, or Kelvin, scale (see Chapter 13).

We now have two relations obeyed by gases, namely, Equation (12.2) at constant temperature, and

$$v = \text{constant} \times T \qquad \textbf{(12.3)}$$

at constant pressure, where T in Equation (12.3) is the absolute temperature; it is hereafter understood that the symbol T refers to absolute temperature on the Kelvin scale.

Boyle's law and Gay-Lussac's law can be combined into a single formula,

$$Pv = RT \qquad \text{or} \qquad PV = nRT \tag{12.4}$$

where v is, as before, the molar volume, and V is the volume occupied by n moles. The constant R is known as the *universal gas constant* and has the experimentally measured value

$$R = 8.31\ \text{J/mol·K} = 2.0\ \text{cal/mol·K} \tag{12.5}$$

Equation (12.4) is the equation of state for an ideal gas, the *ideal gas law*. It is sometimes written

$$Pv = N_A kT \qquad \text{or} \qquad PV = nN_A kT = NkT \tag{12.6}$$

where N is the total number of gas molecules in the volume V and $k = R/N_A$ is called *Boltzmann's constant*, and has the value

$$k = 1.38 \times 10^{-23}\ \text{J/K}$$

● ***Example 12.2*** In a low-temperature experiment, a small sample volume is filled with helium gas to a pressure of 0.5 atm at a temperature of 20 K (the temperature of liquid hydrogen at atmospheric pressure). If at the end of the experiment, the apparatus is allowed to warm to room temperature before the helium gas is pumped from the sample chamber, what will the pressure in that chamber be?

We assume that we can neglect the small change in volume due to thermal expansion of the material of which the sample chamber is constructed. We also assume that we can use the ideal gas law as the equation of state of helium under the stated conditions. Since no helium gas escapes from the chamber, n = constant in Equation (12.4). Hence, the pressures and temperatures are related by

$$\frac{P_1}{P_2} = \frac{T_1}{T_2}$$

or

$$P_2 = P_1 \frac{T_2}{T_1}$$

Setting $T_2 = 293$ K (20 °C), one finds that

$$P(20\ ^\circ\text{C}) = \frac{0.5 \times 293}{20} = 7.3\ \text{atm}$$

This is a fairly high pressure, equal to that about 63 m (210 ft) below the surface of a lake, and enough to rupture delicate, thin-walled tubing. ●

An important practical consequence of the ideal gas law, which you may recall from elementary chemistry, is that at a given temperature and pressure the molar volume is the same for all ideal gases. Thus, to the extent that these gases approximate ideal gases, 2 g of H_2, 20 g of Ne, 44 g of CO_2, and 211 g of Rn, all occupy 22.4 L at 0 °C = 273 K and 1 atm pressure, the condition usually referred to as *STP, standard temperature and pressure*.

Equation (12.4) or (12.6) is an empirical statement of the observed behavior of dilute gases and is a useful predictor for such systems. By concentrating attention on the macroscopic state functions, however, we have precluded derivation of any relations between the state functions and the dynamical variables of the constituent molecules. Nevertheless, we can already draw some important and suggestive conclusions from these relations.

The left-hand side of (12.6) has the dimension

$$[P][V] = \frac{[F]}{[A]}[V] = [F][L] = [\text{energy}]$$

where F, A, V, and L stand for force, area, volume, and length. Consequently, NkT must also have the dimension of energy. One knows that simply lowering a container of gas through some height has no perceptible influence on its temperature or pressure, so that gravitational PE apparently plays no significant role. Moreover, since Equation (12.6) approximates the behavior of real gases most closely when they are very dilute, it must also be true that the mutual interaction energies of molecules do not contribute to the energy kT. Thus one is almost inevitably forced to the conclusion that kT is in some way directly related to the *average KE* of a molecule. Herein lies the profound significance of the concept of absolute temperature in classical physics!

The connection between microscopic dynamical variables of the molecules and the macroscopic state functions is, however, not confirmed by such qualitative arguments, suggestive though they may be. The fundamental, unanswered question remains: What is the relation between temperature and molecular KE, assuming that such a relation exists?

The answer to this question is given by the kinetic theory of gases, to which Boltzmann made immense contributions and to which we now turn our attention.

12.4 Kinetic Theory

We consider N molecules of an ideal gas, in thermal equilibrium with the walls of a container of volume V. Occasionally, a given particle will collide elastically with another, changing the energies and momenta of both, subject to conservation of total energy and momentum. One can show that when a fast-moving particle collides with a slow one, the most probable result is that the fast one loses energy and the slow one gains energy. Intermolecular collisions thus tend toward an equilibrium arrangement in which the particles of the gas acquire and maintain *comparable* kinetic energies and speeds. That does not mean, however, that all particles assume *identical* speeds; far from it. It does suggest that there is some *average* speed about which the particle speeds will cluster. Using sophisticated techniques, one can deduce the distribution of speeds within a large collection of particles. James Clerk Maxwell and Ludwig Boltzmann first carried through such a calculation, with results shown graphically in Figure 12.5.

Next, since there is no net flow of gas in a stationary container, the vector sum of the momenta of all the particles must be zero, and so must the net velocity. That is, *on the average,* as many particles move to the right as to the left, up as down.

We now ask what happens when a particle strikes the wall of the container. The simplest assumption is that it rebounds elastically, like a perfect Ping-Pong ball, and we shall proceed on that assumption. First, however, we must ask how good that assumption is, and if it proves to be a poor one, can it introduce important errors in our calculation.

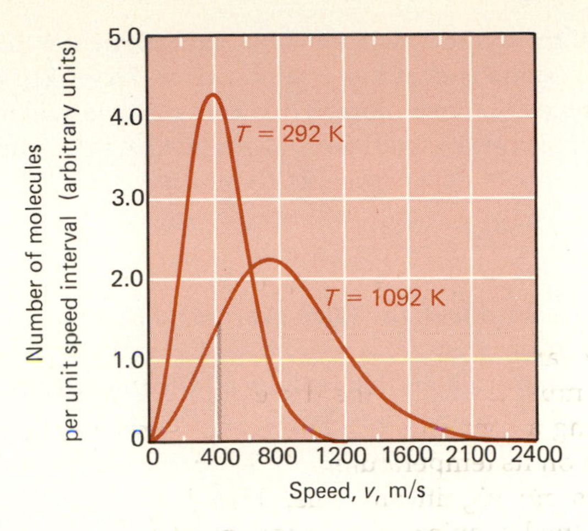

Figure 12.5 The Maxwell-Boltzmann distribution of molecular speed of O_2 in oxygen gas at two different temperatures. The number of molecules in some small range of speeds, for example between 420 m/s and 440 m/s, equals the area of the narrow strip under the curve. In this figure that strip is shown shaded for the higher-temperature curve. The total number of molecules in the gas is equal to the area under the curve.

Suppose a particle on striking the wall adheres to it. In coming to rest, it must release its KE to the wall, and if that process continues for some time and for many particles, significant energy will be transferred to the wall. This energy transfer implies that the container and the gas within it are not in thermal equilibrium, in contradiction to our initial assumption. What happens in a real gas is that a molecule may adhere to the wall briefly, but then absorb energy from the wall and move off into the volume of the container. On the average, the energy and momentum lost by molecules on adhesion is regained as they reenter the volume. Consequently, the assumption that all wall collisions are elastic, though unrealistic, will not lead us astray so long as our principal concern is with average properties and not the specific behavior of a particular gas molecule.

To simplify the formal calculation, we assume that the ideal gas is inside a cubic box of sides L, volume L^3. We focus attention on just one of the N particles, which happens to be traveling toward the wall at $x = L$ with velocity $\mathbf{v}_1$, with components v_{1x}, v_{1y}, v_{1z}. We mark time $t = 0$ the instant the particle strikes that wall. If it is assumed that the collision at the wall is elastic, the velocity afterward has components $-v_{1x}$, v_{1y}, v_{1z}. Hence, the change in the particle's momentum as a result of colliding with the wall is

$$\Delta p_{1x} = m(-v_{1x}) - mv_{1x} = -2mv_{1x} \qquad \Delta p_{1y} = \Delta p_{1z} = 0. \tag{12.7}$$

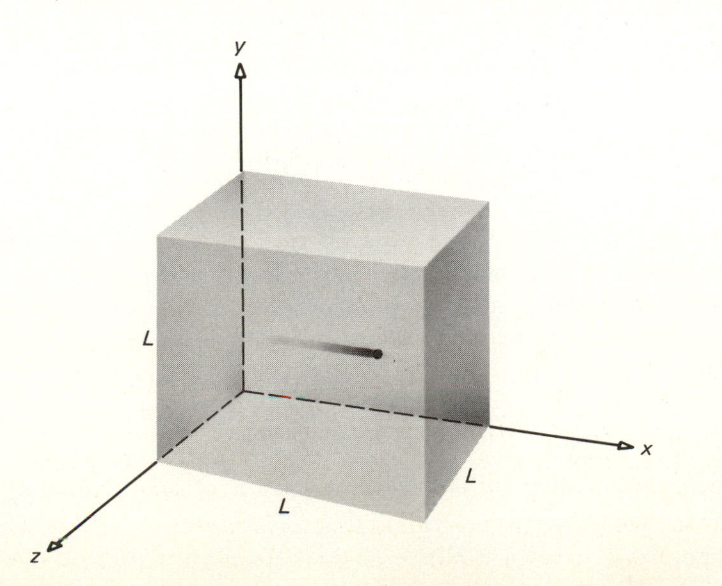

Figure 12.6 A gas is confined to a cubic volume of side length L. One molecule of the gas is shown here. It is assumed that whenever this molecule strikes a wall of the container, it rebounds elastically.

After a brief time interval $\delta t = L/v_{1x}$, this particle will reach and collide with the opposite wall at $x = 0$; at $t = \Delta t = 2\delta t = 2L/v_{1x}$ it rebounds from the wall at $x = L$ once more. Thus this particle will spend the time $2L/v_{1x}$ between successive collisions at the wall $x = L$. The average force exerted *by this wall on that particle* is therefore

$$F_1 = \frac{\Delta p_1}{\Delta t} = \frac{-2mv_{1x}}{2L/v_{1x}} = \frac{-mv_{1x}^2}{L} \qquad \textbf{(12.8)}$$

and by Newton's third law, the force exerted *by this particle on that wall* is just mv_{1x}^2/L.

The total force exerted on that wall by *all* the particles is the sum of the forces due to the N independent particles, that is,

$$F = \frac{mv_{1x}^2}{L} + \frac{mv_{2x}^2}{L} + \cdots + \frac{mv_{Nx}^2}{L} = \frac{m}{L}\sum_{i=1}^{N} v_{ix}^2$$

and by definition, the pressure on that wall is

$$P_x = \frac{F}{A} = \frac{1}{L^2} \times \frac{m}{L}\sum_{i=1}^{N} v_{ix}^2 = \frac{m}{V}\sum_{i=1}^{N} v_{ix}^2 \qquad \textbf{(12.9)}$$

Similarly, the pressure on the walls perpendicular to the y and z directions is

$$P_y = \frac{m}{V}\sum_{i=1}^{N} v_{iy}^2 \qquad P_z = \frac{m}{V}\sum_{i=1}^{N} v_{iz}^2 \qquad \textbf{(12.9a)}$$

The system has cubic symmetry; nothing distinguishes the x axis from the y or z axes. Hence, $P_x = P_y = P_z = P$; that is, the pressure is the same on all the walls of the container. If we add the three equations of (12.9), we obtain

$$3P = \frac{m}{V}\sum_{i=1}^{N} (v_{ix}^2 + v_{iy}^2 + v_{iz}^2) = \frac{m}{V}\sum_{i=1}^{N} v_i^2$$

But $mv_i^2 = 2(\text{KE}_i)$ is just twice the translational KE of the ith particle; and on summing over all N particles, we have $3P = (2/V)(\text{KE})$, or

$$PV = \tfrac{2}{3}(\text{KE}) = \tfrac{2}{3}N\,\langle \epsilon_t \rangle \qquad \textbf{(12.10)}$$

Here, the symbol KE stands for the total translational kinetic energy of all N particles, and $\langle \epsilon_t \rangle$ is the average translational kinetic energy of one particle.

Equation (12.10), which was deduced without any specific assumption about the velocity distribution, is general and is valid for any ideal gas. It bears a striking resemblance to Equations (12.4) and (12.6). As we had surmised from a strictly dimensional argument, the product kT is indeed proportional to the average translational KE per molecule. To be precise,

$$\tfrac{1}{2}kT = \tfrac{1}{3}\,\langle \epsilon_t \rangle \qquad \langle \epsilon_t \rangle = \tfrac{3}{2}kT \qquad \textbf{(12.11)}$$

This is a special instance of a very general and extremely important result, which we now state without proof.

1. *In thermal equilibrium, the average KE of a molecule is ν times $\frac{1}{2}kT$, where ν is the number of degrees of freedom of the molecule.*

2. *The average KE per molecule is equally distributed among the available degrees of freedom.*

The second statement is known as the *law of equipartition of energy.*

A *degree of freedom* is any independent coordinate of the system. For example, a point mass in space has three degrees of freedom; three quantities—in cartesian coordinates, x, y, and z—are needed to locate the mass. If the mass is constrained to move on a surface, it has only two degrees of freedom. If the surface is the x-y plane, its position is given by the x and y coordinates; if it is constrained to move on a spherical surface, like a ship crossing the ocean, its position is best given by two angles, the longitude and the latitude. A diatomic molecule—two point masses held together by a cohesive force—has six degrees of freedom, since three coordinates must be given for each of the two atoms. These six degrees of freedom could equally well be the three coordinates of the center of mass, the two angles that identify the orientation of the line joining the two atoms, and one distance, the separation between the two atoms.

The results obtained so far, especially the equipartition theorem, have several important consequences.

First, the concept of absolute temperature, previously based on extrapolation of ideal gas behavior, now acquires new meaning and physical significance. The temperature of a system is simply a measure of the average molecular KE, $\frac{1}{2}\nu kT$, where ν is the number of degrees of freedom per molecule. For a gas of N independent point particles, the total energy, $\frac{3}{2}NkT$, is called the *internal energy* of that gas and is denoted by the symbol U. In polyatomic gases, the interaction between the atoms of a molecule generally contributes some PE, and the internal energy of a polyatomic gas includes both the potential-energy and the kinetic-energy contributions.

Second, *the average KE per degree of freedom does not depend on the mass of the molecule.* It follows from this that at a given temperature, the average speed of a molecule will be greater the smaller its mass. This result has important practical consequences. For example, at 273 K, the average speed of O_2 is 461 m/s, that of He is 1305 m/s, and that of H_2 is 1844 m/s. The much greater *average* speed of the lighter molecules means that a significant fraction of the molecules of such gases will have speeds over 11,200 m/s, the escape velocity from the gravitational field of the earth. That is why our atmosphere contains no hydrogen gas, even though it must have been abundant at an early stage of geological time; the gas simply escaped into space. Helium is present in the atmosphere only because it is constantly replenished by decay of some radioactive minerals.

A corollary of this kinetic theory result is that in a volume containing two different gases in thermal equilibrium, the molecules of the gas with the lower molecular weight have greater average speed. Since diffusion of gases depends on the average molecular speeds, the concomitant difference in diffusion rates may be used to separate components of a gas that differ only in mass. This was the method used during World War II to separate ^{235}U from ^{238}U.

Third, since the number of degrees of freedom increases with molecular complexity, the molar KE at a given temperature must also increase proportionately. Hence, we can expect that increasing the temperature of a mole of a polyatomic gas by one degree will require more energy than a similar temperature change in one mole of a monatomic gas. That is, we expect the molar heat capacity to increase with increasing molecular complexity.

● ***Example 12.3*** Show that at 273 K, the average speed of an O_2 molecule is 461 m/s and of an H_2 molecule, 1844 m/s.

According to Equation (12.11),

$$\tfrac{1}{2}m\,\langle v^2\rangle = \tfrac{3}{2}kT \qquad \langle v^2\rangle = \frac{3kT}{m}$$

The mass of one oxygen molecule is 32 times the mass of a hydrogen atom, i.e., $32(1.66 \times 10^{-27}$ kg). Thus we have

$$v_{\text{av}} = \sqrt{\frac{3(1.38 \times 10^{-23}\ \text{J/K})(273\ \text{K})}{32(1.66 \times 10^{-27}\ \text{kg})}} = 461\ \text{m/s}$$

for oxygen.

From Equation (12.11), it follows that at a given temperature, the average of v^2 is inversely proportional to the molecular mass. Hence, we can write

$$\frac{v_{\text{av}}(H_2)}{v_{\text{av}}(O_2)} = \sqrt{\frac{m(O_2)}{m(H_2)}}$$

$$v_{\text{av}}(H_2) = \sqrt{16}(461\ \text{m/s})$$

$$= 1844\ \text{m/s}$$

●

The average speed calculated in Example 12.3 is known as the *root-mean-square speed* because it is the square *root* of the *mean* of the *squared* speed. According to Equation (12.11), the average of v^2 equals $3kT/m$ for an ideal gas, so the quantity we have calculated is indeed the root-mean-square speed, denoted by v_{rms}.

12.5 Heat Capacities of Gases

Gases, like liquids and solids, have a heat capacity. The heat capacity of gases, however, in contrast with liquids and solids, depends critically on the experimental conditions under which it is measured.

Consider the arrangement in Figure 12.7. One mole of an ideal gas is confined in a cylindrical volume fitted with a movable piston. We can either clamp the piston in position, maintaining a fixed volume, or apply a constant pressure over the surface of the piston, allowing it to move

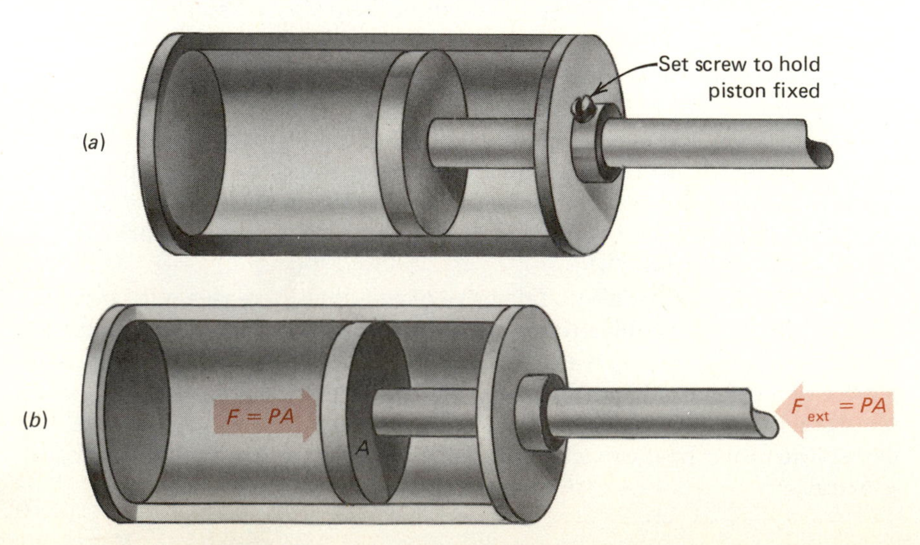

Figure 12.7 *Schematic drawing, showing the arrangement for making measurements (a) at constant volume, and (b) at constant pressure. (a) A set screw holds the piston in fixed position so that the volume occupied by the gas within the cylinder cannot change. (b) A constant force is applied to the rod attached to the piston. The piston is in mechanical equilibrium when the magnitude of this external force equals the product PA, where P is the pressure exerted by the gas in the cylinder and A is the area of the piston.*

back and forth until the externally applied pressure is balanced by the pressure the confined gas exerts on the piston. In either case, we know that at thermal equilibrium,

$$Pv = RT \tag{12.4}$$

At the temperature T, the total molar KE of the gas molecules is

$$\text{KE} = (\nu N_A)(\tfrac{1}{2}kT) = \tfrac{1}{2}\nu RT$$

where ν denotes the number of degrees of freedom per molecule.

12.5 (a) Monatomic Gases

If the gas is monatomic, then $\nu = 3$, and there are no contributions to the total molecular energy of the system other than the kinetic energy of translation of the individual atoms. Thus the total internal energy of the gas is then just

$$U = \tfrac{3}{2}RT \text{ per mole (monatomic gas)} \tag{12.12}$$

Suppose we supply some heat, say ΔQ cal, to one mole of a monatomic gas. The energy ΔQ supplied to the gas as heat can have two effects. It may cause the gas to do some mechanical work, like pushing the piston outward against an external force, or it could increase the internal energy U of the gas, manifesting itself in a corresponding increase in temperature. Or the gas could do some work and also change its temperature.

If the piston in Figure 12.7 is clamped, no mechanical work can be done; we recall that $\Delta W = F\,\Delta s$, and since $\Delta s = 0$, $\Delta W = 0$ also. Hence all the thermal energy ΔQ must go into increasing the internal energy of the molecules. Thus,

$$\Delta Q = \Delta U = \tfrac{3}{2}R\,\Delta T \text{ per mole} \tag{12.13}$$

and consequently, the molar heat capacity of the monatomic gas is

$$c_v = \frac{\Delta Q}{\Delta T} = \frac{3}{2}R \tag{12.14}$$

where the subscript denotes that this is the heat capacity at constant volume.

Note that the *molar* heat capacity is the same for all monatomic gases, such as helium, neon, and argon, although their specific heat capacities (heat capacity per gram) differ greatly (see Table 12.1).

Suppose that we do not clamp the piston of Figure 12.7 but permit it to move until no net force acts on it, that is, until

$$PA = F_{ext} \tag{12.15}$$

where P is the pressure of the gas, A the area of the cylinder, and F_{ext} the external force applied to the piston. Let's assume that our 1 mol of gas has come to thermal equilibrium at some temperature T and that its pressure is P. If we supply some heat, the temperature of the gas will increase and so will the average KE of each atom. Consequently, the pressure exerted by the gas against the piston increases, and the piston tends to move out, increasing the volume occupied by the gas until Equation (12.15) is once again satisfied. Thus, as heat ΔQ is supplied to the gas at constant pressure, the result is not only an increase in the temperature of the gas but also the performance of mechanical work by the gas. The work done is $\Delta W = F\,\Delta s = PA\,\Delta s$, where Δs is the distance moved by the piston. But

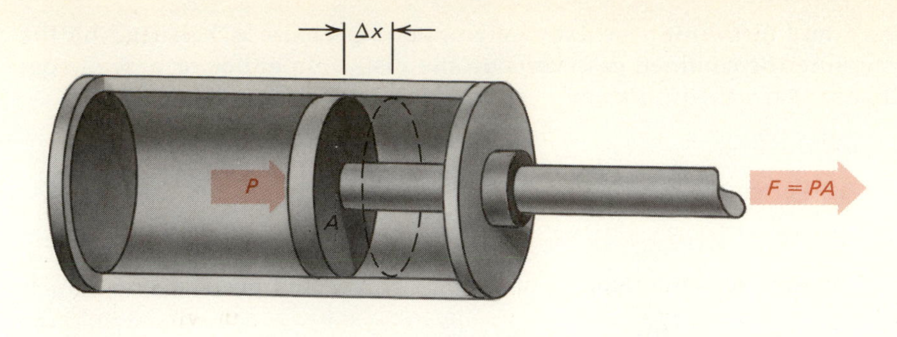

Figure 12.8 The mechanical work done by the molecules of the gas within the cylinder in pushing the piston a distance Δx is $F\ \Delta x = PA\ \Delta x = P\ \Delta V$.

$A\ \Delta s = \Delta v$ is just the change in molar volume. We can therefore write

$$\Delta W = P\ \Delta v \tag{12.16}$$

From conservation of energy, the heat input must equal the increase in internal energy plus the work done by the gas; that is,

$$\Delta Q = \Delta U + \Delta W \tag{12.17}$$

where ΔW denotes the work done *by* the system against external forces and ΔQ is the energy supplied to the system as heat.

Conservation of energy, as written in Equation (12.17), is known as the *first law of thermodynamics.* The broad generality of this law, and of other thermodynamic laws, is responsible for the great versatility of thermodynamics and the commensurate generality of results deduced from thermodynamic reasoning. The next chapter gives a brief introduction to thermodynamics and we, therefore, defer further discussion of this law.

From (12.16) and (12.17), we have

$$\Delta Q = \Delta U + P\ \Delta v \tag{12.18}$$

Table 12.1 Molar heat capacities of gases at 20 °C and atmospheric pressure

	Gas	c_p, cal/mol·K	c_v, cal/mol·K	$\frac{c_p - c_v}{R}$	$\gamma = \frac{c_p}{c_v}$
Monatomic gases	He	4.98	3.00	1.00	1.66
	Ne	4.97	3.03	0.98	1.64
	Ar	5.00	3.01	1.00	1.67
	Kr	4.97	2.96	1.02	1.68
	Xe	4.97	3.00	1.00	1.66
Diatomic gases	H_2	6.89	4.90	1.00	1.41
	N_2	6.96	4.97	1.00	1.40
	O_2	7.02	5.03	1.00	1.40
	CO	6.95	4.96	1.00	1.40
	NO	6.99	5.00	1.00	1.40
Triatomic gases	CO_2	8.74	6.70	1.03	1.30
	N_2O	8.82	6.78	1.03	1.30
	CH_2	8.48	6.47	1.01	1.31
	H_2S	8.64	6.55	1.05	1.32
Polyatomic gases	C_2H_2	9.97	7.92	1.03	1.26
	C_2H_6	11.60	9.51	1.05	1.22

But from the ideal gas law, we also know that if P is constant, $P\,\Delta v = R\,\Delta T$. Thus, for our 1 mol of *monatomic gas,* we obtain, using Equation (12.12),

$$\Delta Q = \tfrac{3}{2}R\,\Delta T + R\,\Delta T = \tfrac{5}{2}R\,\Delta T \qquad \textbf{(12.19)}$$

and

$$c_p = \tfrac{5}{2}R$$

We see that the molar heat capacity of a monatomic gas is much greater at constant pressure than at constant volume! But why should the molar heat capacity of a substance depend on the experimental conditions under which it is measured? It is not difficult to find the answer to this question. Indeed, it is placed in evidence by Equations (12.18) and (12.19). If the *volume* of the gas is fixed, all the heat that is supplied is used to increase its internal energy, that is, to raise its temperature. If the *pressure* is kept constant and the gas allowed to expand, however, some of the heat is used to do mechanical work (the second term on the right-hand side of (12.18) and (12.19)). In that case, the temperature of the gas will rise less than if the volume is constant, and the ratio $\Delta Q/\Delta T$ is therefore greater.

The difference between the molar heat capacities at constant pressure and constant volume is R cal/mol·K. This result does not depend on the molecular structure of the gas, provided that the ideal gas model is a valid approximation. That is, $c_p - c_v = R$ for any gas, whether it is monatomic or polyatomic, if it obeys Equation (12.4).

12.5 (b) Diatomic Gases

The data in Table 12.1 show that for the diatomic gases the molar heat capacity is almost exactly 2 cal/mol·K greater than for the monatomic gases. That result is readily explained by the equipartition theorem and a simplified model of a diatomic molecule.

Suppose a diatomic molecule, like H_2, behaves like a rigid dumbbell, that is, two atoms separated by a fixed distance. Such a molecule has five degrees of freedom, not six, because the last degree of freedom mentioned earlier, the separation between the two atoms, is not "free" in this model.

The molar internal energy is then

$$U = N_A(5)(\tfrac{1}{2}kT) = \tfrac{5}{2}RT$$

and the molar heat capacity at constant volume is

$$c_v = \frac{\Delta U}{\Delta T} = \frac{5}{2}R \simeq 5 \text{ cal/mol·K}$$

Calculation of the molar heat capacity at constant pressure proceeds along the same lines as that for the monatomic gas. Again, one finds that the heat capacity is greater at constant pressure than at constant volume by R cal/mol·K, provided that the ideal gas law is valid. It follows that

$$c_p = \tfrac{7}{2}R \simeq 7 \text{ cal/mol·K}$$

The ratio of the heat capacities

$$\gamma = \frac{c_p}{c_v} \qquad \textbf{(12.20)}$$

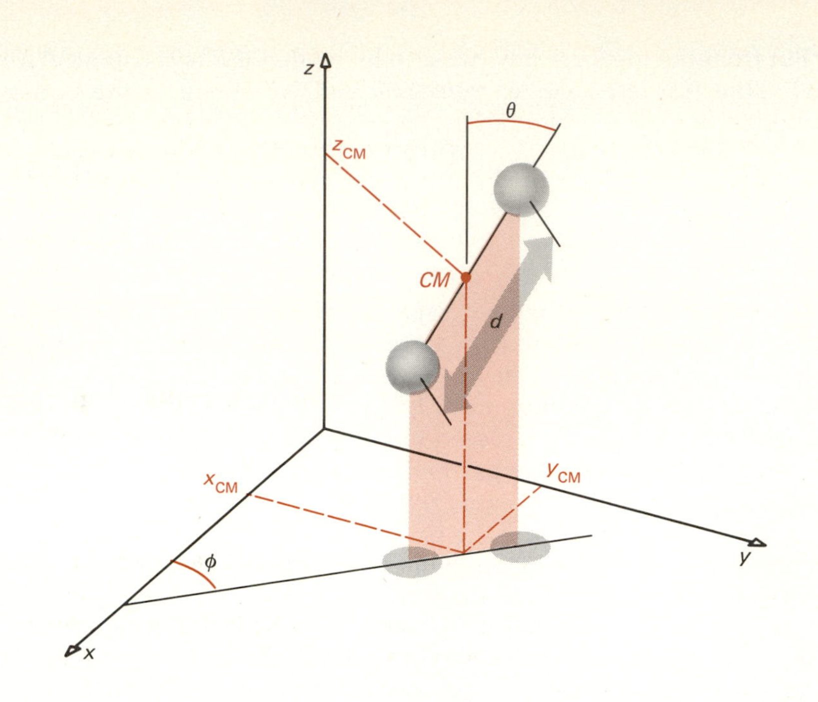

Figure 12.9 Rigid dumbbell model of a diatomic molecule. The five degrees of freedom of the molecule are the three coordinates of its center of mass, x_{CM}, y_{CM}, and z_{CM}, and the two angles θ and ϕ that specify the orientation of the line joining the two atoms.

can be a good indicator of the molecular structure of a dilute gas. For monatomic gases, $\gamma = \frac{5}{2}/\frac{3}{2} = \frac{5}{3} = 1.67$; for diatomic gases $\gamma = \frac{7}{5} = 1.4$; for triatomic gases (for example, CO_2) $\gamma = \frac{9}{7} = 1.29$; and so forth (see Table 12.1).

*12.6 Mixture of Gases

We have earlier restricted discussion to the equation of state and heat capacities of a one-component ideal gas. Generally, however, systems of interest, like air, have more than one gaseous component.

Suppose the volume holds n moles of a monatomic gas. In that case the total internal energy will be $U = nN_A \times \frac{3}{2}kT = \frac{3}{2}nRT$. Similarly, if the gas is diatomic, $U = \frac{5}{2}nRT$. That is, the total internal energy of n moles is simply n times the internal energy per mole, and the heat capacity of n moles is n times the molar heat capacity.

Suppose that our system consists of a mixture of two or more ideal gases in thermal equilibrium in a container of volume V. What is then the

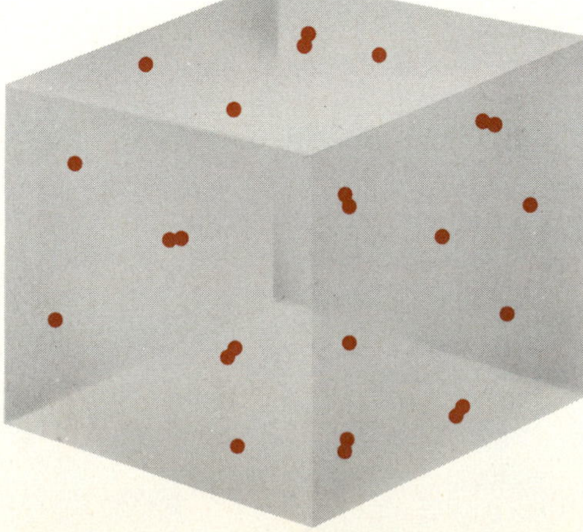
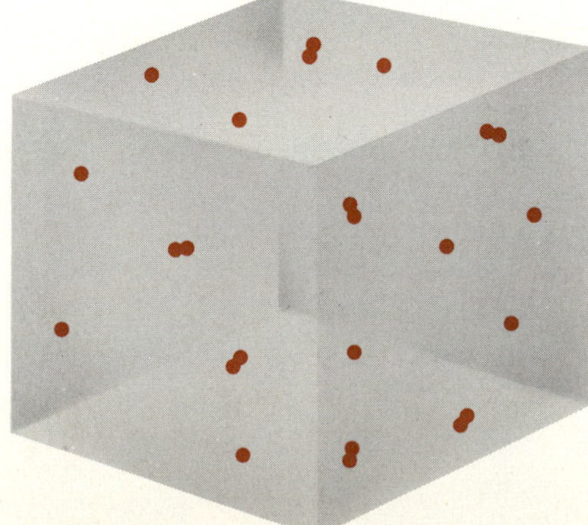

Figure 12.10 The total pressure exerted by the mixture of two gases, a monatomic and a diatomic, within the cubic volume is the sum of the pressures due to each group of molecules separately. The pressure that each group exerts is called the partial pressure due to that part of the mixture.

total pressure against the walls of the container, and how does each component of the gas mixture contribute to that pressure?

It is assumed that each component behaves as an ideal gas, and moreover, that the constituent molecules of different gases do not react with each other—for example, no chemical combinations are formed. Therefore the molecules of each gas move about the container at their normal thermal speeds, striking the walls and exerting a pressure given by the ideal gas law applied to that group of molecules. The total pressure exerted by all the molecules of the mixture is then

$$P = P_A + P_B + \cdots \qquad \textbf{(12.21)}$$

where $P_A, P_B, \ldots$ are the *partial pressures* of the constituent gases and are given by

$$P_A = \frac{n_A RT}{V} \qquad P_B = \frac{n_B RT}{V} \qquad \text{and so on} \qquad \textbf{(12.22)}$$

Here $n_A, n_B, \ldots$ are the number of moles of gas A, gas B, etc., confined to the volume V.

Ordinary air is principally a mixture of 80% N_2 and 20% O_2. Since the total atmospheric pressure is 1 atm at sea level, the partial pressure of N_2 and of O_2 are 0.8 atm and 0.2 atm respectively.

● ***Example 12.4*** To avoid the dangers of nitrogen narcosis, divers are supplied with a gaseous mixture of O_2 and He. What relative composition by weight should a mixture of this kind have so that when it is supplied to a scuba diver at a depth of 100 m, the partial pressure of O_2 will be 1 atm? (Oxygen at pressures above about 1 atm is toxic.)

The total pressure of the gas supplied to the diver must balance the external pressure on his body to prevent collapse of the rib cage. At a depth of 100 m, the total pressure on the diver is 11 atm (1 due to atmospheric pressure, 10 due to the water above him). Hence the total pressure of the gas supplied must also be 11 atm, and the partial pressures must be

$$P_{O_2} = 1 \text{ atm} \qquad P_{He} = 10 \text{ atm}$$

Hence the mixture should contain 10 mol of He for every mole of O_2. Since 1 mol of He has a mass of 4 g and 1 mol of O_2 has a mass of 32 g, the composition of the mixture, by weight, should be 40 g of He and 32 g of O_2, or about 56% He and 44% O_2. ●

12.6 (a) Scuba Diving and Aquatic Mammals

Although all aquatic mammals breathe through lungs, like terrestrial mammals, life in the aquatic environment imposes special demands; the most obvious is a highly efficient use of oxygen to support extended dives. There are, however, several further problems besides adequate oxygen supply that must also be solved, most of them encountered also by scuba divers and occasionally even skin divers. The most familiar is the syndrome known as the bends, diver's disease, or aeroembolism. It generally appears when a diver surfaces after prolonged submersion at depths in excess of about 20 m, and it is more severe with greater depth and diving time. The bends is caused by gas bubbles in the joints (extremely painful), in muscle and fatty tissues, and in the blood. Gas bubbles in the blood are by far the most dangerous because such bubbles can block small blood vessels with fatal consequences if these happen to serve the central nervous system.

How and why are these gas bubbles formed? Although we do not think of a liquid, such as water or blood, as containing dissolved gases, oxygen and nitrogen are normally dissolved in all liquids that are in contact with the atmosphere. Ordinarily, the partial pressures of these two gases in the liquid are 0.2 and 0.8 atm, just as in the surrounding air. If the gaseous partial pressures in the liquid were less, atmospheric pressure would gradually force more gas into solution in the liquid; if it were greater, the gases would come out of solution and form bubbles in the liquid that would then rise to the surface. If water or blood is placed in a tank and the air pressure above the liquid is increased, more oxygen and nitrogen gas will gradually dissolve in these liquids until pressure equilibrium is again established.

Hydrostatic pressure increases about 1 atm for every 10 m of depth. Thus a scuba diver, descending to 20 m, will be subjected to a *total* pressure of 3 atm. To prevent collapse of the lungs under this greater pressure, the diver must breathe air from tanks at a pressure of 3 atm; thus, after some time, the partial pressures of O_2 and N_2 in the diver's blood will be 0.6 atm and 2.4 atm respectively. Equilibration to these higher pressures does not occur immediately but gradually, and this is why the danger of the bends and its severity increases with duration of submersion as well as depth.

The blood and tissues of a diver who returns quickly to normal atmospheric pressure contain much more dissolved gas than can be accommodated under equilibrium conditions; the body liquids are said to be supersaturated with gas. This excess gas is then released in much the same manner that carbon dioxide dissolved in carbonated drinks under pressure forms small bubbles as soon as the cap is removed and the excess pressure released. Moreover, just as shaking a pop bottle before opening increases the rate of bubble formation, so does movement of joints and muscular exertion by the diver. The principal culprit in causing bends is nitrogen, because oxygen is rapidly consumed by the metabolic action of the body cells and is quickly dissipated.

To avoid the bends, a diver must decompress slowly, that is, return to the surface gradually, allowing the slow release of excess dissolved nitrogen through the lungs. Divers are warned to adhere strictly to so-called staging tables during ascent. If accident or some other emergency requires rapid surfacing, it is imperative that the person be quickly transferred to a pressure chamber, and the air pressure brought to the value corresponding to the diving depth and then very gradually reduced to atmospheric.

Even a skin diver, descending repeatedly for brief periods of one or two minutes without auxiliary equipment, can develop aeroembolisms. There is a well-documented instance of a Danish physician who developed the bends during an extended study of underwater escape techniques. After some five hours of repeated two-minute dives he complained of pains in the joints and other symptoms of diver's disease. He was placed into a compression chamber, whereupon these symptoms disappeared, and he was then gradually staged to atmospheric pressure with no ill effects.

Bends can be a hazard even in air travel if a nonpressurized plane ascends rapidly to about 6000 m (18,000 ft), where the pressure is only half an atmosphere, or if a pressurized cabin is suddenly decompressed at high altitude.

For many years it was a source of amazement that seals and whales could dive repeatedly for extended periods, often surfacing only briefly before descending once again to great depths. (Whales are known to feed at depths below 1000 m!) How do these animals avoid the bends, and also, how do they survive submersion for such lengthy time intervals?

One might guess offhand that to store enough oxygen, seals and whales must have much greater relative lung capacities than man. Quite the contrary is true. In man, about 7 percent of the body volume is taken up by the filled lungs, whereas in the bottlenose dolphin the fraction is only about 1 percent. Even more surprising, perhaps, is that seals (and probably also whales) do not fill their lungs before diving but exhale instead.

As the animal dives to great depth, the enormous external hydrostatic pressure tends to collapse the lung, forcing the little remaining air out of the lungs. With no air remaining in the lung, nitrogen cannot diffuse into the bloodstream. Moreover, during dives, blood flow to and from the lungs is apparently interrupted, so that even at modest depths, before complete lung collapse, what little nitrogen might diffuse into the blood is not transported to other parts of the body.

That still leaves us with the puzzle of the sufficiency of oxygen for metabolism. Part of the answer seems to be that the blood and muscle tissues are capable of storing considerably more oxygen per unit volume in aquatic mammals than in man. Also, the amount of blood per kilogram of body mass is about twice as much in the seal as it is in man. It seems that aquatic mammals do have substantial capacity for storing oxygen without recourse to gaseous storage in the lung.

Nevertheless, that storage capacity of blood and tissues is still insufficient to account for the seal's ability to remain submerged for ten minutes or more, and for the whale's much more extended dives. The physiological adaptation that makes such long submersion possible is a vasoconstriction mechanism that controls blood circulation, restricting the flow to all but the most critical organs, such as the brain and heart, during a dive. This drastic reduction in total volume of blood flow also permits an equally dramatic reduction in heart rate without a commensurate drop in blood pressure. The heart of a harbor seal swimming on the surface beats about 140 times a minute; the rate drops almost instantly to less than 10 beats a minute as the animal dives.

Thus aquatic mammals avoid aeroembolism by emptying their lungs before a dive; they store large quantities of oxygen in their blood and muscle tissue, and carefully conserve this stored oxygen by restricting blood flow during dives to the most critical organs. Of course, at the end

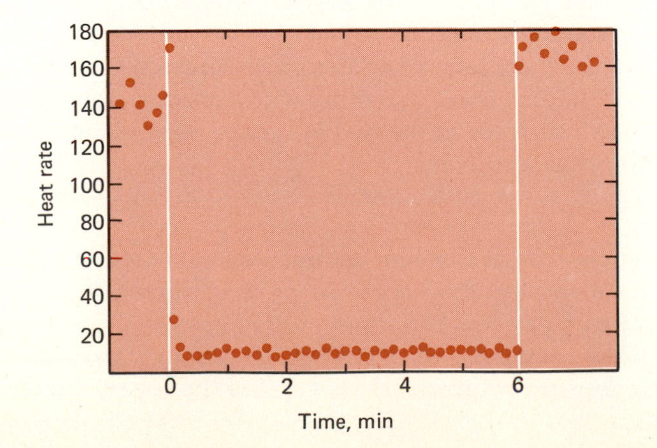

Figure 12.11 Heart rate of a harbor seal just before, during, and immediately after a dive lasting six minutes. Note that the change in heart rate is almost instantaneous.

of a dive the oxygen deficiency in the body must be made good by rapid gas exchange.

Summary

It is often convenient to express physical quantities on a molar basis. One *mole* is the amount of material whose mass in grams equals the molecular weight of the molecules that compose the material. For practical purposes, the molecular weight of atomic hydrogen is 1.

An *ideal gas* is constituted of point particles that collide elastically but exert no long-range forces on each other. Dilute gases at temperatures well above the condensation temperature (boiling point) closely simulate an ideal gas.

Boyle's law,

$$PV = \text{constant} \qquad (T \text{ constant})$$

and Charles's law (also discovered by Gay-Lussac),

$$V = \text{constant} \times T \qquad (P \text{ constant})$$

can be combined into the *equation of state* of the ideal gas,

$$Pv = RT \qquad PV = nRT$$

where v represents the *molar volume*, V is the total volume, R is the *universal gas constant*, whose value is 8.31 J/mol·K = 2 cal/mol·K, and n is the number of moles of the gas within the volume V. Here P is the pressure, and T is the absolute temperature. The gas constant R may be written

$$R = N_A k$$

where N_A is Avogadro's number, 6.02×10^{23} molecules/mol, and k is *Boltzmann's constant* and has the value 1.38×10^{-23} J/K.

Kinetic theory leads to the result that for the ideal gas,

$$Pv = \tfrac{2}{3} N_A \langle \epsilon_t \rangle$$

where $\langle \epsilon_t \rangle$ is the average translational KE of a molecule of the gas. This result is a special case of a more general conclusion of kinetic theory, namely, the law of equipartition of energy.

The average KE per molecule is equally distributed among the available degrees of freedom, with the average KE per degree of freedom equal to $\frac{1}{2}kT$. The total average KE per molecule is $\frac{1}{2}\nu kT$, where ν denotes the number of degrees of freedom.

The molar heat capacity of an ideal gas depends on whether the heat capacity is obtained under conditions of constant volume or constant pressure. The difference in the two heat capacities, c_p and c_v, equals the gas constant; that is,

$$c_p - c_v = R$$

The heat capacity at constant volume depends on the molecular structure of the constituent molecules. At or near room temperture, c_v is $\frac{3}{2}R$ for a monatomic gas and $\frac{5}{2}R$ for a diatomic gas.

If a volume contains a mixture of gases, the total pressure exerted on the walls is the sum of the *partial pressures* of the several gases; the partial pressure of each gas is given by the ideal gas law equation of state, provided that the ideal gas is a satisfactory approximation.

Scuba divers inhale air (or some other mixture of gases) at pressures substantially greater than atmospheric; the additional pressure above atmospheric is 1 atm for each 10-m depth. If a diver ascends from a dive too quickly, the excess gas dissolved in his blood and tissues is released, forming bubbles that cause pain in the joints, a syndrome known as the bends, or divers' disease.

Multiple Choice Questions

12.1 The average KE of a gas molecule can be determined by knowing
(a) the number of molecules in the gas only.
(b) the pressure of the gas only.
(c) the temperature of the gas only.
(d) None of the above is enough by itself.

12.2 A steel tank holds 0.2 m^3 of air at a pressure of 5 atm. The volume this air would occupy at the same temperature but at a pressure of 1 atm is
(a) 0.2 m^3
(b) 1m^3
(c) 5m^3
(d) 10^5 m^3

12.3 If both the temperature and the volume of an ideal gas are doubled, the pressure
(a) increases by a factor of 4.
(b) is also doubled.
(c) remains unchanged.
(d) is diminished by a factor $\frac{1}{4}$.

12.4 An ideal gas undergoes a process that has the net effect of doubling its temperature and doubling its pressure. If V_1 was the initial volume of the gas, the final volume V_2 is
(a) $V_2 = 4V_1$
(b) $V_2 = 2V_1$
(c) $V_2 = V_1$
(d) $V_2 = V_1/4$

12.5 A container holds equal amounts of helium and argon gas by weight at 20 °C.
(a) The partial pressures exerted by the two gases are the same.
(b) There are equal numbers of Ar and He molecules in the container.
(c) A He atom has the same average speed as an Ar atom.
(d) None of the above statements is correct.

12.6 A container holds 0.1 mol of H_2 and 0.1 mol of O_2. If the gases are in thermal equilibrium,
(a) an O_2 and an H_2 molecule have the same average KE.
(b) the average speeds of the O_2 and H_2 molecules are the same.
(c) the two gases contribute equally to the specific heat at constant pressure of the system.
(d) (a) and (c) are both correct statements.

12.7 A container holds a mixture of He and H_2 in thermal equilibrium at room temperature. The partial pressures exerted by these two gases will be the same if
(a) $M(\text{He}) = 2M(\text{H}_2)$
(b) $M(\text{He}) = M(\text{H}_2)$
(c) $M(\text{He}) = \frac{1}{2}M(\text{H}_2)$
(d) $M(\text{He}) = 3M(\text{H}_2)/5$

where $M(\text{He})$ and $M(\text{H}_2)$ are the masses of He and H_2 in this container.

12.8 Two containers are filled each with a different gas. The two containers are at the same temperature. Suppose that the molecular weights of the two gases are M_A and M_B. The average momenta (in magnitude) of the molecules are related by
(a) $p_A = p_B$
(b) $p_A = (M_B/M_A)p_B$
(c) $p_A = (M_B/M_A)^{1/2}p_B$
(d) $p_A = (M_A/M_B)^{1/2}p_B$

12.9 The molar heat capacity of a gas measured at constant pressure differs from that measured at constant volume by R. From this one may conclude that
(a) the gas is a monatomic ideal gas.
(b) the gas is a diatomic ideal gas.
(c) the gas obeys the ideal gas equation of state, but may be monatomic or polyatomic.
(d) the gas is monatomic, and may be an ideal or a real gas.

Problems

Section 12.2

12.1 A block of copper has a mass of 15 g. How many copper atoms are in that block?

12.2 Calculate the atomic volumes of copper and of lead. By *atomic volume,* we mean that amount of space associated with one atom in the solid. The densities of copper and lead are given in Table 9.2.

12.3 Water, H_2O, can be dissociated into H_2 and O_2 gas by electrolysis. If 1 L of water is so dissociated, how many moles of H_2 and O_2 are produced?

12.4 What is the mass of 2 mol of carbon tetrachloride, CCl_4?

• **12.5** Most wine contains 12% ethyl alcohol, C_2H_5OH, by volume. The specific gravity of ethyl alcohol is 0.79. How many moles of ethyl alcohol are in an 0.8-L bottle of wine?

Section 12.3

12.6 Calculate the molar volume of an ideal gas at atmospheric pressure and 273 K and 373 K.

• **12.7** A full cylinder of helium gas contains 42 L of the gas at 180 atm and 300 K. If this gas is liquefied, how many liters of liquid helium will be obtained? The specific gravity of liquid helium is 0.125.

• **12.8** A tire is inflated to a gauge pressure of 2 atm. (Gauge pressure is the actual pressure minus 1 atm.) As the car is driven, the temperature of the tire increases from the initial value of 10 °C to 60 °C, and simultaneously, the volume of the tire increases by 10%. What is the gauge pressure in the tire at the higher temperature?

• **12.9** A plastic balloon has a volume of 4 L and is filled with air at 20 °C. It is then pulled to a depth of 15 m below the surface of a lake, at which point its volume has diminished to 1.6 L. What is the temperature of the water at this depth?

• **12.10** A 10-L container holds oxygen gas at a gauge pressure of 4 atm and at 300 K. How many moles of O_2 are in this container?

• **12.11** A moderately good laboratory vacuum system will achieve a pressure of 10^{-6} torr (1 torr = 1 mm Hg = $\frac{1}{760}$ atm) at room temperature (20 °C). Under these conditions, what is the concentration of air molecules in the system? Give your answer in molecules per cubic centimeter. (The "average air molecule" has a molecular weight of $0.8 \times 28 + 0.2 \times 32 = 28.8$.)

• **12.12** On page 257, it is stated that 1 cm^3 of a gas contains 2.7×10^{19} molecules at STP. Verify that statement.

• **12.13** A diving bell, in the shape of a right circular cylinder 3.5 m tall and 2 m in diameter, closed at the top and open at the bottom, is lowered to the bottom of a freshwater lake that is 80 m deep. Find the pressure at which air must be pumped into the bell. Also determine the level of the water in the bell if no additional air is pumped into the bell during its descent.

• **12.14** An air bubble released from a scuba diver at a depth of 25 m rises to the surface. If the temperature of the water is 7 °C at the 25-m depth and 16 °C near the surface, what is the ratio of the air bubble's volume just below the surface of the water to the volume at the depth of 25 m?

• **12.15** A nitrogen storage tank has a volume of 50 L. If the gauge pressure of the tank at 20 °C reads 120 atm, what is the mass of nitrogen gas, N_2, in the tank? How many moles of N_2 is that?

• **12.16** What is the average separation between molecules in air at *STP?*

• •**12.17** When jam is preserved in Mason jars, the jar is nearly filled with the jam, leaving an air space of only about 10 cm^3. The jar is closed with a 6-cm-diameter rubber-gasketed cover that allows gas to escape but prevents gas from entering the jar, and the jar is then placed into boiling water for half an hour. How much force is needed to lift the cover off this jar after it has cooled to room temperature, 20 °C? Neglect thermal expansion of the jam and of the jar.

Section 12.4

12.18 Small smoke particles have masses of about 10^{-13} g. What are their rms speeds when they are in thermal equilibrium with air at 20 °C?

12.19 In the enrichment of ^{235}U by diffusion fractionation, the uranium is combined with fluorine to form the gas uranium hexafluoride, UF_6. Determine the percentage difference between the rms speeds of the two kinds of gas molecules, $^{235}UF_6$ and $^{238}UF_6$.

• **12.20** At what temperature would the rms speed of O_2 molecules equal 520 m/s?

• **12.21** Interstellar space is permeated with atomic hydrogen at a concentration of about one atom of hydrogen per cubic centimeter. The temperature of interstellar space is about 4 K. What pressure does this gas exert, and what is the average speed of an interstellar H atom?

• **12.22** The escape velocity at the surface of the earth is 11,200 m/s. At what temperatures would the rms velocity of O, H, O_2, and H_2 equal 10 percent of this escape velocity?

• **12.23** A small amount of hydrogen sulfide, H_2S (a noxious, toxic gas, of the odor of rotten eggs), is released at one end of a 50-m-long corridor in a chemistry building. What is the minimum time before the odor of H_2S can be detected at the other end of the corridor? Why is the time for diffusion of the gas to the other end of the hall considerably longer?

• •**12.24** Since our atmosphere is devoid of H_2 but does hold N_2, O_2, Ar, and CO_2, one might conclude that if the average molecular speed is less than about 4% of the escape velocity from a planet's surface, that planet can contain such gases in its atmosphere. Using this criterion, determine the minimum molecular weight of gases that can be expected to persist in the atmosphere of Mars. Would O_2 and H_2O be among those gases? (To do this problem you will need to refer to earlier chapters; in Chapter 7 the acceleration of gravity on Mars was calculated, and in Chapter 11 you will find information about radiative balance, from which a planet's surface temperature can be estimated.)

• **12.25** A tank contains 30 mol of N_2 gas at 9 atm. If the volume of the tank is 50 L, what is the rms speed of the molecules?

• • **12.26** How many moles of air are contained in a room that is 6 m × 4 m × 2.5 m? The temperature of the room is 20 °C, and the pressure is atmospheric. What is the total kinetic energy of the air molecules in that room? If all this energy could be converted to mechanical energy to fire a 10-kg projectile, what would the muzzle velocity of that projectile be?

Section 12.5

• **12.27** Show that if the molar heat capacity of a gas at constant volume is bR cal/mol·K, the heat capacity at constant pressure is $(b + 1)R$ cal/mol·K provided that the equation of state of the gas is the ideal gas law.

• **12.28** How much heat must be supplied to raise the temperature of 60 L of CO_2 at atmospheric pressure from 20 °C to 40 °C?

• **12.29** An electric heater is used in a 5-m × 3.5-m × 2.5-m room. If the heater draws 1000 W, and if heat losses can be neglected, how long will it take to raise the temperature of the room from 12 °C to 18 °C?

• **12.30** Estimate the number of moles of air in a classroom that seats about 40 students. Assume a ceiling height of 13 ft (4 m) and allow room at the front for the instructor's desk. What is the corresponding mass of air in the classroom? Approximately how much heat would be needed to raise the temperature of the room from 18 °C to 20 °C?

• • **12.31** Nitrogen gas (N_2) is confined to a cylinder of 0.03-m^2 cross section. When the gas is at 20 °C, the force compressing the gas is the sum of the force due to the atmosphere, the weight of the piston (20 N), and an additional mass of 40 kg placed on the piston. Under these conditions, the piston is at rest 1 m above the bottom of the cylinder. Find the number of moles of N_2 in the cylinder. How much heat must be supplied to the gas to raise its temperature to 100 °C? Assume that the piston is free to slide up or down in the cylinder.

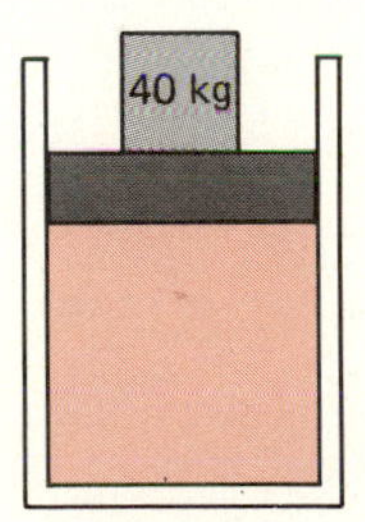

Figure 12.12

Section 12.6

• **12.32** A 30-L container is filled with O_2, H_2, and He giving a total pressure at 0 °C of 1 atm. The partial pressures of these gases are equal. What masses of O_2, H_2, and He does this container hold?

• **12.33** A container holds 1 g of H_2 and 1 g of Ar at STP. What are the partial pressures of the two gases, and what is the volume of the container?

• **12.34** A scuba diver's tanks are filled with air at a pressure of 8 atm. To what depth could this diver descend? At that depth, what would the partial pressure of oxygen be in his system under equilibrium? If a partial pressure in excess of 0.9 atm is considered dangerous, how far can the diver descend safely?

• **12.35** A tank contains a mixture of O_2, H_2, and Ne gases at a total pressure of 2 atm and 20 °C. What is the volume of the tank if the mass of each gas that is present is 10 g? What are the partial pressures of the three gases?

• **12.36** What should the relative fractions of O_2 and Ne gas be so that each gas in the mixture contributes the same amount to the internal energy of the mixture?

• • **12.37** A sealed cylinder of 10-L volume holds a mixture of H_2 and N_2 at STP. The amounts of N_2 and H_2 are such that their partial pressures are 0.5 atm each. How much heat must be supplied to this container to raise its temperature to 80 °C? (Neglect the heat capacity of the container.)

• • **12.38** A container holds 0.1 mol of He and 0.1 mol of O_2 at 80 °C and a total pressure of 2 atm. Find the volume of the container and determine the average speeds of the He atoms and O_2 molecules. What fraction of the total internal energy of the system is attributable to the O_2 gas?

•13•

Thermodynamics

A theory is the more impressive the greater the simplicity of its premises, the more diverse the things it relates, and the more extended its area of applicability. Therefore the deep impression which classical thermodynamics made upon me. It is the only physical theory of universal content which, I am convinced, . . . will never be overthrown.

ALBERT EINSTEIN

13.1 Introduction

Thermodynamics, as the name suggests, pertains to the study of heat and motion, the relations between heat and mechanical, electrical, and other forms of energy or work. This topic is unique in that it most nearly approaches the Aristotelian ideal of an abstract, internally consistent system of logic based on a few "self-evident" axioms. Securely based on a few fundamental postulates, thermodynamics succeeds by careful, logical construction in building a strong yet remarkably versatile edifice. Thermodynamic arguments are deceptively simple, demanding no knowledge of the microscopic constitution or other details of the system under study. The conclusions, therefore, also have the immense virtue of wide generality.

Classical thermodynamics has, however, one serious drawback. It is strictly valid only when applied to systems in thermal equilibrium. Consequently, in analyzing the effects of variations in temperature or pressure on a system, one must assume that these changes occur infinitely slowly. The system is presumed to pass almost imperceptibly from one equilibrium state to a neighboring one, and so on, ultimately arriving at the new state.

In the real world, events take place with finite speed, often very rapidly, and the idealized equilibrium state cannot be maintained during this process. If we supply some heat or do some work on a system, it will pass through a series of complicated transient states when it is not in thermal equilibrium and no temperature can be assigned to it. For example, if a piston is pushed into a cylinder filled with gas, the region of the gas near the piston is briefly at a higher density than the more distant portions of the gas, and some short but finite time elapses before uniform density is reestablished. The analysis of such transient conditions, the approach to thermal equilibrium which is an *irreversible* process, is simply beyond the scope of thermodynamics.*

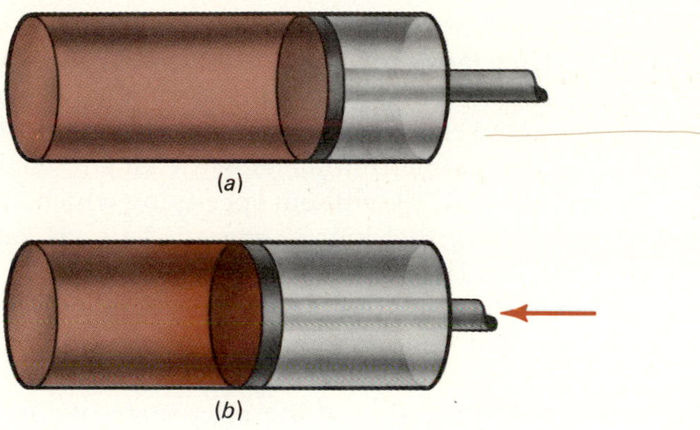

Figure 13.1 *During compression, a gas is not in equilibrium. (a) The piston is stationary; the density and pressure of the gas are uniform. (b) The piston is in motion, compressing the gas in the cylinder; the density and pressure of the gas just ahead of the piston are slightly greater than in the rest of the cylinder. (The effect is here greatly exaggerated.)*

Thermodynamics neither draws on nor provides information about the behavior of the molecular components of a system. Yet as we have seen, it is ultimately the dynamical response of these molecules that are detected as changes in pressure and temperature. Thermodynamics may be compared to a pointillist painting, whose very fine brushstrokes, responsible for the appearance of the total effect, are not apparent when the canvas is viewed as a whole. Only if a small portion of it is scrutinized do the individual brush strokes become evident, but then we may no longer able to tell whether we are looking at a landscape or a portrait.

The connection between the painting and the brushstrokes, between the thermodynamic variables and the dynamic behavior of the molecules, is provided by statistical mechanics.

13.2 Statistical Arguments

We have found that at STP an ideal gas contains 2.7×10^{19} molecules within each cubic centimeter. One cubic centimeter is a relatively small volume, but 2.7×10^{19} is a very large number. Large though it may be, it is still true that knowing (as we do) the forces that these molecules exert on each other, we could apply Newton's laws of motion and calculate in full detail the trajectory of each individual molecule for all time, given

* In the thirties, Lars Onsager succeeded in extending traditional thermodynamics to encompass also nonequilibrium states. This development of irreversible thermodynamics won him the Nobel Prize in 1968.

only the initial conditions. But even if somehow this feat could be accomplished using modern high-speed computers, what would be achieved thereby? What useful information could possibly be extracted from that inconceivably vast computer printout?

But if this mechanistic approach is futile, how should one analyze the bulk properties of fluids? The answer is contained in a branch of physics known as statistical mechanics, which, as the name suggests, concerns itself with the mechanics of systems that must be treated by statistical methods, i.e., with systems composed of a very large number of constituent particles, so that one can apply the laws of probability, the actuarial approach.

The basic idea is that in such a system there is an equilibrium state, corresponding to some distribution of particles that is the most probable by an overwhelming margin, and which the system will approach if left undisturbed. To carry the discussion forward, it is convenient to define two new terms, *microstate* and *macrostate*. The microstate is a description of the system that specifies the position and momentum of each individual *particle*. The characterization of the macrostate is less precise; it gives only the *number* of particles whose properties fall within a given range. For instance, one could characterize a macrostate by requiring that there be 10^{12} molecules per cubic centimeter with kinetic energies between 1.0×10^{-19} J and 1.1×10^{-19} J, without specifying which molecules of the more than 2×10^{19} have that kinetic energy. To each macrostate there generally correspond a large number of microstates.

We can clarify these ideas best with a few simple examples. Suppose we have a box containing some number N of identical coins. We shake the box and then look to see which coins come up tails and which are heads. To satisfy the condition of the microstate, we must number or otherwise identify each coin and tell whether it is heads or tails. The macrostate, however, is completely specified by giving only the total number of heads and tails. For example, if $N = 2$, there are four possible microstates, identified in Table 13.1. There are, however, only three distinct macrostates: two heads, two tails, and one head and one tail. The last of these macrostates comprises two microstates.

Table 13.1 *The four microstates and three macrostates for two identical coins*

Microstate	*1*	*2*	*3*	*4*
Coin 1	H	T	H	T
Coin 2	H	H	T	T
Macrostate	*1*	*2*		*3*

● ***Example 13.1*** Enumerate the macrostates for a system of four identical coins and state how many microstates are included in each.

The simplest procedure, without resorting to the mathematics of probability theory, is to prepare a table like Table 13.2. Each coin is identified by a number from 1 to 4, and each microstate is identified by specifying which side is up for each of the four coins.

The sixteen microstates group into five macrostates, as indicated, comprising one, four, six, four, and one microstates, respectively. ●

Table 13.2 Microstates and macrostates for four identical coins

Microstate	1	2	3	4	5	6	7	8	9	10	11	12	13	14	15	16
Coin 1	H	H	H	H	T	H	H	H	T	T	T	H	T	T	T	T
Coin 2	H	H	H	T	H	H	T	T	H	H	T	T	H	T	T	T
Coin 3	H	H	T	H	H	T	H	T	H	T	H	T	T	H	T	T
Coin 4	H	T	H	H	H	T	T	H	T	H	H	T	T	T	H	T
Macrostate	1	2				3						4				5

We see from Example 13.1 that when the number of coins is increased from two to four, the number of macrostates is increased by two, but the number of microstates increases from four to sixteen. In general, for a system of N coins, the following rules apply.

1. The total number of microstates is 2^N.
2. The total number of macrostates is $N + 1$.
3. The number of microstates contained in the macrostate that corresponds to n heads and $(N - n)$ tails is given by

$$\text{Number of microstates} = \frac{N!}{n!\,(N-n)!}$$

where the symbol "!" denotes the factorial, $1 \cdot 2 \cdot 3 \cdots (n-1) \cdot n \equiv n!$

These simple rules show that as N becomes larger, the number of microstates increases far more rapidly than the number of macrostates. It follows that the number of microstates composing a macrostate must, on the average, also increase dramatically as N increases; this can be verified from rule 3. For instance, if N is increased from 4 to 6, the number of microstates for an equal number of heads and tails jumps from 6 to 20, though the macrostate of all heads still contains only one microstate, as before. For $N = 60$, the numbers of microstates for a few macrostates are listed in Table 13.3, and although N is still small compared with the number of gas molecules in 1 cm^3 at STP the number of microstates is already very large for some macrostates.

Table 13.3 Number of microstates within four specified macrostates for $N = 60$

Macrostate n	$(N - n)$	Number of microstates
60	0	1
50	10	7.54×10^{10}
40	20	4.19×10^{15}
30	30	1.18×10^{17}

The fundamental, and plausible, postulate of statistical mechanics is that *in a system of noninteracting identical particles, every microstate will appear with equal probability*. It then follows that the most probable macrostate is the one that encompasses the largest number of microstates; in our system of identical coins, it is evident from Table 13.3 that the most probable macrostate is that with an equal number of heads and tails. *The equilibrium state of a many-particle system is now identified as the most probable macrostate.*

Let us consider one further example that bears directly on the behavior of gases. Suppose that we have a dilute gas of 60 molecules confined to a fixed volume, as shown in Figure 13.2. We now ask what is the probability that all 60 molecules will be in the left-hand half of this volume

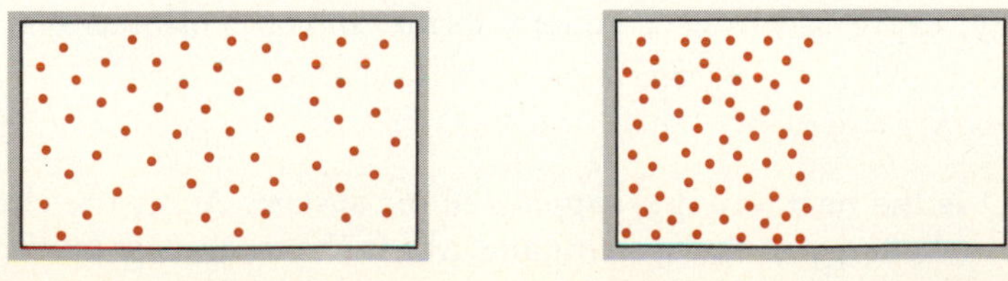

(b)

Figure 13.2 (a) Sixty molecules of a gas are confined in the volume shown. The probability that at some time all 60 will, of their own accord and by virtue of random fluctuations, be found only in the left half of this volume, as in (b), is only 8.5×10^{-18} compared with the probability of finding 30 molecules in each half of the container.

compared with the probability of an equal number in each half. The answer, in analogy with the box of 60 coins, is obtained by associating heads with location in the left half and tails with location in the right half. According to Table 13.3, the probability that all 60 molecules will, at some time, be in the left half compared with the probability of an equal number in each half is $1:1.18 \times 10^{17}$, or almost infinitesimal. A simple extension of these arguments leads to the conclusion that the equilibrium state of a gas of noninteracting molecules corresponds to a uniform distribution within the container. (If interactions are present, one must take them into account. They could, for instance, lead to condensation of the gas into a drop of liquid, and then the distribution would be highly nonuniform.)

In most of our work, we shall be concerned with equilibrium states. As we have seen, it is not necessary to know the dynamical behavior of every particle to characterize the macrostate that corresponds to the equilibrium state. The parameters, such as pressure, volume, and temperature, that identify the equilibrium state are the state functions already introduced in the preceding chapter.

13.3 The Zeroth and First Laws of Thermodynamics

Thermodynamics asks, and generally answers, questions about relations between state functions of a system, for example, pressure, volume, temperature, internal energy, and others such as magnetization. The system may be as simple as a well-insulated enclosure containing an ideal, monatomic gas, or it may comprise a complicated arrangement of pistons, boilers, turbines, photoelectric cells, and the like. But whatever is included in the system, the thermodynamicist insists that its constituents must be at the same temperature, i.e., in thermal equilibrium. Moreover, the system itself is usually assumed to be in one of two ideal conditions. Either it is perfectly isolated thermally from the rest of the world, unable to release energy to or accept energy from an external heat sink or source; or it is in intimate thermal contact with an infinitely large heat reservoir at a specified temperature, so that it exchanges energy but remains at a fixed temperature.

The zeroth law of thermodynamics states that two systems, each in thermal equilibrium with a third, are in thermal equilibrium with each other.

Figure 13.3 The thermometer is in thermal contact with the water in this beaker containing ice and water. If the thermometer is in thermal equilibrium with the water, it measures the temperature of the water. If the water is in thermal equilibrium with the ice, the thermometer also registers the temperature of the ice.

For example, the two systems might be water and ice, the third a thermometer that measures the temperature of the water. The zeroth law says that if the water and ice are in thermal equilibrium and the thermometer is in equilibrium with the water, the temperature of the ice is also the temperature read by the thermometer.

The above statement is, in a sense, no more than a tautology. If we redefine our system to include all three component subsystems, then the requirement of thermal equilibrium of the system as a whole requires that all its parts are at the same temperature.

The *first law of thermodynamics* was given already in Chapter 12. As was pointed out then, it is a restatement of the principle of conservation of energy, expressed in very general terms. In conventional symbols, it reads

$$\Delta Q = \Delta U + \Delta W \qquad \textbf{(13.1)}$$

Here ΔQ is the heat (energy) *supplied to the system*, ΔU is the change in the internal energy of the system, and ΔW is the *work done by the system* against external forces. In any application of the first law, it is crucial to adhere carefully to these definitions.

We have so far concentrated attention largely on the kinetic energy of the constituent molecules. Generally, however, the internal energy U includes contributions from interatomic forces (chemical energy) and even nuclear forces. Similarly, the term ΔW need not be restricted to mechanical work; it may include work done, for example, by electric currents. Thermodynamics is just as useful in analyzing the performance of steam engines as of solar cells.

The changes to which a system may be subjected are of various kinds. For instance, a system may absorb heat while it is kept at constant volume, constant pressure, or constant temperature. Alternatively, a system could be thermally isolated, unable to exchange heat, while its temperature, volume, or pressure is altered. The following nomenclature is widely used in physics and also chemistry and biology.

Isothermal. A process is said to be isothermal if throughout the process the *temperature* of the system remains constant.
Isochoric. A process is said to be isochoric if throughout the process the *volume* of the system remains constant.
Isobaric. A process is said to be isobaric if throughout the process the *pressure* of the system remains constant.
Adiabatic. A process is said to be adiabatic if throughout the process the system exchanges *no heat* with the outside world.

The different variables—temperature, volume, pressure, and heat—must remain fixed *throughout* a process; just because a process has the same temperature at its start and end does not guarantee that it is isothermal.

In any of the foregoing processes, the internal energy U of the system may change, i.e., $\Delta U \neq 0$; or some work may be done by or on the system, $\Delta W \neq 0$; or both terms may be nonvanishing. For the ideal gas, as shown in Section 12.5, U depends only on temperature, and consequently, $\Delta U = 0$ during an isothermal process. But in a real gas, U is a function of both temperature and volume. The most striking example of a drastic change in internal energy at constant temperature is a phase change, for example, a solid at its melting point. As the solid melts at constant pressure, the temperature of the system remains fixed; yet substantial heat, the latent heat of fusion, must be supplied to effect the phase change because the liquid has a much greater internal energy than the solid at the same temperature. Even if there is no phase change, the internal energy of a real gas depends on volume as well as temperature, although the volume dependence is generally very slight.

In Chapter 12 we showed that the mechanical work ΔW due to a volume change ΔV is given by $\Delta W = P\,\Delta V$. For any process, we can always plot a curve of the pressure P as a function of the volume of the system, as in Figure 13.4. The curve of Figure 13.4(*a*) corresponds to an isobaric process—the pressure is constant. In this case, the total work done by the system as its volume changes from V_1 and V_2 is just the sum of the small contributions $\Delta W = P\,\Delta V$, as indicated; hence, the work done by the system in this isobaric process is

$$W_{12} = P(V_2 - V_1) \tag{13.2}$$

The work done is just the area of the shaded rectangle of Figure 13.4(*a*).

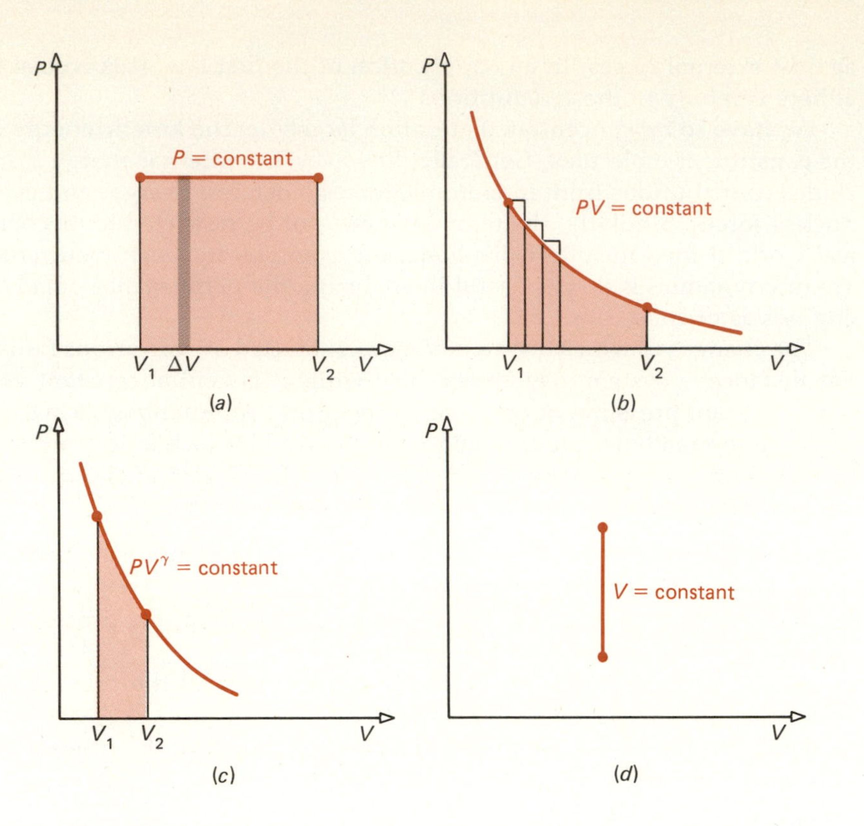

Figure 13.4 Pressure-volume curves for various thermodynamic processes. (a) An isobaric process. Here the pressure is constant as both volume and temperature of the gas change. The work done by the gas as it expands from volume V_1 to volume V_2 equals the shaded rectangular area under the P = constant line. (b) An isothermal process of an ideal gas. The curve obeys the equation PV = constant. The work done by the gas as it expands from V_1 to V_2 is also equal to the area under the P-V curve, shown shaded here. (c) An adiabatic process for an ideal gas. Here the curve obeys the relation PV^{γ} = constant. Again, as the gas expands, the work done by the gas equals the area under the curve. (d) An isochoric process. The volume of the gas remains unchanged as pressure and temperature change. In this case no work is done by or on the gas, and the area under the curve is zero.

Figure 13.4(*b*) is the *P-V* curve of an ideal gas subjected to an isothermal process; that is, the curve satisfies the equation PV = constant. In this case, the pressure is not constant but changes as the volume varies. We can, however, imagine the expansion of the gas from volume V_1 to volume V_2 as a succession of minute isobaric events as shown. The work done during each small isobaric expansion is $\Delta W_i = P_i\,\Delta V$, and the total work done is, as before, calculated by summing the small increments of work. That is,

$$W_{12} = \sum \Delta W_i$$

and this is, once again, just equal to the shaded area under the curve of Figure 13.4(*b*).

Using elementary calculus, one can show that the work done by an ideal gas during an isothermal expansion from the volume V_i to the volume V_f is given by

$$W = nRT \ln \frac{V_f}{V_i} \qquad \textbf{(13.3)}$$

If a gas undergoes an adiabatic process, the work done by the gas as it expands from volume V_1 to volume V_2 also equals the area under the *P-V* curve. That curve is different from the one for the isothermal case for the following reason.

When a gas is compressed, work is done on it. If this is done adiabatically, no heat can escape to the outside. Hence all this work must go into raising the internal energy of the gas, resulting in a temperature increase. Although for the ideal gas, the relation between pressure, volume, and temperature, $PV = nRT$, remains valid throughout such a reversible adiabatic compression (and its reverse, an adiabatic expansion), the curve representing the process on the *P-V* diagram is not one corresponding to PV = constant, since the *temperature varies as the adiabatic change progresses.* Using elementary calculus, one can show that in an adiabatic process,

$$PV^{\gamma} = \text{constant} \tag{13.4}$$

where $\gamma = c_p/c_v$ is the ratio of specific heats at constant pressure and constant volume. This relation is valid for a real gas, not just the ideal gas, provided only that the specific heats do not depend on temperature over the temperature range of interest.

Since no heat is supplied or removed during an adiabatic process, the work done by the system must equal the decrease in its internal energy. That is,

$$\Delta W = -\Delta U$$

For the ideal gas, the work the gas does during an adiabatic expansion is

$$W = -\Delta U = nc_v(T_i - T_f) \tag{13.5}$$

A few *isotherms* and *adiabats* for an ideal monatomic gas are shown in Figure 13.5. Note that the adiabats have steeper slopes than the isotherms.

● ***Example 13.2*** One mole of a monatomic ideal gas at 300 K is kept in a container at constant volume. It is then placed in contact with a heat reservoir at 250 K. How much heat is removed from the gas as it cools to the lower temperature?

Since the process is isochoric (constant-volume), no work is done on or by the gas. Consequently, Equation (13.1) reduces to

$$\Delta Q = \Delta U$$

But $\Delta Q = C_v\,\Delta T$, where C_v is the heat capacity of the gas at constant volume. In this case, $C_v = \frac{3}{2}R$. Thus

$$\Delta Q = \Delta U = \tfrac{3}{2}R\,\Delta T = 1.5(2\text{ cal/mol·K})(250\text{ K} - 300\text{ K}) = -150\text{ cal}$$

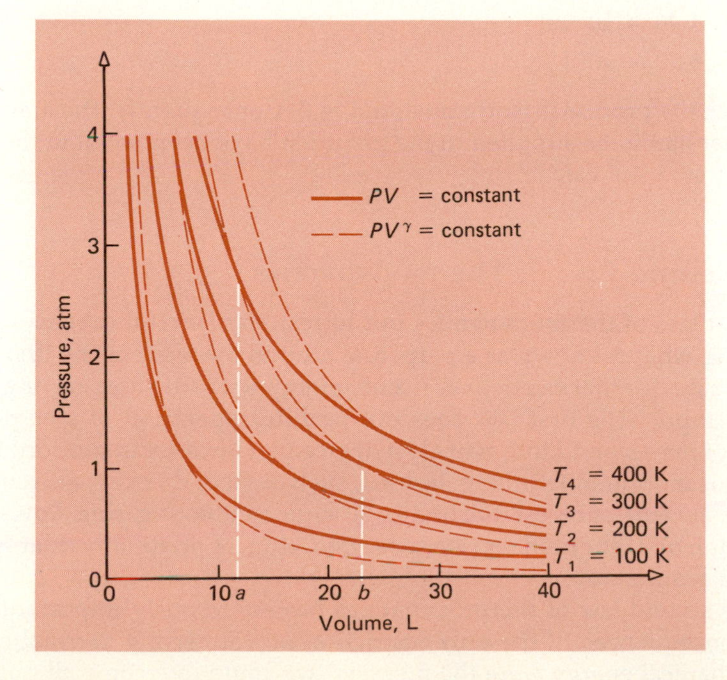

Figure 13.5 *A family of isotherms and adiabats for a diatomic ideal gas. Note that the adiabats have steeper slopes than the isotherms. Consequently, during an adiabatic expansion, for example from a to b, the temperature must always decrease.*

That is, -150 cal is *supplied to* the gas, or 150 cal is *removed from* the gas, as it cools from 300 K to 250 K.

Notice that this is also the amount by which the internal energy of the gas changes in this situation. In an isochoric process, no mechanical work is done by or on the gas and so the change in internal energy equals $C_v \, \Delta T$. ●

● ***Example 13.3*** An ideal gas in contact with a heat bath at 300 K is compressed from an initial volume of 20 L to a final volume of 12 L. In this process, 3 J of energy is expended by the external mechanism that pushes the piston. How much heat is exchanged between the system and the heat bath, and what is the direction of heat flow?

Here we have an isothermal process. Since the temperature remains constant, and since the internal energy of a fixed amount of an ideal gas depends only on temperature, U must remain unchanged, i.e., $\Delta U = 0$. Hence, for this case we can write

$$\Delta Q = \Delta W = -3 \text{ J}$$

The negative sign appears here because work was done on the gas, not by the gas. Similarly, by definition ΔQ is the amount of heat supplied to the gas, and so the result shows that 3 J of thermal energy must be removed from the gas, to be absorbed by the heat reservoir. The direction of heat flow is therefore from the gas to the heat bath. ●

● ***Example 13.4*** An ideal, diatomic gas at a temperature of 60 °C is allowed to expand isothermally until it occupies three times its initial volume. If the amount of gas is 2 mol, how much work was done during the process and how much heat had to be supplied to the gas?

The work done is given by Equation (13.3).

$$\begin{aligned} W &= nRT \ln \frac{V_f}{V_i} \\ &= (2 \text{ mol})(2 \text{ cal/mol·K})(333 \text{ K})(\ln 3) \\ &= 1.46 \times 10^3 \text{ cal} \\ &= 6.12 \times 10^3 \text{ J} \end{aligned}$$

Since the process is isothermal, $\Delta U = 0$. Consequently, the amount of heat that had to be supplied to the gas must have been equal to the work done, 1.46×10^3 cal. ●

13.4 Second Law of Thermodynamics

The first law of thermodynamics put to rest any fanciful dreams of constructing what is known as a perpetual motion machine of the first kind. This is a device that can deliver useful energy without an equal or greater energy input. The first law does not prohibit operation of a *perpetuum mobile* of the second kind, namely a device that once set in motion, maintains this motion indefinitely. Indeed, such devices do exist; a circulating electric current once established in a superconducting ring flows with undiminished strength as long as this ring is maintained at a low enough temperature.

The second law of thermodynamics has enormously important practical consequences. If the only restriction on converting thermal energy to mechanical energy were the first law, we could overcome all our troublesome energy problems without resorting to coal, oil, or nuclear fuels. There is, after all, a vast storehouse of thermal energy in the oceans. Why not extract some of this thermal energy and use it to drive generators, heat cities, propel ships? To be sure, the seas would cool slightly, but by an insignificant amount.

We know that mechanical energy can be converted to heat. Indeed, Joule measured the heat generated by a given amount of mechanical energy. So why don't we simply reverse the process and get mechanical energy from thermal energy directly?

The answer is that there is something peculiarly perverse in nature that prevents this apparently simple reversal. None of the fundamental laws of nature encountered so far—conservation of energy, of momentum, or of angular momentum—precludes such an operation. Yet the process appears to be unattainable. It is the second law which, in its own formal language, tells us that such events are literally unnatural.

Unlike other laws of physics so far encountered, the second law of thermodynamics finds mathematical expression as an *inequality*, not an equality (except in a very special case). It is perhaps for this reason that several, apparently disparate statements of the second law exist. They can be shown to be equivalent, and we shall demonstrate this equivalence for two of these formulations (Section 13.5).

We depart from the usual practice of deriving results and then stating the conclusion. Instead, we present the three best-known formulations of the second law at the outset and then indicate some of the steps that led to these insights. We shall not try to prove the second law; the arguments for the proof are probably more sophisticated and abstract than any in classical physics. Here we concentrate on plausibility arguments instead of logical proofs.

Three statements of the second law most often quoted are:

1. *Heat does not, of itself, flow from a cooler to a hotter body.*

2. *It is impossible to take heat from a reservoir and convert it completely to work if no other changes are taking place in the system or its environment.*

3. *In any process taking place in an isolated system, the entropy of the system either remains fixed (a reversible process) or increases. The entropy of an isolated system and of the entire universe tends toward a maximum.*

The last of these formulations is included here for completeness. We shall define the new term *entropy* in Section 13.5 and there demonstrate the equivalence of the first and third of the above statements.

By a curious quirk of history, the second law actually preceded the first by a few years, although notes of Sadi Carnot, published after his untimely death at age 36, suggest that he was also cognizant of the energy conservation principle. Carnot did not set out to "discover" the second law, or the first law. What troubled him, as a professional engineer, was that "notwithstanding the work of all kinds done by steam engines, notwithstanding the satisfactory condition to which they have been brought today, their theory is very little understood." Carnot therefore constructed a theoretical model of a heat engine so that he could identify and study the parameters that characterized its performance. Here we outline the steps that on elaboration by Lord Kelvin and by Clausius led to the second law.

Figure 13.6 *Nicolas Leonard Sadi Carnot (1796–1832).*

How does one extract mechanical energy from a heat engine? Generally, the process is cyclic; that is, the system undergoes some changes, such that at the end of a series of steps the heat engine is again in its initial state. During one cycle, a net amount of heat is supplied to the engine

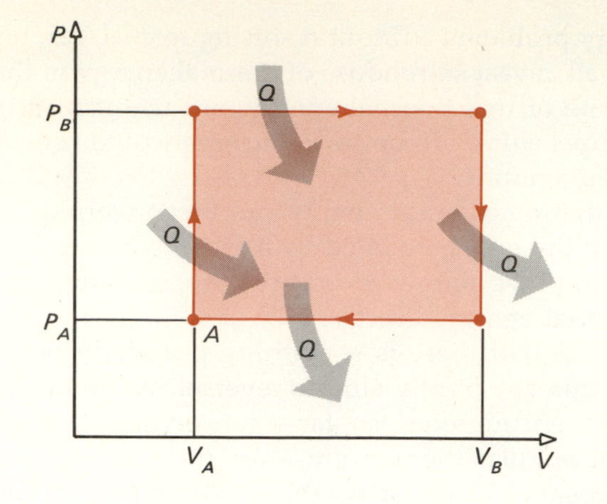

Figure 13.7 One possible cycle for a heat engine. The cycle starts at pressure P_A and volume V_A and proceeds clockwise, as indicated. The wide arrows indicate those portions of the cycle where heat must be supplied and removed from the engine. The mechanical work done by this engine during one cycle equals the area of the rectangle shown shaded in this figure.

and a net amount of mechanical work has been done by the engine. The P-V diagram for one such cyclic process is shown in Figure 13.7.

We begin the cycle at point A, with the gas in the cylinder at pressure P_A and occupying a volume V_A. We next perform an isochoric change, supplying heat to the gas. As a result, the temperature and the pressure of the gas increase until the pressure P_B is reached. At that pressure we now permit the piston to move outward, maintaining the pressure constant; this process requires further supply of heat to the gas. When the volume has increased to V_B, we again clamp the piston in position and bring the cylinder into contact with a heat reservoir at a lower temperature, removing heat from the system. When the pressure has dropped to P_A, we release the piston and isobarically compress the gas to its initial volume V_A; in this process, we shall have to cool the gas further and extract more heat from the system. The system will have returned to its original configuration, and we have completed one full cycle.

Over the isochoric portions of this cycle, $\Delta W = 0$. During the isobaric expansion at pressure P_B from volume V_A to V_B, the gas does work on the outside of $P_B(V_B - V_A)$ J; during the isobaric compression at pressure P_A, the outside does work on the gas of $P_A(V_B - V_A)$ J [that is, the system does work $P_A(V_A - V_B)$ J]. Thus the net work the engine does during one complete cycle is $\Delta W = (P_B - P_A)\ (V_B - V_A)$; the work done by the system equals the area of the rectangle enclosed by the curve on the P-V diagram.

Note that during a portion of the cycle, heat had to be supplied to the gas, and during another portion, heat was released by the gas to the outside, to some heat reservoir. The absorption and subsequent release of

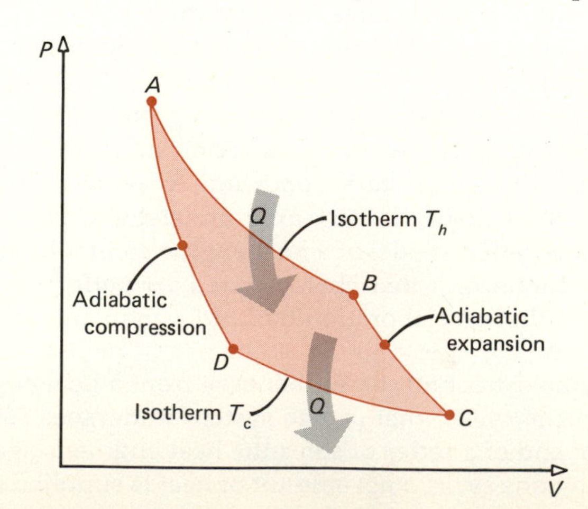

Figure 13.8 Pressure-volume curves for a Carnot engine. The Carnot cycle has two isothermal and two adiabatic stages. Beginning at A, we have an isothermal expansion, with the gas in contact with a heat reservoir at the temperature T_h. Heat must be supplied to the engine during this stage. At B there commences an adiabatic expansion to C. No heat is absorbed or released by the gas between B and C, but the temperature must decrease. (See Figure 13.4.) This stage is now followed by an isothermal compression; the gas is in thermal contact with a heat reservoir at the temperature T_c. During this stage the engine releases heat to the low-temperature heat reservoir. The initial state is regained by an adiabatic compression, from D to A. The temperature of the gas is raised to T_h during this last stage, but no heat is absorbed or released by the gas. The mechanical work done by the engine during one cycle equals the area enclosed by the curve for the cyclic process, shown shaded in the figure.

heat by the working medium is a characteristic of all heat engines. The one Carnot studied involves a cycle, shown in Figure 13.8, in which successive stages alternate between adiabatic and isothermal, rather than between isochoric and isobaric processes. In this *Carnot cycle,* heat is absorbed or released only during the isothermal portions at the fixed temperatures T_h and T_c, greatly simplifying the thermodynamic analysis of this engine.

We define the efficiency in the usual way, as the ratio of useful output divided by the energy input per cycle. That is,

$$\eta = \frac{\text{output per cycle}}{\text{energy input per cycle}} \tag{13.6}$$

Since the system returns to its initial configuration at the end of a complete cycle, $\Delta U = 0$ over a full cycle. (We recall that for the ideal gas, for example, U depends on temperature only; even for a real gas or other medium, U does not depend on past history—only on instantaneous state variables such as pressure, temperature, etc.). Consequently, by the first law, the work done per cycle must equal the difference between the heat input and the heat released during a cycle; that is,

$$\Delta W = Q_{in} - Q_{exh} \tag{13.7}$$

where Q_{in} is the heat input at the temperature T_h and Q_{exh} is the heat energy released by the exhaust at the temperature T_c. Substituting Equation (13.7) into (13.6), we have

$$\eta = \frac{Q_{in} - Q_{exh}}{Q_{in}} = 1 - \frac{Q_{exh}}{Q_{in}} \tag{13.8}$$

Equation (13.8) is the expression for the efficiency of the Carnot engine. This result is independent of the working substance of the engine; it could be anything—a gas, a liquid, or a solid. Consequently, the ratio Q_{exh}/Q_{in} cannot involve the properties of this unspecified medium and can depend only on the temperatures of the two heat reservoirs. Kelvin showed that one could use Equation (13.8) to define a thermodynamic temperature scale such that

$$\frac{T_c}{T_h} = \frac{Q_{exh}}{Q_{in}} \tag{13.9}$$

This temperature scale is *absolute* in the sense that it is independent of the working substance of the Carnot engine. However, if the working substance of the Carnot engine is an ideal gas, one can calculate its efficiency from the ideal gas equation of state, with the result that

$$\eta = 1 - \frac{T_c^g}{T_h^g} \tag{13.10}$$

Here T_c^g and T_h^g are the temperatures of the cold and hot reservoirs measured on a scale defined by the ideal gas law (see Section 12.3). Since Equation (13.10) is just the result that follows if (13.9) is substituted into Equation (13.8), the absolute temperature of the thermodynamic (Kelvin) scale agrees with the absolute temperature determined from the ideal gas law.

The thermodynamic scale's absolute zero can be defined as the temperature of the heat sink of a Carnot engine whose efficiency is unity. It is the same as the absolute zero of the ideal gas law, -273.16 °C.

In practice, the low-temperature heat reservoir for a heat engine is usually the surrounding air or the waters of a lake or the sea. Thus, T_{exh} is generally near 300 K. High efficiency therefore requires that T_{in}, the temperature of the source from which the engine draws its heat, be as high as possible, consistent with material constraints (for instance, we don't want to risk melting engine components). This is the fundamental reason why engineers have pushed designs toward ever higher operating temperatures, and why metallurgists have continued to search for strong and yet machinable high-temperature alloys. The limitation on engine efficiency imposed by friction losses is not nearly so severe as that dictated by the second law.

13.5 Entropy and the Second Law

At the end of a complete cycle, the Carnot engine, the isobaric-isochoric engine, and for that matter, any cyclic system revert to their original thermodynamic state—same pressure, volume, and temperature. During each cycle, a net amount of heat is absorbed and is converted to mechanical energy. It follows that though the internal energy U of the system is a true state function—a function that depends only on the state variables and not on past history—the same cannot be said for the quantity we have identified as the heat Q. Put another way, the system of Figure 13.9 can attain the thermodynamic state characterized by P_C, V_C, starting from the initial state characterized by P_A, V_A, by various possible routes. One of these could be the isothermal-adiabatic path A-B-C; another could be the adiabatic-isothermal path A-D-C. The amount of heat the system absorbs depends not only on the initial and end points *but also on the choice of path.* Using the nomenclature of Chapter 4, we would say that the heat Q *is not a conserved quantity.*

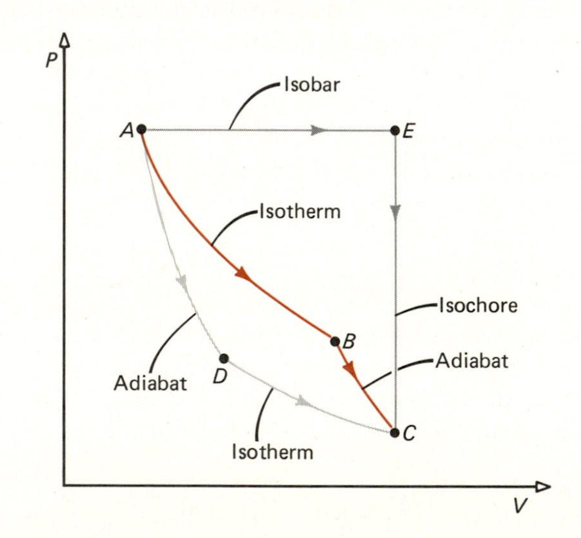

Figure 13.9 *A gas, in the initial state characterized by A on the P-V diagram, can be brought to the state C by various routes. For example, one can begin with an isothermal expansion to B, followed by an adiabatic expansion to C. Or one can start with an adiabatic expansion to D, followed by an isothermal expansion to C. Alternatively, one could use a combination of isobaric and isochoric processes, as in A → E → C. The work done by the gas in following these alternate routes is different, even though the final and initial thermodynamic states are the same. It follows, therefore, that in this case W is not a conserved quantity, and consequently, neither is Q.*

The work of Carnot and Kelvin suggested that in a reversible isothermal process, the quantity Q/T *is* conserved. That is, as the system passes from state A to state C in Figure 13.9, $Q_1/T_1 = Q_2/T_2$, where Q_1 and Q_2 are the amounts of heat the system absorbs during the isothermal processes at temperatures T_1 ($D \rightarrow C$) and T_2 ($A \rightarrow B$), respectively. Pursuing this line of argument, Clausius was able to prove that in any closed system undergoing any *reversible* thermodynamic process,

$$\Delta S = \frac{\Delta Q}{T} = 0 \qquad \textbf{(13.11)}$$

Clausius gave the name *entropy* to this function S, which in contrast to Q is a true state function. Clausius also succeeded in demonstrating that if a closed system suffers an *irreversible* change, the value of S *must increase*. Hence the rule:

The entropy of an isolated system can never decrease; S either remains constant (reversible process) or increases (irreversible process).

Figure 13.10 Rudolf Julius Emanuel Clausius (1822–1888), who showed that the function $S = Q/T$ is conserved in a reversible thermodynamic process. That is, in Figure 13.9, the change in S in going from state A to state C is the same regardless of the path followed. Clausius gave the name entropy to this new state function.

● ***Example 13.5*** A dish containing 10 g of water at 0 °C is placed in a thermally insulated box in which the air is at −10 °C. After equilibrium has been established, all the water has solidified and the ice and air are both at 0 °C. Determine the change in entropy of the water and estimate the entropy change of the air. Is this process reversible or irreversible?

Since the water does not change its temperature, the change in entropy is simply the heat released as it solidifies. The change in heat of the water is

$$\Delta Q = -(80\ \text{cal/g})(10\ \text{g}) = -800\ \text{cal}$$

The entropy change of the water is therefore

$$\Delta S_\text{w} = \frac{\Delta Q}{T} = \frac{-800\ \text{cal}}{273\ \text{K}} = -2.93\ \text{cal/K} = -12.25\ \text{J/K}$$

The entropy of the water has decreased! This does not violate the second law, because as the entropy of the water decreases, that of the air increases. Since no work is done (we neglect the change in the volume of the water on freezing), and since the enclosure is a thermal insulator permitting no heat exchange with the outside, the 800 cal released by the water is absorbed by the air and its temperature is raised by 10 °C. A good approximation of the entropy change of the air is therefore

$$\Delta S_\text{a} = \frac{\Delta Q}{\langle T \rangle} = \frac{800\ \text{cal}}{268\ \text{K}} = 2.985\ \text{cal/K} = 12.48\ \text{J/K}$$

Note that as expected, the total entropy change is positive; i.e.,

$$\Delta S = \Delta S_\text{w} + \Delta S_\text{a} = 0.23\ \text{J/K} > 0$$

The process is clearly irreversible. The inverse would be an event in which we started with an isolated enclosure containing ice and air at 0 °C and wound up with water at 0 °C and air at −10 °C! ●

At the beginning of Section 13.4, we mentioned that the various statements of the second law, though seemingly unrelated, are really equivalent. We now demonstrate that the first statement, "heat does not, of itself, flow from a cooler to a hotter body," is equivalent to the requirement that in an irreversible process the entropy of the system must increase.

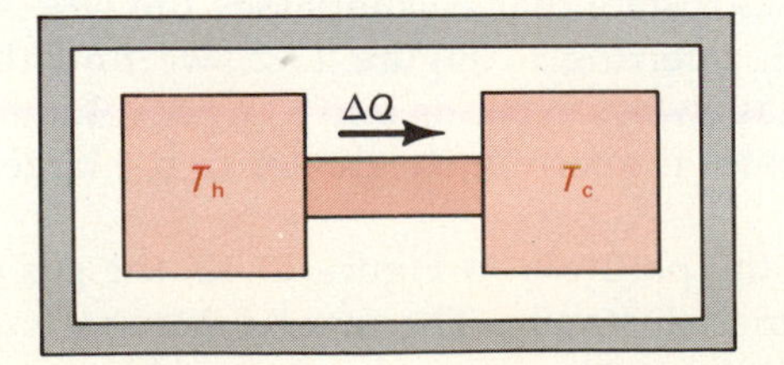

Figure 13.11 Two bodies within an adiabatic enclosure are in thermal contact by means of a heat conductor. As a small amount of heat Q is transported from the hot body, at temperature T_h, to the cold body, at temperature T_c, the entropy of the hot body decreases and the entropy of the cold body increases. The increase in entropy of the cold body exceeds the decrease in entropy of the hot one, and hence, the entropy of the entire system increases. It follows that heat conduction must be irreversible ($\Delta S > 0$); the reverse process cannot occur in nature because it would violate the second law.

We consider two bodies, one at the temperature T_C and the other at the temperature T_H ($T_H > T_C$), connected thermally through a heat conductor. The entire system is in an enclosure that isolates it from the surroundings. The question we are trying to answer is, What change, if any, occurs in the entropy of this system as a small amount of heat ΔQ flows from the hotter to the colder body?

In trying to answer this question, we are immediately faced with a paradoxical situation. Since heat has only been known to flow from a hotter to a cooler body, never in the other direction, thermal conduction is presumably irreversible. We shall soon demonstrate this conclusion using the entropy theorem. But thermodynamics, strictly speaking, allows us to make numerical calculations only for reversible processes.

We can overcome this dilemma by the following artifice. We consider an amount of heat transfer ΔQ so small that the temperature change of either body is negligibly small. If necessary, we can go to the limit as $\Delta Q \rightarrow 0$.

Using this procedure, we can now calculate the infinitesimal entropy change of each body. The hot body's entropy changes by the amount $-\Delta Q/T_H$ and the cold body's entropy increases by $\Delta Q/T_C$ as the infinitesimal quantity of heat ΔQ is transferred from the hot to the cold body. The change in entropy of the closed system is then

$$\Delta S = \frac{\Delta Q}{T_C} - \frac{\Delta Q}{T_H} = \Delta Q \left(\frac{T_H - T_C}{T_H T_C} \right)$$

We see that since $T_H > T_C$, *the entropy change must be positive.* We can therefore conclude that whenever an isolated system that is not in thermal equilibrium approaches thermal equilibrium, the entropy of that system must increase. Conversely, an isolated system will not spontaneously depart from thermal equilibrium, one portion becoming warmer as another cools, because this would violate the second law. Moreover, we see that since ΔS is not zero during a process of heat conduction, that process is indeed irreversible; heat will not, of itself, flow from a cooler to a hotter body. It all ties together quite nicely.

It must be emphasized that Equation (13.11) refers to the *total* entropy of an isolated system. It is conceivable that the entropy of a *portion* of a closed system might decrease, provided that the entropy of another part increases by at least the same amount. The preceding calculation of the entropy change during heat conduction is a good example.

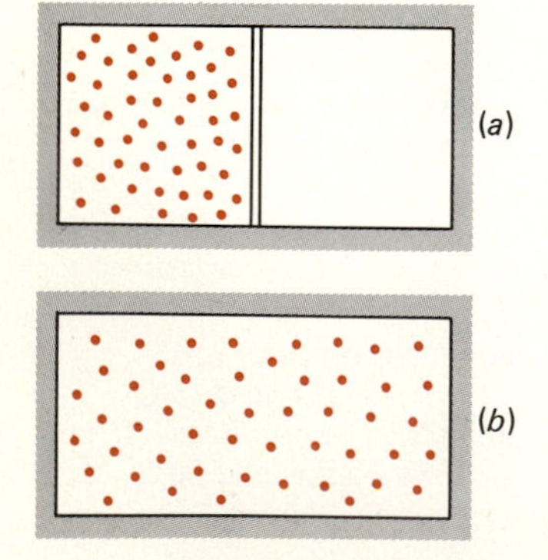

Figure 13.12 *Free expansion of a gas. (a) The gas is confined to half the total volume by a thin partition; the other half of the volume is a perfect vacuum. (b) Once the partition is removed, the gas expands to fill the entire volume.*

*13.6 Statistical Interpretation of Entropy; Heat Death

Earlier (Section 13.2) we treated the fundamental question: What distinguishes the thermal equilibrium state from all other states of a many-particle system? The answer, obtained by appealing to plausibility, was that this is the macrostate that encompasses the largest number of microstates. Indeed, referring to Figure 13.2, we find the answer to the question why a gas always expands into a larger volume—the number of available microstates is enormously greater in the larger volume than in the smaller.

If we remove the partition in Figure 13.12, the gas expands into the larger volume almost instantly. The reverse process has never been observed. It could happen; the gas molecules could organize their motion so that at some moment, all would be in only half the available space. No fundamental dynamical law of nature would be violated, not energy conservation, not momentum conservation. It could happen, but it doesn't!

The basic reason why mechanical energy can be completely converted to thermal energy, but the reverse is forbidden by the second law, is that mechanical energy inevitably corresponds to a highly correlated motion of individual molecules, whereas thermal energy represents random, uncorrelated motion. If I drop a book on the table, the kinetic energy associated with the *correlated* downward motion of the molecules of the book is then shared among the molecules of the table and book as thermal energy, and is manifested as a slight increase in the temperature of book and tabletop. No law of mechanics prevents the molecules of the tabletop from organizing their motion so that they suddenly impart an upward impulse to the book and propel it to my waiting hand. Still, even if I were to stand with my hand outstretched all my life, that miracle would not occur.

Such arguments suggest that irreversibility, associated with the approach to equilibrium, is intimately related to the statistical interpretation of the equilibrium state discussed in Section 13.2. It is to this statistical interpretation of entropy, which lends precise meaning to an otherwise elusive concept, that Boltzmann dedicated his life.

What Boltzmann succeeded in proving was that the entropy function of Clausius is related to the probability of the macrostate. The exact relation he derived is

$$S = k \ln W \tag{13.12}$$

where k is Boltzmann's constant, which we have already encountered in the kinetic theory of gases, and W denotes the number of microstates composing the particular macrostate.

Generally, a highly ordered macrostate has a much lower probability than a more random, disordered one. For instance, the probability of finding the relatively more disordered state of two heads and two tails is six times greater in a random toss of four coins than having all come up heads, a highly ordered state (see Example 13.1). The statistically "natural" tendency, therefore, is toward the greatest degree of disorder.

It is this observation that lends the second law its profound significance, because it introduces the totally novel concept of a *temporal direction* in natural processes. Newton's equations of motion are time-reversible; the progression of a nondissipative (friction-free) dynamical system "forward" in time is as consistent with these equations as the "time-reversed" motion. The operator of a planetarium can reproduce the configuration of stars and planets on the birthday of Tutankhamen with as much confidence as that with which he predicts the exact times of future solar and lunar eclipses. Nor can we, by watching a film of a swinging pendulum, say whether the film is running forward or reversed in the projector unless the time of observation is long enough to tell whether the amplitude of each swing increases or decreases as time proceeds. Moreover, forward and time-reversed motions are consistent not only with Newton's laws but also with relativistic mechanics and even with the modern quantum mechanical description of nature. Yet time proceeds inexorably in one sense only; we get older, gray, and wrinkled, and much as we may wish to invert this unhappy sequence and defer its ultimate culmination, it cannot be. To "grow younger" is as "unnatural" as it is for the book on the table to levitate.

This trend toward general disorder, which, as Eddington said, "sets the arrow of time," holds for the entire universe. Gradually, the highly

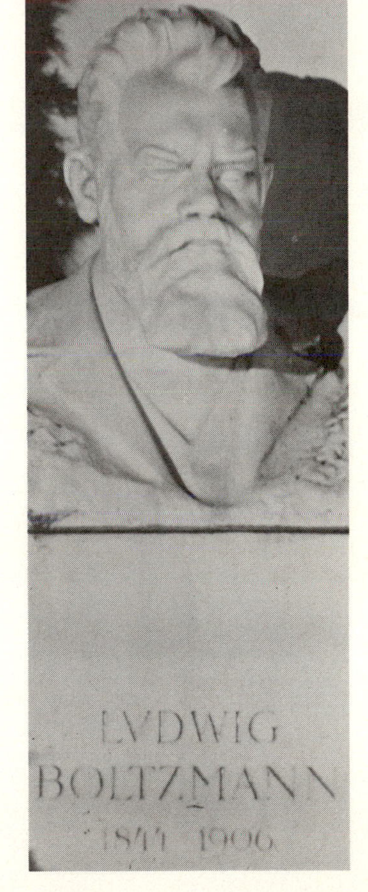

Figure 13.13 *Ludwig Boltzmann (1844–1906). Bust on his grave in Vienna.*

ordered state, in which millions of very hot stars are distributed in a vast expanse of cosmic dust, gases, and occasional planets, must give way to the less ordered state in which the energy of the universe is more uniformly distributed. This is the basis for the prediction of the "heat death," not merely of our sun, but of the entire universe.

Then star nor sun shall waken,
Nor any change of light:
Nor sound of waters shaken,
Nor any sound or sight:
Not wintry leaves nor vernal
Nor days nor things diurnal;
Only the sleep eternal
In an eternal night.

Algernon Charles Swinburne

Summary

Classical thermodynamics applies only to systems in thermal equilibrium and to systems that change so slowly that they may be regarded as making gradual transitions from one equilibrium state to a nearby equilibrium state. Despite these restrictions, many important results can be derived using thermodynamic arguments.

The following nomenclature is used to characterize thermodynamic processes:

An *isothermal* process is one in which the *temperature* remains fixed.
An *isochoric* process is one in which the *volume* remains fixed.
An *isobaric* process is one in which the *pressure* remains fixed.
An *adiabatic* process is one in which *not heat exchange* takes place.

The so-called *zeroth law of thermodynamics* states that two systems, each in thermal equilibrium with a third, are in thermal equilibrium with each other.

The *first law of thermodynamics* expresses the principle of conservation of energy. The formal statement reads

$$\Delta Q = \Delta U + \Delta W$$

where

ΔQ = heat supplied *to* the system
ΔU = change in internal energy of the system
ΔW = work done *by* the system

A heat engine is a device that in a cyclic process, extracts heat from one or more reservoirs at high temperature, releases heat to one or more reservoirs at low temperature, and converts the difference into mechanical (or other nonthermal) energy. The optimum efficiency of a heat engine operating between two heat reservoirs at temperatures T_h (hot) and T_c (cold) is the Carnot efficiency,

$$\eta = 1 - \frac{T_c}{T_h}$$

The *second law of thermodynamics* states that the entropy of an isolated system either remains constant (reversible process) or increases (irreversible process). Alternative statements of the second law are equivalent to the above.

Entropy is a state function. The change in entropy of a system is given by

$$\Delta S = \frac{\Delta Q}{T}$$

where ΔQ is the heat supplied to the system.

The entropy of a state is related to W, the number of ways by which that state can be formed, through the expression derived by Boltzmann,

$$S = k \ln W$$

where k is Boltzmann's constant and W is the number of microstates that make up the macrostate of entropy S.

Since a disordered, or random, arrangement of constituents of a physical system comprises many more microstates than a highly ordered one, the second law prescribes that the degree of disorder of an isolated system must remain constant or increase with time. In other words, all isolated systems, and the universe as a whole, tend toward maximum disorder. This tendency toward disorder "sets the arrow of time."

Multiple Choice Questions

13.1 If a system is subjected to an isochoric process,

(a) no mechanical work is done by the system.

(b) the internal energy of the system remains constant.

(c) the entropy of the system remains constant.

(d) the pressure of the system remains constant.

13.2 Identify the correct statement.

(a) It is possible for a system to undergo an adiabatic change and simultaneously also to change its entropy.

(b) It is possible for a system to undergo an isothermal change and simultaneously also to change its internal energy.

(c) It is possible for an isolated system to suffer a reversible change and simultaneously also to increase its entropy.

(d) None of the above statements is correct.

13.3 In the equation $\Delta Q = \Delta U + \Delta W$, which expresses the first law of thermodynamics, the quantities ΔW and ΔQ represent

(a) the work done on the system and the heat supplied to the system.

(b) the work done by the system and the heat supplied to the system.

(c) the work done on the system and the heat supplied by the system.

(d) the work done by the system and the heat supplied by the system.

13.4 An ideal gas in the state characterized by P_i, V_i, undergoes a change such that the final pressure and volume become P_f, V_f, where $V_f > V_i$. The mass of ideal gas is exactly 1 mol.

(a) The amount of work done by the gas during the process is determined by the values P_i, V_i, P_f, and V_f.

(b) The change in the internal energy of the gas during the process is determined by the values P_i, V_i, P_f, and V_f.

(c) The heat supplied to the gas during the process is determined by the values P_i, V_i, P_f, and V_f.

(d) None of the above statements is true.

13.5 When on a humid day water vapor condenses on the surface of a cold object, the entropy of the water in the process of condensation

(a) increases.

(b) decreases.

(c) remains unchanged.

(d) there is not enough information to decide between (a) and (c).

13.6 A gas changes its state reversibly from P_i, V_i, to P_f, V_f, as indicated in Figure 13.14. The work done by the gas is

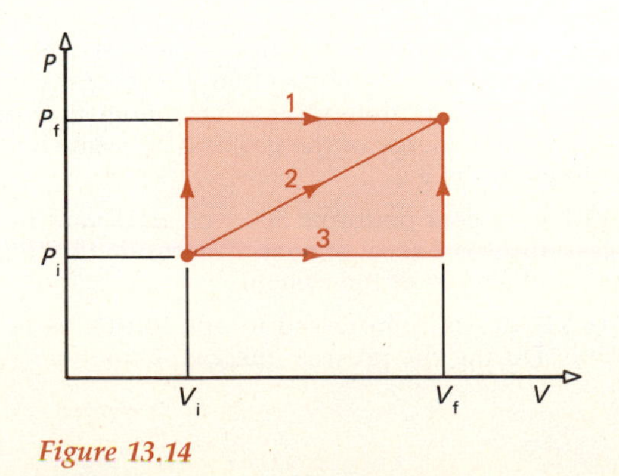

Figure 13.14

(a) greatest for path 1.

(b) least for path 2.

(c) the same for paths 1 and 3.

(d) equal for all three paths.

13.7 The heat capacity of a gas at constant volume is

(a) $\Delta U/\Delta T$.

(b) $\Delta U/\Delta T$ only if the gas is an ideal gas.

(c) $\frac{3}{2}R$ per mole.

(d) always greater than the heat capacity at constant pressure.

13.8 A piece of hot metal is placed in an evacuated, isolated enclosure. If by "system" we mean the metal and its enclosure, then as the metal cools by radiation,

(a) the entropies of the metal and of the system increase.

(b) the entropies of the metal and of the system remain constant.

(c) the entropy of the metal decreases, that of the system increases.

(d) the entropy of the metal decreases, that of the system remains constant.

13.9 During an adiabatic expansion of a gas,

(a) the internal energy of the gas remains constant.

(b) the temperature of the gas remains constant.

(c) no heat is supplied to or removed from the gas.

(d) the gas does no work nor is work done on it.

13.10 In an isothermal compression of an ideal gas,

(a) the work done on the gas is zero.

(b) heat must be supplied to the gas.

(c) the energy stored in the gas remains constant.

(d) None of the above is correct.

Problems

Section 13.2

13.1 What is the probability that in a random toss of six identical coins three will come up heads compared to having all six come up heads?

13.2 What is the probability that in a random toss of six identical coins two will come up heads as opposed to having all six come up heads?

• **13.3** When two dice are thrown, what is the most likely total of the throw? How many microstates does this macrostate comprise?

• • **13.4** A box contains three red and three green marbles. What is the probability a blindfolded person would pick (a) two red marbles? (b) one red and one green marble?

• • **13.5** A box is divided into two equal regions by a partition that has a hole large enough to permit marbles to pass through. The box is given a good shaking. Compare the likelihood that after the shaking of the total of 16 marbles, 8 will be on each side compared to 7 on one side and 9 on the other.

Section 13.3

13.6 An ideal gas is compressed from 12 L to 6 L at constant temperature. In the process, a total of 60 cal is released by the gas. Assuming that the process was reversibly performed, determine what change took place in the internal energy of the gas and how much work was done on the gas.

13.7 A system performs 500 J of work and in the process absorbs 300 cal of heat. Determine the change in internal energy of the system.

• **13.8** A gas is compressed to one-fourth its initial volume. During the process, the gas is maintained at the same temperature, and a total of 80 J of energy is expended. Determine the amount of heat supplied to the gas and the change in the internal energy of the gas.

• **13.9** Repeat Problem 13.8, assuming that the compression is adiabatic instead of isothermal.

• **13.10** An ideal gas is subjected to the isobaric-isochoric cycle shown in Figure 13.15. How much work is done by this gas during one cycle?

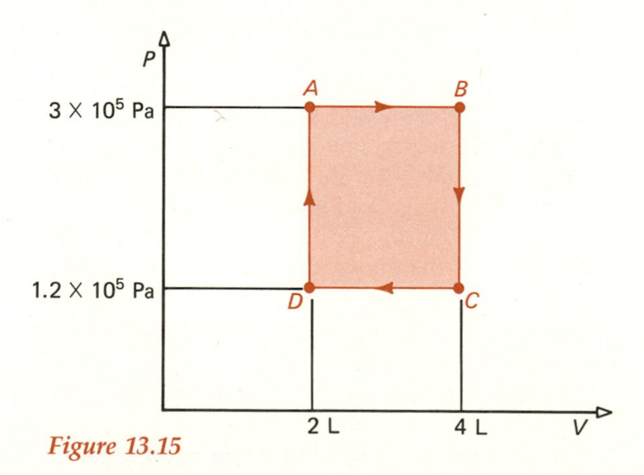

Figure 13.15

• **13.11** Argon gas, initially at 20 °C and atmospheric pressure, is adiabatically compressed to one-tenth its initial volume. Determine the pressure and temperature for the compressed state.

• • **13.12** An ideal diatomic gas is confined to a volume of 14 L at 30 °C and atmospheric pressure. Some heat is then slowly added to this gas, which is allowed to expand at constant pressure until its volume is 21 L. Determine the number of moles of gas in the container, the final temperature, the amount of work done by the gas, and the amount of heat that was supplied to the gas.

• • **13.13** A monatomic gas containing 0.4 mol is subjected to the change from state *A* to state *B* on the *P-V* diagram of Figure 13.16. Determine the work done by the gas in following the isochoric-isobaric path *A-C-B*. Calculate also the change in internal energy of the gas during that process and the heat supplied to the gas.

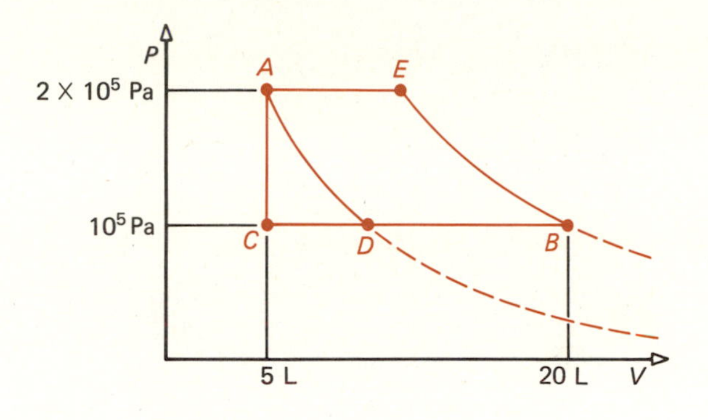

Figure 13.16

• • **13.14** Repeat Problem 13.13 for the case in which the gas attains the state *B* by following the isothermal-isobaric path *A-D-B*.

• • **13.15** Repeat Problem 13.13 for the isobaric-isothermal path *A-E-B*.

• • **13.16** If the heat engine of Problem 13.10 contains 0.1 mol of ideal gas, what are the temperatures at points *A, B, C, D* of Figure 13.15?

Sections 13.4, 13.5

13.17 In a steam turbine, steam heated to 750 °C enters the turbine and is exhausted at 110 °C. What is the maximum efficiency that this turbine can attain?

13.18 A Carnot engine's working substance is water. It uses steam at 240 °C, and this condenses to water at 40 °C. What is the maximum theoretical efficiency of this engine?

13.19 A Carnot engine operates at an efficiency of 26 percent. The exhaust temperature of the working substance is 20 °C. Find the temperature of the hot reservoir with which the working substance must come in contact during the other isothermal part of the cycle.

• **13.20** An ice tray holds 180 mL of water. It is filled with water at 8 °C and placed into the ice compartment of the refrigerator, which is maintained at −5 °C. Estimate the change in entropy of the water as it comes to equilibrium with its surroundings.

13.21 The exhaust temperature of a Carnot engine is 70 °C. If that engine has an efficiency of 35 percent, what must the temperature of the high-temperature reservoir be?

• **13.22** An ideal refrigerator is a Carnot engine operated in a reverse cycle. That is, the amount of heat Q_1 is removed from a heat reservoir at the temperature T_1, and an amount of heat Q_2 is released to a heat reservoir at the higher temperature T_2. Show that to run this refrigerator the work that must be done during one cycle is

$$W = Q_2\left(1 - \frac{T_1}{T_2}\right)$$

• **13.23** A refrigerator is driven by a small motor whose useful output of mechanical power is 100 W. If we assume that this refrigerator is operating as an ideal Carnot refrigerator, and that the hot and cold temperatures of the heat reservoirs are 20 °C and −5 °C, how much ice will this refrigerator make in 1 h if water at 10 °C is put into it?

• **13.24** A closed container holds 1 L of water at 20 °C. The container has a freely movable piston whose external surface is open to the air. (a) How much heat must be supplied to transform this mass of water into vapor at 100 °C and 1 atm? (b) If we assume that water vapor at 100 °C and 1 atm can be approximated by an ideal gas, what is the volume that vapor occupies? (c) How much mechanical work was done in moving the piston, and what was the change in the internal energy of the water?

• • **13.25** Since entropy is a state function, it is possible to describe the changes of an ideal gas on a plot showing entropy versus temperature similar to the more familiar plot of pressure versus volume. Sketch the Carnot cycle as it might appear on a plot of *S* versus *T*.

• • **13.26** Half a mole of helium gas at 300 K is subjected to an isobaric process that reduces the volume the gas occupies from 10 L to 4 L. Determine the initial and final pressures of the gas, the initial and final temperatures, and the change in entropy of the gas, assuming that the process was carried out slowly (that is, reversibly).

• • **13.27** The gas of Problem 13.16 is a diatomic gas. Calculate the efficiency of this heat engine and compare that efficiency with the Carnot efficiency of an engine operating between the highest and lowest temperatures reached by the engine of Problem 13.16.

• • **13.28** Calculate the efficiency of the heat engine that uses the monatomic ideal gas of Problem 13.13 and traces the path *A-E-B-D-A*. Compare this efficiency with the efficiency of a Carnot engine operating between the same two temperature reservoirs.

• • **13.29** A 10-g copper calorimeter can contains 50 g of water at 20 °C. A 20-g block of aluminum, whose temperature is 40 °C, is dropped into the water of the calorimeter. Determine the final temperature of the system and estimate the entropy change.

·14·

Oscillatory Motion

O, it sets my heart a-clickin' like the tickin' of a clock . . .

James Whitcomb Riley, *When the Frost Is on the Punkin*

14.1 Introduction

In this chapter we turn to the last of the three principal forms of motion found in nature. Its distinguishing feature is a repetitive pattern; the system assumes at some later time the same configuration it displayed earlier.

Periodic, repetitive behavior is perhaps even more ubiquitous than translation or rotation (though rotation is, in a sense, also periodic motion, since the system returns to its initial configuration after each full revolution). The seasons, night and day, the phases of the moon, the tides, breathing, and heartbeat are all periodic events. We communicate by means of vibrations, generating pressure oscillations in the air with our vocal cords, and these pressure oscillations are then sensed by an eardrum, whose vibrations ultimately excite well-defined responses of the nervous system. We measure the passage of time by counting the swings of a pendulum, the ticks of a wristwatch, or the number of complete cycles of an electronic oscillator.

Oscillatory behavior is so common largely because it is the natural response of almost any system that is perturbed from stable equilibrium. However, not every system at equilibrium is necessarily in stable equilibrium. Our first task in studying oscillatory motion is therefore to analyze equilibrium itself.

14.2 Stable, Unstable, and Neutral Equilibrium

Suppose a particle is constrained to move along the friction-free track shown in Figure 14.1. We know that if it is at rest anywhere on the horizontal portion between A and B, it will remain so unless acted on by a force. If it is at rest at point D, it will also remain there; and if it is carefully balanced at point C it will not move from that spot. All these locations are therefore equilibrium positions, but their characters are essentially different.

If the particle is in motion along the track at point E, between A and B, it will continue its motion unchanged over this section. It is in equilibrium anywhere over that part, preferring no one point. This condition is called *neutral equilibrium*.

Let us next examine the equilibrium at point C. If the particle is even slightly displaced to the right or the left, it will slide down the track. The point C is therefore referred to as a position of *unstable equilibrium*.

By contrast, if the particle is moved slightly to the right or the left of point D and released, it will return toward D. The combination of gravitational force and reaction force of the track act on the particle in such a way that it is pulled back to the equilibrium position. We say that in the vicinity of D, the particle experiences a *restoring force;* the point D is said to be a point of *stable equilibrium*.

We know from experience that if the particle is displaced slightly from D, it will not return and come to rest at D immediately. Instead it moves toward D, overshoots, and climbs the track on the opposite side; there it comes to rest, turns about, and overshoots the equilibrium point once more. In other words, it oscillates about the equilibrium position.

It is instructive to look into the nature of the forces that lead to stable and unstable equilibrium. Figure 14.2 shows the dependence of the force on the location of the particle on the track of Figure 14.1. Here a positive force signifies a force in the positive x direction—that is, to the right—and a negative force is one that points to the left.

Between A and B no net force acts on the particle. We can define neutral equilibrium, then, as a condition in which no net force acts on a system over a region of finite extent.

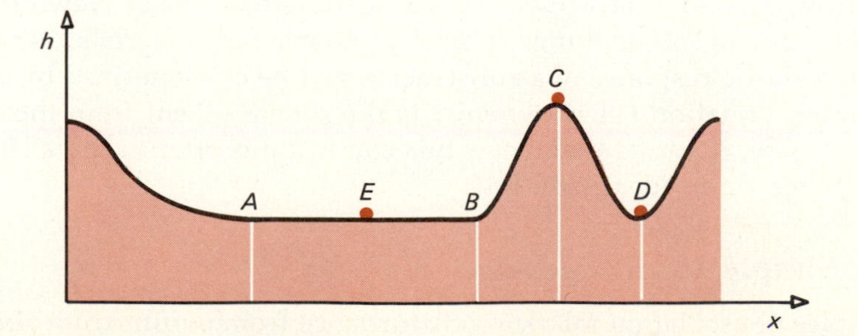

Figure 14.1 Examples of the three types of equilibrium. A particle constrained to move on the track shown in this figure is in equilibrium anywhere between points A and B, as at E; the region between A and B is a region of neutral equilibrium. At C, the particle is in unstable equilibrium; if disturbed slightly, it will move away from C. At D the particle is in stable equilibrium; if displaced slightly from D, it will tend to return to D.

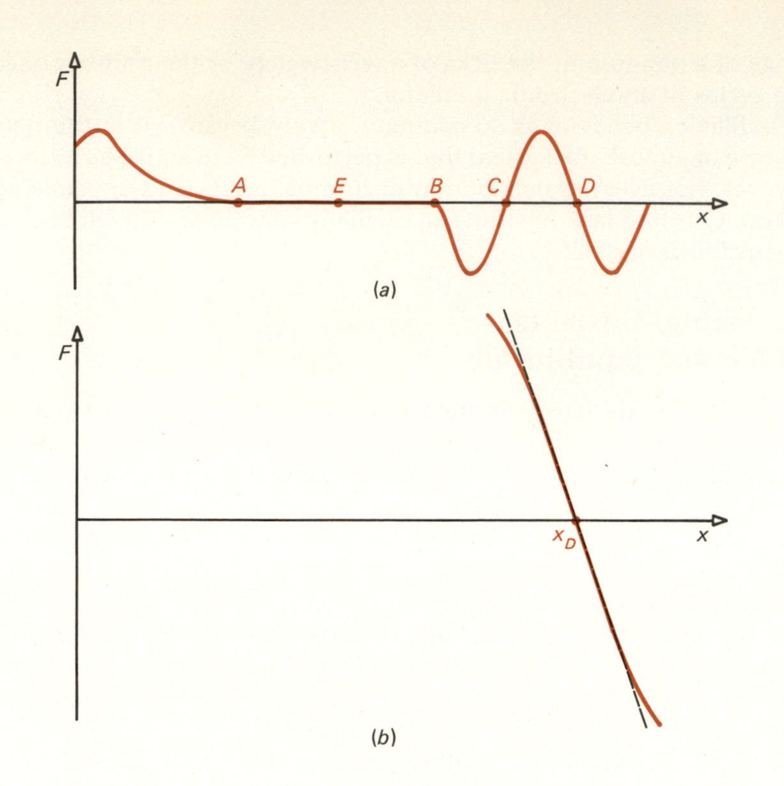

Figure 14.2 *The force acting on a particle on the track of Figure 14.1. (a) At all three equilibrium positions, E, C, and D, the force is zero. Only at D, however, is that force given by an expression of the form $F = -kx$, where x is the displacement from D. (b) The region near $x = x_D$ on an expanded scale. Note that in the immediate neighborhood of x_D the tangent to the curve is a good approximation of the F vs. x curve.*

At C the force is again zero. To the right of C, however, the force points to the right, and to the left of C it points to the left. Consequently, if the particle is displaced slightly from C, it experiences an acceleration in the *same direction as the displacement* and therefore moves farther from C.

At D the force is also zero. There, however, a positive displacement leads to a negatively directed force, and vice versa. The force therefore induces a return to equilibrium; it is a restoring force.

Although the dependence of the force on the displacement from the point of stability need not be linear, *in the immediate vicinity of the position of stable equilibrium it generally has the form*

$$F = -kx \tag{14.1}$$

to a first approximation. *Here x is the displacement from equilibrium.* For the situation just discussed, we have

$$F = -k(x - x_D)$$

We can see that this must be so, because close to D the smooth curve of Figure 14.2 can be approximated by a straight line tangent to this curve at that point.

A force that obeys an expression of the form (14.1) is called a *Hooke's law* force, named after Robert Hooke, a contemporary of Newton, who studied, among other things, elastic properties of materials. He found that the elastic response of a substance could be characterized by a force satisfying Equation (14.1), where x is the displacement from the undeformed state. A spring that meets this condition is often called a Hooke's law spring, or an ideal spring.

14.3 Simple Harmonic Motion

Examples of oscillation following disturbance from equilibrium abound. The swinging pendulum, the vibration of a plucked string, water sloshing back and forth in a bathtub, the gentle swaying of willows in the wind, the fluttering of aspen leaves in a light breeze, the rocking of a boat at anchor are but a few. Generally, these periodic motions are complex, as is, for instance, the repetitive pattern of a normal electrocardiogram. Our attention will be concentrated on a special case of oscillatory behavior, known as *simple harmonic motion,* abbreviated SHM.

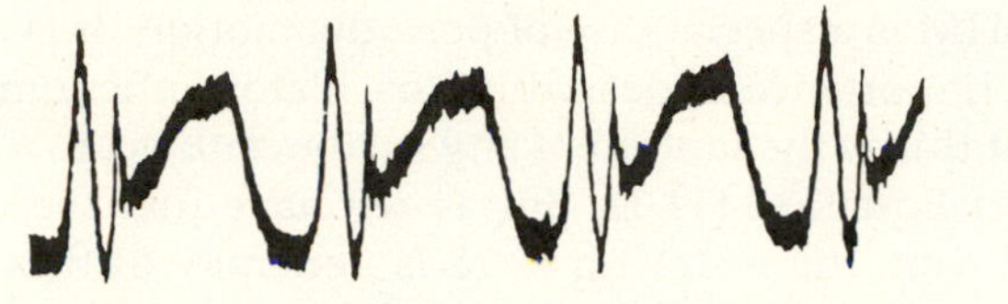

Figure 14.3 A periodic pattern—a normal electrocardiogram trace.

Before considering SHM in detail, we must first define some useful terminology. The dominant characteristic of all periodic events is a repetitive pattern in time. The *period* T is defined as the time interval between successive identical configurations of the oscillating system; it is measured in seconds.

The *frequency of oscillation* f is the number of oscillations, or *cycles,* during a unit time interval, generally one second. Since T is the time required for one cycle, the number of cycles in one second is then

$$f = \frac{1}{T} \tag{14.2}$$

Frequencies are expressed in *hertz,* abbreviated Hz. One hertz equals one cycle per second, and the dimension of f is $[T]^{-1}$.

An oscillating system performs excursions about some average position; for example, a pendulum swings about the vertical. The maximum excursion of the system from its average position is called the *amplitude,* generally designated A.

The characteristic of a system undergoing SHM is that the time dependence of its dynamical variables is of the form

$$A \sin (2\pi ft + \phi) = A \sin (\omega t + \phi) \tag{14.3}$$

The product $2\pi f$ is assigned the symbol ω and is called the *angular frequency;* its unit is the radian per second, the same unit as for angular velocity.

The parameter ϕ is the *phase angle.* Its value depends on the instant that one selects as the zero on the time scale. We shall usually set $\phi = 0$.

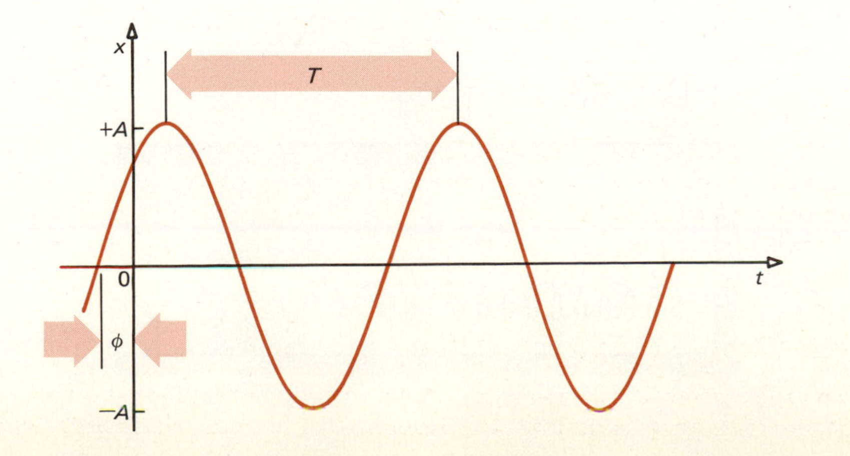

Figure 14.4 The displacement as a function of time of an object executing simple harmonic motion. The period T is the time interval between successive crests. The amplitude A is the maximum excursion from the mean position. The phase angle ϕ depends on the choice of zero for the time scale.

Note that when the time increases by one period T, the expression in Equation (14.3) becomes

$$A \sin [2\pi f(t + T) + \phi] = A \sin (2\pi ft + 2\pi + \phi)$$
$$= A \sin (2\pi ft + \phi)$$

That is, Equation (14.3) displays the periodic pattern of simple harmonic motion with period T.

Though SHM is a special case of periodic motion, it is nevertheless of fundamental importance in describing oscillatory phenomena. First, one can show that this is the motion of any system subject to a restoring force of the form of Equation (14.1). But as we have just seen, right around stable equilibrium the restoring force is generally of this form. Second, even complex periodic patterns of motion can be synthesized from combinations of SHM of different frequencies, amplitudes, and phases, and consequently, they can be analyzed in terms of their constituent sinusoidal components.

14.4 Mass-and-Spring System

The force exerted by an ideal elastic spring clamped at one end is given by

$$F = -kx \tag{14.1}$$

where x is the displacement of the other end of the spring from its equilibrium position. The constant of proportionality k is the *spring constant.* The negative sign indicates that the force F is directed opposite to the displacement from equilibrium. In Figure 14.5, if $x < 0$, as in (*b*), the force is positive (to the right), whereas if $x > 0$, as in (*c*), the force is negative, or points to the left.

Suppose we now attach a mass M that is free to move along the x direction to the unclamped end of the spring. It follows from the preceding discussion that if this mass is pulled a distance x_0 to the right and released, it will oscillate with SHM about the point $x = 0$ with amplitude

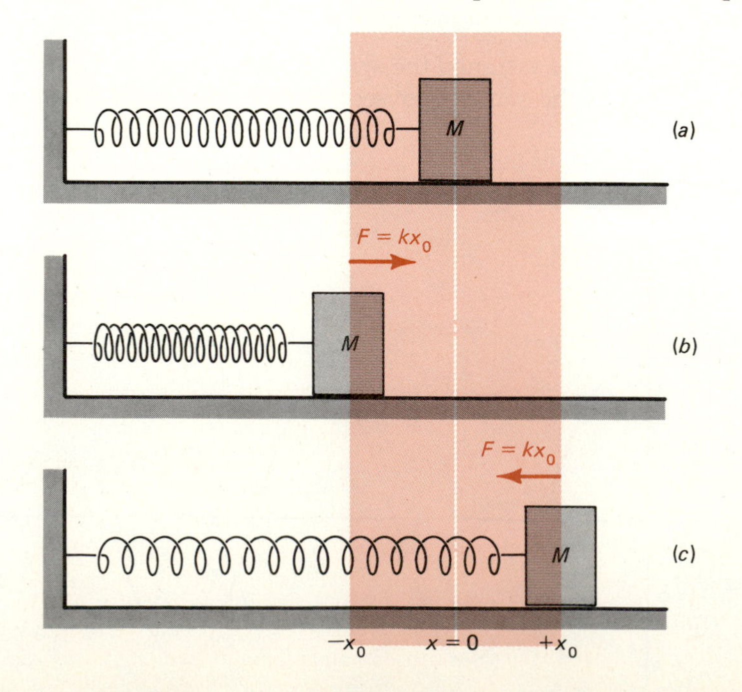

Figure 14.5 A spring exerts a restoring force on the mass M, which slides on a friction-free horizontal surface. (a) M is at the equilibrium position; **F** *= 0. (b) M has been moved to the left. The spring is compressed and the force on M is to the right. (c) M has been moved to the right. The spring is stretched and the force on M is to the left.*

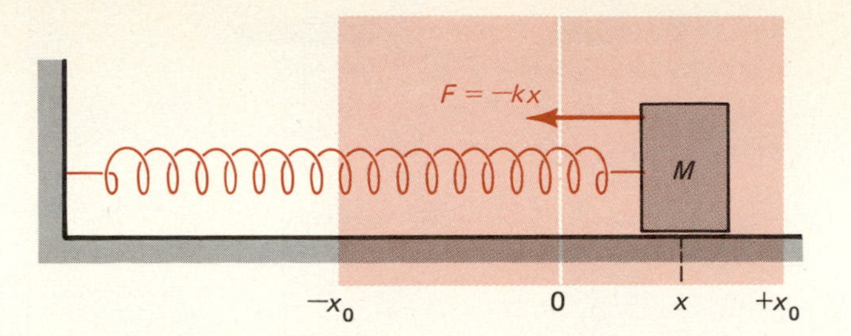

Figure 14.6 An example of a SHM system—a mass M attached to an elastic spring.

x_0. What we now want to determine is the frequency of this oscillation and its dependence on the physical parameters.

Figure 14.6 shows the system at some instant. The mass is at location x, and hence the force exerted on it by the spring is $F = -kx$. Applying Newton's second law, we have

$$Ma = -kx \qquad x = -\frac{M}{k}a \tag{14.4}$$

The apparent simplicity of this equation of motion is deceptive. The difficulty arises from the fact that the acceleration a is neither constant nor dependent on time in some predetermined manner. Instead, a depends on the location of M, and as x changes so does a.*

Still, we can make some progress using only algebra and dimensional analysis. The ratio $M/k = |x/a|$ has the dimension

$$\frac{x}{a} = \frac{[L]}{([L]/[T]^2)} = [T]^2 \tag{14.5}$$

We conclude, therefore, that any temporal parameter such as the period T must be given by an expression of the form

$$T = \text{constant} \times \sqrt{\frac{M}{k}}$$

Dimensional arguments cannot tell us the value of the constant; solving Equation (14.4) leads to the results

$$T = 2\pi\sqrt{\frac{M}{k}} \qquad f = \frac{1}{2\pi}\sqrt{\frac{k}{M}} \qquad \omega = \sqrt{\frac{k}{M}} \tag{14.6}$$

The frequency given by this expression is called the *natural frequency*, or the *resonance frequency*, of the mass-spring system. Equation (14.6) agrees with the qualitative observation that the frequency of such a system is greater the stiffer the spring (greater k) and the smaller the mass.

● ***Example 14.1*** A mass of 0.2 kg is attached to a vertical spring and supported so that the spring exerts no force on the mass. When the support is suddenly removed, the mass oscillates vertically with an amplitude of 10 cm. Find the spring constant and the frequency of oscillation.

At the start of the oscillations, the mass is momentarily at rest. It is therefore at maximum displacement from equilibrium at that instant. Since the vibration amplitude is 10 cm, the *equilibrium position* must be 10 cm below the starting point. Hence,

* Equation (14.4) is a relatively simple *differential equation*, but solving it calls for differential and integral calculus.

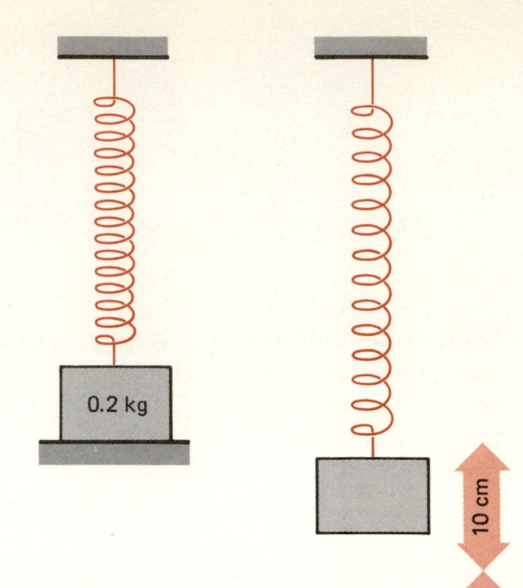

Figure 14.7 The 0.2-kg mass is initially supported so that the spring exerts no force on the mass. When the support is suddenly removed, the mass oscillates as indicated.

$$Mg = (0.1\ \text{m})k$$

$$k = \frac{(0.2\ \text{kg})(9.8\ \text{m/s}^2)}{0.1\ \text{m}} = 19.6\ \text{N/m}$$

From Equation (14.6), we obtain

$$f = \frac{1}{2\pi}\sqrt{\frac{19.6\ \text{N/m}}{0.2\ \text{kg}}} = 1.56\ \text{Hz}$$

14.5 Kinematic Equations of Simple Harmonic Motion

The mathematical expression for the position of the mass of the vibrating system of Figure 14.6 is

$$x(t) = A \sin(\omega t) \tag{14.7}$$

where ω is given by (14.6), and A is the amplitude of vibration, and we have set $\phi = 0$. A plot of x as a function of time is shown in Figure 14.8(a).

Recalling that the instantaneous velocity is the slope of the curve of $x(t)$ (see page 19), we see that $v = 0$ when x is at its greatest excursions, and v is greatest in magnitude when $x = 0$. This is consistent with what we would expect from physical argument. At the point of maximum displacement, the mass changes direction of motion and is instantaneously at rest. Between that point and the instant the mass passes $x = 0$, it experiences an acceleration toward the point of equilibrium, but that acceleration changes direction as soon as the mass passes to the other side of $x = 0$; consequently, $|v|$ must be greatest at $x = 0$.

For a given amplitude, the speed must also be proportional to the frequency. If the frequency is increased and the period correspondingly reduced, the time interval between successive maximum positive and negative displacements is shortened. Hence, the average speed during

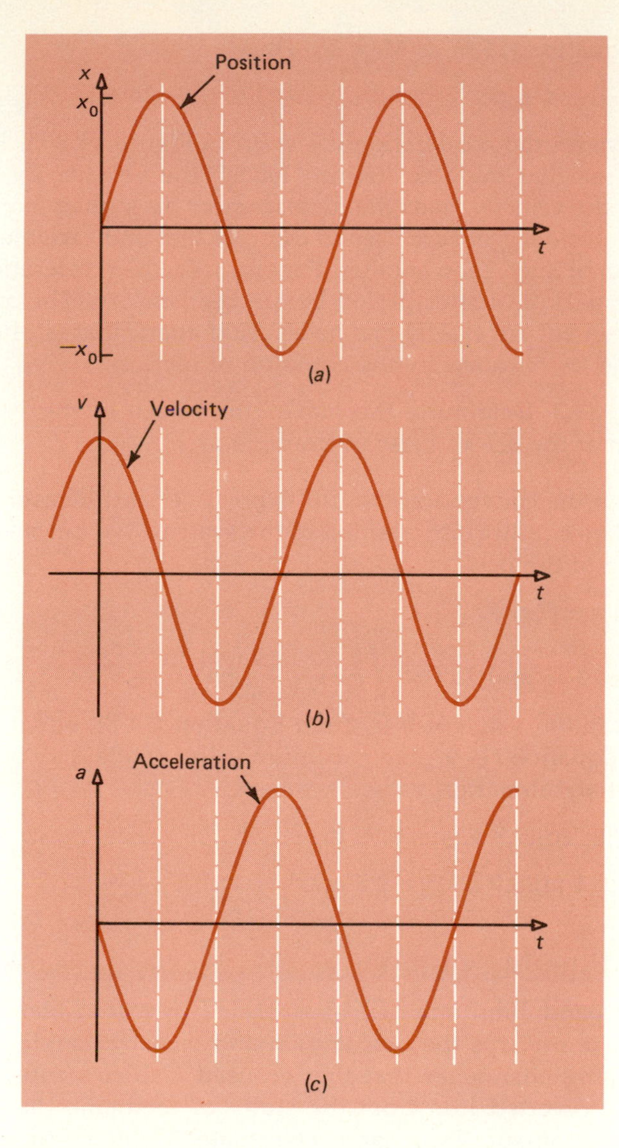

Figure 14.8 The dependence of dynamical variables of a harmonic oscillator on time. (*a*) Position of the mass of Figure 14.6 as a function of time. (*b*) Velocity of this mass as a function of time. Note that the velocity has its greatest magnitude when the mass is at equilibrium, and that the velocity is zero when the displacement is x_0. (*c*) Acceleration of the mass as a function of time. The acceleration is exactly 180° out of phase with the displacement, as required by Equation (14.4).

that half period

$$\langle v \rangle = \frac{2A}{T/2} = 4Af$$

is proportional to the frequency f.

It can be shown, using elementary calculus, that

$$v(t) = A\omega \cos(\omega t) = A\omega \sin\left(\omega t + \frac{\pi}{2}\right) \qquad \textbf{(14.8)}$$

The curve corresponding to Equation (14.8) is shown in Figure 14.8(*b*). It looks just like the curve for $x(t)$ but is displaced in time by one quarter-period. As indicated by Equations (14.7) and (14.8), the two curves differ in phase by $\pi/2$ rad (90°).

Lastly, the curve for the acceleration can be deduced from Figure 14.8(*b*) by plotting its slope, just as Figure 14.8(*b*) could be derived from Figure 14.8(*a*). A more direct approach, however, is to use Equation (14.4),

$$a(t) = -\frac{k}{M}\,x(t) = -\frac{k}{M}\,A \sin(\omega t)$$

But from Equation (14.6), $k/M = \omega^2$. Thus

$$a(t) = -A\omega^2 \sin(\omega t) \tag{14.9}$$

The curve for the acceleration is shown in Figure 14.8(*c*). From these curves we see that the acceleration "leads" the velocity by a quarter-period and the velocity leads the displacement by a quarter-period. The terms *leading* and *lagging* are used to indicate that one variable attains its peak sooner, or later, than another. For example, the acceleration reaches its positive peak a quarter-period before the velocity. The acceleration and displacement are exactly one half-period apart; that is, they are 180° out of phase, as required by the equation of motion.

● ***Example 14.2*** Determine the maximum values of the velocity and acceleration of the 0.2-kg mass of Example 14.1.

We adopt the convention that up is positive, and measure displacement from the equilibrium position of the system. We can then write for the position of the mass

$$x(t) = A \sin(2\pi f t)$$

where $A = 0.1$ m and $2\pi f = \sqrt{\dfrac{k}{M}} = \sqrt{\dfrac{19.6 \text{ N/m}}{0.2 \text{ kg}}} = 9.9 \text{ s}^{-1}$.

The maximum value of v is, from Equation (14.8), equal to $A\omega$; according to Equation (14.9), the maximum value of a is $A\omega^2$. Substituting the appropriate numerical values, we have

$$v_{\text{max}} = (0.1 \text{ m})(9.9 \text{ s}^{-1}) = 0.99 \text{ m/s}$$

$$a_{\text{max}} = (0.1 \text{ m})(9.9 \text{ s}^{-1})^2 = 9.8 \text{ m/s}^2$$ ●

14.6 The Simple Pendulum

The simple pendulum, a mass M suspended by a thin filament from a fixed point, is perhaps the best-known and oldest example of SHM. It was a swaying chandelier that first aroused Galileo's interest in mechanics and diverted him from the study of medicine that his father wanted him to pursue. Two features fascinated Galileo—that the period appeared independent of the amplitude of the swing and that it was also independent of the mass of the bob.

By careful measurements Galileo found that the period depended on the length of the supporting string, increasing as the square root thereof. This dependence has been used for centuries in adjusting pendulum clocks.

Figure 14.9 *A pendulum clock is adjusted by changing the position of the bob on the pendulum's rod. If the bob is lowered, the clock will slow down.*

Figure 14.10 shows a simple pendulum in three positions: (*a*) at the start of its swing; (*b*) at its lowest point; and (*c*) at some intermediate position on the opposite side of its swing. In the accompanying free-body diagrams, the forces acting on the bob are also shown; these are the tension **T** in the string and the weight $M\mathbf{g}$ of the bob.

The string constrains the motion of M to a circle of radius L centered at O, the point of support. The motion of this system is therefore most conveniently analyzed by applying the results of Chapter 8. That is, we specify the position of the pendulum by means of the angle θ that the pendulum makes with the vertical. The equation of motion is

$$\tau = I\alpha \tag{8.11}$$

From Figure 14.11, it is apparent that the tangential component of the weight Mg, or $Mg \sin\theta$, is the only force that contributes to the torque

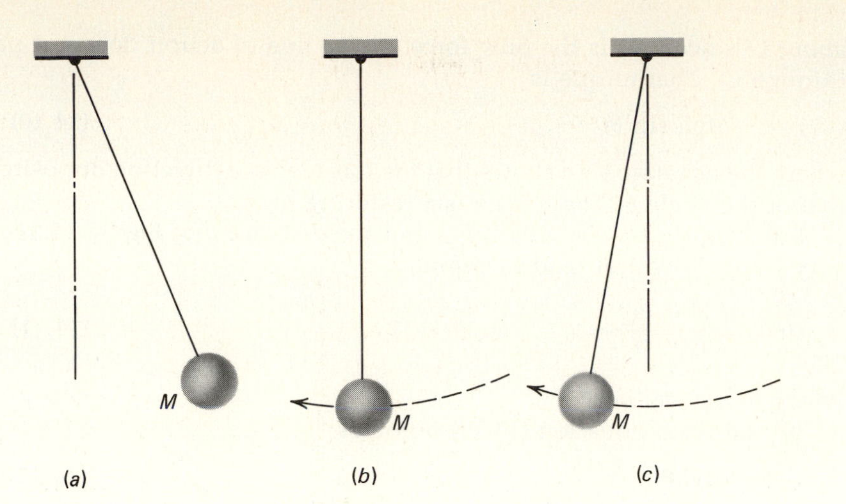

Figure 14.10 *A simple pendulum in three positions: (a) at the start of its swing; (b) at its lowest point; and (c) at some intermediate position on the opposite side of its swing. The free-body diagrams for these three situations are shown in (d), (e), and (f). Note that the tension in the string is not at constant magnitude. (Why does the tension vary as the pendulum swings?)*

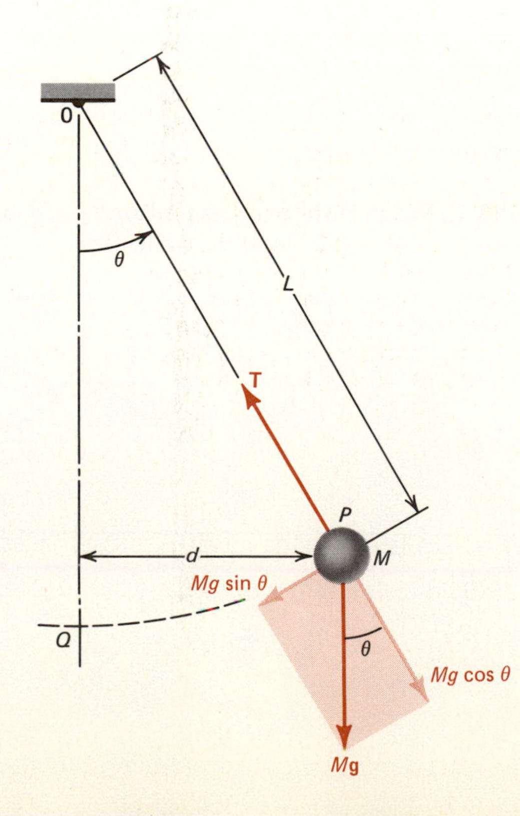

Figure 14.11 *Free-body diagram for a simple pendulum that is executing small-amplitude oscillations about its equilibrium position.*

about O since that is the only force whose line of action does not pass through O. That torque is

$$\tau = -MgL \sin\theta \tag{14.10}$$

where the negative sign shows that the torque has a direction opposite to that of the angle θ. The torque is a restoring torque.

If θ is small, the arc length PQ and the distance d of Figure 14.11 are very nearly identical, and therefore,

$$\sin\theta = \frac{d}{L} \simeq \frac{PQ}{L} = \theta \qquad \text{for } \theta \ll 1 \tag{14.11}$$

where θ is in radians.

In that case, Equation (14.10) becomes

$$\tau = -MgL\theta \tag{14.12}$$

The moment of inertia of the system about point O is

$$I = ML^2 \tag{14.13}$$

Substituting (14.12) and (14.13) into (8.11) gives

$$MgL\theta = -ML^2\alpha \tag{14.14}$$

$$\theta = -\frac{L}{g}\alpha \tag{14.15}$$

Equation (14.15) is identical to Equation (14.4), except that angular displacement and acceleration have replaced the linear variables and L/g appears in place of M/k. It follows that

$$\theta = A \sin(2\pi ft + \phi)$$

where the resonant frequency and period are given by

$$f = \frac{1}{2\pi}\sqrt{\frac{g}{L}} \qquad T = 2\pi\sqrt{\frac{L}{g}} \tag{14.16}$$

● ***Example 14.3*** Figure 14.12 shows the angular displacement of a pendulum as a function of time. Find an expression of the form of Equation (14.3) that represents this motion. What is the length of this pendulum?

From Figure 14.12 the amplitude of oscillation is $\pi/6$ rad. The time interval between successive peak positive deflections is 0.4 s. Thus, $f =$

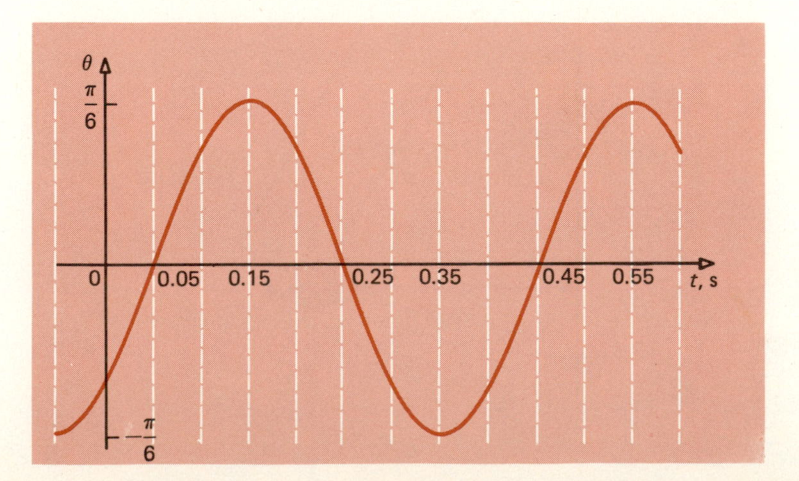

Figure 14.12 The angular displacement of the pendulum of Example 14.3.

$1/T = 1/0.4 = 2.5$ Hz. Finally, at $t = 0$, the sine function is still 0.05 s, or one-eighth of a period from the instant at which its argument is zero; that is, not until $2\pi ft = \pi/4$ does $(2\pi ft + \phi) = 0$. Consequently, $\phi = -\pi/4$. The complete expression for $\theta(t)$ is therefore

$$\theta(t) = \frac{\pi}{6} \sin\left(5\pi t - \frac{\pi}{4}\right) \text{ rad}$$

We determine the length of the pendulum by solving Equation (14.16) for L:

$$L = \frac{g}{4\pi^2 f^2} = \frac{9.8 \text{ m/s}^2}{4\pi^2(2.5 \text{ s}^{-1})^2} = 0.0397 \text{ m} = 3.97 \text{ cm}$$

●

14.7 Equivalence Principle Revisited

In considering whether inertial and gravitational masses might be different (Chapter 3), we stated that experiments had demonstrated their equivalence. We did not then mention what experiments led to this profoundly significant result.

In Equation (14.14), the quantity M on the left-hand side is the gravitational mass; that is, Mg is the gravitational force acting on the object of mass M. On the right-hand side of Equation (14.4), however, M refers to the inertial mass. We canceled these masses in arriving at Equation (14.15), but that step is predicated on the validity of the equivalence principle. We can now see that one method by which the equivalence of inertial and gravitational masses can be established is through precise measurement of the mass dependence of the period of a pendulum.

Careful measurements of that kind were performed already by Newton, who found that gravitational and inertial masses could differ by no more than one part in a thousand. Modern experiments have confirmed the equivalence to within one part in 10^{12}.

14.8 Energy of an Oscillating System

When the mass of the system of Figure 14.5 passes the point $x = 0$, its speed and hence its KE is greatest; the KE then has the value

$$\text{KE}_{\text{max}} = \tfrac{1}{2}Mv_{\text{max}}^2 = \tfrac{1}{2}MA^2\omega^2 \qquad \textbf{(14.17)}$$

At $x = \pm A$, $v = 0$ and KE $= 0$. What has happened to this KE? The answer is that it has been transformed into elastic PE that is now stored in the spring. In compressing or extending the spring by an amount A, the mass M has done work *on* the spring, which resists deformation, according to Equation (14.1). For a spring obeying Equation (14.1), the elastic energy stored in the spring is the product of the *average* force exerted on the spring and the displacement.

The average force that acts on the spring in causing a displacement x is

$$\langle F \rangle = \tfrac{1}{2}kx$$

and hence

$$\text{PE} = (\tfrac{1}{2}kx)x = \tfrac{1}{2}kx^2 \qquad \textbf{(14.18)}$$

At the extreme excursions of the mass, the PE of the spring is

$$\text{PE}_{\text{max}} = \tfrac{1}{2}kA^2 \qquad \textbf{(14.19)}$$

Figure 14.13 Multiple-exposure stroboscopic photograph of a swinging pendulum. The separation between successive positions of the bob indicates the speed of motion. The pendulum is instantaneously at rest at the two ends of its swing; it is moving most rapidly, and has the greatest kinetic energy, at the bottom of its swing.

From Equation (14.6), $k = M\omega^2$, so that Equation (14.19) can be rewritten

$$\text{PE}_{\text{max}} = \tfrac{1}{2}MA^2\omega^2$$

This is identical to (14.17), the expression for the maximum KE, which occurs when the PE is zero.

It is not difficult to show that throughout this periodic motion, energy is constantly exchanged between the spring (PE) and the mass (KE) such that the total energy is always conserved.

The total energy is

$$E = \text{PE} + \text{KE}$$

where PE is given by Equation (14.18), the KE is $\frac{1}{2}Mv^2$, and the displacement and velocity are given by Equations (14.7) and (14.8). Thus the total energy is

$$\begin{aligned} E &= \tfrac{1}{2}kx^2 + \tfrac{1}{2}Mv^2 \\ &= \tfrac{1}{2}k[A\sin(\omega t)]^2 + \tfrac{1}{2}M[A\omega\cos(\omega t)]^2 \\ &= \tfrac{1}{2}kA^2[\sin^2(\omega t) + \cos^2(\omega t)] \\ &= \tfrac{1}{2}kA^2 = \tfrac{1}{2}MA^2\omega^2 \end{aligned}$$

where we have used the trigonometric identity $\sin^2\theta + \cos^2\theta = 1$, and the last equality follows from the relation between k, M, and ω, Equation (14.6). Although both PE and KE depend on time, the total energy is independent of time.

It is left as a problem to show that the motion of the simple pendulum involves a similar transfer of energy, in that case between KE and gravitational PE of the bob. Periodic transformation of energy is a fundamental characteristic of all oscillatory systems, be they mechanical or electrical or of some other type.

● ***Example 14.4*** A mass of 0.5 kg is attached to a spring whose spring constant is $k = 1000$ N/m. What is the energy of the system when it oscillates with an amplitude of 0.1 m? Suppose this mass is replaced by a 1-kg mass and the system is again set into oscillation with the same vibrational amplitude. What is then the maximum KE?

Here is a problem for which we are given more information than we need for the solution. At the displacement extrema, that is, when $x = \pm A = \pm 0.1$ m, $v = 0$, and all the energy is stored as elastic energy in the spring. Consequently, the total energy of the system is

$$E = PE_{max} = \tfrac{1}{2}kA^2 = \tfrac{1}{2}(1000\ \text{N/m})(0.1\ \text{m})^2 = 5\ \text{J}$$

This energy is independent of the mass that is attached to the spring and must also equal the maximum KE of that mass. Therefore, the maximum KE is 5 J for both masses. ●

*14.9 Forced Oscillations; Damping and Resonance

Suppose the mass of a mass-spring system, initially at rest, is given a small brief impulse to initiate SHM. If this initial impulse is followed at regular time intervals of one period by further small impulses, the amplitude of oscillation of the mass will grow larger with each cycle, just as the amplitude of vibration of a child on a garden swing is increased with each push. In principle, we can achieve arbitrarily large vibrational amplitude and vibrational energy by *driving* the system for a sufficiently long time in this manner. This technique of imparting great energy by a series of carefully timed impulses is employed in cyclotrons and other particle accelerators.

In real situations we can never eliminate all frictional or other sources of energy dissipation, and the amplitude of a real oscillating system will, under such periodic driving impulses, attain some maximum amplitude. This maximum amplitude is reached when the energy imparted to the system during one cycle just equals the loss due to dissipative effects. Also, if such a system is set in oscillation and the driving force is removed, the amplitude will diminish, until ultimately, the system comes to rest at equilibrium.

This approach to maximum amplitude, and alternatively to zero amplitude in the absence of a driving force, is *exponential;* that is, the time dependence is of the form

$$A(t) = A_{max}(1 - e^{-t/\tau}) \qquad \text{approach to } A_{max} \text{ starting with } A = 0 \text{ at } t = 0 \tag{14.20}$$

$$A(t) = A_{max}e^{-t/\tau} \qquad \text{approach to zero amplitude starting with } A = A_{max} \text{ at } t = 0 \tag{14.21}$$

The parameter τ, which appears in Equations (14.20) and (14.21) and which has the dimension of time, is called the *time constant.* Its value de-

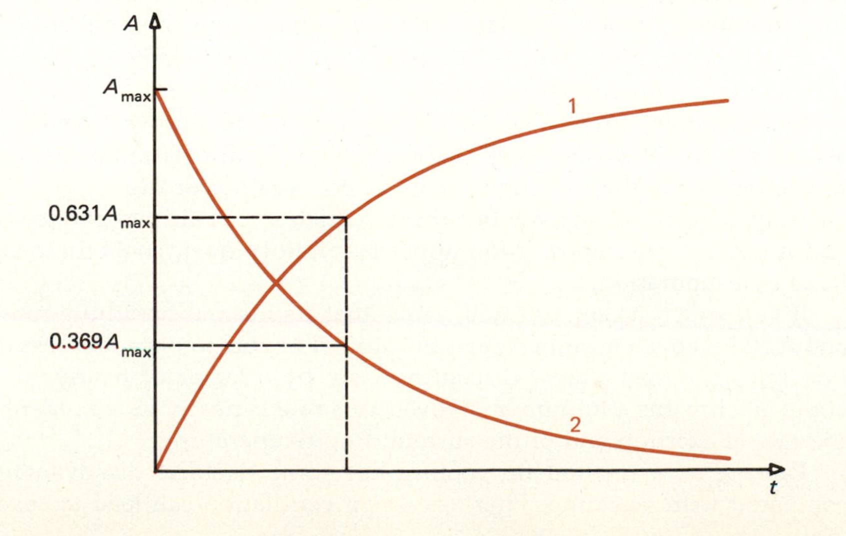

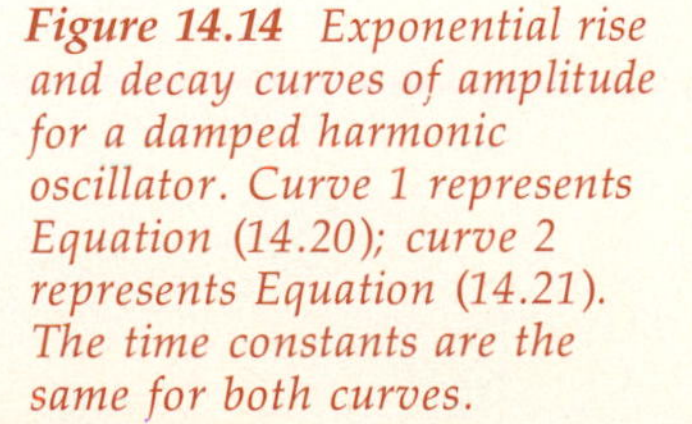

Figure 14.14 Exponential rise and decay curves of amplitude for a damped harmonic oscillator. Curve 1 represents Equation (14.20); curve 2 represents Equation (14.21). The time constants are the same for both curves.

pends on the magnitude of frictional or other *damping* forces. The time constant is a measure of the speed with which a system approaches its ultimate steady-state condition.

If an oscillating system is lossless, $\tau = \infty$, and the amplitude will be constant in the absence of external forces. For finite τ, the decay of $A(t)$ is given by Equation (14.21); the amplitude decreases by a factor of $1/e = 1/2.72 = 0.369$ with each time constant. The curves for $A(t)$ corresponding to (14.20) and (14.21) are shown in Figure 14.14 and the oscillations of a damped system driven at resonance are shown in Figure 14.15.

We have here used the term *resonance* again. The gradual increase in amplitude of an oscillator, shown in Figure 14.15, will occur only if the frequency of the periodic driving force equals the natural frequency of the oscillating system. When that condition prevails, the driving force and the oscillating system are *in resonance*.

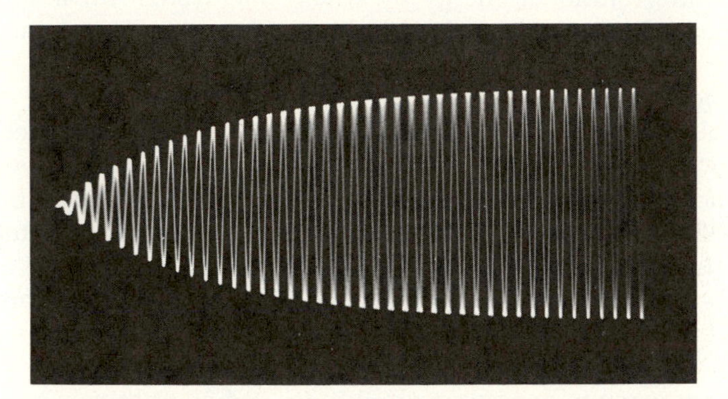

Figure 14.15 *Displacement of a damped harmonic oscillator driven at its natural resonant frequency. The envelope of this curve should be compared with curve 1 of Figure 14.14.*

Physical and also biological systems frequently manifest damped oscillations and resonance. For example, the frame of a car is supported on its wheels by springs. With no shock absorbers, a car oscillates violently after it passes over a bump in the road. Shock absorbers provide the viscous loss to damp these vibrations.

When a voltmeter is placed across the terminals of a battery, the torque that is impressed on the meter movement causes the needle to rotate. There is a spring in the movement, so that the equilibrium position of the needle is proportional to this torque, which in turn is proportional to the voltage. If no damping were incorporated in the meter's construction, the needle would oscillate for a long time about the final equilibrium position. The system is damped, allowing the needle to reach its ultimate position relatively quickly.

One example of resonance in a biological system that we have all observed but do not always recognize as a natural manifestation of mechanical resonance is the panting of dogs in hot weather or during physical exertion. This rapid shallow breathing, which forces air past the tongue and mucous membranes of the upper respiratory tract, cools these surfaces by evaporation.

If you watch a dog you will notice that its normal breathing rate is about 30 breaths a minute. When the animal begins to pant, the respiration rate increases almost discontinuously by a factor of nearly 10, to about 300 breaths a minute; moreover, this rate is nearly independent of the rate of exertion and of the surrounding temperature.

Panting as a method for cooling has some decided disadvantages compared with sweating. First, excessive ventilation can lead to severe depletion of carbon dioxide from the lungs. Since the CO_2 in the lungs is derived from the blood, the blood CO_2 is thereby lowered drastically, leading to an abnormal condition called alkalosis. Second, since an increase in the breathing rate requires muscular effort, the additional exertion and resulting heat production tends to counteract the cooling effect of evaporation.

Alkalosis is reduced by the shallowness of breathing during panting, thereby limiting the additional ventilation largely to mouth, nose, and upper respiratory tract. The extra heat load due to the muscular effort demanded by the greatly accelerated breathing rate would generate more body heat than could be dissipated by evaporation if it were not for the animal's reliance on the natural resonance of the respiratory cavity.

In rough approximation, the chest cavity and respiratory tract can be likened to a hollow chamber, supported by the ribs and surrounded by elastic membranes (skin and tissue), which can be expanded by muscular action. This anatomical system has a natural frequency that is determined by its volume, the mass associated with the supporting structure, and the elastic properties of the membranes, tendons, and muscles surrounding the cavity. This system will attain relatively large vibrational amplitudes with little expenditure of energy per cycle if it is driven at resonance. For medium-sized dogs, the respiratory system's resonance frequency is of the order of 5 Hz, that is, 300 cycles a minute. By switching to that high frequency, the dog greatly increases air flow in and out of the upper respiratory tract with very little muscular effort.

Summary

Equilibrium is of three kinds. A system is in *neutral equilibrium* if it experiences no force when displaced slightly from the equilibrium position. The equilibrium is *unstable* if a slight displacement from equilibrium produces a force on the system that drives it further from the equilibrium position. If a small displacement from equilibrium produces a *restoring force* on the system, the equilibrium is *stable*.

Any physical system executes *simple harmonic motion* (SHM) if it is disturbed slightly from stable equilibrium. The time dependence of all dynamical variables of a system in SHM is of the form

$$\text{Variable} = A \sin (2\pi ft + \phi)$$

where f is the *frequency* of the oscillation, measured in hertz, and ϕ is the *phase constant*. The *period* of the oscillation T is the time interval between successive maxima; T is the reciprocal of f. The quantity A is called the *amplitude* of oscillation.

Two common examples of mechanical systems exhibiting SHM are a mass attached to an elastic spring and a simple pendulum. The periods and frequencies for these systems are given by Equations (14.6) and (14.16). The velocities and accelerations of systems undergoing SHM depend on the amplitude and frequency according to Equations (14.8) and (14.9).

In SHM the system is continually transforming energy between kinetic and potential. The total energy, in the absence of damping, remains constant.

Real oscillating systems always lose some energy to friction or other loss mechanisms. This energy loss results in *damping* of the oscillations.

Generally, the damping is *exponential;* the amplitude of free vibrations of a damped harmonic oscillator has a time dependence of the form

$$A(t) = A_{\max}e^{-t/\tau}$$

where τ is the *time constant.* A damped harmonic oscillator attains maximum amplitude if it is driven at its resonance frequency.

Multiple Choice Questions

14.1 A periodic SHM is given by $y = 3 \sin (\pi t)$ m, where t is in seconds. The period of the system is

(a) 2 s
(b) 2 Hz
(c) 3 m
(d) 0.5 s

14.2 The energy of a vibrating mass-and-spring system is proportional to

(a) the amplitude of vibration only.
(b) the square of the mass.
(c) the square of the frequency.
(d) the square of the product of amplitude and spring constant.

14.3-5 Pendulum A has a mass M_A and length L_A. Pendulum B has a mass M_B and length L_B.

14.3 If $L_A = L_B$ and $M_A = 2M_B$ and the vibrational amplitudes are equal, then

(a) $T_A = T_B$ and the energies of the pendulums are equal.
(b) $T_A = \frac{1}{2}T_B$ and the energies of the pendulums are equal.
(c) $T_A = T_B$ and A has greater energy than B.
(d) $T_A = T_B$ and A has less energy than B.

14.4 If $L_A = 2L_B$, and $M_A = M_B$, and the two pendulums have equal vibrational energies, then

(a) their amplitudes of angular motion are equal.
(b) their periods of motion are equal.
(c) B has a greater angular amplitude than A.
(d) None of the above is correct.

14.5 If pendulum A has a period twice that of B, then

(a) $L_A = 2L_B$ and $M_A = 2M_B$.
(b) $L_A = 2L_B$ regardless of the masses.
(c) $L_A = 2L_B$ and $M_A = M_B/2$.
(d) None of the above is correct.

14.6 The sketches of Figure 14.16 show spheres and hemispheres in contact with various supporting surfaces that are not frictionless. Of these, the following correspond to configurations of stable equilibrium:

(a) A, B, C, and E only.
(b) C and E only.

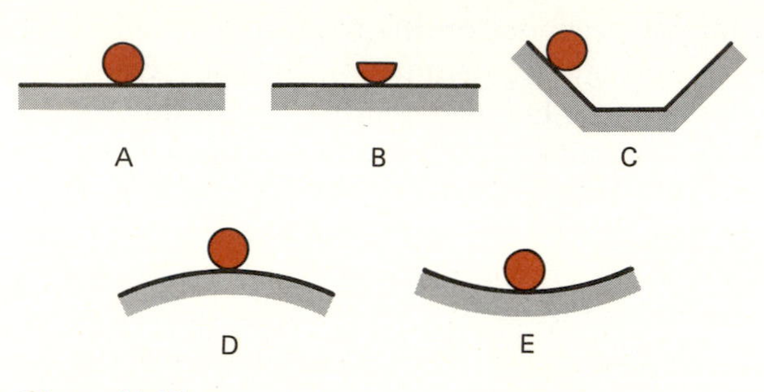

Figure 14.16

(c) B and E only.
(d) B, C, and E only.

14.7 A mass-and-spring system with mass M vibrates with an energy of 4 J when the amplitude of vibration is 5 cm. If the mass is replaced by another of mass $M/2$ and the system is again set in vibration with an amplitude of 5 cm, the energy will be

(a) 1 J
(b) 2 J
(c) 4 J
(d) None of the above

Problems

(Sections 14.2, 14.3)

14.1 A system oscillates at a frequency of 60 Hz with an amplitude of 2 cm. Write an expression for the displacement y as a function of time for the case when (a) $y(0) = 0$ cm; (b) $y(0) = 2$ cm; (c) $y(0) = -2$ cm.

14.2 The displacement of an object is given by

$$y(t) = 0.02 \sin (30\pi t) \text{ m}.$$

What are the amplitude, frequency, and period of the oscillations?

14.3 A system performs simple harmonic motion along the x direction. The amplitude of the oscillations is 3 cm and the period is 0.6 s. Write an expression for the displacement as a function of time for the case that the displacement maximum occurs when (a) $t = 0$ s; (b) $t = 0.15$ s; (c) $t = 0.2$ s.

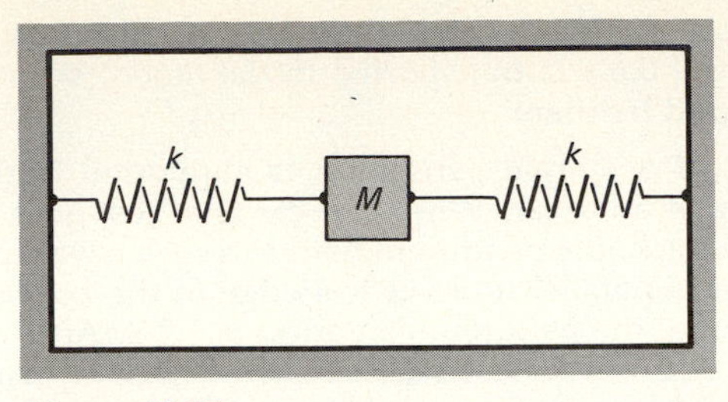

Figure 14.17

• **14.4** An object performs simple harmonic motion at a frequency of 8 Hz. The amplitude of the displacement is 0.06 m. Write an expression for the displacement as a function of time for the case in which (a) the displacement is zero at $t = 0$ and increasing; (b) the displacement is zero at $t = 0$ and decreasing, i.e., going negative; (c) the displacement is 0.03 m and increasing; (d) the displacement is 0.06 m.

(Section 14.4)

14.5 A spring stretches by 0.1 m when a mass of 1 kg is suspended from its end. What mass should be attached to this spring so that the natural vibration frequency of the system will be 5 Hz?

14.6 A mass is attached to a spring of spring constant $k = 400$ N/m. If this mass is displaced from equilibrium by 4 cm and released at $t = 0$, it oscillates at a frequency of 15.6 Hz. Write an expression for the displacement, velocity, and acceleration of this mass as a function of time. Also, determine the value of the mass.

• **14.7** A mass of 0.5 kg is attached to a spring. The mass is then displaced from its equilibrium position by 4 cm and released. Its speed as it passes the equilibrium position is 24 cm/s. Find the spring constant of the spring.

• **14.8** A 0.4-kg mass is supported by a spring. The system is set in vibration at its natural frequency of 2 Hz with an amplitude of 5 cm. Find the spring constant of the spring and the maximum speed of the mass.

• **14.9** The mass of 0.4 kg of Problem 14.8 is augmented by an additional 0.2 kg. At what amplitude should the system vibrate so that the maximum speed of the mass is the same as in Problem 14.8?

• **14.10** Many electronic components are subject to severe vibration to test their endurance under repeated extremes of acceleration. If the testing table has a maximum vibration amplitude of 5 cm, at what frequency must it be operated to confirm a manufacturer's claim that a product can withstand repeated accelerations of $32g$?

• **14.11** An object is vibrating along a straight line in SHM. When it is 6 cm from its mean position, it has an acceleration of 0.2 m/s^2. Find the frequency of oscillation.

• **14.12** A mass of 0.5 kg is supported on a frictionless table and is attached to two identical springs, as shown in Figure 14.17. If the mass is pulled along the line joining the fixed ends of the two springs and then released, it oscillates with a frequency of 12 Hz. Find the spring constant of the springs.

• **14.13** A spring stretches 5 cm when a mass of 0.4 kg is suspended from its free end. If this mass is then pulled 3 cm beyond the equilibrium position and released at $t = 0$, what expression gives the displacement of this mass as a function of time? Be sure to state which is the assumed positive direction, up or down.

• **14.14** A spring, whose unstretched length is 20 cm, is stretched to 24 cm when a 0.2-kg mass is suspended from its free end. Find the spring constant of the spring and the frequency at which the system would vibrate if displaced from equilibrium.

• **14.15** A mass is attached to a spring, as in Figure 14.6. If the spring is stretched 8 cm beyond its normal length and then released, the system vibrates at a frequency of 2.5 Hz. If the initial mass is replaced by another that is 0.05 kg greater, the system then vibrates at 1.8 Hz. Find the value of the initial mass and the vibrational energy of that system.

• **14.16** A 0.5-kg mass slides on a frictionless horizontal track. The mass is attached to one end of a spring whose spring constant is $k = 800$ N/m and whose other end is fixed. When the mass moves past its equilibrium position, it does so with a speed of 5 m/s. Find the frequency and amplitude of the oscillation and the maximum force the spring exerts on the mass.

• **14.17** Figure 14.18 shows various ways in which two identical masses and two identical springs can be

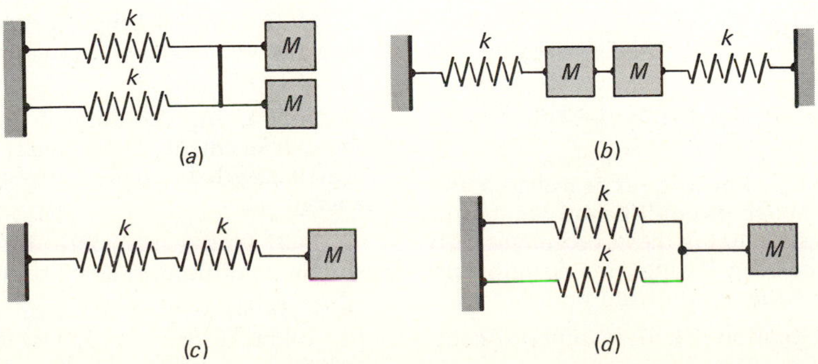

(a) (b) (c) (d)

Figure 14.18

connected. For each arrangement, give the expression for the resonance frequency for motion along the line of the springs.

• • **14.18** A mass of 0.5 kg is attached to a spring and supported on a level friction-free air track. The system is initially at rest. A second mass of 0.7 kg makes an inelastic collision with the first, after which the two move together. After the collision, the system oscillates at a frequency of 0.6 Hz with an amplitude of 20 cm. Find the spring constant of the spring and the speed of the 0.7-kg mass before the collision.

• • **14.19** A mass of 0.5 kg is attached to a spring whose $k = 600$ N/m. The system rests on a level friction-free air track and is initially at rest. A second mass makes an elastic head-on collision with the mass attached to the spring; thereafter the oscillating system vibrates with an amplitude of 0.3 m. If the incident mass is 0.2 kg, what was its incident speed?

• • **14.20** A 1000-kg car, whose shock absorbers are in deplorable condition, vibrates up and down at a frequency of 2 Hz after passing over a bump. When a passenger of 100 kg enters the car, by how much does the car settle?

(Sections 14.5, 14.6)

14.21 The position of an object relative to its equilibrium location is given by $x(t) = 0.6 \cos(7.85t)$, where x is in meters. What are the amplitude of oscillation, the frequency, and the angular frequency? What is the velocity of the object at $t = 0$? What is its acceleration at $t = 0$?

• **14.22** The periodic displacement of a mass from equilibrium is given by $x = 0.3 \sin(47t + \pi/3)$ m. (a) For $t = 0$, find the displacement, velocity, and acceleration of the mass. (b) Find the maximum values of the displacement, velocity, and acceleration, and the values of t for which these maximum values occur. (c) What are the frequency, period, and amplitude of the motion?

• **14.23** A pendulum is released from rest when its supporting string makes an angle of 15° with the vertical. The length of the pendulum is 0.6 m. Write an expression for the angular velocity of the pendulum as a function of time.

• **14.24** The acceleration of gravity varies slightly over the surface of the earth. If a pendulum has a period of 3.0000 s at a location where $g = 9.803$ m/s^2 and a period of 3.0014 s at another location, what is g at this new location?

• **14.25** What should be the length of a simple pendulum whose period is 1 s?

• **14.26** A pendulum is 4 m long. It is pulled aside until its bob is 0.2 m above its equilibrium point, and then released from rest. What is the speed of the bob when the bob is (a) passing through the equilibrium point? (b) 0.1 m above the equilibrium point?

• **14.27** Show that Equation (14.16) is dimensionally correct.

• **14.28** A simple pendulum that oscillates with a period of 0.6 s is transported to the moon. What will its period be there?

• • **14.29** A simple pendulum is supported from the ceiling of a boxcar. When the car is at rest on a horizontal track, the pendulum hangs in such a way that at rest, it is parallel to a vertical edge of the boxcar. At that time, the pendulum's period is 1.2 s. After some time, an observer notices that the pendulum, when it is motionless in the car, hangs so that it makes an angle of 6° with the vertical edges of the car; also, the observer is leaning forward to maintain balance.

This observation could be the result of the car's acceleration on a horizontal track or of the car's climb up a 6° grade at constant speed. Is there any experiment that the observer in the car can perform to choose between these two possibilities? Verify with a suitable numerical calculation.

• **14.30** A pendulum consists of a bob of mass 1.41 kg and a string of length L. What should the value of L be so that the period of the pendulum is 2 s?

• **14.31** A pendulum is made of a massive bob supported by a fine wire of length L. If as a result of thermal expansion the length of the wire increases by 0.02% as the temperature is raised by 50 °C, what corresponding percentage change will there be in the frequency of the pendulum?

• • **14.32** A homogeneous steel sphere of 2-cm diameter rolls, without slipping, on the inner surface of a spherical surface of 1-m diameter. Derive an expression for the period of this SHM for small displacement of the sphere from its equilibrium position.

• • **14.33** A block of wood of mass 2.4 kg is suspended by a massless string 0.8 m long. A bullet is fired into this block and comes to rest in it. The system is observed to vibrate with an amplitude of 15°. Determine the momentum of the bullet, assuming that its mass is negligible compared with that of the wooden block.

• • **14.34** Two equal masses of 1.2 kg each are connected by identical rigid rods of negligible mass; the rods make an angle of 90°, as shown in Figure 14.19. The system can rotate about axis AA' or BB'. Determine the vibration frequencies for these two vibrations, assuming that the angular displacement from equilibrium is small.

(Sections 14.8, 14.9)

14.35 Show that the expressions for the PE and KE of a mass-and-spring system are dimensionally correct.

• **14.36** Using the data of Problem 14.8, determine the displacement of the mass for which the total energy is divided equally between kinetic and potential energy.

• **14.37** Find expressions for the PE and KE of a simple pendulum in small oscillation.

• **14.38** Show that the total energy of a swinging pendulum is conserved. Use the results of Problem 14.37.

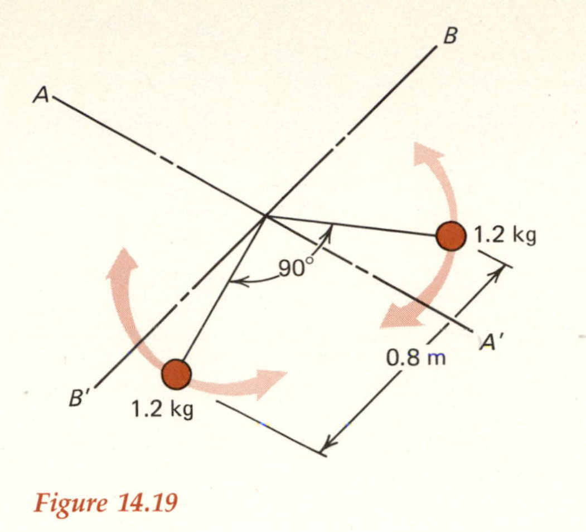

Figure 14.19

• 14.39 Two equal masses are suspended from two different springs. Spring A extends 5 cm, spring B extends 8 cm under the load of the mass. If both systems are set in oscillation with an amplitude of 4 cm, what is the ratio of the energy of the two systems?

• • 14.40 Two equal masses of 0.5 kg each are connected to one end of a spring of spring constant $k = 120$ N/m, as shown in Figure 14.20. The masses slide on a frictionless horizontal surface. The system is set in vibration by displacing the masses a distance of 0.8 m and then releasing them.

Some time after the system has been vibrating, one of the two masses breaks free of the spring.

Find the initial vibration frequency and the energy of the vibrating system; the frequency of vibration after one of the masses has become disengaged from the spring; the amplitude and vibrational energy after the mass has broken free if this occurred (a) as the system passed through its equilibrium position, and (b) as the system experienced maximum acceleration.

• • 14.41 A mother pushes her child on a swing in a playground. The mass of child and swing is 50 kg. The length of the swing is 6 m, and the mother keeps the swing going so that it swings with an amplitude of 23°. In the process she does work at an average rate of 4 W. If she stops pushing the child, by what fraction does the amplitude of the swing decrease with each period?

• • 14.42 A vibrating system of frequency 100 Hz has a time constant of 2 s. If at $t = 0$ the vibration amplitude is 6 cm, and the energy of the system is then 40 J, what is the amplitude of vibration at the end of 2 s and at the end of 4 s? What energy is dissipated by the damping mechanism during each of the two 2-s intervals?

• • 14.43 A 0.5-kg mass rests on a level surface and is attached to one end of a spring as in Figure 14.6. The system has a resonance frequency of 3 Hz. It is set in oscillation and maintained in vibration by means of regular impulses imparted to the mass. At what rate must work be done on the system to maintain an amplitude of vibration of 0.2 m if the coefficient of kinetic friction between mass and surface is 0.2? Compare the energy given the system during one period with the energy of vibration.

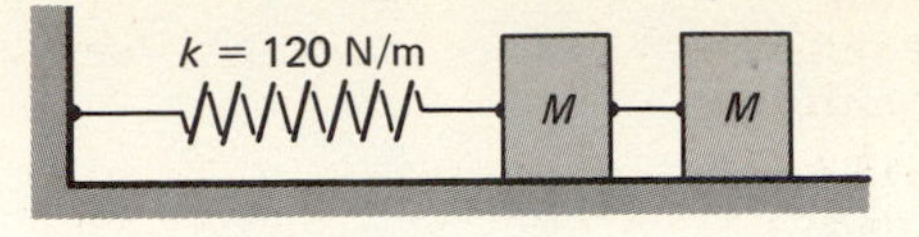

Figure 14.20

• 14.44 Relative to Problem 14.43, what will be the vibration amplitude if the power input is reduced by half?

• • 14.45 The hydrometer of Figure 10.17 (see page 200) has a mass of 25 g and floats in a liquid of 1.025 specific gravity. The stem of the hydrometer is a plastic cylinder, 0.8 cm in diameter, whose specific gravity is 0.7. The hydrometer is now lifted slightly out of the liquid and then released. Show that the ensuing motion is simple harmonic and find the period of this SHM. Assume that the liquid is inviscid.

• • 14.46 Robert Boyle entitled his work *Experiments on the Spring and Weight of Air*. Suppose a 2-kg piston that is free to move rests 1.5 m above the bottom of a cylinder filled with a polyatomic ideal gas at 300 K (Fig. 14.21). This piston is now raised by 20 cm and released. Assume that the number of atoms per molecule is so great that the heat capacities at constant pressure and constant volume are practically identical and so large that the temperature of the gas remains unchanged as the piston moves.

Let y denote the displacement of the piston from its equilibrium position. Find an expression for the force the gas exerts on the piston as a function of y, and determine the net force on the piston as a function of y. Also calculate the frequency of oscillation of this system. What is the "spring constant" of this gas?

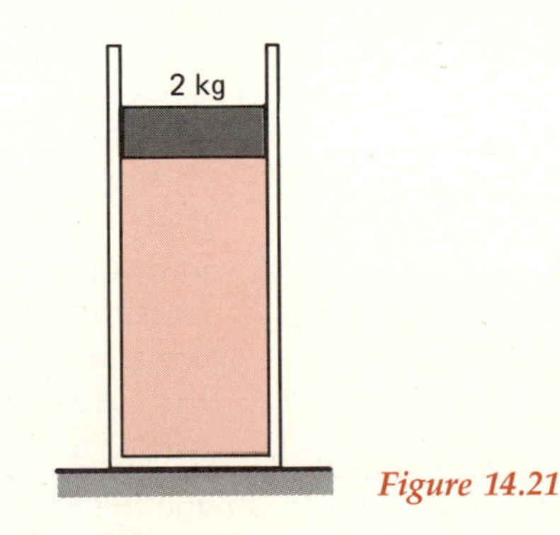

Figure 14.21

•15•

Mechanical Waves

Music, when soft voices die,
Vibrates in the memory

Percy Bysshe Shelley

15.1 Introduction

In this and later chapters we shall be concerned principally with waves, their description, propagation, generation, and detection, and with the various phenomena characteristic of waves, such as interference and diffraction. First, we consider *mechanical vibrations* and *sound waves.* Later we shall examine *electromagnetic waves,* and subsequently (Chapter 30) the evidence that objects we normally regard as particles, such as atomic nuclei and electrons, display wave characteristics and are represented mathematically by "wave functions."

What, then, is a wave? We think we know the answer, having observed waves on the surface of water and watched the circular ripples spread outward as a raindrop strikes a puddle. So we have in mind some sort of undulation of a medium, traveling at some speed, giving rise to a regular, periodic up-and-down motion such as that of a piece of wood floating on a lake. Its displacement is approximated by an expression of the form of Equation (14.3):

$$y(t) = A \sin (2\pi ft + \phi) \tag{14.3}$$

Although we shall, in fact, examine almost exclusively such oscillatory behavior, this is not really necessary for wave propagation. For example, with a flick of the finger, one can initiate a pulse at one end of an array of dominoes that travels down the line and soon tumbles the last. Or I could initiate on a stretched string a single pulse that propagates down the string to the other end; there, if you were holding the end, you could not but move your hand in response to this pulse.

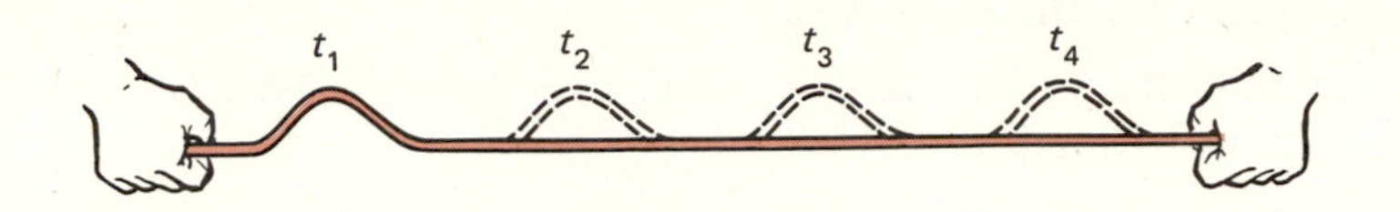

Figure 15.1 *A pulse propagates along a stretched string at constant speed. When it has traveled the full length and reached the end at the right, the hand of the person holding the string will respond to the arrival of the pulse with an upward motion.*

These simple examples focus attention on a very important and fundamental aspect of wave propagation. *It is a mechanism for transmitting energy between two points in space which does not require the physical translocation of material.* If I throw a ball at you and you catch it, your hand recoils because it must absorb the energy and momentum I have imparted to the ball. This energy is carried by the ball as it moves between the two points. But when your hand reacts to the pulse on a string, the section at the other end that I flicked to initiate the pulse never left my hand. It is this remarkable feature of wave propagation that lends it such tremendous importance in our lives. Wave propagation allows us to communicate with each other very rapidly, using for short distances sound waves that travel at a speed of about 330 m/s in air, and for long distances, electromagnetic waves that travel at 300,000,000 m/s.

15.2 Sinusoidal Waves

Although a wave need not have the shape of a sine function, we shall nevertheless restrict our attention mostly to sine waves. This may seem an arbitrary, limiting choice. In fact it is not, for once we know how sine waves behave and propagate, we can predict the behavior of waves of any shape.

Figure 15.2 shows a continuous sine wave at a certain instant. To describe such a wave and discuss its properties, we recall some terms defined already and introduce a few more.

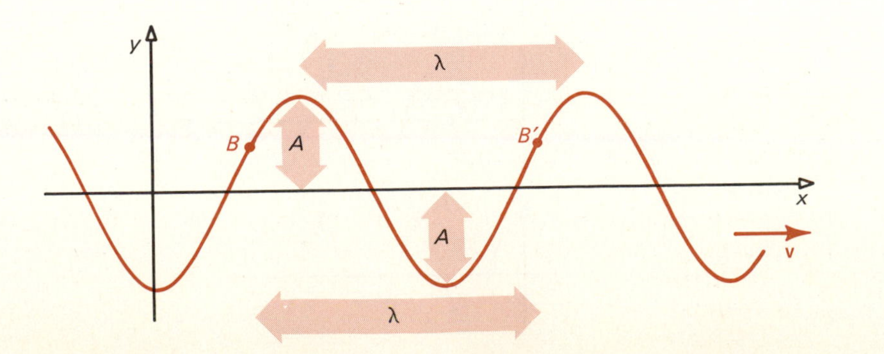

Figure 15.2 *A sinusoidal wave propagating with velocity* **v** *in the x direction. The wave's amplitude A and wavelength λ are indicated.*

DEFINITION: The *amplitude A* of a sine wave is the maximum excursion of the wave from the mean. The *wavelength* λ (Greek letter *lambda*) is the shortest distance between two points on the wave that have identical characteristics.

The amplitude and wavelength are indicated in Figure 15.2. The wavelength is the distance between two adjacent crests, or troughs, or any two equivalent points such as B and B'.

If we focus attention on a particular point in space, such as point B, then as the wave progresses to the right, we note that this point first moves down, reaches a trough, moves up, reaches a peak, then back down, and so on. One complete periodic motion constitutes one *cycle*.

DEFINITION: The frequency f is the number of cycles per unit time. The period T is the time per cycle.

Frequency and period are each other's reciprocals. That is, frequency and period are related by

$$T = \frac{1}{f} \qquad f = \frac{1}{T} \tag{15.1}$$

In Figure 15.3, we show a wave as it appears at successive intervals of time. At time $t = 0$, it appears as in Figure 15.3(*a*). If this wave propagates to the right, in the positive x direction, then slightly later it will be in the position of Figure 15.3(*b*), and so on, until it attains the configuration of Figure 15.3(*i*), which is indistinguishable from the one in Figure 15.3(*a*). During this time, some point such as B has completed one cycle of its oscillatory motion. Consequently, the time elapsed while the wave moved from the position of Figure 15.3(*a*) to that of Figure 15.3(*i*) was one period, T. In that time T, the wave has advanced one full wavelength λ. The *speed of propagation v* of the wave is the distance traveled by the

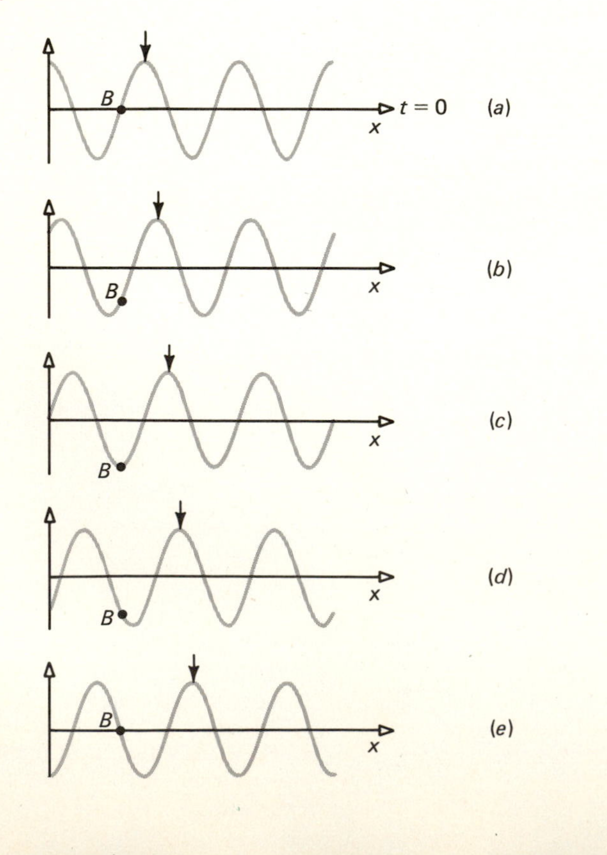

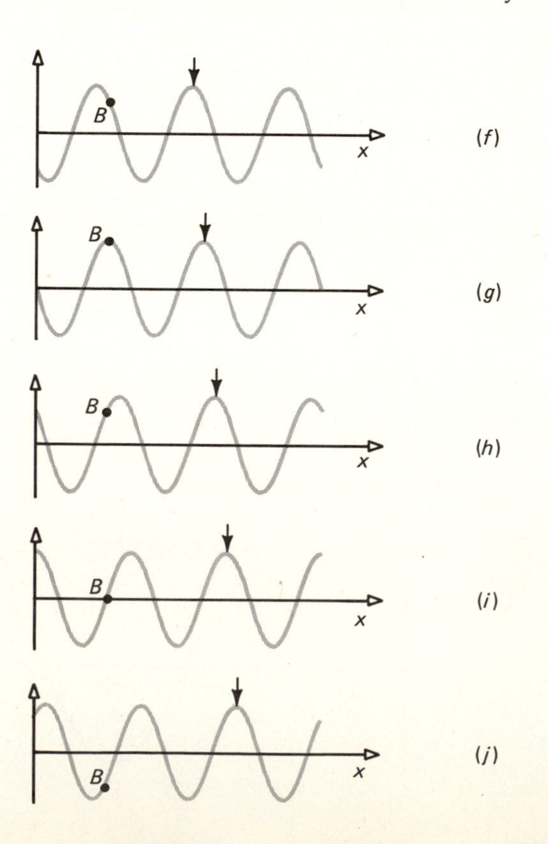

Figure 15.3 *Series of "snapshots" showing a sinusoidal wave propagating to the right. Note that the crest, marked by an arrow, shifts its position as time passes; it moves exactly one wavelength while point B moves through a complete cycle—first down to a minimum, then back up through zero to a maximum, and back to zero displacement—as shown in (i).*

wave divided by the time required for this displacement. Hence

$$v = \frac{\lambda}{T} = f\lambda \qquad (15.2)$$

This simple relation between wave velocity, wavelength, and frequency is valid for all kinds of waves regardless of their physical nature and the mechanism by which they are generated or propagated.

● ***Example 15.1*** What is the wavelength of a sound wave in air at 0 °C if the frequency of the sound is 440 Hz (middle A)?

The speed of sound in air at 0 °C is 330 m/s. From Equation (15.2), we find

$$\lambda = \frac{v}{f} = \frac{330 \text{ m/s}}{440 \text{ s}^{-1}} = 0.75 \text{ m}$$ ●

In Chapter 14 we described the oscillatory motion of a simple pendulum and of a mass supported by a spring; you may have noticed that point B of Figures 15.2 and 15.3 has a pattern of motion just like that of the mass of Figure 14.6. We can therefore use the same mathematical formalism we employed in Chapter 14, and express the displacement of a point of the medium by

$$y = A \sin\left(2\pi\frac{t}{T} + \phi\right)$$
$$= A \sin(2\pi ft + \phi) = A \sin(\omega t + \phi) \qquad (15.3)$$

where A is the amplitude of the wave, ω the angular frequency, and ϕ a phase factor that depends on the choice of zero for the time t.

The spatial dependence of the displacement at a given moment is represented by

$$y = A \sin\left(2\pi\frac{x}{\lambda} + \beta\right) = A \sin(kx + \beta) \qquad (15.4)$$

where $k = 2\pi/\lambda$ is called the *wave number*. The phase angle β now depends on the choice of spatial origin. Notice that whereas the displacement progressed through one complete cycle as the time increased by one period, it now progresses through one complete cycle as x increases by one wavelength.

The mathematical expression that describes the displacement due to a traveling wave is, however, neither Equation (15.3) nor (15.4), which are functions of time only or position only. The displacement of a traveling wave is a function of time *and* position.

To see what the form of that expression must be, we consider a disturbance, represented at $t = 0$ by $y = f(x)$, which propagates in the positive x direction at the speed v. At some later time t, this disturbance, initially centered about $x = 0$, though retaining its shape, has shifted its center to $x = vt$, as shown in Figure 15.4. Consequently, if $y = f(x)$ at $t = 0$, then at the later time t, $y = f(x - vt)$. If $f(x)$ is a sine function, and if we set $v = f\lambda$, we obtain as the general expression for a traveling sine wave

$$y(x, t) = A \sin\left[2\pi\left(\frac{x}{\lambda} - \frac{t}{T}\right)\right] = A \sin(kx - \omega t) \qquad (15.5)$$

If the wave propagates in the negative x direction,

$$y(x, t) = A \sin(kx + \omega t) \qquad (15.5a)$$

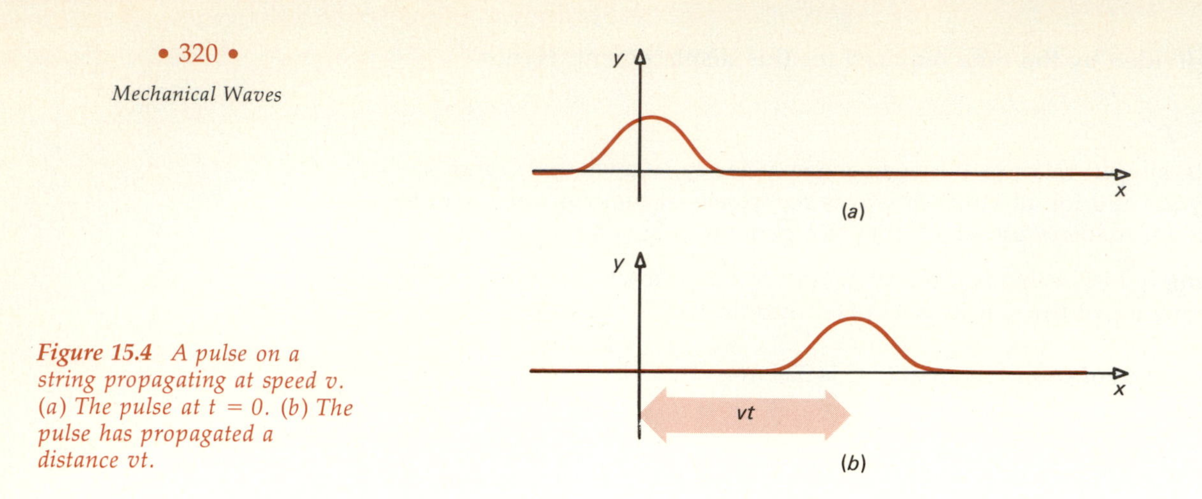

Figure 15.4 A pulse on a string propagating at speed v. (a) The pulse at $t = 0$. (b) The pulse has propagated a distance vt.

● ***Example 15.2*** What are the amplitude, propagation velocity, and frequency of a wave whose displacement is given by

$$y = 0.2 \sin (0.4\pi x + 14\pi t) \text{ m}$$

Comparing this expression with Equation (15.5), we see that the wave amplitude is 0.2 m; the wave number is

$$k = 0.4\pi = \frac{2\pi}{\lambda}$$

Hence,

$$\lambda = \frac{2\pi}{0.4\pi \text{ m}^{-1}} = 5 \text{ m}$$

Similarly,

$$\omega = 2\pi f = 14\pi \text{ s}^{-1} \qquad f = 7 \text{ Hz}$$

Last, the velocity of propagation is given by $v = f\lambda$,

$$v = (7 \text{ Hz})(5 \text{ m}) = 35 \text{ m/s}$$

and is in the negative x direction. ●

Do amplitude, frequency, and wavelength really render a full description of a wave? Not quite. We must specify not only the direction and speed of propagation but also the direction in space along which the vibrations occur. For example, you can take a string stretched along the x direction and wiggle it up and down, or horizontally, or in some arbitrary direction in the plane perpendicular to x. Or you can generate a wave on a Slinky by moving the end sideways, as with the string; but you could also generate a wave by moving the end back and forth in the direction of the Slinky. In that case you will be generating compressional waves.

We use the term *polarization* to characterize the direction of the displacement of the medium. In the case of waves on a string, the wave is necessarily *transversely polarized*—the vibration must be in a plane perpendicular (transverse) to the direction of propagation. The Slinky, however, can also support *longitudinally polarized* waves, commonly called longitudinal, or compressional, waves.

For longitudinal waves, the direction of polarization is specified by the direction of propagation. For transverse waves, on the other hand, an

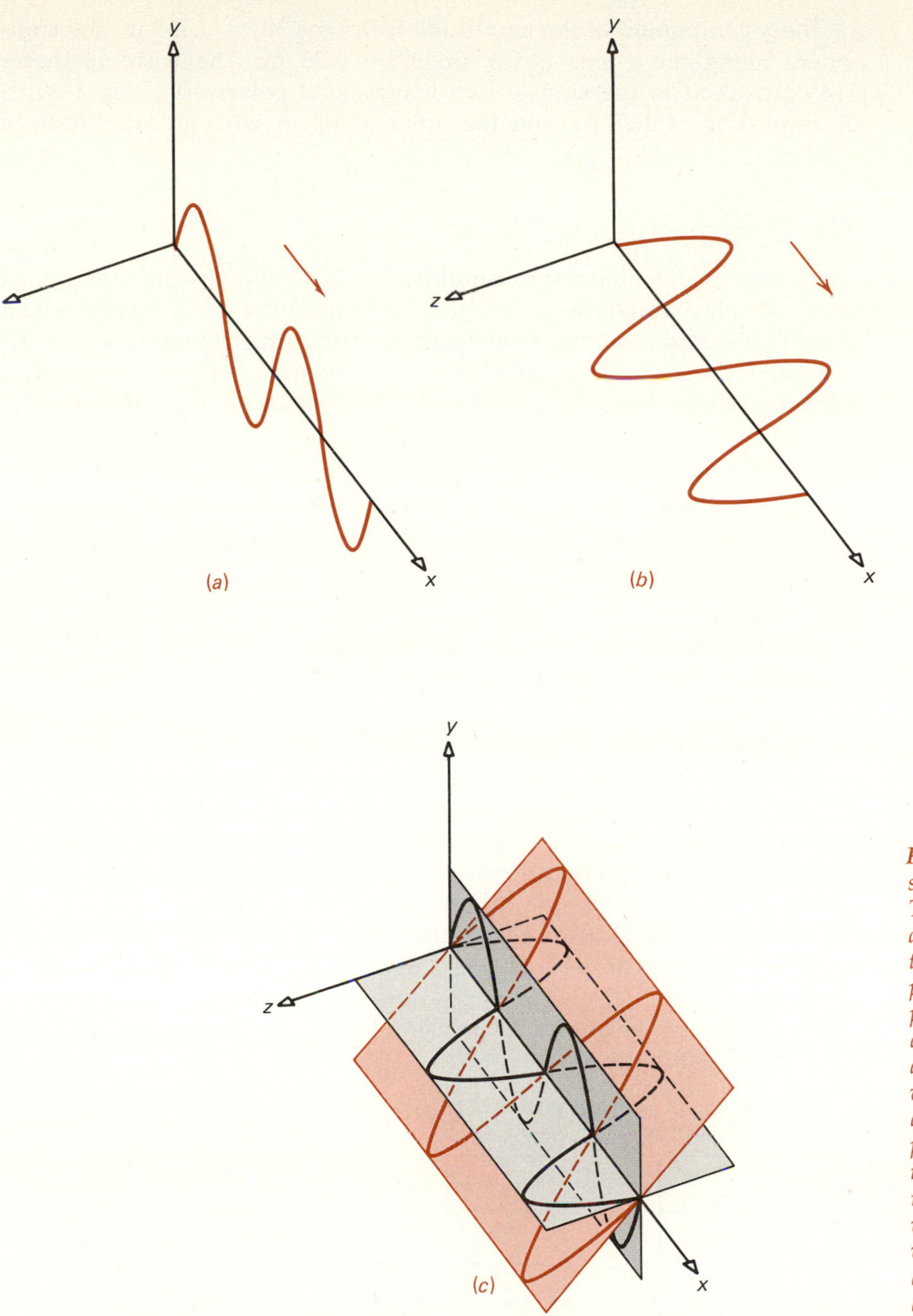

Figure 15.5 A wave on a string is a transverse wave. That is, the direction of the displacement is at right angles to the direction of wave propagation. (a) The wave propagates along the x direction and is polarized along the y direction. (b) The wave is polarized along the z axis. (c) A transverse wave polarized along any direction in the y-z plane can be represented as a sum of two waves of the same wavelength, one polarized along y, the other polarized along z.

infinite number of polarization directions are possible. However, just as an arbitrary vector in a plane can be reduced to a sum of two components, so a transverse wave with arbitrary polarization direction can be represented as a sum of two transverse waves of proper amplitudes and identical frequencies, polarized along two mutually perpendicular directions.

● ***Example 15.3*** A wave on a string stretched along the x direction has an amplitude of 0.5 m. The vibrations make an angle of 30° with the y axis. Decompose this transverse wave into two components oscillating along the y and z axes.

The y component of the amplitude is $0.5 \cos 30° = 0.433$ m; the component along the z axis is $0.5 \sin 30° = 0.25$ m. The wave is therefore equivalent to the sum of two waves, one polarized along z, with an amplitude of 0.25 m, and the other along y, with an amplitude of 0.433 m. ●

15.3 Superposition of Waves

Every day we are subjected to a multitude of waves—several musical instruments play simultaneously, emitting a variety of *sound* waves; a host of radio, TV, and radar transmitters fill the air with electromagnetic radiation at all times. To describe the resulting displacements of the medium due to these simultaneous disturbances, we use the *principle of superposition*.

SUPERPOSITION PRINCIPLE: The displacement at a given point in space and time due to the simultaneous influence of two waves is the vector sum of the displacements due to each wave acting independently.

15.3 (a) Superposition of Two Waves of Identical Wavelengths and Amplitudes

The result of superposition of two waves of identical wavelengths and amplitudes depends on their relative phase. Figure 15.6(a) shows two identical sine waves that are in phase; the result of their superposition is another sine wave of the same wavelength with twice the amplitude. At the other extreme, Figure 15.6(d), the same two waves exactly cancel each other because their phase difference is 180°—they are out of phase with each other. When the phase difference is intermediate between 0° and 180°, as in Figures 15.6(b) and (c), the resultant wave still has the same wavelength, but its amplitude is less than $2A$ and depends on the phase difference.

Figure 15.6 *Superposition of two waves of equal amplitude. The two waves are shown in light color and grey. The resultant, shown in color, depends on the phase difference between the two component waves. (a) The waves are in phase. (b) The phase difference is 90°. (c) The phase difference is 120°. (d) The phase difference is 180°.*

We can derive the analytic expression for the resultant wave using the trigonometric identity

$$\sin \alpha + \sin \beta = 2 \sin [\tfrac{1}{2}(\alpha + \beta)] \cos [\tfrac{1}{2}(\alpha - \beta)] \qquad \textbf{(15.6)}$$

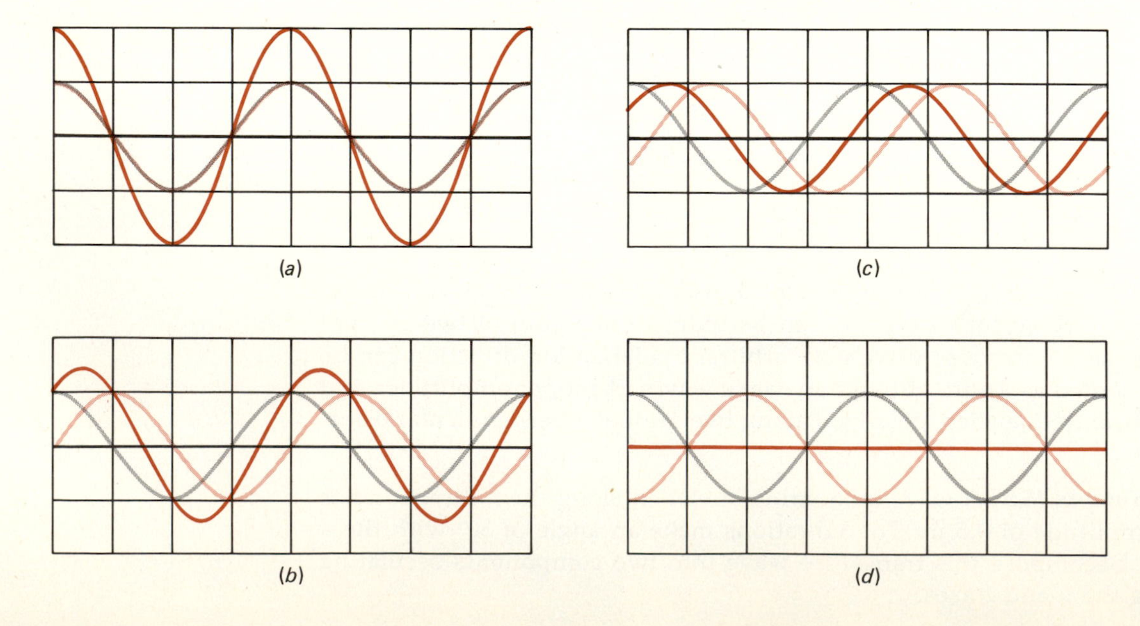

The resultant of two waves of equal amplitude, wavelength, and frequency, given by

$$y_1 = A \sin (kx - \omega t) \qquad \text{and} \qquad y_2 = A \sin (kx - \omega t - \phi)$$

is therefore

$$y = y_1 + y_2 = 2A \sin \left(kx - \omega t - \frac{\phi}{2}\right) \cos \frac{\phi}{2} \tag{15.7}$$

As expected, the resultant wave has the same wavelength and frequency as the two constituents; its amplitude is $2A \cos(\phi/2)$ and depends on the phase difference between the two waves (see Figure 15.6).

15.3 (b) Superposition of Waves of Different Wavelengths and Amplitudes

Waves need not and generally do not have the form of sine functions. Complex waveforms are, however, constituted of a number of sine waves and can be decomposed into sine waves of suitable amplitudes, wavelengths, and phases. The response of a system such as an amplifier, a loudspeaker, or a human ear to stimulus by a complex waveform can be analyzed from its response to the constituent sine waves.

For example, the superposition of traveling sine waves whose wavelengths are λ_0, $\lambda_0/3$, $\lambda_0/5$, etc. (frequencies f_0, $3f_0$, $5f_0$, . . . , where $f_0 = v/\lambda_0$), and whose amplitudes are A_0, $A_0/3$, $A_0/5$, . . . , combine to yield a square wave, as indicated in Figure 15.7. The phase relation between the constituent waves is critical, as demonstrated in Figure 15.8; this complex waveform is the result of the superposition of the first three waves that contribute to the square wave of Figure 15.7, with a 90° phase displacement between the fundamental and the third harmonic and a 180° phase displacement between the fundamental and fifth harmonic.

The two new terms—*fundamental* and *harmonic*—are commonly used in describing temporally periodic patterns. The *fundamental frequency* is defined as the lowest frequency characteristic of a repetitive temporal pattern; that is, if the pattern of displacement first repeats itself exactly after an interval of T_0 s, the fundamental frequency is $f_0 = 1/T_0$ Hz. The *nth harmonic* is the wave whose frequency is nf_0. Thus, the square wave contains only odd harmonics.

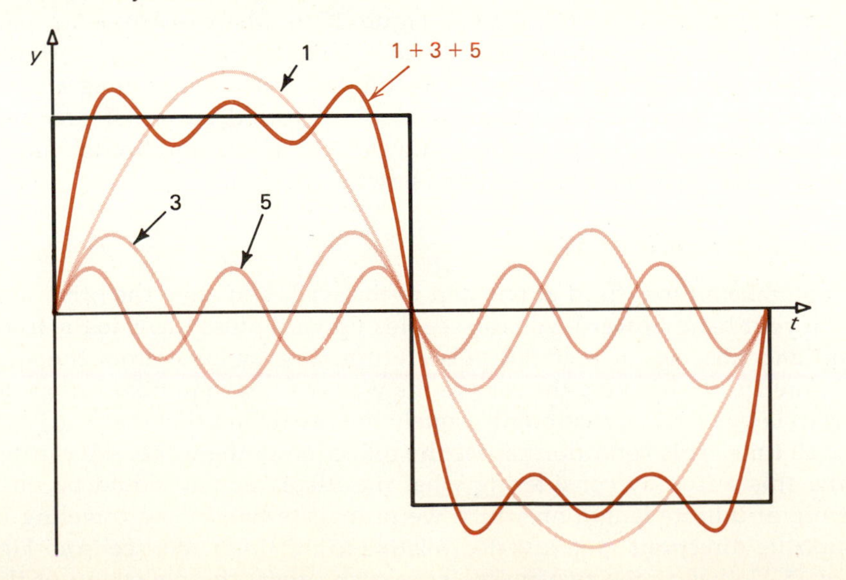

Figure 15.7 A square wave and the first three odd harmonics whose superposition approximates the square wave. Use of higher harmonics improves this synthesis.

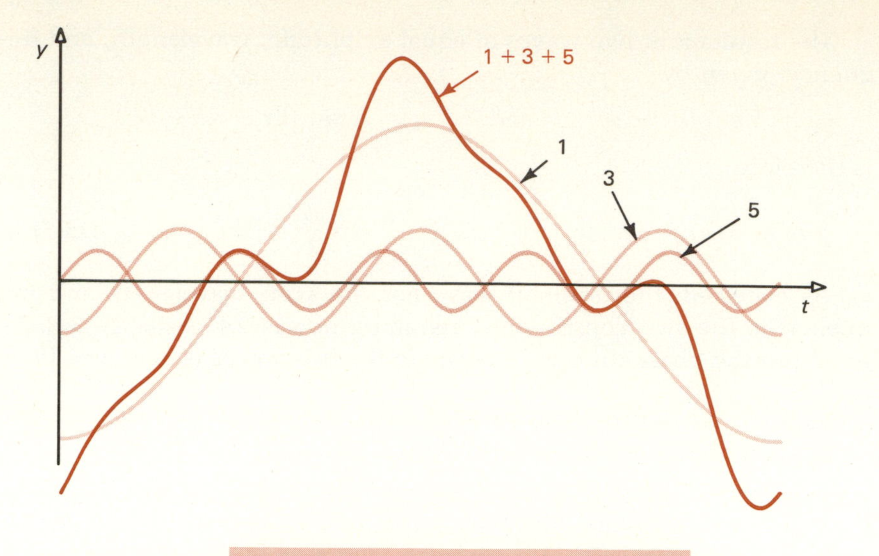

Figure 15.8 The three harmonics of Figure 15.7, with different phase relations, combine to form a wave with a distinctly different shape.

(a)
(b)
(c)
(d)

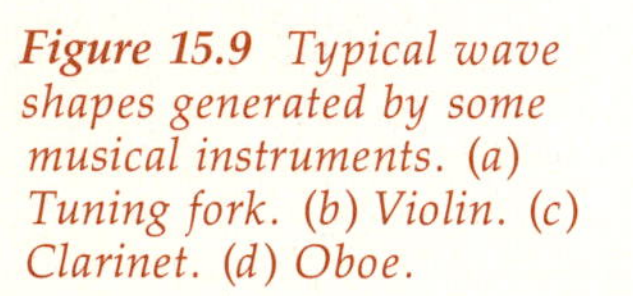

Figure 15.9 Typical wave shapes generated by some musical instruments. (a) Tuning fork. (b) Violin. (c) Clarinet. (d) Oboe.

It is the relative amplitude and phase of the various harmonic components of sound waves produced by different musical instruments—violins, flutes, bassoons, clarinets—that give each instrument its characteristic tone quality and allows us to identify the instrument even when each plays a tone of identical pitch (Figure 15.9). Many instruments, particularly percussion instruments, produce sounds that contain overtones whose frequencies are higher than the fundamental but not integral multiples of it. These instruments produce not sound waves of precisely periodic character but complex sound pulses, whose amplitude usually rises rapidly and then decays more slowly with time.

15.4 Reflection of Waves

If you take a rope, fixed at one end to the wall, and snap the other end with your hand upward, you can see this upward pulse travel to the fixed end and observe an inverted pulse return to your hand from the wall (Figure 15.10). By tying the rope to the wall, you have imposed a *boundary condition* on it at that point, namely the restriction that there $y(t) = 0$ for all time. This condition causes the reflection of the wave. We can see how this arises by considering what the displacement would be on a string of infinite length on which we propagate two pulses traveling in opposite directions, one inverted relative to the other. We see from Figure 15.11 that as the two pulses pass each other, the deflection of the

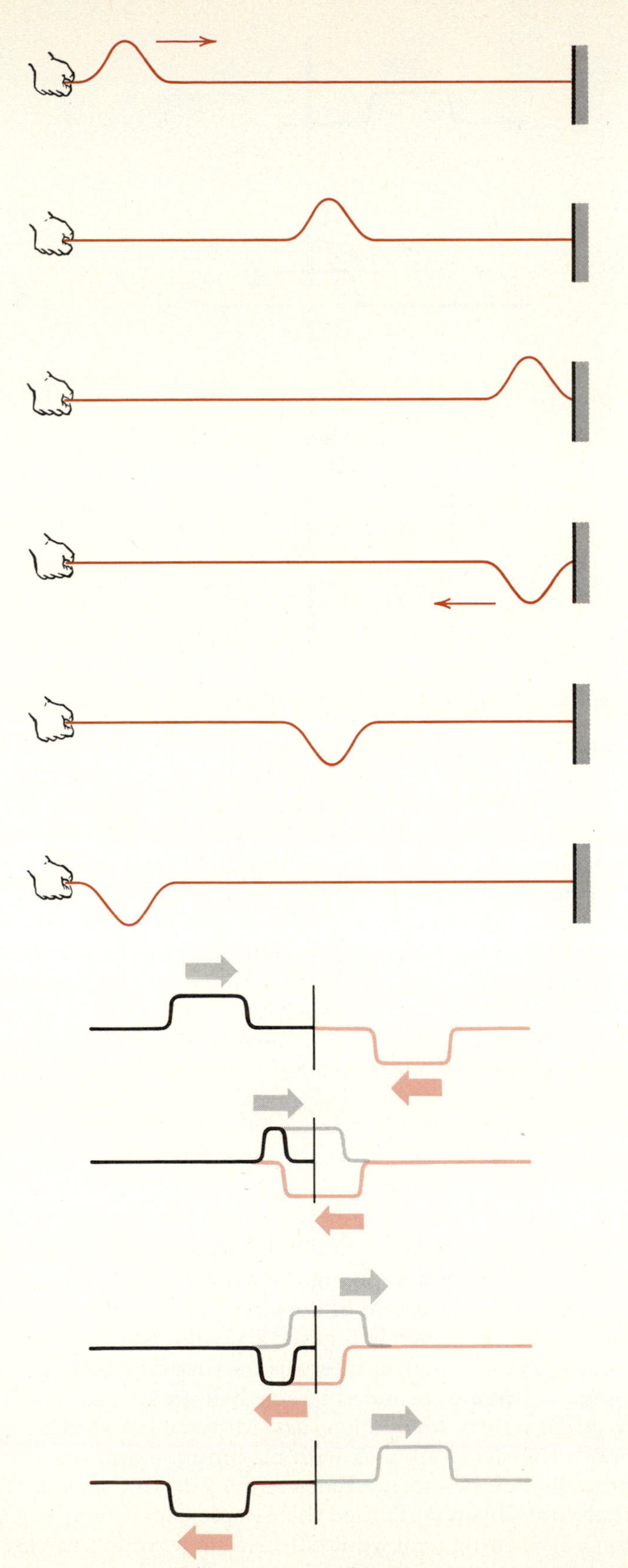

Figure 15.10 A pulse propagating along a string is reflected at the fixed end. The reflected pulse is inverted relative to the original pulse.

Figure 15.11 Reflection of a wave pulse at a fixed point can be represented by the superposition of two traveling pulses of opposite polarity that are propagating along the string in opposite directions.

string at point A is always zero. Thus the effect on wave propagation of fixing point A is equivalent to the presence of the reflected wave on the infinite string.

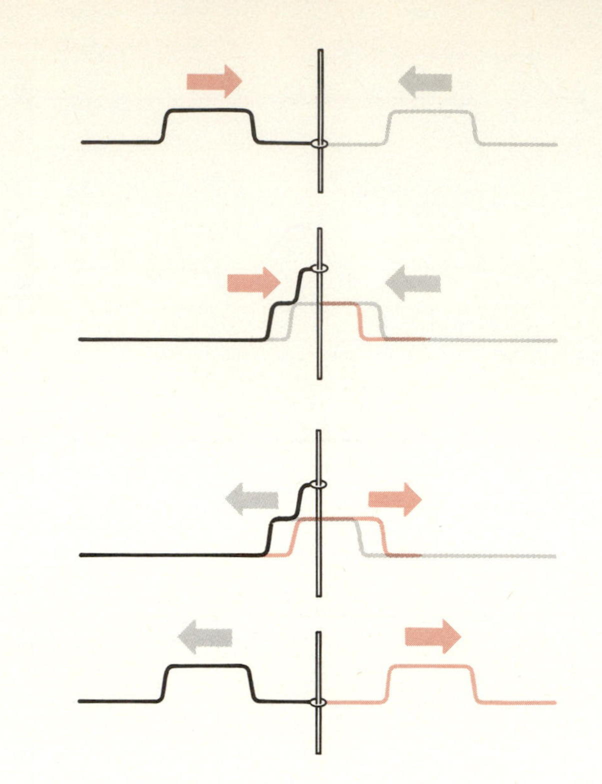

Figure 15.12 *If a string of finite length is terminated so that its end is free to move transversely, a wave pulse is again reflected at this termination. In this case, however, the reflected pulse is not inverted but adds to the displacement at the termination; consequently, the displacement is twice the pulse amplitude at this point. Again, reflection can be represented by the superposition of two waves traveling in opposite directions along an infinitely long string.*

Although this particular boundary condition leads to an inverted reflected wave, inversion is not the general rule. If we terminate the rope with a light ring that is free to slide on a vertical smooth post (Figure 15.12), that end of the string is no longer constrained in its vertical motion. Indeed, it is even free of the vertical pull that would act on it if the rope were extended beyond that point in the normal manner. This termination produces a reflected wave that is in phase with the incident, and the displacement at the "free" end is then twice the amplitude of the incident wave. The situation is analogous to that encountered by surface waves on water that strike a wall, such as a breakwater or steep cliff. At that wall the crests and troughs are almost twice as great as those of the waves further out at sea, a feature familiar to anyone who has lived near the seashore.

15.5 Standing Waves

Suppose that, instead of a single pulse, we send a continuous sine wave down the string of Figure 15.11. At point A we shall again generate an inverted reflected wave, and applying the superposition principle we can determine the net displacement of the string at any point and any time. Such a construction is shown in Figure 15.13, and from this it is apparent that not only at A but at regular intervals the displacement is *always* zero. These points, which are separated by one half-wavelength on the string, are called nodal points, or simply *nodes.* Midway between the nodes, at the *antinodes,* the string vibrates with maximum amplitude.

Note that the nodes and antinodes remain in fixed location along the string. Therefore, if you performed this simple experiment and observed the motion of the string, you would discern no traveling waves. The two traveling waves, the incident and the reflected, have superposed, thereby creating an oscillation of the string such that the wave pattern remains fixed in space. We call this pattern a *standing wave.*

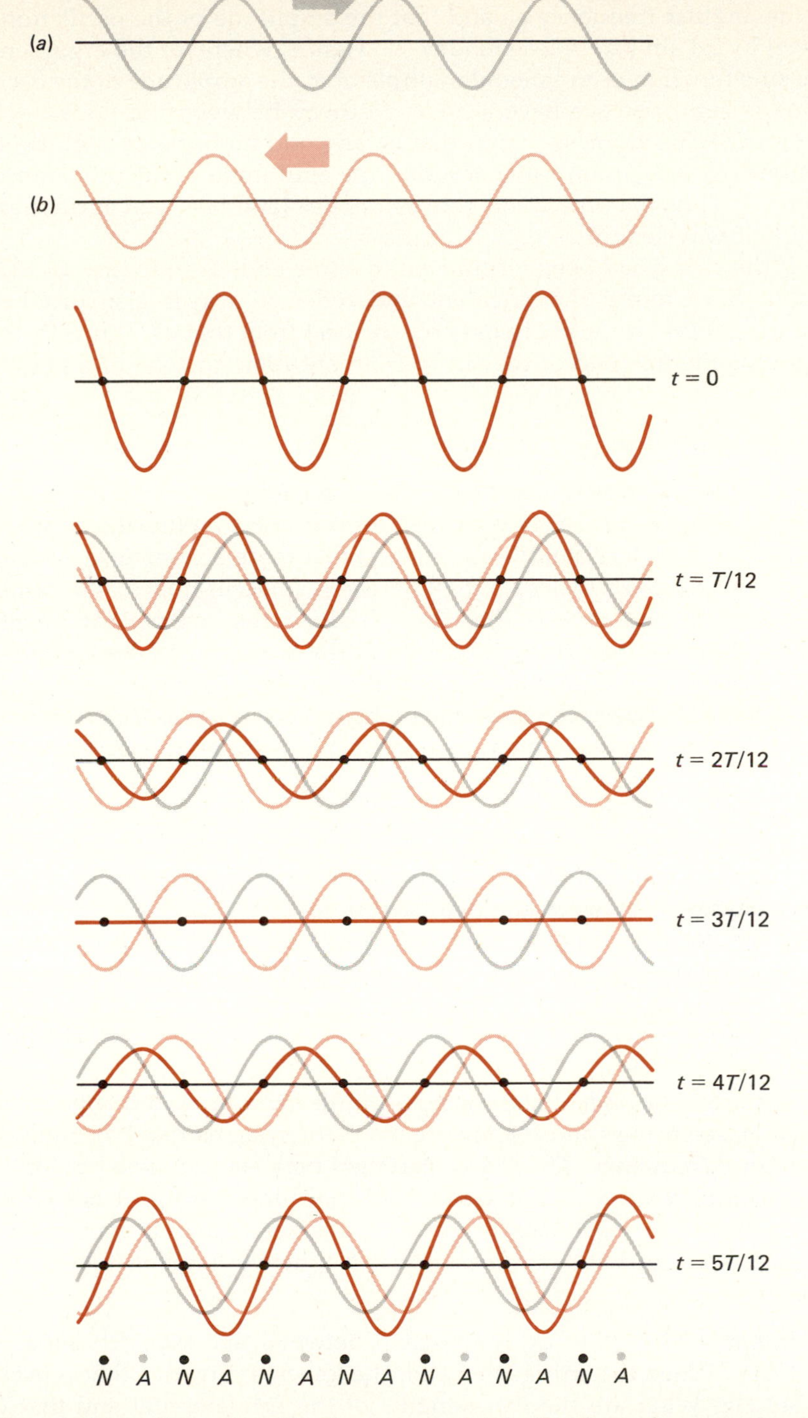

Figure 15.13 Two sine waves of equal amplitude and wavelength, (a) and (b), traveling in opposite directions superpose to give a standing wave. At $t = 0$, the two traveling waves are not momentarily in phase but displaced by one-eighth of a wavelength. Note that the distance between adjacent nodes of the standing wave is exactly $\lambda/2$, where λ is the wavelength of the traveling waves. The letters N and A mark the nodal and antinodal points.

By applying Equation (15.6), we can obtain the formal expression for the standing wave produced by two waves of identical wavelength but propagating in opposite direction. The two waves whose resultant we want are

$$y_1 = A \sin (kx - \omega t) \qquad \text{and} \qquad y_2 = A \sin (kx + \omega t)$$

corresponding to waves propagating in the $+x$ and $-x$ directions, respectively. The resultant is therefore

$$y = y_1 + y_2 = 2A \sin (kx) \cos (\omega t) \tag{15.8}$$

We can see from Equation (15.8) that the displacement oscillates in time at the angular frequency ω, and that the amplitude of the oscillations is given by $2A \sin(kx) = 2A \sin(2\pi x/\lambda)$. Hence, wherever the argument of the sine function is an integral multiple of π, the amplitude of the oscillations is zero, and we have a node. Midway between the nodes, when $x/\lambda = \pi/4, 3\pi/4, 5\pi/4, \ldots$, that is, any odd multiple of $\pi/4$, the sine assumes its maximum value of 1, and the amplitude of the oscillations is then $2A$. Thus the separation between nodes (and between antinodes) is one half-wavelength.

If the string had been terminated in a free end, as in Figure 15.12, we would have found that incident and reflected waves also produce a standing wave, differing in only one respect from that of Figure 15.13. In that case the free end of the string is an antinode instead of a node.

15.6 Resonances

Let us now take a string of length L and fix *both* ends. If we try to establish standing waves on this string, for example by plucking it at some point, the boundary conditions demand that there be a node at each end. But as we just saw, the separation between adjacent nodes of a standing wave is $\lambda/2$, where λ is the wavelength of the component traveling waves. To meet our new boundary conditions, we must therefore be able to fit exactly an integral number of half-wavelengths into the distance L, the distance between the fixed ends. The traveling waves that combine to give the standing wave must therefore have wavelengths given by

$$\lambda_n = \frac{2L}{n} \qquad n \text{ an integer} \tag{15.9}$$

Only if Equation (15.9) is satisfied will an integral number of half-wavelengths fit between the two fixed points.

If the velocity of propagation of a wave on the string is v, then the possible wavelengths given by Equation (15.9) correspond to frequencies

$$f_n = \frac{v}{\lambda_n} = \frac{nv}{2L} = nf_0 \qquad f_0 = \frac{v}{2L} \tag{15.10}$$

These are the *resonance frequencies* of this system. A simple pendulum and a mass-spring combination are resonant systems that have only one resonance frequency. The string stretched between two fixed points is a more complicated resonant system but more interesting; it has an infinite number of resonance frequencies that constitute a harmonic sequence, starting with the *fundamental* frequency f_0 given by Equation (15.10).

● ***Example 15.4*** A string is stretched between two supports separated by 40 cm. When the string is bowed, the lowest frequency that is excited is 240 Hz. What are the wavelengths of the fundamental and first few overtones, and what is the velocity of propagation of transverse waves on the string?

The two ends of the string are fixed and are therefore nodes. Hence, the wavelength of the fundamental oscillation is

$$\lambda_0 = 2L = 80 \text{ cm}$$

The overtones, in this case, are all the harmonics whose wavelengths are

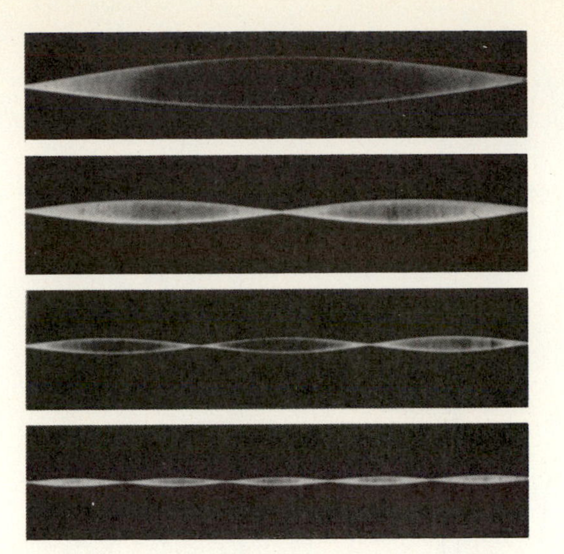

Figure 15.14 Standing waves on a string of fixed length. The fundamental corresponds to the superposition of traveling waves whose wavelength is twice the distance between the fixed ends of the string.

$$\lambda_n = \frac{\lambda_0}{n} = \frac{80 \text{ cm}}{n} = 40 \text{ cm } (n = 2), 26.7 \text{ cm } (n = 3), \cdots$$

and whose frequencies are related to the fundamental frequency f_0 by Equation (15.10). Since the fundamental frequency is 240 Hz, the propagation velocity of waves is

$$v = f\lambda = (240 \text{ Hz})(80 \text{ cm}) = 192 \text{ m/s}$$

15.7 Beats

Musicians who play instruments that must be tuned frequently—violin, cello, guitar, harpsichord—are familiar with the phenomenon of *beating*, the slow, regular variation in intensity when two sound sources of nearly the same frequency radiate simultaneously. Beats are another example of wave superposition that we can readily understand.

In Figure 15.15, we show two waves, of equal amplitude and nearly the same wavelength, and their resultant. At A, the two waves are in phase and the resultant wave has an amplitude that is the sum of the amplitudes of the two waves. However, some distance farther along, the two waves are no longer in phase, because their frequencies and wavelengths differ slightly. At B they are out of phase, and consequently the resultant amplitude is minimum, thereafter rising again to another maximum at C, and so on.

We represent the displacement of the two constituent waves at some fixed point, say at $x = 0$, by

$$y_1 = A \sin \omega_1 t \qquad \text{and} \qquad y_2 = A \sin \omega_2 t$$

From Equation (15.6) we obtain

$$y = 2A \sin \left[\left(\frac{\omega_1 + \omega_2}{2}\right) t\right] \cos \left[\left(\frac{\omega_1 - \omega_2}{2}\right) t\right]$$

If $\omega_1 \simeq \omega_2$, the displacement oscillates at an average frequency $\langle\omega\rangle = (\omega_1 + \omega_2)/2$ with an amplitude that is time-dependent and varies as

$$\cos \left[\left(\frac{\omega_1 - \omega_2}{2}\right) t\right]$$

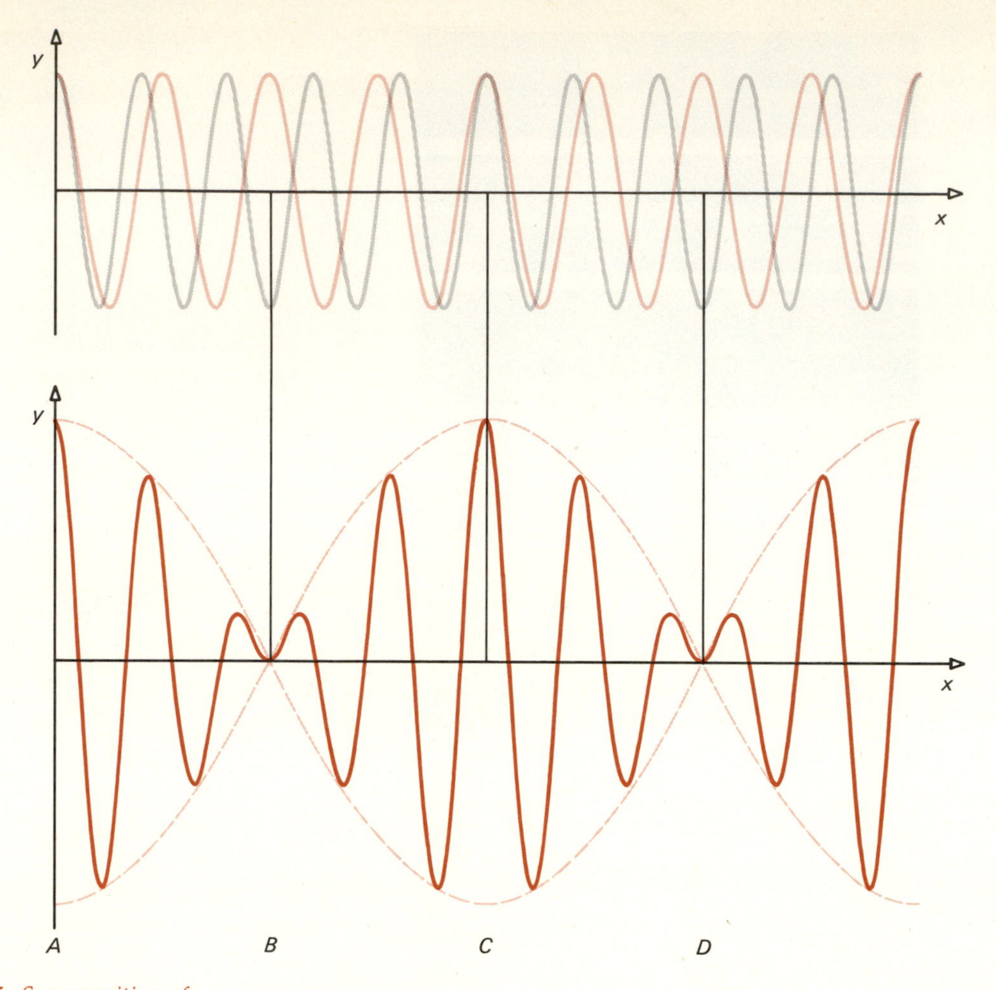

Figure 15.15 Superposition of two waves of equal amplitude but slightly different wavelengths. At B, the two waves are 180° out of phase and cancel on superposition. At A and C, the waves are in phase and the resultant amplitude is twice the amplitude of the individual component waves. At D, the waves cancel again. The result is a wave of wavelength equal to the average wavelength of the two component waves, whose amplitude slowly and regularly oscillates between a maximum and zero.

where this amplitude variation is slow compared to the average frequency. The oscillation at the average frequency will appear with maximum amplitude whenever the cosine function is $+1$ *or* -1. Hence the beat frequency is

$$f_{\text{beat}} = 2\left[2\pi\left(\frac{\omega_1 - \omega_2}{2}\right)\right] = 2\pi\omega_1 - 2\pi\omega_2 = f_1 - f_2$$

Consequently, as this wave pattern travels through the air and strikes our eardrums, we hear a sound whose frequency is the average of the two and whose intensity oscillates at the frequency $f_1 - f_2$.

Piano tuners listen (or did, before electronic tuning aids) for beats—or rather, their absence—when tuning the three strings that should vibrate in almost perfect unison when struck by the hammer. Violinists and cellists rely on fainter beats in tuning their instruments to exact fifths.*

* If a tone has a frequency f, another tone whose frequency is $1.5f$ is said to be one fifth higher in pitch. If f_0 is the frequency of a particular note, the note one octave higher has the frequency $2f_0$, and the note of frequency $3f_0$ is a fifth above that. Hence the third harmonic of the tone f_0 and the second harmonic of the fifth above it have the same frequency. If the two tones are not exactly one fifth apart, these harmonics will generate beats that can be perceived without much effort.

The simple relations between musical harmonics and mathematics was first recognized by the Greek philosopher Pythagoras, who is perhaps better remembered for his famous theorem on right triangles.

15.8 Speed of Propagation of Waves on Strings

We have said nothing yet about the speed of propagation of waves on strings. What physical parameters determine this property?

We can reasonably expect that the mass of the string, its length, and the tension applied to it will affect the speed of wave propagation. Assuming that these are the critical parameters, we now derive an expression for the velocity by dimensional analysis.

We seek an expression of the form

$$[M]^m[L]^\ell[F]^f$$

whose dimension is that of velocity, that is, $[L]/[T]$; here $[F]$ stands for force, and m, ℓ, and f are exponents to be determined.

The dimension of $[F]$ is given by Newton's law, namely,

$$[F] = \frac{[M][L]}{[T]^2}.$$

Thus, the equation that we must satisfy is

$$\frac{[M]^{m+f}[L]^{\ell+f}}{[T]^{2f}} = \frac{[L]}{[T]}$$

It follows that

$$m + f = 0 \qquad \ell + f = 1 \qquad 2f = 1$$

and that therefore,

$$f = \tfrac{1}{2} \qquad \ell = \tfrac{1}{2} \qquad m = -\tfrac{1}{2}$$

Consequently, the expression for the velocity of propagation of a wave on a string must be

$$v = c\sqrt{\frac{FL}{M}} = c\sqrt{\frac{F}{\mu}}$$

where c is an unknown dimensionless constant and μ is the mass per unit length. The constant c, which cannot be determined by dimensional argument, is unity, and the velocity of propagation is

$$v = \sqrt{\frac{F}{\mu}} \qquad \textbf{(15.11)}$$

● ***Example 15.5*** The A string of a violin is found to be flat, resonating at a fundamental frequency of 430 Hz when under a tension of 400 N. How should the tension be adjusted to achieve the correct frequency of 440 Hz?

Since the active length of the string remains fixed as the instrument is tuned, so does the wavelength of the fundamental resonance frequency. To increase the frequency, it is, therefore, necessary to increase the speed of propagation. From Equation (15.10), we can write

$$\frac{v_2}{v_1} = \frac{f_2}{f_1}$$

where v_1 and v_2 are the initial and final wave velocities and f_1 and f_2 the corresponding resonant frequencies. Thus,

$$\frac{v_2}{v_1} = \frac{440}{430}$$

But according to Equation (15.11), the tension F is proportional to the square of the velocity. Consequently,

$$F_2 = F_1 \left(\frac{v_2}{v_1}\right)^2 = (400 \text{ N}) \left(\frac{4.4}{4.3}\right)^2 = 419 \text{ N}$$

● ***Example 15.6*** A popular toy "telephone" consists of two empty tin cans connected with a thin wire. The children using this phone pull the wire taut and set it in vibration by speaking into one can. The vibrations set up at that end are then propagated along the wire to the other end, causing the other tin can to vibrate and generate sound waves. If the wire that is used to connect the cans is 10 m long and weighs 0.01 N, with what tension must the children pull on the cans so that the speed of transmission of sound is as fast along the telephone as in the air?

At room temperature, the speed of sound in air is about 340 m/s. To determine the proper tension, we must first calculate the mass per unit length of the connecting wire. From the data given, we have

$$\begin{aligned}\mu &= \frac{M}{L} \\ &= \frac{(0.01 \text{ N})/(9.8 \text{ m/s}^2)}{10 \text{ m}} \\ &= 1.02 \times 10^{-4} \text{ kg/m}\end{aligned}$$

The proper tension is then, from Equation (15.11),

$$\begin{aligned}F &= \mu v^2 \\ &= (1.02 \times 10^{-4} \text{ kg/m})(340 \text{ m/s})^2 \\ &= 11.8 \text{ N}\end{aligned}$$

a force a bit greater than 2.5 lb.

15.9 Energy of a Vibrating String

We stated at the beginning of the chapter that it is the transmission of energy that lends wave motion its paramount importance and utility. How much energy is associated with the vibrations of a string?

A vibrating string does have more energy than a stationary one under equal tension. We know this from the simple observation that we must do work in plucking or bowing the string to initiate vibrations, and that the string, once plucked, gradually returns to its stationary configuration.

To determine this energy of vibration, we consider a small length Δx of the string. The transverse oscillating motion imparts to Δx a KE given by

$$\Delta\text{KE} = \tfrac{1}{2}mv_y^2 = \tfrac{1}{2}(\mu\ \Delta x)v_y^2 \tag{15.12}$$

where v_y is the instantaneous transverse velocity of the short segment of length Δx and mass $\mu\ \Delta x$. Of course, as with other vibrating systems (see Chapter 14), the KE of the segment Δx is not constant in time but varies between zero and some maximum value. The total energy, however, remains constant (neglecting frictional losses), the system continually transforming PE to KE and vice versa.

Let us consider, then, a section of the string one wavelength long. During one complete period, as the wave progresses a distance λ, each point on the string executes one complete cycle of transverse motion, from zero displacement to the amplitude A, back to zero, to $-A$, and back to zero once again. The total transverse distance traversed by a given point in one period is therefore $4A$. Hence the average speed of motion in the y direction is $\langle v_y \rangle = 4A/T$. Since this is true for every point of the string, an estimate of the average KE of the length λ is

$$\langle \text{KE} \rangle = \tfrac{1}{2}(\mu\lambda)\langle v_y^2 \rangle \simeq \tfrac{1}{2}(\mu\lambda)\langle v_y \rangle^2$$
$$= \frac{8\mu\lambda A^2}{T^2} = 8\mu v f A^2 \qquad \textbf{(15.13)}$$

Equation (15.13) is only approximate; the approximation consisted in setting $\langle v_y^2 \rangle$ equal to $\langle v_y \rangle^2$. This is true for a square wave such as that of Figure 15.7, but not for a sine wave. For a sine wave, an exact calculation leads to the correct result

$$\langle \text{KE} \rangle = \pi^2 \mu v f A^2 \qquad \textbf{(15.14)}$$

which is not vastly different from our approximate result.

For the vibrating string, as for the oscillating mass-spring system, the average KEs and PEs are equal, $\langle \text{PE} \rangle = \langle \text{KE} \rangle$. Hence the total energy of a segment of one wavelength is

$$E = \langle \text{PE} \rangle + \langle \text{KE} \rangle = 2\langle \text{KE} \rangle = 2\pi^2 \mu f v A^2$$

In one second, f wavelengths pass a given point on the string as the wave propagates at the speed v. Therefore, the energy transmitted per second, that is, the power carried by the wave, is f times the energy content in one wavelength:

$$P = fE = 2\pi^2 \mu v f^2 A^2 \qquad \textbf{(15.15)}$$

This has been a somewhat lengthy derivation. Let us see if the result is intuitively reasonable. First, the rate at which energy is transported along the string might be expected to depend on the propagation velocity, and so it does. Second, this energy should depend also on the mass per unit length because, other things being equal, the greater this mass, the greater the average KE. Third, the product of frequency and amplitude must be related to the average transverse speed of vibration: the greater the amplitude the more the transverse distance covered in one cycle; the greater the frequency the faster this distance is traversed. Since the KE depends on the square of the transverse velocity, the appearance of the product f^2A^2 is also understandable.

Our reason for devoting this effort to deriving Equation (15.15) is that the qualitative aspects of this relation retain their validity even when applied to other mechanical waves, specifically sound waves in air. In that case, too, the power transported is proportional to the square of the amplitude, the square of the frequency, and to the wave velocity.

● ***Example 15.7*** A long rope with a mass of 0.2 kg/m is kept under a tension of 80 N. If one end oscillates at a rate of 4 Hz with an amplitude of 15 cm, at what rate is energy transmitted along the rope?

According to Equation (15.11), the velocity of the waves is

$$v = \sqrt{\frac{F}{\mu}} = \sqrt{\frac{80\ \text{N}}{0.2\ \text{kg/m}}} = \sqrt{400\ \text{m}^2/\text{s}^2} = 20\ \text{m/s}$$

The power transmitted by this wave is therefore

$$P = 2\pi^2(0.2\ \text{kg/m})(20\ \text{m/s})(4\ \text{Hz})^2(0.15\ \text{m})^2 = 28.4\ \text{W}$$

That's a fair amount of power; if you try it, you will soon find yourself working up quite a sweat.

Summary

A *wave* is a disturbance in a medium that transports energy without causing a significant translocation of mass as the wave propagates. The shape of a wave is arbitrary. By the superposition principle, any wave form can be synthesized from a sum of sinusoidal waves.

A sine wave is characterized by its *amplitude* A, *frequency* f, *wavelength* λ, and *direction of polarization*. The polarization direction is that direction along which the medium vibrates as the wave moves past, and waves are classified as longitudinal and transverse accordingly.

The expression for a sine wave propagating along the x direction is

$$y = A \sin\left(2\pi\frac{x}{\lambda} - 2\pi ft\right) = A \sin(kx - \omega t)$$

The speed of propagation of a wave is related to its frequency and wavelength by

$$v = f\lambda$$

The phase factor in the mathematical expression for a wave provides information on the relative temporal or spatial displacement between two waves.

The superposition principle states that the displacement of the medium due to the simultaneous influence of two waves is the vector sum of the displacements due to the two waves acting independently.

An *overtone* is any resonance frequency that is higher than the fundamental (lowest) resonance frequency of a system. A *harmonic* is an overtone that is an integral multiple of the fundamental frequency. The nth harmonic is that vibration whose frequency is nf_0, where f_0 is the fundamental frequency.

Standing waves result from the superposition of two waves of identical frequency and wavelength propagating in opposite directions. The position on a standing wave where the displacement is zero is called a *node;* that where the displacement is a maximum is called an *antinode*.

The resonant vibrations on a string whose ends are fixed are standing waves that have nodes at each end of the string. The resonant wavelengths and frequencies of a stretched string of length L are

$$\lambda_n = \frac{2L}{n} \qquad f_n = \frac{nv}{2L} \qquad n \text{ an integer}$$

When two waves of slightly different frequencies superpose, a *beat pattern* is formed. These beats are slow, regular changes in the intensity of the vibration. The *beat frequency* is $|f_1 - f_2|$, where f_1 and f_2 are the frequencies of the constituent waves.

The speed of propagation of transverse waves on a string is given by

$$v = \sqrt{\frac{F}{\mu}}$$

where F is the tension in the string and μ is the mass per unit length of the string.

The power carried by a mechanical wave is proportional to the wave velocity and the square of the frequency and of the amplitude.

Multiple Choice Questions

15.1 An instrument has two strings. Both have the same length and are uniform. One of the strings is under twice as much tension and has twice the mass of the other.

(a) The more massive string has a resonant frequency that equals $\sqrt{2}$ times the frequency of the other.

(b) Both strings vibrate at the same frequency, but the more massive string has a longer wavelength.

(c) The more massive string has a shorter wavelength and a higher vibration frequency.

(d) The frequencies and wavelengths on the two strings are equal, but the velocity of the wave is less on the more massive string than on the other.

(e) Both strings vibrate with the same frequency and have the same wavelength and the same wave velocity.

15.2 The relation between wavelength λ, frequency f, and propagation velocity v of a wave is

(a) $v = f/\lambda$

(b) $f = v/\lambda$

(c) $\lambda = f/v$

(d) $v = \lambda f^2$

(e) None of the above

15.3 A traveling wave passes a point of observation. At this point the time interval between successive crests is 0.2 s.

(a) The wavelength is 5 m.

(b) The frequency is 5 Hz.

(c) The velocity of propagation is 5 m/s.

(d) The wavelength is 0.2 m.

(e) There is not enough information to justify any of these statements.

15.4 The tension in a string stretched between two fixed points is increased. It follows that

(a) the frequency of the fundamental resonance and of all harmonics is increased.

(b) the frequency is unchanged for the fundamental but is increased for all higher harmonics.

(c) the frequency of the fundamental and of all harmonics is decreased.

(d) The wavelength on the string of the fundamental and all harmonics is decreased.

(e) None of the above is a correct statement.

15.5 A wave is described by the equation

$$y = 8 \sin\left[2\pi\left(\frac{x}{20} + \frac{t}{2}\right)\right]$$

where all distances are measured in centimeters and the time is measured in seconds. One can conclude that

(a) the amplitude is 4 cm.

(b) the wavelength is $10/\pi$ cm.

(c) the period is π s.

(d) the frequency is 2 Hz.

(e) the wave is traveling to the left (in the negative x direction).

15.6 A piano tuner compares the fundamental frequency generated by one of the piano strings with the frequency from his tuning fork and hears 4 beats per second. He then tightens the piano string just a bit and hears only 3 beats per second. He concludes that

(a) the string has a higher frequency than the tuning fork and must be tightened more to make the frequencies equal.

(b) the string has a lower frequency than the tuning fork and must be tightened more to make the frequencies equal.

(c) the string has a higher frequency than the tuning fork and must be loosened to make the frequencies equal.

(d) the string has a lower frequency than the tuning fork and must be loosened to make the frequencies equal.

(e) This one observation on the effect of tightening the string does not tell the tuner enough to allow any of these conclusions.

15.7 A sound is produced under water and then propagates to the surface of the water, and some of the sound is transmitted into the air. The velocity of the sound in water is 1450 m/s, the velocity of the sound in air is 330 m/s. As the sound passes from water to air, the effect on the frequency f and the wavelength λ is as follows:

(a) f and λ remain unchanged.

(b) f remains unchanged but λ increases.

(c) f remains unchanged but λ decreases.

(d) f increases but λ decreases.

(e) f decreases but λ increases.

15.8 A string, fixed at both ends, resonates at a fundamental frequency of 120 Hz. What single adjustment, or combination of adjustments, would have the effect of reducing the fundamental frequency to 60 Hz?

(a) Doubling the tension and doubling the length.

(b) Halving the tension and keeping the length fixed.

(c) Halving the tension and doubling the length.

(d) Holding the tension fixed and halving the length.

(e) None of the above.

15.9 A transverse wave is traveling on a string of total mass M, length L, and tension T.

(a) The wavelength of the wave is proportional to L.

(b) The wave velocity depends on M, L, and T.

(c) The frequency of the wave is proportional to the wavelength.

(d) The energy of the wave is proportional to the square root of the amplitude of the wave.

(e) The speed of motion of a point on the string is the same as the velocity of propagation of the wave.

Problems

(Section 15.2)

15.1 A wave has a frequency of 50 Hz and a wavelength of 1 m. What are the period and the velocity of propagation?

15.2 A traveling wave with frequency 10 Hz moves to the right. How long does it take for one complete wavelength to pass a given point?

15.3 As you stand on a pier, you notice that the crest of a wave passes every 2.5 s. If the distance between crests is 3 m, what is the speed of the surface waves?

15.4 Sound waves in air propagate at 330 m/s. What are the wavelengths of sound waves of frequency 20 Hz, 1000 Hz, and 12,000 Hz?

15.5 Standing on a cliff, you watch the ocean waves and notice that a new crest strikes the cliff every 4 s. Comparing the distance between adjacent crests with the length of a 8-m sailboat, you estimate the wavelength of the waves at 10 m. What is the speed of surface waves on the water?

15.6 The average adult can perceive sounds whose frequencies range between 17 Hz and 17,000 Hz. What are the corresponding wavelengths, and how do these compare to dimensions of common objects such as ordinary boxes and pop bottles?

• **15.7** What is the expression for a sine wave that has an amplitude of 5 cm and a wavelength of 0.4 m and that is propagating at 25 m/s along a rope stretched along the z direction? Assume that the direction of propagation is in the positive z direction and that the displacement is in the x direction; the displacement is zero at $t = 0$, $z = 0$, and increases with time at that point.

• **15.8** Write an expression for a sine wave that propagates in the negative z direction at 20 m/s and has a frequency of 12 Hz and an amplitude of 4 cm; the displacement is in the y direction. If at $t = 0$ and $z = 0$ the displacement is 2 cm, what value or values does the phase angle have?

• **15.9** Show that the following expression represents a traveling sine wave, moving in the positive x direction:

$$y = 2.5 \sin \left[2\pi \left(\frac{x}{5} - \frac{t}{4} \right) \right]$$

where x and y are in centimeters and t is in seconds. Sketch y as a function of x for $t = 0$ and $t = 1$ s. What are the wavelength, frequency, period, amplitude, and velocity of this wave?

(Section 15.3)

• **15.10** A traveling sine wave is the result of the superposition of two waves of equal amplitude. If the amplitude of the resultant is 8 cm and that of each of the component waves is 6 cm, what is the phase difference between the two component waves?

• **15.11** Sketch two waves of the same amplitude and wavelength that differ in phase by (a) 60°; (b) 90°; (c) 150°. Using superposition, determine the relative amplitude of the resultant of these waves of amplitude A.

• • **15.12** A taut wire is pulled to one side at its midpoint so that it has the shape of an isosceles triangle. Sketch the first three sine waves whose superposition approximates the shape of the wire.

• • **15.13** Sketch three sine waves whose sum approximates a sawtooth wave.

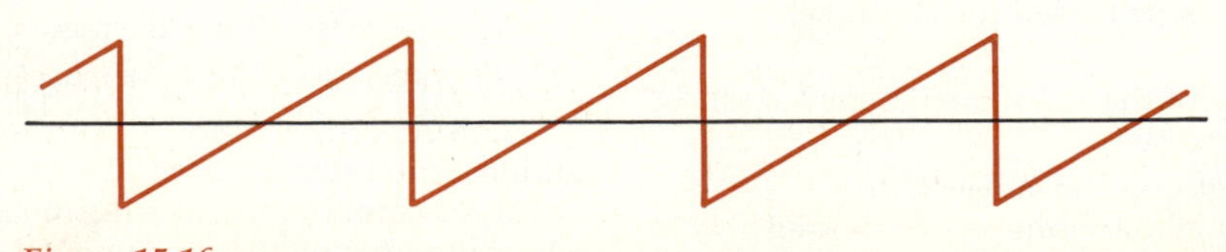

Figure 15.16

(Sections 15.4–15.6)

15.14 Give the expression for the standing wave formed by the traveling wave of Example 15.2 and its reflected wave.

15.15 A standing wave is given by the expression

$y = 0.2 \cos 8.47t \sin 12.8x$

where x and y are in meters and t is in seconds. What are the wavelength and frequency of the two waves that give rise to this standing wave?

• **15.16** A string vibrating at a frequency of 400 Hz has four nodes, including the two at the fixed ends. If the length of the string is 90 cm, what is the speed of a traveling wave on the string?

• **15.17** A string 2 m long is stretched between two posts. If the tension in the string is adjusted so that the lowest resonance frequency is 24 Hz, what is the velocity of wave propagation along the string?

• **15.18** One string of a bass fiddle is 1.4 m long and has a fundamental frequency of 81 Hz. (a) What is the speed of propagation of traveling waves on this string? (b) Where on the fingerboard should you press your finger to produce a sound whose frequency is 109 Hz?

• • **15.19** A string fixed between two points 40 cm apart is plucked at its midpoint; that is, the midpoint is pulled up so that before the string is released, it has the shape of two sides of an isosceles triangle. What is the wavelength of the fundamental frequency of vibration? If the velocity of propagation of waves on this string is 20 m/s, what is the frequency of the fundamental? What is the frequency of the first overtone that is generated by this method of plucking?

(Sections 15.7, 15.8)

15.20 A string instrument is tuned by adjusting the tension of the strings. Suppose that violin X has its A string tuned to 440 Hz and violin Y has its A string tuned to a slightly lower frequency. What is that frequency if beats of frequency 1.5 Hz are heard when both instruments are bowed on the A string?

• **15.21** What should be the percentage change in tension of the A string of violin Y of Problem 15.20 to bring it in tune with violin X?

• **15.22** The E string of a violin is normally tuned to 660 Hz. One of two violins is tuned properly, and as it is played together with the other, a beat of 3 Hz is heard when the two E strings are bowed. What then is the frequency of vibration of the E string of the other instrument if a small reduction in the tension of its E string brings it into tune with the first? What should the percentage reduction in tension be to reduce the beat frequency to zero?

• **15.23** A string is fixed between two supports separated by 0.6 m. If the string is under tension of 25 N, what should its linear density be so that the lowest resonance frequency is 400 Hz?

• **15.24** A guy wire supporting the mast of a sailboat has a linear density of 20 g/m and is 15 m long. If this guy wire vibrates at a resonance frequency of 8 Hz (fundamental frequency), what is the tension in the wire?

• **15.25** A piano tuner strikes a key on the piano and then hits a tuning fork whose frequency of vibration is 328 Hz. He hears a beat whose frequency is 2 Hz, and a slight increase in tension of the string increases the beat frequency to 3 Hz. What was the frequency of the piano string when it was first struck?

• **15.26** Two identical steel wires are stretched between posts 1.2 m apart. When the same tension is applied to both, their fundamental frequency of vibration is 500 Hz. If the tension on one is now increased by 2 percent, what beat frequency will be heard when both wires are set into vibration?

• **15.27** The string of Problem 15.17 has a mass of 0.05 kg. What is the tension in the string?

• **15.28** A piano wire 0.8 m long has a mass of 6 g. The wire is stretched between the two posts on the piano with a tension of 500 N. What is the frequency of the fundamental vibration? If this wire has a diameter of 0.07 mm, what is the diameter of a wire of the same material and length which resonates at half that frequency when under the same tension?

• **15.29** One string of a bass fiddle is 1.8 m long and has a fundamental frequency of resonance of 81 Hz when it is under a tension of 120 N. If the total length of the string (including the part wound about the tuning peg) is 2.1 m, what is the mass of the string? If the tension is increased to 140 N, what will the new resonant frequency be?

• • **15.30** The strings of a violin are tuned to the notes G, D, A, and E. These are all one fifth apart; that is, D has a frequency 1.5 times that of G, A, 1.5 times that of D, and E, 1.5 times that of A. If the G string has a mass per unit length of μ_G kg/m, what should the linear densities of the other strings be so that all strings are under the same tension. (Why is this desirable?)

• • **15.31** You are given a 1.5-m length of wire and you want to know its weight but have no accurate balance. You do have a 3-kg bag of potatoes and a 1-m-long board handy. You fix one end of the wire to one end of the board with a nail, let the wire drape over the other end, and hang the bag of potatoes from it. If you then pluck the wire, you can hear a sound, and having been blessed with absolute pitch, you recognize the sound as an octave below middle A, or 220 Hz. What is the mass of the 1.5-m length of wire?

• • **15.32** A steel wire is fixed at one end, and the other end is looped over a small pulley, supporting a mass of 5 kg. At 20 °C, the resonant frequency of vibration of the system is 300 Hz. What is the change in frequency if the temperature of the wire increases to 30 °C and the distance between the fixed end and pulley remains unaffected by the temperature change?

• • **15.33** Repeat Problem 15.32, assuming that the support for the wire and pulley system is also made of steel.

(Section 15.9)

• **15.34** A long rubber hose of linear density 0.04 kg/m is kept under a tension of 20 N. One end of the hose is attached to a device that prevents the reflection of any energy to the other end of the hose, which is forced to vibrate at a frequency of 4 Hz. What should the amplitude of that vibration be so that energy is transmitted at a rate of 12 W?

• • **15.35** A string whose linear density is 0.01 kg/m is maintained under a tension so that traveling waves propagate at 40 m/s. The ends of the string, whose length is 2 m, are clamped. If a standing wave, of amplitude 10 cm, is established on this string with nodes only at the clamped ends, how much energy is stored in this vibrating system? How much power is transmitted along the string?

• • **15.36** Repeat Problem 15.35 for a standing wave having two nodes between the two fixed ends.

•16•

Sound

Take care of the sense, and the sounds will take care of themselves.

LEWIS CARROLL, *Alice in Wonderland*

16.1 Introduction

Sound is commonly defined as a wave disturbance that evokes the sensation of hearing. Sound waves in air and in liquids are longitudinal compressional waves. More generally, physicists tend to extend the definition of "sound" to include any mechanical wave in a medium, including solids, within a wide frequency range extending from a fraction of a hertz to many megahertz. Sound vibrations with frequencies less than about 20 Hz, the lower limit of human bearing, are called *infrasonics;* vibrations with frequencies above about 20,000 Hz, the upper limit of human hearing, are called *ultrasonics*.

Until recently, most interest in sound outside the audible range centered on ultrasonics because of their wide practical applications in areas such as industrial cleaning, nondestructive testing, and diagnostic and surgical techniques. During the past decade, it was found that infrasonic vibrations can be quite harmful, causing frictional heating of internal organs and irritation of the nervous system. Machinery in a workshop can set up such waves, and if the room acts as a resonant cavity, the amplitudes of the waves can be substantial. That these sound waves cannot be heard makes them particularly insidious and all the more dangerous.

Sound waves can propagate through solids, liquids, and gases. In fact, solids and liquids are better conductors of sound than gases, propagating these waves at higher velocities and often with less attenuation than gases. Since normally, however, the medium through which sound propagates in everyday experience is air, we shall concentrate attention on the properties of compressional waves in gases.

16.2 Generation of Sound

Sound waves are created by the vibration of some object—the diaphragm of a loudspeaker or the vocal cords of the larynx—that causes the air molecules immediately adjacent to that object to move in rhythm with these vibrations.

When the diaphragm of Figure 16.1 moves to the right, it pushes the adjacent air to the right, increasing the air pressure immediately in front of the membrane. Half a cycle later, the diaphragm moves to the left, creating a partial vacuum on its right side. These pressure fluctuations propagate as a wave with a speed that depends on the type of gas and its temperature.

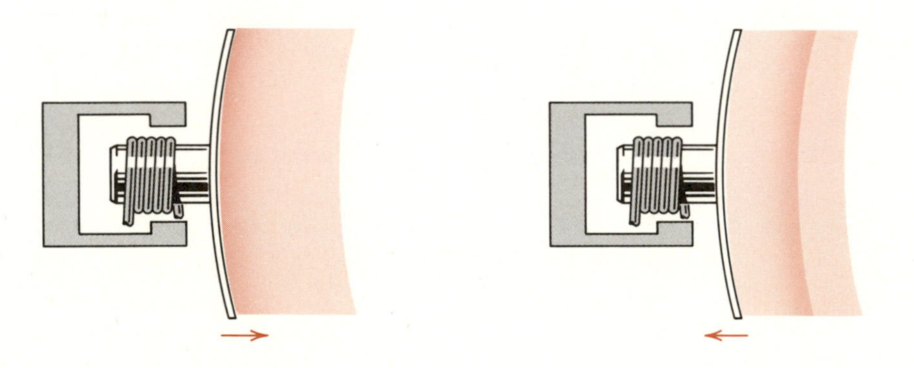

Figure 16.1 The generation of pressure waves in air by a vibrating membrane. As the membrane moves forward, it compresses the air immediately in front of it. A half-cycle later, when the membrane is moving back, the density of the air in front of it is reduced.

16.3 Speed of Sound

The speed with which information of local pressure fluctuations can be transmitted to a distant point in a gas must depend in some way on the properties of the gas. Since the gas molecules are, on the average, very far apart compared to their size, it is only through occasional random collisions between them that one portion of the gas becomes aware, so to speak, of what is going on elsewhere. We expect, therefore, that the speed of sound cannot exceed the average speed of molecular motion; actually the two speeds are nearly the same.

We saw in Chapter 12 that the average speed of molecular motion is related to the absolute temperature of the gas by

$$\tfrac{1}{2}mv_{\mathrm{rms}}^2 = \tfrac{3}{2}kT \qquad \textbf{(16.1)}$$

where v_{rms} is the root-mean-square velocity, m the mass of the gas molecule, k Boltzmann's constant ($k = 1.38 \times 10^{-23}$ J/K), and T the absolute temperature in K.

The speed of sound in a gas is proportional to v_{rms} and is given by

$$v = \sqrt{\frac{\gamma}{3}}\, v_{\text{rms}} \qquad \textbf{(16.2)}$$

Here γ is the ratio of the specific heat of the gas at constant pressure to the specific heat at constant volume (see Chapter 12); it is largest for a monatomic gas and is nearly unity for polyatomic gases, as indicated in Table 12.1.

Substituting the expression for v_{rms} from Equation (16.1) into Equation (16.2) gives

$$v = \sqrt{\frac{\gamma k T}{m}} \qquad \textbf{(16.3)}$$

Equation (16.3) tells us that the *speed of sound in a gas is proportional to the square root of the absolute temperature, and for gases of the same molecular structure, inversely proportional to the square root of the molecular weight of the gas.* Contrary, perhaps, to intuition, *the speed of sound is independent of the pressure of the gas.*

Speeds of sound in various gases are listed in Table 16.1.

● ***Example 16.1*** What is the speed of sound in argon at 25 °C?

We shall answer this question using two approaches. First, we shall calculate $v_A(25\ °\text{C})$ by substituting the appropriate numerical values into Equation (16.3). Second, we shall again employ Equation (16.3) but determine $v_A(25\ °\text{C})$ by comparison with the known speed of sound in oxygen at 0 °C.

Table 16.1 *Speed of sound in representative gases, liquids, and solids*

Substance	*Temperature, °C*	*Speed, m/s*
Gases		
Carbon dioxide	0	259
Oxygen	0	316
Air	0	331
Air	20	343
Nitrogen	0	334
Helium	0	965
Liquids		
Mercury	25	1450
Water	25	1498
Seawater	25	1531
Solids		
Rubber	—	1800
Lead	—	2100
Lucite	—	2700
Gold	—	3000
Iron	—	5000–6000
Glass	—	5000–6000
Granite	—	6000

Argon is a monatomic gas. Consequently, $\gamma_A = 1.67$. The mass of an argon molecule is that of an argon atom. That is,

$$m_A = \frac{40 \text{ g/mol}}{6.02 \times 10^{23} \text{ atoms/mol}} = 6.64 \times 10^{-23} \text{ g/atom}$$

In applying Equation (16.3), we must remember to express the temperature in kelvin; that is, T is the absolute temperature. We then obtain

$$v_A(298 \text{ K}) = \sqrt{\frac{1.67(1.38 \times 10^{-23} \text{ J/K})(298 \text{ K})}{6.64 \times 10^{-26} \text{ kg}}} = 322 \text{ m/s}$$

Alternatively, one could proceed as follows. The ratio of the sound velocities in argon and in oxygen is

$$\frac{v_A(298 \text{ K})}{v_{O_2}(273 \text{ K})} = \sqrt{\frac{\gamma_A k \times 298/m_A}{\gamma_{O_2} k \times 273/m_{O_2}}} = \sqrt{\frac{\gamma_A \times 298 \times m_{O_2}}{\gamma_{O_2} \times 273 \times m_A}}$$

The ratio of the molecular masses just equals the ratio of the molecular weights, namely, 32 for O_2 and 40 for A. The sound velocity in O_2 at 273 K is given in Table 16.1 as 316 m/s. Thus the sound velocity in argon at 298 K is

$$v_A(298 \text{ K}) = (316 \text{ m/s}) \sqrt{\frac{1.67(298 \text{ K})(32)}{1.4(273 \text{ K})(40)}} = 322 \text{ m/s}$$

as before. ●

In liquids, and especially in solids, the effects of local pressure fluctuations are transmitted far more quickly than in gases because the displacement of one atom is felt by its neighbors through their mutual interatomic forces and not as a result of occasional random collisions. Sound velocities in some common liquids and solids are also listed in Table 16.1.

16.4 Relation between Pressure Fluctuation and Average Displacement of Gas Molecules

We now show that the existence of a pressure wave implies the existence also of a wave of average displacement of gas molecules. To understand the relation between pressure and average displacement, consider Figure 16.2.

In Figure 16.2(*a*), the regions marked *A* correspond to high pressure and high density, those marked *B* to low pressure and low density. As a

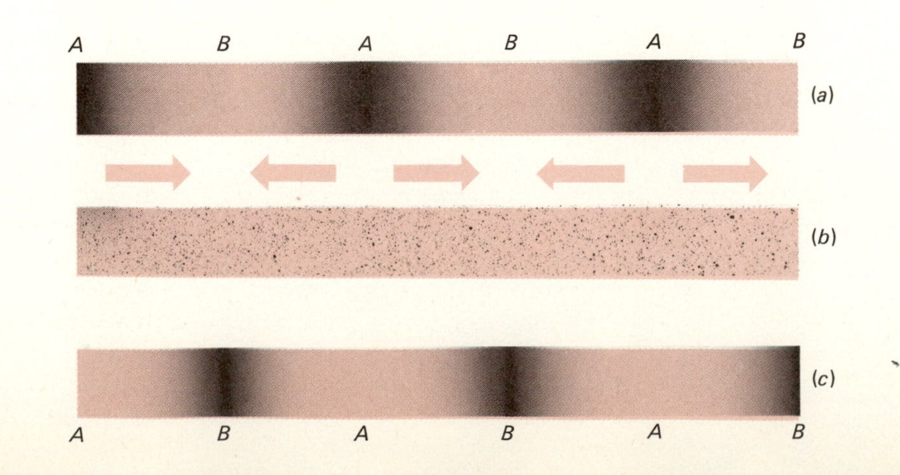

Figure 16.2 *Pressure wave in a column of gas. (a) The molecular density and pressure are above average at A and below average at B. This pressure variation leads to a redistribution, with an average molecular motion as indicated by arrows. (b) The molecular distribution becomes momentarily uniform. (c) Inertia has carried the molecules beyond the equilibrium distribution, so that half a cycle after (a), there is maximum pressure at B and minimum pressure at A.*

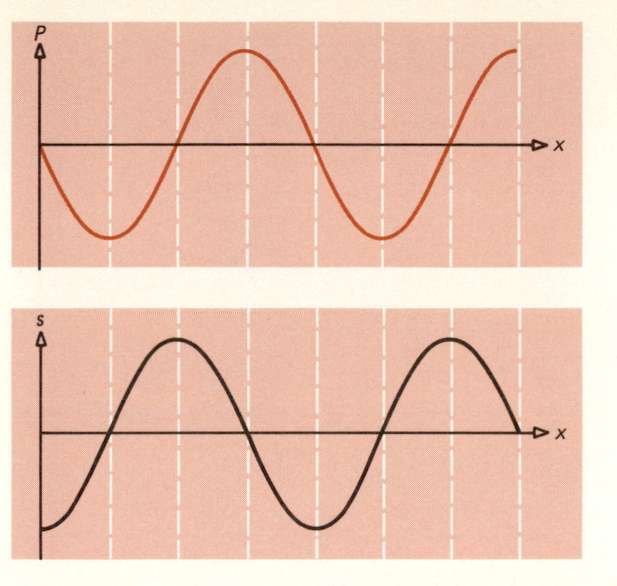

Figure 16.3 As suggested by Figure 16.2, the pressure and displacement waves are displaced in phase by 90°.

result of the excess pressure at A, gas molecules tend to move away from that region, as indicated by the arrows, to fill the partial voids at B. Due to their inertia, however, the gas molecules overshoot the equilibrium condition shown in Figure 16.2(b). The dynamics of the gas is analogous to the mass-spring system in which the mass also overshoots its equilibrium position and oscillates back and forth. Thus, very shortly we shall find the situation shown in Figure 16.2(c), where now A designates low-pressure and low-density regions, and B the high-pressure and high-density regions.

Note that at A and B, where the pressure excursions from the mean are greatest, the net *average* velocity of the air molecules vanishes, equal numbers moving to the right and to the left. On the other hand, midway between A and B, where the pressure maintains its average ambient value, the displacements are largest. Consequently, if we plot pressure and average displacement as functions of position, the curves appear as in Figure 16.3, similar in shape but displaced in phase by 90° relative to each other.

16.5 Acoustic Resonances of Pipes

Just as a cord fixed at two points is constrained to vibrate with certain well-defined wavelengths (see Equation (15.9)), so sound waves in a cavity of given dimensions are similarly restricted by boundary conditions. For complicated geometrical shapes, it is very difficult to compute these *resonant modes of vibration*, and they are best determined experimentally. We consider here only the special case of cavities of cylindrical geometry whose lengths are much greater than their diameters.

If a narrow pipe of length L is closed at both ends, an obvious restriction is thereby imposed on the motion of air molecules at each end. They cannot move back and forth. The two *closed ends must be displacement nodes*, just as the fixed ends of a string in Section 15.6 were nodal points. By exactly the same arguments as those that led to Equations (15.9) and (15.10), the wavelengths for the resonance vibrations in such a pipe are given by

$$\lambda_n = \frac{2L}{n} \qquad \textbf{(16.4)}$$

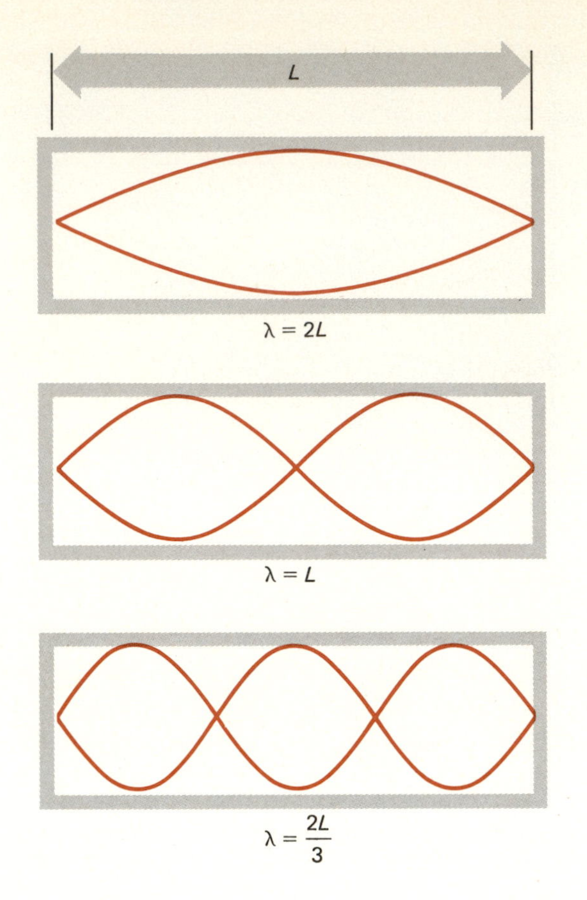

Figure 16.4 Displacement standing waves in a pipe closed at both ends. The possible standing waves must have displacement nodes at the closed ends of the pipe. The three lowest-frequency resonant modes are shown.

and the frequencies by

$$f_n = \frac{nv}{2L} \qquad \textbf{(16.5)}$$

where n is an integer and v is the speed of sound in the gas within the pipe.

Note that since the two ends are displacement nodes, they must be pressure *antinodes* in light of Figure 16.3. At the two ends, the pressure fluctuations at resonance are greatest.

Let us consider next the case of a pipe of length L that is closed at one end but open at the other. Again, the closed end will be a pressure antinode. But at the open end there is no constraint on the displacement of gas molecules, and the pressure there is the ambient air pressure. At resonance, the open end will be at a displacement antinode and a pressure node. The open end of the pipe is analogous to the end of the string terminated by a smooth ring.

Figure 16.5 shows the displacement and pressure patterns of three standing waves in such a pipe. The longest-wavelength, lowest-frequency standing wave that satisfies the stated boundary conditions is a wave whose wavelength is four times the length of the pipe; that is, we must fit one quarter-wavelength into the distance L. The next resonance mode that satisfies the boundary conditions is one whose wavelength is only one-third as long; here the length of the pipe is three-quarters of a wavelength. For this pipe, the resonance condition is given by

$$\lambda_m = \frac{4L}{m} \qquad m = 1, 3, 5, 7, \ldots \qquad \textbf{(16.6)}$$

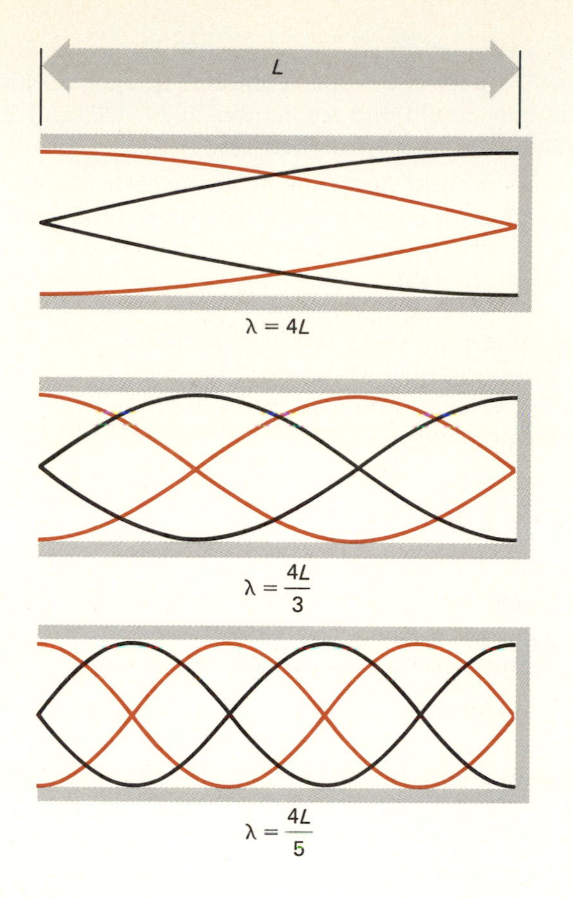

Figure 16.5 Displacement and pressure standing waves for a pipe closed at the right and open at the left end. The displacement is shown in color, the pressure in black. There must be a displacement node at the closed end of the pipe and a pressure node (displacement antinode) at the open end. Consequently, the resonances must meet the condition that an odd number of quarter wavelengths fit into the length of the pipe.

and the frequencies are

$$f_m = \frac{mv}{4L} \qquad \textbf{(16.7)}$$

where m is an *odd* integer.

Lastly, if both ends of the pipe are open, the situation is the same as when both ends are closed, except that the two ends are now *pressure* nodes instead of displacement nodes. Consequently, the resonant wavelengths and frequencies are again given by Equations (16.4) and (16.5).

Note that the open organ pipe (both ends open) will support all harmonics of the fundamental frequency, but the closed one (one end open, the other closed) can resonate only at the fundamental and its odd harmonics. Hence the tone quality of these pipes will be distinctly different.

● ***Example 16.2*** A small loudspeaker is placed above the open end of a pipe and as the frequency of the sound is slowly varied, a resonance is found at 280 Hz and the next one at 360 Hz. The length of the pipe is 2 m. Find the velocity of sound in the pipe and decide whether the other end of the pipe is open or closed.

Let us assume that both ends of the pipe are open. In that case, from Equation (16.5) we have 280 Hz $= nv/4$ m and 360 Hz $= (n + 1)v/4$ m. Thus, 80 Hz $= v/4$ m and $v = 320$ m/s. Is this a reasonable result? Certainly the magnitude of the velocity is right. But if we now solve for n, we find that $n = 4 \times \frac{280}{320} = \frac{7}{2}$, not an integer. Consequently, the initial assumption that the other end of the pipe is open must be incorrect.

You can quickly convince yourself that starting with Equation (16.7) you will again obtain the same value for the velocity, $v = 320$ m/s.

Solving for m in Equation (16.7), where 280 Hz $= mv/4L = m \times 40\ \text{s}^{-1}$, we find $m = 7$, a perfectly respectable odd integer. Lastly, we can also conclude that one should find resonances at 40, 120, and 200 Hz in addition to resonances at higher frequencies.

Example 16.3 A child's toy pipe normally emits a sound of frequency 336 Hz. When this pipe is connected to a tank of an unknown gas and that gas is blown into the pipe, a sound of 980 Hz is produced. What gas was contained in the tank?

Since the same pipe was used as a resonator, the *wavelength* of the sound *in the pipe* remained unchanged. From Equation (16.5) or (16.7), it follows that the velocity of sound in the unknown gas must have been higher than in air. Thus

$$v_{\text{gas}} = v_{\text{air}} \times \left(\frac{980}{336}\right) = 2.92 v_{\text{air}}$$

Looking at Table 16.1, we see that $v_{\text{He}}/v_{\text{air}} = \frac{965}{331} = 2.92$, and we can conclude that the tank must have been filled with helium. You should verify that another light gas, namely hydrogen, would produce a sound of substantially different pitch.

Resonance is an extremely important as well as ubiquitous phenomenon that appears in so many guises that we are often unaware of it. As already mentioned, tuning and playing of musical instruments relies on resonance. But we also call on resonance when we tune a radio or a TV set so that it is particularly sensitive to that narrow portion of the electromagnetic frequency spectrum in which the station of our choice is operating.

Resonance has also been important in biological evolution. For example, the vibration of insect wings is determined by the resonance frequency of this system, which is crudely analogous to a thin steel plate held firmly to the edge of a table top and struck at its free end. In Chapter 14 we described another biological resonance, a dog's panting. Most birds also control their body temperature by panting at the resonance frequency of their respiratory system.

Sometimes resonances can be extremely damaging, however. For example, if the structural members of an airplane resonate at a frequency generated by the plane's engines, the consequent large-amplitude deformations will soon produce cracks and cause failures. A particularly dramatic demonstration of an undesirable resonance occurred on November 7, 1942, when the newly constructed Tacoma Narrows bridge collapsed during a moderate windstorm.

Figure 16.6 *Tacoma Narrows Bridge shortly before its collapse on November 7, 1942.*

16.6 Sound Intensity

DEFINITION: The *intensity* I of a wave is the amount of energy crossing unit area in unit time; intensity is the power that crosses unit area.

Intensities are measured in watts per square meter in the SI.

In Section 15.9 we found that the power transmitted by a traveling wave on a string is proportional to the square of the wave amplitude. This relation holds also for other mechanical waves, such as sound waves in gases, liquids, and solids. It does not matter whether we focus our attention on the displacement or on pressure variations, the intensity is proportional to the square of the amplitude of either. Since it is much simpler to measure pressure fluctuations than average displacements of gas molecules, it is the former that are of greater interest.

These pressure fluctuations are normally very small. For example, the amplitude of the pressure wave that corresponds to a sound intensity of normal conversation is only about 3×10^{-2} Pa, compared with atmospheric pressure of about 10^5 Pa.

As we recede from a sound source of constant strength, one that emits sound energy at constant rate, the sound intensity diminishes. Two factors are responsible for this decrease. First, the absorption of sound energy by the medium—the gradual conversion of this energy to heat—and second, while the *total* power emitted remains constant, the *intensity* diminishes because that power is distributed over a larger surface area.

We consider a source that emits at a rate of P watts uniformly in all directions. To see how the intensity depends on distance from the source, we construct an imaginary sphere of radius R meters with the source at its center. Since the surface area of a sphere is $4\pi R^2$ and the power radiated by the source must pass through the entire spherical surface, the intensity at the distance R will be

$$I(R) = \frac{P}{A} = \frac{P}{4\pi R^2} \qquad \textbf{(16.8)}$$

provided that absorption can be neglected.

We see that the intensity $I(R)$ falls off as $1/R^2$ with increasing distance from the source. This *inverse square law* is valid whenever a conserved quantity, such as energy, charge, or mass, spreads uniformly from a point source. It is, however, valid *only if the source radiates uniformly in all directions in three dimensions.* For instance, for circularly diverging surface waves generated by dropping a pebble into a pond, similar reasoning leads to the conclusion that the intensity is now proportional to $1/R$.

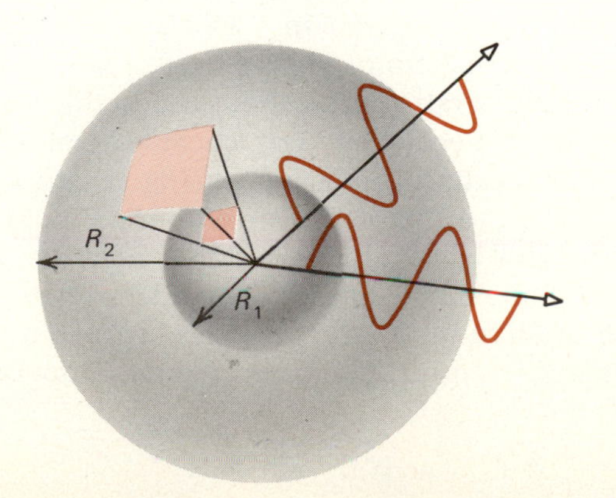

Figure 16.7 *A sound source at $r = 0$ radiates uniformly in all directions. The energy per unit time that passes through the entire surface area of a sphere surrounding the source is the same regardless of the sphere's radius. The intensity of the sound, defined as the power per unit area, therefore diminishes as r increases. Since the surface area of a sphere is proportional to r^2, the intensity decreases as $1/r^2$, if we neglect sound absorption.*

● *Example 16.4* At a distance of 20 m from a large jet engine, the sound intensity is fifty times greater than that which can cause permanent damage to hearing. How far from the engine should you be so that the sound intensity is one-fiftieth that which causes permanent hearing impairment?

We should move to such a distance that the intensity is reduced by a factor of $50 \times 50 = 2500$. The ratio of the distances is therefore $\sqrt{2500} = 50$. One should move to a distance of $50 \times 20\ \text{m} = 1000\ \text{m} = 1\ \text{km}$. ●

16.6 (a) Intensity Levels; Decibels

The ear is an extremely sensitive detector of pressure fluctuations. It is, moreover, capable of responding to a phenomenal range of sound intensities; the strongest intensity most people can endure for brief periods without physical pain is about 10^{12} (a million million) times as great as that which just gives a faintly perceptible sound. (It is important to recognize, however, that *extended* exposure to sound intensities well below the threshold of pain can cause permanent hearing impairment. Most rock concerts, construction equipment (air hammers, etc), old subway trains, and large diesel trucks generate noise levels deleterious to hearing.)

Although the measured sound intensity of a forte passage at a symphony performance may be one hundred thousand to a million times as great as the intensity of a pianissimo passage, most of us, if asked to describe the relative intensities, would say that one is perhaps five or ten times as loud as the other. For reasons that are not fully understood, the physiological response and psychological perception of hearing is roughly logarithmic, not linear. It is therefore convenient to specify sound levels on a logarithmic scale, using the *bel* as the unit. Two sounds differ in intensity by one bel if the ratio of their intensities is 10. Normally a smaller unit, the *decibel* (dB), one tenth of a bel, is used. It is common practice to specify sound intensities in decibels according to the relation

$$\beta = 10 \log_{10} \frac{I}{I_0} \qquad \textbf{(16.9)}$$

where β is the intensity level in decibels and I_0 is the sound intensity at the threshold of audibility, $10^{-12}\ \text{W/m}^2$. Intensity levels of some common sounds are given in Table 16.2.

The logarithmic response of the ear creates some minor problems in sound reproduction. Since sound intensity is proportional to the square of the wave amplitude, a range in "volume" of one million, corresponding roughly to the ratio of fortissimo to pianissimo passages at a concert, would require a ratio of one thousand in the amplitude of vibration of the phonograph needle, an impracticable demand on both phonograph records and pickup. Therefore, during recordings the sound is "compressed"; that is, soft passages are amplified more than loud ones. To compensate for this compression in dynamic range, very good audio amplifiers contain adjustable sound "expanders," which automatically increase amplification of the louder passages compared with softer ones. Without expanders, recordings of symphonies appear somewhat flat

Table 16.2 Typical sound levels and intensities

Sound level, dB	*Intensity, W/m²*	*Sound*
0	10^{-12}	Threshold of hearing
10	10^{-11}	Rustle of leaves
20	10^{-10}	Whisper (1 m away)
30	10^{-9}	Quiet home
40	10^{-8}	Average home, quiet office
50	10^{-7}	Average office
60	10^{-6}	Normal conversation, average traffic
70	10^{-5}	Noisy office
80	10^{-4}	Busy traffic, inside car in traffic
90	10^{-3}	Inside subway train
100	10^{-2}	Machine shop
120	10^{0}	Pneumatic chipper (2 m away), threshold of pain
140	10^{2}	Jet airplane (30 m away)

compared with live performances, partly because of the restricted intensity range.

● ***Example 16.5*** At a distance of 20 m from a source of sound that radiates uniformly in all directions, the sound intensity is 80 dB. At what distance from the source will the intensity have dropped by 30 dB to 50 dB, and what is the intensity at these two locations in W/m²?

A change in intensity by 30 dB connotes a reduction in intensity by a factor of 10^3. Since for this source the intensity is inversely proportional to the square of the distance from the source,

$$\frac{R^2}{400} = 10^3$$

$$R = (20 \text{ m})\sqrt{1000} = 632 \text{ m}$$

The intensity of the sound at these two locations is determined from Table 16.2. At 20 m, the intensity is 80 dB, corresponding to a power per unit area of 10^{-4} W/m². At 50 dB, the intensity is 10^{-7} W/m². ●

16.7 Doppler Effect

Most of us have heard the change in pitch of a train whistle as we were standing at a grade crossing and watching the train pass. As a train approaches, the pitch is high; it drops quickly to a noticeably lower value as the engine passes and recedes.

We denote the frequency of the source by f and assume that at $t = 0$ the source is at A and the wave generated by the source is at a crest. During one period, this crest travels a distance $vT = \lambda$ toward the observer. In this same time interval T, the source has also moved some distance $v_s T$ toward the observer, where v_s is the speed of approach of the source. Consequently, at the time $t = T$, when the next crest is generated by the source, the distance between it and the preceding crest is not λ but

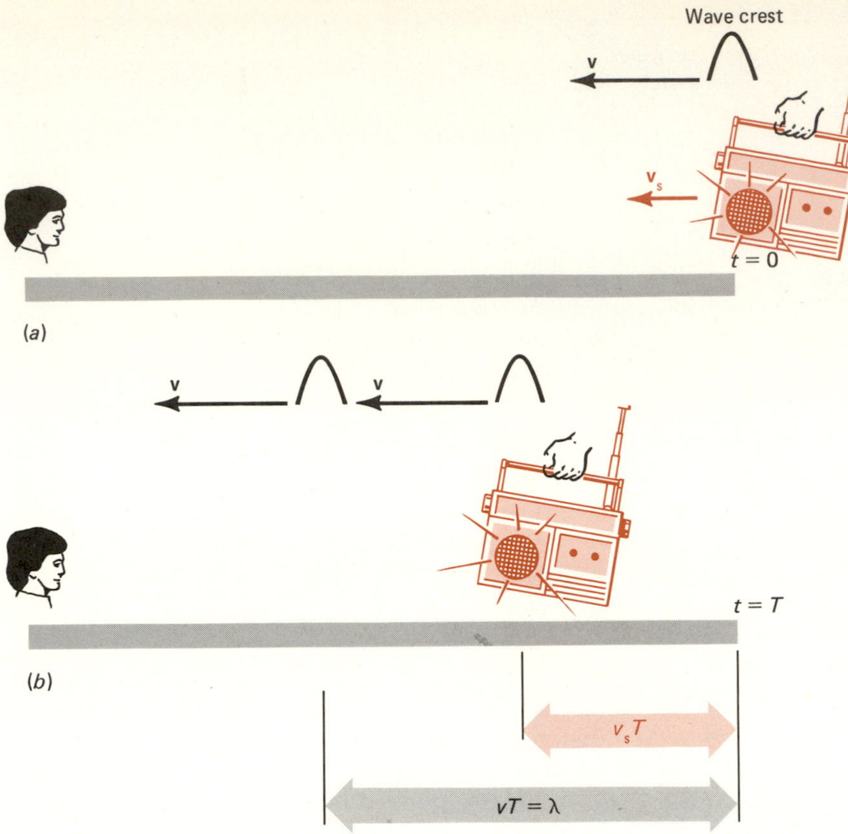

Figure 16.8 Perceived frequency from a moving source. If the source is moving toward the observer with a velocity $\mathbf{v}_s$ relative to the medium (still air) in which the sound velocity is $\mathbf{v}$, the wavelength in the medium is reduced from $\lambda = vT = v/f$, corresponding to the stationary source, to $\lambda' = (v - v_s)T = (v - v_s)/f$. The perceived frequency, as heard by a stationary observer, is then $f' = v/\lambda' = f/(1 - v_s/v)$. (a) Source and observer at $t = 0$. (b) Source and observer at $t = T$, where T is the period of the sound generated by the moving source.

$\lambda - v_s T$. Thus the wavelength of the wave that is propagating toward the observer is $\lambda' = \lambda - v_s T$. It follows that the frequency, i.e. the number of wave crests per second that pass by the observer, is

$$f_s' = \frac{v}{\lambda'}, = \frac{v}{\lambda - v_s T} = \frac{v}{\lambda}\left[\frac{1}{1 - v_s(T/\lambda)}\right]$$

$$f_s' = f\left(\frac{1}{1 - v_s/v}\right) \qquad \textbf{(16.10)}$$

The frequency heard by the observer when the source is approaching is thus greater than f, the frequency of the sound emitted by the source (Figure 16.8).

Similarly, if the source is receding from a stationary observer at a speed v_s, the perceived frequency is lower than f and is given by

$$f_s' = f\left[\frac{1}{1 + v_s/v}\right] \qquad \textbf{(16.10a)}$$

A pitch change also occurs if the source is stationary but the observer is moving toward or away from it with velocity v_0. Suppose the observer is approaching the source; in this case more wave crests per second strike his ear than if he were at rest. The wavelength of the sound in the transmitting medium is unchanged but the relative velocity of the wave (the velocity of the wave relative to the observer) is now $v + v_0$. Therefore, the perceived frequency is

$$f_0' = \frac{v + v_0}{\lambda} = f\left(1 + \frac{v_0}{v}\right) \qquad \textbf{(16.11)}$$

If the observer is receding from the source with velocity v_0,

Figure 16.9 Spherical waves are generated by a source moving with velocity $\mathbf{v}_s$ through a ripple tank. Note that the wavelength is shortened ahead of the source and lengthened behind the source.

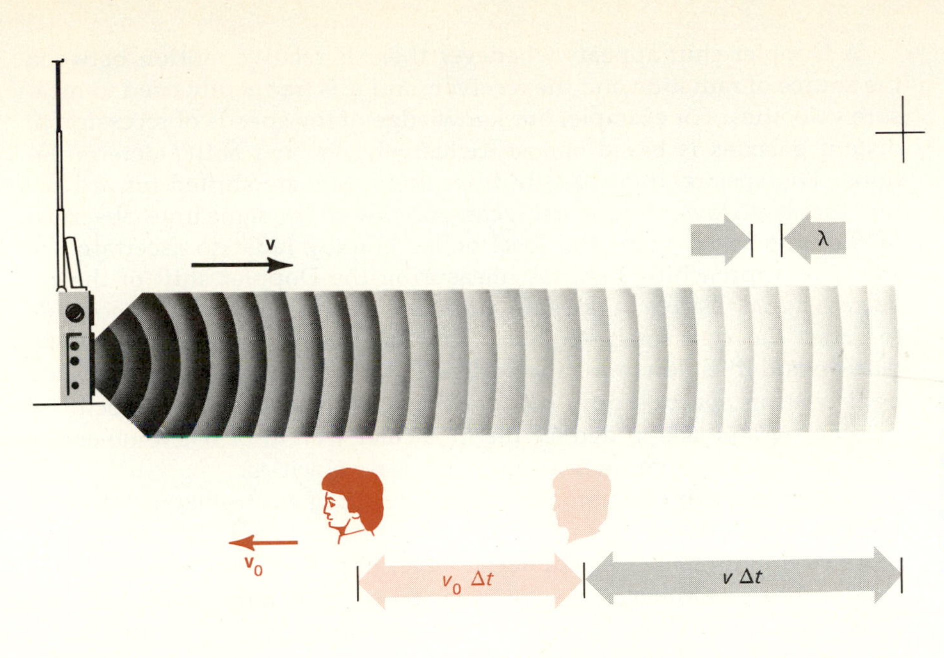

Figure 16.10 An observer approaching a source of sound at velocity $\mathbf{v}_o$ will pass $[(v + v_o)/\lambda]\,\Delta t$ wave crests in a time Δt. Consequently, the frequency of the sound measured by the moving observer will be $f' = f(1 + v_o/v)$.

$$f_o' = f\left(1 - \frac{v_o}{v}\right) \qquad \textbf{(16.11a)}$$

Equations (16.10) and (16.11) can be combined into a single expression for f'

$$f' = f\left(\frac{1 \pm v_o/v}{1 \mp v_s/v}\right) \qquad \textbf{(16.12)}$$

Here the upper sign applies if source and observer approach each other, the lower sign if they recede from each other.

We see from Equations (16.10)–(16.12) that the perceived frequency is greater if source and observer are approaching each other and less if they are receding from each other. The exact frequency shift, $f' - f = \Delta f$, depends, however, not just on the relative velocities but on whether source or observer is moving relative to the transmitting medium.

In many cases speeds v_s and v_o are much less than the speed of sound v. The fractional change $\Delta f/f$ due to the Doppler effect is then given to good approximation by a simple expression. If

$$\frac{v_o}{v} \ll 1 \qquad \text{and} \qquad \frac{v_s}{v} \ll 1$$

we can use Equation (A.11) and expand Equation (16.12) to obtain

$$f' \simeq f(1 \pm v_o/v)(1 \pm v_s/v) \simeq f\left(1 \pm \frac{v_o + v_s}{v}\right)$$

Then

$$\frac{f' - f}{f} = \frac{\Delta f}{f} = \frac{v_r}{v} \qquad \textbf{(16.13)}$$

where

$$v_r = v_o + v_s \qquad \textbf{(16.14)}$$

is the relative speed of approach of source and observer; if source and observer are receding from each other, v_r is negative.

A Doppler shift appears whenever there is relative motion between the source of radiation and the receiver, and it is frequently used to measure velocities. For example, our knowledge of the speeds of recession of distant galaxies is based almost exclusively on "red shift" determinations. The spectral lines of light from their stars are shifted toward the red, that is, to lower frequencies, compared with the same lines observed in the laboratory. State and local police employ radar to ascertain the speed of approaching vehicles, measuring the Doppler shift of the reflected signal relative to the emitted one. Using lasers it is now possible to detect the very small frequency shift of light scattered by a moving particle and thus determine the particle's velocity. This technique, called light-beating spectroscopy, has been used to measure the velocity of blood flow and also to deduce the molecular weight of macromolecules such as proteins from their average thermal velocities.

The correct expression for the Doppler shift of electromagnetic waves cannot, however, be any of Equations (16.10)–(16.12). We know that these equations cannot be applicable here, because if they retained their validity, we would have an experimental method at hand with which we could decide whether it is some distant galaxy or the earth that is moving. But as we shall discuss more fully in Chapter 29, the fundamental hypothesis of the special theory of relativity strictly precludes such experiments. The relativistic Doppler equation is more complicated than Equations (16.10)–(16.12), but it too reduces to the simple form

$$\frac{\Delta f}{f} = \frac{v_r}{c} \tag{16.13}$$

if $v_r \ll c$, where $c = 3 \times 10^8$ m/s is the speed of light.

● ***Example 16.6*** The frequency of a train whistle, as heard by the engineer in the engine, is 800 Hz. If a person at an unguarded crossing hears the whistle at a frequency of 760 Hz, what is the speed of the train? Is the train approaching or receding? Assume that the velocity of sound in air is 340 m/s.

We shall first apply the approximate formula, Equation (16.13), and then recalculate the speed of the train, using the exact expression for the Doppler shift due to a moving source.

From (16.13), we have

$$v_r = (340 \text{ m/s})\left(\frac{40}{800}\right) = 17.0 \text{ m/s} = 61.2 \text{ km/h}$$

Since the perceived frequency is lower than the frequency emitted by the train whistle, the train is receding from the observer.

The exact expression is given by Equation (16.10a), which is equivalent to $\Delta f/f_s = v_s/v$. Hence

$$v_s = (340 \text{ m/s})\left(\frac{40}{760}\right) = 17.9 \text{ m/s} = 64.4 \text{ km/h}$$ ●

16.8 Scattering of Waves

When a wave on the surface of a lake reaches a large pontoon, one can observe a scattered wave radiating outward from the pontoon. A similar wave striking a small boat riding at anchor not only is scattered but also visibly rocks the boat. If, however, we watch a small piece of wood floating on the surface of the lake, we see no scattered wave; the piece of wood simply rises and falls as it is lifted and lowered by the passing wave.

Scattering of waves is of great importance in nature. We see nonluminous objects only because when they are illuminated, they scatter some light toward our eyes. A beam of light in a dustfree room is invisible; we can, however, see such a beam by blowing smoke in its path, because the smoke particles scatter some of the light toward us.

The ability of an object to scatter a wave depends among other factors on the size of the object compared to the wavelength of the radiation. If the object is much smaller than the wavelength, it will scatter the wave only weakly; the wave will largely "go around it," or like the piece of wood on the lake, the object just accommodates to the wave without disturbing it significantly. In the range in which the linear dimension of the scatterer is comparable to or less than the wavelength, the relative intensity of the scattered wave depends strongly on size; for spherical particles, it is roughly proportional to the fourth power of the radius. Conversely, if waves of different wavelength impinge on a small object, the shorter waves will be scattered more effectively than the longer ones. (We shall see one important practical application of this relation in the next section and shall return to it in a later chapter.)

*16.9 Navigation of Bats

It is well known that bats depend on sound rather than light signals to guide them in flight. Although the pattern of sounds emitted and the manner of sound generation vary with individual species, bats generally emit brief, high-intensity ultrasonic pulses, and ascertain the position, distance, motion, and size of an obstacle or prey from the echo reflected to their large and mobile ears. Some fruitbats, in contrast to the majority of bats, which generate the ultrasound in a highly specialized larynx, create the signal by motion of the tongue. Also, most bats emit the sound through their mouths; horseshoe bats, however, use the nose as a highly directional transmitter, beaming the sound by adjusting the horseshoe-shaped folds on each side of the nose. In all species, the emitted sound intensities are extremely high, with pressure amplitudes comparable with those produced by a compressed-air hammer. Generally, bats can sense obstacles and prey at distances to about 4 m.

Figure 16.11 *Horseshoe bat in flight.*

Several species emit short pulses of about 3-ms duration, at a repetition rate of 10 to 30 pulses per second. Each pulse is frequency-modulated, starting at a high frequency of about 100 kHz and ending at about 30 kHz. As the animal approaches its prey, the repetition rate increases to 200 pulses per second and the pulse duration is reduced to achieve better resolution. Direction to the reflecting object is probably determined by comparing the time of arrival of the echo at the two ears, or the relative intensities of the reflections sensed by the ears, or both. It is also probable that the animal can estimate the size of the object by the change in reflected intensity with frequency. Since a scatterer with dimension of a few millimeters, the size of typical prey, scatters 100-kHz waves much more than 30-kHz waves, frequency modulation of the pulse provides a mechanism for estimating the size of objects of interest as food.

A favorite, juicy morsel of bats is the moth. Nature has given these nocturnal insects two defenses against their predators. First, the furry covering is an effective sound absorber, an acoustic camouflage. Second, moths' antennae are sensitive to ultrasonic vibrations, and moths can hear the hunting cries of bats. If the moth is struck by such a pulse, it folds up its wings and drops quickly to the ground.

*16.10 Physiology of Hearing

The mammalian ear, which translates pressure waves into electrical signals (nerve impulses), is crudely analogous to a microphone. However, compared with the ordinary microphone, the ear is an infinitely more refined and sensitive instrument. Despite extensive research by physiologists, psychologists, and physicists, we still do not fully understand how the complex system of ear, auditory nerve, and auditory cortex functions. Much of our "knowledge" is really only conjecture, based on the details of the ear's anatomy.

The ear is divided into three principal regions—the outer ear, the middle ear, and the inner ear.

16.10 (a) Outer Ear

The external, visible part of the outer ear is the *auricle,* or *pinna,* a wrinkled, cartilaginous appendage protruding from the side of the head. The ear canal leading into the head channels the sound waves to the *tympanic membrane,* the eardrum. The tympanic membrane has an area of about 60 mm^2 and a thickness of about 0.1 mm, and vibrates back and forth in response to pressure fluctuations at the inner end of the ear canal.

The ear canal is about 2.5 cm long and very crudely approximates a pipe open at one end and closed at the other. According to Equation (16.7), such a pipe has a resonant fundamental frequency $f = v/4L$; substituting for v the value of the velocity of sound in air, we find that the resonant frequency of the ear canal is about 3300 Hz. The next resonance should come at the third harmonic, that is, at about 10,000 Hz. A plot of

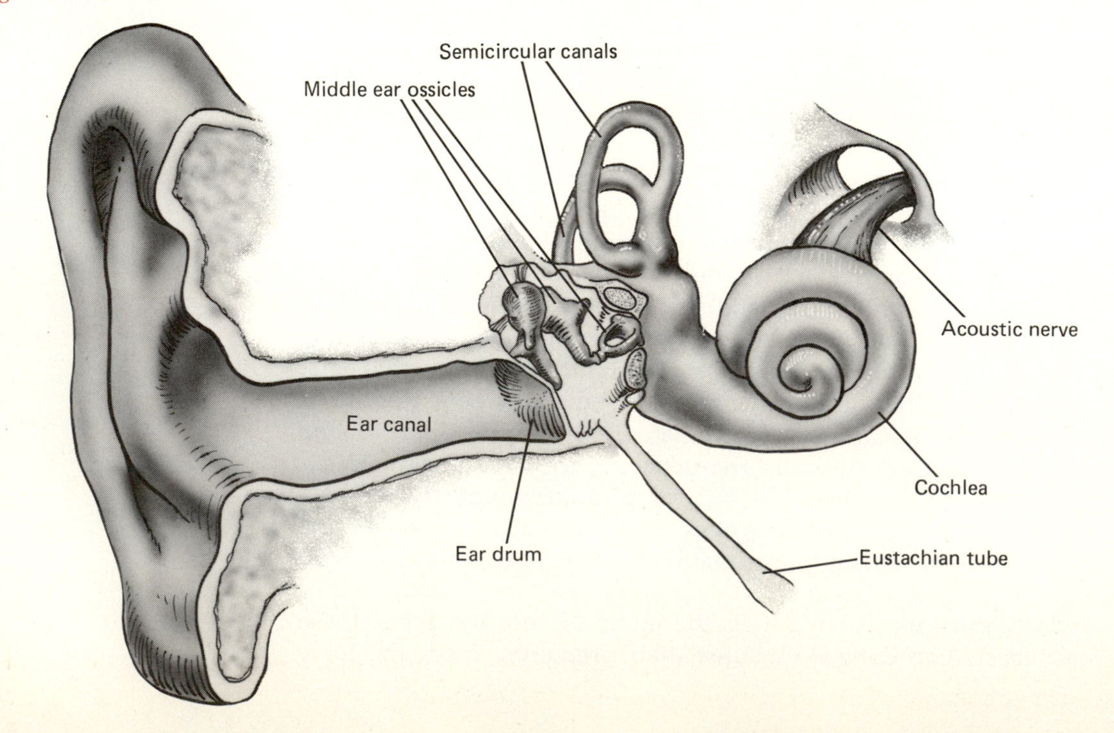

Figure 16.12 *The human ear. The drawing is not to scale; the inner ear is greatly enlarged to show details.*

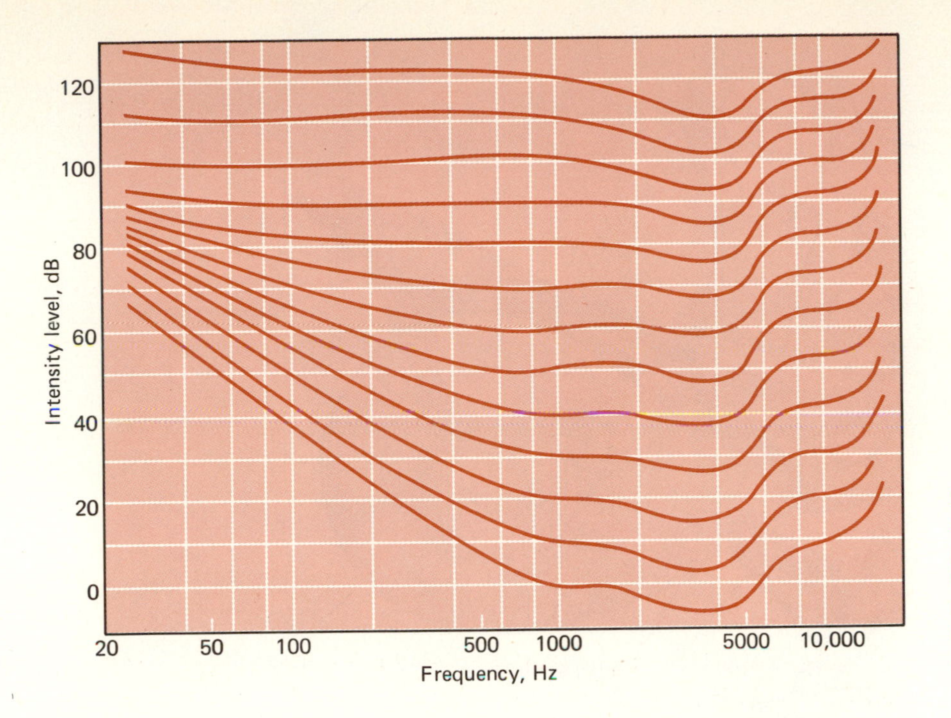

Figure 16.13 Curves showing sound intensity versus frequency for sounds judged by the average listener to be equally loud. The ear's sensitivity is greatest at about 3300 Hz.

the frequency response of the normal human ear clearly reflects these resonances (Figure 16.13). Evidently, the ear is most sensitive at about 3300 Hz; moreover, the next resonance also has an effect, causing a flattening in the upward trend of the response curve between 8000 and 10,000 Hz.

16.10 (b) Middle Ear

The middle ear is an air-filled cavity in the skull, on the internal side of the tympanic membrane; it contains the *auditory ossicles*—the hammer (*malleus*), the anvil (*incus*), and the stirrup (*stapes*). These small bones provide the mechanical linkage between the tympanic membrane and another membrane which covers the *oval window* of the *cochlea,* a tube about 5 mm long and coiled like a snail's shell. The middle ear is also connected to the pharynx through the *Eustachian tube,* whose function is to maintain a pressure balance between the middle ear and the ambient air, preventing damage to the tympanic membrane. The Eustachian tube is normally closed but opens during swallowing or yawning. Even relatively small pressure differences between outer and inner ear can be quite painful, as any air traveler knows, and chewing gum (thereby swallowing saliva) relieves this pain.

The leverage system of the ossicular chain operates so as to reduce the vibration amplitude at the oval window with a corresponding increase in the transmitted pressure (force) amplitude. The muscles holding the ossicles in the middle ear also act as an automatic volume control; as the sound intensity increases, they twist these bones slightly, rotating the stirrup away from the oval window. Simultaneously, other muscles stiffen the eardrum to prevent damage to this thin membrane. These normal autonomous responses require some time, however, so a sudden loud boom or crack is far more likely to be injurious than a sound of equal intensity that has built up more gradually.

The oval window has an area of only about 3 mm^2. The relative sizes of the tympanic membrane and oval window, the pressure-enhancing mechanical advantage of the ossicular chain, and the resonant length of

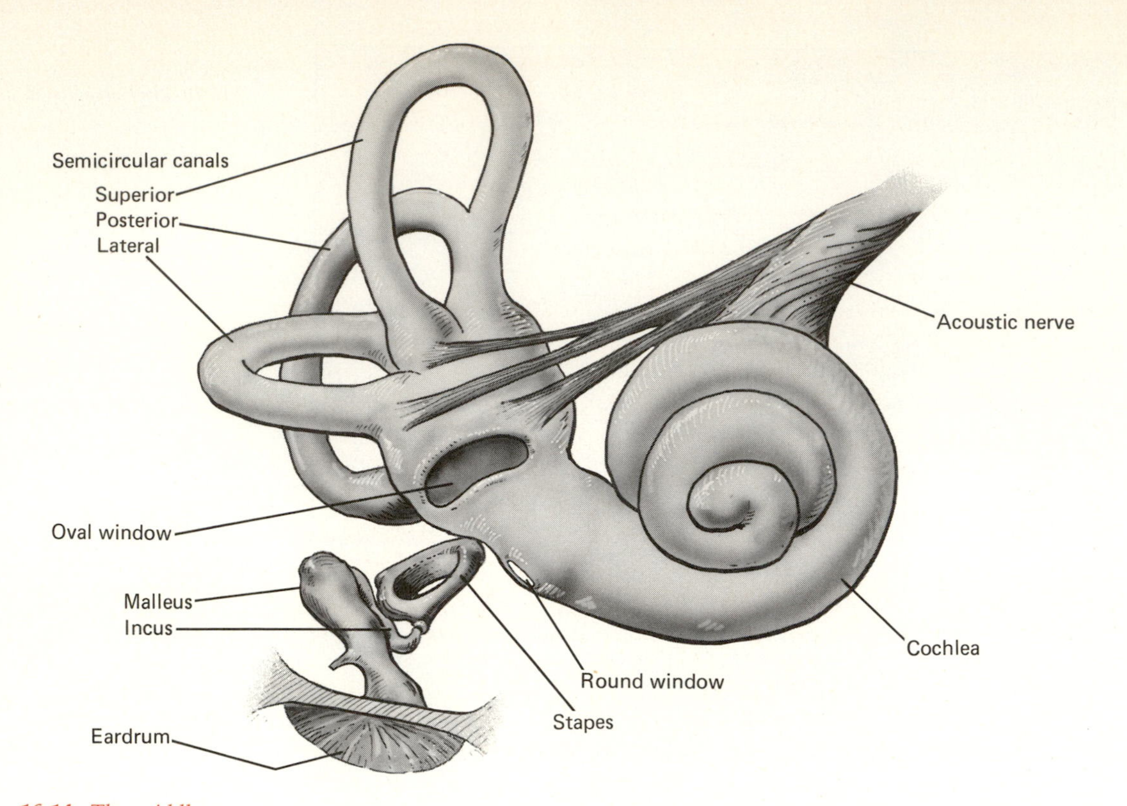

Figure 16.14 *The middle ear ossicles (incus, malleus, and stapes) form the mechanical linkage between the eardrum and the oval window of the cochlea.*

the ear canal combine to yield a substantial sound amplification and partly account for the remarkable sensitivity of the ear. At a frequency near 3000 Hz, the pressure amplitude at the eardrum, the closed end of the ear canal (a pressure antinode at resonance), is roughly twice the external pressure amplitude (see Section 15.4). From Pascal's principle (Section 10.3) it follows that the ratio of the pressure amplitudes at the oval window and at the eardrum is inversely proportional to their areas, giving an amplification of about 20. There is a further amplification in pressure of about 2, produced by the leverage system of the ossicles. Thus, the overall pressure amplification in the ear is about 80. Since the sound intensities are proportional to the square of the pressure amplitudes, the intensity is amplified by a factor of about 6400. These features help to explain how the ear can detect a sound intensity as low as 10^{-12} W/m^2, corresponding to a sinusoidal oscillation in pressure whose amplitude is only 3×10^{-5} Pa. By comparison, atmospheric pressure is about 10^5 Pa, nearly 10^{10} times greater. This sensitivity of the ear is close to the practical limit; pressure fluctuations due to random motion of air molecules, so-called Brownian motion, is about 5×10^{-6} Pa, so that the random noise due to Brownian motion is about 15 dB below the threshold of hearing. You can now also appreciate why a relatively small change in atmospheric pressure such as that experienced in riding a fast elevator can be painful.

16.10 (c) Inner Ear

The inner ear contains several organs, of which only one, the *cochlea*, is active in auditory perception. A cross section of the cochlea reveals three separate tubular regions, the *vestibular* and *tympanic canals* and the *cochlear duct*. The *basilar membrane* separates the tympanic canal from the other two, and *Reisner's* membrane forms the boundary between the vestibular canal and the cochlear duct. At the near end of the vestibular canal is the *oval window*, and at the near end of the tympanic canal is the *round window*, covered by another small, thin membrane. The two canals are connected at the far end of the cochlea by a small opening, the *heliotrema*, and are filled with lymphatic fluid. Attached to the basilar membrane on the side facing the cochlear duct is the *organ of Corti*, the physiological apparatus that transduces mechanical vibrations to electrical nerve impulses. The organ of Corti contains many fine sensory cells, the *hair cells*, that are connected to the *auditory nerve*.

A pressure wave carried to the oval window by the stirrup sets up a sound wave that travels the length of the vestibular canal, through the heliotrema, and returns through the tympanic canal to the round window; the round window is primarily an acoustic load, absorbing the residual sound energy and reducing reflection. The wave set up in the cochlea results in displacements of the basilar membrane and this, in turn, bends the hair cells, exciting the auditory nerve fibers.

The detailed mechanism of pitch perception is still a controversial subject. Von Bekesy has found that the displacement of the basilar membrane is greatest at the far end under low-frequency excitation, and that

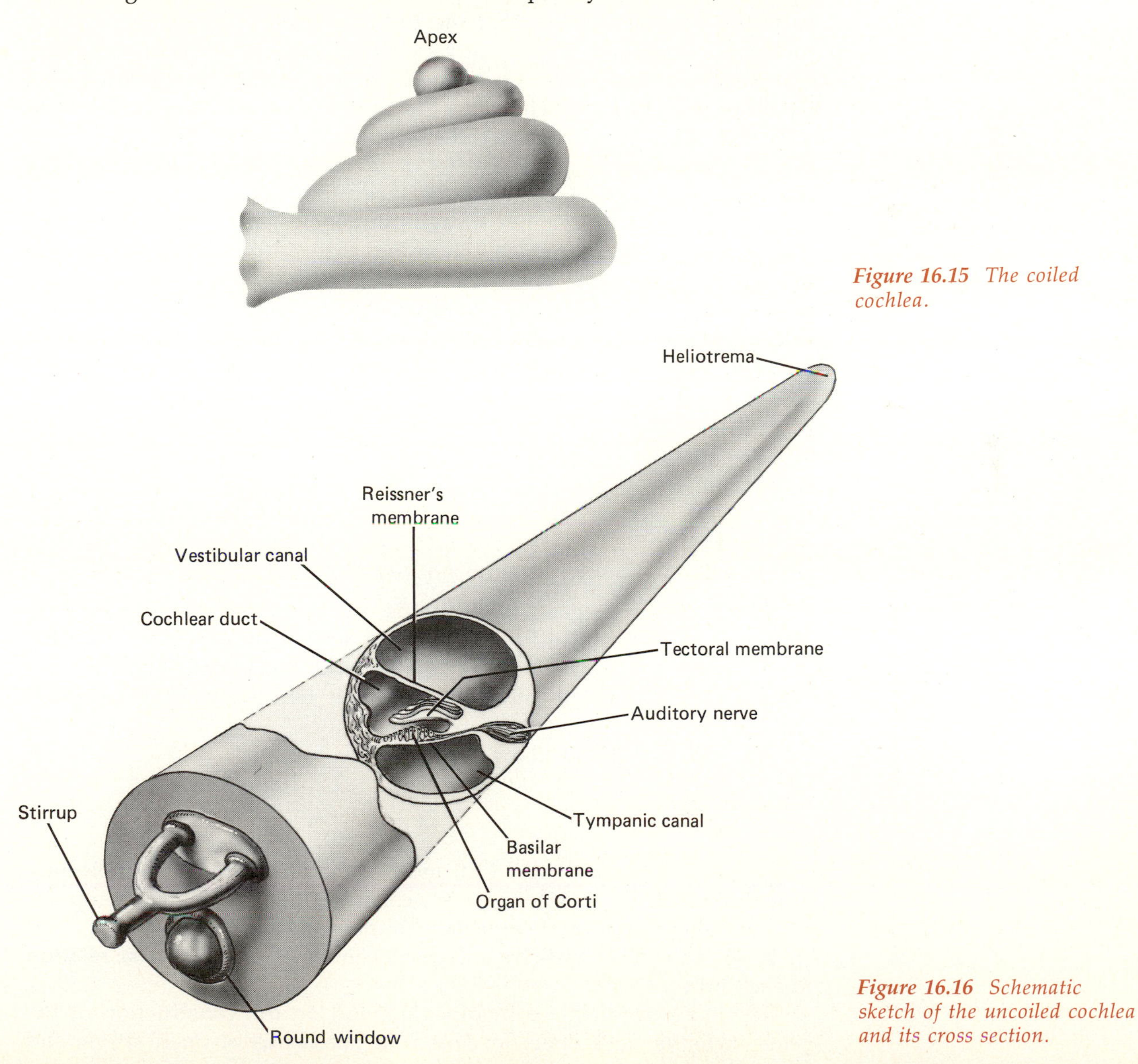

Figure 16.15 The coiled cochlea.

Figure 16.16 Schematic sketch of the uncoiled cochlea and its cross section.

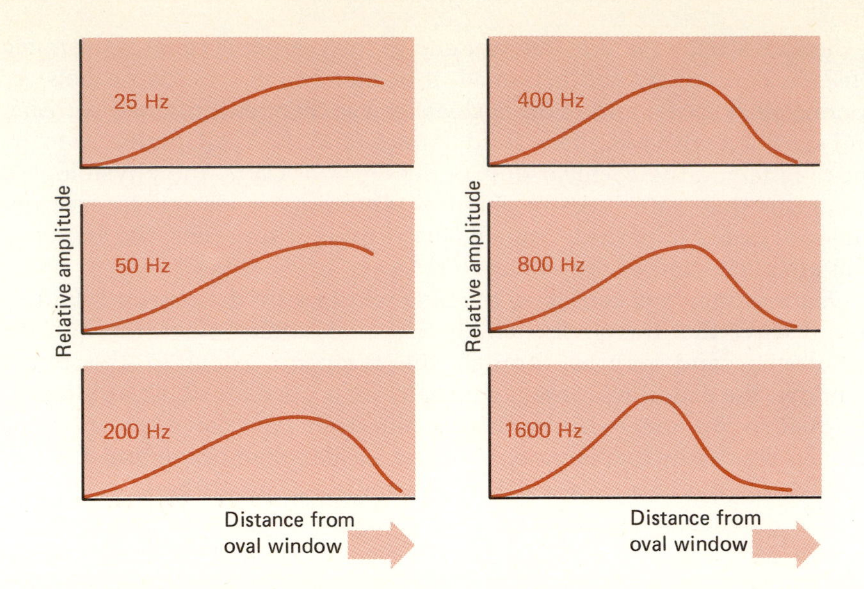

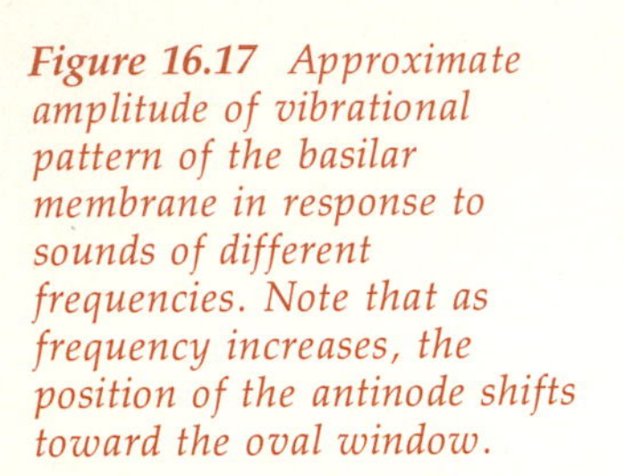
Figure 16.17 *Approximate amplitude of vibrational pattern of the basilar membrane in response to sounds of different frequencies. Note that as frequency increases, the position of the antinode shifts toward the oval window.*

as the frequency is increased, the peak in basilar displacement shifts toward the near end. This observation is consistent with the structural features of the basilar membrane, which is relatively heavy and under small tension near the heliotrema but thin and under considerable tension near the oval window. From the resonant properties of the string, which carry over qualitatively to membranes, we would expect the thicker part to respond to low frequencies and the thin, taut section to be excited in a resonant manner at high frequencies. Thus the distortion of the basilar membrane, which determines the group of hair cells that are stimulated, reflects the pitch of the sound wave.

Hearing is by no means explained by these mechanistic analogies. The auditory cortex of the brain plays an active part, selecting from among a profusion of sounds those that are meaningful and significant. In a noisy environment, one soon finds that the noise is no longer consciously perceived. This feedback between brain and ear is not yet properly understood.

Summary

Sound waves in gases and liquids are compressional waves. In a solid, sound waves may be transverse as well as longitudinal. Audible sound comprises the frequency range between 20 Hz and 20,000 Hz. *Infrasonics* are sounds of lower than audible frequency. *Ultrasonics* are sounds of frequency higher than audible.

The speed of sound in a gas is given by

$$v = \sqrt{\frac{\gamma kT}{m}}$$

where γ is a constant between 1.0 and 1.67 that depends on the molecular structure of the gas, k is Boltzmann's constant, T is the absolute temperature of the gas, and m is the mass of a molecule of the gas.

In sound waves, pressure and displacement bear a 90° phase relation to each other.

Resonance frequencies in pipes depend on the construction of the pipe—whether it is open (or closed) at both ends or open at one and closed at the other end—on the length of the pipe, and on the velocity of sound in the gas within the pipe. For a pipe closed, or open, at both ends the resonance frequencies are given by

$$f_n = \frac{nv}{2L} \qquad n \text{ an integer}$$

where L is the length of the pipe. For pipes open at one end and closed at the other,

$$f_m = \frac{mv}{4L} \qquad m \text{ an } odd \text{ integer}$$

The *intensity* of a wave is defined as the power that crosses a unit area. The intensity of sound at a distance R from a source radiating uniformly in all directions is proportional to $1/R^2$.

Sound intensities are usually measured in decibels (dB). The intensity, in decibels, is given by

$$\beta = 10 \log \frac{I}{I_0} \qquad I_0 = 10^{-12}\ \mathrm{W/m^2}$$

The *Doppler effect* is the change in frequency of sound as perceived by an observer in relative motion with respect to the source of sound. The perceived frequency f' is

$$f' = f\left(\frac{1 \pm v_o/v}{1 \mp v_s/v}\right)$$

where f is the frequency of the sound emitted by the source, v_o and v_s are the speeds of the observer and source when moving toward (upper sign) or away from (lower sign) each other. If the speed of the relative motion is small compared to the speed of wave propagation, this frequency shift is given by

$$\frac{\Delta f}{f} = \frac{v_r}{v}$$

where v_r is the relative speed of observer and source and is taken positive if source and observer are approaching, negative if they are receding.

Provided that v_r is much less than the speed of light, $c = 3 \times 10^8$ m/s, the same expression is also valid for electromagnetic radiation.

The efficacy of an object in scattering waves depends on its size compared to the wavelength.

Multiple Choice Questions

16.1 A truck and a motorcycle are traveling on a highway in the same direction. The truck travels at twice the speed of the motorcycle and passes the cycle. After the truck has passed, the cyclist sounds his horn to signal the truck driver that he may return to the outside lane. The frequency of the motorcycle horn is 400 Hz. Mark the correct statement.

(a) The truck driver and cyclist hear the same frequency of 400 Hz.

(b) The truck driver hears a frequency higher than that heard by the cyclist.

(c) The cyclist hears a frequency of 400 Hz; the truck driver hears a frequency less than 400 Hz.

(d) The cyclist hears a frequency of 400 Hz; the truck driver hears a frequency higher than 400 Hz.

(e) Both cyclist and driver hear the same frequency, which is lower than 400 Hz.

16.2 The velocity of propagation of sound waves in a diatomic gas at 30 °C

(a) is greater the greater the molecular weight of the gas.

(b) is greater the smaller the molecular weight of the gas.

(c) is independent of the molecular weight of the gas.

(d) is less than the speed of sound in a triatomic gas of the same molecular weight.

(e) may be (a) or (b) or (c) depending on the frequency of the sound wave.

16.3 Consider the following two situations: (A) A source of sound, vibrating at 400 Hz, approaches a stationary observer at a speed equal to half the speed of sound in still air. (B) The same source of sound is stationary in still air, but the observer is approaching the source at a speed equal to half the speed of sound.

(a) In cases A and B, the observer hears a sound of the same frequency, which is greater than 400 Hz.

(b) In cases A and B, the observer hears a sound of frequency higher than 400 Hz, but the frequency is slightly higher in case B than in case A.

(c) In cases A and B, the observer hears a sound of frequency higher than 400 Hz, but the frequency is slightly lower in case B than in case A.

(d) In case A, the observer hears a sound of frequency less than 400 Hz.

(e) In case B, the observer hears a sound of frequency less than 400 Hz.

16.4 A resonant system has a fundamental frequency of 100 Hz. If the next-higher frequencies that give a resonance are 300 Hz and 500 Hz, the system could be

(a) a pipe open at both ends.

(b) a pipe closed at both ends.

(c) a string vibrating between two fixed points.

(d) a pipe open at one end and closed at the other.

(e) none of the above.

16.5 The sensitivity of the human ear is greatest near 3000 Hz. This result is consistent with the fact that

(a) the ear canal is a resonant cavity with a fundamental resonant frequency near 3000 Hz.

(b) at a frequency of 3000 Hz, the wavelength of the sound is approximately equal to the separation between the two ears.

(c) nerve impulses travel most rapidly along axons if the sound frequency is about 3000 Hz.

(d) 3000 Hz is the frequency at which the entire skull resonates.

(e) the average frequency of the adult person's heartbeat is 1500 Hz.

16.6 Consider two experiments: (A) A source of sound recedes from a stationary observer at a speed greater than the speed of sound in air. (B) An observer recedes from a stationary source of sound at a speed greater than the speed of sound in air.

(a) In both cases the observer will hear no sound at all.

(b) In case A, the observer will hear a sound of frequency considerably lower than that emitted by the stationary source, but in case B he will hear no sound at all.

(c) In case B, the observer will hear a sound of frequency considerably lower than that emitted by the stationary source, but in case A he will hear no sound at all.

(d) In case A, the observer hears a sound of higher pitch than in case B, but he does hear a sound in both cases.

(e) In case A the observer hears a sound of lower pitch than in case B, but he does hear a sound in both cases.

16.7 A tube that is open at both ends is initially filled with air. Which of the following will decrease the frequency of the fundamental resonance in the tube?

(a) Replacing the air with helium gas, He.

(b) Replacing the air with hydrogen gas, H_2.

(c) Closing one end of the tube.

(d) (a) or (b), but not (c).

(e) (a) or (c), but not (b).

16.8 The speed of sound in an ideal monatomic gas is

(a) proportional to the pressure and inversely proportional to the temperature of the gas.

(b) independent of the temperature and proportional to the pressure of the gas.

(c) independent of the temperature and inversely proportional to the pressure of the gas.

(d) independent of the pressure and proportional to the temperature of the gas.

(e) None of the above is correct.

Problems

Unless it has been specifically stated otherwise, assume that the speed of sound in air at normal temperature is 340 m/s.

(Sections 16.1–16.3)

16.1 Dolphins as well as bats emit ultrasonics. What is the wavelength of a sound of frequency 2×10^5 Hz in water?

16.2 Calculate the speed of sound in helium gas at STP.

16.3 At what temperature will the speed of sound in air equal 300 m/s? 365 m/s?

16.4 The speed of sound in a diatomic gas is measured at 20 °C and found to be 215 m/s. What is the molecular weight of this gas? What is the most probable composition of this gas?

16.5 Compute the speed of sound in hydrogen gas at 0 °C.

• 16.6 The expression for the speed of compressional waves in a solid is $v = \sqrt{B/\rho}$, where B is the bulk modulus of the material (see Chapter 9) and ρ is the mass density. Show that this result is dimensionally correct.

• 16.7 At what temperature is the speed of sound in methane, CH_4, equal to the speed of sound in air at room temperature?

(Section 16.5)

16.8 A straight pipe is closed at one end. What is the shortest length of pipe that will be resonant at a frequency of 70 Hz on a day when the speed of sound in air is 345 m/s?

• 16.9 When the compressor for an organ breaks down, the enterprising organist quickly connects a high-pressure tank of nitrogen to the system. What will the resonance frequency be if the pipe normally emits middle C (264 Hz)?

• 16.10 What is the length of the shortest organ pipe that can give a sound whose fundamental frequency is 42 Hz? What are the frequencies of the first two overtones from this pipe?

16.11 An organ pipe open at both ends is 0.6 m long. What is the frequency of the fundamental resonance and of the first overtone for this pipe at normal temperature?

16.12 Repeat Problem 16.11 for an organ pipe that is open at one end and closed at the other.

• 16.13 You are asked to construct a pipe that will resonate at room temperature at the following frequencies: 170 Hz and 510 Hz, and no other frequencies between 0 and 600 Hz. Describe the pipe and give its length. If this pipe is used when the air temperature has dropped to 10 °C from room temperature (20 °C), will this affect the frequencies, and if so, how?

16.14 A pipe, closed at one end and open at the other, has a fundamental frequency in air of 1360 Hz. Find the length of the pipe and the frequencies of the first few overtones.

• • 16.15 One way of measuring the velocity of sound in air is by listening to resonances when a tuning fork is struck above a tube partially filled with water as the water level is changed. Suppose that a tuning fork that vibrates at 500 Hz is used in this experiment, and resonances are heard when the water level in the tube changes by 35 cm. What is then the velocity of sound in air? What is the shortest tube length that will give a resonance under these conditions?

• • 16.16 A flute can be thought of as a pipe that is closed at one end and open at the other. If such a flute is made of an alloy whose expansion coefficient is vanishingly small, and the instrument is tuned properly at 20 °C, what, if anything, will happen if during a concert the temperature of the air in the hall increases to 28 °C? Give your answer in terms of the ratio of the frequencies of the sounds produced by the instrument at the two temperatures.

(Section 16.6)

16.17 Two sounds have intensities of 10^{-4} W/m^2 and 10^{-7} W/m^2. What is the difference in the decibel levels of the two sounds?

16.18 Two sounds differ in intensity by 20 dB. What is the ratio of the pressure amplitudes of the two sounds, if it is assumed that their frequencies are the same?

16.19 The pressure amplitude at the threshold of hearing (10^{-12} W/m^2) is 3×10^{-5} Pa. What is the pressure amplitude at an intensity of 60 dB? 120 dB?

• 16.20 A sound of frequency 200 Hz and one of 800 Hz have the same intensity at a given point. What is the ratio of the pressure amplitudes of the two sounds at that point?

• 16.21 Most good hi-fi amplifiers claim to deliver 25 W or more of undistorted power. Suppose that you turn up the volume control so 15 W is delivered to the speaker, and that the conversion efficiency of the speaker is 10 percent. Estimate the sound intensity that would reach your ear if you were standing 3 m from the speaker. Assume that the speaker radiates uniformly into the forward hemisphere, and that the average frequency of the sound is 500 Hz.

• 16.22 At a certain distance from a sound source, the intensity is 80 dB. What is the intensity of the sound in watts per square meter, and how much energy falls on a 2-m^2 area during 10 s?

• • 16.23 The inverse-square law assumes that the source of sound radiates uniformly in all directions and that the transmitting medium absorbs no sound. Absorption by air amounts to about 8 dB/km. If the sound intensity at a distance of 100 m from a jet engine is 130 dB, what is the intensity at 2 km? Assume that the sound radiates uniformly.

• 16.24 Repeat Problem 16.23, neglecting absorption of sound by the air.

• • 16.25 When a violinist pulls the bow across a string, the force with which he pulls is not very great, perhaps 0.5 N. Suppose the bow travels across the open A string at 0.5 m/s. A listener 10 m from the violinist hears a sound whose intensity is 80 dB. With what efficiency is the mechanical energy of bowing converted to sound energy?

• • 16.26 At 20 m, the intensity from a uniformly radiating sound source is 60 dB. At what distance from this source will its sound be just barely perceptible if absorption of sound by air is neglected?

(Section 16.7)

• **16.27** Two trains are moving toward each other at the same speed, one-tenth of the speed of sound in air. If a whistle having a frequency of 330 Hz is sounded by one of the trains, what is the frequency of the sound that a passenger on the other train would hear?

• • **16.28** A siren whose frequency is 400 Hz is suspended from a helium balloon. The balloon is carried along by an 80-km/h wind that blows west to east. What frequency will be apparent to a stationary observer due east of the balloon? a stationary observer due west of the balloon?

• **16.29** A motorist traveling at 50 mph approaches a vertical cliff. He sounds his horn and notes that the pitch of the reflected echo corresponds to a frequency of 440 Hz (the driver is a professional violinist). Find the frequency of the sound emitted by the horn.

• • **16.30** A bat flying toward an obstacle at 15 m/s emits sound pulses at a repetition rate of 120 pulses per second. What is the time interval between the echo pulses?

• **16.31** A fog horn emits a sound whose frequency is 200 Hz. A stationary observer maintains that the perceived frequency is not constant but depends on whether it is calm or a gale is blowing. Could that be correct? Calculate the perceived frequency if a wind of 80 km/h is blowing from the fog horn toward the observer.

• **16.32** A whistle emits a sound of 700 Hz. A passenger in a car traveling away from this sound source hears a frequency of 650 Hz. What is the speed of the car?

• • **16.33** A police radar uses a transmitter that generates electromagnetic waves whose wavelength is 1 cm. The speed of propagation of these waves is 3×10^8 m/s. Suppose that a car approaches the radar at a speed of 120 km/h. What is then the difference in frequency between the transmitted signal and that reflected to the radar by the approaching car?

• • **16.34** If in a lecture hall a tuning fork is struck and moved toward the blackboard, students hear a well-defined beat. If the frequency of the tuning fork is 440 Hz, and the beat frequency is 3 Hz, with what speed is the tuning fork approaching the blackboard? (*Note:* The sound heard by students is the superposition of two sounds—one due to the moving tuning fork radiating toward the students, the other the sound from the moving tuning fork that is reflected toward the students by the blackboard.)

(Section 16.10)

16.35 What intensity level must a 100-Hz sound have so that it will appear to be as loud as a 1000-Hz sound whose intensity is 40 dB?

• **16.36** When the headphones placed to the ears of an individual are excited with currents of 300, 400, and 500 Hz superposed on each other, the listener perceives a sound whose frequency is 100 Hz. Explain this observation.

•17•

Electrostatics I: Charges and Interactions Between Charges

Hence have arisen some new terms among us. We say B (and bodies like circumstanced) is electrized positively; *A,* negatively. *Or rather, B is electrized* plus; *A,* minus. *And we daily in our experiments electrize* plus *or* minus, *as we think proper.*

BENJAMIN FRANKLIN

17.1 Introduction

The word *electricity* derives from ηλεκτρον, the Greek word for "amber." As this etymology implies, the early Greeks were familiar with the effects of static electrification manifested by a piece of amber that has been rubbed with fur or wool cloth. Over the centuries it was discovered that other materials also develop "electricities" when rubbed with suitable substances. But careful and thoughtful investigations of electrical effects commenced only toward the end of the seventeenth century. A crucial discovery was made by Charles François de Cisternay du Fay, who, in 1734, wrote to Charles, Duke of Richmond and Lenox, as follows:

> . . . chance has thrown my way another principle . . which casts new light on the subject of electricity. This principle is that there are two distinct electricities, very different from each other: one of these I call *vitreous electricity;* the other, *resinous*

> *electricity.** The first is that of [rubbed] glass, rock crystal, precious stones, hair of animals, wool, and many other bodies. The second is that of [rubbed] amber, copal, gum lac, silk, thread, paper, and a vast number of other substances.
>
> The characteristic of these two electricities is that a body of, say, the *vitreous electricity* repels all such as are of the same electricity; and on the contrary, attracts all those of the *resinous electricity.*

Today we use a different terminology. We now know that the observed effects result from excess charges on the glass and amber; we talk of charges, not "electricities," positive and negative rather than vitreous and resinous, but the basic ideas are much the same.

17.2 Origin of Electric Charges

One of the first questions that comes to mind is, What is the origin of electric charges? This question went unanswered until fairly late in the study of electricity, although Davy, Faraday, and others were convinced that charges were associated with chemical elements. Still, the unequivocal identification of the fundamental charged object, responsible for electrostatic, electrodynamic, and magnetic phenomena, came only toward the close of the nineteenth century with J. J. Thomson's discovery of the electron and the subsequent clarification of atomic structure by Rutherford and Bohr. Today, high-school students learn about electrons, protons, neutrons, and nuclei, and it is perhaps difficult to appreciate the tortuous path that has led to our current understanding of atomic and nuclear phenomena.

Every atom consists of a very small, relatively massive, *positively charged nucleus,* and one or more very much lighter, *negatively charged electrons.* The electrons may be envisaged as occupying a roughly spherical region about the nucleus, and are sometimes depicted as circling the nucleus as planets circle the sun. The analogy to the solar system is sometimes a helpful model. But it is no more than a model; there are, as we shall see later, important fundamental differences between the atomic system of nucleus and electrons and the solar system.

The positive charge of the nucleus attracts the negatively charged electrons, holding them in stable orbits. Those electrons that are fairly

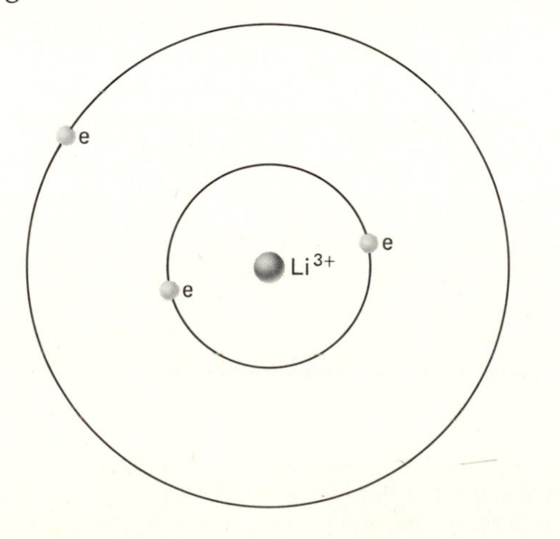

Figure 17.1 Model of the lithium atom. The nucleus contains three protons, each with a charge equal in magnitude to that of an electron but of positive sign. This nucleus is surrounded by three negatively charged electrons, and the entire system is electrically neutral. Two of the electrons are held in orbits fairly close to the nucleus; these electrons are tightly bound to the nucleus and are called core electrons. The third electron is in an orbit with a much greater average radius, and it is much less tightly bound to the nucleus; this electron is called a valence electron.

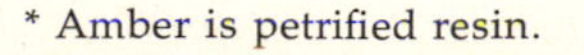

* Amber is petrified resin.

close to the nucleus are tightly bound; that is, they are not easily pulled away; those in more extended orbits can be dislodged more readily. For some materials, such as glass, electrons on atoms near the surface can be transferred to silk by mechanical rubbing. Similarly, electrons from atoms of fur or wool tend to migrate to ebonite or amber. As a result of this charge transfer, ebonite acquires excess negative charge when rubbed with fur, whereas a glass rod becomes electron-deficient and thus positively charged when rubbed with silk.

Numerous experiments have confirmed certain fundamental attributes of charge; charge is *quantized,* charge is *conserved,* and charge is *invariant.*

When we say that charge is quantized, we mean that it can appear only as *integral multiples of a fundamental indivisible unit.* That unit is the electronic charge e.* A certain macroscopic charge of negative sign simply means that an excess number of electrons have been deposited on the object carrying this negative charge; conversely, a net positive charge implies prior removal from this object of a finite, integral number of electrons.

By conservation of charge one means that the total charge in an isolated system must remain constant. That does not restrict the number of electrons or positive charges that may be present; as we shall see later, it is possible to create charged particles by a sufficient expenditure of energy, and the inverse process, the annihilation of charged particles with a concomitant release of energy, is also common. However, in every such process, an equal amount of positive and of negative charge must be created or destroyed, the total, or net, charge remaining constant.

Finally, the concept of invariance refers to the behavior of systems under relativistic conditions, that is, at speeds approaching the speed of light, $c = 3 \times 10^8$ m/s. As discussed in Chapter 28, mass, length, and time are not relativistically invariant. For example, the inertial mass of an object depends on its speed in the reference frame in which the mass is measured. By contrast, charge is invariant; the charge of an electron, a proton, or any other particle remains the same whatever its speed of motion.

These three fundamental properties of charge—its quantized nature, conservation, and invariance—have far-reaching theoretical implications in elementary-particle physics.

How many electrons are transferred from a glass rod to a silk cloth under normal circumstances? The answer depends on various factors, such as number of strokes, contact pressure, and temperature; but typically about 10^9, that is, one billion electrons are involved. That may seem a huge number, but in fact it is quite small when we remember that the total number of atoms in an even relatively small rod of about 100 g is of order 10^{24}. That is also the number of outermost atomic electrons, those most readily dislodged. Thus, only a very small fraction, about $1/10^{15}$, of the atoms are left electron-deficient.

* Some years ago M. Gell-Mann suggested that protons and neutrons are complexes of subnuclear particles that carry charges of $\frac{2}{3}$ and $\frac{1}{3}$ the electronic charge. There followed a period of intensive, though unsuccessful, search for these so-called *quarks.* Evidence has been accumulating that such entities, with fractional electronic charges do exist within a nucleus, though not in isolation. The existence of quarks does not violate the concept of charge quantization; it means only that the basic unit of charge may have to be reduced by a factor of 3.

17.3 Conductors, Insulators, and Semiconductors

It is well known that some materials, such as copper, aluminum, and other metals conduct electricity, while a different group, including glass, rubber, and most plastics are widely used as electrical insulators.

What distinguishes these two types of substances as regards their electrical characteristics is that in conductors the outermost, so-called valence electrons are relatively free to move about within the material, though they cannot readily escape from it; in insulators, by contrast, even valence electrons are tightly bound to their respective nuclei. The precise description of electrons in solids is a subtle matter, requiring quantum theory and the solution of differential equations by sophisticated mathematical techniques. Still, the broad generalization given above is a fair approximation of the real situation.

There is a third class of substances, which includes the elements germanium and silicon and also numerous compounds such as indium antimonide and gallium arsenide, that are neither conductors nor insulators. In these substances, the number of "free" electrons is only a small fraction of those in a metal; moreover, the number depends on temperature and on the concentration and the type of impurity that may be present. These *semiconductors* do conduct electricity, but resistance to charge flow in them is much greater than in metals, though much less than in insulators. It is our ability to control the electrical properties of semiconductors by adjusting the type and concentration of impurities that accounts for their immense technological importance. These materials form the matrices for transistors and for a multitude of minute and highly elaborate integrated solid state circuits that have completely replaced the older vacuum tubes in virtually all electronic devices.

17.4 Interaction between Charges; Coulomb's Law

If a glass or ebonite rod is charged by rubbing, some of that charge may be transferred to another object, such as a light pith ball, by physical contact. By this procedure, one can, in a series of simple experiments diagrammed in Figure 17.2, demonstrate that:

1. Like charges repel each other.
2. Unlike charges attract each other.

From these experiments two basic facts emerge. First, the electrostatic interaction differs in an essential way from the only other interaction-at-a-distance encountered so far, the gravitational interaction; the latter is always attractive, the former may be attractive or repulsive. It is this sign reversal that hides the second important distinction; the electrostatic interaction is enormous compared to the gravitational. This tremendous difference can be appreciated from the following consideration. If two 1-kg spheres of aluminum are separated by 1 m, and if one electron from every billionth atom is transferred from one sphere to the other, leaving one negatively charged and the other positively charged, the force of attraction between the spheres will be more than 100,000 N, or more than ten thousand times the weight of each sphere. We are generally unaware of the great strength of electrostatic forces because objects

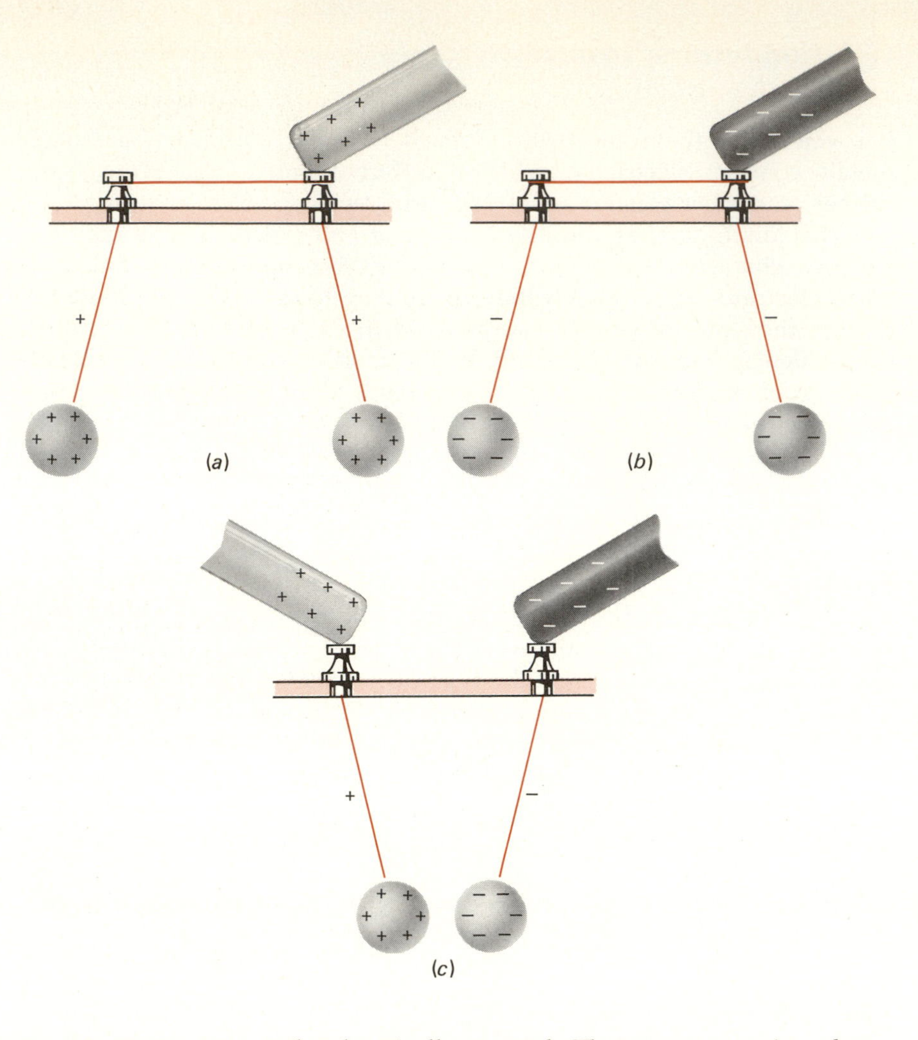

Figure 17.2 Two pith balls that have been painted with aluminum paint are suspended by a metal thread from metallic supports set in an insulating horizontal bar. In (a) and (b) a metal wire connects the two supports. When (a) a positively charged glass rod or (b) a negatively charged ebonite rod is touched to the metal support, the charge distributes itself over the pith balls. The observed effect is a repulsion of the two balls in each case. (c) When the metallic connection between the pith balls is removed, and one is charged positively and the other negatively, the two pith balls attract.

we encounter are nearly electrically neutral. They may contain a huge number of charges, but half of them will be positive, the other half negative and of equal magnitude, producing no net charge. Some small charge imbalance can be achieved and maintained. If the imbalance becomes too large, the object will discharge spontaneously, often with a great display of sparks.

To probe electrostatic forces and deduce quantitative relationships, one needs a device for measuring charge. One of the earliest such instruments is the gold-leaf electroscope, shown in Figure 17.3. It consists of a hollow cylinder in which a thin gold leaf is attached at its upper end to a vertical metal rod. The rod is set in an insulating sleeve and passes out through the top of the chamber; the sides of the chamber are made of glass to permit viewing the disposition of the gold foil.

If, for example, a negative charge is deposited on the small brass sphere atop the metal rod, the many million excess electrons do not remain fixed there. They are able to move about within and on the surface of the metal, and since they repel each other, they will redistribute themselves so as to keep out of each other's way as much as possible. Thus, some of that negative charge will appear on the gold leaf and metal rod; and since like charges repel, the flexible gold foil will diverge from the rod, the angle between them depending on the amount of charge deposited on the electroscope.

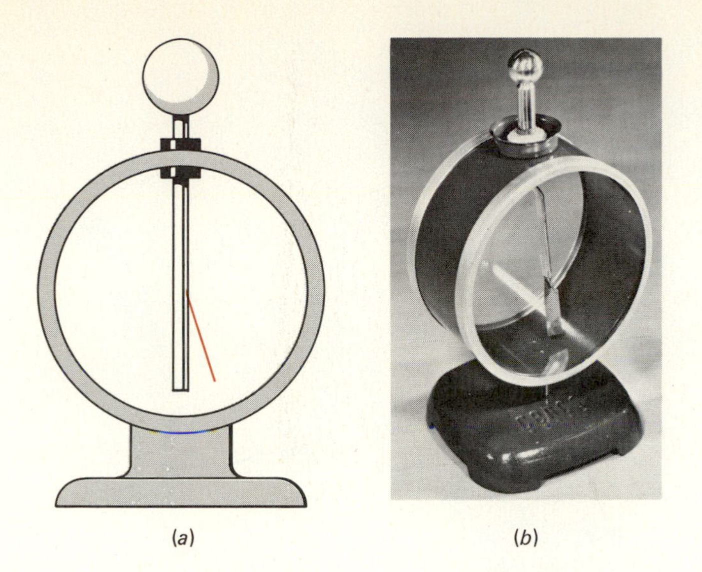

Figure 17.3 (a) The gold-leaf electroscope consists of a flat metal rod to which a thin gold leaf has been soldered. The rod is brought out of the enclosure through an insulating plug and is usually surmounted by a small metal sphere. (b) Photograph of a gold-leaf electroscope.

Although such instruments do not measure charge accurately, and have been replaced by more precise electronic meters, they are still used in lecture demonstrations.

The mathematical relation that governs the interaction between like and unlike charges was enunciated by Charles Coulomb in 1785. Coulomb measured the strength of the forces as functions of the distance between the charges and their magnitudes, using a sensitive torsion balance of the sort employed by Cavendish (see Section 7.3). Coulomb concluded that the magnitude of the force between two charges in air is proportional to the product of the two charges, q_1 and q_2, and inversely proportional to the square of the distance between them. That is,

$$F = k\frac{q_1q_2}{r^2} \tag{17.1}$$

where k is a constant. The direction of the force on the charges is along the line joining them. It is to be understood that a positive force in Equa-

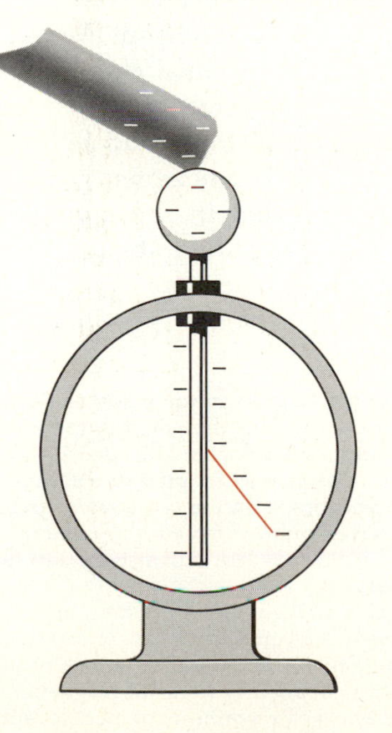

Figure 17.4 Negative charge has been deposited on the sphere atop the electroscope. The charge distributed itself over the entire metallic surface connected to the sphere. As a result, the repulsion of like charges causes deflection of the gold leaf.

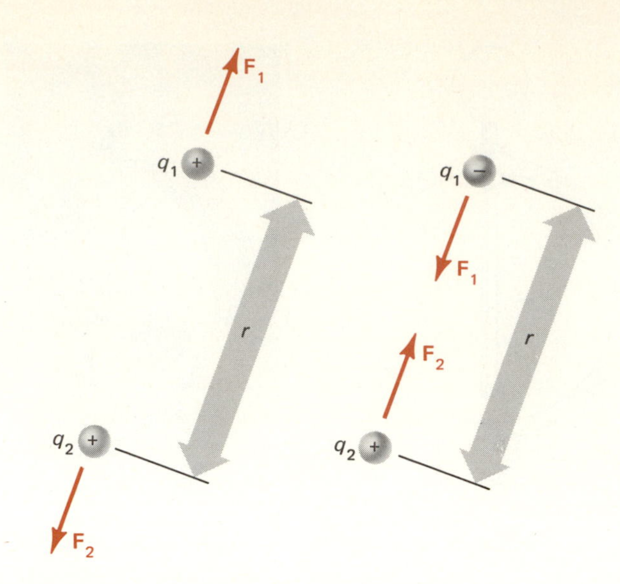

Figure 17.5 Two like charges repel, two unlike charges attract each other. The forces between the charges depend on the magnitude of the charges, q_1 and q_2, and on the separation r between them, according to Coulomb's law, Equation (17.1).

tion (17.1) implies repulsion, a negative, attraction. Since force is a vector quantity, the adjectives *positive* and *negative* are meaningless in any other context.

17.5 Units

The constant of proportionality in Coulomb's law, the factor k in Equation (17.1), depends on our choice of units; we must now decide on a consistent set of units for force, charge, and length.*

The SI unit of charge is the coulomb, abbreviated C. It is a unit derived from the unit of current, the ampere, which is defined in terms of the forces that act between two current-carrying conductors in vacuum (see Chapter 22). In the SI system, the constant k in (17.1) takes the value

$$k = 8.9874 \times 10^9 \text{ N}\cdot\text{m}^2/\text{C}^2 \simeq 9 \times 10^9 \text{ N}\cdot\text{m}^2/\text{C}^2 \qquad \textbf{(17.2)}$$

One disadvantage of the SI system is that the coulomb is an inconveniently large unit. This is apparent from the fact that the force exerted by a charge of one C on another equal charge at a distance of 1 m is 9×10^9 N, roughly equal to the force of gravity experienced by a mass of 10^9 kg at the earth's surface! In a good lightning bolt, a charge of the order of 1–10 C is transferred between ground and the thundercloud, but typical charges generated in the laboratory range between picocoulombs (10^{-12} C) and microcoulombs (10^{-6} C).

There is a natural unit of charge—the charge carried by a proton, the nucleus of a hydrogen atom. This charge is positive and equal in

Figure 17.6 A lightning discharge.

* One procedure, adopted early on in the study of electrostatics, was to set the constant k equal to unity. This then defines the unit of charge as that amount which results in unit force between two identical charges separated by unit length. When this scheme was adopted, the system of units the scientific community used was based on the centimeter, gram, and second, and the corresponding derived units of force (dyne), energy (erg), etc. The unit of charge so defined is now known as one electrostatic unit (esu) of charge; it is also called one statcoulomb.
That scheme worked well enough for some time. Then, in the nineteenth century, Faraday and others discovered the intimate connection between electric currents and magnetic phenomena, and another system of units came into use, in which the unit of current is defined in terms of magnetic interactions. All this took place well before Maxwell's synthesis of electric and magnetic phenomena, and

magnitude to the negative charge carried by an electron. It is designated by the letter e and has the value

$$e = (1.6021892 \pm 0.0000029) \times 10^{-19}\ \text{C} \simeq 1.6 \times 10^{-19}\ \text{C} \qquad \textbf{(17.3)}$$

For the description of macroscopic electrostatic events this unit is as inconveniently small as the coulomb is large. However, in calculations of atomic events, e proves to be a natural as well as a convenient choice.

● ***Example 17.1*** A charge Q experiences a force of 6 N toward the east when a charge of $+5\ \mu\text{C}$ is placed 20 cm to the east of Q. Determine the charge Q.

From Equation (17.1), we can solve for Q_1 knowing Q_2 and F:

$$Q_1 = \frac{Fr^2}{kQ_2} = \frac{(6\ \text{N})(0.2\ \text{m})^2}{(9 \times 10^9\ \text{N}\cdot\text{m}^2/\text{C}^2)(5 \times 10^{-6}\ \text{C})} = 5.33 \times 10^{-6}\ \text{C}$$

We must now decide whether Q_1 is positive or negative. Since the force on that charge is *toward* the positive charge of $5\ \mu\text{C}$, the two charges attract each other. We can therefore conclude that the unknown charge is $Q = -5.33\ \mu\text{C}$. ●

17.6 Grounding and Charging by Induction

Before a tanker truck pumps fuel into the wing tanks of an airplane, both the plane and the truck are electrically connected to a ground post by conducting cables. The purpose is to provide a path to ground for any charges that might have accumulated on the truck or plane, whose rubber tires insulate them from the pavement. If a substantial charge had accumulated, then as the driver prepared to fill the plane's tanks, it might discharge in a spark that jumps between nozzle and fuselage, thereby starting a fire or explosion.

Let us examine this grounding process more carefully. Suppose that we place some negative charge on a metal sphere, as in Figure 17.7(a).

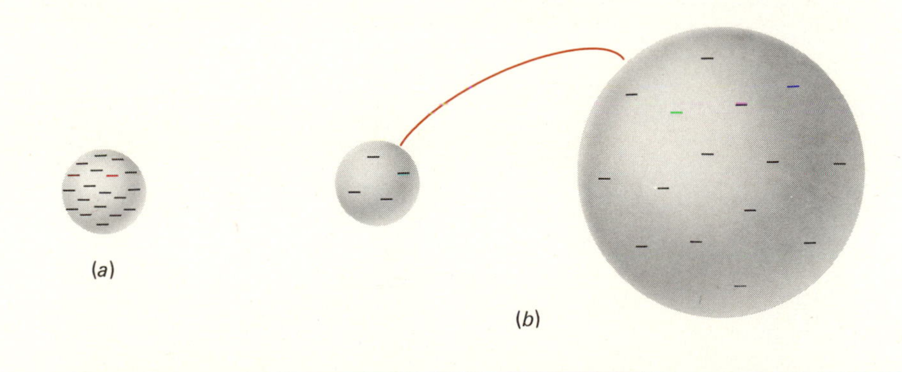

Figure 17.7 (a) *The negative charge on the small metal sphere is distributed uniformly over the sphere.* (b) *If this sphere is electrically connected to another, larger metal sphere, most of the negative charge will move over to the larger sphere as a result of the mutual repulsion between like charges.*

the consequent recognition that the two systems of units, the electrostatic (esu) and the electromagnetic (emu), were not independent. Still, physicists used esu and emu units during the first half of this century largely according to personal preference and depending on whether the problem at hand was most conveniently expressed in one or the other system. What made matters even more exasperating for the unfortunate student of that generation was that the commonplace, practical units, such as the volt and ampere, which differed from their esu and emu counterparts by various conversion factors, proved so much more convenient than the statvolt (esu volt) or abampere (emu ampere) that they, too, infiltrated the scientific literature.
The SI units are an enormous improvement over the near chaos that characterized the use of units in electricity and magnetism until fairly recently.

Since the charges repel, they tend to move as far from each other as possible, but as long as the sphere is electrically insulated from its surroundings, the best the charges can do is to distribute themselves uniformly over the surface. If, however, we now connect this sphere by a wire to a much larger conductor, as in Figure 17.7(*b*), much of the charge will shift to the larger one as a result of the mutual repulsion. If the larger sphere is the entire earth, which does conduct electricity, all the charge on the metal sphere will discharge to ground.

Charge can be conducted *to* ground or *from* ground onto a conductor. Consider, for instance, the situation shown in the sequence of Figures 17.8(*a*), (*b*), and (*c*). When the positively charged glass rod is brought close to the metal sphere, which has been connected to ground by a conducting wire, negative charges, electrons, are attracted toward the posi-

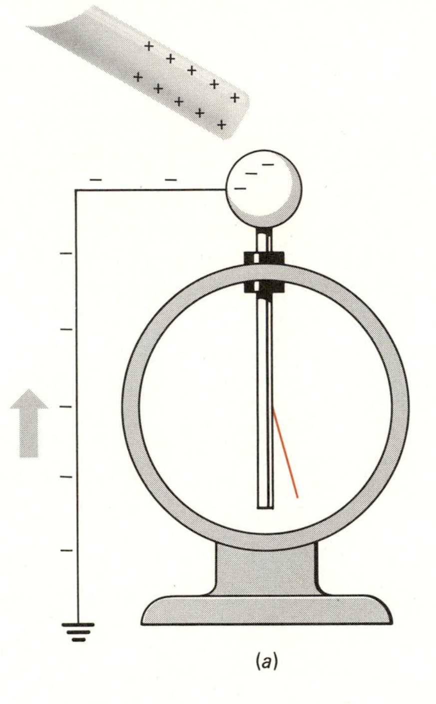

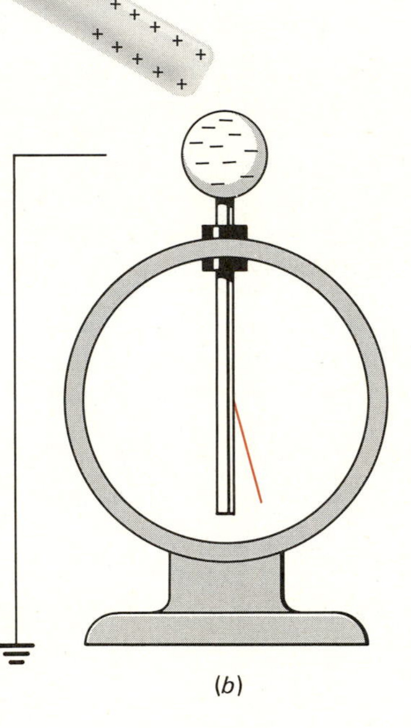

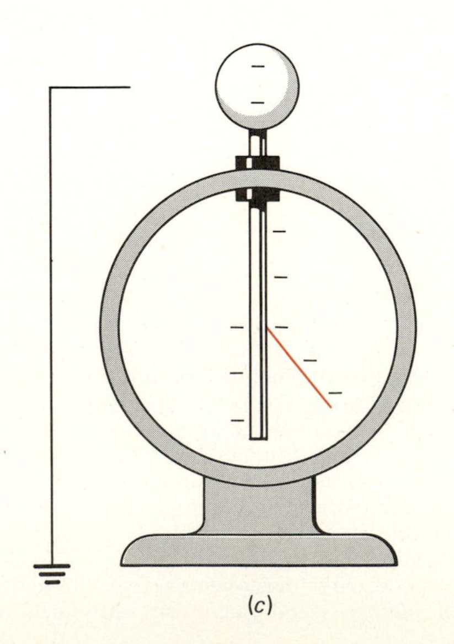

Figure 17.8 Charging by induction. (*a*) The electroscope is connected to ground. As the glass rod is brought near the metal sphere, negative charges are attracted to the sphere from the ground. (*b*) The ground connection is removed while the glass rod is kept in position. (*c*) If the glass rod is then withdrawn, the negative charge on the electroscope, which redistributes itself over this region, causes a deflection of the gold leaf, showing that there is a net charge on the instrument.

tively charged glass and flow from ground to the surface of the sphere. If we then disconnect the wire from the sphere *without moving the glass rod,* whatever charge has been drawn to the sphere from ground is trapped and will remain on the sphere even after the glass rod has been withdrawn.

This method of charging a conducting object is called *charging by induction.* The charge is induced on the conductor by the proximity of another charged object, not transferred to it directly as in charging by conduction. Note that when a conductor is charged by induction, its charge is of a sign opposite to that on the charging object. Note also the convenience of charging by induction; the charging object is not depleted of its charge, and the process can be repeated innumerable times without the need of renewed electrification by friction, that is, rubbing with a silk cloth.

17.7 Induced and Permanent Dipoles

If the gold leaf of an electroscope is watched closely during the charging process depicted in Figure 17.4, one observes that just *before* the glass (or ebonite) rod touches the brass sphere, the gold leaf already swings slightly away from the metal support rod as though in anticipation of events to come. What happens is that as the positively charged glass rod approaches the brass sphere atop the electroscope, it pulls some of the mobile electrons toward the top of the instrument, leaving the lower portion, including the gold leaf, electron-deficient, i.e., positively charged. These positive charges repel each other, and so the gold foil moves away from the electroscope rod.

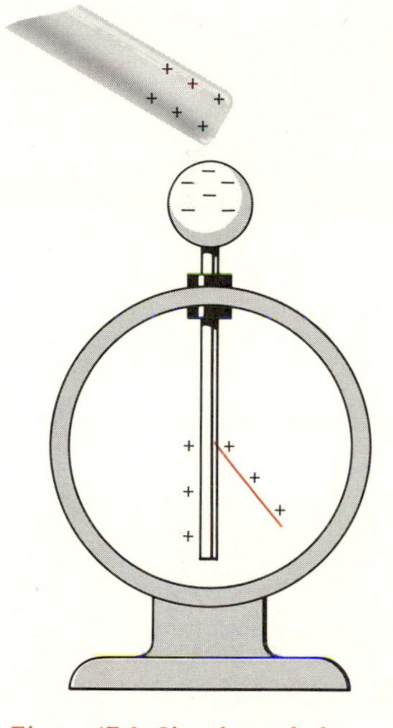

Figure 17.9 *If a charged glass rod is brought near an ungrounded electroscope, there is also a deflection of the gold leaf. The presence of the charges on the glass rod induces a relative displacement of negative and positive charges in the electroscope. The result of this induced dipole is a deflection of the gold leaf.*

Notice that before actual contact with the glass rod, *there is no net charge on the electroscope.* The *proximity* of the charged rod already produces a nonuniform charge distribution on the electroscope, and it is this nonuniformity that is responsible for the observed effect.

A distribution of charges such that equal positive and negative charges are separated by some distance is called an *electric dipole,* or an electric dipole moment. A dipole is a *vector quantity,* in contrast to a charge, which is a scalar. The *strength,* or *magnitude,* of the dipole is defined as the product of the separated charges and the distance between the positive and negative charge. The *direction* of the electric dipole is defined as the direction from the negative to the positive charge.

In the case described above, the presence of the positively charged glass rod *induces a dipole* on the metallic conductor of the electroscope. If the glass rod is removed without touching the electroscope, the charges on the instrument relax to their uniform distribution, and the dipole vanishes.

Not all dipoles are induced. Many molecules carry permanent dipoles, though each molecule is electrically neutral. For example, a simple molecular dipole is formed when sodium and chlorine atoms combine to form a molecule of ordinary table salt, NaCl. The sodium atom gives up one of its electrons, the valence electron, which shifts to the chlorine atom; as a result, the sodium atom, now an ion, is positively charged, and the chlorine ion is negatively charged. The most ubiquitous polar molecule is water, and its very large dipole accounts for many important properties of this liquid, notably its ability to serve as a solvent for a very wide range of chemicals.

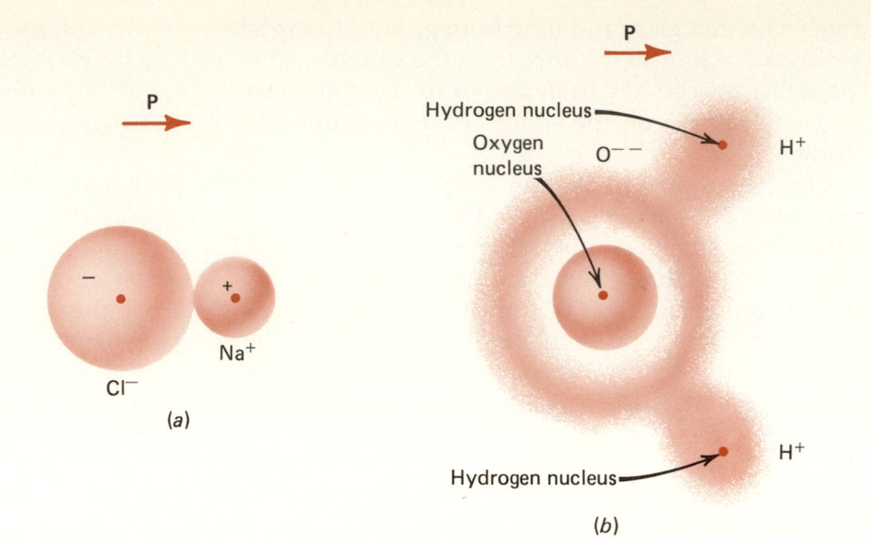

Figure 17.10 *Many molecules are permanent electric dipoles. (a) The NaCl molecule, a combination of an Na^+ and Cl^- ion, is such a dipole. (b) The asymmetric disposition of the two protons (hydrogen ions) about the oxygen ion in the water molecule gives rise to a very large dipole moment for this molecule.*

Formation of a dipole produces a net attraction between a charged object and a neutral one. If a charge $+Q$ is brought near a neutral metal sphere, the mobile electrons will redistribute themselves so that the part nearer the charge Q carries a net negative charge, while the part far from Q carries a net positive charge. Although the metal sphere as a whole remains neutral, this induced dipole and the charge Q attract each other because the negatively charged part of the metal sphere is nearer $+Q$ than the positive.

Even a nonconducting object—one made of wood, for example—will be attracted to the charge Q because some of the molecules of the wood, for instance water, carry permanent dipole moments. Although these dipoles are normally oriented at random, they reorient themselves so that a majority have their negative ends pointing toward the positive charge Q. The result is again a macroscopic dipole and a consequent attractive force between the wood and the charge Q.

If the isolated charge were negative instead of positive, the interaction would still be attractive. In that case, the induced dipole would be oppositely directed.

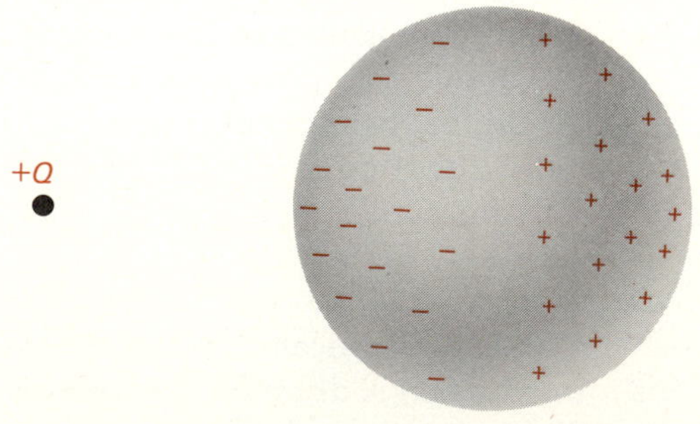

Figure 17.11 *In the vicinity of a point charge $+Q$, the mobile charges on a metal sphere redistribute themselves. The result is a dipole which is attracted to the positive charge.*

A substance in which molecular dipoles are oriented is said to be *polarized.* Insulators can be polarized even if they contain no permanent molecular dipoles. In such materials, the atomic or molecular charge distribution is normally spherically symmetric, but the electrostatic force due to some external unbalanced charge causes a small distortion of the electronic orbits and consequently a small induced dipole. By studying

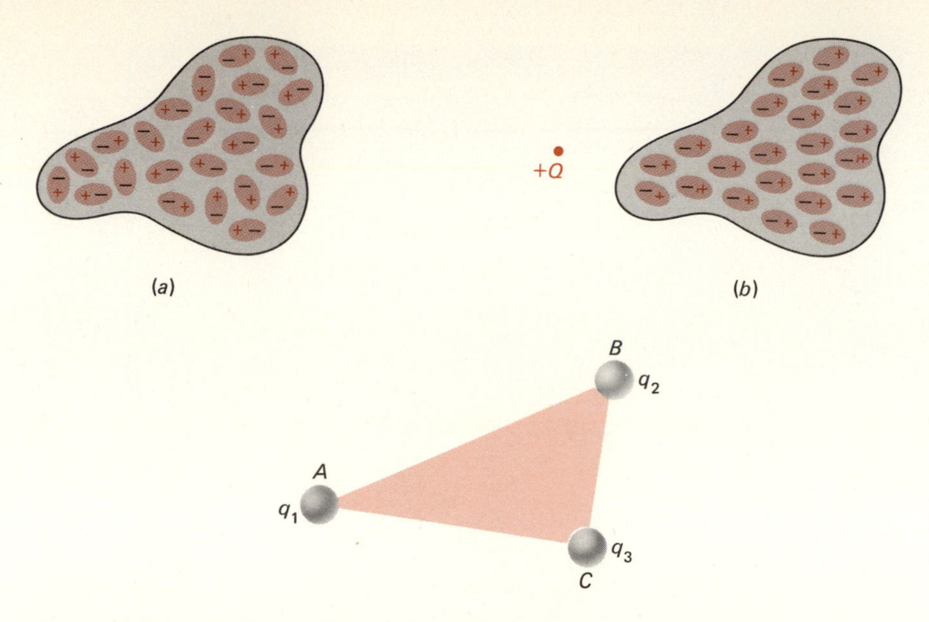

Figure 17.12 (*a*) *In a nonconducting material which contains molecules with permanent dipoles, these dipoles are normally oriented randomly.* (*b*) *In the vicinity of a point charge, the dipoles within this material assume a preferred orientation. As a result of this molecular rearrangement, the material is attracted to the point charge.*

Figure 17.13 *Charges are placed at positions A, B, and C. The presence of the charge q_3 at C does not alter the interaction between charges q_1 and q_2.*

the temperature dependence of the polarization of a substance, one can tell whether polarization is due to the alignment of permanent dipoles or to induced dipoles; such information provides valuable insight into the structure of a material at the atomic level.

17.8 Electrostatic Forces due to a Distribution of Charges

Suppose two charges q_1 and q_2 are located as in Figure 17.13. If a third charge q_3 is then positioned at point C, what is the force that acts on it, and what are the forces on q_1 and q_2?

The simplest and most plausible assumption one might make is that the force on q_3 is the vector sum of the forces due to q_1 and q_2 acting independently on q_3. Similarly, we might assume that the force acting on q_1 is the vector sum of the forces due to q_2 and q_3 acting independently, and likewise for the force on q_2. In other words, we postulate the validity of the superposition principle. Accordingly, the presence of other charges does not influence the interaction between any pair such as q_1 and q_2. Hence the total electrostatic interaction can be expressed as a vector superposition of pairwise interactions. Plausible though this is, the assumption must nevertheless be verified experimentally; when this is done, one does find that this superposition principle holds.* We conclude therefore that

The net force acting on any charge is the vector sum of the forces due to each of the remaining charges of a given distribution.

● ***Example 17.2*** In Figure 17.14, $q_1 = 3\ \mu C$, $q_2 = -6\ \mu C$, and $q_3 = 2\ \mu C$. What are the electrostatic forces on each of the three charges?

The forces due to the pairwise interactions are shown in Figure 17.14. The first subscript indicates the charge that gives rise to the force acting

* There have been some suggestions that this assumption may fail when extremely strong electrostatic forces come into play, and experiments designed to detect such departures have been performed. However, to date there is no experimental evidence to support that suggestion.

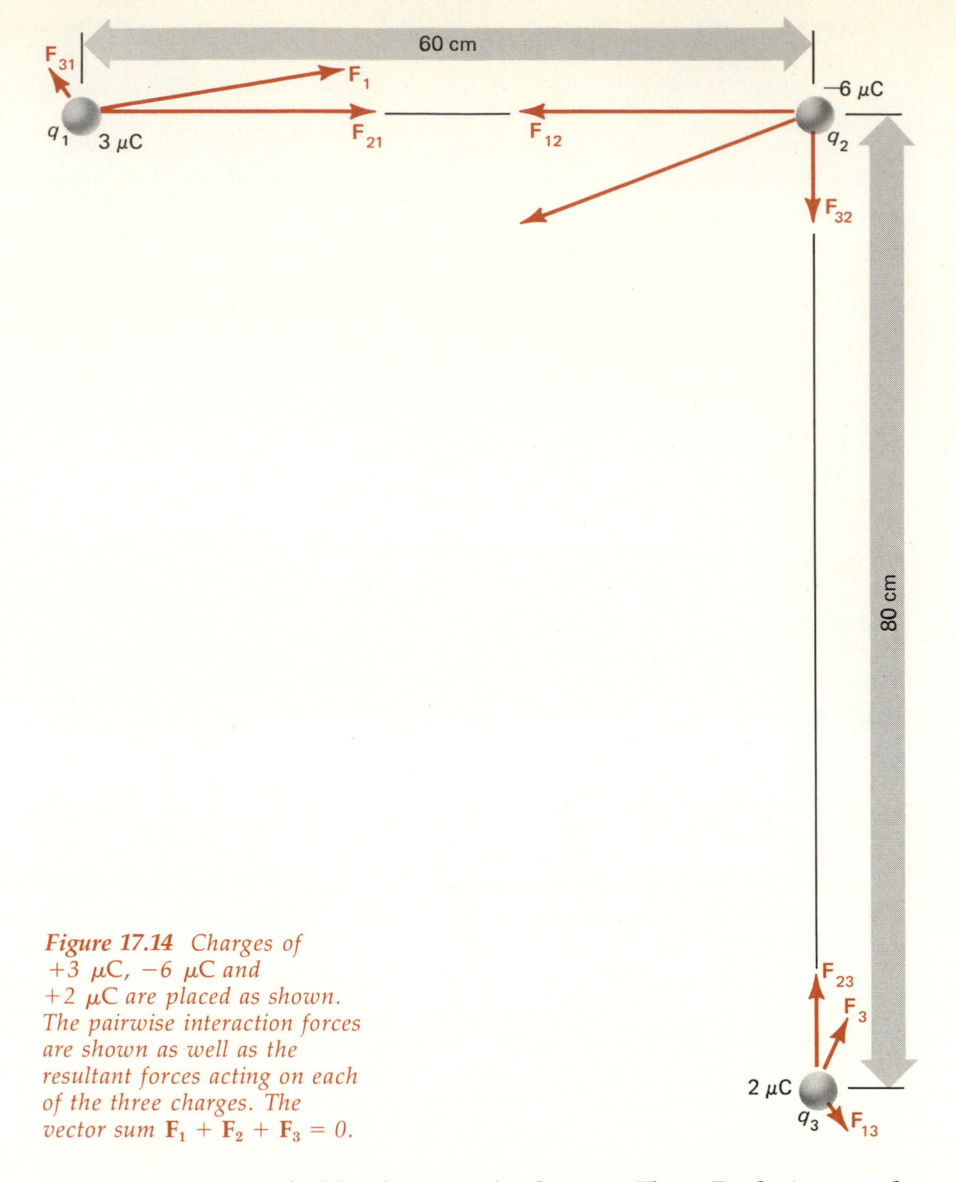

Figure 17.14 Charges of $+3\ \mu C$, $-6\ \mu C$ and $+2\ \mu C$ are placed as shown. The pairwise interaction forces are shown as well as the resultant forces acting on each of the three charges. The vector sum $\mathbf{F}_1 + \mathbf{F}_2 + \mathbf{F}_3 = 0$.

on the charge identified by the second subscript. Thus, $\mathbf{F}_{23}$ designates the force due to the action of charge q_2 on charge q_3. Since q_2 and q_3 are of opposite sign, this force is attractive, as shown. We now list the magnitudes of the pairwise forces between the three charges. Each is obtained by applying Equations (17.1) and (17.2). Thus

$$F_{12} = F_{21} = (9 \times 10^9\ \text{N}\cdot\text{m}^2/\text{C}^2)\frac{(3 \times 10^{-6}\ \text{C})(6 \times 10^{-6}\ \text{C})}{(0.6\ \text{m})^2}$$
$$= 4.5 \times 10^{-1}\ \text{N}$$

$$F_{23} = F_{32} = (9 \times 10^9\ \text{N}\cdot\text{m}^2/\text{C}^2)\frac{(6 \times 10^{-6}\ \text{C})(2 \times 10^{-6}\ \text{C})}{(0.8\ \text{m})^2}$$
$$= 1.69 \times 10^{-1}\ \text{N}$$

$$F_{31} = F_{13} = (9 \times 10^9\ \text{N}\cdot\text{m}^2/\text{C}^2)\frac{(2 \times 10^{-6}\ \text{C})(3 \times 10^{-6}\ \text{C})}{(1.0\ \text{m})^2}$$
$$= 0.54 \times 10^{-1}\ \text{N}$$

According to the superposition principle, the force $\mathbf{F}_1$ acting on q_1 is the sum of the forces due to the presence of q_2 and of q_3. That is,

$$\mathbf{F}_1 = \mathbf{F}_{21} + \mathbf{F}_{31}$$

and similarly for the other two charges.

To perform the vector additions, we follow the procedure discussed in Chapter 1. We take components along x and y directions, add components algebraically, and then find the resultant. The forces $\mathbf{F}_{12}$, $\mathbf{F}_{21}$, $\mathbf{F}_{32}$, and $\mathbf{F}_{23}$ have components only along x or y directions. The force $\mathbf{F}_{13}$ has components along both axes, and these are

$$F_{13,x} = (0.054 \text{ N})(\cos 53°) = (0.054 \text{ N})(0.6) = 0.0324 \text{ N}$$

$$F_{13,y} = -(0.054 \text{ N})(\sin 53°) = -(0.054 \text{ N})(0.8) = -0.0432 \text{ N}$$

Since $\mathbf{F}_{31} = -\mathbf{F}_{13}$, the components of $\mathbf{F}_{31}$ are just the negatives of the components of $\mathbf{F}_{13}$.

The components of the resultant forces acting on q_1, q_2, and q_3 are

$$F_{1x} = (0.45 - 0.0324) \text{ N} = 0.4176 \text{ N} \qquad F_{1y} = 0.0432 \text{ N}$$

$$F_1 = 0.42 \text{ N}$$

$$F_{2x} = -0.45 \text{ N} \qquad F_{2y} = -0.169 \text{ N} \qquad F_2 = 0.48 \text{ N}$$

$$F_{3x} = 0.0324 \text{ N} \qquad F_{3y} = (0.169 - 0.0432) \text{ N} = 0.1258 \text{ N}$$

$$F_3 = 0.13 \text{ N}$$

These forces are shown in Figure 17.14; and though it is not obvious by just looking at the numerical results, it is evident from the action and reaction pairs that the sum of the three forces is zero: $\mathbf{F}_1 + \mathbf{F}_2 + \mathbf{F}_3 = 0$. This must be so because no external forces act on the system of these three charges. ●

Summary

An atom consists of a positively charged nucleus and one or more negatively charged electrons. The number of electrons is just sufficient to cancel exactly the nuclear charge. In general, therefore, a substance is electrically neutral.

Conductors are materials in which some electrons are relatively mobile; these electrons are referred to as *free electrons*. In *insulators* there are no free electrons. In *semiconductors* there are relatively few free electrons and their number depends sensitively on specimen purity and on temperature.

The removal of some electrons leaves an object positively charged; the addition of some excess electrons leaves an object negatively charged. Charged bodies exert *electrostatic forces* on each other. If both bodies carry a net charge of the same sign, they repel each other; if the charges on the two bodies are of opposite sign, the bodies attract each other.

The electrostatic interaction obeys *Coulomb's law:*

$$F = k\frac{q_1q_2}{r^2}$$

Here F is the force between the two charges q_1 and q_2, which are separated by the distance r; k is a constant.

The SI *unit of charge* is the *coulomb*, abbreviated C. In the SI the constant k takes the value

$$k = 9 \times 10^9\ \text{N·m}^2/\text{C}^2$$

The charge of a proton is

$$e = 1.6 \times 10^{-19}\ \text{C}$$

The charge on an electron is $-e$.

Objects can be *charged by conduction* or *induction*.

If an object is electrically neutral but the charge is so distributed that the centers of positive and negative charges do not coincide, that charge distribution is said to possess an *electric dipole moment*. A dipole is a *vector* quantity, directed from the center of the negative to the center of the positive charge distribution; the magnitude of the dipole is the product of the charges and the distance separating the charge centers.

Electric dipoles can be permanent or induced. Many molecules, for example, NaCl and H_2O, have permanent electric dipole moments. Water has a very large dipole moment. Even charge distributions that do not exhibit a permanent dipole may have one induced under the action of a nearby charge.

Electrostatic forces obey the superposition principle.

Multiple Choice Questions

17.1 Figures 17.15(*a*)–(*e*) illustrate a sequence of events, progressing in time from (*a*) to (*e*). Which is the correct statement?

(a) At (*c*) and at (*e*) the sphere carries a net positive charge.

(b) At (*c*) the sphere is uncharged, but at (*e*) it is positively charged.

(c) At (*c*) and at (*e*) the sphere is uncharged.

(d) At (*c*) and at (*e*) the sphere is negatively charged.

(e) At (*c*) the sphere is negatively charged; at (*e*) it is positively charged.

17.2 In Figure 17.16, switch S is initially closed. While charge $+Q$ is in the position shown, S is opened. The charge $+Q$ is then removed. The metallic object A is now

(a) uncharged.

(b) positively charged.

(c) negatively charged.

(d) Any of the above may be correct depending on the charge on A before $+Q$ was brought nearby.

17.3 Good insulators do not carry electric current well because

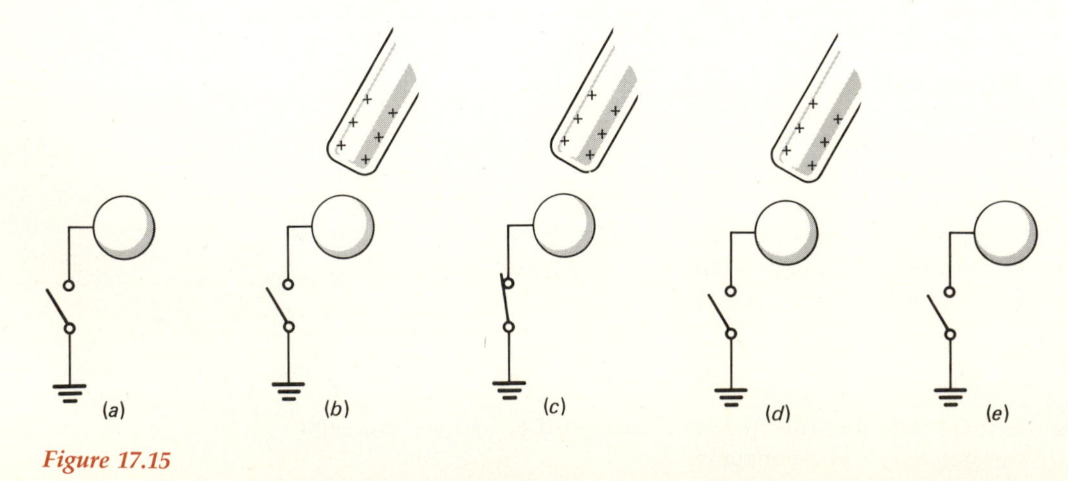

Figure 17.15

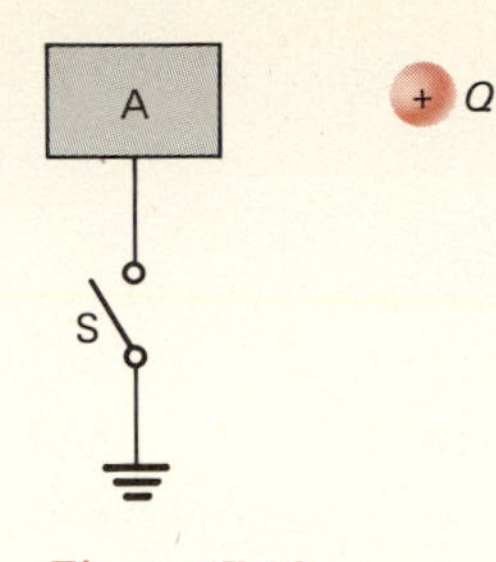

Figure 17.16

(a) the atoms of which they are constituted have no electrons.

(b) the electrons of the constituent atoms are tightly bound to these atoms.

(c) the atoms are not arranged on a regular crystal lattice.

(d) None of the above is the reason.

17.4 A positive charge would experience no net force if placed at

(a) A

(b) B

(c) A or C

(d) A, B, or C

in Figure 17.17.

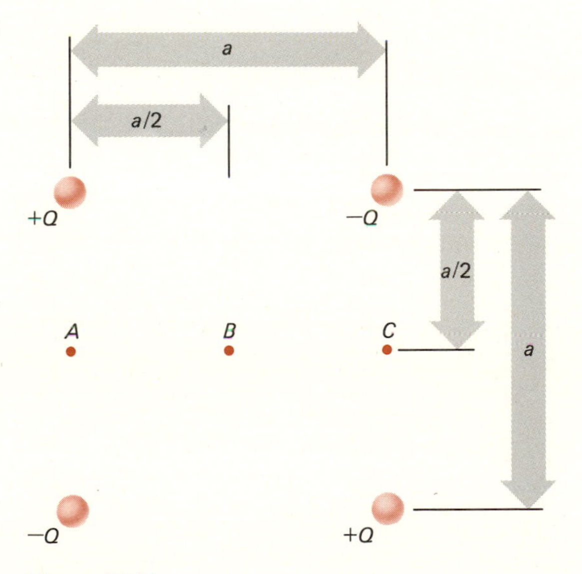

Figure 17.17

17.5 Which of the following statements is true?

(a) A positive charge experiences no electrostatic force near a neutral insulator.

(b) A positive charge experiences an attractive electrostatic force toward a nearby neutral insulator.

(c) A positive charge experiences a repulsive force, away from a nearby neutral insulator.

(d) Whatever the force on a positive charge near a neutral insulator, the force on a negative charge is then oppositely directed.

17.6 Two charged particles attract each other with a force F. If the charge of one of the particles is doubled and the distance between them is also doubled, then the force will be

(a) F

(b) $2F$

(c) $F/2$

(d) $F/4$

17.7 A *charged* insulator and an *uncharged* metal

(a) exert no electrostatic force on each other.

(b) always repel each other electrostatically.

(c) always attract each other electrostatically.

(d) may attract or repel, depending on the sign of the charge on the insulator.

17.8 Three charges are arranged as in Figure 17.18. The magnitude of the force on q_2 due to charge q_1 is F_{21}. The magnitude of the force on q_2 due to charge q_3 is F_{23}. The ratio F_{21}/F_{23} is

(a) 2/3

(b) 4/3

(c) 2

(d) 4

(e) 6

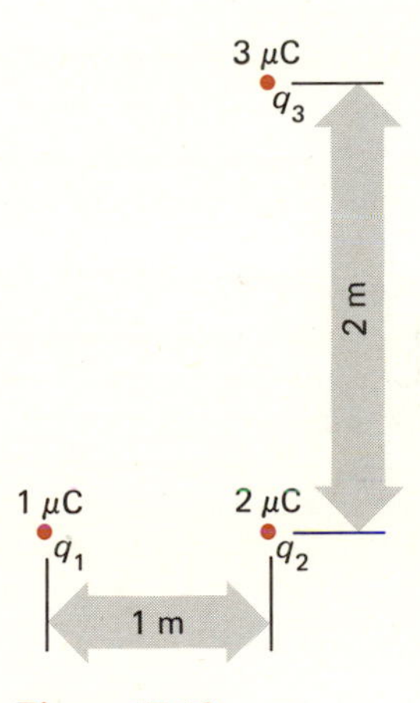

Figure 17.18

17.9 A negatively charged rod is brought near a conductor as in Figure 17.19(*a*). If the switch S, grounding the conductor, is now closed, as in (*b*),

(a) positive charge will briefly flow to the right, leaving the conductor with a net positive charge.

(b) positive charge will briefly flow to the left, leaving the conductor with a net positive charge.

(c) negative charge will briefly flow to the left, leaving the conductor with a net negative charge.

(d) the charge will flow so that the conductor will be left without any net charge.

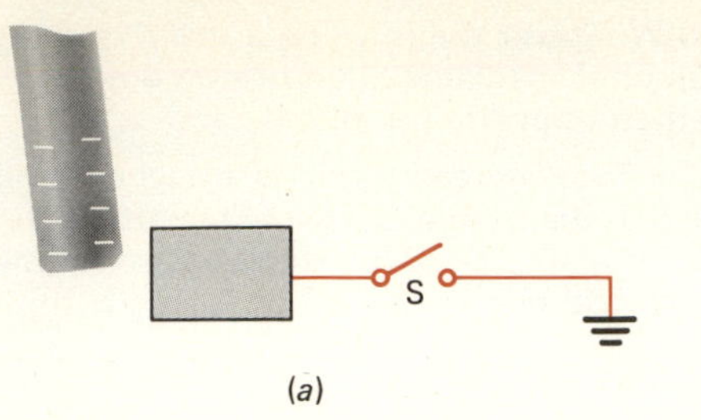

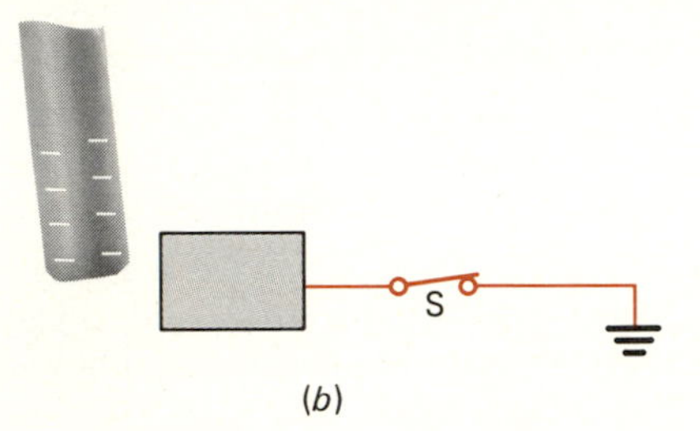

Figure 17.19

17.10 A pith ball is suspended from a thin nylon fiber. If a negatively charged ebonite rod is brought near the pith ball but does not touch it,

(a) the pith ball will become charged by induction.

(b) the pith ball will become polarized.

(c) the pith ball will become charged by conduction.

(d) the pith ball will be repelled from the ebonite rod.

17.11 Two charged objects attract each other with a force F. If the charges on both objects are doubled and the separation between them is also doubled, the force between them is then

(a) $16F$
(b) $4F$
(c) F
(d) $F/2$
(e) $F/4$

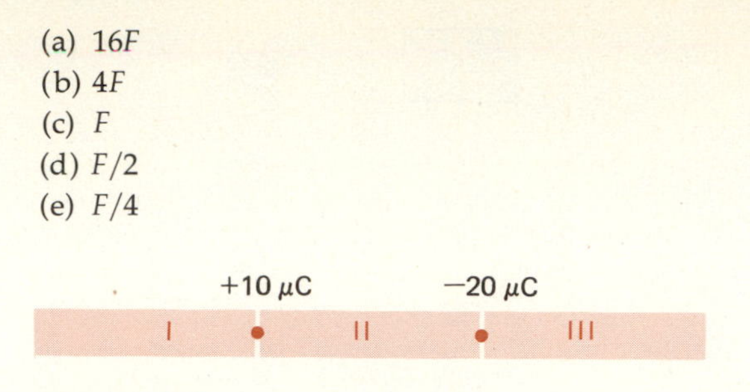

Figure 17.20

17.12 Charges of $+10\ \mu C$ and $-20\ \mu C$ are placed as in Figure 17.20. The force on a $-5\text{-}\mu C$ charge is directed to the right everywhere

(a) in region I.
(b) in region II.
(c) in region III.
(d) in regions I and III.
(e) in regions II and III.

17.13 Three equal charges are placed on three corners of a square. If the force between q_1 and q_2 is F_{12} and that between q_1 and q_3 is F_{13}, the ratio of the magnitudes F_{12}/F_{13} is

(a) $1/2$
(b) 2
(c) $1/\sqrt{2}$
(d) $\sqrt{2}$

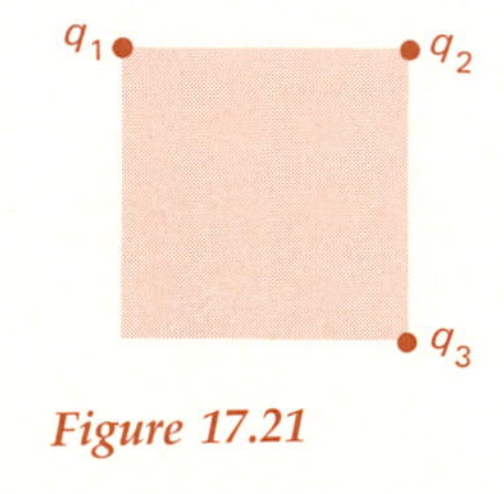

Figure 17.21

Problems

17.1 Two point charges of $+2 \times 10^{-2}$ C and -6×10^{-2} C are 3 m apart. What is the magnitude and the nature of the force between them?

17.2 What should be the separation between two protons so that the electrostatic force on each equals the weight of the proton?

17.3 A *faraday* (F) is the amount of charge of 1 mol of singly ionized material. How many coulombs is one faraday?

17.4 How many hydrogen ions are needed to make a total charge of 1 C? What is the mass of that assembly of ions?

17.5 Two charges repel each other with a force of 0.08 N. What will the repulsive force between them be if their separation is reduced to one-fourth its initial value?

17.6 At what separation is the force between two stationary electrons 4×10^{-12} N?

17.7 The average separation between a neon nucleus and the innermost electron is about 5×10^{-12} m. The charge on the neon nucleus is $+10e$. Calculate the force of attraction between the two charged objects.

• **17.8** A charge of $+3\ \mu C$ is at $x = 0$. A second charge is located at $x = 2$ m. If a charge of $1\ \mu C$ experiences no force when at $x = 4$ m, what is the magnitude and sign of the charge at $x = 2$ m?

• **17.9** The radius of a helium nucleus is about 2×10^{-15} m. Assume that this is the separation between the two protons in that nucleus. What is the electrostatic force between these two particles? Compare this force with the weight of 1 mol of helium, that is, the force of gravity on 6×10^{23} nuclei of helium.

• **17.10** The Bohr model of the hydrogen atom in its normal (ground) state is an electron in a stable circular orbit of radius 0.53×10^{-10} m about an almost stationary proton. What is the force between these two charged particles? What must the speed of the electron be if it is indeed in a stable orbit?

• **17.11** Two small spheres carry positive charges, the total charge being 8 μC. If the separation between the spheres is 20 cm and the force of repulsion between them is 1.8 N, what is the charge on each of the two spheres?

• **17.12** Two spheres carry equal charges. When the two spheres are separated by 10 cm, the force one exerts on the other is 0.024 N. Determine the magnitude of the charge on each sphere. Can you tell the sign of the charge?

• •**17.13** Two small pith balls, each of mass of 0.3 g, are suspended by 30-cm-long fine strings from the same point. The angle between each string and the vertical is 8°. Find the magnitude of the charge on each pith ball, assuming that they are equally charged.

• •**17.14** Repeat Problem 17.13 under the assumption that the charges on the two pith balls are in the ratio of 2:1.

• •**17.15** Calculate the electrostatic force on each of the charges of Figure 17.22.

• •**17.16** Is there any point at which the force on a 1-μC charge vanishes for the charge distribution shown in Figure 17.23? If so, locate that point.

• •**17.17** Repeat Problem 17.16 for the condition that the +2-μC charge is replaced by a −2-μC charge.

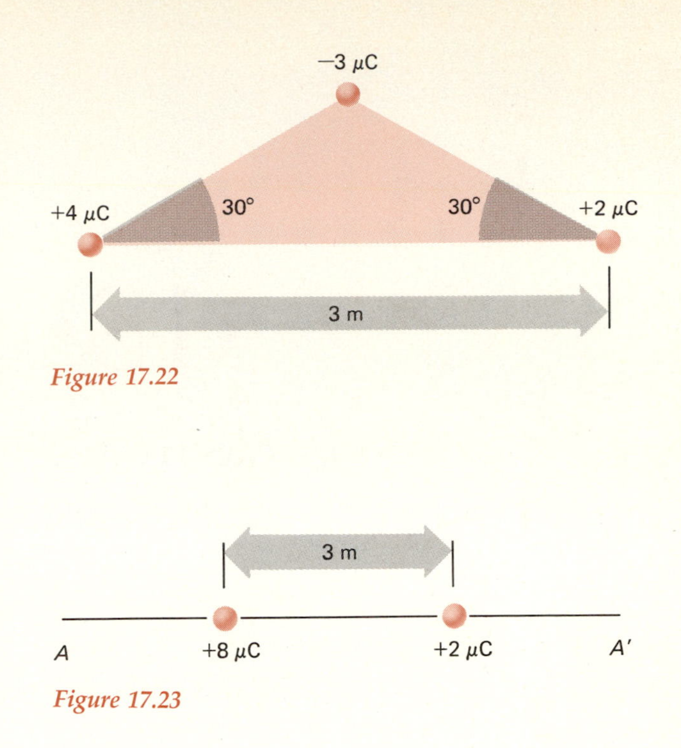

Figure 17.22

Figure 17.23

• •**17.18** At the point or points at which the force on a 1-μC charge is zero, that charge is in equilibrium. If the 1-μC charge is restricted to moving along the line AA′ (Figure 17.23), is that equilibrium stable, neutral, or unstable for Problems 17.16 and 17.17?

• •**17.19** Point charges of 0.3 μC are placed at three corners of a square whose sides are 0.5 m long. A charge of −0.4 μC is placed at the fourth corner. Calculate the total force acting on each of the four charges.

• •**17.20** Charges of 4 μC and −3 μC are separated by 24 cm. Where should a charge of +2 μC be placed so that it experiences no net force? Is that charge then in a stable equilibrium position?

•18•

Electrostatics II: Electric Fields and Potentials

To fields where flies no sharp and sided hail.

GERARD MANLEY HOPKINS

18.1 The Electric Field

DEFINITION: *The electric field* **E**(**r**), *at the point identified by the position vector* **r**, *is the force that a unit positive charge would experience if located at that point and if its placement there did not alter the distribution of any charges in space.*

As we have seen in the preceding chapter, the condition that a unit test charge not distort an existing charge distribution is unrealistic; that test charge will induce electric dipoles and orient permanent dipoles in matter. This is all the more evident when we recall that a unit charge, one coulomb, represents an enormous amount of charge. However, since the force on a given charge at some point in space is, by Coulomb's law, proportional to the magnitude of the charge itself, one can, in principle, make the test charge as minute as desired. Ideally, it could be made so small that its effect on the surroundings would be negligible. One could still determine the electric field by measuring, with some exquisitely sensitive instrument, the force on this very small charge.

In other words, we can redefine the electric field $\mathbf{E}(\mathbf{r})$ as the electrostatic force per unit charge:

$$\mathbf{E}(\mathbf{r}) = \frac{\mathbf{F}_q(\mathbf{r})}{q} \qquad (18.1)$$

where $\mathbf{F}_q(\mathbf{r})$ is the force that acts on a *small* charge q when placed at the point $\mathbf{r}$. Conversely, once $\mathbf{E}(\mathbf{r})$ is known, the force acting on a charge q is given by

$$\mathbf{F}_q(\mathbf{r}) = q\mathbf{E}(\mathbf{r}) \qquad (18.2)$$

Note that in the preceding sentences we have omitted the adjective *positive*. If the force on a positive charge q is $\mathbf{F}$, the force acting on a negative charge of equal magnitude at the same point will be $-\mathbf{F}$. The usefulness of the field concept is precisely that $\mathbf{E}$ fully characterizes the force on any charge throughout the region in which the field is known.

From Equation (18.1) it follows that the unit of electric field is the newton per coulomb, N/C. We shall later find that the electric field can also be expressed in the more conventional unit of volt per meter, V/m.

From Equation (18.1), it also follows that the electric field is a *vector field;* that is, we associate with every point in space a vector $\mathbf{E}(\mathbf{r})$, whose magnitude and direction are functions of the position $\mathbf{r}$.

● ***Example 18.1*** What are the electric fields in the regions near an isolated positive charge Q and in the region near an isolated negative charge of magnitude Q?

According to Equation (17.1), the force on a positive charge q due to another positive charge Q has the magnitude

$$F_q(r) = k\frac{qQ}{r^2}$$

and points away from the charge Q. By definition, the electric field $\mathbf{E}(\mathbf{r})$ is therefore a vector that also points radially outward from the location of the charge Q and has a magnitude given by

$$E(r) = \frac{F_q(r)}{q} = \frac{kQ}{r^2} \qquad (18.3)$$

If the charge Q is negative, the electric field is again given by (18.3), but the direction of $E(r)$ is now reversed; that is, $\mathbf{E}(\mathbf{r})$ points radially inward toward the negative charge because that is the direction of the force that would act on a positive test charge. ●

18.2 Electric Field Lines

A very useful concept for visual representation of an electric field configuration is that of *electric field lines,* or lines of force, first introduced by Michael Faraday. A precise graphic representation of an electric field requires drawing a vector of proper length and direction at every point of space, and such a map would be messy, to say the least. For example, the field map, in two dimensions, of the field due to a single positive point charge would look somewhat like Figure 18.1. A much simpler visualization is afforded by drawing a group of lines that obey the following conditions:

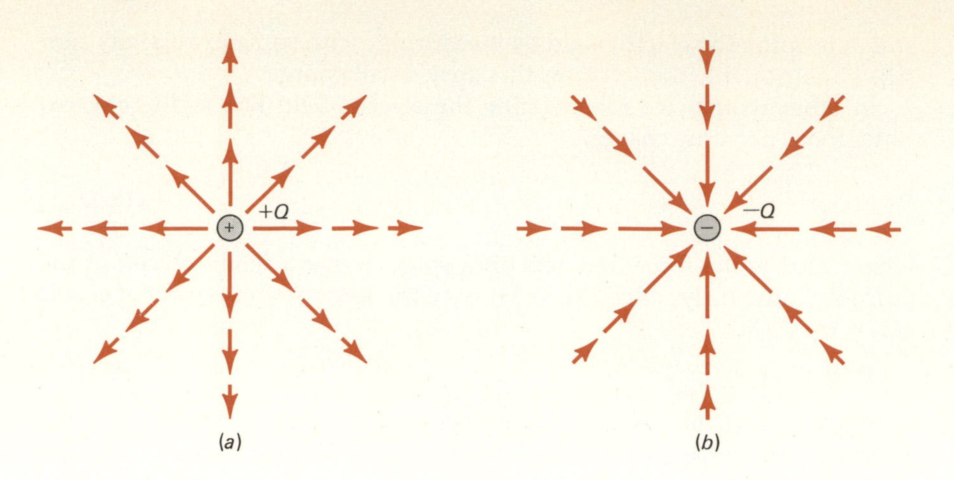

Figure 18.1 A representation of the vector field due to (a) a positive and (b) a negative point charge of magnitude Q. Since the magnitude of $\mathbf{E}$ is proportional to $1/r^2$, the lengths of the vectors must diminish with increasing distance. Very close to the point charge, these vectors would be excessively long; at some greater distance from the charge, they would be so short as to be almost pointlike objects.

1. The lines are directed, pointing away from positive and toward negative charges, so that at any point, the tangent to a line is along the direction of the electric field at that point.

2. The number of lines that are drawn emerging from or terminating on a charge is proportional to the magnitude of the charge.

In drawing the field pattern, a few elementary rules are helpful.

1. Every field line must originate at a positive charge and terminate on a negative charge.

1a. A corollary of the above is the requirement that in a charge-free region the field lines must be continuous.

2. In the immediate vicinity of a point charge, the field lines are radially directed.

3. Field lines do not intersect in a charge-free region.

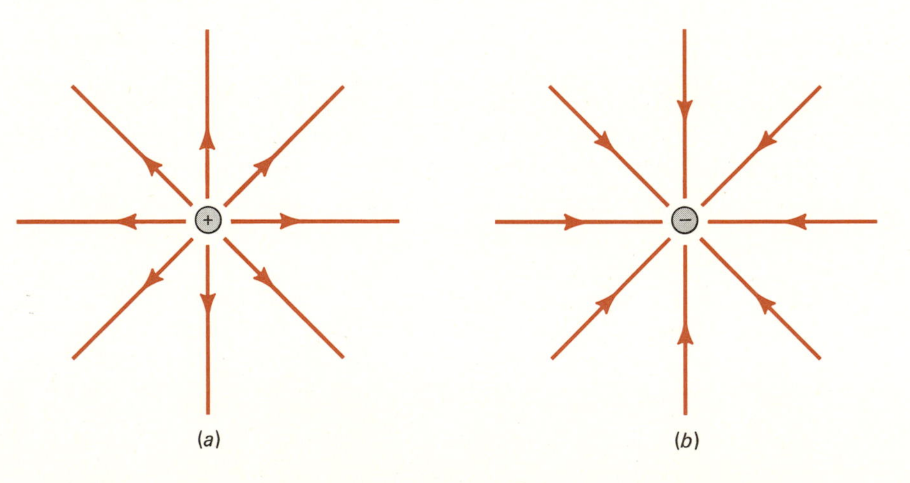

Figure 18.2 Electric-field lines are an alternative way of representing the electric field. Shown here are the field lines due to (a) a positive and (b) a negative point charge. In this method of portraying an electric field, the strength of the field is taken as proportional to the density of the field lines, and the number of field lines emanating from or terminating on a charge is proportional to the charge. Note that the density of field lines, that is, the number of lines crossing unit area, is greatest nearest the point charges, and diminishes as $1/r^2$ since the surface area of a sphere is proportional to r^2.

The last of these rules reflects the fact that the electric field vector is unique at every point in space. If two field lines did intersect, this would mean that at the point of intersection there are two equally valid directions (and possibly magnitudes) of the $\mathbf{E}$ vector and, therefore, of the net force acting on a small test charge.

A few examples will help to clarify the use of these concepts in visualizing the electric field.

● ***Example 18.2*** A charge of $+2\ \mu C$ is located at $x = 0$ and a charge of $-2\ \mu C$ at $x = 2$ cm. Sketch the electric field of this charge distribution, using lines of force. On the basis of this sketch, what is the direction of the electric field anywhere along the perpendicular bisector of the line joining the two charges? Where on this bisector is the field strongest?

We adopt the arbitrary convention of drawing 12 field lines per microcoulomb of charge. Hence, 24 lines emanate from the $+2$-μC charge and terminate on the -2-μC charge.

To start, we note that in the *immediate* vicinity of a point charge, the field is almost exclusively due to that charge; the perturbing effects of more distant charges are negligibly small. Hence, very near each of these point charges, the field lines are radially directed, pointing outward from the positive charge and inward to the negative.

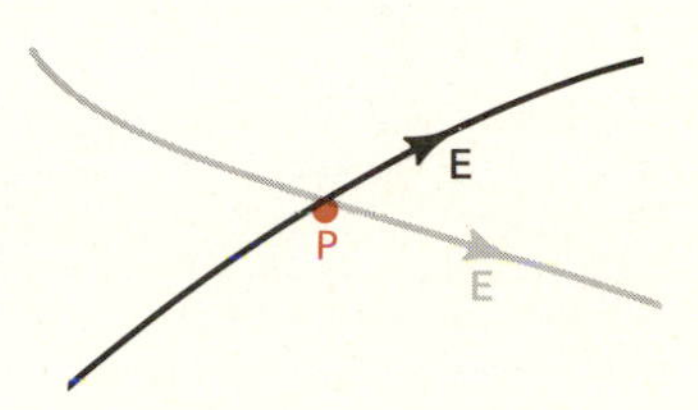

Figure 18.3 Electric-field lines cannot intersect. If they did, as in this figure, a positive charge at the point of intersection P would experience schizophrenia, trying to decide in which direction it should move.

To complete the pattern of field lines throughout the remaining region, we recall the three rules cited above. Since the total charge is zero, the 24 lines emanating from the positive charge must terminate on the negative charge; the lines must be continuous, and they must not intersect. The pattern of connections that meets these conditions is shown in dashed lines in Figure 18.4(a). This is the field pattern of a dipole.

From this field pattern, it is apparent that the electric field at any place along the bisector is parallel to the x axis and points in the positive x direction. Moreover, the density, that is, the concentration of field lines along the bisector line A–A', is greatest at the point P, midway between the two charges. It is at this point on the line A–A' that the electric field has its greatest magnitude. As we shall see presently, the strength of **E** and the density of field lines are proportional to each other. ●

● ***Example 18.3*** A charge of $+3\ \mu$C is located at $x = 0$ and a charge of $-1\ \mu$C at $x = 2$ cm. Sketch the distribution of the electric field in the neighborhood of this charge distribution.

We retain the convention of drawing 12 field lines per microcoulomb of charge. Thus a total of 36 lines must emanate from the $+3$-μC charge, and 12 of these lines must terminate on the -1-μC charge. Very near the $+3$-μC charge, the 36 field lines are uniformly distributed about that

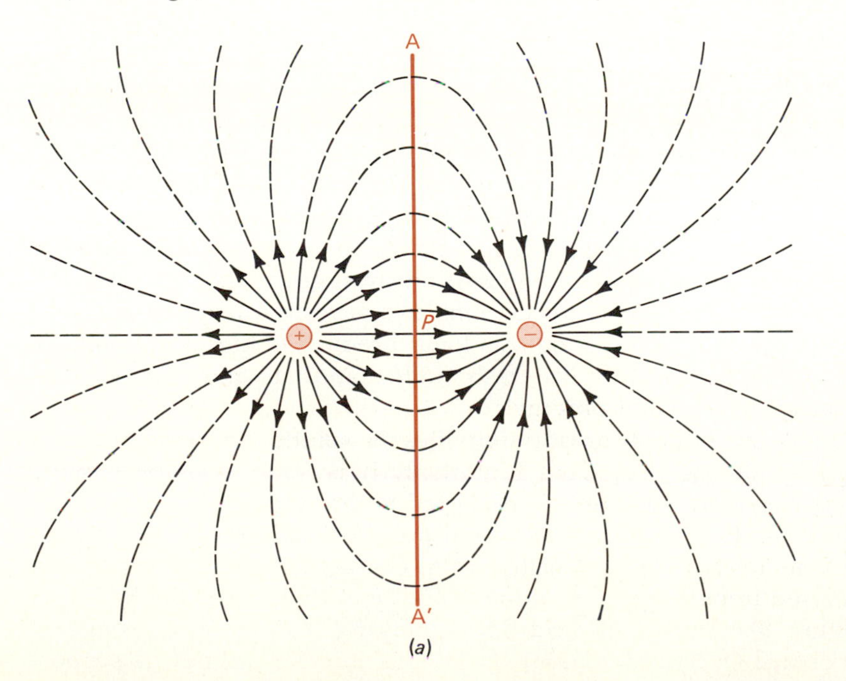

(a)

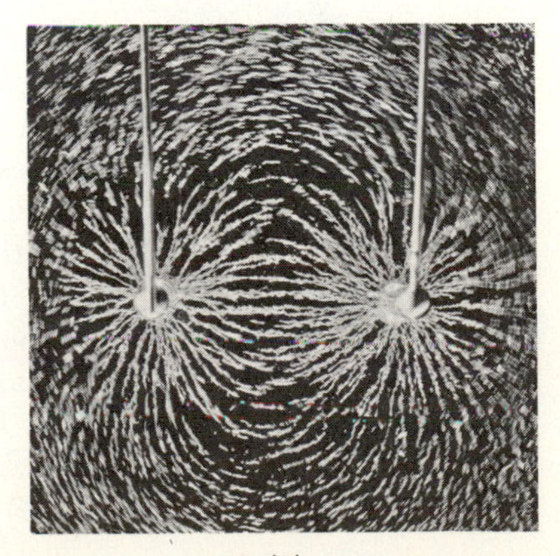

(b)

Figure 18.4 (a) Electric-field lines near an electric dipole of $+2\ \mu$C and $-2\ \mu$C. We have adopted the convention of drawing 12 field lines per microcoulomb of charge. The field lines in the immediate vicinity of each of the charges is shown by solid lines that point radially to or from the charge. These lines must be connected so that they do not cross, as shown by the dotted continuations. (b) The electric field can be exhibited by suspending small bits of fine thread in oil. The field aligns the molecular dipoles of the threads, and they therefore align along the field direction.

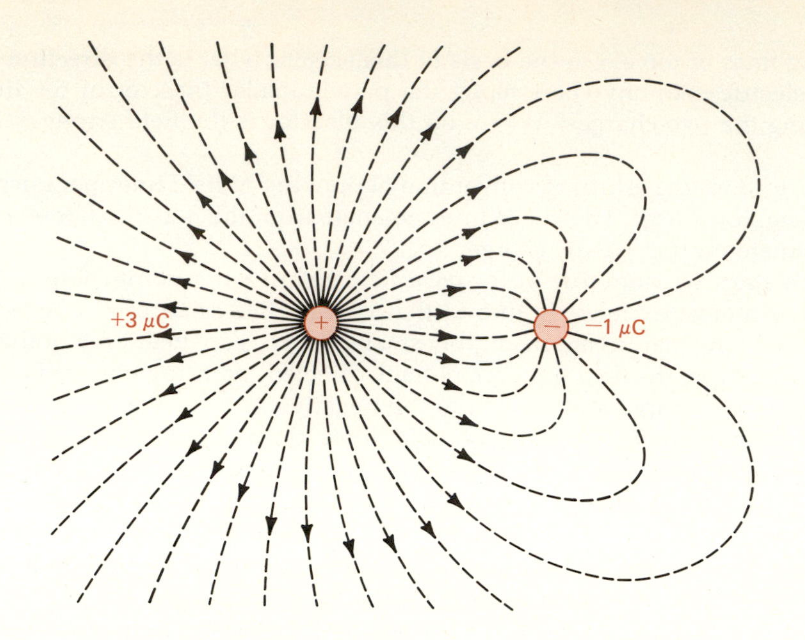

Figure 18.5 Electric-field pattern near a charge of $+3\ \mu C$ and $-1\ \mu C$. Note that as in Figure 18.4, once the field lines near each charge have been drawn, there is no reasonable alternative way that these lines can be connected without crossing each other. Twenty-four field lines extend to infinity. These represent the net charge of $+2\ \mu C$.

charge and point radially outward, whereas near the -1-μC charge, the field lines number 12 and point radially inward toward this charge.

To complete the pattern of field lines, we connect 12 of the lines emanating from the positive charge to the 12 field lines terminating on the negative charge in such a way that no lines intersect. The connecting links are sketched in dashed lines in Figure 18.5, and the field pattern now becomes fairly clear. In particular, we see that near the negative charge, a positive charge would experience a force toward the negative charge. Far from this charge distribution, however, a positive charge experiences a repulsive force no matter what its position; at a sufficiently large distance from this charge distribution, the field pattern is the same as that due to a net charge of $+2\ \mu C$.

18.3 Gauss's Law

Karl Friedrich Gauss (1777–1855) was remarkable even among that small, select group endowed with the gift of mathematical genius. His exceptional talent surfaced when he was but three years old, and before the age of 20 he had made several original and important contributions that clearly established him as an intellectual partner of Archimedes, Newton, and Euler. Gauss applied his efforts to a tremendous range of problems, from the theory of numbers, abstract geometry, and probability theory to astronomical calculations and terrestrial magnetism. In all these spheres, and throughout his long, productive life, he brought to every problem new insight, fresh understanding, and the personal enthusiasm of the child prodigy.

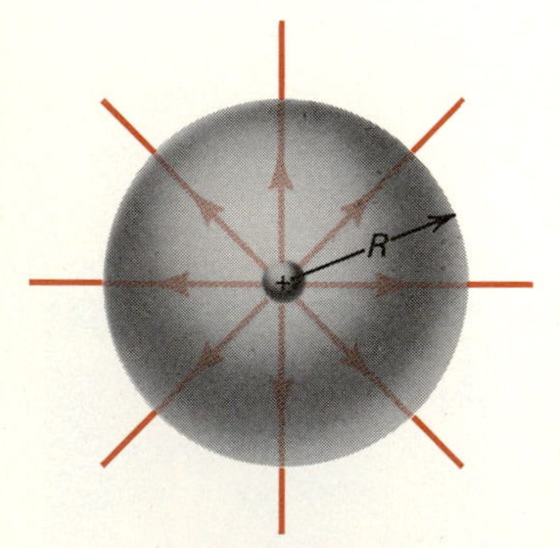

Figure 18.6 A positive point charge Q at the center of a sphere of radius R. The radially directed field lines penetrate this spherical surface. Their density at the spherical surface is inversely proportional to the area of the sphere and consequently, proportional to $1/R^2$.

Gauss's law not only greatly simplifies the solution of numerous electrostatic problems, but more important, it is valid also for moving charges, whereas Coulomb's is restricted to the static case.

Here we do not attempt a general proof of Gauss's law. We do establish the relation for an isolated point charge; its general validity can be inferred from the superposition property of the electric field.

Figure 18.6 shows the field lines emerging radially from a positive point charge Q. By convention, N, the total number of lines, is proportional to Q. Now imagine a sphere of radius R centered on Q. Since the surface area of that sphere is $4\pi R^2$, the density of field lines, $\mathcal{N}(R)$, that is, the number of lines that penetrate the surface per unit area, is given by $\mathcal{N}(R) = N/4\pi R^2$. The important point to notice here is that $\mathcal{N}(R)$ is proportional to $1/R^2$ as the electric field E is. Moreover, since N is proportional to Q and, according to Equation (16.3) E is also proportional to Q, *the density of field lines is proportional to the strength of the field.*

If we now form the product of $E(R)$ and the spherical surface area $4\pi R^2$, we have

$$4\pi R^2 E(R) = 4\pi R^2 \frac{kQ}{R^2} = 4\pi kQ \tag{18.4}$$

Equation (18.4) is a special statement of Gauss's law.

Suppose the surface enclosing the charge Q is not a sphere but has a more general shape, as in Figure 18.7. In that case, the field lines from the charge Q are not everywhere perpendicular to the enclosing surface. It is nevertheless still true that the total number of lines penetrating this arbitrary closed surface is N. Whereas previously the field lines were all perpendicular to the surrounding surface, now a field line has, in general, a tangential and a perpendicular component. Only the perpendicular component penetrates the surface. The general statement of Gauss's law is therefore as follows.

Gauss's Law: The perpendicular component of the electric field summed over any closed surface equals $4\pi k$ times the net charge enclosed within that surface.

In symbolic form, Gauss's law reads

$$\Sigma(E_\perp \, \Delta A) = \frac{Q}{\epsilon_0} \tag{18.5}$$

where $E_\perp$ is the component of the electric field at the small element of area ΔA perpendicular to this surface element, and

$$\epsilon_0 = \frac{1}{4\pi k} \tag{18.6}$$

is known as *the permittivity of free space*. In Equation (18.5) it is understood that when the normal component of the electric field points outward, it is considered positive, and when it points inward, it is considered negative.

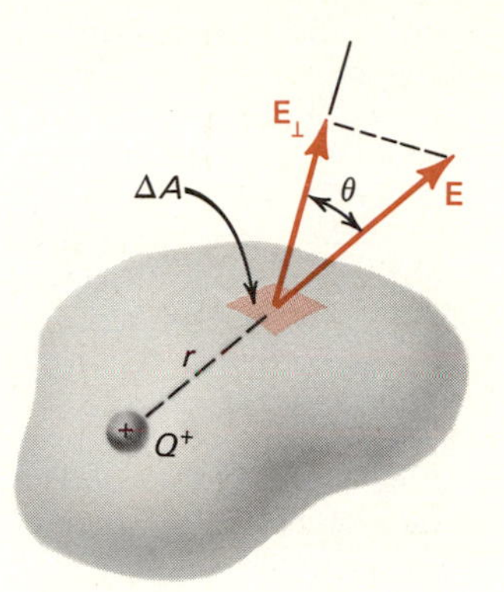

Figure 18.7 A point charge Q enclosed within an irregularly shaped surface. The electric field **E** is not perpendicular to the area element ΔA. In this case, one must use only the normal component of **E**, that is, $\mathbf{E}_\perp$ in the application of Gauss's law.

18.4 Electric Field Inside a Conductor; Shielding

We consider the following situation. A block of metal is placed in a region previously permeated by a uniform electric field. The question we want to answer is this: What is the electric field inside the metal block?

Under the influence of the external electric field, the free electrons of the metal experience a force in a direction opposite to $\mathbf{E}_{ext}$ (why opposite?) and consequently tend to move to the left (Figure 18.8(*a*)). Thus negative charge accumulates on the left surface, and a net positive charge on the right surface of the metal. This charge distribution, induced by the external field, establishes its own electric field, and the total field at any point in space is the vector sum of the external **E** field and the field due to the induced charge on the metal. The latter, as we can see, tends to cancel the former within the metal; indeed, inside the metal, the cancellation is exact! If it were not, and a net **E** field did exist inside the metal,

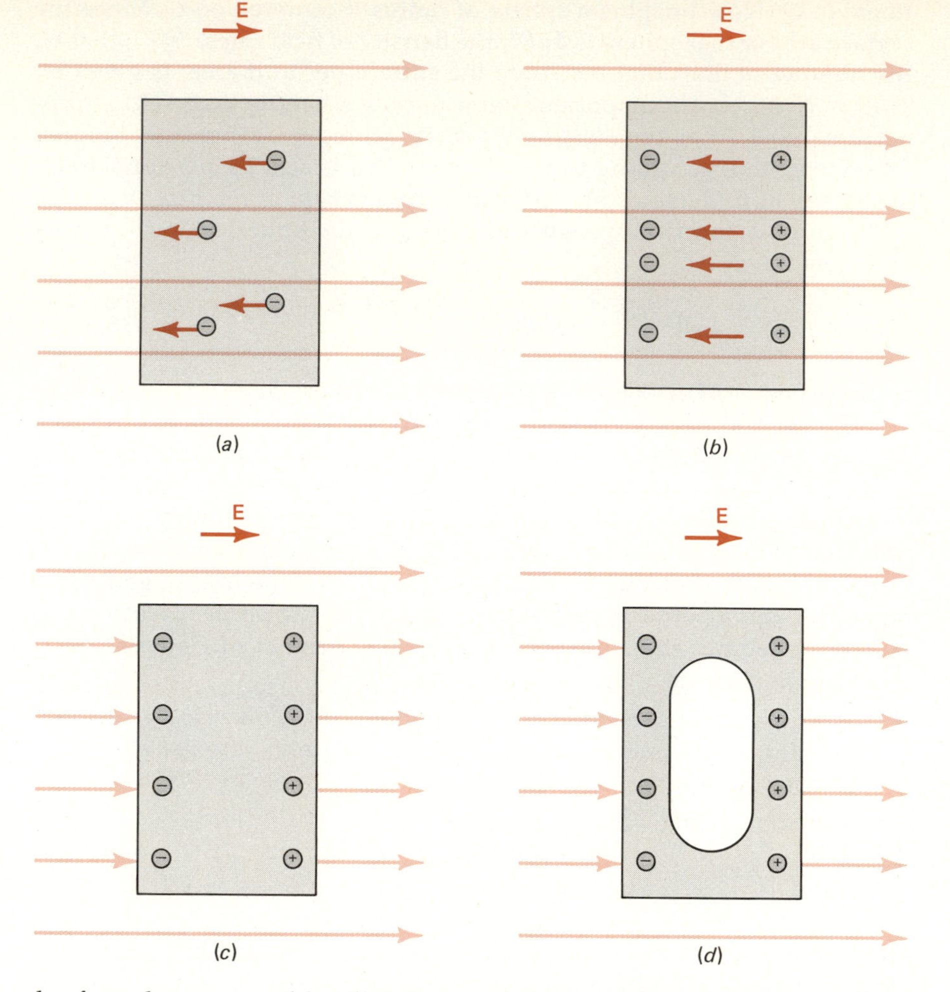

Figure 18.8 *A block of metal is placed into a region of uniform electric field* **E**. *(a) As a result of the presence of the field, forces in the direction indicated act on the mobile, negatively charged electrons in the metal. (b) The electrons will therefore move to the left, and so there will be a net negative charge on one surface, and a net positive charge on the opposite surface of the metal. This charge distribution generates its own field, which is in a direction opposite to that of the applied electric field. (c) The vector sum of these two fields is exactly zero. (d) Since the field inside the metal is zero and there are no unbalanced charges within the metal, one could make a cavity within the block without in any way changing the field configuration. We can therefore conclude that within a closed metallic surface that contains no free charges, the electric field must vanish exactly.*

the free charges would still experience forces and continue to readjust their positions until the forces vanished, that is, until the field vanished. We can therefore conclude that

The electrostatic field inside a conductor is always zero.

From this statement it also follows that there can be no static *unbalanced* charges inside a conductor. This is most readily established by Gauss's law.

Imagine a conductor, as in Figure 18.9, and contained entirely within this conductor a closed "gaussian surface," as shown by the dashed line. Since the **E** field over this surface is everywhere zero, it follows from Equation (18.5) that the net charge enclosed within that surface must vanish. That conclusion is true whatever the shape of this surface.

We can also prove by Gauss's law another important consequence of the absence of an electrostatic field in a conductor, namely,

If there are any unbalanced, static charges on a conductor, they must reside on the surface of the conductor.

Again, we can imagine a closed surface within the conducting body, which carries a net charge Q. As long as the gaussian surface is *within* the conductor, the enclosed charge must be zero. It follows, then, that if a net charge does reside on the body, it can be distributed only over the surface layer of that body.

Since the interior of the conducting block of Figure 18.8(c) contains no unbalanced charges, we could scoop out some of the material, leaving a

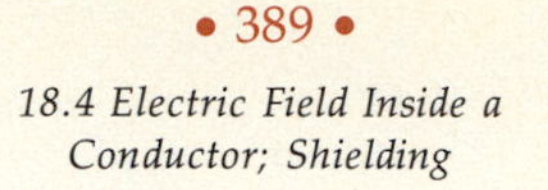

hollow cavity, without altering in any way the charge distribution or electric field anywhere. But we have just concluded that the field **E** within the conductor is zero. It follows that

The electrostatic field in a volume containing no free charges and bounded by a closed conducting surface is everywhere zero.

This result has immensely important practical implications. It tells us that we can shield an object from the influence of electrostatic fields by simply enclosing it within a conducting sheath, for example, some aluminum foil. The electrostatic shield need not even be a solid conductor; a wire mesh will do. The electric field will penetrate the wire mesh ever so slightly, but only to a depth that is a fraction of the diameter of the mesh opening.*

We have seen that a conducting enclosure shields the interior from external electrostatic fields. The next question is obvious: Will such a metal shell shield the *exterior* region from fields that might be generated by charges placed within the metallic enclosure?

Figure 18.10 shows a spherical metallic shell surrounding a positive charge Q placed at its center. If the metallic shell were not present, there would be a radially directed **E** field in the region occupied by the metal shell (see Figure 18.10(*b*)). We know that the electric field inside a conductor must vanish, and so $\mathbf{E} = 0$ within the annular metallic shell. That is, electrons in the metal will be attracted toward the inner surface,

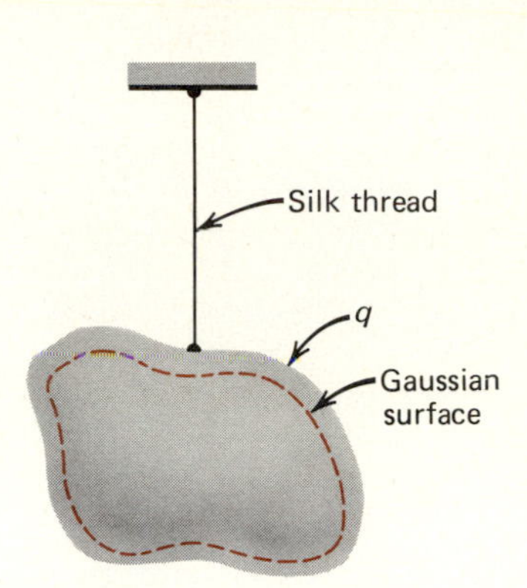

Figure 18.9 *An insulated metallic body. The "gaussian surface" just inside the surface of the body is indicated here by a dashed line.*

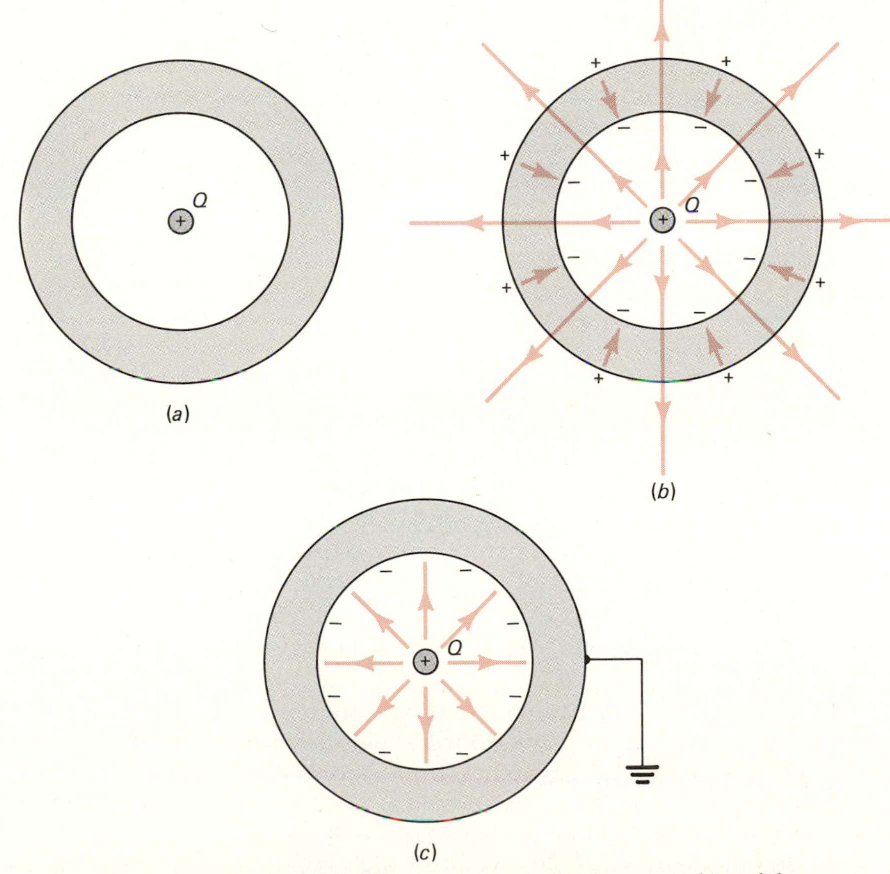

Figure 18.10 *(a) A charge Q is placed in the center of a spherical metallic shell. (b) The electric fields due to this charge and to the charges induced on the surfaces of the shell. The charges induced on the inner and outer surfaces of the shell are such that within the metal the resulting electric field exactly cancels that due to the charge at the center. Note that the field outside the shell, however, does not vanish. (c) If the metallic shell is grounded, no charge remains on the outer surface, and the field in the external region vanishes.*

* The idea of an antigravity "shield," at times a popular *deus ex machina* of the science fiction community, is a natural extension of electrostatic shielding, especially in view of the similarity between gravitational and electrostatic forces. To fashion the antigravity shield, we need only discover and isolate objects of negative gravitational mass.

leaving a positive surface charge over the exterior surface, and this charge separation will proceed until the **E** field within the metal shell vanishes. Using Gauss's law, one can show that the total charge on the inner surface will be exactly $-Q$, so that the total net charge enclosed within the imaginary spherical surface shown dashed in Figure 18.10(*b*) is zero. This leaves a net positive charge $+Q$ distributed over the exterior surface. The electric field outside this shell is then not zero, but is exactly the same as though the metal were removed. *An ungrounded metal shell does not shield the exterior from electric fields that are due to charges placed inside the shell.*

If, however, the metal shell is grounded, as in Figure 18.10(*c*), the charge on the exterior surface will be neutralized by charges attracted to it from ground, and the field due to the interior charge is then indeed shielded from the external world.

18.5 Electrostatic Potential

The concept of potential energy proved very useful in describing the dynamic behavior of massive objects in a gravitational field. The existence of a unique potential energy (relative to some arbitrary zero) is predicated on the fact that the gravitational force is conservative. This is a characteristic of all so-called *central forces,* that is, forces that act radially toward or away from the source of the force field. The electrostatic force evidently meets this criterion, and it is another example of a force from which a potential energy can be defined.

DEFINITION: The electrostatic potential energy difference of a charge q between two points in space is the negative of the work done by the electrostatic force in transporting this charge from the initial to the final position.

DEFINITION: The electrostatic potential difference V_{AB} between points A and B is the difference in electrostatic potential energy per unit positive charge between these two points.

That is,

$$V_{AB} = V_B - V_A \tag{18.7}$$

where V_A and V_B are the electrostatic potentials of points A and B relative to some arbitrary reference point that has been assigned zero potential.

As with the gravitational potential energy, so also in the electrostatic case, the only meaningful quantity is the *difference* in potential between two points; the choice of zero is completely arbitrary. By convention, zero potential is taken to be the potential at a point infinitely far removed from any unbalanced charges. The phrase "absolute potential" is sometimes used to designate the electrostatic potential relative to this zero reference. In practice, however, one often simply assigns zero potential to "ground."*

By definition, the electrostatic potential difference between points A and B is the *negative* of the work done *by* the electrostatic field on a unit *positive* charge in moving that charge from A to B. Alternatively, the potential difference V_{AB} is the work that must be done *against* the electrostatic field in moving a unit positive charge from point A to point B. As with the gravitational field, we need not specify the path followed in moving the charge.

* Strictly speaking, this is an unwarranted assumption. Under normal circumstances the earth carries a net negative charge that is balanced by an equal net positive charge in the clouds and atmosphere above the earth's surface.

In contrast to the electrostatic field, the potential is a scalar function of position.* It is this feature of the potential that makes it such a convenient property for many problems.

The potential throughout space constitutes a *scalar field,* and the value $V(\mathbf{r})$ of that field at some point $\mathbf{r}$ depends on the distribution of charges in space. The potential obeys the superposition principle. That is, the potential at a point $\mathbf{r}$ is the algebraic sum of the potentials due to each individual charge of the total charge distribution.

The unit of potential, or of potential difference, is the *volt* (V), so named in honor of Alessandro Volta (1745–1827).† From the definition of potential difference, it follows that dimensionally

$$[V] = \frac{[W]}{[Q]} \tag{18.8}$$

where we have used the letter W instead of E to denote energy or work, to avoid confusion with the electric field. Since the joule, the unit of work, is one newton meter, one volt equals one newton meter per coulomb; that is, $1\ \text{V} = 1\ \text{N}\cdot\text{m/C}$. But the unit for electric field is newton per coulomb. Consequently,

$$[V] = \text{N}\cdot\text{m/C} = [E][L] \qquad \text{and} \qquad [E] = \frac{[V]}{[L]}$$

In other words, a newton per coulomb equals a volt per meter.

18.6 Electrostatic Potential and the Energy of Charged Bodies

The electrostatic potential $V(\mathbf{r})$ was defined in terms of the work that must be done to transport a unit positive charge from a location of zero potential to the position $\mathbf{r}$. Just as raising a mass through a height h imparts to this mass a PE equal to the work done in the process, so the PE of a charge Q is changed if it is moved through a potential difference V. From the definition of potential, the change in PE is given by

$$\text{PE}_{AB} = QV_{AB} \tag{18.9}$$

where PE_{AB} is the change in PE of the charge Q as it passes from point A to point B through a potential difference V_{AB}.

If a *positive* charge is moved from a location of low potential to one of higher potential, its PE is increased. Such a charge, if left only to the influence of the existing electric field, will seek and move toward the point of lowest potential. As it does, the positive charge acquires KE equal to the loss in PE if it is assumed that there are no losses due to friction or other dissipative mechanisms. In analogy to the gravitational

* We shall omit the adjective *electrostatic* hereafter.

† Volta, whose most famous invention was the voltaic pile, the forerunner of modern batteries, taught physics in his native Como from 1764 to 1777. He gained an international reputation, lecturing in France and in England, where he was elected a member of the Royal Society. In 1815 he was named director of the Faculty of Philosophy of Padua, simultaneously holding the chair of physics at the University of Pavia, which he had occupied since 1779. Volta was especially admired by Napoleon, who had a medal struck in his honor. In 1804, when Volta announced his intention of retiring from Pavia, Napoleon wrote: "I cannot consent to the retirement of Volta; if his duties as professor are fatiguing, his work must be lightened. Let him have, if desired, no more than one lecture to deliver each year; but the University of Pavia would never forgive me if I permitted so illustrious a name to be struck off the list of its members."

case, we then speak of a charge as "dropping" through a potential difference of so many volts.

If the charge that is moved through the potential difference V_{AB} is *negative*, the change in PE of that charge is still given correctly by

$$PE_{AB} = QV_{AB}$$

In that case, however, when Q is negative, the PE of the charge will *decrease* as the charge moves from a location of low potential to a place of higher potential. Hence, a *negative charge* under the influence of the local electrostatic field will seek and move toward the point of *highest potential.* In that case one still speaks of the charge as dropping through a certain potential difference and acquiring, in the process, a KE equal to $Q\,\Delta V$, with the understanding that this negative charge "drops" from a point of low electrostatic potential to a point of higher potential.

18.7 Potentials Associated with a Uniform Electric Field and with the Field from a Point Charge

Figure 18.11 shows a region in space permeated by a uniform electric field **E**. We want to know the potential difference V_{AB} between two arbitrary points A and B.

We determine this potential difference by calculating the work done by the field on a unit positive charge in transporting it from A to B. Since throughout this region a charge q experiences a constant force

$$\mathbf{F} = q\mathbf{E}$$

the work done by that force is

$$W_{AB} = Fs_{AB} \cos \theta$$

where θ is the angle between the force **F** and the displacement $\mathbf{s}_{AB}$ over which this force acts.

In the present instance, the product of the displacement and cos θ is just the distance AB' of Figure 18.11. Since **F** acts in the direction AB', the definition of potential difference gives

$$V_{AB} = -Fs_{AB'}/q = -Es_{AB'} \tag{18.10}$$

In a uniform electrostatic field, the potential difference V_{AB} is the negative of the product of the electric field E and the component of the displacement from point A to point B in the direction of the field.

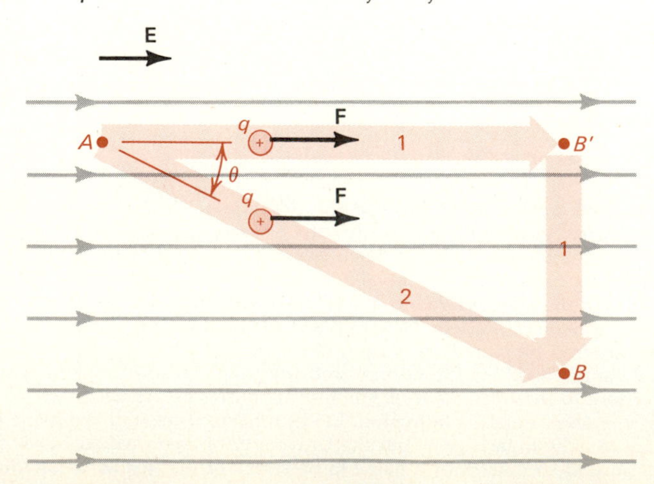

Figure 18.11 A charge of $+q$ coulombs is to be moved from point A to point B. The uniform electric field **E** exerts a force q**E** on this charge. The work done by this force **F** in taking the charge from A to B equals the decrease in electrostatic PE as a result of that displacement. The work is the same whether path 1 or path 2 is followed. The change in electrostatic potential V_{AB} is the change in electrostatic PE divided by the charge q.

Notice that instead of transporting the charge q along the line AB, we could have moved it from A to B' and hence to B. In that case, the force **F** would have done no work on the charge during the latter portion since the line $B'B$ is perpendicular to **F**. It follows that the potential everywhere along $B'B$ must have the same value.

The line $B'B$ is called an *equipotential line*. In three dimensions we could move a charge anywhere on a *plane* perpendicular to the uniform field **E** without doing any work. Such a plane is an *equipotential surface*.

Equipotential surfaces need not be planes; in fact, they are parallel planes only in the very special case of the uniform field considered here. Equipotential surfaces have generally more complicated shapes, but whatever their shapes, these surfaces must be everywhere perpendicular to the electric field. That requirement is demanded by the condition that the potential difference between two neighboring points be zero.

If the electric field is not uniform but is, instead, the Coulomb field due to a single positive point charge, the definition of potential difference is not changed, but its calculation becomes more difficult. The difficulty arises because, as indicated in Figure 18.12, the electric field, and hence the force on a unit positive charge, depends on position. The magnitude of the field **E** at a distance r from a point charge Q is given by

$$E = \frac{kQ}{r^2} \tag{18.3}$$

and increases greatly as the charge Q is approached. Thus, in Figure 18.12, the work done in transporting a test charge against this force field through a displacement Δs along the radial direction depends on where Δs happens to be relative to the location of the charge Q that is responsible for the field.

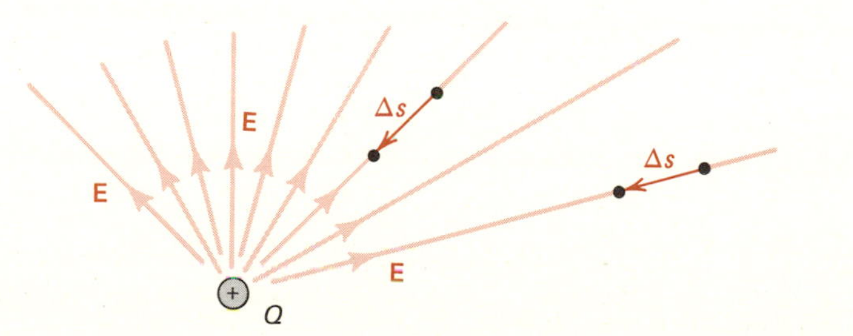

Figure 18.12 The electric field of a point charge is not uniform but increases as the charge is approached. Consequently, the change in potential as one approaches the point charge by an amount Δs depends on the distance of this small displacement from the charge.

To evaluate the work done in moving a unit charge in this field, one must use elementary integral calculus. The result is that

$$V(r) = k\frac{Q}{r} \tag{18.11}$$

where we have adopted the convention that $V(r) \to 0$ as $r \to \infty$.

The equipotential surfaces in the neighborhood of an isolated point charge are shown in Figure 18.13. These surfaces are spheres centered at the charge Q. Everywhere on these surfaces, the electric field, pointing along the radial direction, is perpendicular to the equipotential surface. Whereas in the uniform electric field the spatial separation between equipotential surfaces differing by a fixed voltage is everywhere the same, this is not true here. For a given potential difference, the equipotential surfaces are crowded together as the charge Q is approached. The reason for this becomes evident if one plots $V(r)$ as given by Equation (18.11) and as shown in Figure 18.14 for equal positive or negative charges. As r approaches zero, $V(r)$ diverges in the positive or negative

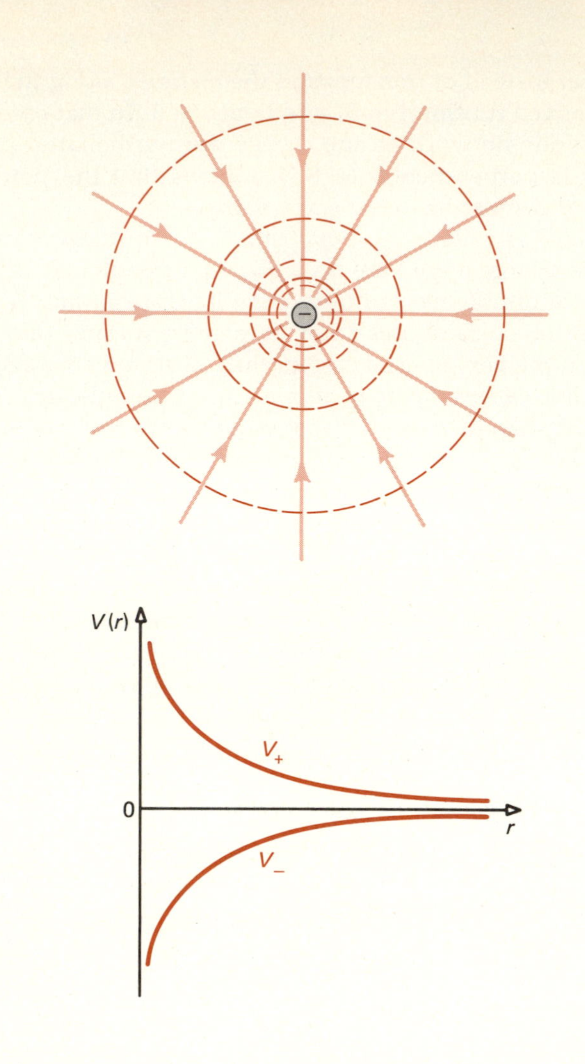

Figure 18.13 *The electric field lines and equipotential surfaces near a negative point charge. Note that the equipotential surfaces are always perpendicular to the electric-field lines. Note, also, that as the charge is approached and the electric field increases, the separation between equipotential surfaces decreases if the difference in potential between neighboring surfaces is kept fixed. Figure 18.13 should be compared with Figure 18.14.*

Figure 18.14 *The absolute electrostatic potential at a distance r from a positive point charge and a negative one. The potential diverges, that is, approaches infinity, as r approaches zero. Note that the potential due to a negative charge is negative.*

direction. A potential difference ΔV will therefore correspond to a much smaller radial separation δr if r is small.

● ***Example 18.4*** A uniform electric field has been established in a region of space. When a charge of -2×10^{-4} C is placed in this region, a force of 3×10^{-2} N pointing to the right must be applied to that charge to maintain it at rest.

1. What is the magnitude and direction of the electric field?
2. If the charge is located at a point at which the potential relative to ground is 2 V, what is the potential relative to ground at a point 5 cm to the left of this position?
3. In which direction will this charge move if the external force is removed?

Since an external force of 3×10^{-2} N to the right is needed to maintain equilibrium, the force due to the electric field must be 3×10^{-2} N to the left. From the definition of electric field (18.1), we have

$$E = \frac{F}{q} = \frac{3 \times 10^{-2}\ \text{N}}{2 \times 10^{-4}\ \text{C}} = 150\ \text{V/m}$$

Since the force which the field exerts on the *negative* charge is to the left, **E** must point to the *right*.

With **E** pointing to the right, work would have to be done on a positive charge against the field to move it to the left. The work done on a *unit* positive charge in shifting its position by 5 cm to the left is $Ed = 150 \times 0.05 = 7.5$ J, and hence, the potential at that point is 7.5 V greater than at the reference point; if the latter is at 2 V, the potential at the new location is 9.5 V relative to ground.

Finally, if no external force is applied to the negative charge, it will move opposite to the direction of **E**; that is, it will move to the left. ●

● ***Example 18.5*** A singly ionized silver atom (one valence electron has been removed from the neutral silver atom) is initially at rest in the uniform field of Example 18.4. If the ion is free to move, what are the magnitude and direction of the acceleration, and what is its velocity after it has traveled 5 cm?

The charge on the silver ion is $q = +1.6 \times 10^{-19}$ C. Hence, the force on this charge in the field $E = 150$ V/m is $(150 \text{ V/m})(1.6 \times 10^{-19} \text{ C}) = 2.4 \times 10^{-17}$ N, directed to the right, that is, in the direction of **E**. The acceleration of the ion is, by Newton's law,

$$\mathbf{a} = \mathbf{F}/m = \frac{2.4 \times 10^{-17} \text{ N}}{108 \times 1.66 \times 10^{-27} \text{ kg}} = 1.34 \times 10^{8} \text{ m/s}^2$$

also directed to the right. (The mass of the silver ion is the product of the atomic mass number of silver, 108, and the mass of one atomic mass unit, $1 \text{ u} = 1.66 \times 10^{-27}$ kg.)

To calculate the velocity after the ion has traveled 5 cm, we can take two different routes. We can use Equation (2.13e) or conservation of energy. We shall use both approaches for illustrative purposes.

From (2.13e) we have

$$v^2 = 2as = 2(1.34 \times 10^{8} \text{ m/s}^2)(5 \times 10^{-2} \text{ m}) = 1.34 \times 10^{7} \text{ m}^2/\text{s}^2$$

and

$$v = 3.66 \times 10^{3} \text{ m/s}$$

This is by far the quickest route once the acceleration has been determined. To use conservation of energy, we would proceed as follows.

In dropping through a potential of $\Delta V = (150 \text{ V/m})(0.05 \text{ m}) = 7.5$ V, the ion loses PE and gains KE of $q\,\Delta V = (1.6 \times 10^{-19} \text{ C})(7.5 \text{ V}) = 1.2 \times 10^{-18}$ J. Thus if the ion started from rest, its kinetic energy at that point must be $1.2 \times 10^{-18} \text{ J} = \frac{1}{2}mv^2$; solving for v, we obtain the same result as before. ●

18.8 The Electron Volt

We remarked earlier that in atomic physics the charge of an electron, though small on a macroscopic scale, is a convenient unit. Example 18.5 is typical of the situation often encountered in atomic physics. Potential differences may range from a few volts to several million volts, but the corresponding potential energy changes are only between 10^{-18} and 10^{-12} joules. Yet, because the masses of atoms and of electrons are very small, when these PE changes convert to KE, the corresponding speeds of the particles are high, approaching the speed of light, 3×10^{8} m/s.

In all such situations, the charge of the particle that falls or is raised through a potential difference of V volts is either one or a small integral number of electronic charges and may be positive or negative. Thus, when an electron or an O_2^- ion moves from a location where $V = 3$ volts to one where $V = 8$ volts, its potential energy decreases, changing by

$$\Delta PE = q\,\Delta V = (-1.6 \times 10^{-19}\text{ C})(5\text{ V}) = -8 \times 10^{-19}\text{ J}$$

If a Ca^{2+} ion is moved between the same two locations, its potential energy will *increase* by

$$\Delta PE = q\,\Delta V = 2(1.6 \times 10^{-19}\text{ C})(5\text{ V}) = 16 \times 10^{-19}\text{ J}$$

In each case, the resulting change in potential energy is the product of a potential difference, expressed in volts, and a small integral multiple of $e = 1.6 \times 10^{-19}$ C. It is therefore convenient and also common practice to express energy differences in atomic events by a different *unit of energy*, the *electron volt*, abbreviated eV.

DEFINITION: One electron volt (eV) is the increase in PE of a charge of e coulombs when raised through a potential difference of 1 V.

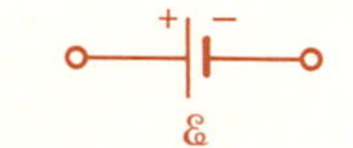

Figure 18.15 *The internationally accepted symbol for a battery. The positive terminal of the battery is designated by the longer, thinner line, the negative by the shorter, heavier line.*

Since

1 joule = 1 coulomb volt

it follows that

$$1\text{ eV} = 1.6 \times 10^{-19}\text{ C}\cdot\text{V} = 1.6 \times 10^{-19}\text{ J} \qquad \textbf{(18.12)}$$

The convenience of this unit of energy in physics and chemistry is best appreciated from a few typical examples. The ionization energies of atoms (the energies needed to remove an electron from a neutral atom) generally range between about 1 and 20 eV. The binding energies of stable molecules (energies needed to dissociate a molecule into its atomic constituents) are usually between about 0.1 and 3 eV. The average KE of translation of molecules, $\frac{3}{2}kT$, is 0.038 eV at room temperature. The energy associated with a single quantum of visible light is in the neighborhood of 3 eV. One could cite many more examples. In each case, the same energy could be expressed in the more conventional unit, the joule, but we would then have to carry along factors of order 10^{-19}. Moreover, the unit electron volt lends itself to a simple model for visualizing these energies. An ion of silver with KE of 6 eV can be produced in the laboratory by attaching the source of silver ions to the positive terminal of a 6-V battery as in Figure 18.16 and allowing the singly charged ion to fall through this potential difference.* When the ion strikes the negative terminal, its KE will be 6 eV. (It is assumed here that the ion, in falling through this potential difference, does not collide with other particles; that is, the region between the two battery terminals must be evacuated.)

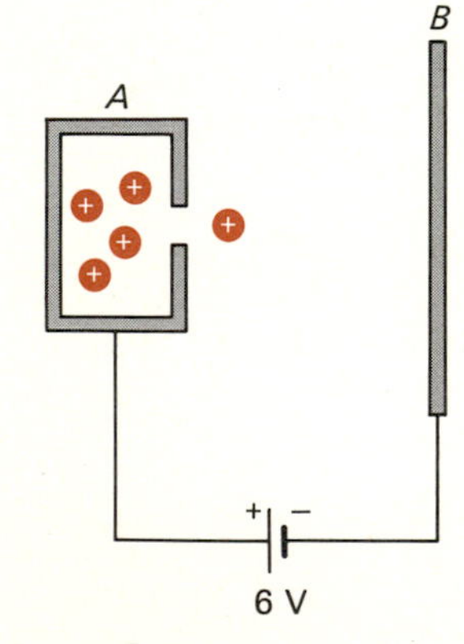

Figure 18.16 *Silver ions, with very little KE, escape through the small aperture in the box A. They are accelerated by the force due to the electric field between A and B created by connecting these two metallic surfaces to the 6-V battery as shown. As the ions drop through this potential difference of 6 V, their KE's increase by 6 eV.*

It must be emphasized again that *the electron volt is a unit of energy;* it is *not* a unit of potential difference or of voltage.

● ***Example 18.6*** An electron is liberated from the metal plate A of Figure 18.17 with vanishingly small KE. If plates A and B are connected by a 12-V battery as shown, what is the kinetic energy of the electron just be-

* In Chapters 20 and 23 we shall describe the basic principles of operation of batteries and other devices that generate and maintain potential differences. For our present purposes, it suffices to accept that a battery exists, and that it is a device that maintains a specified potential difference between its two terminals. The conventional symbol for a battery is shown in Figure 18.15—a pair of parallel lines, one long and thin, the other short and heavy. The long one denotes the positive terminal of the battery; the short and heavy one denotes the negative terminal.

fore it strikes the plate B? What is its speed? What would happen if the two plates were connected to the battery with opposite polarity, that is, if plate B were connected to the negative and plate A to the positive terminal?

Since an electron carries a *negative* charge $-e$, its PE decreases as it moves from a region of low potential to a region of higher potential. In this instance, its PE decreases by 12 eV (the potential difference is 12 V), and since it started with effectively zero KE, its KE just as it strikes the plate connected to the +12-V terminal will be 12 eV.

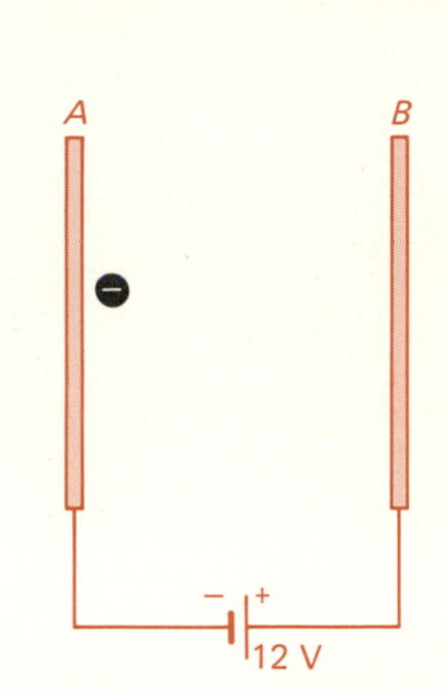

Figure 18.17 An electron leaves plate A with zero KE. When it strikes plate B, its KE is 12 eV. Note that the electron "falls" through a potential increase.

To calculate its speed, we convert to SI units and apply the relation KE $= \frac{1}{2}mv^2$. Thus

$$v = \sqrt{2 \times \text{KE}/m}$$

$$= \sqrt{\frac{2(12\ \text{eV})(1.6 \times 10^{-19}\ \text{J/eV})}{9.1 \times 10^{-31}\ \text{kg}}}$$

$$= 2.05 \times 10^{6}\ \text{m/s}$$

If the battery were connected to the two plates with reversed polarity, the electron could never even approach the other plate. The force due to the electric field would push it back toward plate A instead of pulling it toward plate B. ●

● ***Example 18.7*** A stream of doubly charged 200-eV magnesium ions, Mg^{2+}, enters the shaded region of Figure 18.18 through the hole in metal plate A. With what KE will the ions emerge through the small hole in plate B? Repeat for the ions Ag^{+}, Na^{+}, and O_2^{-}, assuming that they all enter the region with a KE of 200 eV.

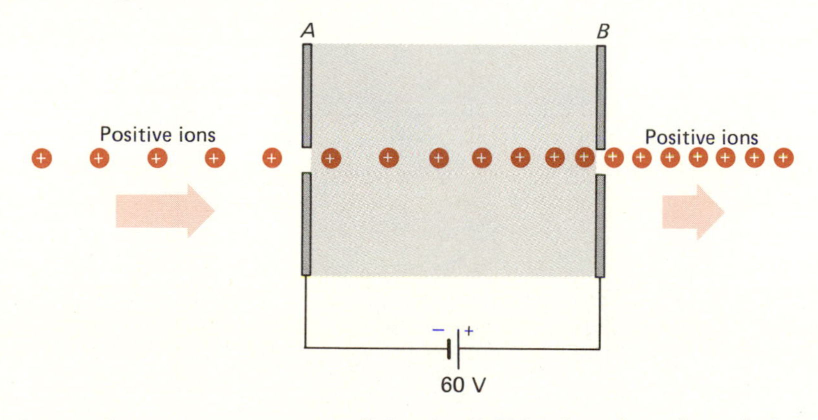

Figure 18.18 Various ions enter the shaded region between plates A and B through the small aperture in plate A. The region between the two plates contains an electric field created by the 60-V battery that is connected to plates A and B. When the ions emerge from the aperture in plate B, their kinetic energies will have changed by amounts that depend on the ionic charge.

Since plate B is at a potential that is 60 V higher than that of plate A, the PE of a positive ion is greater at B than at A. In the case of the doubly ionized ion Mg^{2+}, the increase in PE is $2 \times 60 = 120$ eV, and the KE of the ion must therefore decrease by the same amount as the ion passes from A to B. Thus, as the magnesium ions leave the shaded region, their KE will be only $(200 - 120)$ eV $= 80$ eV. In the case of a singly ionized silver ion, the energy change is only 60 eV, again a reduction in KE because the PE of the positive ion increases as it passes from A to B. As the silver ion leaves the hole through the plate B, its KE will be 140 eV. If the ion is singly ionized sodium, Na^{+}, its KE as it emerges from the hole in B is again 140 eV. Finally, if the ion is O_2^{-}, its PE will *decrease* by 60 eV as it passes from A to B, and so its KE at B will be $(200 + 60)$ eV $= 260$ eV.

Although the silver and sodium ions have exactly the *same energies* at B, their *speeds* will be substantially *different*, the lighter sodium ion moving faster than the more massive silver ion. Similarly, the speeds of the ions at A will differ even though all of them pass through this hole with exactly the same KE of 200 eV. ●

18.9 Potentials due to Charge Distributions

From the superposition principle, it follows that the potential due to several charges is simply the algebraic sum of the potentials due to the individual charges. In calculating the potential for a given charge distribution, one only needs to calculate the potential at the point of interest due to the separate charges and then add these potentials. It is important, however, to remember the sign of the charge, which also determines the sign of the corresponding potential. That is, according to Equation (18.11), the potential due to a negative charge is negative, as shown in Figure 18.14.

● ***Example 18.8*** Equal charges of $+6\ \mu\text{C}$ are positioned at the corners of a square of sides 4 cm long. What are the field and the potential at the center of the square?

With this, as with any problem, it is wise first to sketch the fields arising from the separate charges and think about the problem briefly instead of blindly plugging numbers into formulae. The **E** fields at the center of the square are shown in Figure 18.19, and it is evident from the symmetry of this pattern that the vector sum of the four field vectors will be zero. That is, $\mathbf{E} = 0$ at the center of the square.

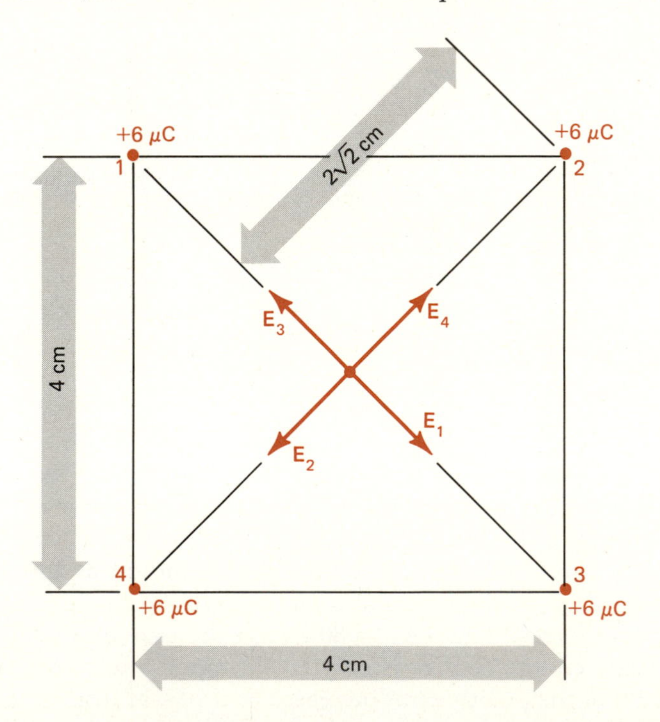

Figure 18.19 *Equal positive charges are located at the corners of a square. The electric field at the center of the square is zero, the vector sum of the four fields due to the four point charges. The potential at that point, however, is not zero.*

That does not mean, however, that the potential at that point also vanishes. There the potential due to each of the charges is the same, namely

$$V_i = \frac{(9 \times 10^9\ \text{N}\cdot\text{m}^2/\text{C}^2)(6 \times 10^{-6}\ \text{C})}{2\sqrt{2} \times 10^{-2}\ \text{m}} = 1.91 \times 10^6\ \text{V}$$

and the potential due to all four identical charges is then four times as much; i.e.,

$$V = 7.64 \times 10^6 \text{ V}$$

a sizable potential.

Example 18.9 Suppose that three of the $+6\text{-}\mu\text{C}$ charges of Example 18.8 are replaced by charges of $-2\ \mu\text{C}$, as shown in Figure 18.20. Find the field and potential at the center of the square.

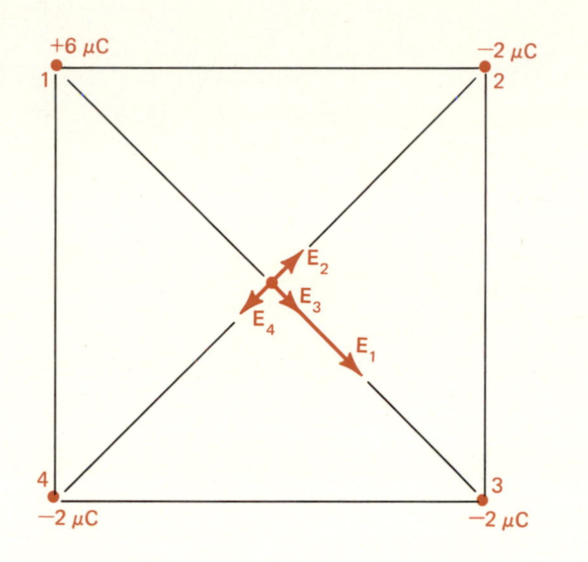

Figure 18.20 *Three charges of $-2\ \mu\text{C}$ and one charge of $+6\ \mu\text{C}$ are placed at the four corners of a square. The electric field at the center of the square does not vanish, but has a magnitude equal to $E_1 + E_3$ and a direction along $\mathbf{E}_1$ and $\mathbf{E}_3$. The potential, however, is now zero at the center of the square.*

This time the field at the center will not vanish. The contributions to the field from the $-2\text{-}\mu\text{C}$ charges on opposite corners do cancel, but that still leaves the contributions from the remaining $-2\text{-}\mu\text{C}$ charge and the $+6\text{-}\mu\text{C}$ charge on the other pair of diagonally opposite corners. The resulting field is then

$$\mathbf{E} = (9 \times 10^9 \text{ N}\cdot\text{m}^2/\text{C}^2)\left[\frac{(2 \times 10^{-6}\text{ C})}{(8 \times 10^{-4}\text{ m}^2)} + \frac{(6 \times 10^{-6}\text{ C})}{(8 \times 10^{-4}\text{ m}^2)}\right]$$
$$= 9 \times 10^7 \text{ V/m}$$

directed toward the $-2\text{-}\mu\text{C}$ charge.

Although in this case the field does not vanish, the potential at the center of the square is zero. We can quickly see this on taking the sum of the potentials due to the four charges, all equidistant from the point of interest, with the total charge adding to zero.

From Examples 18.8 and 18.9, we see that **E** can be zero at a point at which $V \neq 0$, and V can be zero where $\mathbf{E} \neq 0$. By the gravitational analogy, the electric field, proportional to the force that acts on a charge, is akin to the net downward pull experienced by an object, and the electrostatic potential is akin to the gravitational potential. The first of the preceding examples corresponds to the peak of a tall mountain; on a peak, where locally the ground is flat, an object feels no net downward pull and remains at rest if undisturbed. Its gravitational potential is, however, well above that at sea level. On the other hand, on a steeply sloping coast line, an object placed at sea level (zero gravitational potential) will still slide down into the water; that is, it feels a force with a net downward component. Example 18.9 corresponds to this latter situation.

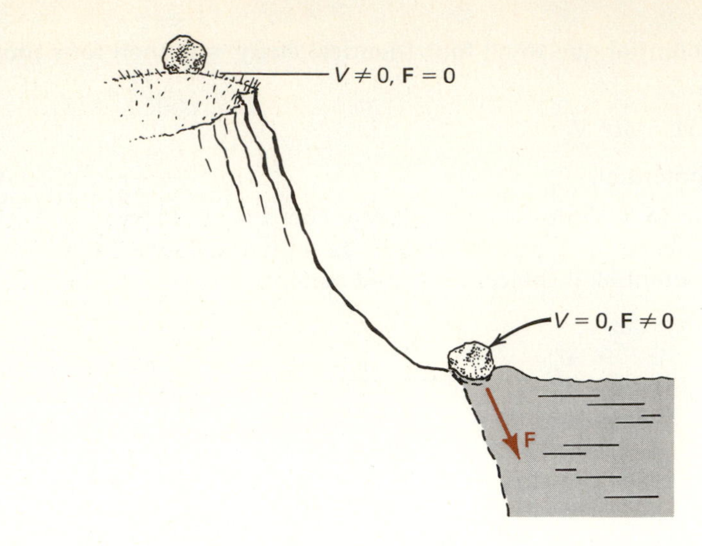

Figure 18.21 *In the earth's gravitational field, an object may be at a point of large potential and yet feel no net force. An example is a rock at the top of a hill above the sea. Yet another rock at sea level, that is, at a location of zero potential, may feel a net unbalanced force that pulls it down into the sea, to regions of negative potential.*

18.10 Equipotential Surfaces and Electric Field Lines

We have established that within a conductor the electrostatic field must vanish. From Gauss's law it follows that under static conditions, there can be no unbalanced charges within a metal. Any charges a metallic object carries must appear on its surface.

Another far-reaching consequence of the conducting property of a metal and its inability to support a steady electric field is that

Under electrostatic conditions, all points in and on a connected conducting object must be at the same potential.

This statement follows from the fact that a potential difference between two points can exist only if somewhere along any *arbitrary* path connecting the two points the electric field does not vanish. If the two points can be connected by a path, for example, the interior of a conducting region, such that $\mathbf{E} = 0$ *everywhere* along this path, the potential difference between these two points *must* be zero.

We therefore conclude that

Any metallic surface must be an electrostatic equipotential surface.

Since the electric field is everywhere perpendicular to an equipotential surface, we now see how one might go about constructing an electric-field configuration that meets any desired specifications, consistent with the basic demand that electric-field lines do not intersect and are continuous in a charge-free region. Given the field configuration, one first draws the corresponding equipotential surfaces. If one then shapes thin conducting sheets to match these surfaces and connects them to batteries of the proper terminal voltages, the correct equipotential surfaces will have been established, and consequently, the specified electric field will exist in the region.

In a region in which the electric field is strong, two equipotential surfaces differing by a potential of one volt must be very close together. Conversely, in regions of weak electric field, the equipotential surfaces are well separated. We have already encountered one example of this in our discussion of the equipotentials surrounding an isolated point charge.

It may be helpful to compare equipotential maps for the electrostatic case with gravitational equipotential maps, namely the contour maps

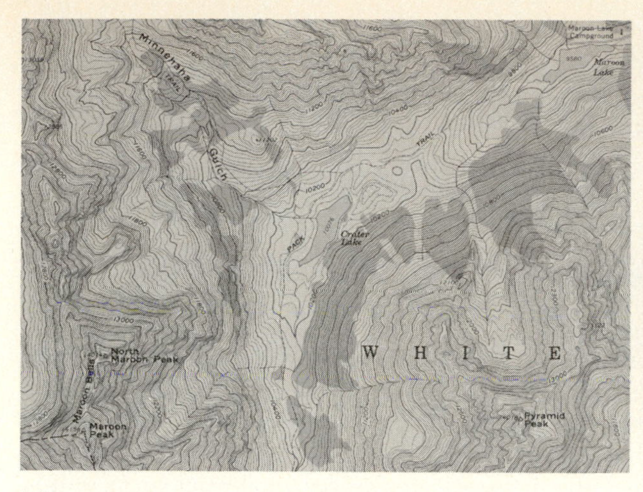

(*a*)

(*b*)

Figure 18.22 *(a) Part of the* USC&GS *survey map Maroon Bells Quadrangle of the White River National Forest. Compare the equipotential lines with (b) the photograph of the Maroon Bells as seen from Maroon Lake at the upper right-hand corner in (a).*

published by the U.S. Coast and Geodetic Survey. On the contour maps, the contour lines, showing lines of constant elevation above sea level, are the intersections of the surfaces of constant gravitational potential (spheres centered on the earth's center) with various protuberances such as mountains on the earth. Closely packed contour lines indicate a region of very steeply sloping ground, like a cliff, and here the net downward pull on an object is greatest.

● ***Example 18.10*** Devise a technique for generating a uniform electric field in a limited region of space.

For this purpose two large, flat metallic plates are placed parallel to each other and are connected to the terminals of a battery. The potential difference between the two plates is then the same as that between the terminals of the battery. Both metal plates are equipotential surfaces, and so the field lines must be perpendicular to the two plates in the immediate vicinity of the metal surfaces. Except near the edges, there is no reason for the field lines to bend one way or another, and they point directly from one plate to the other. Since the field lines are parallel in the intervening region, the field strength **E**, which is proportional to the density of the field lines, is constant in the region between the plates. ●

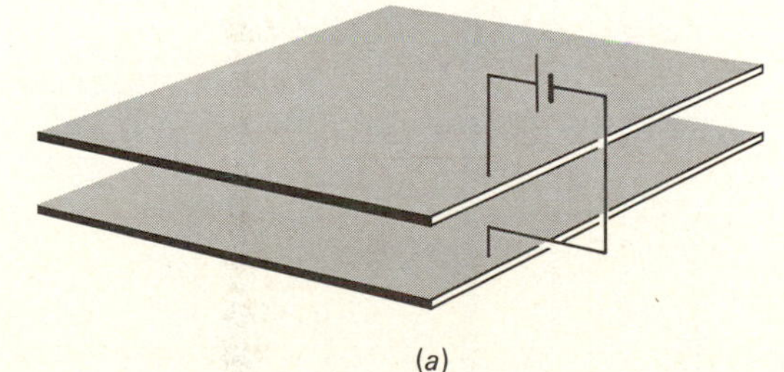

(*a*)

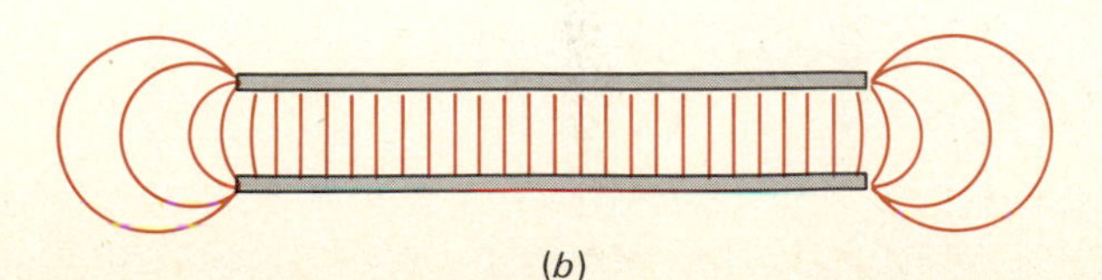

(*b*)

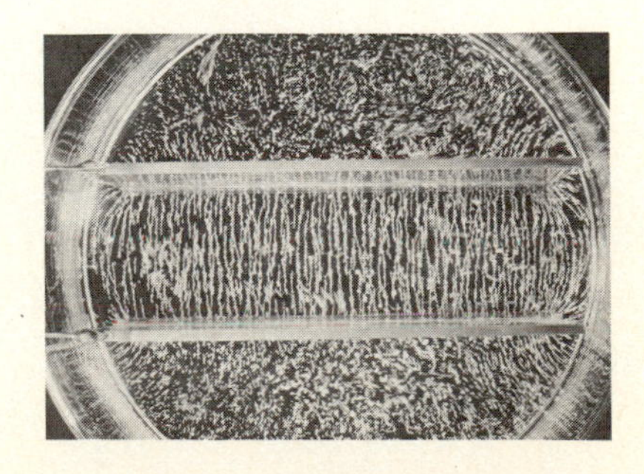

(*c*)

Figure 18.23 *(a) If two large, parallel metal plates are connected to a voltage source, the region between the plates is permeated by a uniform electric field, as shown in (b). The field can be visualized by means of small threads suspended in oil. (c) Two metal plates have been connected to the terminals of a battery and the field between them is uniform except near the edges.*

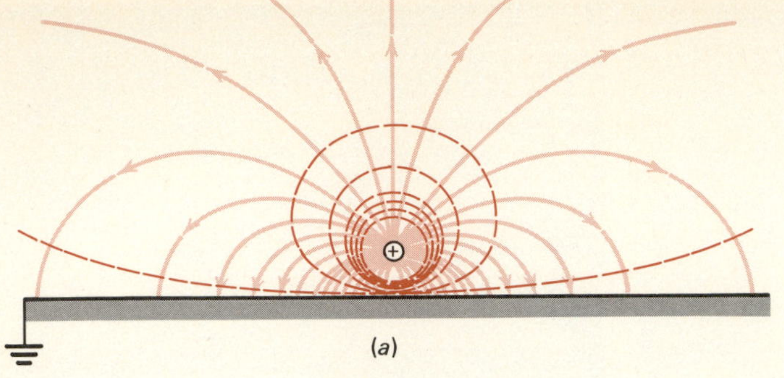

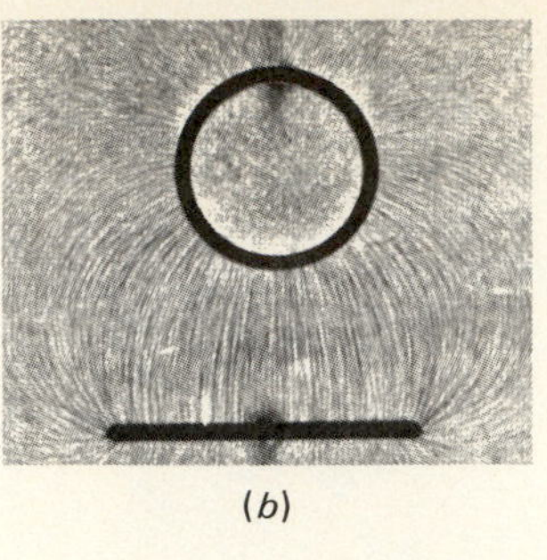

Figure 18.24 (*a*) *Electric-field lines and equipotential surfaces for a positive point charge placed above a very large, flat conducting sheet connected to ground. Note that immediately above the metal sheet the field lines are perpendicular to the metal surface; very close to the point charge, the field lines are directed radially outward from the charge. Close to the conducting sheet, the equipotential surfaces are approximately flat; close to the point charge, the equipotential surfaces approximate spheres. You should compare Figure 18.25*(*a*) *with Figure 18.4*(*a*). (*b*) *Field lines between a circular ring and a metal bar that have been connected to the terminals of a battery. The visual display is obtained by the same technique as in Figure 18.4*(*b*). *Note the similarity to the pattern of* (*a*).

● ***Example 18.11*** Draw the field lines and equipotential surfaces corresponding to a single positive charge Q placed above a grounded metallic plate of infinite extent.

Since the plate is metallic, it constitutes an equipotential surface; it is grounded, and so the potential over the surface of the plate must be zero.

To draw the field lines, we recall the rules for field lines, especially that they must be continuous and not cross in a charge-free region. In the immediate neighborhood of the charge Q, the field lines point radially out from this charge. Just above the metallic plate, the field lines must be perpendicular to the metal plate; that is, they must be vertical. There is only one reasonable way in which these lines can be connected, and that is shown in Figure 18.24(*a*); the corresponding equipotential surfaces are also shown. Note that near the point charge, these surfaces approximate spheres; near the plate, they approximate planes. Comparison of the field and potential configuration obtained here with Figure 18.4 shows that in the region between the grounded metal plate and the point charge the field is identical to that due to a dipole formed by placing a charge $-Q$ at the mirror image of Q and removing the metallic "mirror." Here we have an example of a field configuration achieved by the judicious location of a metallic equipotential surface. ●

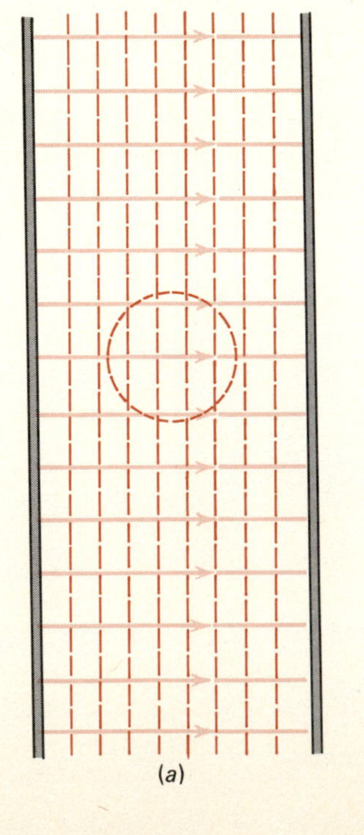

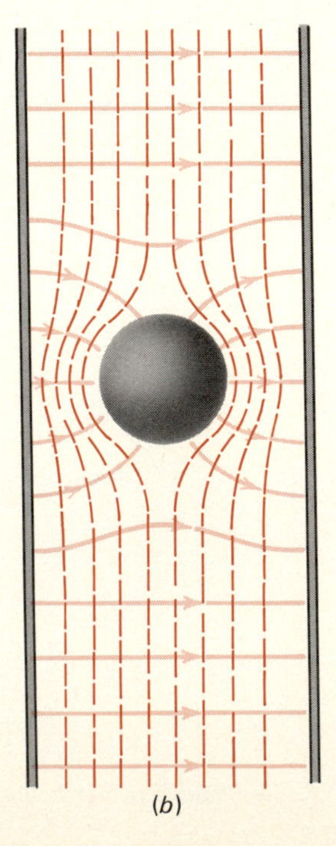

Figure 18.25 (*a*) *A region of uniform electric field is established between two metal plates. When a metallic sphere is placed in the position shown by the dashed circle, the field lines and equipotential surfaces must assume a new shape to satisfy the conditions that* (1) **E** *is always perpendicular to the surface of a metal and* (2) *every metallic surface must be an equipotential.* (*b*) *The new field and potential configuration.*

● ***Example 18.12*** A brass sphere is placed between the two metal plates of Figure 18.25(*a*). Sketch the field lines and equipotential surfaces under the assumption that the brass sphere was uncharged and electrically insulated from the two plates.

Since the brass sphere carries no net charge, as many field lines must emanate from it as terminate on it. We know that inserting this metallic sphere will distort the pattern of field lines and equipotential surfaces, because the entire sphere must be at the same potential. The field lines must come in toward this sphere at normal incidence.

If the sphere is midway between the two plates, its potential will be midway between the potentials of the two plates. Thus, one arrives at the configuration of equipotentials and field lines shown in Figure 18.25(*b*). ●

Summary

The *electric field* at some point in space is defined as the *force per unit positive charge at that point.* For any charge q, positive or negative, the force on that charge is given by

$$\mathbf{F}(\mathbf{r}) = q\mathbf{E}(\mathbf{r})$$

where $\mathbf{E}(\mathbf{r})$ is the electrostatic field at the point $\mathbf{r}$. If q is positive, $\mathbf{F}$ and $\mathbf{E}$ are parallel; if q is negative, $\mathbf{F}$ and $\mathbf{E}$ are antiparallel.

The electric field in the vicinity of an isolated point charge Q has a magnitude given by

$$E(r) = \frac{kQ}{r^2}$$

and points radially away from Q if Q is positive, and toward Q if Q is negative. Here the charge Q is presumed at the origin of coordinates.

The electric field is a *vector field.* It is conveniently visualized in terms of *field lines* that point in the direction of $\mathbf{E}$ and have a density proportional to the strength (magnitude) of $\mathbf{E}$. Field lines must originate at a positive charge and terminate on a negative. In a field-free region, field lines are continuous and cannot intersect.

Gauss's law states:
The perpendicular component of the electric field summed over any closed surface equals the net charge enclosed within that surface divided by the permittivity of free space.

$$\Sigma(E_\perp \, \Delta A) = \frac{Q}{\epsilon_0}$$

where $\epsilon_0 = 1/4\pi k$ is the permittivity of free space.

Using Gauss's law, one can show that within a charge-free region surrounded by a closed conducting surface, the electrostatic field must vanish.

If a conductor carries a net charge, this charge must be distributed over its surface.

The *electrostatic potential difference* V_{AB} between points A and B is defined as the negative of the work done by the electrostatic field in transporting a unit positive charge from B to A. The electrostatic field is a *con-*

servative field; V_{AB} does not depend on the path followed in transporting the unit positive charge from B to A.

By designating some location as that of zero potential, one can speak of *absolute potentials* relative to this zero. By convention, zero potential is the potential at a point infinitely far removed from any charge distribution. In terms of absolute potentials,

$$V_{AB} = V_B - V_A$$

The SI unit of potential is the *volt* (V).

The most widely used SI unit for electric field is the volt per meter, which is equivalent to the newton per coulomb.

In a *uniform* electric field, the potential difference V_{AB} equals the product of the field $\mathbf{E}$ and the component of $\mathbf{s}_{AB}$ in the direction of $-\mathbf{E}$, where $\mathbf{s}_{AB}$ is the displacement from A to B.

The potential in the neighborhood of an isolated point charge Q is given by

$$V(r) = k\frac{Q}{r}$$

The potential is positive if Q is positive and negative if Q is negative.

In the description of atomic events, the electron volt is a convenient unit of energy. One *electron volt* equals the PE change of a charge e in moving through a potential difference of 1 V.

Regions in space that are at the same potential are known as *equipotentials.* Since the electrostatic field must vanish everywhere within a conducting region, all parts of a connected metallic object compose an equipotential region. In particular, the surface of a connected metallic object is an *electrostatic equipotential surface.*

The electric field is locally perpendicular to the equipotential surface.

Multiple Choice Questions

18.1 In a certain region of space the electric field is zero. It follows that in this region

(a) the potential is proportional to r.

(b) the potential is constant.

(c) the potential is inversely proportional to r.

(d) the potential is zero.

18.2 At a distance 10 cm from a point charge, the electric field is 5 V/m and points toward the charge. At a distance 50 cm from this point charge, the field

(a) is 0.2 V/m and points toward the charge.

(b) is 1 V/m and points toward the charge.

(c) is 1 V/m and points away from the charge.

(d) is zero.

18.3 Which of the following is *not* a unit of energy?

(a) Newton meter

(b) Calorie

(c) Volt

(d) Watt second

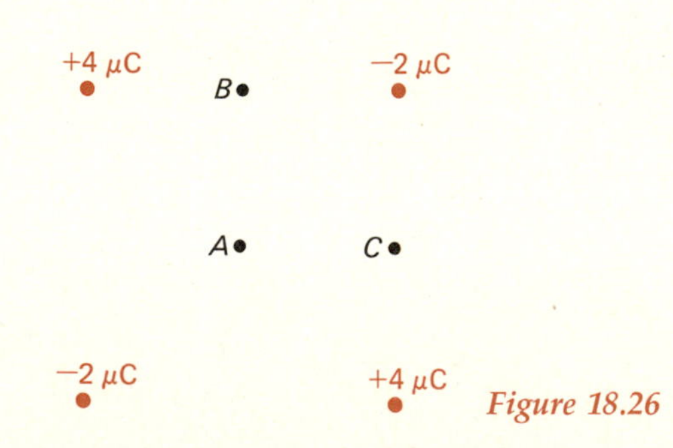

Figure 18.26

18.4 These statements refer to Figure 18.26. Mark the correct one.

(a) Work must be done on a positive charge to move it from A to B.

(b) Work must be done on a negative charge to move it from A to B.

(c) No work is done in moving a positive charge from A to C.

(d) None of the above statements is correct.

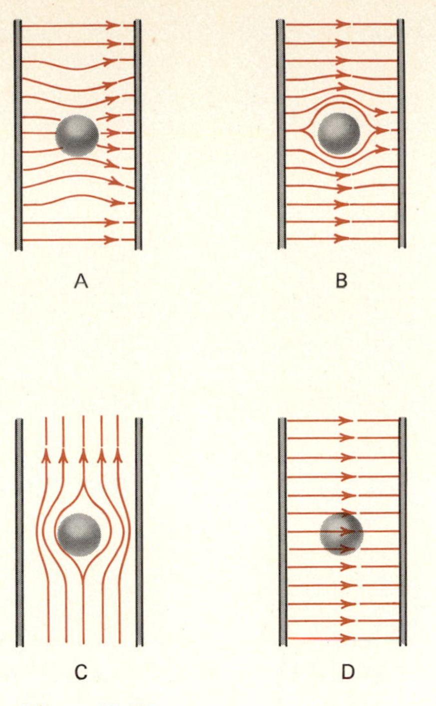

Figure 18.27

18.5 Figures 18.27A–D show electric field lines in the region between two metal plates. A metal sphere has been placed between the two plates. The correct field configuration is

(a) A
(b) B
(c) C
(d) D

18.6 A positive point charge Q is located at the center of a spherical metallic shell shown in Figure 18.28. The following statement is true.

(a) The **E** field vanishes everywhere.
(b) $\mathbf{E} = 0$ for $r > R_0$; $\mathbf{E} \neq 0$ for $r < R_i$.
(c) $\mathbf{E} = 0$ for $r < R_i$; $\mathbf{E} \neq 0$ for $r > R_0$.
(d) $\mathbf{E} \neq 0$ for $r < R_i$ and $r > R_0$.

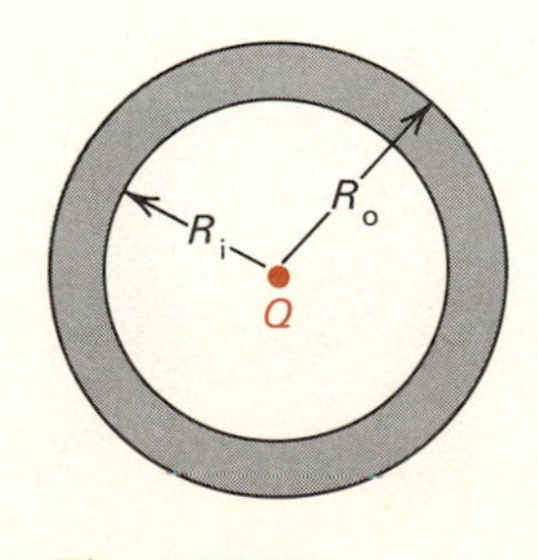

Figure 18.28

18.7 Figure 18.29 shows a group of equipotential lines resulting from a charge distribution in the plane

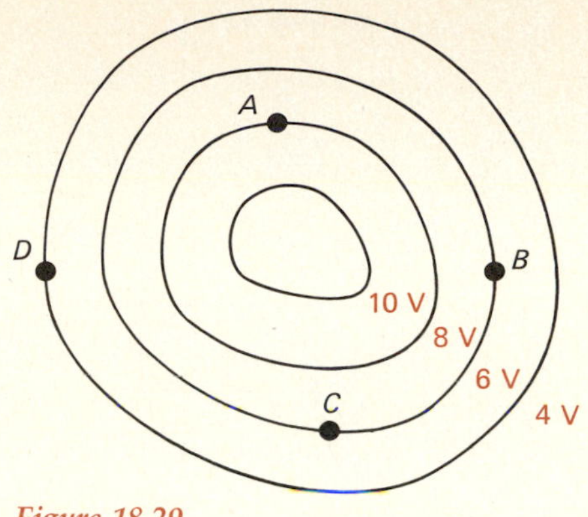

Figure 18.29

of the paper. At which location will an electron ($q = -1.6 \times 10^{-19}$ C) experience a force directed toward the bottom of the page?

(a) A
(b) B
(c) C
(d) D

18.8 In Figure 18.30 the electric field is zero

(a) at some point in region a only.
(b) at some point in region b only.
(c) at some point in region c only.
(d) at some point in region a and some point in region c.
(e) at no point along the x axis.

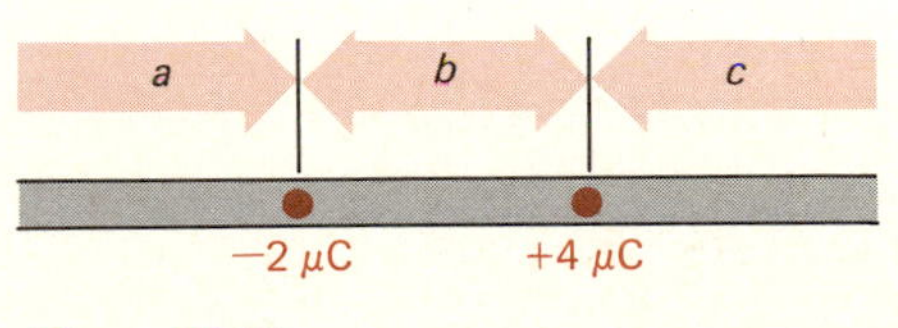

Figure 18.30

18.9 In Figure 18.30 the absolute potential is zero

(a) at some point in region a only.
(b) at some point in region b only.
(c) at some point in region c only.
(d) at some point in region a and also at some point in region b.
(e) at some point in region a and also at some point in region c.

18.10 Two charges q_1 and q_2 are placed as shown in Figure 18.31. The electric field at point A is in the direction indicated. The charge q_2 is related to charge q_1 according to

(a) $q_2 = q_1$
(b) $q_2 = -q_1$
(c) $q_2 = \sqrt{2}\, q_1$
(d) $q_2 = -\sqrt{2}\, q_1$

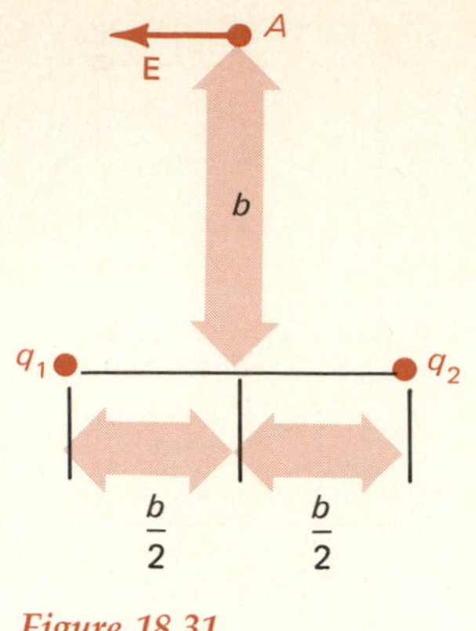

Figure 18.31

18.11 The sketches in Figure 18.32 show some patterns for equipotential lines in the plane of the paper. The one most closely corresponding to the equipotentials due to two equal positive charges is

(a) A
(b) B
(c) C
(d) D

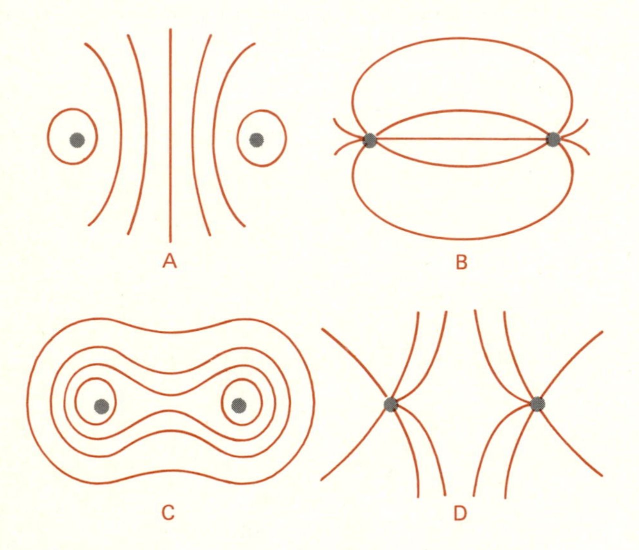

Figure 18.32

18.12 Which of these statements is true for a metallic conductor?

(a) It cannot carry a net charge.
(b) If it carries a net charge, that charge must be distributed uniformly throughout its volume.
(c) If it carries a net charge, that charge must be distributed over its surface.
(d) It must be at zero absolute potential.

18.13 Figures 18.33(A)–(D) represent equipotential lines such that the potential difference between adjacent lines is the same. The figure that most closely represents the potential in the neighborhood of an

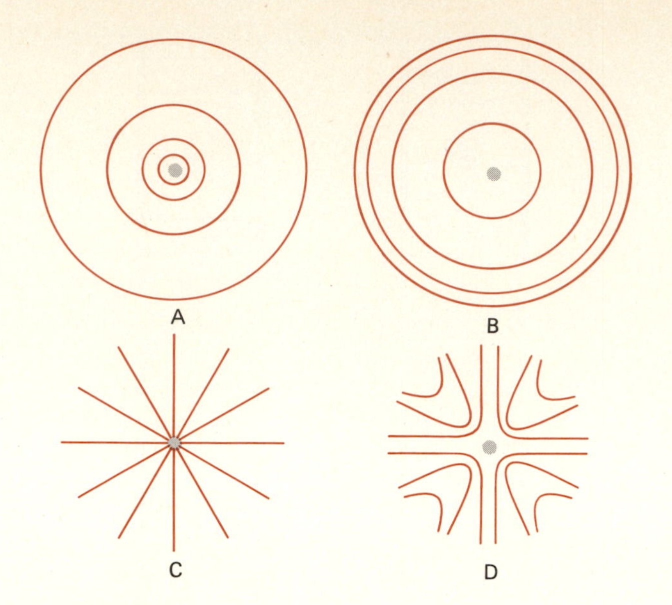

Figure 18.33

isolated point charge is

(a) A
(b) B
(c) C
(d) D

18.14 A charge $-Q$ is brought near an initially uncharged metallic shell as shown, but the charge is not brought into contact with the shell. Which of the following statements is correct?

(a) A net negative charge is induced on the inside surface and a positive charge of equal magnitude appears on the outside surface of the shell.
(b) A net positive charge is induced on the inside and a larger negative charge appears on the outside surface of the shell.
(c) A net positive charge is induced on the outside and a smaller negative charge appears on the inside surface of the shell.
(d) None of the above statements is correct.

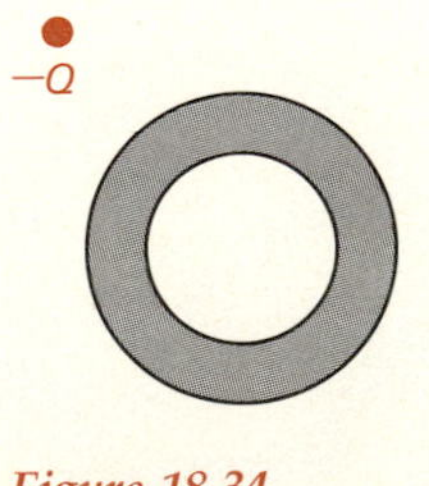

Figure 18.34

18.15 The electric field lines drawn in Figure 18.35 correspond to the field configuration of

(a) two equal positive charges.

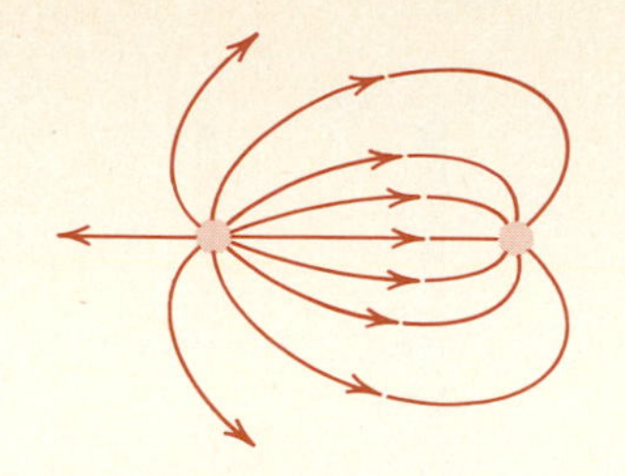

Figure 18.35

(b) two charges Q and $-Q$.

(c) two unequal charges, with the positive the larger.

(d) two unequal charges, with the negative the larger.

18.16 A point charge $-Q$ is placed at the center of a conducting shell as shown in Figure 18.36. Consider these statements.

A. The electric field inside the conducting shell (that is, within the shaded region) is zero.
B. The electric field outside the shell points radially toward the center of the sphere.
C. The field between the charge and the inner surface of the shell is constant.
D. The potential outside the sphere varies as $-kQ/r$.

(a) A only is correct.
(b) A and B are both correct
(c) A, B, and C are correct.
(d) A, B, and D are correct.

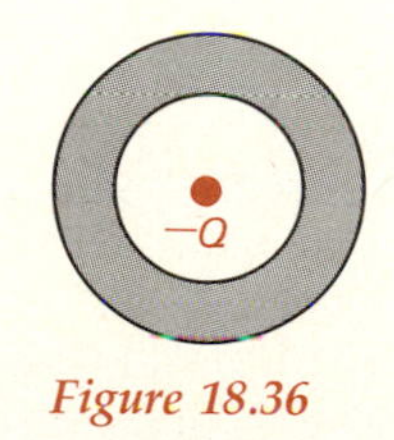

Figure 18.36

18.17 In an electrostatic situation, which of the three statements A, B, and C are *always* true?

A. The electric field just outside a metallic surface is parallel to the surface.
B. The electric field inside a metal is zero.
C. The electric lines of force are perpendicular to equipotential surfaces.

(a) A and B only.
(b) B and C only.
(c) A and C only.
(d) A and B and C.

18.18 The charges of Figure 18.37 are arranged on the four corners of a square. The electric field is zero at the following point or points:

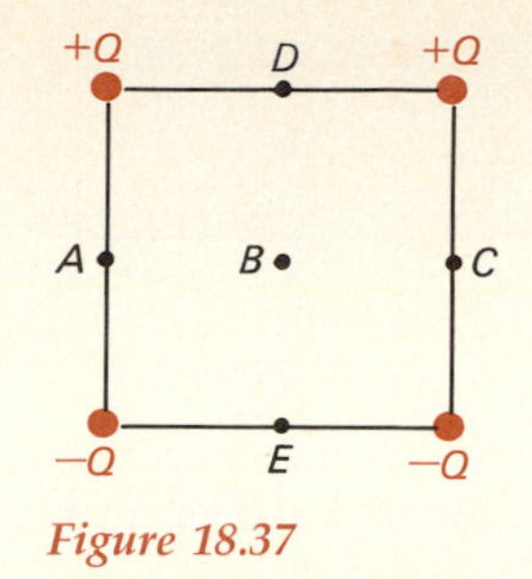

Figure 18.37

(a) A and C
(b) D and E
(c) B
(d) None of the above.

18.19 For the arrangement of charges of Figure 18.37, the potential is zero at

(a) A, B, and C.
(b) D, B, and E.
(c) B only.
(d) A, B, C, D, and E.

18.20 A negative charge is placed at the center of a hollow conducting sphere that is initially uncharged. The electric field lines are most correctly represented by which of the diagrams of Figure 18.38A–E?

(a) A
(b) B
(c) C
(d) D
(e) E

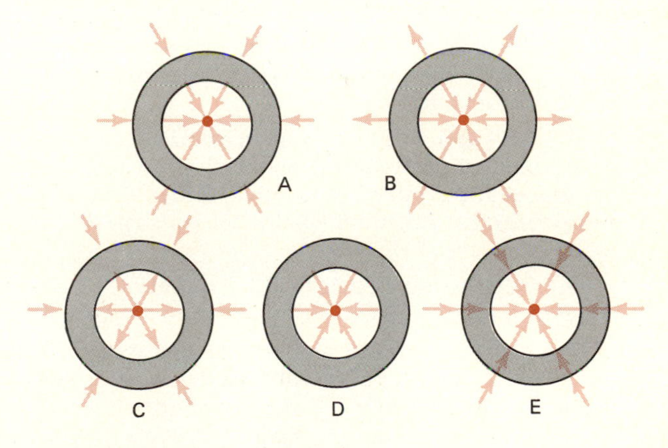

Figure 18.38

18.21 In Figure 18.39, a charged pellet is suspended between two charged horizontal metal plates. The lower plate is charged positively, the upper negatively. Which of the following statements is *not* true?

(a) The electric field between the plates points up.

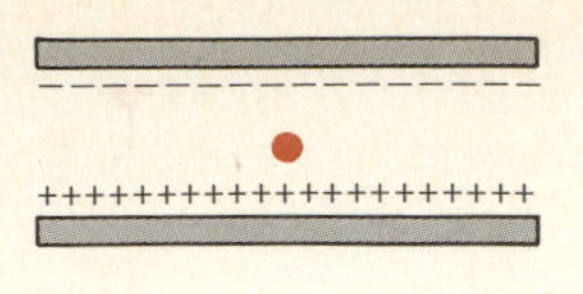

Figure 18.39

(b) The pellet is charged negatively.

(c) The magnitude of the electrostatic force on the pellet is equal to mg.

(d) The plates are at different potentials.

18.22 A proton (mass = m_p, charge = e) and a deuteron (mass = m_d = $2m_p$, charge = e) are accelerated from rest by the same field and fall through the same potential difference ΔV, as indicated in Figure 18.40. Both particles start from the same position just in front of the positive plate. Which of the following correctly describes their properties as the particles pass through the hole in the negative plate?

(a) The KE is less for the proton than for the deuteron.

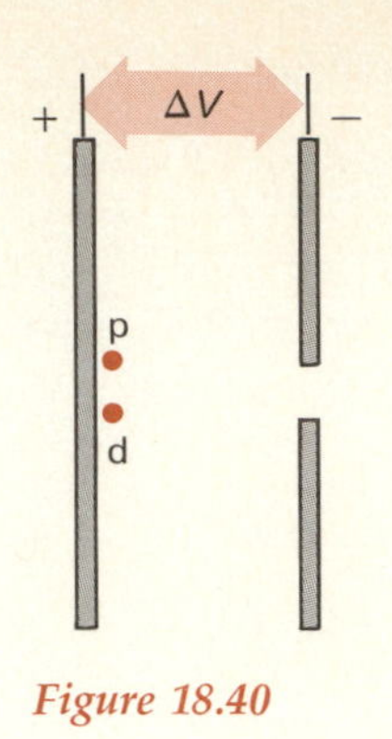

Figure 18.40

(b) The KE is the same for the proton and the deuteron.

(c) The proton has the same speed as the deuteron.

(d) The proton has the same momentum as the deuteron.

Problems

Sections 18.1–18.2

18.1 A small sphere carries a charge of 20 μC. It is in a constant electric field whose magnitude is 10,000 N/C = 10,000 V/m. Under these conditions, the sphere is in equilibrium. What is the mass of the sphere and the direction of the electric field?

18.2 A charge of -2.4 μC experiences an upward force of 3 N. What is the magnitude and direction of the electric field at the charge?

18.3 An electron is located in an electric field of 600 V/m. What is the force acting on the electron and what is its acceleration if it is free to move? If the field points in the positive x direction, what is the direction of the acceleration on the electron?

18.4 What is the electric field 1.5 m below a point charge of -6 μC?

18.5 A point charge of 4 nC is located at x = 30 cm, y = 40 cm. At which point will the electric field be 72 V/m and point in the negative y direction?

18.6 In Figure 18.41, the field at point A is zero. Find the charge Q_1.

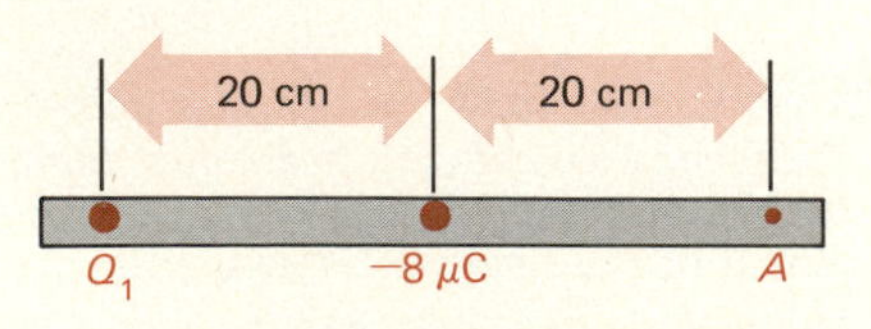

Figure 18.41

• **18.7** A charge of 0.2 μC is placed at x = 0, y = 20 cm and a charge of -0.4 μC is placed at x = 20 cm, y = 0. Determine the electric field at the origin.

• **18.8** A small water droplet has a diameter of 6 μm. If it carries a charge of 24 excess electrons, what is the strength and direction of the uniform electric field that will just balance the gravitational force acting on the droplet? If the magnitude of the electric field is then increased to three times its previous value, how will the droplet behave compared with its behavior in the total absence of the electric field?

• • **18.9** For the water droplet of Problem 18.8, calculate the terminal velocities in the absence of an electric field and in an electric field whose magnitude is 4 times that needed to just balance the force of gravity on the droplet.

• • **18.10** A charge of $+3$ μC is located at x = 0 and a charge of -1 μC at x = 20 cm. Calculate the electric field at (a) x = 5 cm, y = 0; (b) x = 15 cm, y = 0 cm; (c) x = 5 cm, y = 5 cm; (d) x = 15 cm, y = 5 cm. Compare your results with Figure 18.5.

Section 18.3

• **18.11** A metal sphere of 10-cm diameter carries a total charge of 0.4 μC. What is the magnitude of the electric field just outside this sphere?

• **18.12** How many electrons must be removed from an isolated metallic sphere of 10-cm radius so that the electric field just outside the sphere is 600 V/m?

• • **18.13** A positive charge of 2 C is distributed with uniform density throughout a spherical volume of radius 4 m. Taking the origin of coordinates at the center of this spherically symmetric charge distribution, derive expressions for the electric field as a function of r for $r > 4$ m and for $r < 4$ m. Use Gauss's law.

• • **18.14** A coaxial cable consists of an inner solid cylindrical conductor and an outer, cylindrical metallic conductor. Assume that the space between the two conductors is air. The radius of the inner conductor is 0.5 cm, that of the outer conductor is 5 cm (inside radius). Suppose that the inner conductor carries a charge of λ C/m and that the outer conductor is grounded. Use Gauss's law to find the electric field at any point between the two conductors. What do the equipotential surfaces look like? How would your result change if the outer conductor had not been grounded?

• • **18.15** Imagine that a hole 500 km deep is drilled toward the center of the earth. Calculate the weight of a 1-kg mass at the bottom of this hole, assuming that the density of the earth is uniform. If such an experiment could be conducted, would you expect the result to give a weight greater or smaller than the one calculated here?

Sections 18.5–18.8

18.16 Two charges, $+10\ \mu$C and $-20\ \mu$C, are separated by 1.4 m. Determine the electric field at a point midway between the two charges.

18.17 An α particle (helium nucleus) is accelerated in a cyclotron to an energy of 40 MeV (4×10^7 eV). Calculate the speed of this particle. What would be the speed of a proton whose energy is 40 MeV?

18.18 Charges of 6 μC and 24 μC are separated by 0.6 m. What is the electric field and the potential midway between the two charges?

• **18.19** According to the Bohr model of atomic hydrogen, the electron is in a stable orbit about the proton with an orbit radius of 5.3×10^{-11} m. What is the coulomb force on the electron and what is the potential energy of the electron, using the standard convention that $V = 0$ at infinity?

• **18.20** Calculate the electric field and potential at the center of the square of Problem 17.19.

• **18.21** Charges of $+2\ \mu$C, $-2\ \mu$C, and $+4\ \mu$C are placed at the vertices of a square of side 0.4 m, as shown in Figure 18.42. What is the electric field at the fourth corner?

• **18.22** What is the potential at the center and at the lower left-hand corner of the square of Figure 18.42?

• **18.23** An isosceles triangle has charges at its vertices as shown in Figure 18.43. Find the field and absolute potential at point A.

• **18.24** For Problem 18.6, calculate the work done in moving a charge of 10 μC from point A to the point midway between the two charges.

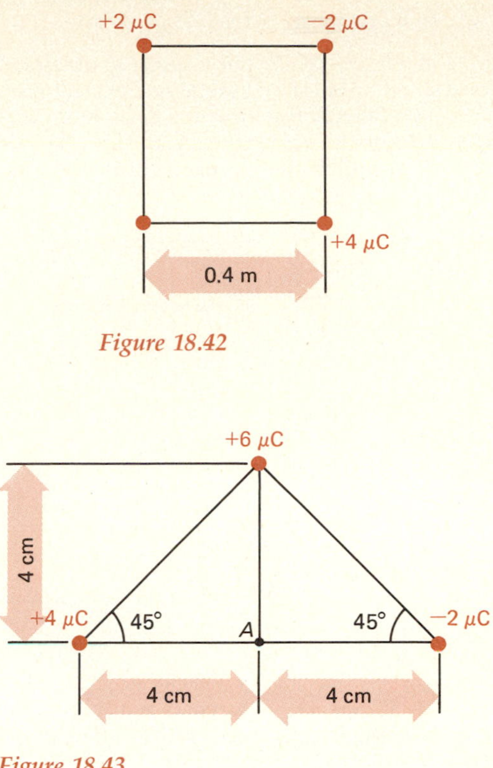

Figure 18.42

Figure 18.43

• • **18.25** A proton is released from rest in a uniform electric field of 1000 V/m. What is the acceleration of this particle and how far from its starting point will it be when its speed is 1000 km/s?

• • **18.26** A pair of parallel plates, separated by 0.05 m, are connected to a 45-V battery, as shown in Figure 18.44. An electron released from rest at the negative plate emerges through a small hole in the positive plate. Find the electric field between the plates, the force acting on the electron, and the change in PE of the electron as it passes from the negative plate to the positive. Also determine the speed of the electron as it emerges from the hole and the time it spends between the two plates.

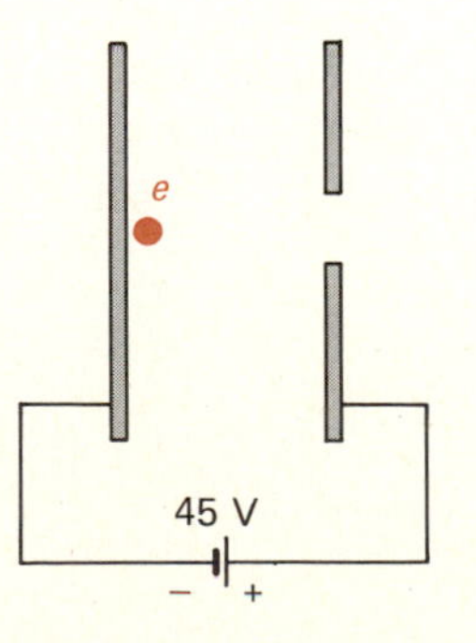

Figure 18.44

Sections 18.9–18.10

18.27 In Figure 18.45, the various geometrical figures portrayed are made of metal and are at potentials as indicated. Sketch the equipotential lines and the electric field lines in each case, and mark the location or locations at which the electric field is strongest.

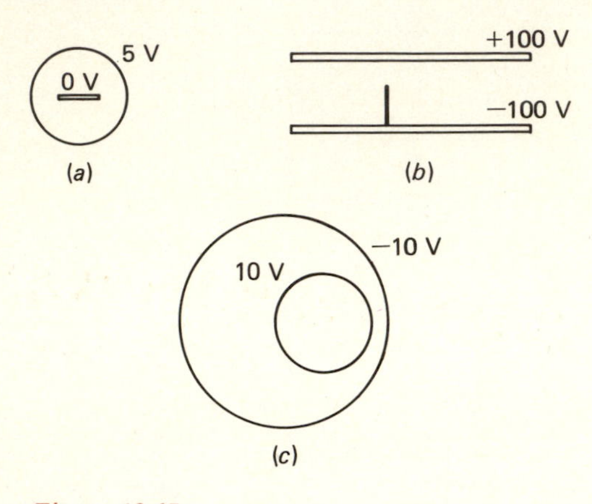

Figure 18.45

18.28 In Figure 18.46, several equipotential patterns are shown. In each case indicate how such patterns might be obtained in practice.

• •**18.29** A charge of $+5\ \mu C$ is placed at $x = 10$ cm and $y = 0$; a second charge is placed at $x = 0$ and $y = 20$ cm. Measurement of the potential at the origin, $x = 0$, $y = 0$, shows that the potential is zero at that point. Find the magnitude and sign of the charge at $x = 0$, $y = 20$ cm, the electric field at the origin, and sketch the equipotential corresponding to $V = 0$ volts.

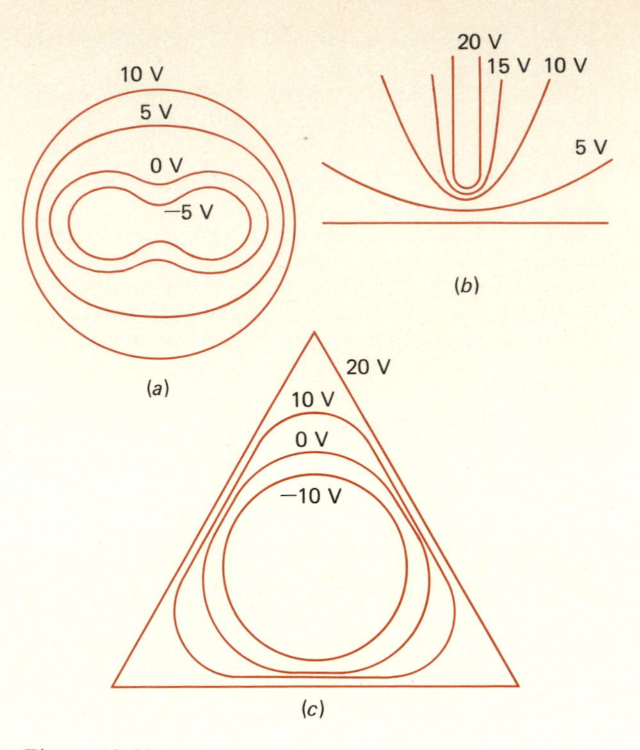

Figure 18.46

• •**18.30** A charge of $50\ \mu C$ is 10 cm above a very large, grounded metal plate. Determine the electrostatic force acting on that charge.

•19•

Capacitors

Striking the electric chain wherewith we are darkly bound.

Childe Harold GEORGE GORDON, LORD BYRON

19.1 Introduction

In 1746, M. Réaumur, a French scientist of some renown, received a letter from Professor Musschenbroek of the University of Leyden in Holland, in which the professor described a "new but terrible experiment which I advise you not to try yourself." Shortly thereafter, the Abbé Nollet received a similar communication from another resident of Leyden, M. Allaman, who reported on this "new experiment which we have made here and which is very remarkable." Following a description of the device, which soon became known as a *Leyden jar,* Allaman concluded:

> You will receive a prodigious shock, which will affect all your arm and even all your body; it is a stroke of lightning: the first time that I received it I was affected to such a degree that for a few moments I could not breathe.

What was this remarkable apparatus, capable of delivering such potent electric shocks? It was nothing more than a thin-walled glass jar,

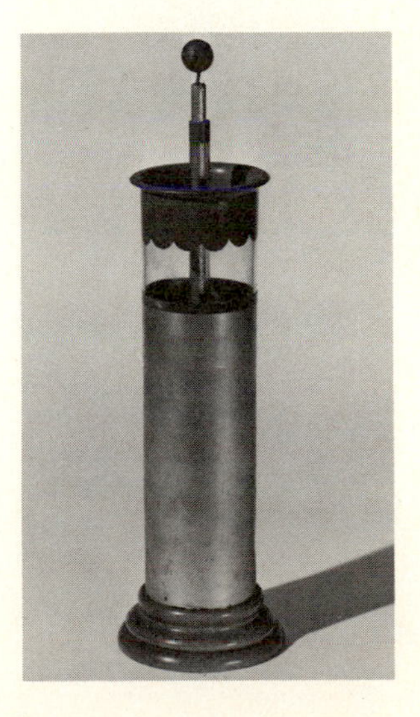

Figure 19.1 A Leyden jar. The earliest capacitor was a thin-walled glass vessel whose inside and outside surfaces were coated with a metal film.

coated on the inside as well as outside surface with conducting materials that were not, however, continuous across the lip of the jar and were therefore not in electrical contact. The present-day name for this and similar arrangements of two conducting surfaces separated by an insulator is *capacitor*. As we shall see presently, such a device can maintain a substantial charge separation. When this charge imbalance is released by flowing through the body of an unwary experimenter, the electric shock can be most unpleasant and even fatal. Large capacitors should be treated with due respect and caution.

19.2 Capacitors and Capacitance

The simplest capacitor, and the one we shall examine in detail, consists of two flat, parallel metal plates separated by a small distance d. The region between the plates may be vacuum (or air, which has practically the same electrical properties) or some other nonconducting material such as oil, glass, or mica. Suppose we take such an air capacitor, which carries no charge on either plate, and connect it to a battery as shown in Figure 19.2. Before the switch is closed, the two capacitor plates are at the same potential and the region between them is field-free. If we now close switch S, the potential difference between the two plates, which we denote by V, must assume the same value as that across the terminals of the battery.*

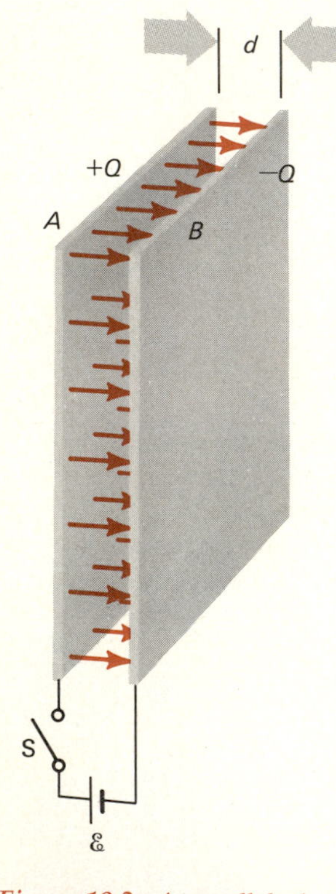

Figure 19.2 *A parallel plate capacitor with air as the insulating substance. After switch S is closed, the potential V across the capacitor will equal the potential across the battery terminals, here designated by ℰ. The magnitude of the electric field in the space between the two plates, separated a distance d, is then* $E = V/d$, *with* **E** *pointing as shown.*

If there is a potential difference V between plates A and B, there must exist an electric field **E** in the region between the plates. In Section 18.7 we found that the magnitude of the field between two parallel equipotential surfaces is

$$E = \frac{V}{d} \tag{19.1}$$

where V is the potential difference between the surfaces and d their separation.

The electric field lines between the two plates are perpendicular to the plates (see Figure 18.23) and of uniform density. These field lines must originate from positive and terminate on negative charges. Consequently, the existence of the voltage V and field **E** demand that a positive charge $+Q$ reside on the surface of plate A and an equal negative charge $-Q$ on the surface of plate B. The magnitude of these charges is proportional to the density of field lines, that is, to the strength of the field **E**; hence $E \propto Q$. Since $V = Ed$, the charge Q must also be proportional to the potential difference V.

DEFINITION: The *capacitance C* is the proportionality constant that relates Q to V:

$$Q = CV \qquad \text{or} \qquad C = \frac{Q}{V} \tag{19.2}$$

In Equation (19.2), V is the voltage across the capacitor, and Q is the charge on *one* plate (not the net charge on the two plates, which would be zero in the case just considered).

The unit of capacitance, the coulomb per volt, has been given the name *farad* (F), in honor of the British scientist Michael Faraday

* In this chapter we shall use V to indicate not an absolute potential but the potential difference or voltage across the capacitor.

(1791–1867), who introduced the concept of capacitance and made numerous other contributions to electricity and magnetism. Since one volt equals one joule per coulomb, the dimension of the farad is

$$[C] = \frac{[Q]}{[V]} = \frac{[Q]^2}{[W]} = \frac{[Q]^2[T]^2}{[M][L]^2}$$

where $[C]$ denotes the dimensions of capacitance, $[Q]$ charge, $[V]$ potential, and $[W]$ energy (work), respectively.

We know that one coulomb is a very large amount of charge by laboratory standards. Since a capacitor of 1 F accumulates a charge of 1 C when a potential of only 1 V is impressed across its plates, it is not surprising that typical laboratory capacitors have capacitances ranging between fractions of a picofarad (1 pF $= 10^{-12}$ F), and for relatively large units, a few thousand microfarads (1 μF $= 10^{-6}$ F).

We might expect the capacitance of a parallel-plate capacitor to depend on three parameters: the area A of the plates, the separation d between the plates, and the nature of the insulating material between the plates. Indeed, as the following calculation shows, these factors do enter in the final result.

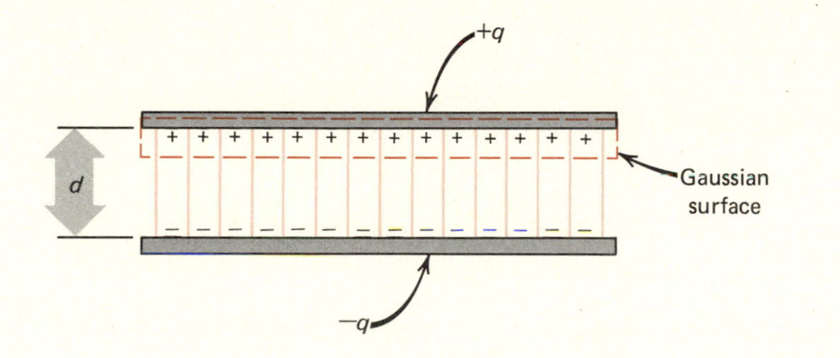

Figure 19.3 Cross section of a parallel-plate capacitor, showing the positive and negative surface charges, electric-field lines, and gaussian surface (dashed).

To calculate the capacitance of the system of Figure 19.3, we turn to Gauss's law (see Section 18.3). The gaussian surface that we use is in the shape of a flat parallelepiped, whose large area equals the area of the capacitor plate, shown end-on in Figure 19.3. One area of this parallelepiped is entirely within the metal of one of the plates, and the other large area is in the region between the plates. If we assume that the fringing field near the edges of the capacitor can be neglected, the only portion of the gaussian surface penetrated by electric field lines is the area between the two plates. (The **E** field in a conductor vanishes under static conditions.)

For this case, Gauss's law, Equation (18.5), becomes

$$\epsilon_0 EA = Q \tag{19.3}$$

where A is the area of the capacitor plate. Using the definition, Equation (19.2), and Equation (19.1), one now obtains

$$C = \frac{Q}{V} = \frac{Q}{Ed} = \epsilon_0 \frac{A}{d} \tag{19.4}$$

where $\epsilon_0 = 8.85 \times 10^{-12}$ F/m is the permittivity of free space.

Note that C is proportional to the area A and inversely proportional to the separation d. These dependences make good sense. If the voltage V and separation d are kept constant, **E** is fixed; the total charge Q is then proportional to the area A, and hence, $C = Q/V$ must also be proportional to A. If on the other hand, the area A and the charge Q are held constant, **E** is again fixed according to Gauss's law; the voltage V is then proportional to the separation d, and consequently, $C = Q/V$ is inversely proportional to d. It follows that to attain a large capacitance we should make the plate area as large as possible and reduce the plate separation to a minimum consistent with mechanical and electrical requirements.

Air capacitors are normally used only as variable tuning capacitors in radio receivers and transmitters. These capacitors have a group of fixed parallel metal plates, electrically connected together, and another set of interconnected plates, insulated from the fixed plates, and mounted on a rotatable shaft. The movable group of plates mesh into the spaces between the fixed plates, and the effective capacitor area can be adjusted by rotating the shaft.

Other capacitors are made of two thin metallic foils, separated by a thin sheet of insulating, or *dielectric,* material. For example, one inexpensive method of construction is that of the tubular capacitor, illustrated in Figure 19.4(b).

The dielectric insulator used in the tubular and other capacitors has two advantages over air. First, it allows a rigid, compact construction with a very small separation d between the two conducting surfaces. Second, the electric polarization induced in the dielectric by the field between the plates greatly increases the capacitance over its value if air or vacuum furnished the insulation.

To see how this increase in capacitance arises, we consider a parallel-plate air capacitor with charges $+Q$ and $-Q$ on its plates, as in Figure 19.6. The potential between the plates is $V = Q/C_0$, and the electric field between the plates is $E = V/d$; the subscript "$_0$" indicates that we are referring to the value of C when the insulating material is air.

Suppose we now insert a polarizable dielectric into the space between the two plates. As described in Chapter 17 and illustrated in Figure 19.5, the electric field between the plates will tend to orient permanent dipoles as well as induce dipoles. The result of that polarization is a new charge configuration, as shown in Figure 19.5(c). Notice that as shown in Figure 19.6(b) the end result of the alignment or induction of dipoles is a net negative charge on the surface of the dielectric facing the positive capacitor plate, and a positive charge on that surface of the dielectric facing the negatively charged capacitor plate. Notice also that inserting the insulating dielectric has *not* altered the charge on the capacitor plates (these plates are *not* connected to a battery). What has changed, however, is the electric field **E** between the plates. The surface charges on the dielectric slab establish an electric field within the slab that is opposed to the field that caused the polarization.

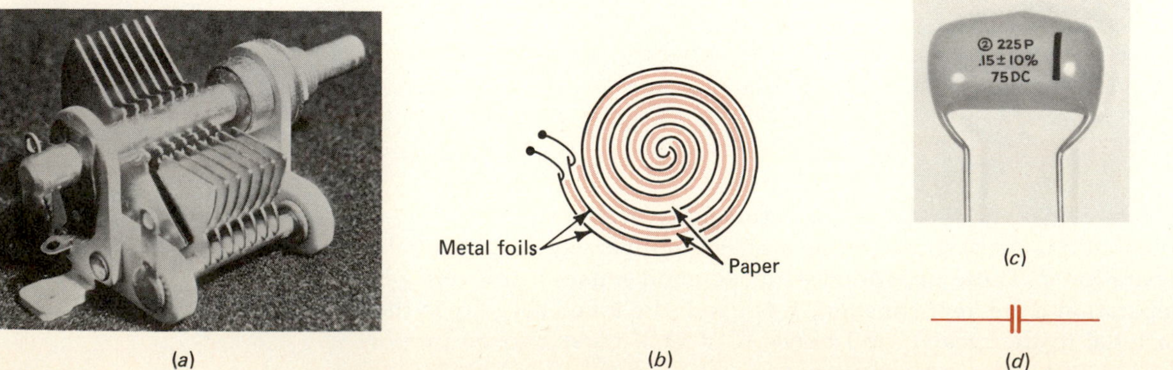

Figure 19.4 *(a) A variable tuning capacitor used in some radio receivers. (b) Schematic drawing of a tubular paper capacitor. The two strips of metal foil are separated by long thin strips of oil-impregnated paper. The metal foils protrude past the paper on opposite ends, and the protruding portions are joined and soldered to connecting wires. The entire compact cylindrical capacitor is then enclosed in a cardboard tube, and the electrical characteristics printed thereon. (c) Photograph of a commercial polyester film capacitor. (d) Circuit symbol for a fixed capacitor.*

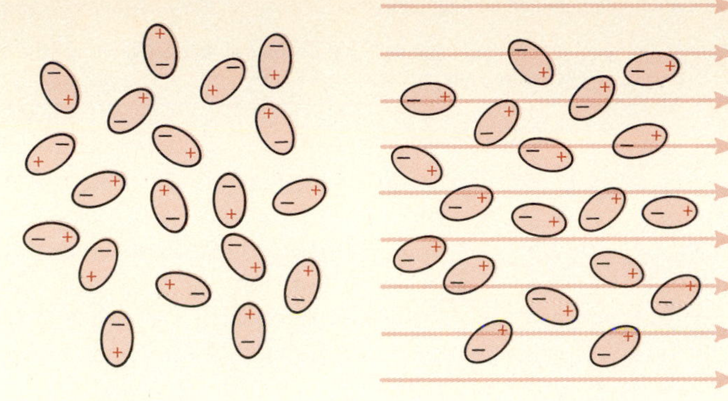

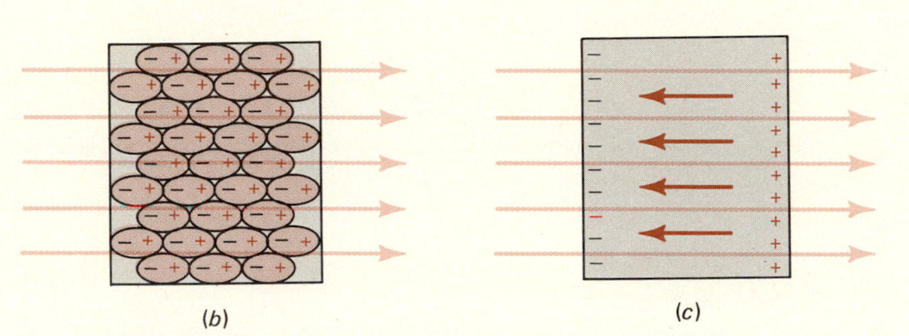

Figure 19.5 (a) In the absence of an electric field, the permanent dipoles of a dielectric are arranged randomly. When an external field is applied, here pointing left to right, these dipoles assume a more ordered configuration, aligning at least partly with the field. (b) In a nonpolar dielectric, an electric field induces dipoles, as indicated schematically in this figure. (c) Whether because of alignment of permanent dipoles or because of induced dipoles, the result is a net negative surface charge on the left surface and a net positive surface charge on the right. The electric field due to these surface charges is directed opposite to the external field. The net field within the dielectric is therefore less than the external field.

With fixed charges on the capacitor, the field between the plates is reduced by the insertion of the dielectric. The ratio of the electric field in the evacuated space to the electric field in the dielectric,

$$\frac{E_0}{E_\kappa} = \kappa \tag{19.5}$$

defines the *dielectric constant* κ. The dielectric constants of a variety of insulating solids and liquids are listed in Table 19.1.

If the dielectric completely fills the space between the capacitor plates, the ratio of the voltages across the capacitor without and with the dielectric in place is

$$\frac{V_0}{V_\kappa} = \frac{E_0 d}{E_\kappa d} = \kappa$$

and consequently,

$$\frac{C_0}{C_\kappa} = \frac{Q/V_0}{Q/V_\kappa} = \frac{1}{\kappa}$$

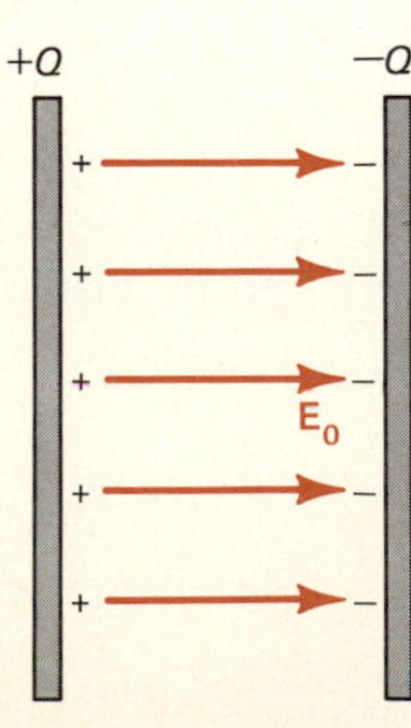

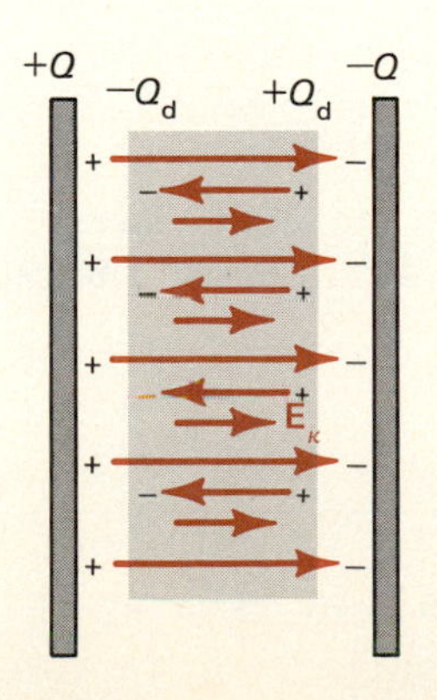

Figure 19.6 Electric field in a parallel plate capacitor with charge Q on its plates (a) before, and (b) after insertion of a polarizable dielectric material. The polarization of the dielectric results in the appearance of charges $+Q_d$ and $-Q_d$ on the surfaces of the dielectric. The net electric field in the region between the capacitor plates is then the vector sum of the field due to two charge distributions $+Q$ and $-Q$ on the capacitor plates and $-Q_d$ and $+Q_d$ on the adjoining dielectric surfaces. Hence, the electric field is less after insertion of the dielectric.

Table 19.1 Dielectric constant κ and dielectric strength of various substances

Material	Dielectric constant κ	Dielectric strength, kV/mm
Vacuum	1.00000	∞
Dry air	1.00054	0.8
Water	78	—
Transformer oil	4.5	12
Mica	5.4	10–150
Paper	3.5	14
Pyrex	4.5	13
Bakelite	4.8	12
Polyethylene	2.3	50
Polystyrene	2.6	25
Teflon	2.1	60
Neoprene	6.9	12
Titanium dioxide	100	6

NOTE: The values given are at approximately room temperature and for steady electric fields.

That is,

$$C_\kappa = \kappa C_0 \tag{19.6}$$

The replacement of vacuum, or air, by a dielectric such as transformer oil, Bakelite, or mica significantly increases the capacitance of a capacitor.

But that is not the only benefit attendant to the dielectric. There is a limit on the electric field strength that can be maintained in a region of insulating material. In dry air, for example, the maximum electric field is about 800,000 V/m. If E exceeds this value, the air becomes temporarily conducting, and a spark jumps across the region. This spark carries charge which tends to neutralize the charge imbalance responsible for the large field that caused the spark.

The maximum field that can be maintained in a dielectric without such breakdown is known as its *dielectric strength.* Dielectric strengths also are listed in Table 19.1. Most commercial dielectrics have a dielectric strength one to two orders of magnitude greater than that of air. Thus, a paper capacitor not only holds more charge for a given potential difference but also can sustain voltages nearly twenty times as great as those that could be applied to an air capacitor of similar dimensions.

● ***Example 19.1*** A parallel-plate capacitor is constructed of two sheets of metal, 7 m by 15 m, and separated by a layer of air 0.5 cm thick. What is the capacitance of that system, and how much voltage must be applied if this capacitor is to hold a charge of 0.01 C?

Note that the metal plates are about the size of a normal classroom for about 30 to 40 students. The capacitance is given by Equation (19.4):

$$C = \frac{\epsilon_0 A}{d}$$

$$= \frac{(8.85 \times 10^{-12}\ \text{F/m})(7\ \text{m})(15\ \text{m})}{5 \times 10^{-3}\ \text{m}}$$

$$= 1.86 \times 10^{-7}\ \text{F} \simeq 0.2\ \mu\text{F}$$

To hold a charge of 0.01 C, one would have to apply a voltage of

$$V = \frac{Q}{C} = \frac{10^{-2}\ \mathrm{C}}{2 \times 10^{-7}\ \mathrm{F}} = 50{,}000\ \mathrm{V}$$

The electric field between the plates would then be 10^7 V/m, exceeding the breakdown field for air. In other words, this capacitor cannot hold a charge of 0.01 C. ●

● ***Example 19.2*** Suppose that the two metal plates of Example 19.1 were separated by a slab of Bakelite of 0.5-cm thickness. What would the capacitance of the system be, and what is the maximum charge this capacitor could store?

Bakelite has a dielectric constant of 4.8. Hence the capacitance of the system is

$$C_\kappa = 4.8(1.86 \times 10^{-7}\ \mathrm{F}) = 8.9 \times 10^{-7}\ \mathrm{F}$$

The dielectric strength of Bakelite is 12 kV/mm = 1.2×10^7 V/m. The maximum potential that can be placed across the capacitor is now

$$V_{\mathrm{max}} = (12\ \mathrm{kV/mm})(5\ \mathrm{mm}) = 60\ \mathrm{kV} = 60{,}000\ \mathrm{V}$$

The maximum charge that can be stored by this capacitor is

$$\begin{aligned} Q_{\mathrm{max}} &= CV_{\mathrm{max}} \\ &= (8.9 \times 10^{-7}\ \mathrm{F})(6 \times 10^4\ \mathrm{V}) \\ &= 5.34 \times 10^{-2}\ \mathrm{C} \simeq 0.05\ \mathrm{C} \end{aligned}$$

●

19.3 Electrostatic Energy of a Capacitor; Energy Stored in an Electric Field

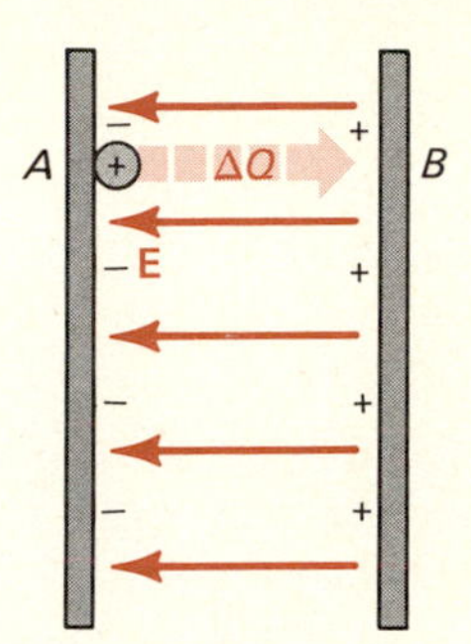

Figure 19.7 *Charging an initially uncharged capacitor involves transferring charge from one plate to the other. As positive charge ΔQ is moved from plate A to plate B, the field* **E**, *pointing from B to A is established. It is against the force due to this field that work must be done to move additional charge from A to B.*

A potential difference is generated across a capacitor by transporting charges from one plate to the other. Initially, with both plates neutral and at the same potential, the first, minute amount of charge ΔQ can be transferred from plate A to plate B of Figure 19.7 with nearly no expenditure of energy. However, as this is done, a small potential difference $\Delta V = \Delta Q/C$ appears across the capacitor, and further transport of charge requires that work be done in raising this charge through the existing, although small potential. As more and more charge accumulates on the capacitor plates, the voltage across the plates increases proportionately, as shown in Figure 19.8. If having attained a potential V, we now wish to carry some more charge ΔQ from the negative to the positive plate, we

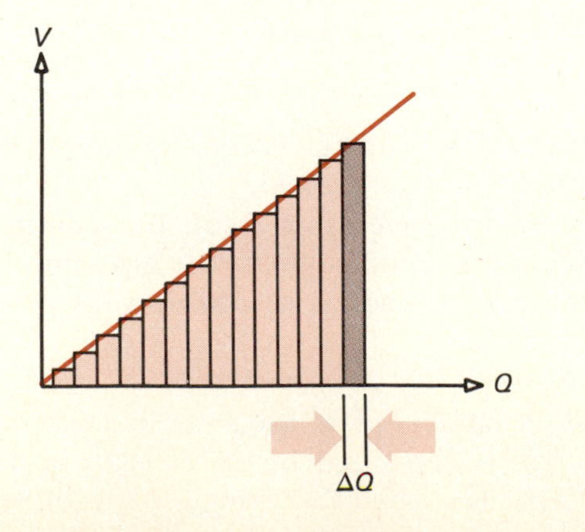

Figure 19.8 *For a capacitor, $V = Q/C$. Hence, a plot of V versus Q is a straight line of slope $1/C$. The work ΔW that must be done to bring a charge ΔQ across the capacitor when a voltage V exists between the plates is $\Delta W = V\,\Delta Q$ (see the heavily shaded rectangle). The total work performed in charging the capacitor to the voltage V is therefore given by the area under the straight line, namely, $\frac{1}{2}QV$.*

must expend an amount of energy equal to

$$\Delta W = V\,\Delta Q \tag{19.7}$$

to overcome the electrostatic force that pushes the charge back toward the negative plate.

From Figure 19.8, we see that the total amount of work done in charging a capacitor to a voltage V is given by the area of the triangle under the V versus Q curve. The area of a triangle equals half the product of base and altitude. Thus

$$W = \tfrac{1}{2}QV \tag{19.8}$$

Since this work can be performed without frictional losses, and since the electrostatic field itself is conservative, it follows from conservation of energy that the energy stored in the capacitor must equal W. *The potential energy stored in a charged capacitor is given by*

$$\mathrm{PE} = \frac{1}{2}\,QV = \frac{1}{2}\,CV^2 = \frac{1}{2}\frac{Q^2}{C} \tag{19.9}$$

The three expressions are completely equivalent and are obtained from each other by applying Equation (19.2). Whichever is used is a matter of convenience, as illustrated in subsequent examples.

While establishing an electric field requires displacement of charges and a concomitant consumption of energy, it is often useful to think of the stored energy as associated with the very existence of the electrostatic field itself, without regard to how that field was created. To obtain a relation between stored energy and the electric field, we turn once more to the parallel-plate capacitor.

If a parallel-plate capacitor, whose capacitance is

$$C = \frac{\kappa\epsilon_0 A}{d}$$

is charged to a voltage V, the magnitude of the electric field $\mathbf{E}$ in the region between its plates is

$$E = \frac{V}{d}$$

The energy stored in the capacitor is given by Equation (19.9),

$$\mathrm{PE} = \frac{1}{2}\,CV^2 = \frac{1}{2}\,\kappa\epsilon_0\left(\frac{A}{d}\right)(Ed)^2 = \frac{1}{2}\,\kappa\epsilon_0 E^2(Ad). \tag{19.10}$$

The product Ad is just the volume within which the electric field is (ideally) confined. Thus we are led to the result

$$\text{Electrostatic energy per unit volume} = \tfrac{1}{2}\,\kappa\epsilon_0 E^2 \tag{19.11}$$

The energy per unit volume given by Equation (19.11) is often called the *energy density* of the electric field.

Although we have derived this result by examining a special field configuration, it can be demonstrated that Equation (19.11) is generally valid, even when $\mathbf{E}$ is a function of position, as in the region about a point charge.

● ***Example 19.3*** A parallel-plate capacitor is constructed of two metal disks of 10-cm radius separated by an air gap of 1 mm. What is the charge on this capacitor when the voltage between the plates is 100 V? How much energy is then stored in this capacitor?

From Equation (19.2) we have $Q = CV$. To determine Q, we must first calculate C. The capacitance is given by Equation (19.4).

$$C = \frac{\epsilon_0 A}{d}$$

$$= \frac{(8.85 \times 10^{-12}\ \text{F/m})(0.1\ \text{m})^2\pi}{10^{-3}\ \text{m}}$$

$$= 2.78 \times 10^{-10}\ \text{F} = 278\ \text{pF}$$

The charge on the capacitor is therefore

$$Q = CV = (2.78 \times 10^{-10}\ \text{F})(10^2\ \text{V}) = 2.78 \times 10^{-8}\ \text{C} = 0.0278\ \mu\text{C}$$

The stored energy is

$$\text{PE} = \tfrac{1}{2}QV = \frac{(2.78 \times 10^{-8}\ \text{C})(10^2\ \text{V})}{2} = 1.39\ \mu\text{J}$$

Example 19.4 The capacitor of Example 19.3 is disconnected from the 100-V battery after it is fully charged. If the separation between the plates is then increased to 2 mm in such a way that no charge can leak off the plates, what is the potential across the capacitor? Also determine the energy stored in the capacitor before and after the plates are separated by the additional 1 mm.

From (19.4) we see that an increase of d from 1 mm to 2 mm will reduce C by a factor of 2. Denoting the quantities after the increase in d with a prime, we have

$$d' = 2d \qquad A' = A \qquad \text{hence} \qquad C' = \frac{C}{2}$$

But we know that $Q' = Q$ because we have not allowed any charge to leak off the plates. Rewriting Equation (19.2), $V = Q/C$, we see that $V' = Q'/C' = 2Q/C = 2V$. Thus, as the plates are separated to 2 mm, the potential difference across the capacitor increases to 200 V.

To calculate the stored energy, we use one of the three equivalent forms of (19.9); the first or last is most convenient since these two involve the charge Q, which in this case remains unchanged.

As calculated in Example 19.3, the initial stored energy is PE = 1.39 μJ.

After the separation between the plates has been increased to $d' =$ 2 mm, $Q' = Q$; but $V' = 2V$, as we have just found. Hence,

$$\text{PE}' = \tfrac{1}{2}Q'V' = \tfrac{1}{2}Q(2V) = 2\text{PE} = 2.78\ \mu\text{J}$$

This result may seem puzzling at first. We have done something that has *decreased* the capacitance of the system, and yet the stored energy has *increased!* Though the formula $\text{PE} = \tfrac{1}{2}Q^2/C$ tells us that PE must increase if we keep Q fixed and decrease C, this really only begs the question. To understand what is happening we should ask, If the energy stored in the capacitor is increased, where does this extra energy come from? As soon as the matter is put in these terms, into the language of energy conservation, it is not difficult to find the answer.

The charges $+Q$ and $-Q$ on the two capacitor plates attract each other, as do all charges of opposite sign. Hence, work must have been done against the electrostatic force of attraction by an external force to separate the plates to a greater distance. That work by the external force is recovered and appears as stored energy in the capacitor.

● ***Example 19.5*** A 6-μF, parallel-plate air capacitor is connected across a 100-V battery. After the capacitor is fully charged, the battery is removed and a slab of dielectric that completely fills the space between the plates is inserted. If the dielectric constant of this material is $\kappa = 8.0$, what is the potential across the capacitor after the slab is inserted? Is it necessary to do work either to insert or to remove the slab? If so, which of these processes requires work to be done on the system, and how much?

Let us first be certain we understand the physical situation. Once the capacitor is charged and the battery has been disconnected, there can be no change in the charge on the capacitor plates. However, the potential difference across the capacitor can change. We shall first find the charge on the capacitor and then determine the capacitance after the dielectric has been inserted. Using Equation (19.2), and keeping the charge Q constant, we can find the voltage V_κ across the plates after insertion of the dielectric. To answer the last question, we shall calculate the energy stored in the capacitor before and after the slab is inserted.

The charge on the air capacitor is

$$Q_0 = C_0V = (6 \times 10^{-6}\ \text{F})(10^2\ \text{V}) = 6 \times 10^{-4}\ \text{C}$$

If the space between the plates is filled with dielectric, the capacitance is increased to

$$C_\kappa = \kappa C_0 = 8(6\ \mu\text{F}) = 48\ \mu\text{F}$$

Since the charge has not changed, the voltage across this larger capacitor is

$$V_\kappa = \frac{Q}{C_\kappa} = \frac{6 \times 10^{-4}}{48 \times 10^{-6}} = 12.5\ \text{V}$$

Before the dielectric is inserted, the stored energy is

$$\text{PE} = \frac{1}{2}QV = \frac{(6 \times 10^{-4}\ \text{C})(10^2\ \text{V})}{2} = 3 \times 10^{-2}\ \text{J}$$

After the dielectric has been inserted, the stored energy is

$$\text{PE}_\kappa = \frac{1}{2}QV_\kappa = \frac{(6 \times 10^{-4}\ \text{C})(12.5\ \text{V})}{2} = 0.375 \times 10^{-2}\ \text{J} = \frac{\text{PE}}{8}$$

Since the stored energy in the capacitor is *less* if the insulator is the dielectric, it follows that we shall have to *do work on the system to remove the dielectric slab,* thereby increasing the stored energy. Every system normally seeks the lowest energy configuration. In this case, the dielectric will be pulled into the space between the capacitor plates, and the energy of the system thereby reduced.

The work needed to remove the dielectric is the difference between PE and PE_κ; i.e.

$$W = \text{PE} - \frac{\text{PE}}{8} = \frac{7\text{PE}}{8} = 2.625 \times 10^{-2}\ \text{J}$$

●

19.4 Capacitors in Series and Parallel Combinations

Two or more capacitors are said to be connected *in series* if the total voltage across the combination is the algebraic sum of the potential dif-

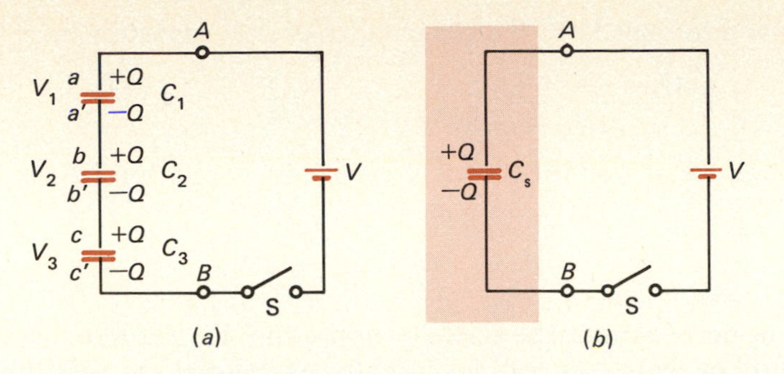

Figure 19.9 (*a*) *A series combination of three capacitors.* (*b*) *The equivalent capacitance of this combination.*

ferences across the individual capacitors of the combination. Three capacitors connected in series are shown in Figure 19.9.

Two or more capacitors are said to be connected in parallel if their plates are connected so as to form one pair of equipotential surfaces. A parallel combination of three capacitors is shown in Figure 19.10. In a parallel combination, the potential difference is the same across the combination as across each capacitor.

For any combination of capacitors, it is always possible to find an *equivalent capacitance,* that is, a single capacitor of proper capacitance so that the voltage-versus-charge relation at the terminals of the combination is the same as the voltage-versus-charge relation of the equivalent capacitor. The equivalent capacitor responds to an applied voltage as the combination does, and stores the same energy for a given potential difference as the combination. Since it is much simpler to use a single element in computations than several capacitances of various values, it is useful to know how to reduce series and parallel combinations to their equivalents.

Of the two, the parallel combination is the simpler, and we therefore consider it first.

In Figure 19.10(*a*) the plates connected to the positive terminal of the battery are connected to each other; the plates connected to the negative terminal of the battery are similarly connected to each other. It follows that

$$V_1 = V_2 = V_3 = V$$

The charge on each capacitor is given by Equation (19.2):

$$Q_1 = C_1V \qquad Q_2 = C_2V \qquad Q_3 = C_3V$$

If we want to replace these three capacitors by a single capacitor that will store the same total charge $Q = Q_1 + Q_2 + Q_3$ when the potential

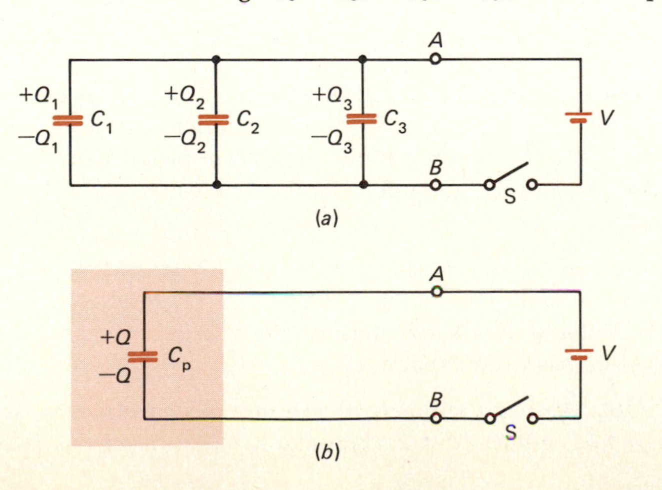

Figure 19.10 (*a*) *A parallel combination of three capacitors.* (*b*) *The equivalent capacitance of this combination.*

across its terminals is V, the capacitance of that capacitor must be

$$C_p = \frac{Q}{V} = \frac{Q_1 + Q_2 + Q_3}{V} = C_1 + C_2 + C_3$$

In general, for a combination of N capacitors connected in parallel,

$$C_p = \sum_{i=1}^{N} C_i \qquad (19.12)$$

Let us now examine the series combination. We assume that the three capacitors of Figure 19.9(a) are initially uncharged, so that the voltage across each, as well as across the combination, is zero before switch S is closed. Again, we want to determine what single capacitor could be connected between terminals A and B of Figure 19.9(b) so that the battery delivers the same charge to this equivalent capacitor as it delivers to the series combination of Figure 19.9(a).

Suppose the battery deposits a charge $+Q$ on the plate a of the capacitor C_1 after the switch has been closed. This charge will induce an equal but opposite charge, $-Q$, on plate a' of C_1. Since, however, the total charge on any electrically insulated surface must remain constant, the total charge on the equipotential surface that comprises plate a' of C_1 and plate b of C_2 must be zero. Consequently, a positive charge $+Q$ must appear on plate b of C_2. Following the same reasoning, we see that the charge on capacitor C_3 must also be Q.

Thus, the condition that must be satisfied by the series combination is

$$Q_1 = Q_2 = Q_3 = Q$$

The total voltage between terminals A and B is

$$V = V_1 + V_2 + V_3 = \frac{Q_1}{C_1} + \frac{Q_2}{C_2} + \frac{Q_3}{C_3} = \frac{Q}{C_1} + \frac{Q}{C_2} + \frac{Q}{C_3} = \frac{Q}{C_s}$$

The single capacitor that would store a charge Q when a potential difference of V is impressed across its plates must meet the condition

$$\frac{1}{C_s} = \frac{1}{C_1} + \frac{1}{C_2} + \frac{1}{C_3}$$

In general, for N capacitors in a series combination, the equivalent capacitance is given by

$$\frac{1}{C_s} = \sum_{i=1}^{N} \frac{1}{C_i} \qquad (19.13)$$

In this case, although the charge on each of the series capacitors is the same regardless of the value of the capacitance, the total voltage is divided among the series constituents in *inverse* proportion to their capacitances; the largest voltage appears across the capacitor with the smallest capacitance.

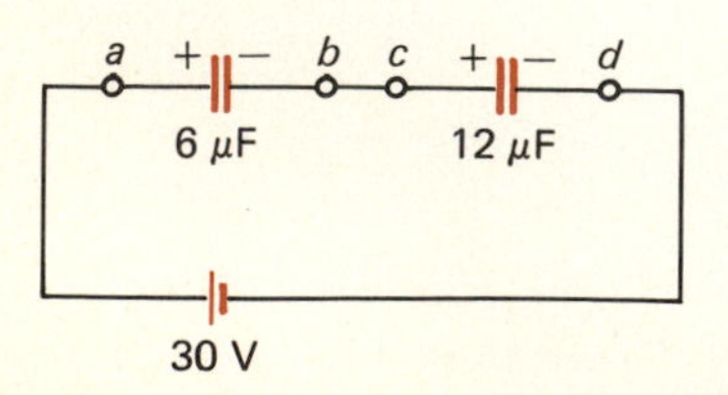

Figure 19.11 A 12-μF and a 6-μF capacitor in series across a 30-V battery.

● ***Example 19.6*** A 12-μF capacitor and a 6-μF capacitor are connected in series across a 30-V battery, as in Figure 19.11. Find the equivalent capacitance of that series combination, the charge on each capacitor, and the voltage across each capacitor.

The equivalent capacitance of the series combination is given by Equation 19.13, where $N = 2$, namely,

$$\frac{1}{C_s} = \frac{1}{C_1} + \frac{1}{C_2} = \frac{1}{12\ \mu\text{F}} + \frac{1}{6\ \mu\text{F}} = \frac{1}{4\ \mu\text{F}}$$

The equivalent capacitance of this series combination of 12 μF and 6 μF is a capacitor of 4 μF.

The charge that would be deposited on this equivalent capacitor by the 30-V battery is given by Equation (19.2),

$$Q = CV = (4 \times 10^{-6}\ \text{F})(30\ \text{V}) = 120\ \mu\text{C}$$

That is also the charge that flows onto each of the two series capacitors; the corresponding voltages are then given by

$$V(12\ \mu\text{F}) = \frac{120\ \mu\text{C}}{12\ \mu\text{F}} = 10\ \text{V}$$

$$V(6\ \mu\text{F}) = \frac{120\ \mu\text{C}}{6\ \mu\text{F}} = 20\ \text{V}$$

Note that the total voltage across the combination divides in inverse proportion to the capacitance; that is, for the combination of two capacitors in series,

$$\frac{V_1}{V_2} = \frac{C_2}{C_1}$$

Summary

A *capacitor* consists of two conducting bodies separated by an insulator. The *capacitance* of a capacitor is defined as the ratio

$$C = \frac{Q}{V}$$

where Q is the charge on one conducting surface and V is the potential difference between the conducting surfaces.

The capacitance of a *parallel-plate* capacitor is given by

$$C = \kappa\epsilon_0 \frac{A}{d}$$

where κ is the *dielectric constant* of the insulating material between the metallic plates, ϵ_0 the *permittivity of free space,* A the area of each of the capacitor plates, and d the separation between the plates. The dielectric constant of air is unity (to three significant figures); the dielectric constants of several insulating liquids and solids are given in Table 19.1.

Energy must be expended to charge an initially uncharged capacitor. That energy is stored in the capacitor as electrostatic potential energy, and its value is given by

$$\text{PE} = \frac{1}{2} QV = \frac{1}{2} CV^2 = \frac{1}{2}\frac{Q^2}{C}$$

Capacitors are sometimes connected electrically in *series* or in *parallel combinations.* The capacitance that is equivalent to N capacitors connected in parallel is given by

$$C_p = \sum_{i=1}^{N} C_i$$

The equivalent capacitance of a series combination is given by

$$\frac{1}{C_s} = \sum_{i=1}^{N} \frac{1}{C_i}$$

Multiple Choice Questions

19.1 Two unequal capacitors, initially uncharged, are connected in series across a battery. Which of the following is true?

(a) The potential across each is the same.

(b) The charge on each is the same.

(c) The energy stored in each is the same.

(d) The equivalent capacitance is the sum of the two capacitances.

19.2 A capacitor is charged by placing charges $+Q$ and $-Q$ on its plates. The following statement is correct.

(a) The energy stored in the capacitor is $Q^2C/2$.

(b) The potential across the capacitor is QC.

(c) The potential across the capacitor is C/Q.

(d) The energy stored in the capacitor is $QV/2$.

19.3 Three capacitors, with capacitances $C_a < C_b < C_c$, are connected in series. It follows that

(a) the equivalent capacitance is greater than C_c.

(b) the equivalent capacitance is greater than C_a but less than C_c.

(c) the equivalent capacitance is less than C_a.

(d) None of the above is necessarily true.

19.4 The capacitance of a parallel-plate capacitor is

(a) inversely proportional to the separation between the plates and inversely proportional to the area of the plates.

(b) inversely proportional to the separation between the plates and proportional to the area of the plates.

(c) proportional to the separation between the plates and proportional to the area of the plates.

(d) proportional to the separation between the plates and inversely proportional to the area of the plates.

19.5 An air capacitor with parallel plates carries a charge Q. When a dielectric with $\kappa = 3$ is inserted between the plates,

(a) the voltage across the capacitor increases by a factor of 3.

(b) the voltage across the capacitor decreases by a factor of 3.

(c) the charge on the plates increases by a factor of 3.

(d) the charge on the plates decreases by a factor of 3.

(e) None of the above effects occurs.

19.6 A parallel-plate air capacitor is connected to a battery that maintains a voltage of 12 V between its terminals. Under these conditions, the charge on the capacitor is Q. If the separation between the plates is tripled while the capacitor is kept connected to the battery,

(a) the energy stored in the capacitor is tripled.

(b) the energy stored in the capacitor is reduced by a factor of 3.

(c) the energy stored in the capacitor is kept constant.

(d) the energy stored in the capacitor is reduced by a factor of 9.

19.7 A parallel-plate air capacitor carries a charge Q. If a dielectric slab with dielectric constant $\kappa = 2$ is slipped between the plates,

(a) the stored energy remains unchanged.

(b) the stored energy is increased by a factor of 2.

(c) the stored energy is reduced by a factor of 2.

(d) None of the above is correct.

19.8 Three identical capacitors, each with capacitance C, are combined as shown in Figures 19.12*A–D*. If the same voltage V is applied to each combination, the one that stores the greatest energy is

(a) A (b) B (c) C (d) D

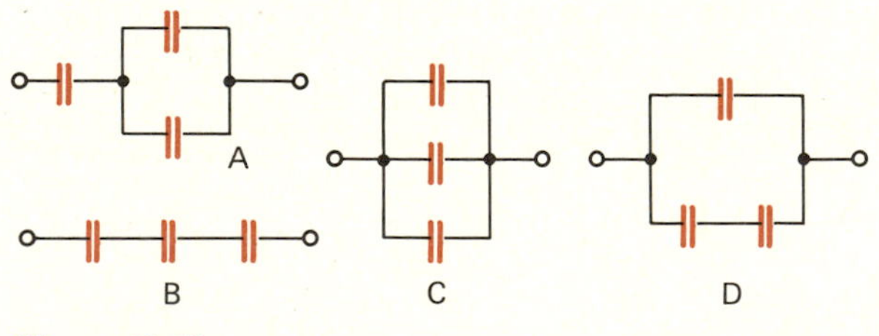

Figure 19.12

19.9 A parallel-plate capacitor is connected to a battery that has a constant terminal voltage. If the capacitor plates are pulled apart,

(a) the electric field decreases and the charge on the plates also decreases.

(b) the electric field remains constant but the charge on the plates decreases.

(c) the electric field remains constant but the charge on the plates increases.

(d) the electric field increases but the charge on the plates decreases.

Problems

(Section 19.2)

19.1 A parallel-plate air capacitor has plates of area 100 cm^2. What is the separation between the plates if the capacitance is 40 pF?

19.2 The inside and outside surfaces of a Pyrex beaker are painted with aluminum paint. The thickness of the glass is 1 mm and the beaker has a diameter of 8 cm and a height of 18 cm. Calculate the capacitance of this Leyden jar.

19.3 An uncharged 4.0-μF capacitor is connected to a 12-V battery. How much charge is drawn from the battery?

19.4 What is the area of the plates of a parallel-plate air capacitor of 0.3-μF capacitance if the separation between the plates is 1 mm?

19.5 When 900 μC is transferred from one plate of an uncharged capacitor to the other, the voltage across the capacitor becomes 600 V. Calculate the capacitance of the capacitor.

19.6 The potential difference between the interior and exterior of a cell is typically 100 mV. The insulating cell membrane that separates the external ionic solution from the cell plasma, also a conducting medium, is made of an organic material (lipid), like a soap film, that has a dielectric constant of about 4.0. Long before cell membranes were observed by electron microscopy, their thickness had been reliably established by measuring the capacitance of a cell. The results clustered about the value of 1 μF per square centimeter of cell surface. Estimate the typical membrane thickness and the electric-field strength within the membrane.

19.7 A parallel-plate capacitor has a plate separation of 0.3 mm, a plate area of 100 cm^2, and a capacitance of 500 pF. Determine the dielectric constant of the insulating material between the plates.

• **19.8** "Suppose we take such an air capacitor, which carries no charge on either plate, and connect it to a battery as shown in Figure 19.2. Before the switch is closed, the two capacitor plates are at the same potential and the region between them is field-free" p. 412.

Prove that the statement of the second sentence is correct.

• **19.9** Immediately following Equation (19.1) is the following statement. "The electric field lines between the two plates are perpendicular to the plates and of uniform density."

Prove that the field lines are indeed of uniform density in the situation discussed there.

• **19.10** A capacitor is constructed by rolling two sheets of aluminum foil between sheets of paper that are 0.07 mm thick. What must the total area be of each foil if the capacitance is 0.01 μF? What maximum charge can this capacitor hold, and what is the voltage across the capacitor under these conditions?

• **19.11** A capacitor is constructed using three rectangular metal plates, each with sides 8 cm and 10 cm. The two external plates that surround the central plate are connected, as shown in Figure 19.13. If the air gap between the plates is 1 mm, what is the capacitance of this system?

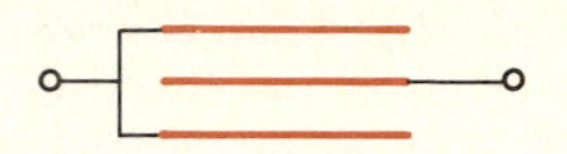

Figure 19.13

• • **19.12** A parallel-plate air capacitor is constructed of plates of 50 cm^2 area separated by 6 mm. Another metal plate, not connected to either capacitor plate, whose thickness is 4 mm, can be inserted between the capacitor plates so that a 1-mm gap remains on either side. Calculate the capacitance of the system before and after the insertion of the insulated metal plate.

• • **19.13** A capacitor is made of two concentric spheres. Their radii are R_a and R_b, where $R_b > R_a$ and $R_b - R_a \ll R_a$. Find an expression for the capacitance of this system if the insulating material between the two metal spheres is air.

• • **19.14** Four metal plates of identical shape are connected together, as shown in Figure 19.14. If the area of each plate is A and the separation between the plates is d, what are the capacitances for the two arrangements shown?

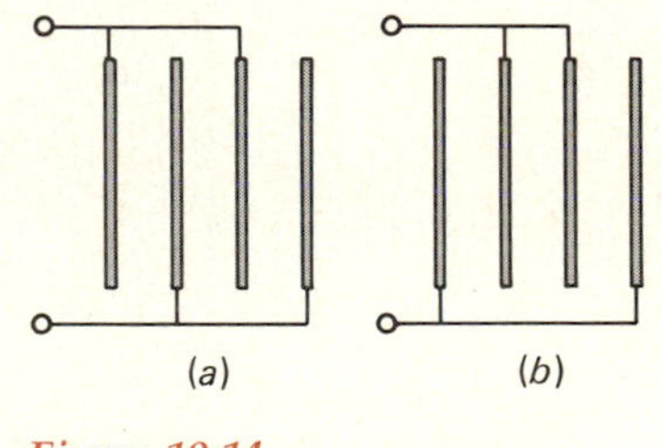

Figure 19.14

(Section 19.3)

• **19.15** A dielectric slab, whose dielectric constant is 4.5, is inserted between the plates of a parallel-plate capacitor so that it just fills the space between the plates. The capacitor is then charged until the stored energy is 0.3 mJ and the battery is disconnected. How much work must be done to remove the dielectric slab?

• **19.16** A parallel-plate air capacitor has a capacitance of 15 pF. What is the potential across the capacitor and the energy stored when the capacitor is charged to 0.6 μC? If without permitting the charge to escape, one increases the separation between the plates to three times its original value, what are the new values for C, V, and the stored energy? How much work was done by or on the capacitor during the plate separation?

• **19.17** A parallel-plate air capacitor carries a charge of 0.1 μC when a potential difference of 400 V exists between its plates. For an air gap between the plates of 0.4 mm, find the area of the plates, the capacitance of the capacitor, and the energy stored by the capacitor.

• **19.18** A 200-pF air capacitor is connected to a 500-V battery and then removed from the battery. The capacitor is then submerged in transformer oil. What is the energy stored in the capacitor before and after submersion in the oil?

• **19.19** A 500-pF air capacitor is connected to a 400-V battery. The capacitor is then submerged in transformer oil. How much energy is provided by (or to) the battery as the capacitor is lowered into the oil?

• **19.20** A 300-pF parallel-plate air capacitor is connected to a 500-V battery. If a tight-fitting sheet of Bakelite is inserted between the capacitor plates, how much charge flows from (or to) the battery?

• **19.21** A parallel-plate air capacitor has a plate separation of 1.4 mm. What is the maximum charge per unit area that can be stored on the capacitor plates?

• **19.22** A variable, parallel-plate air capacitor, similar to that shown in Figure 19.4, has a capacitance that varies between 35 pF and 400 pF as the movable plates are rotated through 180°. With the plates fully meshed, the capacitor is connected to a 200-V battery. The connection to the battery is then broken. What is the potential across the capacitor if the shaft holding the variable plates is rotated through 180°? Determine the energy stored in the capacitor in the initial and in the final configuration.

• • **19.23** In Example 19.5 we saw that work must be done to remove a dielectric slab from the space between the plates of a charged parallel-plate capacitor. It follows, then, that a force must be applied to the slab to pull it out. Determine how the average force depends on the charge on the capacitor plates. The charged capacitor is not connected to a battery.

• • **19.24** If a total of 0.2 mJ must be expended to transfer 0.6 μC from one plate of a 2-μF capacitor to the other, what charge was residing on each plate before the additional charge was transferred?

• • **19.25** A 40-μF capacitor is charged to a voltage of 120 V. How much energy must be expended to transfer 0.4 μC from the negative plate to the positive?

(Section 19.4)

• **19.26** Determine the equivalent capacitance of the combination of three capacitors in Figure 19.15.

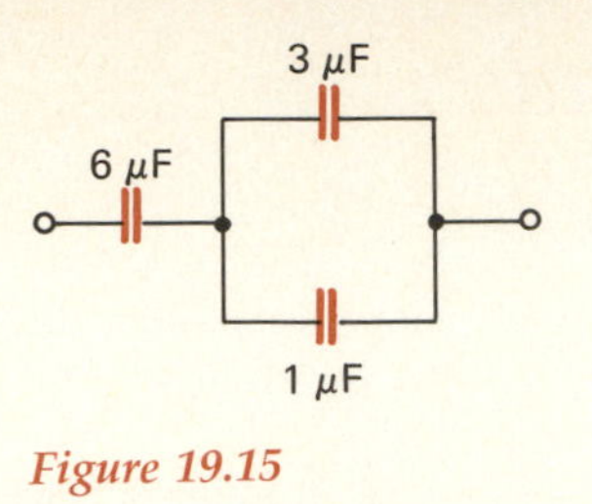

Figure 19.15

• **19.27** If a battery with terminal voltage of 12 V is connected across the combination of capacitors of Figure 19.15, how much energy will be stored in each of the three capacitors after steady state has been attained?

• **19.28** In Figure 19.16, the capacitance of each of the capacitors is 4 μF. Calculate the charge on each capacitor and the energy stored by each.

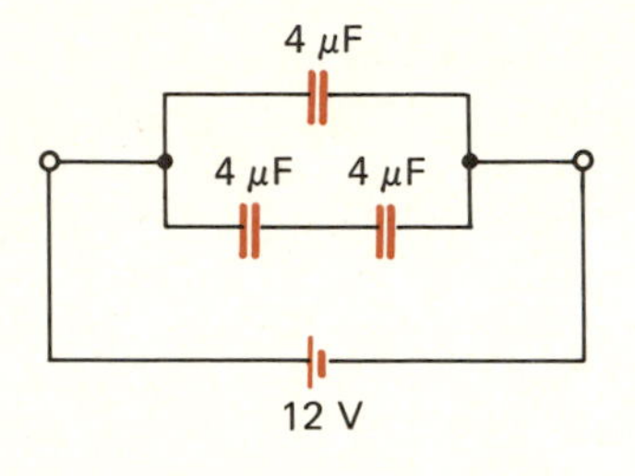

Figure 19.16

• **19.29** Twenty identical capacitors, each with capacitance of 20 μF, are connected in parallel across a 120-V battery. They are then disconnected from the battery and reconnected as shown in Figure 19.17(b). (a) What is the equivalent capacitance of the parallel combination, and what is the energy stored in the system after the battery has been connected to the parallel combination? (b) What is the equivalent capacitance of the series combination of Figure 19.17(b), and what is the charge on that equivalent capacitance? (c)

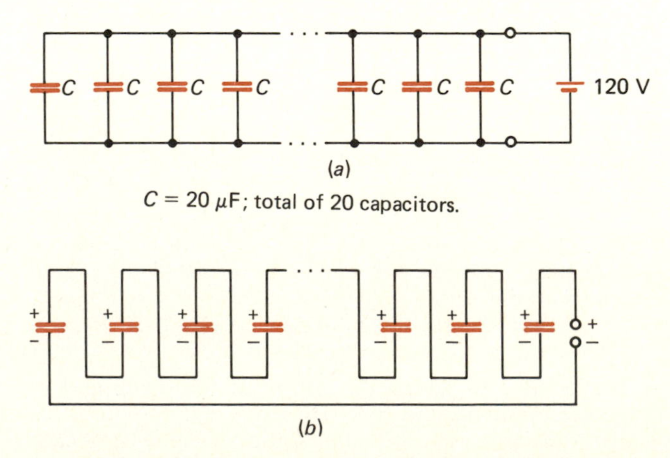

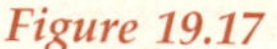
Figure 19.17

Has there been a change in the stored energy as a result of the reconnection of the capacitors? If so, how did that come about? (d) If a spark jumps across the combination of Figure 19.17(*b*), discharging the capacitors in 3 μs, at what rate is energy released by the spark discharge?

• • **19.30** A 40-μF capacitor is charged by connecting it to a 16-V battery. The battery is then disconnected and the charged capacitor is next connected across a 24-μF capacitor that was initially uncharged. After steady state has been reached, what is the charge on each capacitor and the voltage across each?

• • **19.31** In Figure 19.18, the charge on each capacitor is 24 μC. Find the capacitance of C_1 and the total energy stored in the system.

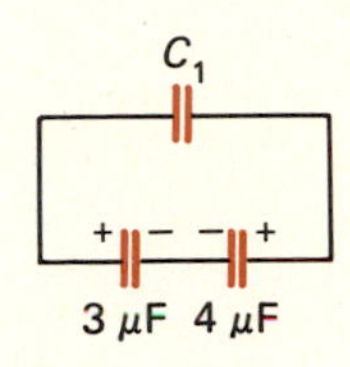

Figure 19.18

• • **19.32** Determine the capacitance of the capacitor network shown in Figure 19.19.

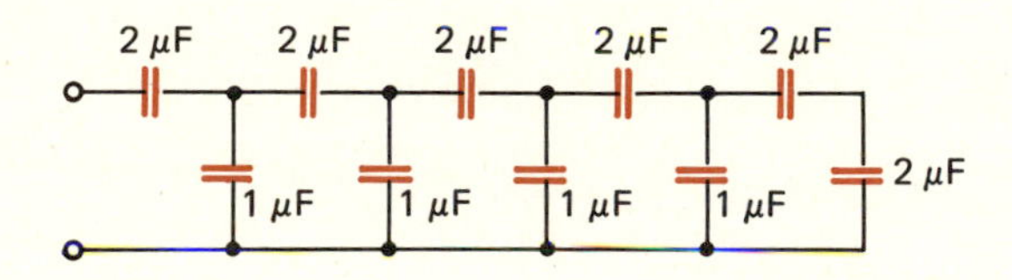

Figure 19.19

• • **19.33** An 8-μF capacitor is charged to 100 V. It is then connected to another capacitor, charged to 40 V, positive plate to positive, negative plate to negative. After the connection has been completed, the voltage across the capacitor plates is 60 V. Determine the capacitance of the other capacitor.

• • **19.34** A rectangular parallel-plate capacitor is constructed so that a dielectric slab can be inserted between the plates, as indicated in Figure 19.20. The capacitor is connected to a battery of voltage V, with the slab fully inserted. The slab is then withdrawn while the battery remains connected to the capacitor plates. Let C_0 be the capacitance when the dielectric slab of dielectric constant κ is completely withdrawn. Find expressions for

(a) The capacitance as a function of x.

(b) The stored energy as a function of x.

(c) The force that must be applied to remove the slab once it is fully inserted. Does this force depend on x?

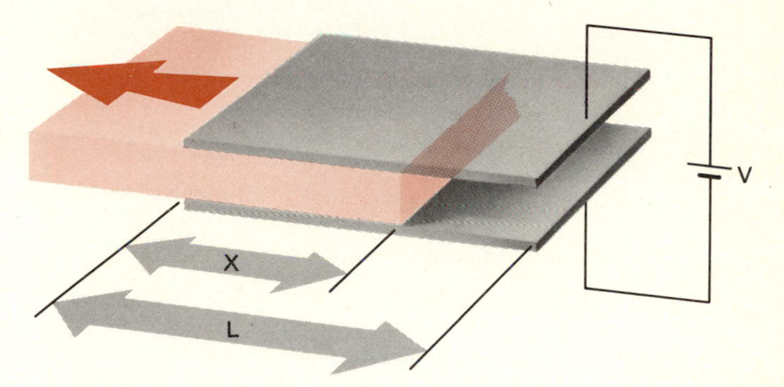

Figure 19.20

20 Steady Electric Currents

"We must take the current when it serves,
Or lose our ventures."

William Shakespeare, *Julius Caesar*

20.1 Sources of Electromotive Force; Batteries

A battery was earlier defined (Section 18.8) as a device that maintains a specified potential difference between its two terminals. In Figure 20.1 such a device is shown in a purely schematic representation.

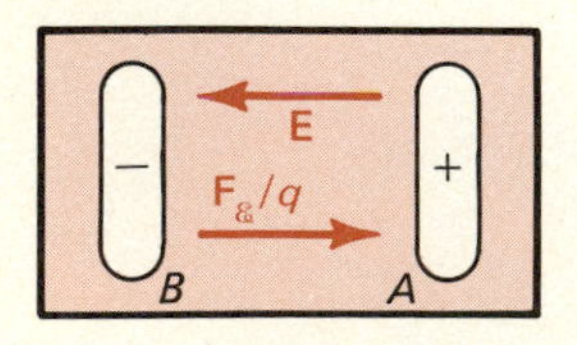

Figure 20.1 Schematic representation of a battery. The device is able to maintain a steady potential difference V_{AB} between its terminals.

In the region between the positive (+) terminal A and the negative (−) terminal B there must exist an electric field **E** that points as indicated, from right to left. In this field, mobile charges would flow toward the terminal of opposite sign (negatively charged electrons would flow toward the positive terminal) and neutralize the charge imbalance that must be at the root of the electric field, and the voltage between the terminals would therefore decay to zero. In other words, the energy of the system, which includes energy stored in the electrostatic field (see Equation (19.11)) could be lowered by suitable charge motion; without any other forces, such motion is to be expected.

But this does not happen. Evidently, some force within the system counteracts the electrostatic force $q\mathbf{E}$. This countervailing force, which we denote by $\mathbf{F}_{\mathcal{E}}$, must be equal to $q\mathbf{E}$ in magnitude and opposite in direction; i.e., $\mathbf{F}_{\mathcal{E}} = -q\mathbf{E}$. The potential difference V_{AB} between the battery terminals is defined as the work that would have to be done by an external force to move a unit positive charge from A to B against the electrostatic field (see Section 18.5). For the battery, that "external" force is this force $\mathbf{F}_{\mathcal{E}}$, and it is common practice to speak of the work done by this force on unit charge as the *electromotive force* of the battery. The term is really a misnomer since the electromotive force, abbreviated "emf" in written and oral communication, has the dimension and unit of electrostatic potential (volt = joule per coulomb), that is, *energy per unit charge, not force*. However, the terminology is now firmly established. The symbol for emf is the script capital letter $\mathcal{E}$, and the unit is the volt.

The source of energy that produces an emf may be mechanical, thermal, chemical, or electromagnetic. In an earlier chapter (Section 11.4) we mentioned the use of thermocouples for measuring temperature differences; these devices are examples of the conversion of thermal to electrical energy. Generators use mechanical energy to produce an emf. The best-known source of emf is the battery, which converts chemical to electrical energy.

The battery is based on the observation by Alessandro Volta that a potential difference appears between two dissimilar metals when they are partially immersed in an acid solution. When a metal is placed into an acid, some metal ions go into solution. These ions carry a positive charge, and thus leave the metal negatively charged. An electric field is established in the acid solution that tends to pull the positive ions back toward the metal, and thus limits the number of ions that go into solution. If the potential of the acid relative to the metal were measured, the acid would be at a positive potential, as indicated in Figure 20.2.

The solubilities of metal ions in an acid differ for different metals. Consequently, some metals will become more negative, and some less,

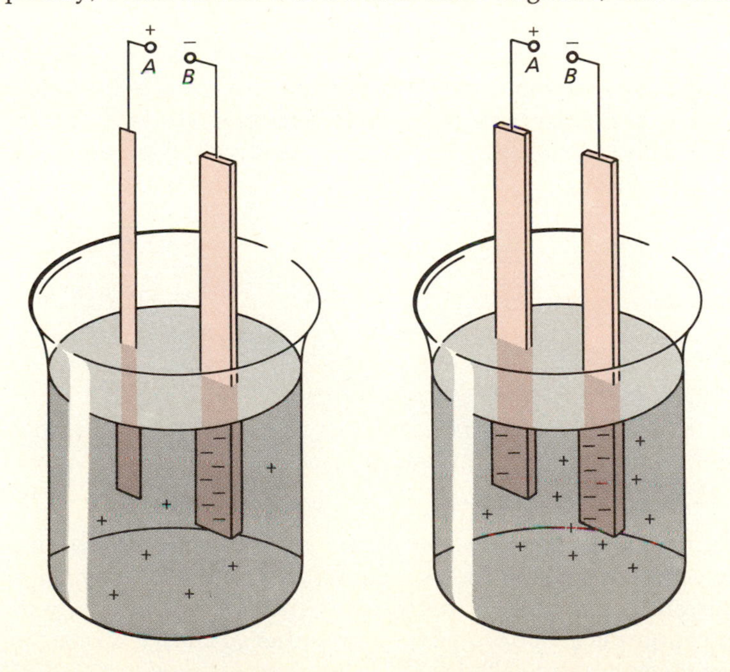

Figure 20.2 *(a) As positive ions of a metal go into solution in an acid, the metal acquires a net negative charge, the acid a net positive charge. If a conductor, impervious to the acid, were placed into the solution, a potential difference V_{AB} would appear, with the metal the negative terminal and the impervious conductor the positive terminal. (b) The solubilities of two dissimilar metals in a given acid are generally different. Here, metal A has a lower solubility than metal B. A meter that measures potential differences would show A positive relative to B.*

relative to the acid solution in which they have been immersed. Hence, when two dissimilar metals are inserted into the same acid bath, as in Figure 20.2(*b*), a potential difference appears between them, which is just the difference between the potentials developed by the two metals relative to the common acid.

The magnitude of this potential difference depends on the metals used and on the acid solution. In some batteries, one of the *electrodes*, as the conducting rods of batteries are called, is not a metal but a metal oxide; this is the case in the ordinary car battery, which has lead and lead oxide electrodes. Zinc and zinc oxide are the electrodes in so-called dry cells (which are not truly dry). The standard symbol for a battery, shown earlier, is reproduced in Figure 20.3.

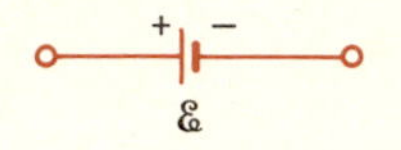

Figure 20.3 *The symbol for a source of emf. The long, narrow line designates the positive terminal of the source, the short heavy line the negative.*

20.2 Electric Current

An *electric current* is said to exist whenever there is a net flow of charge. Most commonly, the moving charges are confined to a limited region, for example, the interior of a metallic wire, the cylindrical volume of a neon tube, or the cross section of the collimated electron beam in a TV picture tube. The amount of current is defined as the amount of charge that passes in unit time through an area perpendicular to the flow. That is,

$$\langle I \rangle = \frac{\Delta Q}{\Delta t} \tag{20.1}$$

Here ΔQ is the charge that, during the time Δt, crosses an area perpendicular to the charge flow. The instantaneous current is obtained by the now familiar limiting process,

$$I = \lim_{\Delta t \to 0} \left(\frac{\Delta Q}{\Delta t}\right) \tag{20.2}$$

The unit of current, one coulomb per second, is called the ampere (A), named for André Marie Ampère (1775–1836), a French scientist who made many contributions to current electricity.

By definition, *the direction of current is that of positive charge flow.* If it is negative charges that move, the current direction is then opposite to the velocity of the negative charges. This latter circumstance is by far the more common; in metallic conductors, negatively charged electrons are the charge carriers.

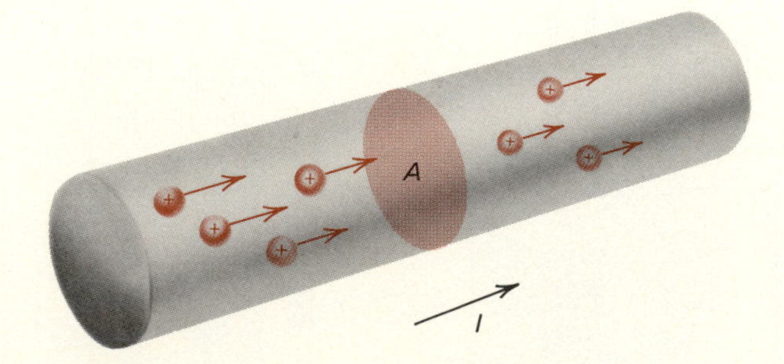

Figure 20.4 *The current in this cylindrical conductor is defined as the positive charge that crosses the area A per unit time.*

In a metal, mobile *conduction electrons* are in continuous motion, like the molecules of a gas. If the average velocity of the electrons is zero, the total current is also zero. A macroscopic current resulting from a net motion of charges is proportional to the number of charge carriers and their average velocity, also called their *drift velocity*.

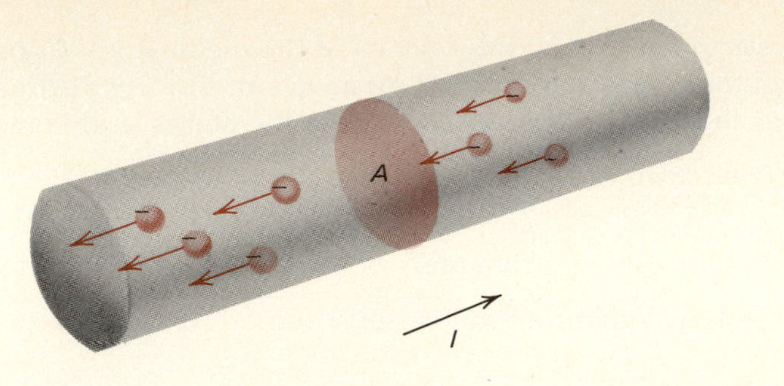

Figure 20.5 When the charge carriers are negatively charged, the velocity of the charge carriers and the current are oppositely directed.

Consider a beam of particles of uniform density, n/m^3, each carrying a charge of q coulombs. In a time Δt, a charge that travels with a speed v_d covers a distance $\Delta \ell = v_d \, \Delta t$. The total charge within the volume $\Delta V = A \, \Delta \ell$ of Figure 20.6 is $nq \, \Delta V = \Delta Q$, and this is the amount of charge that sweeps past the cross-sectional area A during the time interval Δt.

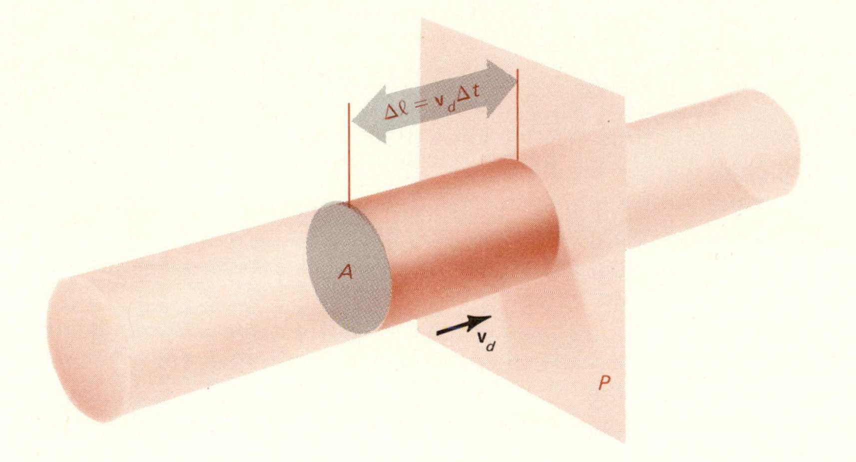

Figure 20.6 The cylindrical region contains particles that move with an average drift velocity $\mathbf{v}_d$. The density of the particles is n per m^3, and each particle carries a charge of q C. In the time interval Δt, all the charged particles within the volume $v_d \, \Delta t \, A$ will have passed across the reference plane P.

If we now substitute into Equation (20.2), we have

$$I = \frac{\Delta Q}{\Delta t} = \frac{nqAv_d \, \Delta t}{\Delta t} = nqv_dA = JA \qquad \textbf{(20.3)}$$

Here J is the *current density*, measured in amperes per square meter, and is given by

$$J = nqv_d \qquad \textbf{(20.4)}$$

● ***Example 20.1*** A current of 2 A flows in a copper wire of 1-mm² cross section. What is the drift velocity of electrons in that wire? How long does it take an electron to travel 10 cm (about the length of an incandescent bulb filament) in this wire under these circumstances?

We know that the current density is

$$J = \frac{I}{A} = \frac{2 \text{ A}}{10^{-6} \text{ m}^2} = 2 \times 10^6 \text{ A/m}^2$$

The charge q on each carrier is -1.6×10^{-19} C, because the carriers are electrons. To determine v_d, we need n, the density of the charge carriers in the conductor.

We make the assumption, which is generally correct for most metals, that the number of conduction electrons per atom equals the number of valence electrons, that is, the valency of the metal atom. Copper, silver, gold, and the alkali metals are all monovalent. The conduction electrons in copper then have the same density as the atoms in this metal. Thus,

$$n = \frac{(1 \text{ electron/atom})(6 \times 10^{23} \text{ atom/mol})(8.92 \text{ g/cm}^3)}{63.5 \text{ g/mol}}$$
$$= 0.843 \times 10^{23} \text{ electrons/cm}^3 = 0.843 \times 10^{29} \text{ electrons/m}^3$$

We can now substitute the various numerical values into (20.4) and obtain

$$v_d = \frac{2 \times 10^6 \text{ A/m}^2}{(-1.6 \times 10^{-19} \text{ C/elect})(0.843 \times 10^{29} \text{ elect/m}^3)}$$
$$= -1.48 \times 10^{-4} \text{ m/s}$$

a leisurely rate of -0.148 mm/s (the negative sign appears because the electronic charge is negative, and v_d is therefore directed opposite to J).

To traverse 10 cm at this speed will require a time

$$t = \frac{10^{-1} \text{ m}}{1.48 \times 10^{-4} \text{ m/s}} = 0.674 \times 10^3 \text{ s} = 11 \text{ min } 14 \text{ s}$$

That is a long time. Yet we know that as soon as we close the proper switch, charge flows through a circuit and lamps light up. We need not wait several minutes, not even seconds, to witness the effect of current in a circuit, and there appears to be no observable dependence on the distance between the wall switch and light fixture, a distance generally considerably greater than 10 cm.

The point is that one does not have to wait until a *particular* electron at the battery terminal reaches the lamp for the lamp filament to respond to the current. When the switch is closed, the entire charge distribution within the conductor is set in motion almost instantaneously, much as water starts to flow in a long pipe as soon as we open a faucet. ●

20.3 Resistance and Resistivity

If a wire is connected between the terminals of a battery, positive charge flows through this *external circuit*, from the positive to the negative terminal, that is, from the point of higher to the point of lower potential, as shown in Figure 20.7. Within the battery, the flow of positive charge is from the negative terminal to the positive, opposite to the electric field; the charge flow is driven not by the electrostatic field but by the chemical reaction of the battery. In the external circuit, the charge flow is driven by the field **E**. A common analogy to the flow of charge in an electric circuit is the flow of water in a hydraulic system. In a gravitational field, water always flows downhill; there are, however, devices—pumps—that can force the uphill flow of water by drawing on some other source of energy.

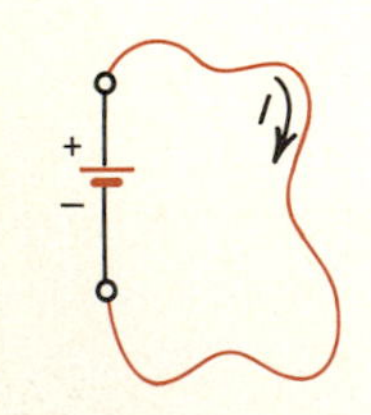

Figure 20.7 *If the two terminals of a battery are connected by a wire, current flows in that wire as indicated. In the external circuit, positive charge flows from the positive to the negative terminal. Inside the battery, positive charge flows from the negative to the positive terminal, driven by the emf of the battery against the electrostatic field.*

If the wire between the battery terminals were an ideal, perfect conductor, with mobile charges that feel no force except the externally imposed electrostatic field, these charges would experience a constant acceleration as a result of a field **E**. The average velocity of the charge carriers would therefore continually increase with time, and so would the current. This does not happen in practice, however. Instead, the current very quickly reaches a steady value that is proportional to the potential difference between the ends of the wire. It is because the wire offers some *resistance* to the flow of charge carriers that a steady state is attained.

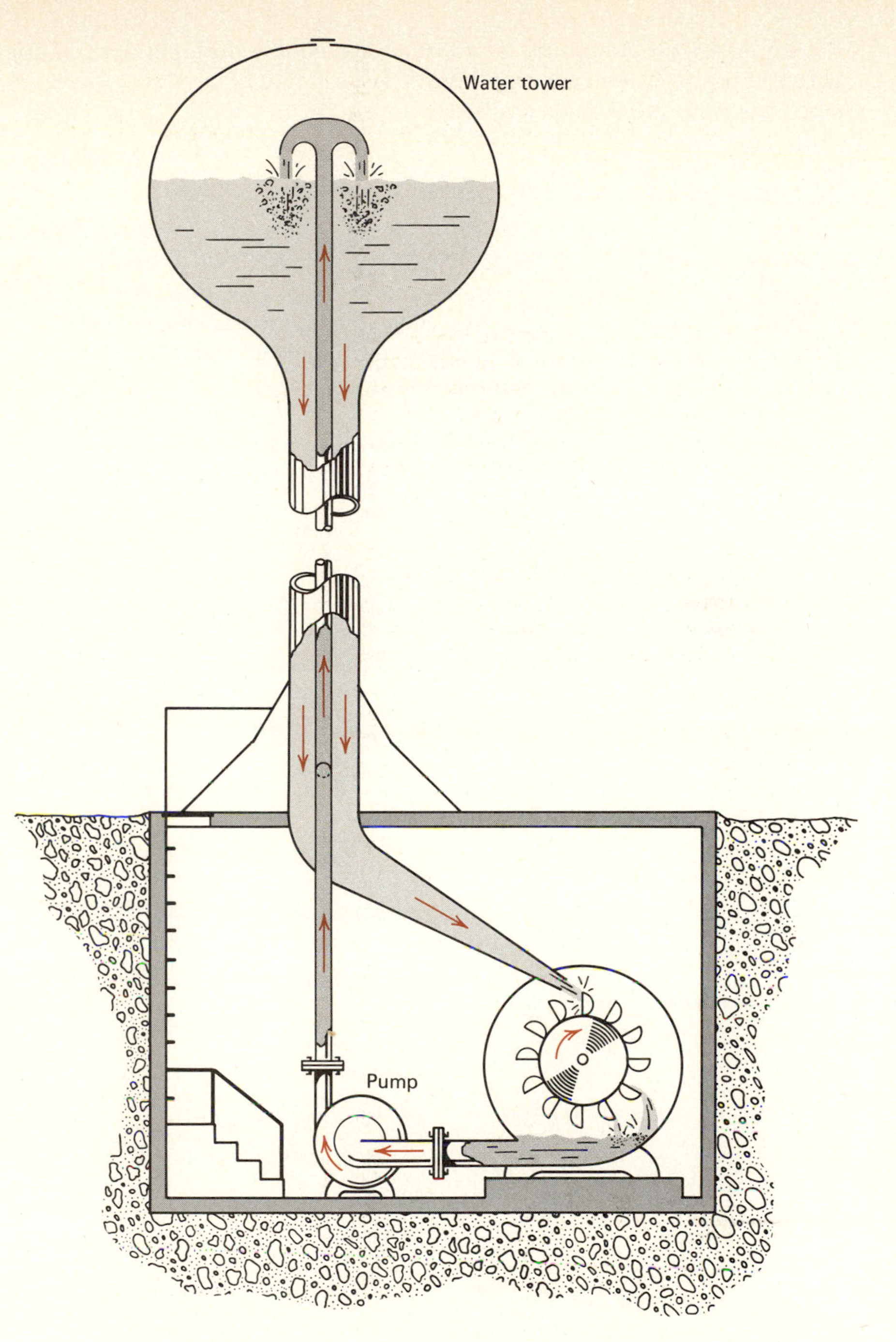

Figure 20.8 Hydraulic equivalent of an electric circuit. Water flows from the high potential point (water tower) to a place of lower potential, performing work in the process (the water wheel). The liquid is returned to the high potential position by the pump, which draws on another form of energy to raise the potential energy of the water.

The resistance of the wire is defined as the ratio of voltage to current; that is,

$$R = \frac{V}{I} \tag{20.5}$$

where R is the resistance, I is the current that flows through this resistance, and V is the *potential drop* across the resistance; that is, V is the potential difference between the two ends of the resistive element in the presence of the current I. The unit of resistance is the *ohm* (Ω), named after Georg Simon Ohm (1787–1854). One ohm equals one volt per ampere. Any circuit component that introduces only resistance in a circuit is called a (pure) *resistor*.

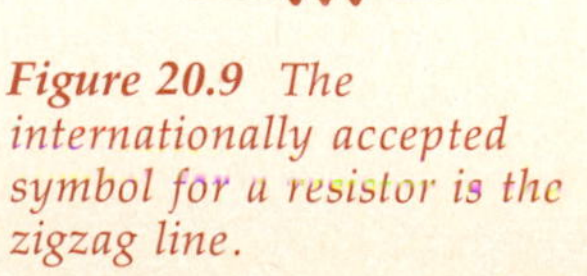

Figure 20.9 The internationally accepted symbol for a resistor is the zigzag line.

Very often the resistance of a circuit element is independent of the current through it, at least over a fairly wide range of currents. The relation, Equation (20.5) or its equivalent,

$$V = IR \tag{20.6}$$

where R is taken as constant, is known as *Ohm's law*.

Ohm's law is not a basic law of nature in contrast to Newton's laws of motion, the second law of thermodynamics, or the laws of conservation of energy and momentum. There are many important resistive systems in which Ohm's law is not obeyed. These play a central role in solid state electronics. For most simple circuit elements, wires, electric heaters and such, however, Ohm's law is either valid, or at the very least, a good approximation.

The resistance of a conductor depends on its length ℓ and cross section A and on an intrinsic material property, its *resistivity*. The relation between the resistance R and the resistivity ρ is

$$R = \frac{\ell}{A}\rho \tag{20.7}$$

The unit of resistivity is the ohm meter.

Room-temperature resistivities of substances span an enormous range, from the low values for ultrapure metals, such as copper and silver, to the very large resistivities of good insulators, such as glass, Teflon, and Mylar. Resistivity values at 20 °C are listed in Table 20.1 for some pure metals, and a few alloys, semiconductors, and insulators. These values range over 25 decades.

Resistance to charge flow in a conductor arises because the charge carriers encounter various obstacles that tend to impede their motion. As they collide with these obstacles, the carriers lose momentum and energy gained from the electric field since the previous collision. As we shall see presently, this loss of energy causes heating of the resistor through which the current flows.

Table 20.1 Resistivities and temperature coefficients of resistivity at 20 °C

Substance	*Resistivity ρ, $\Omega \cdot m$*	*Temperature coefficient α_r, 1/C° or 1/K*
Copper	1.69×10^{-8}	3.9×10^{-3}
Silver	1.59×10^{-8}	3.8×10^{-3}
Gold	2.44×10^{-8}	3.4×10^{-3}
Aluminum	2.83×10^{-8}	4.0×10^{-3}
Tungsten	5.33×10^{-8}	4.6×10^{-3}
Platinum	10.4×10^{-8}	3.9×10^{-3}
Manganin	48.2×10^{-8}	0
Constantan	48.9×10^{-8}	2×10^{-6}
Nichrome	100×10^{-8}	4×10^{-4}
Carbon	3.5×10^{-5}	-5×10^{-4}
Germanium	~ 0.5	$\sim -5 \times 10^{-2}$
Silicon	~ 1000	$\sim -7 \times 10^{-2}$
Wood	10^{8}–10^{14}	—
Glass	10^{10}–10^{14}	—
Fused quartz	5×10^{17}	—

The obstacles that scatter the conduction electrons may be impurities dissolved in a pure metal, for the attainment of an ideally pure metal is impossible. Even the ideally pure metal crystal (except a superconductor at sufficiently low temperatures) would have a finite resistance because the thermal motion of the atoms in a metal crystal scatters the free electrons. Since the amplitude of this vibrational motion in a crystal increases with increasing temperature, it is to be expected that the scattering of free electrons, and hence the resistivity of a metallic conductor, will also increase as the temperature is raised. That is, indeed, what one observes.

The temperature coefficient of resistivity is defined like the coefficient of thermal expansion:

$$\alpha_r = \frac{1}{\rho}\frac{\Delta\rho}{\Delta T} \tag{20.8}$$

Values of α_r for several pure metals, a few alloys, and carbon, a nonmetallic conductor, are given in Table 20.1.

The resistivity of a semiconductor is much less predictable, depending sensitively not only on temperature but also on the type and concentration of impurities dissolved in the host material. (See Chapter 31.)

● ***Example 20.2*** A-10 m length of 1-mm diameter copper wire is wound on a spool. What is the resistance of this length of wire at room temperature?

According to Table 20.1, the resistivity of copper at 20 °C is 1.69×10^{-8} $\Omega\cdot$m. To find R we only need to substitute the appropriate numerical values into Equation (20.7). The area of the wire is

$$A = \frac{\pi d^2}{4} = \frac{\pi}{4} \times 10^{-6}\ \text{m}^2$$

Thus

$$R = \frac{(10\ \text{m})(1.69 \times 10^{-8}\ \Omega\cdot\text{m})}{(\pi/4) \times 10^{-6}\ \text{m}^2} = 0.215\ \Omega$$ ●

● ***Example 20.3*** If the temperature of the wire of Example 20.2 increases to 80 °C, what is its resistance at this new temperature?

We calculate ΔR from (20.8), using the value of α_r from Table 20.1. In this calculation we generally neglect the very small correction due to thermal expansion; that is, we presume that $\Delta R/R = \Delta\rho/\rho$ (see Problem 20.10). Thus

$$\Delta R = \alpha_r R \Delta T = (3.9 \times 10^{-3}\ \text{K}^{-1})(0.215\ \Omega)(60\ \text{K}) = 0.050\ \Omega$$

$$R(80\ ^\circ\text{C}) = R(20\ ^\circ\text{C}) + \Delta R = 0.215\ \Omega + 0.050\ \Omega = 0.265\ \Omega$$ ●

20.4 Current, Voltage, and Power Dissipation in a Simple Circuit

Figure 20.10 shows the simplest possible direct-current circuit. It consists of an ideal battery with emf $\mathcal{E}$ and an external resistor R, the *load*. The wires that connect the load to the battery terminals are assumed to be ideal, resistance-free conductors.

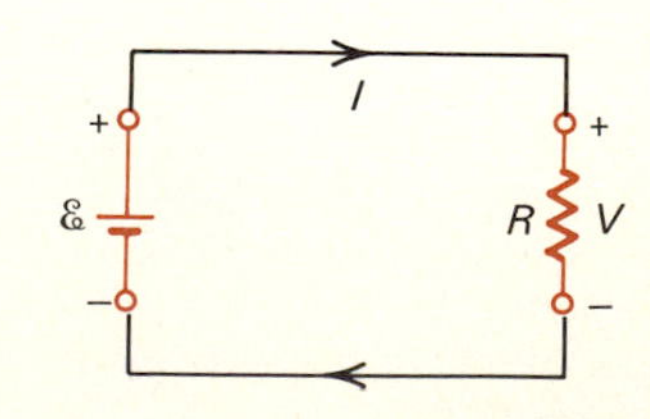

Figure 20.10 *A very simple circuit. An ideal source of emf is connected to a load resistor R. The direction of current and the polarities of the potentials are indicated.*

The load may be a toaster, a light bulb, or any other device that dissipates electrical energy. For the present we need not inquire further into its construction.

The direction of current in the external circuit is as indicated, from the positive terminal to the negative. The potential drop across the resistor R is in the direction of current and is given by

$$V = IR \qquad \textbf{(20.6)}$$

The positive side of R and the positive battery terminal are connected together, and so are the negative side of R and the negative battery terminal. It follows that

$$\mathcal{E} = V \qquad \text{or} \qquad \mathcal{E} = IR \qquad \textbf{(20.9)}$$

for this simple circuit.

In the circuit of Figure 20.10, whenever some charge $\Delta Q = I\,\Delta t$ is transported from the positive end of the load resistor to the negative, that charge loses potential energy in the amount of

$$\Delta \text{PE} = V\,\Delta Q = VI\,\Delta t$$

Conservation of energy now requires that this potential energy surface as energy in some other form: heat, mechanical energy, chemical energy, etc. In each case, the rate at which electrical energy is dissipated in the load is

$$P = \frac{\Delta W}{\Delta t} = VI = I^2R = \frac{V^2}{R} \qquad \textbf{(20.10)}$$

where the alternative forms are obtained by making use of Equation (20.6). Because Joule used the heat produced by a current passing through a resistor in one of the more precise measurements of the mechanical equivalent of heat, the I^2R heating rate in a resistor is often called *Joule heating.*

At the battery, the charge ΔQ is *raised* through this same potential difference by the emf $\mathcal{E}$. Hence the battery delivers energy in the amount $W = \mathcal{E}\Delta Q = \mathcal{E}I\,\Delta t$ to the charge ΔQ. The rate at which the battery does work, that is, its power output is therefore

$$P = \mathcal{E}I \qquad \textbf{(20.11)}$$

From Equation (20.9), we see that the power the battery delivers exactly equals the power dissipation in the load; this must be so since there is no other source or sink of energy in the system.

20.5 Resistors in Series and Parallel Combinations

Every practical electrical circuit is more complicated than the one shown in Figure 20.10. For example, a typical Christmas tree decoration consists of ten or a dozen small lamps, placed in series in a circuit, as shown in Figure 20.11; switches connect the battery of an automobile to the starter motor, horn, headlights, heater fan, and radio, as shown schematically in

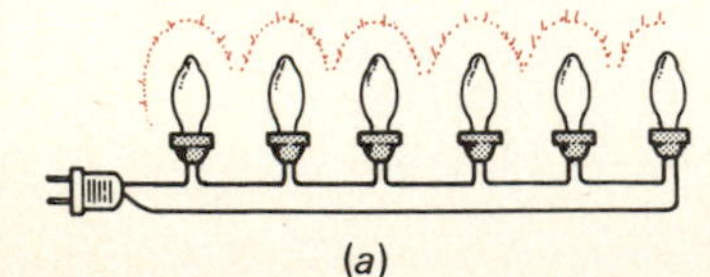
(*a*)

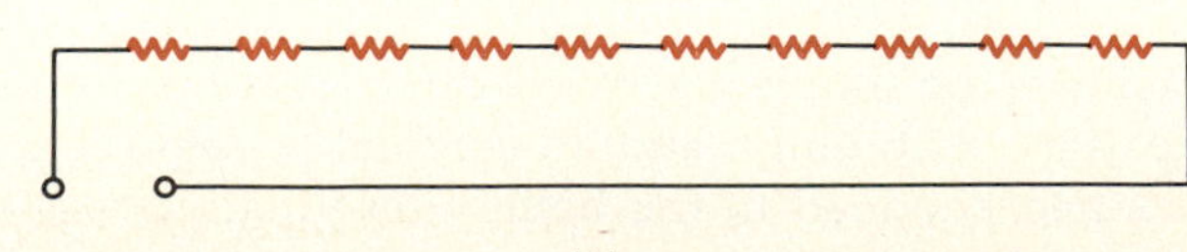
(*b*)

Figure 20.11 *(a) A string of Christmas tree lights connected in series. (Today, Christmas tree lights are usually connected in parallel.) (b) The string of lights is represented by an equal number of resistors connected in series. In this case, the resistances of the resistors are all identical.*

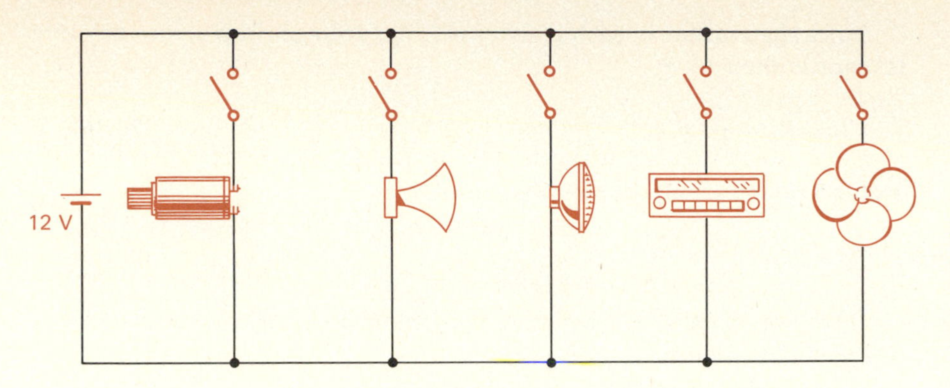

Figure 20.12 On a car, the starter motor, horn, headlights, radio, heater fan, and other electrical equipment are all connected in parallel to the terminals of the battery.

Figure 20.12. Figures 20.11 and 20.12 are examples of series and parallel connections.

20.5 (a) Resistors in Series

When two or more resistors are connected in series, as in Figure 20.13, the same current flows in each resistor of the circuit. As in Chapter 19, where we considered series and parallel combinations of capacitors, we now want to find the equivalent resistor that can replace the series combination. That equivalent resistor should draw the same current from the battery as the series combination.

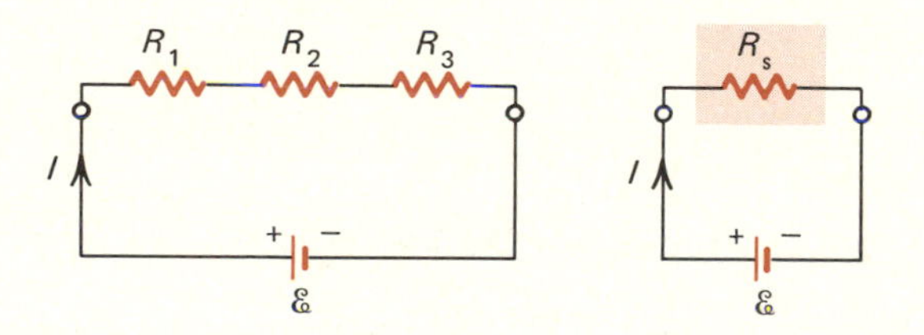

Figure 20.13 Three resistors connected in series across a battery. The equivalent resistance of the series combination, R_s, is that of a resistor which will draw the same current when placed across the terminals of the same battery.

To find the value of R_s (which one can sense intuitively), we note that when a current I flows through these resistors of the series combination, the individual potential drops, V_1, V_2, and V_3, are in the same direction. Consequently, V_{AB} is the sum of these potential drops, and V_{AB} must also equal the emf $\mathcal{E}$ of the battery. Thus,

$$\mathcal{E} = V_1 + V_2 + V_3 = IR_1 + IR_2 + IR_3 = I(R_1 + R_2 + R_3) \qquad \textbf{(20.12)}$$

and solving for the current, we find

$$I = \frac{\mathcal{E}}{R_1 + R_2 + R_3} \qquad \textbf{(20.13)}$$

It follows from (20.13) that if we want to replace the series combination of R_1, R_2, and R_3 by a single resistor that draws the same current from the battery, that equivalent resistor must have a resistance

$$R_s = R_1 + R_2 + R_3 \qquad \textbf{(20.14)}$$

By induction, the generalization of (20.14) for any number N of resistors in series is

$$R_s = \sum_{i=1}^{N} R_i \qquad \textbf{(20.15)}$$

where R_s is the resistance of the equivalent resistor of the series combination of the N resistors of resistances R_i.

Since the current in each resistor is I, the voltage drop across any one resistor is then

$$V_i = IR_i = \frac{\mathcal{E}}{R_s} \times R_i \tag{20.16}$$

and the power dissipated by that resistor is

$$P_i = IV_i = \frac{\mathcal{E}}{R_s} \times \frac{\mathcal{E}}{R_s} \times R_i = \left(\frac{\mathcal{E}}{R_s}\right)^2 R_i \tag{20.17}$$

If we sum over all the resistors, we obtain the total power dissipated in the circuit, namely,

$$P = \sum_{i=1}^{N} \left(\frac{\mathcal{E}}{R_s}\right)^2 R_i = \left(\frac{\mathcal{E}}{R_s}\right)^2 R_s = \frac{\mathcal{E}^2}{R_s} \tag{20.18}$$

which is just the power delivered by the emf $\mathcal{E}$ to a load resistor of resistance R_s.

● ***Example 20.4*** A chain of twelve identical Christmas tree lights connected in series is to be placed across a 120-V source of emf. Each light should dissipate 15 W. What resistance must each of the light bulbs have?

Since each light dissipates 15 W, the total energy delivered by the voltage source must be

$$P = 12(15 \text{ W}) = 180 \text{ W}$$

and therefore the current in the circuit must be

$$I = \frac{P}{\mathcal{E}} = \frac{180 \text{ W}}{120 \text{ V}} = 1.5 \text{ A}$$

The total potential drop across the twelve bulbs is 120 V; since the bulbs are identical, the potential drop across each must be (120 V)/12 = 10 V. Consequently, the resistance of each light bulb is

$$R = \frac{10 \text{ V}}{1.5 \text{ A}} = 6.67 \ \Omega$$ ●

● ***Example 20.5*** Two resistors are connected in series across a 20-V battery. The resistance of one of the resistors is 16 Ω. What must the resistance of the other be if the potential across the 16-Ω resistor is 8 V? What is the power dissipated by this second resistor?

Since the potential drop across the 16-Ω resistor is 8 V, the current in that resistor must be $I = (8 \text{ V})/(16 \ \Omega) = 0.5$ A. The sum of the potential drops across the two series resistors must be 20 V, the emf $\mathcal{E}$ of the battery. Hence, the potential drop across the unknown resistor must be $V = (20 - 8) \text{ V} = 12$ V. But the current is the same in this resistor as in the other resistor of the series combination, namely, 0.5 A. Therefore, the value of the unknown resistance is $R = (12 \text{ V})/(0.5 \text{ A}) = 24 \ \Omega$. The power dissipated by this 24-Ω resistor is $I^2R = (0.5 \text{ A})^2(24 \ \Omega) = 6$ W. ●

Example 20.5 demonstrates an important general principle. In a series circuit, the potential drops across the component resistors stand in the same ratio as the resistances of the resistors. It is left as a problem to show that the same rule also holds for the power dissipation in each of the component resistors.

20.5 (b) Resistors in Parallel

Figure 20.14 shows three resistors connected in parallel across a battery. Again, we want to find the equivalent resistance R_p of this parallel combination. If a resistor with resistance R_p is connected across the terminals of the battery, the battery should deliver the same current as it does to the parallel combination and consequently supply the same amount of total power. We must therefore find the total current delivered by the battery.

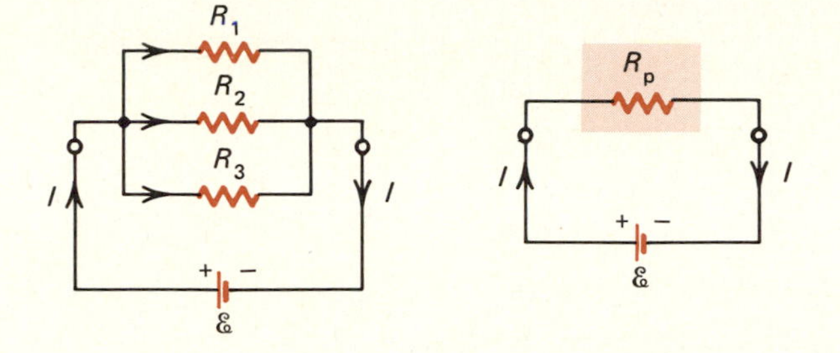

Figure 20.14 *Three resistors connected in parallel across a battery. The equivalent resistance of the parallel combination R_p is that of a resistor which, when placed across the terminals of the same battery, draws the same current as the total current drawn by the parallel combination.*

In this circuit, the total current delivered by the battery divides into three parts, I_1, I_2, and I_3, which flow through the three resistors of the parallel combination. Whereas in this parallel combination the currents in the three resistors need not be the same, it is apparent from the circuit that the potential drop across each resistor must be the same. Specifically,

$$V_1 = V_2 = V_3 = I_1R_1 = I_2R_2 = I_3R_3 = \mathcal{E} \tag{20.19}$$

and

$$I_1 = \frac{\mathcal{E}}{R_1} \qquad I_2 = \frac{\mathcal{E}}{R_2} \qquad I_3 = \frac{\mathcal{E}}{R_3}$$

Since

$$I = I_1 + I_2 + I_3 = \mathcal{E}\left(\frac{1}{R_1} + \frac{1}{R_2} + \frac{1}{R_3}\right) = \frac{\mathcal{E}}{R_p}$$

it follows that the equivalent resistance is given by

$$\frac{1}{R_p} = \frac{1}{R_1} + \frac{1}{R_2} + \frac{1}{R_3} \tag{20.20}$$

The above is readily generalized for any number of resistors in a parallel combination:

$$\frac{1}{R_p} = \sum_{i=1}^{N} \frac{1}{R_i} \tag{20.21}$$

where R_p is the equivalent resistance of the N resistances R_i in the parallel combination.

Since the voltage across the resistors of the combination is the same, the power dissipated in each is

$$P_i = \frac{V^2}{R_i} = \frac{\mathcal{E}^2}{R_i} \tag{20.22}$$

Summing over all elements of the parallel combination, we see that the total power dissipated in the resistors is equal to

$$P = \frac{\mathcal{E}^2}{R_p} = \mathcal{E} I \tag{20.23}$$

which is the power delivered by the battery.

Note, as a general rule, that *the resistance of a series combination is always greater than the largest of its components; the resistance of a parallel combination is always less than the smallest of its components.*

Whereas in the series combination the voltages divide in proportion to the resistances, in the parallel combination the currents divide in proportion to the reciprocal of the resistances.

Table 20.2 Relations for series and parallel combinations of resistances

Series combinations	*Parallel combinations*
$V_s = V_1 + V_2 + V_3 + \cdots$	$I_p = I_1 + I_2 + I_3 + \cdots$
$I_s = I_1 = I_2 = I_3 = \cdots$	$V_p = V_1 = V_2 = V_3 = \cdots$
$R_s = R_1 + R_2 + R_3 + \cdots$	$\frac{1}{R_p} = \frac{1}{R_1} + \frac{1}{R_2} + \frac{1}{R_3} + \cdots$
$R_s > R_1, R_2, R_3, \ldots$	$R_p < R_1, R_2, R_3, \ldots$

● ***Example 20.6*** An 800-W toaster, a 200-W lamp, and a 300-W electric blanket are connected across 120 V. What are the resistances of the individual components and what is the equivalent resistance of this combination?

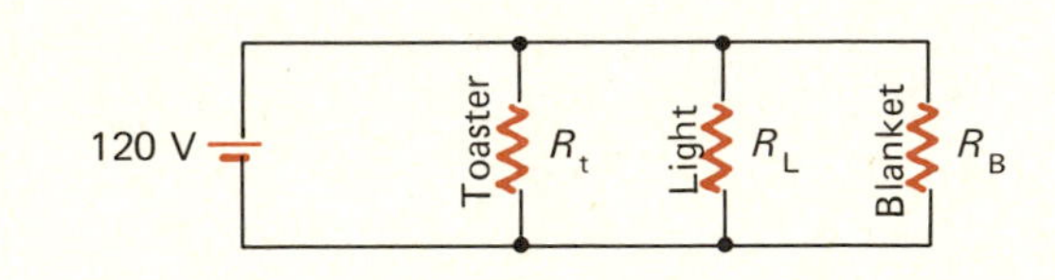

Figure 20.15 A toaster, a lamp, and an electric blanket connected in parallel across a 120-V source.

Each component is connected across 120 V when it dissipates the stated power. Hence the resistances can be obtained from Equation (20.10):

$$R_t = \frac{(120\ \text{V})^2}{800\ \text{W}} = 18\ \Omega \qquad R_L = \frac{(120\ \text{V})^2}{200\ \text{W}} = 72\ \Omega$$

$$R_b = \frac{(120\ \text{V})^2}{300\ \text{W}} = 48\ \Omega$$

The equivalent resistance of the combination is

$$R_p = \left[\frac{1}{18\ \Omega} + \frac{1}{72\ \Omega} + \frac{1}{48\ \Omega}\right]^{-1} = 11.1\ \Omega$$

In this case, we could have obtained R_p more simply by the relation

$$P = \frac{\mathcal{E}^2}{R_p}$$

with $P = 800 + 200 + 300 = 1300$ W, the total power delivered by the voltage source. ●

20.6 The Real Battery; Internal Resistance

We have hitherto referred to a battery as a source of emf. While it does serve this function, a real battery is not an ideal source. An ideal source would be able to maintain the same potential difference between its terminals regardless of the amount of current drawn, and no energy would be dissipated in heat within the ideal battery. No real voltage source meets these conditions.

A real battery produces a potential difference of $\mathcal{E}$ volts between its terminals when no current flows, that is, on *open circuit*. However, if the external circuit is completed and current is drawn from the battery, the potential difference between the battery terminals decreases by an amount roughly proportional to the current. In the limit, when a heavy copper bar of nearly zero resistance connects the battery terminals, the current in this *short circuit* is not enormous, as it would be for an ideal battery, but has some large but finite value.

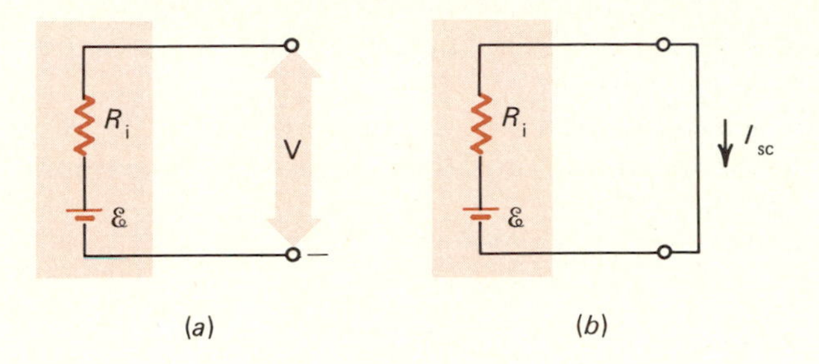

Figure 20.16 The real battery may be simulated by an ideal battery in series with an internal resistance R_i. (*a*) The open-circuit voltage between the battery terminals is equal to the emf of the ideal battery. (*b*) The short-circuit current I_{sc} is related to the internal resistance by $I_{sc} = \mathcal{E}/R_i$, where $\mathcal{E}$ is the open circuit emf.

This behavior can be simulated by the combination of an ideal battery in series with a resistance, the *internal resistance* of the real battery. The emf of the ideal battery is then the open-circuit voltage across the terminals of the real battery, and the internal resistance is the ratio of this open-circuit voltage to the short-circuit current.

At least in a broad, qualitative way, it is not difficult to understand the origin of this internal resistance and the reason for the limitation on the short-circuit current. In any battery, the electrical energy is derived from a chemical reaction; within the battery itself, the charge is carried by ions moving in an electrolyte, an ionic conductor, generally a liquid. The flow of ions in the electrolyte is subject to resistive mechanisms, and moreover, the chemical reactions at the battery plates proceed at finite rates. This latter restriction ultimately places an upper limit on the amount of current that a battery with a given plate area can deliver. The greater the area of the plates in contact with the electrolyte, the greater the short-circuit current. Furthermore, the temperature of the battery is also important, because chemical reaction rates are generally sensitive to temperature, and reactions are slower at lower temperatures. It is largely for this reason that a battery, able to start a car at normal temperature, may fail to do so if the temperature drops to a chilly −20 °C. It sometimes happens that if near-short-circuit current is drawn from such a battery for about ten seconds in the first abortive attempt to start the car, and another attempt is made half a minute later, the battery almost miraculously succeeds in delivering enough power to start the engine. The reason it works the second time is that the energy dissipated in forcing current through the internal resistance of the battery during the first attempt heated the electrolyte so that a short time later the chemical reaction rate is somewhat greater and a larger current can be drawn for a brief interval.

What is sometimes of concern is how much power a battery or some other source of electrical power can deliver to an external load, and under what conditions that maximum power transfer is achieved.

To answer these questions, we first sketch the equivalent circuit, as in Figure 20.17. The circuit shows the real battery, consisting of the ideal

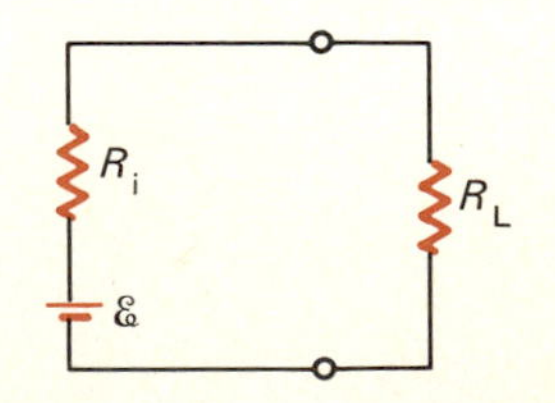

Figure 20.17 The load resistor R_L is connected to the terminals of a real battery, whose open-circuit voltage is $\mathcal{E}$ and whose internal resistance is R_i. The power delivered to the load depends on the load resistance.

voltage source and an internal resistance, and the external load resistor R_L. We recognize this circuit as two resistors in series with an ideal emf. Hence, the power dissipated in the load resistor can be obtained from Equation (20.17), where $R_s = R_L + R_i$. Thus,

$$P_L = \mathcal{E}^2 \frac{R_L}{(R_L + R_i)^2} \tag{20.24}$$

If the load resistance R_L is very large compared with the internal resistance of the battery, then the denominator in Equation (20.24) is approximately equal to R_L^2, and (20.24) can be written $P_L \simeq \mathcal{E}^2/R_L$. Under these conditions, P_L increases as R_L is reduced. However, if R_L is reduced so that its value is substantially less than R_i, the exact formula (20.24) then reduces to $P_L \simeq \mathcal{E}^2 R_L/R_i^2$; now, evidently, decreasing R_L diminishes P_L. Thus, for some value of R_L in the neighborhood of R_i, the power dissipation must have a maximum.

A plot of P_L as a function of R_L is shown in Figure 20.18. As expected, there is indeed a maximum, and the condition for maximum power dissipation is that $R_L = R_i$.

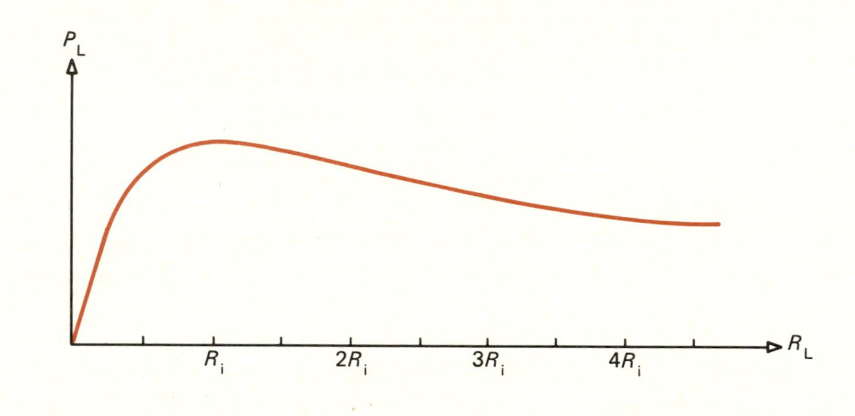

Figure 20.18 *The power dissipated in the load resistor R_L as a function of load resistance. Maximum power dissipation is attained when $R_L = R_i$, where R_i is the internal resistance of the voltage source.*

Maximum transfer of electrical power from a source with internal resistance R_i to a load occurs when the load resistance equals the internal resistance of the source; that is, when $R_L = R_i$.

Matching of load and source resistances is often a critical design consideration in electrical and electronic instrumentation. For example, the final amplification stage of an audio amplifier is designed so that the effective source resistance matches the resistance of the loudspeakers. Moreover, analogous considerations apply to the transfer of energy by mechanical linkages, as, for example, in the human ear, where a complicated lever system of bones transmits vibrations of the eardrum to the sensing apparatus of the inner ear (see Section 16.10).

● ***Example 20.7*** A battery has an open-circuit terminal voltage of 15 V. When a load resistance of 2 Ω is connected to the battery terminals, 72 W is dissipated in the load. Find the internal resistance of the battery and the power dissipated in the battery when the 2-Ω resistor is connected. Also determine the power that would be dissipated in the battery under short-circuit conditions and the maximum power that this battery can deliver to a load.

Since 72 W is dissipated in a 2-Ω resistor, we can use the expression $P = V^2/R$ to find the voltage drop across the load; its value in this case is also the potential difference between the terminals of the real battery. We have

$$72\ \text{W} = \frac{V^2}{2\ \Omega} \qquad V^2 = 144\ \text{V}^2 \qquad V = 12\ \text{V}$$

The current in the circuit is then

$$I = \frac{V}{R} = \frac{12\ \text{V}}{2\ \Omega} = 6\ \text{A}$$

That current of 6 A, in flowing in the internal resistance of the battery, results in a voltage drop at the terminals from 15 V at open circuit to the 12 V with the load attached. Thus,

$$IR_i = 3\ \text{V} \qquad \text{and} \qquad R_i = \frac{3\ \text{V}}{6\ \text{A}} = 0.5\ \Omega$$

The power dissipated in the battery when the current of 6 A is flowing is given by

$$P = I^2R_i = (6\ \text{A})^2(0.5\ \Omega) = 18\ \text{W}$$

Under short-circuit conditions, we have

$$I_{sc} = \frac{\mathcal{E}}{R_i} = \frac{15\ \text{V}}{0.5\ \Omega} = 30\ \text{A}$$

and the power dissipated within the battery is then

$$P_{sc} = I_{sc}^2R_i = 450\ \text{W}$$

Maximum power will be delivered under matched load conditions, that is, when $R_L = R_i = 0.5\ \Omega$. The power delivered to the load is then given by Equation (20.24):

$$P_L = \frac{(15\ \text{V})^2(0.5\ \Omega)}{(1.0\ \Omega)^2} = 112.5\ \text{W}$$

Summary

A *source of electromotive force* (emf) forces the flow of charge around a closed circuit. The emf $\mathcal{E}$ of the source is equal in magnitude to the potential difference between the terminals of the source under *open-circuit* conditions. The internal resistance R_i of the source is the ratio of its emf to the *short-circuit* current I_{sc}.

Current is the flow of charge. The unit of current is the *ampere* (A). One ampere corresponds to the flow of one coulomb across an area in one second. The dimension of current is

$$[I] = \frac{[Q]}{[T]}$$

The average current is given by

$$\langle I \rangle = \frac{\Delta Q}{\Delta t}$$

and the instantaneous current is

$$I = \lim_{\Delta t \to 0} \frac{\Delta Q}{\Delta t}$$

The average velocity of the charge carriers responsible for the flow of charge is called the *drift velocity* $\mathbf{v}_d$ of the carriers. The current per unit area is the *current density* $\mathbf{J}$. Current density and drift velocity are related by

$$\mathbf{J} = nq\mathbf{v}_d$$

where n is the density of charge carriers, that is, the number of carriers per unit volume, and q is the charge per carrier.

The current density has the same direction as the drift velocity if the carriers are positively charged; $\mathbf{J}$ is directed opposite to $\mathbf{v}_d$ if the carriers are negatively charged.

The *resistance* of a *resistor* is defined as the ratio of the potential difference between the two terminals of the resistor to the current. That is,

$$R = \frac{V}{I}$$

where V is the *voltage drop* across the resistor in the presence of the current I. If R is independent of I, the circuit element is said to obey *Ohm's law,*

$$V = IR$$

with R constant.

The resistance of a wire is given by

$$R = \frac{\ell}{A}\rho$$

where ℓ is the length and A the cross-sectional area of the wire and ρ is the *resistivity* of the material of which the wire is made.

The unit of resistance is the *ohm* (Ω); the unit of resistivity is the ohm meter ($\Omega \cdot$m).

The resistivity of a metal depends on its temperature. The temperature coefficient of resistivity is defined by

$$\alpha_r = \frac{1}{\rho}\frac{\Delta\rho}{\Delta T}$$

The power dissipated when a current I passes through a resistor R is given by

$$P = I^2R = VI = \frac{V^2}{R}$$

This power dissipation causes heating of the resistor and is often referred to as *Joule heating.*

The equivalent resistance of a series combinaton of N resistors is given by

$$R_s = \sum_{i=1}^{N} R_i$$

The equivalent resistance of a parallel combination of N resistors is given by

$$\frac{1}{R_p} = \sum_{i=1}^{N} \frac{1}{R_i}$$

A source of emf with internal resistance R_i delivers maximum power to a load resistor when the resistance of the load equals the internal resistance of the source. When $R_L = R_i$, the load is said to be matched to the source.

Multiple Choice Questions

20.1 Two metal rods have exactly the same resistance. Rod A has a length L_A and diameter D_A. The length L_B and diameter D_B of rod B are related to L_A and D_A by $L_B = 2L_A$ and $D_B = 2D_A$. It follows that rod A has a resistivity related to that of rod B by

(a) $\rho_A = \frac{1}{4}\rho_B$
(b) $\rho_A = \frac{1}{2}\rho_B$
(c) $\rho_A = \rho_B$
(d) $\rho_A = 2\rho_B$
(e) $\rho_A = 4\rho_B$

20.2 In Figures 20.19 (a) and (b), the battery emf's are the same, and the resistors have the same resistances. The current in (a) is I_a. The current I_b is given by

(a) $I_b = I_a$
(b) $I_b = 2I_a$
(c) $I_b = 4I_a$
(d) $I_b = 16I_a$
(e) $I_b = 32I_a$

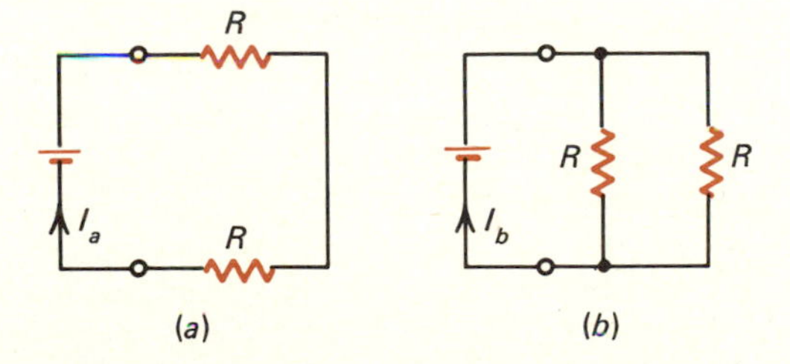

Figure 20.19

20.3 Two resistors, R_1 and R_2, are connected in parallel across the terminals of a battery. If $R_1 < R_2$, and R_p denotes the resistance of the equivalent resistor of the parallel combination, then

(a) the power dissipated by R_2 is greater than the power dissipated by R_1, and R_p is less than either R_1 or R_2.

(b) the power dissipated by R_2 is less than the power dissipated by R_1, and R_p is less than either R_1 or R_2.

(c) The resistance R_p is greater than either R_1 or R_2.

(d) The resistance R_p is the geometrical mean of R_1 and R_2.

(e) None of the above statements is true.

20.4 In the circuit of Figure 20.20, the current flowing through the battery is related to the current through the 4-Ω resistor by

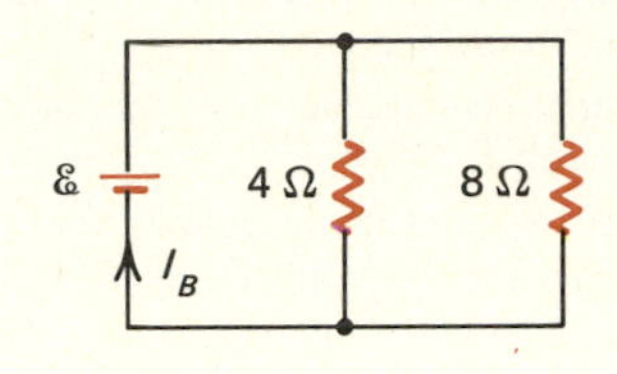

Figure 20.20

(a) $I_B = I_4/3$
(b) $I_B = 1.5I_4$
(c) $I_B = 2I_4$
(d) $I_B = 3I_4$
(e) $I_B = 12I_4$

20.5 When two identical resistors are connected in series across a battery, the power dissipated by them is 20 W. If these resistors are connected in parallel across the same battery, the total power dissipated will be

(a) 5 W
(b) 10 W
(c) 20 W
(d) 40 W
(e) 80 W

20.6 If in the accompanying circuit, the resistance of R_2 is decreased,

(a) the current through R_1 increases.
(b) the current through R_1 is constant.
(c) the voltage drop across R_2 decreases.
(d) the power dissipated by R_2 decreases.
(e) Both (b) and (d) are correct.

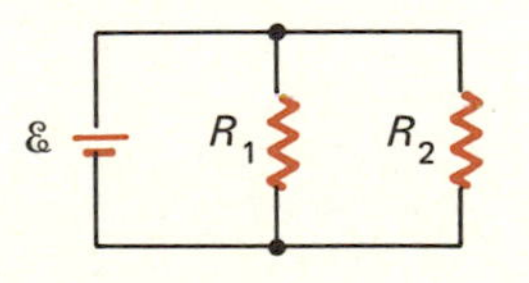

Figure 20.21

20.7 Two metals, A and B, are partially submerged in a dilute acid. When these metals are connected by a copper wire, current flows in that wire from metal B to metal A. Identify the correct statement.

(a) The number of A ions in solution is greater than the number of B ions, and the net flow of ions in solution is from A to B.

(b) The number of A ions in solution is less than the number of B ions, and the net flow of ions is from A to B in the acid.

(c) The number of A ions in solution is greater than the number of B ions, and the net flow of ions in solution is from B to A.

(d) The number of A ions in solution is less than the number of B ions, and terminal A is positive relative to B.

20.8 The resistivity of a metal generally increases with increasing temperature because

(a) more impurities dissolve in the metal as the temperature is raised.

(b) fewer and fewer electrons can move about within the solid.

(c) thermal expansion decreases the likelihood that an electron will jump from one atom to its neighbor.

(d) thermal agitation of the ions scatters electrons out of their paths along the direction of the electric field lines.

20.9 When a current flows in a metal, heat is produced at a rate given by I^2R. The rate at which heat is generated per unit volume of the metal is

(a) VJ

(b) V^2/ρ

(c) $J^2\rho$

(d) J/ρ

20.10 When a load is matched to the source,

(a) $IR_L = \mathcal{E}$

(b) $P_L = \mathcal{E}I/2$

(c) $P_L = \mathcal{E}I$

(d) $I = \mathcal{E}/R_i$

where $\mathcal{E}$ denotes the source emf, I the current, R_i the internal resistance of the source, R_L the load resistance, and P_L the power dissipated in the load.

Problems

(Sections 20.1, 20.2)

20.1 A wire carries a current of 2.5 A. How many electrons cross a given area of the wire in 1 s?

20.2 A battery charger delivers a current of 4 A for 5 h to a 12-V storage battery. What is the total charge that passed through the battery?

20.3 The beam current in a TV tube is 6 μA. How many electrons strike the screen per second?

• **20.4** What is the drift velocity of electrons in an 0.4-mm diameter silver wire that carries a current of 12 A?

• **20.5** There are approximately 2×10^{23} free electrons per cubic centimeter in aluminum. Calculate the drift velocity of electrons in a piece of aluminum wire that has a 1.5-mm diameter and carries a current of 2.5 A.

• • **20.6** The proton (H^+ ion) beam of a 40-MeV cyclotron is 0.5 μA. How many protons strike the target per second, and how much energy is deposited on the target by this beam in 10 s if the protons are stopped within the target? If the target is a 10-g block of lead, initially at room temperature, what would its temperature be at the end of 10 s if none of the energy deposited by the beam was removed?

(Section 20.3)

20.7 An 18-Ω resistor is connected to a 12-V battery. How much current does the battery deliver?

20.8 A resistor is connected to a 6-V battery. If the current flowing from the battery is 3 A, what is the resistance of the resistor?

20.9 A 15-Ω resistor is connected to a voltage source. If a current of 4.5 A flows in the resistor, what is the potential difference across the terminals of the voltage source?

• **20.10** The resistance of a spool of 0.05-mm-diameter copper wire is 40 Ω at room temperature. How long is the wire?

• **20.11** A 30-m length of aluminum wire has a resistance of 1.4 Ω. What is its diameter?

• **20.12** A wire-wound resistor is made from manganin. What length of manganin wire of 0.03-mm diameter is required to wind a 400-Ω resistor?

• **20.13** A 4-m length of wire of 0.2-mm diameter has a resistance of 8 Ω. The wire is then pulled through a set of dies until its diameter has been reduced to 0.1 mm. Assuming that there is no loss of material and that the drawing process does not change the wire's resistivity, determine the resistance of the wire after the drawing.

• **20.14** At what temperature will a given length of copper wire have the same resistance as a piece of aluminum wire of the same length and diameter kept at room temperature?

• **20.15** Find the dimension of resistivity in terms of $[M]$, $[L]$, $[T]$, and $[Q]$, where $[Q]$ stands for the dimension of charge (coulomb).

• **20.16** A 47-Ω resistor is to be made by winding a coil of platinum wire. Just 1.8 g of platinum is to be used for this purpose. What should the diameter of the platinum wire be?

• **20.17** A lamp has a tungsten filament whose resistance at room temperature is 100 Ω. What is its resistance at the operating temperature of 1800 °C?

• **20.18** A two-wire power line carries a current of 150 A. If the conductors are made of copper and the voltage drop per kilometer of line must not exceed 0.2 V, what should the diameter of each wire be? What is the mass of this power line per 150-m length (the approximate distance between towers)?

• **20.19** Resistance thermometers are used for some applications. Suppose a wire-wound resistor is made from a 100-m length of 0.2-mm-diameter aluminum wire. What is its resistance at room temperature? If this resistor is to be used as a thermometer, to what precision must one measure its resistance to determine temperature changes to within 0.2 °C?

• **20.20** Tungsten wire is frequently used for incandescent lamps. If the diameter of the wire is 0.08 mm, what length should the filament be so that the resistance of the lamp is 11 Ω at 1800 K?

• **20.21** Electrical wire is made of either copper or aluminum. With what ratio of wire diameters for the two materials will equal lengths have equal resistances? What is the ratio of the masses of these two types of wire?

• • **20.22** The temperature coefficient of resistivity $\alpha_r = \frac{1}{\rho}\frac{\Delta\rho}{\Delta T}$ equals $\frac{1}{R}\frac{\Delta R}{\Delta T}$ only if the dimensions of the conductor do not change with temperature. Show that if the linear thermal expansion coefficient is α_ℓ and the conductor expands isotropically, then

$$\frac{1}{R}\frac{\Delta R}{\Delta T} = \alpha_r - \alpha_\ell$$

How important numerically is the correction due to thermal expansion in calculating the change in resistance of an aluminum wire?

(Section 20.4)

20.23 Show that I^2R has the dimension of power.

20.24 Estimate the average power consumption in your home. Check your estimate by consulting a recent bill from the local power company.

20.25 A small transistor radio is rated at 0.27 W. What is the current supplied by its 9-V battery?

20.26 When a 6-Ω resistor is connected to an ideal battery, 18 W is dissipated in this resistor. Find the current in this resistor and the voltage drop across it.

• **20.27** Small immersion heaters are widely used to prepare hot drinks. What rating and resistance should such a heater, intended for use in a car (12-V source), have if it is to raise the temperature of 0.18 L of water from room temperature to 90 °C in 3 min? (Neglect heat loss to the surroundings.)

• **20.28** An electric iron is rated at 1500 W. How much current does this appliance draw from a 110-V source? What is the resistance of the heating element? If the heating element is made of Nichrome ribbon 1.6 mm × 0.3 mm in cross section, what must the length of this ribbon be? Nichrome is a nickel-chromium alloy extensively used in toasters, electric space heaters, irons, and the like.

• **20.29** A 1200-W toaster is to be used with a source of 110 V. How much current will this toaster draw and what should its resistance be? Suppose that a portion of the heating coils short together, reducing the resistance to 80 percent of its former value. What will the change be, if any, in its power consumption when connected to 110 V as before?

• **20.30** Resistors used in radios and TV sets are rated according to their ability to dissipate power. What is the maximum voltage that can be applied across a 1/8 watt, 230-Ω resistor without damaging the resistor?

• **20.31** An electric teakettle can bring 0.5 L of water from 20 °C to 100 °C in 2 min when the heating element is connected across a 120-V source. Neglecting heat losses, calculate the resistance of the heating element of this teakettle.

• **20.32** Nichrome ribbon 0.2 mm thick is used to wind a resistive element of a 1500-W heater that is to be used at a rated voltage of 120 V. If the width of the ribbon is 3 mm, how long should the ribbon be?

• • **20.33** Many large electrical systems are water-cooled. A large electromagnet draws 200 A when 100 V is applied across its coils. At what rate should cooling water circulate through the coils so that the temperature rise is no greater than 35 °C at peak current?

• **20.34** A self-cleaning oven has heating coils whose total power consumption is 4 kW. The oven is connected to a 240-V source. What is the total current that flows into the unit during the self-cleaning? If self-cleaning involves the continuous operation of these coils for 2.5 h, what is the cost of the cleaning in a region where the electric rate is 7¢/kW·h?

• **20.35** One of the most common sources of fire is excessive heating of wires carrying greater-than-rated current. What diameter copper wire can carry a current of 20 A and generate no more than 2 W of heat per meter of conduit?

• **20.36** How much energy was stored in the battery of Problem 20.2, if the charging is assumed to be 90 percent efficient?

• • **20.37** If the heater of Problem 20.32 attains a temperature of 500 °C, what is the percent change in resistance and power dissipation relative to the values at 20 °C?

• • **20.38** An incandescent lamp uses a tungsten filament. At its operating temperature of 1730 K, the lamp draws a current of 1.2 A when connected to a 110-V source. Find the current drawn when the lamp is initially connected to the 110-V source, and the power consumed initially and under normal operating conditions.

(Section 20.5)

20.39 A wire of resistance R is cut into four equal lengths. The four sections are then joined at the ends to make one braided conductor one-fourth as long as the original wire. What is the resistance of the braided conductor?

• **20.40** Find the equivalent resistance of 3 Ω, 6 Ω, and 8 Ω connected in parallel.

• **20.41** Determine the resistance between terminals A and B of the combination of resistors shown in Figure 20.22.

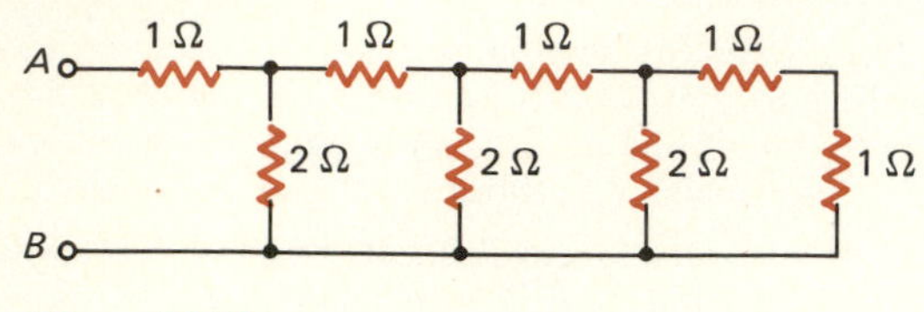

Figure 20.22

• **20.42** How many different resistance values can be obtained using three identical 16-Ω resistors? What is the equivalent resistance of each of the combinations?

• **20.43** Resistors of 2 Ω, 3 Ω, and 6 Ω are connected in parallel. The current through the 6-Ω resistor is 3 A. Find the currents in the other two resistors.

• **20.44** You are given three resistors whose resistances are 2 Ω, 3 Ω, and 4 Ω. How many different values of resistance can be obtained by suitable combinations of these three elements? What are these values?

• **20.45** Two resistors of 8 Ω each are connected in parallel. What should the resistance of a third resistor be so that when it is also connected in parallel with the other two, the total resistance of the combination is 2 Ω?

• • **20.46** Find the equivalent resistances that could replace the networks between the terminals for Figures 20.23(*a*)–(*d*).

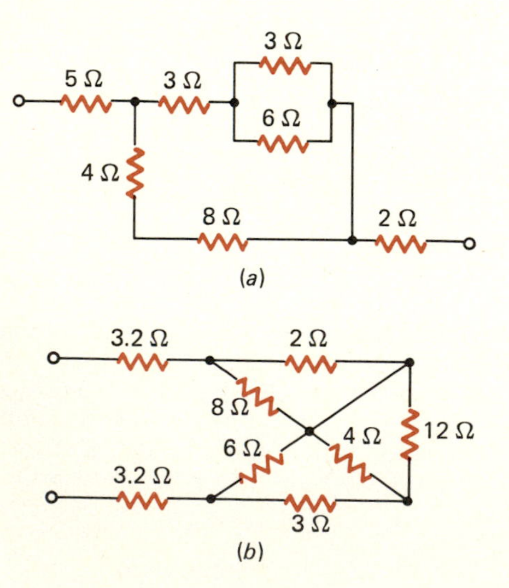

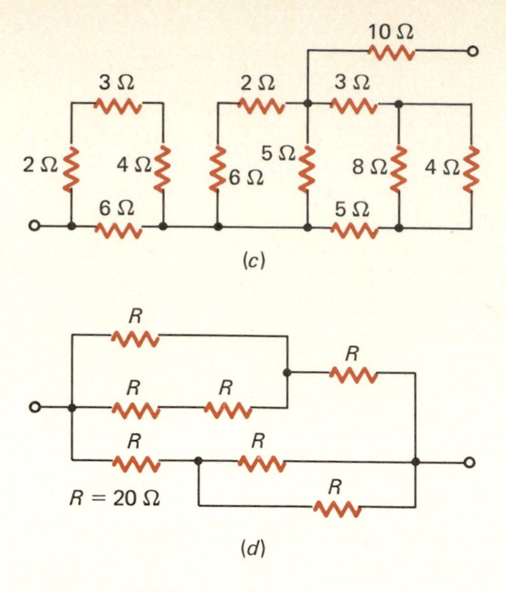

Figure 20.23

• • **20.47** A 12-Ω and an 8-Ω resistor are connected in series. What resistance should a third resistor, connected in parallel with the 12-Ω resistor, have so that the equivalent resistance of the entire combination is 12 Ω?

(Section 20.6)

• **20.48** A 12-V storage battery has an internal resistance of 0.04 Ω. A charging current of 5 A is sent through this battery for 3 h. How much energy was stored in the battery during this time?

• • **20.49** When a 25-Ω resistor is connected across a battery, the current in the circuit is 2 A. If a 100-Ω resistor is placed in parallel with the 25-Ω resistor, the total current drawn from the battery is 2.25 A. Find the emf and the internal resistance of the battery.

• **20.50** A battery has an emf of 45 V and an internal resistance of 2.5 Ω. Find the current in a 20-Ω resistor connected across the battery terminals, and the power dissipated by it.

• **20.51** What is the maximum power that the battery of Problem 20.50 can supply to a load? What should the load resistance be under these conditions?

• **20.52** A variable resistor is connected to a battery, and the power dissipated in this resistor is measured as the value of the resistance is adjusted. When R = 1.8 Ω, this power attains a maximum value of 405 W. Find the emf and the internal resistance of the battery.

• **20.53** A 12-V car battery has an internal resistance of 0.018 Ω. The starter draws a current of 120 A. What then is the potential across the battery terminals?

•21•

Direct-Current Circuits

Across the wires the electric message came:
'He is no better, he is much the same.'

ALFRED AUSTIN, *On the Illness of the Prince of Wales*

21.1 Introduction

Electric circuits in many practical applications often involve a network of resistors and incorporate one or more sources of emf. In the following pages we give a prescription for "solving" such networks, that is, calculating the currents in each of the resistors of the network. The method, which is generally valid and applies to situations far more complicated than any we shall be concerned with, is based on two rules, first clearly stated by Gustav Kirchhoff. These rules are restatements of conservation of charge and of energy, with special reference to electric circuits.

Frequently, a circuit can be solved without recourse to Kirchhoff's rules. This simple situation prevails when the circuit can be reduced to an equivalent resistance in series with a battery and is exemplified by the first of several examples in this chapter. This is followed by a presentation and discussion of Kirchhoff's rules and examples illustrating their applications.

We next turn our attention to the measurement of current, voltage, and resistance, and the limitations encountered in making such measurements. A few networks, known as bridges, that permit more precise determination of voltages or resistances than ordinary meters are briefly described.

21.2 Solution of Simple Direct-Current Circuits

Suppose a 22-V battery, with internal resistance of 0.5 Ω, is connected to the resistance network shown in Figure 21.1(*a*). Some current will then flow in each of the seven resistors of the network. The question we want to answer is, What are the magnitudes and directions of these currents?

In this instance, the resistance network can be reduced to a single equivalent resistance using Equations (20.15) and (20.21). Therefore, the circuit can be solved simply by applying Ohm's law. Let us see how that procedure works for the circuit of Figure 21.1(*a*). The resistive network of Figure 21.1(*a*) can be simplified in several steps, outlined by the diagrams in Figure 21.1, to a single resistor of 5 Ω. Thus, as far as the battery is concerned, the seven resistors between terminals *A* and *B* of Figure 21.1(*a*) are exactly equivalent to the 5-Ω resistor of Figure 21.1(*e*). The current drawn from the battery is, by Ohm's law,

$$I_1 = \frac{22\text{ V}}{(5.0 + 0.5)\ \Omega} = \frac{22\text{ V}}{5.5\ \Omega} = 4\text{ A}$$

This is the current in the 1.4-Ω and the 3.6-Ω equivalent resistor of Figure 21.1(*d*). It follows that the voltage drop across the 3.6-Ω resistor of Figure 21.1(*d*) is

$$V_{3.6} = (4\text{ A})(3.6\ \Omega) = 14.4\text{ V}$$

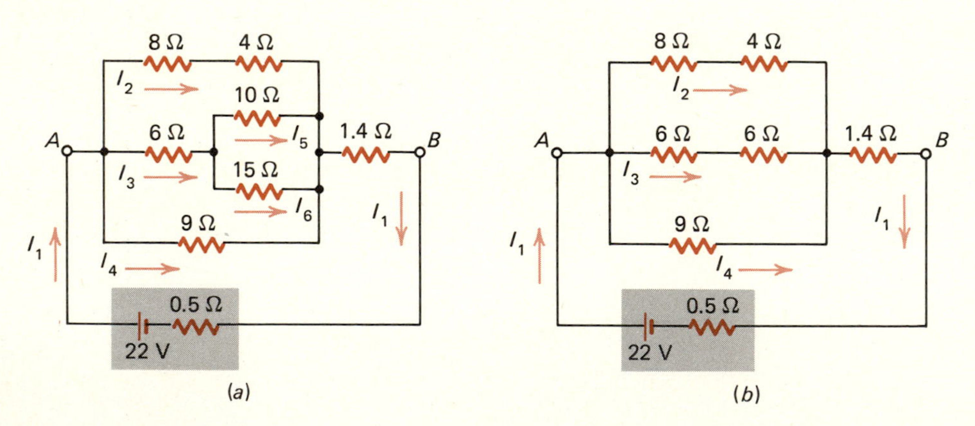

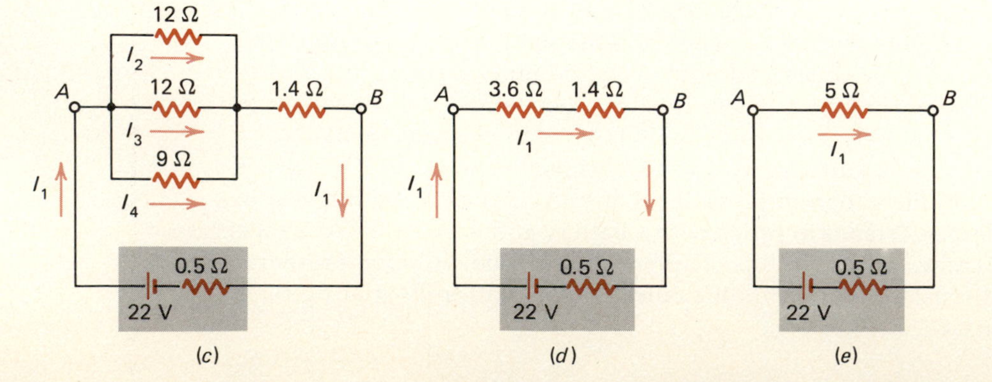

Figure 21.1 *A resistive network connected to a source of emf. The currents in each resistor can be calculated by first reducing the network to its equivalent resistance, as indicated in the sequence (a)–(e), and then working backward to the original network.*

This is the potential drop across each of the three resistors in the parallel combination of Figure 21.1(c). Consequently, the currents in the two 12-Ω resistors and the 9-Ω resistor of Figure 21.1(c) are

$$I_{12} = \frac{V}{R} = \frac{14.4\ \text{V}}{12\ \Omega} = 1.2\ \text{A}$$

$$I_9 = \frac{14.4\ \text{V}}{9\ \Omega} = 1.6\ \text{A}$$

These are the currents I_2, I_3, and I_4 in Figures 21.1(a) and (b). The current of 1.2 A in the 6-Ω resistor of Figure 21.1(b), which is the equivalent of the parallel combination of 15-Ω and 10-Ω resistors of Figure 21.1(a), results in a voltage drop across this combination of

$$V_6 = (1.2\ \text{A})(6\ \Omega) = 7.2\ \text{V}$$

That voltage drop must then appear across the 15-Ω and the 10-Ω resistors of the original circuit. Consequently, the currents in these resistors, designated I_5 and I_6 in Figure 21.1(a), are

$$I_5 = \frac{7.2\ \text{V}}{10\ \Omega} = 0.72\ \text{A}$$

$$I_6 = \frac{7.2\ \text{V}}{15\ \Omega} = 0.48\ \text{A}$$

Identification of all the unknown currents completes the solution of this circuit.

21.3 Kirchhoff's Rules

Not all circuits are amenable to solution by reduction to an equivalent resistance in series with a source of emf. Though it looks deceptively simple, the circuit of Figure 21.2, for example, cannot be further simplified. The presence of batteries in two of the three branches precludes any parallel combinations that would otherwise be possible.

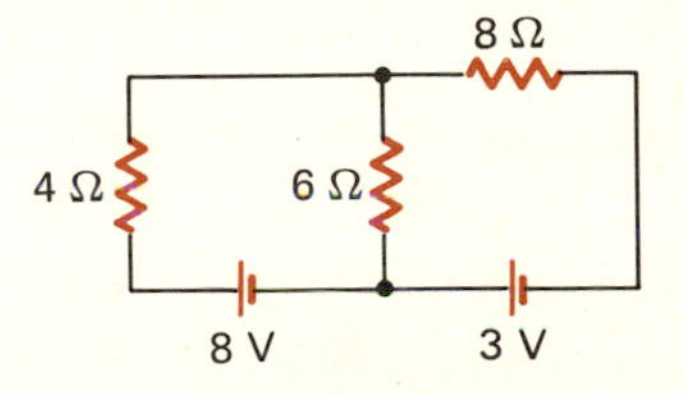

Figure 21.2 *A relatively simple circuit which cannot be reduced to an equivalent resistance in series with an emf.*

Another circuit that cannot be reduced to an equivalent resistance by series and parallel combinations, though it contains only one source of emf, is that of Figure 21.3. This is one of the simplest bridge circuits, and we shall return to it in a later section.

To solve for the currents in the branches of the circuits of Figure 21.2 and 21.3, one turns to the rules named after Gustav Kirchhoff. Kirchhoff's rule I (KR-I) states:

The net current flowing into a junction must vanish. (KR-I)

An alternative, equivalent statement is that the current flowing into a junction must equal the current flowing out of that junction. The mathematical formalism for KR-I is

$$\Sigma I = 0 \quad \textbf{(21.1)}$$

where currents directed toward the junction (sometimes called a branch point) are taken as positive, and currents directed away from the junction are taken as negative.

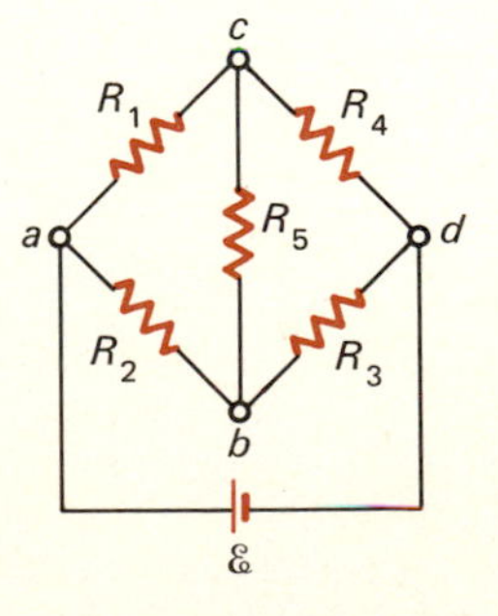

Figure 21.3 *A resistive network connected to a single source of emf. An equivalent resistance cannot be found using the techniques of series and parallel combinations.*

The second of Kirchhoff's rules (KR-II) states:

The algebraic sum of the emf's around any closed loop must equal the algebraic sum of the potential drops (IR drops) around the same closed loop. (KR-II)

In mathematical terms, KR-II is written

$$\Sigma\,\mathcal{E} = \Sigma\, IR \qquad \textbf{(21.2)}$$

The first rule is a statement of continuity of charge flow. If this rule were violated, either charge would have to accumulate at a junction, thereby raising its potential to large values, or the law of conservation of charge, believed to be a fundamental law of nature, would be violated. The second rule rephrases conservation of energy in the context of charge flow. In traversing a closed loop, a positive charge may gain electrostatic energy $q\mathcal{E}$ from a source of emf and may lose electrostatic energy in moving through a resistor with potential drop IR between its terminals. When this charge has returned to its starting point, however, the total energy change must be zero. Put another way, every point in a circuit is at a well-defined potential relative to some arbitrary reference; therefore, when a complete loop is traversed, the total change in potential must be zero.

The application of Kirchhoff's rules is best illustrated with an example. We shall solve for the currents in the circuit of Figure 21.2. First, however, we give a few general guidelines you are advised to adopt in these cases.

1. Draw a clear circuit diagram.

2. Indicate on this circuit diagram all known quantities: emf's (including the sign), resistance values, voltage drops (including directions), currents (including directions).

3. Select suitable symbols for the unknown quantities to be calculated. Show these on the circuit diagram. Generally, directions of unknown currents and potential drops must be assumed.

It does not matter what direction is selected. If, for example, the true direction of a current is opposite to what has been assumed, the calculated current will have a negative value. *It is, however, imperative that once the directions have been assigned, these directions are maintained throughout the solution.*

4. If the circuit permits reduction by series and parallel combination of resistors, it should be so simplified. It is generally sound practice to redraw the circuit as these reductions are carried forward.

5. If the circuit cannot be further simplified by such combination, or if such combination leads only to trivial reductions in the circuit's complexity, the linear equations corresponding to application of KR-I and KR-II to the circuit should be written. The voltage drops, emf's, and current equations must be consistent with the assumed directions for the unknown quantities.

6. Solve the set of linear equations.

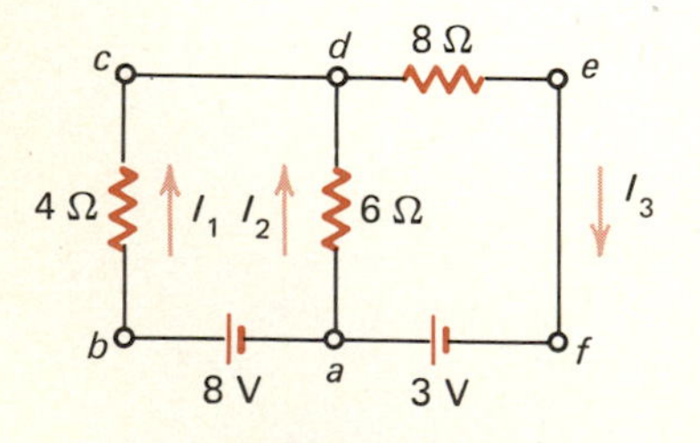

Figure 21.4 A simple circuit that cannot be solved without application of Kirchhoff's rules. The current directions shown are those assumed for the purpose of writing the circuit equations. They are not necessarily the correct directions of the true currents in the circuit.

● ***Example 21.1*** We now apply Kirchhoff's rules by solving the circuit of Figure 21.2, which we have redrawn here for ready reference. The three unknown currents, I_1, I_2, and I_3, are shown on this diagram together with arrows indicating their assumed directions.

We begin by writing equations for the unknown currents based on KR-I. In the circuit of Figure 21.4, there are two junction points, at a and d. Currents I_1 and I_2 are directed toward d, and I_3 is directed away from that junction. Hence,

$$I_1 + I_2 - I_3 = 0 \qquad \textbf{(21.3)}$$

We could write a corresponding relation for junction a but the equation would be exactly the same as (21.3) except for a sign reversal of every term. That equation would therefore convey no additional information.

We next apply KR-II. Since we have three unknown quantities, the three currents, we shall need three independent equations to effect a solution. One of these equations is Equation (21.3). We must therefore write two equations based on KR-II before we can complete the solution.

The first loop we consider is the loop a-b-c-d-a. Staring at a, we have an emf of 8 V, and this is the only emf in the loop. Following along the path of this loop, we have the following IR drops: a voltage drop of $4I_1$ from b to c, no voltage drop from c to d, and finally a voltage drop of $-6I_2$ from d to a. *Note the negative sign in this voltage drop.* It appears here because the assumed direction of current is opposite to the direction in which we are traversing the loop. If I_2 has this direction, point d will be at a lower potential than a, and going from d to a results in an *increase* of potential, or a *negative potential drop.* Thus, KR-II gives

$$8 = 4I_1 - 6I_2 \quad \textbf{(21.4)}$$

The third equation is obtained by considering the loop a-d-e-f-a. This loop contains one emf of 3 V, and traversing the loop in the stated direction, we are also going in the direction of the assumed currents. The equation for this loop is therefore

$$3 = 6I_2 + 8I_3 \quad \textbf{(21.5)}$$

There are certain prescriptions for solving N linear equations in N unknowns. With only three unknowns, the simplest procedure is to use the current equation to eliminate one of the unknowns from the remaining equations; this reduces the problem to solving two equations in two unknowns, which is readily accomplished (see Appendix A).

From (21.3), we have

$$I_3 = I_1 + I_2 \quad \textbf{(21.3a)}$$

and substituting into (21.5) gives

$$3 = 6I_2 + 8I_1 + 8I_2 = 8I_1 + 14I_2 \quad \textbf{(21.5a)}$$

We can now eliminate the unknown I_1 by multiplying Equation (21.4) by 2 and subtracting the resulting equation from (21.5a). Following these steps, one obtains

$$-13 = 26I_2$$

$$I_2 = -0.5 \text{ A}$$

The negative sign tells us that the direction of the current is opposite to that assumed; the current in the 6-Ω resistor flows from d to a, not as shown in Figure 21.4.

Next we find I_1, using either (21.5a) or (21.4), with the result

$$I_1 = 1.25 \text{ A}$$

and obtain I_3 from (21.3):

$$I_3 = 1.25 \text{ A} - 0.5 \text{ A} = 0.75 \text{ A}$$

Lastly, it is wise to check the solution by inserting the calculated currents into one of the original loop equations, for instance, (21.4). ●

In Example 21.1, we could write only one independent current summation equation although the circuit had two junctions. We then wrote two loop equations to get the necessary three independent equations for solving the problem. The question naturally arises, How many current and loop equations can be written? The answer is that the number of independent current equations is one fewer than the number of junctions. Having written these equations based on KR-I, one must then write enough loop equations so that the total number of linear relations between the unknowns equals the number of unknowns.

For example, the bridge circuit of Figure 21.3 has four junctions, *a*, *b*, *c*, and *d*. If it is assumed that the resistance values are given and the emf of the battery is known, the number of unknown currents is six, namely, the currents in each of the five resistors and the current through the battery. One would therefore write $4 - 1 = 3$ current equations and $6 - 3 = 3$ loop equations, giving the six equations needed to determine the values of the six unknowns. A simple-looking circuit may not be so simple to solve.

● ***Example 21.2*** The circuit of this example is very much like the one of Figure 21.4. The important difference between this example and the previous is that, in the present instance, some currents, potential drops, and power dissipations are known while some emf's and resistances are among the unknown parameters. The problem is to find the values of the unknown resistances and of the emf of the second battery.

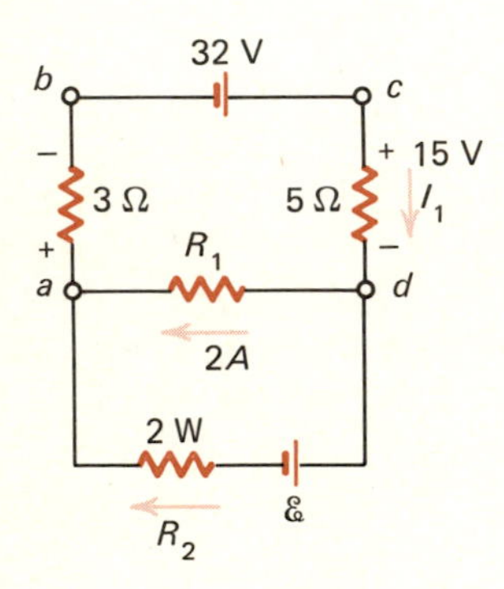

Figure 21.5 Although this circuit is similar to the one in Figure 21.4, we can determine the values of the unknown quantities without resort to Kirchhoff's rules. The solution is given in the text.

In this case one could start by writing the various KR-I and KR-II equations, but a better procedure is to look for a circuit element for which more than one parameter is given. If there is such an element, the third of the triplet of current-voltage-resistance is quickly determined, and one can then proceed from that point.

Since the potential drop across the 5-Ω resistor is 15 V with polarity as shown, we can immediately conclude that the current flowing in this resistor is $I_1 = (15\text{ V})/(5\ \Omega) = 3$ A. It is also the current in the 3-Ω resistor and through the 32-V battery.

Next, summing the potential changes from *a-b-c-d*, namely,

$$-(3\text{ A})(3\ \Omega) + 32\text{ V} - 15\text{ V} = 8\text{ V} = (2\text{ A})R_1$$

we see that R_1 must have the value 4 Ω.

Since 3 A enters the junction at *d* of which 2 A flows through the resistor $R_1 = 4\ \Omega$, a current of 1 A must flow through the battery of emf $\mathcal{E}$ and the unknown resistor R_2 in the direction shown; that is, the 32-V battery is effectively charging the unknown battery with a charging current of 1 A. But we know that this current of 1 A dissipates 2 W of power in the resistor R_2; consequently, R_2 must have a resistance of 2 Ω.

Finally, we can determine the unknown emf by recalling that the voltage V_{ad} is 8 V. That must also be the voltage across the resistor R_2, namely 2 V, plus the emf $\mathcal{E}$ of the battery. Thus, $\mathcal{E} = 6$ V. ●

21.4 Measurement of Current and Voltage

Knowing how to calculate currents, voltages, resistances, and power dissipation in various circuit components often has great practical value, allowing, for instance, the identification of a faulty circuit component. Before one can put this knowledge to good use, one must know how to measure voltages and currents. The devices used to measure these are

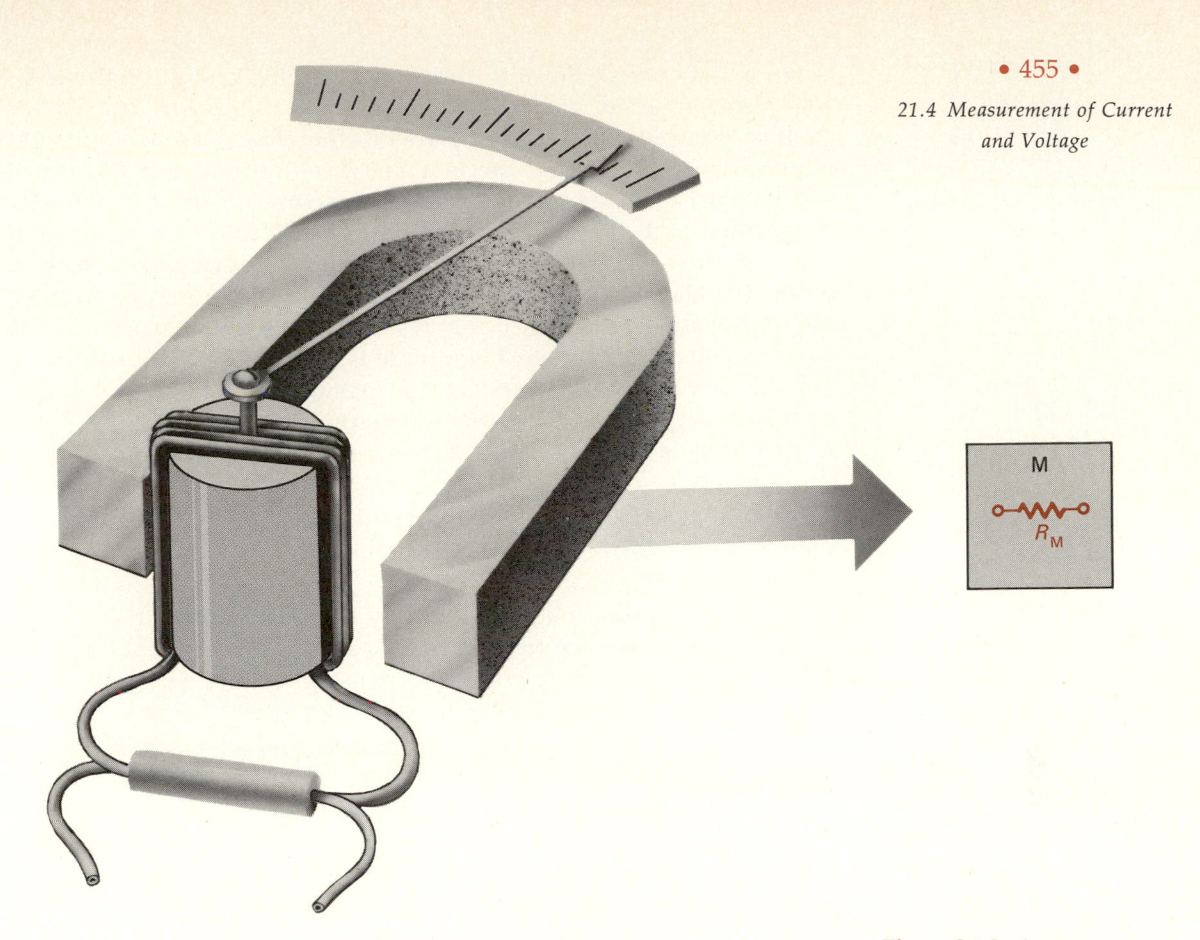

Figure 21.6 A meter movement may be thought of as a device that has some resistance R_M and an indicator (needle) that deflects full scale when a certain current flows through the movement. Generally, the current needed for full-scale deflection of the meter movement is quite small, typically between a few microamperes and a few tenths of a milliampere.

known as *voltmeters* and *ammeters*. Later we shall examine the fundamental operating principles of such instruments in detail. Here we simply assume that one can construct a meter that has the following properties. There are two terminals on the meter, and between those terminals is a resistance R_M (which may be about 100 Ω). If a small current I_M passes through this resistance, a needle on the meter shows full-scale deflection.

In the following pages we shall see how this basic instrument, which is referred to as a *galvanometer*, can be incorporated into a simple network and used to measure a wide range of currents and voltages. We shall also examine the effect of introducing voltmeters and ammeters into a circuit. It is often tacitly assumed that currents and voltages registered by meters are the currents that flow and voltages that appear in the circuit in the absence of these meters. We shall see that this is not so and shall examine the conditions that must be satisfied if the placement of meters in a circuit is to have minimal disturbing influence.

21.4 (a) Ammeters

We assume that we have a galvanometer at hand with the following characteristics. The resistance of the meter is 100 Ω, and full-scale deflection requires a current through the galvanometer of 2×10^{-3} A. As it stands, this meter can be used as a milliammeter with a full-scale reading of 2 mA, or as a voltmeter with full-scale reading of 0.2 V, since this is the potential difference that will appear across the meter terminals when full-scale current flows through the meter.

Suppose, however, that we need to measure much larger currents. Let's assume that we want a meter that will read full-scale deflection

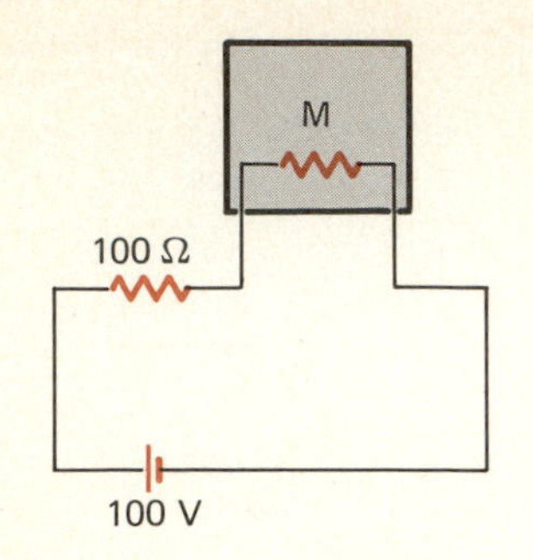

Figure 21.7 *If a meter movement with full-scale current rating of a fraction of a milliampere is connected directly into a circuit in which a much larger current flows, the meter will be badly damaged.*

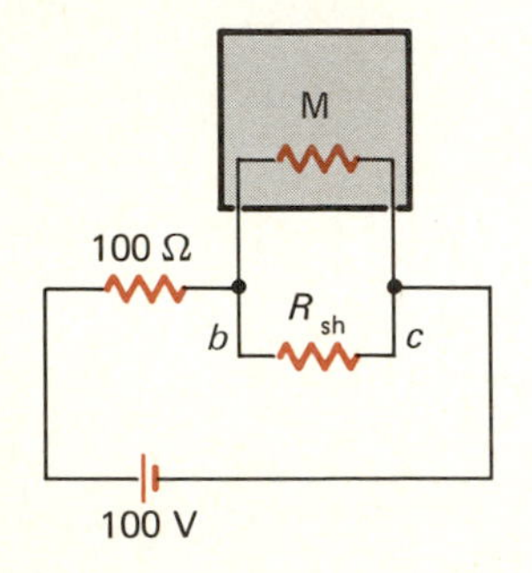

Figure 21.8 *To avoid damage to the meter movement and yet make it possible to use this movement to measure large currents, one must place a shunt R_{sh} in parallel with the meter movement. This shunt resistance is generally small, and through this shunt flows most of the current in the circuit.*

when a 1-A current flows in the circuit. Such a circuit might well be a 100-V emf in series with a 100-Ω load resistor.

If we simply placed this meter into the circuit, as in Figure 21.7, it would have the following effects. First, the additional resistance of 100 Ω of the meter movement would reduce the current from 1 A to 0.5 A. But more important, that current of 500 mA is 250 times the rated full-scale current of the meter, and if allowed to flow through this instrument, it will surely damage the movement (most probably cause overheating and melting of the fine wires in the device). So, is the answer a trip to the store to buy a 1-A ammeter? No; there is a much less expensive solution. We simply place a small resistance in parallel with the meter movement, so that when 1 A flows into the junction b of Figure 21.8, 998 mA flows through this so-called *shunt resistor* and only 2 mA through the meter movement.

Now that we know how the scheme should work, all we need do is determine the correct value for this shunt resistance R_{sh}. We know that the meter movement gives full-scale deflection when 0.2 V appears across its terminals, that is, when 2 mA passes through its 100-Ω resistance. Since the same voltage must also appear across the shunt, but for a shunt current of 998 mA, the shunt resistance must be

$$R_{sh} = \frac{V_{sh}}{I_{sh}} = \frac{0.2\ \text{V}}{(998 \times 10^{-3}\ \text{A})} = 0.2004\ \Omega \simeq \frac{0.2\ \text{V}}{1\ \text{A}} = 0.2\ \Omega$$

If we want to change to an ammeter with full-scale deflection for 50 mA, the proper shunt resistor would be one with resistance of $R_{sh} = 0.2/0.048 = 4.17\ \Omega$; note that in this case, with the total current only 25 times the meter current (not 500 times, as above), the true shunt current cannot be approximated by the total current.

21.4 (b) Voltmeters

Suppose we wanted to use this galvanometer as part of a voltmeter capable of measuring voltages up to 100 V. If we simply connected the basic meter movement across a 100-V battery, the result would also be disastrous. A current of 1 A would flow through a device designed to carry a current of 2 mA; the needle would swing violently across the meter face and wrap itself about the post at the end of the scale, and the very large current would heat and melt the fine wire of the meter unless corrective measures were taken quickly.

In this case a shunt resistor is of no value. All a shunt would do would be to increase the total current drawn from the battery without changing the overload current in the meter. The meter would still burn out.

What must be done is to insert a *resistance in series with the meter* so that when a voltage of 100 V is impressed across the series combination of meter and series resistor, the current in the meter is the rated full-scale current of 2 mA. Referring to Figure 21.9, we see that we must have

$$R_M + R_{se} = \frac{100\ \text{V}}{(2 \times 10^{-3}\ \text{A})} = 50{,}000\ \Omega$$

The series resistor must therefore have the value

$$R_{se} = 49{,}900\ \Omega$$

The multimeters in wide use consist of a meter movement and a switching circuit whereby different shunt resistors can be placed in parallel with the meter for current measurement, and different resistors can be placed in series with the meter movement for voltage measurement.

21.4 (c) Influence of Meters in a Circuit

Both the ammeter and the voltmeter dissipate some, though relatively little power. That power must derive from some source, and that source is the circuit whose electrical properties are being measured. Consequently, the presence of meters in a circuit inevitably disturbs the circuit, and the meter readings may not represent the currents and voltages without the meters. To illustrate the dilemma, we consider the following problem. Given a resistor R, find the value of its resistance.

We assume that we have a suitable ammeter and voltmeter at hand, as well as a battery and a second resistor to limit the current drawn from the battery.* We can now construct a simple circuit, such as that of Figure 21.10(a), which includes both ammeter and voltmeter; one measures the current in the unknown resistor, the other the voltage drop across it. Suppose the meter readings are I_M and V_M. Is the resistance of the unknown resistor really given by $R = V_M/I_M$?

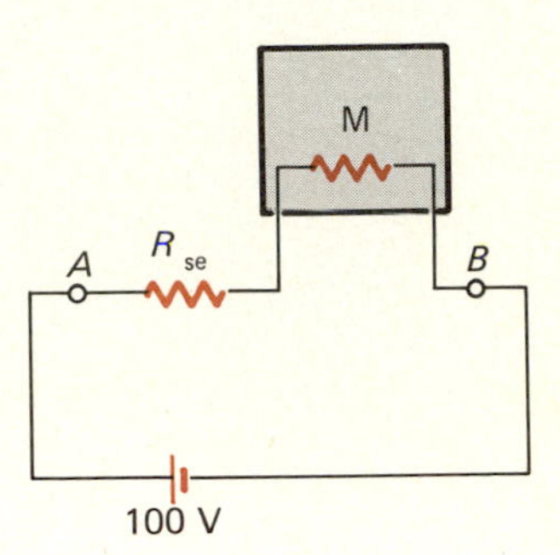

Figure 21.9 *A meter movement can also be used to measure voltages. In that case, a large resistance R_{se} is placed in series with the meter movement, so that when a fairly large voltage is applied across the terminals A–B, the current that flows through the meter is small enough to give a suitable needle deflection.*

In the circuit of Figure 21.10(a), the ammeter reads the true current in the unknown resistor; however, the voltmeter reads the potential drop *across the series combination of the resistor R and the ammeter.* Since the ammeter has a finite resistance (approximately equal to that of its shunt resistance), the voltmeter reading will be greater than the voltage across the resistor R alone. Hence, the ratio V_M/I_M will be greater than the true value of R.

Why not arrange the circuit so that the voltmeter does read the true voltage across the unknown resistor, as in the circuit of Figure 21.10(b)? In that case, the ammeter reading gives the current through the resistor *plus the current through the voltmeter.* Hence, the current read by that meter is not the true current through the unknown. For this circuit, the ratio V_M/I_M will be smaller than R.

From these simple examples we can draw two important conclusions. The first applies specifically to voltmeters and ammeters and the most desirable characteristics for such instruments. The ideal voltmeter is one with infinite resistance (drawing no current from the circuit); the ideal ammeter is one with zero resistance (no voltage across the meter itself). These ideals cannot be achieved, but they can be closely approached. As a general rule, then, we can say that ammeters should have a very low resistance, and voltmeters should have a very high resistance.

The second important conclusion is that performing a measurement on a circuit disturbs the circuit itself. This is but a special example of a

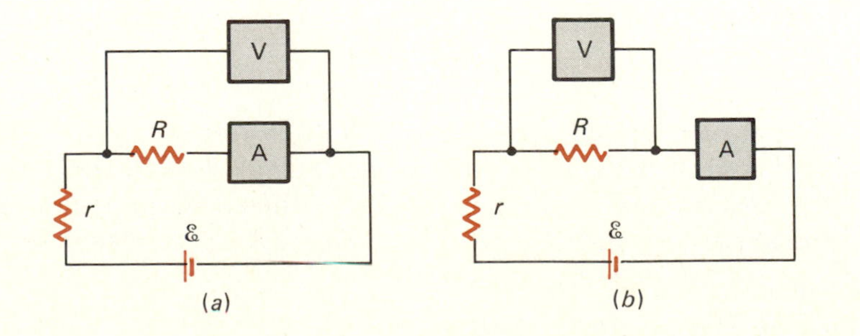

Figure 21.10 *A battery, a current-limiting resistor r, a voltmeter, and an ammeter may be used to determine the resistance R of an unknown resistor. However, the ratio of the voltmeter-to-ammeter readings is not equal to the value of R. (a) In this circuit, the voltmeter reads the potential across the series combination of R and the ammeter. (b) In this circuit, the ammeter reads the sum of the currents flowing in R and in the voltmeter.*

* The use of a current-limiting resistor is essential to avoid damaging the ammeter. If the resistance of the unknown resistor is very low, and this resistor is connected directly across the terminals of the battery, a correspondingly very large current will flow in the circuit. That current could well exceed the current rating of the ammeter. Moreover, this current could conceivably be so great as to burn out the unknown resistor, bringing the experiment to a premature but embarrassing conclusion.

basic concept in physics that has troubled philosophers for centuries. Every process of measurement on a system modifies that system so that the result of the measurement does not truly indicate the state of the system in the absence of the measuring instrument. But since we have no way of obtaining data on a system without performing some measurements, what this appears to tell us is that our ability to probe the structure and constitution of a physical system is limited. The limitation exists no matter what kind of probe we use, whether electrostatic, electrodynamic, mechanical, or optical (light waves), and no matter how sophisticated the instrumentation.

● ***Example 21.3*** A voltmeter with internal resistance of 10,000 Ω and an ammeter with internal resistance of 1.0 Ω are used to measure the resistance of an unknown resistor. The instruments and the resistor are connected as in Figure 21.10(*a*). Under those conditions, the voltmeter and ammeter readings are 48 V and 8 A. Find the true value of the unknown resistance.

The potential difference of 48 V appears across the unknown resistor and the ammeter, connected in series. Hence,

$$48\text{ V} = (8\text{ A})[(R + 1.0)\Omega]$$
$$R(8\text{ A}) = (48 - 8)\text{ V}$$
$$R = \frac{40\text{ V}}{8\text{ A}} = 5\ \Omega$$

If we had simply taken the ratio of the voltmeter and ammeter readings, we should have arrived at an incorrect result, $R = 6\ \Omega$.

A better circuit in this case is that of Figure 21.10(*b*). If we adjust the resistor r until the ammeter again reads 8 A, the reading of the voltmeter can be found from the following argument.

The ammeter current branches into the current I_R through the resistor and the current I_V through the voltmeter. The ratio of the two currents is the inverse of the ratio of the resistances of the two paths; that is,

$$\frac{I_V}{I_R} = \frac{R}{R_V} = \frac{5}{10^4} \qquad \text{and} \qquad I_V + I_R = I_A = 8\text{ A}$$

Thus,

$$I_R(1 + 5 \times 10^{-4}) = 8\text{ A}$$
$$I_R = 7.996\text{ A}$$

and

$$V_R = RI_R = 39.98\text{ V} = V_V$$

Note that with this arrangement, the ratio $V_V/I_A = 4.9975\ \Omega$ is very nearly the correct value of 5.0 Ω. It is worth the effort to analyze these two situations carefully and understand why the two different voltmeter and ammeter arrangements give such widely divergent results. ●

*21.5 The Wheatstone Bridge

The importance of bridge circuits is that they can be used with null instruments; that is, when the bridge is properly adjusted, the sensing instrument, whether it is a voltmeter or an ammeter, will read zero. But when a meter reads zero, no current flows through the meter movement, and the meter therefore dissipates no power.

The Wheatstone bridge is the arrangement of resistors presented earlier, in Figure 21.3, as an example of a relatively simple circuit not amenable to reduction to an equivalent resistor. The circuit is reproduced in Figure 21.11.

Suppose the switch S is open. In that case the two branches, *a-b-c* and *a-d-c* with resistors $R_1 + R_x$ and $R_2 + R_3$, respectively, are in parallel. The current flowing in R_1 and R_x is

$$I_1 = \frac{\mathcal{E}}{R_1 + R_x} \tag{21.6}$$

and that in the other branch is

$$I_2 = \frac{\mathcal{E}}{R_2 + R_3} \tag{21.7}$$

If $I_1 R_1 = I_2 R_2$, the potentials at b and d will be identical and the potential difference $V_{bd} = 0$. If we then close switch S, no current will flow through the ammeter A; that is, the circuit is unaffected by closing or opening the switch. Under these conditions, the Wheatstone bridge is said to be in balance.

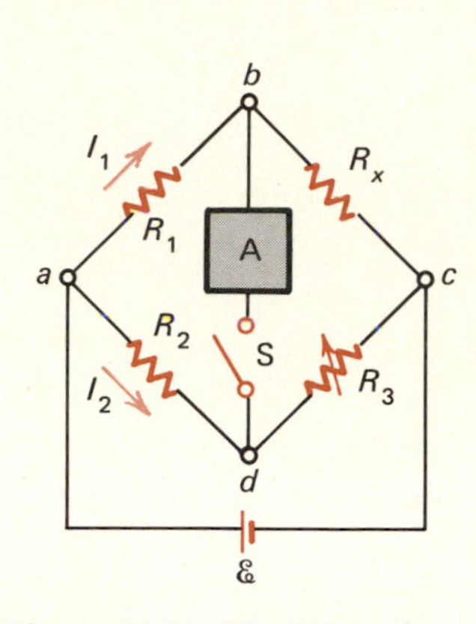

Figure 21.11 The Wheatstone bridge. This bridge circuit can be used to determine the value of an unknown resistance R_x. Here R_1 and R_2 are fixed resistors of known value, and R_3 is a precisely calibrated adjustable resistor. The latter is adjusted until one observes no deflection of the galvanometer when switch S is closed. The bridge is then said to be balanced, and R_x can be determined from the condition $R_x : R_3 = R_1 : R_2$.

The balance condition is now easily expressed in terms of the four resistances. We have

$$\left(\frac{\mathcal{E}}{R_1 + R_x}\right) R_1 = \left(\frac{\mathcal{E}}{R_2 + R_3}\right) R_2 \tag{21.8}$$

But if $V_{ab} = V_{ad}$, then $V_{bc} = V_{dc}$; i.e., $I_1 R_x = I_2 R_3$, and therefore,

$$\left(\frac{\mathcal{E}}{R_1 + R_x}\right) R_x = \left(\frac{\mathcal{E}}{R_2 + R_3}\right) R_3 \tag{21.9}$$

If we divide Equation (21.9) by (21.8), we obtain the desired result

$$\frac{R_x}{R_1} = \frac{R_3}{R_2} \qquad R_x = \frac{R_1 R_3}{R_2} \tag{21.10}$$

If R_x is an unknown resistance, its value can be determined by placing it in this bridge circuit and adjusting one of the resistances of the other arms of the bridge until balance is achieved. When the meter needle shows no movement on closing and opening switch S, the bridge is balanced and the unknown resistance can be calculated from Equation (21.10).

Since at balance no current flows in the measuring instrument, our earlier assertion that every measurement disturbs the system being measured seems to be contradicted. There is no contradiction here, however, because to determine the condition of balance, one must move away from balance by increasing and decreasing the resistance of R_3, while observing the meter deflection. True balance can only be approximated, because even the best meter requires some small, nonzero current to give an observable deflection; that is, even at an apparent balance, a minute current, below the limit of meter sensitivity, may yet flow in the galvanometer.

Summary

Given the values of resistances and of the emf's in a direct current (dc) network, one can determine the currents in those resistors by solving a set of linear equations deduced by applying *Kirchhoff's rules*, KR-I and

KR-II, to the network. The first of these rules states:

The net current flowing into a junction must vanish. (KR-I)

The second states:

The algebraic sum of the emf's around any closed loop must equal the algebraic sum of the potential drops (IR drops) around the same closed loop. (KR-II)

Circuits can always be solved by applying these two rules.

Currents and voltages are measured with *ammeters* and *voltmeters.* The basic meter movement for both instruments is the same; it is a device that registers full-scale deflection when a relatively small current passes through the instrument, which has a moderate internal resistance, of the order of 100 Ω.

To convert such an instrument to an *ammeter,* a suitable *shunt resistor* is placed in parallel with the meter movement. To convert the same instrument to a *voltmeter,* a suitable *series resistor* is connected in series with the meter movement.

The presence of ammeters and voltmeters in a circuit disturbs the circuit slightly.

Multiple Choice Questions

21.1 For the circuit of Figure 21.12, which of the following equations is correct?

(a) $I_3 + I_6 = I_5$
(b) $I_2 + I_1 = I_3$
(c) $I_1 + I_4 = I_5$
(d) $I_1 + I_2 = I_5 + I_6$

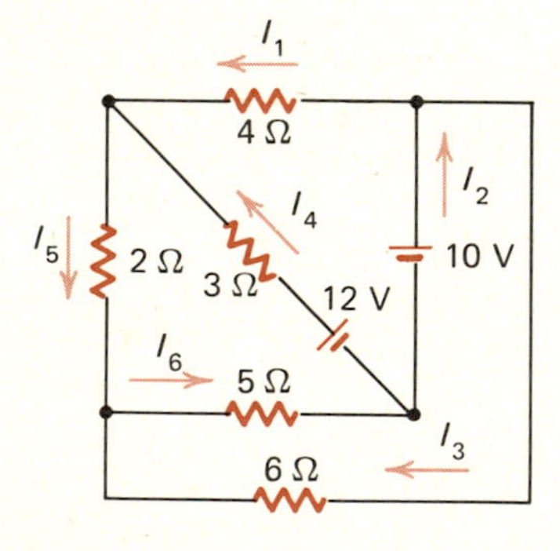

Figure 21.12

21.2 For the circuit of Figure 21.12, which of the following equations is correct?

(a) $4I_1 + 2I_5 + 6I_3 = 10$
(b) $3I_4 - 4I_1 = 2$
(c) $3I_4 + 2I_5 - 5I_6 = 12$
(d) $4I_1 + 2I_5 + 6I_3 = 0$

21.3 A 10-mA ammeter with internal resistance of 2 Ω is to be used to read currents up to 50 mA. One should

(a) use a shunt resistor whose resistance is less than 2 Ω.
(b) use a shunt resistor whose resistance is greater than 2 Ω.
(c) use a series resistor whose resistance is greater than 2 Ω.
(d) use a series resistor whose resistance is less than 2 Ω.

21.4 The 10-mA ammeter of Question 21.3 is to be used as a voltmeter that reads full scale for a voltage of 25 V across its terminals. One should use a

(a) shunt resistor whose resistance is less than 2 Ω.
(b) shunt resistor whose resistance is greater than 2 Ω.
(c) series resistor whose resistance is greater than 2 Ω.
(d) series resistor whose resistance is less than 2 Ω.

21.5 A meter has an internal resistance R and deflects full scale with a voltage of 50 mV across its terminals. To convert this to a 20 V voltmeter, one should connect

(a) a resistor much smaller than R in parallel with the meter.
(b) a resistor much larger than R in parallel with the meter.
(c) a resistor much smaller than R in series with the meter.
(d) a resistor much larger than R in series with the meter.

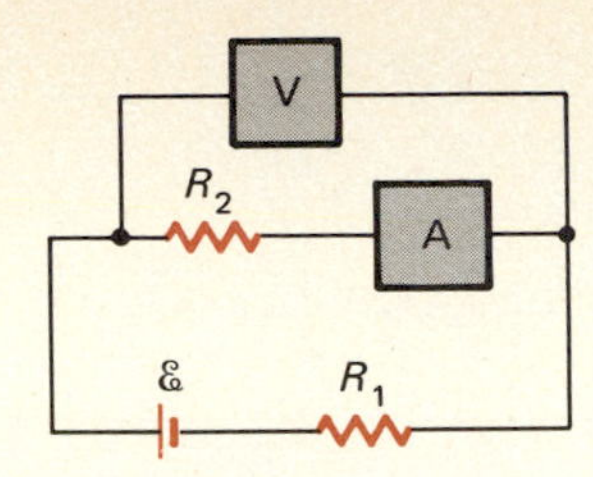

Figure 21.13

21.6 If V_M and I_M are the readings of the voltmeter and ammeter of the circuit of Figure 21.13, the value of the resistance of resistor R_2

(a) equals V_M/I_M.
(b) is greater than V_M/I_M.
(c) is less than V_M/I_M.
(d) can be any of the above, depending on the ratio R_2/R_1.

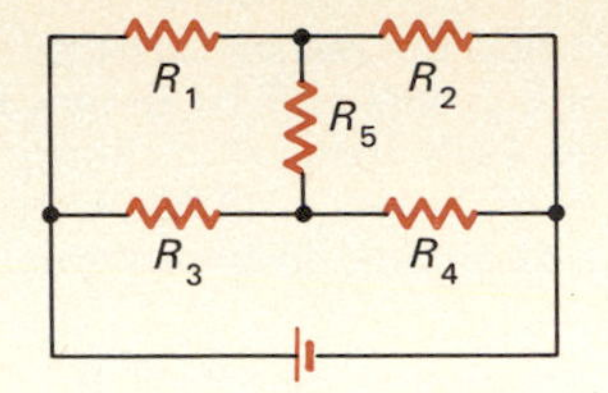

Figure 21.14

21.7 In the circuit of Figure 21.14, the current in the resistor R_5 will be zero if

(a) $(R_1/R_2) = (R_3/R_4)$
(b) $(R_1/R_2) = (R_4/R_3)$
(c) $R_1R_2 = R_3R_4$
(d) (a) and (c) are both correct.

Problems

(Sections 21.2–21.3)

• **21.1** A rheostat is a wire-wound resistor with a slide-wire contact. Such a rheostat can be used as a *voltage divider*, as indicated in Figure 21.15. That is, the voltage across the resistor R_L depends on the position of the sliding contact on the rheostat. Suppose that the rheostat is made of wire of uniform resistance per unit length, and that the resistance between the terminals of the rheostat resistor is 10,000 Ω. Find the potential across the load resistor R_L when the sliding contact is at the point b, where $a\text{-}b = (a\text{-}c)/3$, if $R_L = 2000$ Ω and if $R_L = 200{,}000$ Ω. What general conclusions, if any, can you draw from these results?

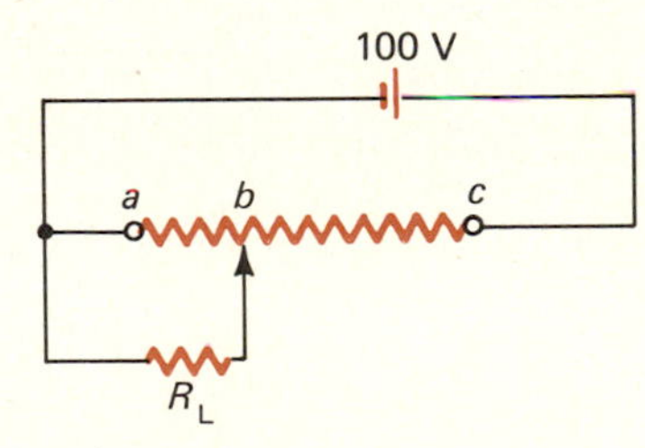

Figure 21.15

• **21.2** The resistors of the simple network of Figure 21.16 are identical, and each can dissipate no more than 20 W. What is the maximum power that can be dissipated by this network?

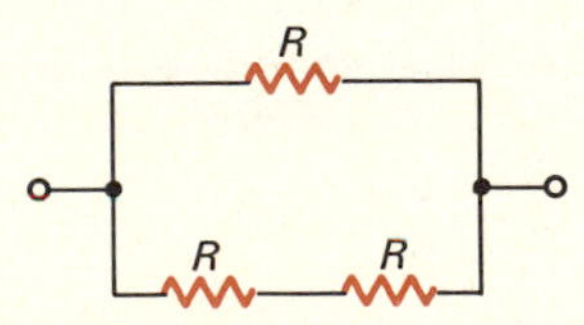

Figure 21.16

• **21.3** For the circuit of Figure 21.17, find the current in the 5-Ω resistor, the power dissipated by the 4-Ω resistor, and the voltage drop across the 2-Ω resistor.

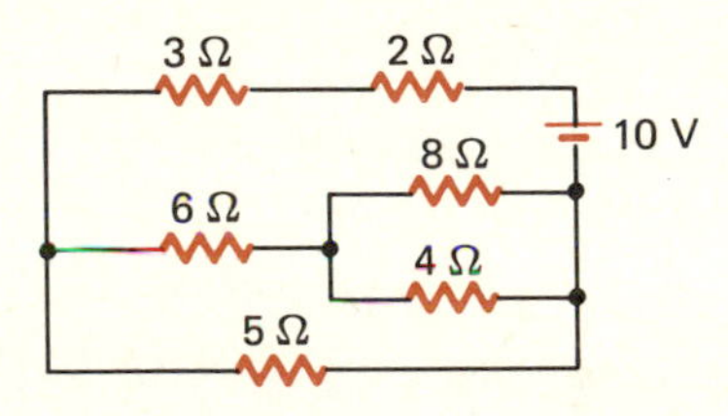

Figure 21.17

• **21.4** Find the currents in all the resistors of the circuit of Figure 21.18.

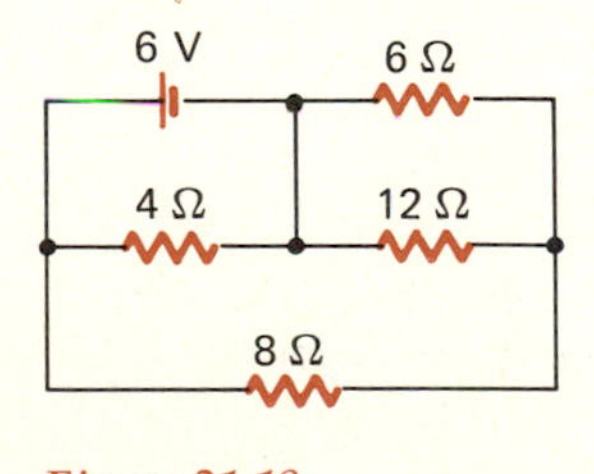

Figure 21.18

• **21.5** The internal resistance of a car battery depends sensitively on temperature, increasing with diminishing temperature. Suppose that the internal resistance of the battery is 0.012 Ω at 5 °C and 0.14 Ω at −15 °C, and that the resistance of the starter motor is 0.2 Ω. What currents will be drawn by the starter motor from this 12-V battery on mornings when the temperature is 5 °C and −15 °C. In each instance, what power is dissipated in the internal resistance of the battery?

• **21.6** Find the power delivered by the battery in Figure 21.19.

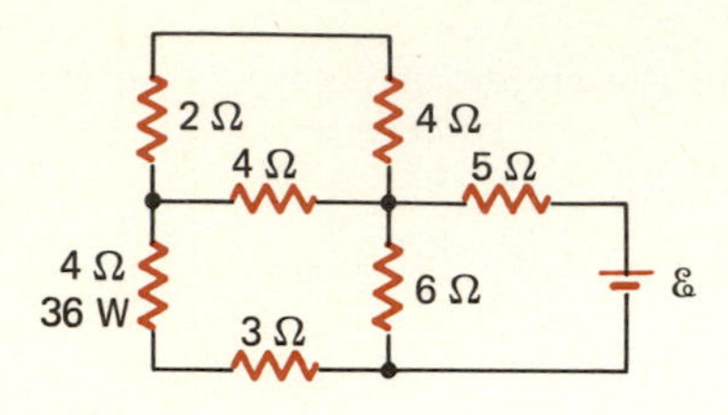

Figure 21.19

• **21.7** Find the currents in the resistors of the network of Figure 21.20(*a*) and (*b*).

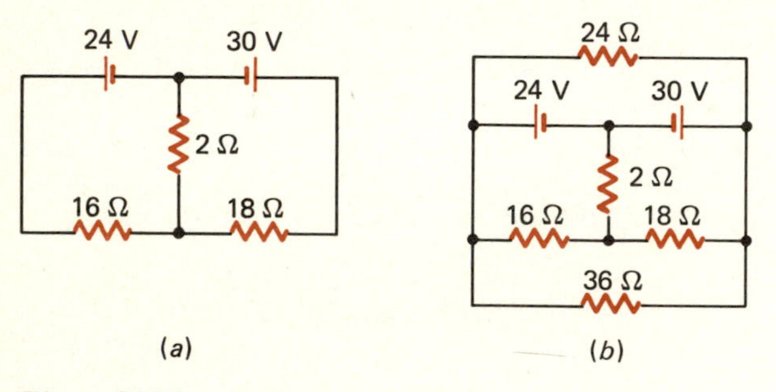

Figure 21.20

• **21.8** Find the unknown values of the resistances in the circuit of Figure 21.21.

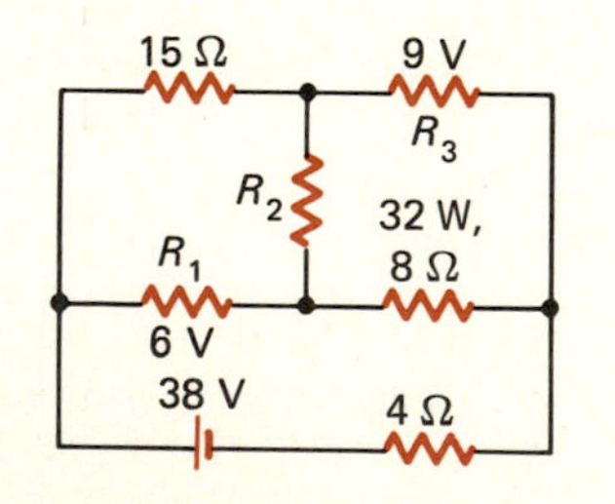

Figure 21.21

• • **21.9** Find current drawn from the battery by the circuit of Figure 21.22 when switch S is open and when switch S is closed. How much current flows in the wire connecting terminals *A* and *B* when S is closed? What is the equivalent resistance between *D* and *C* under the two conditions?

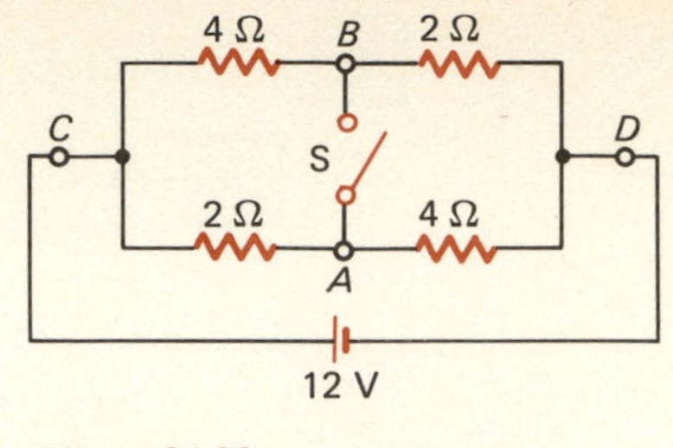

Figure 21.22

• • **21.10** Find the currents in all the resistors of the circuit of Figure 21.23.

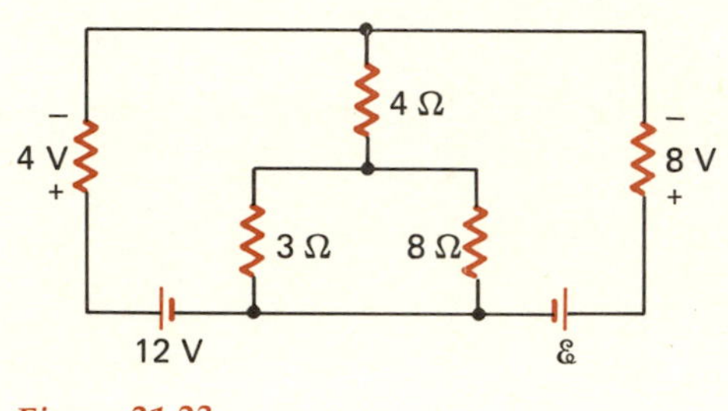

Figure 21.23

• • **21.11** Find the current in each of the resistors of the network of Figure 21.24.

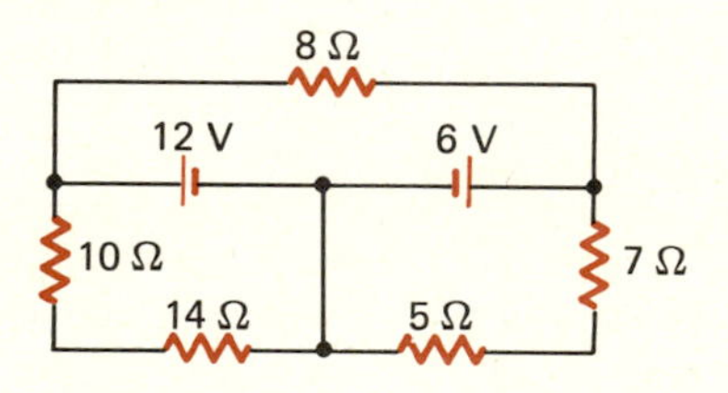

Figure 21.24

• • **21.12** Determine the current flowing in each of the resistors of Figure 21.25.

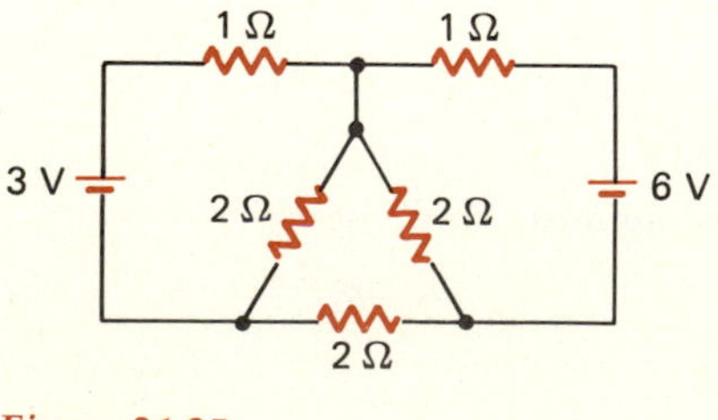

Figure 21.25

• • **21.13** Calculate the currents in each of the resistors of Figure 21.26.

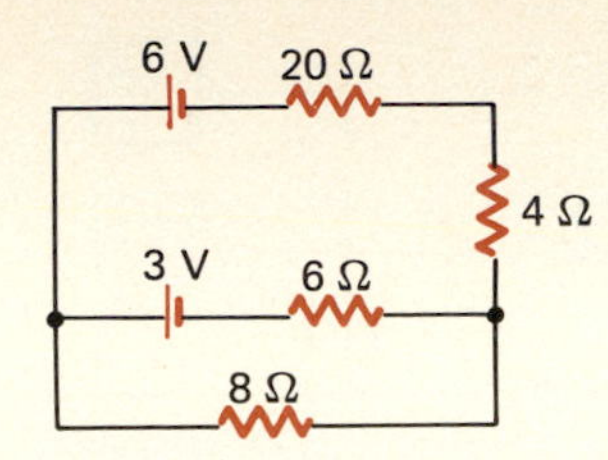

Figure 21.26

• •**21.14** Calculate the currents in each of the resistors of Figure 21.27.

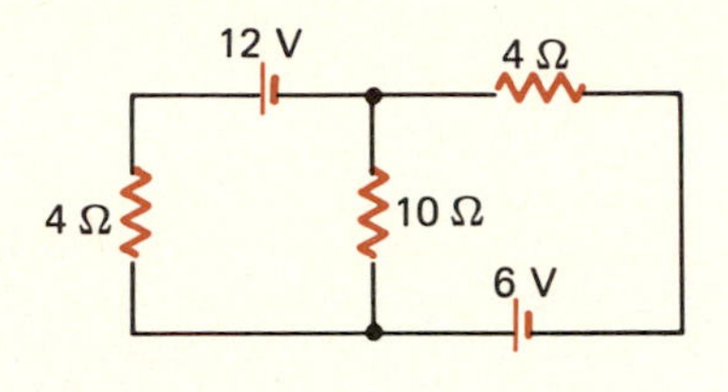

Figure 21.27

(Sections 21.4–21.5)

• **21.15** A meter has an internal resistance of 20 Ω and gives full-scale deflection when the current in the meter is 50 μA. Find the resistances that should be used to transform this instrument into a meter that will read full scale for (a) a current of 1 A; (b) a current of 200 μA; (c) a voltage of 1 V; (d) a voltage of 10 mV. State how these resistances should be connected to the meter.

• **21.16** An ammeter reading full scale for a current of 0.5 A and having an internal resistance of 0.2 Ω, and a voltmeter reading 10 V full scale and having an internal resistance of 5000 Ω, are used to measure an unknown resistance R with the circuit of Figure 21.28(*a*). What is the value of R if these meters read 0.400 A and 6.10 V? By what fraction does the true value of R differ from the ratio $V_M/I_M = 6.10/0.400$? What would the voltmeter read if the meters were reconnected as in Figure 21.28(*b*) and the current-limiting resistor r readjusted so that $I_M = 0.400$ as before? For which arrangement is V_M/I_M more nearly equal to R, and why is one circuit better than the other?

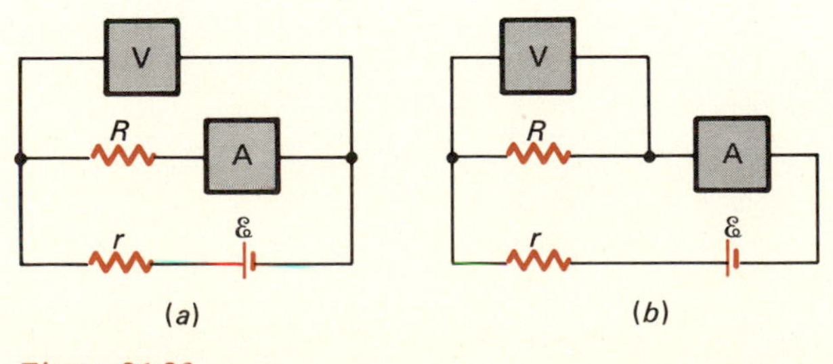

Figure 21.28

• **21.17** The bridge of Figure 21.29 is balanced when the value of the variable resistor is 132 Ω. What is the value of the unknown resistor R_x, and how much power is dissipated in this unknown at balance?

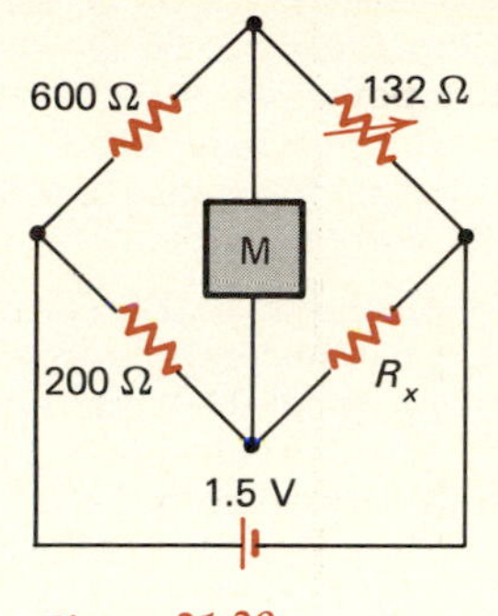

Figure 21.29

• **21.18** An alternative to the Wheatstone bridge of Figure 21.11 is the "slide-wire Wheatstone bridge" of Figure 21.30. If $R_1 = 2000$ Ω and the current in the meter is zero when the slide wire contact is one-third the distance along the slide wire as indicated, what is the value of the unknown resistor R_x?

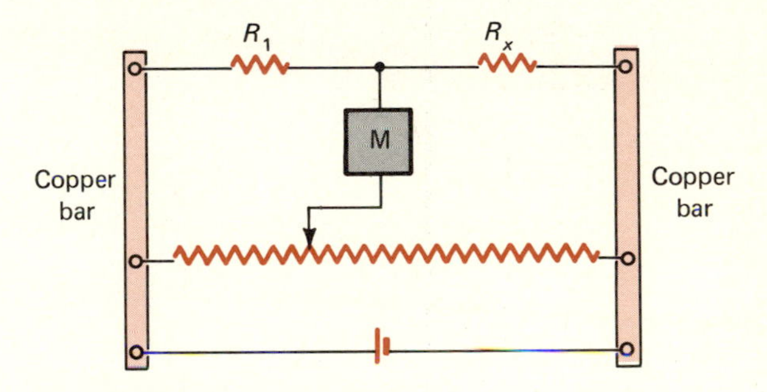

Figure 21.30

• •**21.19** A meter has an internal resistance of 4 Ω and gives full-scale deflection when a current of 100 μA passes through the instrument. Show a circuit that incorporates this meter and allows it to be used as a multirange ammeter with full-scale deflections for 0.1 A, 1.0 A, and 5.0 A.

• •**21.20** A meter has an internal resistance of 10 Ω and gives full-scale deflection when the current in the meter is 0.100 mA. This meter is to be the sensing instrument of a multiscale voltmeter as shown in Figure 21.31, with full-scale readings for voltages of 5 V, 25 V, and 100 V. What should the values be of the unknown resistors shown in Figure 21.31?

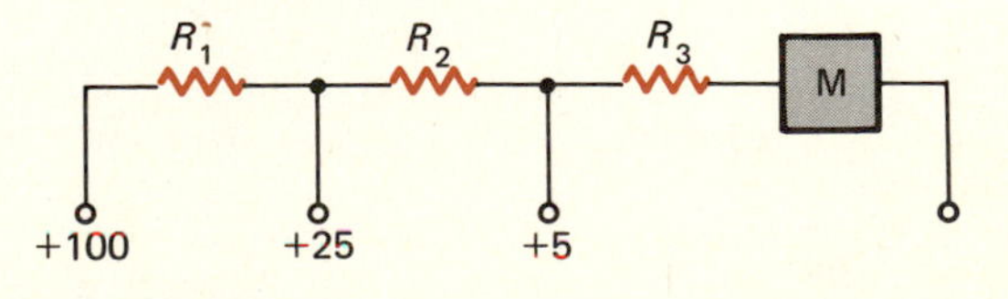

Figure 21.31

• • **21.21** A 300-Ω resistor and a 200-Ω resistor are connected in series across the terminals of an ideal 9-V battery. A voltmeter connected across the 300-Ω resistor reads 5.0 V. What is the internal resistance of the voltmeter?

• • **21.22** If a voltmeter of 50,000-Ω internal resistance is connected across the 100,000-Ω resistor of Figure 21.32, what will the voltmeter read? If another voltmeter, with an internal resistance of 500,000 Ω is used, what will that meter read?

• • **21.23** A 4000-Ω and a 12,000-Ω resistor are connected in series, and the combination is connected to the terminals of an ideal 12-V battery. A voltmeter with an internal resistance of 40,000 Ω is used to measure the potentials across the resistors. What will the voltmeter read when it is connected across the 4000-Ω resistor and when it is connected across the 12,000-Ω resistor?

• • **21.24** You have a 12-V battery and an ammeter and a voltmeter. The internal resistance of the battery is 0.1 Ω and the two meters have resistances 0.15 Ω and 5000 Ω, respectively. You are given two resistors, R_A and R_B, and are to measure their resistances. If you know that R_A has a resistance of about 1.0 Ω, and R_B a resistance of about 100,000 Ω, which of the two circuits, Figures 21.10(*a*) or 21.10(*b*), should be used in each case so that the ratio of voltmeter reading to ammeter reading closely approximates the true resistance? Justify your answer.

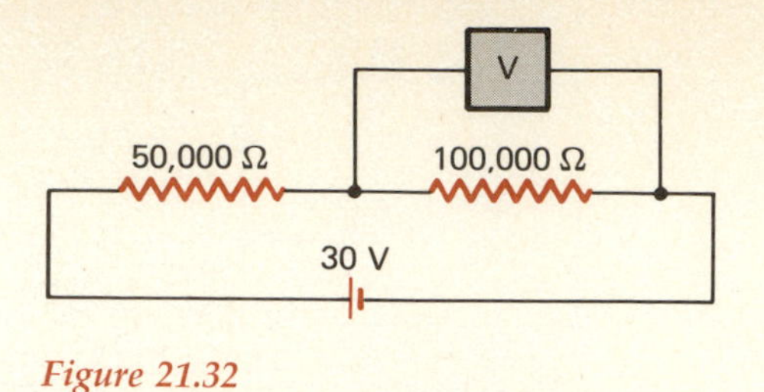

Figure 21.32

• • **21.25** An ammeter with an internal resistance of 0.02 Ω is connected in series with a variable resistance and a battery. When the resistance value is 0.8 Ω, the meter reads 10 A; when the resistance value is 0.5 Ω, the meter reads 15 A. What are the open-circuit voltage and the internal resistance of the battery?

•22•

Magnetism

But ere the first wild alarm could get out abroad among the crew, the old man with a rigid laugh exclaimed, "I have it! It has happened before. Mr. Starbuck, last night's thunder turned our compasses—that's all. Thou hast before now heard of such a thing, I take it."

"Aye; but never before has it happened to me, Sir," said the pale mate, gloomily.

Here, it must needs be said, that accidents like this have in more than one case occurred to ships in violent storms. The magnetic energy, as developed in the mariner's needle, is, as all know, essentially one with the electricity beheld in heaven; hence it is not to be much marvelled at, that such things should be.

HERMAN MELVILLE, *Moby Dick*

22.1 Introduction

In the summer of 1820, Hans Christian Oersted wrote the following:

> The first experiments on the subject which I undertake to illustrate were set on foot in the classes for electricity, galvanism, and magnetism, which were held by me in the winter just past. By these experiments it seemed to be shown that the magnetic needle was moved from its position by the help of a galvanic apparatus.

Oersted's discovery marked the beginning of modern understanding of magnetism. The mysterious power of certain stones, most commonly found in the region of Magnesia in Asia Minor, to draw to themselves small bit of iron was known in antiquity, probably long before the electrostatic effects of rubbed amber had been noted. By about 100 B.C., the Chinese recognized that these stones, when suspended by a thin string, tended to orient in a particular direction. During the eleventh and twelfth centuries, crude magnetic compasses began to appear aboard ships.

In Oersted's day, "electricity, galvanism, and magnetism" were regarded as three quite distinct areas for scientific study. Electricity, concerned with effects such as those studied by Cavendish, Coulomb, and others, centered on the interactions between stationary charges. Galvanism, as the name suggests, originated with the observation of Luigi Galvani that a frog's sciatic nerve responds to electrical stimulation. That soon led to further investigations by Alessandro Volta and the discovery of electrochemical reactions that could be used to construct "Voltaic piles," that is, batteries, capable of maintaining a continuous current in a closed circuit. The study of these currents and their chemical and other effects came to be known as galvanism.* Magnetism was the study of properties of magnets, their interactions and their effects, if any, on other objects.

Some properties of magnets were already well known by the time of Oersted: If two bar magnets are placed so that their north-seeking ends, their north poles, are in close proximity, they repel each other, and similarly for two south poles; on the other hand, a north pole attracts a south pole. The similarity between the behavior of magnets and electric charges did not escape the attention of natural philosophers; but until Oersted's momentous discovery, no one had been able to detect any connection between magnetic and electric effects.

Figure 22.1 Hans Christian Oersted (1777–1851).

Little wonder, then, that Oersted invited several distinguished men to impeccable reputation, a professor of medicine, a knight of the Danish Order, a professor of natural history, the minister of justice, and one or two others to witness a repetition of these experiments, which he describes as follows:

> Let the opposite poles of the galvanic apparatus [*battery*] be joined by a metallic wire. To the effect which takes place in this conductor and surrounding space, we will give the name of electric conflict.
>
> Let the rectilinear part of this wire be placed in a horizontal position over the magnetic needle duly suspended, and parallel to it. . . . These things being thus arranged, the magnetic needle will be moved, and indeed, under that part of the joining wire which receives electricity most immediately from the negative end of the galvanic apparatus, will incline toward the west. . . .
>
> The effects of the joining wire on the magnetic needle pass through glass, metal, wood, resin, earthenware, stones. The passing of the effects through all these materials in electricity and galvanism has never been observed. The effects, therefore, which take place in electric conflict are as different as possible from the effects of one electric force on another.

In this last paragraph, Oersted stresses that although a current cannot pass through an insulator such as wood or resin, and the electrostatic forces are shielded by metallic sheets, the magnetic effect of the current, which he calls "electric conflict," is not influenced by the interposition of such materials.

* The name has survived to this day in our reference to zinc-plated steel as "galvanized" even though electroplating is now rarely used.

22.2 The Magnetic Field

In 1820, the scientific community was more than ready for this discovery, which sparked a period of hectic activity. Within months, Ampère, Arago, and Biot in France, Faraday in England, and others elaborated and extended these investigations. Ampère, in particular, recognized that if a current exerts a force on a magnet, then by Newton's third law, a magnet must exert a force on a current. The experiment was performed without delay and gave the expected result—a current-carrying wire experienced a force near a magnet. In analogy to the gravitational and electrostatic cases, one can envisage a *magnetic field* that permeates space and acts on the current. This magnetic field, designated by **B**, can be defined in terms of the force on a current, just as the electric field was defined in terms of the force on a charge. Since a current is the result of charges in motion, the definition we shall use rests on the force experienced by an isolated charge moving through a magnetic field.

The experimental observations on which this definition of the magnetic field is based are as follows.

1. In a region in which a compass needle experiences a torque that tends to orient it along some spatial direction, a moving charge q experiences a force that depends on:
(a) The magnitude and sign of the charge.
(b) The *velocity* of the charge, that is, the speed as well as the direction of its motion.

2. This velocity-dependent force is in addition to and independent of whatever velocity-independent electrostatic force may act on the charge.

We have encountered velocity-dependent forces before. For example, viscous forces, resulting in a terminal velocity during free fall, are velocity-dependent. Those forces act along the direction of motion.

In the present instance, the velocity-dependent force has a different character. At every point in a region where **B** does not vanish, a charge q, moving with velocity **v**, experiences a force **F** whose *magnitude* is given by

$$F = Bqv \sin \theta \tag{22.1}$$

where θ is the angle between the magnetic field vector **B** and the velocity vector **v**. The *direction* of the force vector is obtained by applying Right-Hand Rule No. 1:

RHR-I: Place the right hand so that the thumb points along **v**, the direction of motion of *positive* charge, and let the fingers point in the direction of the component of **B** perpendicular to **v**. The palm then faces in the direction of the force **F**, which acts on the positive charge.

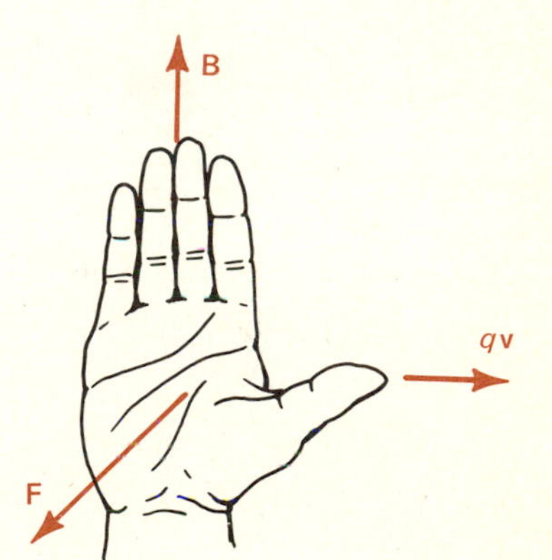

Figure 22.2 The direction of the force acting on a moving charge in a magnetic field can be established by applying Right-Hand Rule 1, RHR-I, illustrated here.

If the moving charge is *negative* instead of positive, the direction of the force acting on it is opposite to the force prescribed by RHR-I. One way of looking at this matter is to use RHR-I so the thumb points in the direction of $q\mathbf{v}$ instead of **v**. If q is negative, we must, keeping the direction of the fingers the same, reverse the direction of the thumb so that it points opposite to **v**. The hand is rotated through 180°, interchanging palm and back of hand.

This force on a moving charge in a magnetic field is called the Lorentz force, in honor of the Netherlands physicist H. Lorentz (1853–1928). This Lorentz force is the basis of numerous devices, including electric motors, ammeters, voltmeters, TV picture tubes, blood flowmeters, electromagnetic pumps, "magnetic bottles" for plasma confinement in controlled fusion experiments, mass spectrometers, and cyclotrons and other large particle accelerators.

The Lorentz force is not a central force. Consequently, it is not possible to relate the magnetic field to a scalar potential function. Since the force due to the field **B** is at all times perpendicular to the direction of motion of the charge, the work done by this force must be exactly zero. Hence, a charge moving in a time-independent magnetic field experiences no change in its potential or kinetic energy.

In the electrostatic case, the magnitude and direction of the electric field was defined in terms of the force acting on a stationary positive charge (we did not then require the charge to be at rest because we had not yet come to consider magnetic fields, and did not want to complicate the definition unnecessarily). Similarly, *we now define the magnetic field* **B** *in terms of the force that acts on a positive charge q that is moving with velocity* **v**.

According to this prescription, measuring **B** involves the following procedure. We first determine the force acting on the charge when it is at rest, and then subtract that force from those forces subsequently measured. We then find the force on the charge *that arises from its motion*. We may find that there is no velocity-dependent force whatever and conclude that in this region of space, $\mathbf{B} = 0$. We may find a velocity-dependent force; in that case we must identify the direction of the velocity for which this force attains a maximum, $\mathbf{F}_{max}$. The *magnitude* of **B** is then given by

$$B = F_{max}/qv \qquad \textbf{(22.1a)}$$

and the direction of **B** is obtained from RHR-I.

With the direction of **B** defined by RHR-I, the north-seeking pole of a compass needle points in the direction of the field. Since a compass needle placed between the poles of a magnet points toward the south pole, the direction of the magnetic field in the air space between the pole faces of a magnet such as that shown in Figure 22.3 points from the north to the south pole face.

Figure 22.3 *An electromagnet, such as the 12-in. Varian magnet shown here, can produce fields of about 3 T in the air gap between its pole faces. In a typical laboratory configuration, this air gap may be between 4 and 10 cm. The coils that energize the magnet are water-cooled, and the unit dissipates about 20 kW.*

The SI unit of magnetic field is the *tesla* (T). Another unit that is still widely used is the *gauss* (G). The latter has the advantage that one gauss is about the strength of the magnetic field at the earth's surface. The tesla is a much larger unit:

$$1 \text{ T} = 10{,}000 \text{ G} = 10^4 \text{ G}$$

The field produced between the pole faces of large laboratory electromagnets is typically in the range between 1 and 3 T. The field near the pole faces of a good permanent magnet is about 0.4 T.

As you perhaps know already, one can create an excellent visual image of a magnetic field in a plane by scattering small iron filings over a piece of paper. Such filings behave like tiny dipoles, orienting themselves parallel to the field direction. Placed above an ordinary bar magnet, they show the field pattern of Figure 22.4, which is strikingly similar to Figure 18.4. It is because the field pattern of a bar magnet is the pattern of a dipole that one speaks of *magnetic dipoles,* even though the isolation of separate magnetic north and south poles appears impossible.

As in the description of the electric field, *magnetic field lines,* or *lines of flux* as they are called, are very useful. As was true for the **E** field, the **B** field has a strength proportional to the density of these flux lines, and the phrase *magnetic flux density* is often used for **B**. If in some region the flux density **B** is uniform, the total *flux* ϕ that passes through a given area is

$$\phi = B_{\perp} A \text{ Wb} \tag{22.2}$$

where $B_{\perp}$ is the component of **B** perpendicular to the area A, and Wb is the abbreviation for *weber,* the SI unit of magnetic flux. The concept of

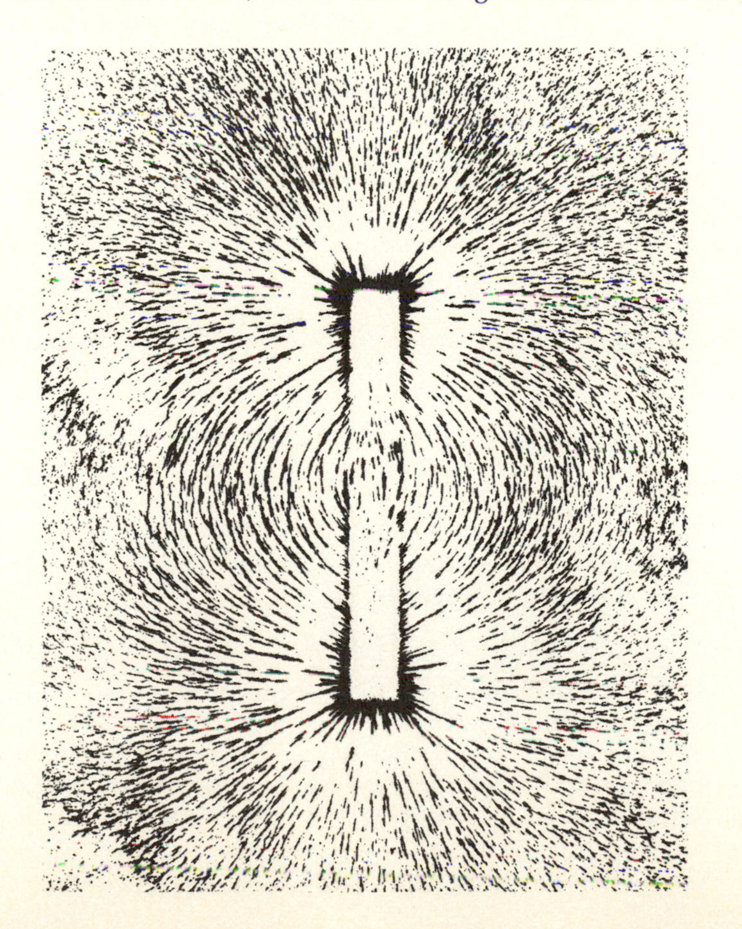

Figure 22.4 *The field pattern in the neighborhood of a small bar magnet can be portrayed by means of iron filings, which serve as very small magnets. The filings align themselves along the field lines. The pattern of this figure should be compared with the pattern of the electric dipole shown in Figure 18.4.*

flux will prove especially valuable in discussing electromagnetic induction (Chapter 23).

● ***Example 22.1*** What is the magnitude of the Lorentz force on an electron whose velocity is 2×10^6 m/s in a magnetic field of 2×10^{-2} T (200 G) directed perpendicular to the velocity? How strong an electric field would be required to produce a force of the same strength?

The Lorentz force is given by Equation (22.1). In this instance, $\theta = 90°$. Hence,

$$F = (2 \times 10^{-2}\ \text{T})(1.6 \times 10^{-19}\ \text{C})(2 \times 10^6\ \text{m/s}) = 6.4 \times 10^{-15}\ \text{N}$$

An electric field will provide the same force if **E** is at right angles to both **v** and **B** and has a magnitude given by

$$E = \frac{F}{q} = \frac{6.4 \times 10^{-15}\ \text{N}}{1.6 \times 10^{-19}\ \text{C}} = 4 \times 10^4\ \text{V/m}$$

In general, the force on a charge q, moving at velocity **v**, due to a field **B** perpendicular to **v**, will be the same as the force due to a field **E** perpendicular to both **v** and **B** if

$$Bqv = qE \qquad \text{or} \qquad E = Bv$$

●

22.3 Force on a Current in a Magnetic Field

In Chapter 20 we saw that a distribution of particles with a density of n per m^3, each carrying a charge q, and moving with an average drift velocity $\mathbf{v}_d$, constitutes a current density $\mathbf{J} = nq\mathbf{v}_d$ A/m^2. The current through an area A is the product $JA = nqv_dA$, where A is the cross-sectional area perpendicular to the current density **J**.

Suppose that a wire of length ℓ is placed in a region of uniform magnetic field and that a current I is flowing in that wire. Since the charges moving within the wire experience a force due to the field **B** but are constrained to remain within the wire, the wire itself exhibits this force to the observer, as Ampère first discovered.

We can now calculate this force, using Equations (22.1) and (20.3). The total number of charge carriers in the length ℓ and cross section A is

$$N = nV = nA\ell$$

The force on any one of these charge carriers, moving at the velocity $\mathbf{v}_d$, is

$$F_q = Bqv_d \sin\theta$$

where θ is the angle between **B** and $\mathbf{v}_d$. Since the charges in the wire are constrained to move in the direction of the wire, the force on the conducting wire of length ℓ is given by

$$F = NF_q = nA\ell Bqv_d \sin\theta = BI\ell \sin\theta \qquad \textbf{(22.3)}$$

Here θ is the angle between **B** and $\boldsymbol{\ell}$, where the vector $\boldsymbol{\ell}$ is parallel to the wire and points in the direction of the current I.

The direction of the force **F** that acts on the wire is obtained by invoking the same right-hand rule that gave the force on a moving positive charge, RHR-I. The only difference is that now the thumb must be placed so that it points in the direction of the current.

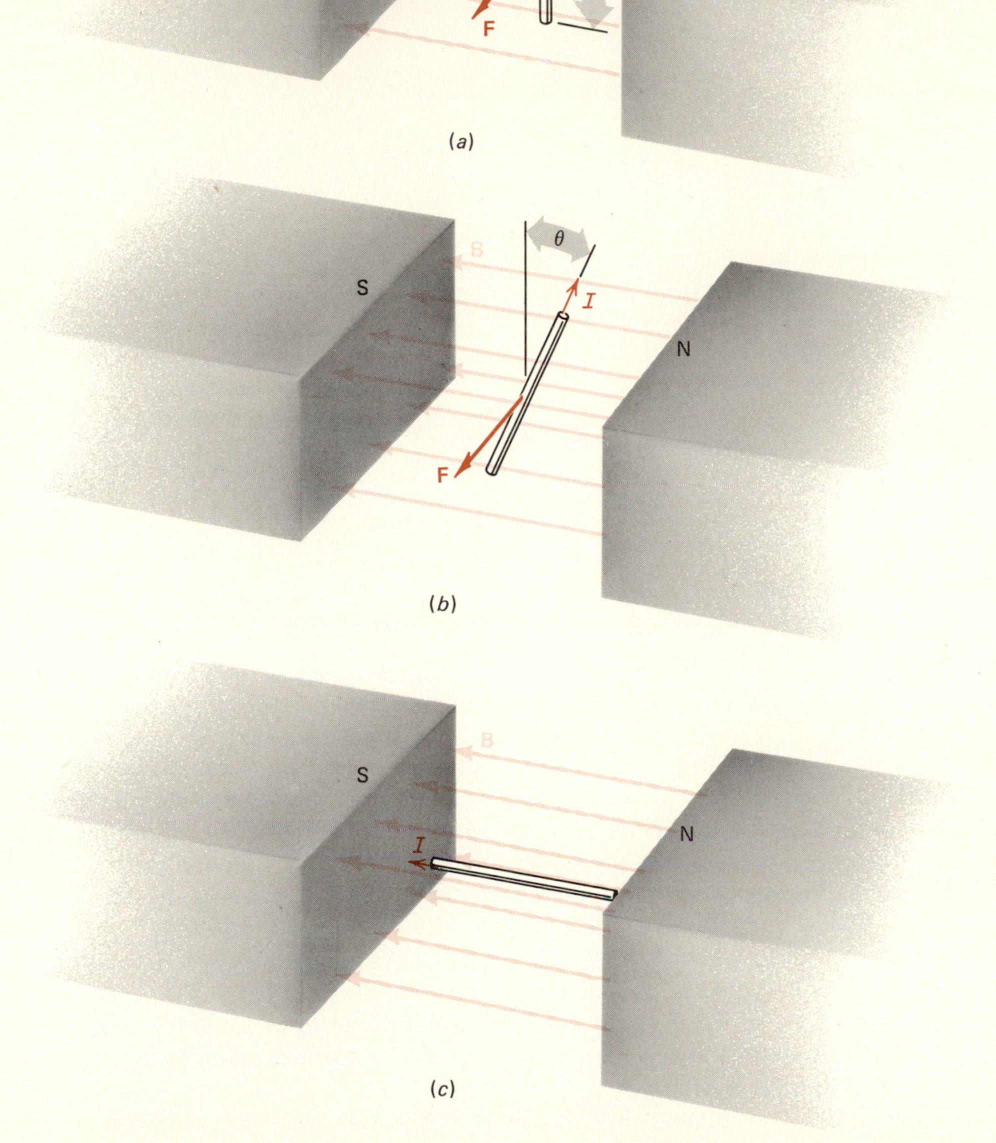

Figure 22.5 A wire of length ℓ carrying a current **I** in a magnetic field **B** experiences a force **F**, given by Equation (22.3). That force, for a given **I** and **B**, is greatest when ℓ and **B** are at right angles, as in (a). If the angle between ℓ and **B** is not 90°, **F** is proportional to the component of the field perpendicular to the wire, as illustrated in (b). If ℓ and **B** are collinear, as in (c), **F** = 0. Note that in every case the force **F** is perpendicular to both ℓ and **B**, that is, **F** points along a line normal to the plane formed by ℓ and **B**.

● ***Example 22.2*** A horizontal wire carries a current from east to west. What is the direction of the force on this current if we assume that at this location the magnetic field of the earth points due north?

When the thumb of the right hand points west and the fingers point north, the palm faces down; hence the force on this wire will be down. ●

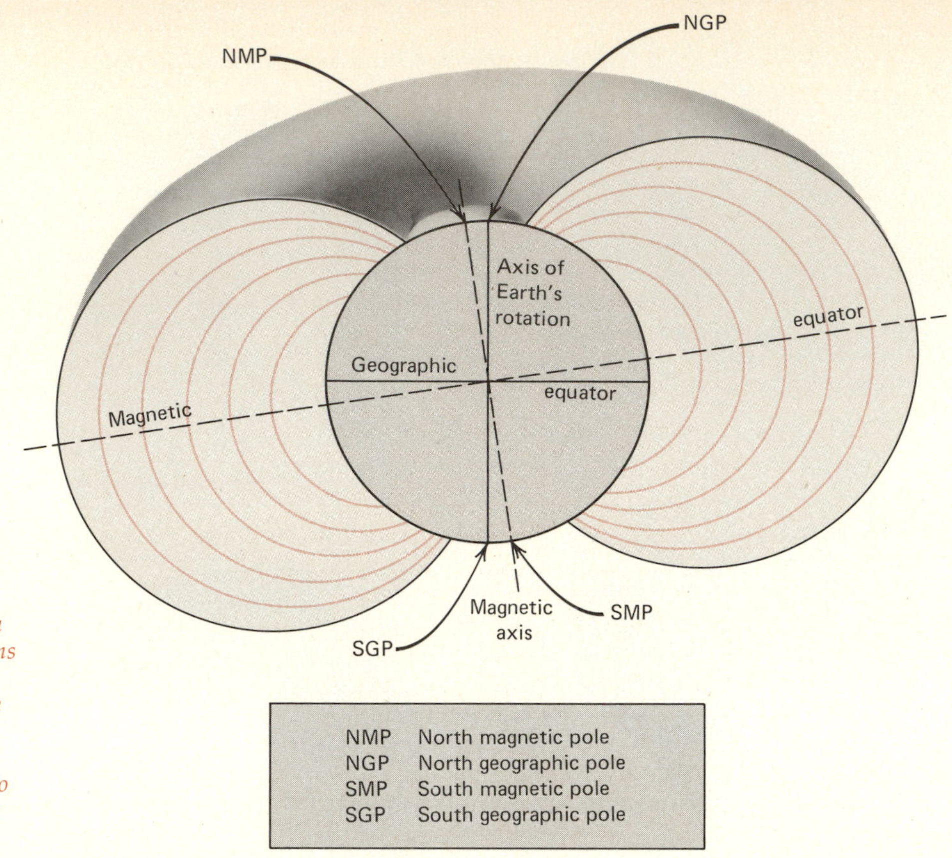

Figure 22.6 *Schematic drawing of the earth's magnetic field. The field pattern is similar to that of a magnetic dipole. The locations of the geographic north and south poles (which lie on the earth's rotation axis) do not coincide with the magnetic north and south poles, nor do the geographic and magnetic equators overlap.*

*22.4 The Earth's Magnetic Field

Example 22.2 reminds us that we live in a region of weak magnetic field. The earth's magnetic field resembles that of a dipole, as indicated in Figure 22.6.

Numerous puzzling aspects of the earth's magnetic field remain unresolved. The pattern of field lines is continually shifting, the magnetic north and south poles moving significantly in times that are short by geological reckoning. Correspondingly, the *magnetic declination,* as the difference between the true, geographic north and the magnetic north is called, also changes year by year in amounts of order 10 minutes of arc. Hence, over a decade or two, these changes can amount to corrections of one degree or more. Even more mysterious, there is evidence that every few hundred thousand years the earth's magnetic field reverses completely, though the earth's rotational or orbital motions seem to suffer no corresponding perturbation.

As Figure 22.6 suggests, the magnetic field is not parallel to the earth's surface except at the magnetic equator (which does not coincide with the geographic equator). At other locations, a magnetic needle free to move in a horizontal *and* vertical plane would point down as well as toward the north in the northern hemisphere. The angle the needle makes with the horizontal plane is called the *dip angle,* or *magnetic inclination.*

● ***Example 22.3*** A vertical wire carries a current of 4 A straight up in a region in which the geomagnetic field has a downward inclination of 60°, points north and has a magnitude of 1.2 G. Find the direction of the force on this wire due to the current, and compare its magnitude with that of the gravitational force on the wire, assuming its mass per unit length is 20 g/m.

By RHR-I, the direction of the force on the wire is to the west; its magnitude per unit length is given by dividing Equation (22.3) by ℓ:

$$F = \frac{BI\ell \sin \theta}{\ell} = BI \sin \theta$$
$$= (1.2 \times 10^{-4}\ \mathrm{T})(4\ \mathrm{A})(\sin 150^\circ)$$
$$= 2.4 \times 10^{-4}\ \mathrm{N/m}.$$

The gravitational force on that wire is

$$W = mg = (20 \times 10^{-3}\ \mathrm{kg/m})(9.8\ \mathrm{m/s^2}) = 0.196\ \mathrm{N/m}$$

about a thousand times the magnetic force.

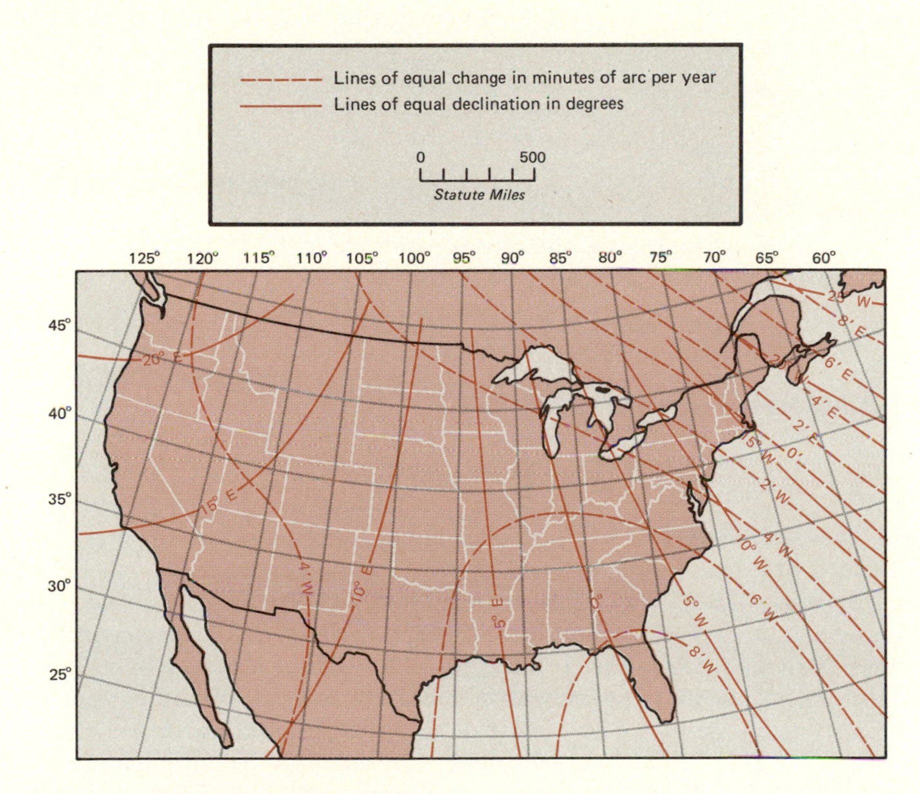

Figure 22.7 The magnetic declination in the United States. These deviations of the magnetic north from the geographic north do not remain fixed but change by a few minutes of arc per year.

22.5 Trajectory of Moving Charges in a Magnetic Field

Suppose that a particle of mass m and charge q is traveling with velocity $\mathbf{v}$ in a region of uniform magnetic field $\mathbf{B}$, where $\mathbf{B}$ is perpendicular to $\mathbf{v}$, as indicated in Figure 22.8. This particle then experiences a Lorentz force, given by (22.3), whose direction is perpendicular to both $\mathbf{B}$ and $\mathbf{v}$. If q is positive, the force is directed upward at point A in Figure 22.8. Under the influence of that force, the particle will be deflected from its rectilinear path, but as the velocity vector changes direction in response to the acceleration, so does the force, which must always remain at right angles to $\mathbf{v}$.

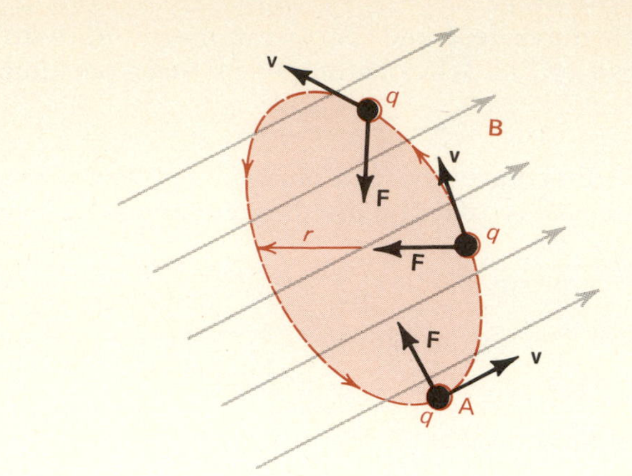

Figure 22.8 *A charged particle moving with velocity* **v** *perpendicular to a uniform magnetic field* **B** *experiences a force that is always perpendicular to* **v** *and to* **B** *and is of constant magnitude. Although the particle is continually accelerated, the acceleration is always at right angle to* **v**, *and so the speed of the particle remains constant. The particle's trajectory is a circle.*

This is reminiscent of a situation encountered earlier (Chapter 6). We saw then that the force needed to maintain uniform circular motion is of constant magnitude and perpendicular to the velocity. This centripetal force F_c is given by

$$F_c = \frac{mv^2}{r} \qquad \textbf{(6.17)}$$

where r is the radius of the circular orbit and v the tangential speed of the particle.

In the present instance, the centripetal force is the Lorentz force, and the particle will trace a circular trajectory so that

$$F_c = Bqv = \frac{mv^2}{r}$$

Solving for the orbit radius, we find

$$r = \frac{mv}{Bq} \qquad \textbf{(22.4)}$$

This simple relation has provided the key to many treasure chests of modern physics.

Equation (22.4) tells us that in a region of uniform magnetic field **B**, a particle of known charge and traveling perpendicular to **B** with speed v, will follow a circular orbit whose radius is a direct measure of its momentum, mv. Since practically all elementary particles are either neutral or carry a charge of $-e$ or $e = 1.6 \times 10^{-19}$ C, observation of a charged particle's trajectory in a magnetic field is the standard technique for measuring momentum in particle physics. Equation (22.4) retains its validity also at relativistic velocities, provided that the relativistic momentum is used in place of mv.

Energetic charged particles are most frequently observed by allowing them to pass through a *bubble chamber,* a volume containing a superheated liquid. Tiny bubbles, the forerunners of more violent boiling, form along the path of an energetic charged particle, because as this particle traverses the liquid, it ionizes some of the molecules along its path, and these ions become nucleation sites for bubbles. A typical bubble-chamber photograph, showing a variety of particles and some nuclear collisions, is reproduced in Figure 22.9.

You will notice that some particles in Figure 22.9 describe clockwise paths, and some, counterclockwise paths. The sense of rotation in a magnetic field of given direction depends on the sign of the charge; a nega-

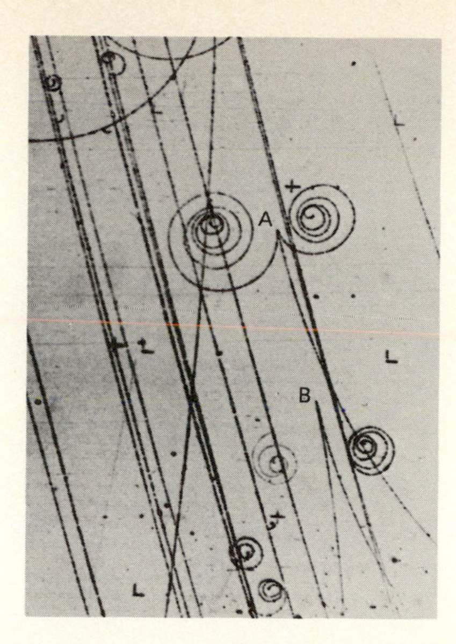

Figure 22.9 *A bubble chamber photograph, taken at the Fermi National Accelerator Laboratory, which illustrates several "events." The practically straight tracks are protons from the primary beam of the accelerator. At A, a collision of a proton with a nucleus of the liquid results in the creation of several new particles; some are positively charged, some negatively charged, and some neutral. One of these neutral particles quickly decays, creating a positive and a negative particle at B. The spiral tracks are made by electrons and positrons of only moderate energy as they lose energy and momentum in ionizing molecules of the liquid in the chamber.*

tively charged particle moving with velocity **v** at point A in Figure 22.8 would experience a downward force and therefore rotate clockwise. Thus, from photographs such as that of Figure 22.9, it is possible to identify the sign of the charge on the particle as well as its momentum. Moreover, by studying the density of the track, that is, the amount of ionization the particle produces in passing through the liquid, one can deduce additional information from which its mass can be inferred.

Next we note that the radius of the orbit of a particle of mass m in a fixed uniform field **B** is proportional to its speed v. Since the circumference of a complete orbit is $2\pi r$, the time for one complete orbit is $T = 2\pi r/v = 2\pi rm/Bqr = 2\pi m/Bq$; correspondingly, the number of complete revolutions per second is given by

$$f_c = \frac{1}{T} = \frac{Bq}{2\pi m} \qquad \textbf{(22.5)}$$

A very important feature of Equation (22.5) is that f_c does not depend on the speed of the particle. This constancy of f_c is used to advantage in the *cyclotron*, and f_c is therefore known as the *cyclotron frequency* of the particle in the field **B**.

22.5 (a) The Cyclotron

The essential elements of a cyclotron are shown in Figure 22.10. The particles, which most commonly are protons, deuterons, and α particles (^{4}He nuclei), and occasionally heavier ions, are released near the center of the machine by the ion source S. Each "Dee" in Figure 22.10 is half a hollow metallic pillbox, and the two Dees are connected to a radio frequency (rf) oscillator capable of generating voltages of 10,000–100,000 V. Thus potential differences appear across the narrow gap between the two Dees and establish in this region a strong electric field that reverses direction at regular intervals. Within the Dees, the electric field vanishes, because the pillboxes are made of copper, and, as we know, a good conductor is an equipotential surface. The Dees are enclosed within a vacuum chamber (not shown in Figure 22.10), which is placed between the pole faces of a large electromagnet. The magnetic field between the pole faces can be controlled by adjusting the current in the coils of the electromagnet.

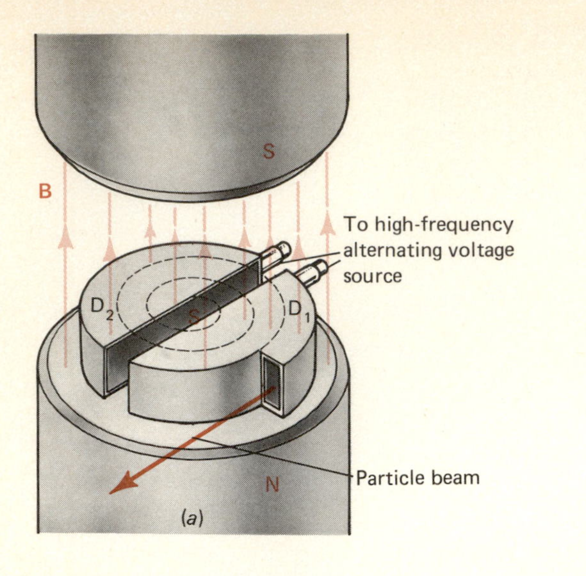

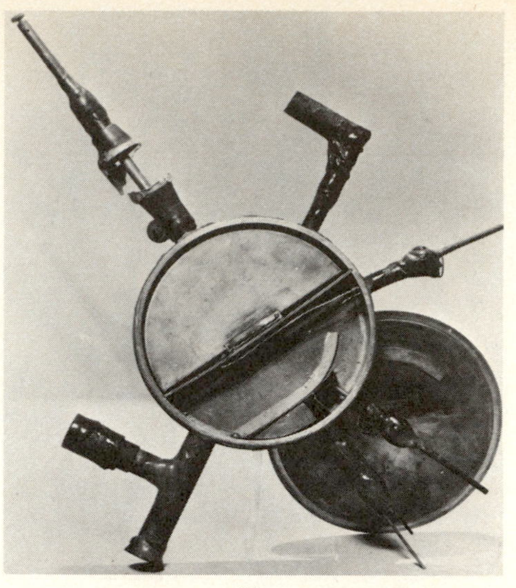
(b)

(c)

Figure 22.10 (a) *Essential elements of a cyclotron. Positive ions generated by the ion source S are accelerated as they cross the gap between the two halves of the copper pillbox, known as Dees. The Dees are connected to an alternating voltage source whose frequency matches the cyclotron frequency of the ions in the constant magnetic field* **B** *of the large electromagnet. As the ions gain energy with each traversal of the gap, they spiral outward and are ultimately extracted, emerging as an energetic beam of particles. The Dees and the "beam line" are maintained in a very good vacuum to reduce the probability that an ion will collide with an atom or a molecule and thereby be scattered out of its orbit.* (b) *The first cyclotron constructed by D. O. Lawrence. The Dees are only a few centimeters in diameter.* (c) *Aerial view of the large accelerator of the Fermi National Laboratory, Batavia, IL. The diameter of the accelerator is 2,000 m. The main administration building (at lower left) is 16 stories high.*

Suppose the magnetic field has been set so that the cyclotron frequency f_c of a charged particle emitted by the ion source S at the center of the magnet just matches the frequency of the rf oscillator connected to the Dees. In that case, a positive ion released by the ion source just when the Dee, marked D_1, is at a peak negative potential will be accelerated toward this Dee by the electric field in the gap. While inside the Dee, it describes a circular orbit under the influence of the magnetic field and returns to the gap at the end of half a period. By then, the potential produced by the oscillator has also completed half a cycle, and D_1 is now at peak-positive potential, D_2 at peak-negative potential. Consequently, as the particle crosses this gap, it is accelerated once again, gaining energy qV_0, where q is the charge on the particle and V_0 is the maximum potential difference generated by the oscillator between the two Dees.

Each time the particle crosses the gap, it gains energy, its speed increases, and the radius of its circular orbit also increases. As the particle nears the extremity of one of the Dees, it is pulled from its circular path by a large electric field due to a voltage pulse applied to a deflecting plate inserted in that Dee.

A photograph of the first cyclotron, constructed by D. O. Lawrence, with Dees having a diameter of only a few centimeters, is shown in Figure 22.10(*b*). It is but a miniature compared with modern accelerators. Kinetic energies of tens and hundreds of million electron volts (MeV) can be imparted to ions accelerated in cyclotrons. Other machines, such as the enormous circular accelerators at CERN in Geneva, the Fermi Laboratory outside Chicago, and elsewhere can achieve energies a thousand times greater still.

22.5 (b) Hall Effect

Suppose we place a conducting slab in a uniform magnetic field, as in Figure 22.11, and pass a current I through this slab. In a metal this current is carried by negatively charged electrons, which therefore move with a drift velocity directed opposite to I. As they move, they experience a Lorentz force $F_B = Bev$ toward the bottom of the slab, as may be verified with RHR-I. Since there is no external conducting path connecting the top to the bottom surface, these electrons gradually accumulate on the lower surface of the slab, with the upper surface then carrying a net positive charge. This charge separation sets up an electric field, the so-called Hall field $\mathbf{E}_{\mathrm{H}}$, within the material, and this field exerts an electrostatic force on the electrons that opposes the Lorentz force. When

$$F_B = Bev = eE_{\mathrm{H}} \qquad E_{\mathrm{H}} = Bv \tag{22.6}$$

no further charge separation occurs, and the electrons then move undeviated across the slab.

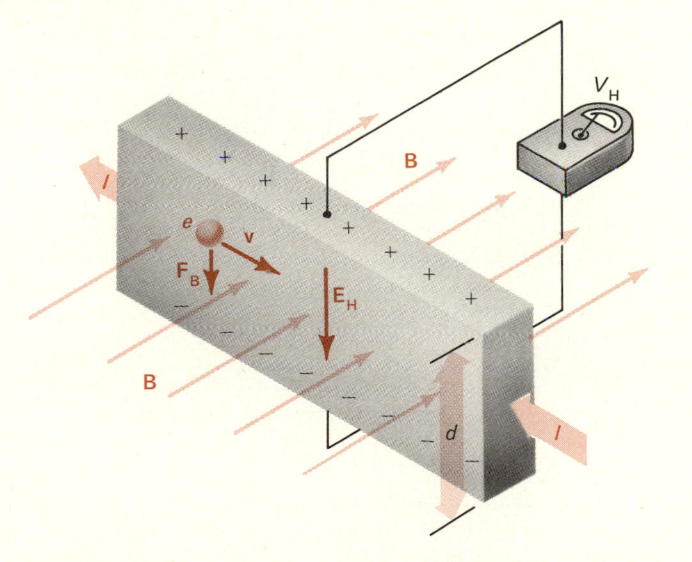

Figure 22.11 *The Hall effect. A current I flows in a conducting slab of width d. It is here assumed that the current is carried by electrons, which therefore travel from left to right, opposite to I. In a magnetic field* **B**, *directed as shown, these electrons experience a Lorentz force toward the bottom of the slab and therefore deviate in that direction. However, as negative charge accumulates on the lower surface, and positive on the upper, the electric field* $\mathbf{E}_{\mathrm{H}}$ *(called the Hall field) due to this charge separation quickly becomes large enough to just balance the Lorentz force on the negative charges. The potential between the two surfaces, which can be measured with a good voltmeter, is the Hall voltage* $V_{\mathrm{H}} = E_{\mathrm{H}}d$.

This Hall field results in a potential difference $V_{\mathrm{H}} = E_{\mathrm{H}}d$ between the two sides of the slab. That potential, which can be measured with a good voltmeter, depends on B, and through v, on the current I. Moreover, as an examination using RHR-I will confirm, the polarity of V_{H} for given current and field directions depends on the sign of the moving charges. The Hall effect is therefore a convenient indicator for determining whether positive or negative charges are responsible for current flow. Hall probes are also common sensors for measuring magnetic fields. The Hall field generated by an ionic liquid, such as blood, flowing through a magnetic field provides a measure of the flow rate; electromagnetic flowmeters now have wide medical application.

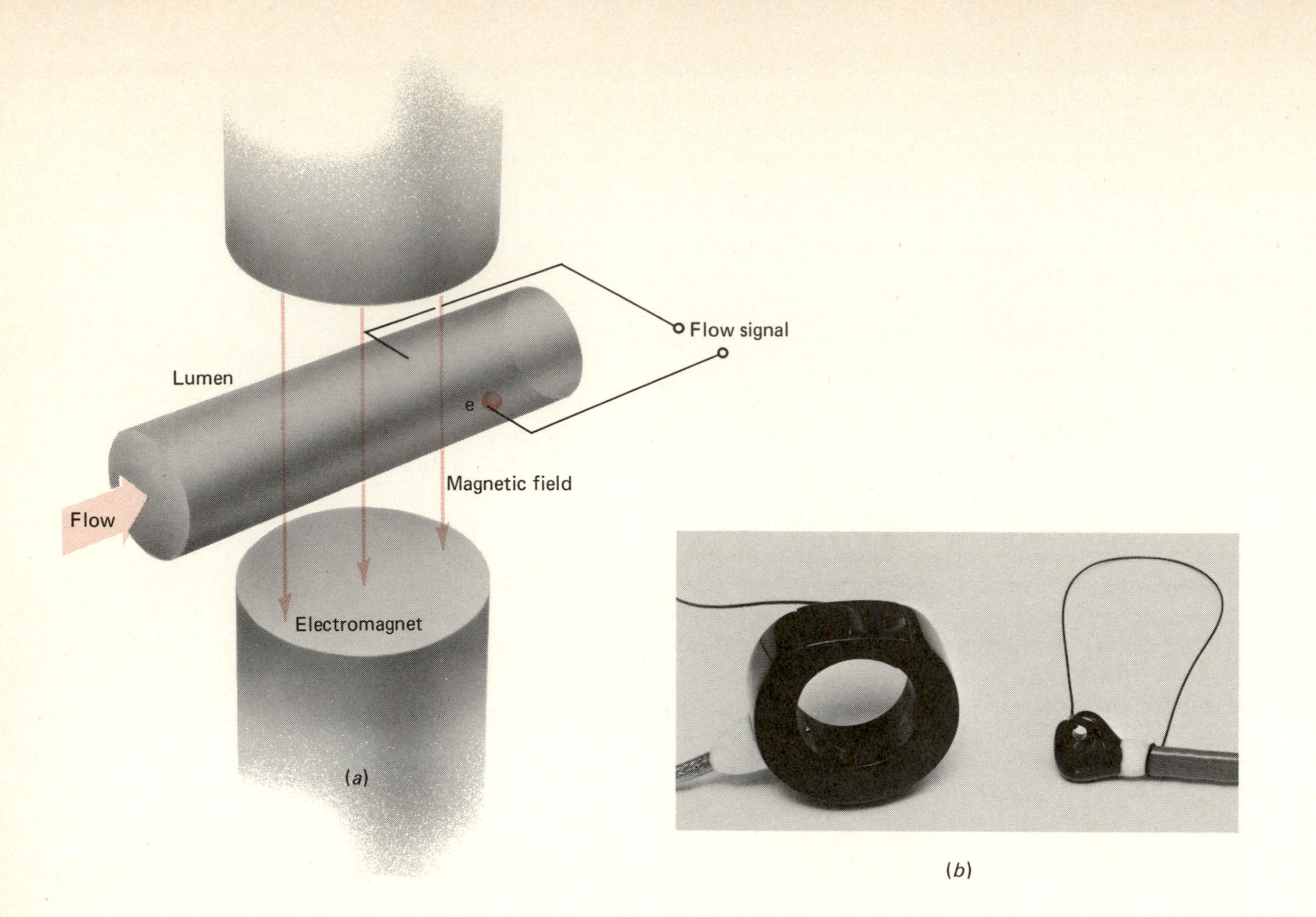

Figure 22.12 (*a*) Blood flow can be measured without inserting a catheter by detecting the Hall voltage generated across the blood vessel by motion of ions in the blood as they pass through a magnetic field. (*b*) Cuff-type probes for measuring flow rates in medical applications.

Example 22.4 What is the kinetic energy of protons whose orbit diameter is 2 m in a magnetic field of 1.6 T?

From Equation (22.4), we have

$$mv = p = Ber$$

The kinetic energy of the particle is given by

$$\text{KE} = \frac{p^2}{2m} = \frac{(Ber)^2}{2m}\ \text{J} = \frac{(Ber)^2}{2me}\ \text{eV} = \frac{B^2r^2e}{2m}\ \text{eV}$$

Hence,

$$\text{KE} = \frac{(1.6\ \text{T})^2(2\ \text{m})^2(1.6\times 10^{-19}\ \text{C})}{2(1.67\times 10^{-27}\ \text{kg})} = 4.9\times 10^8\ \text{eV}$$

$$= 490\ \text{MeV}$$

Example 22.5 The active element of a Hall-effect gaussmeter consists of a slab of semiconductor 1 mm thick, 6 mm wide, and 1.5 cm long. A current of 0.1 A flows the length of the slab, which is placed so that its flat surface is perpendicular to the magnetic field to be measured. The density of electrons in the semiconductor is $10^{24}\ \text{m}^{-3}$. Find the potential difference between the two sides of the slab for a magnetic field of 1.2 T.

The Hall voltage across the width of the slab is $V_H = E_H d$, where E_H is given by Equation (22.6). To evaluate E_H, we must know the velocity of the electrons. From Equation (18.6), we have

$$v = \frac{J}{ne} = \frac{I}{tdne}$$

where t is the thickness of the slab and td its cross-sectional area. Thus,

$$\begin{aligned} V_H &= E_H d \\ &= Bvd = \frac{BI}{tne} \\ &= \frac{(1.2\text{ T})(0.1\text{ A})}{(10^{-3}\text{ m})(10^{24}\text{ m}^{-3})(1.6 \times 10^{-19}\text{ C})} = 0.75\text{ mV} \end{aligned}$$

22.6 Generation of Magnetic Fields

22.6 (a) Magnetic Field Due to an Infinitely Long Straight Current

As Oersted noted, a compass placed under a wire directed north-south will deflect and point in the east-west direction when a current passes in the wire. Moreover, the compass reverses direction if it is placed above this wire rather than below. A complete survey of the field due to a long straight current-carrying wire shows that the field lines form concentric circles about the wire, as in Figure 22.13.

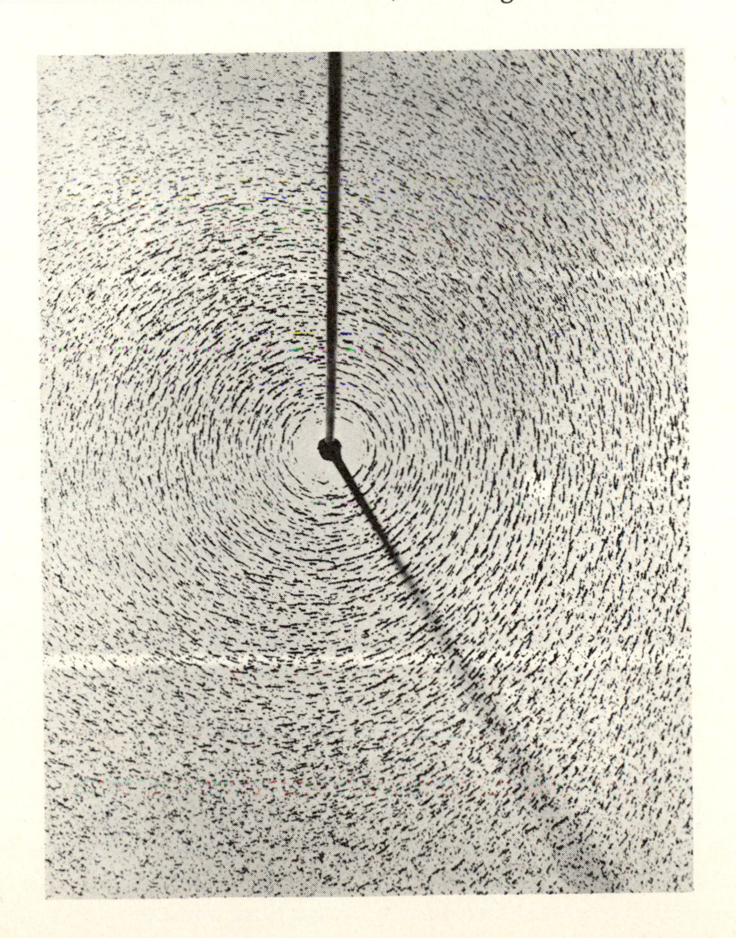

Figure 22.13 *The magnetic field lines due to a long straight current form concentric circles about the current.*

The magnetic field points neither toward nor away from the source of the field, the current, but always at right angles to it! The *direction* of the magnetic field, which can be determined by observing the orientation of a compass needle, obeys Right-Hand Rule No. 2:

RHR-II: If the current is grasped in the right hand so that the thumb points in the direction of the current, the fingers of the right hand encircle the current in the direction of the magnetic field.

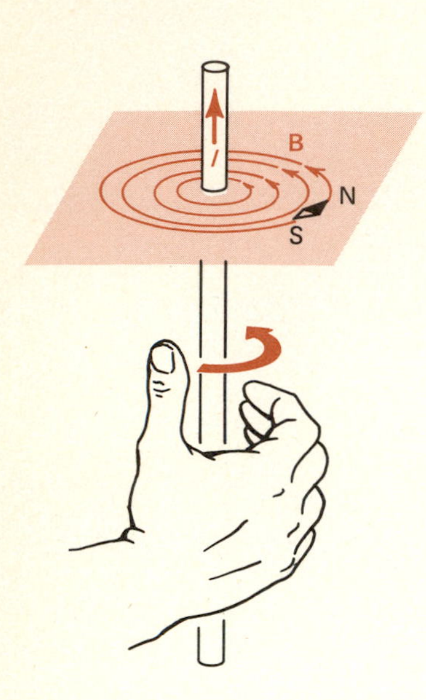

Figure 22.14 The direction of the field due to a long straight current can be established by exploring the surrounding region with a compass. One then finds that the direction of the field is consistent with Right-Hand Rule No. 2, RHR-II, illustrated here.

● ***Example 22.6*** A horizontal wire carries a current from east to west. What is the direction of the magnetic field due to this current directly above and below the wire?

Referring to Figure 22.15, applying RHR-II shows that directly above the wire the **B** field points north; directly below, it points south. ●

The *strength* of the magnetic field a distance r from a long straight wire carrying a current I is found to be proportional to the current and inversely proportional to the distance r. That is,

$$B = \text{constant} \times \frac{I}{r} \qquad \textbf{(22.7)}$$

By definition, the value of the constant in Equation (22.7) is 2×10^{-7} if the surrounding medium is vacuum and SI units are used. For reasons that will be apparent shortly, Equation (22.7) is generally written

$$B = \frac{\mu_0}{2\pi}\frac{I}{r} \qquad \textbf{(22.8)}$$

where

$$\mu_0 = 4\pi \times 10^{-7}\ \text{T·m/A} \qquad \textbf{(22.8a)}$$

is known as the *permeability of free space*.

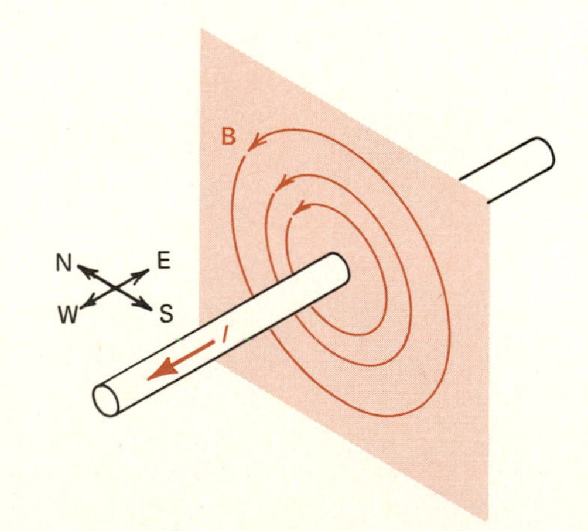

Figure 22.15 The field lines about a wire carrying current from east to west.

22.6 (b) Force between Two Currents; Definition of the Ampere

Since a current in a long straight wire generates a magnetic field, it follows from Equation (22.3) that a second current-carrying wire placed near the first may experience a force. If two long parallel wires, separated

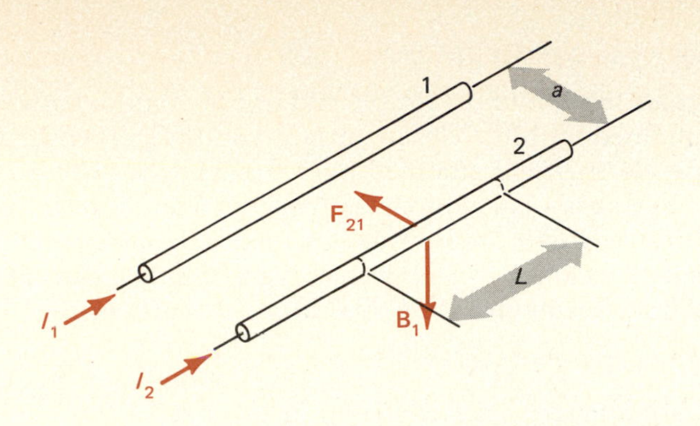

Figure 22.16 *Two long, parallel wires, carrying currents I_1 and I_2, exert forces on each other. Current I_1 produces a magnetic field* $\mathbf{B}_1$ *at the position of current I_2. If I_1 and I_2 are parallel, wire 2 will feel a force* $\mathbf{F}_{21}$ *directed toward I_1 due to current I_2 and field* $\mathbf{B}_1$*. Wire 1 will then feel an equal but opposite force. The two wires, therefore, attract each other under these conditions.*

by a distance a, carry currents I_1 and I_2, as shown in Figure 22.16, one exerts a force on the other, which we now calculate.

Consider a length L of wire 2, with current I_2. This wire is in a region of magnetic field $\mathbf{B}_1$, generated by current I_1. If we denote the force on I_2 due to I_1 by $\mathbf{F}_{21}$, then according to Equation (22.3),

$$F_{21} = B_1(2)I_2L \sin \theta \tag{22.9}$$

where $\mathbf{B}_1(2)$ designates the magnetic field due to current I_1 at the location of I_2.

The magnitude of this field is given by Equation (22.8), and its direction by RHR-II. It follows from the latter that $\mathbf{B}_1(2)$ points down, as shown in Figure 22.16, and is therefore perpendicular to the wire carrying I_2. The direction of the force $\mathbf{F}_{21}$ is determined by applying RHR-I, where the thumb of the right hand points along the direction of I_2. The force $\mathbf{F}_{21}$ therefore points in the direction shown in Figure 22.16. By Newton's third law, wire 1 must then feel a force $\mathbf{F}_{12} = -\mathbf{F}_{21}$ directed toward wire 2. Thus, *two long parallel wires, carrying currents in the same direction, attract each other.*

The magnitude of this attractive force is now readily determined by substituting Equation (22.8) into (22.9):

$$\begin{aligned} |F_{12}| &= |F_{21}| \\ &= B_1(2)I_2L \sin \theta \\ &= \frac{\mu_0}{2\pi}\frac{I_1}{a} I_2L \sin \theta \\ &= \frac{\mu_0}{2\pi}\frac{I_1I_2L}{a} \end{aligned} \tag{22.10}$$

Equation (22.10) is the defining relation for the unit of current. The exact definition of the ampere is as follows.

DEFINITION: The ampere is that constant current which if maintained in two straight parallel conductors of infinite length, of negligible circular cross section, and placed one meter apart in a vacuum, would produce between these conductors a force equal to 2×10^{-7} newton per meter of length.

Although the ampere is, by international agreement, the fourth fundamental unit (the other three are the kilogram, the meter, and the second), it is more convenient for dimensional analysis to express all electric and magnetic quantities in terms of the unit of charge, the coulomb. The coulomb is really a derived unit and is defined as the quantity of charge that crosses an area when a current of one ampere passes through this area for one second.

22.6 (c) Ampère's Law

In Chapter 18 we found a simple and useful relation between the total charge contained within a closed surface and the perpendicular component of the electric field summed over this surface. There is a relation analogous to Gauss's law, known as *Ampère's law,* that can be employed in calculating magnetic fields. This law relates the tangential component of the magnetic field around a closed curve to the total current that passes through the surface bounded by this curve. Its mathematical statement is

$$\Sigma B_{\parallel}\, \Delta \ell = \mu_0 I \tag{22.11}$$

where $B_{\parallel}$ is the tangential component of the **B** field along the small element of length $\Delta\ell$ of the closed curve that bounds the area penetrated by the current I.

The simplest application of Ampère's law is the derivation of the magnetic field due to a long straight current. Imagine this current penetrating a circular disk, as indicated in Figure 22.17. By symmetry, the magnitude of $B_{\parallel}$ must be the same everywhere over the circumference of this disk. Hence, Equation (22.11) reduces to

$$B_{\parallel}\Sigma\, \Delta \ell = B(2\pi r) = \mu_0 I \tag{22.12}$$

where r is the radius of the disk. If we divide both sides of (22.12) by $2\pi r$, we arrive at the expression for the field due to an infinitely long current, Equation (22.8).

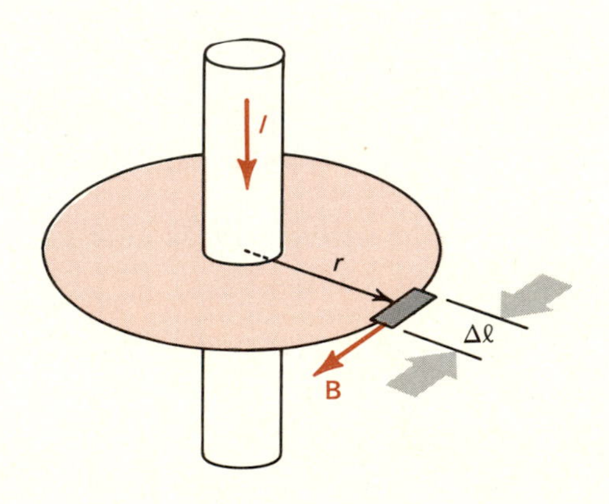

Figure 22.17 *Ampère's law as it applies to the field near a current I.*

22.6 (d) Current Loops, Solenoids, and Toroids

Ampère not only observed and measured the forces between two currents but he also recognized that a simple current loop establishes a magnetic field that bears a striking similarity to that of a small bar magnet. With astonishingly accurate, almost prophetic intuition, Ampère suggested that the magnetism of iron and other materials should be attributed to minute current loops of atomic dimensions:

> I conceive the phenomena presented by magnets, by considering them as if they were assemblages of electric currents in very small circuits about their particles.

Today it is generally accepted that all magnetic effects can be traced to two primary sources: (1) currents, that is, relative motion of electric charges, positive or negative, and (2) magnetic dipole moments of fundamental subatomic particles, such as electrons, protons, and neutrons, that are related to their intrinsic spin angular momenta. Here we limit discussion to magnetic effects associated with various current configurations.

Let us now see if we can understand what led Ampère to his remarkable conclusion. In Figure 22.18 we show a single loop of wire that carries a current I as indicated. If we apply RHR-II, that is, place the right hand near the top of the loop, with thumb pointing in the current direction, we see that the **B** field due to this portion of the loop points out of the paper below and into the paper above that segment of the wire. If we repeat this application of RHR-II at any portion of the loop, we see that the counterclockwise current produces a magnetic field that points out of the page inside the loop and into the page outside the loop. This field pattern is shown in perspective in Figure 22.19(a); in Figure 22.19(b) we see the same pattern portrayed by iron filings.

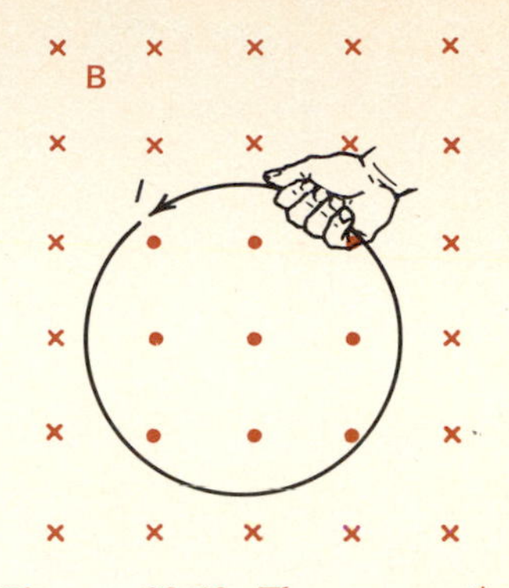

Figure 22.18 The magnetic field due to a single loop of current. Applying Right-Hand Rule 1, we see that for the assumed direction of current, the magnetic field inside the loop points out of the page and the field outside the loop points into the page in the plane of the paper.

Comparison of Figure 22.19 with Figure 22.4 suggests that the fields from a current loop and from a squat, small bar magnet are much alike. In fact, operationally—that is to say, in terms of their effects on the external environment—one simply cannot distinguish between a small current loop and a small bar magnet.

A corollary of this inability to distinguish between current loops and magnets as regards their effects on surrounding space is that their response to an external field also be identical. In particular, we know that a bar magnet experiences a torque in a uniform magnetic field that tends to align it with the field lines. Will such a torque also act on a current loop? Let's see.

It is convenient to alter the shape of the loop from circular to rectangular, as in Figure 22.20. The rectangular loop of width a and length b is located in a uniform magnetic field **B**, as shown. To determine the torque,

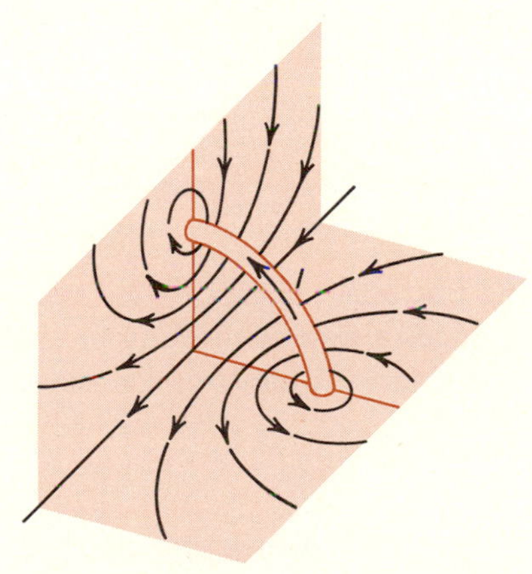

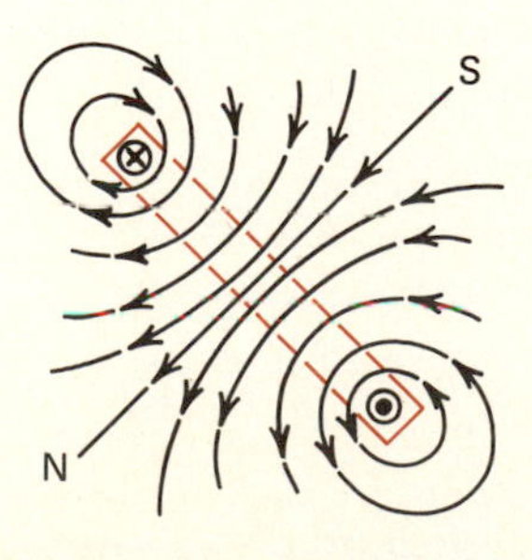

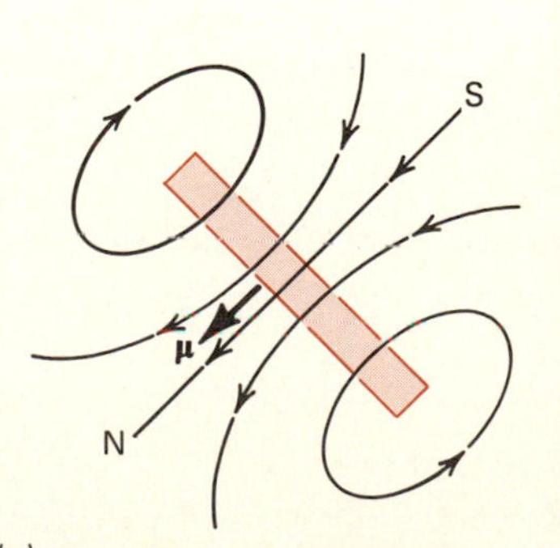

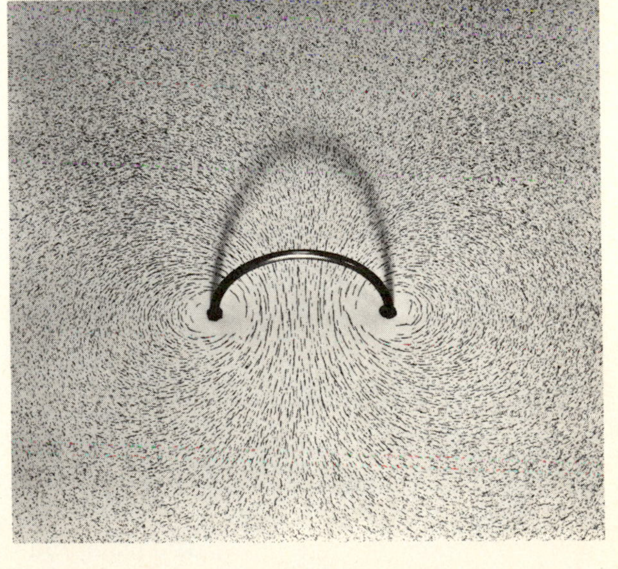

Figure 22.19 (a) Perspective drawing, showing the magnetic field near a current loop. (b) This field pattern is illustrated by the iron filings near the current loop. (c) The field patterns due to a current loop and to a squat bar magnet are indistinguishable, justifying Ampère's hypothesis.

if any, on this loop, we calculate the forces acting on each side of the rectangle when a current I flows in the loop.

The vertical and horizontal forces $\mathbf{F}_v$ and $\mathbf{F}_h$ are shown in Figure 22.20. The sum of these forces vanishes; there is no net force on the coil and it will not, as a whole, accelerate in any direction. There is, however, a torque on the rectangular loop about the vertical axis. This torque is given by

$$\tau = F_h\left(\frac{a}{2}\right)\sin\phi + F_h\left(\frac{a}{2}\right)\sin\phi = F_h a \sin\phi$$

where F_h is the force experienced by the wire of length b on the side of the rectangular loop and $(a/2)\sin\phi$ is the moment arm of this force about the vertical axis through the center of the loop.

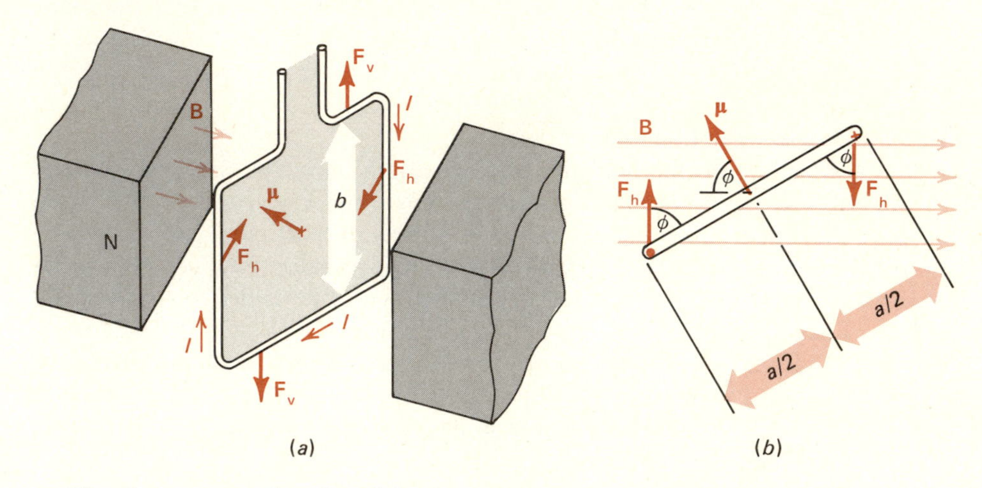

Figure 22.20 *A current-carrying loop of wire in a uniform magnetic field experiences a torque. The torque is such as to align the equivalent bar magnet, indicated by the magnetic moment vector* $\boldsymbol{\mu}$, *with the magnetic field* **B**. *(a) Perspective view. (b) Top view.*

The force on each side of the loop is given by (22.3) and is

$$F_h = BIb$$

The torque is therefore

$$\tau = BIab \sin\phi \qquad \textbf{(22.13)}$$

As expected, the current loop in the uniform magnetic field behaves just like a bar magnet; that is, it experiences a torque tending to align it with the field. It is convenient to define a vector quantity, the magnetic moment $\boldsymbol{\mu}$, whose magnitude is given by

$$\mu = IA \qquad \textbf{(22.14)}$$

where A is the area of the loop (in this case $A = ab$), and whose direction is that of the magnetic field *due to the current loop at its center.* The magnetic moment when placed in an external magnetic field will then experience a torque that tends to align $\boldsymbol{\mu}$ and **B**. The direction of the magnetic-moment vector is most readily established by means of Right-Hand Rule No. 3:

RHR-III: If the right hand encircles the current loop or coil so that the fingers point in the direction of the current, then the thumb will point in the direction of the magnetic moment and also in the direction of the magnetic field inside the loop or coil due to its current.

The magnitude of the magnetic field due to a single loop of current at some arbitrary point in space cannot be calculated without resort to cal-

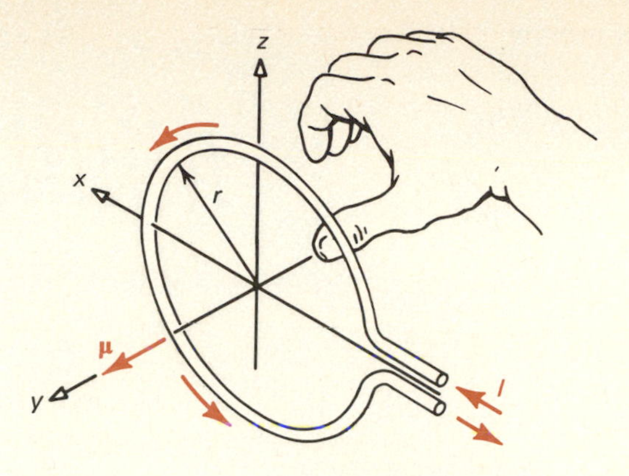

Figure 22.21 *The direction of the magnetic moment of a current loop can be determined by applying Right-Hand Rule 3, RHR-III, illustrated here. Note that the magnetic field at the center of the loop has the same direction as the magnetic moment* $\boldsymbol{\mu}$.

culus. At the center of the loop, the field is

$$B = \mu_0 \frac{I}{2r} \qquad \text{(single loop)} \qquad \textbf{(22.15)}$$

where r is the radius of the circular loop.

A *solenoid* (the word derives from Greek and was introduced by Ampère) is a long cylindrical coil with many loops placed next to each other. The magnetic field inside a rather loosely wound and relatively short solenoid is shown in Figures 22.22(*a*) and (*b*). Not surprisingly, this field is much like the field of a longer bar magnet, since such a magnet can be thought of as composed of many short magnets, placed end to end with opposite poles facing each other. The individual loops of the solenoid correspond to short magnets as we saw in the preceding pages.

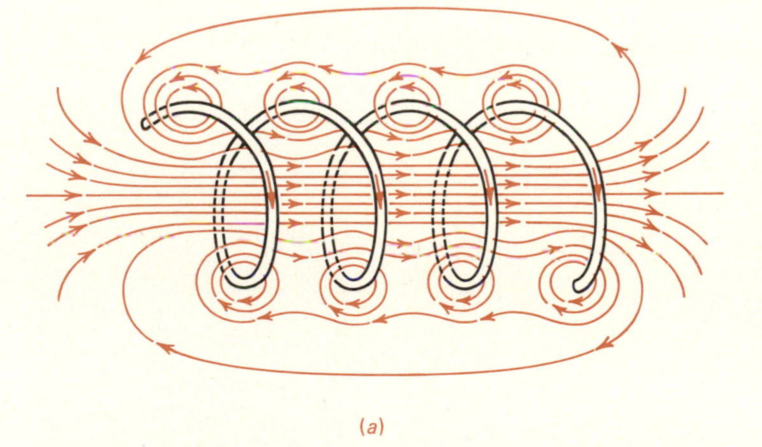

(*a*)

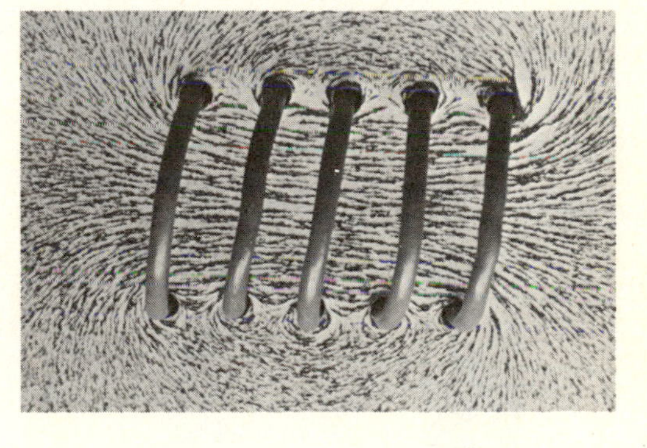

(*b*)

Figure 22.22 *A series of current loops constitutes a solenoid. The magnetic field inside and outside a loosely wound solenoid is sketched in (a) and displayed in (b). Note that near the solenoid most of the field is confined within the coil and that inside the solenoid the field is relatively uniform.*

From these figures it appears that we can expect the field inside a long, tightly wound solenoid to be essentially uniform, and so it is. The magnitude of the field is given by

$$B = \mu_0 n I \qquad \text{(solenoid)} \qquad \textbf{(22.16)}$$

where n is the *number of turns per unit length,* not the total number of turns. Note that well within the solenoid, the field is independent of position and its value does not depend on the diameter of the cylindrical coil. (These statements are not valid for the regions near the ends of the solenoid.)

A *toroid,* shown in Figure 22.24, is really a solenoid that has been bent into a complete circle. Whereas the magnetic field lines of a solenoid

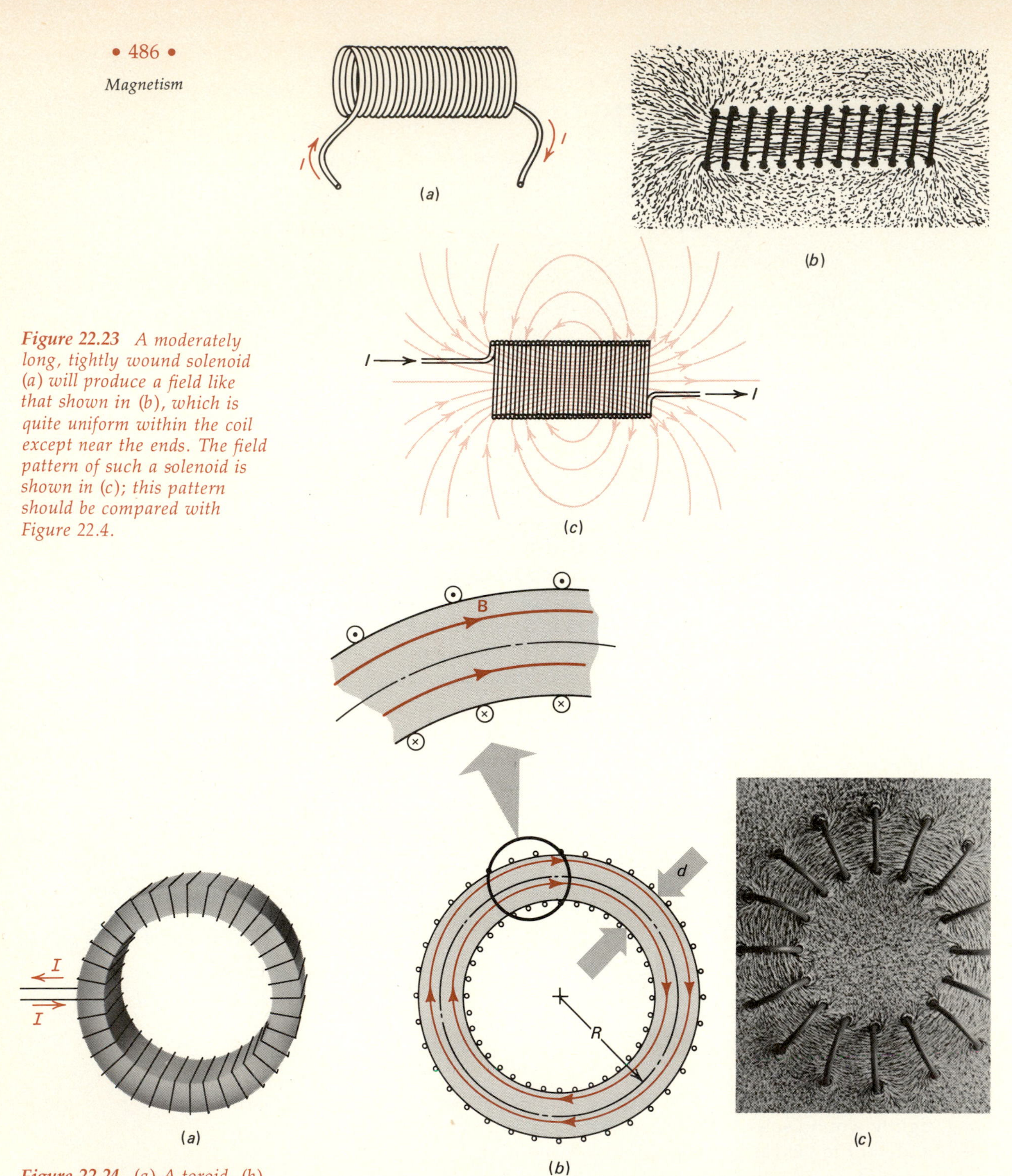

Figure 22.23 A moderately long, tightly wound solenoid (a) will produce a field like that shown in (b), which is quite uniform within the coil except near the ends. The field pattern of such a solenoid is shown in (c); this pattern should be compared with Figure 22.4.

Figure 22.24 (a) A toroid. (b) Cross section of a toroid. (c) The field of a toroidal coil is almost completely confined to the annular region.

must close upon themselves in the region external to the coil, the flux of the toroid is almost entirely confined to the annular region inside the coil.

The magnetic field within the toroid can be calculated by applying Ampère's law. Imagine a circular path of radius R, shown dashed in Figure 22.24(b), which shows the toroid of Figure 22.24(a) in cross section. The current which penetrates the surface bounded by that circle is a downward current of value NI, where I is the current in the toroid's winding and N is the number of turns on the toroid. The direction of the **B** field due to this current is given by RHR-II and is indicated in Figure 22.24(*b*).

Since the circumference of the circle of radius R is $2\pi R$ and the magnetic field inside the toroid must have circular symmetry, we can conclude from Equation (22.11) or (22.12) that the field inside the toroid is

$$B = \frac{\mu_0 NI}{2\pi R} \quad \text{(toroid)} \tag{22.17}$$

Note that the magnetic field inside the toroid is not quite uniform; it approaches uniformity if the diameter of the coil winding is small compared to the radius of the toroid. In that case, Equation (22.17) is equivalent to the expression for the solenoid, Equation (22.16), since $N/2\pi\langle R\rangle = n$, where $\langle R\rangle$ is the average radius of the toroid and n the number of turns per unit length.

● *Example 22.7* A solenoid is to be wound on a cylindrical form, 20 cm long and 5 cm in diameter, which will produce a field of 0.2 T in the central region with a current of 1.5 A. How many turns should be wound on this form?

From Equation (22.16), we have

$$n = \frac{B}{\mu_0 I} = \frac{0.2\ \text{T}}{(4\pi \times 10^{-7}\ \text{T·m/A})(1.5\ \text{A})} = 1.06 \times 10^5\ \text{turns/m}$$

Hence,

$$N = (1.06 \times 10^5\ \text{turns/m})(0.2\ \text{m}) = 21{,}200\ \text{turns}$$ ●

*22.7 Meters

Oersted quickly recognized that the deflection of a compass needle could provide a measure of the current in a circuit. Indeed, the earliest galvanometers were of this type; the indicator was a compass needle mounted between coils that carried the unknown current (Figure 22.25).

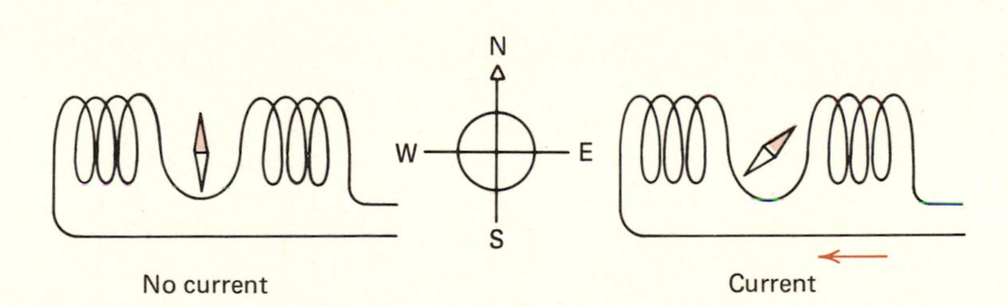

Figure 22.25 A primitive galvanometer. A small current flowing through the coils establishes a magnetic field in the region of the compass needle, which then deflects from the north-south orientation. The angle the needle makes with the N-S direction is a measure of the magnetic field due to the current in the coils.

Nondigital meters in use today are almost all based on the D'Arsonval design, in which the roles of magnet and coil are interchanged; that is, the magnet is stationary and the coil through which the current passes is allowed to rotate about an axis. The principal parts of the instrument consist of a permanent magnet with suitably shaped pole faces, a light, rectangular coil supported on a fine suspension, and a soft iron cylinder. The cylinder assures that the magnetic field in the air gap within which the coil is free to rotate is radially directed and nearly uniform over most of the region.

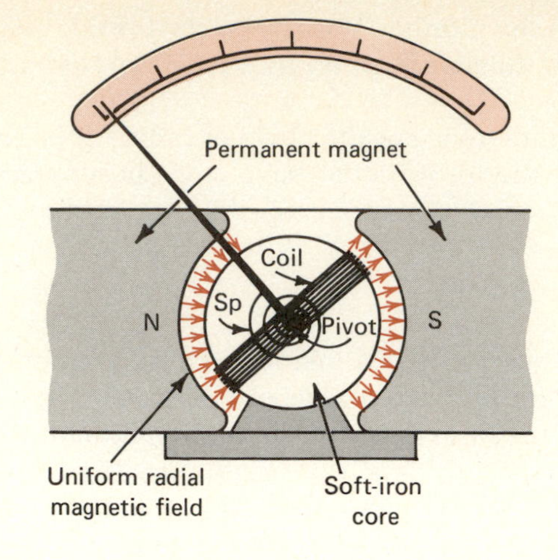

Figure 22.26 *Drawing of a typical mèter movement, based on the D'Arsonval design. The magnetic field lines in the air gap are indicated.*

Summary

A *magnetic field* **B** is said to exist in any region in which a moving charge experiences a force that depends on its velocity **v**. If **B** and **v** make an angle θ, the force on the moving charge is given by

$$F = Bqv \sin \theta$$

The *direction* of **B** can be determined by means of *Right-Hand Rule 1*, illustrated in Figure 22.2.

A force acts on a current in a magnetic field. The force on a length ℓ of wire carrying current I due to a magnetic field **B** that makes an angle θ with the current direction is given by

$$F = BI\ell \sin \theta$$

The direction of the force is given by Right-Hand Rule 1.

A charged particle moving with a velocity **v** in the plane perpendicular to a magnetic field describes a circular trajectory. The radius of this circular path is given by

$$r = \frac{mv}{Bq}$$

where m is the mass of the charged particle. The number of revolutions made by this particle per second is known as the *cyclotron frequency*,

$$f_c = \frac{Bq}{2\pi m}$$

The *Hall effect* is the appearance of a potential difference across a conductor when it is placed in a magnetic field that has a component perpendicular to the current in the conductor. The potential difference is in a direction that is perpendicular to the plane formed by the current and magnetic field directions and is proportional to the current and to the magnetic field.

A current gives rise to a magnetic field. The **B** field in the region about a very long straight current is in the form of concentric circles in the plane perpendicular to the current. According to *Ampère's law*,

$$\Sigma B_{\parallel} \, \Delta \ell = \mu_0 I$$

where $B_{\parallel}$ is the component of **B** tangent to the small segment of line $\Delta\ell$, and the totality of these segments forms a closed curve. The current that passes through the surface bounded by this closed curve is represented by I.

Application of Ampere's law to an infinitely long straight current gives

$$B = \frac{\mu_0}{2\pi}\frac{I}{r}$$

where r is the distance from the current I. Here and in Ampère's law, the quantity μ_0 is the *permeability of free space,* whose value is by definition

$$\mu_0 = 4\pi \times 10^{-7}\ \mathrm{T{\cdot}m/A}$$

The direction of the magnetic field near a current I is established by means of Right-Hand Rule 2, illustrated in Figure 22.14.

Two currents flowing in parallel directions attract each other as a result of the magnetic fields generated by them. This force is used to define the unit of current, one ampere.

A single loop of current creates a magnetic field identical to that of a magnetic dipole. The *magnetic moment* of a single loop of current I is

$$\mu = IA$$

where A is the area of the current loop. The direction of $\boldsymbol{\mu}$ is that of the magnetic field at the center of the coil due to the current flowing in it, and can be determined by using Right-Hand Rule 3, illustrated in Figure 22.21.

A *solenoid* is a long cylindrical coil, usually having many turns of wire. The field near the center of a solenoid is given by

$$B = \mu_0 n I$$

where n is the number of turns per unit length.

The field inside a *toroidal coil* is given by

$$B = \frac{\mu_0 N I}{2\pi R}$$

where R is the distance from the center of the toroid and N is the total number of turns wound on the toroid.

Multiple Choice Questions

22.1 Four long parallel wires are arranged so that they are at the four corners of a square when viewed end-on. Each wire carries the same amount of current, and the current directions are as indicated. The magnetic field at the center of the square

(a) points toward the top of the page.
(b) points to the bottom of the page.
(c) points to the right.
(d) points to the left.
(e) is zero.

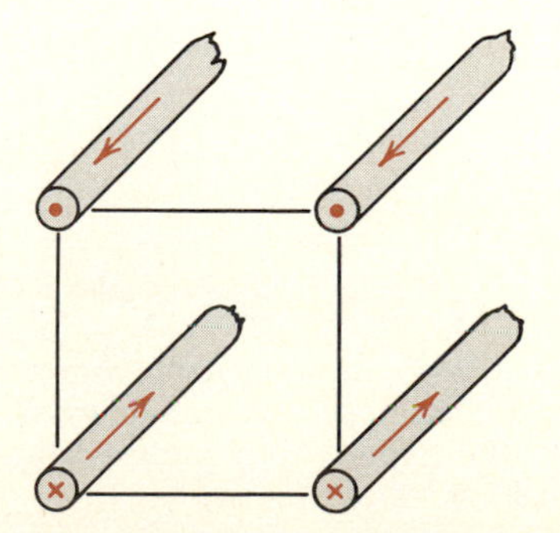

Figure 22.27

22.2 The force on an electric charge moving in a magnetic field is

(a) independent of the speed of the charge.
(b) inversely proportional to the charge.

(c) directed perpendicular to the velocity of the charge.

(d) directed along the magnetic field.

(e) (c) and (d) are both correct.

22.3 Current I_1 flows in the wire parallel to the x axis (Figure 22.28). This wire is placed just above a wire that carries a current I_2 in the y direction. The magnetic interaction force will tend to

(a) push I_2 into the paper.

(b) pull I_2 out of the paper.

(c) turn I_2 clockwise.

(d) turn I_2 counterclockwise.

(e) turn I_2 about the x axis so that the arrow comes out of the paper.

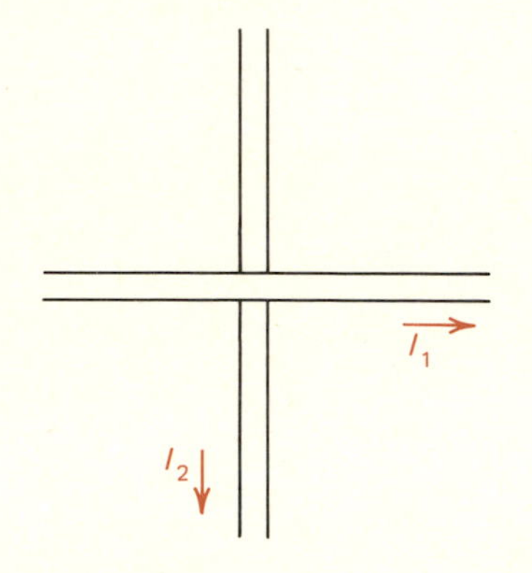

Figure 22.28

22.4 Two long straight wires carry the same amount of current in the directions indicated. The wires cross each other in the plane of the paper. The field $\mathbf{B}_A$ at A and the field $\mathbf{B}_B$ at B are related as follows.

(a) $\mathbf{B}_A = -\mathbf{B}_B$

(b) $\mathbf{B}_A = \mathbf{B}_B$

(c) $|\mathbf{B}_A| < |\mathbf{B}_B|$

(d) $|\mathbf{B}_A| > |\mathbf{B}_B|$

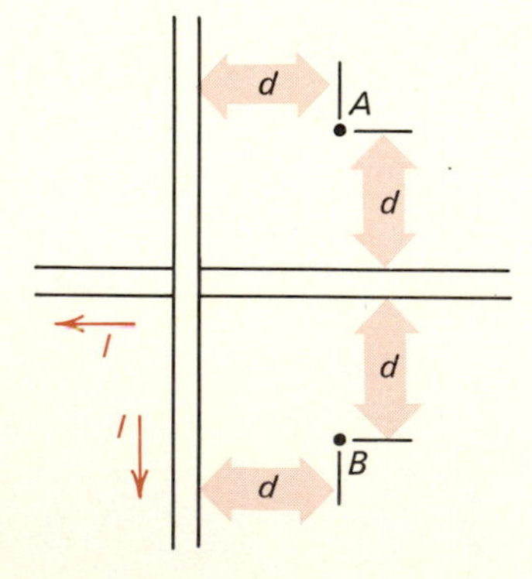

Figure 22.29

22.5 An electron moving with velocity $\mathbf{v}$ to the right enters a region of uniform magnetic field that points out of the paper. After the electron enters this region, it will be

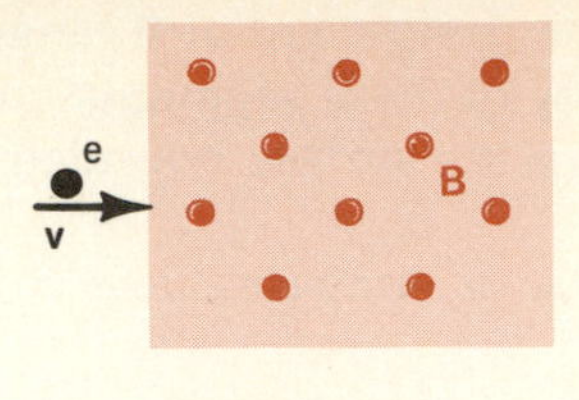

Figure 22.30

(a) deflected out of the plane of the paper.

(b) deflected into the plane of the paper.

(c) deflected upward.

(d) deflected downward.

(e) undeviated in its motion.

22.6 A charged particle moves through a constant magnetic field. The Lorentz force on this particle

(a) increases its kinetic energy whatever the sign of the charge.

(b) increases the KE of the particle if its charge is positive and decreases the KE of the particle if its charge is negative.

(c) does not influence the particle's kinetic energy.

(d) is in the direction of motion of the particle.

(e) is independent of the speed of the particle whatever the direction of motion.

22.7 A pair of electrical conductors are in the shape of very long thin metallic strips. The conductors, shown in cross section, carry equal currents in opposite directions. The currents are uniformly distributed inside the thin strips. The magnetic field at the point A, midway between the strips is

(a) directed up.

(b) directed down.

(c) directed to the right.

(d) directed to the left.

(e) zero.

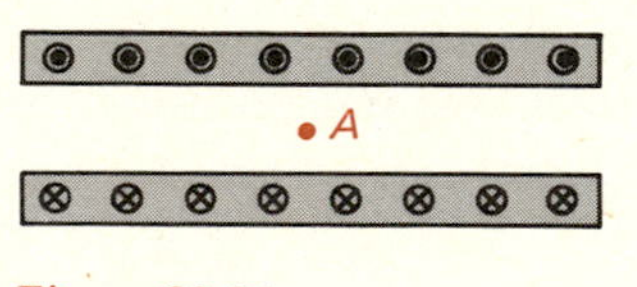

Figure 22.31

22.8 A rectangular loop, carrying a small current I, is in the plane of the paper. A magnetic field due to a large magnet points out of the plane of the paper.

(a) The torque on the coil is such as to push side *1* into and side *3* out of the paper.

(b) The total field inside the coil is greater than that outside the coil.

(c) The torque on the coil is such as to push side *3* into the paper and side *1* out of the paper.

(d) Sides *1* and *3* of the coil are both pulled out of the plane of the paper.

(e) None of the above is correct.

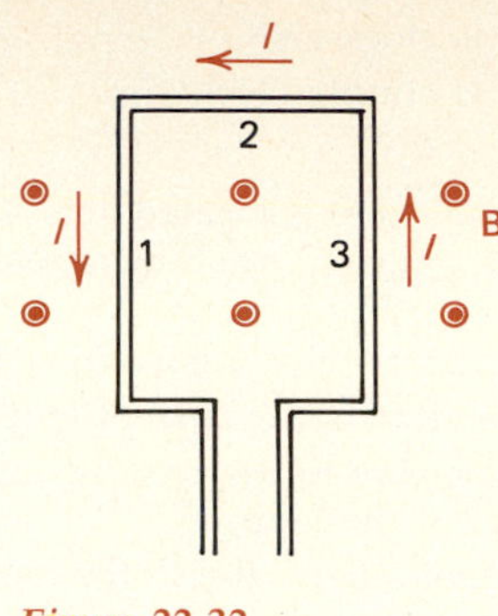

Figure 22.32

22.9 A proton (positively charged hydrogen nucleus) has an initial north-directed velocity but is observed to curve toward the east as a result of a magnetic field **B**. The direction of **B** is

(a) to the east.
(b) to the west.
(c) upward, out of the page.
(d) downward, into the page.

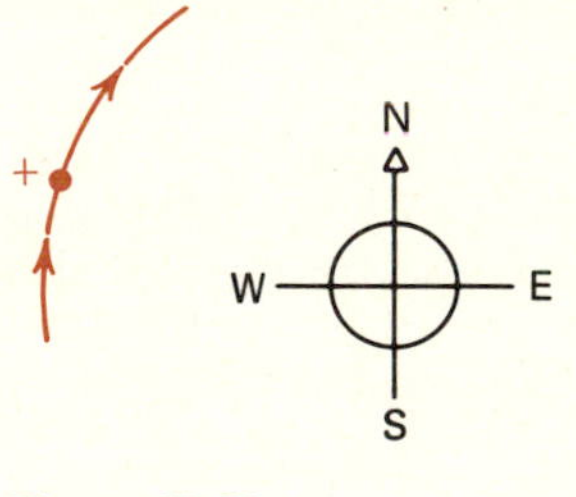

Figure 22.33

22.10 The net force on a circular coil carrying a current I in a uniform magnetic field that is directed perpendicular to the plane of the coil is

(a) zero.
(b) proportional to the area of the coil.
(c) proportional to the magnetic field as well as the area of the coil.
(d) proportional to the magnetic field, area of the coil, and number of turns on the coil.
(e) proportional to the product of current, number of turns, and magnetic field.

22.11 Two long straight wires carry currents of 20 A and 10 A as shown. The wires cross at 90°. The figure that most nearly represents the magnetic field in the plane formed by the two wires is

(a) *A*
(b) *B*
(c) *C*
(d) *D*
(e) *E*

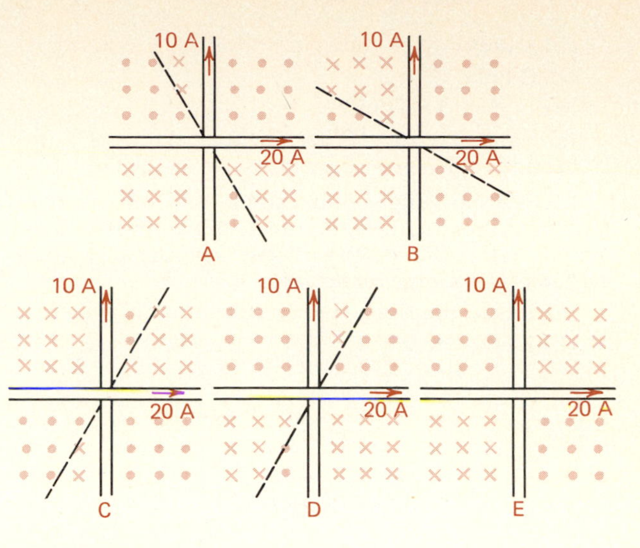

Figure 22.34

22.12 The charged particles describing circular paths (Figure 22.35) in the uniform magnetic field **B** have equal masses and equal kinetic energies. Select the correct relation from the choices given below.

(a) $q_1 = +, q_2 = -; |q_1| > |q_2|$
(b) $q_1 = +, q_2 = -; |q_1| = |q_2|$
(c) $q_1 = +, q_2 = -; |q_1| < |q_2|$
(d) $q_1 = -, q_2 = +; |q_1| > |q_2|$
(e) $q_1 = -, q_2 = +; |q_1| < |q_2|$

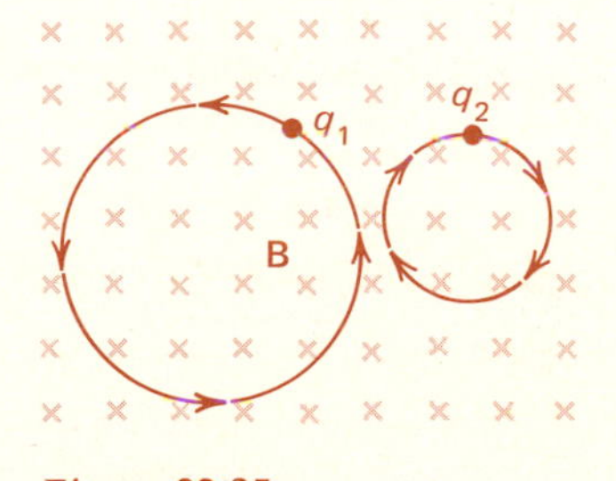

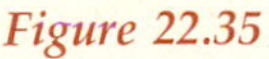

Figure 22.35

22.13 Two parallel wires carry currents of 100 A and 50 A in opposite directions as shown in Figure 22.36. In which region can the magnetic field vanish?

(a) *A* only.

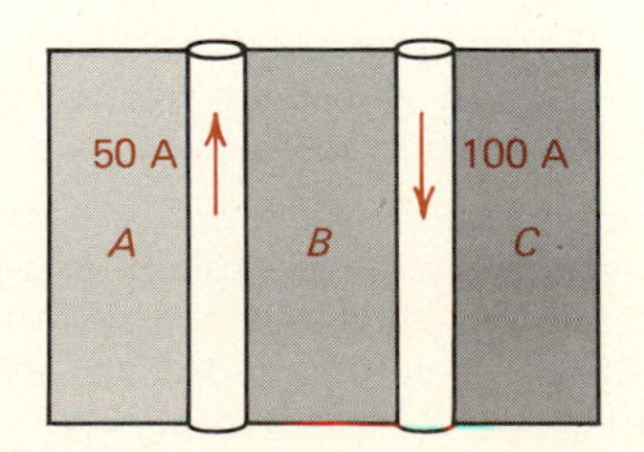

Figure 22.36

(b) B only.

(c) C only.

(d) In A and also in C.

(e) Since the currents are unequal, the field cannot vanish anywhere.

22.14 A bar magnet and a single-turn current loop (shown in cross section, Figure 22.37) are oriented as shown. Select the statement that correctly describes the force and torque *on the loop*.

(a) F is to the right, $\tau = 0$.

(b) F is to the left, $\tau = 0$.

(c) $F = 0$, τ is clockwise.

(d) $F = 0$, τ is counterclockwise.

(e) $F = 0$, $\tau = 0$.

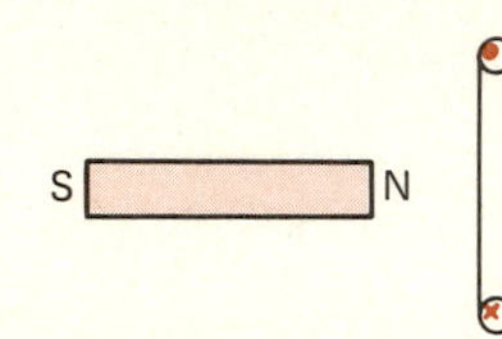

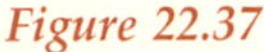

Figure 22.37

22.15 A current loop is located near a long straight wire that carries a current as shown. Select the correct statement.

(a) The loop exerts a repulsive force on the wire.

(b) The wire attracts the current loop.

(c) The **B** field due to the wire is out of the page in the region of the loop.

(d) The magnetic field due to the loop is zero at the location of the wire.

(e) The loop and wire exert no net forces on each other.

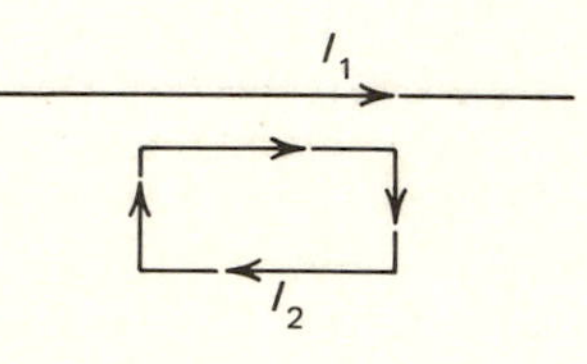

Figure 22.38

Problems

(Sections 22.2, 22.3)

22.1 An electron traveling at a velocity of 5×10^6 m/s east enters a region of uniform magnetic field of 10^{-2} T directed up. Find the magnitude and direction of the force that acts on this electron.

22.2 (a) A charged particle experiences a force due east when moving downward in a magnetic field pointing due south. What is the sign of the charge on the particle?

(b) A beam of electrons with velocity due south is to be deflected upward. What direction of the magnetic field would occasion that deflection?

• **22.3** A wire carries a current of 15 A directed straight up. If a horseshoe magnet that produces a B field of 0.4 T in its air gap is placed about the wire so that the wire is between the north and south poles of the magnet, with the north pole to the east of the wire, what is then the magnitude and direction of the force per unit length on this wire in the region of the field?

• **22.4** How fast should a proton travel at right angles to a magnetic field of 10^{-4} T so that the magnetic force just balances the gravitational force? If the direction of motion of the proton is to the north, what should the direction of the magnetic field be?

• **22.5** A beam of 2000-eV protons is directed north to south in the northern hemisphere at a point where the earth's magnetic field is pointing due north and where its dip angle is 53°. If the magnitude of the earth's field is 6×10^{-5} T at that location, what is the magnitude and direction of the force that acts on a proton?

• • **22.6** If the rectangular coil in Question 22.15 has a length of 25 cm and a width of 8 cm, and if the nearer side of the coil is 10 cm from the infinitely long straight wire, what force does the coil exert on the wire if $I_1 = 8$ A and $I_2 = 16$ A.

(Section 22.5)

• **22.7** In Figure 22.9 numerous spiral tracks are visible. What is the characteristic feature of particles that leave such inward spiraling tracks?

Figure 22.39 is a schematic drawing of a mass spectrometer, an instrument used to measure atomic masses. Ions produced by the source S *are accelerated to a known speed by the potential difference applied between the cathode* C *and anode* A. *The collimated beam then enters a region of uniform magnetic field, where its trajectory describes a circle of radius* R. *Problems 22.8 through 22.12 relate to this instrument.*

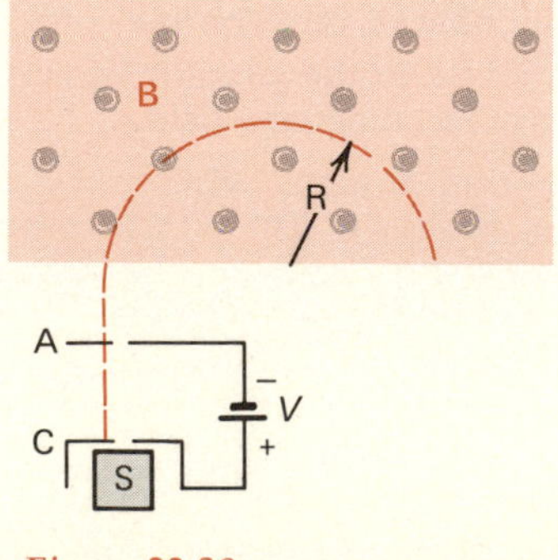

Figure 22.39

• **22.8** If the cathode-to-anode voltage is fixed at 1000 V, what is the radius of the circular path of singly ionized nitrogen atoms if the magnetic field is 0.7 T?

• **22.9** If the ion source provides doubly ionized nitrogen atoms, what will their radius of curvature be if the accelerating voltage and magnetic field of the instrument are left as in Problem 22.8?

• **22.10** Lithium has two *isotopes*, ^{6}Li and ^{7}Li. These two species differ only in atomic mass, one having a mass of 6 u, the other a mass of 7 u, where 1 u = 1.66×10^{-27} kg is known as an atomic mass unit. If a mixture of singly ionized ^{6}Li and ^{7}Li ions is accelerated through a potential difference of 800 V, what is the strength of the magnetic field so that the ^{6}Li ions' path is circular with a radius of 12 cm? What will be the radius of the trajectory of the ^{7}Li ions?

• **22.11** Singly ionized atoms that have been accelerated through a potential of 800 V describe a circular trajectory of 16-cm radius in a magnetic field of 0.2 T. What is the mass of these atoms?

• **22.12** If in Problem 22.11 the atoms had been doubly ionized on emerging from the ion source, what would their mass be?

• **22.13** A 20-keV beam of electrons is directed due south. What should the orientation and strength of the magnetic field be so the electrons of the beam describe a circular path of 40-cm diameter in the vertical plane with a clockwise sense when viewed from the west?

• **22.14** The magnetic field of a cyclotron magnet is 1.2 T. The accelerated particles are extracted when at a radius of 60 cm. What is the energy of the particles if they are protons? if they are deuterons? (Both have charge e, and $m_p = 1.67 \times 10^{-27}$ kg and $m_D = 3.34 \times 10^{-27}$ kg.)

• **22.15** An ion source produces a mixture of protons and deuterons of negligible kinetic energy. These particles are then accelerated by an electric field. After they have dropped through a potential of 2000 V, they enter a region of uniform magnetic field that is perpendicular to the velocity of the particles. Determine the radii of the circular paths of these particles in that field of magnitude 0.8 T.

• • **22.16** The apparatus shown in Figure 22.40 consists of a pair of parallel plates (shown on edge) and a large magnet (not shown). The field of the magnet is perpendicular to the electric field between the plates and directed into the plane of the paper. Charged particles enter this region through a slit at the left and emerge through the exit slit at the right. What direction must the **E** field have so that positively charged particles entering this region from the left may traverse to the exit slit undeviated? Will this same orientation of **B** and **E** fields also allow negatively charged particles to pass undeviated across this region? Show that this system can be used as a velocity selector for charged particles; that is, show that for a particular value of **B** and **E**, particles of a specific velocity will cross from entrance to exit slit regardless of their charge or mass.

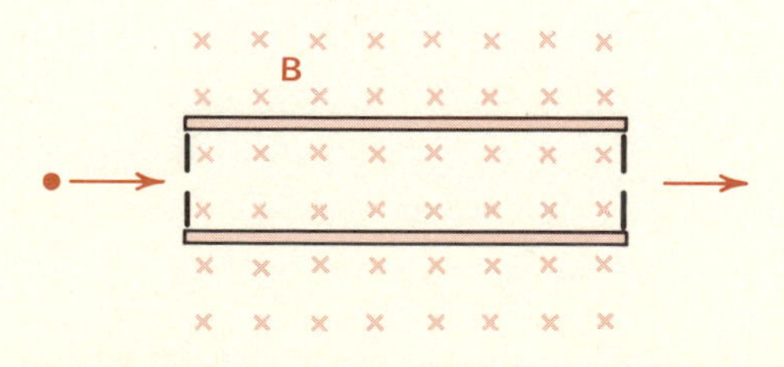

Figure 22.40

• • **22.17** The concentration of free electrons in silver is 5.85×10^{28} m^{-3}. If a sample of 0.2-mm thickness and 2-cm width is used to measure the Hall voltage in a magnetic field of 1.6 T, what current must be used to obtain a Hall voltage across the sample of 10 μV?

(Section 22.6)

22.18 A long wire carries a current of 50 A. At what distance from this wire does the magnetic field due to that current equal the earth's field, which has the value 5×10^{-5} T?

22.19 The rectangular coil shown in Figure 22.41 has 200 turns of wire and carries a current of 0.2 A. What is the magnetic moment of this coil?

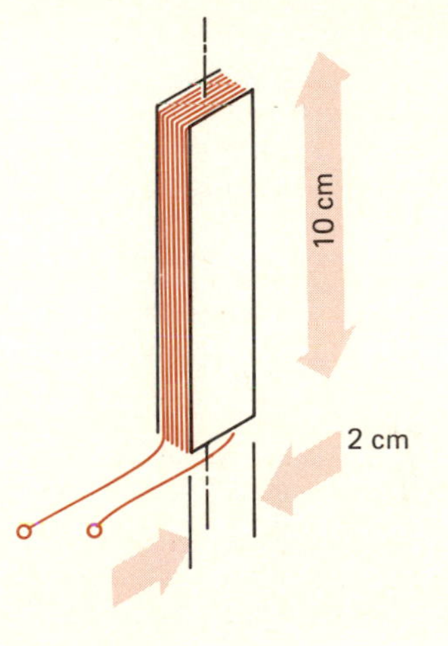

Figure 22.41

22.20 A 10-cm-long, 3-cm-diameter solenoid is wound with 4000 turns of wire. What current is required to produce a field of 0.3 T at the center of this solenoid?

• **22.21** If the coil of Problem 22.19 is located in a magnetic field of 1.2 T that is perpendicular to the axis of rotation of the coil (indicated in Figure 22.41) and makes an angle of 30° with the plane of the coil, what is then the torque that acts on this coil?

• **22.22** An east-west transmission line carries a current of 150 A. The two wires are separated by 1.2 m, and the northern wire carries current to the east. What is the magnetic field at a point midway between the two wires, and also at a point 2 m below this point?

• **22.23** You are given 200 m of copper wire that can carry a current of 0.4 A without excessive heating. This wire is to be used to make a solenoid of 1.5-cm diameter and 30-cm length. Determine the maximum magnetic field that can be achieved at the center of this solenoid.

• **22.24** The same wire as in Problem 22.23 is to be used to wind a toroidal coil. The average radius of the toroid is to be 10 cm, and the cross-sectional diameter of the coil is 1.4 cm. What is the maximum magnetic field at the average toroidal radius?

• **22.25** A circular coil of 8-cm radius has 160 turns and carries a current of 0.4 A. What is the magnetic field strength at the center of the coil?

• **22.26** A solenoid should produce a magnetic field of 0.2 T at its center and have a diameter of 4 cm. Its length shall be 30 cm. If the current in the solenoid must not exceed 8 A, how many turns must be wound on the cylindrical form?

• • **22.27** A solid cylindrical conductor of radius R carries a current I so that the current density is constant over its cross section. Assuming that the length of the conductor is much greater than its radius, show by Ampère's law that at a distance $r < R$ from the cylinder axis the magnetic field is given by

$$B = \frac{\mu_0 I r}{2\pi R^2}$$

• • **22.28** Figure 22.42 represents two very long straight wires separated by 0.8 m, each carrying a current of 40 A. The directions of the currents are indicated in the usual way. Find the magnitude and direction of the magnetic field at points A, B, and C.

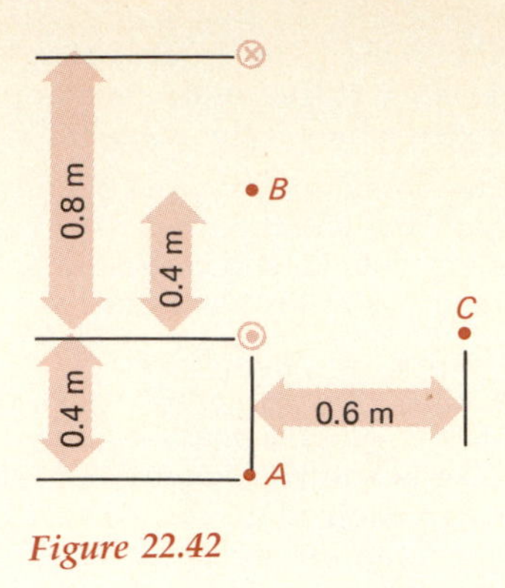

Figure 22.42

• • **22.29** Two long straight wires are suspended as shown in Figure 22.43. Each wire has a mass of 30 g/m. Both wires carry equal currents in opposite directions. If the angle between the suspensions is 8°, what is the current in each wire?

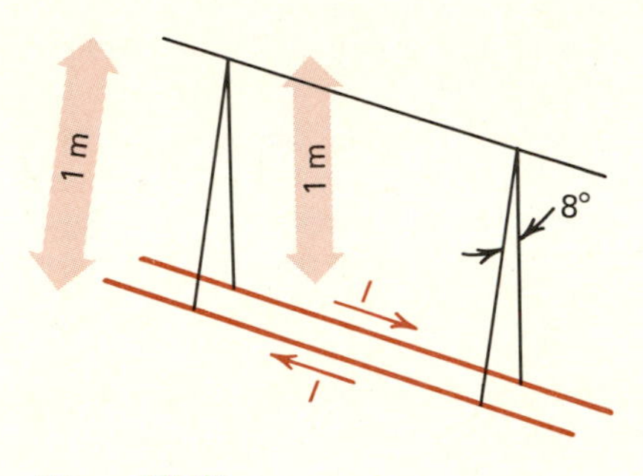

Figure 22.43

•23•

Electromagnetic Induction

I know not, but I wager that one day your government will tax it.

MICHAEL FARADAY, responding to Sir Robert Peel, prime minister, who asked what possible use electromagnetic induction might have.

23.1 Introduction

There could be no better model of the self-educated and self-motivated scientist than Michael Faraday. The son of an English blacksmith, Faraday was apprenticed at age 14 to a bookbinder, in whose establishment "there were plenty of books, and I read them." He not only read widely, but set about the task of self-education with great determination and persistence, actively corresponding with friends for the express purpose of perfecting his writing skills.

Natural philosophy was, from the start, a source of great fascination for Faraday, and he bent every effort to this interest. After attending a series of public lectures by Humphrey Davy and taking careful notes, Faraday bound and presented them to Davy with a request for a position at the Royal Institution, however menial. By chance, one of Davy's assistants became involved in a brawl about that time and was fired, and Faraday was engaged in his place. Thus began the career of perhaps the most remarkable and inventive experimental scientist, if not of all time, then surely of that century.

Although Faraday is best remembered for his seminal work in electromagnetism, he began his scientific career as assistant (and sometime valet) to Sir Humphrey, and in that capacity he turned his attention to chemical investigations. During the decade following Oersted's discovery, when Arago, Biot, Ampère, and others were concentrating their efforts on electricity and magnetism, Faraday could do so only in his "spare time." In those years he succeeded in liquefying carbon dioxide, hydrogen sulfide, and chlorine, discovered tetrachloroethylene and benzene, and was even engaged in such mundane tasks as searching for ways to improve the manufacture of optical glass and testing samples of oatmeal for impurities. His frustration at being thus occupied is evident in the following passage, taken from a letter to Ampère, written November 17, 1825.

Figure 23.1 Michael Faraday (1791–1867).

> Every letter you write me states how busily you are engaged and I cannot wish it otherwise knowing how well your time is spent. Much of mine is unfortunately occupied in very common place employment, and this I may offer as an excuse (for want of a better) for the little I do in original research.

23.2 Induced EMFs; Faraday's Law and Lenz's Law

The nagging puzzle to which Faraday's mind kept turning was this. Once Oersted had shown that an electric current generates a magnetic field, the question arose, Should not "good conductors of electricity, when placed within the sphere of this action [magnetic fields] . . . have current induced through them"?

To test this hypothesis, Faraday wound two large solenoids on the same cylindrical form so that one would be under the magnetic influence of the other whenever current passed through either. Repeated attempts to observe any effects met with failure. But then, in 1831, Faraday, who was a keen observer, noticed "the slight deflection of the [galvanometer] needle . . . at the moment of completing the connexion, always in one direction, and the equally slight deflection produced when the contact was broken, in the other direction."

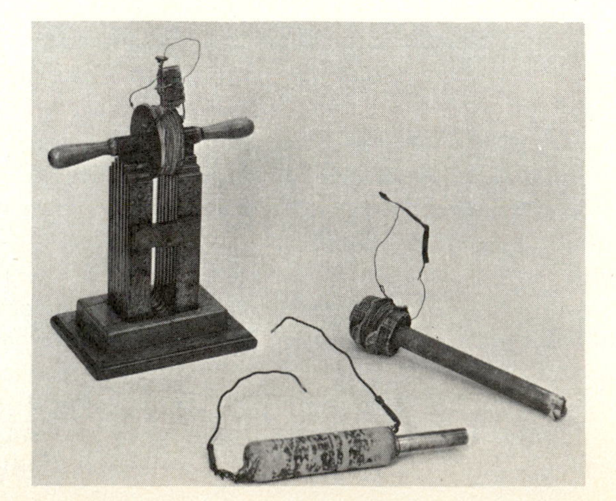

Figure 23.2 Some of the coils used by Faraday during his experiments on electromagnetic induction.

At this point, Faraday, with characteristic inventiveness, solved a nasty technical problem by replacing the clumsy galvanometer with a small steel needle.

> The results . . . led me to believe that the battery current through one wire, did, in reality, induce a similar current through the other wire, but that it continued for an instant only, . . . and, therefore, might magnetise a steel needle, although it scarcely affected the galvanometer.
>
> This expectation was confirmed: for on substituting a small hollow helix, formed round a glass tube, for the galvanometer, introducing a steel needle, making contact as before between the battery and the inducing wire, and then removing the needle before the battery contact was broken, it was found magnetised.
>
> When the battery contact was first made, then an unmagnetised needle introduced into the small indicating helix, and lastly the battery contact broken, the needle was found magnetised to an equal degree apparently as before; but the poles were of the contrary kind.

The beauty and impeccable logic of the argument, and the care and precision with which the conclusions were tested, are characteristic of Faraday's work.

Within a relatively short span he had determined that whenever there was a temporal change in the magnetic flux through a closed circuit, an emf was induced in that circuit. The magnitude of this induced emf is proportional to the rate at which the flux is changed, though the manner by which that change is brought about is of no consequence. For instance, an emf can be induced in a loop of wire by any of several ways:

1. Placing it near another loop or coil and changing the current in that second coil. This was the procedure Faraday used initially (Figure 23.3(*a*)).

2. Keeping the current in the second coil steady, but moving the two coils relative to each other (Figure 23.3(*b*)).

3. Moving a permanent magnet into or out of the loop (Figure 23.3(*c*)).

4. Rotating, or otherwise moving the loop in a steady magnetic field to change the flux that passes through it (Figure 23.3(*d*)).

5. Changing the shape of the loop in a steady field so that its area changes with time (Figure 23.3(*e*)).

The crucial observation of Faraday was that the emf induced in a single loop was related to the *time variation* of the magnetic flux that penetrated the loop. This can be summarized in one simple mathematical relation, known as *Faraday's law of induction:*

$$\mathcal{E} = -\frac{\Delta\phi}{\Delta t} \tag{23.1}$$

where $\mathcal{E}$ is the emf induced in a single closed loop when the magnetic flux penetrating that loop changes with time at the rate $\Delta\phi/\Delta t$. If N coils of wire, wound in the same sense, are connected, as for example in a solenoid, the induced emf is given by

$$\mathcal{E} = -N\frac{\Delta\phi}{\Delta t} \tag{23.2}$$

which is the usual form in which Faraday's law is written.

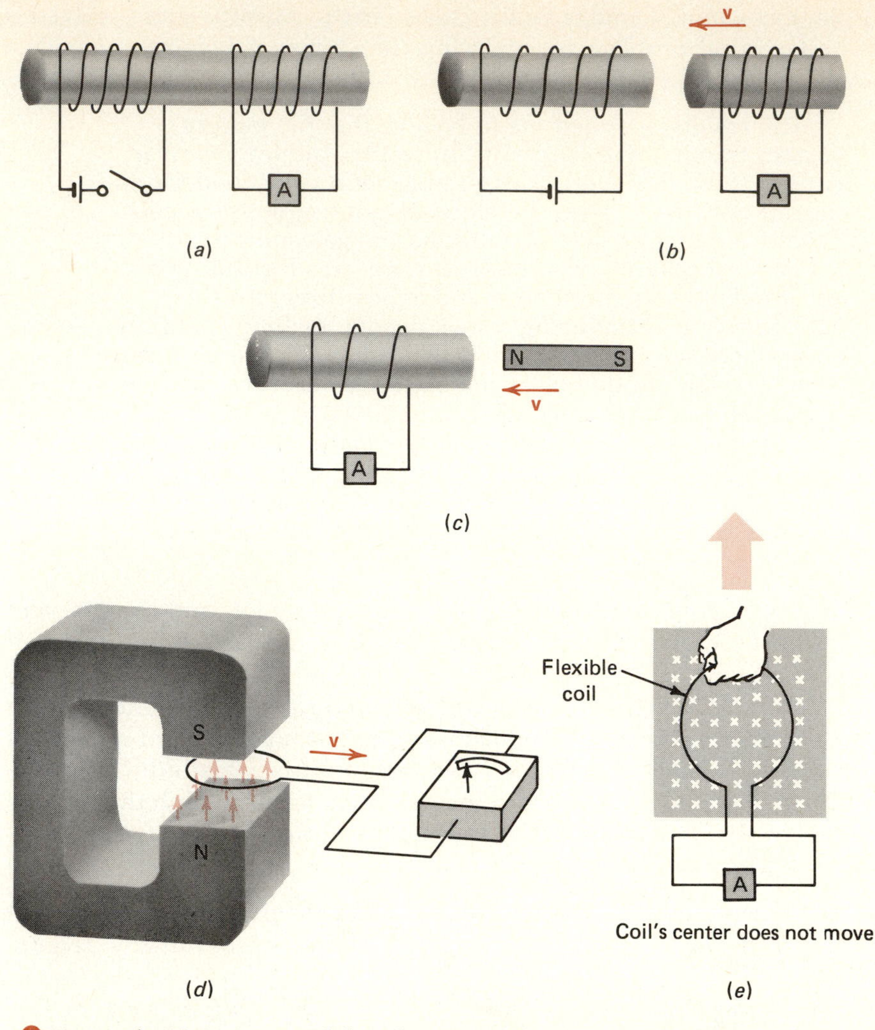

Figure 23.3 *Illustrations of several processes that cause an induced emf in the coil connected to the ammeter* (A).

● ***Example 23.1*** A toroidal coil is wound with 4000 turns of wire. The average radius of the toroid is 10 cm and the diameter of the coils is 1.5 cm. A second coil of 400 turns is wound over the first. What emf is induced in this second coil if the current in the 4000-turn coil is changed at a rate of 25 A/s?

The induced emf in the 400-turn coil is given by Faraday's law, Equation (23.2). To determine this emf, we must therefore find the flux through the coil.

The magnetic field inside a toroid is given by Equation (22.17). Since the diameter of the coils is relatively small compared with the radius of the toroid, the assumption that B is constant within the toroid will not introduce a large error. The flux linking the second coil then equals the flux density B times the cross-sectional area of the coils; i.e.,

$$\phi = BA = \frac{\mu_0 NI}{2\pi R}\frac{\pi d^2}{4} = \frac{(2 \times 10^{-7}\ \mathrm{T{\cdot}m/A})(4000)(I)(\pi)(0.015\ \mathrm{m})^2}{4(0.1\ \mathrm{m})}$$

$$= 1.41 \times 10^{-6}\, I\ \mathrm{Wb}$$

The flux change $\Delta\phi/\Delta t$ is therefore given by

$$\frac{\Delta\phi}{\Delta t} = 1.41 \times 10^{-6}\frac{\Delta I}{\Delta t} = (1.41 \times 10^{-6}\ \mathrm{Wb/A})(25\ \mathrm{A/s})$$

$$= 3.53 \times 10^{-5}\ \mathrm{Wb/s}$$

and the magnitude of the induced emf is

$$\mathcal{E} = (400)(3.53 \times 10^{-5}) = 0.0141 \text{ V}$$

The negative sign appears in Equations (23.1) and (23.2) as a reminder that the sense of the induced emf obeys Lenz's law. This law states:

LL-I: The sense of the emf induced in a loop is such that the current that would flow if the circuit were completed opposes the change in flux through the loop.

Note particularly the words "change in flux." It is not the flux itself that matters; it is the change in flux that determines the magnitude as well as sense of the induced emf.

If this change in flux results from the motion of the loop relative to a magnetic field, another, equivalent statement of Lenz's law is often convenient:

LL-II: If a change in flux through a loop arises as a result of motion of the loop relative to a constant magnetic field, the sense of the induced emf is such that the current which would flow in the completed circuit will resist the relative motion.

Both LL-I and LL-II are cumbersome statements, and it is perhaps best to see their application first before we consider the underlying meaning of Lenz's law.

Example 23.2 When the switch in Figure 23.4 is closed, which terminal will be momentarily at the higher potential, A or B?

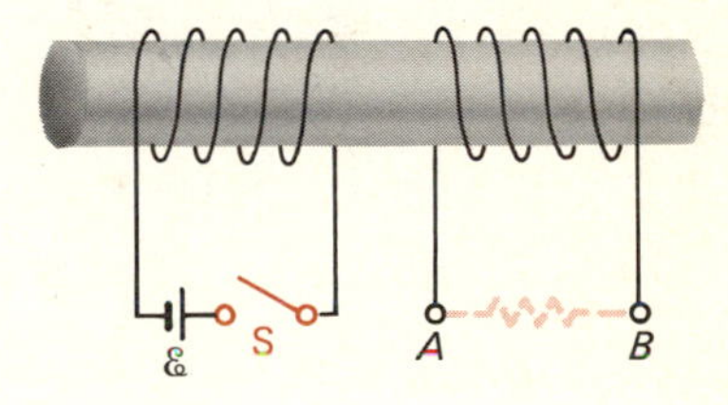

Figure 23.4 *When switch S is closed, an emf is induced in the other coil. The induced emf is such as to make A positive relative to B for a short time.*

To answer the question it is simplest to imagine a low resistance placed between A and B, thus completing the circuit. The coil terminating at A and B is to be viewed as the *source of emf,* like a battery operating for a brief interval, and the resistor, shown dashed, as the external load.

Before the switch is closed, there is no flux in either coil. Some time after the switch has been closed, current in the left-hand coil will have established a magnetic field, which points to the right, according to RHR-III. Therefore, during a brief time, as this current was building up from zero to its steady value, the flux through both coils increased from zero to some nonzero value, pointing to the right.

According to LL-I, the induced emf in the other coil must send a current through the external resistor and the coil that will generate a magnetic flux opposing the change, thus creating flux lines *that point to the left.* Such a magnetic field is produced if a current flows from B to A *through the coil,* and therefore *from A to B through the external resistor.* Consequently, A will be at a higher potential than B; A will be positive relative to B for a brief time after switch S is closed.

Example 23.3 The rectangular loop of copper, located partially in a region of uniform magnetic field as shown, is pulled to the right. In what direction does the induced current flow in this loop?

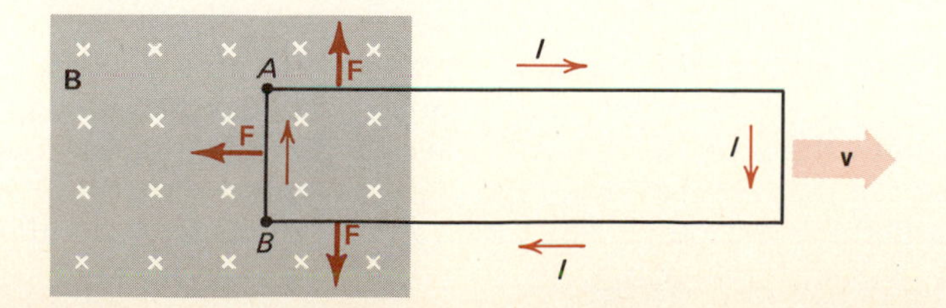

Figure 23.5 *As the rectangular conducting loop is pulled to the right with velocity* **v**, *a current is induced in this loop. This current whose direction is shown, causes a force on the portion AB which opposes the motion.*

Here LL-II is most convenient, although LL-I is also readily applied; we shall use both approaches.

LL-I: As the loop is pulled to the right, the magnetic flux within the loop, directed into the page, is diminished. Consequently, the induced current should flow so as to augment that flux, and from RHR-III, this means a current clockwise around the loop.

LL-II: If a current flows, that portion of the conductor in the magnetic field will experience a force. The force must be perpendicular to the conductor, that is, vertical on the horizontal arms and either to the right or the left between A and B. To oppose the motion, the force must point to the left; by RHR-II this means a current flowing up in this section, from B to A, consistent with the preceding conclusion. ●

Now that we have seen how Lenz's law is applied in a few situations, it is almost obvious why an induced current, if one exists, *must* be in the direction the law dictates. Consider, for instance, Example 23.3. If the induced current flowed counterclockwise, from A to B, the magnetic force on the segment AB would be to the right, in the direction of motion. The loop would therefore accelerate, inducing an even larger current, which would only increase the force and enhance this acceleration. Thus, starting almost from rest, the loop could acquire substantial kinetic energy before leaving the field region, in violation of energy conservation.

23.3 Motional EMF

Scrutiny of the conditions depicted in the last of the Lenz's law examples shows that the emf that drives the current around the loop is induced entirely as a result of motion of segment AB. This conclusion is also consistent with the fact that the magnetic forces on the other portions of the loop do not result in the performance of any work since those forces are perpendicular to the direction of motion. Whatever energy is expended in pulling the loop out of the magnetic field and thereby inducing the current flow arises from the opposing force acting on segment AB. One can also demonstrate this by modifying the experimental arrangement slightly. A copper bar, in sliding contact with a fixed U-shaped conductor, is pulled to the right. It is now simple to show that a clockwise current is induced in this conducting loop, and that that current must derive just from the lateral motion of the bar. Thus we conclude that an emf is induced by virtue of motion of a conductor in a magnetic field, or more correctly, by the relative motion of conductor and magnetic field.

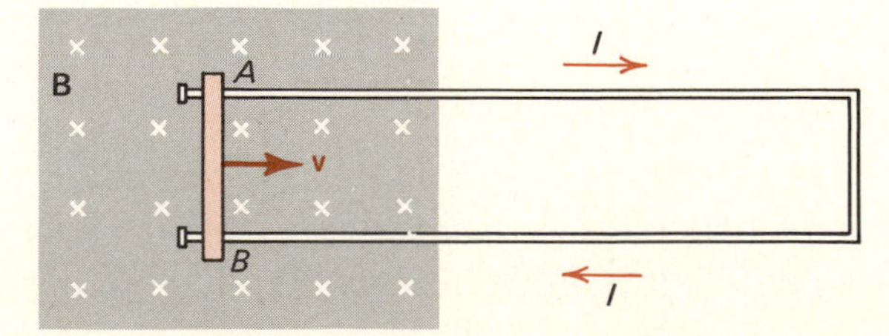

Figure 23.6 The emf induced in the loop of Figure 23.5 is due entirely to the translation of segment AB. The same emf will be induced in AB if the bar is made to slide to the right with velocity **v**.

Motional emf, as that term is conventionally used, refers to the arrangement in which mobile charges are caused to move through a magnetic field by the brute-force technique of bodily shifting the conductor. The result of that maneuver is intimately related to a situation encountered earlier, the Hall effect. The latter can be viewed as the motional emf

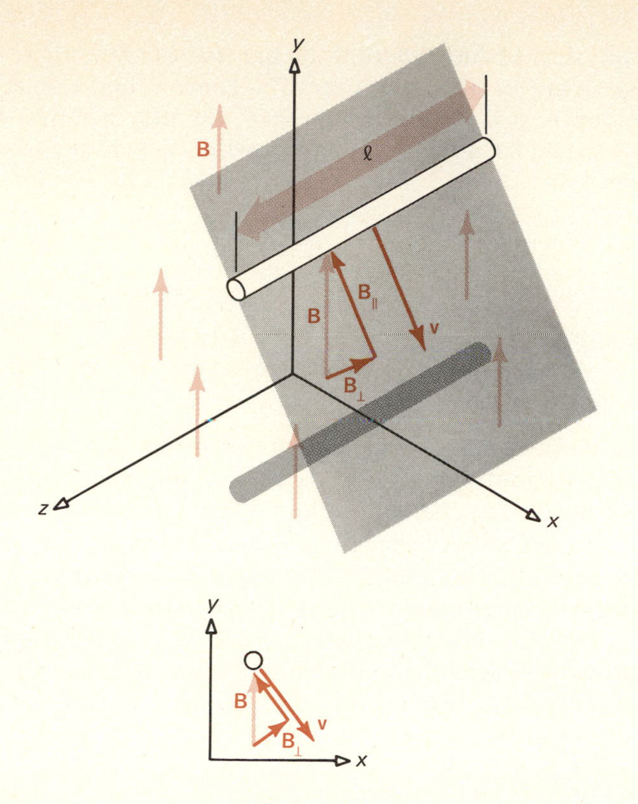

Figure 23.7 The motional emf induced in the conductor of length ℓ and moving with velocity **v** through the uniform magnetic field **B** is given by $B_\perp \ell v$. The direction of the induced emf is determined by applying RHR-II

resulting from the translation of charges in a magnetic field. In the generation of the Hall emf, the conductor is stationary, and the charge carriers are caused to move with some drift velocity by connecting the ends of the conducting slab to the terminals of a battery.

The electric field that arises from the charge separation due to the Lorentz force on the charge carriers is (see Section 22.5(b))

$$E = B_\perp v \tag{23.3}$$

The induced emf is therefore

$$\mathcal{E} = B_\perp v \ell \tag{23.4}$$

where ℓ is the length of the conductor along the direction of the electric field, that is, in the direction perpendicular to the plane formed by **B** and **v**. Here $B_\perp$ is the component of the magnetic field strength that is perpendicular to **v**.

With the aid of Figure 23.8, we can quickly convince ourselves that Equation (23.4) is equivalent to Equation (23.1) and does not represent

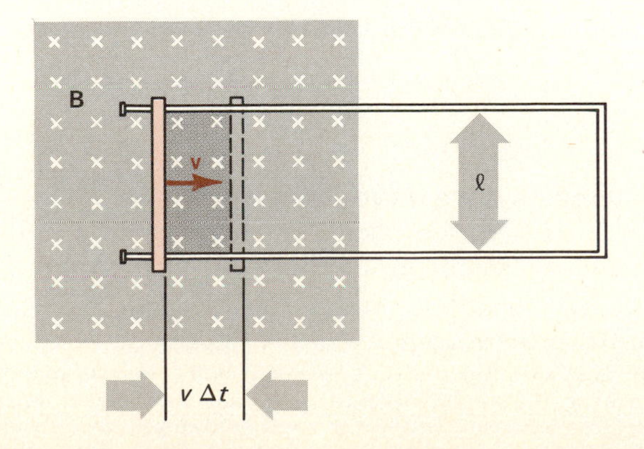

Figure 23.8 As the conductor moves to the right, a motional emf is induced in it. This emf can also be calculated using Faraday's Law, since this motion changes the flux that links the loop by diminishing the area of the loop.

some new physical phenomenon. If the bar moves to the right at velocity v, its displacement in time Δt is $v\ \Delta t$. The concomitant reduction in the area of the loop is $\ell v\ \Delta t$, and consequently, the flux enclosed within the loop is diminished by $\Delta\phi = B\ell v\ \Delta t$. According to Faraday's law, the induced emf is

$$\mathcal{E} = -\frac{\Delta\phi}{\Delta t} = \frac{B\ell v\ \Delta t}{\Delta t} = B\ell v \tag{23.4}$$

23.4 Mutual Inductance and Self-Inductance

23.4 (a) Mutual Inductance

Faraday, in his original experiments on induction, employed two coils in close proximity. He found that when the current in one of the coils was changed, that change brought about an emf across the terminals of the second coil. Ultimately, he discovered that it was not the change in current but the change in the magnetic flux associated with the current that caused the induced emf. In practice, however, it is often the relation between the time variation of the current and the induced emf that is of primary interest. One therefore defines a quantity, the *mutual inductance*, that relates the induced emf directly to the change in current. The defining relation is

$$\mathcal{E}_2 = -M_{21}\frac{\Delta I_1}{\Delta t} \tag{23.5}$$

where $\mathcal{E}_2$ is the emf induced in coil 2 as a result of a change of current ΔI_1 in time Δt in coil 1. The unit of mutual inductance is the *henry* (H), named in honor of Joseph Henry, who independently discovered electromagnetic induction and first identified the self-inductance property of a coil.*

● ***Example 23.4*** Determine the mutual inductance of the two superimposed toroidal coils of Example 23.1.

We have previously established that the flux linking each of the 400 turns of the outer coil is related to the current in the 4000-turn coil by

$$\phi = 1.41 \times 10^{-6} I \text{ Wb}$$

Consequently, if we write

$$\mathcal{E}_2 = -N_2\frac{\Delta\phi}{\Delta t} = -(400)(1.41 \times 10^{-6})\frac{\Delta I}{\Delta t} = -M_{21}\frac{\Delta I_1}{\Delta t}$$

we see that M_{21} must have the value

$$M_{21} = 400(1.41 \times 10^{-6}) = 5.64 \times 10^{-4} \text{ H}$$ ●

23.4 (b) Self-Inductance

In all the foregoing, we invariably assumed that the induced emf is a consequence of flux change caused by some external influence, for example, a changing current in another coil, or shifting of a permanent

* Henry's life paralleled Faraday's not only in professional interest; in other ways, their fortunes also mirrored each other's. Henry, too, was apprenticed to a trade at an early age (13), in his instance to a master watchmaker. Michael Faraday was appointed director, and was offered, but refused, the presidency of the Royal Institution, founded in part through the efforts of the expatriate American, Count Rumford (Benjamin Thompson); Joseph Henry became the

magnet. There is, however, nothing in electromagnetic induction that demands that the flux change must originate in this manner.

Consider the circuit of Figure 23.9. Before switch S is closed, no current flows in the coil, and hence, no flux links the turns of this solenoid. Some time after closing switch S, a steady current $I = \mathcal{E}/R$ will flow in the circuit, where R is the resistance of the wire of which the coil is constructed. Now some flux links the turns of this coil. Thus, during the intervening time there must have been a change in the flux and according to Faraday's law, an emf must have been induced in that coil.

This induced emf will have been in a direction to oppose the change in flux, that is, to oppose the buildup of flux due to the current developed in the coil. Since this current is forced through the coil by the battery, the induced emf, in this instance, must be opposite to the emf of the battery; it is therefore called a *back-emf*.

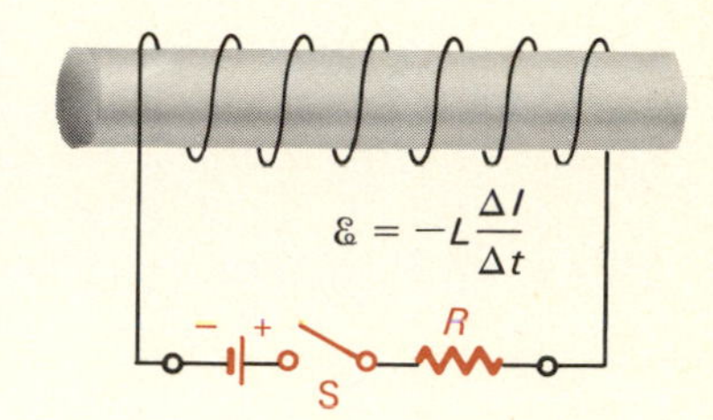

Figure 23.9 An emf is induced in the coil if the current through the coil changes with time. The constant of proportionality relating this induced emf to the rate of current change is the self-inductance of the coil L. The induced emf opposes the current change produced by some external influence (e.g., a battery), and for this reason is called a back-emf.

It is here convenient to define a proportionality constant, like the mutual inductance, that relates this back-emf directly to the time rate of change of current. The relation is

$$\mathcal{E} = -L\frac{\Delta I}{\Delta t} \tag{23.6}$$

where L is called the *self-inductance* of the coil and $\mathcal{E}$ is the emf induced in that coil by the change in current flowing through it.

The circuit symbol for inductance is shown in Figure 23.10(*a*). This symbol represents an ideal inductance, made of a conductor with zero resistance. A characteristic feature of the ideal inductance is that a current, once established in a circuit that contains ideal inductance only, will continue to flow indefinitely without decay; such a circuit dissipates no energy, and no power has to be supplied to maintain the current.

Normally, a practical inductor, wound with copper wire, has some resistance as well as inductance. Even though one cannot separate its inductance and resistance physically, it is a useful abstraction to do so in theory. A real, common, garden-variety inductor is therefore usually represented as a series combination of ideal inductance and resistance, as in Figure 23.10(*c*), just as the real battery was represented as a series combination of ideal emf in series with an internal resistance.

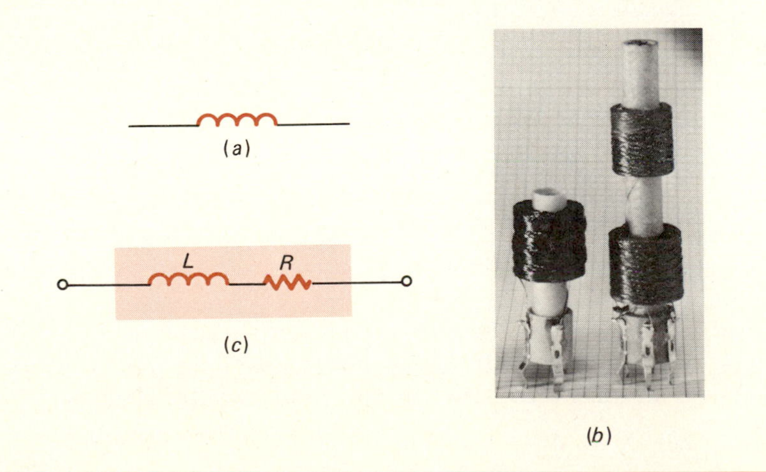

Figure 23.10 (*a*) Circuit symbol for an ideal inductor. Every real inductor (see (*b*)), except those made of superconducting wire and maintained at low temperature, has some resistance as well as inductance. It is electrically equivalent to an ideal inductance L in series with a resistance R, as shown in (*c*).

first secretary of the Smithsonian Institution, which owes its origin to an endowment by the Englishman James Smithson. Faraday, during his early years as assistant to Davy, lived in quarters in the Royal Institution; Henry, after he left Albany for Washington, lived in a small apartment in the Smithsonian, and there he occasionally entertained President Lincoln, who was fond of Henry and would walk across the mall to spend a few relaxing moments with his friend.

It is, however, possible to fashion an ideal inductance by winding the coil with niobium, lead, or other superconducting wire, and operating the coil at a sufficiently low temperature.

The self-inductance of an air-core coil depends on the number of turns and on its dimensions and shape. For example, the multi-layer coils on the forms of Figure 23.10 have self-inductances that will differ somewhat from the self-inductance of a single-layer solenoid of equal total turns, and therefore greater length, wound on a form of the same diameter.

The self-inductance of a coil is therefore not easily calculated precisely, although it is usually not too difficult to make reasonable estimates. Calculation is simplest for a toroidal coil.

Suppose that the toroid is of average radius R and that the coil has a diameter d and carries N turns. The magnetic field within the toroid is

$$B = \frac{\mu_0 NI}{2\pi R} \qquad \textbf{(22.17)}$$

and the flux linking each turn is therefore

$$\phi = BA = B\left(\frac{\pi d^2}{4}\right) = \frac{\mu_0 NId^2}{8R}$$

If the current in this coil changes at the rate $\Delta I/\Delta t$, the corresponding flux change is

$$\frac{\Delta\phi}{\Delta t} = \frac{\mu_0 Nd^2}{8R}\frac{\Delta I}{\Delta t}$$

Consequently, the emf induced between the terminals of this toroidal coil under the stated conditions of current change is

$$\mathcal{E} = -N\frac{\Delta\phi}{\Delta t} = -\frac{\mu_0 N^2 d^2}{8R}\frac{\Delta I}{\Delta t}$$

From comparison of the above expression with the defining equation for self-inductance, Equation (23.6), we see that the self-inductance of the toroidal coil is given by

$$L = \frac{\mu_0 N^2 d^2}{8R} \qquad \textbf{(23.7)}$$

23.5 Energy Stored in an Inductance

The inductance of Figure 23.11 is an ideal inductance. Immediately after switch S is closed, the battery will attempt to force charge flow at an infinite rate through the circuit—infinite rate, because with zero resistance in the circuit, there is no limit on the current supplied by the ideal battery.

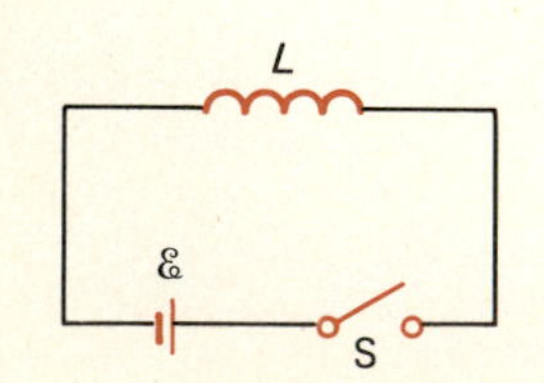

Figure 23.11 This circuit contains only an ideal battery and an ideal inductance. When the switch S is closed, there is no energy dissipated in the circuit; all the energy supplied by the battery is stored in the inductance.

For this circuit, as for any other, Kirchhoff's rules must be satisfied at all times. In particular, KR-II (see page 451) must hold. With only an ideal battery and an ideal inductance in the circuit, KR-II becomes

$$\mathcal{E} - L\frac{\Delta I}{\Delta t} = 0 \qquad \frac{\Delta I}{\Delta t} = \frac{\mathcal{E}}{L} \qquad \textbf{(23.8)}$$

which immediately tells us that the current in this circuit cannot assume some large value an instant after the switch is closed. The rate at which the current can change in this circuit is limited by the inductance L!

The amount of energy delivered by a battery during a time interval Δt equals the product of the battery emf and the charge ΔQ forced through the external circuit in this time interval. Thus

$$\Delta W = \mathcal{E}\Delta Q = L\frac{\Delta I}{\Delta t}\Delta Q = L\frac{\Delta I}{\Delta t} I\,\Delta t = LI\,\Delta I \qquad \textbf{(23.9)}$$

We interpret Equation (23.9) as follows. The ideal inductance, without any resistance, will support a flow of current without an additional driving force; that is, no emf and no power are needed to maintain a steady current. However, to increase this current by ΔI, one must supply $LI\,\Delta I$ joules of energy.

In Figure 23.12, we show the quantity LI plotted against I; the curve is simply a straight line with slope L. The product $LI\,\Delta I$, shown cross-hatched in Figure 23.12, is the energy required to increase the current through this inductance from I to $I + \Delta I$. It is evident from the geometrical construction that starting from $I = 0$, the total energy expended to achieve a final current I is just the area of the shaded triangle; that is,

$$\text{Energy} = \tfrac{1}{2}LI^2 \qquad \textbf{(23.10)}$$

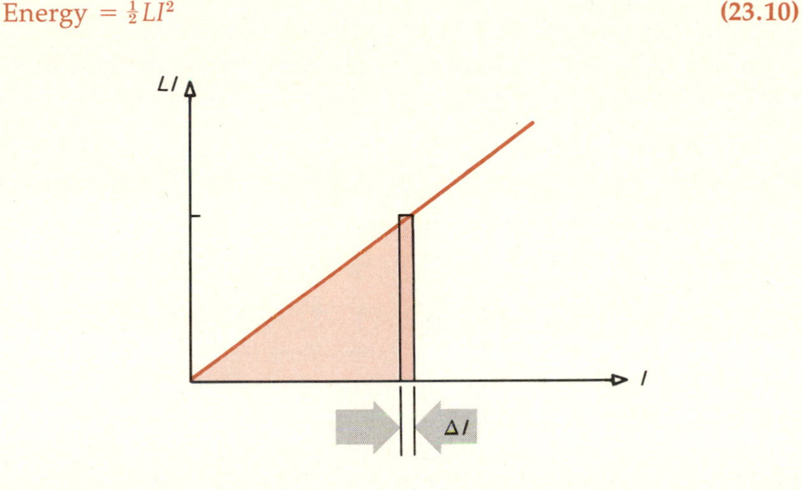

Figure 23.12 The additional energy needed to increase the current through the ideal inductance L of Figure 23.11 from I to $I + \Delta I$ is $LI\,\Delta I$. This is represented by the area of the thin, dark-colored rectangle. The total energy expended by the battery in bringing the current in the inductor from zero to the value I is given by the area of the shaded triangle, which is equal to $\frac{1}{2}LI^2$.

This is the energy supplied by the battery in Figure 23.11, and since there is no loss of energy in the circuit, this must then be the energy that is stored in the inductor. Equation (23.10) is similar in form to Equation (19.8) for the energy stored in a capacitor. In that case, we argued that this energy can be viewed as stored in the electric field between the plates. Similarly, one can show that the energy stored in an inductance is stored in the magnetic field resulting from the current in the inductor's windings.

● ***Example 23.5*** An air-core solenoid consists of 45,000 turns of superconducting wire wound on a form that is 30 cm long and has in inner diameter of 5 cm. The mass of the unit is 6 kg, and it can carry a maximum current of 40 A. We want to determine approximate values of:

1. Its self-inductance.
2. The magnetic field strength at the center of the coil under peak current conditions.
3. The energy stored in this coil at peak current.

Before we do the calculations, we should comment on the adjective *approximate*. A practical solenoid, such as the one described in this example, consists of many layers of windings. In the present case, there might be perhaps 30 layers with about 1500 turns per layer. The ideal solenoid generates a field of constant strength inside the air core and produces a small field, in the opposite direction, in the region just outside the solenoid (see Figures 22.22 and 22.23). In this more realistically designed solenoidal magnet, the field, though uniform near the center of the air core, falls off gradually as one proceeds through the layers of the coil to its outer radius. The flux that links the turns of the various layers of the coil is therefore not constant, and a calculation that assumes it is must be in error. The magnitude of the error will depend on construction details.

We apply the simple solenoid formula and obtain for the field in the central region

$$B = \mu_0 nI = (4\pi \times 10^{-7}\ \text{T}\cdot\text{m/A})\left(\frac{45{,}000}{0.3\ \text{m}}\right) I = 6\pi \times 10^{-2} I\ \text{T}$$

The total flux through the core of the solenoid is

$$\phi = BA = (6\pi \times 10^{-2} I\ \text{T})\pi(0.025\ \text{m})^2 = 3.75\pi^2 \times 10^{-5} I\ \text{Wb}$$

If we now assume (and this is not a valid assumption for the real coil) that this flux links all 45,000 turns of the solenoid, the induced emf resulting from a change in coil current with time is given by

$$\mathcal{E} = -N\frac{\Delta\phi}{\Delta t} = -(4.5 \times 10^4)(3.75\pi^2 \times 10^{-5})\frac{\Delta I}{\Delta t} = -16.7\frac{\Delta I}{\Delta t} = -L\frac{\Delta I}{\Delta t}$$

The inductance of this coil is therefore about 16.7 H. The true inductance is likely to be substantially less because the real coil is a layered structure.

The field at the center of the solenoid when $I = 40$ A is

$$B = (6\pi \times 10^{-2})40 = 7.54\ \text{T}$$

That field value is fairly reliable; that the coil is made of many layers does not affect that part of the calculation, and to the extent that end effects can be neglected, the above value of B in the central region is correct.

The energy stored in the coil is approximately

$$W = \tfrac{1}{2}LI^2 = \tfrac{1}{2}(16.7\ \text{H})(40\ \text{A})^2 = 13{,}360\ \text{J}$$

a substantial amount of energy. If all that energy were converted into kinetic energy of the coil, for example, this 6-kg mass would acquire a speed of

$$v = \sqrt{\frac{2(13{,}360\ \text{J})}{6\ \text{kg}}} = 66.7\ \text{m/s} = 150\ \text{mph}$$

*23.6 Superconductivity and Applications

We have mentioned superconductors at various times and have just used a superconducting coil in Example 23.5. A brief digression, a quick look at the unusual phenomenon of superconductivity is in order.

As so often happens, the story begins with a curious quirk of history. At the beginning of this century, physicists were engaged in mild controversy over the theoretically predicted behavior for the resistivity of metals. You will recall (Chapter 20) that the resistivity of metals at room temperature decreases with decreasing temperature, whereas that of carbon and semiconductors increases as the temperature is reduced. At

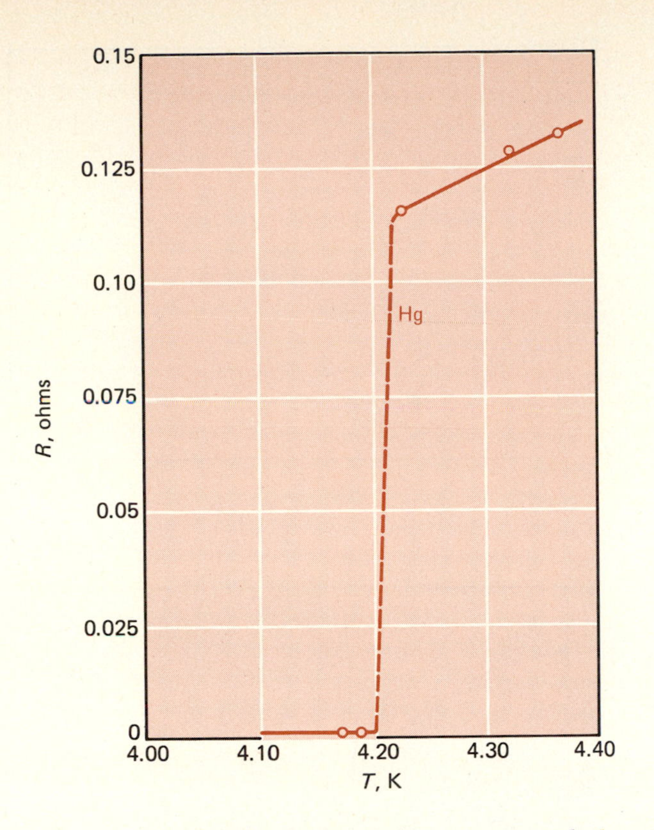

Figure 23.13 *The resistance of a sample of pure mercury at low temperatures, as first reported by Kamerlingh-Onnes.*

the time, one group of physicists argued that at extremely low temperatures the resistivity of *all* materials must rise, approaching infinity as T approaches 0 K. Another group maintained that metals should display a gradually diminishing resistivity with decreasing temperature, the limiting value depending on the purity of the sample, the resistivity of the ideally pure metal approaching zero as T approaches 0 K.

In 1908, Kamerlingh-Onnes succeeded in liquefying helium, and shortly thereafter initiated experiments on the low-temperature resistivity of metals. After some initial work with platinum and gold, it became apparent that significant results could be obtained only for a metal of exceptional purity. Mercury, which is liquid at room temperature, can be highly purified by repeated distillation, and Kamerlingh-Onnes therefore turned to mercury for his next sample.

The results obtained in 1911, and subsequently refined and confirmed in 1912 and 1913, revealed that the resistivity of mercury vanished suddenly when the temperature dropped below 4.17 K, a value just slightly below the temperature of liquid helium at atmospheric pressure, 4.2 K. The importance of this result was not immediately appreciated by Kamerlingh-Onnes, who at first unsuccessfully tried to fit it into the existing theoretical framework. Within a few years, however, it became apparent that this was a new, unexplained and seemingly inexplicable natural phenomenon, and numerous research groups throughout the world embarked on studies of superconductivity as soon as their laboratories managed to produce liquid helium. More than 40 years passed before Bardeen, Cooper, and Schrieffer formulated a sound theory of superconductivity.

Today, the number of metals and alloys known to be superconducting is legion; in fact, nonsuperconducting metals are in the minority. Yet, despite years of research aimed at discovering commercially useful materials that remain superconducting at temperatures above 21 K (the tem-

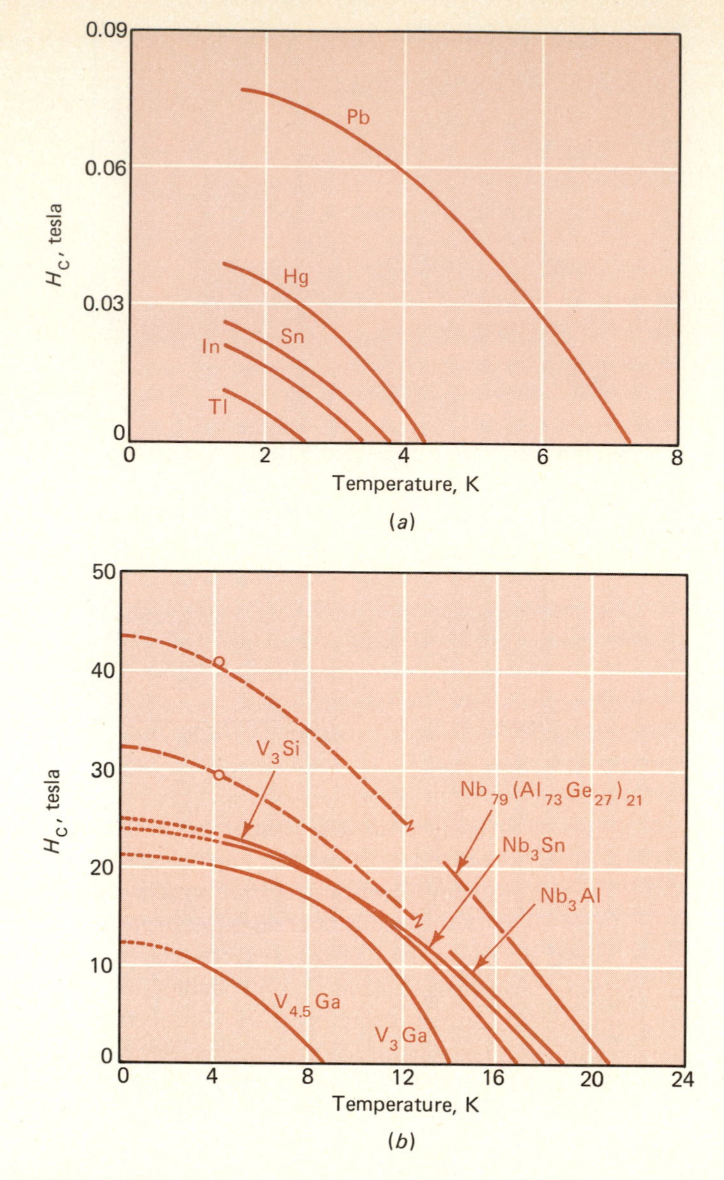

Figure 23.14 *(a) Curves of critical field versus temperature for several superconducting metals. Note the similarity of these curves; this agrees with theory, which predicts a universal curve, if the data are properly scaled. The metal is superconducting at fields and temperatures below the critical curve, and it behaves like a normal metal above that curve. Note that these critical fields are rather low; these metals could not be used to wind superconducting magnets. (b) Curves of critical field versus temperature for some high-field superconducting alloys. Niobium-tin, Nb_3Sn, is one of several alloys used in the manufacture of high-field superconducting solenoids.*

perature of liquid H_2 at atmospheric pressure is 20.4 K, and hydrogen is much cheaper and far more abundant than helium), none has been found to date.*

The following are two essential and important properties of all superconductors. First, below a certain critical temperature T_c, characteristic of a particular metal, the resistivity of the metal is *exactly* zero. Second, in a sufficiently strong magnetic field, superconductivity is destroyed, and the metal assumes a resistivity expected of normal metals. The critical magnetic field for destruction of superconductivity varies from metal to metal but is always greatest at $T = 0$ K, decreasing with increasing temperature, and approaching zero as T approaches T_c.

Although the existence of this critical field places a limit on the magnetic fields that can be achieved by sending large currents through solenoids made with superconducting windings, the accessible fields are still high when compared to those that can be produced with reasonable

* $Nb_{79}(Al_{73}Ge_{27})_{21}$, whose critical temperature is just above the boiling point of hydrogen, is very brittle and cannot be machined or drawn into wires.

effort using conventional electromagnets. Superconducting solenoids capable of producing fields in excess of 10 T are now commercially available; by contrast, fairly large laboratory electromagnets generally produce fields of only 2–3 T in comparable volumes, and do this only with the consumption of tens of kilowatts. Moreover, the superconducting solenoid, once energized, can be operated in its persistent mode indefinitely without further power input provided that its temperature is maintained near 4 K. Thus, although the initial investment involving cost of wire and low-temperature facilities can be very great, so can the ultimate saving in energy. As a result, large superconducting magnets are now common in research laboratories.

Much current development work is aimed at using superconductors for large-scale electric-power transmission. The initial investment costs are also staggering, but so may be the saving of otherwise wasted energy.

One application with almost magical overtones is magnetic levitation. To understand the effect, consider what might happen if we gradually

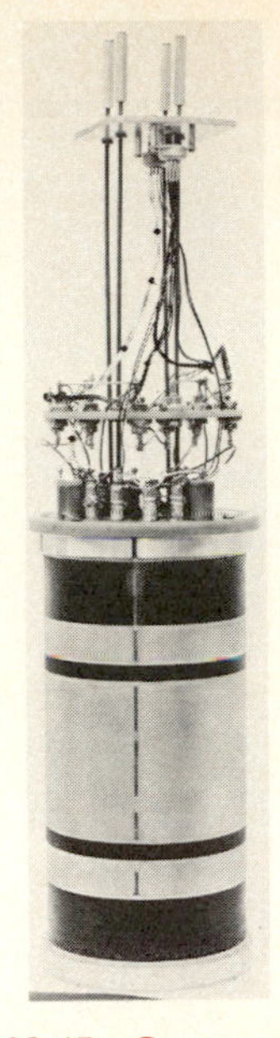

Figure 23.15 Commercial superconducting solenoid.

Figure 23.16 Very large superconducting magnet nearing final assembly. The magnet is used at the Argonne National Laboratory in conjunction with a liquid hydrogen bubble chamber.

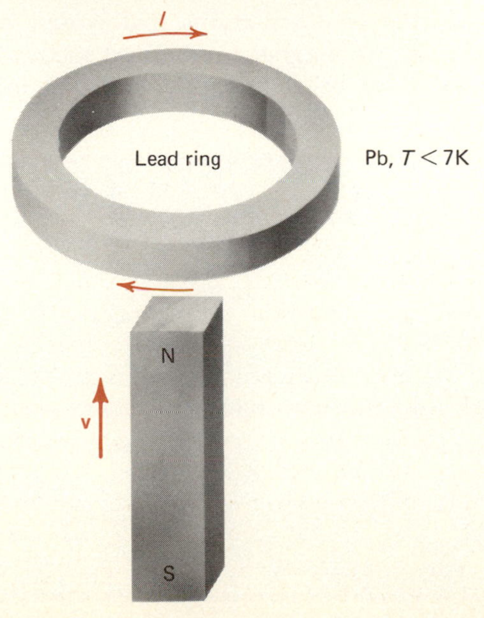

Figure 23.17 As the bar magnet is raised toward the superconducting lead ring, a current is induced in the ring, such that no flux links the ring. This induced current gives the ring a downward-directed magnetic moment. Thus ring and magnet repel each other. As the magnet is raised and brought closer to the ring, the circulating current increases until the repulsive force between magnet and ring balances the gravitational force on the ring, and the ring levitates.

Figure 23.18 *Instead of levitating a superconducting ring above a magnet, one can levitate the magnet above a superconductor. Here a small magnet is levitating above a pan of tin. The temperature of the liquid helium is which the pan has been submerged has been reduced to about 3.2K.*

raise a magnet toward a superconducting ring. As the flux from the magnet tries to pass through the ring, it induces a circulating current in it that, by Lenz's law, opposes the motion of the magnet. This simply means that the magnetic moment of the induced current repels the approaching magnet. Ultimately, the repulsive force between magnet and ring will be enough to counterbalance the weight of the ring, and the ring will float, or levitate, above the magnet. Raising the magnet further will only raise the ring.

In Japan, Germany, and elsewhere, plans are now under way to construct magnetically levitated trains that would carry superconducting coils and ride above a magnetic rail. Short sections of such magnetically levitated trains have been operating in Japan for some years in pilot studies.

Figure 23.19 *A practical application of magnetic levitation, the Japanese Supertrain.*

Finally, we must mention the numerous, extremely sensitive modern instruments that use superconducting devices. It is now possible to construct voltmeters capable of measuring potentials as small as 10^{-14} V, and magnetometers that can sense less than 10^{-10} times the earth's field. These magnetometers have already found application in geology, archeology, and medical technology; for instance, in many situations magnetocardiography and magnetoencephalography prove vastly superior to their electrical counterparts, because the source of the magnetic disturbance can often be located much more precisely than the source of the corresponding electrical signal.

23.7 Generators

When the then prime minister, Sir Robert Peel, on visiting Faraday at the Royal Society, saw his crude dynamo and asked what possible use that might be, Faraday is reputed to have replied: "I know not, but I wager that one day your government will tax it." Perhaps the story is true; Faraday had a fine sense of humor. But clearly, he also did know the purpose of the generators and motors he designed and constructed.

If the terminals of a rotating coil placed in a uniform magnetic field are connected to slip rings, as in Figure 23.20(*a*), a sinusoidal emf is induced between terminals A and B. To see this, let us look at the coil of Figure 23.20 and calculate the change in flux $\Delta\phi$ in a short time interval Δt, during which the coil changes its orientation from θ to $\theta + \Delta\theta$.

The total flux through the coil at any moment is the product of the magnetic field B and the projection of A, the area of the coil, on the plane perpendicular to **B**. (Maximum flux links the coil when the coil is perpendicular to **B**; no flux links the coil when it is parallel to **B**). The projection of the coil area on this plane is just $A \cos \theta$.

During the time Δt, the angle changes by an amount $\Delta\theta = \omega \, \Delta t$, where ω is the angular velocity of the coil (see Chapter 6). The change in

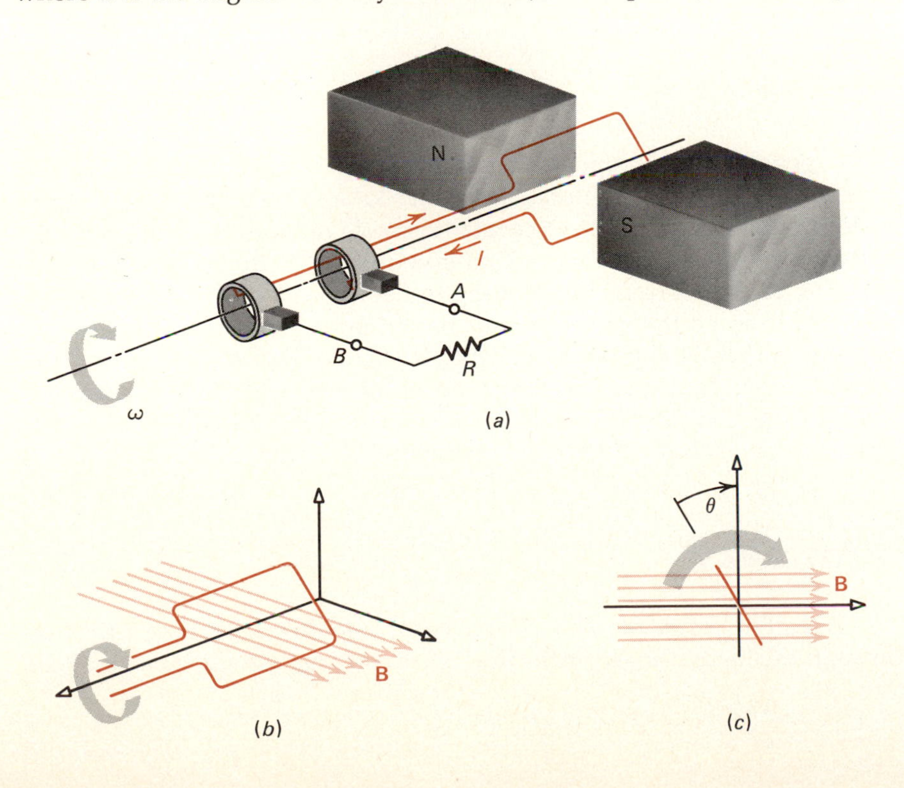

Figure 23.20 If a coil rotating between the pole faces of a magnet is connected to slip rings, as in (*a*), the polarity of V_{AB} and the current direction in the load resistor reverse every time the coil completes half a revolution. The flux through the coil depends on the position of the coil, as is shown in (*b*) and in cross section, in (*c*). As derived in the text, the induced emf is a sinusoidal function of time; the peak value of V_{AB} depends on the strength of the magnetic field, the area of the coil and the number of turns on it, and the angular velocity of rotation.

flux in this brief interval is

$$\begin{aligned}\Delta\phi &= AB[\cos(\theta + \Delta\theta) - \cos\theta]\\ &= AB[\cos\theta\cos\Delta\theta - \sin\theta\sin\Delta\theta - \cos\theta]\\ &= -AB\,\Delta\theta\sin\theta = -AB\omega\,\Delta t\sin\theta\end{aligned} \tag{23.11}$$

In arriving at this result, we first applied the trigonometric identity $\cos(a + b) = \cos a \cos b - \sin a \sin b$. In the next step, we recognized that $\Delta\theta$ is infinitesimally small and that therefore $\cos \Delta\theta = 1$ and $\sin \Delta\theta = \Delta\theta$. Lastly, we replaced $\Delta\theta$ by $\omega\,\Delta t$.

Applying Faraday's law, we obtain

$$\mathcal{E} = -N\frac{\Delta\phi}{\Delta t} = NAB\omega\sin\omega t = V_0\sin\omega t \tag{23.12}$$

A plot of Equation (23.12) is just a sine curve whose period is $T = 2\pi/\omega$ and whose amplitude is $NAB\omega$. The induced emf oscillates, or alternates, between positive and negative values at a frequency $f = \omega/2\pi$ Hz, that is, at a frequency equal to the number of revolutions per second of the rectangular coil. The frequency of the alternating voltage delivered by this generator is synchronous with the rotational speed of its armature.

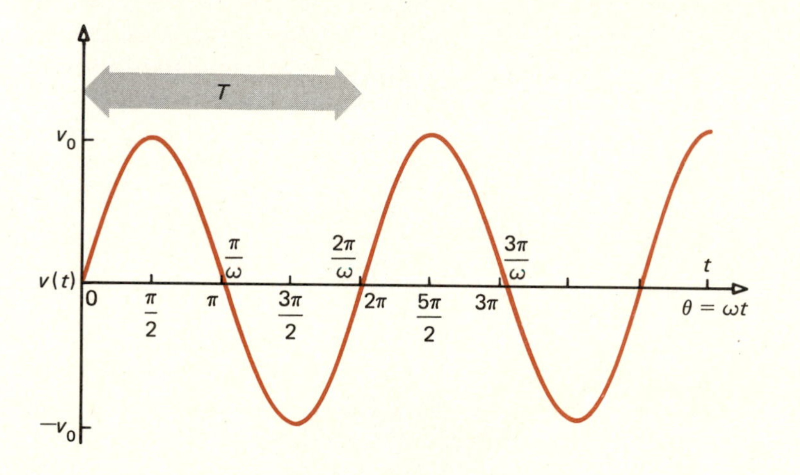

Figure 23.21 The time dependence of the emf induced in the coil of Figure 23.20(a).

A generator, on open circuit with no load connected to its terminals—and with frictional and other losses neglected—once set in motion continues to rotate at constant speed, generating induced emf's at the terminals. If, however, a load is connected to the generator terminals, the current direction, according to Lenz's law, will oppose the motion that induced the emf, and thus the armature will soon come to rest unless a torque is applied to the shaft by some device. The greater the current drawn from the generator, the greater the electrical power it delivers to the load, and therefore the greater the power that must be supplied to the machine to keep it turning.

Summary

An emf is induced in a closed loop whenever the magnetic flux through the loop changes with time. That emf is given by *Faraday's law,*

$$\mathcal{E} = -\frac{\Delta\phi}{\Delta t}$$

For a coil with N turns, wound in the same sense, Faraday's law of induction becomes

$$\mathcal{E} = -N\frac{\Delta\phi}{\Delta t}$$

The method by which the time-varying flux is achieved is immaterial.

The negative sign in Faraday's law is a reminder that the sense of the induced emf obeys *Lenz's law:*

The sense of the emf induced in a loop is such that the current that would flow if the circuit were completed opposes the change in flux through the loop.

An emf is also induced whenever a conductor moves through a magnetic field with a velocity that has a component perpendicular to **B**. One then refers to the induced emf as a *motional emf*. The sign of this motional emf is given by a variant of Lenz's law:

The sense of the motional emf is such that the current that would flow if the circuit were completed will resist the motion.

When a current change in one coil induces an emf in a second coil, the two coils are said to be coupled through a *mutual inductance*. This mutual inductance, $M_{12} = M_{21}$, is defined by

$$\mathcal{E}_2 = -M_{21}\frac{\Delta I_1}{\Delta t}$$

The *self-inductance* L of a coil is defined by

$$\mathcal{E} = -L\frac{\Delta I}{\Delta t}$$

where $\mathcal{E}$ is the induced emf when the current in the coil changes with time at the rate $\Delta I/\Delta t$.

The *energy stored in a coil* with self-inductance L is

$$E = \tfrac{1}{2}LI^2$$

A *superconductor* is a material whose electrical resistance vanishes below some temperature, known as the *critical temperature for superconductivity*. In sufficiently strong magnetic fields, superconductivity is suppressed. This *critical field* for the suppression of superconductivity depends on temperature and on the material.

Multiple Choice Questions

23.1 If the current I in the long wire increases with time, the current induced in the loop

(a) circulates clockwise.

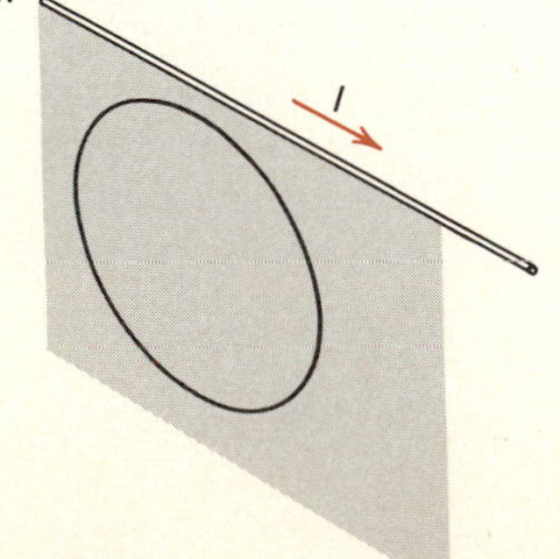

Figure 23.22

(b) is zero.

(c) circulates counterclockwise.

(d) has a direction that depends on the dimensions of the loop.

23.2 A circular loop of wire is rotated at constant angular speed about an axis that makes an angle θ with a constant magnetic field, as shown in Figure 23.23. The emf induced in the coil will be zero at all times if

(a) $\theta = 0°$

(b) $\theta = 90°$

(c) neither (a) nor (b).

(d) either (a) or (b).

23.3 The bar magnet is moved to the right, away from the coil. The current flowing through the meter

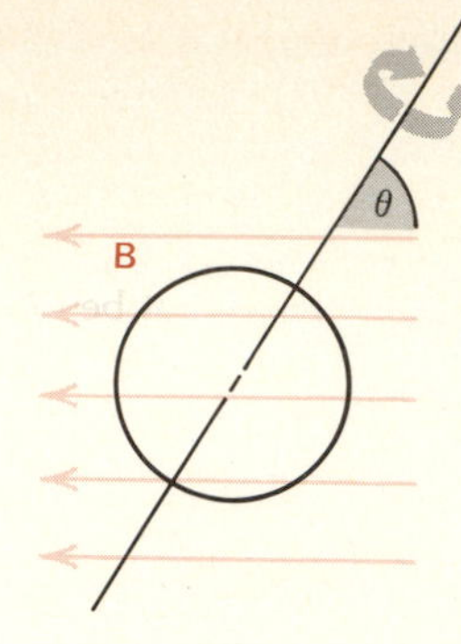

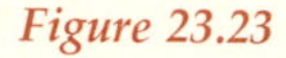

Figure 23.23

[A] is

(a) directed from a to b.

(b) directed from b to a.

(c) zero.

(d) in a direction that cannot be determined from the information provided.

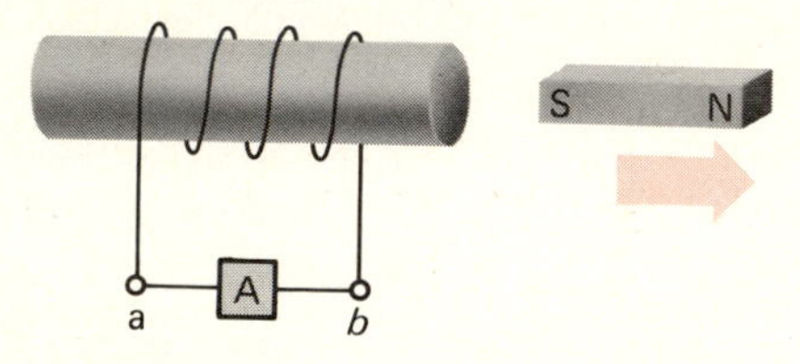

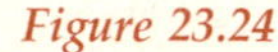

Figure 23.24

23.4 The current in a solenoid is changing at a constant rate. The solenoid is wound with n_1 turns per meter. A second coil of N_2 turns, wound around the first solenoid has an emf $\mathcal{E}_0$ induced in it. If now both n_1 and N_2 are doubled, and the rate of change of current maintained the same as before, the emf induced in the second coil will be

(a) $\mathcal{E}_0/4$

(b) $\mathcal{E}_0$

(c) $4\mathcal{E}_0$

(d) $16\mathcal{E}_0$

23.5 Figure 23.25 shows a bar magnet and a single coil of wire in cross section. The induced current in that coil is as indicated (into the wire at the bottom, out of the wire at the top). From that information one can conclude

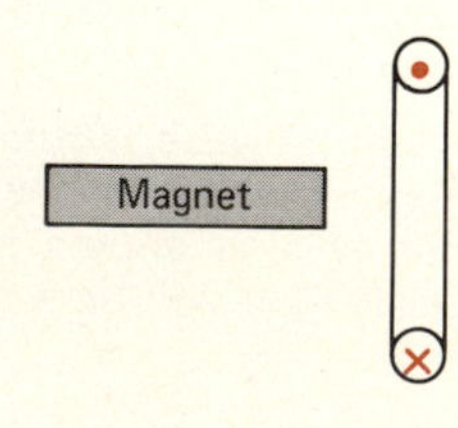

Figure 23.25

(a) that the magnet is moving to the right.

(b) that the magnet is moving to the left.

(c) that either (a) or (b) is true, depending on which end of the magnet is the north pole.

(d) nothing about the motion of the magnet relative to the coil; it could be at rest.

23.6 Two different wire loops are concentric and lie in the same plane. The current in the outer loop is clockwise and increasing with time. The induced current in the inner loop then

(a) is clockwise.

(b) is zero.

(c) is counterclockwise.

(d) has a direction that depends on the ratio of the loop radii.

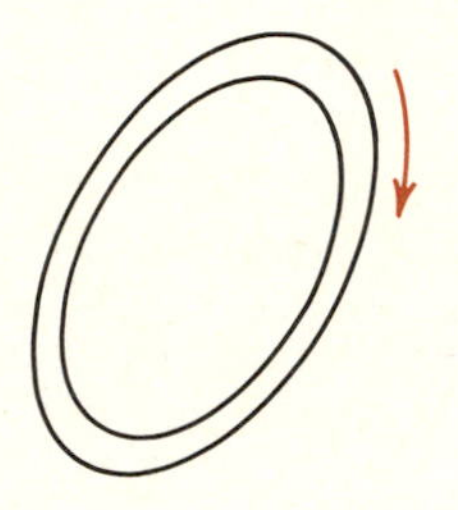

Figure 23.26

23.7 A wire loop is rotated in a uniform magnetic field about an axis perpendicular to the field. The direction of the current induced in the loop reverses once each

(a) quarter revolution.

(b) half revolution.

(c) full revolution.

(d) two revolutions.

23.8 A uniform magnetic field **B** permeates the shaded region of Figure 23.27. The direction of **B** is into the page. Which of the wire loops, moving with velocity **v**, experiences a force toward the top of the page?

(a) A

(b) B

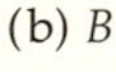

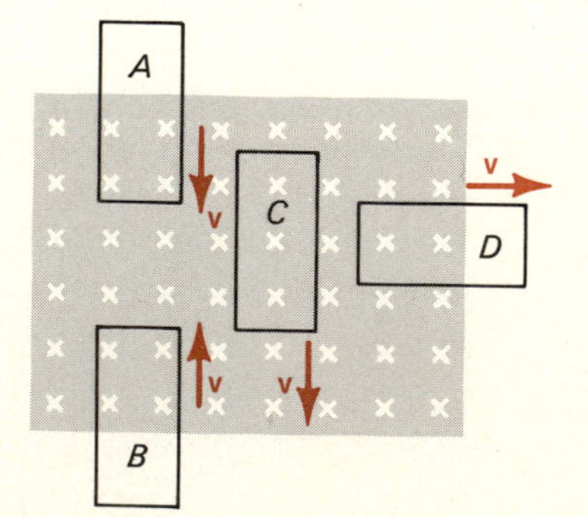

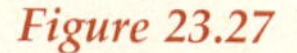

Figure 23.27

(c) C

(d) D

(e) A and C

23.9 Two coils are wound on identical cardboard forms of cylindrical shape. One coil is a solenoid with 1000 turns and a length of 20 cm. The other coil has 500 turns over the same length of 20 cm. If the coil with 1000 turns has a self-inductance of 10 mH, the self-inductance of the other coil will be

(a) 5 mH.

(b) 2.5 mH.

(c) 1.25 mH.

(d) of some value that cannot be determined from the information given.

Problems

(Section 23.2)

23.1 The circular loop shown in Figure 23.28 is in a uniform magnetic field **B** that points into the page. The diameter of the loop is 50 cm. If the field **B** increases at a rate of 0.4 T/s, what are the magnitude and sense of the induced emf between points A and B?

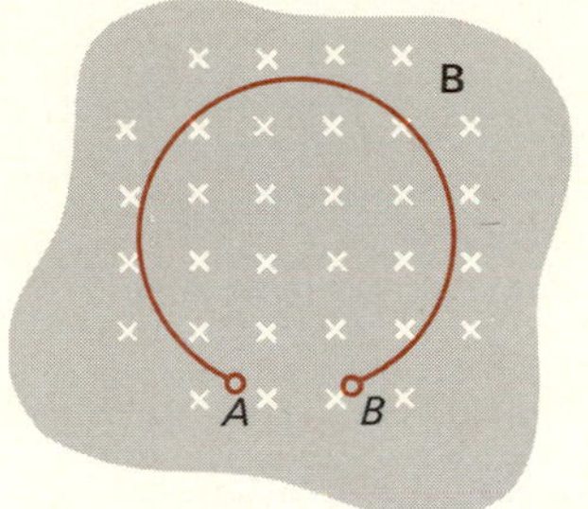

Figure 23.28

23.2 A rectangular coil is located in the plane of the paper. The dimensions of the coil are 5 cm × 3 cm, and the coil has 30 turns. In going from terminal A to B, one must pass through these coils clockwise. If initially there is a magnetic field pointing out of the plane of the paper of magnitude 2 T, and the field then goes to zero at a uniform rate in a time interval of 4 s, what is the emf (magnitude and sense) induced in the coil?

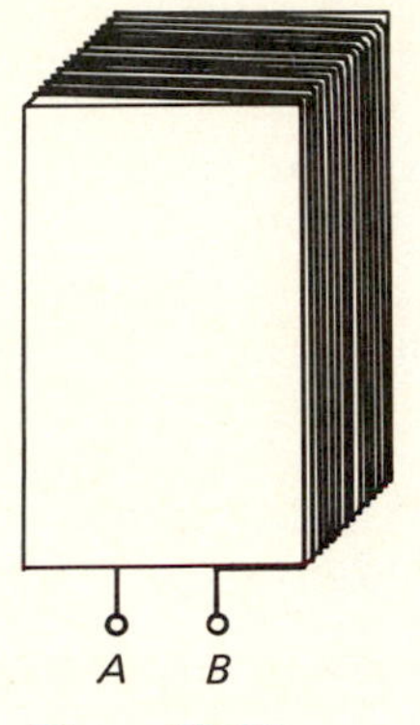

Figure 23.29

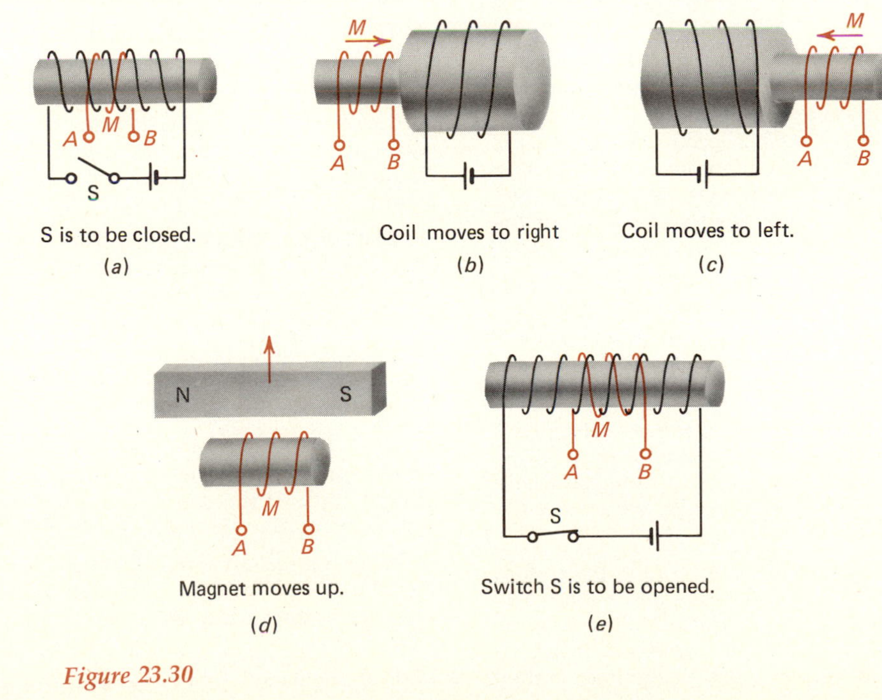

Figure 23.30

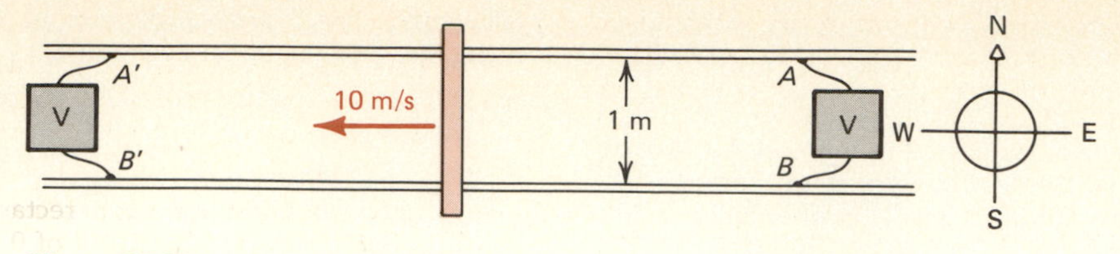

Figure 23.31

• **23.3** A closed circular loop of diameter 40 cm is made of a length of copper wire whose resistance is 0.012 Ω. The current in the loop due to a changing magnetic field is 0.04 A. What is the rate of change of B, that is, what is the value of $\Delta B/\Delta t$, if

(a) **B** is uniform and perpendicular to the plane of the loop?

(b) **B** is uniform and makes an angle of 30° with the plane of the loop?

• **23.4** For the conditions indicated in Figure 23.30, determine the sense of the induced emf in coil M.

(Section 23.3)

• **23.5** The set of horizontal, parallel rails shown in Figure 23.31 is located in a region of uniform magnetic field **B** of magnitude 0.05 T, directed along the vertical and pointing up. When a metal bar slides along the rails to the west (left) with a speed of 10 m/s, what will the voltmeter connected between terminals A and B read? between terminals A' and B'?

• **23.6** Repeat Problem 23.5, assuming that **B** has a component of 0.04 T pointing north and a component of 0.03 T pointing up.

• **23.7** Repeat Problem 23.5, assuming that **B** has a component of 0.04 T pointing west and a component of 0.03 T pointing down.

• • **23.8** In Figure 23.31 the voltmeter connected to terminals A'-B' is replaced by a 6-V battery in series with a 30-Ω resistor, and **B** = 0.05 T, pointing up. What is the equilibrium speed with which the bar slides along the track? If the bar slides to the left, which terminal, A' or B', is connected to the positive terminal of the battery? Neglect friction.

• • **23.9** Suppose that in Problem 23.8 the battery is replaced by a 0.1-Ω resistor and the 30-Ω resistor is removed. Find the power required to move the rod along the track at 1.5 m/s, assuming that the resistance of the track and rod is negligibly small.

• • **23.10** The bar AB of Figure 23.6 has resistance R and the track has negligible resistance. Show that the power dissipated in pushing the bar through the field at a speed v is given by $P = B^2L^2v^2/R$, where L is the length of the bar AB.

• **23.11** The rectangular closed loop of copper wire, shown in Figure 23.32, is 50 cm long and 8 cm wide, and has a mass of 0.005 kg. Initially, the loop is at rest between poles of an electromagnet that produces a horizontal **B** field of 2.0 T. If the loop is then allowed to fall under its weight, what is the direction of the current induced in this loop?

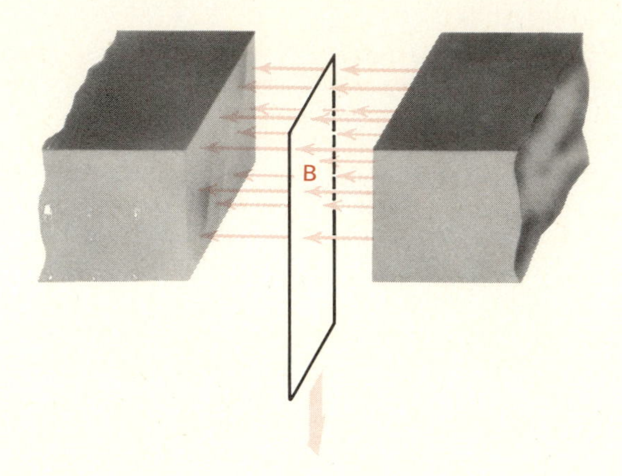

Figure 23.32

• • **23.12** If after release, the loop of Problem 23.11 acquires a terminal speed of 0.2 m/s, what is the resistance of the loop?

(Sections 23.4–23.5)

• **23.13** Derive an expression for the self-inductance of an ideal solenoid, assuming that the field inside the coil is uniform and neglecting end corrections.

23.14 The measured self-inductance of an air-core solenoid of 8-cm length and 2-cm diameter is 3 mH. Determine the approximate number of turns of this solenoid.

23.15 A toroidal coil has 10,000 turns and is wound on a form whose average toroidal radius is 18 cm. The measured self-inductance of this coil is 0.8 H. Determine the diameter of the coils.

23.16 A single-layer solenoid is 10 cm long and 4 cm in diameter and has a total of 800 turns. Find the self-inductance of this coil.

23.17 A coil with a 5-mH self-inductance is to be wound on an air-core cylinder of 3-cm diameter and 80-cm length. If the coil is in the form of a solenoid, how many turns will have to be wound on the form?

• **23.18** If a second coil of 120 turns is wound directly over the coil of Problem 23.17, what is the mutual inductance between the two coils?

23.19 How large a current must flow through a 0.2-H inductance so that the energy stored is 8 J?

• **23.20** The current flowing through the solenoid of Problem 23.16 produces a magnetic field of 0.1 T within the solenoid. How much energy is stored in this coil? What is the value of the current?

• **23.21** A 20-cm-long solenoid is wound with a total of 800 turns on a cylindrical cardboard form of 4-cm diameter. Determine the self-inductance of this coil and the energy stored when it carries a current of 1.2 A, assuming that the field within the solenoid is uniform.

• **23.22** If a second coil of 200 turns and a total length of 5 cm is wound over the coil of Problem 23.21 near the central region, what is the average emf induced in the 200-turn coil if the 1.2-A current in the solenoid changes direction in 0.04 s?

• **23.23** What is the mutual inductance of the two coils of Problem 23.22?

• • **23.24** Two coils are wound on the same toroidal form. The first coil has a total of 1200 turns, the second 400 turns. The diameter of the coils is 2 cm; the radius of the toroidal form is 8 cm at its middle. The first coil is connected to a 12-V battery through a switch. The second coil is connected to a 10-Ω resistor. Assuming that when the switch is closed, the current increases at an average rate of 8 A/s, determine how much current will flow through the 10-Ω resistor.

• • **23.25** Using the result of Problem 23.13 and Equation (22.16), show that the energy stored in the solenoid, $\frac{1}{2}LI^2$, can also be written $\frac{1}{2}(B^2/\mu_0)V$. Consequently, one can regard a region with a magnetic field **B** as having an energy density of $\frac{1}{2}B^2/\mu_0$.

(Section 23.7)

23.26 A rectangular coil 8 cm wide and 12 cm long has 80 turns. At what speed must it rotate in a field of 0.4 T so that the maximum induced voltage will be 80 V?

23.27 An ac generator has a rectangular loop of 40 turns rotating in a uniform field of 0.3 T at 1800 rpm. The loop is 8 cm long and 6 cm wide. Find the maximum value of the induced emf in the loop.

• **23.28** A simple dc generator has 80 rectangular loops on its armature. Each loop is 12 cm long and 8 cm wide and the armature rotates at 1200 rpm about an axis parallel to the long side. If the average field through which the loops' sides move is 0.4 T, what will the output voltage of this generator be?

• **23.29** A 20-cm × 10-cm rectangular coil rotates about an axis through its center and parallel to one of the long sides. The speed of rotation is 30 rps. There are 50 turns on this coil and it is located in a uniform magnetic field of 0.8 T, directed perpendicular to the axis of rotation. Write an expression for the emf induced in this coil and find the maximum instantaneous value of that emf.

• • **23.30** A square coil with 80 turns is wound on a 18-cm × 18-cm form. The coil rotates about an axis through the diagonal of the square. It is located in a uniform magnetic field of 0.6 T. If **B** makes an angle of 45° with the axis of rotation, what must the angular speed of the coil be so that the maximum instantaneous induced emf is 14 V?

Time Dependent Currents and Voltages; AC Circuits

Civilization depends on the supply of materials and of energy as its two necessities. Electrical energy is the only form which is economically suited for general energy transmission and distribution.

CHARLES PROTEUS STEINMETZ

24.1 Introduction

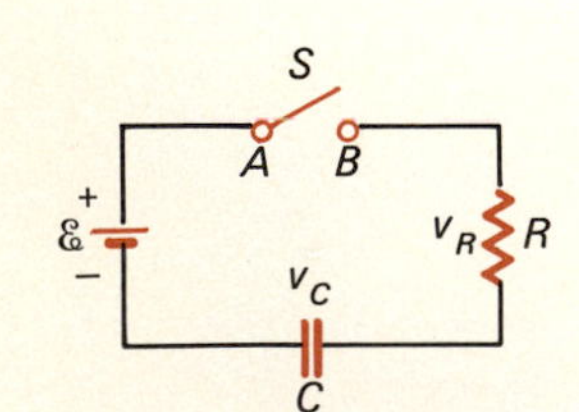

Figure 24.1 *A simple RC circuit. Before switch S is closed, the capacitor C is uncharged. After the switch is closed, the potentials across the capacitor and resistor display the time dependence shown in Figure 24.2.*

In preceding chapters, we restricted discussion almost exclusively to steady-state conditions: electrostatics, direct steady currents, and steady magnetic fields. Explicit time dependence appeared as an essential feature only in electromagnetic induction.

In this chapter, the temporal variation of fields and currents plays a central role. Although we have now firmly established the fundamental relations between charges, fields, and currents, generalizing these relations to time-varying conditions poses some conceptual as well as mathematical problems. We shall stress the concepts and try to make the mathematical results and functional dependences plausible without the burden of formal derivations.

We first try to answer the following questions: If a capacitor is connected to a battery through a resistor and a switch (Figure 24.1), how does the charge on the capacitor change with time after the switch is closed? Suppose that after this capacitor has been fully charged, it is removed from the circuit and allowed to discharge by connecting a resistor across its plates. In what manner will the charge on the capacitor decay to zero? If an inductor and a resistor are in series, how will the current in this circuit change with time when it is connected to a battery? How will this circuit respond if we disconnect the battery, interrupting the current?

Having considered the response of simple circuits to a discontinuous change in applied potential, we then turn to the response of these same circuit elements to sinusoidal alternating voltages. Whereas under steady, direct-current (dc) conditions, an ideal capacitor functions like an open circuit and an ideal inductor like a short circuit, these components exhibit dramatically different behavior when subjected to alternating voltages and currents. In particular, we shall see that a simple series circuit consisting of capacitance, inductance, and resistance displays resonance properties analogous to those of mechanical systems (considered in Chapter 14).

24.2 Response of Simple Circuits to Current or Voltage Discontinuities

24.2 (a) RC Circuits

We assume that before switch S of Figure 24.1 is closed, the capacitor C is uncharged, and shall mark time from the instant the switch is closed. For $t < 0$, no current flows in the circuit, and the voltage across the resistor is therefore zero; since the capacitor is initially uncharged, the potential across it is zero for $t < 0$. Consequently, for $t < 0$, the entire emf of the battery appears across the terminals A and B of the switch.

As soon as the switch is closed, $V_{AB} = 0$, and to satisfy Kirchhoff's Rule 2 (KR-II, p. 451), $v_R + v_C$ must equal $\mathcal{E}$.* Thus,

$$v_R + v_C = iR + \frac{q}{C} = \mathcal{E} \tag{24.1}$$

must hold at all times.

The problem posed in attempting to solve this equation is as follows. Since charge cannot pass across the insulating material separating the capacitor plates, a current through the resistor must result in the deposition of charge on the capacitor plates. Thus the two terms on the left-hand side of Equation (24.1) are not independent but intimately related. Still, we can see, at least qualitatively, the expected course of events for this circuit. Just after S is closed, there is no charge on C and hence $v_C = 0$ at $t = 0$. All the emf $\mathcal{E}$ of the battery must therefore appear as iR drop across the resistor, and so the magnitude of the resistance determines the value of the initial current; i.e.,

$$i(0) = \frac{\mathcal{E}}{R} \tag{24.2}$$

As charge continues to flow, equal but opposite charges accumulate on the two capacitor plates, and v_C increases. So that Equation (24.1) will

* We adopt the common convention of designating *time-dependent* variables, such as voltages, currents, or charges, with lowercase letters.

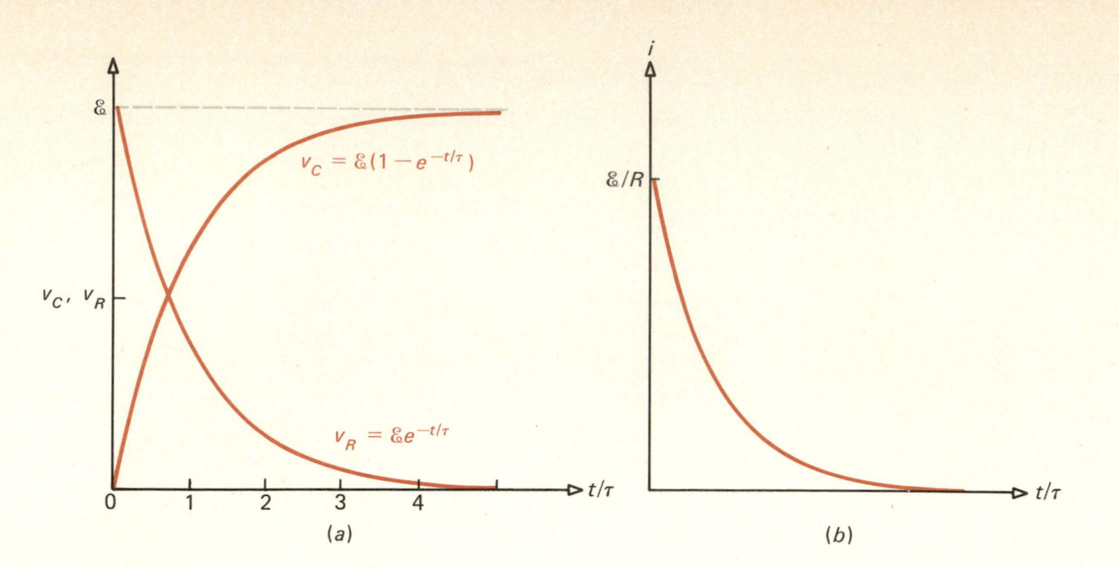

Figure 24.2 (a) The potentials across the capacitor and resistor of Figure 24.1 following closing of switch S at time $t = 0$. Note that the sum of the two voltages equals the battery emf ℰ at all times, as required by Kirchhoff's Rule 2. (b) Current in the circuit of Figure 24.1. When switch S is closed at $t = 0$, the current i jumps discontinuously from $i = 0$ to $i = ℰ/R$ and thereafter decays to zero exponentially. The time scale for the curves is in units of the time constant $\tau = RC$.

be satisfied, the gradual increase in v_C must be offset by an equal decrease of v_R. Hence the current i gradually diminishes with time as more and more charge appears on C. In the limit as $t \to \infty$, we find $v_C \to ℰ$ and $i \to 0$.

Using elementary calculus and the definition of the instantaneous current, $i = \lim_{\Delta t \to 0}(\Delta q/\Delta t)$, one can show that the asymptotic approach of v_C to its limiting value ℰ is exponential, following a curve like that of Figure 24.2. The mathematical relations that express the time dependences of v_C, v_R, and i in this particular case are

$$v_C(t) = ℰ(1 - e^{-t/\tau}) \qquad \textbf{(24.3a)}$$

$$v_R(t) = ℰe^{-t/\tau} \qquad \textbf{(24.3b)}$$

$$i(t) = \frac{v_R(t)}{R} = \frac{ℰ}{R}e^{-t/\tau} \qquad \textbf{(24.3c)}$$

where τ, a parameter with the dimension of time, is called the *time constant*.

The curves for v_C, v_R, and i are shown in Figure 24.2. Since, from Equation (24.1), $v_R = ℰ - v_C$, v_R must decrease from its peak value of ℰ at $t = 0$ to zero, as v_C grows from zero to ℰ. Thus, the two curves of Figure 24.2(a) are mirror images. Lastly, since $i = v_R/R$, the time course of i mimics that of v_R.

The time constant τ characterizes the rate at which the various observables, such as voltages and currents, approach their steady-state values. For a circuit consisting of a capacitance C and resistance R in series, the time constant is given by

$$\tau = RC \qquad \textbf{(24.4)}$$

It is reasonable that the values of resistance and capacitance should determine the rate at which the capacitor approaches its asymptotic voltage, and that this process should be slower—have a longer time constant—the greater R or C. If R is increased in the circuit of Figure 24.1, then for any value of v_R, the corresponding current i is reduced proportionately, and so is the rate at which charge is deposited on the capacitor plates. Consequently, an increase in R must increase the time required to charge the capacitor. If C is increased, this increases the amount of charge that must be placed on the capacitor plates to bring the capacitor to within a given fraction of its asymptotic potential. If, by holding R fixed, we do not allow an increase in current, the only way this greater charge can be transported is if the current flows for a longer time.

If a charged capacitor is discharged through a resistor R, as in Figure 24.3, the charge does not vanish from the plates instantaneously but decays to zero exponentially with the time constant RC.

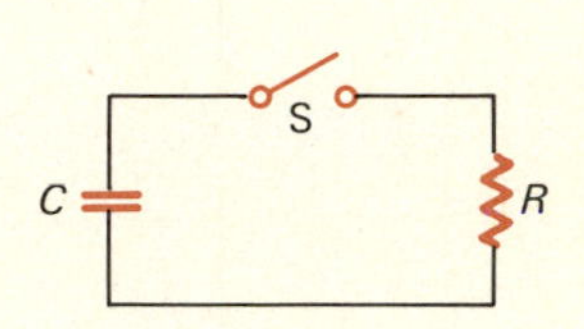

Figure 24.3 Discharge of a charged capacitor through a resistor. The charge on the capacitor flows through the resistor R after switch S is closed. The decay time for this process is also given by $\tau = RC$.

One of the characteristic features of the exponential function $e^{-t/\tau}$ is that the *fractional change* of the function during a given time interval is always the same. That is, if at some instant the *difference* between the function $f(t)$ and its asymptotic value $f(\infty)$ is $F(t) = f(t) - f(\infty)$, then after one time constant has elapsed, that difference will have been reduced by a factor of $1/e = 0.37$:

$$f(t + \tau) - f(\infty) = F(t + \tau) = 0.37F(t) \quad \textbf{(24.5)}$$

After two time constants have elapsed,

$$F(t + 2\tau) = 0.37F(t + \tau) = 0.37^2F(t)$$

and more generally,

$$F(t + N\tau) = 0.37^N F(t) \quad \textbf{(24.6)}$$

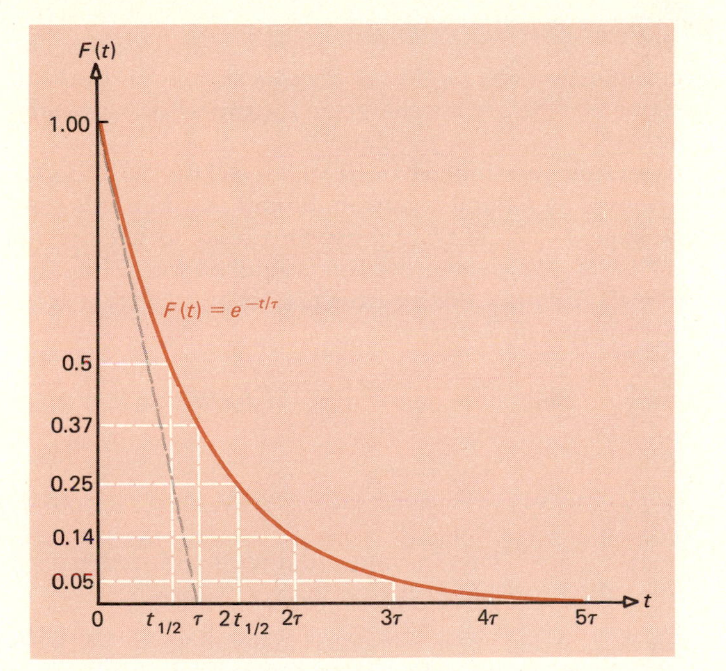

Figure 24.4 The function $F(t) = e^{-t/\tau}$ as a function of time. After time $t = \tau$ has elapsed, the value of the function has dropped from one to $1/e = 0.37$; at $t = 2\tau$, the function is $(1/e)^2 = 0.14$; and at $t = 3\tau$, $F(3\tau) = (1/e)^3 = 0.05$. The decay time is also the time that would be needed for a linear decay from 1 to 0 if that linear decay had the same slope as the initial slope of the function $F(t)$. The time marked $t_{1/2}$ is the time needed for the function $F(t)$ to diminish by a factor of $\frac{1}{2}$; $t_{1/2} = 0.693\tau$. Thus $F(t_{1/2}) = \frac{1}{2}$, and $F(2t_{1/2}) = (\frac{1}{2})^2 = \frac{1}{4}$, and $F(3t_{1/2}) = (\frac{1}{2})^3 = \frac{1}{8}$.

● ***Example 24.1*** A 0.1-μF capacitor is connected to a 45-V battery through a 3-Ω resistor. If the capacitor is initially uncharged, how much charge will have accumulated on it after 0.3 s, 0.6 s, and 1.5 s? How large is the current at these times?

The time constant for this RC circuit is

$$\tau = (0.1 \times 10^{-6}\ \text{F})(3 \times 10^{6}\ \Omega) = 0.3\ \text{s}$$

The initial value of the charge on the capacitor is zero, and the final, asymptotic value is given by $Q = CV = (10^{-7}\ \text{F})(45\ \text{V}) = 4.5 \times 10^{-6}$ C. According to the exponential law, the difference between the initial and final values of charge will be reduced by a factor of 0.37 during each time constant. Hence, starting with $q(0) = 0$ and an initial difference in charge of 4.5 μC, this difference will drop by $t = 0.3$ s to $0.37(4.5\ \mu C) = 1.665\ \mu C$, and the charge on the capacitor will then be $(4.5 - 1.665)\mu C = 2.835\ \mu C$.

Alternatively, we have, from (24.3a),

$$q = Cv_C = C\mathcal{E}[1 - e^{-t/\tau}] = 4.5 \times (1 - e^{-t/0.3})\ \mu C \tag{24.7}$$

which, for $t = 0.3$ s, gives the same result as before. After two time constants, the charge is $(4.5\ \mu C)(1 - 0.37^2) = 3.884\ \mu C$, and after 1.5 s $= 5\tau$ has elapsed, $q(1.5) = (4.5\ \mu C)(1 - 0.37^5) = 4.47\ \mu C$, a charge within 1 percent of the limiting value of 4.5 μC.

The current through the resistor when the circuit is first completed is

$$i(0) = \frac{\mathcal{E}}{R} = \frac{45\ V}{3 \times 10^6\ \Omega} = 15\ \mu A$$

It then decays to zero according to (24.3c). Following the same arguments as before, we have

$$i(0.3) = 0.37(15\ \mu A) = 5.55\ \mu A$$

$$i(0.6) = 0.37^2(15\ \mu A) = 2.05\ \mu A$$

$$i(1.5) = 0.37^5(15\ \mu A) = 0.104\ \mu A$$

●

Another question one might ask about a system of this kind is, How long must one wait before the charge on the capacitor has reached half its final value?

To answer this question, we need only recall the definition of the natural logarithm. The natural logarithm of any number N is the power to which the base $e = 2.718\ldots$ must be raised to get the number N. That is, $e^{\ln N} = N$. We denote the time required for $q(t)$ to change from $q = 0$ to $q = C\mathcal{E}/2$ by $t_{1/2}$. We see from (24.3a) or (24.7) that at $t = t_{1/2}$, the value of $e^{-t_{1/2}/\tau}$ must be $\frac{1}{2}$; that is,

$$e^{-t_{1/2}/\tau} = \frac{1}{2} \quad \text{and} \quad \ln\left(\tfrac{1}{2}\right) = \frac{-t_{1/2}}{\tau} = -0.693$$

Hence,

$$t_{1/2} = 0.693\tau \tag{24.8}$$

We shall meet $t_{1/2}$ again later in connection with radioactive decay, where $t_{1/2}$ is the so-called half-life of the radioactive substance.

● ***Example 24.2*** A capacitor is charged to a voltage of 100 V. When it is connected to a voltmeter whose internal resistance is 10 MΩ, the meter reads 100 V for an instant, then slowly reads ever diminishing voltages. After 6 s, the meter reading has dropped to 50 V. What is the value of the capacitance of the capacitor? How long after the initial connection to the voltmeter will the meter read 25 V?

From the statement of the problem, we know that the time required for reducing the voltage by $\frac{1}{2}$ is 6 s. That is, $t_{1/2} = 6$ s. From Equation (24.8), we have

$$\tau = RC = \frac{t_{1/2}}{0.693} = 8.66\ s$$

$$C = \frac{8.66\ s}{10^7\ \Omega} = 0.866\ \mu F$$

Since for the exponential function the fractional change during a given time interval is always the same, we can write a relation for $t_{1/2}$ equivalent to Equation (24.5), namely,

$$F(t + Nt_{1/2}) = (\tfrac{1}{2})^N F(t)$$

The decay in capacitor voltage from 100 V to 25 V corresponds to a fractional change by a factor of $\frac{25}{100} = \frac{1}{4} = (\frac{1}{2})^2$. Hence, the voltage will reach that value after $2t_{1/2} = 12$ s. ●

● ***Example 24.3*** Simple *RC* circuits have many important applications. We shall describe a number of them in a later section of this chapter; here we consider one, that of providing a *linear time base* for cathode ray oscilloscopes (CRO) and TV picture tubes.

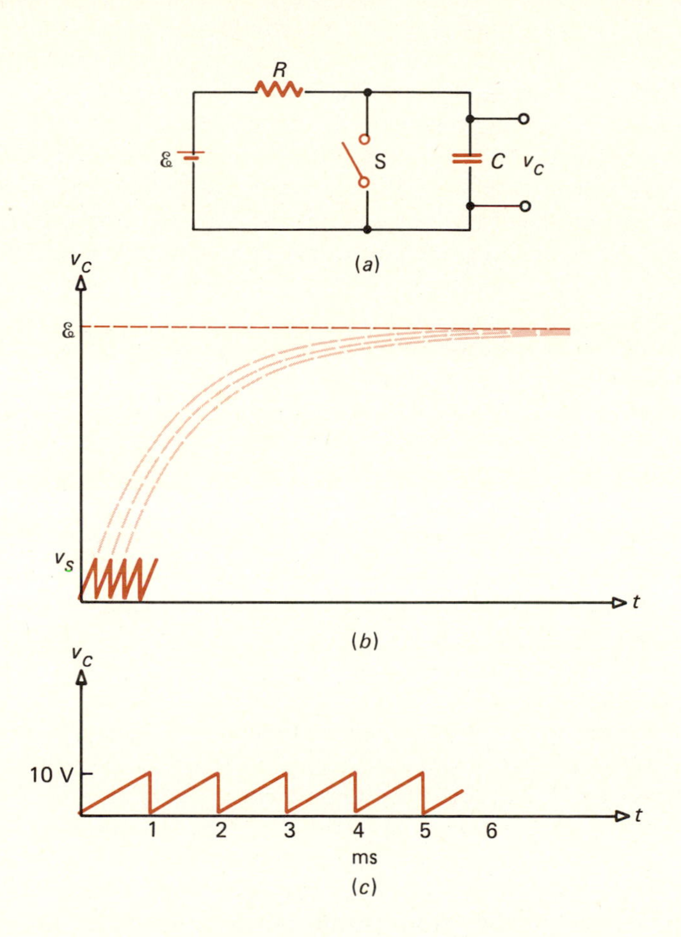

Figure 24.5 *(a) The basic circuit for generating a linear time base. The switch S is activated by the voltage v_C across the capacitor C; when v_C reaches a preset value V_S, generally a small fraction of ℰ, S automatically closes for a very brief interval, discharging the capacitor, which then recharges again, and the cycle repeats. (b) The voltage v_C as a function of time for the circuit of (a). (c) The sawtooth pattern of voltage v_C shown on an expanded time scale. The time intervals of 1 ms and the peak voltage of 10 V are for the circuit parameters used in the text example.*

Figure 24.5 shows this so-called sweep circuit, which controls the uniform horizontal motion of the electron beam across the face of the tube, in its most rudimentary form. The circuit consists of a steady voltage source ℰ, a resistance *R*, capacitance *C*, and a switch S. That switch, which is really an intricate electronic circuit, has the property that it automatically closes when the potential between its terminals exceeds some preset value, but remains closed only very briefly, perhaps 1–10 picoseconds (1 ps = 10^{-12} s). To see how this circuit functions as a time base, let us insert some typical values for ℰ, *R*, *C*, and the critical switch voltage V_S, and calculate the voltage v_C as a function of time.

We assume that ℰ = 1000 V, $R = 10$ MΩ, $C = 0.01$ μF, and $V_S =$ 10 V. Moreover, we further assume that the resistance between the capacitor terminals is 10 $\mu\Omega$ when S is closed (a reasonable value for a good switch), and that this switch remains closed for only 1 ps.

We begin with $v_C = 0$ and follow v_C as time progresses. Since the switch is placed across the capacitor, the potential between its terminals is also v_C. The switch is therefore initially open, and v_C will increase with time, following an exponential curve with time constant $\tau = RC = 0.1$ s. As soon as v_C reaches 10 V, the switch closes briefly, allowing the capacitor voltage to drop drastically. While S is closed, the resistance in series with C, through which the capacitor must discharge, is 10 $\mu\Omega$, so that the time constant for the decay of v_C is $\tau' = (10^{-5}\ \Omega)(10^{-8}\ \text{F}) = 10^{-13}$ s. If the switch stays closed for 1 ps, it remains closed for $10\tau'$; consequently, we conclude from Equation (24.6) that in this brief interval v_C decays from 10 V to $(0.37)^{10}(10\ \text{V}) = 4.8 \times 10^{-4}$ V, effectively zero. (It is left as a problem to show that the current flowing through this switch resistance of 10 $\mu\Omega$ due to the 1000-V source can be neglected.) As soon as the switch opens, v_C again rises along the exponential curve, now displaced in time, until $v_C = V_S = 10$ V, when the switch closes briefly once again. Hence, the voltage v_C follows the sawtooth pattern of Figure 24.5.

Since the portion of the exponential curve that is traced by v_C during its rise to 10 V is only a small part of the complete exponential, the time dependence of v_C is nearly linear. We can see that this will be true as follows.

When $v_C = 0$, the current i is $\mathcal{E}/R = (1000\ \text{V})/(10^7\ \Omega) = 0.1$ mA. Just before S closes, $v_C = 10$ V and the current is then $i = (\mathcal{E} - v_C)/R = (990\ \text{V})/(10^7\ \Omega) = 0.099$ mA. We see that the current in the circuit changes by only 1 percent during this time; that is, the rate at which charge is deposited on the capacitor and its voltage increases is constant within 1 percent.

Since the linear approximation is a good one here, we can quickly determine the time required for one complete period of this repetitive pattern. When $v_C = 10$ V, the charge on the capacitor is $q = Cv_C = 0.1\ \mu$C. Since the charging current is 0.1 mA = 100 μA, the time needed to accumulate a charge of 0.1 μC on the capacitor is $T = 0.1\ \mu\text{C}/100\ \mu\text{A} = 1$ ms. Since this time is very long compared with the 1 ps during which the capacitor discharges, the latter can be neglected. Thus, the sweep time for this circuit is 1 ms.

A voltage of 10 V applied to the deflection plates of a TV tube is not enough to deflect the beam substantially. The sawtooth voltage generated by the sweep circuit is therefore amplified before being impressed on the horizontal deflection plates. The spot on the tube then travels across the face of the tube from left to right at the predetermined sweep rate and returns to the left side almost instantaneously before repeating its trajectory.

In TV sets, the horizontal sweep rate is set to synchronize with the transmitted signal. Laboratory CROs have provision for adjusting the sweep rate by substituting different resistors or capacitors, or both, in the basic RC circuit and by varying, within limits, the potential V_S at which the shorting switch closes. ●

Before turning to other circuits, let us briefly consider energy changes during charging or discharging of a capacitor. For instance, in Example 24.2, the energy stored in the capacitor changes from its initial value of

$$\tfrac{1}{2}Cv_C^2(0) = \tfrac{1}{2}(8.66 \times 10^{-7}\ \text{F})(100\ \text{V})^2 = 4.33 \times 10^{-3}\ \text{J}$$

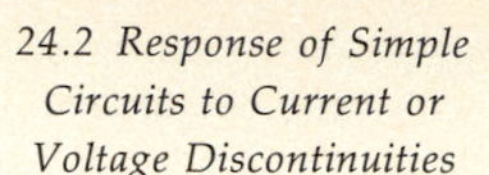

to

$$\tfrac{1}{2}C\, v_C^2(6) = \tfrac{1}{2}(8.66 \times 10^{-7}\ \text{F})(50\ \text{V})^2 = 1.08 \times 10^{-3}\ \text{J}$$

What happened to the energy difference? It is dissipated as heat in the resistor through which the charge must flow as the capacitor is discharged. Similarly, when a capacitor is charged, as in Figure 24.1, not all the energy supplied by the battery is stored in the capacitor. In fact, the amount of energy stored in the capacitor exactly equals the energy dissipated as heat by the resistor through which the charging current must flow.

24.2 (b) RL Circuits

We next examine the response of a circuit (Figure 24.6) consisting of an inductance and a resistance in series with a battery and switch. Before the switch is closed, no current flows in this circuit. Long after the switch has been closed, the current must have the value $i(\infty) = \mathcal{E}/R$, since the ideal inductance L does not impede the flow of a steady current. The question is, Does this current attain the value $\mathcal{E}/R$ instantaneously, or have we here another example of asymptotic approach?

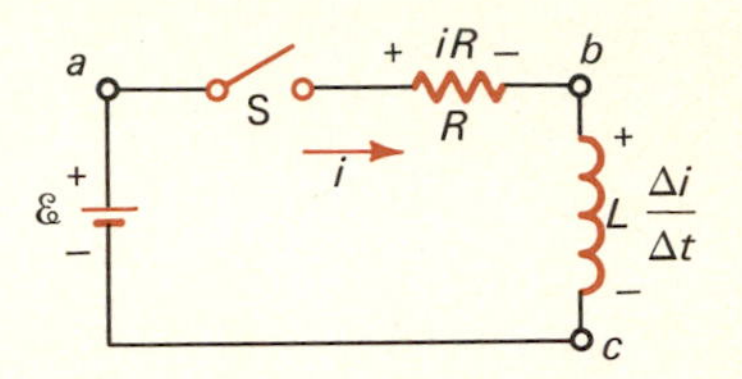

Figure 24.6 A simple RL circuit. After the switch S is closed, the current increases gradually, approaching the asymptotic value $\mathcal{E}/R$.

We can see quickly that the current cannot change suddenly in this circuit. A change in current induces a back-emf, that is, a voltage drop, in the inductance L of magnitude $v_L = L(\Delta i/\Delta t)$. Applying KR-II to this circuit, we obtain

$$iR + L\frac{\Delta i}{\Delta t} = \mathcal{E} \qquad \textbf{(24.9)}$$

This equation can be solved like Equation (24.1), and not surprisingly, leads to an exponential dependence of the variables. We say "not surprisingly," because if we substitute $i = \Delta q/\Delta t$ into the first term of Equation (24.1), we see that this equation has assumed the same form as (24.9), except that the variable has been changed from i to q.

For the RL circuit, one obtains the following relations:

$$i(t) = \frac{\mathcal{E}}{R}(1 - e^{-t/\tau}) \qquad \textbf{(24.10a)}$$

$$v_R(t) = \mathcal{E}(1 - e^{-t/\tau}) \qquad \textbf{(24.10b)}$$

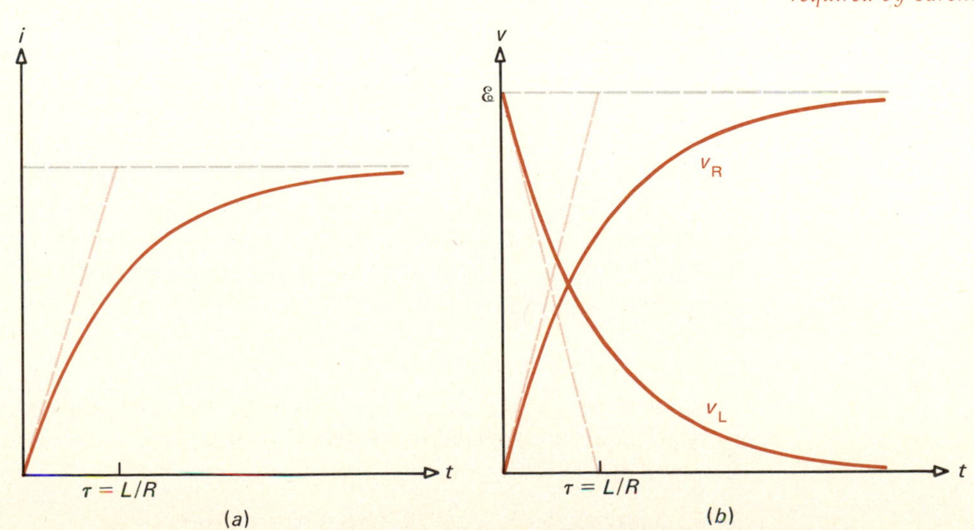

Figure 24.7 (a) The current in the circuit of Figure 24.6 as a function of time following closure of switch S. (b) The potentials across the resistor and the inductor of Figure 24.6 as functions of time after closing of switch S at $t = 0$. Note that v_L and v_R add to the battery emf $\mathcal{E}$ at all times, as required by Kirchhoff's Rule 2.

$$v_L(t) = \mathcal{E}e^{-t/\tau} \tag{24.10c}$$

where now the time constant τ is given by

$$\tau = \frac{L}{R} \tag{24.11}$$

A simple dimensional argument shows that this combination of inductance and resistance must be correct. Both iR and $L(\Delta i/\Delta t)$ have the dimension of voltage. Therefore,

$$[I][R] = \frac{[\mathcal{L}][I]}{[T]}$$

where $[\mathcal{L}]$ stands for the dimension of inductance, not length. From the above, it follows directly that dimensionally

$$\frac{[\mathcal{L}]}{[R]} = [T]$$

It is left as a more challenging exercise to present a plausibility argument for this dependence of the time constant on the circuit parameters.

Example 24.4 In the circuit shown, the coil represents the field coils of a large laboratory electromagnet. The self-inductance of the field coils is 12 H, and the rated maximum current for this magnet is 160 A. The resistance of the magnet coils is 1.5 Ω. When the magnet is operating and is to be deenergized quickly, a 13.5-Ω resistor is connected across the terminals of the magnet coils just before the power supply is disconnected.

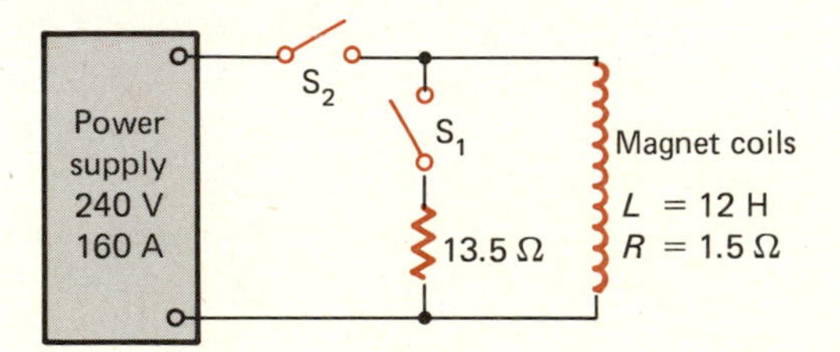

Figure 24.8

How much time elapses before the magnet current drops to 5 A? Why is it not practical to shorten the decay time by using a much larger shunt resistor, for example, a 600-Ω resistor or one even larger? And why is it important to connect a shunt resistor across the coils *before* disconnecting the power supply? What would happen if switch S_2 were opened before S_1 is closed?

We shall assume that the closure of S_1 precedes the opening of S_2 by a small fraction of a second only, and shall count time from that moment of closure of S_1. Thus, at $t = 0$, $i = 160$ A. This current decays exponentially to zero, with a time constant given by Equation (24.11). The total resistance in the circuit is $R = 13.5 + 1.5 = 15\ \Omega$. Hence the time constant of this RL circuit is

$$\tau = \frac{L}{R} = \frac{12\ \text{H}}{15\ \Omega} = 0.8\ \text{s}$$

Since the ratio of final to initial current is

$$\tfrac{5}{160} = \tfrac{1}{32} = (\tfrac{1}{2})^5$$

the interval of time required for the decay to 5 A is $5t_{1/2}$, where $t_{1/2}$ is defined by Equation (24.8) and has the value

$$t_{1/2} = 0.693(0.8 \text{ s}) = 0.554 \text{ s}$$

The time needed before the current has dropped to 5 A is 5(0.554 s) = 2.77 s.

To see why it would not be desirable to shorten the decay time, we need only bear in mind that at $t = 0$, the entire magnet current of 160 A must pass through the shunt resistor. Thus, the potential between the magnet terminals the moment switch S_2 is opened is iR = (160 A)(13.5 Ω) = 2160 V. If that resistor were replaced by one of 600 Ω, the potential across the coils of the magnet at that moment would be 96,000 V. The likelihood of insulation failure is then high.

We can now also understand why it is important that switch S_1 is closed *before* switch S_2 is opened. If S_2 were opened without provision for an alternative current path, the attempt to stop the flow of current in an inductor almost instantaneously would result in an enormous induced back-emf. The associated arcing and insulation breakdowns would almost certainly cause severe damage. ●

Although in general appearance the behavior of *RC* and *RL* circuits show great similarity, there is one important and revealing difference. In the *RC* circuit, the *voltage across the capacitor* cannot change discontinuously, but the current in the circuit can change to a new value instantaneously. In the *RL* circuit, the *current* cannot change discontinuously although the voltage across the inductance can. It is instructive to inquire into the fundamental reason for this difference.

The central issue is the way in which energy is stored in each circuit. In the first, the energy is proportional to the square of the *voltage across the capacitor;* in the *RL* circuit, the stored energy is proportional to the square of the *current in the inductance.* Energy conservation demands that if the energy stored in a circuit element changes, that energy must be supplied from or absorbed by some source or sink. Any energy change that occurs "instantaneously"—in an infinitesimal time interval—implies that energy is supplied or dissipated at an infinite rate, a practical impossibility. Finite power requirements demand that the current in an inductor and the voltage across a capacitor change only gradually; quickly, perhaps, but not discontinuously.

● ***Example 24.5*** A capacitor of 5 μF, charged to a potential of 90 V, is discharged by connecting a 100,000-Ω resistor across its terminals. How much energy is dissipated in the resistor during the first second after the circuit is completed?

We have here again a simple *RC* circuit. Its time constant is RC = 0.5 s. Hence one second corresponds to two time constants, and thus at the end of 1 s the potential across the capacitor will have decreased from 90 V to $(0.37)^2(90 \text{ V}) = 12.3$ V.

During that second, the energy stored in the capacitor drops from

$$\tfrac{1}{2}Cv_C^2(0) = \tfrac{1}{2}(5 \times 10^{-6} \text{ F})(90 \text{ V})^2 = 2.025 \times 10^{-2} \text{ J}$$

to

$$\tfrac{1}{2}(5 \times 10^{-6} \text{ F})(12.3 \text{ V})^2 = 0.038 \times 10^{-2} \text{ J}$$

That energy change of 1.99×10^{-2} J must be dissipated in the resistor. ●

24.3 Alternating Currents and Voltages

We have seen how a sinusoidally varying emf can be generated (Chapter 23). Such alternating voltages and currents are almost universally used throughout the world in residential and industrial electrical power consumption. Here we shall give only a cursory explanation of alternating circuits, just enough so that you can appreciate a few of the diverse applications in which capacitors and self and mutual inductances play a crucial role.

The expression for an alternating voltage or current is of the form

$$v(t) = V_0 \sin(\omega t + \alpha) \qquad i(t) = I_0 \sin(\omega t + \beta) \tag{24.12}$$

where V_0 and I_0 are the voltage and current amplitudes. (Sometimes the term *peak voltage*, or *peak current*, is used.) The angular frequency ω is related to the frequency of the alternating signal by

$$\omega = 2\pi f \tag{24.13}$$

and α and β in Equation (24.12) are phase factors.

Suppose that a resistor is connected across the terminals of an ideal alternating voltage source. The voltage across this resistor is then as shown in Figure 24.9, and the current follows the same pattern, since at every instant

$$i(t) = \frac{v(t)}{R}$$

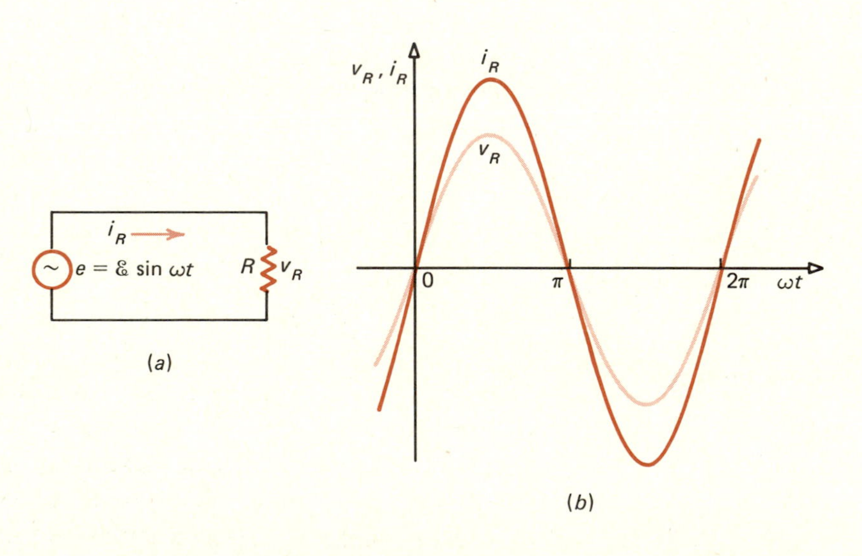

Figure 24.9 *(a) An ac circuit containing only a resistor R. The symbol ⊙ is conventionally used to designate a sinusoidal voltage source. (b) In this circuit, both current i_R and voltage v_R are sine functions and are in phase.*

Looking at these curves, we notice that while there may be a current in the resistor at any particular instant, the average current and the average voltage over one or more complete cycles is exactly zero. Does this mean that no power is dissipated in the resistor? Surely not; we know this, because a toaster, an electric stove element, and the like, when connected to the ac line do get hot.

The instantaneous power dissipated in the resistor is given by

$$P(t) = i^2(t)R \tag{24.14}$$

Although $i(t)$ changes sign every half-cycle, $i^2(t)$ is positive definite, and the *average* of i^2R is therefore not zero, even though the average of i is zero.

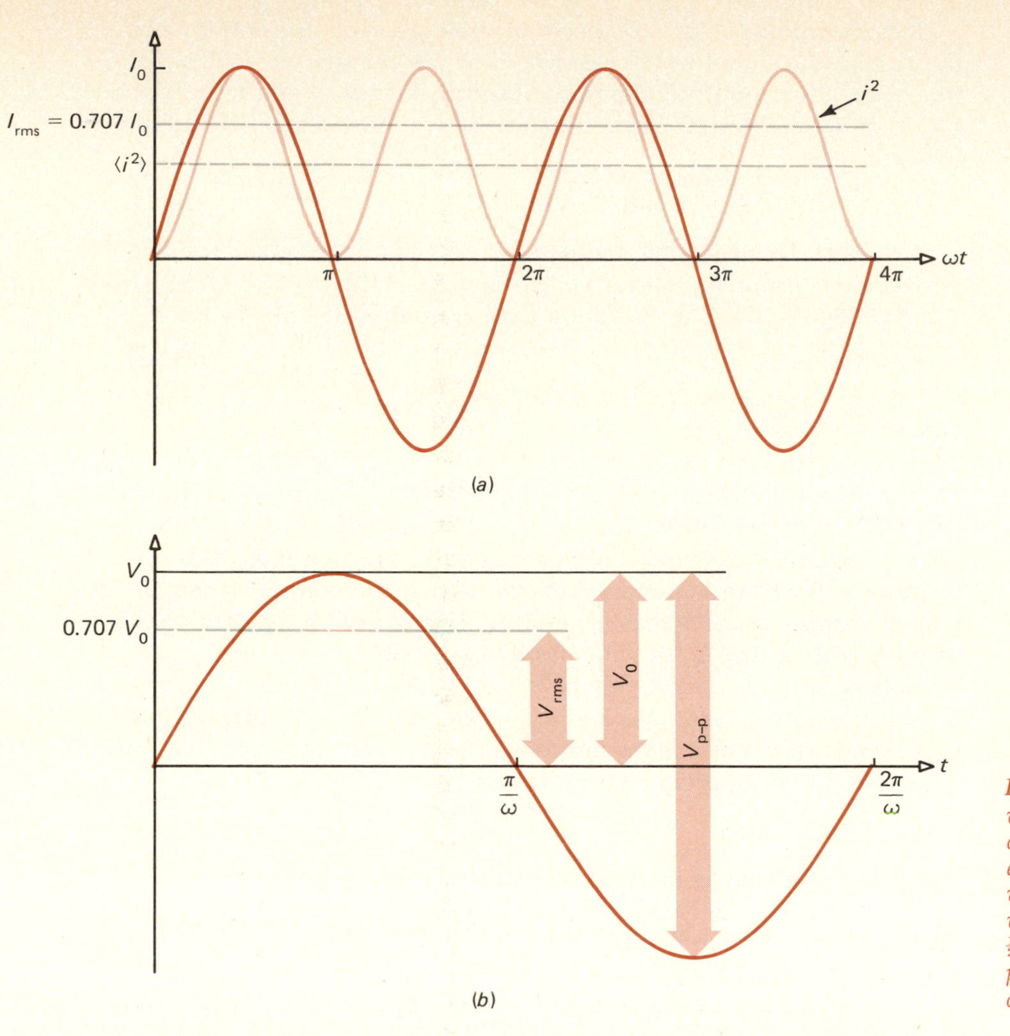

Figure 24.10 (a) The average value of the ac current i over one or more complete cycles is exactly zero. The average value of i^2, however, does not vanish; it is given by $\langle i^2 \rangle = \frac{1}{2}I_0^2$. (b) The amplitude, peak-to-peak and rms values of a sinusoidal voltage.

It is simple to determine the average value of i^2 when i is given by (24.12). The trigonometric identity

$$\sin^2 \theta = \tfrac{1}{2}(1 - \cos 2\theta) \qquad \textbf{(24.15)}$$

is useful here. Since the average value of cos 2θ is zero, the average of $i^2(t)$ is

$$\langle i^2(t) \rangle = \langle I_0^2 \sin^2 \omega t \rangle = I_0^2 \langle \sin^2 \omega t \rangle = \tfrac{1}{2} I_0^2 \qquad \textbf{(24.16)}$$

It follows that the effect of passing an alternating current through a resistor is the same, *in so far as average power dissipation is concerned,* as forcing a steady current whose value is

$$I_{\text{rms}} = \frac{I_0}{\sqrt{2}} = 0.707\ I_0 \qquad \textbf{(24.17)}$$

through that resistor. This current is called the *rms current,* where "rms" stands for "root-mean-square." The designation reminds us that this current is obtained by taking the square root of the mean value of the square of the alternating current. Similarly, the *rms voltage* is given by

$$V_{\text{rms}} = \frac{V_0}{\sqrt{2}} = 0.707 V_0 \qquad \textbf{(24.18)}$$

It is common practice to specify ac voltages and currents by stating their rms values, and in the remainder of the chapter we shall abide by this convention, omitting the rms subscripts.* Thus, when we talk about 115-V ac, we are describing a sinusoidally varying voltage whose amplitude, or peak value, is

$$V_0 = \sqrt{2}(115\text{ V}) = 1.414(115\text{ V}) = 163\text{ V} \tag{24.19}$$

Next, let us see what happens if we place an ideal inductance between the terminals of an ac voltage source. Applying Kirchhoff's Rule 2 and using the defining equation for L, Equation (23.6), we have

$$\mathcal{E}_0 \sin \omega t - v_L = 0 \tag{24.20}$$

where $v_L = L\dfrac{\Delta i}{\Delta t}$

In contrast to the resistive case, it is now not the current but the *slope* $\Delta i/\Delta t$ of the current-time curve that is proportional to the potential v_L across this circuit element. The current and voltage for an inductance are related as shown in Figure 24.11(*b*). Note that the voltage is zero at the precise instant at which the current is at its positive or negative peak because it is then that $\Delta i/\Delta t$ vanishes momentarily.

This qualitative description of the current-voltage relation in the inductive circuit is not complete, since we do not yet know how the magnitudes of current and voltage are related. That relation is obtained from Equation (24.20) by elementary calculus, and leads to the result

$$V_L = \omega L I_L = X_L I_L \qquad X_L = \omega L = 2\pi f L \tag{24.21}$$

where X_L is known as the *inductive reactance*.

Current and voltage obey something akin to Ohm's law, but with some important differences. First, as is evident from Figure 24.11, current and voltage are separated in phase, with the current lagging the voltage by 90°; second, the reactance depends not only on the value of the inductance L but also on the frequency. The higher the frequency, the greater

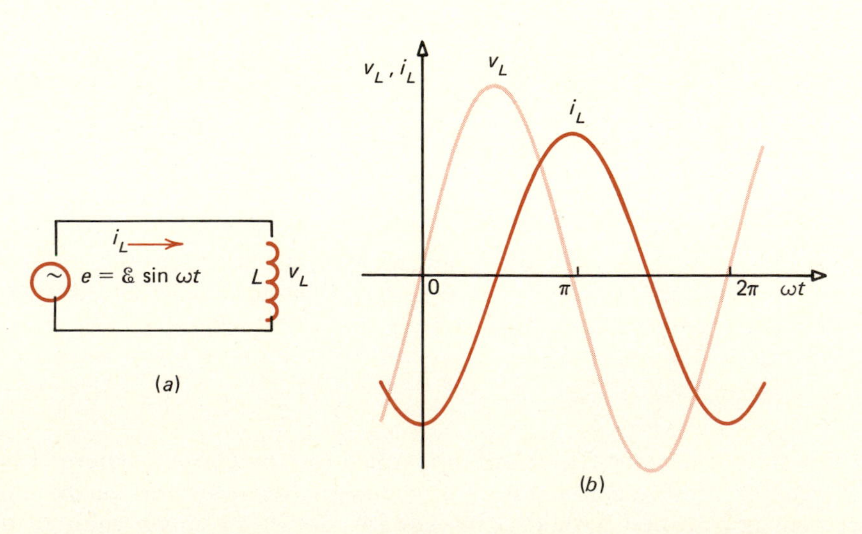

Figure 24.11 (*a*) *An ac circuit containing only an inductance L.* (*b*) *In this circuit the current and voltage sinusoids are displaced in time by one quarter period, the current lagging the voltage.*

* The phrases "ac voltage" and "ac current" are contradictory and redundant, respectively, since they mean, literally "alternating-current voltage" and "alternating-current current." Still, the abbreviation ac to designate an electrical quantity displaying simple harmonic time dependence has become so universally accepted that we are forced to perpetuate this linguistic abomination.

the inductive reactance and the smaller the current for a given impressed ac voltage.

We proceed in a similar manner to arrive at the current-voltage relation in an ac circuit containing only a capacitance C. In this case, the instantaneous voltage v_C is proportional to the charge q on the capacitor. The current, by definition, is given by $i = \Delta q/\Delta t$; replacing q by Cv_C, we have

$$i = C\frac{\Delta v_C}{\Delta t} \tag{24.22}$$

which bears a formal similarity to Equation (24.20).

The temporal relation between voltage and current for the capacitance in an ac circuit is shown in Figure 24.12(b). The current is zero whenever v_C is at a maximum or a minimum. There is again a 90° phase difference between current and voltage, but now the voltage lags the current, or what is equivalent, the current leads the voltage.

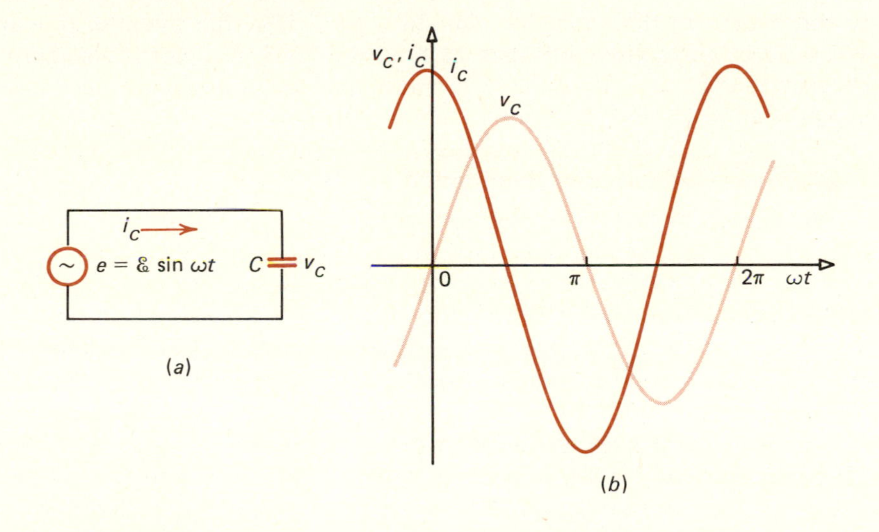

Figure 24.12 (a) An ac circuit containing only a capacitance. (b) In this circuit the current and voltage sinusoids are displaced in time by one quarter period, the current leading the voltage.

A quantitative relation between V_C and I_C can be derived from (24.22) and leads to the result

$$V_C = \frac{I_C}{\omega C} = X_C I_C \qquad X_C = \frac{1}{\omega C} = \frac{1}{2\pi f C} \tag{24.23}$$

The quantity X_C is the *capacitive reactance*. In contrast to the inductive reactance, the capacitive reactance is inversely proportional to the frequency and becomes infinite as f approaches zero. This is consistent with earlier conclusions, namely, that a capacitor in a dc circuit acts like an open circuit, preventing the flow of a steady current.

● ***Example 24.6*** A capacitor is connected to the output terminals of an oscillator operating at 1000 Hz with output voltage of 10 V. The current drawn from the oscillator is 63 mA. What is the value of the capacitance?

From Equation (24.23) we have

$$X_C = \frac{V}{I} = \frac{(10.0 \text{ V})}{(0.063 \text{ A})} = 159 \ \Omega$$

and

$$\frac{1}{2\pi fC} = 159\ \Omega$$

$$C = \frac{1}{(2\pi)(1000\text{ Hz})(159\ \Omega)} = 10^{-6}\text{ F} = 1\ \mu\text{F}$$

The instantaneous power delivered by the ac source of Figure 24.12 is

$$p(t) = e(t)i(t) = (\mathcal{E}_0 \sin \omega t)(I_0 \cos \omega t)$$

where I_0 is the amplitude of the alternating current in the circuit. From Equation (A.24) we can write

$$p(t) = \mathcal{E}_0 I_0 \sin 2\omega t$$

which is an oscillating function whose average is zero. We can also see this qualitatively by looking at Figure 24.12. During the first quarter of the period, between $\omega t = 0$ and $\omega t = \pi/4$, both e and i are positive and $p > 0$. During the following quarter period, however, e is positive and i is negative, so that $p < 0$. We see then that the average power delivered by the source to the capacitor vanishes. Similarly, the average power delivered by the source of Figure 24.11 to the ideal inductor is also zero. Therefore, *no power is dissipated by an ideal inductor or an ideal capacitor in an ac circuit.*

24.3 (a) RLC Circuit; Resonance

We have now seen the effect of resistance, capacitance, and inductance as individual circuit elements in an ac circuit. What happens if all three are placed in series between the terminals of an ac voltage source?

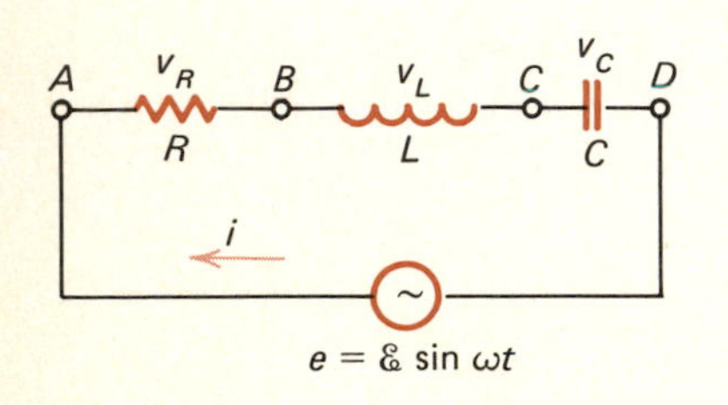

Figure 24.13 A series circuit containing a resistance R, a capacitance C, and an inductance L, connected to an ac voltage source.

The answer is obtained by applying Kirchhoff's Rule 2, remembering, however, that this rule applies to the *instantaneous* values of the potential drops and not to their rms or their peak values. Hence, although the rms voltages across R, L, and C of the circuit of Figure 24.13 are IR, IX_L, and IX_C, the rms voltage across the three circuit elements in series is *not* the arithmetic sum of the three separate voltage drops. Instead, the voltage across that series combination is given by a further generalization of Ohm's law,

$$V = ZI \qquad \textbf{(24.24)}$$

Here Z is the *impedance* of the circuit, and is given by

$$Z = \sqrt{R^2 + (X_L - X_C)^2} \qquad \textbf{(24.25)}$$

Since the inductive and capacitive reactances depend on frequency, the impedance of the circuit of Figure 24.13 (and for that matter, any circuit containing capacitors or inductors, or both) depends on this variable. It follows from Equation (24.25) that the impedance will reach a minimum, and the current, for fixed applied voltage, will reach a maximum when $X_L = X_C$. The circuit is then *in resonance* with the applied voltage source.

At resonance, the impedance of the series RLC circuit just equals the resistive component, the inductive and capacitive reactances canceling each other. For given circuit parameters, this condition obtains at a particular frequency only. The current, at constant voltage, as a function of frequency is shown in Figure 24.14 for several values of R. The resonance condition corresponds to that of maximum current. Note that the resonance becomes more pronounced and narrower as the resistance diminishes.

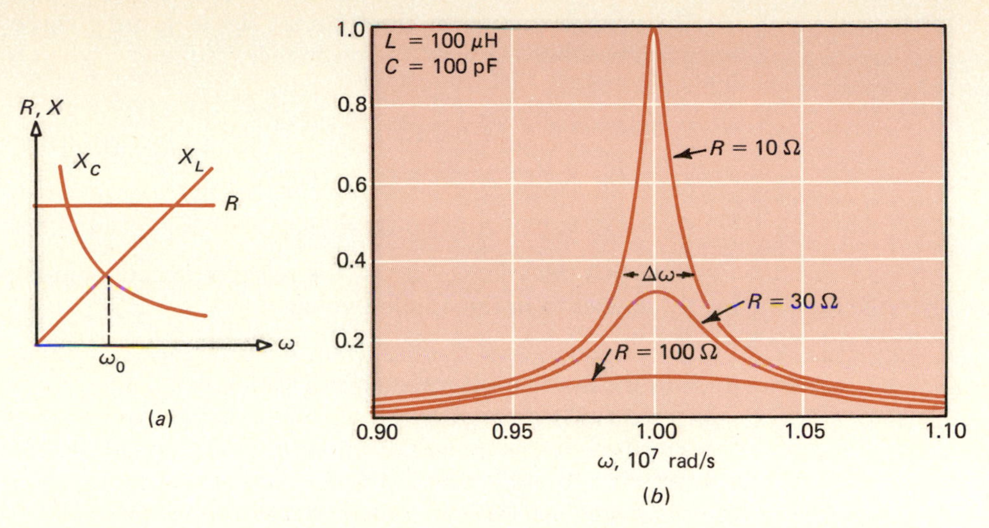

Figure 24.14 (a) The frequency dependence of the reactance and resistance of the three circuit elements of Figure 24.13. The resonance frequency is that for which $X_L = X_C$, and is indicated in the figure. (b) The current in a circuit containing a 100-μH inductance and a 100-μF capacitance in series with a resistance R, in the neighborhood of the resonance frequency, $\omega_0 = 10^7$ rad/s. The resonance becomes sharper and the current at resonance larger as the resistance decreases. Here Δω is the full width at half-maximum, usually called the half-width of the resonance.

These results suggest that an *LC* circuit (neglecting for the moment the resistance) should have its own natural frequency of oscillation, just as the pendulum and the mass-spring systems do (see Chapter 14). The natural frequency of these electrical oscillations is determined by the resonance condition

$$X_L = X_C \qquad \omega L = \frac{1}{\omega C} \tag{24.26}$$

Solving for the resonance frequency, we obtain

$$\omega_0 = \frac{1}{\sqrt{LC}} \qquad f_0 = \frac{1}{2\pi\sqrt{LC}} \tag{24.27}$$

● ***Example 24.7*** A series circuit consists of a 10-Ω resistor, a 0.004-μF capacitor, and a 36-mH inductor. This series circuit is connected to the terminals of an ac oscillator whose output is 50 V and whose frequency is adjustable over a wide range.

What is the resonance frequency of this circuit? How much current flows in this circuit at resonance? What are the voltages across each of the three circuit components?

From Equation (24.27), we have

$$f_0 = \frac{1}{2\pi\sqrt{(0.004 \times 10^{-6}\ \text{F})(36 \times 10^{-3}\ \text{H})}} = 13.3\ \text{kHz}$$

At that frequency, $X_L = X_C$, and consequently,

$$Z = R = 100\ \Omega \qquad (f = f_0)$$

The current in the circuit is then

$$I = \frac{V}{Z} = \frac{V}{R} = \frac{50\ \text{V}}{100\ \Omega} = 0.5\ \text{A}$$

To answer the remaining questions, we must calculate the values of X_L and X_C at the frequency f_0. These reactances are given by Equations (24.21) and (24.23). We have

$$X_L = (2\pi)(13.3 \times 10^3\ \text{Hz})(36 \times 10^{-3}\ \text{H}) = 3000\ \Omega$$
$$X_C = X_L = 3000\ \Omega \text{ when } f = f_0$$

The voltages across the three circuit elements are given by the generalization of Ohm's law. At the resonance frequency, when the current is 0.5 A,

$$V_R = IR = (0.5\text{ A})(100\ \Omega) = 50\text{ V}$$

$$V_L = IX_L = (0.5\text{ A})(3000\ \Omega) = 1500\text{ V}$$

$$V_C = IX_C = (0.5\text{ A})(3000\ \Omega) = 1500\text{ V}$$

Note that the voltages across the reactive circuit elements are significantly greater than the applied voltage. ●

This resonance of the inductance-capacitance combination is analogous to the mechanical resonance of the mass-spring system. In both instances there is a periodic transfer of energy between the two elements that compose the resonant system. In the mass-spring system, the kinetic energy of the mass is zero at the displacement extrema, when all the energy is stored as elastic potential energy in the spring. As this potential energy is transferred to the mass, it acquires speed and kinetic energy. Its inertia carries it beyond the equilibrium point, and it now transfers more and more of its kinetic energy back to the spring, which stores it once again as potential energy. The process then repeats, with a period given by Equation (14.6).

In the inductance-capacitance circuit, electrical energy is stored instead of mechanical. At one extreme of the cycle, the current in the circuit of Figure 24.15 is zero, and the capacitance is fully charged. It now starts to discharge through the inductance, but as we have seen earlier, the current in an inductance cannot change discontinuously, nor the voltage across a capacitance. Thus the current builds gradually to a maximum, which comes just at the instant when the capacitance is completely discharged. At that moment, all energy resides in the magnetic field generated by the current in the inductance. Just as the inertia of the mass carries it past the equilibrium point, so the requirement that the current in an inductance cannot change discontinuously precludes a sudden cessation of current in this circuit even though at that instant there is no potential driving this current. As the current continues to flow, it now charges the capacitance in the sense opposite to the previous voltage. Gradually, more and more of the magnetically stored energy is transferred to the capacitance and stored as energy in its electric field, until the current momentarily ceases and then reverses direction.

The inductance is analogous to the mass, and the capacitance to the compliance, $1/k$, of the spring of the mechanical oscillator. The resistance of the electrical circuit, which provides a mechanism for gradually dissipating energy, has its counterpart in the frictional loss between the surface of the table and the mass of Figure 14.6. This one-to-one matching of electrical and mechanical parameters has some useful practical consequences. It frequently happens that one must know the likely resonance characteristics of a mechanical system because the presence of such resonances may well cause serious structural failures (for example, the Tacoma Narrows bridge, and more recently, fatigue failures of the Electra turbo-prop airplane). Such resonances, if they exist, can often be damped or totally eliminated by proper design modifications. The necessary design analysis is greatly facilitated by using electrical analogs of the mechanical system.

Resonant circuits are also used in many everyday applications. Each radio and television station is assigned a particular, relatively narrow range of frequencies, and the station transmitter incorporates numerous

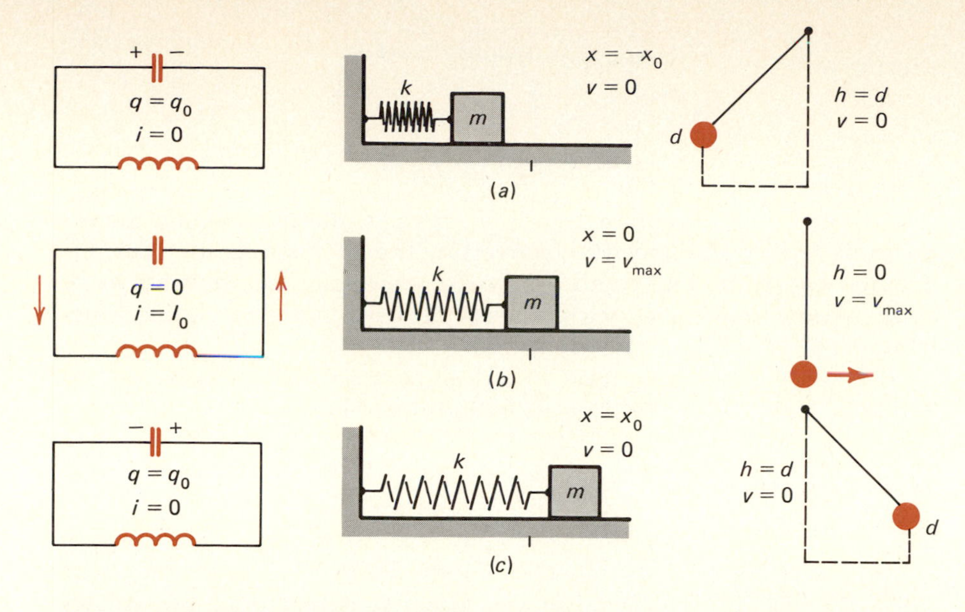

Figure 24.15 *The resonance of an LC circuit is analogous to the mechanical resonance of a mass-spring system or of a simple pendulum. (a) The capacitor is fully charged with one polarity, and the current in the circuit is momentarily zero. To this correspond the conditions of the mechanical systems as shown. The mass is at rest and the spring at maximum compression; the pendulum bob is at rest and maximum height. (b) The capacitor is completely discharged and the current in the inductance is a maximum. The mechanical analog is that of the spring with its free length and the mass moving at maximum speed; for the pendulum, the equivalent configuration has the bob at its equilibrium point and moving with maximum speed. (c) The current is again momentarily zero and the capacitor once again fully charged, now with polarity opposite to (a). The mass is momentarily at rest, the spring at maximum extension; the pendulum bob is also at rest and at its greatest height at the opposite end of its swing.*

resonant circuits to maintain and control this operating frequency. At the other end, the receiver, "tuning in" a station involves adjusting a variable capacitor or inductor in an *LC* circuit until the natural frequency of that circuit matches the frequency of the transmitted signal. Then the current induced in the antenna circuit of the receiver due to the rather small electrical signal picked up by the antenna becomes especially large. Many electronic digital watches use a resonant circuit as the basic timekeeping element, although the most precise timepieces rely on the electromechanical vibrations of a tiny quartz crystal.

*24.4 Filters

Suppose we place a circuit consisting of an inductor and a capacitor between an ac source and its load resistor, as in Figure 24.16. Since for low frequencies, the capacitive reactance is very large and the inductive reactance very small, the current path of least impedance at low frequencies will be through the inductor and the load resistor. At high frequencies, however, the *LR* branch will present a much higher impedance than the parallel capacitive reactance, and so most of the high-frequency current will flow through the capacitor and very little through the load resistor.

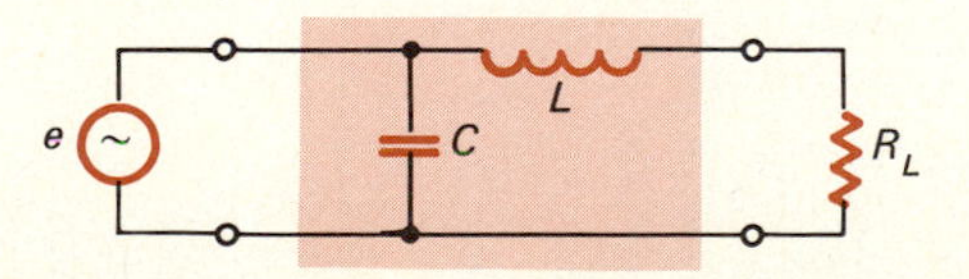

Figure 24.16 *An inductance-capacitance combination can serve as a low-pass filter. The filter (enclosed within the shaded rectangle) when inserted in the circuit between an ac source and its load will largely shunt high-frequency signals through the capacitance, while low-frequency signals will produce a current mostly through the inductance-and-load resistance branch with little current through the capacitive branch.*

We see that this *LC* network, inserted between source and load, allows low-frequency signals to pass through the load but blocks and bypasses the high-frequency signals from the source. Such an *LC* network is called a *low-pass filter*.

Interchanging the positions of inductor and capacitor converts the low-pass filter to a *high-pass filter*. In the circuit of Figure 24.17, low-frequency current is effectively blocked from passing through the load resistor by the capacitor, whereas high-frequency current will pass more readily through capacitor and load resistor than through the inductor.

Figure 24.17 Interchange of capacitance and inductance transforms the low-pass filter of Figure 24.16 into a high-pass filter.

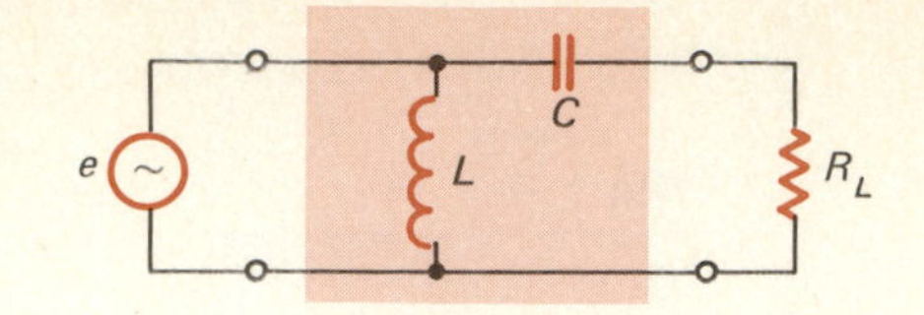

Although the *LC* combination is effective in filtering, usually the inductor is replaced by a resistor, converting the circuit to an RC filter. Inductors tend to be bulky and expensive, especially for applications at relatively low frequencies, such as the 60 Hz used in normal power transmission.

Filters are extensively used in almost all electronic devices, for example, in bass and treble controls in audio amplifiers.

24.5 Transformers

On your travels you may have seen a power substation on the outskirts of a town, such as the one shown in Figure 24.18. The metal containers house oil-filled transformers, whose function is to reduce the voltage from some high value, perhaps 230,000 V or more, to about 12,000 V for distribution throughout the city on overhead or underground lines. Smaller transformers, often visible mounted just below the high-voltage lines on the utility poles, change the potential to 230 V for residential use.

In this section we consider two questions. First, how is this transformation of potential achieved and, second, why go through all that voltage conversion in the first place? We shall see that it is this ease of increasing and decreasing voltages that proves to be one of the major advantages of ac over dc in power distribution.

A power transformer is nothing more than two coils wound close together on the same laminated soft iron core. Thus, whatever flux links one coil also links the other, and any flux change will induce the same emf in each turn of either coil. One coil, called the *primary*, is connected to an ac source and carries an alternating current, which generates the flux in the core. This ac flux induces a back-emf

Figure 24.18 Suburban substation, showing a large bank of transformers, with their tall ceramic insulators, and their terminals connected to conductors coming from a high-tension transmission line.

$$e_1 = -L\frac{\Delta i}{\Delta t} = -N_1\frac{\Delta \phi}{\Delta t}$$

in the primary coil and a corresponding emf

$$e_2 = -M_{21}\frac{\Delta i}{\Delta t} = -N_2\frac{\Delta \phi}{\Delta t}$$

in the secondary coil. Hence, the ratio of the ac voltages at the terminals of the primary and secondary windings is

$$\frac{\mathcal{E}_2}{\mathcal{E}_1} = \frac{N_2}{N_1} \quad \text{or} \quad \mathcal{E}_2 = \frac{N_2}{N_1}\mathcal{E}_1 \tag{24.28}$$

The turns ratio, N_2/N_1, determines the ratio of secondary to primary voltage; whether a transformer is a step-up or a step-down transformer is determined by its use rather than by its construction. For example, if $N_2/N_1 = 1/20$, an ac voltage of 120 V applied to the side with the larger number of turns will produce an ac voltage of 6 V at the terminals of the low-turn coil; on the other hand, if one applied 24 V to the coil with the fewer turns, a voltage of 480 V would appear at the other pair of terminals.

Figure 24.19 Small transformer—a bell transformer. The laminated core holding primary and secondary windings is shown.

Transformers can alter the potential at which electrical power is delivered but do not increase or decrease the amount of power (assuming the losses in the transformer can be neglected). That is, for the ideal transformer,*

$$P_1 = P_2 \quad \text{i.e.,} \quad \mathcal{E}_1 I_1 = \mathcal{E}_2 I_2 \tag{24.29}$$

where P_1 is the power delivered to the primary and P_2 the power delivered by the secondary. Step-up in voltage, therefore, means a commensurate reduction in current.

But why use transformers? Efficient though they may be, some energy is still wasted whenever such a unit is inserted in a distribution system. Why not simply transmit electrical power from the generating station to the residential or industrial user at the potential of 230 V?

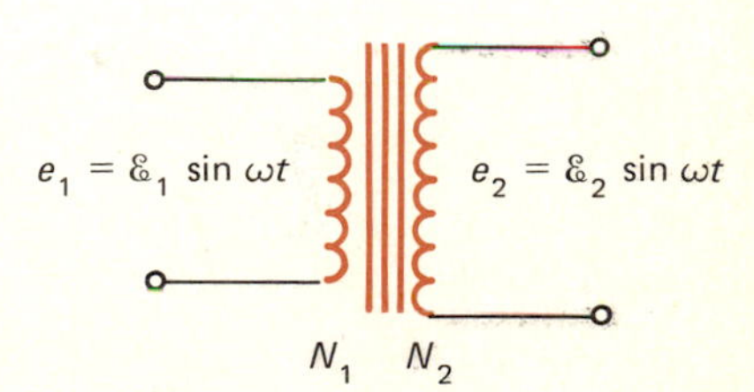

Figure 24.20 Symbolic diagram of a transformer. An ac voltage e_1 applied to the terminals of the primary winding produces an emf e_2 at the terminals of the secondary. The ratio of the rms voltages, $\mathcal{E}_1/\mathcal{E}_2$, equals the ratio of the number of turns, N_1/N_2, on the two windings. Two or three straight lines between the two coils indicate the presence of a laminated soft-iron core; the absence of such lines means that the transformer is an air-core unit, generally used in radios and television sets when the frequencies are high.

The answer to this question is apparent once we make a quick order-of-magnitude estimate of the current that would have to be carried by a transmission line that serves a community of 100,000 inhabitants. To estimate the average power consumption per resident, we shall assume that each light bulb, TV set, hi-fi unit, and other small appliance takes 100 W, and that every large one—toaster, each hot plate of the electric stove, clothes dryer, air conditioner, hair dryer—consumes 1000 W. Since there are numerous commercial demands (advertising, store lighting, open freezers in food stores, operation of industrial machinery, etc.) during the daylight hours, it is not unreasonable to assume that every resident will probably consume on the average 1 kW of power.† Thus, this average community uses about 100,000 kW. If that power is delivered at a voltage of 230 V, the transmission line carrying the current from the generating plant, possibly a hydroelectric plant some 100 miles from the city, must be capable of handling about 100,000,000 W/230 V = 435,000 A, an immense current. To avoid excessive power loss in the

* Transformers are extremely efficient electrical devices. Large, oil-filled power transformers used in substation distribution systems have an efficiency over 98.5 percent.

† Although close, the estimate is on the low side; U.S. electrical power consumption in 1978 was 1.2 kW per capita.

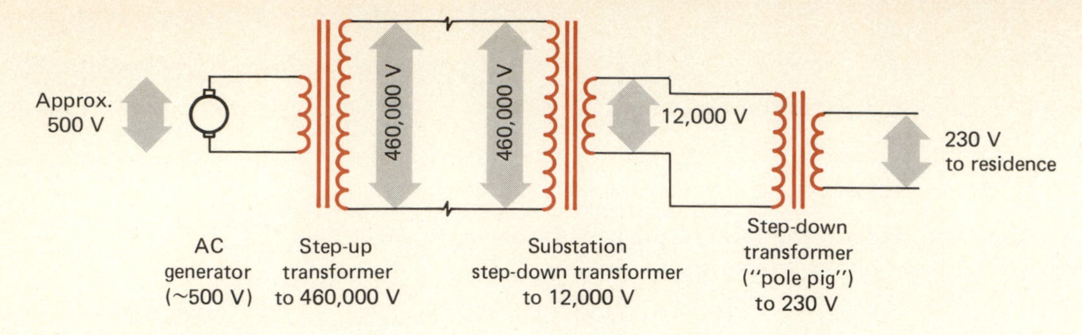

Figure 24.21 *Schematic of an ac power-transmission system. At the generating plant, the output from the generator is fed to the primary of a step-up transformer, which raises the voltage to 460,000 V. Power is transmitted at this very high tension to the outskirts of a city. There, at a substation, transformers reduce the voltage to about 12,000 V for distribution throughout the city on overhead power lines or underground cables. Before power is brought to a residence, the voltage is further reduced to 230 V by a transformer (called a* pole pig*) attached to the utility pole.*

transmission lines, the "wires" would have to be copper or aluminum cables nearly as thick as the pine poles that support the power lines in many residential areas today. Even then the I^2R losses in transmission would be substantial.

By transmitting power over long distances at 460,000 V, the current can be reduced to 217 A—still large, but manageable with reasonably sized cables that can be supported on steel towers. To bring power at that voltage to the center of the city would, however, pose prohibitive safety hazards. The voltage level is therefore reduced to about 12,000 V with transformers at a substation outside the high-population region, and the power is then distributed within the city at this lower potential.

● ***Example 24.8*** An ideal transformer has 4000 turns on its primary winding. The transformer is used to supply power to a 6-V chime with the primary connected to the 120-V house circuit. How many turns should be on its secondary winding? If the chime draws a current of 0.2 A, how much current flows in the primary of the transformer? What is the impedance of the load? What is the apparent impedance of the load as seen from the primary side of the transformer?

Since the ratio of primary to secondary voltage is 120/6 = 20, this must also be the turns ratio, N_1/N_2. Hence,

$$N_2 = \tfrac{4000}{20} = 200 \text{ turns}$$

Since the transformer is 100 percent efficient, we can apply Equation (24.29) and obtain

$$I_1 = \tfrac{6}{120} I_2 = (\tfrac{1}{20})0.2 \text{ A} = 0.01 \text{ A}$$

The impedance of the chime is given by Equation (24.24) as

$$Z_2 = \frac{V_2}{I_2} = \frac{6 \text{ V}}{0.2 \text{ A}} = 30\ \Omega$$

If we did not know that the transformer was in this circuit and simply measured the ratio of voltage to current in the 120 V circuit, we would conclude that the impedance of the load connected to the 120-V line is

$$Z_1 = \frac{120 \text{ V}}{0.01 \text{ A}} = 12{,}000\ \Omega$$ ●

As we have just seen in Example 24.8, a transformer is a device not only for changing voltages and currents, but also for modifying the load impedance as seen by the source. In Example 24.8, the actual load impedance was 30 Ω; however, seen from the primary side of a transformer of 20:1 turns ratio, this impedance is transformed to 12,000 Ω. It is left as a problem to prove that the ratio of primary to secondary impedance is given by

$$\frac{Z_1}{Z_2} = \left(\frac{N_1}{N_2}\right)^2 \tag{24.30}$$

Impedance transformation is another important application of transformers. We saw in Chapter 20 that for maximum power transfer from source to load, the load resistance, in a dc circuit, should equal the internal resistance of the battery or other source of emf. The corresponding condition, with impedance taking the place of resistance, also applies to ac circuits, and transformers are widely used for impedance matching in electronic systems.

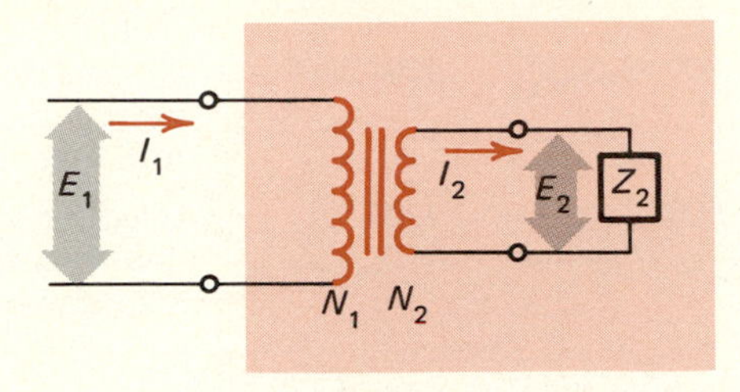

Figure 24.22 *Transformers are also often used for "impedance matching." The impedance Z_2 connected to the terminals of the secondary winding of the transformer draws a current I_2 when the secondary voltage is E_2, where $I_2 = E_2/Z_2$. The corresponding primary current and voltage are given by Equations (24.28) and (24.29). The effective impedance presented to the primary circuit is then $Z_1 = E_1/I_1$, and is related to Z_2 by $Z_1 = (N_1/N_2)^2 Z_2$. That is, if the entire circuit within the shaded region of this figure were enclosed within a box, ac measurements on the terminals of that box would imply that it contains an impedance Z_1, not Z_2.*

*24.6 Physiological Effects of Electric Currents

Most of us tend to treat an electrical circuit with greater respect the greater the voltage, and in a broad sense this is perfectly reasonable and justified. Yet it is really not the potential difference that is harmful but the current that flows through the body. Often, especially in winter, when humidity indoors is low, one accumulates a significant charge on the body in walking across a nylon carpet, and on touching a doorknob or other metallic object, experiences an unpleasant shock. Sometimes a spark of a few millimeters is seen to jump from the finger, showing that the potential difference between body and ground was several thousand volts. Still, except for a momentary unpleasantness, there is no lasting harm. In other circumstances, a dangerous level of voltage may be as little as 30–50 V.

Tables 24.1 and 24.2 summarize the physiological effects of 60-Hz alternating current on the average person. Note that the critical parameter is the amount of current that flows through the torso, not the voltage between the two points of contact. Most of the resistance between, say, the left hand and the right is due to the layer of skin, provided that it is relatively dry. The internal body fluids have various ions in solution and are moderately good conductors on the whole. That is why so many fatal electrical accidents occur in the bathroom; grasping a hair drier with an insulation fault on stepping out of a shower can be the last conscious act of a person.

An electric current passing through the body is so damaging because diverse parts of the body communicate through the transmission of electrical pulses. Currents flowing through tissues as a result of contact with an external source interfere with normal nerve and muscle impulses, causing muscle spasms. At somewhat larger currents, rhythmic contractions of the heart are affected. The consequent spasmodic and rapid un-

Table 24.1 Physiological effects of a 60-Hz AC current

Current	Voltage for Stated Current Assuming Body Resistance of 10,000 Ω	1,000 Ω	Physiological effects
1 mA	10	1	Threshold for sensation.
5 mA	50	5	Considered maximum harmless current.
10–20 mA	100–200	10–20	Discomfort to onset of uncontrolled muscular contractions; "can't let go."
50 mA	500	50	Considerable pain, but usually no cardiac or respiratory malfunction.
100–300 mA	1,000–3,000	100–300	Normal respiratory function, but ventricular fibrillation initiated.
5 A	50,000	5000	Used for ventricular defibrillation. Arrest of respiratory function.

NOTE: Electrical contact is assumed with surface of skin.

Table 24.2 Typical body resistance for various skin contact conditions

Skin condition and type of contact		Typical resistance
Light touch, finger	Dry	50 kΩ–1 MΩ
	Wet	5 kΩ–20 kΩ
Holding wire in hand	Dry	10 kΩ–50 kΩ
	Wet	2 kΩ–6 kΩ
Holding uninsulated electric drill	Dry	0.5 kΩ–3 kΩ
	Wet	0.2 kΩ–1 kΩ
Hand immersed in tap water		200 Ω–500 Ω

controlled cardiac contractions, known as fibrillations, can lead to death in a short time. The most effective method today for reestablishing normal heart function is to subject the heart to a severe electrical shock, forcing a brief overall muscular contraction, after which the heart often resumes its normal, coordinated pattern, pumping blood through the body.

Summary

Simple circuits, consisting of a resistance and capacitance or a resistance and inductance, respond exponentially to a discontinuous change in applied voltage. The approach to the steady state is governed by the *time constant* τ of the circuit, which has the value

$$\tau = RC \qquad \text{or} \qquad \tau = \frac{L}{R}$$

for the resistance-capacitance or the resistance-inductance circuit.

The *halving time* $t_{1/2}$, which is the time interval during which the difference between the value of voltage (or current) and its steady-state value is reduced by a factor of 2, equals

$$t_{1/2} = 0.693\tau$$

In an alternating-current (ac) circuit, the current and voltage vary sinusoidally with time:

$$i(t) = I_0 \sin(\omega t + \alpha)$$

$$v(t) = V_0 \sin(\omega t + \beta)$$

where ω is the angular frequency, which is related to the frequency by

$$\omega = 2\pi f$$

The root-mean-square (rms) values of current or voltage are

$$I_{\text{rms}} = \frac{I_0}{\sqrt{2}} = 0.707 I_0$$

$$V_{\text{rms}} = \frac{V_0}{\sqrt{2}} = 0.707 V_0$$

Normally, when ac currents and voltages are given, their rms values are stated and the subscript "rms" is omitted.

The current-voltage relation for ac circuits containing resistance only is Ohm's law,

$$V_R = RI_R$$

When an ac voltage is applied across a capacitor, the current and voltage are displaced in phase by 90°, the current leading the voltage. The values of current and voltage are related by

$$V_C = X_C I_C$$

where X_C is called the *capacitive reactance* and is given by

$$X_C = \frac{1}{\omega C}$$

When an ac voltage is applied across an inductor, the current and voltage are displaced in phase by 90°, the current lagging the voltage. The values of current and voltage are related by

$$V_L = X_L I_L$$

where X_L is called the *inductive reactance* and is given by

$$X_L = \omega L$$

A series circuit that contains inductance and capacitance as well as resistance has an *impedance* Z, defined by

$$Z = \frac{V}{I}$$

which depends on frequency. This impedance is given by

$$Z = \sqrt{R^2 + (X_L - X_C)^2}$$

When the angular frequency ω has the value

$$\omega_0 = \frac{1}{\sqrt{LC}}$$

the impedance attains a minimum. At that *resonance frequency*, $X_L = X_C$ and the impedance Z equals the resistance R.

Inductors, capacitors, and resistors can be connected to serve as *filters* that pass only currents of either high or low frequency.

Two coils, wound so that the flux produced by the current in one links the turns of the other, constitute a *transformer*. The ratio of the emf's at the terminals of the transformer windings is

$$\frac{\mathcal{E}_1}{\mathcal{E}_2} = \frac{N_1}{N_2}$$

where N_1 and N_2 are the number of turns on the primary and secondary windings. The currents in these windings stand in the ratio

$$\frac{I_1}{I_2} = \frac{N_2}{N_1}$$

Transformers are widely used to change the voltage levels in electric-power transmission, and also for impedance matching.

Multiple Choice Questions

24.1 The switch in the circuit shown is connected first to A, and then to B. Before the connection to A is made, the capacitor is uncharged. Which of the curves in Figure 24.23 shows how the current I varies with time?

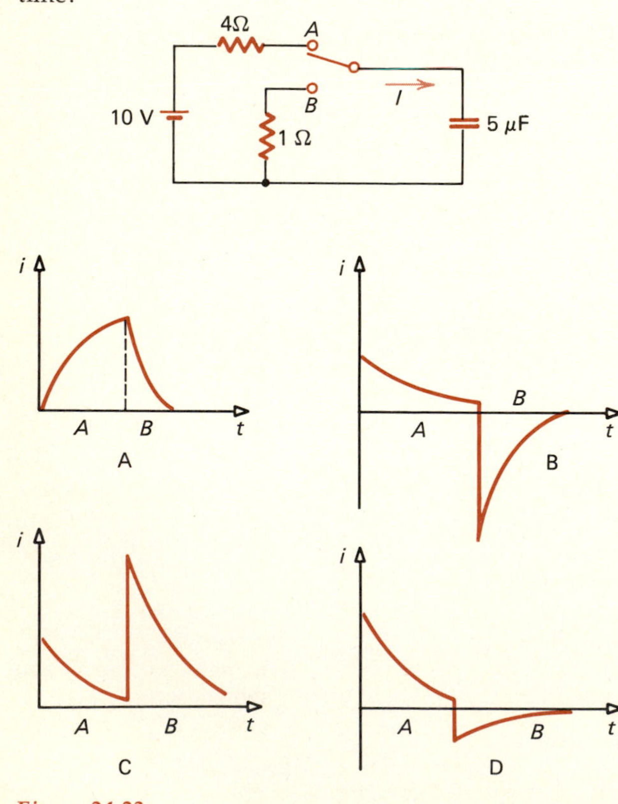

Figure 24.23

(a) A
(b) B
(c) C
(d) D

24.2 A capacitor is connected across a source of ac voltage, as shown in Figure 24.24. If the frequency of the voltage source is increased while the peak value of the voltage is kept fixed,

(a) the rms current decreases.
(b) the rms current increases.
(c) the phase relation between current and voltage changes.
(d) None of the above is correct, since no current flows in this circuit.

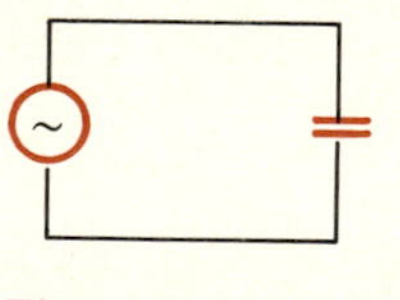
Figure 24.24

24.3 A generator produces a voltage that is given by $v = 240 \sin 120t$ V, where t is in seconds. The frequency and rms voltage are

(a) 60 Hz and 240 V.
(b) 19 Hz and 120 V.
(c) 19 Hz and 170 V.
(d) 754 Hz and 170 V.

24.4 The switch in the circuit shown is first connected to B. It is then turned to A, and then back to B.

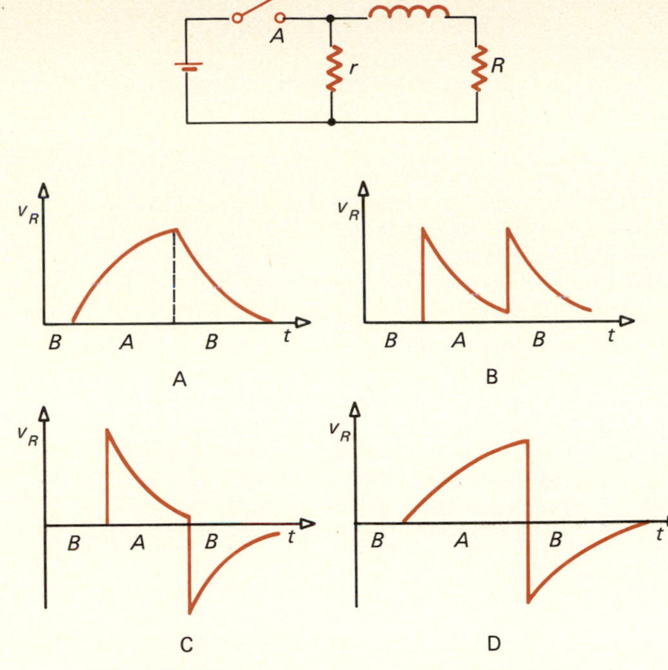

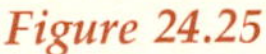

Figure 24.25

The voltage across the resistor R is best described by the curve

(a) A
(b) B
(c) C
(d) D

24.5 For the circuit shown in Figure 24.26, which of the following statements is *false?*

(a) The rms voltage across the resistor is R times the rms current.
(b) A decrease in the frequency f increases the voltage across the capacitor.
(c) An increase in the capacitance C increases the voltage across it.
(d) The power dissipated in the circuit equals the product of the rms voltage across the resistor and the rms current in the circuit.

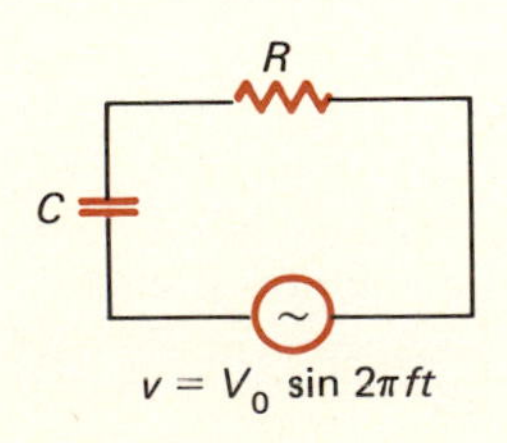

Figure 24.26

24.6 The critical quantity in causing injury due to electric shock is

(a) the current.
(b) the potential difference.
(c) the polarity of the potential.
(d) the induced emf.

24.7 In the circuit of Figure 24.27, the power dissipation at very high frequencies will be

(a) mostly in R_1.
(b) mostly in R_2.
(c) the same in R_1 and R_2.
(d) zero.

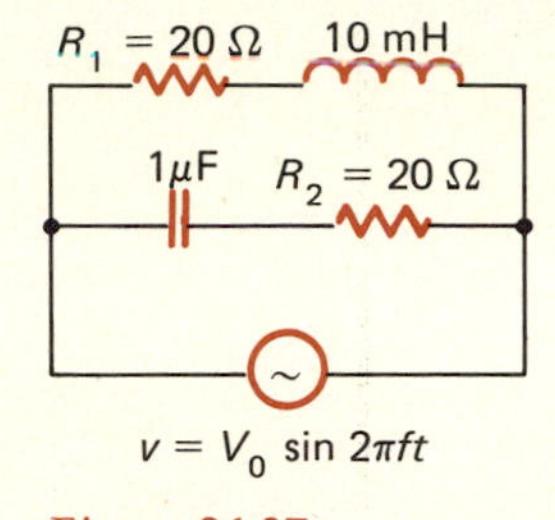

Figure 24.27

24.8 For the series R-L-C circuit of Figure 24.28,

(a) the power dissipated is I^2R no matter what the frequency.
(b) the power dissipated is I^2R only at resonance.
(c) the power dissipated is a minimum at resonance.
(d) None of the above statements is correct.

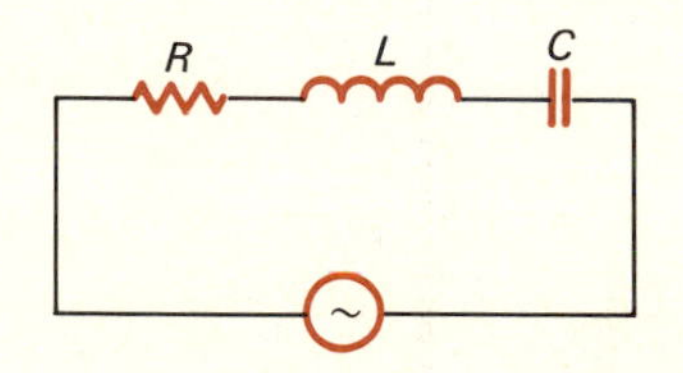

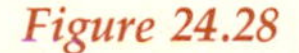

Figure 24.28

24.9 A transformer has N_1 turns on its primary winding and N_2 turns on its secondary. A load resistance r is connected across the secondary, and the power delivered by the secondary is P_2 when the voltage across the load resistance is V_2. Under these conditions, the current in the primary is given by

(a) $I_1 = P_2/V_2$
(b) $I_1 = (P_2/V_2)(N_1/N_2)$
(c) $I_1 = (P_2/V_2)(N_2/N_1)$
(d) $I_1 = (P_2/V_2)(N_2/N_1)^2$

24.10 An inductance and a resistance are connected in series across a battery, as shown in Figure 24.29. At the instant just after switch S is closed,

(a) the voltage across R equals $\mathcal{E}$.
(b) the current in R is zero.
(c) the voltage across L is zero.
(d) the current is $\mathcal{E}/R$.

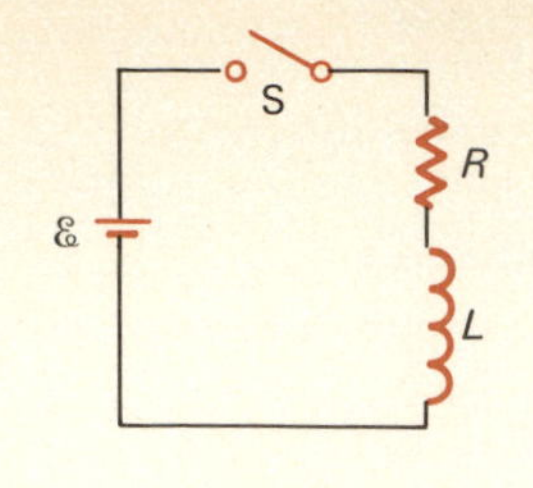

Figure 24.29

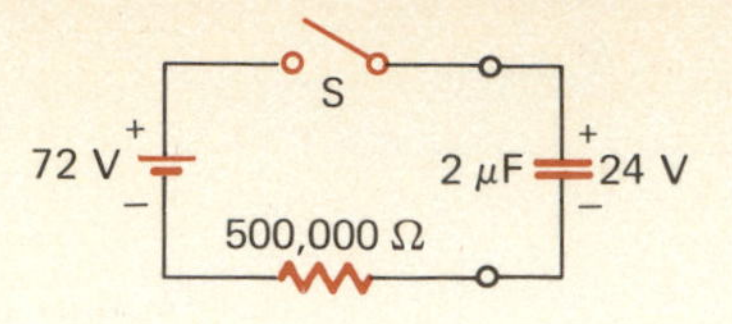

Figure 24.30

Problems

(Section 24.2)

24.1 What resistance should be placed in series with an 0.4-μF capacitor so that the time constant of the circuit is 0.8 s?

24.2 A 25-mH inductor has an internal resistance of 30 Ω. How large a resistance should be put in series with this inductor to reduce the time constant to 0.2 ms?

24.3 What is the time constant of a series circuit consisting of a 25-pF capacitor and a 5-MΩ resistor?

24.4 Modify the sweep circuit of Example 24.3 so that the sweep time is 20 ms, (a) using the same capacitor; (b) using the same resistor.

• **24.5** A resistor and a capacitor are connected in series across a 12-V battery. The capacitor is initially uncharged; 0.4 s after the connections are completed, the energy stored in the capacitor is 0.25 W (∞), where W (∞) is the energy stored at $t = \infty$. If the resistor has a resistance of 2.5 MΩ, what is the value of the capacitance and the charge on the capacitor at the end of the 0.4 s?

• **24.6** A capacitor with a leaky dielectric acts like a resistor in parallel with an ideal capacitor. If this capacitor is connected to a 12-V battery, the energy stored in the capacitor is 480 μJ. If the battery is disconnected from the capacitor at $t = 0$, the energy stored drops to 30 μJ at $t = 12$ s. Determine the values of C and R.

• **24.7** An inductor of 15 H and internal resistance of 40 Ω is connected in series to an ideal 12-V source through a switch. What is the rate of increase of current at the instant the switch is closed? How long after the switch is closed will the current have the value 0.225 A?

• • **24.8** A 2-μF capacitor is charged to a voltage of 24 V. It is then placed in the circuit in Figure 24.30, with polarity as indicated. At $t = 0$, switch S is closed. Sketch the voltage across C as a function of time. At what time will the voltage across the resistor equal 24 V?

• • **24.9** Repeat Problem 24.8, assuming that the 2-μF capacitor, charged to 24 V, is placed in the circuit of Figure 24.30 with inverted polarity.

• **24.10** The coils of a large electromagnet have inductance of 25 H and resistance of 8 Ω. If at $t = 0$, this magnet is connected to a dc source of 200 V, how much time will elapse before the current in the coils is (a) 12.5 A? (b) 24 A?

• **24.11** After the magnet of Problem 24.10 has been connected to the 200-V source for a long time, it is disconnected from the power source. What is the minimum shunt resistance that should be placed across the magnet terminals before opening the switch S, if the voltage across the terminals of S should not exceed 5000 V? How much time will then elapse before the magnet current is reduced to 0.1 percent of its initial value?

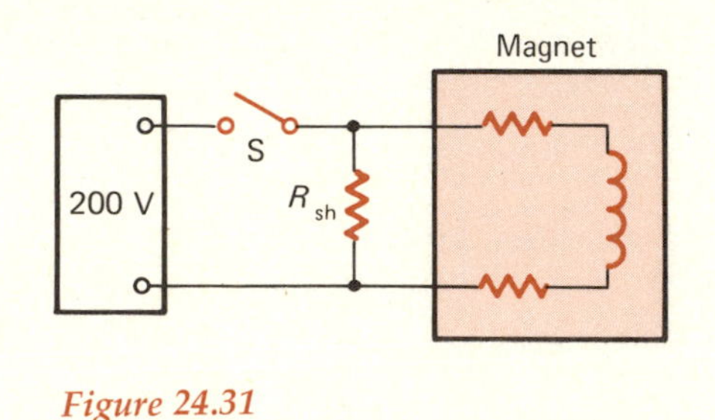

Figure 24.31

• **24.12** A 100,000-Ω resistor and a 5-μF capacitor are connected in series to a 12-V battery. Before the switch S is closed, the capacitor is uncharged.

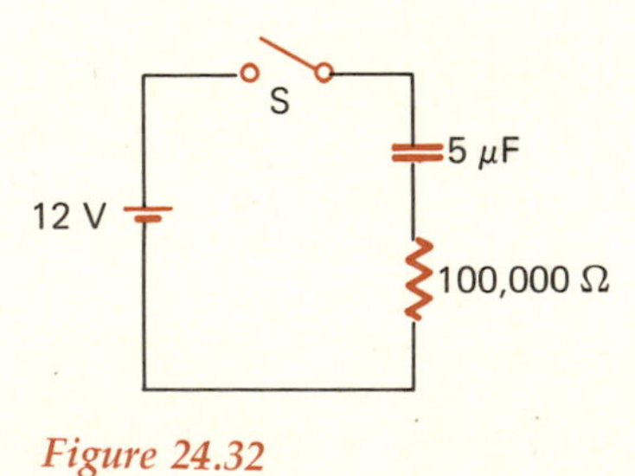

Figure 24.32

(a) Sketch the voltages across C and R as functions of time after S is closed.

(b) What is the current in the circuit immediately after S is closed?

(c) At what instant, following closure of S, will the voltages across R and C be equal?

(d) What is the voltage across the capacitor 2 s after S has been closed?

• • **24.13** A 15-μF capacitor is charged to a potential of 100 V. At time $t = 0$, it is discharged by connecting a copper wire of 0.01-Ω resistance across its terminals.

(a) At what time t will the voltage across C be 50 V?

(b) What is the initial current in the circuit after the connection across the capacitor terminals has been made?

(c) How much energy is dissipated in the resistor during the first halving time $t_{1/2}$? At what average power is energy dissipated in the resistor during this time interval?

• • **24.14** Repeat Problem 24.13 with the 0.01-Ω copper wire replaced by a 400-Ω resistor.

(Section 24.3)

24.15 An inductance has a reactance of 235 Ω at a frequency of 50 Hz. What is the value of this inductance?

24.16 If the inductance of Problem 24.15 is connected to a 150-Hz source, what is its reactance?

24.17 A 150-mH inductor has a resistance of 25 Ω. What are its reactance and impedance at the frequencies of 0 Hz (dc), 60 Hz, and 500 Hz?

24.18 A 0.1-μF capacitor is connected across the terminals of a variable-frequency oscillator. At what frequency is its reactance 84 Ω? What will the reactance be if the frequency is then tripled?

24.19 An inductor of 250 mH has a reactance of 1450 Ω. What is the frequency of the current in the inductor?

24.20 At what frequency will the reactance of a 0.3-μF capacitor be 200 Ω?

24.21 An inductor has a reactance of 20 Ω and resistance of 25 Ω at a frequency of 60 Hz. What is the inductance of this coil?

24.22 The current through a circuit is given by $i(t) = 56.4 \sin(754t)$ A, where t is in seconds. What is the rms current, and what is the frequency of the ac source?

• **24.23** A 5-mH inductor has an internal resistance of 30 Ω. When this inductor is connected to a 50-V ac source, the average power dissipated in the inductor is 25 W. Find the frequency of the source.

• **24.24** The current flowing in a series *RC* circuit connected to a 120-V, 60-Hz source is 0.2 A. If the capacitance is 6 μF, what is the value of the resistance?

• **24.25** A tuning capacitor, such as the one shown in Figure 19.4a, has a capacitance that varies from 30 pF to 365 pF. What inductance should be used with this capacitor so that the *LC* circuit will be in resonance over the entire AM frequency band from 550 kHz to 1600 kHz?

• **24.26** A series *RC* circuit has an impedance of 30 Ω and a resistance of 15 Ω. Find the reactance of the capacitor. If these impedance and resistance values were obtained for a source frequency of 400 Hz, what is the impedance of this RC circuit at 600 Hz?

• **24.27** A series *RC* circuit has a time constant of 0.1 ms. Its impedance at a frequency of 1 kHz is 180 Ω. Determine the values of R and C.

• **24.28** An inductor has a reactance of 50 Ω and an impedance of 80 Ω at a frequency of 120 Hz. Find the impedance of this inductor at a frequency of 80 Hz.

• **24.29** An inductor has a resistance of 20 Ω and an inductance of 10 mH. At what frequency will the inductor's impedance be 25 Ω? If a 12-V ac source operating at that frequency is connected across this inductor, what is the power dissipation in the inductor?

• **24.30** An ac voltage source of 24 V is connected across a series *RL* circuit. The current in the circuit is 4 A and the power dissipated in the circuit is 32 W. If the frequency of the ac is 60 Hz, what are the resistance and inductance components of this circuit? Calculate the power dissipation in the circuit when the frequency of the source is changed to 40 Hz while the voltage is kept at 24 V.

24.31 A capacitor is connected across the terminals of a 120-V, 60-Hz source. If the current in the circuit is 0.2 A, and the resistance of the circuit is negligible, what is the capacitance of this capacitor?

• **24.32** A capacitor and a resistor of 200-Ω resistance are connected in series across the terminals of a 120-V, 60-Hz source. The current in the circuit is 0.2 A. Find the capacitance of the capacitor.

• **24.33** For the circuits of Problems 24.31 and 24.32, find the power supplied by the source.

• • **24.34** Calculate the current in the circuit of Example 24.7 if the source frequency is reduced to $0.8f_0$, where f_0 is the resonance frequency. What are then the voltages across the three circuit elements? How much power is dissipated in the circuit?

• **24.35** What are the peak voltages across the resistor, capacitor, and inductor of Example 24.7?

• • **24.36** What are the peak voltages across the resistor, capacitor, and inductor of Problem 24.34?

• **24.37** A series *RLC* circuit is constructed using a 5-Ω resistor, an 80-mH inductance, and a 2-μF capacitance. Determine the resonance frequency of this circuit?

• • **24.38** If the series circuit of Problem 24.37 is connected across a 10-V ac source operating at the resonance frequency, what is the current in the circuit? What are the potentials across the three circuit elements?

• • **24.39** A 24-mH inductor with internal resistance of 40 Ω is connected to a source whose voltage is given by $v(t) = 354 \sin(150\pi t)$ V, where t is in seconds. Determine the peak current in the circuit, the maximum and rms voltages across the inductor, the average power dissipation, and the maximum and average energy stored in the inductor.

(Section 24.4–24.5)

24.40 A substation transformer is used to reduce the line voltage from 230,000 V to 4600 V for secondary distribution. What is the turns ratio for this transformer? If a current of 300 A flows in the 4600-V line, what is the current in the 230,000-V line?

24.41 A transformer is used to reduce the voltage from 230 V to 6.5 V. The number of turns on the secondary is 48. How many turns should there be on the primary winding?

24.42 If the transformer of Problem 24.41 were connected with its secondary (48 turn) winding across a 120-V line, what would be the voltage across the terminals of the other winding?

• **24.43** Prove the relation for impedance transformation, Equation (24.30).

• •**24.44** The last stage of amplification of a hi-fi unit has an internal resistance of 8000 Ω. The output from that amplifier is to be used to drive either earphones whose resistance is 2000 Ω or a loudspeaker whose resistance is 4 Ω. Design a simple circuit, incorporating necessary switches and transformers, so that maximum power is transferred from the amplifier to the respective loads. Specify the turns ratios of the transformers.

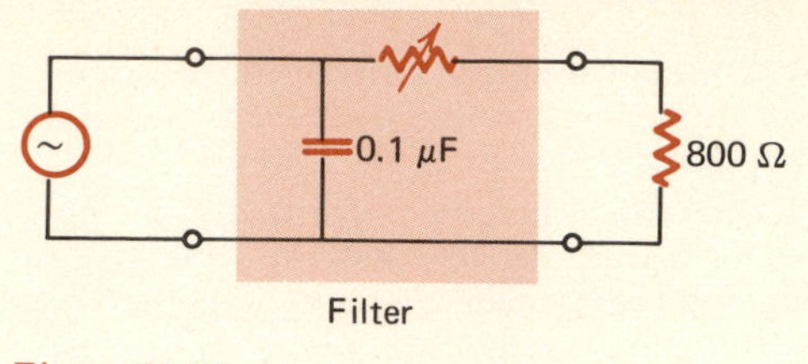

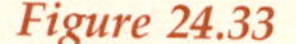

Figure 24.33

• •**24.45** A very crude tone control is a low-pass *RC* filter. Suppose that an adjustable *RC* filter is placed between the source of ac and the load whose resistance is 800 Ω. The filter's resistance can be varied between 200 Ω and 2400 Ω. The capacitance of the capacitor used in the filter is 0.1 μF. At what frequency will as much current flow through the load as through the capacitor when the filter resistance is (a) 200 Ω? (b) 2400 Ω? Will this tone control accentuate the high or the low frequencies if the load is a set of earphones?

•25•

Electromagnetic Waves and the Nature of Light

Before I began the study of electricity I resolved to read no mathematics on the subject till I had first read through Faraday's "Experimental Researches on Electricity."

JAMES CLERK MAXWELL

25.1 Introduction

Michael Faraday and James Clerk Maxwell, the two men who more than any others elucidated electromagnetic phenomena, could hardly have come from more disparate social and educational backgrounds.*

Faraday began his career as a bookbinder's apprentice, with minimal formal education and consequently little feeling for or confidence in mathematical abstraction. James Clerk Maxwell's family had ample financial resources and young James was sent to the best private schools and universities. Six years at the Edinburgh Academy and three at the University of Edinburgh preceded his entrance to Trinity College, Cambridge. Maxwell demonstrated exceptional mathematical abilities at a very early age; he wrote his first serious paper, "The Theory of Rolling Curves," when he was fourteen. On graduation from Cambridge with

* *Maxwell* was his mother's maiden name, adopted, in the Scottish tradition, by his father, John Clerk, on acquiring the Maxwell estate.

high honors at age 24, he was offered and accepted the chair of physics at Marichal College, Aberdeen. He returned to Cambridge in 1871 as the first professor of experimental physics (a curious title for Maxwell, who referred to himself as one who had "not made a single experiment worthy of the name"). The chair had been endowed the year before by the Duke of Devonshire, who also provided funds for a physical laboratory, the now famous Cavendish Laboratory. Maxwell planned and equipped this laboratory and directed it until his untimely death in November 1879, when he was 49 and at the height of his career.

Figure 25.1 James Clerk Maxwell (1831–1879).

25.2 Displacement Current

In formulating mathematical expressions for the physical features embodied in the laws of Coulomb, Ampère, and Faraday, Maxwell became aware of an inconsistency in Ampère's law (see Section 22.6c). This law relates the current through a surface to the magnetic field along the curve that bounds this surface, as shown in Figure 25.2. To see the source of the difficulty, consider the simple circuit consisting of an inductor and capacitor (Figure 25.3).

Suppose we now imagine two surfaces bounded by the same curve, as shown in Figure 25.4. According to Ampère's law, a magnetic field must appear along the curve whenever a current flows in the conductor and penetrates the surface S_1. Yet if we think of this same curve as bounding the surface S_2, we must conclude that the magnetic field will vanish because no current flows across the surface S_2, which passes only through regions of insulating material. Clearly, these conclusions, though consistent with Ampère's law, cannot both be correct. Maxwell, recognizing that verification of Ampère's law rested wholly on experiments performed on "closed circuits," proposed adding another term, namely,

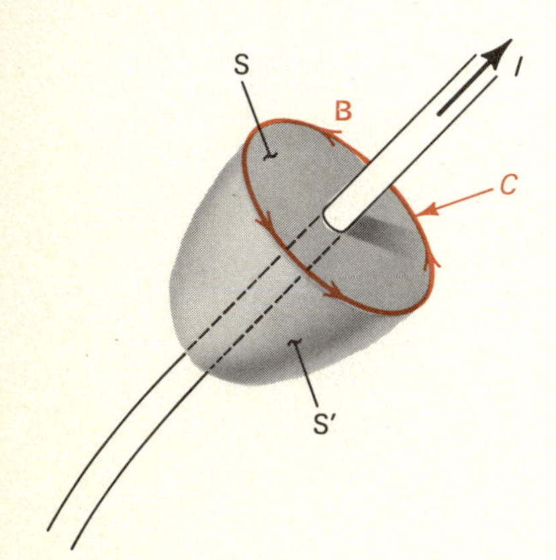

Figure 25.2 Ampère's law states that the magnetic field over a closed curve, such as C, is proportional to the current that penetrates any surface bounded by that curve, for example, surfaces S and S'. Here we have indicated the direction of **B** along C as given by RHR-I.

$$j_d = \epsilon_0 \frac{\Delta\phi_E}{\Delta t} \qquad \textbf{(25.1)}$$

to the current in Ampère's law. Here ϕ_E is the electric flux, defined, in analogy to the magnetic flux, by

$$\Delta\phi_E = E\,\Delta A \qquad \textbf{(25.2)}$$

where E is the electric field and ΔA is the small element of surface area perpendicular to the field vector **E**.

Notice that this additional term, called the *displacement current,* is proportional to the rate of change of electric flux and vanishes under

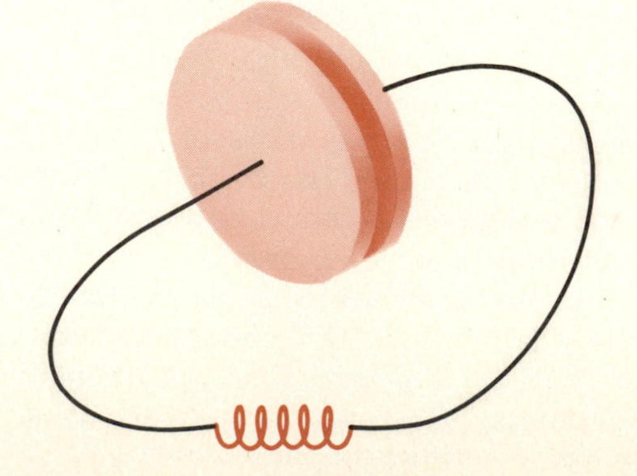

Figure 25.3 Physical configuration of an LC circuit, showing the parallel-plate capacitor and inductor. An oscillating current is easily established in such a circuit.

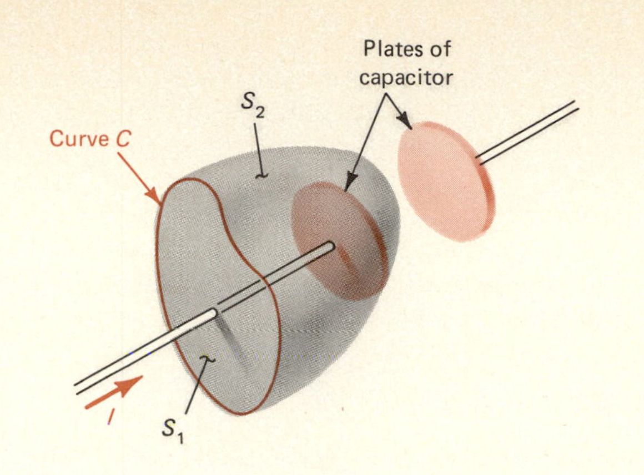

Figure 25.4 The parallel-plate capacitor of Figure 25.3 with two Ampère's law surfaces, S_1 and S_2, bounded by the same closed curve C.

steady-state conditions. Generally, it is also negligibly small in conductors, in which the normal current dominates (you will recall that in a perfect conductor, $\mathbf{E} = 0$). This new term, which relates the magnetic field to the time rate of change of an electric field, restores a symmetry to the description of electromagnetic phenomena heretofore absent. Faraday had demonstrated that a changing magnetic field induces an electric field, but the converse, the induction of a magnetic field by a changing electric field, had been neither observed nor suggested before Maxwell's hypothesis.

Maxwell's postulated displacement current proved, however, far more than a cosmetic device to improve the symmetry between electric and magnetic events. As Maxwell showed in just a few mathematical steps, his equations, when they included the displacement current, led directly and inexorably to electromagnetic waves.

Maxwell's equations, which so beautifully and concisely summarize all that was then and is now known about electromagnetism, are beyond the scope of this text, and so is the formal argument that predicts electromagnetic waves. Here we only summarize the principal features of these waves and try to make their existence and generation plausible.

Maxwell's equations have survived unscathed for over a century, particularly through a period when many of the traditional concepts and theories of physics suffered drastic changes. It would be difficult to exaggerate the impact of Maxwell's work, not only on classical physics but also on modern physics. Contrary to popular myth, it was not the Michelson-Morley experiment (Chapter 28) but Einstein's firm conviction that Maxwell's equations correctly describe nature that motivated the work that led to his theory of special relativity. In a very real sense it is more appropriate to say that Einstein's theory is consistent with Maxwell's equations than the converse, though of course both statements are factually accurate.

25.3 Electromagnetic Waves

Suppose we locate an electric dipole at some point in space which we designate as the origin of our coordinate system. We now know (see Chapter 18) how to calculate the electric field from this simple charge distribution and we also know that it can be represented by field lines like those shown in Figure 25.5. That figure simply tells us that if we were to place a suitable measuring device at point A, it would sense a field $\mathbf{E}$ of a particular magnitude, pointing straight down.

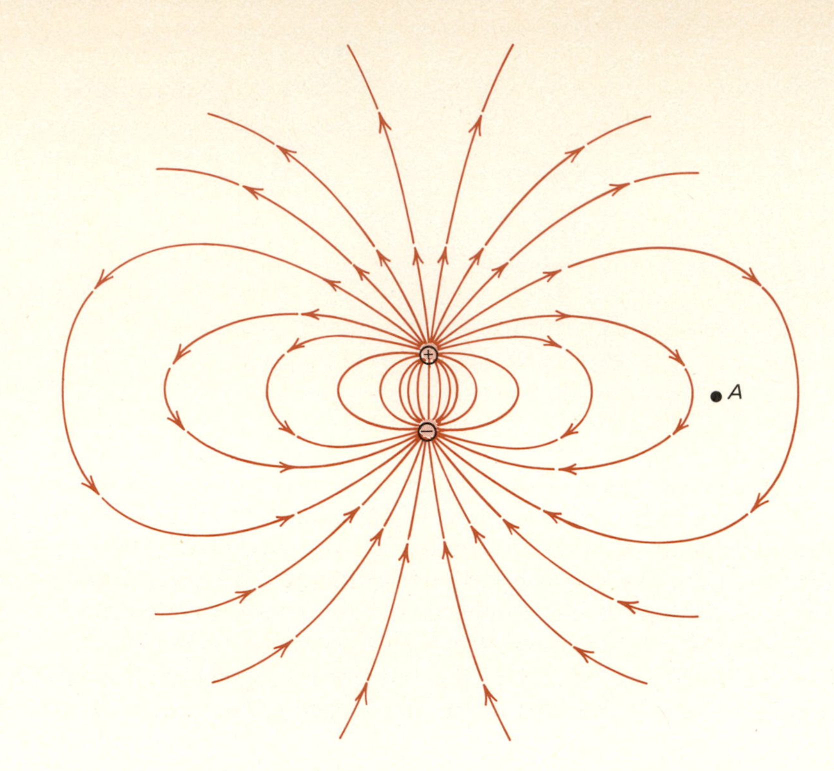

Figure 25.5 The electric field of an electric dipole. A measuring device placed at point A would register an electric field pointing down.

Next, let us assume that starting at time $t = 0$, the charges of the dipole perform simple harmonic motion about the origin at some frequency f, so that after half a period, at $t = 1/2f$, the dipole again has its maximum value but is inverted with respect to its initial orientation. We expect that our measuring device will register a corresponding sinusoidal variation of frequency f in the electric field at point A. We should now ask, and try to answer, two important questions: First, will the oscillations of $\mathbf{E}$ at point A commence at time $t = 0$ or will there be some delay? Second, by what mechanism is this information of the dipole's oscillations communicated to the instrument at A?

One can arrive at the answers to both questions by careful experiment; they are also contained in Maxwell's equations. Information about distant charge and current distributions is provided by their electric and magnetic fields. This information is not communicated instantaneously, however, but propagates outward through empty space at a finite speed, given by

$$v = \frac{1}{\sqrt{\epsilon_0 \mu_0}} \qquad \textbf{(25.3)}$$

where ϵ_0 and μ_0 are the permittivity and permeability of free space.

One has the distinct impression that Maxwell was surprised when he substituted the then measured values of these physical constants into his equations and found that the speed of propagation of these waves was just the speed of light, which, at that time, was already known to better than 0.1 percent accuracy. As Maxwell wrote,

> The velocity of transverse undulations in our hypothetical medium, calculated from the electromagnetic experiments of MM. Kohlrausch and Weber, agrees so exactly with the velocity of light calculated from the optical experiments of M. Fizeau, that we can scarcely avoid the inference that *light*

consists in the transverse undulations of the same medium which is the cause of electric and magnetic phenomena.

Since this speed of light in vacuum is an important fundamental constant of nature, it is designated by a special symbol, the lowercase letter c; its value is

$$c = 299{,}792 \text{ km/s} \simeq 300{,}000 \text{ km/s} = 3 \times 10^8 \text{ m/s} \quad \textbf{(25.4)}$$

Hence, information about a change in the disposition of charges or currents at one point in space is communicated to another by the propagation of electromagnetic waves. To see how these waves arise, and to determine the relative orientations of the **E** and **B** fields, let us now take a more careful look at our dipole.

In Figure 25.5, we show the electric-field pattern just before $t = 0$. Suppose that instead of continued simple harmonic motion of the dipole charges, we prevent further charge motion after just half a period; that is, at time $t = T/2$, with the dipole inverted relative to its orientation at $t = 0$, we suddenly stop the charge oscillations. We then examine the electric field in the region about the origin at some later time, say at $t = 2T$. For distances greater than $2cT$ from the origin, the field will be the same as before, since regions beyond that range cannot yet be affected by the inversion of the dipole during the time between $t = 0$ and $t = T/2$. Within a radial distance of $\frac{3}{2}cT$, the field will be a simple dipolar field, now inverted relative to the earlier pattern because of the inversion of the dipole at the origin.

In the spherical shell of thickness $cT/2$, corresponding to the distance traversed by the wave traveling with velocity c for a time $T/2$, the electric field must have undergone some modification. Although the details can only be worked out using Maxwell's equations, we can visualize the field without recourse to that formalism. We need only recall that in a charge-free region, the electric-field lines must either be closed curves or

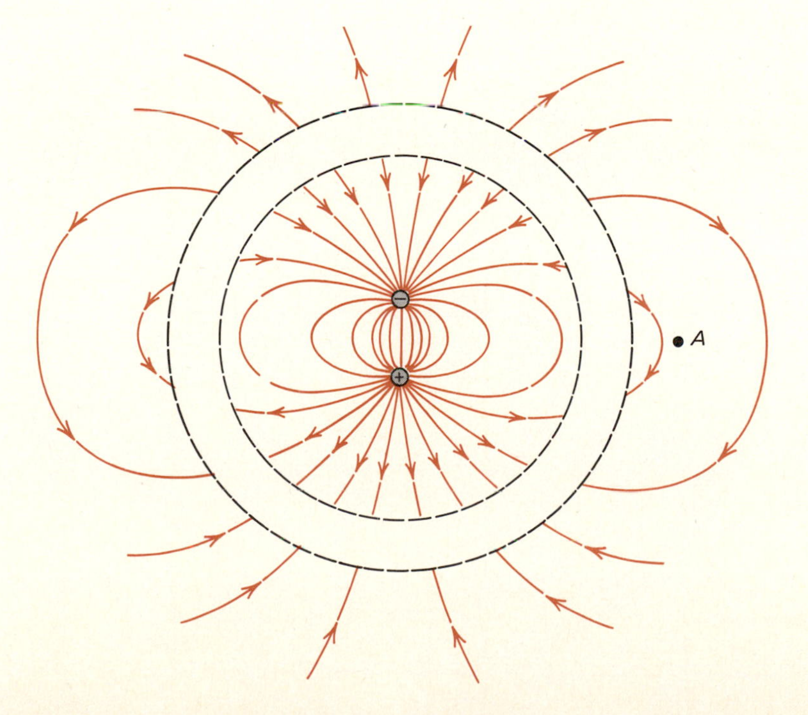

Figure 25.6 *The electric field of the dipole of Figure 25.5 at time 2T after this dipole began a process, lasting a time T/2, during which its polarity reversed.*

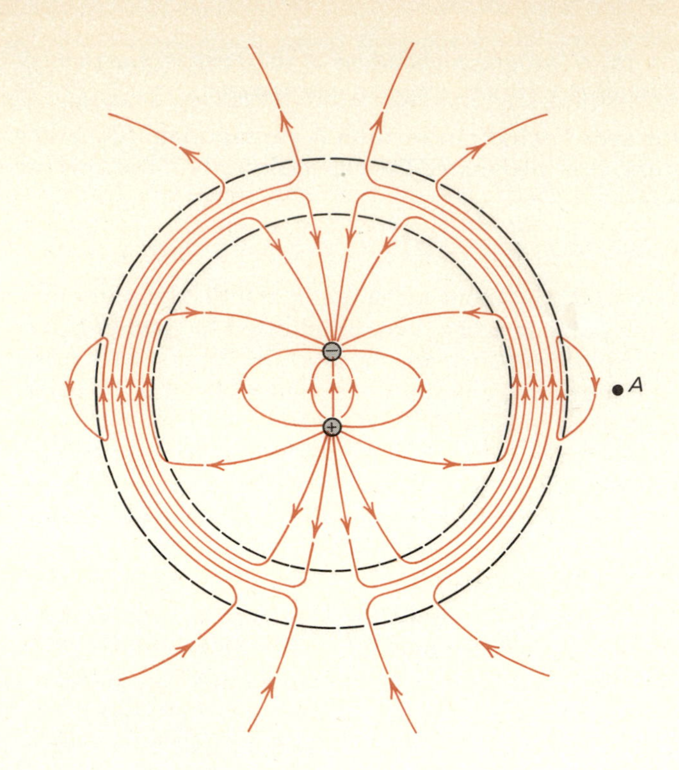

Figure 25.7 The field pattern of Figure 25.6 with the electric-field lines completed according to the rules for electric lines of force (see Section 18.2). These stipulate that lines in a charge-free region must be continuous and lines must not cross.

extend to infinity. Figure 25.7 shows the field lines for $r < \frac{3}{2}cT$ joined in a plausible way to those for $r > 2cT$.

If instead of interrupting the simple harmonic vibrations of the dipole after half a period we permit them to continue, the electric-field configuration would appear as shown in Figure 25.8.

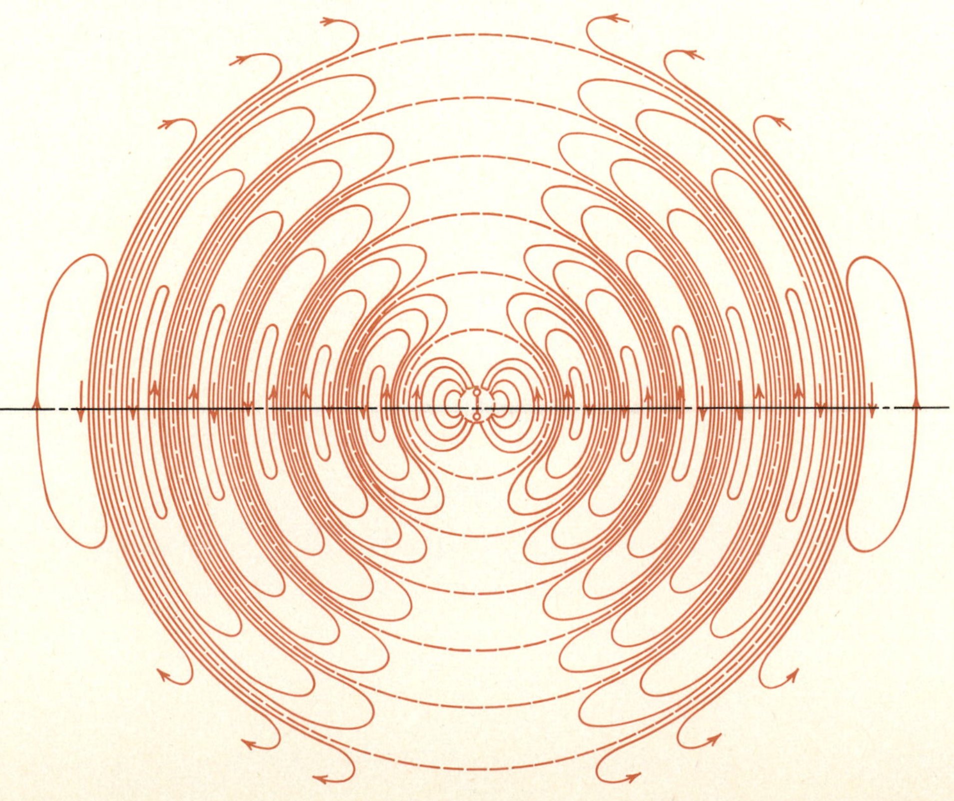

Figure 25.8 The electric field in the vicinity of an oscillating electric dipole. This field pattern propagates radially outward at a speed of $c = 3 \times 10^8$ m/s in vacuum.

A time-varying dipole generates, however, not only an electric field but also a magnetic field. This follows from the fact that a moving charge constitutes a current, and currents give rise to magnetic fields. In this particular instance, the current (motion of positive charge) performs vertical oscillations at the frequency f; an application of Right-Hand Rule 1 (Section 22.2) shows that in the plane perpendicular to the dipole, the **B** field lies in that plane and forms a pattern of concentric circles. Viewed from above, the **B** field of the oscillating dipole is as shown in Figure 25.9, where the field intensity is indicated roughly by the density of field lines. This field pattern will also propagate outward at the speed c in a vacuum. In Figure 25.9, the direction of propagation is outward along radial lines, with **B** tangent to the circles and **E** perpendicular to the plane of the figure, as indicated by the usual symbols ⊙ and ⊗. Note that **E** and **B** are mutually perpendicular and both are everywhere perpendicular to the direction of wave propagation. Electromagnetic waves are transverse waves—transverse undulations, in the words of Maxwell—with both field vectors **E** and **B** perpendicular to the direction of wave propagation.

If we were far from the origin and examined the wave as it passed, the curvature of the **E** and **B** vectors would be so slight that to all intents and purposes, it would be negligible. Such a wave, with a flat wave front, is called a *plane wave*. (Much of our subsequent discussion will focus on the properties of such plane waves.) The electric and magnetic fields of a sinusoidal plane electromagnetic wave are shown in Figure 25.10.

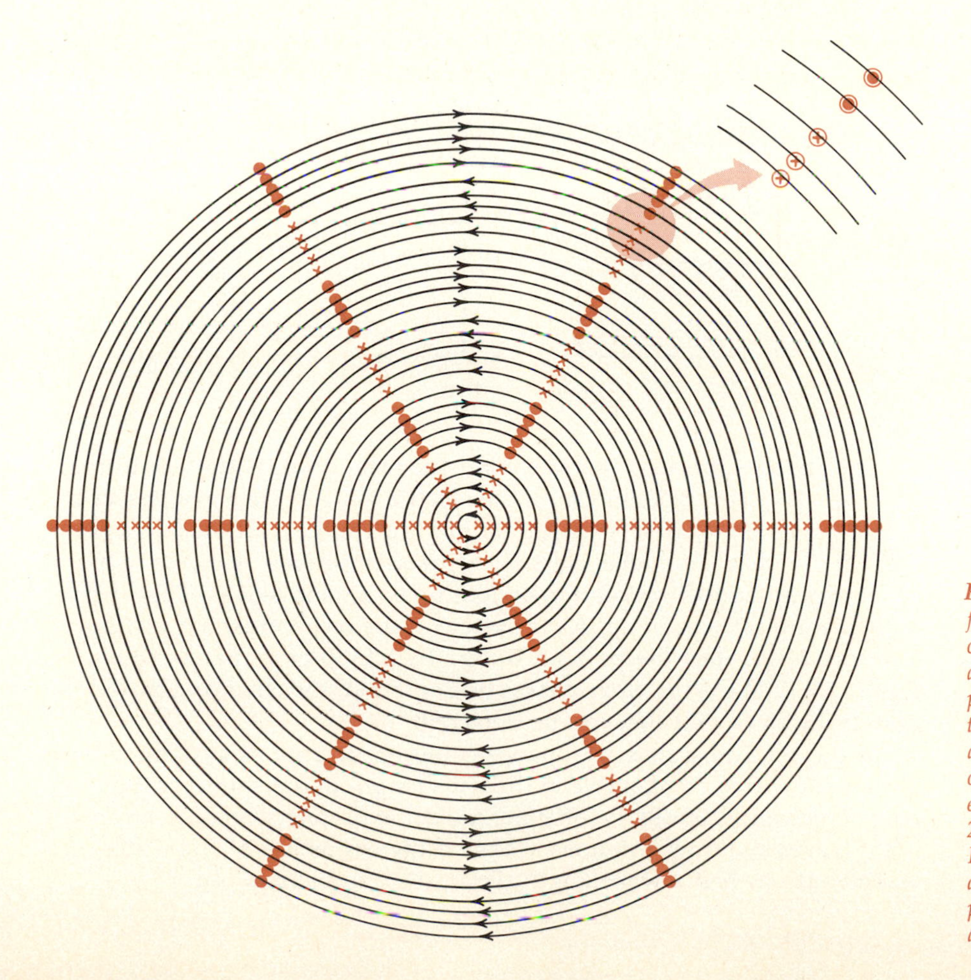

Figure 25.9 *The magnetic field in the vicinity of an oscillating electric dipole. The direction of the dipole is perpendicular to the plane of the paper, and its location is at the center and in the plane of the drawing. The electric-field lines of Figure 25.8, in which the plane of Figure 25.9 is shown by the dot-dash line, intersect this plane, as indicated by the ⊙ and ⊗ symbols.*

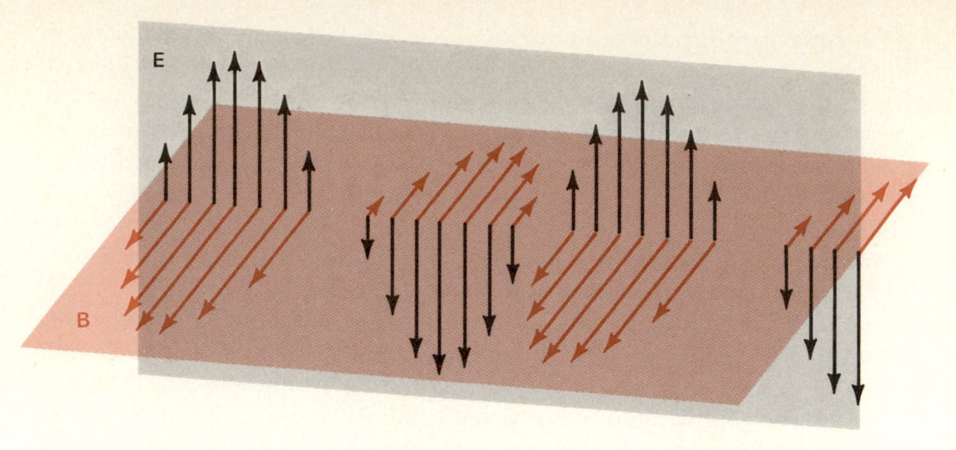

Figure 25.10 Far from the oscillating dipole, the spherical wave front of the electromagnetic wave appears as a plane wave front. The pattern of electric and magnetic fields at one instant is shown here. The direction of propagation of an electromagnetic wave is always the direction in which a right-handed screw would advance if rotated as the vector **E** is rotated through 90° so as to overlap the vector **B**.

We can now see how one might go about generating electromagnetic waves in practice, and how such waves might be detected. One technique is to force an alternating current to flow in a length of wire, as in Figure 25.11(*a*). That wire, called the transmitting antenna, is coupled at its midpoint to an ac generator, the oscillator, by a transformer. Sometimes it is more convenient to use a vertical conductor whose lower end is permanently at ground potential (Figure 25.11(*b*)). In that configuration, the induced current will make the top of the antenna alternately positive and negative with respect to ground. In either case, the time-varying charge and current distributions will establish sinusoidal electromagnetic waves that propagate outward into space.

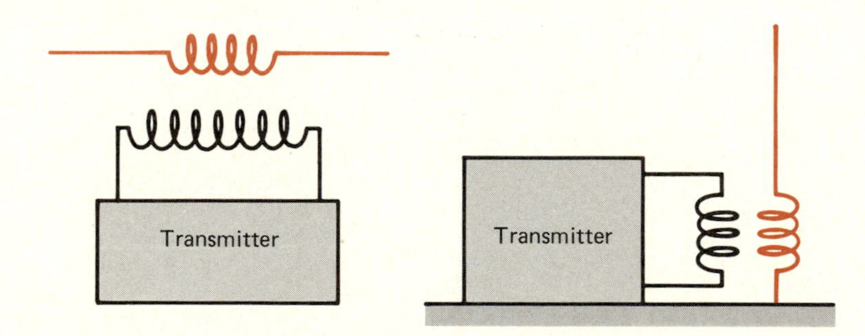

Figure 25.11 (*a*) Schematic of a transmitting antenna driven at its midpoint. (*b*) A vertical antenna grounded and driven near its ground point.

The inverse process, the detection (or reception) of the electromagnetic wave, occurs as follows. As the electric field of the wave passes the receiving antenna, it sets up an alternating charge motion (alternating current) in that conductor, just as a passing surface wave on a pond causes an up-and-down motion of floating twigs and leaves. That current is then coupled to the amplifying system of the receiver by a transformer. One can also detect the passing **B** field by a loop antenna, as shown in Figure 25.12(*b*). As the wave passes this antenna, the changing flux in the loop induces an emf that can be detected.

In the acoustical case, resonance could occur only in systems whose dimensions were comparable to the wavelength of the sound at the resonance frequency. Similarly, efficient radiation or detection of electromagnetic waves requires antennae whose dimensions are of the order of a wavelength or half a wavelength. From the general relation

$$\lambda = \frac{c}{f}$$

we find that at a frequency of 10^6 Hz = 1 MHz, which is in the midrange of the AM radio band, the wavelength is $(3 \times 10^8 \text{ m/s})/(10^6 \text{ Hz}) = 300 \text{ m} \simeq 1000$ ft. AM broadcast antennae are therefore usually long wires, supported between two tall towers, with feeder lines to the transmitter descending from the midpoint. Often the tall tower is itself the vertical antenna, as in Figure 25.13(*b*). FM and TV stations broadcast at much higher frequencies, around 10^8 Hz, so their antennae are correspondingly more compact.

The term *microwaves* refers to electromagnetic waves in the range of wavelengths between 10 cm and 1 mm. A typical radar installation will transmit and receive signals of about 1-cm wavelength, corresponding to a frequency of 3×10^{10} Hz = 30 GHz. Still shorter wavelengths fall into the infrared, followed by the visible and ultraviolet regions of the spectrum. The wavelengths of X rays are comparable to atomic dimensions, and γ rays have wavelengths approaching nuclear dimensions. However, the distinction between ultraviolet (UV), X rays, and γ rays is based not so much on their wavelengths as on how these radiations are gen-

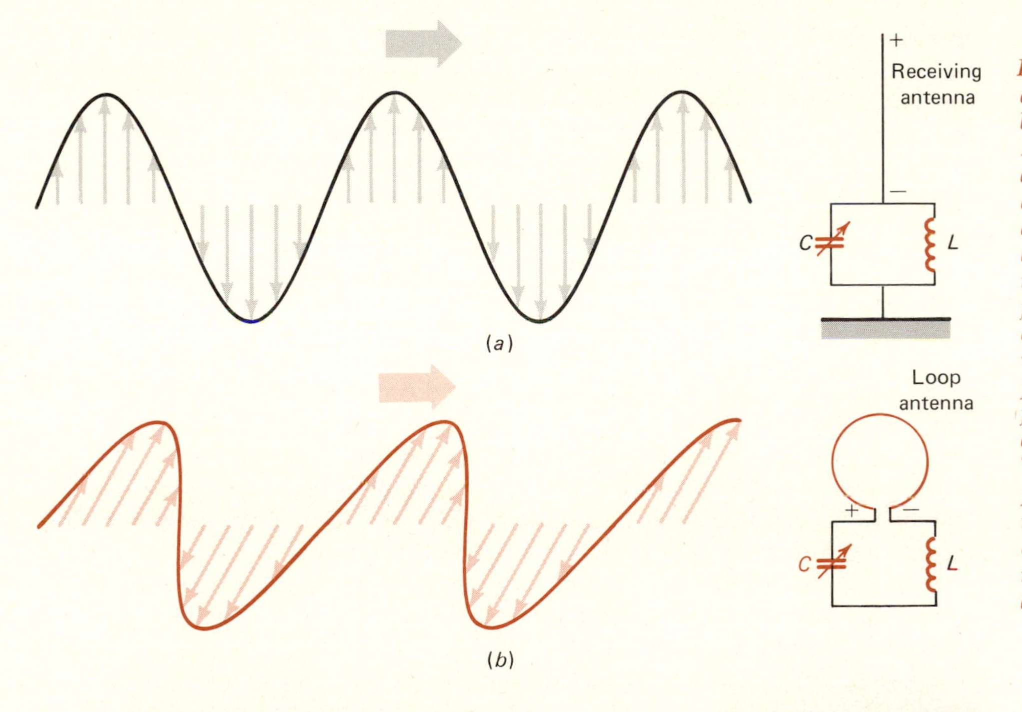

Figure 25.12 The electromagnetic wave radiated by the vertical antenna of Figure 25.11(*b*) can be detected by (*a*) a vertical antenna that responds to the alternating electric field or (*b*) a loop antenna oriented so that the alternating magnetic field of the wave induces an emf in the loop. An LC circuit with an adjustable capacitor for tuning to a particular frequency forms part of the antenna circuit in each case. The loop antenna is often used for direction finding because the signal is sensitive to the orientation of the loop; it is a maximum if the propagation direction is, as in (*b*), in the plane of the loop, and zero if perpendicular to it.

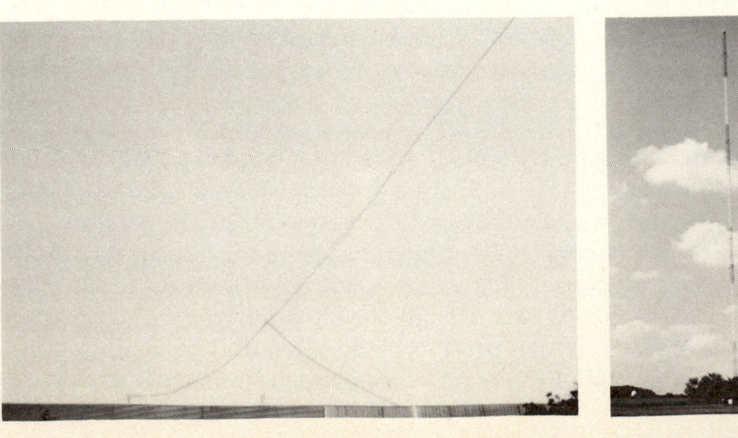

Figure 25.13 (*a*) Photograph of a horizontal wire antenna with feeder lines coming from the midpoint. (*b*) Photograph of a vertical antenna tower.

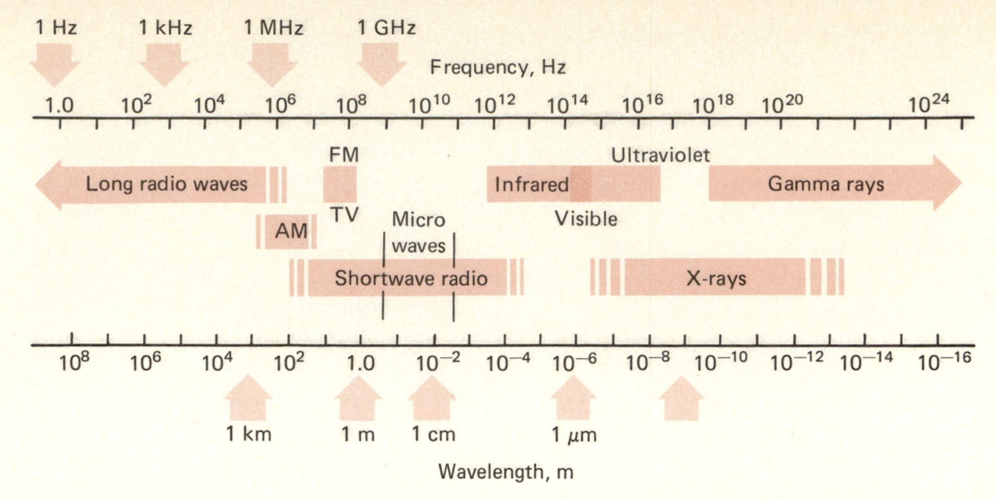

Figure 25.14 The frequencies and wavelengths of the principal divisions of the electromagnetic spectrum on logarithmic scales.

erated, a topic to which we shall return later. For this reason, the spectral regions covered by these three types of electromagnetic waves overlap substantially, as shown in Figure 25.14.

● ***Example 25.1*** A simple loop is connected to a source of high-frequency current. If the frequency of this current is 480,000 Hz, what is the wavelength of the electromagnetic waves radiated by this transmitter? How should a straight rod that is to serve as receiving antenna be oriented if the plane of the transmitting loop is horizontal and the receiving antenna is far from the transmitter but at the same elevation?

The wavelength of the radiation is given by the relation

$$\lambda = \frac{c}{f} = \frac{3 \times 10^8 \text{ m/s}}{4.8 \times 10^5 \text{ Hz}} = 625 \text{ m}$$

In the plane of the loop, the magnetic field produced by the loop current will point either up or down, depending on the sense of the circulating current. In any event, it points in the vertical direction. With the direction of the radiation radially outward from the source, the condition that **E** be perpendicular to **B** and also to the direction of propagation means that **E** must be horizontal and perpendicular to the radius vector from the source. In other words, the radiation pattern from the horizontal loop in the horizontal plane looks just like Figure 25.9, except that the roles of **E** and **B** are interchanged.

So that the radiation can be detected, a straight rod antenna must therefore be placed horizontally, not vertically, and this rod should be at right angles to the line from receiving to transmitting antenna. ●

25.4 Light

Though we now know that visible light is an electromagnetic wave, the nature of light was for many centuries a great mystery, a subject for philosophical argument, speculation, and occasionally, heated dispute. Before about 1800, the generally accepted view, favored a century earlier by Sir Isaac Newton and consequently endowed with the authority of established science, held that light consisted of a stream of tiny particles, traveling in straight lines, which produced the sensation of vision as they impinged on the retina of the eye.

There were indeed sound arguments to support this interpretation. Waves have the well-known property of "bending around" an obstacle and therefore do not cast sharp shadows. They also combine according to the principle of superposition and thus cause effects such as beats and standing waves. Light, however, traveled in straight lines, obstacles did cast sharp shadows, and no beats or standing waves could be produced.

Then, in the early 1800s, Young in England and Fresnel in France demonstrated interference and diffraction phenomena using visible light. Though light was recognized as a wave phenomenon, the nature of light waves remained uncertain until the latter part of that century, when Maxwell predicted electromagnetic waves that propagate at the speed of light.

Maxwell died eight years before Heinrich Hertz succeeded in generating and detecting electromagnetic waves. Hertz then went on to demonstrate that electromagnetic waves showed the same interference, reflection, refraction, and polarization effects as light did. It is perhaps ironic that in the course of these experiments Hertz also discovered the photoelectric effect, which Einstein explained on the assumption that light has a corpuscular character. Today light—and electromagnetic radiation of all wavelengths—is recognized as having both wave and particle properties; which of these attributes dominates is determined by the experiment we perform. These matters will be treated in Chapter 29.

25.5 Speed of Light

Probably Egyptians, Greeks, and others tried to measure the speed of light, but one of the earliest recorded experiments was the one by Galileo, who used a lamp, shutter, and mirror. The mirror was positioned on a hill some considerable distance from the lamp and shutter. Galileo attempted to determine the speed of light by measuring the time interval between removal of the shutter from the lantern and his perception of the light reflected to his eyes by the mirror; Galileo concluded that "if not instantaneous, it is extremely rapid."

In view of the extraordinary speed of light, it is understandable that the earliest measurements of c relied on astronomical observations.*

In 1675 the Danish astronomer Roemer published results of very careful measurements of the period of one of the moons of Jupiter, which had been discovered about half a century earlier by Galileo. Roemer

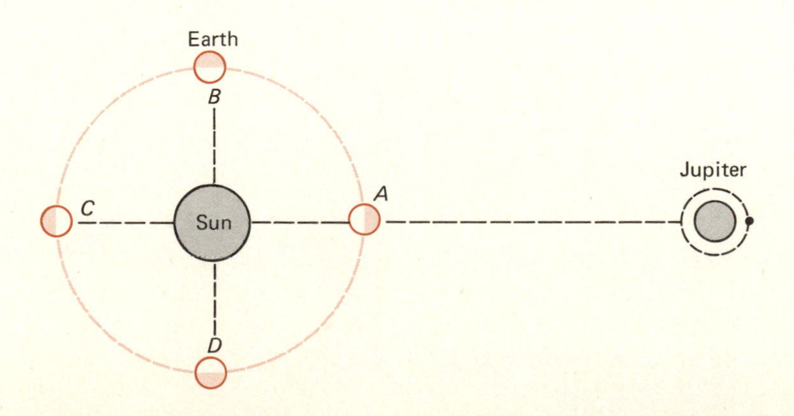

Figure 25.15 *Roemer's method for measuring the speed of light. The period of Jupiter's moon appears to be greatest when Earth, in its orbit about the sun, is at B and moving away from Jupiter; the period as measured by an astronomer on Earth is shortest when Earth is at D and approaching Jupiter.*

* The symbol c denotes the speed of light in vacuum. In air, the speed of light is slightly less. We usually do not concern ourselves with this small difference and shall use c to designate the speed of light in air as well as in vacuum.

found a roughly annual variation, such that the period appeared to be slightly longer when Earth, in its motion about the sun, was receding from Jupiter, and slightly shorter a half-year later when Earth approached Jupiter. Roemer correctly deduced that these small time discrepancies were due to the finite speed of light. Knowing the radius of Earth's orbit and the measured variation in the occultation times of Jupiter's moon, he calculated that $c = 200{,}000$ km/s.

About fifty years later an English astronomer, Bradley, calculated the speed of light from stellar aberration. This aberration is the apparent shift in the position of a star due to the velocity of Earth in its orbital motion about the sun (to which one must also add a small correction due to the rotation of Earth about its own axis). To see how this arises, consider raindrops falling on a calm day. They will drop through a fixed vertical tube. If, however, we move the tube at a velocity **v**, then a raindrop can emerge from the bottom only if the tube is slightly tilted in the direction of **v**. From Figure 25.16(*b*) we see that the angle of tilt with the vertical is given by $\tan^{-1}(v/v_r)$, where $\mathbf{v}_r$ is the velocity of the falling raindrops. Bradley correctly argued from the corpuscular theory that a corresponding inclination of a telescope tube is required to compensate for the motion of Earth, and that over one year this effect would cause stars to shift their apparent position in a circular pattern. From the measured aberration, Bradley concluded that $c = 304{,}000$ km/s.

Over a century elapsed before Fizeau performed the first terrestrial measurement of c, using, in effect, a refined version of the Galilean technique. On one hill in Paris he placed a light source, lenses, and a toothed wheel (his shutter); on another, distant hill a mirror reflected the light to the shutter wheel and the eye of the observer. With the wheel stationary, Fizeau adjusted the optics until he could see the light reflected by the distant mirror. When the wheel was rotating, no light was seen, because in the time required for light to travel from the wheel to the mirror and back, the wheel had turned enough to obstruct the light path to the observer. However, if the wheel spun fast enough, the light would reappear; in that case, the wheel had rotated $1/n$ of a complete revolution (n

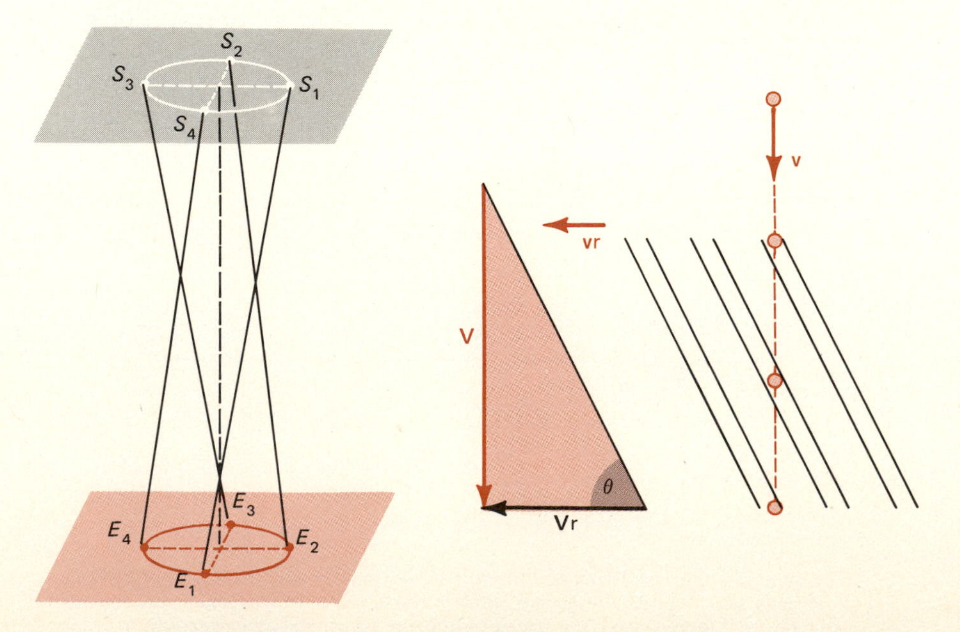

Figure 25.16 *Bradley's determination of the speed of light from observation of stellar aberration. (a) As a result of Earth's orbital motion, a star appears to shift position during a year. (b) Schematic drawing showing that if a long tube is moving with velocity* $\mathbf{v}_r$*, it must be tilted in the direction of* $\mathbf{v}_r$ *to allow a particle falling with velocity* **v** *to traverse the tube without striking its sides. Similarly, a telescope must be tilted to compensate for the motion of the observatory.*

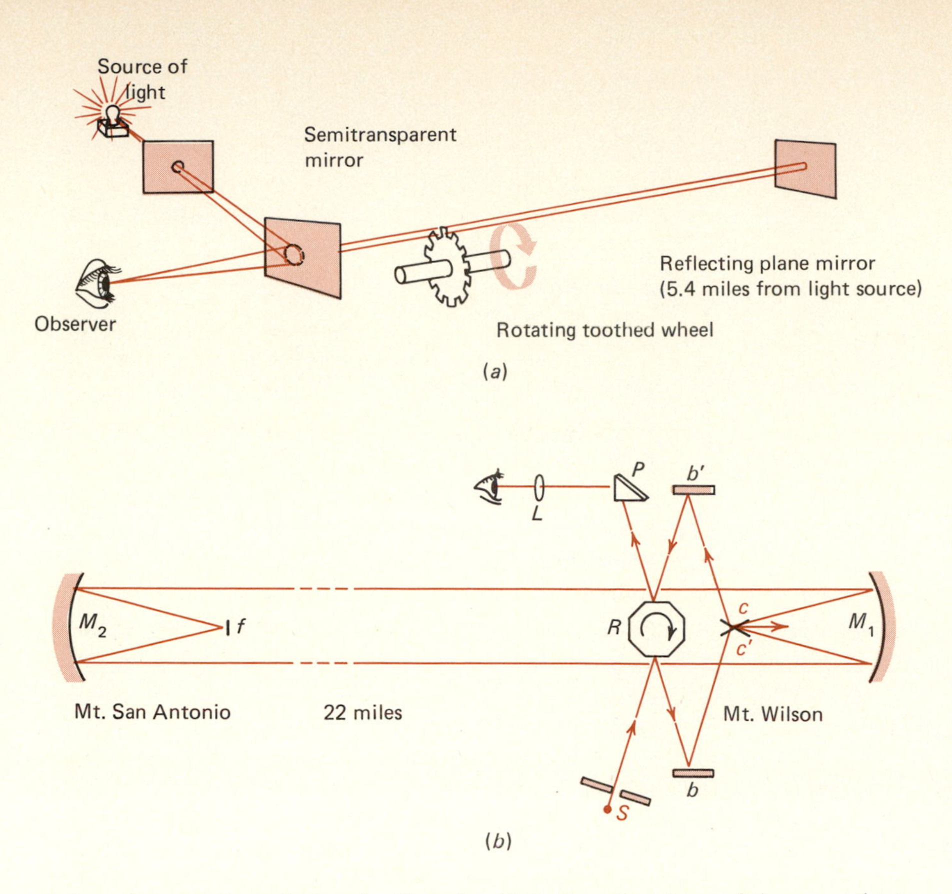

Figure 25.17 (*a*) Fizeau used a rapidly rotating toothed wheel and a long light path (approximately 10.4 miles) to measure the speed of light. If, while the light is traveling from wheel to mirror and back, the wheel turns to place a neighboring slot in the light path, the observer will see a bright spot. (*b*) Foucault's use of a rotating mirror was also the method used by A. A. Michelson in his experiments. His measurement in 1926, using mirrors placed on Mt. Wilson and Mt. San Antonio, was not surpassed in precision for a quarter-century.

being the number of slots in the wheel) during the time of the round trip of light from wheel to mirror and back. From the rotational speed of the wheel and the distance to the mirror, Fizeau calculated that c = 313,300 km/s.

Shortly thereafter another Frenchman, Foucault, improved Fizeau's technique, replacing the slotted wheel by a rotating mirror. This method was also favored by Michelson, who made many precise measurements of c between 1880 and 1930. (His 1926 value $c = (299{,}796 \pm 4)$ km/s remained unchallenged for 25 years.) Foucault also made the first determination of the velocity of light in a medium other than air or vacuum, and found that the velocity of light was substantially less in water than in air. This proved to be a very important result because refraction of light (bending of a light beam at an interface) had been explained by proponents of the corpuscular theory on the assumption that light traveled *faster* in water (or glass) than in air; on the other hand, refraction as interpreted by the wave theory required that light travel *more slowly* in water than in air, which was what Foucault found.

The most recent precise measurement, using not mirrors but laser interferometry, has yielded the value

$$c = (299{,}792.4562 \pm 0.0011) \text{ km/s}$$

It may seem almost pointless to attempt further improvement on such an accurate determination of a physical constant; yet, the greatest source of error in laser-ranging experiments between Earth and its moon is this small uncertainty of about 1 m/s (1 part in three hundred million) in the speed of light!

● ***Example 25.2*** The period of one of the moons of Jupiter is 42.5 h. What is the difference in the apparent period of this moon when Earth is

Table 25.1 Speed of electromagnetic radiation (light) in free space

Date	Experimenter	Country	Method	Speed, km/s	Uncertainty, km/s
1600(?)	Galileo	Italy	Lanterns and shutters	"If not instantaneous, it is extraordinarily rapid."	
1675	Roemer	England	Astronomical	200,000	
1729	Bradley	England	Astronomical	304,000	
1849	Fizeau	France	Toothed wheel	313,300	
1862	Foucault	France	Rotating mirror	298,000	500
1876	Cornu	France	Toothed wheel	299,990	200
1880	Michelson	USA	Rotating mirror	299,910	50
1883	Newcomb	England	Rotating mirror	299,860	30
1883	Michelson	USA	Rotating mirror	299,853	60
1906	Rosa and Dorsey	USA	Electromagnetic theory	299,781	10
1926	Michelson	USA	Rotating mirror	299,796	4
1950	Bergstrand	Sweden	Geodimeter	299,792.7	0.25
1957	Bergstrand	Sweden	Geodimeter	299,792.85	0.16
1958	Froome	England	Microwave	299,792.50	0.10
1967	Grosse	West Germany	Geodimeter	299,792.5	0.05
1967	Simkin, Lukin, Sikora, and Strelenskii	USSR	Microwave interferometer	299,792.56	0.11
1972	Evenson, Wells, Peterson, Danielson, Day, Barger, and Hall	USA	Laser	299,792.4562	0.0011

receding from Jupiter and when it is approaching Jupiter in its motion about the sun?

The radius of Earth's orbit about the sun is 1.5×10^{11} m. (See Appendix B). The orbital speed of Earth is therefore

$$\frac{2\pi(1.5 \times 10^{11}\ \text{m})}{(365.25\ \text{days})(24\ \text{h/day})} = 1.075 \times 10^8\ \text{m/h}$$

Hence, in 42.5 h Earth travels

$$(42.5\ \text{h})(1.075 \times 10^8\ \text{m/h}) = 4.57 \times 10^9\ \text{m}$$

The time required for light to traverse that distance is

$$t = \frac{L}{c} = \frac{4.57 \times 10^9\ \text{m}}{3 \times 10^8\ \text{m/s}} = 15\ \text{s}$$

Consequently, as Earth recedes from Jupiter, the apparent period is about 15 s longer; a half-year later, when Earth approaches Jupiter, the apparent period is about 15 s shorter than the true period. The difference in the apparent periods is therefore half a minute. ●

● *Example 25.3* In Fizeau's experiment, the distance between shutter wheel and mirror was 8633 m (from a home in the suburb of Suresnes to the hill of Montmartre in Paris). The disk had 720 slots around its circumference. To quote Fizeau: "In the circumstances in which the experiment has been tried, the first eclipse occurs with 12.6 turns per second. With twice the velocity the point [of light] shines out again." What was the speed of light as measured by Fizeau?

When the wheel turned at 25.2 rev/s, the light reappeared with full brightness. Since the wheel had 720 slots, the time interval required to shift from one slot to the neighboring slot was the time for one revolution divided by 720, or

$$T = \frac{1}{720}\left(\frac{1}{25.2}\ \text{rev/s}\right) = 5.51 \times 10^{-5}\ \text{s}$$

Hence,

$$c = \frac{L}{T} = \frac{2(8633\ \text{m})}{5.51 \times 10^{-5}\ \text{s}} = 3.133 \times 10^{8}\ \text{m/s}$$

25.6 Index of Refraction

In vacuum, the speed of light is $c = 3 \times 10^8$ m/s. In any other medium, light propagates more slowly; in air, for example, $v_{\text{air}} = c/1.0003$, in water $v_{H_2O} = c/1.33$, and in diamond $v_{\text{dia}} = c/2.42$. The ratio c/v_m, where v_m is the speed of light in some transparent medium m, is known as the *refractive index*, or the *index of refraction*, of that material; it is denoted by the letter n:

$$n = \frac{c}{v} \qquad v = \frac{c}{n} \tag{25.5}$$

Thus for water, $n = 1.33$; the refractive indices of various other substances are listed in Table 25.2.

Whereas in vacuum electromagnetic waves of all wavelengths propagate at the same speed c, in all other matter the speed of light varies with wavelength, i.e., with frequency. This frequency dependence of the speed of propagation, known as *dispersion*, has important consequences. One is the dispersion of white light into the colors of the rainbow as the light passes through a glass prism, a phenomenon that Newton found so pleasing and fascinating. The values of n in Table 25.2 are for yellow light of wavelength 589 nm.

Table 25.2 *Refractive indices at the wavelength of 589 nm*

Material	$c/v = n$
Air*	1.0003
Water	1.33
Ethanol	1.36
Acetone	1.36
Fused quartz	1.46
Benzene	1.50
Lucite or Plexiglas	1.51
Crown glass	1.52
Sodium chloride	1.53
Polystyrene	1.59
Carbon disulfide	1.63
Flint glass	1.66
Methylene iodide	1.74
Diamond	2.42

* Normal temperature and pressure.

Since $v = f\lambda$, it follows that when light enters a medium of refractive index n from air, either f or λ, or both, must change. A little thought shows that f remains constant, and therefore λ decreases from λ_0 to

$$\lambda_n = \frac{\lambda_0}{n} \tag{25.5a}$$

where λ_0 is the wavelength in air (vacuum). Consider the analogous situation of a long string composed of two sections differing in mass per unit length. From Equation (15.11) the propagation velocities of transverse waves on the two sections of the string will be different. If we initiate a wave at the left on the rope of Figure 25.18 and the wave propagates to the right, the number of times that point A travels up and down each second is just given by the frequency of the wave. But that motion at point A is what causes the propagation, although at a lower velocity, of this disturbance in the right portion of the string. The frequency of the distur-

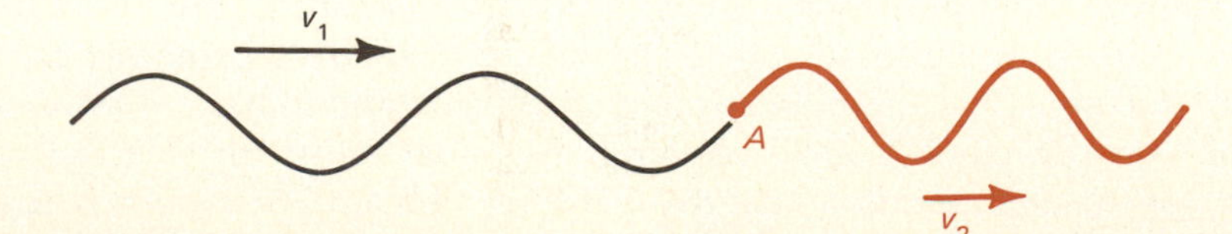

Figure 25.18 *A wave on a rope consisting of two sections of different mass per unit length. The propagation velocity is greater on the lighter section, but the frequencies are identical for both parts of the rope. Hence, the wavelength is greater where the speed of propagation is greater.*

bance is, then, the same as on the left; but since the velocity is less, the distance between successive crests, that is, the wavelength, is correspondingly reduced.

25.7 Waves and Rays

If on a sunny day we make a small opening in a window blind, we can observe a narrow beam—a *ray*—of light in our room. We can, as Newton did, reflect this ray with a mirror, observe its refraction (bending) as it passes through a plate of glass or enters a tank of water, and view the brilliant colors produced by placing a glass prism in its path. The description of light in terms of straight rays is a natural one in the Newtonian corpuscular context. Reflection by mirrors and propagation through lenses is analyzed most conveniently using rays. We shall employ this method, but first we must convince ourselves that this construct of light rays is consistent with the wave theory, whose validity we acknowledge.

Suppose we initiate a pulse at some point in space. A wave will then propagate outward, much as a surface wave expands radially on a pond if we drop a pebble into the water. The locus of all points that have the same displacement for the water wave, or the same magnitude of the **E** field for the electromagnetic wave, is called the *wave front*. For example, the wave front of a water wave might be the curve that connects neighboring points of a given crest. In three dimensions, a wave front is a surface.

Consider a source of light that is radiating electromagnetic waves uniformly in all directions. As the wave fronts travel outward, we can specify their motion by arrows, as in Figure 25.19. These arrows, which are everywhere perpendicular to the wave front, are the light rays of common parlance. At a great distance from the source, the wave front is nearly a plane and the corresponding rays are nearly parallel to each other. A light beam of *parallel rays* thus corresponds to a *plane wave*.

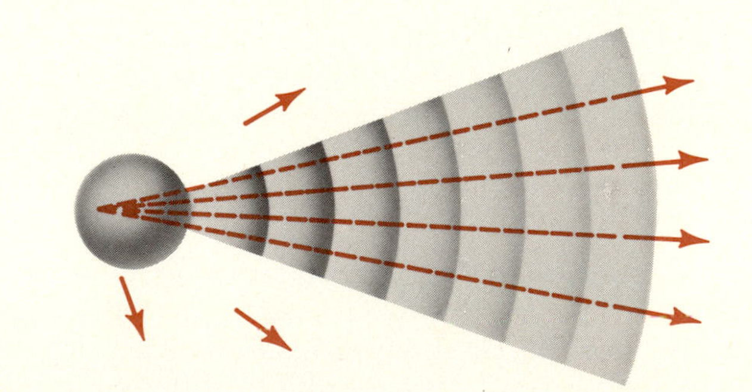

Figure 25.19 *A source emitting waves uniformly in all directions. The wave fronts are spherical, and the direction of propagation of the wave as seen by an observer at some point is perpendicular to the wave front. Very far from the source, the curvature of the wave front is so small locally that the wave front appears to be a plane.*

25.7 (a) Huygens's Principle

Huygens, in his book *Traité de la Lumière*, proposed that each point on a wave front may be regarded as a new source of waves, the so-called Huygens wavelets. The subsequent position of the wave front is then obtained from the superposition of the wavelets emanating from these fictitious sources. In Figure 25.20 we show how Huygens's principle can be applied to spherical and plane wave propagation and to transmission of a plane wave through a small opening. We shall return to this last case later; for the moment, note that the wave front beyond the aperture is not exactly a plane wave—the wave front is slightly curved near each end.

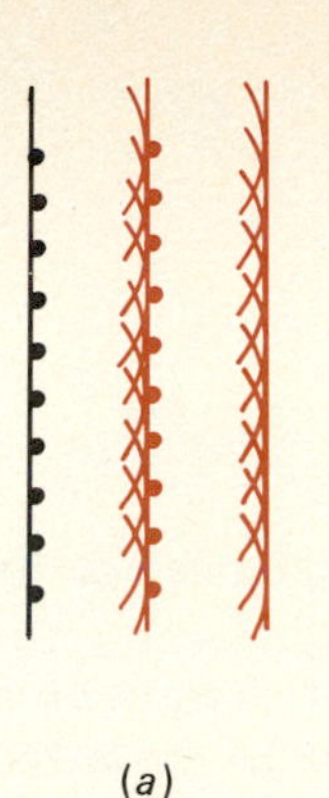

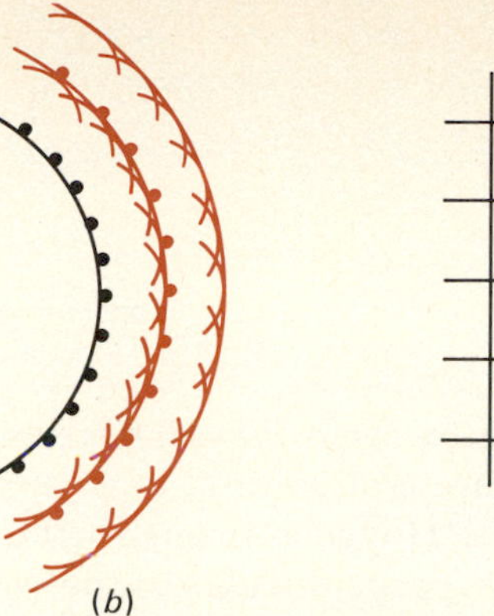

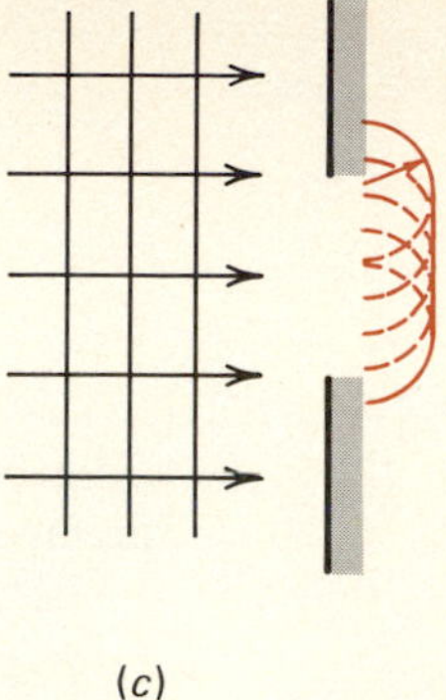

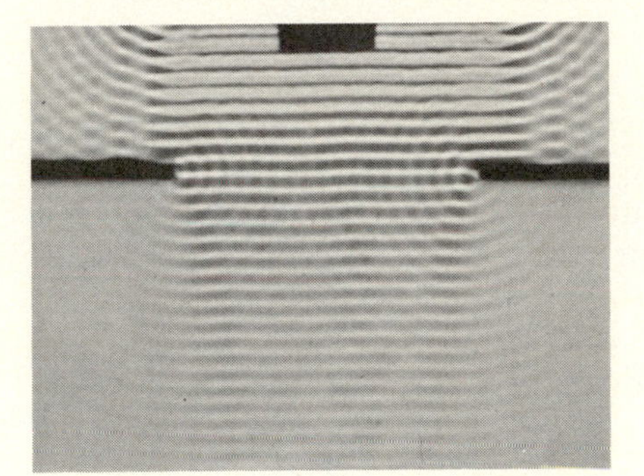

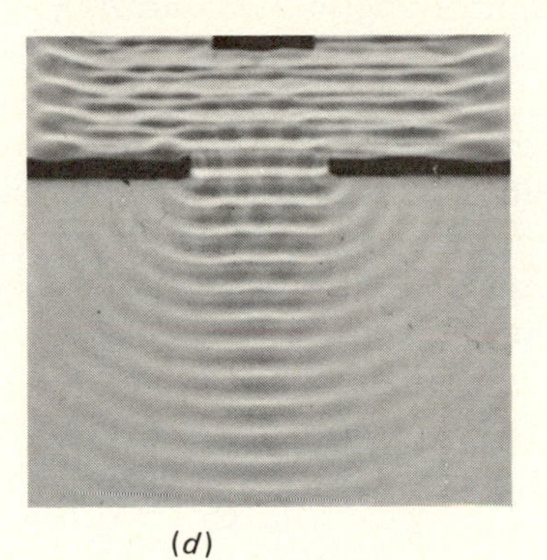

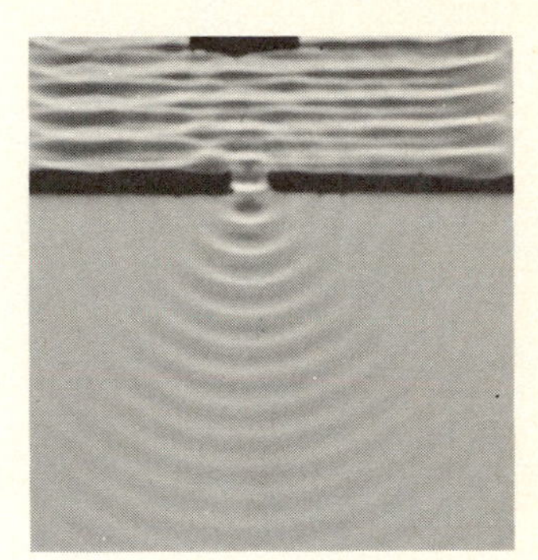

Figure 25.20 Huygens's principle. Any point on a wave front may be viewed as a source of a spherical wave, and the new wave front can be obtained by following these new Huygens wavelets and constructing their envelope. (a) Huygens wavelets propagating outward from a plane wave front generate a new plane wave. (b) Huygens wavelets propagating outward from a spherical wave generate another spherical wave front of larger radius. (c) A plane wave propagating through an aperture produces a wave front that is not a true plane wave; the wave front near the edges is rounded. (d) Plane waves in a ripple tank are incident from above on an aperture. Notice that as the aperture is made narrower, the transmitted wave front departs ever more from that of a plane wave and approaches, instead, that of a point source with its outgoing spherical wave.

This implies that the waves bend around the obstacle and that therefore, an obstacle will not cast a sharp shadow. Failure to observe such *diffraction* effects in Newton's time provided some of the strongest arguments against the wave theory. On the other hand, we shall find Huygens wavelets indispensable for explaining these same effects, which were seen in due time.

25.8 Reflection of Plane Waves

We now use Huygens's principle to prove the *law of specular* (*mirror*) *reflection,* which states:

1. The reflected ray lies in the plane formed by the incident ray and the normal to the reflecting surface.*

2. The angle of reflection equals the angle of incidence, where these angles are defined as the ones the reflected and incident rays make with the normal to the surface.

To prove the law of reflection by means of the Huygens construction, consider a plane wave, represented by the ray **I**, incident on the reflecting surface S of Figure 25.22. At some time $t = 0$, a particular wave front of the incident wave has attained the position shown as F_0 in Figure 25.22. This wave front is in contact with the surface S at point A. A Huygens wave, starting at A at time $t = 0$ will have expanded radially outward a distance $R_1 = c\,\Delta t$ in time Δt. In the same time interval, the in-

* The normal to a surface is the line that is perpendicular to the surface.

Figure 25.21 The angles of incidence and reflection for rays striking a surface are defined as the angles the rays make with the line perpendicular to that surface. For specular (mirror) reflection, the angle of incidence equals the angle of reflection.

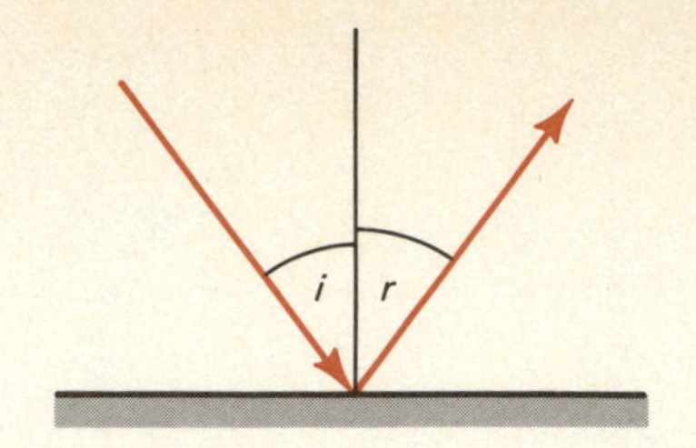

cident wave front will have advanced a distance $c\ \Delta t$ to F_1, contacting the surface S at point B.

By Huygens's construction, the reflected wave front is G and propagates perpendicular to the wave front, or in the direction of the ray **R**. Since $EB = AD = c\ \Delta t$, and the hypotenuse AB is common to the two right triangles ADB and AEB, these triangles are congruent. Hence

$$\alpha_i = \alpha_r \qquad \textbf{(25.6)}$$

where α_i and α_r are the angles between the wave fronts and the surface S. Since ray **I** is perpendicular to AF_0, and **R** is perpendicular to BG, the angles that **I** and **R** make with the normal to S are also α_i and α_r. Thus Equation (25.6) is simply a statement of the law of reflection that we set out to prove.

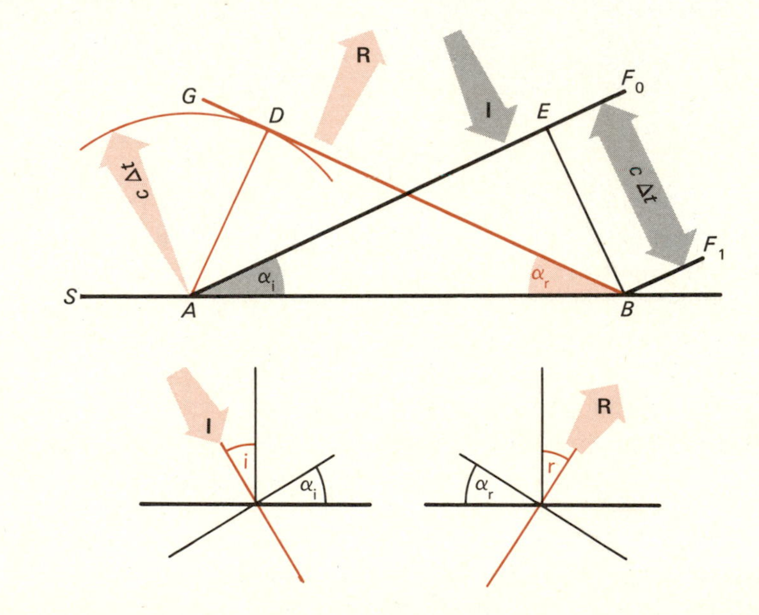

Figure 25.22 The law of reflection proved by means of the Huygens construction. The incident plane wave fronts shown for two successive times are F_0 and F_1. The incident wave is propagating along **I**. The reflected wave front is along BG, and propagating along **R**. The triangles ADB and AEB are congruent, and the angles α_i and α_r are equal. These angles equal the angles of incidence and reflection, respectively, as shown in the inset.

25.9 Refraction

Refraction of light at the interface between two media having different indices of refraction leads to many commonplace phenomena such as the apparent foreshortening of an object standing in water and the displacement of an object viewed obliquely through a thick piece of glass. Figure 25.23 shows refraction and reflection of plane waves in a ripple tank at the interface separating two regions with different velocities of propagation. To find the relation between the angles of refraction and incidence, we again use Huygens's principle.

The angle of refraction is defined as the angle between the refracted ray and the normal to the refracting interface. In Figure 25.24, the line AA' represents the wave front of the incident plane wave propagating in

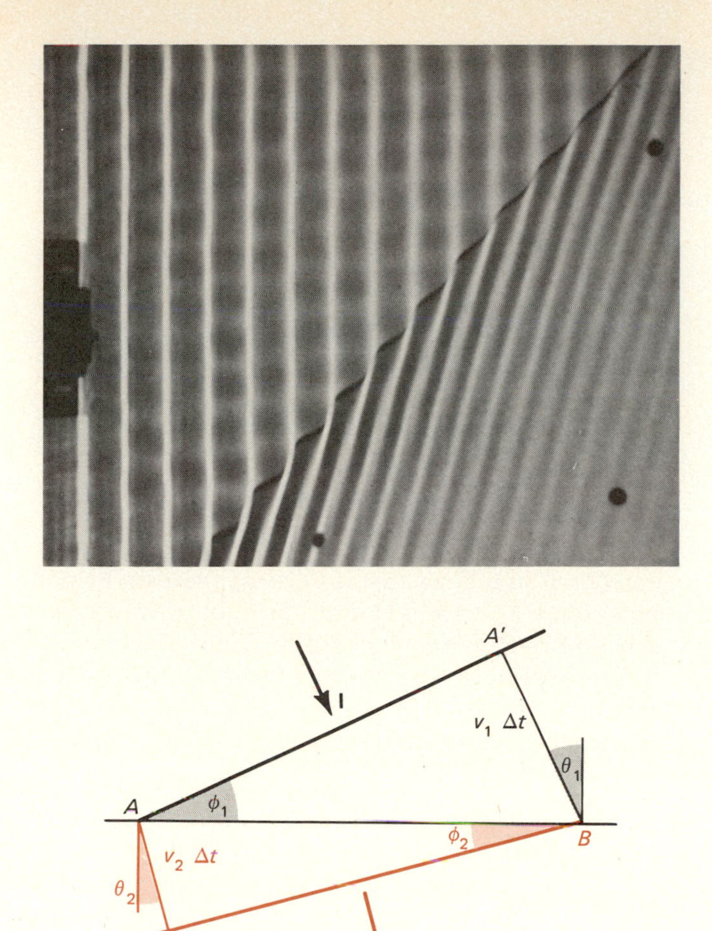

Figure 25.23 *The speed of surface waves on water depends on the depth of the water. At the diagonal interface, the depth in this ripple tank changes, and plane waves incident on this interface are refracted. There is also some reflection visible in the photograph.*

Figure 25.24 *Derivation of Snell's law of refraction using a Huygens construction. During the time in which the incident wave front advances from A' to B, the transmitted wave front propagates from A to B'. Since* **I** *is perpendicular to AA' and* **R** *is perpendicular to B'B, the angles ϕ_1 and ϕ_2, which AA' and B'B make with the interface AB, equal the angles θ_1 and θ_2 between the normal to this interface and* **I** *and* **R**, *respectively.*

medium 1 in the direction indicated by ray **I**. At time $t = 0$, the wave front meets the interface between medium 1 and medium 2 at A, and a spherical Huygens wavelet propagates radially outward from that point. In medium 1, the wave proceeds at the speed v_1, in medium 2 at the speed v_2. In a time $\Delta t = A'B/v_1$, the incident wave front has advanced from A' to B. In the same time interval, the Huygens wave centered at A has advanced a distance $AB' = v_2\,\Delta t$, where v_2 is the speed of propagation in medium 2. Thus the wave front of the plane wave transmitted into medium 2 is BB', and the direction of propagation in medium 2 is represented by the vector **R**, perpendicular to BB'.

From Figure 25.24, we see that

$$AB = \frac{v_1\,\Delta t}{\sin \phi_1} = \frac{v_2\,\Delta t}{\sin \phi_2}$$

Since $\theta_1 = \phi_1$, and $\theta_2 = \phi_2$, we obtain *Snell's law:*

$$\frac{v_1}{v_2} = \frac{\sin \theta_1}{\sin \theta_2} \qquad \textbf{(25.7)}$$

Equation (25.7) is usually written in terms of refractive indices, n_1 and n_2, rather than velocities, v_1 and v_2. Since $v_1 = c/n_1$ and $v_2 = c/n_2$, Snell's law takes the form

$$n_1 \sin \theta_1 = n_2 \sin \theta_2 \qquad \textbf{(25.8)}$$

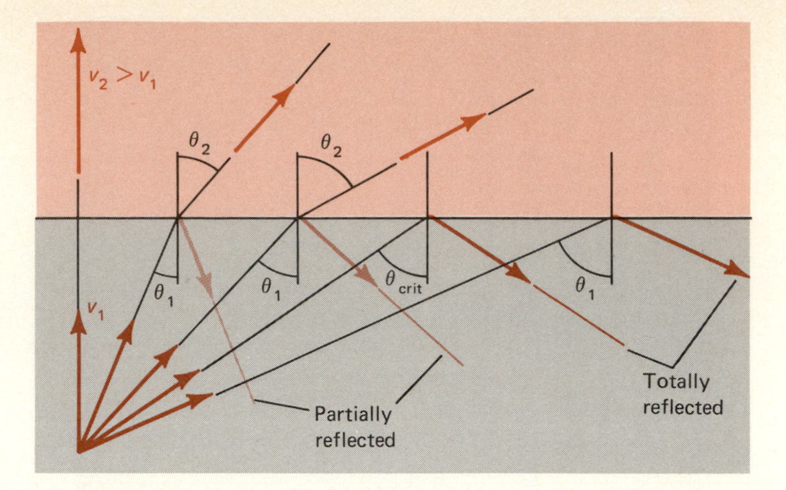

Figure 25.25 Refraction of light as it passes from a medium of high refractive index to a medium of lower refractive index, as from glass to air. When the angle of incidence is small, some of the light is internally reflected, some refracted. When the angle of incidence exceeds a certain critical angle, all the light is internally reflected.

From Equation (25.8), it follows that if $n_2 > n_1$, then $\sin \theta_2 < \sin \theta_1$, and θ_2 will be smaller than θ_1. In other words, as light travels from a medium of a low index of refraction to one of a high index, it is bent toward the normal as shown in Figure 25.24. Conversely, light is refracted away from the normal as it passes from a medium of large refractive index to a medium of a smaller refractive index.

● ***Example 25.4*** A diver looks up and sees the sun at an angle of 30° with the zenith. To an observer above water, what is the angle the sun's rays make with the zenith?

If we designate air medium 1 and water medium 2, then from Equation (25.8) we have

$$1 \times \sin \theta_1 = 1.33 \times \sin 30° = 0.665$$
$$\theta_1 = 41.7°.$$

That is, the sun is 41.7° below the zenith or 48.3° above the horizon. ●

Refraction is not restricted to sharp interfaces between two media but arises even if the change of refractive index is gradual. Although the refractive index of air is nearly unity, it is still slightly larger than 1, and its value depends on the density, and hence on the temperature of the air. On a hot, calm summer day, the layer of air immediately above a black asphalt pavement is substantially hotter than the surrounding air several inches higher. Hence a ray of light approaching the pavement at nearly grazing incidence is refracted away from the normal as it penetrates the hot layer and may reenter the upper layer of air as though it had been reflected by the road (Figure 25.26). To the driver of a car, the effect simu-

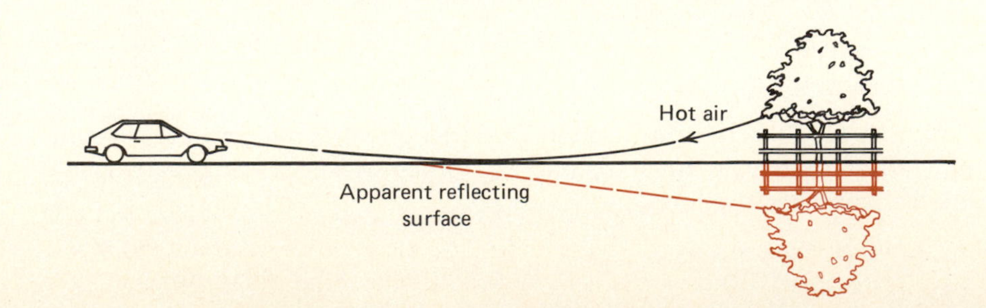

Figure 25.26 Hot air has a lower index of refraction than cooler air, which is more dense. A ray of light entering the region of hot air above a pavement on a hot summer day is refracted as shown in the sketch. To an observer, this refraction simulates the effect of reflection by the pavement.

lates the presence of a layer of water on the highway some distance, perhaps half a mile, ahead of the car. This is but one of many kinds of mirages that can be observed under suitable atmospheric conditions.

25.10 Total Internal Reflection

Consider a beam of light passing from medium 1, with refractive index n_1, into medium 2 with refractive index n_2. From Equation (25.8), we can write

$$\sin \theta_2 = \frac{n_1}{n_2} \sin \theta_1 \tag{25.9}$$

When $n_1 > n_2$, there will be a range of angles $90° > \theta_1 > \theta_c$ such that $\sin \theta_2 > 1$. Under these circumstances, Equation (25.9) cannot be satisfied for any real angle θ_2, and no refracted ray will be propagated into the second medium; all the incident light is then reflected at the interface.

This phenomenon is called *total internal reflection,* since the incident beam is generally inside some material such as glass or plastic and is reflected at the glass-air interface. (For other waves, particularly sound waves, the term *total external reflection* might be more appropriate; see Problem 25.32.)

The critical angle for total internal reflection, θ_c, is that angle for which $(n_1/n_2) \sin \theta_c = 1$; that is,

$$\theta_c = \sin^{-1} \frac{n_2}{n_1} \tag{25.10}$$

If the angle of incidence is equal to θ_c or greater, the light will be totally reflected.

● ***Example 25.5*** A plastic material whose index of refraction is 1.5 is shaped into a cube. If this cube is placed on a table, is it possible to see what is under the cube by looking into the side of the cube?

From Figure 25.27, we see that a light ray can emerge from the side of the cube only if the angle of incidence is less than the critical angle θ_c. For this material, $\theta_c = \sin^{-1}(1/1.5) = 41.8°$. From Figure 25.27 it is evident, however, that a light ray coming from the lower face of the cube so that the angle of incidence with the side face is less than 41.8° must make an angle of at least 48.2° with the normal to the lower face. But we see, by applying Equation (25.8) or (25.9) that even at grazing incidence, for $\theta_1 = 90°$, the angle of refraction is 41.8°. Thus any ray that enters the cube through the lower face must strike the side face at an angle greater than the critical angle for total internal reflection. Consequently, it is impossible ever to see what is below such a cube by looking into a side face. ●

25.11 Dispersion

The speed of propagation of light in all substances depends to some degree on the frequency of the radiation; only in vacuum is the velocity of propagation of electromagnetic waves independent of frequency.

For example, the index of refraction of fused silica (quartz glass) is 1.4636 for blue light of wavelength 480 nm (in air) and 1.4561 for red light of wavelength 670 nm. Consequently, when a ray containing both red and blue light strikes a piece of fused silica at other than normal inci-

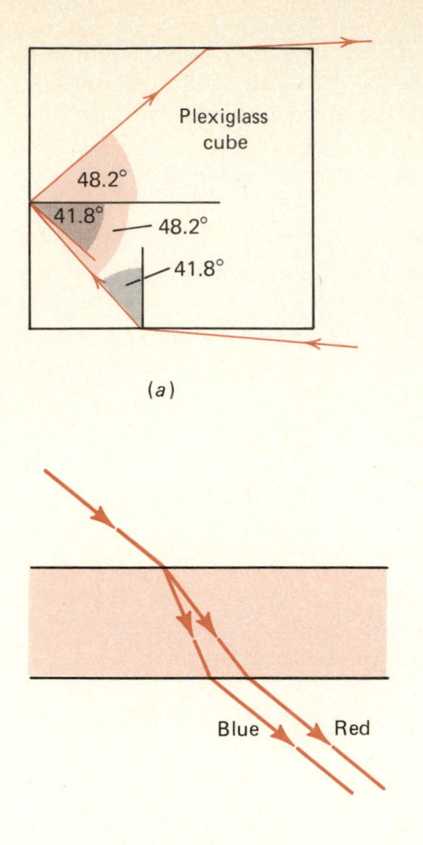

Figure 25.27 To emerge from the side of the cube of plexiglass, a ray of light must make an angle of no more than 41.8° with the horizontal. A ray of light that has entered the cube from below, however, will have been refracted toward the vertical and will make an angle of 41.8° or less with the vertical. Such a ray makes an angle of 48.2° or more with the horizontal and will therefore be reflected internally at the side of the cube.

Figure 25.28 Dispersion by a piece of glass. A beam of light containing both red and blue light is incident on the glass from above. Because the index of refraction of glass is somewhat greater for short-wavelength, blue light than for the longer-wavelength red light, the blue part of the beam is refracted more strongly at the top surface. After refraction at the lower interface, both parts of the beam emerge along parallel directions but are now spatially separated.

dence, there will be two refracted rays in the glass propagating at slightly different angles; the blue ray will be bent toward the normal slightly more than the red one. A beam of white light, which comprises all wavelengths of the visible spectrum, will similarly be *dispersed* into its various spectral components on refraction by glass. Prisms of glass or of other transparent substances are frequently employed in spectroscopes, instruments designed to identify the spectral composition of a light source. How such a prism achieves this separation is shown schematically in Figure 25.29. Notice that the light is dispersed twice, once at the air-glass interface, and once more at the glass-air interface.

Although the "celebrated Phenomena of Colours" had been known at least since Greek times, its true origin was not understood until Newton applied his genius to the problem. The generally accepted explanation at that time was that light, in passing through the prism, is "modified;" that is, it appears first red and then on passing through thicker layers of glass is substantially modified, thereby changing to green and ultimately violet. This "explanation" is remarkable only in that it explains none of the salient features nor answers any relevant question. For example, if indeed the thickness of the glass determines the color of the emerging light, then why are no colors produced when white light passes through an ordinary piece of plate glass, and why is the pattern of colors the same when an incident beam enters the prism near the apex as when one enters near the base?

Much like Einstein, who conceived his most revolutionary ideas during his years of scientific semi-isolation as patent clerk in Bern, so Newton initiated and largely completed his most important work during the plague years of 1665 and 1666 when Cambridge University was closed and Newton lived with his mother in Woolsthorpe. There,

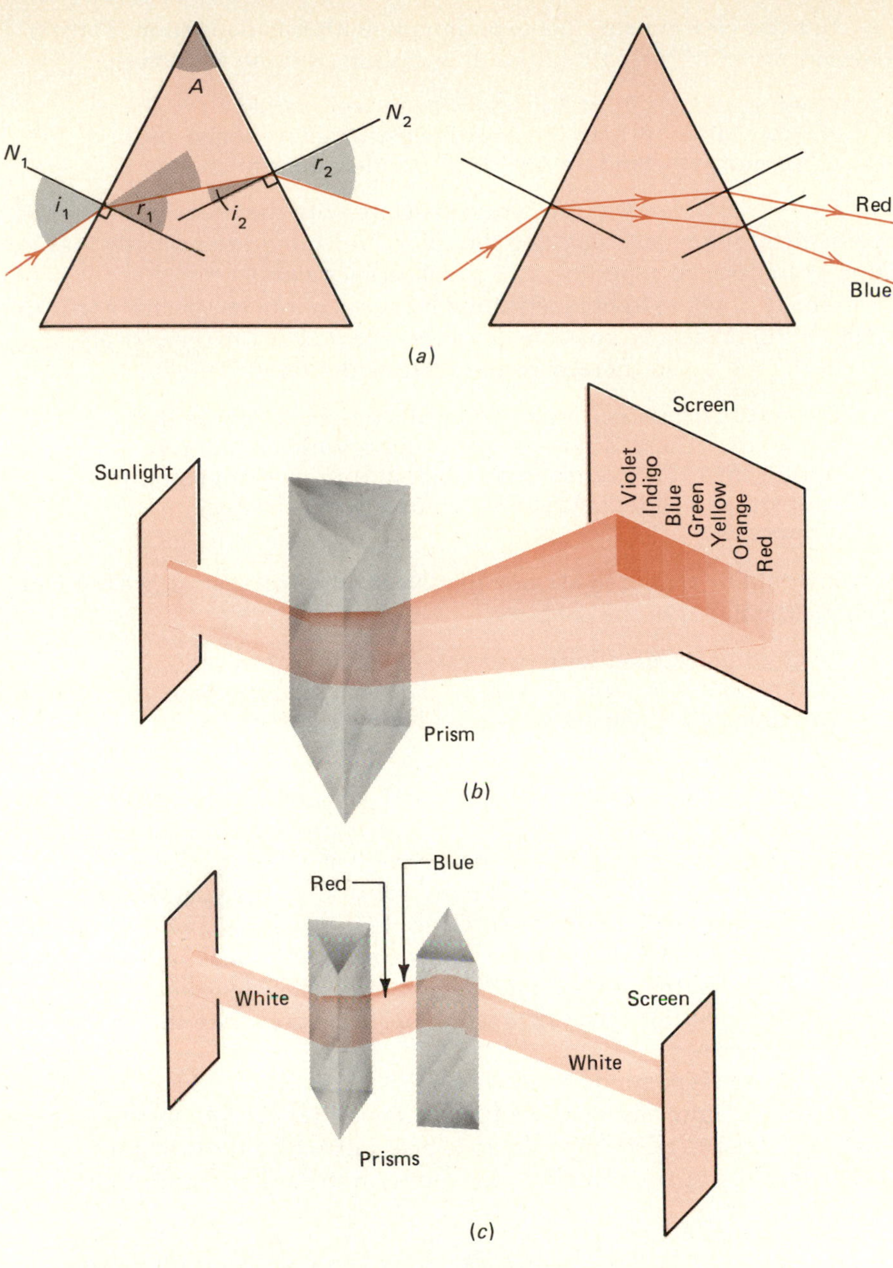

Figure 25.29 Dispersion of white light by a glass prism. (a) Refraction of the red and blue beams at the two interfaces results in propagation along different directions. (b) Sunlight passing through a narrow slit is refracted and dispersed by the prism, yielding the colors of the rainbow on the screen. (c) If a second prism, identical to the first, is interposed in reverse direction in the path of the dispersed beam, the dispersed colors are again combined and yield a narrow beam of white light. Experiments of this kind were performed and described by Newton.

> I procured me a Triangular glass-Prisme, to try therewith the celebrated Phenomena of Colours. . . . It was at first a very pleasing divertisement, to view the vivid and intense colours produced thereby; but after a while applying myself to consider them more circumspectly, I became surprised to see them in *oblong* form; which according to the received laws of Refraction, I expected should have been *circular*.
>
> And I saw that the light, tending to [one] end of the Image, did suffer a Refraction considerably greater than the light tending to the other. And so the true cause of the length of the Image was detected to be . . . that *light* consists of *Rays differently refrangible*. . . .

In other words, Newton proposed that white light is not modified but physically separated into different-colored components as a result of the slight difference in the "refrangibility" of glass to divers colors.

Nor did Newton stop his experiments with this deduction. Placing a second prism in the path of the now dispersed light, he noted:

> When any one sort of Rays hath been well parted from those of other kinds, it hath afterwards obstinately retained its colour notwithstanding my utmost endeavours to change it.

This observation was totally inconsistent with the prevalent view that passage through glass modified the color; Newton quite properly viewed this as his crucial experiment (*Experimentum Crucis*).

Finally, if white light is separated by a prism into its spectral constituents, it should also be possible to reverse the process, to *combine* all colors of the rainbow and thereby reconstitute white light. And so,

> The most surprising, and wonderful composition was that of Whiteness. There is no one sort of Rays which alone can exhibit this. 'Tis ever compounded, and to its composition are requisite all the aforesaid Colours, mixed in due proportion.

It would be wrong to assume that Newton's explanation was universally acclaimed. On the contrary:

> I was so persecuted with discussions arising from the publication of my theory of light that I blamed my own imprudence for parting with so substantial a blessing as my quiet to run after a shadow.

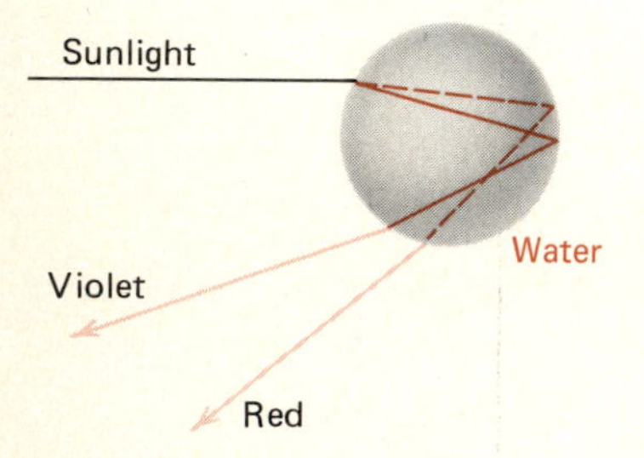

Figure 25.30 Refraction of sunlight by a spherical drop of water, followed by internal reflection and a second refraction. The light that emerges has been dispersed into its constituent colors.

In due course, his theory withstood all attempts to fault it. One remarkable feature of this work is the elegant application of what we now call the scientific method. Having arrived at the hypothesis that the colors were due to dispersion arising from a color dependence of the index of refraction, Newton then deduced (1) that subsequent passage of a particular color through another prism should not cause further dispersion and (2) that the proper combination of the spectral colors should reconstitute white light. He then set about to demonstrate the validity of these inferences.

Dispersion is common to all substances. Anyone who has enjoyed the sight of a rainbow has witnessed dispersion of white light by water. Figure 25.30 shows what happens to a beam of white light incident on a spherical drop of water. From this, and with the aid of Figure 25.31, you

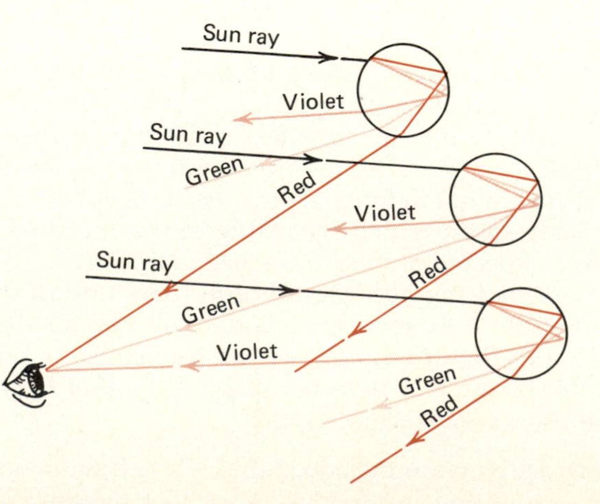

Figure 25.31 Sunlight dispersed to form a rainbow and viewed by an observer. Note that the red part of the primary rainbow is near the top of the rainbow, the blue part near the bottom. Also, the observer is not looking toward the sun but facing away from it.

can now see how the rainbow is formed, and why the violet-blue color appears at the lower part, the red at the upper part, of the primary rainbow.

Summary

In the attempt to fit the laws of electromagnetic phenomena into a consistent mathematical framework, Maxwell found an apparent inconsistency in Ampère's law. To remove this inconsistency, Maxwell introduced the *displacement current,*

$$j_d = \epsilon_0 \frac{\Delta \phi_E}{\Delta t}$$

where ϕ_E is the *electric flux.*

The mathematical equations Maxwell deduced implied the existence of electromagnetic waves. The equations predicted that such waves propagate at the speed of light, and from this Maxwell inferred that light is electromagnetic radiation.

Electromagnetic radiation spanning the range from extremely long wavelengths of several hundred kilometers to wavelengths approaching nuclear dimensions has been observed. Light is electromagnetic radiation whose wavelength falls in the range 400–700 nm.

The propagating electromagnetic disturbance consists of both electric and magnetic fields. These fields are mutually perpendicular, and the direction of propagation of the wave is perpendicular to both **E** and **B**.

An electromagnetic wave is generated whenever there is an alternating current. The most common forms of antennae are a straight wire or a circular loop. The same configuration also serves as a receiving antenna.

In vacuum, light propagates at the speed $c = 3 \times 10^8$ m/s, regardless of its wavelength. The speed of light is less in materials—gases, liquids, and transparent solids—than in vacuum. The ratio of c to the speed of light in the medium is the *refractive index* of the medium, or the *index of refraction,* designated by n. That is, $n = c/v$, where v is the speed of propagation of light in the medium.

The index of refraction, which is always greater than unity, depends on the wavelength of the radiation. This property of a medium is known as *dispersion.*

A *wave front* is the locus of all points that are in phase in a wave. The progression of a wave can be followed by applying *Huygens's principle,* which states that each point on a wave front can be regarded as a new source of spherical waves.

The propagation of light can often be studied effectively by the method of *light rays.* A ray of light has the direction of propagation of the wave front and is perpendicular to the wave front.

At a plane-mirror surface, an incident light wave is reflected so that the angle of incidence equals the angle of reflection.

When a ray of light passes from one medium, with index of refraction n_1, into a second medium, with index of refraction n_2, the direction of propagation generally changes. The angles of incidence and refraction (θ_1 and θ_2) are related by *Snell's law:*

$$n_1 \sin \theta_1 = n_2 \sin \theta_2$$

At an interface between two media with different indices of refraction, light is reflected as well as refracted. If the direction of propagation is from a medium of high refractive index to a medium of lower refractive index, there is a maximum angle of incidence for which a refracted ray appears in the other medium. If the angle of incidence exceeds this critical angle, all the light is *internally reflected.* The *critical angle for internal reflection* is given by

$$\theta_c = \sin^{-1}\frac{n_2}{n_1}$$

Multiple Choice Questions

25.1 An electromagnetic wave is propagating due east. Therefore,

(a) the **E** field may point north-south, the **B** field up-down only.

(b) The **E** field may point up-down, and the **B** field north-south only.

(c) The **E** field may point north-south, the **B** field east-west.

(d) Either (a) or (b) can give a wave propagating due east.

25.2 A vertical antenna radiates electromagnetic waves as a result of an alternating current in the antenna whose frequency is 1.2 MHz. That antenna, if it radiates efficiently, has a height of about

(a) 3 cm

(b) 1 m

(c) 100 m

(d) 500 m

25.3 A vertical antenna in Cleveland, Ohio, radiates electromagnetic waves. The radiation is detected in Toledo, Ohio, by a receiving antenna in the form of a circular loop. Toledo is about 100 km due west of Cleveland. For optimum orientation, the plane of the loop antenna is

(a) horizontal.

(b) vertical and in the north-south plane.

(c) vertical and in the east-west plane.

(d) Either (b) or (c) will do equally well.

25.4 A vertical antenna

(a) radiates uniformly in all directions.

(b) radiates uniformly in the horizontal plane, but most strongly in the vertical direction.

(c) radiates most strongly and uniformly in the horizontal plane.

(d) The answer depends critically on the wavelength of the radiation.

25.5 When a charged capacitor is discharged by placing a small resistor between its terminals,

(a) the displacement current flows through the resistor.

(b) the displacement current exists in the region between the capacitor plates.

(c) no displacement current appears in this case.

(d) Both (a) and (b) are correct.

25.6 A ray of light passes from air into water, striking the surface of the water with an angle of incidence of 45°. Which of the following four quantities change as the light enters the water? [I] Wavelength, [II] Frequency, [III] Speed of propagation, [IV] Direction of propagation.

(a) I and II only.

(b) I, III, and IV only.

(c) II, III, and IV only.

(d) III and IV only.

(e) I, II, III, and IV.

25.7 A beam of light propagating in medium A with index of refraction $n(A)$ passes across an interface into medium B with index of refraction $n(B)$. The angle of incidence is greater than the angle of refraction; $v(A)$ and $v(B)$ denote the speed of light in A and B. Which of the following is true?

(a) $v(A) > v(B)$, and $n(A) > n(B)$

(b) $v(A) > v(B)$, and $n(A) < n(B)$

(c) $v(A) < v(B)$, and $n(A) > n(B)$

(d) $v(A) < v(B)$, and $n(A) < n(B)$

(e) $v(A) < v(B)$, but there is not enough information to state definitively the relation between $n(A)$ and $n(B)$.

25.8 Visible light is unique among other forms of electromagnetic radiation because

(a) it is a transverse wave.

(b) it propagates at 3×10^8 m/s.

(c) it produces a latent image on a photographic plate.

(d) it causes sunburn.

(e) None of the above.

25.9 The wavelength region to which the human eye is sensitive falls in the range of

(a) 10–50 nm.

(b) 400–800 nm.

(c) 2000–4000 nm.

(d) 20,000–50,000 nm.

(e) None of the above.

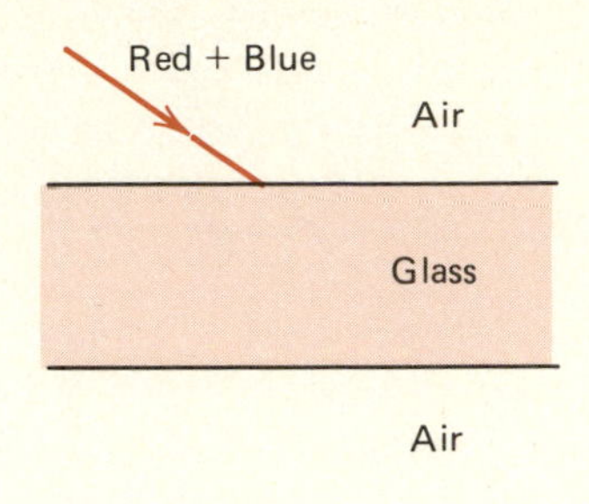

Figure 25.32

25.10 A ray of light composed of blue light and red light is incident on a plate of glass as indicated. When the light emerges from the other side of the plate, the two colored beams will

(a) appear at the same point and propagate in the same direction.

(b) appear at different points and propagate in the same direction.

(c) appear at the same point and propagate in different directions.

(d) appear at different points and propagate in different directions.

(e) The blue beam will not emerge at all because it suffers internal reflection.

25.11 A medium is said to be dispersive if

(a) light of different wavelengths propagates at different speeds.

(b) light of different wavelengths propagates at the same speed but has different frequencies.

(c) light is gradually bent rather than sharply refracted at an interface between the medium and air.

(d) light is never totally reflected internally.

(e) None of the above is correct.

25.12 Of the following relations, (A) $f/\lambda = n$, (B) $v = f/\lambda$, (C) $v = nc$, (D) $v = c/n$, the correct one(s) is (are)

(a) (A)

(b) (B) and (C)

(c) (D)

(d) (A) and (D)

(e) none

Problems

(Sections 25.1–25.3)

25.1 A radio station broadcasts on a frequency of 2.4 MHz. What is the wavelength of the radiation?

25.2 The U.S. Bureau of Standards operates radio stations that broadcast at frequencies of exactly 5 MHz, 10 MHz, and 20 MHz. What are the wavelengths of these radiations?

25.3 X rays are used for analyzing crystal structure. It is then necessary that the wavelength of the radiation be of the same order of magnitude as the separation between atoms in the crystal. Atoms within a crystal are typically between 0.1 and 1 nm apart. What is the frequency of X rays whose wavelength is 0.6 nm?

25.4 A pulse of light is directed at a reflecting surface on the moon. How much time elapses between the transmission of the pulse and the reception of the reflected signal?

25.5 Radar waves of 1-cm wavelength are widely used in police speed traps. What is the frequency of this electromagnetic radiation?

• • **25.6** A parallel-plate air capacitor has a plate area of 30 cm^2. It is part of an LC circuit that oscillates at a frequency of 3 MHz. If the inductance $L = 0.1$ mH and the maximum potential across the capacitor plates is 500 V, what is the average value of the displacement current during the half-period in which the potential changes from $+500$ V to -500 V?

• **25.7** A parallel-plate capacitor is charged to 600 V; its capacitance is 0.5 μF, and the insulating material is air. If the capacitor is discharged through a 1000-Ω resistor, what is the average displacement current during the first time constant?

• **25.8** Figure 25.10 shows the **E** and **B** fields of an electromagnetic wave propagating toward the right. The direction of propagation for that wave is obtained by a right-hand rule: Rotate the **E** vector toward the **B** vector; the direction of propagation is that in which a right-handed screw will advance. Show that this rule is consistent with the representation of **B** and **E** of Figure 25.9.

(Section 25.5)

25.9 The radius of the earth's orbit is about 1.5×10^8 km. How long does it take light to traverse that distance?

• **25.10** Laser-ranging experiments are now widely used, for example, in measuring the distance between a point on the surface of the earth and a reflector on the surface of the moon. Time intervals can now be measured to an accuracy of 0.1 nanosecond (1 ns $= 10^{-9}$ s). If the distance between the transmitter and the reflector on the moon is obtained by such a time-lapse measurement, and the time interval for the round trip of a light pulse is determined to within 0.1 ns, what is the precision of the measurement? Use the best current value for the speed of light in vacuum.

• **25.11** At what speed must the eight-sided mirror of Michelson's apparatus of Figure 25.17(*b*) rotate so that the observer can see the image of the light source? Give the two lowest rotational speeds.

(Section 25.6)

25.12 Yellow light from a sodium vapor lamp has a wavelength of 589 nm. What is the frequency of this electromagnetic wave?

25.13 Determine the speed of light in carbon disulfide.

25.14 What is the speed of light in ethanol?

25.15 What is the wavelength of light in glass of refractive index 1.42 if the wavelength in vacuum is 550 nm?

25.16 The speed of light in ice is 2.3×10^8 m/s. What is the refractive index of ice?

(Section 25.8–25.9)

25.17 Figure 25.33 is a photograph showing a beam of light incident on a thick piece of glass. Using a protractor, estimate the index of refraction of this glass.

Figure 25.33

25.18 A beam of light in air strikes a plate of crown glass at an angle of incidence of 60°. What is the angle of refraction?

25.19 Figure 25.34 shows the path of a ray of light from air into a liquid. Find the index of refraction of the liquid.

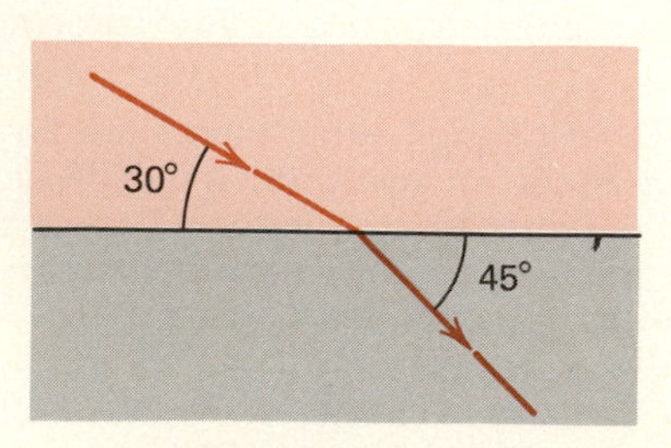

Figure 25.34

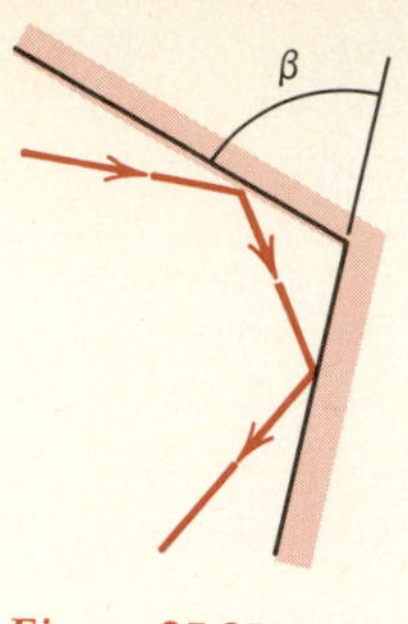

Figure 25.35

• **25.20** Figure 25.35 shows two mirrors that make an angle β. Show that any ray that is successively reflected by both mirrors will make an angle of 2β with the incident ray.

• **25.21** A ray enters one face of a cube such as that of Figure 25.27 at an angle of incidence α. It is reflected internally at the vertical face and emerges from the upper face. Show that the angle the ray makes with the normal of the upper surface when it emerges is α.

• **25.22** An aquarium is filled with water. A ray of light originates from within the aquarium and leaves through one of the vertical glass walls. If the glass has a refractive index of 1.5, and the ray after leaving makes an angle of 37° with the horizontal, what is the direction of the ray in the water? Suppose that the glass were replaced by a transparent substance that has a refractive index of unity. What would the direction of the ray in the water be then?

• **25.23** If a diver below the surface of a freshwater lake observes the sun just as it sets, in what direction and at what angle must he look?

• **25.24** A diver submerged in a liquid looks up and sees the sun when his line of sight makes an angle of 30° with the vertical. An observer above the liquid notes that at that moment the sun is just 45° above the horizon. Determine the index of refraction of the liquid.

• **25.25** A scuba diver looks up toward the calm surface of a freshwater lake and notes that the sun is at an angle of 25° with the vertical. At what angle above the horizon is the sun as seen by someone on dry land?

• • **25.26** A ray of light passes through an isosceles prism whose vertex angle is 45°. The ray's direction inside the prism is parallel to the base of the prism.

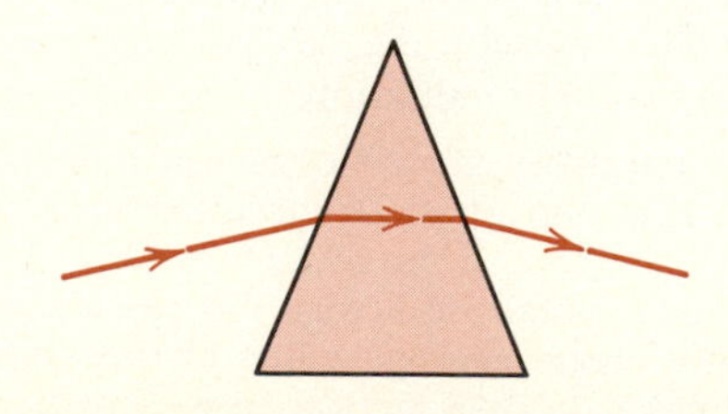

Figure 25.36

What is the angle of incidence of the ray if the prism's index of refraction is $n = 1.5$?

(Sections 25.10–25.11)

25.27 What is the critical angle for total internal reflection at the interface between diamond and air?

25.28 A laser beam is sent upwards through a solution of an organic liquid. When the angle of incidence exceeds 30°, the beam is totally reflected at the interface with the air. Find the index of refraction of this solution.

25.29 A light beam is directed at a semicylindrical piece of glass as shown in Figure 25.37. By rotating the cylinder about its axis, one finds that the critical angle for internal reflection is 37°. What is the index of refraction of the glass?

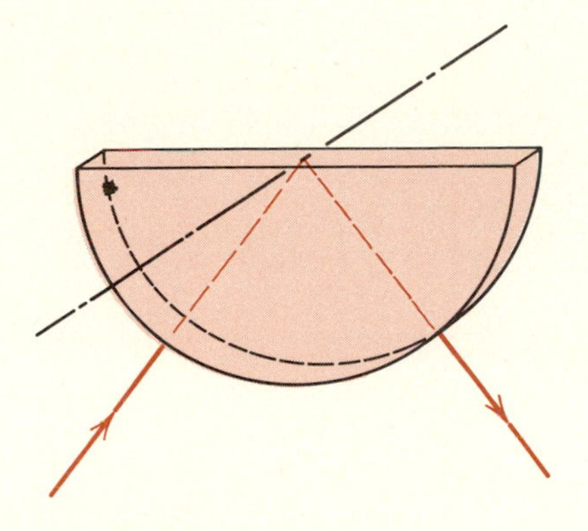

Figure 25.37

• **25.30** If an oil film is spread over the surface of the glass of Figure 25.37, the critical angle for total internal reflection changes to 53°. What is the index of refraction of the oil?

• **25.31** An optical fiber (light pipe) is made of material with $n = 1.7$ and is given a protective coating with another substance whose index of refraction is 1.3. What is the critical angle for total internal reflection?

• **25.32** Two men decide to recover some valuables from the bottom of a shallow lake during the night and before their true owner can claim them. One dons a scuba outfit and dives while the other remains in a parked car, ready to sound the horn in case of trouble. There is trouble, he sounds the horn several times, but the other man does not surface, and later swears he never heard a sound.

Is this possible? Recall that the speed of sound is about 330 m/s in air and about 1450 m/s in water. Is there a critical angle for total reflection of sound at an air-water interface? If so, is it for sound propagating from air to water or from water to air? What is that critical angle?

• **25.33** An oil that has an index of refraction of 1.15 is spread over the surface of water, which has an index of refraction of 1.33. Find the critical angle for total internal reflection at the water-oil interface.

• **25.34** Hot gases emit monochromatic light with wavelength determined by the atomic composition of the gas. One such monochromatic emission appears at 630-nm wavelength when measured in the laboratory. This same emission, observed by an astronomer, and originating from a distant star, appears to have a wavelength of 645 nm. If this shift in wavelength is due to the Doppler effect, at what speed is this star receding from or approaching the earth?

• • **23.35** A solid cylinder is made of Lucite, and the ends of the cylinder are polished and perpendicular to the cylinder axis. Prove that a ray of light that enters the cylinder at one end face at any angle will be reflected internally at the sides of the cylinder until the ray emerges from the other end.

• • **25.36** A ray of light consisting of blue light of wavelength 480 nm and red light of wavelength 670 nm is incident on a 4-cm-thick plate of quartz glass. The indices of refraction for these wavelengths are 1.4636 and 1.4561, respectively. If the angle of incidence is 60°, what is the separation between the two beams when they emerge from the opposite face of the plate?

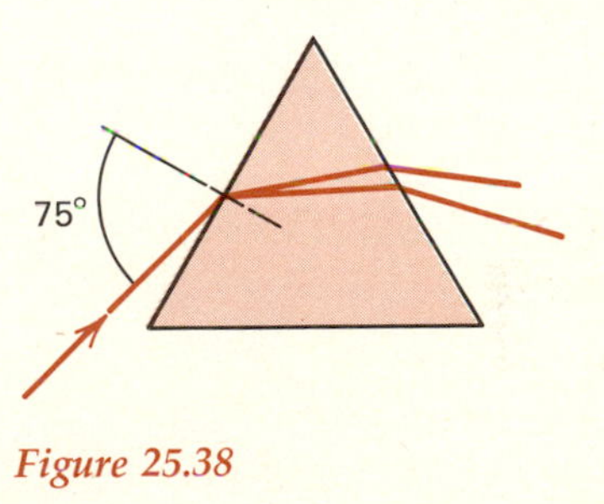

Figure 25.38

• • **25.37** An equilateral prism (all the prism angles are 60°) is made of glass that has an index of refraction $n(\lambda)$ with the following values: at $\lambda = 434$ nm, $n = 1.528$, and at $\lambda = 656$ nm, $n = 1.511$. If a thin ray of light containing these two wavelengths is incident on one face of the prism with an angle of incidence of 75°, what is the angular separation between the emerging beams?

26

Geometrical Optics; Optical Instruments

When you look
into a mirror
it is not
yourself you see,
but a kind
of apish error
posed in fearful
symmetry.

When you look
into a mirror
it is not
yourself you see,
but a kind
of apish error
posed in fearful
symmetry.

JOHN UPDIKE, "Mirror"

26.1 Plane Mirrors

Figure 26.1 shows an object placed in front of a plane mirror. We now use the ray concept to locate the image formed by this mirror. The figure shows three rays that start from the point P of the object and are specularly reflected at O, B, and C. The reflected rays are ultimately sensed by an observer (the eye in the lower left-hand corner of Figure 26.1), and as far as the observer can discern, appear to radiate from P', located behind the mirror.* Thus the plane mirror forms a *virtual image* $P'Q'$ of the *real object* PQ. This image is *upright* and of the same size as the object, and the *image distance* $i = OQ'$ equals the *object distance* $p = OQ$. *The image is virtual* because the rays only *appear* to emanate from the position behind the mirror, a fact we readily accept but which is a cause of some consternation to animals and very young children. We shall see very shortly that

* A similar ray construction can be made for every point of the object PQ.

not all images are virtual; lenses and mirrors can also create *real images,* which can be projected onto a screen.

Although the plane mirror is perhaps the oldest optical instrument known to man, it remains an important element in the modern arsenal of sophisticated optical devices. For example, the earth-moon laser-ranging experiments, initiated in 1969, rely on high-quality reflectors. The distance between the transmitting and receiving station on the earth and a particular location on the moon has been and is being monitored by accurate measurement of the time required for a light pulse to travel from the earth to the reflector on the moon and back to the earth.

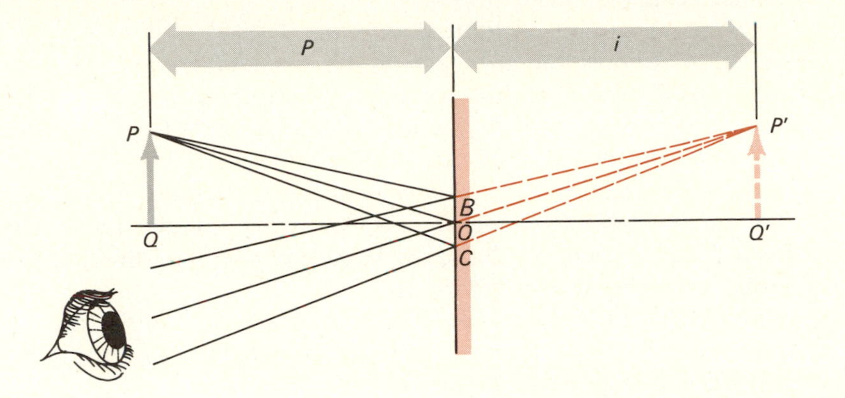

Figure 26.1 A plane mirror forms a virtual, erect image P′Q′ of the object PQ, as shown by the ray construction. The light rays entering the eye appear to originate from the image behind the mirror.

But how, you may well ask, can one possibly orient a mirror to within a minute fraction of a degree so that it will reflect a narrow beam back to the exact spot on the earth from which it came? Even if that feat could be accomplished, the rotation of the earth about its axis would bring the station out of alignment with the reflected beam except for a fleeting moment each day.

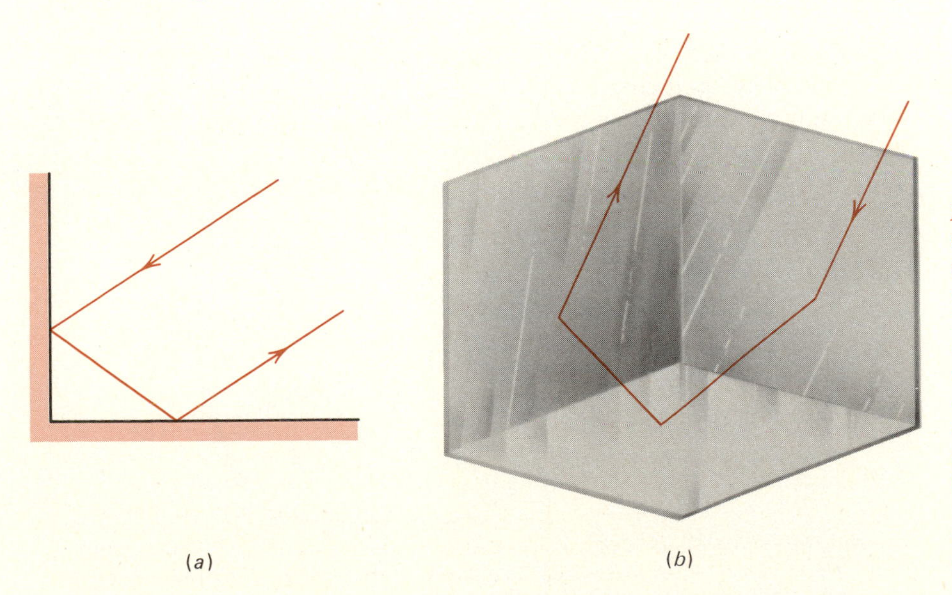

Figure 26.2 The corner reflector, also called a retroreflector. (a) The corner reflector in two dimensions. The two mirrors are at 90° to each other, and the reflected and incident rays are exactly antiparallel whatever the angle of incidence. (b) In three dimensions, the three mirror planes intersect at right angles, and again the reflected and incident rays are exactly antiparallel.

These difficulties are easily overcome by using a *corner reflector,* also called a *retroreflector,* made of three plane mirrors that are mutually perpendicular. A ray intercepted by such a mirror system is reflected through exactly 180° regardless of its initial direction. Once an array of such corner reflectors is placed on the moon's surface, generally facing toward the earth, it is necessary only to aim the laser toward that spot on the moon; the beam will be reflected precisely to its point of origin.*

26.2 Spherical Mirrors

Spherical mirrors are reflecting portions of a spherical surface. To locate the images such mirrors form, we shall use a few rays specially chosen so that their paths can be easily followed.

26.2 (a) Concave Mirrors

We shall locate the image formed by a concave spherical mirror by following the paths of several light rays. One class of rays of particular interest comprises rays that approach the mirror parallel to its axis. If the distance of these rays from the axis is small compared with the mirror's radius of curvature, all these rays will be reflected so as to pass through the same point on the axis, the *focal point* of the mirror.

We locate the focal point of a concave mirror as follows. The concave mirror of Figure 26.3 is a portion of the sphere whose center is at C. Since a radius is always perpendicular to the surface of a sphere, the line CA is perpendicular to the mirror surface at A.

The ray DA, which is parallel to the mirror axis CO and strikes the mirror at A, is reflected according to the law of reflection so that $i = r$, as shown. It then follows that the triangle CFA is isosceles, i.e. $CF = FA$. If, moreover, $OA \ll OC$, the condition mentioned earlier, the angles i and r are small, and $CF = FA \simeq FO \simeq \frac{1}{2}CO$. All such rays parallel to the mirror axis will be reflected through the point F, the focal point of the mirror, located midway between the mirror and its center of curvature. If R is the radius of curvature of the mirror, we have

$$OF = f = \frac{R}{2} \tag{26.1}$$

where f is the *focal length* of the mirror.

Conversely, an incident ray that passes through the focal point before striking the mirror will be reflected so as to emerge parallel to the mirror axis.

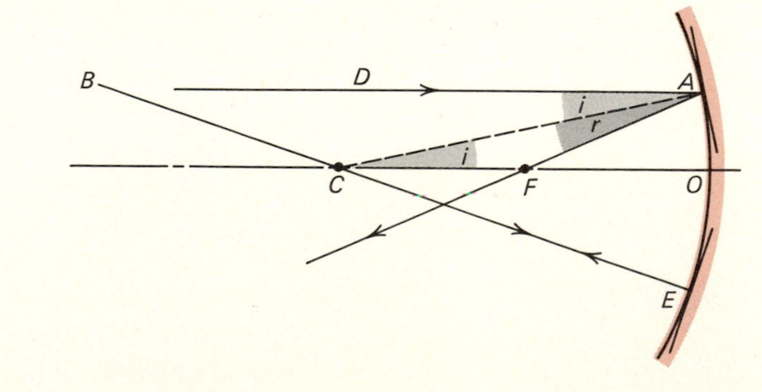

Figure 26.3 Location of the focal point of a concave mirror using the ray construction. The ray DA parallel to the optic axis CO, which strikes the mirror at A, is reflected so that the angles of incidence and reflection i and r are equal. Hence, the triangle CAF must be isosceles. If the angle i is small, AF = FO = ½CO. Thus, the focal distance f = FO equals half the radius of curvature CO of the mirror.

* This marvelous property of corner reflectors has long been known. Shortly after World War II, when Massachusetts highway patrols began to use radar to catch exuberant motorists, two MIT students quickly developed, and sold, a simple device guaranteed to fool any radar. It consisted of four light aluminum corner reflectors, each with an effective reflecting area of about 15 cm^2, mounted windmill fashion at the end of two crossed bars. The center of the device had a small bearing that slipped over a car's radio antenna. As the car moved forward, the wind drove the windmill so that the reflector, which at any instant returned

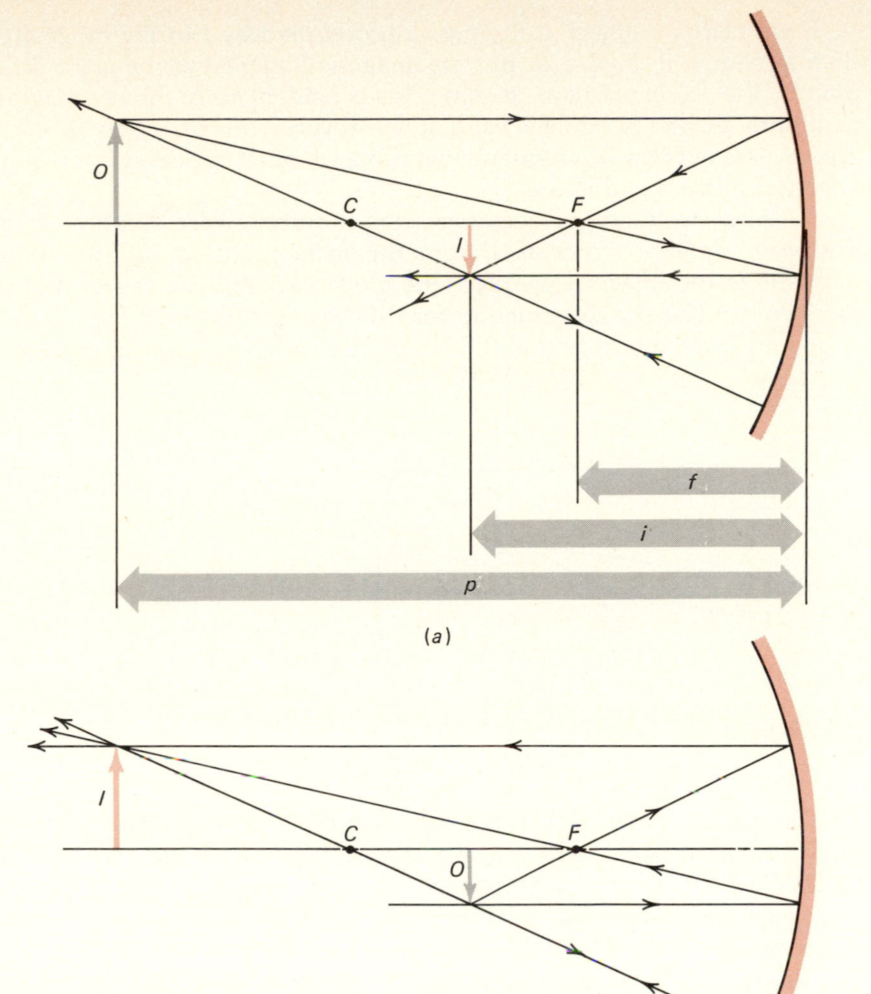

Figure 26.4 Formation of real images by a concave spherical mirror. (*a*) An object *O* placed at the object distance $p > 2f$ gives an inverted, real, and diminished image *I* at the image distance *i*, where $f < i < 2f$. The three principal rays are shown. (*b*) If an object is placed at the position at which the image appeared in (*a*), the mirror will create a real image where the object was placed in (*a*). Note that in (*b*), the rays are just the reverse of those of (*a*).

One last ray is also readily traced; any ray, such as *BCE*, passing through the center of curvature must be reflected on itself because it is directed along a radius of the spherical surface.

We now use these three *principal rays* to locate the image formed by the mirror when an object is placed before it. In Figure 26.4(*a*), the concave mirror forms a real, inverted image *I* of the object *O*. This image is *real*, since the light rays that reach the observer as though they originated from *I* do, in fact, converge at the image before diverging toward the observer's eye. The presence of such a real image is easily verified by placing a ground-glass screen (or piece of thin paper) at the image position, whereupon this image will become visible on the screen.

---

the radar signal to the transmitter, moved *backward* relative to the car. Thus, the net forward velocity of the active reflector was always less than the velocity of the car itself, the more so the greater the car's speed. This device can be quite effective if, as usually happens, the radar is pointed at approaching vehicles. If the police use a hand-held radar and in disbelief of their initial reading, try to confirm it by checking the speed of the receding car. . . .

Reflectors mounted on bicyles and along highways also use arrays of small rectroreflectors.

If we place an object at the position occupied by I in Figure 26.4(a), then as shown in Figure 26.4(b), its image will appear at the place occupied by the object of Figure 26.4(a). This is evident since the ray diagram of Figure 26.4(a) is equally valid if we reverse the direction of every arrow. Reversibility is a fundamental property of all optical systems composed of mirrors and lenses.

As the *object distance* p decreases from infinity toward the focal point, the *image distance* i increases from the focal point to infinity. What happens to the image if we place the object *between* the mirror and its focal point? The ray construction for this case is shown in Figure 26.5, and from it we conclude that the image is now virtual and upright.

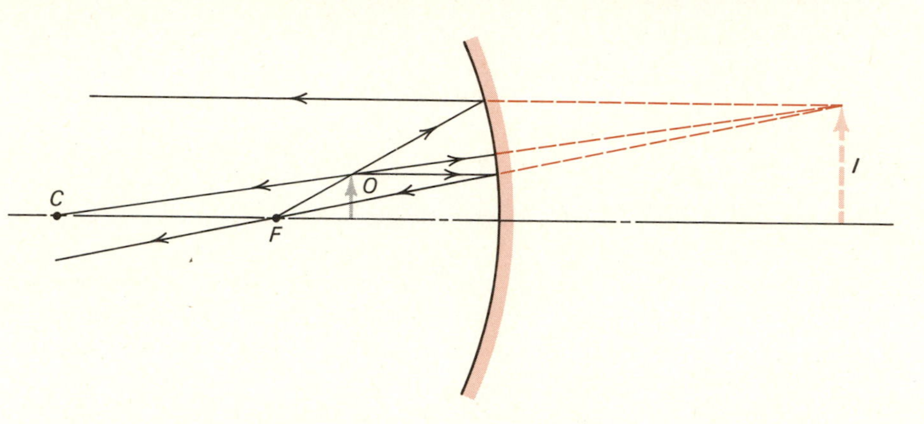

Figure 26.5 *If an object is placed at a distance $p < f$ from a concave spherical mirror, the image formed by the mirror is upright, virtual, and enlarged. The three principal rays, two of which suffice to locate the image, are shown.*

Though it is often convenient and a valuable aid to trace rays and locate images thereby, there is also a simple algebraic expression that relates object distance, image distance, and focal length.

Figure 26.6 shows the ray construction for the concave mirror once more, including the ray AQE from the object A to the center of the mirror Q, whence it is reflected to E. Since the mirror at Q is perpendicular to the line HQ, the law of reflection requires that the angles AQH and EQH be equal. Further, both triangles AQH and EDQ are right triangles. It follows that since the angles of these two right triangles are equal, the two triangles are similar. Consequently,

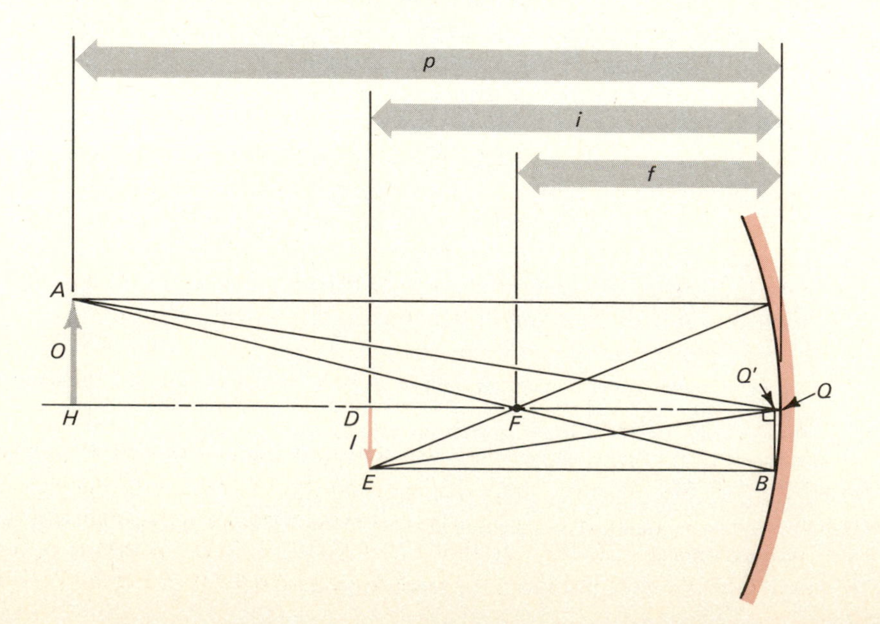

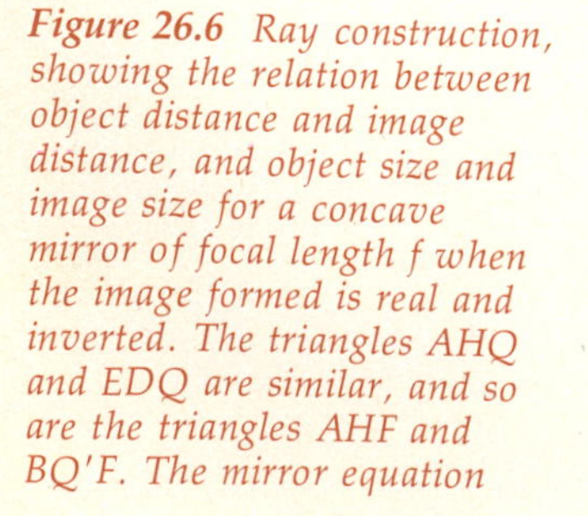

Figure 26.6 *Ray construction, showing the relation between object distance and image distance, and object size and image size for a concave mirror of focal length f when the image formed is real and inverted. The triangles AHQ and EDQ are similar, and so are the triangles AHF and BQ′F. The mirror equation*

$$\frac{1}{p} + \frac{1}{i} = \frac{1}{f}$$

then follows if for the spherical mirror, the object and image sizes are such that $QQ' \ll QF$.

$$\frac{AH}{DE} = \frac{HQ}{DQ} = \frac{p}{i} = \frac{O}{I} \tag{26.2}$$

where O and I represent object and image heights.

Next, consider the two right triangles AHF and $BQ'F$, where BQ' is the line drawn from B perpendicular to HQ. Since the angles HFA and BFQ' are equal, the two triangles are also similar, and we can write

$$\frac{AH}{BQ'} = \frac{HF}{FQ'} = \frac{O}{I} = \frac{p-f}{f} \tag{26.3}$$

In arriving at the last step, we have neglected the distance QQ', which is small compared to the focal length $f = QF$. Combining the above equations, we obtain

$$\frac{p}{i} = \frac{p-f}{f} = \frac{p}{f} - 1$$

and

$$\frac{1}{p} + \frac{1}{i} = \frac{1}{f} \tag{26.4}$$

Equation (26.4) is the *mirror equation,* relating object distance, image distance, and focal length. Note that as expected from reversibility, the mathematical expression is symmetrical with respect to an interchange of object and image distances.

Equation (26.4) agrees with the qualitative observation concerning object and image distances of the preceding discussion, based on the ray diagrams. For $p \to \infty$, $i = f$, and i increases as p approaches f. If $p < f$, we see that i must be *negative* to satisfy Equation (26.4); the magnitude of i is then always greater than p. The negative value of i is a formal indication that the image is *virtual,* formed *behind* rather than in front of the mirror.

The *magnification m* is defined as the ratio of image height to object height; that is, $m = I/O$. The value of m is negative if the image is inverted, positive if it is erect. It follows from Equation (26.2) that

$$m = -\frac{i}{p} \tag{26.5}$$

where i is positive if the image is real and negative if it is virtual.

Let us then summarize our results for the concave mirror.

I. For $p > f$ the image distance i is positive; the image is real and inverted, and formed in front of the mirror.
   If $p > 2f$, $f < i < 2f$ and $|m| < 1$; m is negative.
   If $p = 2f$, $i = 2f$ and $m = -1$.
   If $f < p < 2f$, $i > 2f$ and $|m| > 1$; m is negative.

II. For $p < f$, the image distance is negative; the image is virtual and upright, and formed behind the mirror. Moreover, $|i| > p$ and $m > 1$.

26.2 (b) Convex Mirrors

Again, we begin by drawing a ray diagram for the incident ray DA parallel to the mirror axis. Since AC is normal to the mirror surface at A, the triangle AFC must be isosceles, like the corresponding triangle of Figure 26.3. As before, it follows that $FC = QC/2 = R/2$, where R is the radius of curvature of the mirror. Since the construction of Figure 26.7 is

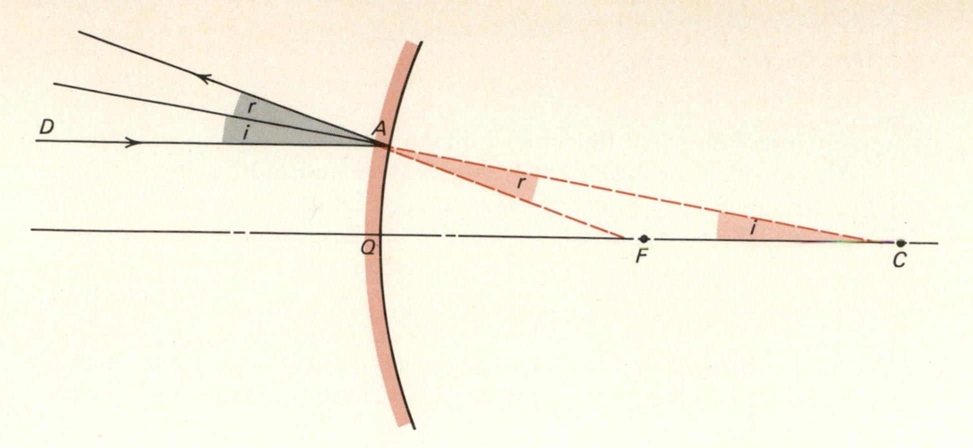

Figure 26.7 The location of the focal point of a convex spherical mirror. The triangle ACF is isosceles, and if the angle of incidence i is small, $AF = FC = \frac{1}{2}QC$. That is, $f = \frac{1}{2}R$, where R is the radius of curvature of the mirror.

valid for all parallel rays, all such rays will be reflected so that they appear to originate from the point F, the focal point of the convex mirror. In this case, however, the focal point is located *behind* the mirror.

We can now draw the principal rays for the convex mirror and use them to locate the image. These principal rays are (1) a ray, collinear with a radius, that strikes the mirror at normal incidence and is therefore reflected on itself; (2) a ray parallel to the mirror axis and reflected as though it originated from the focal point F; and (3) a ray directed toward the focal point and reflected parallel to the mirror axis.

These three rays are shown in Figure 26.8 and help to locate the image I of the object O. Evidently, the image formed by the convex mirror is virtual, erect, and reduced. Inspection will show that for the convex mirror, the image is *always virtual, erect,* and *diminished,* regardless of the position of the real object.

One can again derive a simple expression relating p, i, and f. There is, however, no need to do so. Comparison of Figures 26.5 and 26.8 reveals that 26.8 is identical to 26.5 if we interchange image and object. Consequently, Equation (26.4) must still be valid provided that we make the object distance negative. Rather than impose this requirement, it is an accepted convention to assign a *negative* focal length to a convex mirror. Equation (26.4) then remains valid for the convex mirror with a positive object distance p and yields the correct *negative* image distance; the negative sign reminds us that the image is *virtual* and *upright*.

Thus, *for the convex mirror, f is negative*. The image distance is also *negative;* its magnitude is always less than the focal distance.

Figure 26.8 Image formation by a convex spherical mirror. The image formed by this mirror is virtual, upright, and diminished, whatever the object distance. The three principal rays are shown.

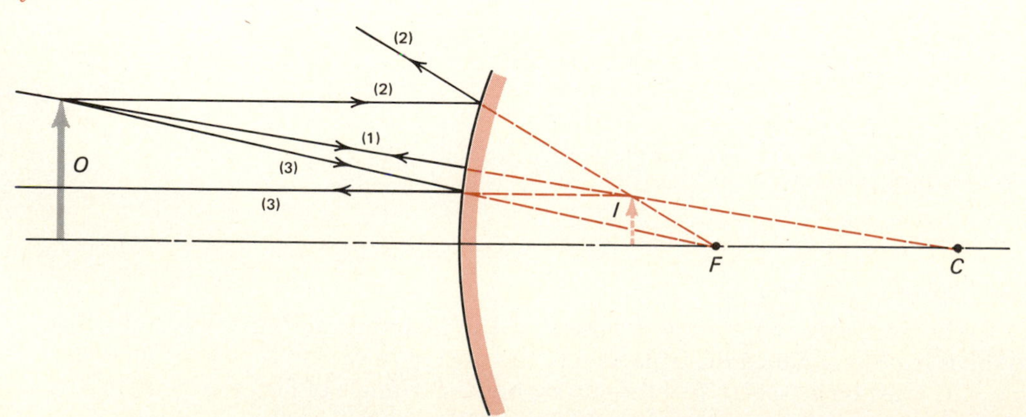

● ***Example 26.1*** You want to buy a shaving mirror that will permit you to see your face, enlarged by 50 percent, 40 cm from your eyes. What kind of mirror should you buy, and what radius of curvature should it have?

Since you want a mirror that will give an enlarged, upright image, it must be concave. To solve this problem it is advisable to sketch the mirror, showing it and the image and object in approximate, scaled dimensions. This procedure generally helps in all problems of geometrical optics and often prevents errors. Such a sketch is shown in Figure 26.9.

We know that the image should be enlarged and upright; hence, it must be a virtual image behind the mirror. Consequently, the object must be placed between the mirror and its focal point. Thus we arrive at the sketch of Figure 26.9.

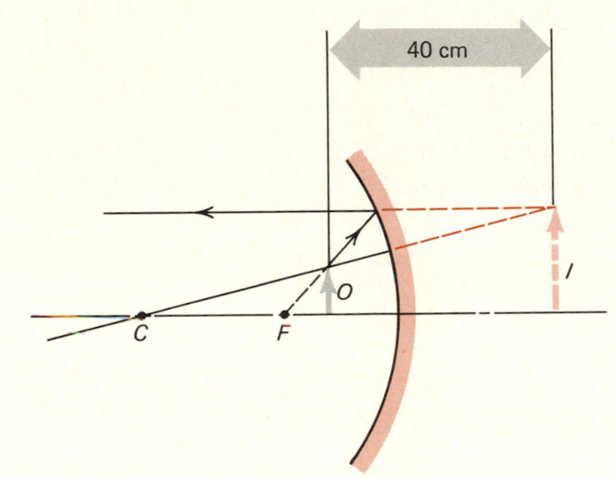

Figure 26.9 *The image of an object in a shaving mirror. The image should be upright and enlarged. Therefore, a shaving mirror must be concave, and the object distance must be less than the focal length.*

From the information given we can now write

$$m = -\frac{i}{p} = 1.5$$

and

$$p + |i| = p - i = 40 \text{ cm}$$

where we have used the fact that i must be negative. Solving these two equations for i and p, one obtains $p = 16$ cm, $i = -24$ cm. Next we insert these values into Equation (26.4):

$$\frac{1}{16 \text{ cm}} - \frac{1}{24 \text{ cm}} = \frac{1}{f} = \frac{1}{48 \text{ cm}} \qquad f = 48 \text{ cm}$$

and from Equation (26.1) we have

$$R = 2f = 96 \text{ cm}$$ ●

● ***Example 26.2*** When an object is placed 50 cm in front of a convex mirror, the image size is half the size of the object. What is the focal length of the mirror?

The sketch for this problem is shown in Figure 26.10. The magnification $m = -i/p = 0.5$. Since the mirror is convex, the image distance must be negative; hence, $i = -0.5p = -25$ cm. From Equation (26.4),

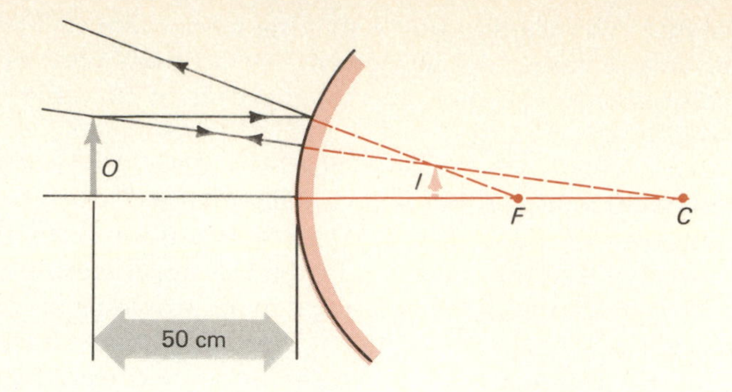

Figure 26.10 Relative positions of object and image for a convex spherical mirror that forms a virtual image half as big as the object.

we then have

$$\frac{1}{f} = \frac{1}{50\text{ cm}} - \frac{1}{25\text{ cm}} = -\frac{1}{50\text{ cm}} \qquad f = -50\text{ cm}$$

26.3 Lenses

Lenses function by refracting light at their interfaces. Consequently, their action depends not only on the shape of the lens surfaces but also on the indices of refraction of the lens material and the surrounding medium. Generally, the medium is air, for which $n = 1$. In some applications, however, one or both surfaces of a lens may be in contact with water or some other liquid; in that case, the refractive power of the lens will be substantially altered.

Figure 26.11 shows light rays passing through a *converging* (double convex) and through a *diverging* (double concave) lens. In the first in-

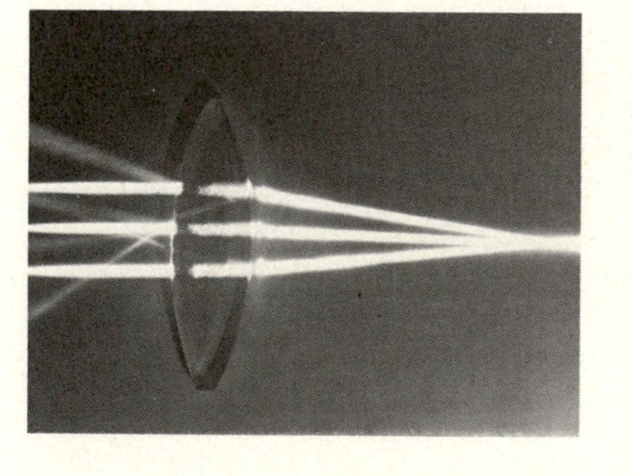

(a)

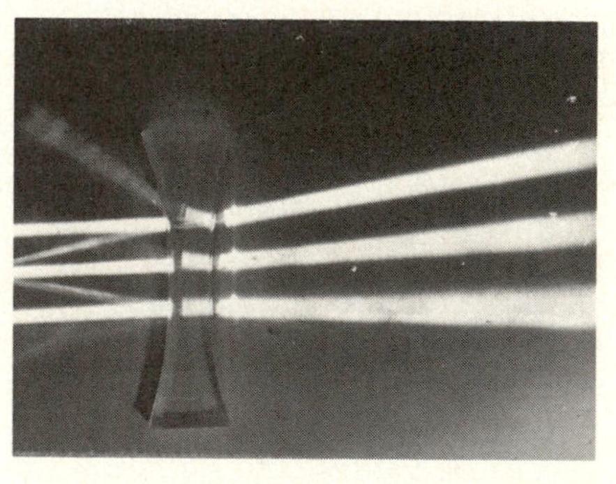

(b)

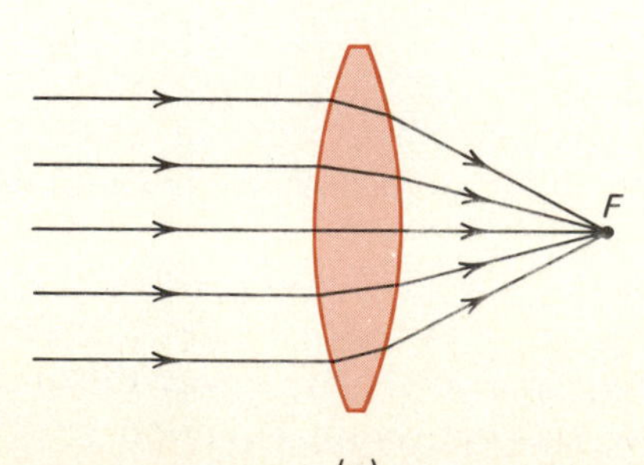

(c)

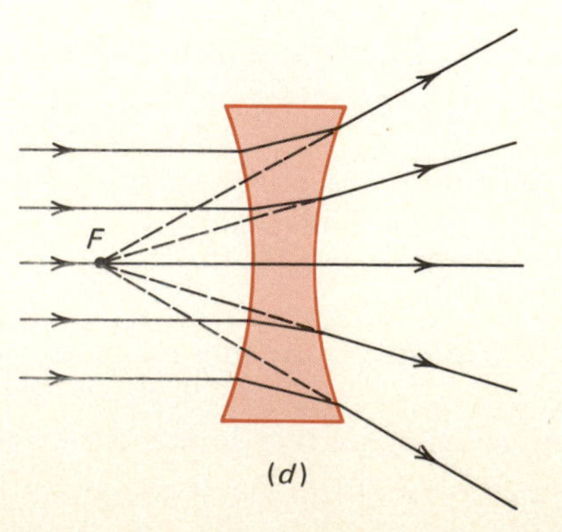

(d)

Figure 26.11 Light rays parallel to the axis of a converging (double convex) lens are brought to a focus by this lens; rays parallel to the axis of a diverging (double concave) lens are diverged by this lens so that they appear to have originated from the focal point of that lens. (a), (b) Photographs showing the pattern of light rays. (c), (d) The path of these parallel rays, shown schematically.

stance, incident rays parallel to the lens axis are refracted so that they *converge* and pass through a point on the axis, the focal point of the lens. In the second, parallel incident rays are *diverged* by the lens and propagate as though they had originated from a point on the opposite side of the lens, the focal point of the diverging lens. To understand better how refraction at the lens surfaces produces these effects, consider how a ray passes through a converging lens, as shown in Figure 26.12.

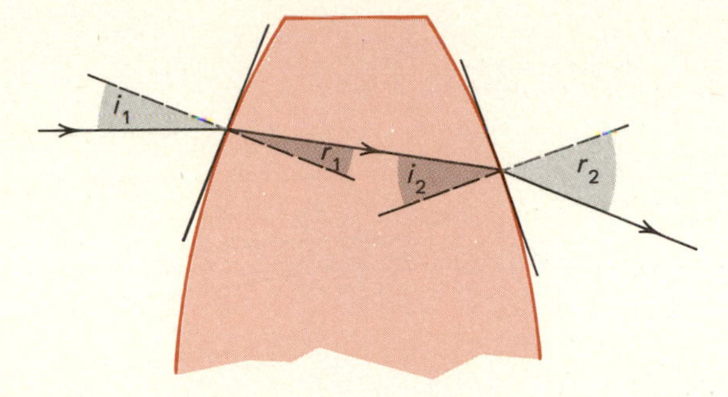

Figure 26.12 *A ray parallel to the lens axis is incident from the left on a double convex, converging lens. At both interfaces, the light is refracted toward the lens axis.*

The incident ray strikes the first surface at an angle of incidence i_1. If we assume that the lens substance has a greater index of refraction than the surrounding medium—the situation that normally prevails—then according to Snell's law, the angle of refraction r_1 will be smaller than i_1; in other words, the ray is bent toward the normal. The ray now passes through the lens and strikes the second surface at the angle i_2. At this interface, the ray is refracted away from the normal (going from an optically dense medium to one less dense), so that r_2 is greater than i_2, as shown. Note that the interior of the lens plays no role, at least to first approximation. In all our discussion we shall restrict ourselves to the *thin-lens approximation,* in which the thickness of the lens is assumed negligible compared with its focal length.

Though we shall not do so here, one can show that a lens whose surfaces are spherical segments will focus parallel rays. Moreover, since it does not matter whether the beam is incident from the left or from the right, a lens has two symmetrically positioned focal points, one on each side of the lens. The distance between the focal point and the lens is called the focal length of the lens, denoted by f.

It is often convenient to use the unit *diopter* (D) to characterize a lens. The power of a lens, in diopters, is simply the reciprocal of its focal length in meters:

$$D = \frac{1}{f} \qquad f \text{ in meters} \tag{26.6}$$

Optometrists use the adjectives *positive* instead of *converging* and *negative* instead of *diverging.*

26.4 Principal Rays for Lenses and Locating Images

The three principal rays used to establish the position of images for converging and diverging lenses are shown in Figure 26.13. Two of the rays pass through the focal points of the lens, or in the case of the diverging lens, appear to do so. An incident ray parallel to the axis of the lens is refracted so as to pass through the focal point (or in a diverging lens, ap-

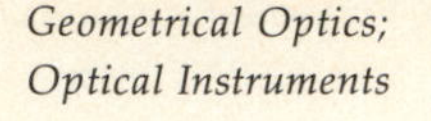

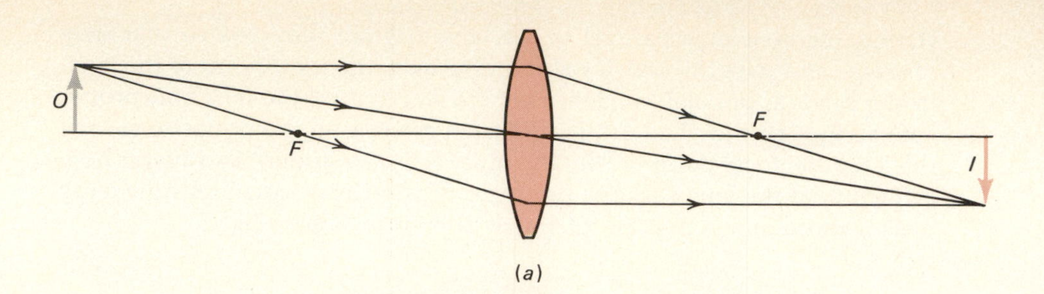

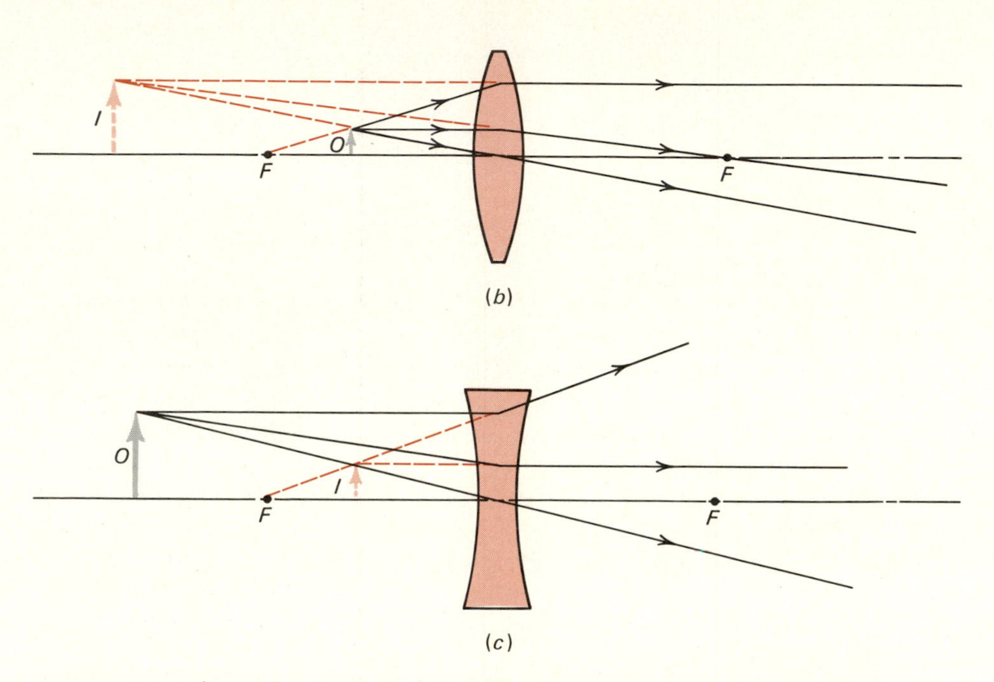

Figure 26.13 The ray constructions for converging and diverging lenses, showing the three principal rays. (*a*) The object distance from the converging lens is greater than the focal length; the image is real and inverted and is formed on the opposite side of the lens at an image distance greater than the focal length. (*b*) The object distance from the converging lens is less than the focal length; the image is virtual and upright and is located on the same side of the lens as the object. In this case, the image is always enlarged. (*c*) The lens is a diverging one, and the image formed by this lens is always virtual, upright, and diminished.

pear to come from the focal point). The other ray passes through (or is directed toward) the focal point and is refracted by the lens to emerge parallel to the lens axis. The third principal ray passes through the center of the lens and in the thin-lens approximation, is undeviated, as though it passed through a very thin piece of plate glass.

From Figure 26.13, we see that an object placed outside the focal point of a converging lens, as in Figure 26.13(*a*), yields a real, inverted image beyond the focal point on the opposite side of the lens. If the object is placed between a converging lens and its focal point, the lens forms a virtual, upright, and enlarged image of the object, as in Figure 26.13(*b*). From Figure 26.13(*c*), it is apparent that a diverging lens always creates a virtual, upright, and diminished image of a real object.

Evidently, there is a close similarity between the patterns of behavior of spherical lenses and mirrors. Indeed, comparing Figures 26.4(*a*) and 26.13(*a*) reveals that one is just the mirror image of the other, as shown in Figure 26.14. Similarly, Figures 26.5 and 26.13(*b*), and Figures 26.8 and 26.13(*c*), are reflections of each other. There is then no need to repeat the derivation of Equation (26.4); as for mirrors, so for lenses

$$\frac{1}{p} + \frac{1}{i} = \frac{1}{f} \qquad \textbf{(26.7)}$$

provided we adopt the following conventions:

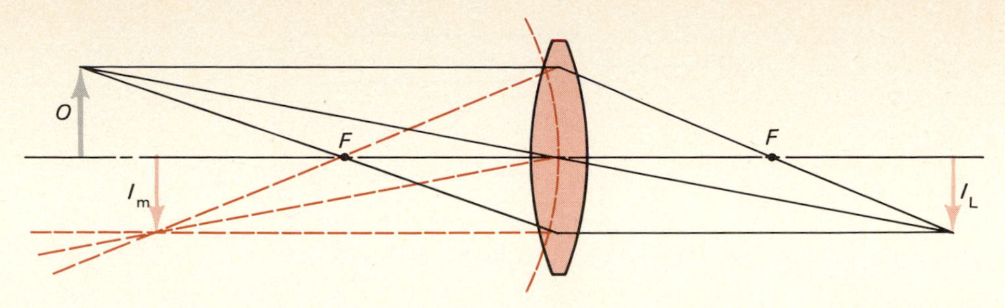

Figure 26.14 The ray construction for a concave spherical mirror is just the mirror image of the ray construction for a converging lens. The concave mirror surface and the reflected rays are shown dashed in this figure.

1. The focal length of a *converging* lens is *positive*. The focal length of a *diverging* lens is *negative*.

2. The object and image distances are *positive* if object and image are on *opposite sides* of the lens. The object distance is positive and the image distance is *negative* if object and image are on the *same side* of the lens.

We find that for converging lenses:

I. For $p > f$, the image distance is positive; the image is real and inverted, and formed on the opposite side of the lens.
   If $p > 2f$, $f < i < 2f$ and $|m| < 1$; m is negative.
   If $p = 2f$, $i = 2f$ and $m = -1$.
   If $f < p < 2f$, $i > 2f$ and $|m| > 1$; m is negative.

II. For $p < f$, the image distance is negative; the image is virtual, upright, and enlarged, and formed on the same side of the lens as the object. $|i| > p$ and $m > 1$.

For diverging lenses, the image of a real object is always virtual, upright, and diminished.

We have noted that the focusing action of a lens derives from the refraction at the lens surfaces, and that the intervening region of glass plays no essential role. In principle, then, a "lens" such as the one shown schematically in Figure 26.15, consisting of a series of concentric circular trapezoids and an end prism, should be just as effective as the more conventional lens shown dashed. Indeed, such *Fresnel lenses* are widely used, for example as condensing lenses for overhead projectors and also in lighthouse and airport beacons. Since the Fresnel lens is thin, it is much lighter than a conventional lens of equal diameter and power and incidentally, also absorbs less light. It does have its drawbacks, evidenced by the faint but clearly visible concentric circles cast by an overhead projector on the screen.

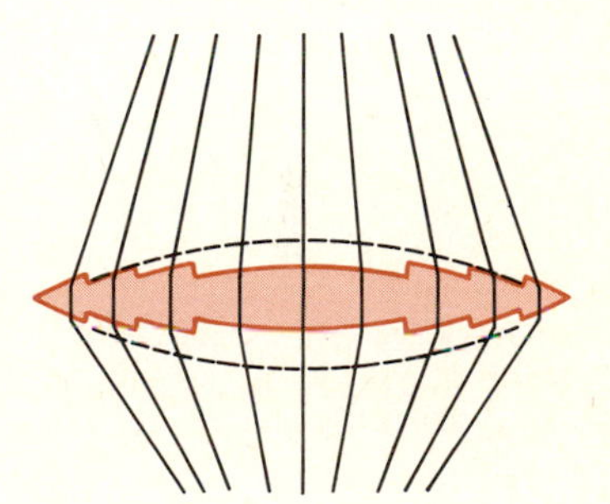

Figure 26.15 A double convex Fresnel lens. The refraction of light at the surfaces of this lens is the same as it would be at the surfaces of the normal, thicker lens, shown dashed. Many plastic Fresnel lenses are made with one side flat, so that the lens can be attached to a flat glass surface such as a mirror or plate glass.

*26.5 Aberrations

In the discussion of image formation, we assumed a perfection that cannot be achieved in practice. Real mirrors and lenses suffer from various intrinsic defects that produce distortions, known as *aberrations*. We consider only two kinds, *chromatic* aberration and *spherical* aberration, although there are a number of others as well.

Spherical aberration is the blurring of an image due to rays that traverse a lens far from its optic axis. The same defect also manifests itself with spherical mirrors, and its cause is immediately evident from Figure 26.16. Parallel rays incident near the mirror axis cross at the focal point F; however, rays relatively far from the axis—for which the small angle approximation is no longer valid—cross the axis at points nearer the

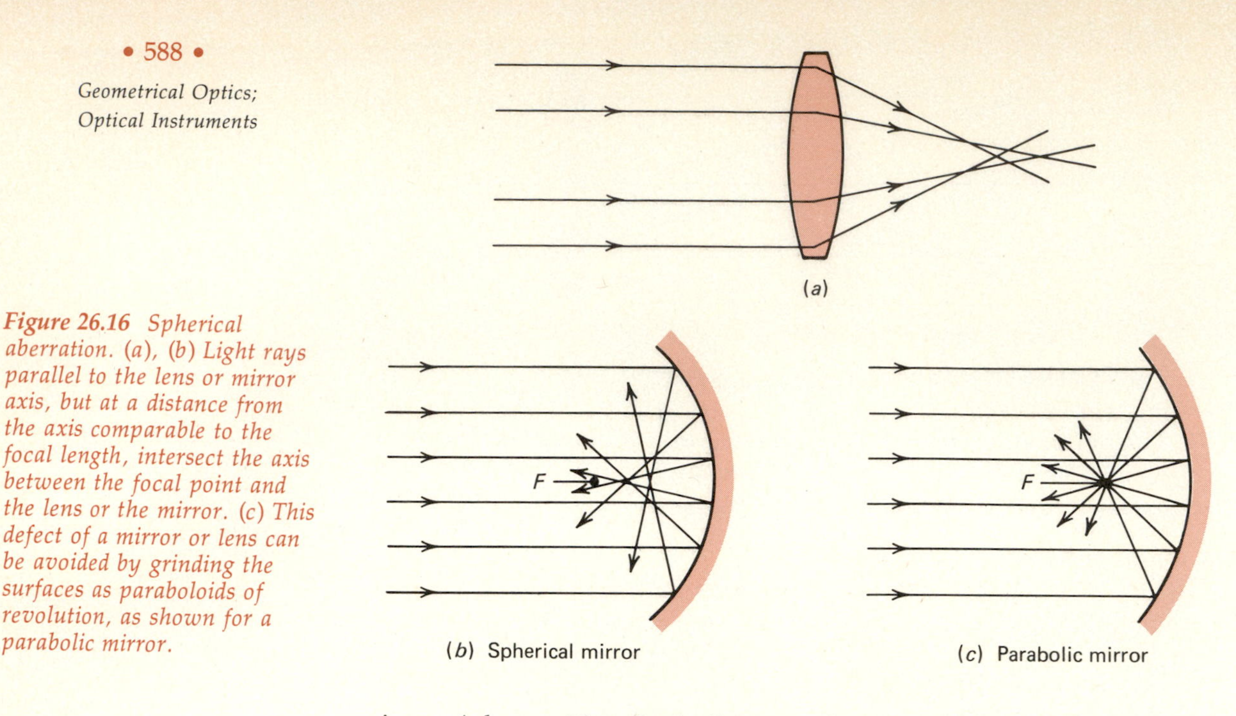

Figure 26.16 Spherical aberration. (a), (b) Light rays parallel to the lens or mirror axis, but at a distance from the axis comparable to the focal length, intersect the axis between the focal point and the lens or the mirror. (c) This defect of a mirror or lens can be avoided by grinding the surfaces as paraboloids of revolution, as shown for a parabolic mirror.

mirror. A lens with spherically ground surfaces shows the same defect, and in both cases the cure is to use parabolic instead of spherical surfaces. Unfortunately, grinding such surfaces is difficult and expensive.

As exemplified by the Fresnel lens, every lens can be viewed as a continuous series of graded, concentric prisms. Since the index of refraction of the glass or other substance of which the lens is made depends somewhat on wavelength (color), it follows that the focal length of the lens will also be a function of wavelength. Thus, if a lens is placed so that an object radiating green light casts a sharp image on a screen, the image will be slightly out of focus if the object is illuminated with red light. Chromatic aberration can be greatly reduced by means of properly constructed com-

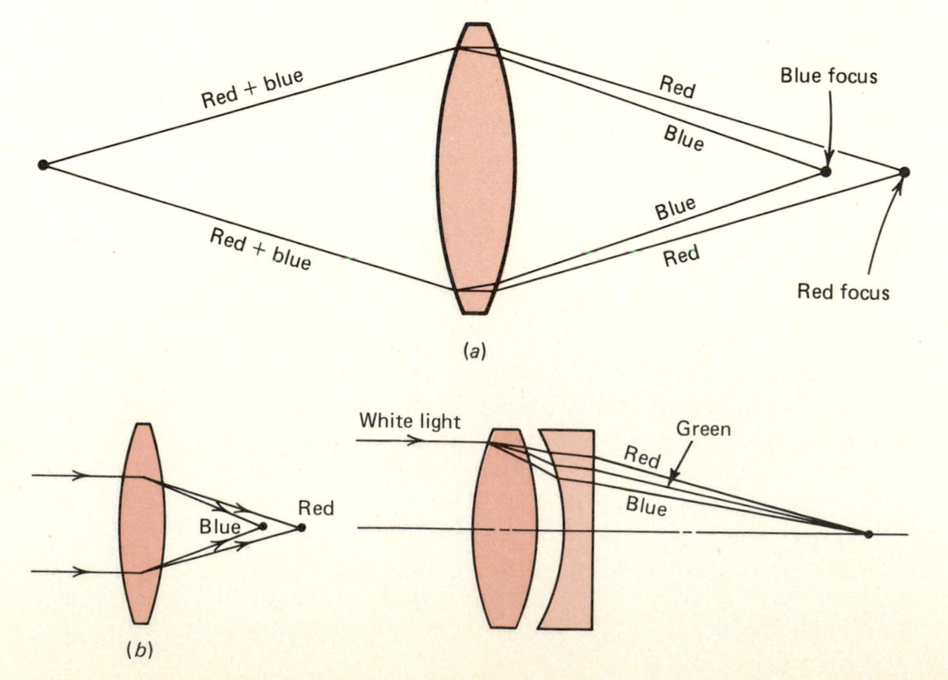

Figure 26.17 Chromatic aberration arises because the index of refraction of a lens varies with wavelength. For glass, the index of refraction is greater at short wavelengths (blue light) than at long wavelengths (red light). Consequently, the focal length for blue light is shorter than for red light. A camera that has a simple converging lens cannot bring a multicolored object to a sharp focus at the film plane. Chromatic aberration can be corrected by using a lens made of two different pieces of glass with different indices of refraction and different dispersive properties. All moderately good cameras and binoculars use such achromatic lenses.

pound lenses. Such *achromatic* lenses are made of glasses that have different average indices of refraction and also different dispersions, and they are always used now in cameras, field glasses, microscopes, and other optical instruments of good quality.

Since the focusing action of mirrors arises from reflection and not refraction, mirrors are free of this defect.

26.6 The Eye

A diagrammatic representation of the human eye in cross section is shown in Figure 26.18. The eye, with refracting surfaces at the cornea and the relatively thick lens, is not a simple optical system amenable to the thin-lens approximation. For many purposes, however, it has optical properties that can be considered equivalent to those of a single lens whose center is situated about 5 mm behind the real lens and 15 mm in front of the retina. One can then apply the simple thin-lens equation to this *reduced eye*.

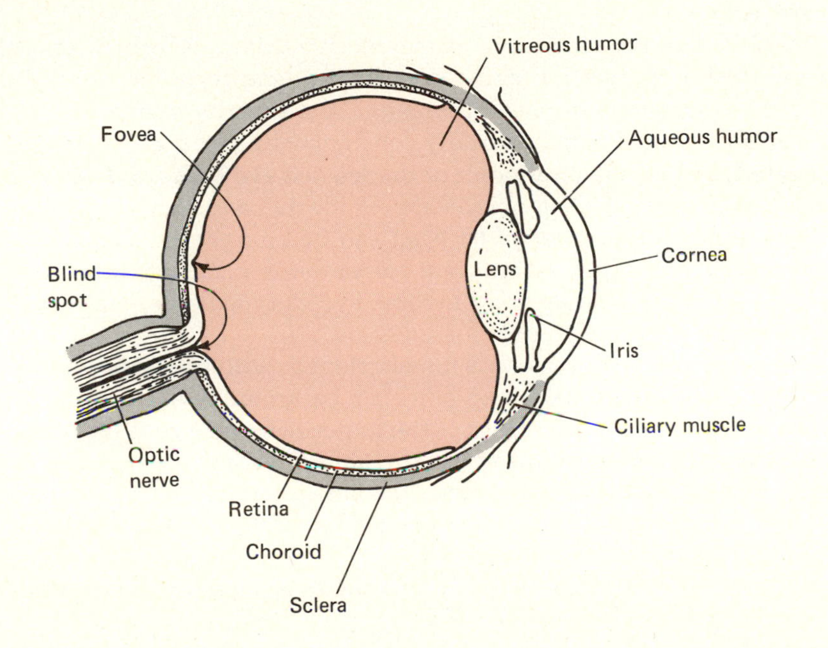

Figure 26.18 Cross section of the human eye (schematic drawing), showing its principal features.

Since the reduced eye, when relaxed, focuses an object at infinity on the retina, its refractive power is about 67 diopters. Most of that refraction occurs at the air-cornea interface (about 45 D). At rest, the lens has a power of about 20 D, but as it accommodates to more nearby objects, its power can increase to 30 D, and even more in young children. Such accommodation, especially in adults, cannot be maintained without excessive strain, and for practical purposes the *normal near point* of the eye is taken as 25 cm, corresponding to an increase in the power of the lens by 4 D.

The most common defects of the eye are *myopia* (nearsightedness), *hypermetropia* (farsightedness), *presbyopia* (lack of accommodation), and *astigmatism* (cylindrical distortion). Fortunately all these defects can be corrected with suitable lenses.

Of these defects, perhaps the most widespread is astigmatism, which commonly results from uneven curvature of the cornea, giving rise to distortion of the image, which can generally be corrected by cylindrically

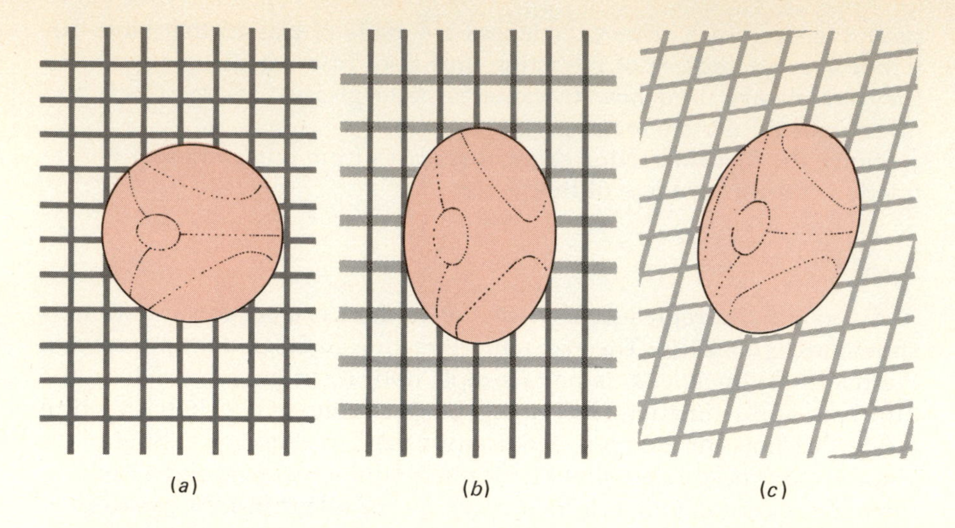

Figure 26.19 *Drawings of a beach ball in front of a square grid. (a) Ball and grid as seen with a nonastigmatic lens. (b) and (c) The same scene as viewed through a lens with severe astigmatism. The pictures simulate the effect that a cornea of corresponding distortion has on the image formed on the retina of the eye.*

ground lenses. Figure 26.19 shows the sort of distortion that arises from severe astigmatism.

Myopia and hypermetropia usually derive from a deformation of the eyeball during maturation and consequently tend to progress until the age of 15–18 years is reached. Thereafter, the severity of myopia or hypermetropia usually remains nearly constant. The myopic eye is slightly elongated, so that the focal point of the relaxed eye is in front of the retina. Consequently, distant objects appear blurred and only objects closer to the eye than its *uncorrected far point* can be brought to sharp focus on the retina. The uncorrected far point corresponds to the maximum object distance for which the unaided myopic eye can focus the image on the retina.

The hypermetropic eyeball is foreshortened, with the result that even at maximum accommodation an object 25 cm from the eye is still not in focus. The *uncorrected near point* of the hypermetropic eye is farther than 25 cm and corresponds to the minimum object distance for which the unaided hypermetropic eye can focus the image on the retina. In severe cases, the hypermetropic individual may be unable to see any object distinctly, no matter what the distance.

Although distortion of the eyeball is the most common cause of myo-

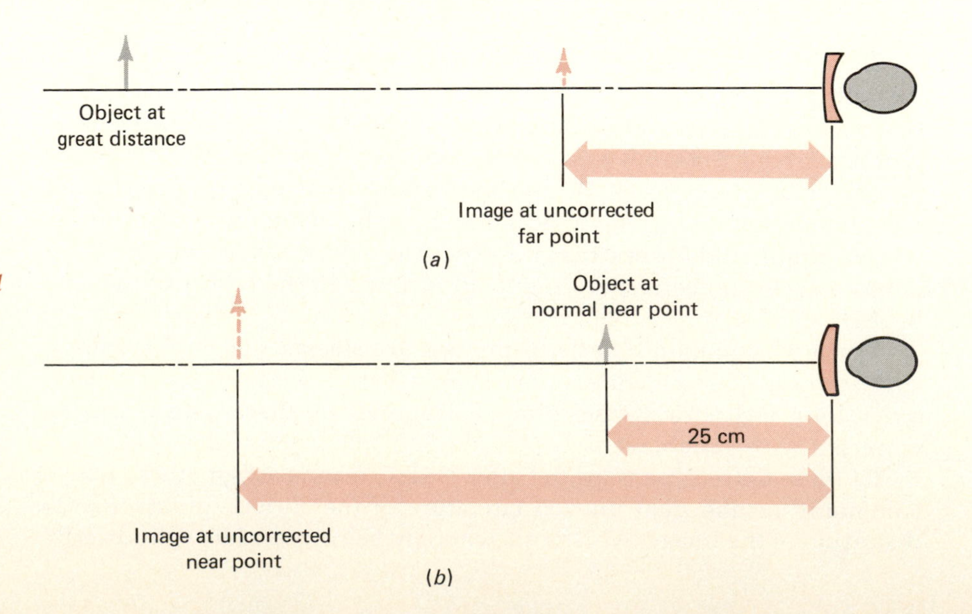

Figure 26.20 *Corrective lenses for myopia and hypermetropia. (a) The diverging lens forms a virtual image of a distant object at a point at which the uncorrected eye can bring it to a sharp focus on the retina. (b) The converging corrective lens forms an image of a nearby object at a point far enough from the eye that the uncorrected eye can focus the image on the retina. (The lenses shown have one concave and one convex surface, the usual practice.)*

pia and hypermetropia, excessive or inadequate curvature of the cornea or some defect of the lens can also cause these visual impairments.

Whatever the origin of these malfunctions, perfect or near-perfect vision can usually be restored by means of an appropriate lens placed before the imperfect eye. The purpose of the lens is to form a virtual image of the real object at a position where the uncorrected eye can see it in sharp focus without strain (see Figure 26.20).

Example 26.3 Corrective Lenses for the Myopic Eye

What is the focal length of the corrective lenses for a myopic individual whose uncorrected far point is 12.5 cm?

The corrective lens must be one that will form, at a point 12.5 cm in front of the lens, a virtual, upright image of an object at infinity. To determine the type and focal length of the proper corrective lens, we make use of the lens equation, Equation (26.7), recalling that the image distance is negative if the object is real and the image virtual.

Thus

$$\frac{1}{\infty} - \frac{1}{0.125\ \text{m}} = \frac{1}{f} = -8\ D$$

Hence, the proper corrective lens is a negative (diverging) lens that has a focal length of −12.5 cm, or a power of −8 D. That a diverging lens is required is apparent from Figure 26.21. Because the eyeball is excessively elongated, the optical apparatus of the eye has a converging power that is too great to focus parallel rays on the retina. The corrective lens compensates for this excessive converging power of cornea and lens by diverging the incident parallel rays before they reach the cornea.

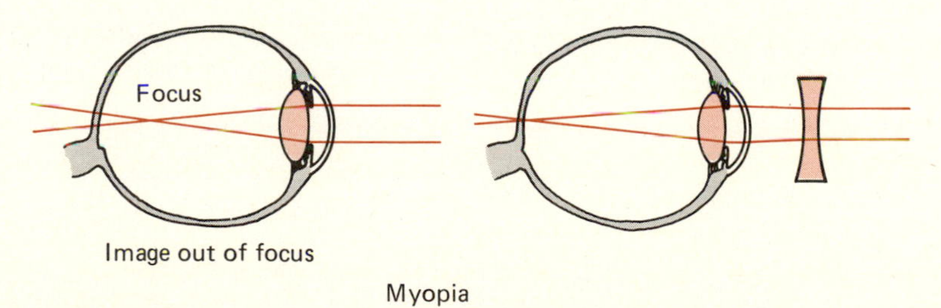

Figure 26.21 A diverging lens corrects for myopia.

Example 26.4 Correction for Hypermetropia

A farsighted woman is unable to focus on objects closer than 2 m from her eye without excessive strain. What is the focal length of corrective lenses that will allow her to focus on reading material held 25 cm in front of her eyes?

The proper corrective lens should form a virtual image at the uncorrected near point when an object is placed 0.25 m from the eye. Again, we establish the focal length of the corrective lens by means of Equation

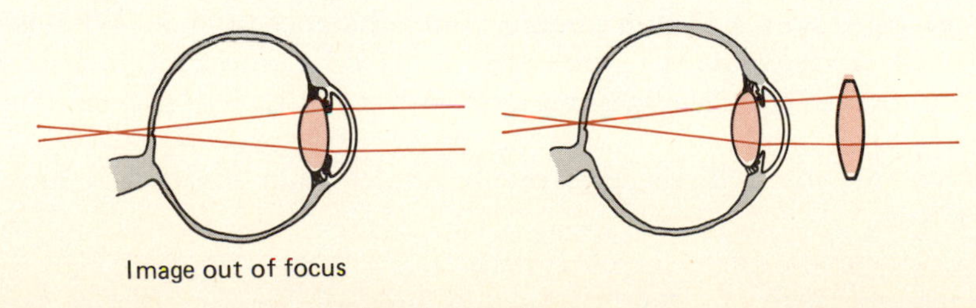

Figure 26.22 A converging lens corrects for hypermetropia.

(26.7), where now $p = 0.25$ m and $i = -2$ m; the negative sign signifies that the image is virtual. Thus

$$\frac{1}{p} + \frac{1}{i} = \frac{1}{0.25\text{ m}} - \frac{1}{2\text{ m}} = \frac{1}{f} = 3.5\text{ D} \qquad f = 28.6\text{ cm}$$

Hence the proper corrective lens is a converging lens of +3.5 D. ●

Before leaving the discussion of vertebrate eyes, we should point out that accommodation by adjusting the power of the lens is not universal in the animal kingdom. In fishes and amphibians, the lens is nearly spherical and of fixed shape. To accommodate to different object distances, this lens is moved along the optic axis as a camera lens is positioned to obtain a sharp image on the film plane.

26.7 Camera and Projection Lantern

The camera is basically a simple instrument, with a lens, of fixed focal length, whose position can be adjusted relative to the film plane. In addition to the lens and film, the camera also contains an adjustable iris and shutter, which determine the exposure of the film.

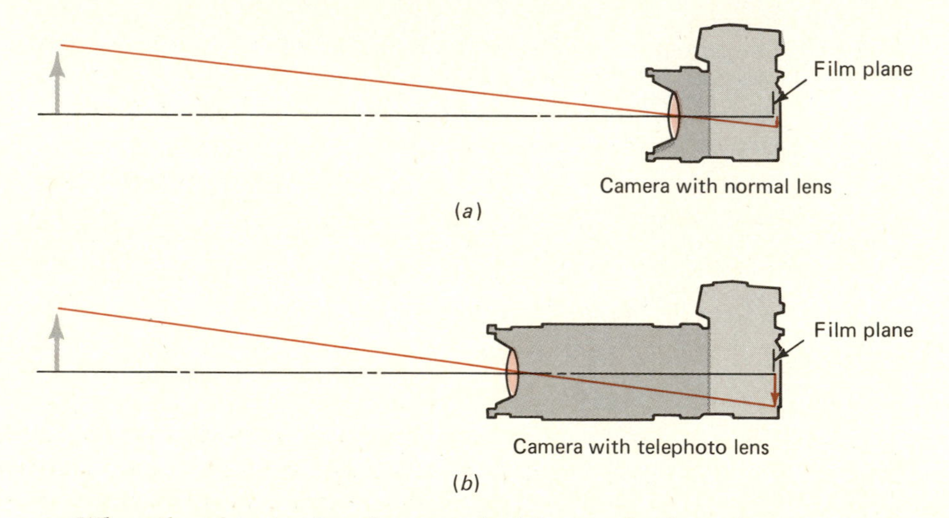

Figure 26.23 Expensive cameras can be fitted with lenses of different focal lengths. (a) With a lens of the usual 50-mm focal length, the image on the film of a distant object is very small. (b) The camera is fitted with a long-focal-length telephoto lens; the image of the same distant object is then significantly larger.

When the object to be photographed is very far from the camera, the film-to-lens distance is the focal length of the lens. As the object distance is reduced, the lens is moved forward, away from the film plane, to allow for the increase in image distance with diminishing object distance. The image that is formed is real and inverted.

Expensive cameras permit the user to select lenses of various focal lengths, ranging from short-focal-length wide-angle lenses to long-focal-length telephoto lenses. The following example demonstrates the advantage of a telephoto lens in nature photography.

● ***Example 26.5*** A deer, approximately 1 m tall and 1.5 m long, is to be photographed with a 35-mm camera from a distance of 20 m. What is the image size if a standard 55-mm lens is used? Compare this with the image obtained by replacing that lens with a 400-mm telephoto lens.

Since the object distance is much greater than the focal length of either lens,

$$\frac{1}{i} = \frac{1}{f} - \frac{1}{p} \simeq \frac{1}{f} \qquad i \simeq f$$

Consequently,

$$m = -\frac{i}{p} \simeq -\frac{f}{p}$$

For the 55-mm lens, the magnification is

$$m = -\frac{5.5 \times 10^{-2}\ \text{m}}{20\ \text{m}} = -2.75 \times 10^{-3}$$

The image of the deer on the film will therefore measure

$$(1\ \text{m})(2.75 \times 10^{-3}) \times (1.5\ \text{m})(2.75 \times 10^{-3}) = (2.75\ \text{mm}) \times (4\ \text{mm})$$

The area occupied by the deer's image is only about 11 mm^2, roughly 1.6% of the (35 mm)(20 mm) = 700 mm^2 of the area of the entire transparency. One will have to look closely even to find the deer in the picture, let alone see much detail.

Suppose we replace the standard lens with a 400-mm telephoto lens. Since f is now greater, let us first determine the true image distance:

$$\frac{1}{i} = \frac{1}{f} - \frac{1}{p} = \frac{1}{0.4\ \text{m}} - \frac{1}{20\ \text{m}} = 24.95\ \text{m}^{-1} \qquad i = 401\ \text{mm}$$

Again we can safely neglect the small difference between i and f. Now the magnification is $-(0.4/20) = -2 \times 10^{-2}$, and the image of the deer measures about 30 mm × 20 mm; it nearly fills the entire area of the transparency. ●

Conversely, to photograph a large cathedral from a distance of one or two city blocks it may be necessary to use a lens of relatively short focal length, perhaps 28 mm or less.

The projection lantern is the optical inverse of the camera. The transparency is illuminated by a light source placed at the focal point of a converging lens or mirror, the *condenser,* so that the slide is illuminated by a nearly uniform beam of parallel rays. The projection lens then forms an enlarged, inverted real image of the transparency on a distant screen. Whereas with the camera, the most favorable focal length of the lens was determined by the size and distance of the object, the focal length of the projection lens is dictated by the size and distance of the screen on which the image is to be projected.

● ***Example 26.6*** What should the focal length of the lens of a slide projector be so that the image of a 35-mm slide will just fill a 2 m wide screen placed 6 m from the projector?

If the 35-mm slide, when projected, is to fill the screen, the magnification of the system must be

$$m = -\frac{2\ \text{m}}{3.5 \times 10^{-2}\ \text{m}} = -57 = -\frac{i}{p}$$

Since the image distance is 6 m, the object distance must be

$$p = \frac{6\ \text{m}}{57} = 0.105\ \text{m} = 10.5\ \text{cm}$$

The focal length of the projector lens is obtained from Equation (26.7):

$$\frac{1}{f} = \frac{1}{p} + \frac{1}{i} = \frac{1}{0.105\ \text{m}} + \frac{1}{6\ \text{m}} = 9.69\ \text{m}^{-1} \qquad f = 10.3\ \text{cm}$$ ●

26.8 Simple Magnifying Glass, or Jeweler's Loupe

Perhaps the simplest optical instrument is the jeweler's loupe. Since the size of the image projected on the retina of the eye is determined by the angle subtended by the rays impinging on the pupil, an object is effectively enlarged if it can be viewed at close range. There is a limit beyond which one cannot go, however, for the eye is unable to accommodate sufficiently to give a sharp image if the object is mucn closer than the uncorrected near point. The magnifying glass overcomes this limitation by forming a virtual image of a nearby object at a point sufficiently far to permit comfortable viewing.

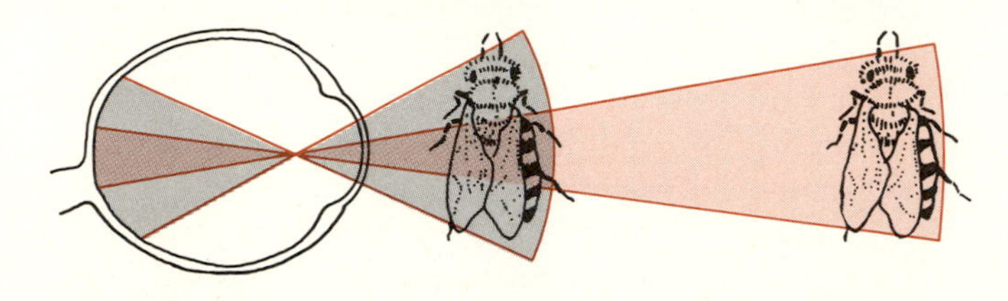

Figure 26.24 *The size of the image on the retina depends on the distance of the object from the eye.*

From what we have just said, it is obvious that the lens must be a converging one, with the object placed just inside its focal point. Suppose we use a lens whose focal length is 2.5 cm. From Equation (26.7), we find that an object placed 2.27 cm from this lens will form a virtual, upright, and enlarged image 25 cm from the lens. As far as the viewer can tell, the object appears to be 11 times as large as it would be if placed 25 cm from the unaided eye and yet is at a convenient distance for comfortable viewing. The lens allows us to bring the object very close to the eye so that the angle subtended by it is large. The magnifying lens refracts the diverging rays from the nearby object, converging them a bit so that in conjunction with the refractive power of the eye, focused to the normal near point, it forms a sharp image on the retina.

The magnification of the object by the loupe is given by Equation (26.5),

$$m = -\frac{i}{p} = -\frac{(-25)}{2.27} = 11$$

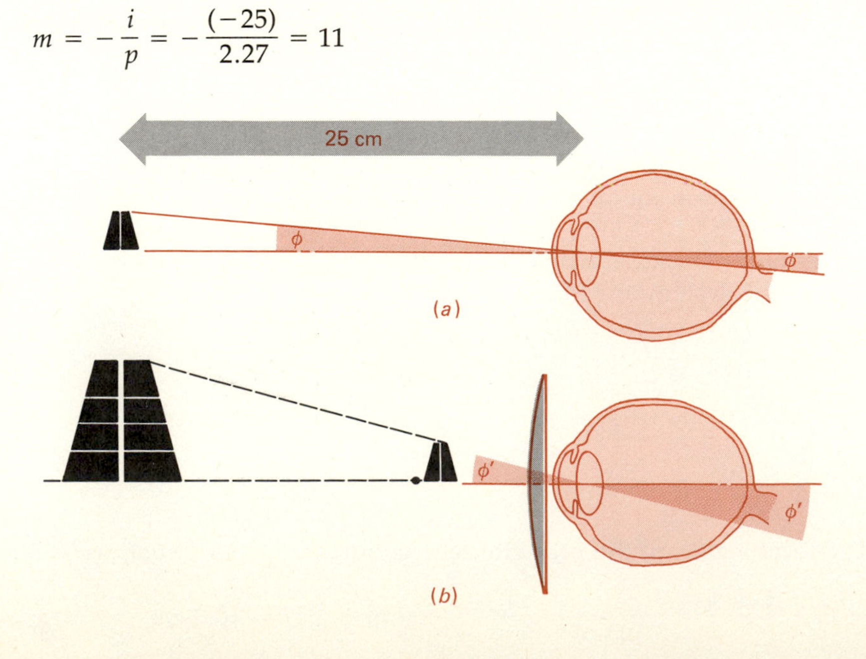

Figure 26.25 *The jeweler's loupe, or magnifying glass, allows the small object under study to be brought very close to the eye. The loupe forms a virtual, erect, and enlarged image of the object at a convenient distance for viewing.*

Since the object is located just inside the focal point of the loupe, the object distance is almost equal to the focal length. Consequently, the expression

$$m \approx \frac{0.25}{f} = \frac{D}{4} \qquad \textbf{(26.8)}$$

is usually a good approximation. For our example, the exact result and the approximate magnification obtained by applying Equation (26.8) differ by 10 percent.

26.9 Microscopes

The compound microscope is a two-lens instrument consisting of a converging *objective* lens of short focal length and, at a convenient distance, a second converging lens of greater focal length—the eyepiece, or *ocular*, which serves as a magnifying glass. A ray diagram for such an instrument is shown in Figure 26.26.

The objective is placed so that the sample to be examined rests just outside its focal point. This lens then forms a real, inverted, and enlarged image of the object some distance from the lens. By proper adjustment of the object distance, this image can be located just inside the focal point of the ocular and now serves as the real object for that lens. The ocular, performing the same function as the loupe described earlier, forms a virtual, enlarged image of its object.

● ***Example 26.7*** A microscope has an objective of 5-mm focal length and an ocular of 3-cm focal length. The distance between the two lenses, fixed by the microscope barrel, is 20 cm. Determine the magnification of this instrument and also the correct object-to-objective distance.

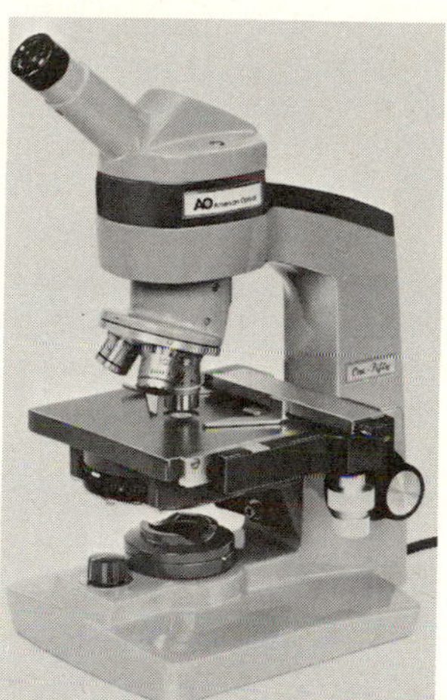

(*a*)

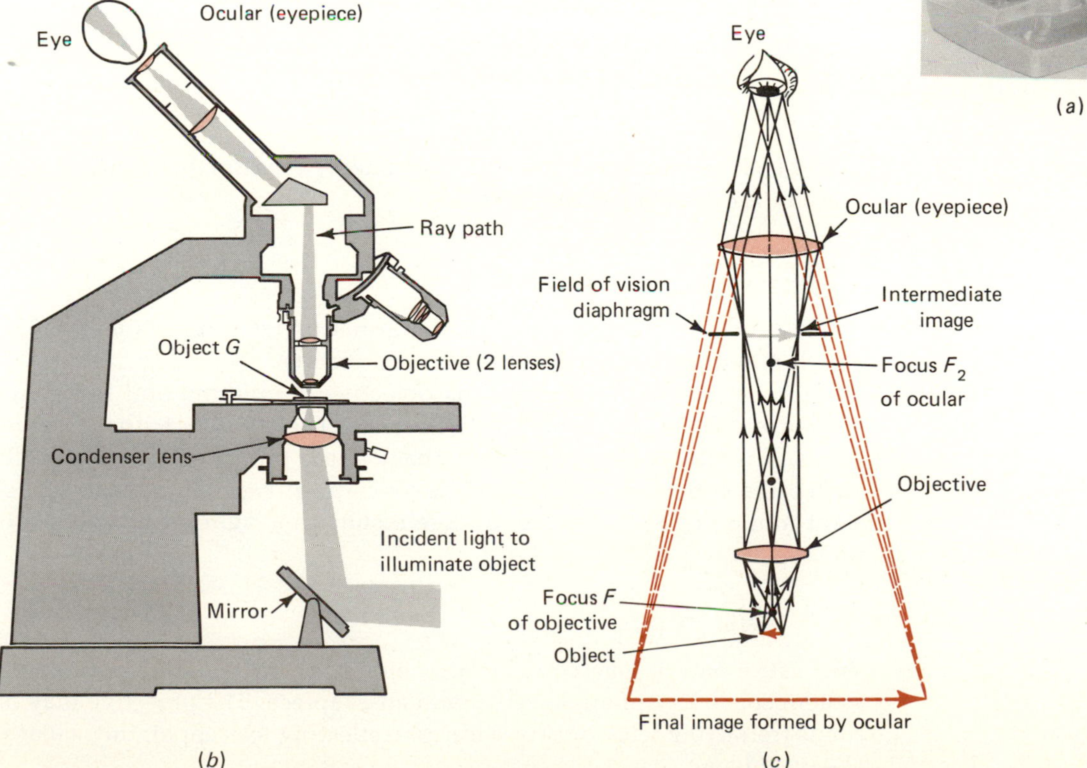

Figure 26.26 *The compound microscope. (a) A monocular microscope showing the turret carrying several objective lenses. (b) Schematic cross section of a compound microscope. (c) Ray diagram, showing image formation in a compound microscope. The object is placed just outside the focal point of the objective lens. The real image formed by this objective lens serves as the object for the ocular, or eyepiece. The ocular forms a virtual enlarged image of its object because the real image of the objective is located just inside the focal point of the ocular.*

In attacking problems of this sort, one proceeds from the final image, working backwards toward the object. We know that for comfortable viewing, the virtual image of the ocular should be at an image distance of −25 cm. (The image is virtual, hence i must be chosen as a negative distance.) Applying the lens equation, we determine the object distance for the eyepiece:

$$\frac{1}{p_e} = \frac{1}{f_e} - \frac{1}{i_e} = \frac{1}{3\text{ cm}} + \frac{1}{25\text{ cm}} = 0.3733\text{ cm}^{-1}$$

$$p_e = 2.68\text{ cm}$$

The object of the eyepiece is also the real image formed by the objective. Since the two lenses are separated by 20 cm, the image distance of the objective must be $i_o = 20 - 2.68 = 17.32$ cm. Making use of the lens equation once more, one obtains

$$\frac{1}{p_o} = \frac{1}{f_o} - \frac{1}{i_o} = \frac{1}{0.5\text{ cm}} - \frac{1}{17.32\text{ cm}} = 1.94\text{ cm}^{-1}$$

$$p_o = 5.15\text{ mm}$$

This object distance is indeed just slightly greater than the focal length of the objective.

The overall magnification is now readily determined; it is simply the product of the individual magnifications of the two lenses; that is,

$$m = m_o \times m_e = \left(-\frac{i_o}{p_o}\right)\left(-\frac{i_e}{p_e}\right)$$

$$= \left(-\frac{17.32}{0.515}\right)\left(-\frac{(-25)}{2.68}\right) = -314$$

Since for compound microscopes,

$$p_o \simeq f_o \qquad p_e \simeq f_e$$

and

$$i_o = L - p_e \simeq L - f_e \simeq L$$

where L is the barrel length of the instrument, a reasonably good approximation for the magnification is

$$m \simeq -\frac{0.25L}{f_o f_e} = \frac{-LD_oD_e}{4} \qquad \textbf{(26.9)}$$

where L is in meters. In our example, Equation (26.9) yields $m = -333$, which is within 10 percent of the correct result.

Since the image formed by the objective is inverted and the image formed by the ocular is upright with respect to its object, the observer sees an inverted, enlarged, virtual image of the preparation on the slide. Hence the common, and at first disconcerting, experience that left-to-right motion of the microscope stage results in a right-to-left motion of the image.

26.10 Telescopes

The astronomical telescope is also a simple two-component optical system consisting of an objective and an eyepiece. The objective may be a lens (refracting telescope) or a mirror (reflecting telescope); the ocular is always a lens.

The function of the objective is twofold. Because in most applications the instrument is aimed at a faint source, it should gather as much of the light as possible. It should also form a real image of high resolution that can be enlarged by the eyepiece. Both goals call for objectives of large diameter, which is why astronomers have clamored for ever larger instruments.

Since the object distance is practically infinite and unknown, we cannot use the standard formula to compute the magnification of the instrument. In discussing the simple magnifying glass, we saw that magnification can be regarded in terms of the angle subtended by the object without and with the aid of the loupe. In considering the telescope's magnification, we again turn to the description of magnification in terms of angles.

Suppose the object subtends an angle of θ_o radians when viewed with the unaided eye. This is then also the angle subtended by the real image at the objective lens (or mirror); that is, $\theta_o = I_o/f_o$, where I_o is the size of the image formed by the objective of focal length f_o (see Figure 26.27). If this real image is located at the focal point of the ocular, the ocular creates a virtual image at infinity; the angle subtended by this virtual image at the eyepiece is $\theta_e = I_o/f_e$, where f_e is the focal length of the eyepiece. The magnification of the telescope can be expressed in terms of these two angles:

$$|m| = \frac{\theta_e}{\theta_o} = \frac{f_o}{f_e} \tag{26.10}$$

Note that in contrast to the compound microscope, the magnification of the telescope is greater the greater the focal length of the objective.

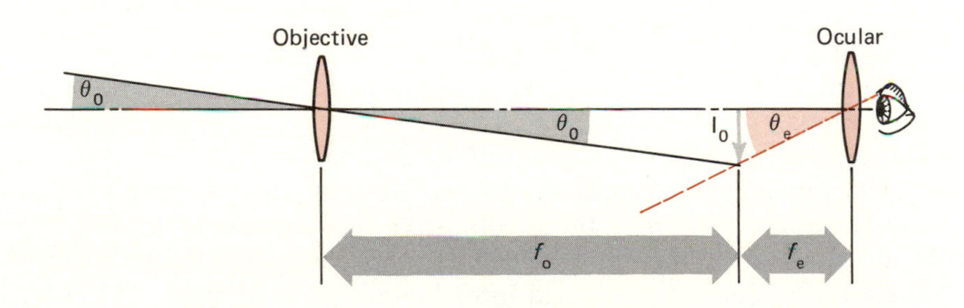

Figure 26.27 The astronomical telescope in schematic form. The objective lens forms a real, inverted image of the distant object at the lens's focal point. This image is then viewed through the ocular. The ocular is placed so that the objective's image is at the ocular's focal point. The magnification of the telescope is the ratio f_o/f_e, where f_o and f_e are the focal lengths of the objective and eyepiece.

Large astronomical telescopes always use reflectors as objectives. Mirrors have many advantages over lenses in this application. A large-diameter lens, even one of long focal length, would be extremely heavy. Since it can be supported only around its periphery, it would be subject to severe and varying strains as the telescope is tilted in the earth's gravitational field; these strains in the glass would lead to considerable distortions. Moreover, lenses must be corrected for chromatic aberration, a defect not present in mirrors. By contrast, a mirror can be supported over its full back surface, thus minimizing strains due to rotation and tilting of the instrument. Also, by using a honeycomb construction a mirror can be made relatively light without reducing its mechanical strength or stability. The major disadvantage of a telescope mirror is that since it is front-surfaced, the reflecting surface is in constant contact with the atmosphere and subject to tarnishing by various pollutants. The reflecting surface must therefore be removed at regular intervals and the mirror resilvered, a tedious and expensive operation.

Figure 26.28 *The 200-inch-diameter mirror of the Hale Observatory at Mt. Palomar before its installation. The honeycomblike support structure of the glass is clearly visible.*

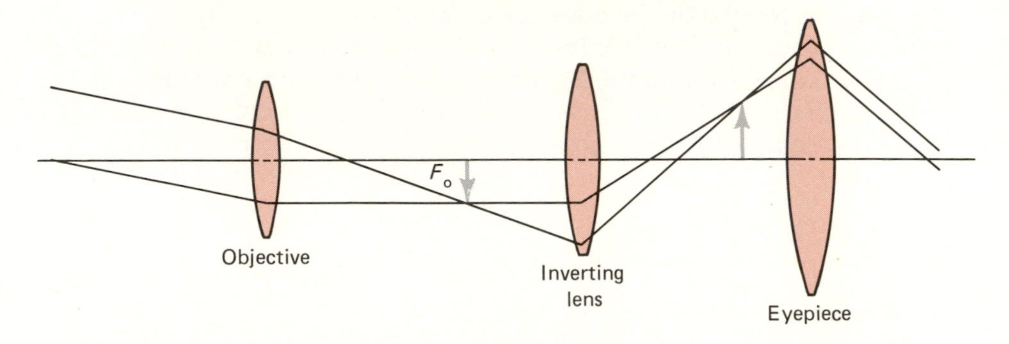

Figure 26.29 *A terrestrial telescope uses an inverting lens between the objective and ocular.*

Like the simple microscope, the simple telescope yields an inverted virtual image. While this inversion is inconsequential in astronomical observations, it does make the instrument unsuitable for terrestrial viewing. The problem is overcome by incorporating an intermediate lens between objective and eyepiece, whose principal function is to reinvert the image formed by the objective. An alternative technique is to use two reflecting prisms; one turns the image right side up, the other reflects in the vertical plane and interchanges left and right. This is the method used in binocular field glasses; it has the advantage of reducing the physical length of the instrument without a concomitant reduction in

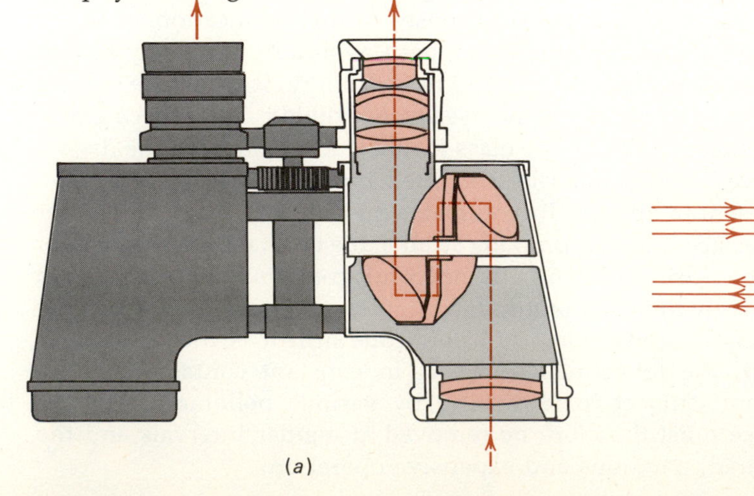

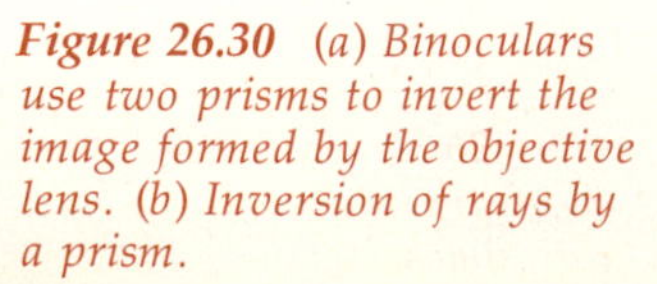

Figure 26.30 *(a) Binoculars use two prisms to invert the image formed by the objective lens. (b) Inversion of rays by a prism.*

the optical length of the telescope barrel. Since the magnification is greater the greater the objective's focal length, a long barrel is generally desirable.

*26.11 Optical Instruments in Medical Diagnosis

26.11 (a) Ophthalmoscope

Anyone who has had a thorough physical examination has endured the discomfort of retinal illumination with an ophthalmoscope, a small, hand-held instrument for visual inspection of the cornea, lens, and retina. The instrument consists of a rechargeable battery powering a small, moderately intense light source, suitable optics for directing this light into the patient's eye, and a selection of small lenses of varying power mounted on a rotatable disk so that any one can be placed in front of the eye of the examining physician.

As shown in the diagram of Figure 26.32, light from the source is collimated and directed through the patient's pupil by a small mirror. The

Figure 26.31 *A modern ophthalmoscope.*

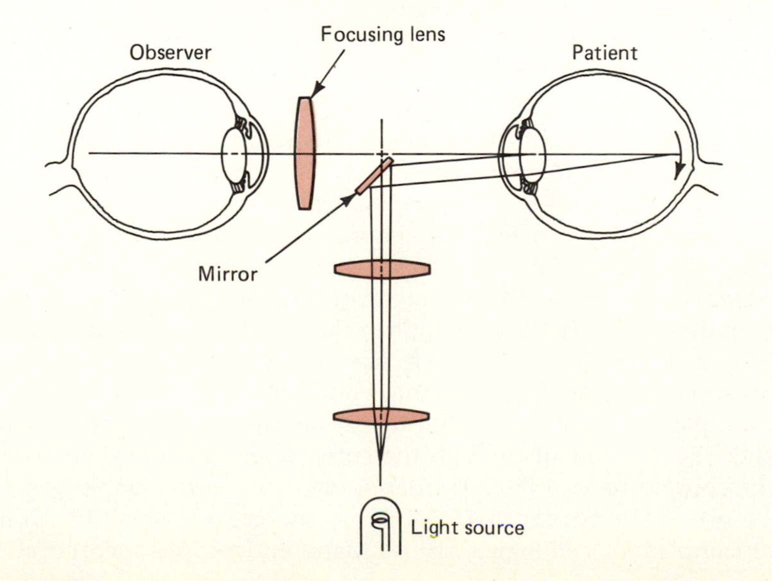

Figure 26.32 *Schematic ray diagram for the ophthalmoscope.*

physician looks over the mirror at the illuminated part of the eye. In retinal examination, the principal function of the instrument is to illuminate the retina; the cornea and the lens of the eye then function as a magnifying glass of about 15-mm focal length through which the physician looks to see an enlarged image of the retina.

*26.11 (b) Endoscopes

Endoscopy, the visual observation of various internal body cavities—such as colon, bronchi, esophagus, and stomach—is an essential part of modern diagnostic practice. Until recently, instruments used for this purpose consisted of long stainless steel tubes containing a system of lenses and at the far end a small, intense light source. These rigid and cumbersome devices have been replaced by much more convenient and versatile flexible fiber-optic endoscopes.

To understand how these instruments work, we recall that a ray incident internally on a glass-air interface will suffer total internal reflection if the angle of incidence is greater than the critical angle of Equation (25.10). Consider, for example, a smooth fiber of clear plastic whose end surfaces are flat, as in Figure 26.33. Light entering at one end will then be

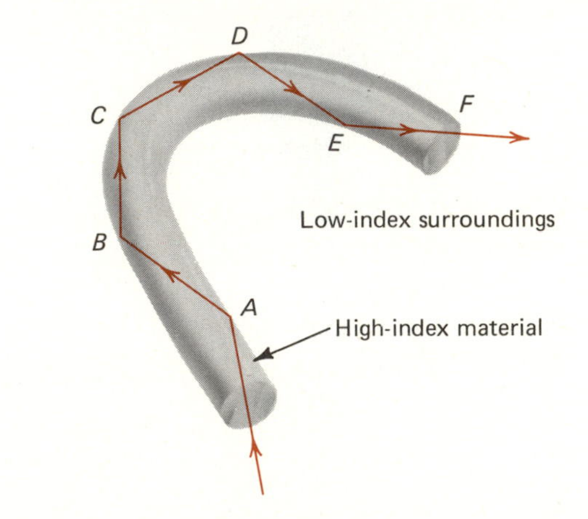

Figure 26.33 A flexible plastic pipe with smooth ends serves as a light pipe, or optical fiber. If the index of refraction of the light pipe is high enough, light entering through one of the flat ends is internally reflected until it emerges from the other end.

constrained to follow the contour of the fiber, suffering multiple internal reflections, until it emerges at the opposite flat end, provided that there are no excessively sharp kinks in the fiber. Thus optical information can be transmitted readily from one point to another even though "straight-line" transmission may be impeded by obstacles. Moreover, one can construct a light pipe that consists not just of one fiber but of a large number of extremely fine fibers, and the pattern of light that emerges from one end is then a replica of what entered the other.

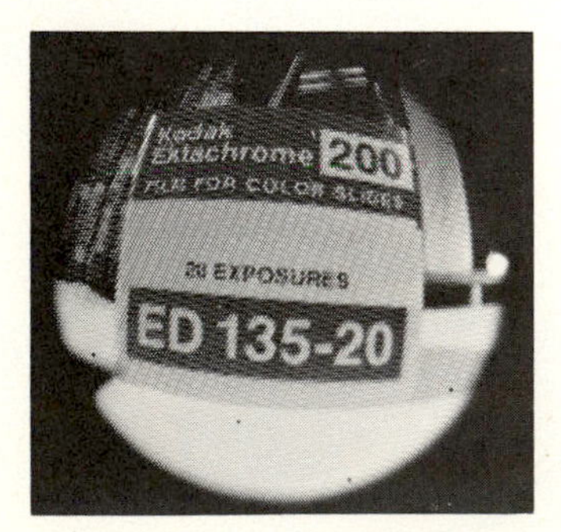

Figure 26.34 A bundle of many small-diameter light pipes can be used to transmit an image. The photograph shows a bundle of 5-μm-diameter fibers.

The fiber-optic bronchoscope, shown in Figure 26.35, includes two features generally incorporated in modern endoscopes to enhance their effectiveness and versatility as diagnostic and surgical instruments. The knob on the side of the instrument handle operates a set of fine wires that control the orientation of the angle section. By proper articulation of the tip, an endoscope can be guided into body passages and around flexures. For example, by carefully controlling the articulated tip, a physician can advance the instrument through the entire length of the colon.

The compactness of flexible fiber optics also frees valuable space that can be used to admit forceps, diathermy snares, or a small brush to retrieve samples for cytological study. Many endoscopes incorporate two

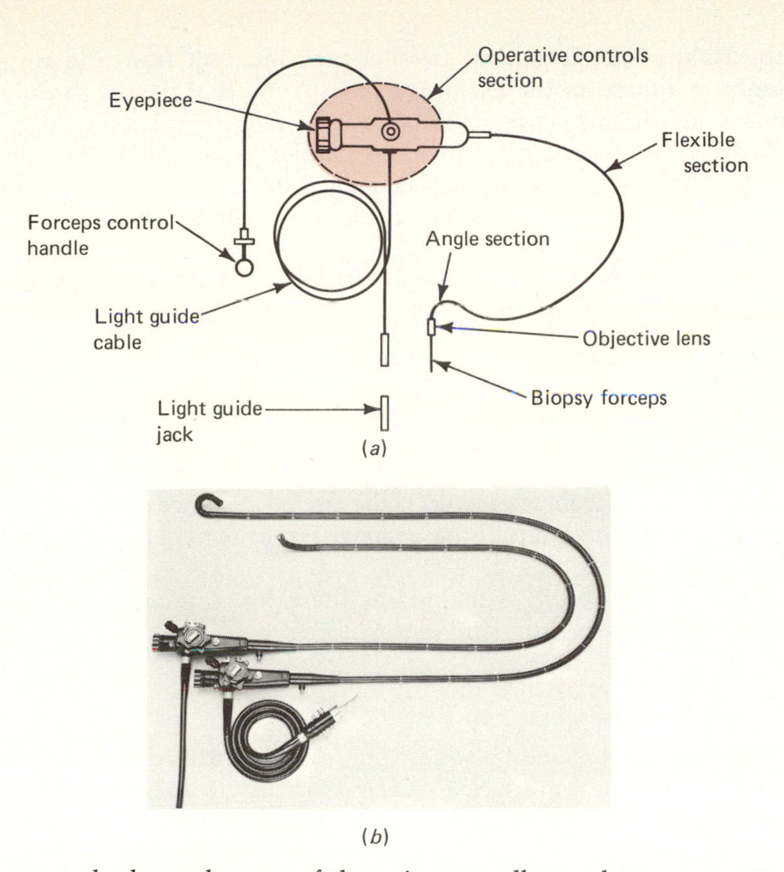

Figure 26.35 (a) Schematic drawing of a fiber-optic endoscope. (b) Photograph of a modern fiber-optic endoscope.

or more such channels; one of these is generally used to carry water to wash the objective lens.

Fiberscopes are also extremely useful in many industrial applications, permitting visual inspection of remote and otherwise inaccessible areas such as the interior of boilers, pumps, and engines.

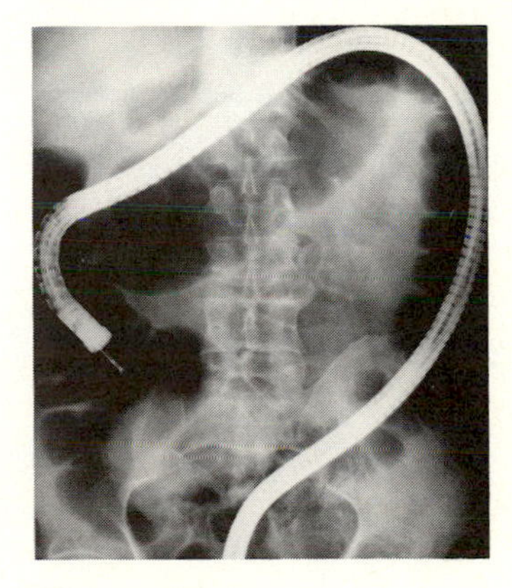

Figure 26.36 An X-ray picture showing a fiber-optic endoscope in the colon. A small forceps extending from the tip of the endoscope is clearly visible.

Summary

The formation of images by mirrors and lenses is most conveniently described in terms of the ray concept.

A plane mirror forms an erect, virtual image of an object. The image size is the same as the size of the object, and the image is located behind the plane mirror. The image distance equals the object distance.

A *concave spherical mirror* brings incident parallel rays to *focus* at the *focal point* of the mirror. The *focal length* f is related to the radius of curvature R of the mirror by

$$f = \frac{R}{2}$$

A *convex spherical mirror* reflects incident parallel rays so that they appear to have originated from a focal point located behind the mirror. The focal length of the convex mirror is also half the radius of curvature. The focal length for this mirror is given a *negative* sign.

A converging lens refracts light rays at its surfaces so that incident rays parallel to the lens axis are brought to a focus on the opposite side of the lens. A lens has two focal points, symmetrically positioned. For the converging lens, parallel rays incident from the left are brought to a focus to the right of the lens, and parallel rays incident from the right are brought to a focus to the left of the lens. A ray that passes through the center of a thin lens proceeds undeviated.

A diverging lens refracts incident parallel rays so that they emerge from the lens in such directions that they appear to originate from the focal point on the side of the lens from which the parallel rays are incident.

For spherical mirrors and lenses, the object distance p, image distance i, and focal length f are related by

$$\frac{1}{p} + \frac{1}{i} = \frac{1}{f}$$

The focal length f is positive for the concave mirror and negative for the convex mirror; f is positive for the converging (convex) lens and negative for the diverging (concave) lens. If the image distance i is positive, the image is real and inverted; if the image distance is negative, the image is virtual and erect.

The magnification m is defined as the ratio of image size to object size. For spherical mirrors and lenses,

$$m = -\frac{i}{p}$$

Spherical mirrors and lenses are subject to various defects, known as *aberrations. Spherical aberration* causes blurring of images because rays parallel to but far from the optic axis are brought to a focus closer to the mirror (or lens) than rays near the optic axis. *Chromatic aberration* arises because the index of refraction of a lens depends on the wavelength of the incident light. Since the focal length of a lens depends on the index of refraction, different colors have different focal lengths.

The human eye's cornea and lens refract incident rays so that they are brought to focus on the retina. In the *myopic* (nearsighted) person, the eyeball is excessively long, and consequently, a sharp image is formed at a point in front of the retina rather than on the retina. The corrective diverging lens forms a virtual, upright, and diminished image of the distant object at an image distance sufficiently close to the eye that the myopic person can see this image clearly. In the *hypermetropic* (farsighted) person, the eyeball is too short. The corrective lens is converging, and it assists the converging optics of the cornea and eye lens to bring the rays from an object to focus on the retina.

A *camera* has a lens of fixed focal length that forms a real, inverted image of an object on the film plane. Focusing is accomplished by moving the lens with respect to the film plane.

A *magnifying glass* is a converging lens. It is so positioned that the lens forms a virtual, upright, and enlarged image of the object to be examined. The magnification is approximately $D/4$, where D is the power of the lens in diopters.

The *compound microscope* consists of an *objective lens* of short focal length and an eyepiece, or *ocular lens.* The magnification of the compound microscope is approximately equal to $LD_oD_e/4$, where L is the length of the microscope barrel in meters and D_o and D_e are the powers of the objective and eyepiece in diopters.

Telescopes are basically two-component systems, containing an objective and an eyepiece. The objective forms a real image of the distant object, which is then viewed with the desired magnification by means of a properly selected ocular lens. Large telescopes use mirrors as objectives instead of lenses.

A *light pipe* is a flexible clear plastic rod that has a high index of refraction. Light is carried from one end of the pipe to the other by total internal reflection at the walls of the pipe. A flexible bundle of very fine light pipes can transmit an image from one point to another around various obstacles.

Multiple Choice Questions

26.1 You are given a concave mirror of 1-m radius of curvature. The image formed by this mirror using a real object

(a) is always real and inverted.
(b) is always virtual and enlarged.
(c) is always virtual.
(d) is always virtual and diminished.
(e) can be real or virtual.

26.2 If a real object is placed just inside the focal point of a diverging lens, the image is

(a) virtual, erect, and diminished.
(b) real, inverted, and enlarged.
(c) real, inverted, and diminished.
(d) virtual, erect, and enlarged.
(e) virtual, inverted, and diminished.

26.3 An object is placed at a distance of $1.5f$ m from a converging lens of focal length f m. The image formed by this lens is

(a) virtual, erect, and larger than the object.
(b) virtual, erect, and smaller than the object.
(c) real, inverted, and larger than the object.
(d) real, inverted, and smaller than the object.
(e) real, inverted, and of the same size as the object.

26.4 As an object is moved from a great distance toward the focal point of a concave mirror, the image moves from

(a) a great distance toward the focal point and is always real.
(b) the focal point toward a great distance from the mirror and is always real.
(c) the focal point to a great distance and is always virtual.
(d) the focal point to a position immediately adjacent to the mirror and is always real.
(e) None of the above is correct.

26.5 The left and right lenses of a pair of eyeglasses have powers of 1.25 diopters and 1.33 diopters, respectively. The left and right lenses are

(a) converging, with focal lengths of 0.8 cm and 0.75 cm, respectively.
(b) converging, with focal lengths of 1.25 cm and 1.33 cm, respectively.
(c) converging, with focal lengths of 80 cm and 75 cm, respectively.
(d) converging, with focal lengths of 1.25 m and 1.33 m, respectively.
(e) none of the above.

26.6 Mark the correct statement.

(a) A concave mirror always forms a virtual image of a real object.
(b) A convex mirror never forms a real image of a real object.
(c) The virtual image formed by a concave mirror is always diminished.
(d) The virtual image formed by a convex mirror is always enlarged.
(e) A concave mirror always forms an enlarged real image of a real object.

26.7 An object is placed 40 cm from a converging lens of 10-cm focal length. The image that is formed will be

(a) real, inverted, reduced.
(b) real, inverted, enlarged.
(c) virtual, upright, enlarged.
(d) virtual, inverted, reduced.
(e) virtual, upright, reduced.

26.8 An object is placed 10 cm from a diverging lens of focal length -20 cm. The image that is formed will be

(a) real, inverted, reduced.
(b) real, inverted, enlarged.
(c) virtual, upright, enlarged.
(d) virtual, inverted, reduced.
(e) virtual, upright, reduced.

26.9 A converging lens has a refractive index of 1.6. When the lens is used in air, its focal length is 20 cm. If it is immersed in water, its focal length is

(a) greater than 20 cm.
(b) unchanged, 20 cm.
(c) less than 20 cm.
(d) negative.
(e) infinite.

26.10 An object is placed 40 cm from a concave mirror, whose radius of curvature is 50 cm. The image will be

(a) real, inverted, enlarged.
(b) real, inverted, diminished.
(c) real, upright, enlarged.

(d) real, upright, diminished.

(e) none of the above.

26.11 Most commonly, the myopic person

(a) has an eyeball that is longer than normal, and a sharp image is formed by the eye in front of the retina.

(b) has an eyeball that is shorter than normal, and a sharp image is formed by the eye in front of the retina.

(c) has an eyeball that is shorter than normal, and a sharp image is formed behind the retina.

(d) has an eyeball that is longer than normal, and a sharp image is formed by the eye behind the retina.

(e) None of the above is true; myopia refers to an asymmetric distortion of the image.

26.12 If an object is placed closer than the near point of a farsighted person,

(a) the image is formed in front of the retina, and a diverging corrective lens will lead to a sharp retinal image.

(b) the image is formed behind the retina, and a diverging lens produces a sharp retinal image.

(c) the image is formed in front of the retina, and a converging lens produces a sharp retinal image.

(d) the image is formed behind the retina, and a converging corrective lens produces a sharp retinal image.

(e) no corrective lens is needed for a sharp retinal image.

26.13 A compound microscope has three objectives (focal lengths f_o) and two eyepieces (focal lengths f_e). For maximum magnification, one should select the combination of objective and eyepiece so that

(a) f_o and f_e are both as large as possible.

(b) f_o is the smallest available, f_e the largest available.

(c) both f_o and f_e are the smallest available.

(d) f_o is the largest available, f_e the smallest available.

(e) f_o is about equal to f_e.

26.14 A projector shows a picture that is in focus but too large for the screen on which it is shown. Rather than move the slide projector closer to the screen, the projectionist may

(a) refocus the projector lens by moving it forward.

(b) refocus the projector lens by moving it backward toward the slide.

(c) replace the projector lens with one of longer focal length.

(d) replace the projector lens with one of shorter focal length.

(e) None of these measures will help; one must get a bigger screen.

Problems

(Section 26.1)

26.1 You stand 0.8 m from a plane mirror, camera in hand, ready to take your own picture. What is the distance to which the camera should be focused?

• **26.2** You have a rectangular mirror. If you hold the mirror 40 cm from your eyes, you can see the entire building that is directly behind you. If the mirror is 12 cm high and the building is 24 m tall, how far are you standing from the building?

• • **26.3** Show, by a ray diagram, that if you look into the vertex of a two-dimensional corner reflector—two vertical mirrors placed at right angles—you will appear in that mirror as you do to other people, that is, without the left-for-right inversion.

(Section 26.2)

26.4 The image of an object viewed in a convex mirror whose focal length is −80 cm appears to be 40 cm behind the mirror. Use a ray diagram to locate the object and to estimate the magnification.

26.5 Locate the image of an object placed (a) 50 cm, (b) 20 cm, from a concave mirror whose focal length is 30 cm.

26.6 Do Problem 26.4 using the mirror equation.

26.7 Do Problem 26.5 using the mirror equation.

• **26.8** If an object is placed 80 cm from a spherical mirror, the image appears upright and has a magnification of 1.5. Characterize the mirror, giving its focal length, and locate the image.

• **26.9** If a spherical mirror forms a real image, at an image distance of 20 cm, of an object placed 40 cm from the mirror, what is the radius of curvature of this mirror?

• **26.10** An object is placed 20 cm in front of a spherical mirror. The image that is formed is twice as large as the object and is erect. Find the image distance and the radius of curvature of the mirror. Is the mirror concave, or is it convex?

• **26.11** A plane mirror always forms a virtual, upright image of the same size as the object. Show that this result is consistent with the mirror equation.

• **26.12** An object is to be placed in front of a concave mirror of focal length 0.5 m so as to give an enlarged upright image. If this image is to be three times the size of the object, what should the object distance be?

• **26.13** A concave mirror has a focal length of 40 cm. An object is located in front of this mirror so that its image is real and its magnification is -2. What is the object distance?

• **26.14** An object is placed in front of a concave mirror of 50-cm radius of curvature so that its magnification is $+1.5$. What are the object distance and image distance?

• • **26.15** A telescope mirror has a focal length of 6 m. What is the diameter of the image of the moon produced by this mirror?

(Sections 26.3, 26.4)

26.16 An object is placed 30 cm from a diverging lens, whose focal length is -20 cm. Use a ray diagram to locate the image and estimate the magnification.

26.17 Do Problem 26.16 using the lens equation.

26.18 An object 10 cm high is placed 30 cm to the right of a lens. The image formed by this lens is inverted and 4 cm high. Find the image distance and the focal length of the lens.

• **26.19** A lens forms an erect image of an object twice the size of the object. The image appears 40 cm from the lens. Determine the object distance and focal length of the lens.

• **26.20** A lens forms an erect image of an object that is half the size of the object. The image appears 20 cm from the lens. Determine the object distance and the focal length of the lens.

• **26.21** A converging lens has a focal length of 40 cm. A 4-cm-tall object is placed 60 cm in front of this lens. Where is the image formed? Is it real or erect? What is its height?

• **26.22** For a thin lens of given geometrical shape, the focal length is proportional to $(n - 1)^{-1}$, where n is the index of refraction of the lens substance. A lens made of fused silica has a focal length of 80.00 cm for light of wavelength 670 nm. What is the focal length of the lens for blue light of wavelength 480 nm? For fused silica, $n(670) = 1.4561$ and $n(480) = 1.4636$.

• • **26.23** A converging lens of 4 diopters is placed 1 m to the left of a plane mirror. An object is positioned 40 cm to the left of the lens. Construct a ray diagram to locate all the images this system forms; then, using the lens and mirror equations, determine the image distances.

• • **26.24** A light bulb and a screen are separated by a distance d. A converging lens of focal length f is placed between bulb and screen. Show that if $d < 4f$, no position of the lens will produce a sharp image on the screen, but if $d > 4f$, there are two positions of the lens for which the bulb will cast a sharp image on the screen. Find the two positions of the lens relative to the location of the bulb.

(Section 26.6)

26.25 A person cannot focus on objects closer than 50 cm from his eyes. What lenses will permit him to read normally?

26.26 An optometrist prescribes glasses with a power of -4 D for his patient. What is the patient's far point without the glasses?

26.27 A myopic person wears corrective lenses of -6.5 D. What is that person's uncorrected far point?

• **26.28** A hypermetropic person uses glasses of 2.5-D power. These bring his near point to 25 cm. As the years pass, he notices that he must now hold reading material at a distance of 35 cm to bring the print into focus. What should be the new prescription for his corrective lenses?

• **26.29** Glasses of $+3$ D restore a girl's vision so that she can focus on objects between infinity and 20 cm. What are her near and far points without these corrective glasses?

• **26.30** A person goes to the optometrist, who prescribes corrective lenses of -40-cm focal length. With the aid of these glasses, that person's far point is at infinity, and the near point is at 20 cm. What are his uncorrected far and near points?

• **26.31** A hypermetropic person wears glasses of $+3.5$-D strength. What is that person's uncorrected near point?

• **26.32** A myopic person who normally wears corrective lenses of -8 D removes his glasses when examining fine print of a document. What is the ratio of the image formed by the letters of the document on the person's retina without glasses and with the corrective glasses?

• • **26.33** A middle-aged woman wears bifocals whose prescriptions are -4.5 D/-3.5 D. Which part refers to the distance part, and which to the reading part of the lenses? What are her far and near points without the corrective lenses?

• • **26.34** An older, myopic man can focus, without glasses, only in the narrow range between 11 cm and 16 cm. Will a single lens correct his vision so that he can see objects at great distances and also focus on objects located just 25 cm from his eyes? If so, how strong should the corrective lenses be? If not, what should be the prescription for his bifocals?

(Sections 26.7–26.8)

26.35 A jeweler's loupe has a focal length of 2 cm. Assuming that the loupe is held so that it forms an image at the normal near point of 25 cm, determine the magnification.

26.36 What is the focal length of a magnifying glass that gives a magnification of 10?

• **26.37** A camera is fitted with a 50-mm lens. Objects within the range from infinity to 1.5 m can be focused on the film plane by properly positioning the lens. Over what distance is the lens moved if the focus is adjusted over the full range?

• **26.38** A 35-mm slide projector is to be used to show slides on a screen 1.5 m wide and placed 6 m from the projector. What should the focal length of the projection lens be so that the 35-mm slide covers the full width of the screen? What is then the slide-to-lens distance?

• **26.39** A 35-mm slide projector has a projection lens whose focal length is 12 cm. If the room in which the projector is to be used is 8 m long, what size screen should be purchased for this projector?

• **26.40** A 35-mm camera is fitted with a standard lens of 55-mm focal length. The focusing adjustment allows focusing on an object between infinity and 1.2 m. When the lens is set for focusing on an object at 1.2 m, what is the distance between lens and film plane?

• **26.41** A camera is fitted with a telephoto lens of 200-mm focal length. If the lens can be moved forward by as much as 16 mm, over what range can one focus this camera?

• • **26.42** Close-up pictures of flowers can be taken using a camera with a standard lens by placing a so-called close-up ring between the camera body and the lens assembly. How thick should the close-up ring for the camera of Problem 26.40 be so that an object 25 cm from the camera will be in focus when the lens assembly is set at the infinity setting? What is then the range of object distance for which a sharp image can be obtained with this lens?

(Sections 26.9–26.10)

• **26.43** A microscope with an 18-cm barrel has an eyepiece whose focal length is 1.2 cm. What focal length of the objective lens would achieve an overall magnification of 360?

• **26.44** A reflecting telescope uses a concave mirror of 16-m radius of curvature. For an overall magnification of 400, what focal length should the eyepiece have?

• **26.45** What is the magnification of an astronomical telescope that has an objective lens of focal length 1.4 m and an eyepiece of 4 D?

• **26.46** A pair of binoculars has a magnification of 8. The effective length of the barrel (distance between objective lens and ocular) is 22 cm. What focal lengths should the objective and ocular lenses have?

• **26.47** A microscope is constructed using two converging lenses of focal lengths 0.4 cm and 1.6 cm, mounted in a tube of 15-cm length. Draw a ray diagram, showing the location of object and final image, and determine the magnification of this instrument.

• • **26.48** A compound microscope has interchangeable objective lenses whose focal lengths are 1.5 mm, 4 mm, and 8 mm. If the separation between objective and eyepiece is 20 cm and the focal length of the eyepiece is 2.5 cm, what are the approximate magnifications of the instrument? What should the object distance be when the 8-mm objective lens is used?

• • **26.49** The telescope Galileo used had a converging objective and a *diverging* lens as the eyepiece. Use a ray diagram to show that by placing the diverging lens between the converging lens and its focal point, the combination of lenses forms a virtual, enlarged, and upright image of a distant object.

•27•

Physical Optics

But soft! what light through yonder window breaks . . .

WILLIAM SHAKESPEARE, *Romeo and Juliet*

27.1 Introduction

Until 1800 the Newtonian corpuscular theory of light was dominant. Then Thomas Young in England, and Augustin Jean Fresnel in France, performed a series of definitive experiments that appeared to settle the issue conclusively in favor of the wave theory. The class of experiments that demonstrate the presence of waves relies on interference of two or more waves as they combine according to the superposition principle. We have already described several effects of superposition of waves in connection with sound waves. We saw then that when two waves of equal amplitude and frequency combine at some point in space, the resultant displacement of the medium may be zero, may be twice that due to each individual wave, or may have any intermediate value, depending on the *phase relation* between the two waves. Superposition of two waves of equal frequency and amplitude and traveling in opposite directions produces a standing wave. From this one is led directly to resonances on strings and in pipes. Finally, beats arise from the superposition of two waves of slightly different frequencies.

None of these effects would be evident if sound energy propagated as a stream of tiny particles that strike the eardrum. For example, two identical particles, traveling in the same direction and arriving at the eardrum simultaneously would simply produce a disturbance twice as great as that caused by one particle. There is no conceivable mechanism by which the two independent, identical particles could interact so as to cancel each other's effect. We intend to show that each of the phenomena mentioned above as well as numerous others we have not considered for sound waves—all a direct consequence of superposition of waves—can be observed with visible light. Such experiments then force us to accept the concept of light as a wave phenomenon.

27.2 Coherence

If the effect of superposing two waves with a well-defined phase difference is to be observed, the two waves not only must be of the same frequency but also *must maintain their phase relation during the time required for observation.* If the phase relation fluctuates in a random manner while we make our measurement, we shall observe only the *average* of effects due to the superposition of two waves. Two sources that maintain the relative phase between the emitted waves are said to be *coherent.*

Light radiating from an ordinary source such as an incandescent, a neon, or a carbon arc lamp results from the transition of the atoms of the source between various energy states (see Chapter 30). A typical source contains many billions of such excited atoms, each capable of generating a brief light pulse lasting about 10^{-8} s. Since these emissions are uncorrelated in time, the light waves from two such separate sources are also uncorrelated in phase, and the phase relation between these waves is constantly changing. Over a time interval long enough to permit visual or photographic examination, any interference effects would have shifted randomly many thousand times, and none would be observed.

To obtain two or more coherent sources in the laboratory, one employs one of two techniques, *wavefront splitting* or *amplitude splitting.* In each, one starts with a single source of light, for example, a narrow slit placed in front of an intense source. Cylindrical waves then propagate from this slit.

To obtain two coherent sources by wavefront splitting, one places a second screen with two narrow slits some distance beyond the first. If these slits are positioned symmetrically as in Figure 27.1, the waves arriving at the slits will be exactly in phase. These two slits now serve as two *coherent sources* of new cylindrical waves. Note that when a second wave train, displaced from the first by some arbitrary phase lag, arrives at the primary slit some 10^{-8} s later, the two secondary slits will still emit as coherent sources.

Instead of two slits, one could have interposed a half-silvered mirror, a flat piece of glass on which a very thin layer of silver or aluminum has been deposited by evaporation. The reflecting layer is so thin that only half the incident light is reflected, the rest being transmitted through the glass. The two beams are coherent, since they derive from the same source. They can then be projected onto the same region by means of one or more mirrors. Commercial instruments that require two coherent beams generally use this amplitude-splitting method.

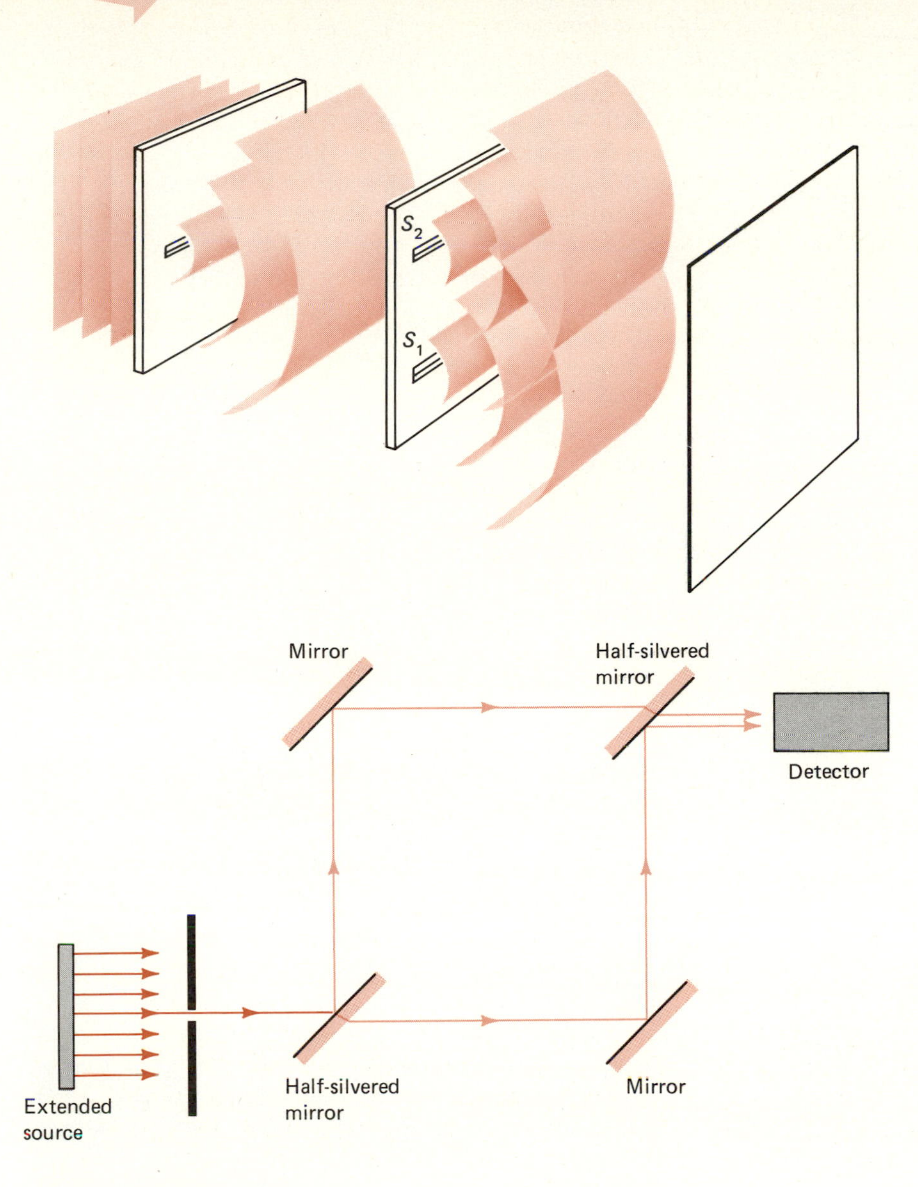

Figure 27.1 *Achieving two coherent sources by wavefront splitting. Plane waves from a distant source are incident on the first slit. Cylindrical waves from this slit impinge on the two slits of the second screen. The waves emanating from those slits are coherent and can interfere with each other.*

Figure 27.2 *Achieving two coherent beams by amplitude splitting. Light from a source is passed through a slit and is incident on a half-silvered mirror. The two beams, the transmitted and the reflected beams coming from that mirror, are now coherent and will interfere when brought together. The device shown here is called a Mach-Zehnder interferometer.*

27.3 Interference

27.3 (a) Double-Slit Interference

Now that we know how coherent sources can be obtained, let us consider the effect of superposition of coherent light from two closely spaced slits S_1 and S_2 falling on a screen placed some considerable distance from these slits, as shown in Figure 27.3. In a typical situation, the separation between the slits, d, might be about 1 mm, and the distance to the screen, D, several meters; Figure 27.3 is therefore a schematic, not an accurate scale drawing.

Since the waves radiating from the two slits are in phase, they will also be in phase when they reach the point O on the screen, which is equidistant from each slit. The resultant amplitude at O will therefore be $2A$, where A is the amplitude of each individual wave. Consider next the amplitude at some off-center point such as P. In traveling to this point,

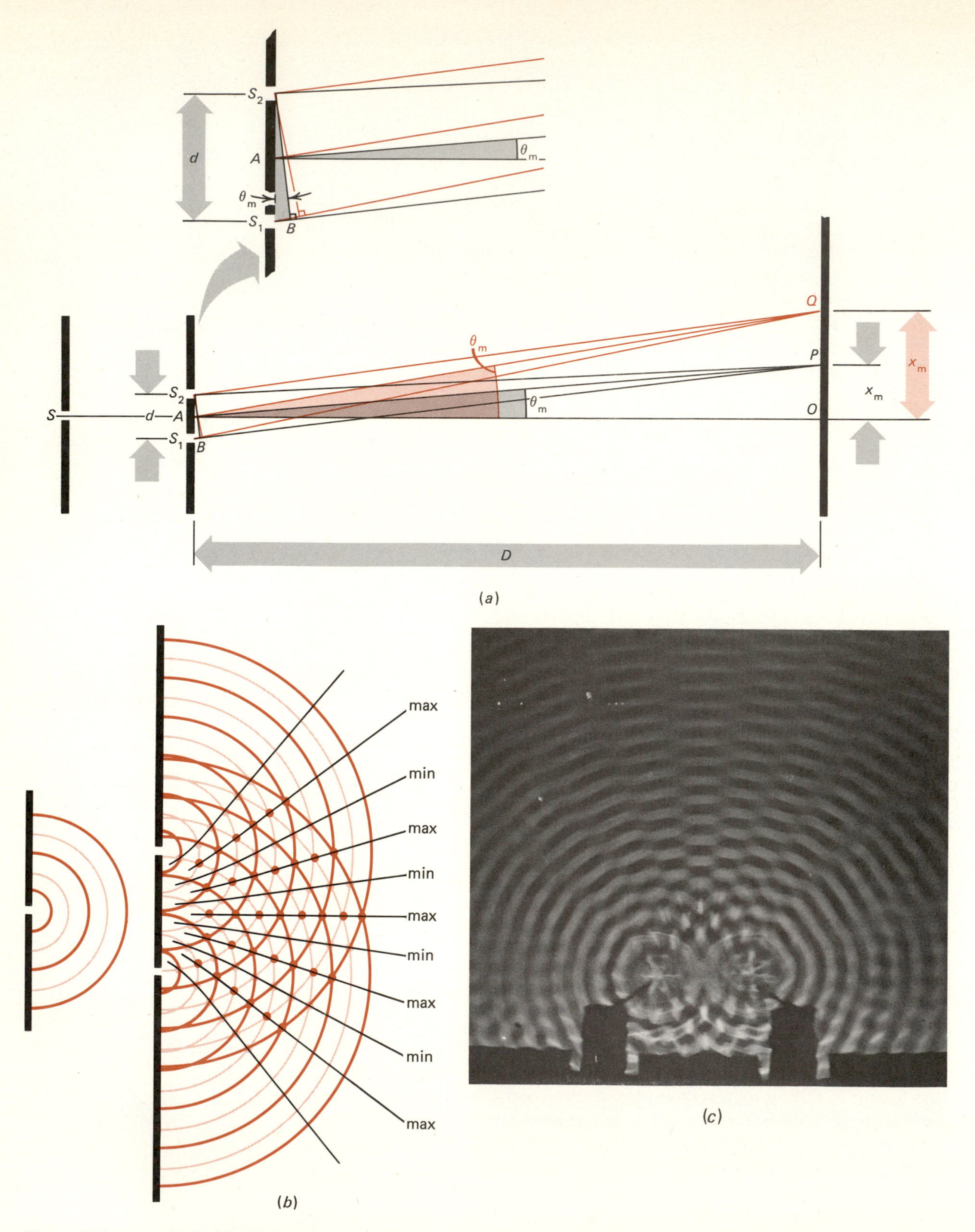

Figure 27.3 Young's double-slit interference experiment. (a) Schematic diagram, showing the path differences from slits S_1 and S_2 to the screen that result in an interference pattern on the distant screen. If the angle θ_m is such that $S_1P - S_2P = (m + \frac{1}{2})\lambda$, the waves from slits S_1 and S_2 interfere destructively at P; if θ_m is such that $S_1Q - S_2Q = m\lambda$, the waves from slits S_1 and S_2 interfere constructively at Q. (b) A diagrammatic picture showing how the Huygens waves from the two slit sources interfere, giving maxima and minima along different directions. (c) The interference patterns from two coherent sources in a ripple tank.

the wave from S_1 has covered a distance slightly greater than that covered by the wave from S_2. If this path difference, BS_1 in Figure 27.3, is exactly one half-wavelength, $\lambda/2$, the two waves arrive at P 180° out of phase, and consequently, the resultant amplitude is zero.

At the point Q, the waves will again be exactly in phase provided the path difference between QS_1 and QS_2 is one complete wavelength; at that point the amplitude will then be $2A$ once again.

At P, the waves interfere *destructively*, at O and Q they interfere *constructively*. The condition for *constructive interference is that the path difference be an integral number of full wavelengths; for destructive interference, the path difference must be an odd number of half-wavelengths.*

Since D is much greater than the slit separation d, the lines AP, S_1P, and S_2P are approximately parallel. The line S_2B, drawn perpendicular to S_1P, is then also perpendicular to AP. Also, as corresponding sides of triangles S_1S_2B and APO are mutually perpendicular, these triangles are similar. Consequently the path difference $S_1P - S_2P = S_1B$ is then given by

$$S_1B = d \sin \theta$$

where θ is the angle subtended by OP at the slits. We therefore obtain the following conditions for constructive and destructive interference:

Constructive Interference

$$\sin \theta_m = \frac{m\lambda}{d} \qquad m = 0, 1, 2, 3, \ldots \tag{27.1}$$

Destructive Interference

$$\sin \theta_m = \frac{(m + \frac{1}{2})\lambda}{d} \qquad m = 0, 1, 2, 3, \ldots \tag{27.2}$$

From Figure 27.3, the positions on the screen at which maxima and minima appear are given by

$$x_m = D \tan \theta_m \tag{27.3}$$

where x_m is the distance measured from the point O, and θ_m is the angle given by Equations (27.1) for maxima and (27.2) for minima.

Finally, if the slit separation is much greater than the wavelength, that is, if $d \gg \lambda$, then $m\lambda/d$ will be very small for small integral values of m. Under these conditions

$$\sin \theta_m \simeq \tan \theta_m$$

and if, moreover, we express the angle θ_m in radians,

$$\sin \theta_m \simeq \tan \theta_m \simeq \theta_m$$

Equations (27.1), (27.2), and (27.3) then give:

Constructive Interference

$$\theta_m = \frac{m\lambda}{d} = \frac{x_m}{D} \qquad x_m = \frac{m\lambda D}{d}$$

$$m = 0, 1, 2, 3, \ldots \tag{27.4}$$

Destructive Interference

$$\theta_m = \frac{(m + \frac{1}{2})\lambda}{d} = \frac{x_m}{D} \qquad x_m = \frac{(m + \frac{1}{2})\lambda D}{d}$$

$$m = 0, 1, 2, 3, \ldots \tag{27.5}$$

where θ_m is in radians.

The interference pattern on the screen will exhibit a series of regularly spaced intensity maxima and minima, as shown in Figure 27.4. This intensity pattern generated by double-slit interference is readily derived. We recall that the intensity of a wave is proportional to the square of its amplitude; we further recall that the amplitude of the resultant of two sinusoidal waves is given by Equation (15.7).

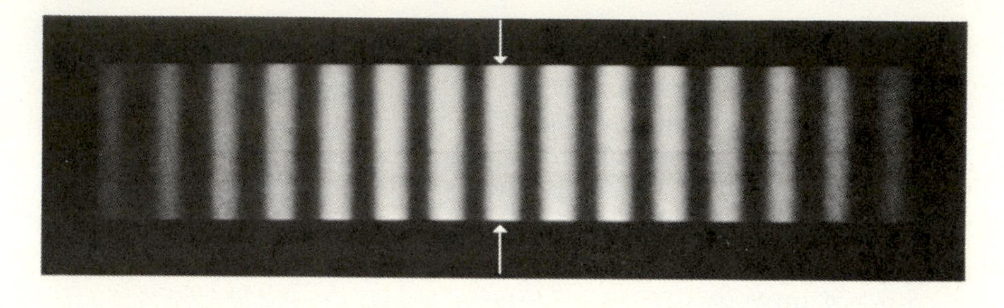

Figure 27.4 *Interference fringes produced by a double slit illuminated with monochromatic coherent light. The arrow marks the central maximum.*

We denote the electric fields of the two waves at the point of interest on the screen by

$$E_1 = E_0 \sin(\omega t) \qquad \text{and} \qquad E_2 = E_0 \sin(\omega t + \phi)$$

The phase difference ϕ arises because of the path difference, as described before. Since a path difference of one wavelength corresponds to a phase difference of 2π radians, we have

$$\phi = \frac{2\pi}{\lambda}(\text{path difference}) = \frac{2\pi}{\lambda} d \sin\theta \tag{27.6}$$

From Equation (15.7), we obtain

$$E = E_1 + E_2 = 2E_0 \cos\frac{\phi}{2} \sin\omega t$$

and for the intensity,

$$I = 4I_0 \cos^2\frac{\phi}{2} \tag{27.7}$$

where I_0 is the intensity of the light at the screen that is due to one slit only and is proportional to E_0^2.

If we again make the small angle approximation, replacing $\sin\theta$ by θ, and use Equation (27.6), we obtain

$$I = 4I_0 \cos^2\left(\frac{\pi d\theta}{\lambda}\right) = 4I_0 \cos^2\left(\frac{\pi dx}{\lambda D}\right) \tag{27.8}$$

This intensity pattern is shown in Figure 27.5. Notice that the average intensity obtained from the interference pattern is the same as that from two *incoherent* slit sources, namely, $2I_0$.

The following qualitative features of double-slit interference deserve emphasis.

1. For a given, fixed slit separation, the distance between adjacent fringes is proportional to the wavelength of the radiation. (See color insert facing p. 624.)

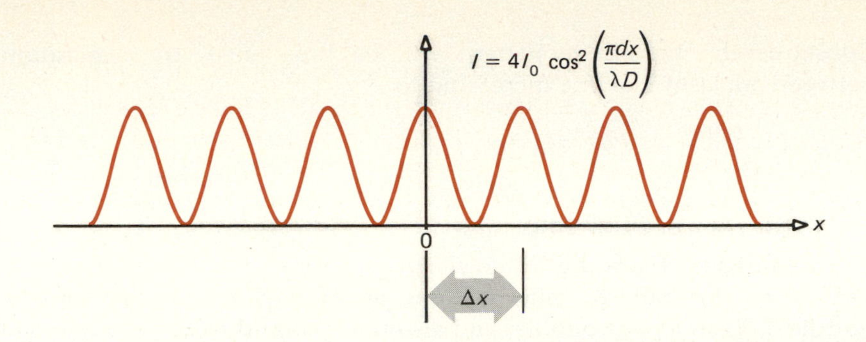

Figure 27.5 The theoretical intensity pattern due to double-slit interference. I_0 is the intensity that would be observed if only one slit were open. Intensity maxima occur whenever $\cos(\pi dx/\lambda D) = \pm 1$; i.e., when $(\pi dx/\lambda D) = m\pi$. Hence, the separation between adjacent maxima is $\Delta x = \lambda D/d$.

2. At constant wavelength, the distance between fringes is inversely proportional to the slit separation; the distance between adjacent fringes increases as the slit separation is reduced.

● *Example 27.1* A pair of slits is illuminated with coherent yellow light from a sodium vapor lamp, $\lambda = 589$ nm. On a screen 1.5 m from the slits, an interference pattern is formed, and measurement of the fringes reveals that the separation between adjacent maxima is 1.6 mm. What is the separation between the two slits? If the mask containing the two slits is rotated about an axis parallel to the slits through an angle of 45°, how will the fringe pattern change?

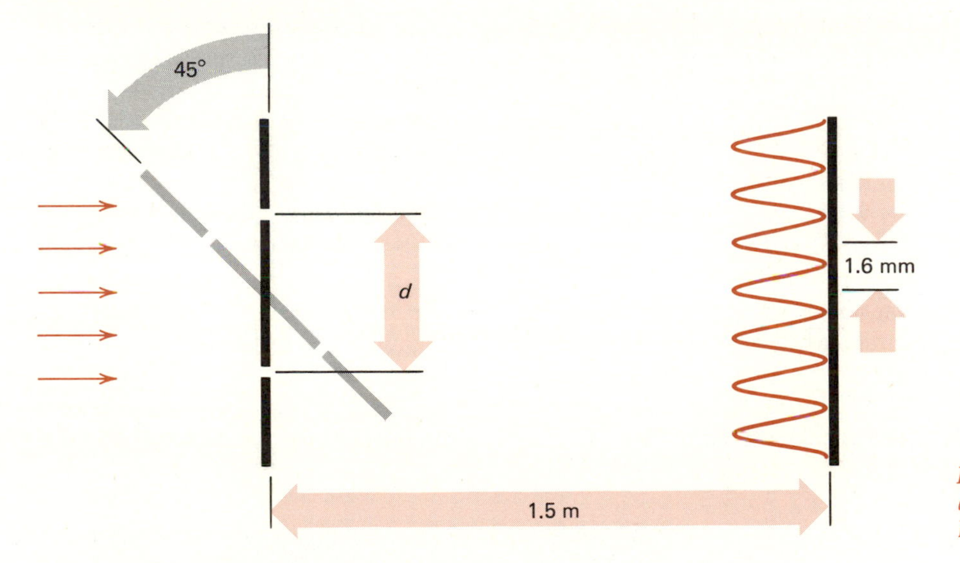

Figure 27.6 Sketch of the double-slit arrangement for Example 27.1.

From Equation (27.4) and also, Figure 27.5, we see that the separation between adjacent maxima is given by $\Delta x = \lambda D/d$, and thus,

$$d = \frac{\lambda D}{\Delta x} = \frac{(5.89 \times 10^{-7}\ \text{m})(1.5\ \text{m})}{1.6 \times 10^{-3}\ \text{m}} = 5.5 \times 10^{-4}\ \text{m}$$

If the mask is then rotated through 45°, the path length of the wave that penetrates the upper slit is increased after it passes through the slit. That increase in path length is exactly compensated by the increase in path length before the mask of that portion of the wavefront that passes through the lower slit. Consequently, the position of the central maximum is not affected by that rotation. The effective slit width, however, is reduced to

$$d = (0.55\ \text{mm})(\cos 45°).$$

Consequently, the fringe pattern will be broadened, the separation between adjacent maxima increasing to

$$\Delta x = \frac{1.6 \text{ mm}}{\cos 45^\circ} = 2.26 \text{ mm}$$

27.3 (b) Thin-Film Interference

Interference effects are by no means restricted to the optics laboratory. The color patterns reflected from an oil slick on the surface of a puddle or from a soap bubble, and also the splendid iridescent colors of hummingbird, pheasant, and mallard feathers and of insect wings, are all manifestations of interference.

To understand how such colors appear, consider the reflection of monochromatic light from a thin film of uniform thickness, such as that shown in Figure 27.7(*a*). We shall simplify matters by assuming that the film is illuminated at nearly normal incidence. Part of the light will be reflected at the front surface; the rest is transmitted to the rear of the film. There a portion will again suffer reflection. Although a fraction of this reflected light is reflected once again on reaching the front interface, we disregard this complication and assume that all the light reflected at the rear surface is transmitted through the front surface.

The two beams, *1* and *2* of Figure 27.7(*a*), are coherent and differ in phase by an amount that depends on the film thickness and its index of refraction. However, before we calculate the phase difference associated with the additional path traversed by beam *2*, we first note one aspect of reflection of light at interfaces separating media of different refractive indices.

We recall (see Section 15.4) that a wave on a string is reflected with inverted phase if the end of the string is fixed, but experiences no phase reversal on reflection at an unsupported end. For light waves, a 180° phase change occurs on reflection if light strikes a medium whose index of refraction is greater than that of the incident medium. However, in the reverse case—internal reflection, as it is sometimes called—the reflected wave is in phase with the incident.

In Figure 27.7(*a*) therefore, beam *1* suffers a 180° phase change on reflection; beam *2* does not. To this fixed phase change, we must now add a

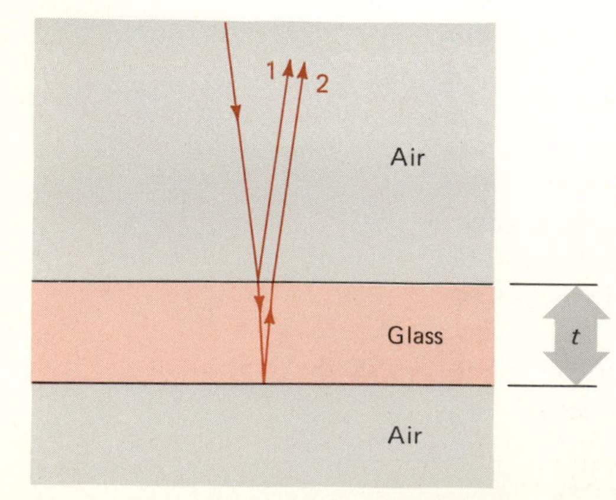

(*a*)

(*b*)

Figure 27.7 *Thin-film interference: an example of interference by amplitude splitting. (a) The part of the incident beam reflected by the top surface of a thin glass plate interferes with the part of the beam reflected by the bottom surface of this plate. (b) Light and dark interference fringes in the light reflected by a soap film. The region near the top of the film is so thin that it appears black; that is, the light reflected from front and back layers interferes destructively.*

change associated with the distance traversed by beam 2 in the medium of refractive index n.

We recall that the wavelength of light in a medium of refractive index n is $\lambda_n = \lambda/n$, where λ is the wavelength in vacuum (or very closely, air). Now in going from the top to the bottom surface and back up, beam *2* traverses a total additional distance $2t$. The number of wavelengths corresponding to this path length is $2t/\lambda_n$. Since a path difference of one wavelength corresponds to a phase change of 2π, this path difference gives a phase change

$$\phi' = 2\pi\left(\frac{2t}{\lambda_n}\right) = \frac{4\pi tn}{\lambda}$$

Thus the *total* phase change between beams *1* and *2*, including the π radians due to reflections, is

$$\phi = \pi\left(1 + \frac{4tn}{\lambda}\right) \tag{27.9}$$

We shall observe constructive interference whenever this total phase shift is an integral multiple of 2π, that is, whenever $tn/\lambda = \frac{1}{4}$ or an odd multiple of $\frac{1}{4}$. Hence,

Constructive Thin Film Interference

$$t = \frac{(m + \frac{1}{2})\lambda}{2n} \qquad m = 0, 1, 2, 3, \ldots \tag{27.10}$$

● ***Example 27.2*** A soap film whose index of refraction is $\frac{4}{3}$ is 300 nm thick. If this film is illuminated at normal incidence with white light, what is the color of the reflected light?

From Equation (27.10), we find that constructive interference on reflection occurs for wavelengths

$$\lambda = \frac{2nt}{(m + \frac{1}{2})} = \frac{2(1.33)(300 \text{ nm})}{(m + \frac{1}{2})} = 1600 \text{ nm}, 533.3 \text{ nm}, 320 \text{ nm}, \ldots$$

Of these wavelengths, only the second falls in the visible range; therefore, this film reflects predominantly green light, and appears green when viewed in white light. ●

An interesting prediction that follows from Equation (27.9) is that if t is small enough, incident light will interfere *destructively* over the entire visible spectrum. Destructive interference is dominant if $t \ll \lambda$ because then the only substantial phase difference between the two beams of Figure 27.7(*a*) is that due to reflection of beam *1* at the front interface. Very thin films, therefore, appear black when viewed by reflected light.

Black lipid films have been the object of intensive study by biophysicists in recent years because the membranes that bound cells and cell organelles are bimolecular lipid films about 5–10 nm thick. It is now generally recognized that these membranes play an essential role in all vital processes, serving as the substrates to which various enzymes and other proteins are attached. The ionic concentration of the cellular (cytoplasmic) fluid is controlled by the selective permeability of the surrounding membrane, and brief changes in this permeability are responsible for the appearance and propagation of nerve impulses that provide the basic communication mechanism in the body. The hope is that study of artificial lipid films, whose composition can be carefully controlled,

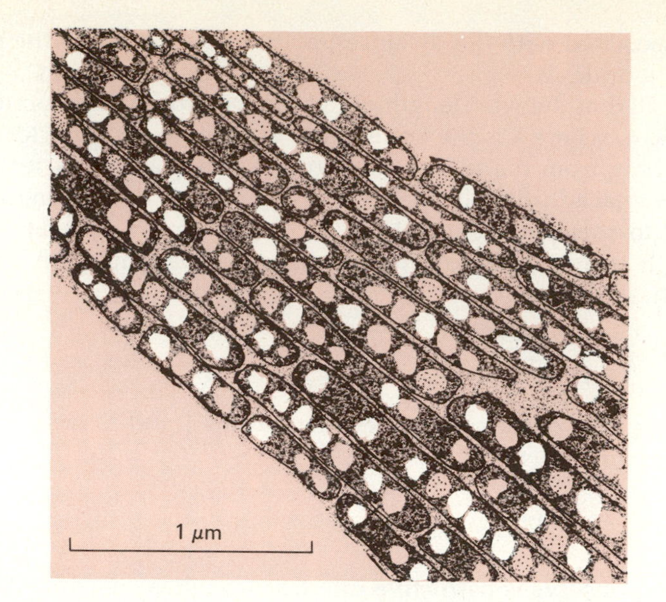

Figure 27.8 *Iridescent colors of bird feathers result from interference similar to the thin-film interference of a soap film. In these feathers, the interference arises from reflection of light by regularly spaced layers of melanin fibers, as shown in the drawing from an electron micrograph of the cross sections of a feather taken from a hummingbird.*

will illuminate the properties of the much more complex biological membranes.

More familiar biological manifestations of interference are the spectacular colors displayed by the plumage of many birds. The cross section of such feathers, when viewed under the electron microscope, reveals a regular arrangement of layers of melanin fibers. Light reflected from successive layers will interfere constructively provided that the proper phase relation is satisfied. The feathers thus appear to be brightly colored although they contain no pigmentation. Moreover, since the condition for constructive interference depends on the angle of incidence, the color pattern changes somewhat with the viewing angle.

The fact that a very thin film appears black on reflection implies that all the incident light is transmitted (except for an insignificant fraction that may be absorbed). It follows that it should be possible to devise a thin-film coating to reduce reflection from a glass surface.

Suppose we coat a sheet of glass with a film of material that has an index of refraction intermediate to that of glass and air. In that case, reflection at *both* film surfaces will give rise to a 180° phase change on reflection. Consequently, the condition for *destructive* interference is then the same as for constructive interference of the unsupported film, i.e. Equation (27.10). Therefore, if the film has a thickness $t = \lambda/4n$, where n is the index of refraction of the coating material, no light is reflected and all the incident light is transmitted into the glass.

It is obviously impossible to make a coating that will appear black over the entire visible spectrum; in practice, a compromise is sought, and both blue and red light are partially reflected. Such *quarter-wave coatings* are now commonly applied to all camera and binocular lenses, giving them their characteristic purple (blue plus red) color.

● ***Example 27.3*** Magnesium fluoride, whose index of refraction is 1.38, is frequently used to coat lenses. How thick should this coating be for maximum transmittance at a wavelength of 530 nm?

The coating thickness should be

$$t = \frac{\lambda}{4n} = \frac{530 \text{ nm}}{4(1.38)} = 96 \text{ nm}$$

●

27.3 (c) Interferometers

Thin-film interference is one of many examples of interference due to amplitude splitting. Most laboratory interferometers—the one Michelson devised is an example—use this technique for obtaining two coherent beams. Figure 27.9 shows this instrument, which consists of two front-surfaced mirrors M_1 and M_2, a half-silvered mirror M_h, and a compensator plate C. The half-silvered mirror is often called a *beam-splitter,* since part of the light from the source slit S is reflected by M_h to M_2, while the rest is transmitted through C to M_1. Some of the light reflected by M_2 is then transmitted through M_h, and part of the light reflected by M_1 is also reflected by M_h. The compensator plate C is of the same thickness and material as the supporting substrate of the half-silvered mirror, so that both beams traverse the same material in this interferometer.

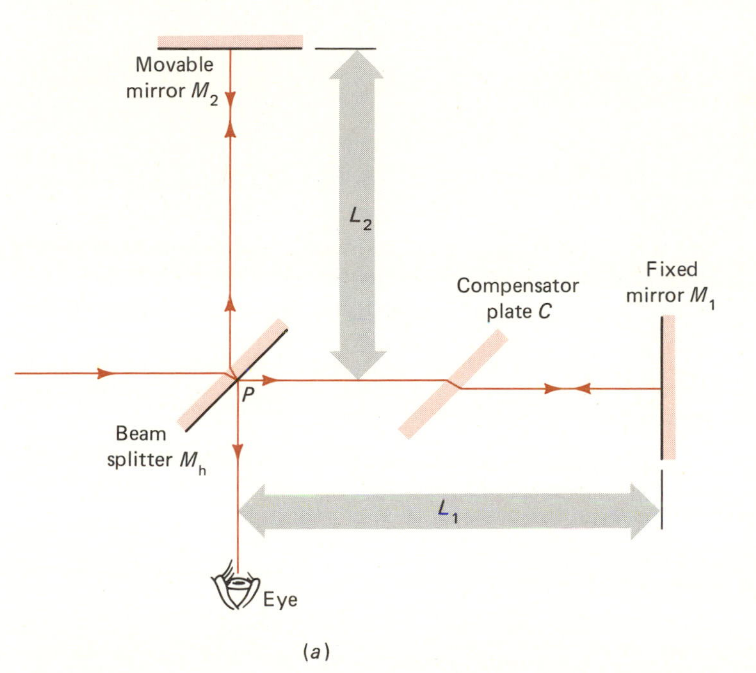

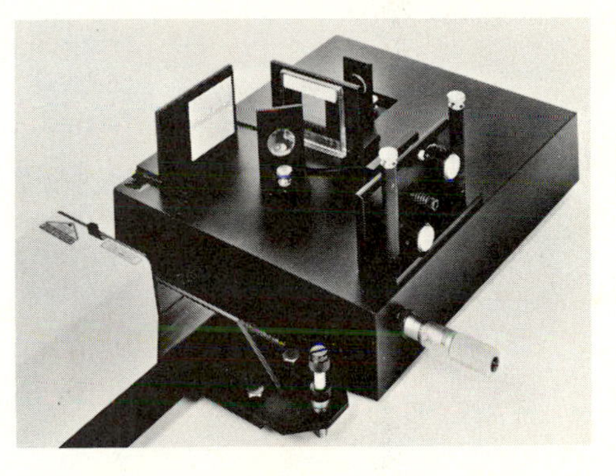

Figure 27.9 *The Michelson interferometer. (a) Schematic diagram, showing the beam splitter, compensator plate, and two mirrors. (b) A photograph of a typical student laboratory interferometer.*

The two beams, which are coherent since they originated from the same source slit, now interfere—constructively or destructively depending on their phase relation. Generally, one of the mirrors, for example M_1, is fixed; the other may be moved by means of a very fine worm gear. A shift in the position of M_2 by half a wavelength changes the path length of beam 2 by one full wavelength, and shifts the observed pattern by one complete fringe, from say a minimum through a maximum to the next minimum. It follows that by the simple expedient of counting fringes as M_2 is slowly moved through a known distance, the wavelength of a monochromatic source can be determined. A more important application of the interferometer, however, is measuring the index of refraction.

Suppose we insert in one path a glass tube which can be evacuated and then slowly filled with gas. Since the index of refraction of the gas is greater than unity, the wavelength of light within the tube of length L will be smaller when it is filled with gas than when it is evacuated.

Suppose the index of refraction of the gas is $n = 1 + \epsilon$, where ϵ is small compared with 1 (for example, for air at atmospheric pressure $\epsilon = 0.0003$). We now have $\lambda_{gas} = \lambda_0/n = \lambda_0/(1 + \epsilon)$, where λ_0 is the

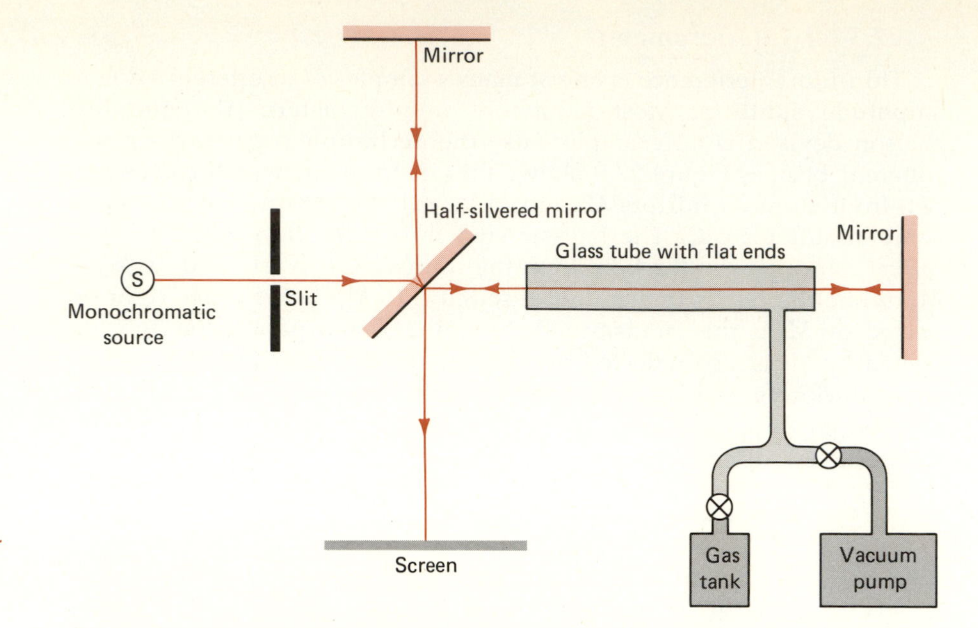

Figure 27.10 *Schematic diagram of an interferometer for measuring the refractive index of a gas.*

wavelength in vacuum. The total path length traversed by the light in the tube is $2L$. If now $2L/\lambda_0 = N$, then $2L/\lambda_{gas} = (1 + \epsilon)N$. Consequently, as the gas is slowly admitted to the tube, one will observe a total of ϵN fringes move across the screen on which the interference pattern is displayed. Since L and λ are known, ϵ and therefore n are readily ascertained.

● ***Example 27.4*** The index of refraction of sulfur dioxide at room temperature and atmospheric pressure is 1.000686 at $\lambda = 589$ nm. If a 30-cm-long glass tube is placed in the fixed arm of a Michelson interferometer and filled with SO_2 at one atmosphere, how many fringes will move across the screen as the tube is slowly evacuated?

In this case

$$\frac{2L}{\lambda} = \frac{6 \times 10^{-1}\text{ m}}{5.89 \times 10^{-7}\text{ m}} = 1.019 \times 10^6$$

and $\epsilon = 6.86 \times 10^{-4}$. Hence a total of $(1.019 \times 10^6)(6.86 \times 10^{-4}) = 699$ fringes will move across the screen. Clearly this technique allows precise determinations of refractive indices of gases. ●

The most celebrated use of the Michelson interferometer was in the historic experiment that put to rest the various "ether theories" of the medium through which light propagates. It demonstrated the absence of the expected "ether wind drift," thus raising the curtain on the special theory of relativity (see Chapter 28).

One further application of the interferometer deserves mention. This is the stellar interferometer, which Michelson proposed initially in 1890; it is shown schematically in Figure 27.11. The purpose of the instrument is to measure stellar diameters and the separation between close double stars.

The two movable mirrors M_1 and M_2 reflect the incident light from a distant source to mirrors M_3 and M_4 and hence through slits S_1 and S_2 mounted in front of the objective of a telescope. The slits S_1 and S_2 generate the normal, Young double-slit interference pattern on the screen at the focal plane of the telescope objective.

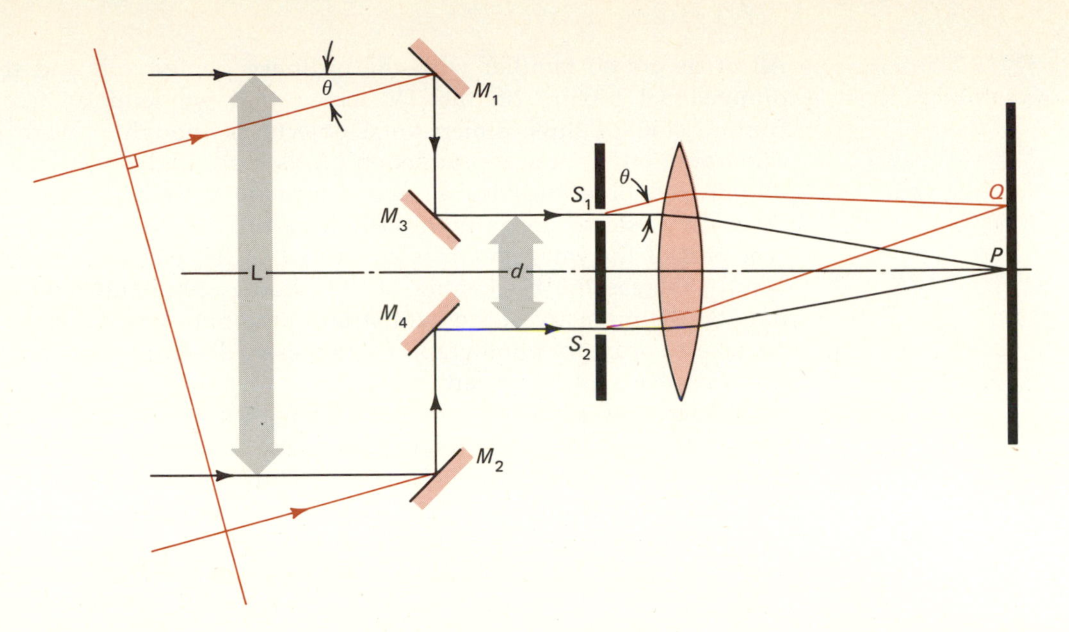

Figure 27.11 The Michelson stellar interferometer.

Suppose the instrument is directed at a double-star system, with two stars of equal luminosity, so that one star is exactly along the instrument axis. Light from that star will generate a fringe pattern such as that of Figure 27.4 with the central maximum located at P. To simplify matters, we assume that the incident light from both stars is monochromatic and of the same wavelength (one can, for example, place a filter in front of the telescope objective). Light rays from the companion star make a small angle θ with the instrument axis, and hence, when the plane waves arrive at mirrors M_1 and M_2, they have a small phase difference, given by $L\theta/\lambda$ (θ is in radians, and we have used the approximation $\sin\theta = \theta$, valid for $\theta \ll 1$). Again the two slits will yield a fringe pattern, now centered not at P but at Q, where $PQ = (L\theta/\lambda)f_o$ and f_o is the focal length of the objective. Since the light from the *two stars* is *incoherent,* these two sources do *not* interfere with each other. The intensity on the screen is therefore simply the sum of the intensities from the two independent fringe patterns.

The operation of the instrument relies on the fact that by changing the mirror separation L, one can shift the two interference patterns relative to each other so that the maxima of one coincide with the minima of the other. When this happens, the observer sees a uniform illumination on the screen, a condition satisfied when $L = \lambda/2\theta$.

Though the mathematical analysis is complicated, one can show that this technique can also be used to determine the angle ϕ subtended by the finite-sized disk of a single star. It was thus that in 1920 ϕ for α-Orion, better-known as Betelgeuse, was measured and found to be 0.047 second of arc. From the known distance of α-Orion (determined from stellar parallax), the diameter of this red giant was calculated to be about 250 million miles, or substantially greater than Earth's orbit about the sun.

Stellar interferometry has advanced greatly since the days of Michelson, largely as a result of new imaginative concepts (correlation interferometry) that cannot be adequately explained here. Suffice it to say that stars with angular diameters of only 0.0005 second of arc (one hundredth of α-Orion's) have now been measured.

*27.4 Holography

All of us are so familiar with the ordinary photograph and the two-dimensional picture on the TV screen that we tend to accept this compression of three-dimensional objects as though it were a basic constraint on optical reproduction. It is not! Light scattered by an illuminated object carries all the information needed to re-create a three-dimensional image if only we make full use of it. The difficulty is that part of the information is contained in the *phase* of the scattered wave, whereas the blackening of the photographic emulsion depends only on the intensity of the light striking the film. Consequently, when we take an ordinary photograph we carelessly discard whatever information the phase might impart.

Holography is a technique that uses both intensity and phase information in recording and reconstructing the image. To do this requires illumination by a coherent light source. Until the development of lasers (see Section 30.7), coherent light sources with useful intensities for photography were almost nonexistent, and holography, though invented in 1947, remained virtually unknown for 15 years. How does the process work?

Figure 27.12 shows one common arrangement. Part of the coherent beam of light, the reference beam, is incident on a mirror and reflected to the emulsion, while the remainder illuminates the object. Light scattered by the object also strikes the emulsion, and at every point on that emulsion, the scattered light and reference beam interfere according to the superposition principle. Thus the intensity of illumination at any given point on the emulsion depends on the relative phases and amplitudes of the scattered and reference waves. The phase of the scattered wave depends, however, on the distance between the object and the emulsion and thus carries the three-dimensional information we seek.

When we develop the exposed film we find that the photograph bears no resemblance to the object whatever. The film has a mottled, gray appearance and under magnification reveals an apparently random distribution of light and dark regions. To view the hologram, we must illuminate the film with coherent light so that the light scattered by the emulsion can interfere and yield the reconstructed image. You can then "see" the object as though it were truly in place at the image position; and moving your head from side to side, you can observe the object shifting relative to the background and revealing parts that were previously obscured. Moreover, if the original object has reasonable depth, you have to adjust the focus of your eye as you look at different portions of the holographic image. Figure 27.14 can give but a very inadequate impression of holographic realism.

27.5 Diffraction

27.5 (a) Diffraction Grating

There is really no fundamental difference between interference and diffraction; both are consequences of superposition of waves. It has become customary, however, to refer to effects arising from superposition of a few waves as interference, and to speak of diffraction when many waves are involved. Thus two slits are said to produce an interference pattern, and a mask containing a large number, say several thousand slits, is called a *diffraction grating*.

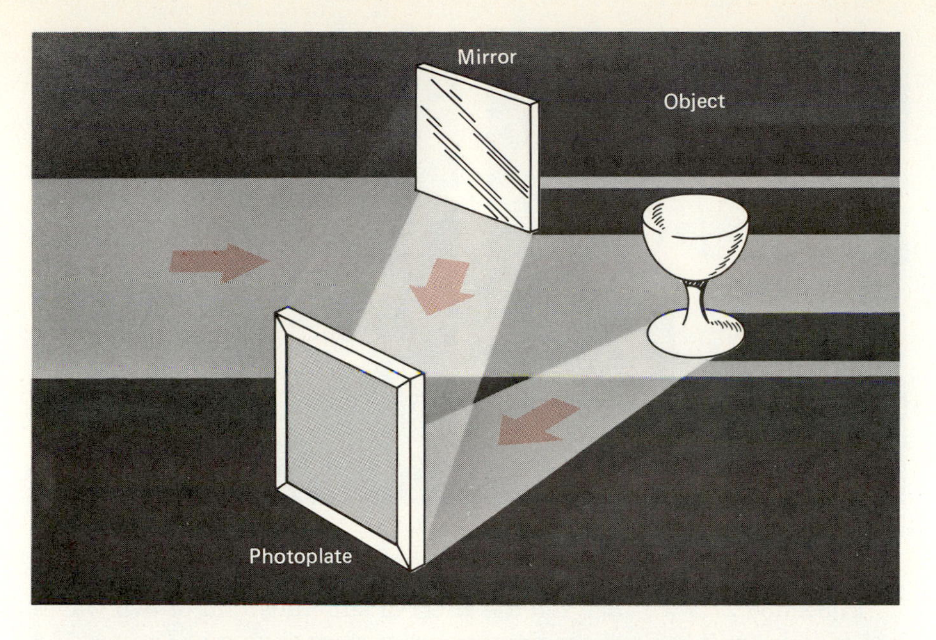

(a)

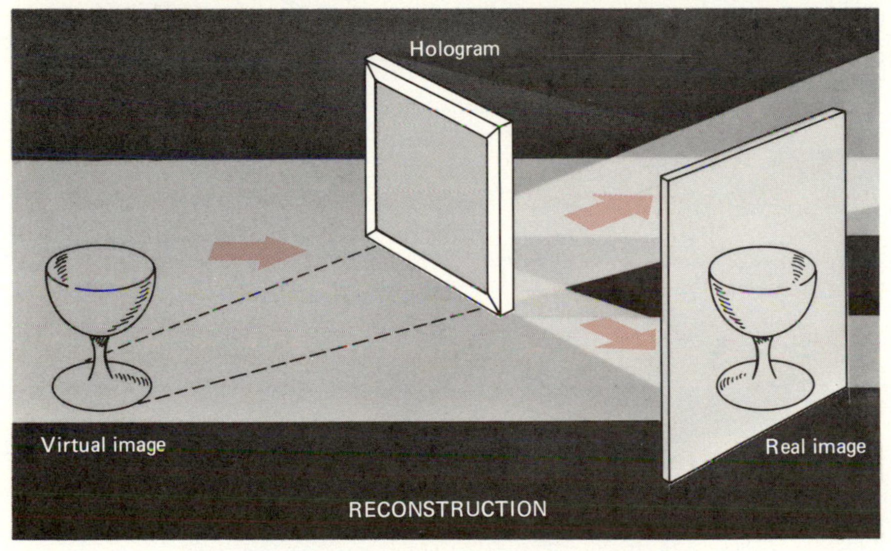

(b)

Figure 27.12 Holography. (a) A holograph is produced on the photographic plate by the interference pattern of the primary beam (reflected to the plate by the mirror) and the light scattered by the object. (b) Illumination of the hologram with coherent light results in the formation of a real and a virtual image of the original object.

Figure 27.13 The appearance of a typical hologram does not suggest the wealth of information it contains.

Figure 27.14 Photographs of the same holographic image taken from two different directions, showing the three-dimensional property of the reconstructed image.

In Section 27.3 we analyzed the effect of superposition of two waves and found that it leads to the interference pattern shown in Figures 27.4 and 27.5. Suppose we used a mask containing not two but a large number of identically spaced narrow slits. If the distance between adjacent slits is d, the individual waves from neighboring slits will again be in phase whenever

$$\sin \theta_m = \frac{m\lambda}{d} \qquad m = 0, 1, 2, 3, \ldots \tag{27.1}$$

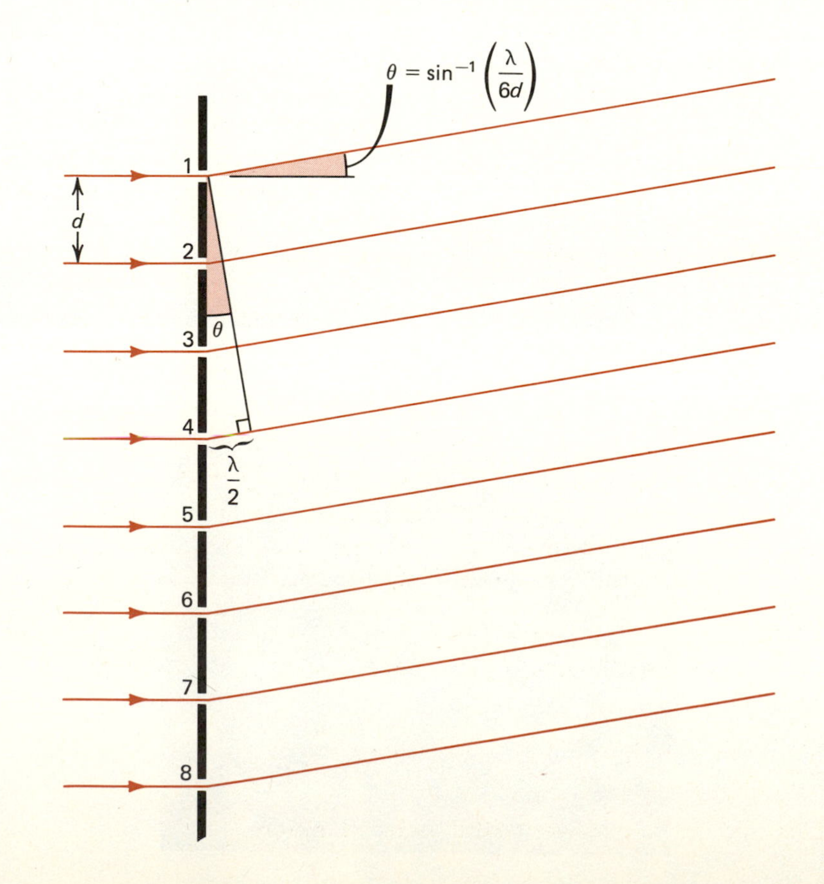

Figure 27.15 Interference of waves from a diffraction grating. For the case shown here, the phase difference between adjacent slits is not 180° but only 60°. Nevertheless, the intensity of light propagating in the direction $\theta = \sin^{-1}(\lambda/6d)$ is zero because the wave emanating from slit 1 interferes destructively with the wave emanating from slit 4. This pairwise cancellation extends over the entire grating. Consequently, the diffraction grating gives, instead of the broad maxima observed in double-slit interference, a series of very sharp maxima whenever the direction of propagation meets the condition $\theta = \sin^{-1}(m\lambda/d)$.

Also, as before, if the path difference from adjacent slits to the screen is $\lambda/2$ or an odd integral multiple thereof, we have perfect destructive interference. But consider now the situation when that path difference is less than $\lambda/2$, say $\lambda/6$. For that angle, given by $\sin\theta = \lambda/6d$, the intensity from just two slits is according to Equation (27.8), $4I_0 \cos^2(\pi/6) = 3I_0$, that is, only slightly less than the maximum intensity.

Though at that angle the path difference to the screen for waves from slits 1 and 2 is only $\lambda/6$, the path difference for waves emanating from slits 1 and 4 is three times as great, i.e. $3(\lambda/6) = \lambda/2$. At that angle, therefore, the wave from the slit labeled "1" in Figure 27.15 is exactly out of phase with the one from slit 4; waves originating from slits 2 and 5 differ in phase by 180° at the screen as do waves from slits 3 and 6, and so on. Thus over the entire grating, waves radiating from various slits will pairwise interfere destructively in that direction, giving zero intensity at the screen, even though *neighboring* slits do not give perfect destructive interference. If the number of regularly spaced slits is large, the intensity pattern under monochromatic illumination consists of a series of extremely sharp, intense maxima. The first noncentral maximum, corresponding to $m = 1$ in Equation (27.1), is referred to as the *first-order* maximum, that for $m = 2$ as the *second-order* maximum, and so on.

As with the double slit, the angle at which the mth-order peak appears for a given grating depends on the wavelength of the incident radiation. Since the peaks are now extremely narrow, illumination by light containing several wavelengths yields several peaks in each order, each peak corresponding to a particular wavelength of the incident light. A grating therefore acts much like a prism, dispersing incident light into its spectral components. Dispersion by a grating, however, is of different character and also different origin from dispersion by a prism.

Dispersion by a prism arises because the index of refraction of the prism depends on wavelength. Dispersion by a grating is a consequence of interference of many coherent sources. A prism deviates red light least, and violet light most. From Equation (27.1), we see that for the grating the angle of deviation is least for violet light (short λ) and greatest for red light (long λ). Last, a prism yields a single spectrum. A grating yields many spectra, one for each order.

The number of spectral orders produced by a grating depends on the spectral region of interest—infrared, visible, ultraviolet—and on the grating spacing d. Since the sine of an angle cannot exceed 1, the order number is limited by Equation (27.1) to

$$\sin\theta_m = \frac{m\lambda}{d} \leq 1 \qquad m \leq \frac{d}{\lambda} \tag{27.11}$$

● ***Example 27.5*** The angle of deviation of light of 400-nm wavelength is 30° in second order. How many lines per centimeter are there on this grating? How many orders of the complete visible spectrum are produced by this grating? Are these visible spectra clearly separated in all orders?

From Equation (27.1), with $m = 2$ and $\lambda = 400$ nm $= 4 \times 10^{-7}$ m, we obtain

$$d = \frac{m\lambda}{\sin\theta_m} = \frac{2(4 \times 10^{-7}\text{ m})}{\sin 30^\circ} = 16 \times 10^{-7}\text{ m} = 16 \times 10^{-5}\text{ cm}$$

The number of slits per centimeter is therefore

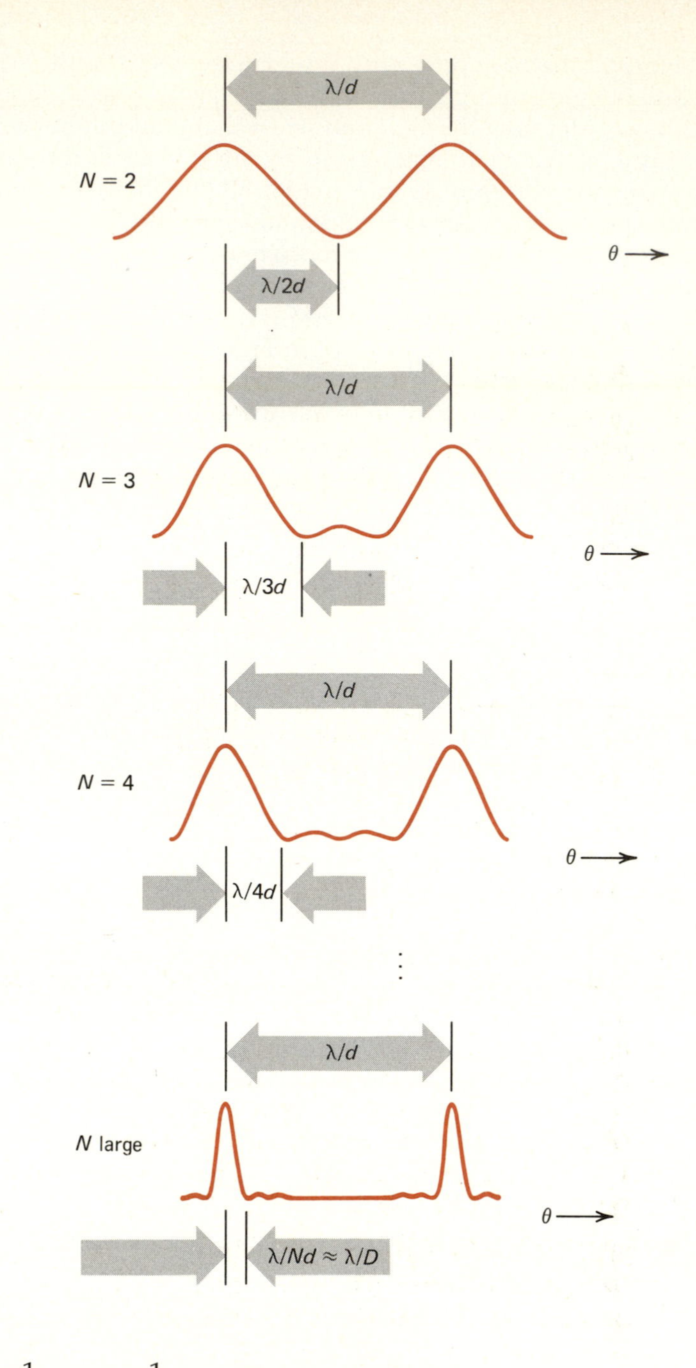

Figure 27.16 The development of sharp maxima as the number of slits of a diffraction grating is increased from two to some large number N. Note that for a grating, the width of the maximum is inversely proportional to the length of the grating Nd; here N is the total number of lines on the grating and d is the separation between adjacent lines.

$$N = \frac{1}{d} = \frac{1}{16 \times 10^{-5}\ \text{cm}} = 6250\ \text{lines/cm}$$

In any order m, the angle of deviation θ_m is greatest for the longest wavelength. From Equation (27.11), it follows that the limitation on the number of orders is imposed by the longest wavelength of the spectral region of interest. For the visible spectrum, this is 700 nm = 7×10^{-7} m; at that wavelength, $d/\lambda = 2.3$. We see that for this grating complete spectra are visible only in first and second order. In third order, the only portion of the visible spectrum that appears is that confined to wavelengths less than $d/3 = 533$ nm.

Spectra overlap in higher orders if the deviation angle for two different wavelengths is the same in two different orders. Thus, for example, the deviation angle for 400-nm light in third order equals the

Diffraction and Interference of Light

Double-Slit Interference Patterns

The spacing between adjacent interference fringes depends on the wavelength of the light. The interference patterns for red and blue light generated by a double-slit arrangement are shown.

Single-Slit Diffraction Patterns

The width of single slit diffraction depends on the wavelength of the light. The single slit diffraction patterns for red, green, and blue light are shown.

Optical Spectra

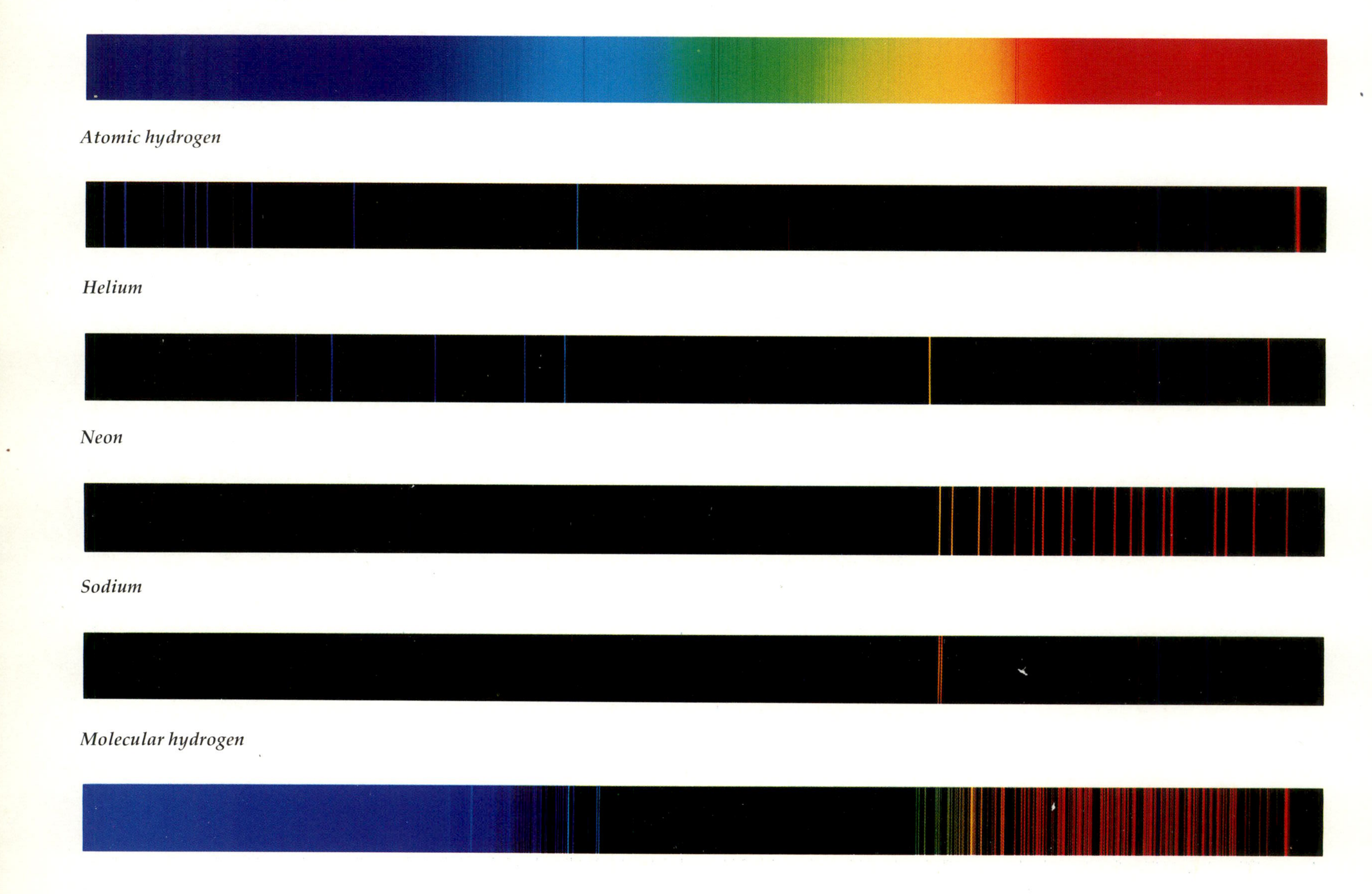

deviation angle for 600-nm light in second order. In our case, the third-order spectrum with wavelengths between 400 nm and 467 nm will overlap the second-order spectrum in the wavelength region between 600 nm and 700 nm.

Note that in these calculations we did not use the small angle approximation. That would have been incorrect since the deviation angles of interest are not small. ●

From this example we see that there is an optimal grating spacing for each wavelength region. Obviously, we must have $d > \lambda$ to obtain constructive interference even in first order. However, if $d \gg \lambda$, the dispersion, that is, the difference in deviation angle for nearby wavelengths, will be very small in first and other low orders. Although the dispersion increases in higher orders, overlapping of spectra from different orders limits the usefulness of higher orders in grating spectroscopy. The grating spacing d is typically somewhere between 3λ and 5λ, where λ is the average wavelength of the spectral region in which the grating is to be used. Thus gratings intended for use in the infrared have far fewer lines per centimeter than gratings used in the visible spectrum.

Gratings have many advantages over prisms as dispersive elements. First, they permit direct measurement of wavelength; since the angle θ_m can be measured quite accurately, knowledge of the grating constant d leads immediately to λ by application of Equation (27.1). In practice this is not a very important feature, since spectroscopists generally introduce into the spectrometer some substance that emits or absorbs light at well-established wavelengths. In an experiment, these lines are then used as fiducial markers. More important, a prism can be used only over a limited wavelength region, that within which the prism is transparent. Prisms made of glass are not suitable for ultraviolet spectroscopy, since glass strongly absorbs uv radiation.

In the preceding, we assumed that the incident light is transmitted through regularly spaced slits. While such *transmission gratings* are in use, *reflection gratings* are far more common. These can be constructed by depositing a thin film of aluminum on an optically flat surface and then removing some of the reflecting metal by cutting regularly spaced lines with a sharp ruling tool. The fabrication of high-quality gratings is an exacting and expensive task, and relatively few are made. Most gratings used in laboratory instruments are *replica gratings,* good plastic castings of high-precision master gratings.

You probably have a reflection grating at home, though you may not be aware of it. If you take an LP record and allow light to strike it at a glancing angle you will see a rainbow of colors from this grating of regularly spaced ridges.

We have so far restricted discussion to one-dimensional gratings, that is, to structures that are periodic in only one dimension. Suppose that we could construct a regular array of reflecting or transmitting units in two or three dimensions. What sort of diffraction pattern might we get?

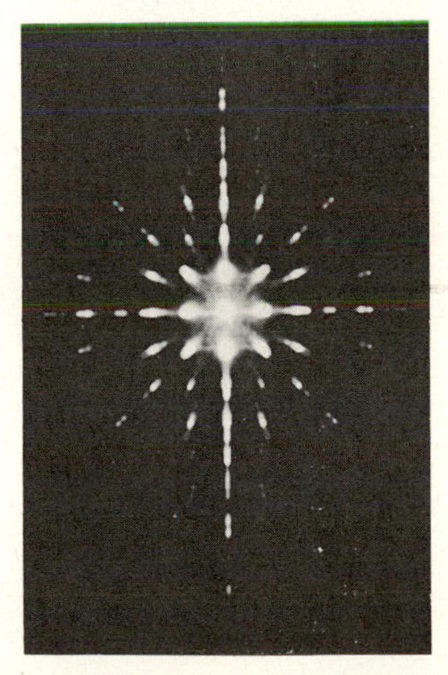

Figure 27.17 The diffraction pattern from a two-dimensional grating. In this case, the pattern is that obtained by allowing light to pass through a rectangular mesh.

To answer that question, you need only look at a small light source (for example, a distant street light) through sheer rayon or nylon, as in an open umbrella. The resulting pattern will be similar to the one shown in Figure 27.17. A three-dimensional regular array of scatterers will also generate a distinctive diffraction pattern, and the atoms of a crystalline solid constitute just such an array. However, since the separation between atoms of a solid is of the order 0.1 nm, we shall not be able to observe diffraction effects if we use visible light whose wavelength is

Figure 27.18 *The diffraction pattern obtained from a three-dimensional grating. The diffraction pattern is that produced by X rays incident on a single crystal of table salt, NaCl.*

several thousand times as long. From the preceding comments, it follows that crystal diffraction can appear only if the radiation has a wavelength shorter than about 0.1 nm; X rays meet this requirement.

If a single crystal of some material is irradiated with X rays, the reflected pattern is like that shown in Figure 27.18. That pattern evidently has the fourfold symmetry of a square, and from this, one can conclude that the crystal under examination must display the same symmetry in its crystal structure. Moreover, the separation between the various spots on the film depends on the distance between atoms just as the separation between various orders of the line grating depends on the grating constant d. Thus X-ray diffraction is a powerful tool in the study of crystal structure; it has also been used to investigate very complicated systems such as proteins. Indeed, X-ray diffraction was the key in revealing the double-helical structure of DNA.

27.5 (b) Diffraction by a Single Aperture

If we illuminate a slit with a beam of light, the intensity pattern we might expect to observe on a screen placed some distance from the slit is that shown in Figure 27.19(*a*); that is what geometrical optics predicts. Instead we observe a pattern like that of Figure 27.19(*b*), consisting of a central bright band, bounded by successive dark and bright bands of diminishing intensity. Moreover, if we *reduce* the width of the slit, the width of the central bright band and the spacing between successive maxima *increase*. This *single-slit diffraction pattern* is yet another manifestation of the wave nature of light and can be understood by applying Huygens principle. (You can observe this pattern by cutting a narrow slit in a piece of cardboard and viewing a distant street light through the slit.)

To see how these maxima and minima arise, consider the interference of the various Huygens wavelets originating at the slit of Figure 27.20(*a*). Here we imagine the light from the single slit to derive from a large number N of Huygens sources evenly distributed over the width of the slit shown in Figure 27.20(*b*).

We assume that the direction of the incident beam is perpendicular to the plane of the slit. All of the Huygens wavelets will then be in phase at

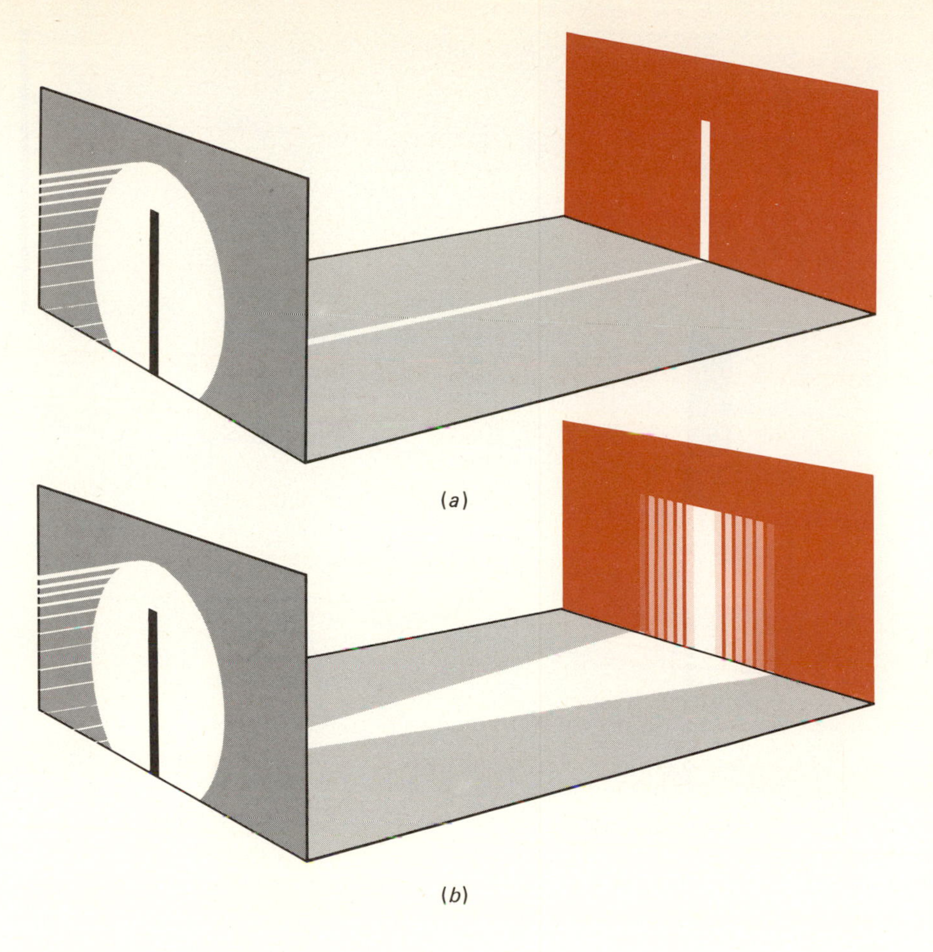

Figure 27.19 Light passing through a narrow slit. (a) Prediction of geometrical optics—a sharply defined image of the slit appears on the screen. (b) Actual observation—a broad diffraction pattern is seen, here greatly exaggerated. The width of the central diffraction maximum increases as the width of the slit is reduced.

the slit. They will therefore also be in phase at the point O on a screen, which we assume to be far from the slit (i.e., $D \ggg d$). At O then, we get the *central maximum.*

Suppose now that the line OP subtends an angle θ_1 at the slit so that

$$\frac{d}{2}\sin\theta_1 = \frac{\lambda}{2} \qquad \text{or} \qquad d\sin\theta_1 = \lambda$$

In that case the wavelet emitted by the Huygens source 1 of Figure 27.21(b) will, on reaching the screen, lead by half a wavelength that emitted by the source $(N/2) + 1$; similarly, waves from sources 2 and $(N/2) + 2$ will be out of phase by 180°; and so on, over the entire slit. Consequently, on the screen at point P, these waves result in pairwise destructive interference. Their superposition will give zero amplitude and intensity. Thus, the location of the first minimum is given by

$$OP = D\tan\theta_1 \simeq D\sin\theta_1 = \frac{D\lambda}{d} \qquad \textbf{(27.12)}$$

where we have made the usual small-angle approximation, which will be valid provided that $d \gg \lambda$.

By the same argument, if $\sin\theta_2 = 2\lambda/d$, wavelets originating from sources 1 and $(N/4) + 1$ interfere destructively on the distant screen, as do sources $(N/2) + 1$ and $(3N/4) + 1$. So viewing the slit as though it were constructed of four adjoining parts, we see that adjacent parts interfere destructively and another minimum will appear at that angle. In

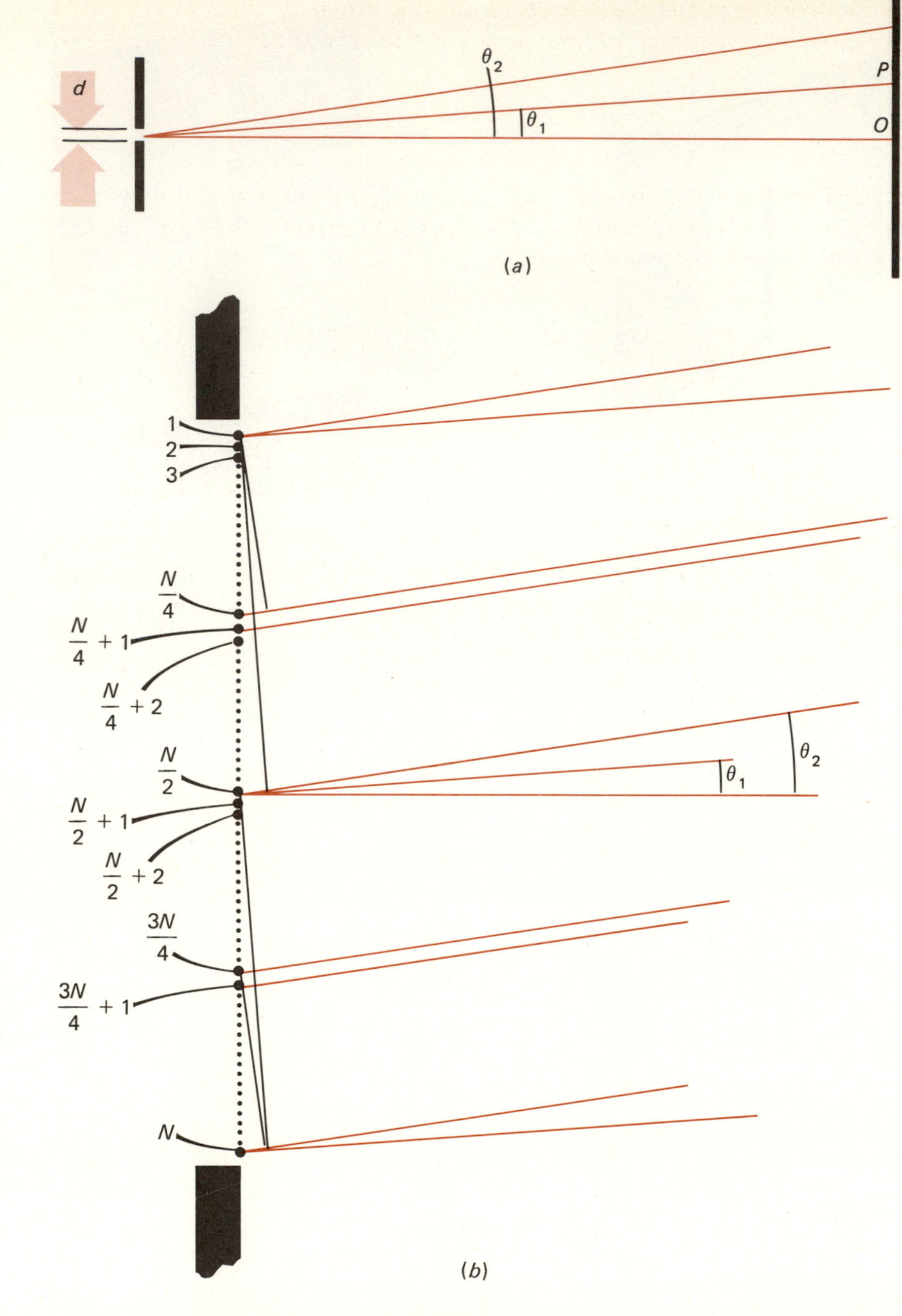

Figure 27.20 *(a) A plane wave of monochromatic light is assumed incident on the slit of width d. On a distant screen, there is a central bright region about the point O. At P, where OP subtends the angle $\theta_1 = \sin^{-1}(\lambda/d)$, the intensity drops to zero. The intensity then rises to another lesser maximum, and at the angle $\theta_2 = \sin^{-1}(2\lambda/d)$, it falls to zero once again. (b) The slit, of width d, is shown greatly enlarged. We imagine the slit divided into a large number N of Huygens sources. If the incident wave is a plane wave propagating perpendicular to the slit, all N Huygens sources radiate in phase. Hence, if the sources designated 1 and (N/2) + 1 are viewed from such an angle that their waves are 180° out of phase, they will cancel. Similar pairwise cancellation applies to all other Huygens sources for that direction. The next zero intensity occurs when the wave from source 1 is canceled by the wave from source (N/4) + 1.*

general, then, if $D \gg d \gg \lambda$ we have

DIFFRACTION MINIMA

$$\theta_m = \frac{m\lambda}{d};\ x_m = \frac{mD\lambda}{d} \qquad \textbf{(27.13)}$$

$$m = 1, 2, 3, \ldots$$

where x_m is the separation between the mth minimum and the central maximum on the screen a distance D from the slit. Note that Equation (27.13) is identical to Equation (27.4). Now, however, it identifies the location of a *minimum* and the $m = 0$ value is missing; for $m = 0$, we obtain not a diffraction minimum but the central maximum.

Between these successive minima, the intensity rises to subsidiary maxima, which appear roughly midway between the minima—at angles given by

$$\phi_m \simeq \sin \phi_m \simeq \frac{(m + \frac{1}{2})\lambda}{d} \qquad m = 1, 2, 3, \ldots \qquad \textbf{(27.14)}$$

Equation (27.14) is only approximate, but a good approximation; the correct coefficients of λ/d are 1.43, 2.46, and 3.47 instead of 1.5, 2.5, and 3.5 for the first three noncentral maxima.

Figure 27.21 *Photograph of a single-slit diffraction pattern. Note that the central maximum is twice as wide as the subsidiary maxima. In contrast to the double-slit interference pattern, the intensities diminish rapidly as one moves away from the central maximum.*

The importance of Equations (27.13) and (27.14) is not so much in their numerical results as in the qualitative predictions, namely:

1. The image cast by a slit is *not* the expected geometrical one. Instead one observes an image such as the one shown in Figure 27.21.

2. For a given wavelength, the *diffraction pattern is wider the narrower the slit.*

3. For a given slit width, the *width of the diffraction pattern is proportional to the wavelength;* that is, it is wider the greater the wavelength. (See the color plates.)

4. The *intensities of the noncentral maxima are much less than the intensity of the central maximum.*

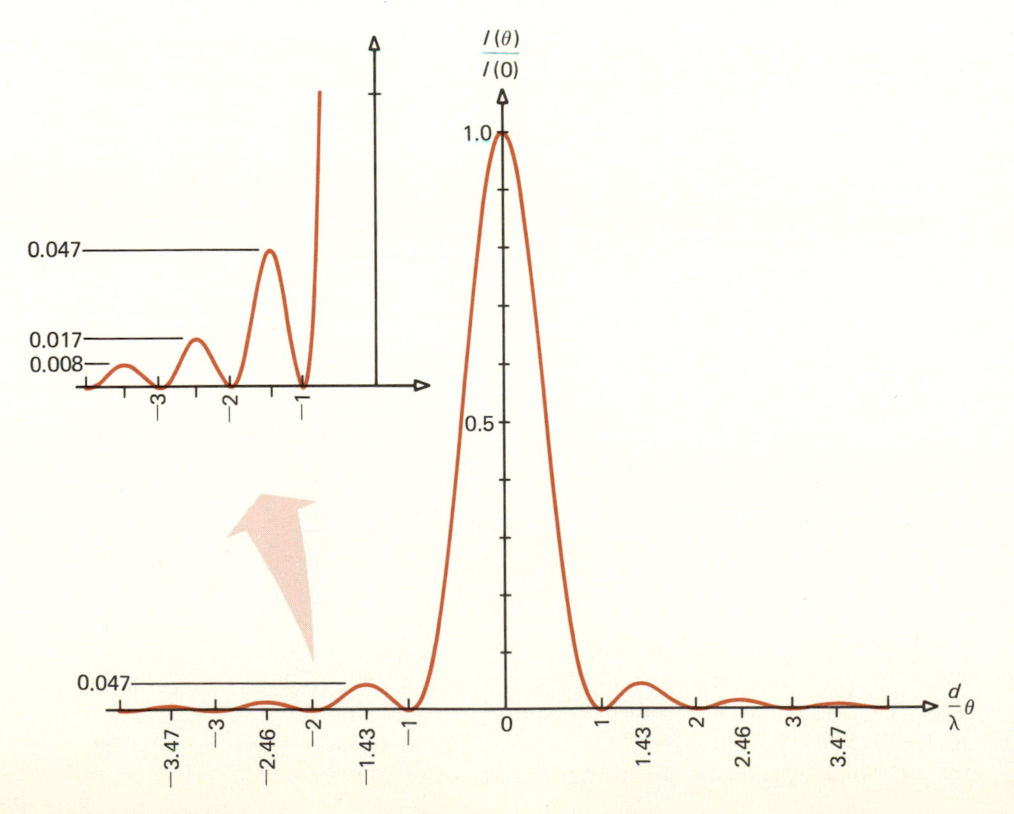

Figure 27.22 *The intensity distribution of the single-slit diffraction pattern.*

5. The *width of the central maximum,* i.e., the angular separation between the bounding minima, *is twice the widths of the noncentral maxima.*

● ***Example 27.6*** A slit 0.1 mm wide is illuminated with monochromatic light of wavelength 500 nm. How wide is the central maximum on a screen 1 m from the slit?

Figure 27.23 *Diffraction from a small circular aperture.*

From Equation (27.13), the separation between the two minima bounding the central maximum is

$$x_1 - x_{-1} = \frac{2D\lambda}{d} = \frac{2(1\text{ m})(5 \times 10^{-7}\text{ m})}{10^{-4}\text{ m}} = 10^{-2}\text{ m} = 1\text{ cm}$$

Thus the central bright band from this narrow slit is about 100 times as wide as the slit itself. ●

We have focused attention on the diffraction pattern from a narrow slit because the geometry is simpler than for the more important circular aperture. However, a small circular aperture also generates a diffraction pattern, in this case a series of concentric bright and dark rings, as shown in Figure 27.23. As for the slit, the width of the pattern varies inversely with the aperture diameter and directly with the wavelength of the illumination.

Diffraction by a single aperture has many important practical consequences. Before we discuss these, let us recall that it was the apparent failure of light to exhibit just these effects that led Newton and other scientists of his day to favor the corpuscular theory of light. Even after the beautiful demonstrations of interference and diffraction by Young and Fresnel, the wave theory did not gain universal acceptance. Scientists, then as now, are rather a conservative lot, at least in their professional life. To air the controversy, and perhaps help resolve it, the French Academy, in 1818, selected this issue as the topic for its annual "competition." The commission of distinguished scientists that heard Fresnel's presentation included Gay-Lussac, Laplace, Poisson, Arago, and Biot. Of these, only Arago, who had sponsored and assisted Fresnel, was sympathetic to

Figure 27.24 *Shadow cast by a circular object. The central bright spot Poisson predicted is clearly visible.*

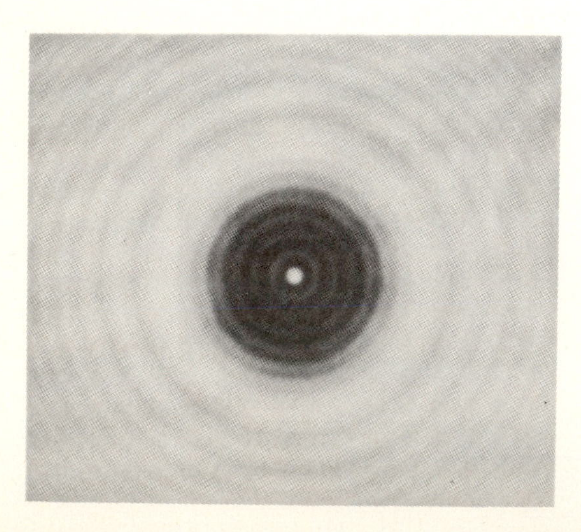

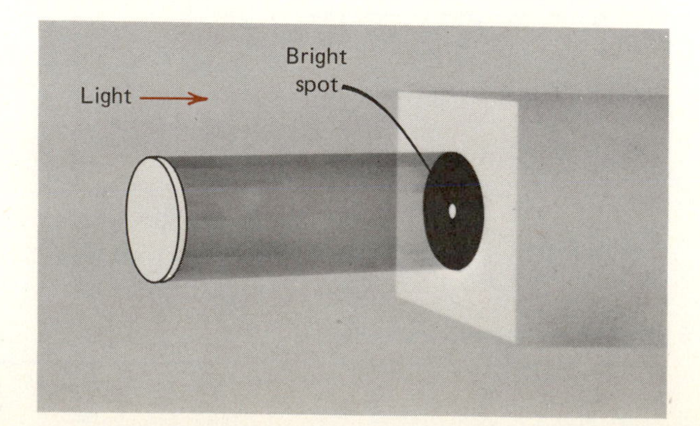

his view. During the session, Poisson, a fine and quick-witted mathematician, called attention to a remarkable consequence of Fresnel's theory. If a screen were placed at a suitable distance from a circular obstacle illuminated by a beam of light, diffraction of waves around this obstacle should cause the appearance of a bright spot at the exact center of the circular shadow. But as Poisson reminded his colleagues, no such bright spot had ever been observed, not even by M. Fresnel. Following the session, Fresnel and Arago returned to their laboratory to perform the experiment. They must have been immensely gratified to see a bright pinpoint of light at the center of the shadow, just as Poisson had predicted. This little irony is by no means unique in the history of science.

27.6 Resolving Power

Suppose we project the image of two adjacent but distant light sources on a screen by passing the light through a small aperture, as in Figure 27.25. The light from each source will create a diffraction pattern on the screen that will be more extended the smaller the aperture. If these diffraction patterns are so broad that their central maxima overlap substantially, it will be impossible to identify the intensity distribution as one produced by two separate sources and not by one, more extended source. Generally, two sources can be resolved only if the central diffraction maximum of one falls at or beyond the first diffraction minimum of the other. This is the *Rayleigh criterion,* and for a slit aperture, it means that the angular separation of the two sources must be greater than λ/d radians, where d is the slit width. That is,

$$\theta_{\text{min}} = \frac{\lambda}{d} \qquad \text{(slit aperture)} \tag{27.15}$$

If the aperture is circular, Rayleigh's criterion is

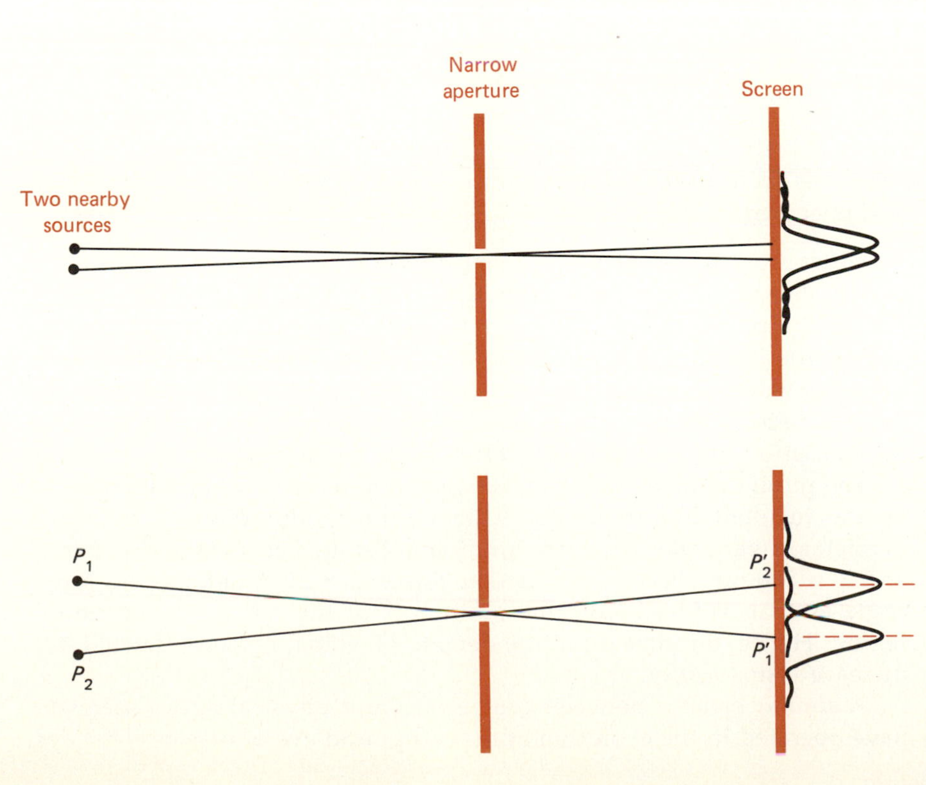

Figure 27.25 *Two nearby sources cannot be resolved if the diffraction patterns overlap substantially. If the two sources are separated by such an angle that when viewed through the aperture, the central maximum from one source falls at least as far as the first minimum of the other source, the two sources can be resolved.*

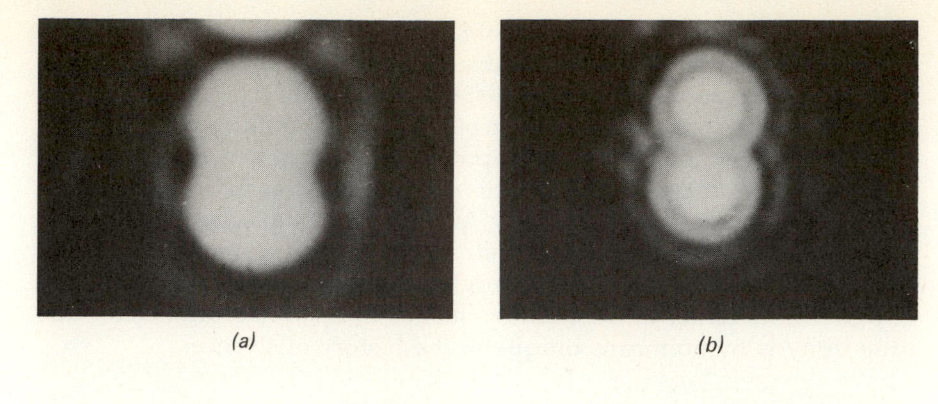

Figure 27.26 *Images from two nearby point sources. (a) The two sources are barely resolved. (b) The two sources are well resolved and their individual diffraction rings are clearly evident.*

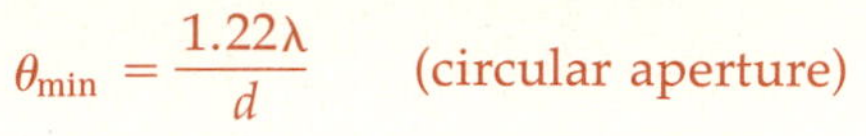

$$\theta_{\min} = \frac{1.22\lambda}{d} \qquad \text{(circular aperture)} \qquad \textbf{(27.15a)}$$

where $\theta_{\min}$ is the smallest angular separation for which two point sources can still be resolved, and d is the diameter of the circular aperture. In Figure 27.26(a), the two sources are just barely resolved; in Figure 27.26(b), they are well resolved.

Clearly, it is advantageous to use as large an aperture as practicable, and this is one reason for the large lenses and mirrors in astronomical instruments. Similarly, the useful magnifying power of a microscope is limited not by the power of its lenses but by the aperture of the objective. From Equation (27.15), it also follows that the resolution can be improved by resorting to short-wavelength radiation whenever possible. Thus, as the technology for generating radio waves of short wavelength—microwaves—progressed during World War II and later, radar antennae could be reduced in size without loss of resolving power.

● ***Example 27.7*** A ground-control radar generates a beam of 3-cm microwaves. How large should the parabolic antenna dish be if the radar is to resolve two planes separated by 20 minutes of arc?

From Equation (27.15a), we have $d = 1.22\lambda/\theta_{\min}$. Twenty minutes of arc equals 0.00582 radian, and so we find that $d = 630$ cm $= 6.3$ m. The radar dish should have a diameter of about 6.5 m. If, however, 1-cm microwaves were used, the diameter of the antenna could be reduced to just over 2 m. ●

Diffraction limitations have been important also in the evolution of vertebrate and insect vision. Were it not for diffraction due to the finite size of the pupil, the resolving power of the eye would be greater the greater the density of visual receptors on the retina, that is, the more "fine grained the film." In the human eye, the density of retinal receptors implies a resolving power of about 2×10^{-4} radian. Let us now estimate the limitation on resolving power due to diffraction effects.

The pupil diameter may vary between 8 mm and 1.5 mm, depending on the incident light intensity. Under normal illumination the pupil diameter is generally between 2 mm and 4 mm. Let us take $d = 3$ mm and $\lambda = 500$ nm. We then find that $\theta_{\min} \simeq 2 \times 10^{-4}$ radian, the same angle as that deduced from the graininess of the retina in the central region. Hence, an increase in the density of retinal receptors would not increase visual acuity.

A similar balance between geometrical and physical optics seems to have operated in the evolution of the compound eye of insects. That eye

Figure 27.27 The compound eye of a horsefly. The striations are the result of an interference pattern.

is a composite of numerous narrow cones, packed together to form a roughly hemispherical body. Each cone, called an *ommatidium,* is covered by a cornea of approximately circular shape; the region below the cornea serves as a light pipe to the photoreceptor, located near the vertex of the cone. The lateral faces of the ommatidium are lined with light-absorbing pigment to prevent radiation in one ommatidium from entering an adjacent one.

Two sources so close to each other that they fall within the acceptance angle of one ommatidium cannot be resolved. For a bee, the diameter of the cornea is about 2×10^{-2} mm, and R, the radius of the compound eye, is about 2 mm. Hence, the acceptance angle is $\phi \simeq 10^{-2}$ radian, giving a resolving power two orders of magnitude below that of the human eye. Could the resolution be improved by reducing the acceptance angle and increasing the number of ommatidia packed into the compound eye?

To answer that question let us determine the diffraction angle of an aperture of 2×10^{-2} mm for visible light, say $\lambda = 500$ nm. From Equation (27.15a), we find $\theta_{\text{min}} \simeq 3 \times 10^{-2}$ radian, which suggests that the resolving power of the bee's eye is already severely diffraction-limited and that nature has given the bee more ommatidia than it should have. In fact, this is not so; the bee's photoreceptors are sensitive to shorter wavelengths than the human rods and cones, well into the ultraviolet, so in our calculation we should have used not 500 nm but a shorter wavelength, perhaps 200 nm. At that wavelength, diffraction and acceptance angles are roughly equal. Evolution seems to have conformed almost perfectly to the laws of physics in achieving optimum performance of the visual apparatus.

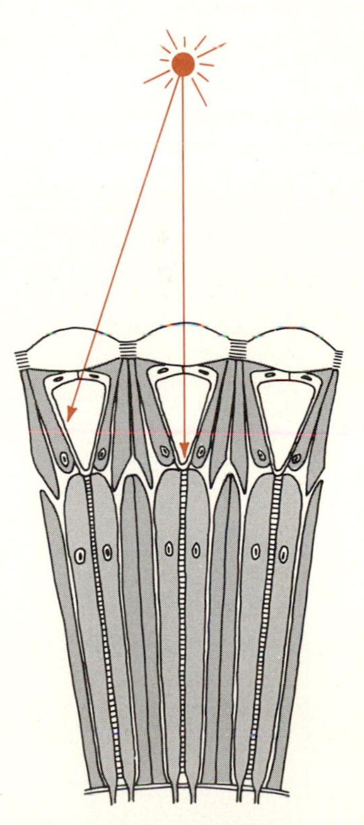

Figure 27.28 Schematic drawing of the ommatidia of the compound eye.

27.7 Polarization

So far, we have studiously avoided mentioning one important characteristic of electromagnetic waves, their polarization. We recall that electromagnetic waves are oscillations, in space and time, of electric and magnetic fields, and that these fields are mutually perpendicular and are both perpendicular to the direction of propagation of the wave. One defines the *direction of polarization* of an electromagnetic wave as the direction of the oscillating *electric* field.

Polarization of a transverse wave evidently implies a rotational asymmetry of the wave about the direction of propagation. For example, a string oscillating in the vertical plane is distinguishable from one oscillating in the horizontal plane. One will freely pass through a vertical slit; the other will not. Yet light from an incandescent lamp or from the sun appears to show no evidence of such asymmetry. The reason is that this light consists of an immense number of waves, each polarized in some random direction in the plane normal to the propagation direction. Such light is called unpolarized, or randomly polarized.

We focus here on three questions (1) How can one tell, given a source of polarized light, that it is polarized? (2) How can one determine the direction of polarization? (3) By what mechanism can polarization be achieved from a source of unpolarized light?

Let us assume that we have a device that transmits light polarized in one direction without attenuation and totally absorbs light polarized perpendicular to this direction. Such an ideal *polarizer* is the electromagnetic analog of the slit of Figure 27.29. Although no such perfect polarizer exists, commercial Polaroid sheets are modest approximations.

If we interpose this ideal polarizer in a beam of unpolarized light of intensity I_0, the intensity of the transmitted light I_1 will be $\frac{1}{2}I_0$ and the light will be *plane-polarized* along the preferred direction. If we now place a second polarizer into the beam behind the first, the intensity I_2 of the light transmitted through both is found to depend on the orientation of the second polarizer relative to the first, according to the *law of Malus*,

$$I_2 = I_1 \cos^2 \theta \tag{27.16}$$

where θ is the angle between the preferred directions of the two polarizers. This relation we now derive from the proportionality between the intensity and the square of the field amplitude.

In Figure 27.30, we show the **E** vector of the polarized wave of intensity I_1, and its decomposition into components parallel and perpendicular to the preferred direction of the second polarizer. The parallel

Figure 27.29 *A thin slit in a solid block acts as a polarizer for a transverse wave on a string. (a) The incident wave is already polarized in the direction of the axis of the polarizer and propagates with undiminished amplitude. (b) The axis of polarization of the polarizer is perpendicular to the polarization of the incident wave; no wave is propagated past the polarizer. (c) The direction of polarization of the incident wave is at some arbitrary angle with the polarizer's direction; the amplitude of the transmitted wave is not zero but is less than the amplitude of the incident wave. The amplitude of the transmitted wave equals the component of the incident amplitude along the polarizer's axis; that is,* $A_{tr} = A_{in} \cos \theta$

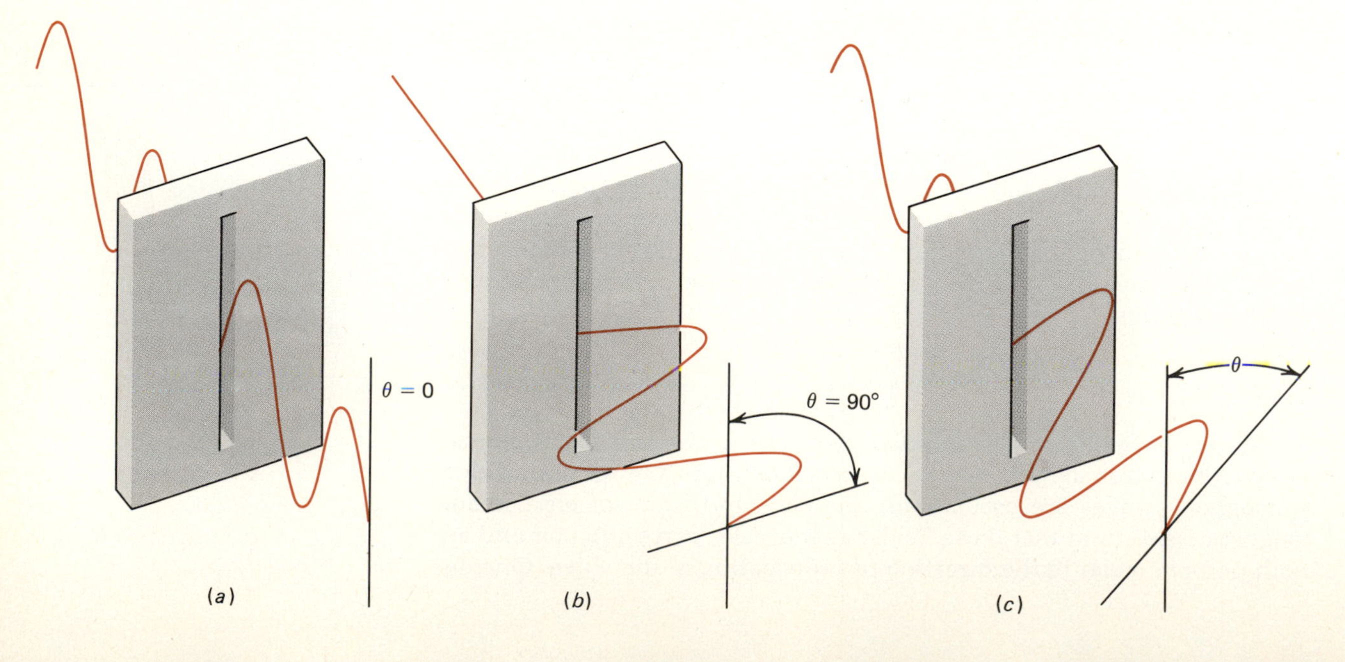

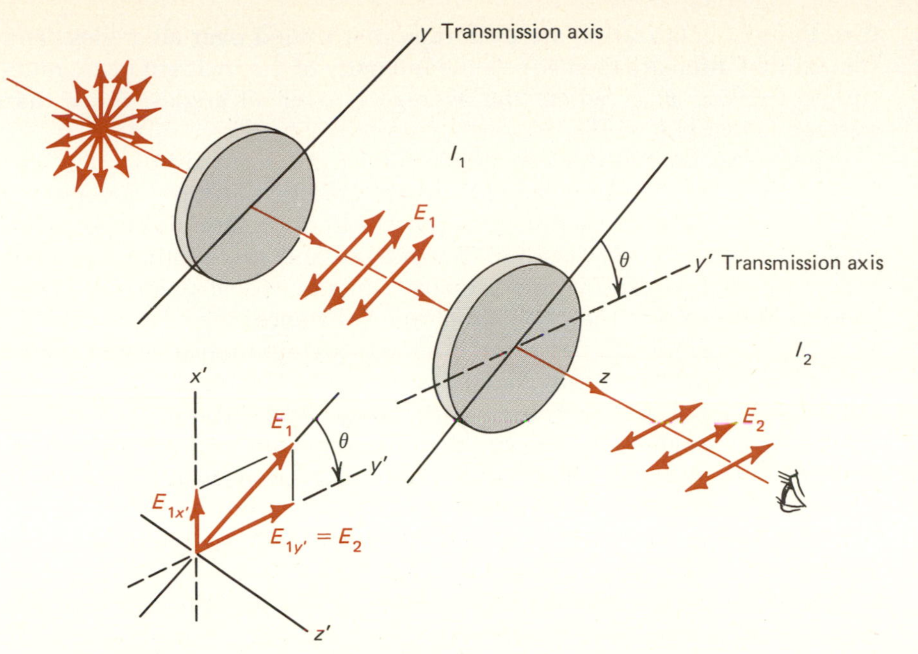

Figure 27.30 *Unpolarized light is incident on the first polarizer, which transmits only those components of the incident electric-field vectors that oscillate in the direction of the polarizer's axis. The transmitted wave is then polarized as indicated and has an intensity I_1. The second polarizer transmits only the portion of this wave corresponding to the component of the electric field oscillating along its axis of polarization. Because the intensity of the electromagnetic wave is proportional to the square of the* **E** *field, the ratio of the intensities I_2/I_1 is given by $I_2/I_1 = \cos^2\theta$*

component is transmitted; the other is totally absorbed by the second polarizing sheet. Since the parallel component is given by

$$E_{\parallel} = E_{y'} = E_1 \cos\theta = E_2 \tag{27.17}$$

it follows that

$$I_2 = kE_2^2 = kE_1^2 \cos^2\theta = I_1 \cos^2\theta \tag{27.16}$$

where k is the constant that relates the intensity to the square of the electric-field amplitude.

● *Example 27.8* A polarizing sheet is placed in the path of a polarized beam of light. The intensity of the transmitted light is half that of the incident light. What is the angle between the polarization of the incident beam and the transmitted beam?

We are given that $I_2/I_1 = 1/2$. Hence, from Equation (27.16), $\cos^2\theta = \frac{1}{2}$, and $\cos\theta = \sqrt{\frac{1}{2}}$. Therefore, $\theta = 45°$. ●

If we want to know whether light from a given source is polarized, partially polarized, or unpolarized, we need only interpose a polarizer (often called an *analyzer*) in the beam and rotate it about an axis parallel to the beam. If the light is plane polarized, the transmitted light intensity will vary according to Equation (27.16); by noting the position of the analyzer that results in maximum intensity of the transmitted light, we obtain the direction of polarization of the incident beam. In practice, it is better to search for the intensity minimum since a null can generally be identified more precisely than a maximum. The polarization direction is then perpendicular to the direction at which the null is found. If the incident beam is only partially polarized, that is, if it is the sum of an unpolarized and a polarized beam, the transmitted intensity will again vary with rotation of the analyzer but never vanish completely. If the beam is not plane-polarized, no intensity variation with angle of the analyzer will be observed.

Finally, we can now understand why the intensity I_1 of Figure 27.30 is half of I_0. The incident, unpolarized beam consists of light waves whose directions of polarization are uniformly distributed over all angles. The transmitted intensity then equals the intensity of the incident beam multiplied by $\langle \cos^2 \theta \rangle_{av}$, where the average is over all angles. Since that average is just $\frac{1}{2}$, $I_1 = \frac{1}{2}I_0$.

There are various devices that produce polarized or partially polarized light from an unpolarized source. One is a *dichroic* material. A substance is said to be dichroic if it absorbs light preferentially for a particular direction of polarization. Tourmaline and many other naturally occurring crystals exhibit this property, and our ideal polarizer of the preceding pages was the perfect dichroic substance.

The most common polarizer in use today is a synthetic dichroic material invented by E. H. Land. *Polaroid* is produced by stretching a sheet of polyvinyl alcohol, thereby aligning the long hydrocarbon chains of the polymeric molecules. The sheet is then impregnated with iodine, which attaches itself to the polymers and results in good electrical conduction along the chains, but allows little conduction perpendicular to them. Electromagnetic waves whose **E** vector is parallel to the polymers are then strongly absorbed by this material.

There are also crystals that are *birefringent*. As the adjective implies, these materials exhibit *two different refractive indices* for polarization along two crystallographic directions. A beam of unpolarized light entering such a crystal will be refracted into two beams, polarized mutually perpendicular, which propagate in slightly different directions. Various methods can be used to divert or eliminate one of the two beams, and the emerging light is then plane-polarized.

The two beams produced by the birefringent crystal are not, as once was thought, different forms of light, but differ only in polarization direction. Still, that difference does have some interesting consequences. For example, the two beams, though they may derive from a coherent source, will when recombined, fail to show an interference pattern because their field vectors are mutually perpendicular. On the other hand, if either beam is split (by amplitude or wave splitting), the two parts can be made to interfere in the usual way. It was this observation that led Fresnel and Arago to conclude that light waves were transverse.

Specular reflection also causes polarization. If a beam of unpolarized light is obliquely incident on a sheet of glass or other dielectric material of a different index of refraction from air, the reflected beam is partially polarized. For a particular angle of incidence, given by

$$\tan \theta_B = \frac{n_r}{n_i} \quad \textbf{(27.18)}$$

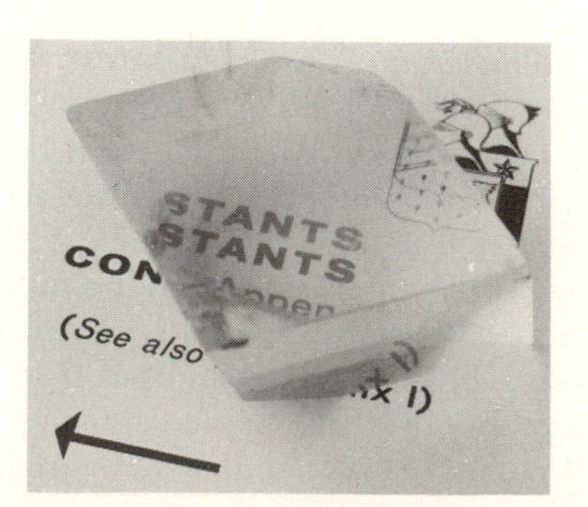

Figure 27.31 Double refraction. The double image produced by a birefringent calcite crystal.

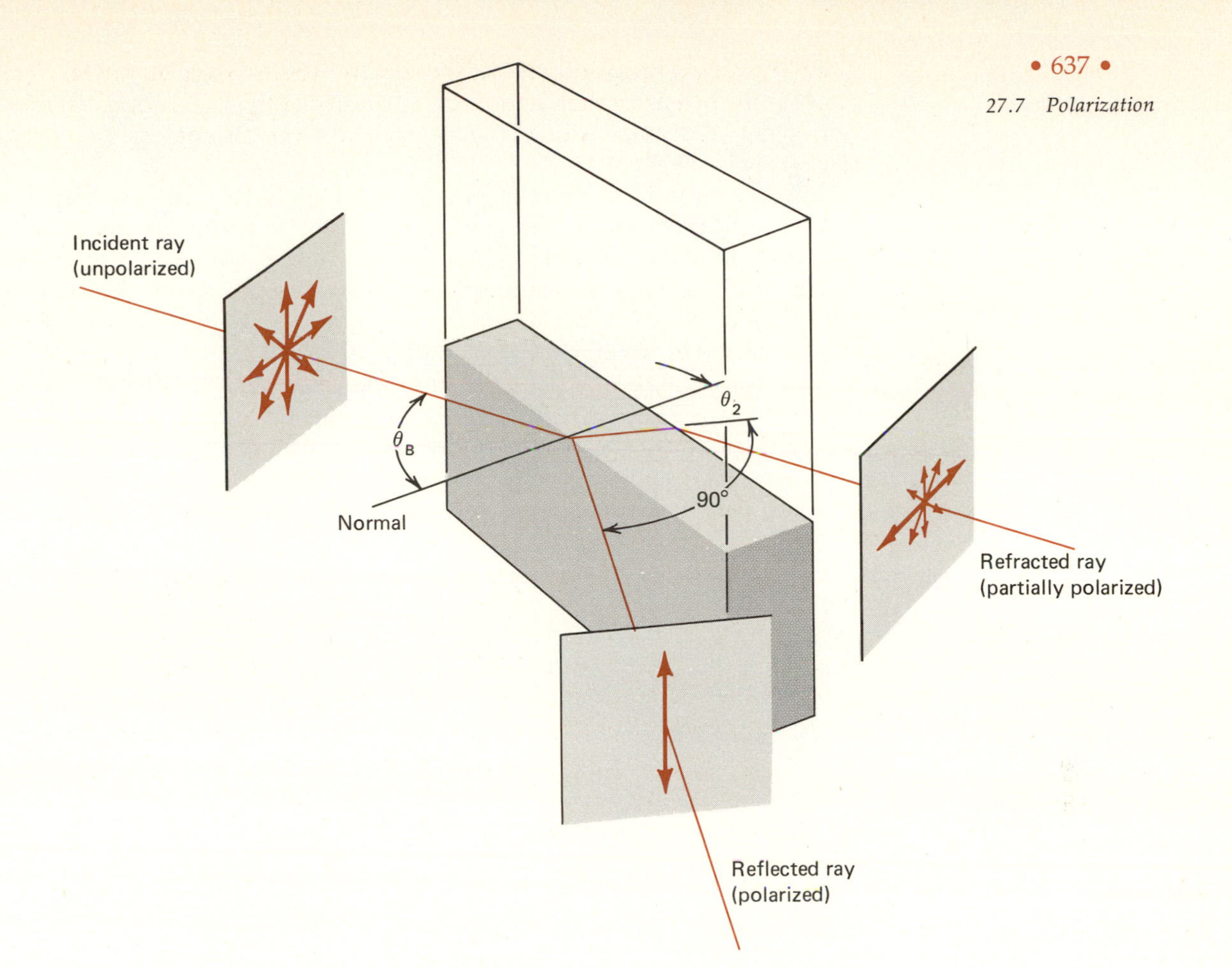

Figure 27.32 *Polarization by reflection. The incident wave is unpolarized and has electric field vectors parallel as well as perpendicular to the plane formed by the incident and reflected rays. If the angle of incidence is such that the reflected and refracted rays are at 90° to one another, the reflected ray will be completely polarized in the direction perpendicular to the plane of incident and reflected rays. The corresponding angle of incidence is known as Brewster's angle. For incidence at angles other than Brewster's angle, the reflected ray is only partially polarized.*

where n_i and n_r are the indices of refraction of the incident and reflecting medium, the reflected beam contains only light polarized parallel to the plane of the reflecting surface. This special angle is known as *Brewster's angle,* after David Brewster, who deduced this relation empirically.*

Polaroids in sunglasses are oriented so as to absorb horizontally polarized light. Since light reflected by the surface of water and other horizontal reflecting surfaces is predominantly polarized in that direction, such glasses are effective in reducing reflected glare.

● ***Example 27.9*** Sunlight is reflected from a calm lake. The reflected light is 100 percent polarized. What is the angle between the sun and the horizon?

The index of refraction of water is 1.33. From Equation (27.18), we have $\tan\theta_B = 1.33/1 = 1.33$, and $\theta_B = 53°$. Since this is the angle of incidence, or the angle that the sun's rays make with the normal to the horizontal surface of the lake, the sun must be $90° - 53° = 37°$ above the horizon. ●

Earlier we pointed out that the ability of an object to scatter incident waves depends on the dimension of the object compared to the wavelength. For air molecules, the probability that light waves will be scattered, though small, depends strongly on wavelength; the shorter the

* David Brewster (1781–1868) was professor of physics at St. Andrews University and is the inventor of the kaleidoscope.

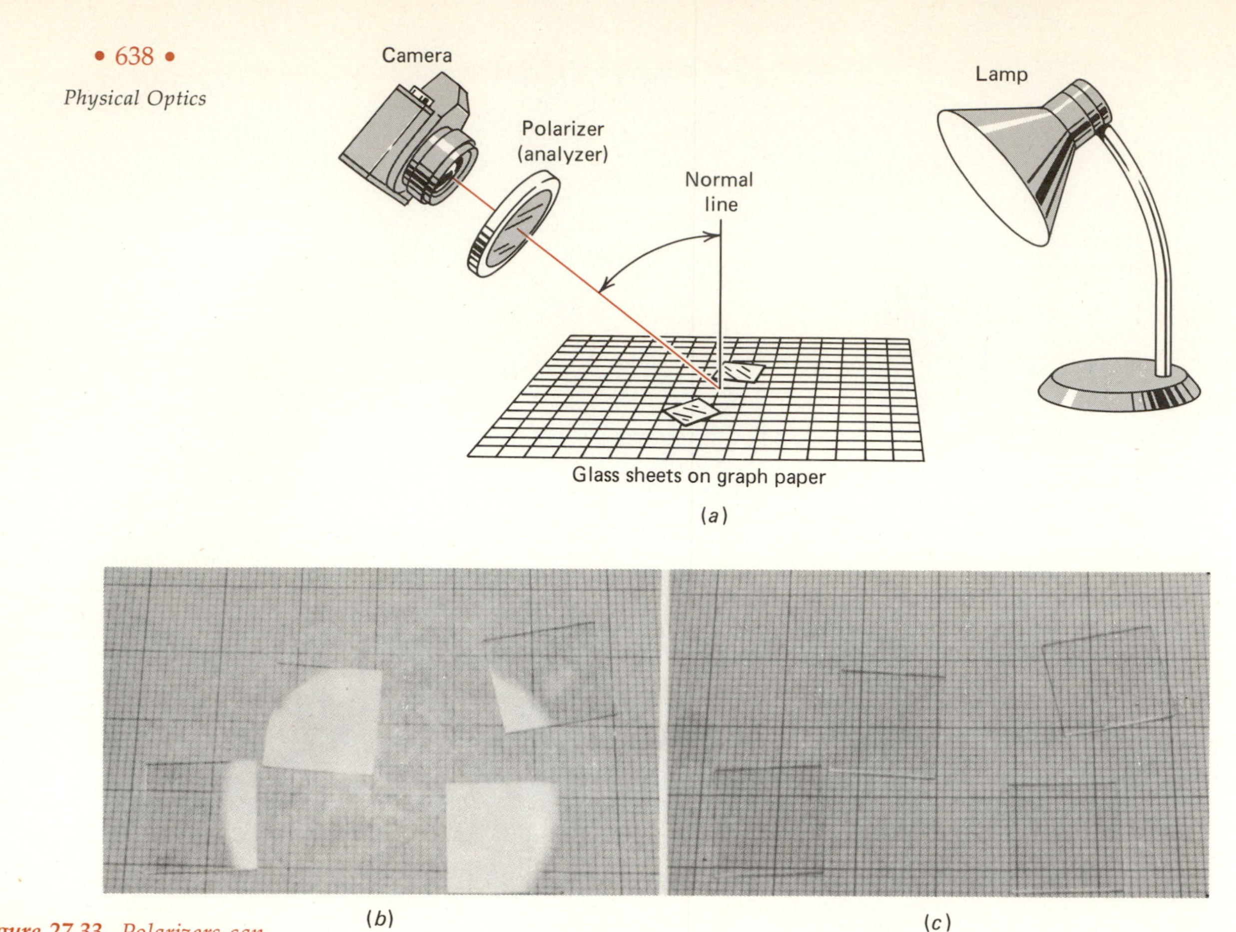

Figure 27.33 *Polarizers can reduce reflected glare. (a) A few microscope cover plates on a sheet of graph paper, taken (b) without and (c) with a polarizing filter.*

wavelength, the greater the probability. Light from the clear sky is really sunlight that has been scattered toward our eyes by molecules in the atmosphere. Since scattering is more pronounced in the blue (short-wavelength) portion of the spectrum, we can now understand why the sky is blue and why the sun appears reddish at sunset or when seen through a humid or smoky, polluted atmosphere.

However, not only is the sky blue, but light coming from it is also partially polarized. You can readily observe this by placing a piece of Polaroid (for example, one lens of a pair of Polaroid sunglasses) in front of your eye and rotating it as you look at the sky on a clear day. You will note a change in light intensity with orientation of the Polaroid. For a certain orientation, a dark band will appear in the sky, showing that light from that region is strongly polarized. Moreover, if you estimate the angle between the lines from your eye to the sun and to the darkest part of the sky, you will find that it is approximately 90°. You can also determine the direction of polarization of the light, since Polaroid sunglasses are constructed so that under normal wear the preferred direction for transmission is vertical. You will then find that the direction of polarization of light from the sky is perpendicular to the plane formed by the three points—you, the sun, and the part of the sky you are viewing.

To see how this polarization arises, consider a single molecule in the path of an unpolarized beam, as shown in Figure 27.34. The oscillating electric field of the incident electromagnetic wave will set the electric charges (electrons) of the molecule in vibration, and these oscillating charges then act as small antennae, radiating at the same frequency as the

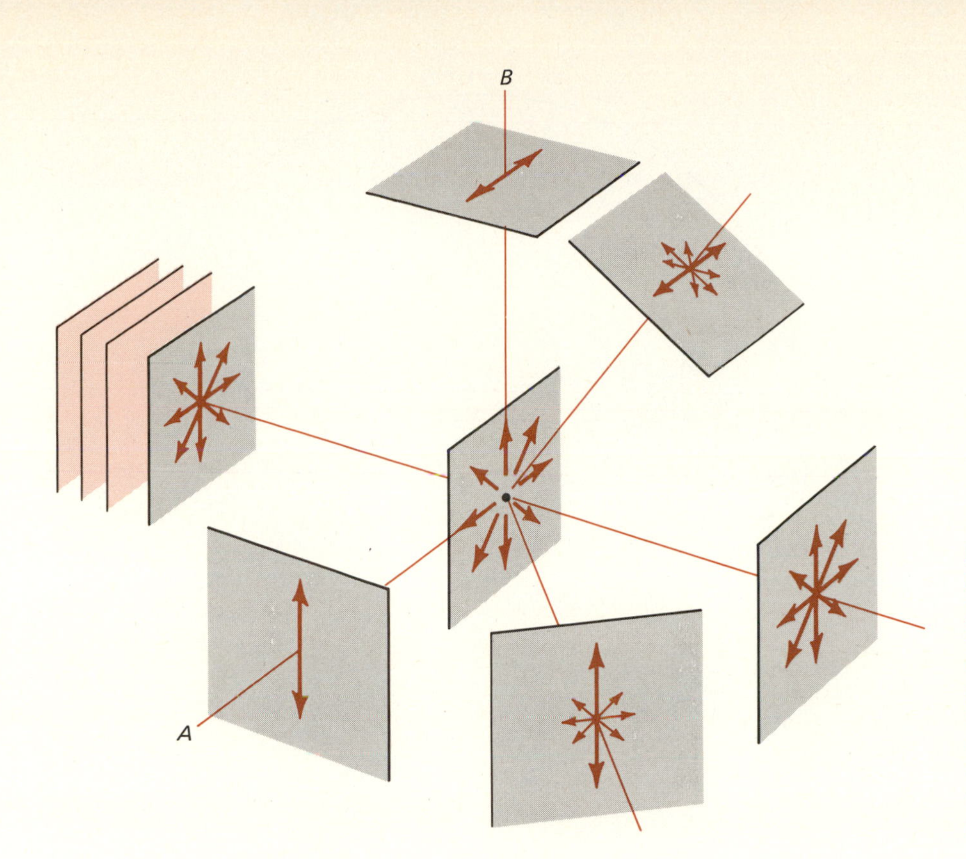

Figure 27.34 Unpolarized light is incident on a molecule from the left. Light scattered in the direction A or B (or any other direction in the plane perpendicular to the direction of the incident ray) is completely polarized as indicated. Light scattered into some arbitrary direction is partially polarized.

incident light. Since the **E** field of the scattered (reradiated) light must be perpendicular to the direction of propagation, it follows that the scattered light, when viewed from A in Figure 27.34 must be vertically polarized; when it is viewed from B, it must be horizontally polarized. In other directions, the scattered light is partially polarized.

Most insects and many birds rely on this polarization of the sky as a compass to direct them in their flight. The cornea of the insect ommatidium is dichroic and serves as an analyzer for sensing polarization. Even the human eye is somewhat sensitive to polarization, this sensitivity varying considerably among individuals.

Navigation by sensing the polarization of the sky light is not limited to birds and insects. Sailors and airline navigators occasionally use a polarizer as a "twilight compass" to locate the sun's position below the horizon before the sky is dark enough to allow stellar observation. It is now believed that the early Vikings employed this technique on their extended voyages. At night they navigated by the stars; by day, by the sun, when it was not obscured by clouds. In the summer at high latitudes, however, the sun is just below the horizon at night, and it illuminates the atmosphere enough to prevent stellar observations. The Vikings' magical "sun stone," which allowed them to locate the sun's position below the horizon with great accuracy was probably a crystal of cordierite, a dichroic mineral common in Scandinavia.

Many molecules have a structure that is intrinsically right-handed or left-handed; that is, the pattern of the individual atoms of the molecule are arranged like a right-handed or left-handed screw. Such substances include most organic molecules of biological interest—monosaccharides, amino acids, and also more complicated polymeric molecules such as proteins, polysaccharides, and phosphoglycerides. These exhibit a prop-

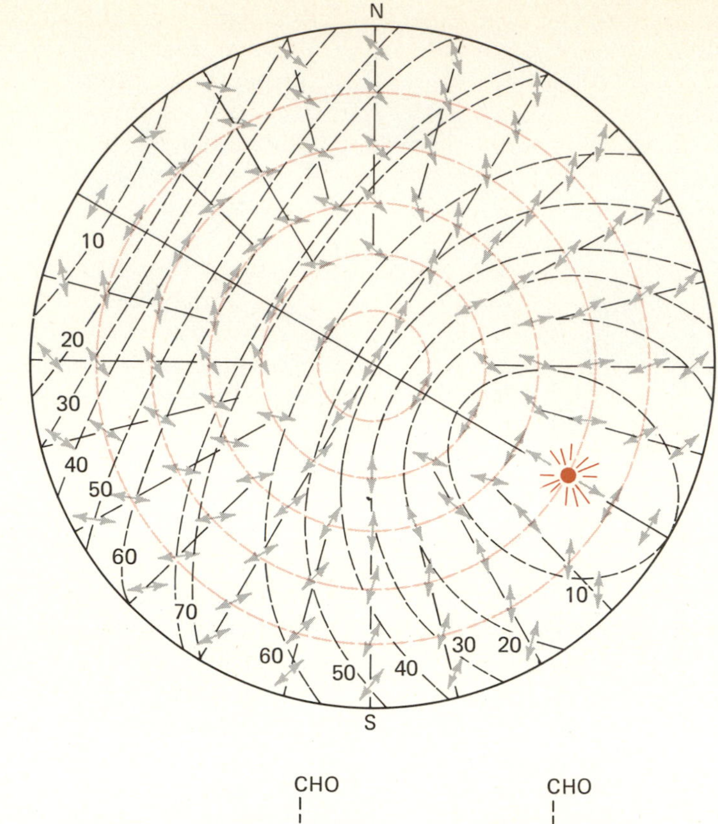

Figure 27.35 Polarization of light from the unclouded atmosphere. The arrows indicate the preferred direction of polarization when the sun is about 30° above the horizon in the ESE direction as shown. The numbers give the percentage of polarization within the corresponding dotted regions. Note that the maximum polarization obtains when the line of sight is at 90° to the line toward the sun's position.

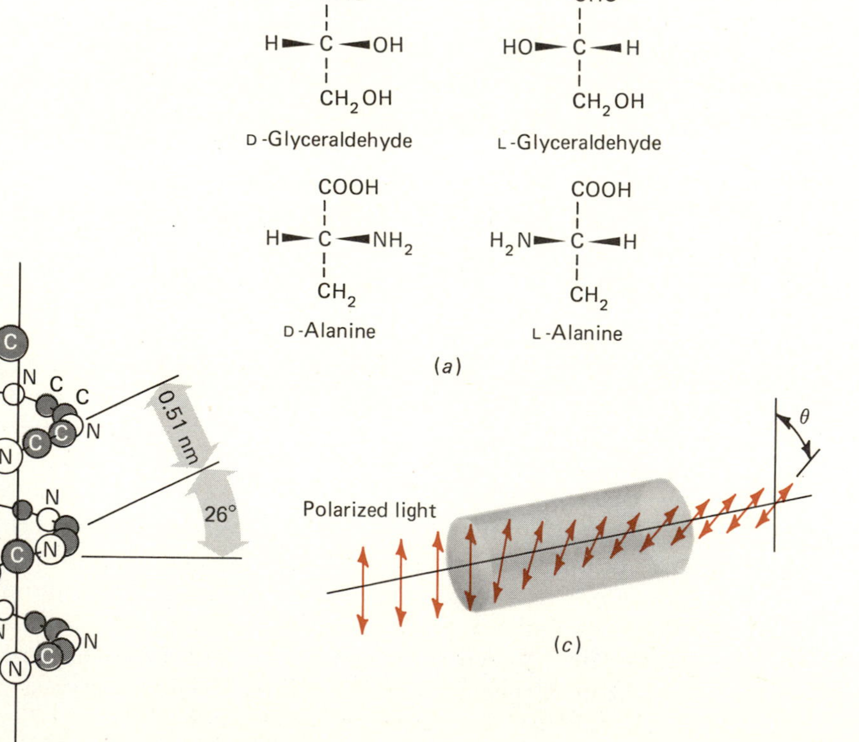

Figure 27.36 (*a*) The chemical configuration of stereoisomers of glyceraldehyde and alanine. Stereoisomers have essentially the same chemical composition; they differ, in that one isomer is a mirror image of the other, changing a left-handed spiral into a right-handed one. (*b*) Proteins are long-chain polymers of amino acids. They combine to form a helical structure, the so-called α helix shown here. (*c*) When polarized light passes through a solution containing molecules with distinctly right-handed or left-handed properties, the direction of polarization of the light is rotated.

erty called *optical activity*. A material is said to be optically active if it causes *rotation of the direction of polarization* of light that passes through it. Table 27.1 lists the *specific rotation* of some amino acids in aqueous solution. The specific rotation is arbitrarily defined as the observed rotation in millidegrees per centimeter of path length for a concentration of 1 g/100 mL. Some amino acids show positive rotation, some negative; they are *dextrorotatory* or *levorotatory*, respectively. When a number of such amino acids link to form a protein, the rotatory power of the protein is generally not the sum of the individual rotatory powers of the constituent amino acids but depends on the structure of the polypeptide configuration. Optical activity is often a useful analytic tool especially in conjunction with other methods.

One can easily observe optical activity by placing a transparent container between two crossed Polaroids. If the container is filled with Karo syrup, you will find that some light is transmitted through the system and that you must rotate the second polarizer to achieve another transmission minimum. Diluting the syrup with water will reduce the rotatory power of the solution.

Table 27.1 Specific rotation of some amino acids in aqueous solutions

Amino acid	Specific rotation
L-Alanine	+1.8
L-Arginine	+12.5
L-Leucine	−11.0
L-Isoleucine	+12.4
L-Phenylalanine	−34.5
L-Glutamic acid	+12.0
L-Histidine	−38.5
L-Aspartic acid	+5.0
L-Methionine	−10.0
L-Lysine	+13.5
L-Serine	−7.5
L-Proline	−86.2
L-Threonine	−28.5
L-Tryptophan	−33.7
L-Valine	+5.6

Summary

The wave nature of light is manifested by *interference* and *diffraction,* phenomena that arise from the superposition of waves. *Interference* refers to the effect of superposition of a few waves; *diffraction* is the term used to describe the effect of superposition of many waves.

Interference and diffraction can be observed only with *coherent* sources. Two sources are said to be coherent if they maintain a constant phase relation between the waves that they emit.

Two closely spaced slits, illuminated by a single monochromatic source, produce an interference pattern on a distant screen. The intensity at the screen is given by

$$I = 4I_0 \cos^2 \frac{\pi d x}{\lambda D}$$

where d is the separation between the slits, x is the distance on the screen from the central maximum, and D is the distance between the slits and the screen. The positions of the interference maxima are given by

$$x_m = \frac{m\lambda D}{d} \qquad m = 0, 1, 2, 3, \ldots$$

The above results are valid only if $D \gg d \gg \lambda$.

Light reflected from the front and back surfaces of a thin film also produces interference effects. For an unsupported film of thickness t and index of refraction n, the condition for constructive interference is

$$t = \frac{(m + \frac{1}{2})\lambda}{2n} \qquad m = 0, 1, 2, \ldots$$

Very thin unsupported films appear black when viewed by reflected light. Lens coatings use thin films whose index of refraction is intermediate between air and glass to minimize reflection of incident light.

An *interferometer* is a device that uses interference between two coherent beams in optical measurements. It can be used to measure wavelengths and refractive indices, and it has also been applied effectively in astronomical studies.

Holography is a technique for producing a three-dimensional image of an object. A *hologram* is a photographic record of the interference pattern generated by a reference beam and the light scattered by the object.

When the hologram is illuminated by a coherent source, a three-dimensional image is formed.

A *diffraction grating* is a mask containing numerous uniformly spaced slits, or reflecting surfaces. When diffraction gratings are illuminated with monochromatic light, they produce a series of very sharp and intense lines. The angles at which these lines appear are given by

$$\sin \theta_m = \frac{m\lambda}{d} \qquad m = 0, 1, 2, \ldots$$

Here m is called the *order* of the diffraction maximum and d is the separation between lines of the grating.

Diffraction gratings are used to disperse polychromatic light into its spectral components. Generally, several spectra are observed, one for each order, and may overlap in higher orders.

A single-slit aperture creates a *diffraction pattern,* with an intense central maximum and alternate dark and bright bands of decreasing intensity bordering that central maximum. The angle at which the *minima* of the diffraction pattern appear is given by

$$\theta_m \simeq \sin \theta_m = \frac{m\lambda}{d} \qquad m = 1, 2, 3, \ldots$$

The resolving power of many optical instruments is limited by diffraction. The minimum angle for resolving two sources is given by the *Rayleigh criterion.* For a circular aperture of diameter d, this is

$$\theta_{\min} = \frac{1.22\lambda}{d}$$

The *polarization direction* of an electromagnetic wave is the direction of vibration of the electric-field vector.

Light from an ordinary source is unpolarized or randomly polarized. Such unpolarized light can be polarized by passing it through a *birefringent* crystal or a *dichroic* material, or by specular reflection.

A polarizing sheet that is used to ascertain the direction of polarization is called an *analyzer.* If polarized light of intensity I_1 passes through an ideal analyzer, the intensity of the transmitted beam is given by

$$I_2 = I_1 \cos^2 \theta$$

where θ is the angle between the direction of the analyzer and the direction of polarization of the incident beam.

When light is scattered by small particles or air molecules, the scattered light is partially polarized. It is this effect that causes the polarization of light from the sky.

Molecules with a spiral structure, which includes many important organic molecules, are *optically active.* Such molecules cause the direction of polarization to rotate as light passes through the optically active material.

Multiple Choice Questions

27.1 Mark the correct statement.

(a) The resolving power of a microscope is improved if blue light is used instead of red.

(b) The resolving power of the human eye is about 3–4 times better than the diffraction limit suggests.

(c) The compound eye of an insect is superior to the mammalian eye in resolving power because each ommatidium has a small acceptance angle.

(d) None of the above statements is correct.

27.2 A double-slit interference experiment is set up in a chamber that can be completely evacuated. With monochromatic light, an interference pattern is observed when the container is open to the air. As the container is evacuated, a careful observer will note that the interference fringes

(a) disappear completely.

(b) move slightly further apart.

(c) move slightly closer together.

(d) do not change at all.

(e) change their color more toward the red.

27.3 A beam of unpolarized light is propagating in the x direction. The electric field vector

(a) may oscillate in any arbitrary direction in space.

(b) must oscillate along the z direction.

(c) must oscillate in the y-z plane.

(d) must oscillate in the x direction.

(e) must have a steady component in the x direction.

27.4 You are walking along a street just before sunset, going due west, next to a modern building with glass walls that reflect the sun's rays into your eyes. You put on a pair of Polaroid sunglasses and find that

(a) reflected glare from the windows is much reduced because the Polaroids absorb light polarized horizontally.

(b) reflected glare is much reduced because the Polaroids normally absorb light that is polarized vertically.

(c) reflected glare is reduced just as much as light from other sources; that is, the entire scene is somewhat darkened, but the reflected light from the windows is not particularly diminished.

(d) reflected glare from the windows is actually enhanced compared with other light because the Polaroids normally absorb light that is polarized vertically.

(e) reflected glare from the windows is actually enhanced compared with other light because the Polaroids normally absorb light that is polarized horizontally.

27.5 In single-slit diffraction,

(a) the intensities of successive maxima are roughly the same, falling off only gradually as one goes away from the central maximum.

(b) the central maximum is about as wide as the other maxima.

(c) the central maximum is about twice as wide as the other maxima.

(d) the central maximum is much narrower than the other maxima.

(e) the slit width must be less than one wavelength for a diffraction pattern to be apparent.

27.6 In which of the following is diffraction *not* exhibited?

(a) Viewing a light source through a small pinhole.

(b) Examining a crystal by X rays.

(c) Using a microscope under maximum magnification.

(d) Resolving two nearby stars with a telescope.

(e) Determining the direction of polarization with a birefringent crystal.

Problems

(Section 27.3)

27.1 Two slits are separated by 0.3 mm and are illuminated with a monochromatic coherent light source whose wavelength is 615 nm. Adjacent fringes on a screen are separated by 1.4 cm. What is the distance between the screen and the slits?

27.2 In a double-slit experiment, the mask containing the slits is 1.4 m from the viewing screen. The slits are illuminated with coherent light of 590-nm wavelength. If the number of bright fringes per centimeter on the viewing screen is 16, what is the slit separation in the mask?

27.3 The interference pattern formed by two slits is such that the separation between adjacent maxima is 1.2 cm when the pattern is viewed on a screen 2.4 m from the slits. If the slit separation is 0.2 mm, what is the wavelength of the light with which the slits are illuminated?

27.4 When a double slit is illuminated with monochromatic light of 600-nm wavelength, the third noncentral maximum appears at an angle of 12° from the incident beam direction. What is the slit separation?

27.5 The slit separation of a double slit is 4×10^{-5} m. When the slits are illuminated with monochromatic light, the separation between adjacent maxima on a screen 2 m from the double slit is 1.2 cm in the neighborhood of the central maximum. What is the wavelength of the light used to illuminate the slits?

• **27.6** A double slit with a slit separation of 2×10^{-5} m is illuminated at normal incidence with light of two wavelengths, 533 nm and 400 nm. Locate the first five interference maxima on a screen 2 m from the slits. Do these maxima overlap for orders other than $m = 0$?

• **27.7** A soap film appears red when viewed at normal incidence in white light. The index of refraction of the soap solution is 1.4. What is the smallest thickness this film could have, if we take 640 nm as the wavelength of red light?

• **27.8** A soap film appears green ($\lambda = 510$ nm) when viewed with white light incident normally to the film. If the index of refraction of the film is 1.37, what is the minimum film thickness?

• **27.9** A film of thickness 5×10^{-7} m has an index of refraction of 1.3. For what wavelengths in the visible part of the spectrum will this film give constructive interference on reflecting light at normal incidence?

• **27.10** The wavelength of a monochromatic source is measured with a Michelson interferometer. When the mirror is moved a distance of 0.26 mm, 878 fringes pass by the viewing screen. What is the wavelength of the light?

• **27.11** What must the mirror displacement be in a Michelson interferometer so that 600 fringes pass across the screen when light of wavelength 640 nm is used?

• **27.12** How much must the mirror in Michelson's interferometer be displaced to cause 800 fringes to pass by the observer's eye when the wavelength of the source is 540 nm?

• **27.13** An evacuated glass tube is placed in one arm of a Michelson interferometer so that the light must pass through the tube, which is 10 cm long. As air is slowly admitted into the tube, fringes pass before the eye of the observer. By the time the air pressure in the tube has reached atmospheric, 150 fringes of 400-nm-wavelength light have been counted. What is the index of refraction of air as determined from this experiment?

• **27.14** A wedge is formed (Figure 27.37) by placing two pieces of flat glass plates one atop the other, with the top plate touching the bottom one along one edge and separated from it by a 0.1-mm-thick sheet of paper at the other edge. The two edges are 4 cm apart. If this arrangement is now illuminated from above with monochromatic light of 620-nm wavelength, one observes a series of parallel bright and dark fringes. What is the separation between the adjacent bright fringes?

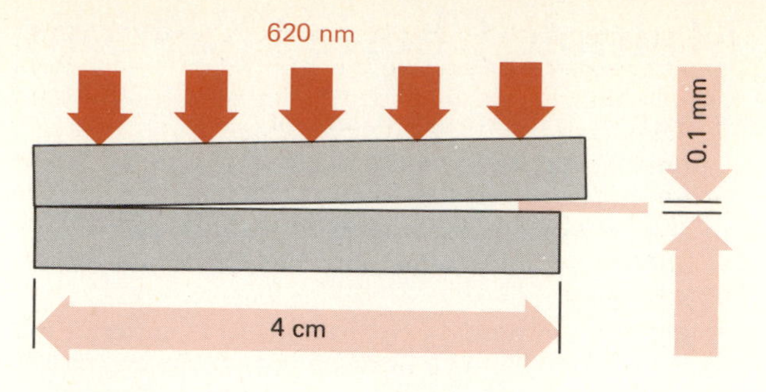

Figure 27.37

• • **27.15** Suppose the wedge of Problem 27.14 is filled with an oil whose index of refraction is 1.8. How will this change the separation between adjacent bright fringes?

• • **27.16** A plano-convex glass lens whose radius of curvature is 0.8 m is placed on a flat plate and illuminated from above with light of 550-nm wavelength. The resulting pattern, known as Newton rings, is a series of concentric light and dark circular fringes, with the central region a dark circle. Why is the central region dark? What are the radii of the first and second bright rings?

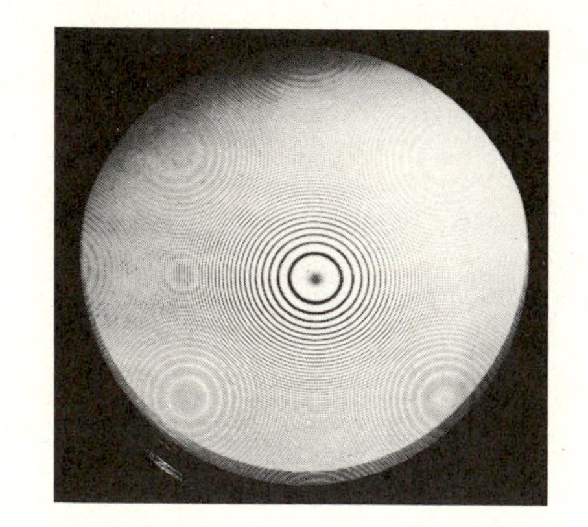

Figure 27.38

• • **27.17** Suppose that before placing the lens on the glass plate of Problem 27.16, a drop of oil with index of refraction 1.8 is deposited on the plate. If the index of refraction of the glass is 1.5, how will the presence of the oil modify the pattern of Newton rings? Will the central region be dark or bright? What will be the radii of the first and second bright circles?

• • **27.18** Derive an expression for the radius of the mth dark Newton ring of Problems 27.16 and 27.17.

• • **27.19** It is sometimes observed that the picture on a TV fades and reappears at regular intervals. This occurs if one is watching the transmission from a distant station while an airplane is flying in the vicinity of the receiver. In this case the line-of-sight wave from the transmitter and that reflected to the receiver from the airplane interfere with one another, constructively or destructively, depending on the location of the plane. Suppose that the transmitter is 40 km from the receiver and is transmitting at a frequency of 88 MHz. If a plane is flying directly above the receiving antenna, radially outward from the transmitting station, at an altitude of about 400 m, what is the precise altitude of the plane if the TV picture fades just as the plane is exactly overhead?

• • **27.20** Suppose that the plane in the Problem 27.19 is flying at 400 km/h. At what rate will the picture on the TV screen fade in and out while the plane is flying over the receiver?

(Section 27.5)

27.21 Light of 540-nm wavelength falls on a 0.02-mm-wide slit. Determine the angular width of the central diffraction maximum. That is, what is the angular separation between the two minima on either side of the central maximum?

27.22 The width of the central maximum produced by a slit 0.016 mm wide is 3°. Determine the wavelength of the incident light.

27.23 The second-order line of wavelength 520 nm propagates at an angle of 18° with the normal to a grating when the grating is illuminated at normal incidence. Determine how many lines are on this grating if the grating is 2 cm wide.

27.24 What is the angular width of the first-order spectrum of white light that is dispersed by a grating with 5000 lines/cm?

• **27.25** How wide is a slit that gives a diffraction pattern such that the angular separation between the first and second minimum is 4° when the slit is illuminated with light of 589-nm wavelength?

• **27.26** A factory whistle emits a steady tone of 1000 Hz on a day when the temperature is near 0 °C. A high wall stands between the factory and the outside. There is an opening in the wall for a narrow door, about 1 m wide. As a girl walks parallel to the wall at a distance of 50 m from it, she hears the whistle most intensely when she is directly opposite the open door; as she walks farther, she notices that the sound diminishes to a distinct minimum and then increases again. What is the distance she walks between the maximum and the minimum positions?

• **27.27** A narrow slit 0.08 mm wide is illuminated with monochromatic light, and a diffraction pattern is observed on a screen 5 m from the slit. The width of the central maximum of this pattern is 4 cm. What is the wavelength of the incident light? How would this pattern change, if at all, if the slit width were increased to 0.16 mm?

• **27.28** A grating gives rise to a third-order diffraction maximum for yellow light of wavelength 590 nm at the same angle as the diffraction maximum of blue light in fourth order. What is the wavelength of the blue light?

• **27.29** Figure 27.39 shows three different intensity patterns. These were observed on a screen 5 m from slit structures that were illuminated with monochromatic light of wavelength 480 nm. What sort of slit arrangement produced each of the patterns, and what were the respective slit separations or slit widths?

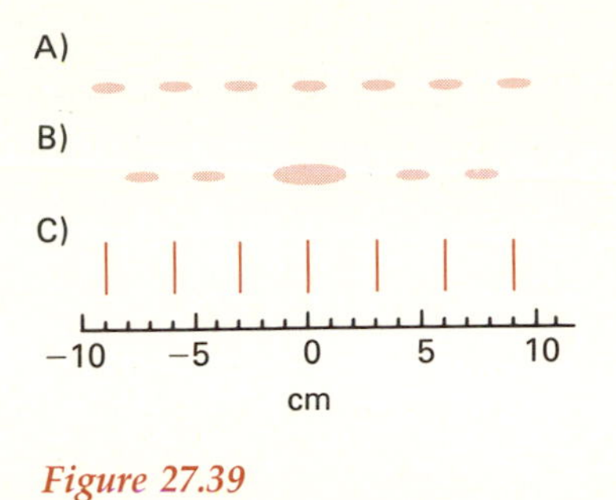

Figure 27.39

• **27.30** Monochromatic light is incident on a transmission grating that has 1000 lines/cm. The pattern on a screen 6 m from the grating shows a series of regularly spaced narrow lines separated by 20 cm. What is the wavelength of the illumination if the light falls on the grating at normal incidence?

• • **27.31** Suppose that in Problem 27.30 the incident beam makes an angle of 30° with the normal to the plane of the grating. What is then the wavelength of the incident light, if the pattern is as described in Problem 27.30 and the screen is normal to the beam?

• **27.32** A diffraction grating has 4000 lines per centimeter and is illuminated with white light at normal incidence. The transmitted light is displayed on a screen 1 m from the grating. Calculate the positions of the first-order and second-order maxima for light of 500-nm wavelength.

• **27.33** For the grating of Problem 27.32, red light of 680-nm wavelength overlaps light of 510-nm wavelength. What is the order of the 680-nm and 510-nm spectrum for which this phenomenon occurs? Is there another part of the visible spectrum that also overlaps at that position?

• **27.34** How many complete spectra can be observed with the grating of Problem 27.32?

• • **27.35** A diffraction grating has 4000 lines/cm. What is the angular separation between the blue (436 nm) and green (546 nm) emission lines from the spectrum of mercury vapor in first and in second order? How many orders can be observed with this transmission grating, in which both lines are visible?

• • **27.36** A grating with 6000 lines/cm is used to study the small spectral region extending from 480 nm to 505 nm. What is the angular separation of this spectral region in first order and in the largest visible order?

• • **27.37** A grating gives no third-order spectrum for any part of the visible region. What is the least number of lines per centimeter for this grating?

• • **27.38** When the interference pattern from a double slit is examined, one finds that the intensity of the maxima gradually diminishes as the order increases; it falls to zero and then gradually increases to a subsidiary maximum, decreases again, and so on. This slow intensity variation is the result of diffraction by the slits. When a double slit is illuminated with 580-nm light and viewed on a screen 2 m from the slits, the interference maxima are separated by 1.8 cm. The intensity of the maxima diminishes with increasing m, and for $m = 8$, the intensity is zero. Determine the separation between the two slits and the width of each slit.

• • **27.39** A transmission grating is used to disperse white light incident perpendicular to the grating. On a screen 10 m from the grating, the wavelength of 500 nm and 600 nm are separated by 8 cm in the second-order spectrum. What is the separation of these wavelengths in the first-order spectrum? Which of the two colors is farther from the central slit image? What is the number of lines per centimeter on this grating?

(Section 27.6)

• **27.40** Railroad tracks are spaced about 1.7 m apart. On a clear day, how high could a plane fly before the pilot is no longer able to resolve the two separate rails? Assume that the diameter of the pupil is about 3 mm and take as the wavelength that to which the eye is most sensitive, namely 530 nm.

• **27.41** Your friend claims that he can resolve two sources of light with his bare eyes just as well as with binoculars. He claims that the binoculars only magnify the object but do not help in defining details. Is he correct in this assessment?

• **27.42** If the pupil diameter of the dark-adapted eye is 4 mm, what is the maximum distance at which two point sources of light that are 3 mm apart can still be resolved? (You might try to confirm this by using a cardboard mask with two small pinholes placed in front of a light source that is completely enclosed.)

• **27.43** Repeat Problem 27.42, if the sensing instrument is not the human eye but the 200-inch-diameter Mt. Palomar telescope.

(Section 27.8)

27.44 What is Brewster's angle for an interface between air and glass of refractive index 1.5?

27.45 If the glass plate of Problem 27.44 is submerged below the surface of water, what is then Brewster's angle?

• **27.46** Unpolarized light is incident from the left on a polarizer that has its polarizing axis vertical. The light, after passing through the first polarizer, then passes through another such unit. If the intensity of the incident, unpolarized beam was I_0 and the intensity of the emerging beam is $I_0/4$, what is the angle between the polarizing axes of the two polarizers?

• **27.47** A person studying the polarization of the light from the sky on a clear day finds that it is most pronounced if he looks in a direction that is 37° above the horizon and due west. Where is the sun when this observation is made?

• **27.48** Two Polaroids are placed into the unpolarized beam of a light source. The angle between the axes of polarization of the two Polaroids is 60°. If the intensity of the incident beam is I_0, what is the intensity of the transmitted light?

• **27.49** A beam of vertically polarized light is passed through an aqueous solution of L-glutamic acid whose concentration is 80 g of glutamic acid per liter. The path of the beam in the solution is 18 cm. How should a second polarizer placed after the container of glutamic acid be oriented so that the transmitted light will be at a minimum?

• **27.50** What is the relative concentration of L-arginine and L-phenylalanine in aqueous solution that gives zero rotation of the plane of polarization?

• • **27.51** Unpolarized light of intensity I_0 is incident on the following arrangements of polarizing sheets. The first polarizer is oriented vertically; the second is oriented so it makes an angle of 30° with the vertical; the third and last polarizer makes an angle of −30° with the vertical. Determine the intensity of the light transmitted through this group of polarizers as a fraction of I_0.

• • **27.52** If in Problem 27.51 the first and second polarizers are interchanged, will that alter the transmitted intensity? If so, how?

•28•

* Relativity

The relativity theory arose from necessity, from serious and deep contradictions in the old theory from which there seemed no escape. The strength of the new theory lies in the consistency and simplicity with which it solves all these difficulties, using only a few very convincing assumptions. . . . The old mechanics is valid for small velocities and forms the limiting case of the new one.

A. EINSTEIN AND L. INFELD, *The Evolution of Physics* (1938)

Although the topic of this chapter is of great interest, it is much more difficult than the material in any of the others. An understanding of relativity is, however, not a prerequisite for the remaining chapters of this textbook. All that is used in Chapters 29–32 is the mass-energy equation, $E = mc^2$, which could be accepted as a law of nature *ad hoc,* without reference to the special theory of relativity. The entire chapter has therefore been designated as optional material.

28.1 Introduction

The first two decades of the twentieth century brought forth a revolution in natural philosophy that many regard as more profound and devastating of accepted dogma than the Copernican-Galilean revolution. Within this brief span of time there surfaced the theory of relativity (actually two theories, special and general relativity) and the quantum theory. The new theories, in sharp contrast to the works of Clausius, Boltzmann, Maxwell, Kelvin, Rayleigh, and others, were not extensions and elaborations of Newtonian mechanics but marked instead a precipitous break with the fundamental postulates and premises of Newton. Many contemporary scientists considered the two new theories so utterly contrary to common sense and experience that they simply could not but reject them.* It is curious, though not altogether surprising, that most scientists found it easier to accept the quantum theory than the theory of special relativity, even though the former really constitutes a much more dramatic and significant departure from Newtonian physics than the latter, which is still in the classical tradition.

These new theories not only were upsetting to older physicists; they shook the very foundations of nineteenth-century philosophy so firmly based on the *a priori* assumption of causality. By causality we simply mean that in a well-defined system, a specific cause must lead to a uniquely predictable effect. But this eminently rational postulate of causality was suddenly called into question by the statistical interpretation given to quantum theory by Max Born, an interpretation currently accepted by the vast majority of physicists. Especially Einstein, who could hardly be accused of an unimaginative or conventional frame of mind, never accepted this probabilistic viewpoint; it was in this context that Einstein made his now famous remark *"Gott würfelt nicht!"* (God does not throw dice!)

In the next five chapters we shall be concerned with these developments of the twentieth century. The light they have shed on the microcosmos of atoms and nuclei and the macrocosmos of stars and galaxies has nurtured the almost miraculous technological progress without which life as we know it today would not only be impossible but even unimaginable.

The theory of special relativity appeared under the guise of the title "On the Electrodynamics of Moving Bodies," one of five articles published by Einstein in Volume 17 of the *Annalen der Physik* in 1905, the *anno mirabilis* of modern physics. That volume also contains the article that explains the origin of Brownian motion and was instrumental in silencing the last opponents of the atomic hypothesis; a paper on the generation and transformation of light which contains, almost as an aside, the explanation of the photoelectric effect that the Nobel Committee cited in November 1922, in making its belated award of the 1921 Nobel Prize to Albert Einstein; and a brief, three-page article entitled "Does the Inertia of a Body Depend on its Energy Content?" in which the now universally recognized equation $E = mc^2$ first appears. Little wonder that today libraries that do have a copy of Volume 17 of *Annalen der Physik*, the "Einstein volume" as it is often called, keep it under lock and key in their special collection along with other rare and valuable books.

These monumental achievements of scientific creativity were, however, not the result of a sudden spurt of activity but the culmination of deep and careful analysis and introspection. In later years, Einstein recounted that on and off, he had pondered the problem treated in the first of the relativity papers for at least a decade, and credited his success

* One is reminded of the words of Max Planck, one of the early architects of quantum physics: "An important scientific innovation rarely makes its way by gradually winning over and converting its opponents. . . . What does happen is that its opponents gradually die out, and that the growing generation is familiarized with the ideas from the beginning."

Figure 28.1 Albert Einstein. (1879–1955)

partly to the tranquility and scientific isolation of his employment in the Swiss patent office. Since the job made no demands on his mental faculties, he could concentrate his efforts on more important matters.

28.2 Classical Relativity

To appreciate the problem Einstein and his contemporaries faced, we must first briefly review the predictions of Newtonian mechanics. In Chapter 2 we considered the descriptions of an object as viewed by observers moving at constant velocity relative to each other. We studied as an example the motion of a canoe crossing a uniformly flowing stream; while to an observer on a raft in the stream the canoe appears to travel at right angles to the river bank, to one at rest ashore the canoe's velocity makes a lesser angle with the shore line.

There is a simple relation between the velocity of an object in one reference frame S (which we may arbitrarily designate the "rest frame") and in another frame S' that has a constant velocity $\mathbf{u}$ relative to S, where $\mathbf{u}$ is assumed in the x direction:

$$v'_x = v_x - u \qquad v'_y = v_y \qquad v'_z = v_z \tag{28.1}$$

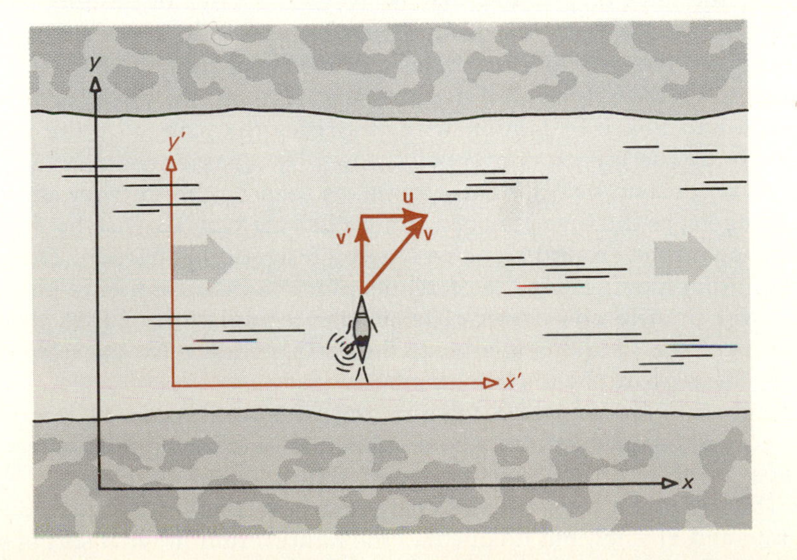

Figure 28.2 In the coordinate system (x', y'), at rest relative to the water of the stream, the canoe is headed at right angles to the shoreline. Viewed from shore, that is, in the coordinate system (x, y), the velocity $\mathbf{v}$ of the canoe makes an angle less than 90° with the shoreline.

It then follows that if in a time t the canoe's displacement is $\mathbf{r}$ with components

$$x = v_x t \qquad y = v_y t \qquad z = v_x t$$

in the S frame, the displacement $\mathbf{r}'$ in the S' frame will have the components

$$\begin{aligned} x' &= v'_x t = (v_x - u)t \\ y' &= v'_y t = v_y t \\ z' &= v'_z t = v_z t \end{aligned} \qquad \textbf{(28.2)}$$

Equations (28.2) are those of a *Galilean transformation* from one inertial frame to another. They are not only formally simple but make eminently good sense. Indeed, all of us use the Galilean transformation constantly, at least subconsciously. In deciding if we can safely cross a road before the oncoming traffic reaches us, if we can pass a car traveling at 60 km/h on a two-lane road without risking a head-on collision with another approaching in the distance, we instinctively apply these coordinate transformations.

In Chapter 2 we also pointed out that there does not seem to be any one coordinate system that can be identified as the "true" rest frame. This was already recognized by Newton who wrote:

> The parts of space cannot be seen, or distinguished from one another by our senses. . . . From the positions and distances of things from any body considered as immovable, we define all places; and then with respect to such places, we estimate all motion. . . . And so, instead of absolute places and motions, we use relative ones.

But, he also wrote,

> The center of the system of the world is immovable. This is acknowledged by all, while some contend that the earth, others that the sun, is fixed in that center.

The formal, mathematical reason that an identification of the true "immovable" center of the world, the absolute rest frame, is impossible is that the laws of physics, that is, Newton's laws of motion, are not changed by a Galilean transformation. Consequently, there simply is no experiment that we can ever perform in reference frame S that can tell us whether we are moving and frame S' is at rest, or vice versa, or whether both S and S' are in motion with velocities that differ by the relative velocity $\mathbf{u}$. The trajectory of a ball thrown by a passenger in an airplane flying at constant velocity is the same, as seen by the thrower and other passengers, as the trajectory of a similar ball thrown likewise by a person standing on the ground and observed by him and his friends. The question, "Am I moving?" is therefore meaningless because nature precludes our ever finding an answer. The precise formulation of that question must be, "Am I moving relative to the earth's surface (or the stern of the ship, the cockpit of the plane, etc)?"

There the matter rested for two centuries until Maxwell formulated his equations and deduced the existence of electromagnetic waves that propagate through space at the speed of light.

That light is a wave phenomenon had been known since the days of Young and Fresnel, but no one had the faintest notion what sort of wave it might be (except that it had to be a transverse wave), or how to describe it in formal mathematical language. Long before Maxwell, physicists struggled to identify the medium that transmits these waves, that oscillates at almost unimaginably high frequencies. All known waves were transmitted by a medium—surface waves on water by a liquid, sound waves by air or an elastic solid. And so scientists and philosophers simply invented the medium that supports light waves, calling it the luminiferous ether, or simply *the ether.*

The trouble was that this ether was apparently endowed with bizarre properties to say the least. Newton had demonstrated that the speed of propagation of a wave in an elastic medium is greater the greater the stiffness of the medium. A medium transmitting a wave at the speed of light would have to be incredibly resilient. Yet, this same medium in no way impeded the motion of objects, and moreover, it resisted all attempts to remove it from a region. For example, while sound cannot propagate through an evacuated space, no vacuum pump was ever able to remove the ether. The ether was the most tauntingly elusive thing ever encountered.

Though unable to reveal its presence and measure its properties, physicists adamantly defended its existence. That defense was based on the unshakable conviction that there could be no true force at a distance, that is, that physical events must be transmitted through the stress or distortion of some medium. To quote Maxwell,

> Whatever difficulties we may have in forming a consistent idea of the constitution of the ether, there can be no doubt that the interplanetary and interstellar spaces are not empty, but are occupied by a material substance or body which is certainly the largest, and probably the most uniform body of which we have any knowledge.

Sooner or later the ether would reveal itself to a really ingenious experimenter. With the work of Maxwell, that opportunity suddenly presented itself. Maxwell's equations predicted electromagnetic waves propagating at the speed $c = 3 \times 10^8$ m/s; but, it was asked, with respect to what? The only sensible, reasonable answer was, *With respect to the luminiferous ether!*

With all space, as Maxwell had stated, permeated by this ether and the earth itself surrounded and embedded in this medium, one had to choose one of two possible options. Either the earth, as it traveled in its orbit about the sun, was moving through the ether—though one could discern no drag or notice any other effects of this relative motion—or the earth, in its motion, dragged the ether along, much as the earth's atmosphere rotates and travels along with the solid globe.

The second possibility had already been eliminated by James Bradley in 1725, when he explained stellar aberration in terms of the finite speed of light and deduced a speed in good agreement with that derived from other astronomical and subsequent terrestrial measurements. Hence the earth must be traveling, at least during most of the year if not at all times, with some finite velocity relative to the ether. At last one could identify that elusive true rest frame, the inertial frame of the ether. All that needed to be done was to measure carefully the speed of light on earth in several directions—for example, parallel and perpendicular to the velocity of the earth in its orbit about the sun—and compare the results.

That was more easily said than done. The average speed of the earth in its orbit is about 30 km/s, or about 1/10,000 of the speed of light. Still, it should be possible. In 1881, Albert A. Michelson did devise a technique capable of the requisite precision.

28.3 The Michelson-Morley Experiment

The schematic diagram of Figure 28.3 will assist in analyzing one of the most celebrated experiments in the history of physics. A narrow beam of light is split into two by a half-silvered mirror M_1. Beam A proceeds to mirror M_2 and is reflected to M_1. Here a portion is transmitted through the half-silvered glass plate and a portion is reflected. Beam B is reflected by mirror M_3, and a portion of this reflected beam is transmitted through M_1 and combines with the reflected portion of beam A. Since beams A and B derive from the same initial beam, they are coherent, and on combining, can interfere according to their relative phase. That phase relation is determined by the difference in the optical path lengths for beams A and B. That difference, we now show, depends on the motion of the apparatus relative to the ether!

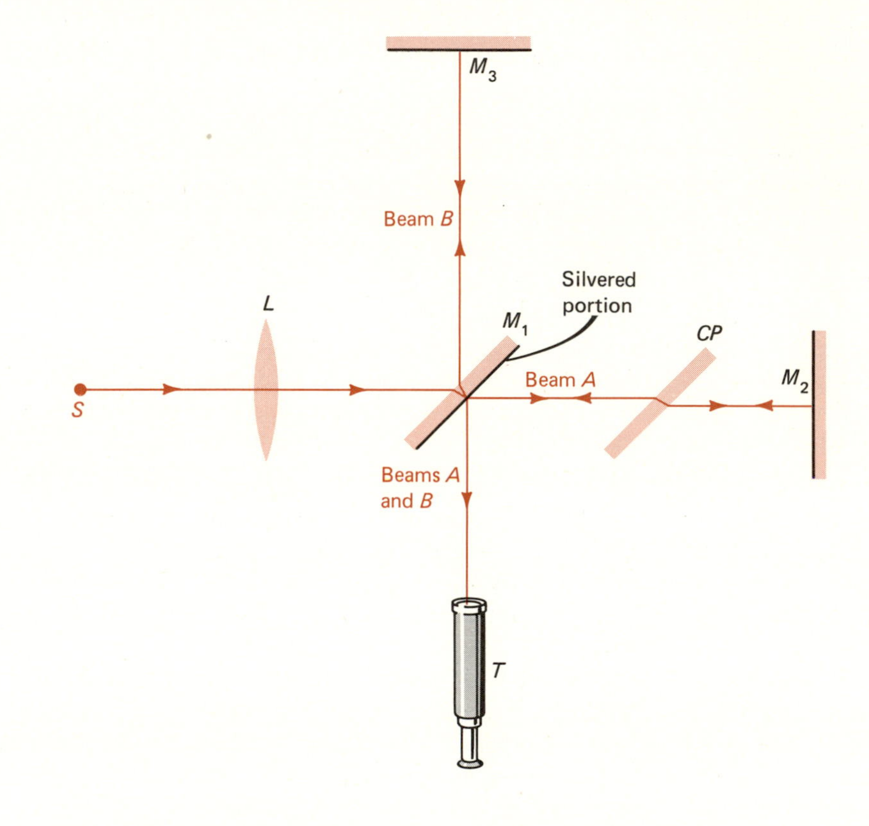

Figure 28.3 *Schematic diagram of the Michelson-Morley apparatus. Light from the source S is collimated by the lens L and then strikes the plate M_1. The back of this plate is partially silvered so that a portion of the incident beam is transmitted, passes through the compensation plate CP and is then reflected by mirror M_2. The portion reflected by the half-silvered surface of mirror M_1 strikes mirror M_3 and is reflected to M_1. At M_1 the beams from M_3 and M_2 are again partly transmitted and partly reflected, respectively, and then interfere. This interference pattern can be observed by means of the telescope T. The compensation plate CP, in light path A, assures that the optical paths of beams A and B are identical.*

Let us assume that at the time of the experiment, the earth moves with velocity $\mathbf{u}$ relative to the ether. During the first part of the light's travel along path A of Figure 28.4, the apparatus is moving in the direction of propagation, and hence, the speed of the light in the inertial frame of the apparatus is $c - u$; the time required for the light to go from M_1 to M_2 is $L/(c - u)$. On the return lap, the time is $L/(c + u)$ since mirror M_1 is approaching the light with velocity $\mathbf{u}$ relative to the ether. Thus for a round trip along A, the elapsed time is

$$t_A = \frac{L}{c-u} + \frac{L}{c+u} = \frac{2Lc}{c^2 - u^2} = \frac{2L}{c}\frac{1}{(1 - u^2/c^2)} \tag{28.3}$$

Let us now calculate the round-trip time for beam B. With the apparatus moving with velocity $\mathbf{u}$ relative to the ether, the path of light in that beam is as shown in Figure 28.4. The path length is therefore given by

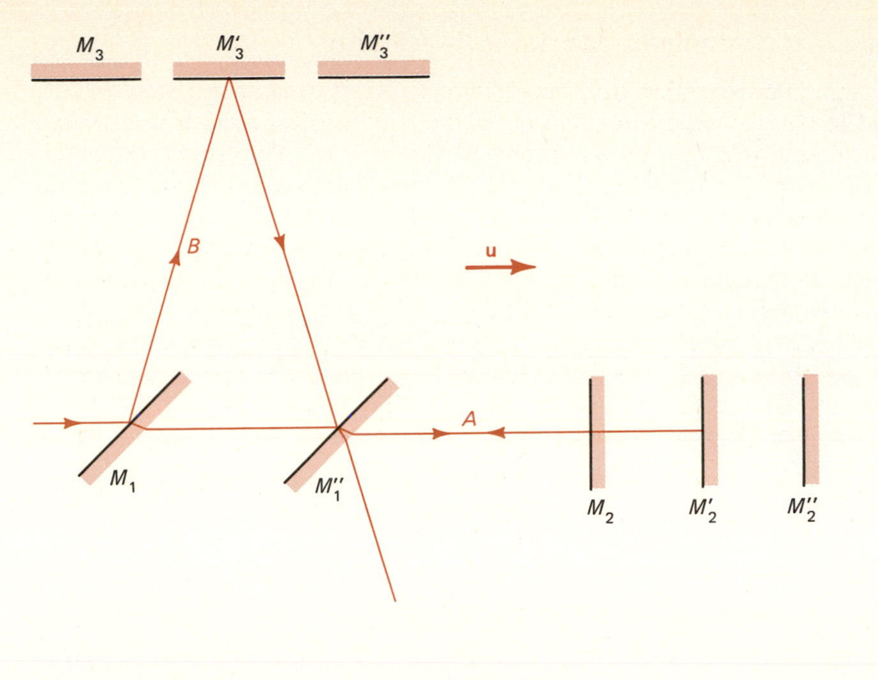

Figure 28.4 *The paths of beams A and B of Figure 29.3 as seen from a reference frame at rest relative to the ether. The motion of the apparatus is enormously exaggerated in this diagram. During the time that the light travels from* M_1 *to* M_2*, the mirror* M_2 *has moved from its initial position to* M_2'*; the reflected beam A meets the oncoming mirror* M_1 *at* M_1''*. Light reflected from* M_1 *to* M_3 *strikes mirror* M_3 *at* M_3'*, and the beam B then meets mirror* M_1 *at position* M_1''*.*

$$2\left[L^2 + \left(\frac{ut_B}{2}\right)^2\right]^{1/2} = ct_B \tag{28.4}$$

which simply states that the distance traveled by light in the rest frame of the ether in time t_B is ct_B. If we factor out L and divide both sides of (28.4) by c, we obtain

$$t_B = \frac{2L/c}{\sqrt{(1 - u^2/c^2)}} \tag{28.5}$$

Thus, as a result of the motion relative to the ether, there is a time difference between the two paths of

$$\Delta t = t_A - t_B = \frac{2L}{c}\left[\frac{1}{(1 - u^2/c^2)} - \frac{1}{\sqrt{(1 - u^2/c^2)}}\right] \tag{28.6}$$

Since $u^2/c^2 \simeq 10^{-8} \ll 1$, this expression can be greatly simplified, using the binomial expansion (Equation A.26). For $x \ll 1$, we get $(1 + x)^{-1} \simeq 1 - x$ and $(1 + x)^{-1/2} \simeq 1 - \frac{1}{2}x$. Equation (28.6) then reduces to

$$\Delta t = \frac{L}{c}\frac{u^2}{c^2} \tag{28.7}$$

If the apparatus is turned through 90°, the roles of paths A and B are interchanged, with the result that then the time difference is the negative of (28.7). Hence the total time shift as a result of such rotation will be twice the value of (28.7); that is,

$$\Delta t_{\text{tot}} = \frac{2L}{c}\left(\frac{u}{c}\right)^2$$

This translates to a phase shift of

$$2\pi\frac{\Delta t_{\text{tot}}}{T} = \frac{2\Delta t_{\text{tot}}\pi c}{\lambda} = \frac{4\pi L}{\lambda}\left(\frac{u}{c}\right)^2 \tag{28.8}$$

where T is the period of the light wave.

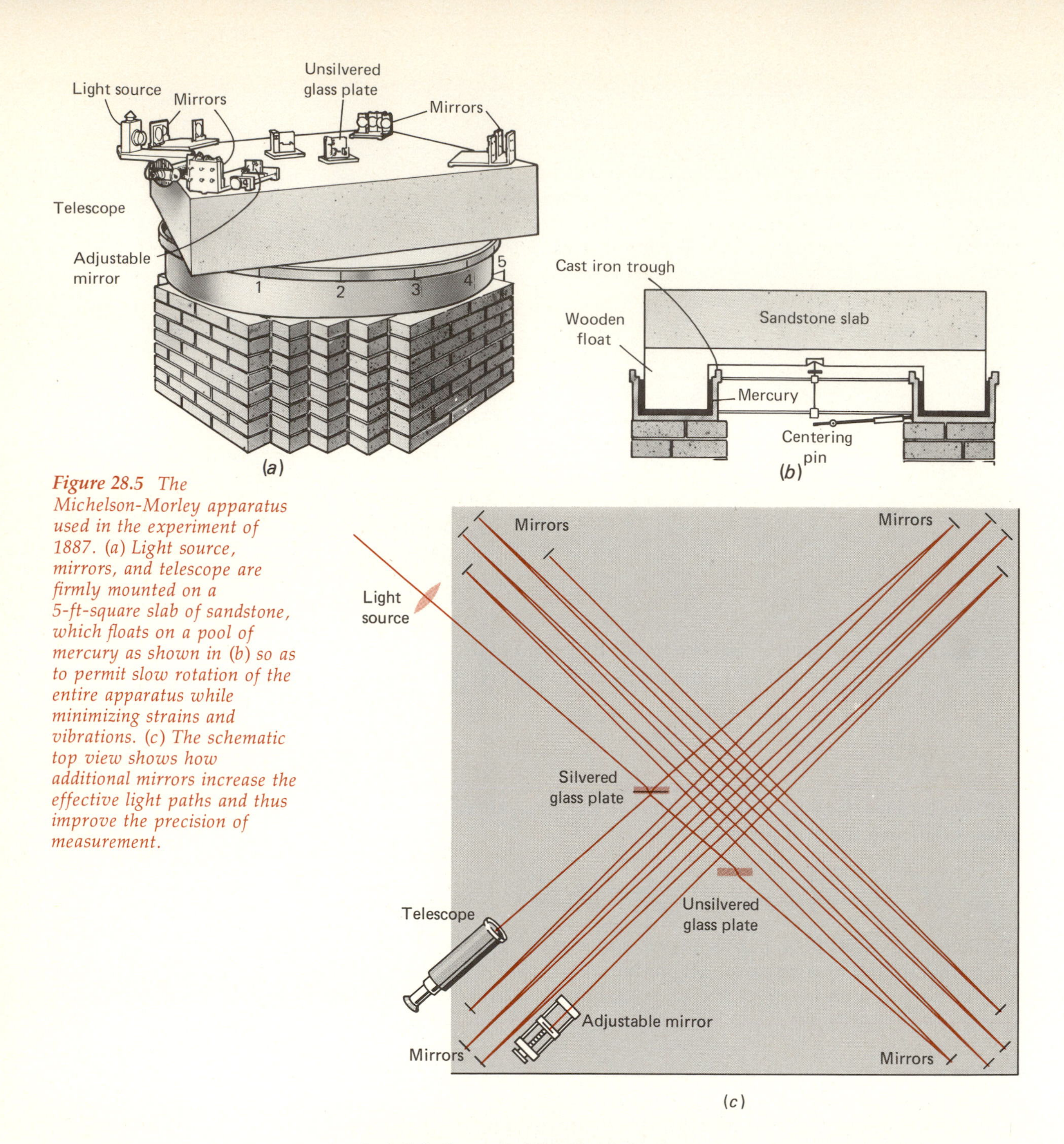

Figure 28.5 *The Michelson-Morley apparatus used in the experiment of 1887. (a) Light source, mirrors, and telescope are firmly mounted on a 5-ft-square slab of sandstone, which floats on a pool of mercury as shown in (b) so as to permit slow rotation of the entire apparatus while minimizing strains and vibrations. (c) The schematic top view shows how additional mirrors increase the effective light paths and thus improve the precision of measurement.*

In Michelson and Morley's final arrangement, shown in Figure 28.5, the optical system, mounted on a large stone slab and floated on mercury, included several additional mirrors to increase the optical path length to 22 m. Even though the factor $(u/c)^2$ is only about 10^{-8}, $2L/\lambda = (22\text{ m})/(5.5 \times 10^{-7}\text{ m}) = 4 \times 10^7$ for the wavelength of yellow light used in that experiment. The expected phase difference as a result of a 90° rotation is therefore, $0.4 \times 2\pi = 0.8\pi$ radians, a significant and readily measurable phase change.

The fruits of their labors seemed most disappointing:

> The displacement to be expected was 0.4 fringe. The actual displacement was certainly less than a twentieth part of this, and probably less than the fortieth part.

There remained one extremely remote possibility: By a quirk of fate, they had performed this experiment at the precise moment when the earth just happened to be nearly at rest relative to the ether. "The experiment," they wrote in their report, "will therefore be repeated at intervals of three months, and thus all uncertainty will be avoided." Michelson never performed these repetitions (which were carried out, with expected null results, by Morley and D. C. Miller). His deep disillusionment with this justly celebrated experiment is evident from the following passage written fifteen years after the event:

> I think it will be admitted that the problem, by leading to the invention of the interferometer more than compensated for the fact that this particular experiment gave a negative result.

That was in 1902, a few years before the appearance of Einstein's relativity theory. It is questionable, however, whether knowledge that his work had confirmed one of the most profound theories of nature would have altered his views. Indeed, we know otherwise, for when Einstein and Michelson met briefly in 1931, Michelson remarked that he regretted that his experiment might have been responsible for giving birth to such a "monster"—referring to the theory of special relativity.*

Following publication of the Michelson-Morley experiment, physicists wracked their brains trying to reconcile this negative result with the unequivocal conclusion from stellar aberration measurements that the earth did, indeed, move relative to the ether. The Irish theorist George F. FitzGerald proposed the following *ad hoc* explanation. As a result of motion through the ether, the linear dimensions of all objects are contracted along the line of relative motion by the fractional amount $\sqrt{(1 - u^2/c^2)}$. Since a meter stick placed alongside any object would suffer the same fractional contraction, there can be no operational procedure for verifying or disproving this assertion of FitzGerald. H. A. Lorentz, who independently had thought of this possibility, justified the hypothesis in terms of a possible change of electromagnetic forces between constituent atoms due to this motion. This length contraction, now known as the Lorentz-FitzGerald contraction, is also a consequence of Einstein's theory, but is not, in this theory, an arbitrary axiom contrived to explain an otherwise incomprehensible observation.

Figure 28.6 *Albert A. Michelson, Albert Einstein, and Robert A. Millikan at the California Institute of Technology in Pasadena, California, in 1931.*

* In fact, this experiment was not the motivation for the 1905 paper. As Einstein recalled, "In my own development, Michelson's result has not had a considerable influence. I even do not remember if I knew of it at all when I wrote my first paper on the subject."

28.4 The Special Theory of Relativity

28.4 (a) Postulates of the Theory

If not the Michelson-Morley experiment, what did motivate Einstein's relativity paper? The answer has to do with the fact that in contrast to Newton's equations, Maxwell's equations are *not* invariant under a Galilean transformation. This implies that there *is* a preferred reference frame in nature, namely the one in which Maxwell's equations are valid. They appear to be valid on earth, and so it might seem that Pope Urban and the Inquisition displayed prophetic foresight in steadfastly maintaining the geocentric theory and forcing Galileo to recant his heretical assertions.

The experimental manifestation of this lack of invariance of the field equations is the apparently different behavior of stationary and moving charges. Suppose we observe two charges separated by some distance d and at rest relative to each other and to us. The force of repulsion between these charges is a measurable quantity and is given by Coulomb's law, Equation (17.1). Suppose these same charges are examined by another observer who is moving with velocity $\mathbf{u}$ perpendicular to the line joining the two charges. That observer will conclude that there are two identical charges, separated a distance d, each traveling with velocity $-\mathbf{u}$. But a moving charge corresponds to a current, so these two charges, according to the second observer, exert not only a Coulomb force on each other but also an attractive force associated with their velocity, not relative to *each other* but relative to the *observer.* In other words, observers in different inertial frames will measure *different forces* between *the same two charges!*

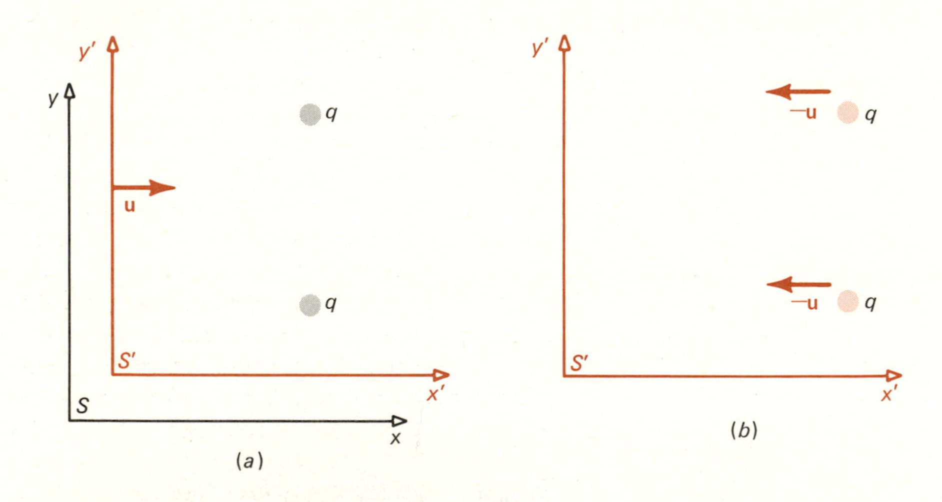

Figure 28.7 *(a) Two identical charges q are at rest in the reference frame S; an observer in S would find that the force between these charges is given by Coulomb's law, Equation (17.1). (b) The same two charges as seen by an observer in the inertial frame S′ that is moving to the right with velocity* **u**. *In that reference frame, the two charges are moving to the left with speed u, and the force between them is therefore not given by Coulomb's law because of the magnetic interaction due to their motion.*

The problem, as Einstein saw it, had to do not so much with ether and ether drift as with the validity of the principle of relativity, which may be stated as follows:

PRINCIPLE OF RELATIVITY: The laws of nature are the same in all inertial reference frames.

This principle must be abandoned, or Maxwell's equations are incorrect—or the transformation equations by which we translate the laws of physics from one inertial frame to another are not given correctly by Equations (28.1) and (28.2). But since those equations do leave Newton's laws invariant, any modification of the transformation rules must mean some modification also of Newton's laws. Einstein proceeded from two fundamental postulates:

POSTULATE I: The principle of relativity is correct and valid for all natural events.

POSTULATE II: The speed of light in vacuum, measured in any inertial reference frame, is c regardless of the motion of the light source with respect to that inertial reference frame.

The second postulate represents a dramatic departure from all previous scientific thought, for it does seem to violate common sense. Suppose light source O emits a pulse of light. An observer in reference frame S, in which O is at rest, measures the speed of the pulse as c. A second observer, approaching the light source O at a speed of, say, $0.5c$, also measures the speed of propagation of the same light pulse. Common sense tells us that this second observer should find that the pulse travels past him at a speed of $1.5c$. Not so, said Einstein; the second observer would also conclude from the readings of *his* instruments, that the light pulse propagates at the speed c.

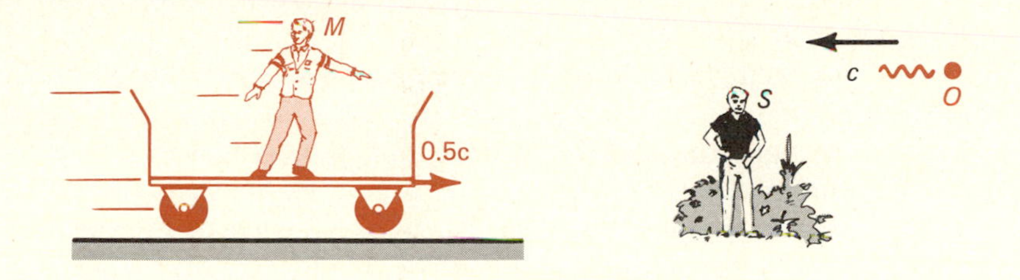

Figure 28.8 Source O emits a light pulse. Observer S, stationary relative to O, measures the speed of this pulse and finds that it travels at speed c. Observer M is moving toward the light source O at a speed of $0.5c$. He, too, measures with his instruments the speed of the pulse as it passes. According to Newtonian theory, observer M should find that the speed of the pulse is $1.5c$. Einstein's second fundamental postulate states that both observers will measure the same speed, c.

The results of the theory of special relativity follow from the two basic postulates in a sequence of logical arguments. What makes this theory such a troublesome obstacle on first encounter is not its mathematical complexity—there is nothing more than elementary algebra here—but the very close and precise logic that inexorably leads to the astounding conclusion "that a balance clock at the equator must go more slowly, by a very small amount, than a precisely similar clock situated at one of the poles under otherwise identical conditions."

Here we shall not derive all the transformation relations. However, because the concept of different time and length scales for observers in different inertial frames seem by far the most troublesome to the student, we shall give a simple derivation of the equations governing the transformations of time and length.

28.4 (b) Simultaneity and Time Dilation

The first, critical step in developing the theory is the recognition that an event occurs not only at some well-defined position in space but also at a well-defined position in time. Since we perceive distant events by light (or radio) signals, propagating at finite, although very high speed, it is no longer obvious that a transformation from one inertial frame to another involves only the spatial coordinates and not the time coordinate. In other words, we must ask, What do we mean when we say that two events are simultaneous?

According to Newton,

> Absolute, true and mathematical time of itself and from its own nature flows equally without relation to anything external.

It is here that Newton and Einstein part company. Einstein recognized that if points A and B are well separated in space, an observer at A can determine the time values of events in close proximity to A, and an observer at B, with his clock, can do likewise relative to events in close proximity to B.

> But it is not possible without further assumption to compare, in respect of time, an event at A with an event at B.

Einstein, for the first time now, invokes his second postulate and establishes "by definition" that the time required for light to travel from A to B equals the time for travel from B to A. One can then answer the question, What is meant by simultaneity?

Suppose we have flashlights at A and B. How can one decide whether both are turned on at the same instant? Clearly, if I stand closer to A than to B, I will see the flash from A first. Einstein carefully defines the process for deciding on the simultaneity of events at two distant spatial locations. He says, Put an observer, at rest in the reference frame in which A and B are at rest, at the exact midpoint between A and B. Then, if the signals from A and B reach this observer at the same instant, the two events are simultaneous.

Next he asks, Will such simultaneous events also appear so to another observer who is moving in the direction AB with velocity $\mathbf{u}$ relative to the stationary observer? And the answer is an unequivocal, no. Suppose these two simultaneous events in S occur at the instant at which observer O' passes O. Some time later, O records the two flashes at the same moment. But O' is approaching light source B and receding from light source A with speed u. Consequently, the signal emitted by B will reach him before the signal from A, and so he concludes that the two events were not simultaneous. Of course, before O' can decide on the matter of simultaneity, he must be certain that he, too, was midway between the two sources when they were turned on. This he can do by placing a long photographic emulsion on the ground in his reference frame; at a later

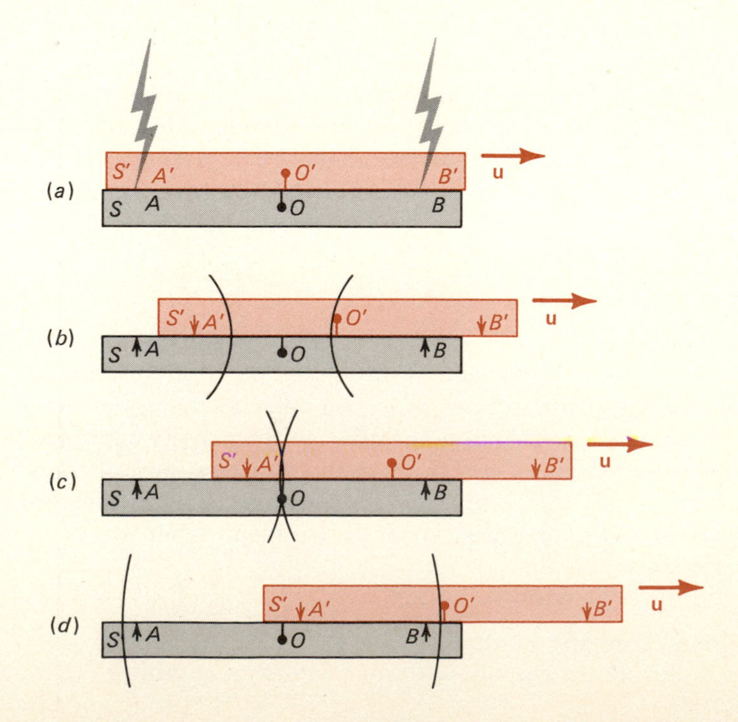

Figure 28.9 *Relativity of simultaneity. Two events occur at A and B. Observer O, stationed midway between A and B, concludes that the two events are simultaneous because the light flashes reach him at the same instant, in (c). Observer O′, who is moving at speed u to the right, concludes that the two events were not simultaneous. In (a) A and A′, B and B′, and O and O′ occupy pairwise the same positions. Light from B, B′ reaches O′ in (b) well before it arrives at O, in (c), and quite some time before light from A, A′ reaches O′, in (d).*

time and at his leisure, he can then pace off the distance between his station and the places where the emulsion was exposed by the light pulses. The result of this *Gedanken* (thought) experiment is that *events that are simultaneous in one inertial frame are not simultaneous when viewed by an observer in another inertial frame.*

Let us now examine one of the best-known consequences of the theory, *time dilation*. By this is meant that, as seen by a "stationary" observer, clocks in a moving inertial frame run more slowly than identical clocks at rest in the reference frame of the stationary observer. For convenience, time as measured by these stationary clocks is referred to as the *proper time* for this stationary observer. (There is, of course, nothing improper about the other clocks.) By clocks we mean any system that may be used to measure time intervals, be they pendulum clocks, digital watches, or processes such as radioactive decay times, the gestation period of a cat fetus, or a normal heartbeat. The startling and at first perplexing conclusion of the theory rests, like everything else, on the two fundamental postulates and nothing more; it has, however, stimulated extended and sometimes heated arguments, particularly about the so-called twin paradox (see Section 28.4(c)).

To measure time intervals, we need a clock, and the clock we devise is both simple and also ideal for our purposes (though hardly practical). It consists of two plane, parallel mirrors, separated by a fixed distance. A very brief pulse of light is initiated at one mirror, and this pulse now bounces back and forth in the space between them. We designate one time unit as the interval required for one complete round trip of the pulse.

We now propose that there are two such identical clocks, one carried by observer O in inertial frame S, the other by O' in S'. Our man O measures time intervals Δt, which are of duration $2d/c$, on his proper clock, and notes that the clock carried by O' measures intervals $\Delta t'$. As seen by O, the clock carried by O' is traveling with velocity $\mathbf{u}$, and the path of the light pulse in this clock, as seen by O, is therefore as shown in Figure 28.11(b). Evidently, this light path is longer than the path for the clock that O carries. We see that the situation is exactly like the one we already encountered in our analysis of the Michelson-Morley experiment. Thus we can immediately write out the result: For the clock in S', as judged by

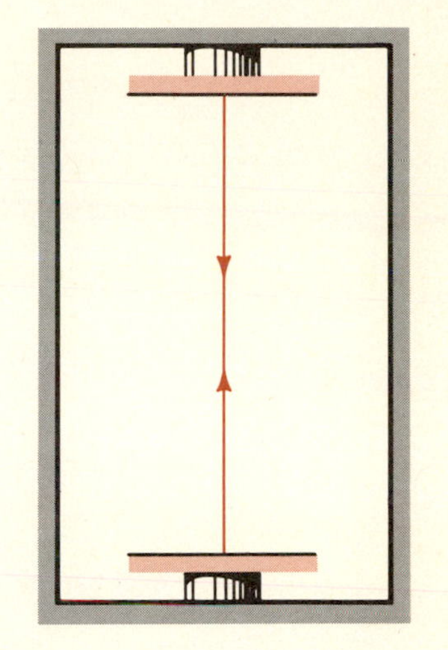

Figure 28.10 The portable light clock. One tick of this clock corresponds to the time interval required for a complete round trip of a light pulse between the two parallel mirrors in the evacuated enclosure.

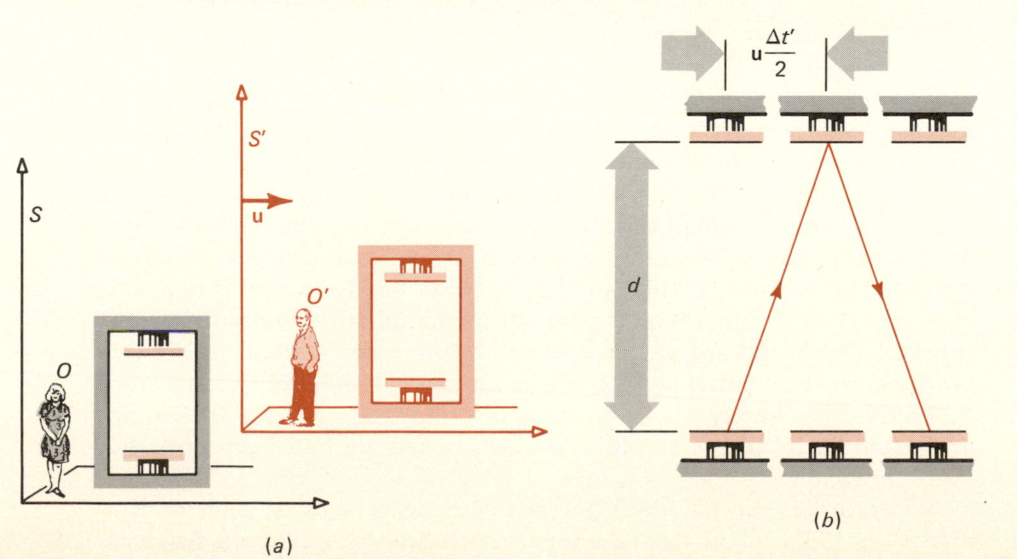

Figure 28.11 Time dilation. (a) Observers O and O′ carry identical light clocks of the type shown in Figure 28.10. Observer O′ is moving with velocity **u** relative to O. (b) The light path followed by a light pulse in the clock of O′, as seen by observer O. That path is evidently longer than 2d, and since the speed of light is the same in all inertial reference frames, the time interval for one tick of the O′ clock, as measured by O, must be longer than the time interval for one tick of the clock carried by O himself. Observer O therefore concludes that the moving clock runs slow.

the observer O in S, the path length is

$$2\sqrt{d^2 + \left(\frac{u\,\Delta t'}{2}\right)^2}$$

But since light travels at constant speed c, whether measured in frame S or S', that distance must equal $c\,\Delta t'$. Solving for $\Delta t'$, one finds

$$\Delta t' = \frac{2d/c}{\sqrt{1 - u^2/c^2}}$$

whereas $\Delta t = 2d/c$. Hence we have

$$\Delta t' = \frac{\Delta t}{\sqrt{1 - u^2/c^2}} \qquad \textbf{(28.9)}$$

Since the denominator in (28.9) is less than 1 (unless $u = 0$), the interval $\Delta t'$ is greater than Δt. That is, the observer O concludes that the mirror clock in S', which is moving relative to his clock, ticks more slowly; *it runs slow.*

"Oh, well," you may say, "so moving clocks are unreliable and we had better be certain that we measure time intervals only by clocks that are at rest. The clock in S is correct; the one in S' is slow."

"Not so," says observer O' with great indignation. "My clock is not slow, it is absolutely accurate. In fact, according to my superior clock, and contrary to what O may have told you, it is *his* clock that is slow compared with mine, not the other way about."

And so it is! To the observer in S', the clock in S' gives the proper time and the one in S is the moving clock and runs slow. So, which one is right? Both are! How can that be? One or the other clock must be slow; each cannot be slow compared with the other. What sort of nonsense is this? Surely, if one clock is slow, the other must be fast, and we can decide which is the slow one by bringing one of the clocks back after a long trip and comparing it with the one that was never moved. This brings us to the infamous twin paradox.

28.4 (c) Twin Paradox

Tim and Jim are identical twins. Tim, who is more adventurous, sets out on a journey to one of the nearest stars, only 35 light years from Earth. His spaceship is remarkable in that it can accelerate quickly and attain a speed of $0.99c$.

When Tim reaches this star, he finds that it is an unpleasant place and quickly turns around and heads for home, again at the phenomenal speed of $0.99c$. As he disembarks on Earth he is greeted by a startling change in scenery; buildings that were brand new when he left are now shabby or have crumbled altogether; the life style of people has changed dramatically, and so have transportation, dress, and customs. When he arrives at his home it is still standing, but only barely so, and the same is true of his twin brother who is now wizened and hard of hearing. Tim is shocked, but it is hard to know which is the more startled of the two brothers. Jim can hardly believe that in the more than seventy years since Tim left he has barely aged ten years, and Tim responds that he cannot understand why Jim is so old since Tim insists he had been gone not quite ten years.

Wherein lies the paradox? If Tim's clock were slow as seen by Jim, then, does it not follow from the theory, that Jim's clock should also have

Figure 28.12 The twin paradox. (a) Tim and Jim, both in their late teens, wave goodbye to each other as Tim sets off on a long journey. (b) On his return Tim is shocked to find his brother gray, hard of hearing, and prematurely aged, while Jim is amazed how young and healthy Tim appears after such a long and strenuous trip.

run slow as seen by Tim? Tim, if he understood relativity, might well have expected Jim to be the youthful one because so far as Tim could tell, Earth was at first rapidly receding and then rapidly approaching his spaceship. Perhaps things just are not as we have described them, and everything will have evened out when they meet. After all, no one has ever tried this silly experiment.

Well, the answer is that the silly experiment has been done, and more of that later. In any event, it is indeed Tim, the adventurous space traveler, who will be by far the younger of the two. There is, after all, something unique that sets Tim's frame of reference apart from that of his stick-in-the-mud brother. Tim's was *not* an inertial reference frame at all times. It could not have been because to return to Earth he had to decelerate, reverse direction, and then accelerate once more. The system is not symmetrical with respect to the two reference frames.

Although time dilation was one of the important results of the 1905 paper, nearly *fifty years later* there were still occasional arguments on the twin paradox in the scientific literature.

● ***Example 28.1*** At exactly noon EST, a distress call is received at Houston from a space station on Neptune. At the time of year in question, the distance between Earth and Neptune is 4.5×10^9 km. The request is for some highly unstable antibiotics, which lose their potency within five hours of removal from culture. At 1 P.M. EST a rocket is ready, and just before blastoff the antibiotics are put aboard. The rocket's speed over nearly all the trip is $0.7c$.

At what time, according to the Houston chronometer, was the distress call sent from Neptune, and at what time will the rocket arrive on Neptune? Will the antibiotics still be potent on arrival at their destination?

Since the call from Neptune must be transmitted by radio, the time required for the signal to travel from Neptune to Earth was

$$\Delta t = \frac{4.5 \times 10^{12}\ \text{m}}{3 \times 10^{8}\ \text{m/s}} = 1.5 \times 10^{4}\ \text{s} = 4\ \text{h, } 10\ \text{min}$$

Consequently, the distress call must have been sent from Neptune at 7:50 A.M. EST.

Since the rocket leaves Houston at 1:00 P.M. EST and travels at a speed of $0.7c$, the return trip takes $15{,}000\ \text{s}/0.7 = 21{,}429$ s, or a bit over 5 h and 57 min, nearly 6 h, as determined by the Houston chronometer. The rocket will therefore arrive on Neptune at 6:57 P.M. EST.

Will the medicine still be of value? If we forgot about time dilation we might conclude the negative. However, according to the clock *on the rocket ship,* the time elapsed during this journey was less than 5 h, 57 min. The rocket clock records a time interval of

$$\sqrt{1 - 0.7^2}\,(21{,}429\ \text{s}) = 15{,}303\ \text{s} = 4\ \text{h, } 15\ \text{min, } 3\ \text{s}$$

Hence, the antibiotics will still be potent for another 45 min. ●

● ***Example 28.2*** Neglecting the time required for acceleration at Earth and the distant star, determine by how much each of the twins, Tim and Jim, has aged during Tim's round trip.

We must first be certain we understand the meaning of *light year.* One light year is the *distance* that light travels in the course of one year. An object traveling at a speed less than c, say pc, where $p < 1$, would take more time to cover the same distance than a pulse of light. Thus, since the round-trip distance of Tim's journey was 70 light years, the time needed for the trip, as measured by Jim's clock, was

$$\frac{70\ \text{yr}}{0.99} = 70.707\ \text{yr} = 70\ \text{yr, } 8\ \text{mo, } 15\ \text{days}$$

Tim's clock, however, is slow compared with Jim's. The time dilation factor is

$$\sqrt{1 - 0.99^2} = 0.141$$

and consequently, Tim will have aged by only

$$0.141(70.7\ \text{yr}) = 9.97\ \text{yr} \simeq 10\ \text{yr}$$

If Tim set out on his twentieth birthday, he will be eleven days short of his thirtieth birthday, whereas Jim will be nearing his ninety-first. ●

28.4 (d) The Lorentz-FitzGerald Contraction

As mentioned earlier, FitzGerald and Lorentz had proposed a length contraction as a *deus ex machina* explanation for the null result obtained by Michelson and Morley. The proposed solution was such that an independent verification of that contraction was clearly out of the question. It seemed as though nature had somehow conspired against man, had devised just the right compensating mechanism that would forever place the determination of the ether rest frame beyond his reach.

The same length contraction is also a consequence of special relativity and is still identified as the *Lorentz-FitzGerald contraction.* However, its origin is not some obscure dependence of interatomic forces on velocity

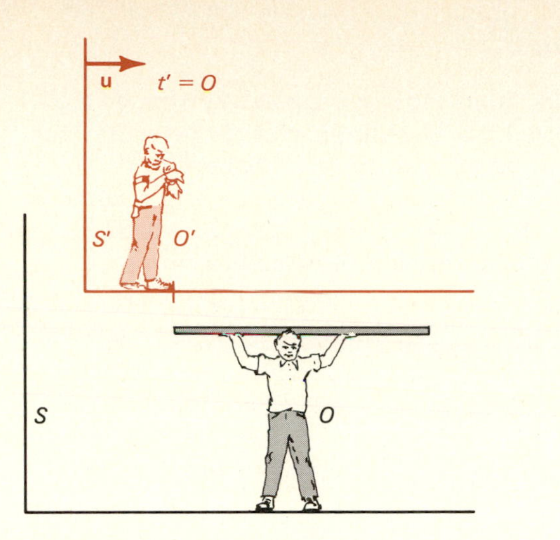

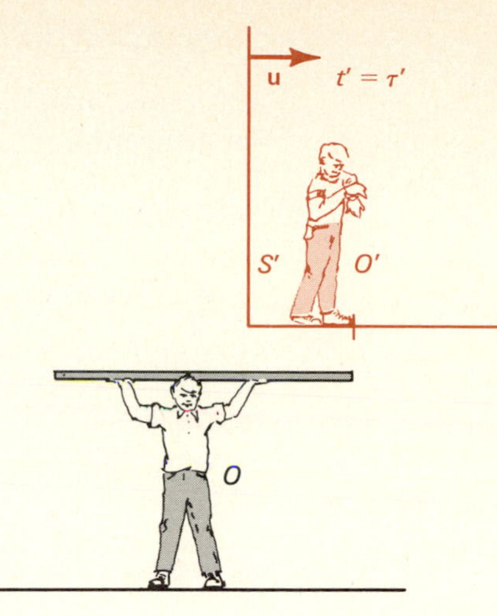

Figure 28.13 Lorentz-FitzGerald contraction. Observer O holds up a long rod of length L so that O', who is whizzing by at velocity **u** can measure its length. Observer O' measures the time τ' taken to pass from one end of the rod to the other, and concludes that the length of the rod is $L' = u\tau'$. However, because the clock carried by O' runs slow, as seen by O, the length L' is less than the length L.

relative to the perennially elusive ether, nor is it introduced as an *ad hoc* assumption to accommodate some troublesome experimental result. Length contraction, like time dilation, is now a consequence of the two basic postulates.

To see how this contraction arises, we suppose that we have a rod of length L, as measured by an observer O in the reference frame S in which L is at rest. This rod, with *proper length* L, is also to be used as a standard of length by observer O', who is traveling with velocity **u** relative to S in a direction parallel to the rod. Observer O' now sets out to measure the length of the rod.

To this end he uses his precise chronometer and simply measures the time interval τ' between the passing of the front and the rear of the rod past a fiducial marker in his reference frame. The length of the rod, as measured by O', is then simply $u\tau' = L'$. But O in reference frame S, who sees O' pass by at speed u, knows that the time interval τ' measured by O' is less than the interval τ that he measures for the passage of the fiducial marker past the front and back of the rod; that is, observer O knows that $\tau' < \tau$, because *moving clocks run slow.* In fact,

$$\tau' = \tau\sqrt{1 - \frac{u^2}{c^2}}$$

Consequently, the length L' measured by O' with his clock is not $L = u\tau$ but

$$L' = u\tau' = L\sqrt{1 - \frac{u^2}{c^2}} \qquad (28.10)$$

This is precisely the Lorentz-FitzGerald contraction.

● ***Example 28.3*** As it appears to Tim during his trip, what is the distance between Earth and the star that was his destination?

In Tim's reference frame, the distance of 35 light years, as measured by an observer on Earth, is diminished by the Lorentz-Fitzgerald contraction to

$$35 \times \sqrt{1 - 0.99^2} = 4.935 \text{ light years}$$

or just a bit under five light years. ●

28.5 Addition of Velocities

With time intervals and lengths subject to such apparently strange transmogrification, it is to be expected that the simple rule for velocity transformation, Equation (28.1), will be invalid. In fact, the second postulate tells us that this must be, if each of two observers, moving relative to each other, concludes that a pulse of light propagates with speed c.

The relativistically correct transformation of velocities is

$$\begin{aligned} v'_x &= \frac{v_x - u}{1 - v_x u/c^2} & \qquad v_x &= \frac{v'_x + u}{1 + v'_x u/c^2} \\ v'_y &= \frac{v_y\sqrt{1 - u^2/c^2}}{1 - v_x u/c^2} & \qquad v_y &= \frac{v'_y\sqrt{1 - u^2/c^2}}{1 + v'_x u/c^2} \\ v'_z &= \frac{v_z\sqrt{1 - u^2/c^2}}{1 - v_x u/c^2} & \qquad v_z &= \frac{v'_z\sqrt{1 - u^2/c^2}}{1 + v'_x u/c^2} \end{aligned} \qquad \textbf{(28.11)}$$

where u is the speed in the x direction of the S' frame relative to the S frame. These equations reduce to the Galilean transformation equations in the limit as $u \to 0$, that is, when $u/c \ll 1$. At normal speeds, those very much less than c, Newtonian mechanics is a valid approximation. Note that in contrast to the Galilean transformation, relative motion of the two reference frames along the x direction influences the y and z components of velocity as well as the x component. The reason is that

$$v'_y = \frac{\Delta y'}{\Delta t'}$$

and although $\Delta y' = \Delta y$, the corresponding time intervals Δt and $\Delta t'$ are related through the transformation that gave us time dilation.

● ***Example 28.4*** An observer on Earth notes that two spaceships are approaching each other from opposite directions. At exactly noon EST, each ship is 600,000 km from Earth, and each is traveling at $0.8c$.

At what time will the two ships collide? What is the speed of one of the ships as measured by the captain of the other? How much time does each captain have before the crash?

To cover a distance of 600,000 km at a speed of $0.8c$ requires

$$\frac{6 \times 10^8 \text{ m}}{(0.8)(3 \times 10^8 \text{ m/s})} = 2.5 \text{ s}$$

Therefore the two ships will collide 2.5 s after noon.

In the reference frame of one of the ships, the speed of the other is given by the first of the equations (28.11), namely,

$$v_x = \frac{v'_x + u}{1 + v'_x u/c^2}$$

In this case $v'_x = 0.8c$ and $u = 0.8c$. Thus

$$v_x = \frac{1.6c}{1.64} = 0.976c$$

Compared with the clock of the observer on Earth, the clocks carried by each of the spaceship captains run slow. Hence the time interval before collision, as deduced by those aboard the ships, is not 2.5 s but

$$(2.5 \text{ s})\sqrt{1 - 0.8^2} = 1.5 \text{ s}.$$ ●

28.6 Mass-Energy Equivalence

We come now to the famous relation $E = mc^2$, derived by Einstein in a gem of a scientific paper, three pages long, beautifully reasoned and devoid of mathematical complexity. Einstein considers radioactive decay (see Chapter 32) in which a nucleus emits a γ ray, that is, electromagnetic radiation of high frequency. For such an isolated system, on which no external forces act, energy and momentum must be conserved in the decay. Using these two conservation conditions and the relativistic expression for momentum, Einstein shows that the mass of the daughter nucleus must be less than the mass of the parent by the amount $\Delta m = E/c^2$, where E is the energy carried off by the radiation. In his own words: "If a body gives off energy E in the form of radiation, its mass diminishes by E/c^2."

This conclusion is valid generally, regardless of how energy is released. For example, in some nuclear reactions, energy is released as kinetic energy of the participating particles, and then, that energy too must be compensated by a change in mass of the system. If a system of mass m_0 were to disappear spontaneously, the energy released would be

$$E = m_0c^2 \qquad \textbf{(28.12)}$$

Correspondingly, it is conceivable that a massive object might be created by the expenditure of sufficient energy, and indeed, such events do occur. That is, Equation (28.12) implies an equivalence between energy and mass that allows reactions in both directions.

This mass-energy equivalence has served as a springboard in the solution of numerous scientific problems, among them one of the oldest facing man, the source of the sun's energy.

● ***Example 28.5*** The sun radiates energy at a rate of 3.8×10^{26} W. At what rate does the mass of the sun diminish? How long would it take for the sun to lose an amount of mass equal to that of Phobos, the larger of the two moons of Mars? What is the fractional change in the sun's mass over one billion years? (The approximate age of the universe is estimated at 15 billion years.) The mass of the sun is 2×10^{30} kg; the mass of Phobos is 1.6×10^{16} kg.

In one second, the energy radiated is 3.8×10^{26} J. The change in the sun's mass is therefore

$$\Delta m = \frac{E}{c^2} = (3.8 \times 10^{26}\ \text{J})/(9 \times 10^{16}\ \text{m}^2/\text{s}^2) = 4.2 \times 10^9\ \text{kg}$$

Since one year has 3.15×10^7 s, the sun loses mass at a rate of $(4.2 \times 10^9\ \text{kg/s})(3.15 \times 10^7\ \text{s/yr}) = 1.32 \times 10^{17}$ kg/yr.

In only

$$\frac{1.6 \times 10^{16}\ \text{kg}}{1.32 \times 10^{17}\ \text{kg/yr}} = 0.121\ \text{yr} = 44\ \text{days}$$

the sun loses as much mass as that of the larger of the two satellites of Mars.

The total mass of the sun is, however, so enormous that the conversion of mass to energy at this fantastic rate has no measurable influence on the sun. During *one billion* years, the change in the sun's mass is

$$(1.32 \times 10^{17}\ \text{kg/yr})(10^9\ \text{yr}) = 1.32 \times 10^{26}\ \text{kg}$$

or

$$\frac{1.32 \times 10^{26}}{2 \times 10^{30}} = 6.6 \times 10^{-5} = 0.0066\%$$

28.7 Experimental Confirmation of Special Relativity

28.7 (a) Mass-Energy Equivalence

Of the various predictions of the theory of special relativity, the first subjected to experimental test was the mass-energy relation. The reason is that while it is extremely difficult to accelerate clocks and meter sticks to speeds beyond a few hundred meters per second—at which relativistic effects are minute—it is easy to produce electrons traveling at speeds approaching c and to measure their mass. For example, an electron that falls from rest through a potential difference of 100,000 V acquires a speed in excess of $0.5c$.

Experiments performed between 1905 and 1910 (and continued for some years thereafter) involved measuring the mass of electrons of known velocities. This was done with a velocity selector using crossed electric and magnetic fields (see Problem 22.16), followed by a region of uniform transverse magnetic field. The electrons, on entering the last-mentioned region, followed a circular trajectory of radius $R = mv/Be$. Here m is the *relativistic mass,* which according to the theory of special relativity, is given by

$$m = \frac{m_0}{\sqrt{1 - v^2/c^2}} \quad \textbf{(28.13)}$$

where m_0 is the *rest mass* of the electron. From the known value of e, and measurements of B, R, and v, the mass m could be calculated. These experiments fully confirmed the predicted dependence of mass on speed.

Further confirmation of mass-energy equivalence comes from observations on the creation and annihilation of particles. This topic is discussed in Section 29.9.

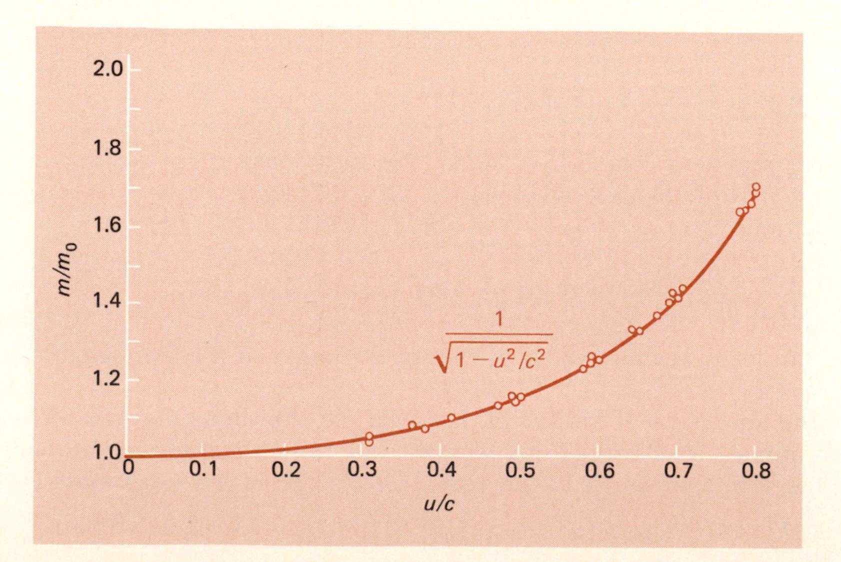

Figure 28.14 *Direct experimental confirmation of the dependence of mass on speed. The data are in excellent agreement with the relativistic expression, Equation (28.13).*

In 1972, two physicists, J. C. Hafele and R. E. Keating, reported the results of an experiment performed the preceding October:

> During October 1971, four cesium beam atomic clocks were flown on regularly scheduled commercial jet flights around the world twice, once eastward and once westward, to test Einstein's theory of relativity with macroscopic clocks. From the actual flight paths of each trip, the theory predicts that the flying clocks, compared with reference clocks at the U.S. Naval Observatory, should have lost 40 ± 23 nanoseconds during the eastward trip, and should have gained 275 ± 21 nanoseconds during the westward trip. . . .
>
> Relative to the atomic time scale of the U.S. Naval Observatory, the flying clocks lost 59 ± 10 nanoseconds during the eastward trip and gained 273 ± 7 nanoseconds during the westward trip, where the errors are the corresponding standard deviations. These results provide an unambiguous empirical resolution of the famous clock "paradox" with macroscopic clocks.

Previous to that experiment, there had been verifications of time dilation, but those results were based on less direct evidence (see below). The large uncertainties in the calculated predictions reflect the complications that arise because numerous corrections must be applied in calculating the final prediction. The clock on earth at the Naval Observatory is not in a true inertial frame because the earth rotates about its axis. During the flights, there were numerous periods of acceleration and deceleration relative to the rotating earth; the flight path was not due east nor due west. The flying clocks were in a weaker gravitational field than the clock at the Naval Observatory because the planes flew at an average altitude of 7000 m; this is important because the theory of general relativity predicts that clock rates are also influenced by acceleration of the reference frame and by gravitational fields.

Before the Hafele-Keating experiment, the best indication of time dilation came from observations of particles with short half-lives. As a result of cosmic radiation that strikes the upper layers of our atmosphere, certain particles, called mu mesons (μ mesons), or muons, are created. These particles are not stable but decay into an electron and two neutrinos.* This decay follows an exponential time dependence just like the pattern displayed by a discharging *RC* circuit (see Section 24.2). The half-life of the muon is known to be 1.5×10^{-6} s. That is, if at some moment we have 100 muons in a container, 1.5 μs later only about 50 will be left, and another 1.5 μs thereafter we can expect to find only about 25.

Now even if the muons traveled toward the surface of the earth at the speed of light—and they nearly do—they would still cover only about $(1.5 \times 10^{-6}\ \text{s})(3 \times 10^{8}\ \text{m/s}) = 450$ m before about half of them decayed, and we should hardly expect to find any at sea level. In fact, they are abundant there.

The critical experiment on muon decay and time dilation was performed in 1941 by Rossi and Hall, who measured the muon flux atop Mt. Washington in New Hampshire, at an altitude of about 2000 m, and

* It is not important to know about this terminology or the decay scheme to understand the arguments that follow.

again at sea level. The ratio of these fluxes "should" be

$$\frac{\mu(2000)}{\mu(0)} = 2^{2000/450} = 2^{4.44} = 21.7$$

The experiment gave a ratio of 1.4!

The resolution of this apparent discrepancy is found in the theory of special relativity. Since $1.4 \simeq \sqrt{2} = 2^{1/2}$, it seems that according to the clock of the muons, only about half of a half-life, or roughly 0.75 μs, elapsed during that trip of 2000 m as measured by an observer on the earth. Since the muon's clock is moving relative to our clock, it is slow, indeed very slow, since it ticks off only 0.75 μs; as measured by the clock carried by the observer at rest on the earth, a particle traveling at the speed of light would require $(2000\text{ m})/(3 \times 10^8\text{ m/s}) = 6.7\ \mu\text{s}$ to cover that distance. The time dilation is therefore $6.7/0.75 = 8.93$. Hence, according to Equation (28.9),

$$\sqrt{1 - \frac{u^2}{c^2}} = \frac{1}{8.93}$$

from which we can now solve for the muon velocity and obtain

$$u = 0.994c$$

The small decay in muon flux in descending the 2000 m can also be viewed, and understood, as it might be by an observer at rest in the inertial frame of the muon. Such an observer would see the ground approaching at terrifying speed, $0.994c$; as seen by this observer, the Lorentz-FitzGerald contraction compresses the height of Mt. Washington to a mere $2000/8.93 = 224$ m, a distance that the approaching earth covers in roughly 0.75 μs.

28.8 General Relativity

During the decade following publication of the theory of special relativity, Einstein worked to extend the theory to encompass all frames of reference, even accelerating ones. Before achieving this feat, he published a paper in 1911 that enunciated the principle of equivalence to which we alluded in Chapter 3.

PRINCIPLE OF EQUIVALENCE: An inertial frame in a uniform gravitational field is equivalent to a reference frame that has a constant acceleration with respect to this inertial frame.

That principle led Einstein to predict that light, on passing through a gravitational field, will be deflected from its otherwise straight trajectory. In particular, he calculated that light from distant stars on passing close to the sun's surface will be deflected by the small but measurable amount of 0.83 second of arc.* A German scientific expedition, dispatched to Russia in 1914 to make astronomical observations during a total solar eclipse, arrived on the scene just as World War I erupted. After a brief period of internment, the expedition members (minus equipment) were returned to their native country.

By then, Einstein had already concluded that his 1911 calculation was incorrect and was immersed in a program of mathematical self-education

* Unknown to Einstein, Johann Georg von Soldner had already made that same prediction a century earlier; his argument was based on the then fashionable corpuscular theory of light.

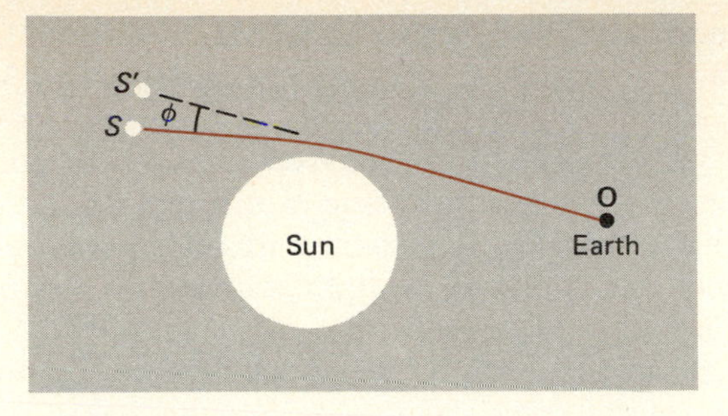

Figure 28.15 *On passing close to the sun, light from a star is deviated by the sun's gravitational field. As a result, the star's apparent position, as seen by an observer on earth, is shifted slightly. Here the deviation is greatly exaggerated; the angle is actually only 1.74 seconds of arc.*

and some highly unconventional speculations. In contrast to his 1905 papers, his 1916 paper on the theory of general relativity is comprehensible only to individuals with a high degree of mathematical sophistication.

The predictions of that theory can, however, be stated in simple language. Bizarre though they may appear at first, they have been verified by a number of dramatic experiments. The theory has also led to several conclusions that have greatly influenced cosmological thinking.

Among the successes of the theory are the explanation of the precession of the elliptical orbit of Mercury by the extremely small but still measurable amount of 43 seconds of arc *per century*, and the bending of light that grazes the sun by 1.74 seconds of arc. The last was first confirmed during the total solar eclipse of 1919 by the British Royal Astronomical Society expeditions to Principe, an island in the Atlantic off West Africa, and to Sobral in Brazil.

Another consequence of the theory is the possible existence of *black holes*, burned-out stars that have collapsed under their own enormous gravitational field to objects of such unimaginably high density that this gravitational field prevents the escape even of light from their surface. Although the idea of a gravitational field inhibiting or modifying the character of light may seem bizarre, this so-called *gravitational red shift* has actually been measured in the laboratory in an experiment of fantastic precision. In 1960, R. V. Pound and G. A. Rebka compared the frequencies of γ rays emitted by radioactive nuclei at ground level and at an altitude of 21 m above ground. The minute change in gravitational field between these two locations should, according to general relativity, cause a frequency change of 2 parts in 10^{15}. The measurements confirmed this prediction to within experimental accuracy of a few percent. The gravitational red shift can also be viewed as a gravitational time dilation, an effect alluded to earlier during the discussion of the Hafele-Keating experiment. Regarded from that standpoint, the frequency shift of 2 parts in 10^{15} is equivalent to a difference in the rates of clocks of 1 second in about 15 million years!

Summary

Before 1905, it was generally believed that light and other electromagnetic waves were supported by a substance called the ether. Maxwell's equations predicted that all electromagnetic waves moving through vacuum (permeated by the ether) travel with a speed $c = 3 \times 10^8$ m/s. It appeared that the true rest frame of the universe, the ether, could be established by a careful measurement of the speed of light. Only in the rest frame of the ether would this speed equal Maxwell's prediction. All such measurements, the most precise due to Michelson and Morley, gave a null result.

The theory of special relativity is based on two fundamental hypotheses: (1) The laws of nature are the same in all inertial reference frames. (2) The speed of light in vacuum, measured in any inertial reference frame, is c regardless of the motion of the light source with respect to the reference frame of the observer.

These two postulates lead to several startling conclusions.

1. Events that appear to be simultaneous in one reference frame are not simultaneous when perceived by an observer in a different inertial reference frame.

2. Moving clocks run slow.

3. If an observer at rest relative to an object measures its length to be L, another observer, moving with speed u in the direction along which L has been measured, will conclude that this object is of length L', where

$$L' = L\sqrt{1 - \left(\frac{u}{c}\right)^2}$$

This reduction in the length of a moving object is known as the Lorentz-FitzGerald contraction.

A further consequence of special relativity is the equivalence between mass and energy, expressed by the relation

$$E = mc^2$$

The theory of general relativity, which removes the restriction that the two reference frames must be inertial frames (moving at constant velocities), rests on the principle of equivalence:

> An inertial frame in a uniform gravitational field is equivalent to a reference frame that has a constant acceleration with respect to this inertial frame.

The predictions of the theory of general relativity include black holes, the bending of light by a gravitational field, the gravitational red shift, and the precession of the orbit of Mercury.

The predictions of special and of general relativity theory have been confirmed by a variety of experiments.

Multiple Choice Questions

28.1 The negative result of the Michelson-Morley experiment was

(a) inconsistent with the view that the speed of light is constant in the ether and the ether is moving along with the earth.

(b) inconsistent with the view that the speed of light is constant in the ether, and that the earth is in motion relative to the ether.

(c) inconsistent with the postulate that the speed of light in vacuum is independent of the motion of source or observer.

(d) inconsistent with the Lorentz-FitzGerald contraction.

28.2 A stick moves past an observer with speed u in a direction parallel to the length of the stick. The observer measures the length of that stick and concludes that it is 2 m. Another observer, at rest relative to the stick, will conclude that the length of the stick is

(a) less than 2 m.

(b) equal to 2 m.

(c) greater than 2 m.

(d) either (a) or (c) depending on whether the first observer made his measurement while the stick was approaching or receding.

28.3 An observer sees a mass-spring system moving past at a speed u, and notes that the period of that system is T. Another observer, who is at rest relative to the mass-spring system, also measures its period. The second observer will find a period that is

(a) greater than T.

(b) equal to T.

(c) less than T.

(d) either (a) or (c) depending on whether the system was approaching or receding from the first observer.

28.4 Reference frame S' is moving in the x direction with speed u. In that reference frame, an object has a velocity v'_y, and no component of velocity in the x' direction. To an observer in the reference frame S, that object has a velocity with

(a) a y component equal to v'_y and an x component equal to u.

(b) a y component less than v'_y and an x component equal to u.

(c) a y component greater than v'_y and an x component of zero.

(d) None of the above.

28.5 A small bomb is made so that it will explode exactly 5 s after the rocket in which it is located is launched. If we assume that this rocket accelerates so that it attains a speed of $0.99c$ within 1 s from liftoff, the explosion will occur

(a) before the rocket has traveled 12×10^8 m.

(b) between 12×10^8 and 15×10^8 m from Earth.

(c) more than 15×10^8 m from Earth.

(d) There is not enough information available to decide.

28.6 Mark the correct statement.

(a) The twin paradox has never been resolved satisfactorily.

(b) Of the two twins, the one who remains on Earth waiting for the sibling will be the younger on reunion.

(c) Of the two twins, the one who remains on Earth waiting for the traveling sibling will be the older on reunion.

(d) Recent experiments using atomic clocks have demonstrated that atomic clocks are influenced only by changes in gravitational fields and are not subject to the time dilation of special relativity.

28.7 Light from a star is noted to be red-shifted. That red shift

(a) is due only to a Doppler effect, indicating that the star is receding.

(b) is due only to a gravitational effect; that is, it is a relativistic red shift predicted by the general theory of relativity.

(c) may be the result of a combination of Doppler and gravitational red shifts.

(d) indicates absorption of light by interstellar dust.

28.8 The following observations support the theory of special relativity.

(a) Bending of light near the sun observed during a total solar eclipse.

(b) The precession of the orbit of Mercury.

(c) The difference in the wavelengths of γ rays emitted by certain nuclei when at ground level and when at an altitude of 20 m.

(d) The observation of μ mesons, created in the upper atmosphere, near sea level.

Problems

(Section 28.4)

• **28.1** Two identical clocks are constructed. Both have a period of 5 s between ticks. One of these clocks is set in motion so that its speed is $0.6c$ relative to the observer, who uses the other clock. What is the time interval between ticks of the moving clock as measured by this observer?

• **28.2** An unstable nuclear particle has a lifetime of 10^{-7} s. If this particle is moving with a speed of $0.98c$ relative to the laboratory, what is its lifetime as measured by an observer in the laboratory?

• **28.3** For the particle of Problem 28.2, find the distance that it will have traveled in the laboratory during its lifetime.

• **28.4** The half life of the K^+ particle is 1.2×10^{-8} s when it is at rest. If K^+ particles travel at a speed of $0.92c$ in the laboratory, what will their half-life be as measured in the laboratory?

• • **28.5** A certain star is at a distance of 30 light years from Earth. At what constant speed must a spaceship travel from Earth to that star so that a passenger aboard will age only 10 years during that trip?

• • **28.6** If the half-life of pions at rest is 2.6×10^{-8} s, at what speed must pions travel so that the half-life, as measured in the laboratory, is 4.2×10^{-8} s?

• • **28.7** Many satellites move in orbits about 600 km above the surface of Earth. If such a satellite stays in that orbit for one year and is then brought back to Earth, what will be the time difference between an atomic clock aboard the satellite and one on Earth, if it is assumed that they were initially synchronized and if effects due to changes in the gravitational fields at the two locations are neglected?

• •**28.8** The half-life of pions (π mesons) at rest is 2.6×10^{-8} s; that is, at the end of 2.6×10^{-8} s, half of the pions initially present will have decayed.

A high-energy beam of pions is generated by an accelerator and the pions are observed to travel 20 m before half have decayed. What is the speed of the pions as measured in the laboratory?

• •**28.9** A stick 5 m long travels parallel to its length at a speed of $0.8c$. How long, according to a stationary observer, does it take for the stick to move past this observer? How long is that stick as measured by the stationary observer?

• •**28.10** Two spaceships pass each other in opposite directions. The speed of spaceship B as measured by a passenger in spaceship A is $0.3c$. The passenger in B concludes that while spaceship B is 30 m long, spaceship A is 40 m long. What are the lengths of the two spaceships as measured by a passenger in A?

(Section 28.5)

• **28.11** Two spaceships travel in exactly opposite directions as seen from Earth. The speed of each spaceship, as measured on Earth, is $0.6c$. What is the relative speed of the two spaceships as measured by a passenger on either?

• •**28.12** The captain of one spaceship observes that he is on a direct collision course with another spaceship, and that their relative speed of approach is $0.87c$. How fast is each spaceship moving according to an observer who sees them approach at equal but oppositely directed velocities?

• **28.13** Show that the velocity transformation equations (29.11) reduce to the Galilean transformation in the limit when $u \ll c$.

• •**28.14** Show that if a light source approaches a stationary observer with speed u_s, the observed light will be Doppler-shifted by exactly the same amount as if the observer were approaching a stationary source with speed $u_o = u_s$.

• •**28.15** A spaceship is traveling in the x direction with a speed of $0.8c$ as measured by an observer on earth. A gunner aboard the spaceship shoots a projectile that has a speed of $0.6c$ relative to the spaceship and is traveling in a direction perpendicular to the motion of the spaceship. What is the speed of the projectile as measured by the observer on earth?

(Section 28.6)

28.16 Estimate the decrease in the mass of the earth due to the annual energy consumption in the United States of about 10^{20} J.

28.17 Suppose it were possible to convert 1 g of mass entirely to energy. How long would this mass-energy conversion yield an output power of 1 MW?

28.18 At what rate must mass be converted to energy to produce 10 MW?

28.19 What is the mass of an electron whose speed is $c/2$.

• **28.20** An electron has a rest energy of 0.51 MeV. If the electron is moving with a speed of $0.8c$, what are its mass, its momentum, and its total energy (including its rest energy)?

• **28.21** An electron is accelerated by letting it fall through a potential difference of 2×10^6 V. What is the velocity of this electron?

• •**28.22** Show that the difference between mc^2 and m_0c^2, where m is given by Equation (29.13), reduces to the classical expression for kinetic energy in the limit $v \ll c$.

• **28.23** At what speed is the mass of a particle three times its rest mass?

• •**28.24** Electrons accelerated in the Stanford Linear Accelerator (SLAC) achieve an energy of 10^4 times their rest energy. If an electron travels at the speed corresponding to this energy for a distance of 1 km in the laboratory, what is the distance traversed as "seen" by the electron in its reference frame?

•29•

Origins of the Quantum Theory

Kurz zusammengefasst, kann ich die ganze Tat nur als einen Akt der Verzweiflung bezeichnen.

(In summary, I can only characterize the whole work as an act of desperation.)

M. PLANCK in a letter to R.W. Wood, July, 1931, referring to his proposal for quantization of energy.

It is the theory which decides what we can observe.

ALBERT EINSTEIN, 1926 (as quoted by Werner Heisenberg)

29.1 Introduction

The theory of relativity was very much the creation of one man, and forever bears the imprimatur of the brilliant, logical mind that conceived and developed that broad new perspective of space and time. By contrast, the quantum theory grew like a giant jigsaw puzzle, to which numerous imaginative minds made diverse and individual contributions. It is a many-splendored thing, and like a kaleidoscope, presents different and startling facets depending on how one twists and turns the formalism.

The new theory touches every aspect of modern physics and most of classical physics. Thermal expansion, specific heats, latent heats, the magnetism of iron, nickel, and similar substances, superconductivity as well as the ordinary electrical properties of metals and semiconductors, spectral lines of atoms and the structure of molecular spectra, lasers,

X rays, radioactivity and practically all nuclear phenomena—the list could go on almost indefinitely—all require an understanding and the application of quantum theory for a satisfactory explanation. No single scientist, however resourceful and tireless, could possibly cope with the enormous diversity of problems literally crying for solution; and each new solution not only strengthened the confidence of the practitioners in the quantum theory but pointed to new directions and potential applications.

Probably the earliest indications that Newtonian mechanics, despite its phenomenal successes, was not the answer to all of nature's puzzles, were found in specific heat data. During the early part of the nineteenth century, measurements of the specific heats of numerous solids showed that the molar heat capacities were invariably near $3R = 6$ cal/mol·degree, a regularity recognized in 1819 by two French scientists. This empirical law of Dulong and Petit found a ready explanation in the statistical arguments of Boltzmann and Maxwell and was at first a testimonial to the validity of that theory. However, more and more departures from this law appeared during the late 1800s when improved techniques for liquefaction of gases permitted measurement of heat capacities at relatively low temperatures, for example, near the temperature of liquid nitrogen (79 K). For reasons that defied understanding, the heat capacities of solids decreased substantially as the temperature was lowered.

There were other indications that all was not well. Among the unsolved problems were the spectral lines of the elements, which Wollaston first saw in 1802 and Fraunhofer measured more precisely in 1814. By 1880 atomic spectroscopy had become one of the "hottest" fields in experimental physics. It was soon appreciated that each element had its own unique spectral signature and that minute amounts could yield a satisfactory spectrum. Spectral analysis gained acceptance as a valuable analytic tool, and the study and identification of spectra assumed great practical importance. But the origin of these bright lines remained a complete mystery.

Then, in 1885, Johann Balmer, an obscure Swiss schoolteacher, found a simple numerical formula, namely

$$\lambda = 364.6 \frac{m^2}{m^2 - 4} \text{ nm} \tag{29.1}$$

that gave an excellent fit to the strong lines in the spectrum of hydrogen. But this was just numerology with no physical justification. The mystery of the spectral lines remained unresolved until Bohr showed how a judicious wedding of the Rutherford model of the atom with the quantum concepts of Planck and Einstein led to an expression of the form Balmer had hit on by trial and error.

The discoveries of X rays and radioactivity just before 1900 also defied classical explanations. But these were, after all, such startlingly new phenomena that physicists could argue that with enough time for proper analysis and reflection, an acceptable explanation would be forthcoming.

The problems of spectral lines and heat capacities could not be dismissed so cavalierly. Though a good many scientists tended to sweep these matters under the rug, others clearly recognized therein a fundamental problem and an exciting challenge. By a curious quirk of history, the clue to solving these difficulties came from an unexpected quarter, from an attempt to find a satisfactory expression for the spectral distribution of blackbody radiation.

29.2 Blackbody Radiation

Radiation is one process by which heat can be transferred. The power radiated per unit area varies with the fourth power of the absolute temperature of the radiating body—this is Stefan's law, Equation (11.11)—and depends on a parameter called the emissivity ϵ. Objects that absorb radiation effectively are also good emitters, and one whose surface is perfectly black—i.e., is a perfect absorber—is also a perfect emitter with emissivity $\epsilon = 1$. Such an object is called a *black body*.

As an object gets hotter, not only does it radiate more strongly but its color changes perceptibly. When a toaster or an electric space heater is first turned on, you can, by placing a hand near it, feel the heat radiated by the element, although you can discern no visible radiation; nearly all the energy is radiated in the infrared, to which our eyes are not sensitive. However, as its temperature increases, the element glows first a dull red, then a brighter red, and its color may even change to orange; some objects, such as the filament of an incandescent lamp, that are raised to a much higher temperature, glow with a bright yellow to white appearance.

The radiation emitted is not confined to one wavelength or narrow spectral region but spans the entire electromagnetic spectrum. As indicated in Figure 29.1, however, the intensity depends on wavelength and

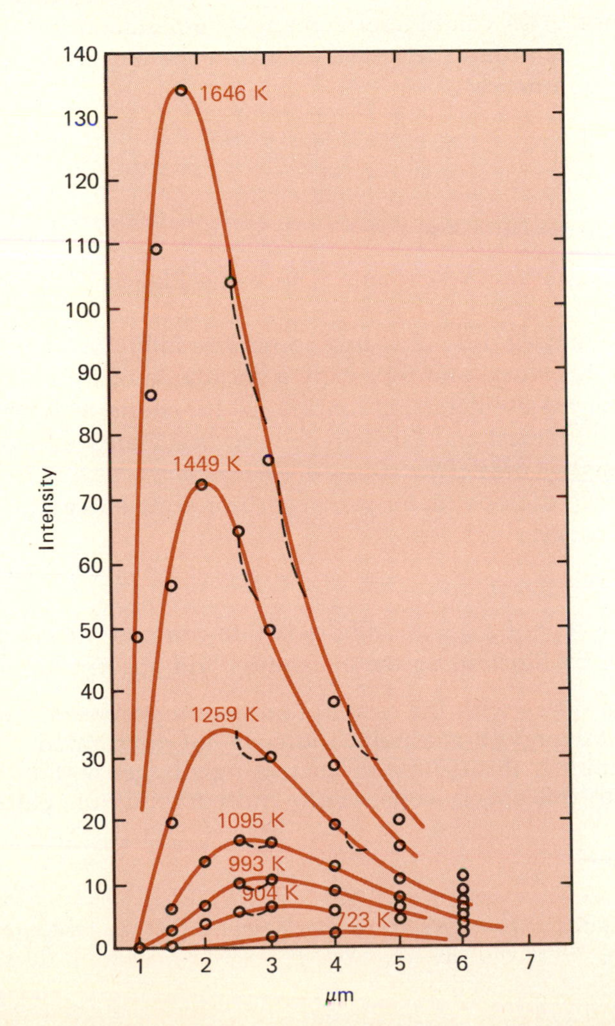

Figure 29.1 *Spectral distribution of blackbody radiation. The lines through the experimental points, the open circles, show the results of Lummer and Pringsheim (the dips in the curves at about 2.7 μm and 4.3 μm reflect absorption by water vapor and carbon dioxide). The theoretical predictions based on Wien's law and his formula for $\phi(\lambda T)$ are shown in color; evidently, Wien's formula does not agree well with the data for longer wavelengths, and the discrepancy is well outside the limits of experimental error.*

temperature, and the most intense part shifts to shorter wavelengths (higher frequencies) as the temperature increases.

The problem that occupied the attention of Max Planck and a few other theorists around 1900 was finding a satisfactory formula for this spectral distribution. In 1893, Wilhelm Wien, using thermodynamic arguments, derived a simple relation between intensity and wavelength; it involved just one arbitrary function of a single variable, $g = \lambda T$, the product of wavelength and temperature. Once this function $\phi(g)$ was known, the radiation at any temperature and wavelength could be predicted.

Wien had suggested a form for $\phi(g)$, and though it seemed to fit moderately well to data then available, later and more precise measurements left no doubt that especially at long wavelengths, theory and experiment were out of joint.

Figure 29.2 Max Planck (1858–1947).

In October 1900, Planck, in much the same way that Balmer had done, proposed a formula which "so far as I can see by quick inspection, represents the . . . data . . . satisfactorily. . . . I therefore feel justified in directing attention to this new formula which . . . I take to be the simplest excepting Wien's."

There still remained the task of justifying this new formula, which called for "a few weeks of the most strenuous work" of Planck's life. Planck constructed a mechanical model, a collection of oscillators or resonators, as he called them, able to absorb and emit radiant energy. He then turned to the combinatorial methods Boltzmann had developed and found that his formula followed only if in the required summations, he used energy intervals*

$$\Delta E = h\nu \tag{29.2}$$

Here

$$h = 6.626 \times 10^{-34}\ \mathrm{J{\cdot}s}$$

is known as *Planck's constant.* (The value Planck deduced was 6.55×10^{-34} J·s).

Planck himself did not at first appreciate fully the far-reaching implications of his work, and for nearly a decade he continued to regard the energies of his oscillators as continuous functions. Within a few years, however, Ehrenfest and Einstein showed that Planck's theory was internally consistent only if *the energy of each oscillator was an integral multiple of $h\nu$.* Hence each oscillator can absorb or emit energy only in quantal units of energy:

$$E = h\nu \tag{29.3}$$

The concept of a quantum of radiation (photon) first appears in 1905 in a paper by Einstein on the generation and absorption of light.

> In accordance with the assumption herein proposed, the radiant energy from a point source is not distributed continuously throughout an increasingly larger region, but, instead, this energy consists of a finite number of spatially

* We introduce here the notation widely employed in this branch of physics, where the symbol for frequency is the Greek letter ν.

localized energy quanta which, moving without subdividing, can only be absorbed and created as whole units.

According to this model, when a beam of light strikes an absorbing material, the light intensity is reduced because photons are removed from the beam. However, only an integral number of photons can be taken out of the beam.

The reason one does not normally notice this quantized, quasicorpuscular character of light is that the number of photons per second striking even a small area of 1 mm^2 is, for normal light intensities, enormous. The density of the "finite number of spatially localized energy quanta" is so great that they appear as a continuous stream. At very low intensities, however, this quantized nature of light can be observed, using commercially available sensitive photon counters.

● ***Example 29.1*** An incandescent 50-W light bulb is viewed from a distance of 1 km. Assume that the efficiency of the bulb is 10 percent (that is, only 5 W of visible radiation is emitted) and that only 1 percent of this radiation is in the wavelength region between 550 and 554 nm. How many photons from that wavelength region penetrate the pupil of the observer's eye? (Assume a pupil diameter of 2 mm).

We must first decide how many photons in the range of interest are emitted by the bulb each second. From the information given, we know that the energy radiated per second within that narrow range is

$$(50\ \text{W})(0.1)(0.01) = 0.05\ \text{W}$$

That must also be the product of the energy per photon multiplied by the number of photons emitted per second in the range 550–554 nm. The energy per photon is

$$\begin{aligned} E &= h\nu = \frac{hc}{\lambda} \\ &= \frac{(6.63 \times 10^{-34}\ \text{J}\cdot\text{s})(3 \times 10^{8}\ \text{m/s})}{5.52 \times 10^{-7}\ \text{m}} \\ &= 3.6 \times 10^{-19}\ \text{J} \end{aligned}$$

where we used the average wavelength of 552 nm. Consequently, the total number of photons emitted per second is

$$N = \frac{5 \times 10^{-2}\ \text{J/s}}{3.6 \times 10^{-19}\ \text{J}} = 1.39 \times 10^{17}\ \text{photons/s}$$

If it is assumed that these photons are radiated uniformly in all directions, the number passing through a hole of 2-mm diameter and 1 km distant equals the ratio of the area of the 2-mm hole to the area of a sphere of 1-km radius multiplied by N. Thus we have that the number of photons entering the observer's eye is

$$\begin{aligned} n &= \frac{(1.39 \times 10^{17}\ \text{photons/s})(\pi \times 10^{-6}\ \text{m}^2)}{4\pi \times 10^{6}\ \text{m}^2} \\ &= 3.5 \times 10^{4}\ \text{photons/s} \end{aligned}$$

Thus even at this considerable distance from a weak light source, the number of photons, within a narrow wavelength range, entering the pupil of an observer's eye is still about 35,000 each second. ●

Using the concept of quantized radiation, Einstein successfully explained some puzzling features of photoelectric emission, in particular certain experimental observations of Phillip Lenard.*

29.3 The Photoelectric Effect

The photoelectric effect relates to the following phenomenon. If a metal surface is illuminated by visible or ultraviolet radiation, electrons are released *provided that the frequency of the radiation exceeds a critical threshold.* Lenard had demonstrated that the number of electrons increases with the intensity of the radiation but their energy does not.

According to classical electrodynamics, the electric field of the incident radiation would set electrons near the surface in motion, and on gaining enough energy, they might be released. In that case, the energy of the emitted electrons would be expected to depend on the light intensity. Furthermore, whereas classical theory predicted that under incident

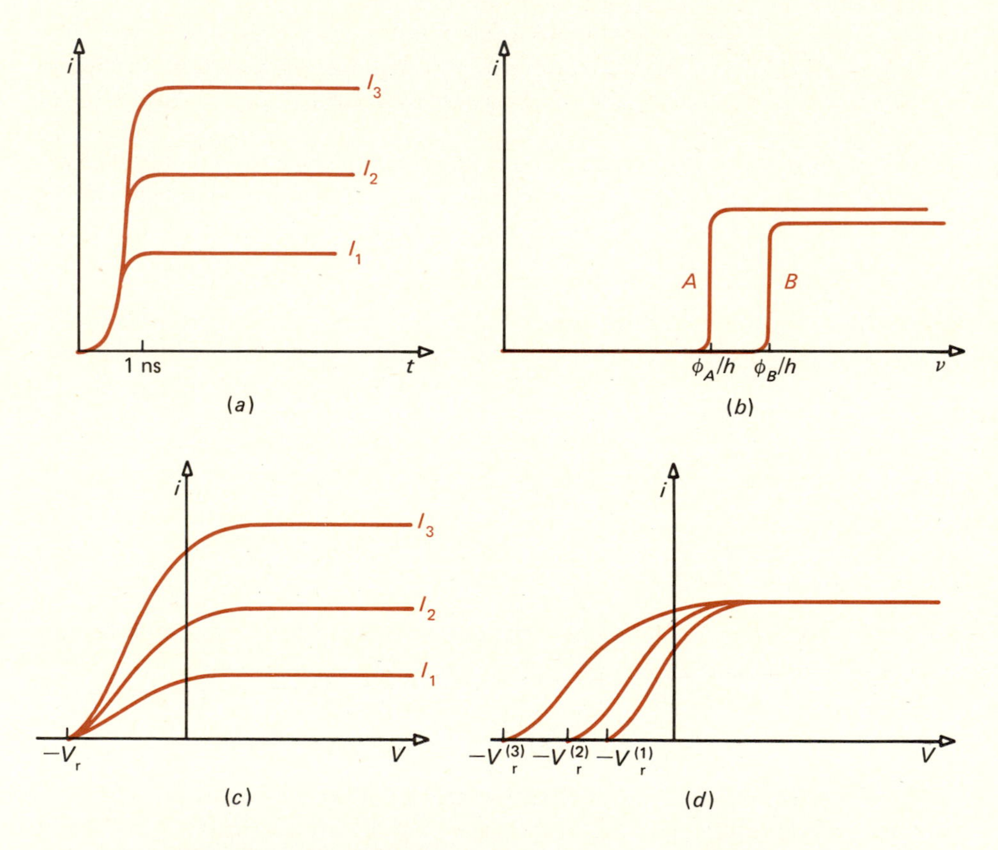

Figure 29.3 Graphical representation of some critical features of photoelectric emission. (*a*) If the emitting surface is illuminated starting at time $t = 0$, the photoelectric current i rises to its saturation value within less than one nanosecond. The saturation value increases with increasing intensity of illumination. (*b*) The photoelectric current as a function of frequency of the incident radiation exhibits a sharp threshold, which depends on the emitting substance; A and B are for two different materials, with work functions ϕ_A and ϕ_B. (*c*) The current, as measured by the ammeter of Figure 29.4, as a function of potential across the phototube for monochromatic illumination at three different intensities. Whatever the intensity, the photocurrent vanishes if the potential is less than $-V_r^c$, where V_r^c is the critical retarding potential. (*d*) The photocurrent as a function of potential at constant photon flux for light of three different frequencies. The critical retarding potential for frequency ν_1 is represented by $V_r^{(1)}$, and here $\nu_3 > \nu_2 > \nu_1$. As the frequency of the incident light increases, so does the critical retarding potential.

* Einstein, in his paper, credited Lenard, referring to his "trailblazing research" (*bahnbrechende Arbeit*). Years later, Lenard emerged as one of the most virulent anti-Semites and enthusiastic supporters of Hitler. His assessment of Einstein's work is a grim reminder of those times:

> The most important example of the dangerous influence of Jewish circles on the study of nature has been provided by Herr Einstein with his mathematically botched-up theories consisting of some ancient knowledge and a few arbitrary additions. This theory now falls to pieces, as is the fate of all products that are estranged from nature. Even scientists who have otherwise done solid work cannot escape the reproach that they allowed the relativity theory to get a foothold in Germany, because they did not see, or did not wish to see, how wrong it is, outside the field of science also, to regard this Jew as a good German.

light of low intensity minutes would be needed before an electron could absorb sufficient energy to escape, experiments showed that emission commenced almost instantly once the light was turned on.

The explanation of these observations is straightforward with Einstein's model. It is assumed that a minimal energy, called the *work function* ϕ, must be imparted to an electron before it can escape from the metal. The work function ϕ is an intrinsic material property, and for most substances is in the range 1–10 eV. No emission occurs unless $h\nu \geq \phi$; but when that condition is satisfied, some of the electrons may leave the metal with kinetic energy (KE) as great as

$$\mathrm{KE}_{\mathrm{max}} = h\nu - \phi \qquad \textbf{(29.4)}$$

The value of $\mathrm{KE}_{\mathrm{max}}$ can be measured with the circuit of Figure 29.4. To reach the anode A in the presence of the retarding voltage V_{r}, an electron must leave the cathode C with a kinetic energy of at least V_{r} eV. The current vanishes altogether when

$$eV_{\mathrm{r}} > eV_{\mathrm{r}}^{\mathrm{c}} = h\nu - \phi = \mathrm{KE}_{\mathrm{max}} \qquad \textbf{(29.5)}$$

because if the retarding potential V_{r} exceeds the critical value $V_{\mathrm{r}}^{\mathrm{c}}$ none of the emitted electrons has enough energy to overcome the electrostatic potential barrier.

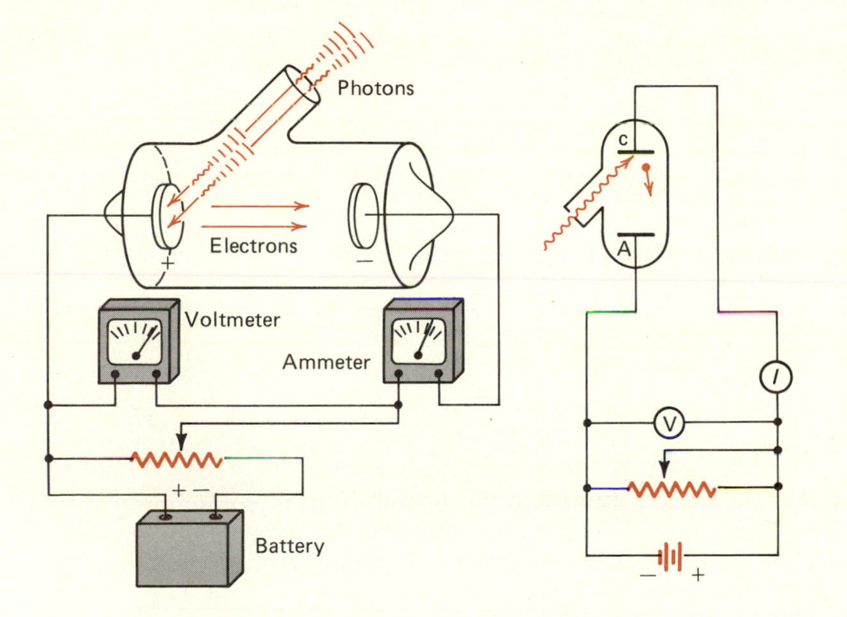

Figure 29.4 Schematic of an experimental arrangement for verifying the Einstein equation, Equation (29.4).

Experiments to verify this prediction were performed in the following decade by Millikan, among others. He found excellent agreement with the Einstein formula, Equation (29.5) and in the process, also measured Planck's constant. (See Figure 29.5.)

● ***Example 29.2*** From the data shown in Figure 29.5, determine the value of Planck's constant and the work function of the substance Millikan used as the electron emitter.

From Equation (29.5), it follows that when the critical retarding voltage goes to zero, the maximum KE of the electrons is also zero and the energy of the incident quanta then just equals the work function of the substance. In this instance, that frequency is 4.4×10^{14} Hz, and the work

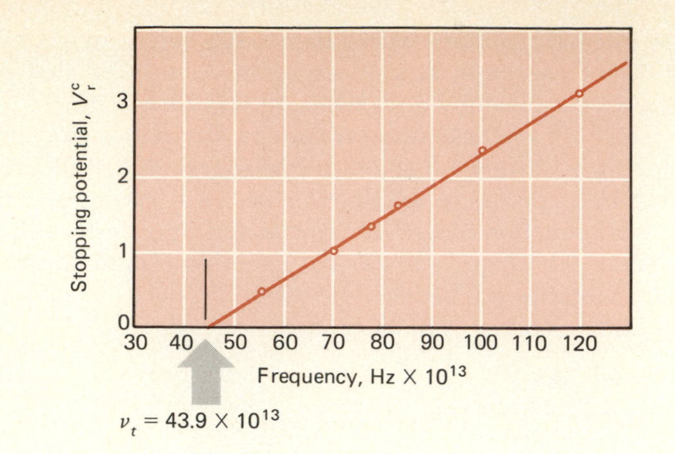

Figure 29.5 Critical retarding potential versus frequency. Data shown here were obtained by R. A. Millikan and published in Physical Review 7, 362 (1916). The points fall on a straight line in agreement with Einstein's prediction.

function is therefore

$$\phi = h\nu_t = (6.63 \times 10^{-34}\ \text{J}\cdot\text{s})(4.4 \times 10^{14}\ \text{Hz})$$
$$= 2.9 \times 10^{-19}\ \text{J} = 1.82\ \text{eV}$$

Referring again to Equation (29.5), we find that the slope of the straight line shown in Figure 29.5 must equal h/e. Reading off the graph gives the stopping potential as 3 V when the frequency of the incident radiation is 11.5×10^{14} Hz. Consequently,

$$\frac{3\ \text{eV}}{(11.5 - 4.4) \times 10^{14}\ \text{Hz}} = 4.23 \times 10^{-15}\ \text{eV}\cdot\text{s}$$

$$h = (4.23 \times 10^{-15}\ \text{eV}\cdot\text{s})(1.6 \times 10^{-19}\ \text{J/eV}) = 6.76 \times 10^{-34}\ \text{J}\cdot\text{s}$$

in moderately good agreement with the accepted value of h.

29.4 Specific Heat of Solids

Einstein based his calculation of the specific heat of solids on the same model that had been used earlier to explain the law of Dulong and Petit. Each atom in the crystal is presumed to vibrate about its equilibrium position with a frequency ν, determined by the atom's mass and the interatomic restoring forces acting on it.

Einstein then calculated the internal energy U of such a model crystal in thermal equilibrium with its surroundings at a temperature T, under the assumption that the energy of each oscillating atom must take on the value $nh\nu$, where n is an integer. He then found that $C_v = \Delta U/\Delta T$ (see Sec-

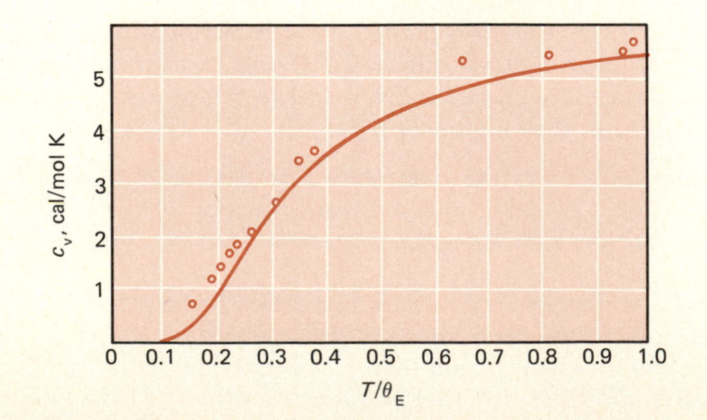

Figure 29.6 The molar heat capacity of diamond as a function of temperature. Einstein's theory predicts that the molar heat capacities of all solids must fall on one universal curve, provided that the data are plotted against a "reduced temperature" T/θ_E. Here $\theta_E = h\nu/k$ and is known as the Einstein temperature.

tion 11.5) was not a constant except in the limit when $T \gg h\nu/k$, but decreased with diminishing temperature in rough agreement with observation (Figure 29.6).

One does not have to go through the details of the calculation to see why the Planck condition should have this consequence. Suppose the crystal is at a very low temperature, so that every atom is in its lowest-energy state, called the *ground state*. In that case, the minimum energy the system can absorb from its surroundings is one quantum $h\nu$. If, however, $kT \ll h\nu$, the probability that the environment can impart that much energy to one of the atoms of the crystal is exceedingly small; the crystal does not absorb much energy and its heat capacity is correspondingly small. The ability of the crystal to absorb energy increases as kT approaches $h\nu$; and when $kT \gg h\nu$, the effects of energy quantization become negligible and the heat capacity approaches the classical value of $3R$.

A few years after Einstein's calculation, Debye improved the model. He relaxed the assumption of a single vibrational frequency and obtained results in excellent accord with experiment for a great variety of crystals and over a wide temperature range. (See Figure 29.7).

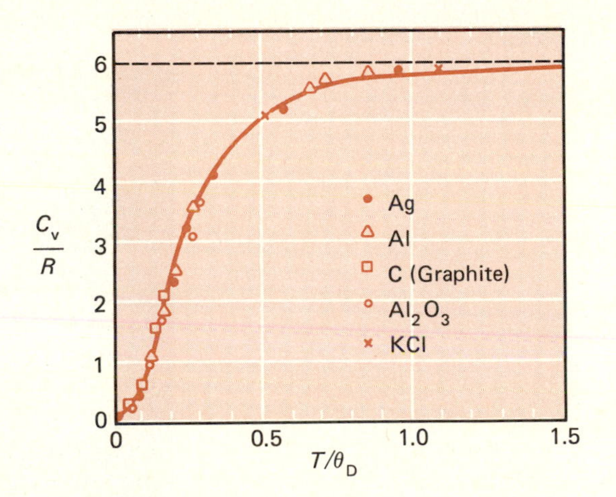

Figure 29.7 *The molar heat capacities of five solids as functions of temperature. When plotted against the reduced temperature T/θ_D, all data fall on the universal curve deduced by Debye. Here θ_D is a characteristic parameter of a particular material and is known as the Debye temperature.*

29.5 X Rays

Toward the end of the nineteenth century, the study of gas discharges was a very active and fashionable field of research. At the University of Würzburg, Professor Wilhelm Roentgen used such a discharge tube, a glass envelope containing gas at low pressure, with two metal electrodes connected to a spark coil to produce the high voltages necessary for the discharge. Late in 1895, he noticed that with the tube covered by "a somewhat closely fitting mantle of black cardboard . . . a paper screen washed with barium-platino-cyanide lights up brilliantly and fluoresces equally well whether the treated side or the other be turned toward the discharge tube." Roentgen concluded that the agency responsible for this fluorescence originated at the glass walls of the tube bombarded by energetic electrons.

In a few weeks of intensive effort, Roentgen determined that this mysterious agency caused fluorescence of many different minerals, fogged photographic film, traveled in straight lines and could not be deflected by a magnetic field, and discharged electrified objects, whether charged positively or negatively. He found also that it was generated whenever energetic electrons struck a solid object, and more effectively if the target was a heavy element, such as tungsten or platinum, than a light one, and that it penetrated, to some degree, practically any material, for example, several centimeters of wood, while 15 mm of aluminum "weakens the effect considerably though it does not destroy fluorescence." Lastly, Roentgen had his wife place her hand over a photographic plate while he energized the discharge, and developed the first anatomical X-ray picture, the outline of a hand, the bones clearly in evidence and also his wife's wedding ring on one finger.

Within six weeks of the announcement of his discovery, which earned Roentgen the first Nobel Prize in physics, physicians used this new scientific marvel in their work. Unfortunately, it was some time before the deleterious effects of X rays on biological systems were properly recognized, and for many years people willingly and unwittingly exposed themselves to hazardous levels of radiation.

Intensive research on X rays began almost instantly throughout Europe and America and continued for several decades. Although it was suspected fairly early that X rays were high-frequency electromagnetic waves, the issue was not settled unequivocally until 1912, when follow-

Figure 29.8 *Wilhelm Konrad Roentgen (1845–1923).*

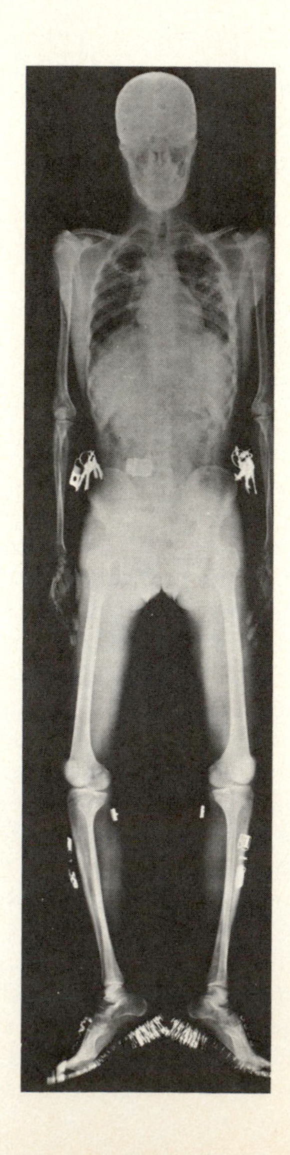

Figure 29.9 *A photograph of historical interest, taken by Roentgen. Radiograph of a clothed man, showing not only his skeletal structure but also his shoes (note the nails in soles and heels), metal clasps of his garters, and keys in his jacket pockets.*

Figure 29.10 A fluoroscope of the kind many shoe stores used about 1930–1950 in Europe and America, before the serious health hazards of such "toys" were properly appreciated. The level of radiation to which the child is subjected is substantially greater than that measured by the technician seated on the right.

ing a brilliant suggestion of Max von Laue, Friedrich and Knipping demonstrated that X rays were diffracted by the regular atomic array in a crystal lattice just as visible light is diffracted by a ruled grating. From the known spacing between atoms in a crystal, one could then calculate the wavelengths of the incident X rays, and these were found to be between about 1 and 5×10^{-11} m. With modern equipment, one can produce X rays with wavelengths between 10^{-9} m and 10^{-12} m.

The work of Laue, Friedrich, and Knipping had two immediate consequences. First, researchers now had a technique, like that used in optics, for dispersing an X-ray beam into its spectral components. X-ray spectroscopy soon led to some of the most perceptive and profound insights in the field of atomic structure and also was instrumental in a critical test of the photon hypothesis.

Figure 29.11 A Laue X-ray photograph of a sodium chloride single crystal. In this instance, the X-ray beam was parallel to one of the cubic axes of the crystal (see Figure 9.1). Note that this picture is symmetrical to rotation through 90°, as is also true of a cube that is rotated about one of its cube axes. Continuous, that is polychromatic, X-ray radiation is used in this work. X-ray diffraction patterns can also be obtained from a polycrystalline sample, as shown in Figure 29.18. To obtain such powder patterns, as these rings are called, one must use monochromatic X rays.

Second, the potential of X rays as probes for determining crystal structure was recognized and exploited, especially by the father-and-son team of W. H. and W. L. Bragg; their impact on the complexion of British science is without parallel in modern times. Since the days of the Braggs, England has emerged as the world's leader in X-ray crystallography. In recent times, X-ray structure analysis has concentrated on complicated organic systems, and it is surely no accident that the DNA puzzle was solved by two biologists in Cambridge, one a young American Ph.D. who had gone there for the express purpose of working in the renowned crystallographic laboratory of Sir Lawrence Bragg.

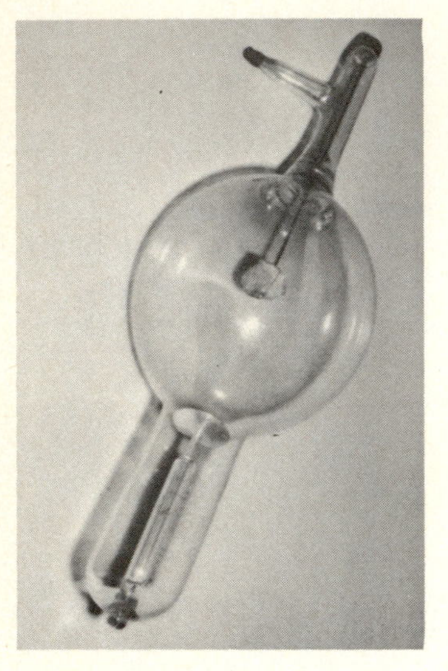

(*a*)

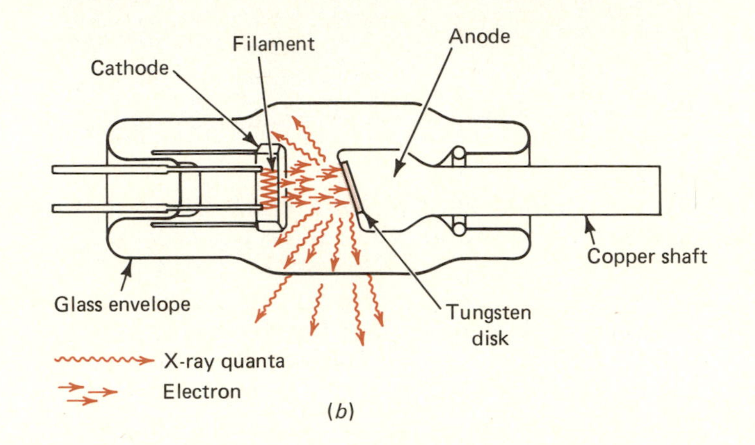

(*b*)

Figure 29.12 (*a*) *An early X-ray tube.* (*b*) *A modern X-ray tube. Electrons emitted by the heated filament are accelerated across the vacuum space by the electric field between cathode and anode. When they strike the target, here a tungsten disk, a small fraction, typically about 1 percent, release their acquired energy in the form of X rays. Most of the energy of the impinging electrons results in raising the temperature of the anode. This heat must be dissipated to avoid melting the anode. Sometimes a copper shaft, as shown here, provides adequate heat conduction, but for higher intensities it is customary to circulate cooling water through the anode. Even then, it may not be possible to avoid melting the target if the same small region is under continuous electron bombardment.*

The spectrum of radiation emitted by an X-ray tube consists of two distinct portions, a continuous part and superimposed on this, a series of sharp and intense lines. The continuous spectrum depends on the voltage across the tube, but does not vary substantially as the target material is changed.* The line spectrum, by contrast, is characteristic of the target material and provided that the accelerating voltage exceeds a certain threshold value, independent of that voltage.

Figure 29.14 shows typical curves of the continuous spectrum for three different applied voltages. Evidently, an increase in anode potential results in an increase in the total energy radiated and also shifts the short-wavelength cutoff toward smaller values of λ.

This continuous radiation was given the name *Bremsstrahlung*, a German word meaning, literally, braking radiation. The term is highly descriptive of the mechanism, namely radiation by a charged object subjected to deceleration. In this particular instance, electrons, on striking the target anode, are quickly brought nearly to rest from high speed. While the classical electrodynamic interpretation is essentially correct, and the word *Bremsstrahlung* has survived in the physics vocabulary, that theory could not explain the spectral distribution, particularly the very sharp short-wavelength threshold and its voltage dependence.

These features are readily understood in terms of the Planck-Einstein postulates. On falling through a potential V_A between cathode and

* The efficiency for production of X rays does depend on atomic number of the target, so a change from copper to tungsten increases the intensity of the continuous radiation but has no effect on its spectral distribution.

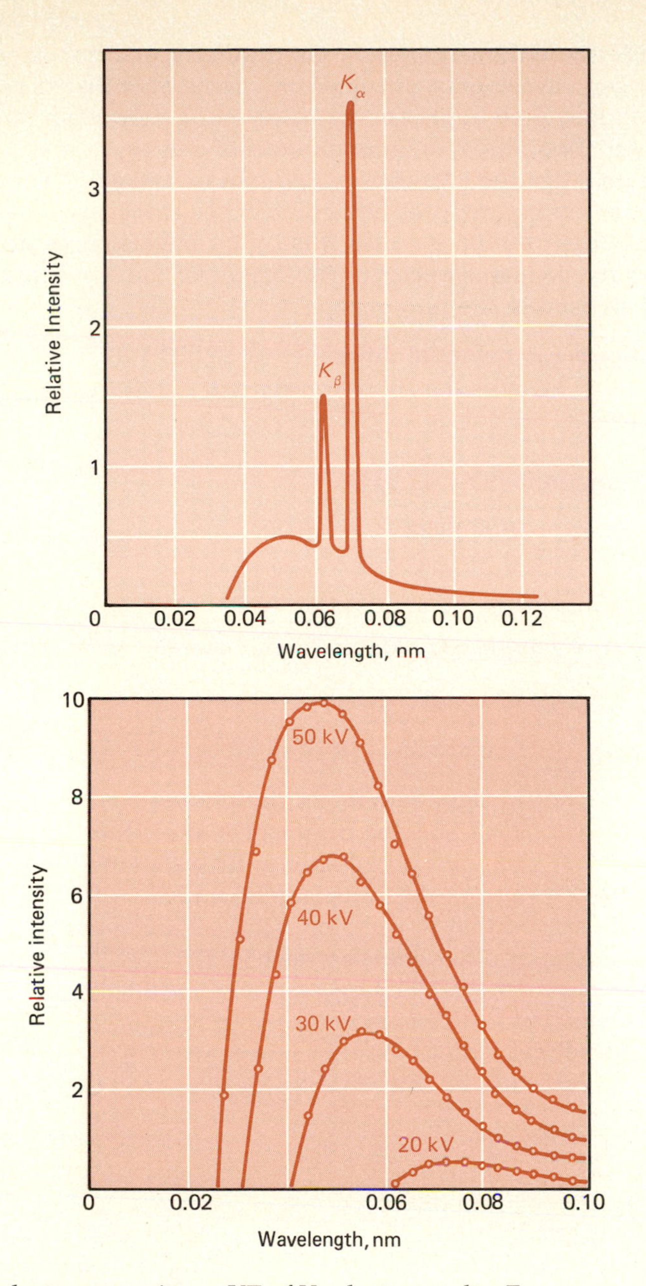

Figure 29.13 Typical X-ray spectrum. In this case, the target material is molybdenum. The characteristic lines, here marked K_α and K_β, superimposed on the continuous background, are even sharper than indicated on this drawing.

Figure 29.14 The spectral distribution of continuous emission of X rays from a tube with a tungsten target. Note that the intensity increases dramatically with increasing anode potential, and that the short wavelength cutoff shifts to the left, that is, to shorter wavelength.

anode, an electron acquires a KE of V_A electron volts. From energy conservation it then follows that the maximum energy of a photon created by the rapid deceleration of this electron cannot exceed that value. Thus

$$h\nu_{\max} = eV_A = \frac{hc}{\lambda_{\min}}$$

from which we obtain

$$\lambda_{\min} = \frac{hc}{eV_A} \qquad (29.6)$$

One of the earliest fairly precise determinations of Planck's constant was based on this simple relation. We have here the inverse of the photoelectric effect; now the maximum energy of the emitted photon is directly related to the energy of the incident electron.

To understand the origin of the continuous spectrum, extending from λ_{min} to longer wavelengths, one need only remember that in decelerating, an electron may give up its energy by emitting more than one quantum of radiation. Two, three, or more photons of energy $h\nu_i$ may be released, provided only that the total energy $\Sigma_i h\nu_i$ not exceed eV_A.* For example, if two photons are emitted, we immediately see that there will be a broad wavelength distribution because it is by no means necessary that both photons have the same energy. (The origin of the X-ray line spectrum will be discussed in the next chapter.)

Example 29.3 A potential difference of 10,000 V is maintained across an X-ray tube. What is the short-wavelength cutoff for X rays produced by this tube?

From Equation (29.6) we have

$$\lambda_{min} = \frac{(6.63 \times 10^{-34}\ \text{J·s})(3 \times 10^{8}\ \text{m/s})}{(1.6 \times 10^{-19}\ \text{C})(10^{4}\ \text{V})} = 1.24 \times 10^{-10}\ \text{m}$$

Hence the shortest wavelength of the X-ray spectrum from this tube is 0.124 nm, a length comparable to the spacing between atoms in a crystal.

29.6 Compton Scattering

Despite the acknowledged success of the photon concept in explaining the photoelectric effect and the origin of X rays, the light-particle hypothesis did not gain wide acceptance until 1923, when A. H. Compton announced his results on the scattering of X rays.

Compton argued that if a photon behaves like a particle, then scattering of photons by other particles, such as electrons, must be governed by the same requirements of energy and momentum conservation as any other collisions. For instance, as indicated in Figure 29.15, when an incident photon collides with an electron at rest, some of the photon's energy and momentum must be imparted to the electron. Consequently, the

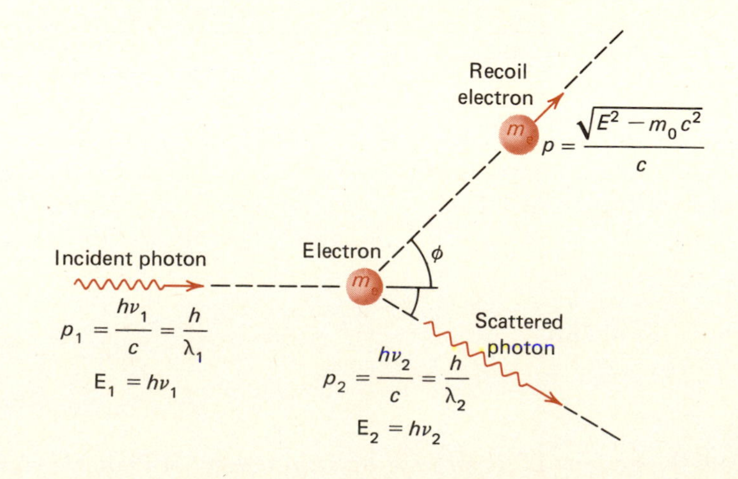

Figure 29.15 *Geometry of Compton scattering. An incident photon collides with an electron nearly at rest (the energy of the electron bound to an atom in the target material—solid, liquid, or gas—is negligible compared with the energy of the incident photon). After the collision, some of the energy and momentum of the incident photon has been transferred to the electron, which recoils with finite energy and momentum. The scattered photon has less energy and therefore a longer wavelength than the incident photon. The difference in wavelength between scattered and incident photons depends on the scattering angle θ.*

* The quantity $\Sigma_i h\nu_i$ may be, and generally is less than eV_A because there are other nonradiative processes by which the incident electron can lose energy.

scattered photon must have less energy than the incident, and according to Planck's relation, a lower frequency, or longer wavelength. Moreover, the change in wavelength must depend on the energy transferred to the electron, and this, in turn, depends on the angle through which the photon is scattered.

Applying the two conservation conditions, one arrives at a relation between the change in wavelength and the scattering angle θ:

$$\Delta\lambda = \lambda' - \lambda = \left(\frac{h}{m_0 c}\right)(1 - \cos\theta) \qquad \textbf{(29.7)}$$

where m_0 is the rest mass of the electron. The quantity $(h/m_0 c)$ is known as the *Compton wavelength*. For electrons, $h/m_0 c = 2.42 \times 10^{-3}$ nm, which is very small compared with the wavelength of visible light. Thus it would be extremely difficult to observe the Compton effect with visible light.

X-ray wavelengths are, however, typically in the neighborhood of 10^{-1} to 10^{-2} nm, and for such radiation, the Compton effect, with wavelength shifts up to 50 percent, is readily observable.

Compton measured wavelength changes that were in complete accord with Equation (29.7). His was the crucial experiment that really convinced nearly all physicists that electromagnetic radiation did, indeed, behave like a stream of particles traveling at the speed of light, as Ein-

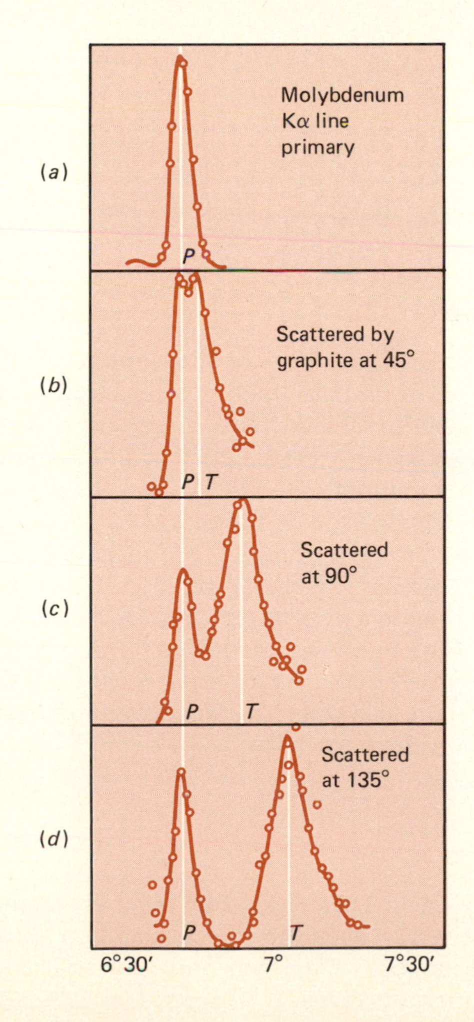

Figure 29.16 *Compton's results, showing the spectral distribution of X rays scattered from a graphite target. The portion labeled P corresponds to so-called Rayleigh-scattered X rays, that is, X rays scattered elastically according to the theory of electromagnetism. The peaks labeled T correspond to X rays scattered inelastically in conformity with the corpuscular model, and this peak shifts to longer wavelengths with increasing scattering angle. (The abscissa is the angle at which the calcite crystal, which served as a diffraction grating, was positioned, and this is a measure of the X-ray wavelength.) These figures demonstrate the wave-particle duality especially vividly, each curve exhibiting both features—classical scattering (P) and Compton scattering (T).*

stein had first proposed almost twenty years earlier. At the same time, Compton recognized the dilemma his measurements had left unanswered:

> The crystal spectrometer measured a wavelike property, the wavelength, and measured it by a characteristically wavelike phenomenon, interference. But the effect of the graphite scatterer on the value of that wavelike property could be understood only in terms of particle-like behavior.

This schizophrenic attribute of electromagnetic waves, the *wave-particle duality* as it came to be known, proved a troublesome conceptual obstacle for some years to even the most creative and resourceful minds, and it continues to be so to almost every student. Perhaps as we go along, the matter will become less confusing and more rational; for the moment, all we can say is that this is the way nature reveals itself to us.

● ***Example 29.4*** In a Compton scattering experiment, one finds that the fractional change in X-ray wavelength is 1 percent when the scattering angle is 120°. What was the wavelength of the X rays used in that experiment?

When the scattering angle is 120°, the change in wavelength is given by

$$\Delta\lambda = (2.42 \times 10^{-3}\ \text{nm})[1 - (-0.5)] = 3.63 \times 10^{-3}\ \text{nm}$$

Since $\Delta\lambda/\lambda = 0.01$, we have

$$\lambda = 3.63 \times 10^{-1}\ \text{nm}$$

a wavelength attainable with an X-ray tube able to maintain a potential difference of 3500 V or more. ●

29.7 Matter Waves

Although by 1924, many physicists had become reconciled with, though not resolved the wave-particle duality of electromagnetic energy, the suggestion contained in the doctoral dissertation of a young French nobleman was greeted with great skepticism and derision. (It is reported that one Parisian humorist referred to his theory as "La Comédie Française.") Louis Victor, Duc de Broglie, who on the death of an older brother became the Prince de Broglie, proposed that particles, such as electrons or nuclei, exhibit wavelike properties. The heuristic arguments that led Prince de Broglie to this hypothesis are based on certain features of Bohr's model of the hydrogen atom (see Section 30.4).

Figure 29.17 *Louis Victor, Prince de Broglie (b. 1892).*

De Broglie proposed that a *matter wave* must be associated with any particle, whether light or heavy. The wavelength of that wave is related to the momentum of the particle by

$$\lambda = \frac{h}{p} \tag{29.8}$$

known as the *de Broglie relation*.

That idea may seem farfetched, but by then many physicists had ceased to be surprised by even the most outlandish notions. At any rate, within a short time, C. J. Davisson and L. H. Germer reported to a meeting of the American Physical Society:

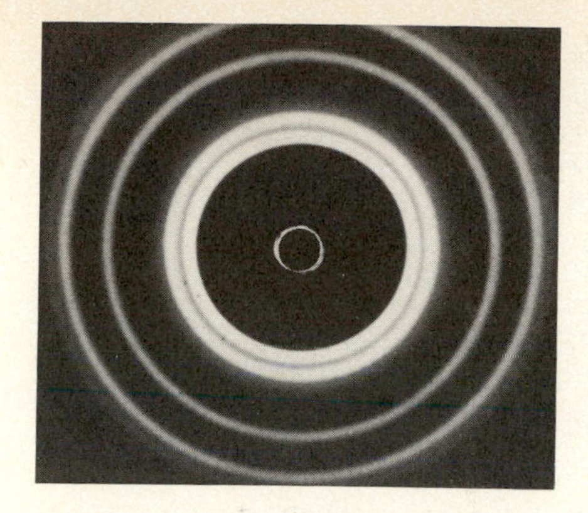

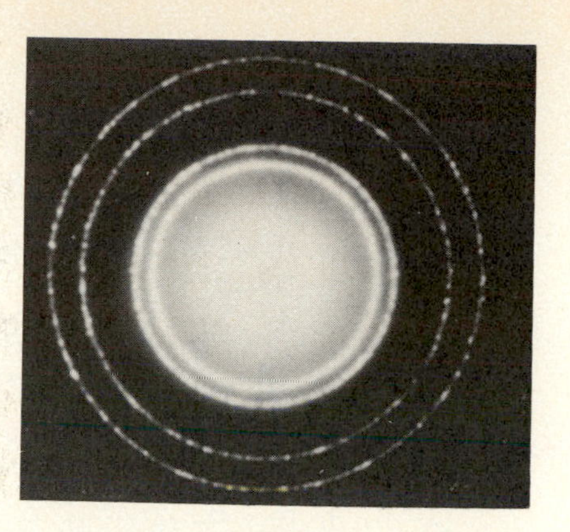

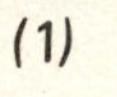

(1) (2)

(a)

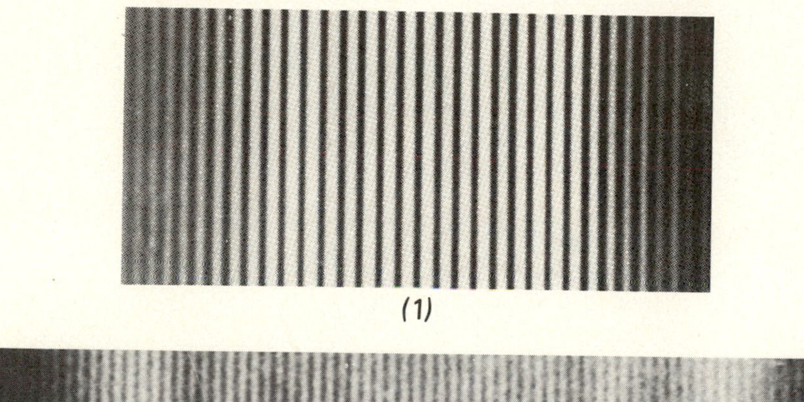

(1)

(2)

(b)

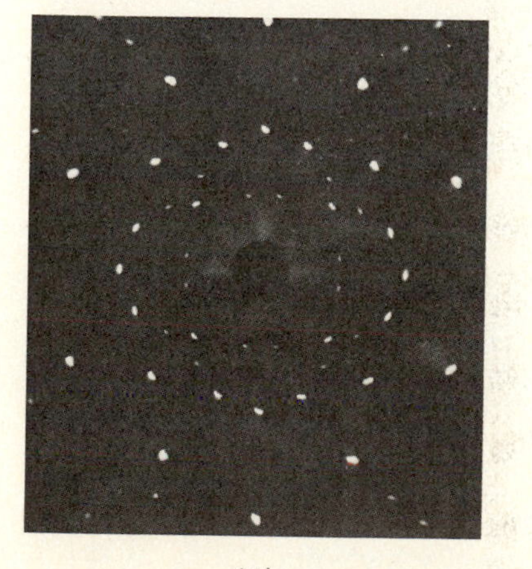

(1) (2)

(c)

Figure 29.18 *Comparison of diffraction effects produced by electromagnetic waves and by particles. (a) X-ray and electron diffraction patterns obtained by passing a monochromatic beam through a thin polycrystalline foil of aluminum: (1) The pattern from X rays with $\lambda = 0.071$ nm. (2) The pattern from electrons with $\lambda = 0.05$ nm (600-eV electrons). The second picture has been enlarged for ease of comparison. (b) The double-slit interference pattern using (1) monochromatic light and (2) monochromatic electrons. (c) X-ray Laue pattern from a sodium chloride crystal (1), and the corresponding diffraction pattern from a beam of neutrons (2).*

> In bombarding a {111} surface of a single crystal of nickel with a beam of electrons of uniform speed it has been found that, for certain definite bombarding speeds, there are beams of scattered electrons leaving the crystal in perfectly sharp directions. . . . The wavelengths of the wave disturbances which could give rise to these beams are quite accurately given by $\lambda = h/mv$, in accordance with the wave mechanics.

● ***Example 29.5*** What is the KE, expressed in electron volts, of an electron whose de Broglie wavelength equals that of a 10,000-eV photon?

In Example 29.3 we calculated the wavelength of a 10,000-eV photon as 0.124 nm.

By the de Broglie relation, Equation (29.8), an electron will have this wavelength if its momentum is

$$p = \frac{h}{\lambda} = \frac{6.63 \times 10^{-34}\ \text{J·s}}{1.24 \times 10^{-10}\ \text{m}}$$
$$= 5.35 \times 10^{-24}\ \text{kg·m/s}$$

The corresponding energy is given by

$$E = \frac{p^2}{2m}$$
$$= \frac{(5.35 \times 10^{-24}\ \text{kg·m/s})^2}{2(9.1 \times 10^{-31}\ \text{kg})}$$
$$= 1.57 \times 10^{-17}\ \text{J} = 98\ \text{eV} \approx 100\ \text{eV}$$ ●

● ***Example 29.6*** What is the wavelength of a 2-g Ping-Pong ball moving with a velocity of (*a*) 1 mm per century? (*b*) 1 m/s?

In (a), the momentum of the ball is

$$p = mv = (2 \times 10^{-3}\ \text{kg}) \left(\frac{10^{-3}\ \text{m}}{(100\ \text{yr})(3.15 \times 10^{7}\ \text{s/yr})}\right)$$
$$= 6.34 \times 10^{-16}\ \text{kg·m/s}$$

and hence,

$$\lambda = \frac{6.63 \times 10^{-34}\ \text{J·s}}{6.34 \times 10^{-16}\ \text{kg·m/s}} = 1.05 \times 10^{-18}\ \text{m}$$

That distance is to be contrasted with typical atomic dimensions of about 10^{-10} m and nuclear sizes of about 10^{-15} m.

If the ball is moving faster, at 1 m/s, its momentum will be correspondingly greater and the wavelength smaller. In that case we have

$$p = mv = 2 \times 10^{-3}\ \text{kg·m/s}$$

and

$$\lambda = \frac{6.63 \times 10^{-34}\ \text{J·s}}{2 \times 10^{-3}\ \text{kg·m/s}} = 3.3 \times 10^{-31}\ \text{m}$$

Evidently, the de Broglie wavelength of macroscopic objects is far too small to be measured, no matter how slow the motion. ●

Verification of the de Broglie relation only deepened the mystery. Though it established a natural symmetry between wavelike and

Figure 29.19 *Shown here, on arrival at the Stockholm railway station, prior to receiving the 1933 Nobel prize, from left to right: Mrs. Heisenberg (the physicist's mother), Mrs. Dirac, Mrs. Schroedinger, P. A. M. Dirac, Werner Heisenberg, and Erwin Schroedinger.*

particle-like manifestations—*complementarity* was the word Bohr favored—it left numerous unanswered questions. What was this field that propagated and behaved like a wave but was intimately associated with the particle? What sort of equations governed the field, equations perhaps analogous to Maxwell's for the electromagnetic field? How should one interpret the amplitude, phase, and intensity of this wave? That is, how are these wave properties to be associated with physical observables such as position, velocity, and kinetic energy?

These were the tough problems challenging the very best theoretical minds of the time. In 1926 two articles appeared in *Annalen der Physik,* one by Erwin Schroedinger, another by Werner Heisenberg, both treating the same problem but in such totally different ways that they seemed to be completely unrelated. Schroedinger, by various ingenious arguments, succeeded in deducing an equation that served as a good formal mathematical framework on which to support de Broglie's matter waves. In this work, Schroedinger was guided by the de Broglie relation between wavelength and momentum, by the classical relation between energy and momentum, by the condition that the equation must lead to wavelike solutions, and finally, by the presumption that the superposition principle holds for these waves as it does for sound and electromagnetic waves. Having deduced his equation, Schroedinger demonstrated that it gave results in excellent agreement with various dynamical properties of electrons. This *Schroedinger equation* remains today the point of departure for most quantum mechanical problems in atomic and molecular physics.

Heisenberg's approach used a formalism based on general properties of mathematical constructs known as matrices. Yet his and Schroedinger's results were the same, and in due time, Schroedinger was able to demonstrate that the two methods were fundamentally identical.

The physical interpretation of particle waves as it has evolved over the years was a source of great controversy, especially between Bohr and Einstein. Students who have a philosophical bent may enjoy some of the historical accounts of the Einstein-Bohr debates.

29.8 The Uncertainty Principle

One of the immediate consequences of quantum mechanics was the *Heisenberg uncertainty principle.* This states that it is *in principle* impossible to simultaneously specify precisely both position and momentum of a particle. Instead, if we denote by Δx and Δp_x the uncertainties in position and momentum along the x direction, then according to this principle of Heisenberg, the product of these uncertainties must always satisfy the inequality

$$\Delta x \, \Delta p_x \gtrsim \frac{\hbar}{2\pi} = h = 1.05 \times 10^{-34} \text{ J} \cdot \text{s} \tag{29.9}$$

with corresponding relations for the other coordinates.

Although it is difficult to reconcile the uncertainty principle with normal everyday observations, it is not difficult to see that it is a necessary consequence of wave mechanics. In Chapter 27, in discussing interference and diffraction, we remarked that such phenomena were the characteristic signatures of waves, and that was exactly the basis for the experimental verification of de Broglie's hypothesis.

If one wanted to localize a particle in, say, the y direction, one way to do this might be to send the particle through a narrow slit, as indicated in Figure 29.20. If a beam of particles strikes the mask, those passing through the slit must have been localized within a region $\Delta y = d$, where d is the slit width. We can draw some conclusions about the momentum in the y direction by observing the particles as they strike a screen some distance from the slit. If these are our ordinary, garden-variety classical particles, they will strike the screen along a thin strip, the projection of the slit on the screen. But if a wave propagates through the slit, then we know that the intensity on the screen will display a diffraction pattern substantially wider than the slit. Moreover, *as the slit width is reduced, the width of the diffraction pattern increases.*

A particle that arrives at the screen some distance from the center must have had a finite y component of momentum Δp_y in addition to its x component p_x, as indicated in Figure 29.20(c). The optical relation, Equation (27.13), between slit width and distance to the first diffraction minimum shows that if we try to delimit the position more narrowly by reducing d, we inevitably increase the diffraction width, that is, increase the uncertainty in p_y. Using the Rayleigh criterion, Equation (27.15),

$$\theta = \frac{\lambda}{d} = \frac{\lambda}{\Delta y} \tag{29.10}$$

the geometrical definition of the angle (for $\theta \ll 1$),

$$\theta = \frac{\Delta p_y}{p_x} \tag{29.11}$$

and the de Broglie relation between wavelength of momentum,

$$\lambda = \frac{h}{p_x} \tag{29.8}$$

we find

$$\frac{\Delta p_y}{p_x} = \frac{\lambda}{\Delta y} = \frac{h}{p_x \, \Delta y}; \qquad \Delta p_y \, \Delta y = h$$

which is consistent with the Heisenberg uncertainty condition.

The uncertainty principle also limits the precision of an energy measurement. If the time during which this measurement is made is Δt, the uncertainty in the energy ΔE is given by an expression similar to (29.8), namely,

$$\Delta E \, \Delta t \gtrsim \hbar \tag{29.12}$$

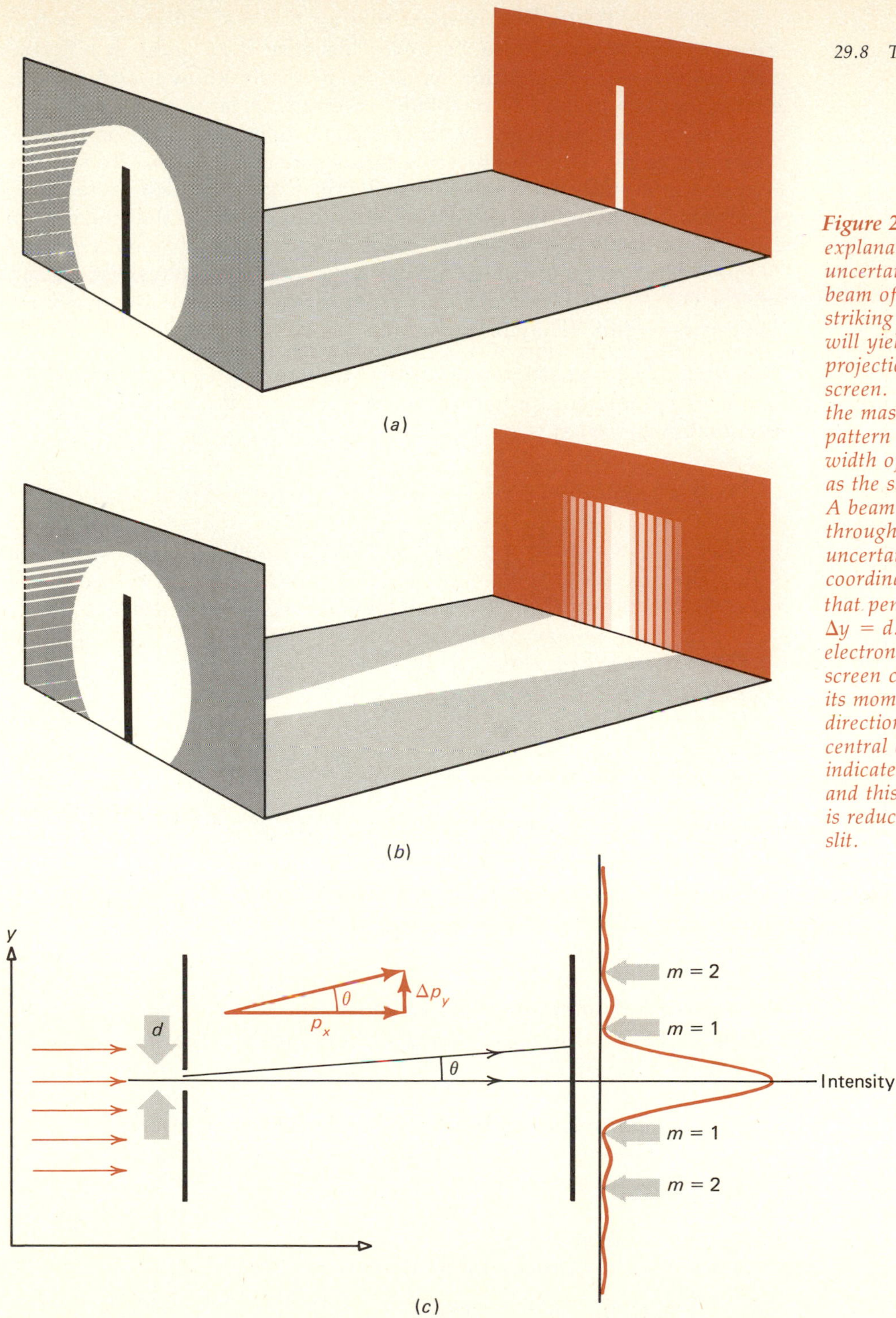

Figure 29.20 Optical explanation of the Heisenberg uncertainty principle. (a) A beam of classical particles striking the mask with a slit will yield a geometrical projection of the slit on the screen. (b) A wave that strikes the mask gives a diffraction pattern on the screen; the width of this pattern increases as the slit width is reduced. (c) A beam of electrons passes through a slit of width d. The uncertainty in the y coordinate of the electrons that penetrate the slit is $\Delta y = d$. The position of the electron on impact at the screen can be used to calculate its momentum in the y direction; the width of the central diffraction maximum indicates the uncertainty Δp_y, and this width increases if Δy is reduced by narrowing the slit.

One reason why the uncertainty principle is of no import in our normal activities is that Planck's constant h is so small that the uncertainties in position and momentum of even a fairly light object, such as a Ping-Pong ball, are far too small to be observed. In the description of atomic processes, however, the typical momenta and displacements are such that Equation (29.8) or (29.12) is critical. We will cite just one example.

Until about 1930 it was thought that as $T \rightarrow 0$ K, all atomic motion ceased. This cannot be true, however, because if an atom in a crystal were truly at rest, its position would be precisely known, and thus its momentum would have to be totally uncertain. A large momentum uncertainty implies a large KE, in contradiction to the presumption that the system's energy must approach a minimum as $T \rightarrow 0$ K. In fact the lowest energy state of a harmonic oscillator is one in which there is still some vibration about the classical equilibrium point, so that the sum of average potential and average KE is a minimum. This so-called *zero-point vibration* is not some fictitious attribute postulated by physicists to satisfy the theory; there are numerous experiments that have unequivocally demonstrated this behavior.

● ***Example 29.7*** An electron is known to exist within a region of 10^{-10} m extent, the diameter of a hydrogen atom. What is the uncertainty of its momentum and what is its approximate kinetic energy?

From the Heisenberg uncertainty principle, we have

$$\Delta p \simeq \frac{\hbar}{\Delta x} = \frac{1.05 \times 10^{-34}\ \text{J·s}}{10^{-10}\ \text{m}} = 1.05 \times 10^{-24}\ \text{kg·m/s}$$

The KE of an electron whose momentum is of this magnitude is

$$\begin{aligned} \text{KE} &= \frac{p^2}{2m} \\ &= \frac{(1.05 \times 10^{-24}\ \text{kg·m/s})^2}{2(9.1 \times 10^{-31}\ \text{kg})} \\ &= 6.1 \times 10^{-19}\ \text{J} = 3.8\ \text{eV} \end{aligned}$$

Though 3.8 eV is a small amount of energy on a macroscopic scale, it is significant on an atomic scale. ●

The uncertainty principle has profoundly affected not only the physics of the twentieth century but philosophy as well. Heretofore it had been a foregone conclusion that given precisely stated initial conditions, for example, position and momentum of a particle at $t = 0$, and knowing the forces acting on the particle thereafter, one could calculate the trajectory for $t > 0$ to arbitrary precision. It must be so, everyone concluded, because it worked that way. The motion of planets, the moon, the comets—all could be given with meticulous accuracy by astronomers, who based their reliable predictions of eclipses and conjunctions on solutions of Newton's laws of motion. To be sure, one had to allow for relativistic effects in a few exceptional cases, such as the orbit of Mercury, but that did not really invalidate the deterministic philosophy that was at the root of classical science.

Along came the new quantum theory, which maintained that since it was *in principle* impossible to measure simultaneously momentum and position to arbitrary precision, it was therefore *in principle* impossible to plot the trajectory of a particle to arbitrary precision. These limitations have nothing whatever to do with instrumental problems, with our failure to read meters accurately, or with imprecise calibrations. Quantum theory was clearly incompatible with any mechanistic determinism in philosophy, and this had to be abandoned.

29.9 Particles and Antiparticles

Despite the enormous success of the Heisenberg-Schroedinger quantum mechanics in explaining observed and in predicting new phenomena in atomic physics, these theories are deficient in one important respect. Neither the Schroedinger wave equation nor Heisenberg's matrix mechanics meets the requirements of relativity theory; that is, the equations change when subjected to a relativistic coordinate transformation.

A relativistically correct quantum theory was developed by Dirac in 1928. One of the most startling predictions of the Dirac theory was the *positron,* the *antiparticle* of the electron. The positron should have the same mass as the electron but carry a *positive* charge. Since conservation of charge seems to be a fundamental law of nature, a positron can be created only with the simultaneous creation of an electron. That process, called *pair production,* in which an electron-positron pair is created, each with rest mass m_0, requires the expenditure of at least $2m_0c^2$ of energy. In modern accelerators and in cosmic rays (where positrons were first observed after Dirac's prediction), electron-positron pairs are created copiously when energetic particles collide with nuclei. In such events, the KE of the colliding system is diminished by $2m_0c^2$ plus the KE that the electron-positron pair acquires in the process. Even an electromagnetic wave of sufficient energy may, on interacting with a charged particle such as a nucleus, cause pair production. Since the rest mass of an electron is equivalent to an energy of 0.51 MeV, pair formation has a lower threshold of 1.02 MeV.

The inverse process, the annihilation of an electron-positron pair with an accompanying release of at least 1.02 MeV of energy in the form of two photons, is also possible and can be observed under suitable experimental conditions. If just before annihilation the CM of the electron-positron pair is at rest, two photons must be created, propagating in opposite directions, so that both energy and momentum can be conserved.

Pair production and annihilation are dramatic demonstrations of the mass-energy equivalence of relativity theory. As more powerful accelerators were constructed in the 1950s and thereafter, proton-antiproton and other particle-antiparticle pairs were created in the laboratory. In every instance, the energetics of the process conformed to the equation $E = mc^2$.

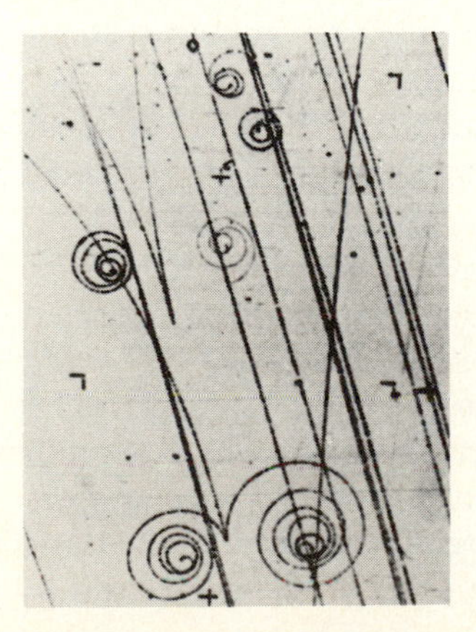

Figure 29.21 *Pair production. High-frequency electromagnetic radiation, that is, a high-energy γ ray, interacts with an electron in the lower portion of this bubble-chamber photograph. An electron-positron pair with relatively little kinetic energy is created, and a recoil electron is seen moving off at high speed. Near the center of the photograph, a γ ray interacts with a nucleus and an electron-positron pair is created. The nuclear recoil is so small that it leaves no visible track in the chamber.*

Summary

In an effort to obtain an analytic expression to match the experimental results on blackbody radiation, Planck found that an oscillator must change its energy by an amount

$$\Delta E = h\nu$$

where ν is the frequency of the oscillator and h is a universal constant, *Planck's constant,* which has the value

$$h = 6.626 \times 10^{-34}\ \text{J·s}$$

That electromagnetic radiation could transmit energy only in finite-sized quanta of energy, called *photons,*

$$E = h\nu$$

was shown to follow from the observations on the *photoelectric effect.* The energy of the emitted electrons ranges between zero KE and a maximum given by the *Einstein relation,*

$$\text{KE}_{\text{max}} = h\nu - \phi$$

where ϕ is the *work function* of the emitting material. If the energy of the incident photon is less than the work function, no electrons are emitted, no matter what the intensity of the radiation.

The low-temperature departure of the specific heats of solids from classical prediction could also be explained by assuming that the atomic vibrations of a crystal lattice are quantized, and that the energy levels differ by an integral number of units of $h\nu$.

When an energetic electron impinges on a solid target and comes to rest, some of its energy may be converted to electromagnetic radiation. If the electron energies are several kilo-electron-volts, X rays are emitted. The minimum wavelength of the emitted radiation is related to the energy of the incident electrons by

$$\lambda_{\text{min}} = \frac{hc}{eV_{\text{A}}}$$

where V_{A} is the anode potential of the X-ray tube; that is, eV_{A} is the energy of the electrons that strike the target of the X-ray tube.

A photon can be scattered by an electron. In this event, known as *Compton scattering,* the electron acquires some of the energy and momentum of the incident photon. As a result, the scattered photon has less energy than the incident and therefore a greater wavelength. The difference between the wavelengths of the scattered and incident photons is given by

$$\Delta\lambda = \lambda' - \lambda = \left(\frac{h}{m_0 c}\right)(1 - \cos\theta)$$

Here m_0 is the rest mass of an electron and θ is the angle between the incident photon and the scattered photon. The quantity $h/m_0c = 0.00243$ nm is known as the electron's *Compton wavelength.*

Prince Louis de Broglie suggested that a particle has a wave character. The wavelength of the particle wave is given by

$$\lambda = \frac{h}{p}$$

known as the *de Broglie relation.*

Heisenberg concluded that the wave nature of particles precluded the simultaneous measurement of a particle's position and momentum with arbitrary precision. The Heisenberg uncertainty principle states that the product of the uncertainties in a particle's position coordinate and corresponding momentum coordinate must be greater than or equal to Planck's constant divided by 2π; that is,

$$\Delta x \, \Delta p_x \geq \frac{h}{2\pi} = \hbar$$

A similar inequality

$$\Delta E \, \Delta t \geq \hbar$$

applies to the measurement of energy and the time during which this energy state exists.

Multiple Choice Questions

29.1 Two sources, A and B, emit light of wavelength 400 nm and 600 nm, respectively. If the power radiated by the two sources is the same, then during one second

(a) source A emits more than twice as many photons as B.

(b) source A emits more photons than B, but fewer than twice as many.

(c) sources A and B emit the same number of photons.

(d) source A emits fewer photons than B but more than half as many.

(e) source A emits less than half as many photons as B.

29.2 An electron has

(a) a KE independent of its wavelength.

(b) a momentum independent of its wavelength.

(c) an energy proportional to its momentum.

(d) an energy proportional to the square of its momentum.

(e) a wavelength proportional to the square of its momentum.

29.3 A proton and an electron are traveling at the same velocity. The proton has a wavelength

(a) greater than that of the electron.

(b) the same as that of the electron.

(c) less than that of the electron.

(d) less than that of the electron only if the proton's KE is less than its rest energy.

(e) greater than that of the electron only if the proton's KE is less than its rest energy.

29.4 The work function of a material is ϕ. The threshold wavelength for emission of photoelectrons is

(a) $\phi/h\nu$

(b) ϕ/e

(c) c/ϕ

(d) $e\phi/h\nu$

(e) hc/ϕ

29.5 It *follows* from the uncertainty principle that the amplitude of atomic vibration in a solid at absolute zero

(a) is much less than kT.

(b) cannot be zero.

(c) is proportional to the uncertainty Δp of the momentum.

(d) depends only on the time during which observations are made.

(e) is proportional to the atomic mass.

29.6 In photoelectric emission, the number of electrons ejected per second

(a) is proportional to the intensity of the light.

(b) is proportional to the wavelength of the light.

(c) is proportional to the frequency of the light.

(d) is proportional to the work function of the material.

(e) None of the above is correct.

29.7 In the Compton effect, a photon of wavelength λ and frequency ν scatters from an electron that is initially at rest. In this process,

(a) the electron gains some of the photon's

energy, and therefore, the scattered photon's wavelength is less than λ.

(b) the photon loses some of its energy, and therefore, the scattered photon's wavelength is greater than λ.

(c) the photon is absorbed, and the electron acquires the momentum of the photon.

(d) the photon's momentum is diminished, and therefore, the frequency of the scattered photon is greater than ν.

(e) None of the above is correct.

29.8 The wavelike character of electrons is most clearly revealed by

(a) the Compton effect.

(b) the photoelectric effect.

(c) diffraction of electrons by a crystal.

(d) emission of photons in electron-positron annihilation.

(e) blackbody radiation.

29.9 The existence of quantized photons (the particle-like nature of light) is most clearly revealed by

(a) emission of radiation in electron-positron annihilation.

(b) the interference pattern in a double-slit experiment.

(c) the diffraction of light by a small aperture.

(d) the Doppler effect.

(e) the Compton effect.

29.10 According to the uncertainty principle,

(a) the uncertainty in the momentum of a particle is always greater than $\hbar$.

(b) the uncertainty in the position of a particle is always greater than $\hbar$.

(c) the product of the uncertainties in position and momentum of a particle is always greater than $\hbar$.

(d) the energy of a particle is always uncertain by an amount greater than $\hbar$.

(e) the product of the uncertainties in energy and momentum of a particle is always greater than $\hbar$.

29.11 If the de Broglie wavelengths of a proton and an electron are equal, it follows that

(a) the proton has a lower velocity than the electron.

(b) the proton has a greater velocity than the electron.

(c) the proton has the same velocity as the electron.

(d) the proton has the same energy as the electron.

(e) the proton has the same angular momentum as the electron.

Problems

(Section 29.2)

29.1 A stiff piece of steel vibrates at a frequency of 18,000 Hz. What is the separation between adjacent energy levels for this oscillating system?

29.2 The atoms in the crystal lithium fluoride vibrate with a resonant frequency of 1.9×10^{13} Hz. What is the energy separation between energy levels for this vibration?

• **29.3** A monochromatic light source radiates 10 W with a wavelength of 590 nm. How many photons are emitted by this source per second?

• **29.4** A light source emits 10^{20} photons per second. If the wavelength of the emitted light is 600 nm, what is the power radiated?

• **29.5** A light source radiates at 25 W. If it emits 10^{20} photons per second, what is the wavelength of the light emitted? Assume the source is monochromatic.

• **29.6** According to Wien's law for blackbody radiation, the product $T\lambda_{max} = C$, where C is a constant and λ_{max} is the wavelength at which the spectral distribution has its maximum intensity. Use Figure 29.1 to verify that relation and to determine the constant C. The intensity of the sun's radiation is a maximum near 500 nm. What is the temperature of the surface of the sun?

• • **29.7** A monochromatic light source radiates at a wavelength of 500 nm uniformly in all directions. At a distance of 100 m from this source, the light flux is 0.1 W/m^2. Find the photon flux for that distance and the power radiated by the source.

• • **29.8** A 25-W, 600-nm light source radiates uniformly in all directions. Determine how far from this source it is still visible, assuming that the eye can perceive a source if at least 80 photons per second pass through its 4-mm-diameter pupil. Neglect absorption of light by the atmosphere.

(Section 29.3)

29.9 Experiment gives a photoelectric-emission threshold of 520 nm for a particular metal. What will the maximum energy of emitted electrons be if this metal is irradiated with light of (a) 620-nm wavelength? (b) 420-nm wavelength?

29.10 When a surface is illuminated with 420-nm light, the maximum KE of the emitted electrons is

0.3 eV. What is the threshold wavelength for photoemission from this surface?

29.11 A metal has a work function of 2.6 eV. What is the threshold wavelength for photoemission from this surface?

• **29.12** A metal whose work function is 2.3 eV is irradiated with monochromatic light. It is found that a retarding potential of 0.8 V must be applied to stop flow of photoelectric current. Determine the wavelength of the light used to irradiate the metal.

• **29.13** When a surface is irradiated with light of 510-nm wavelength, a photocurrent appears, which vanishes if a retarding potential greater than 0.7 V is applied across the phototube. When a different source of light is used, it is found that the critical retarding potential is changed to 1.1 eV. Find the work function of the emitting surface and the wavelength of the second monochromatic source of light.

• **29.14** Determine the maximum speed of electrons emitted from a surface with work function of 3.4 eV by a photon of 310-nm wavelength.

(Sections 29.5–29.6)

29.15 An X-ray tube in a dentist's unit typically uses a high voltage of 20,000 V impressed across the tube. What is the shortest-wavelength X ray that can be obtained with this tube?

29.16 What voltage must be used across an X-ray tube to obtain X rays with a wavelength of 0.12 nm?

29.17 The short-wavelength cutoff from an X-ray tube is 0.62 nm. What voltage is applied across this tube?

• **29.18** The difference in potential between cathode and anode of a TV picture tube is typically 2000 V. What is the minimum wavelength of photons emitted by the electrons as they come to rest at the face of the tube?

• **29.19** Determine the greatest fractional change in energy due to Compton scattering of a photon of 5 eV, 500 eV, and 50,000 eV.

• **29.20** When an incident beam of photons is Compton-scattered through 90°, the wavelength of the scattered photons is found to be 0.12 nm. What was the wavelength of the incident photons?

• •**29.21** In a Compton-scattering event an electron, initially at rest, is scattered in the direction of the incident X-ray beam. If the energy of the recoil electron is 180 eV, what was the energy of the incident X ray? What was the wavelength of the scattered X ray?

• •**29.22** A 50,000-V X-ray tube carries a current of 8 mA. Assuming that only 1 percent of the energy of the electron beam is converted to X rays, and that the average energy of the X rays produced is 0.75 of the maximum X-ray energy, determine (a) the number of X-ray photons emitted per second; (b) the amount of heat that must be removed from the anode to keep it from becoming excessively hot.

(Section 29.7)

29.23 What is the energy of a neutron whose de Broglie wavelength is 0.2 nm?

29.24 An electron has a KE of 100 eV. What is its de Broglie wavelength?

• **29.25** What is the energy of electrons whose de Broglie wavelength equals that of a 1000-eV photon?

• **29.26** An electron and a proton ($m_p = 1830m_e$) have the same kinetic energy, 500 eV. What is the ratio of their de Broglie wavelengths?

• **29.27** What must be the kinetic energy of a proton be so that its de Broglie wavelength is the same as that of a 100-eV electron?

(Section 29.8)

29.28 An atom has a diameter of approximately 0.2 nm. What is the uncertainty in the momentum of an electron confined to a region of that size?

• **29.29** It was once thought that nuclei were made of protons and electrons. The radius of a nucleus is about 2×10^{-15} m. What kinetic energy would an electron confined to this space have? a neutron?

• **29.30** In a TV tube, electrons are accelerated through a potential difference of 10,000 V. The screen of the tube is 30 cm from the electron gun, which has an aperture of 0.2 mm through which the electrons emerge. What is the width of the spot on the screen due to broadening by the uncertainty condition?

• •**29.31** A proton passes through two shutters that define its speed in the x direction to within 1 km/s. If its measured speed is 215 ± 1 km/s, what are the uncertainties in its energy and its momentum in the direction of motion? If the shutter aperture is 1 μm wide in the y direction, what is the uncertainty in the y component of the proton's velocity?

• •**29.32** Many elementary particles have very short lifetimes. The eta particle, a neutral meson, has a mass about 1074 times that of an electron and a lifetime of only 2×10^{-19} s before decaying into two photons. (a) What is the energy of each of the two photons? Assume that the eta was at rest before the decay. (b) Why isn't it possible for the particle to decay to a single photon? (c) What is the uncertainty in the mass of the eta particle?

30

Atomic Structure and Atomic Spectra

When it comes to atoms, language can be used only as in poetry. The poet, too, is not nearly so concerned with describing facts as with creating images.

NIELS BOHR

30.1 Introduction

We have not yet mentioned perhaps the most profound and crucial development of the early years of quantum physics, the synthesis of the Rutherford planetary model of the atom with quantum concepts. This was the monumental achievement of Niels Bohr, and it served as the seed from which our present-day understanding of atomic structure has developed.

By the turn of the century, the electron, as a fundamental particle, was firmly established. In 1897, J. J. Thomson had determined its charge-to-mass ratio. Electrons were known to be liberated in thermionic emission (from a heated metal) and in the photoelectric effect. Before 1900, Becquerel had found that certain elements emitted so-called β rays, which were in fact electrons. Evidently, electrons were pervasive constituents of atoms.

During the early years of the twentieth century, physicists were occupied in devising models of atoms that would be consistent with known

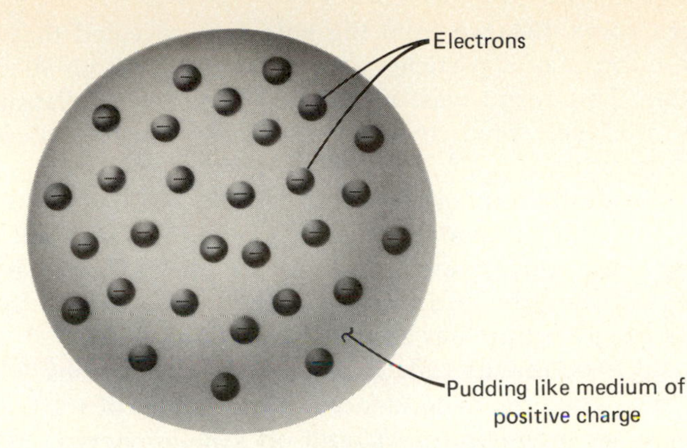

Figure 30.1 The plum-pudding model of the atom. Electrons are embedded within a uniform sphere of positive charge.

physical and chemical data. Such models had to feature (1) electrons as constituents, (2) some positive charge to neutralize the negative charges of the electrons, and (3) some scheme to account for the various atomic weights and the different chemical properties of atoms.

One such model, proposed about 1904 by J. J. Thomson, was dubbed the plum-pudding model. The atom was envisioned as a spherical blob containing N electrons embedded in a uniform positive-charge distribution whose total charge just canceled that of the electrons. The electron "plums" were stuck inside the spherical, positive "pudding."

About this time, in Manchester, England, two young experimenters, H. Geiger and E. Marsden, studied the scattering of a well-collimated beam of α particles by a thin gold foil. Earlier work of Rutherford, by then professor of physics at Manchester, had identified α particles as doubly ionized helium atoms (He^{2+}). The α particles were deflected from their straight path presumably because of electrostatic interaction between these swiftly moving charged objects and the atoms of the gold foil. Geiger and Marsden's results, obtained by the excruciatingly tedious process of counting individual scintillations on a zinc sulfide

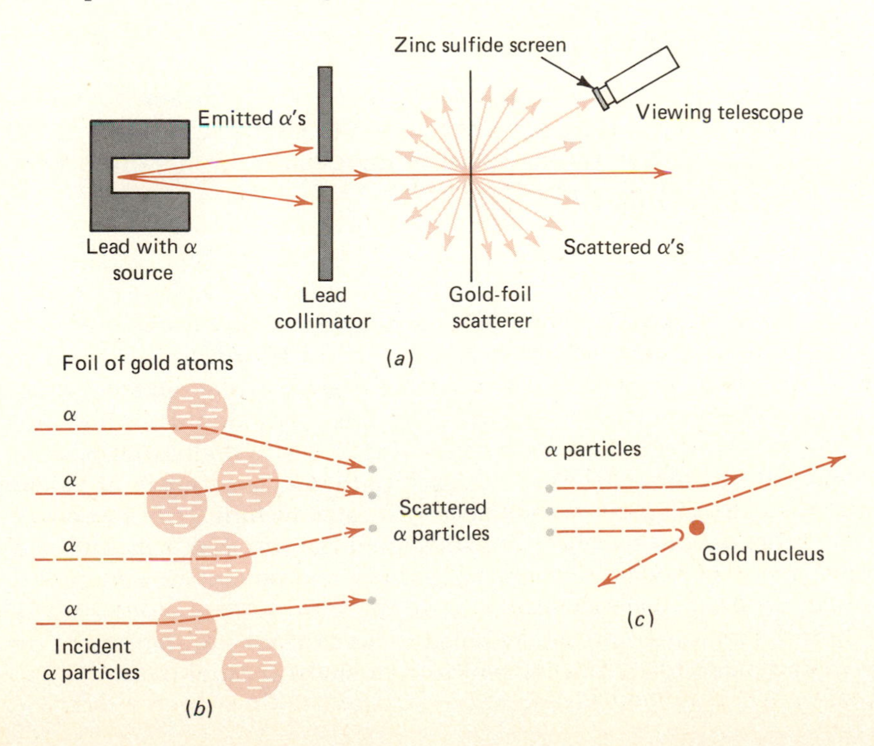

Figure 30.2 The Geiger and Marsden experiment. (a) A collimated beam of α particles impinges on a thin gold foil; the scattered α particles are observed by the scintillations they produce on a zinc sulfide screen. (b) The Thomson plum-pudding model would predict only very small deviations in the trajectory of the particles. (c) Large-angle scattering is possible with the nuclear model if the incident α particle is on a near-collision course with the gold nucleus.

screen placed some distance from the gold foil, revealed that a surprisingly large number of α particles were scattered through 90° and even greater angles. This result was unexpected because calculations, using the Thomson model, predicted a vanishingly small probability for large-angle scattering.

Early in 1911, Rutherford showed that the data of Geiger and Marsden were consistent with an atomic model in which the very light electrons circulate about a highly concentrated, positively charged particle that constitutes practically the entire mass of the atom. This is the *nuclear model* that in broad outline has survived to this day.

Subsequent experiments fully substantiated the various predictions in Rutherford's paper: the dependence of the number of scattered α particles on angle, on atomic number, and on kinetic energy. Yet despite this excellent supporting evidence, the model was badly flawed.

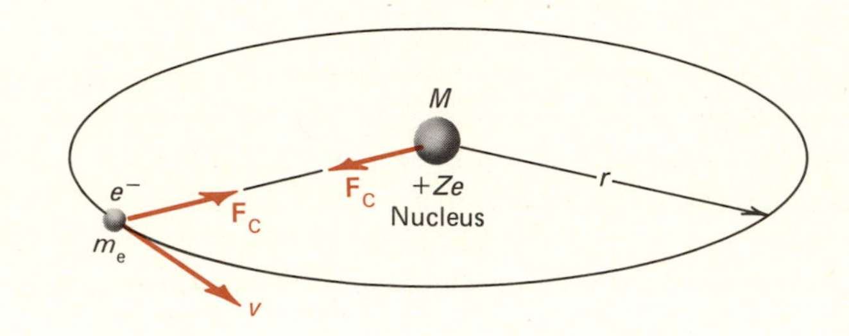

Figure 30.3 *In the Rutherford planetary model, an electron is in a stable orbit about the nucleus if the Coulomb force between electron and nucleus equals the centripetal force needed to maintain the electron in the circular path.*

In Rutherford's model, electrons circle the positive nucleus as planets circle the sun, with the central force of attraction of electrostatic rather than gravitational origin. The electrons' motion is still governed by Newton's laws of motion. For a stable circular orbit, the force of attraction to the massive positively charged nucleus must equal the centripetal force required to maintain that circular motion. Hence

$$k\frac{(Ze)e}{r^2} = \frac{mv^2}{r} \qquad \textbf{(30.1)}$$

where k is Coulomb's constant, $k = 9 \times 10^9\ \mathrm{N \cdot m^2/C^2}$, Ze and $-e$ are the nuclear and electronic charges, and m is the electronic mass; r is the radius of the orbit and v the speed of the electron in that orbit. From (30.1), r and v are related by

$$v^2 = \frac{kZe^2}{mr} \qquad \textbf{(30.2)}$$

Any orbit for which Equation (30.2) is satisfied will be a stable orbit.

The only trouble with this model is that, according to Maxwell's theory of electromagnetism, a charge that is accelerating radiates energy as electromagnetic waves. As a result of this "radiation damping," the electron loses energy and must spiral ever closer to the central positive charge; it thus ultimately must crash into the nucleus, much as Skylab lost energy by air friction and crashed to earth in June 1979. Moreover, during this inward spiraling , which should last less than 1 μs, the electron's angular velocity increases continually and its radiation frequency should therefore also increase.

Figure 30.4 *Sir Ernest Rutherford (1871–1937).*

Thus, the Rutherford model failed to account for the stability of the atom and failed to explain the emission of sharp spectral lines.

30.2 Bohr's Model of the Hydrogen Atom

In the fall of 1911, the 26-year-old Niels Bohr, having just completed his doctoral dissertation, arrived in Cambridge. Disappointed because venerated physicists such as Thomson and Jeans expressed little or no interest in his ideas, Bohr left Cambridge in March 1912 to join Rutherford in Manchester. With that association began one of the most intensely fruitful periods of Bohr's remarkable career, culminating in a series of papers in *The Philosophical Magazine* entitled "On the Constitution of Atoms and Molecules." In the introduction to the first of these, Bohr writes,

> This paper is an attempt to show that the application of the above ideas [the Planck-Einstein quantization conditions] to Rutherford's atom-model affords a basis for a theory of the constitution of atoms.

Figure 30.5 Niels Bohr (1885–1962).

To this end, Bohr made the following assumptions.

1. Electrons can exist only in certain special orbits about the central nucleus. These Bohr called *stationary orbits,* or *stationary states.*

2. The dynamical equilibrium of the system in the stationary states is governed by Newtonian mechanics.

3. Transitions of the system between different stationary states are accompanied by the emission or absorption of radiation, whose frequency ν is given by the Planck formula

$$h\nu = \Delta E \tag{30.3}$$

where ΔE is the difference in energy between the two stationary states.

The first of the assumptions is simply an arbitrary, *ad hoc* escape from the dilemma of radiation damping. That it presented a drastic, revolutionary departure from accepted theory was fully appreciated by Bohr, Rutherford, and all knowledgeable physicists. Its justification would have to rest on the success of the model.

The second assumption posed no problem. It stipulated that despite other strange features of the model, stationary orbits must still conform with ordinary mechanics.

The third assumption caused perhaps the greatest consternation. It was by no means clear just how the electron made a transition from one stationary state to another. Was it a sudden jump or a gradual change? The answers came only much later, with the development of the quantum theory by Heisenberg and Schroedinger.*

According to the Bohr model, the hydrogen atom consists of a relatively heavy nucleus, the *proton,* which carries a positive charge e, and an electron of charge $-e$ in a circular orbit about the nucleus. The equilibrium condition Equation (30.2) must still apply, with $Z = 1$, since we are dealing with a singly charged nucleus. So far the model is that of Rutherford. However, Bohr introduced new restrictions, equivalent to the con-

* Heisenberg recalls that in 1926, thirteen years after Bohr's paper, Schroedinger, then on a visit to Copenhagen, debated vehemently with Bohr, and finally summed up his argument with, "The whole idea of quantum jumps is sheer fantasy!" To which Bohr replied: "What you say is absolutely correct. But it does not prove that there are no quantum jumps. It only proves that we cannot imagine them."

dition that only those orbits can exist for which the angular momentum of the electron round the nucleus equals an integral multiple of $h/2\pi = \hbar$, where h is Planck's constant. Symbolically, we require that the angular momentum L is

$$L = n\hbar \tag{30.4}$$

The angular momentum of a mass m in a circular orbit of radius r is (see Equations (8.12) and (8.15))

$$L = I\omega = mr^2\omega$$

so that Bohr's condition requires

$$mr_n^2\omega = mvr_n = n\hbar \qquad n = 1, 2, 3, \ldots \tag{30.5}$$

where r_n is the orbit radius corresponding to the integer n.

Solving Equation (30.5) for v, we have

$$v = \frac{n\hbar}{mr_n}$$

and squaring gives

$$v^2 = \left(\frac{n\hbar}{mr_n}\right)^2$$

If we substitute the above into the left-hand side of Equation (30.2) with $Z = 1$ and solve for the orbit radius, we find

$$r_n = \frac{n^2\hbar^2}{mke^2} \tag{30.6}$$

The radii given by Equation (30.6) are, according to Bohr, the only possible orbit radii for an electron bound to a proton. Each integer n identifies a particular stationary state of the hydrogen atom. Note that the smallest orbit corresponds to $n = 1$; the value r_1 is now called the *Bohr radius* and is denoted by the special symbol a_0. According to Equation (30.6), the radii of other stationary states are related to a_0 by

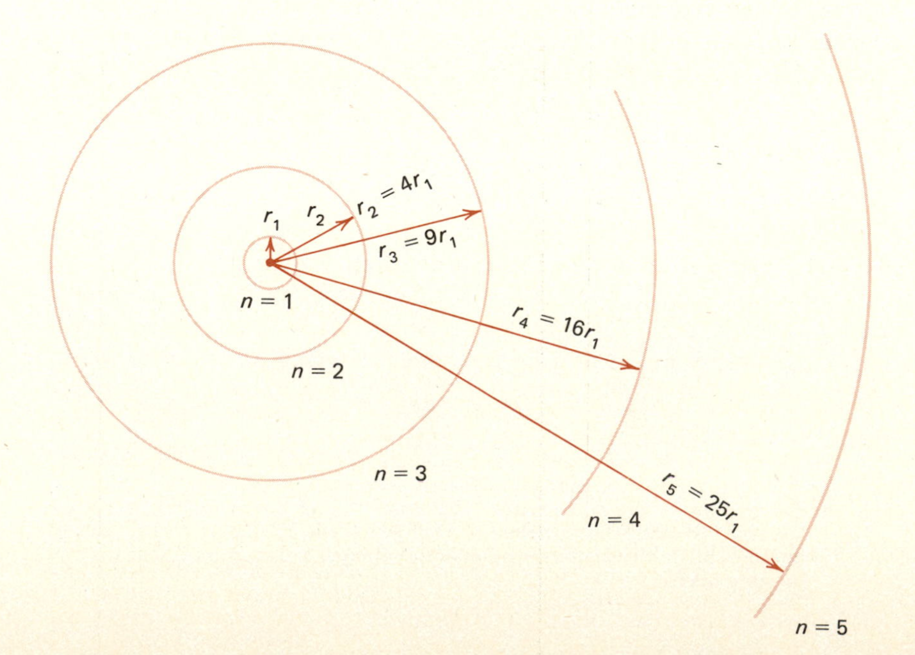

Figure 30.6 Circular orbits of the lower stationary states of the hydrogen atom, according to the Bohr model. Note the relative sizes of the various orbits.

$$r_n = n^2 a_0 \tag{30.7}$$

where

$$a_0 = \frac{\hbar^2}{mke^2} = 5.29 \times 10^{-11}\ \text{m} \tag{30.8}$$

One can now calculate the energy of the hydrogen atom for a given value of n. The total energy contains two contributions, the kinetic energy KE and the potential energy PE of the electron in the electrostatic field of the proton.

The KE is just

$$\text{KE} = \frac{1}{2}mv^2 = \frac{ke^2}{2r_n} \tag{30.9}$$

where we have used Equation (30.2) to obtain the second form.

The PE is, by definition, the product of the electronic charge and the absolute potential at the distance r_n from the positively charged proton. According to Equation (18.11), we have

$$\text{PE} = -eV(r_n) = \frac{-ke^2}{r_n} \tag{30.10}$$

The PE is negative because we have used the standard convention, which assigns zero PE to charges at infinite separation.

Adding potential and kinetic energies, we obtain the total energy

$$E_n = \frac{ke^2}{2r_n} - \frac{ke^2}{r_n} = -\frac{ke^2}{2r_n} \tag{30.11}$$

and substituting for r_n from Equation (30.7) gives*

$$E_n = -\frac{(mk^2e^4/2\hbar^2)}{n^2} = -\frac{E_0}{n^2} \tag{30.12}$$

where

$$E_0 = 2.2 \times 10^{-18}\ \text{J} = 13.6\ \text{eV}$$

The negative sign in Equation (30.12) means that the energy of the atom in the state E_n is less than the energy of the ionized atom, for which the proton and electron are infinitely far apart.

● ***Example 30.1*** Determine the radius and energy of the hydrogen atom in the state $n = 3$.

According to Equation (30.7),

$$r_3 = 3^2 a_0 = 9(5.29 \times 10^{-11}\ \text{m}) = 4.77 \times 10^{-10}\ \text{m}$$

The energy of the atom in that stationary state is given by Equation (30.12):

$$E_3 = \frac{-E_0}{3^2} = \frac{-(13.6\ \text{eV})}{9} = -1.51\ \text{eV}$$

●

* When Bohr performed these same calculations, using the then accepted values for the various physical constants, he obtained $a_0 = 5.15 \times 10^{-11}$ m and $E_0 = 13$ eV. He was greatly encouraged because "these values are of the same order of magnitude as the linear dimensions of the atom . . . and the ionization potentials."

30.3 Emission and Absorption of Radiation; Energy-Level Diagram

According to the last of Bohr's postulates, whenever a hydrogen atom changes from one stationary state to another, energy is either emitted or absorbed. If that energy is in the form of electromagnetic radiation, its frequency is related, through Planck's formula, to the difference between the energies of the stationary states. That is, if E_i and E_f are the energies of the hydrogen atom in the initial and final stationary states, then

$$h\nu = |E_f - E_i| \qquad \textbf{(30.13)}$$

where ν is the frequency of the radiation. If $E_i > E_f$, conservation of energy requires that a photon of energy $h\nu$ be emitted. The inverse process is also possible; a hydrogen atom in a stationary state of low energy can be raised to a state of higher energy by the absorption of a photon whose energy $h\nu$ equals the energy difference between the two stationary states.

The *energy-level diagram* is an excellent visual aid in understanding these absorption and emission processes. Such a diagram for hydrogen is shown in Figure 30.7. Here the vertical axis is used to represent energy, and the horizontal lines denote the sharply defined energies of the various stationary states. Since these states all have negative energies, the line representing zero energy appears at the top of the diagram. The state for $n = 1$, with energy $E_1 = -13.6$ eV, the lowest-lying stationary state, is called the *ground state* of hydrogen.

The energy levels for $n = 2, 3, 4, \ldots$ are also shown in Figure 30.7. Since levels with larger n have higher energies, the transition from a state of large n to a state of smaller n must be accompanied by the emission of energy, whereas transition from a state of small n to a state of larger n requires absorption of energy by the atom. It is common practice to indicate atomic transitions with arrows from the initial to the final state.

If we now substitute Equation (30.12) into (30.13) and divide by Planck's constant, we obtain the frequency of the transition radiation:

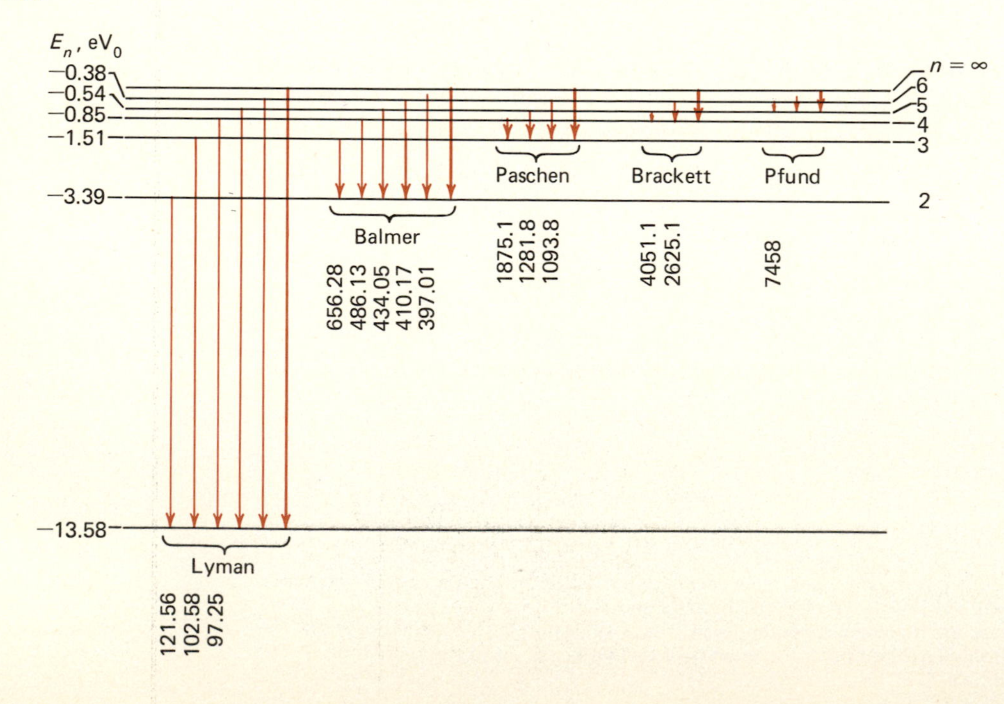

Figure 30.7 *Energy-level diagram for hydrogen. The energy E = 0 corresponds to the ionized state of hydrogen with the free electron at rest. The arrows indicate the atomic transitions corresponding to emission of electromagnetic radiation with wavelengths in nanometers, as indicated. The heavy arrows show the high-frequency, or short-wavelength, limit for each group of lines, called a spectral series.*

$$\nu = \frac{|E_f - E_i|}{h} = \frac{E_0}{h}\left|\left(\frac{1}{n_f^2}\right) - \left(\frac{1}{n_i^2}\right)\right| \quad (30.14)$$

The corresponding wavelength of the radiation is given by

$$\frac{1}{\lambda} = \frac{\nu}{c} = \frac{E_0}{hc}\left|\left(\frac{1}{n_f^2}\right) - \left(\frac{1}{n_i^2}\right)\right| = R\left|\left(\frac{1}{n_f^2}\right) - \left(\frac{1}{n_i^2}\right)\right| \quad (30.15)$$

where

$$R = \frac{E_0}{hc} = 1.1 \times 10^7\ \text{m}^{-1} \quad (30.15a)$$

is known as the *Rydberg constant.*

If $n_f = 2$, the expression for the wavelength is exactly that deduced empirically by Balmer, which gave such excellent agreement with experiment. These spectral lines from hydrogen fall into the visible region of the electromagnetic spectrum and were therefore the first to be observed and measured with good precision. They are now known as the *Balmer lines.*

Transitions from states $n_i = 4, 5, \ldots$ to $n_f = 3$ fall in the near-infrared part of the spectrum, and by 1913, had been studied by Paschen. The remaining spectral series had not yet been observed but, as Bohr wrote:

> If we put t_2 [n_f] = 1 and 4, 5, . . . , we get series respectively in the extreme ultraviolet and the extreme ultra-red, which are not observed, but the existence of which may be expected.

Indeed, within a few years the Lyman series ($n_f = 1$) as well as the Brackett ($n_f = 4$) and longer-wavelength series were identified and measured.

● ***Example 30.2*** What are the wavelengths of the longest-wavelength line of the Balmer and of the Paschen series? What are the short-wavelength limits approached by these series?

When λ is greatest, $1/\lambda$ will attain its smallest value. From Equation (30.15), it follows that $1/\lambda$ will be least when $(1/n_i^2) - (1/n_f^2)$ is smallest. For the Balmer series, $n_f = 2$, and so the longest-wavelength line will arise from the transition $n_i = 3$ to $n_f = 2$. Hence,

$$\text{Balmer series: } \lambda_{max} = \frac{1}{R}\left(\frac{1}{4} - \frac{1}{9}\right)^{-1} = 6.545 \times 10^{-7}\ \text{m} = 654.5\ \text{nm}$$

The final state for the Paschen series is the state $n_f = 3$. The maximum wavelength for that series is therefore that which arises from the $n_i = 4$ to $n_f = 3$ transition. Consequently,

$$\text{Paschen series: } \lambda_{max} = \frac{1}{R}\left(\frac{1}{9} - \frac{1}{16}\right)^{-1} = 1870\ \text{nm}$$

As n_i is increased, the energy difference between the initial and final states increases for both series. However, as n becomes larger, the energy levels E_n become more closely spaced, as is evident from Figure 30.7. In the limit as $n_i \to \infty$, E_n approaches zero (see Equation 30.12). Therefore, the greatest energy difference and correspondingly, the shortest wavelength, will be associated with transitions from states with very large n to $n_f = 2$ (Balmer series) and $n_f = 3$ (Paschen series). These short-wavelength limits are

$$\text{Balmer series:} \quad \lambda_{min} = \frac{4}{R} = 363.6 \text{ nm}$$

$$\text{Paschen series:} \quad \lambda_{min} = \frac{9}{R} = 818.2 \text{ nm}$$

Since the visible region of the spectrum encompasses the range 385–770 nm, we see that even the shortest wavelength of the Paschen series falls just beyond the visible part of the spectrum. All but the most energetic photons of the Balmer series can be sensed by the human eye. ●

Bohr, in the same series of papers, also noted that certain lines, first observed by Pickering in the spectrum of the star ζ-Puppis and subsequently seen in the laboratory using a gas discharge containing a mixture of helium and hydrogen, can be ascribed to transitions between levels of singly ionized helium atoms. Singly ionized helium differs from neutral hydrogen only in that the nuclear charge is $2e$ instead of e. The Bohr model is readily generalized to the case of a single electron in orbit about a positive charge Ze, with the result that the expressions for the radius r_n and the stationary-state energy E_n are

$$r_n = \frac{n^2\hbar^2}{mkZe^2} = \frac{n^2 a_0}{Z} \tag{30.6a}$$

$$E_n = -\frac{Z^2 E_0}{n^2} \tag{30.12a}$$

Notice that the radii are diminished by a factor Z and the magnitudes of the energies increased by the factor Z^2. In the stronger electrostatic field of the greater nuclear charge, the electron describes a more compact orbit, and the amount of energy needed to extract it from that orbit is increased by the factor Z^2. That is, the electron is more tightly bound, and the ionization energy is correspondingly increased.

● ***Example 30.3*** Show that for singly ionized helium, the emission lines for transitions terminating on the $n = 4$ Bohr orbit and originating from states with $n = 6, 8, \ldots$ have the same wavelengths as the lines of the Balmer series. How do the orbit radii of the $n = 2, 3, 4, \ldots$ state of hydrogen compare with the orbit radii of the $n = 4, 6, 8, \ldots$ states of ionized helium?

From Equations (30.12a) and (30.15) we have, for $Z = 2$,

$$\frac{1}{\lambda} = 4R \left| \left(\frac{1}{n_f^2}\right) - \left(\frac{1}{n_i^2}\right) \right|$$

If $n_f = 4$ and $n_i = 6, 8, \ldots$, this expression can be factored by writing $n_f^2 = 4 \times 2^2$ and $n_i^2 = 4 \times 3^2, 4 \times 4^2, \ldots$, to give

$$\frac{1}{\lambda} = R \left| \left(\frac{1}{4}\right) - \left(\frac{1}{n_i'^2}\right) \right|$$

where $n_i' = n_i/2 = 3, 4, \ldots$. The above is just the formula for the Balmer series.

To compare the radii of the various orbits, we turn to Equation (30.6a). For example, for hydrogen we have $r_2(\mathrm{H}) = 4a_0$; the state of ionized helium that has the same energy is $n = 4$, and its radius is $r_4(\mathrm{He}^+) = 8a_0$. Similarly, the radius of the state for $n = 3$ of hydrogen is $9a_0$; the state of the same energy in ionized helium has $n = 6$, and its radius is $18a_0$. States of equal energy have radii that are in the ratio 1:2. ●

30.4 De Broglie Waves and the Bohr Model

The Bohr model not only clarified the relations between wavelengths of atomic spectral lines but also was a fertile soil for de Broglie's ideas. In Chapters 15 and 16 we saw that when the spatial extension of a system bears some simple relation to a wavelength, resonance conditions arise that are characteristic of the system's natural frequencies of vibration. The examples that we studied in detail concerned standing waves on strings and in organ pipes.

If as de Broglie envisioned, a wave of some kind should be associated with an electron, would it not be reasonable to expect some sort of *resonance condition* to give rise to Bohr's stationary states? Standing waves on a string, for example, give the appearance of stationary configurations.

A plausible geometric condition for such a resonance is that the circumference of a stationary-state orbit should equal an integral number of electron wavelengths. That condition would be consistent with standing waves as depicted in Figure 30.8. It is also consistent with Bohr's angular-momentum quantum condition, as we now demonstrate.

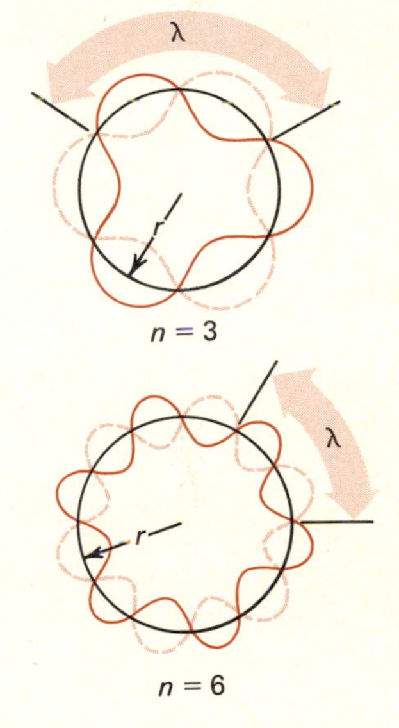

Figure 30.8 Standing de Broglie waves around the circumference of a Bohr orbit. Resonances for $n = 3$ and for $n = 6$ are shown.

The de Broglie resonance requirement may be written

$$2\pi r_n = g\lambda_n \qquad (30.16)$$

where r_n is the radius of the nth stationary state, λ_n the de Broglie wavelength of the electron in that stationary state, and g an integer.

The de Broglie wavelength is related to the electron's momentum through Equation (29.8), so that we can rewrite (30.16),

$$2\pi r_n = \frac{gh}{p} = \frac{gh}{mv}$$

Multiplying by mv and dividing by 2π, we obtain

$$mvr_n = L = g\hbar \qquad (30.5)$$

which is just the Bohr condition provided we identify the integer g with the quantum number n. This last requirement confers a new significance on the quantum number n; it is the number of de Broglie wavelengths that can be fitted into the circumference of that stationary-state orbit.

The concept of a stationary, or at least relatively long-lived, state of a system as a resonance has proved fruitful, especially in nuclear and elementary-particle physics.

30.5 Atomic Wave Functions, Quantum Numbers, and Atomic Structure

According to the quantum theory, the complete description of the electron's wave function requires not one but three quantum numbers. In this quantum theory, one quantum number n, called the *principal quantum number*, is the integer of the Bohr model. Now, however, n is related to the average distance between the electron and proton but is not directly related to the orbital angular momentum.

The need for three quantum numbers is consistent with classical mechanics, where the dynamics of a particle involves the specification of three coordinates. The reason the remaining two quantum numbers did not appear in the Bohr model is that the energy of hydrogen depends on n only, not on the other two quantum numbers, at least to a very good approximation.

The physical meaning of the remaining two quantum numbers of hydrogen, designated by the letters l and m, is as follows. *The quantum number l specifies the magnitude of the orbital angular momentum* of the stationary state; *the quantum number m specifies the component of this angular momentum* along some special direction in space. To be precise, the relation between these quantum numbers and the angular momentum is

$$L = \sqrt{l(l+1)}\,\hbar \tag{30.17}$$

$$L_z = m\hbar \tag{30.18}$$

where z is a particular direction in space.

What is that special direction? In the isolated hydrogen atom, there is no preferred direction. One can, however, deliberately select and identify a spatial direction, for example, by placing the atom between the pole faces of an electromagnet. The orientation of the magnetic field then serves as the direction for space quantization. Since this was the original association of that quantum number, m is sometimes referred to as the magnetic quantum number.

The wave equation for hydrogen has solutions only if the three quantum numbers meet the following conditions.

1. n is a positive integer greater than or equal to 1; $n \geq 1$.

2. l is zero or a positive integer whose value is less than or equal to $n - 1$;

$$l \leq n - 1. \tag{30.19}$$

3. m is a positive or negative integer, or zero, satisfying the condition

$$|m| \leq l. \tag{30.20}$$

Thus, m may range between $+l$ and $-l$.

These quantization rules are a direct consequence of the Schroedinger equation and have no classical analogs. They apply even when the force does not obey Coulomb's law, provided only that it is a central force.

30.5 (a) Electron Spin

There is yet one more quantum number. It relates not to a dynamical variable, such as energy or angular momentum, but to an *intrinsic* property of the electron, its *spin.* The hypothesis of an intrinsic spin angular momentum of an electron was advanced in 1925 by two young Dutch physicists, G. Uhlenbeck and S. A. Goudsmit, who showed that by assigning a spin quantum number of $\frac{1}{2}$ to the electron, many hitherto inexplicable features of atomic spectra could be understood. The spin quantum numbers $s = \frac{1}{2}$ (for the electron) implies an intrinsic spin angular momentum

$$S = \sqrt{s(s+1)}\,\hbar = \sqrt{\tfrac{3}{4}}\,\hbar$$

whose component along a given direction can assume only two possible values

$$S_z = +\tfrac{1}{2}\hbar \qquad \text{or} \qquad S_z = -\tfrac{1}{2}\hbar$$

30.5 (b) Many-Electron Atom and the Periodic Table

It was already apparent when J. J. Thomson, and later Rutherford, proposed their atomic models that atoms are distinguished primarily by their nuclear charge and corresponding number of electrons needed to neutralize that charge. Since all electrons are alike, the different chemical properties of the elements must arise from the different atomic configurations, that is to say, the different arrangements of electrons in orbits about the nucleus.

Once the Bohr model had been advanced and its success demonstrated for the simplest systems, hydrogen and ionized helium, the question arose how the numerous electrons of the heavier atoms were organized in the Bohr scheme. To this question, the Schroedinger equation gave no satisfactory answer.

Since the $n = 1$ level would always be the lowest energy level, one might expect, on classical arguments, that under equilibrium conditions at low temperatures, all electrons would fall into this level. Experimental results by spectroscopists could, however, not be explained with that assumption. The crucial key to the puzzle was provided by Wolfgang Pauli, a rotund Austrian theorist who spent most of his professional life in Zürich. He proposed what is now known as the *Pauli exclusion principle:*

PAULI EXCLUSION PRINCIPLE: No two electrons can occupy the same quantum state.

That is, in any atom, no two electrons can have the same four quantum numbers—n, l, m, m_s.

The rules relating the quantum numbers, Equations (30.19) and (30.20), are mathematical consequences of the wave equation. The basis and justification for the exclusion principle, however, was empirical—its application to all problems relating to electronic structure produced results in complete accord with observation.

With the aid of Equations (30.19) and (30.20) and the Pauli principle, one can now construct a logical model for the electronic structure of atoms. In the ground state, the electrons should occupy the lowest possible energy levels consistent with the Pauli principle. Thus for hydrogen, the ground state will have the quantum numbers

$$\text{H:} \quad n = 1, \quad l = 0, \quad m = 0, \quad m_s = +\tfrac{1}{2} \text{ or } -\tfrac{1}{2}$$

with the two spin orientations giving the same energy in the absence of a magnetic field.

The next atom in the periodic table is helium, with nuclear charge $2e$, holding two electrons in orbits. The ground-state configuration of helium is

$$\text{He:} \quad n = 1, \quad l = 0, \quad m = 0, \quad m_s = +\tfrac{1}{2}$$

$$\phantom{\text{He:}} \quad n = 1, \quad l = 0, \quad m = 0, \quad m_s = -\tfrac{1}{2}$$

The element with atomic number 3 is lithium. According to the Pauli principle and Equations (30.19) and (30.20), only two of its three electrons can be accommodated in states with principal quantum number $n = 1$; the third electron must occupy a state with $n = 2$, with a more extended orbit.

States with $l = 0$ or 1 can exist for $n = 2$. It turns out that states with small values of angular momentum, that is, small values of l, have the lowest energies for a given value of n. Hence the two states with $n = 2$, $l = 0$, will be filled before states with $l = 1$ are occupied. Thus, the ground state of lithium is

$$\text{Li:}\quad \begin{array}{llll} n = 1, & l = 0, & m = 0, & m_s = +\frac{1}{2} \\ n = 1, & l = 0, & m = 0, & m_s = -\frac{1}{2} \\ n = 2, & l = 0, & m = 0, & m_s = \pm\frac{1}{2} \end{array}$$

And in this way one can construct the ground-state configuration of atoms with increasing nuclear charge. In that process, a tabulation of electronic states, Table 30.1, is helpful. In this table we also identify the principal quantum number and angular-momentum states using the accepted spectroscopic notation. The totality of the orbits associated with a given principal quantum number is thought of as constituting a *shell*, and these shells are designated by capital letters: K ($n = 1$), L ($n = 2$), M ($n = 3$), ... The angular-momentum states are also designated with a

Table 30.1 *Tabulation of n, ℓ, m quantum states*

Shell	n	ℓ	m	Spectroscopic designation	Total number of available states: ℓ-subshell	n-shell
K	1	0	0	1s	2	2
	2	0	0	2s	2	
	2	1	−1			
L	2	1	0	2p	6	8
	2	1	1			
	3	0	0	3s	2	
	3	1	−1			
	3	1	0	3p	6	
	3	1	1			
M	3	2	−2			18
	3	2	−1			
	3	2	0	3d	10	
	3	2	1			
	3	2	2			
	4	0	0	4s	2	
	4	1	−1			
	4	1	0	4p	6	
	4	1	1			
	4	2	−2			
	4	2	−1			
	4	2	0	4d	10	
N	4	2	1			32
	4	2	2			
	4	3	−3			
	4	3	−2			
	4	3	−1			
	4	3	0	4f	14	
	4	3	1			
	4	3	2			
	4	3	3			

Table 30.2 Ground-state electronic configurations of the first eighteen elements

Element	*Symbol*	*Atomic number, Z*	*Electronic configuration*
Hydrogen	H	1	1s
Helium	He	2	$1s^2$
Lithium	Li	3	$1s^22s$
Beryllium	Be	4	$1s^22s^2$
Boron	B	5	$1s^22s^22p$
Carbon	C	6	$1s^22s^22p^2$
Nitrogen	N	7	$1s^22s^22p^3$
Oxygen	O	8	$1s^22s^22p^4$
Fluorine	F	9	$1s^22s^22p^5$
Neon	Ne	10	$1s^22s^22p^6$
Sodium	Na	11	$1s^22s^22p^63s$
Magnesium	Mg	12	$1s^22s^22p^63s^2$
Aluminum	Al	13	$1s^22s^22p^63s^23p$
Silicon	Si	14	$1s^22s^22p^63s^23p^2$
Phosphorus	P	15	$1s^22s^22p^63s^23p^3$
Sulfur	S	16	$1s^22s^22p^63s^23p^4$
Chlorine	Cl	17	$1s^22s^22p^63s^23p^5$
Argon	Ar	18	$1s^22s^22p^63s^23p^6$

letter, though the sequence is in no alphabetic order:* s ($l = 0$), p ($l = 1$), d ($l = 2$), f ($l = 3$), ... The group of states of given l and n is referred to as a *subshell*.

The symbol (1s) signifies an $n = 1$, $l = 0$, state, the symbol (3p) an $n = 3$, $l = 1$, state. If the electronic configuration of an atom consists of two electrons in the $n = 1$ state, two electrons in the $n = 2$, $l = 0$, state, and four electrons in the $n = 2$, $l = 1$, state, the complete structure would be written $(1s)^2(2s)^2(2p)^4$. The complete electronic ground-state configurations of the lighter atoms are shown in Table 30.2.

Electrons in closed shells or closed subshells are tightly bound to the atom and therefore do not normally participate in chemical or electrical phenomena. Except for the rare gases—helium, neon, argon, krypton, xenon, and radon—the closed-shell configurations are often not listed in giving the electronic structure. Atoms that have one electron outside a filled subshell are the alkali metals (lithium, sodium, potassium, rubidium, and cesium); the halides (fluorine, chlorine, bromine, and iodine) are one electron short of a closed subshell structure.

One can now understand the regularity in the chemical properties of the elements, formalized by the periodic table. In the heavier atoms, electrons whose principal quantum numbers are small ($n = 1, 2$) occupy orbits close to the nucleus, and their energies are low. That is, they are so tightly bound that they can make transitions to an unoccupied state only if given considerable energy. By contrast, electrons in the outermost orbits of the atom are weakly bound, and it is these electrons that participate in chemical reactions with other atoms. Consequently, atoms with

* The designation derives from early spectroscopy. The letters *s*, *p*, and *d* stand for *s*harp, *p*rincipal, and *d*iffuse. The dominant (principal) spectral lines derived from transitions involving the $l = 1$ levels; those that originated from $l = 0$ states proved to be exceptionally sharp, whereas transitions from $l = 2$ states appeared to give diffuse lines.

*Table 30.3 Periodic table of the elements**

Symbol—**Cl** 17—Atomic number
Atomic mass†—35.453
$3p^5$ —Electron configuration

Group I	Group II	Transition elements										Group III	Group IV	Group V	Group VI	Group VII	Group 0
H 1 1.0080 $1s^1$																	**He** 2 4.0026 $1s^2$
Li 3 6.94 $2s^1$	**Be** 4 9.012 $2s^2$											**B** 5 10.81 $2p^1$	**C** 6 12.011 $2p^2$	**N** 7 14.007 $2p^3$	**O** 8 15.999 $2p^4$	**F** 9 18.998 $2p^5$	**Ne** 10 20.18 $2p^6$
Na 11 22.99 $3s^1$	**Mg** 12 24.31 $3s^2$											**Al** 13 26.98 $3p^1$	**Si** 14 28.09 $3p^2$	**P** 15 30.97 $3p^3$	**S** 16 32.06 $3p^4$	**Cl** 17 35.453 $3p^5$	**Ar** 18 39.948 $3p^6$
K 19 39.102 $4s^1$	**Ca** 20 40.08 $4s^2$	**Sc** 21 44.96 $3d^14s^2$	**Ti** 22 47.90 $3d^24s^2$	**V** 23 50.94 $3d^34s^2$	**Cr** 24 51.996 $3d^54s^1$	**Mn** 25 54.94 $3d^54s^2$	**Fe** 26 55.85 $3d^64s^2$	**Co** 27 58.93 $3d^74s^2$	**Ni** 28 58.71 $3d^84s^2$	**Cu** 29 63.54 $3d^{10}4s^1$	**Zn** 30 65.37 $3d^{10}4s^2$	**Ga** 31 69.72 $4p^1$	**Ge** 32 72.59 $4p^2$	**As** 33 74.92 $4p^3$	**Se** 34 78.96 $4p^4$	**Br** 35 79.91 $4p^5$	**Kr** 36 83.80 $4p^6$
Rb 37 85.47 $5s^1$	**Sr** 38 87.62 $5s^2$	**Y** 39 88.906 $4d^15s^2$	**Zr** 40 91.22 $4d^25s^2$	**Nb** 41 92.91 $4d^45s^1$	**Mo** 42 95.94 $4d^55s^1$	**Tc** 43 (99) $4d^55s^2$	**Ru** 44 101.1 $4d^75s^1$	**Rh** 45 102.91 $4d^85s^1$	**Pd** 46 106.4 $4d^{10}5s^6$	**Ag** 47 107.87 $4d^{10}5s^1$	**Cd** 48 112.40 $4d^{10}5s^2$	**In** 49 114.82 $5p^1$	**Sn** 50 118.69 $5p^2$	**Sb** 51 121.75 $5p^3$	**Te** 52 127.60 $5p^4$	**I** 53 126.90 $5p^5$	**Xe** 54 131.30 $5p^6$
Cs 55 132.91 $6s^1$	**Ba** 56 137.34 $6s^2$	57– 71§	**Hf** 72 178.49 $5d^26s^2$	**Ta** 73 180.95 $5d^36s^2$	**W** 74 183.85 $5d^46s^2$	**Re** 75 186.2 $5d^56s^2$	**Os** 76 190.2 $5d^66s^2$	**Ir** 77 192.2 $5d^76s^2$	**Pt** 78 195.09 $5d^06s^1$	**Au** 79 196.97 $5d^{10}6s^1$	**Hg** 80 200.59 $5d^{10}6s^2$	**Tl** 81 204.37 $6p^1$	**Pb** 82 207.2 $6p^2$	**Bi** 83 208.98 $6p^3$	**Po** 84 (210) $6p^4$	**At** 85 (218) $6p^5$	**Rn** 86 (222) $6p^6$
Fr 87 (223) $7s^1$	**Ra** 88 (226) $7s^2$	89–103‖	**Rf** 104 (261) $6d^27s^2$	**Ha** 105 (262) $6d^37s^2$	106 (263)	107 (261)											

§ Lanthanide series	**La** 57 138.91 $5d^16s^2$	**Ce** 58 140.12 $5d^14f^16s^2$	**Pr** 59 140.91 $4f^36s^2$	**Nd** 60 144.24 $4f^46s^2$	**Pm** 61 (147) $4f^66s^2$	**Sm** 62 150.4 $4f^66s^2$	**Eu** 63 152.0 $4f^76s^2$	**Gd** 64 157.25 $5d^14f^76s^2$	**Tb** 65 158.92 $5d^14f^86s^2$	**Dy** 66 162.50 $4f^{10}6s^2$	**Ho** 67 164.93 $4f^{11}6s^2$	**Er** 68 167.26 $4f^{12}6s^2$	**Tm** 69 168.93 $4f^{13}6s^2$	**Yb** 70 173.04 $4f^{14}6s^2$	**Lu** 71 174.97 $5d^14f^{14}6s^2$
‖ Actinide series	**Ac** 89 (227) $6d^17s^2$	**Th** 90 (232) $6d^27s^2$	**Pa** 91 (231) $5f^26d^17s^2$	**U** 92 (238) $5f^36d^17s^2$	**Np** 93 (239) $5f^36d^07s^2$	**Pu** 94 (239) $5f^66d^07s^2$	**Am** 95 (243) $5f^76d^07s^2$	**Cm** 96 (245) $5f^76d^17s^2$	**Bk** 97 (247) $5f^86d^17s^2$	**Cf** 98 (249) $5f^{10}6d^07s^2$	**Es** 99 (254) $5f^{11}6d^07s^2$	**Fm** 100 (253) $5f^{12}6d^07s^2$	**Md** 101 (255) $5f^{13}6d^07s^2$	**No** 102 (255) $6d^07s^2$	**Lr** 103 (257) $6d^17s^2$

Notes * Atomic mass values given are averaged over isotopes in the percentages in which they exist in nature.
† For an unstable element, mass number of the most stable known isotope is given in parentheses.

similar structure of the outermost electrons, the *valence electrons,* have similar chemical properties.

Thus, the formal results of quantum theory, when combined with the Pauli exclusion principle, give meaning to the previously mysterious pattern of the periodic table discovered by Mendeleev more than half a century earlier.

● ***Example 30.4*** What are the ground-state electronic configurations of sodium, $Z = 11$, and of selenium, $Z = 34$?

From Table 30.1, we see that 10 electrons just fill the K and M shells. Hence the last electron of the sodium atom must be the first to occupy an $n = 3$ state. Since the state of lowest energy is that for $l = 0$, the valence electron configuration will be (3s), and the complete electronic structure of sodium is $(1s)^2(2s)^2(2p)^6(3s)$.

Referring again to Table 30.1, we see that the K, L, and M shells together hold a total of 28 electrons. Thus six more electrons must go into states with $n = 4$ for selenium. Two of those 6 electrons will be in s states, leaving 4 electrons to occupy 4 of the six 4p orbits. Thus, selenium has 4 electrons in an unfilled subshell; that is, there are 4 electrons that can participate in chemical reactions, and the chemical valence of selenium is therefore 4. The valence electron configuration of selenium is $(4p)^4$; the complete electronic configuration is $(1s)^2(2s)^2(2p)^6(3s)^2(3p)^6(3d)^{10}(4s)^2(4p)^4$. ●

30.6 Characteristic X-Ray Lines

In Chapter 29 we saw how generation of the continuous X-ray spectrum could be explained on the basis of the Planck-Einstein hypothesis. Using Bohr's atomic model, one can now also understand the origin of the characteristic X-ray lines (see Figure 29.13). The fact that their wavelength depends only on the target material implies that these lines are as much a unique feature of the target's atomic structure as the more readily observed optical spectral lines.

Characteristic X-ray lines arise as follows. In a heavy atom such as zinc, niobium, or tungsten, the two electrons in the K shell are very strongly bound. Each of these two electrons is in the Coulomb field of a nucleus of charge Z that is only partly screened by the other electron. Thus, each of the two electrons in the K shell feels, if not the full nuclear charge, then an effective charge that is approximately $(Z - 1)e$.

For instance, according to Equation (30.12a), the energy of either of the two electrons in the $n = 1$ state of niobium is approximately

$$\begin{aligned} E_1 &= -(Z')^2E_0 \simeq -(Z - 1)^2E_0 \\ &= -(40)^2(13.6 \text{ eV}) = -21{,}760 \text{ eV} \end{aligned}$$

Here Z' is the effective nuclear charge experienced by the electron and Z the bare nuclear charge.

When the niobium anode of an X-ray tube is bombarded by very energetic electrons, say of 50,000 eV, an atomic electron may be knocked out of its tightly bound $n = 1$ state by such a collision. That leaves the atom with an unoccupied low-lying level. Typically, within less than a nanosecond, an electron from a higher level, from the L ($n = 2$) or M ($n = 3$) shell, will drop into the lower energy state. Energy is conserved by the emission of a photon of wavelength $\lambda = hc/\Delta E$, where ΔE is the difference in the energies of the initial and final stationary states. Since these energies are well defined, so is the wavelength of the emitted radiation.

● ***Example 30.5*** Estimate the wavelength of the X ray emitted from a tantalum target when an electron from an $n = 4$ state makes a transition to an empty $n = 1$ state. Estimate the minimum bombarding energy required to make that transition possible.

The atomic number of tantalum is 73; that is, the nuclear charge is $73e$ and there are 73 extranuclear electrons in the neutral tantalum atom. Of these, two occupy the $n = 1$, eight the $n = 2$, and eighteen the $n = 3$ states. To create the condition necessary for the transition, one of the $n = 1$ electrons must be ejected from this stationary state. We begin by calculating the approximate binding energy of the electron in that state.

The effective nuclear charge that acts on that electron is $Z' \simeq Z - 1 = 72$. Hence, the energy of the electron in this state is approximately

$$E_1 = -(72)^2(13.6 \text{ eV}) = -70{,}500 \text{ eV}$$

Since all stationary states except the outermost are occupied by electrons, the $n = 1$ level can be depleted of one electron only if that electron is given enough energy to reach one of the excited valence electron levels or be completely freed. The excited valence levels are, however, no more than a few electron volts below the ionization energy—negligible compared with the more than 70 keV of binding energy of the $n = 1$ electron. Thus, the answer to the second part of the problem is that for the transition to be possible, the bombarding energy must be at least 70.5 keV.

The X ray of interest is emitted when an electron from an $n = 4$ state falls into the empty $n = 1$ state. The effective nuclear charge seen by an $n = 4$ electron depends on the other quantum numbers, l and m; we can assume, as a rough approximation, that this electron is in the field of the nucleus shielded by the 27 electrons in the $n = 1$, $n = 2$, and $n = 3$ states (the $n = 1$ state has now only 1 electron). Consequently, the initial energy for the transition is

$$E_i = -\frac{(Z')^2E_0}{n^2} = -\frac{(73 - 27)^2(13.6 \text{ eV})}{4^2} = -1800 \text{ eV}$$

and the final energy is that calculated earlier, $-70{,}500$ eV. Consequently,

$$h\nu = (70{,}500 - 1800) \text{ eV} = 68{,}700 \text{ eV}$$

$$\lambda = \frac{hc}{\Delta E} = 0.018 \text{ nm}$$ ●

* 30.7 Masers and Lasers

One of the most exciting developments of the past two decades, which also quickly spawned a multimillion-dollar industry, is the *laser* (an acronym for *l*ight *a*mplification by *s*timulated *e*mission of *r*adiation). The laser is the successor of the *maser*, a device based on the same principles but used for detecting and amplifying microwave radiation. By far the most sensitive microwave detector now known, masers have been widely applied in radio astronomy. They also played a crucial role in the detection of the faint microwave radiation believed to be the remnant of the primordial "Big Bang" explosion of the universe about 18 billion years ago, work that earned Arno Penzias and Robert Wilson the 1978 Nobel Prize in physics.

To understand the operation of these devices, we must summarize some information concerning transitions between stationary states. Bohr's postulate of quantum jumps, combined with the Planck relation, explained in a beautifully simple manner the intricate pattern of wavelengths of atomic absorption and emission lines. But in solving one puzzle, this model raised several new questions. How did an atom in an excited state, of moderately high energy, know to which of the numerous lower states it should decay? What determines the relative probabilities of these various possible transitions? How long would one have to wait before these transitions take place? These questions were answered by the quantum theories of Schroedinger and Heisenberg. However, even before those theories were developed, Einstein had considered the emission and absorption of radiation involving two energy states separated by ΔE, using only general thermodynamic arguments.

Einstein identified three distinct processes.

1. *Absorption of radiation*, by which the system can be raised from the lower to the higher level, provided that $h\nu = \Delta E$.
2. *Spontaneous emission of radiation* of frequency $\nu = \Delta E/h$, with the simultaneous transition of the system from the higher to the lower state.
3. *Stimulated emission of radiation*, whereby the system is induced to decay from the higher to the lower state as a result of exposure to radiation of frequency $\nu = \Delta E/h$.

It is the last of these processes that is so critical to laser (and maser) operation. Suppose we can construct an atomic system that has just three energy levels: a ground state of energy E_0, an excited state of energy E′, and another excited state of energy E_m, as shown in the energy-level diagram of Figure 30.9.* An atom in an excited state, such as E', will normally decay in a relatively short time of about 10^{-8} s to a lower energy level. When there are two or more states of lower energy, as in this case, decay to any of these states is possible. The relative probabilities of the various decay schemes can be calculated using the quantum theory; they can also be determined experimentally.

There often are excited states that are *metastable*, that is, states in which an atom will exist for a long time before decaying to a lower level, because the probability for *spontaneous* decay to the ground state is very small. However, if an atom in such a metastable state is subjected to radiation of the proper frequency, the transition to the ground state may be

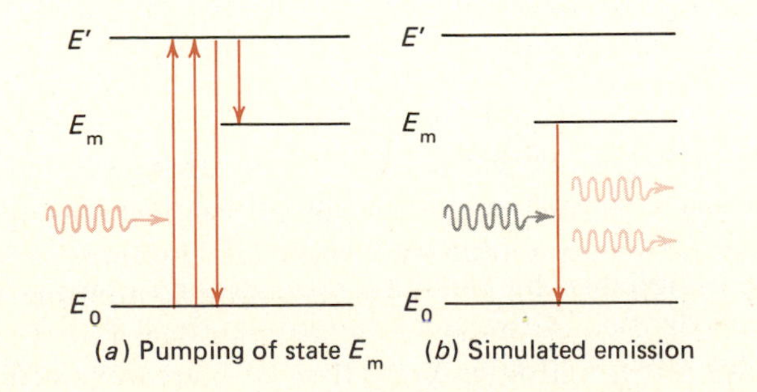

Figure 30.9 (a) The energy levels of a three-level laser. The excited state E_m is a long-lived metastable state. Optical pumping raises atoms from the ground state E_0 to the excited state E′; they then decay either to state E_m or back to the ground state. Those in the state E_m remain there; those that return to the ground state are once again pumped up to the excited state E′. After a short time, the metastable state is strongly populated. (b) An incident photon of energy $h\nu = E_m - E_0$ will stimulate transitions from the metastable excited state to the ground state.

* No system exists that has just three energy levels; real systems have an infinite number of energy levels. However, only three, or sometimes four, levels actively participate in laser operation, so this idealized model is simple yet adequate for our purposes.

stimulated and will then occur with high probability and therefore quickly.

Suppose that E_m in Figure 30.9 is such a metastable state. Suppose also, for convenience of argument (though this is not essential), that the probabilities for spontaneous transitions from E' to E_0 and from E' to E_m are equal. If we now irradiate the system containing a large number of these atoms in their ground state with photons of energy $h\nu_p = E' - E_0$, many atoms will absorb a photon and be excited to the state E'. Once in that state, half will decay to E_0, the other half to E_m. The ones in the state E_m remain there; those that have returned to E_0 will again be "pumped" up to E', half of them decaying once more to E_m. Soon, then, the ground state is depleted by this process and the metastable state highly populated, producing what is called a *population inversion*.

If, at this stage, photons of energy $E_L = h\nu_L = E_m - E_0$ are admitted into the system, it is highly probable that they will *stimulate* one or more atoms to make the transition from E_m to E_0, thereby releasing additional photons of the same frequency. In a laser, the active substance is contained in a cylindrical region fitted with optically flat mirrors at the ends, which confine the radiation within the cylindrical volume so that it can stimulate further transitions. One of the end mirrors is only partially silvered, allowing some of the radiation to escape in a highly collimated, monochromatic beam. The laser tube is the optical analog of the resonant pipe in acoustics, and the distance between the two end mirrors must be carefully adjusted to be exactly an integral number of half-wavelengths. Some modern lasers are designed to operate at several wavelengths or over a range of wavelengths, and the mirror adjustment tunes the laser to the desired wavelength.

The laser or maser is an extremely sensitive detector of monochromatic radiation, since a few photons, even just one, at the resonant frequency can stimulate an avalanche that releases billions and trillions of photons. Although the maser is used primarily to detect microwaves, lasers are used almost exclusively to generate narrow, high-intensity coherent beams of monochromatic radiation.

The extreme collimation of laser beams is achieved by multiple reflection between front and rear mirrors, so that photons not coaxial with the laser tube to within a very small angle are lost from the beam. Generally, beam divergence is close to the diffraction angle for the laser aperture (see Chapter 27). This very small divergence angle means that the power of the beam remains confined to a small cross-sectional area; that is, the intensity remains high over long distances, as evidenced in Figure 30.10. Extreme collimation combined with the development of high-intensity pulsed lasers made possible the laser-ranging experiments discussed in Chapter 25.

The other unique feature of laser emission is the coherence of the radiation. In Chapter 27, we defined the meaning of coherence; in the present context it simply means that over the entire aperture of the laser tube, the oscillations of the electromagnetic field are in phase. That this is true can be established by placing a transmission grating in front of the laser; a well-defined diffraction pattern is then visible on a distant screen. If the same experiment is tried using a similar-sized aperture with light from a sodium vapor lamp or other nearly monochromatic source, no interference pattern will be visible because of lack of coherence in the incident radiation.

Figure 30.10. A well-collimated beam from a 15 W argon ion laser traverses the night sky over the city of Lyon, France.

Because laser beams can easily produce harmful lesions in the retina, they present a serious hazard and should be treated with proper respect and caution. Lasers with rated outputs of 10 mW or more must be considered dangerous. At no time should an even apparently faint laser beam be permitted to fall on the eye.

Summary

Results on the scattering of α particles by thin foils of gold led Rutherford to propose the *nuclear model* of the atom. The inherent classical instability of this model was obviated by Bohr's postulate that electrons in atoms can exist only in certain *stationary states*, for which the angular momentum about the nucleus is given by

$$L = n\hbar$$

where n is an integer and $\hbar = h/2\pi$. For circular orbits, this condition is equivalent to the requirement that the orbit circumference be an integral multiple of the electron's de Broglie wavelength.

For the hydrogen atom, the Bohr quantization condition leads to a set of circular orbits whose radii are

$$r_n = \frac{n^2\hbar^2}{mke^2} = n^2 a_0 \qquad a_0 = 5.29 \times 10^{-11}\ \text{m}$$

and to an energy of

$$E_n = -\frac{mk^2e^4}{2\hbar^2 n^2} = -\frac{E_0}{n^2} \qquad E_0 = 13.6\ \text{eV}$$

In these expressions, n is an integer (1, 2, . . .) and k is the Coulomb constant. If the atomic system consists of a nucleus of charge Ze and a single electron, the orbit radii and energies are given by

$$r_n = \frac{n^2 a_0}{Z} \qquad \text{and} \qquad E_n = -\frac{E_0 Z^2}{n^2}$$

Energy-conserving transitions between these allowed levels account for the emission and absorption lines of atomic hydrogen with frequencies given by the Planck-Einstein relation

$$h\nu = \Delta E = |E_f - E_i|$$

where the subscripts "f" and "i" denote the final and initial state of the atom. This expression can be restated, using the preceding equation for the allowed energy levels, to give

$$\frac{1}{\lambda} = \frac{\nu}{c} = R\left|\left(\frac{1}{n_f^2}\right) - \left(\frac{1}{n_i^2}\right)\right| \qquad R = \frac{E_0}{hc} = 1.1 \times 10^7\ \text{m}^{-1}$$

The quantity R is called the *Rydberg constant.*

The complete description of an electron in an atomic orbit requires four quantum numbers: n, l, m, and m_s. Here n is the *principal quantum number;* l is related to the angular momentum of the electron in its orbit by

$$L = \sqrt{l(l+1)}\,\hbar$$

The component of the angular momentum along a direction of quantization, generally identified as the z direction, is

$$L_z = m\hbar$$

The quantum number m_s specifies the component of the spin angular momentum of the electron along the direction of quantization; that is,

$$S_z = m_s\hbar$$

These quantum numbers must satisfy the following conditions.

$$n \geq 1$$
$$n - 1 \geq l \geq 0$$
$$|m| \leq l$$
$$m_s = \pm\tfrac{1}{2}$$

The *Pauli exclusion principle* states:

Two electrons may not occupy the same quantum state.

Bombardment with energetic electrons may dislodge an atomic electron from a low-lying stationary state in a heavy atom. If an atomic electron from a higher energy state ($n = 3, 4, \ldots$) then drops into the available low-lying ($n = 1, 2$) state, energy is conserved by the emission of a photon. Generally, the energy of the emitted photon is in the range of kilo-electron-volts, which places the wavelength in the region of X rays. These transitions are responsible for the *characteristic X-ray lines.*

Multiple Choice Questions

30.1 In Rutherford's model of the atom,

(a) positive and negative charges were uniformly distributed in a spherical volume.

(b) the electrons were restricted to a few circular orbits with well-defined radii.

(c) electrons would be expected to spiral into the nucleus while radiating electromagnetic waves.

(d) the electrons were assigned a spin angular momentum.

30.2 Bohr's quantum condition on electron orbits

(a) required that the energies of an electron in a hydrogen atom be equal to nE_0, where E_0 is a constant energy and n is an integer.

(b) required that the angular momentum of the electron about the hydrogen nucleus equal $n\hbar$.

(c) required that no more than one electron occupy a given stationary state.

(d) required none of the above.

30.3 If the voltage across an X-ray tube is doubled, the energies of the characteristic X rays emitted by this tube will

(a) double.
(b) quadruple.
(c) halve.
(d) remain constant.

30.4 Bohr's model for the hydrogen atom led to the conclusion that the radii of the stationary-state orbits are

(a) proportional to n^2,
(b) proportional to n,
(c) independent of n,
(d) proportional to $1/n$,
(e) proportional to $1/n^2$,

where n is the principal quantum number.

30.5 A gas of atomic hydrogen is bombarded with electrons of sufficient energy to excite the hydrogen atoms to the $n = 4$ level. The total number of emission lines in the resulting spectrum is

(a) 1
(b) 3
(c) 6
(d) 9
(e) 16

30.6 Bohr's model for the hydrogen atom led to the conclusion that the magnitude of the energies of the electron in the stationary orbits about the proton are proportional to

(a) n^2
(b) n
(c) a constant and independent of n
(d) $1/n$
(e) $1/n^2$

where n is the principal quantum number.

30.7 Which of the following is *not* a possible configuration of electrons in an atom?

(a) $(1s)^2(2s)^2(2p)^5$
(b) $(1s)(2s)^2(2p)^6(3s)$
(c) $(1s)^2(1p)^6(2s)$
(d) $(1s)^2(2s)^2(2p)^6(3s)^2(3p)^2(3d)^4$
(e) All the above are possible configurations.

30.8 The number of electrons that can be accommodated in an $l = 2$ subshell is

(a) 2
(b) 6
(c) 8
(d) 10
(e) 18

30.9 The radius of the electron in the $n = 1$ orbit in triply ionized beryllium ($Z = 4$) is

(a) $4a_0$
(b) $3a_0$
(c) $a_0/3$
(d) $a_0/4$
(e) $a_0/16$

where a_0 is the Bohr radius.

30.10 The number of electrons that can be accommodated in the $n = 4$ shell is

(a) 8
(b) 18
(c) 32
(d) 36
(e) infinite

Problems

(Sections 30.1–30.4)

30.1 The bright yellow lines emitted by sodium have wavelengths near 590 nm. What is the energy of a quantum of radiation of that wavelength?

30.2 Determine the wavelengths of the Balmer line of shortest wavelength.

30.3 What is the wavelength of the Balmer line that corresponds to the $n = 5$ to $n = 2$ transition?

30.4 Determine the longest wavelength of the Lyman series of lines.

30.5 Compute the ionization energy of doubly ionized lithium.

30.6 Sketch the energy-level diagram for doubly ionized lithium and identify some transitions that would lead to emission of visible radiation.

• **30.7** Light of 1091-nm wavelength is emitted from a hydrogen discharge. Identify the transition that produces this emission.

• **30.8** A hydrogen discharge emits light whose wavelength is 396 nm. What is the transition that produces this emission?

• **30.9** Light of wavelength 409 nm is observed from a hydrogen discharge. What transition produces this emission?

• **30.10** Derive equations (30.6a) and (30.12a).

• **30.11** A hydrogen atom makes a transition from the $n = 7$ to the $n = 1$ state. What is the fractional change in the mass of the hydrogen atom as a result of this transition?

• **30.12** Determine the potential energy of an electron in the $n = 2$ stationary state of hydrogen.

• **30.13** Using the hydrogenic model, estimate the ionization energy of neutral lithium, that is, the energy required to remove the least tightly bound of the three atomic electrons of lithium. Remember that the other two electrons screen some of the charge of the lithium nucleus. The observed ionization energy is 5.39 eV.

• **30.14** How much energy would be needed to remove the 1s electron from palladium? from platinum?

• **30.15** A gas of hydrogen atoms is excited such that electrons are raised from the $n = 1$ (ground) state to the $n = 4$ (excited) state. The atoms return to the ground state, emitting radiation in the process. Calculate the wavelengths of all the emission lines that can be observed.

• **30.16** Calculate the approximate minimum potential necessary to observe the $n = 2$ to $n = 1$ characteristic X ray of zinc ($Z = 30$).

• • **30.17** The Bohr model contains no correction for relativistic effects, such as a mass change of the electron as it circulates about the proton. Is that neglect justified? Explain.

• • **30.18** Relativistic correction can be safely neglected in determining the energy levels of hydrogen. Is it a good approximation likewise to neglect these corrections in calculating the energy of the 1s electrons in palladium? in platinum? Justify your answer.

• • **30.19** What is the magnetic field at the proton due to the orbital motion of the electron in the Bohr model for the $n = 2$ state?

(Sections 30.5–30.7)

30.20 What are the possible angular momentum quantum numbers of an electron whose principal quantum number is 5?

• **30.21** Without referring to Table 30.2, write the ground-state configuration of nitrogen.

• **30.22** An X-ray tube has an anode made of palladium. What potential difference is needed between cathode and anode if one wishes to observe all the characteristic X-ray lines emitted by the anode?

• **30.23** An X-ray tube has an anode (target) made of tin. What is the approximate wavelength of the most energetic characteristic X ray that will be emitted by this X-ray tube? If the voltage across the tube is 20,000 V, will one observe this most energetic characteristic X ray? What is the shortest wavelength of the X rays that can be observed with a potential difference of 20,000 V?

• • **30.24** The most energetic characteristic X ray emitted from a certain X-ray tube has a wavelength of 0.212 nm. What is the most likely target material?

• • **30.25** A small laser emits pulses of radiation of 642-nm wavelength. Each pulse lasts 20 ms, and the laser emits one pulse each second. If the average energy output of the laser is 10 mW, what is the power radiated during each pulse? How many photons are released with each pulse?

• • **30.26** Suppose the laser beam of Problem 30.25 comes out of a 7-mm-diameter half-silvered mirror. How wide will the beam be after it has traveled 10 km? Neglecting absorption by the atmosphere, determine the intensity, in photons per square centimeter per second, at the surface of the moon if the laser is aimed in that direction.

• • **30.27** The wavelength of the characteristic X ray due to the $n = 3$ to $n = 1$ transition in aluminum ($Z = 13$) is about 1 nm. Estimate the wavelength of the X ray due to the corresponding transition in iron ($Z = 26$) and in tellurium ($Z = 52$).

• 31 •

Aggregates of Atoms; Molecules and Solids

A three-element electronic device which utilizes a newly discovered principle involving a semiconductor as the basic element is described. It may be employed as an amplifier, oscillator, and for other purposes for which vacuum tubes are ordinarily used.

J. BARDEEN AND W. H. BRATTAIN,
Physical Review 74, 230 (1948).

(*Abstract of the article that announced the development of the transistor.*)

31.1 Introduction

The preceding chapter was concerned with the properties of single atoms, treated as separate entities. Atoms rarely appear in such isolation on earth. In fact, with the exception of the inert gases, elements generally form atomic aggregates. Some, such as the noble metals, often appear naturally in the pure state in the crystalline phase. Mostly, however, elements form chemical compounds, sometimes with atoms of the same element but more often with atoms of other elements. Thus hydrogen, oxygen, nitrogen, and carbon form the diatomic molecules H_2, N_2, O_2, NO, and CO, and a myriad of polyatomic molecules such as H_2O, NH_3, NO_2, HNO_3, CH_4, $C_6H_{12}O_6$ (glucose), $NH_2(CH_2)_2COOH$ (β-alanine), and $CH_3(CH_2)_{16}COOH$ (stearic acid).

In this chapter we first give a brief overview of the forces that hold molecules together. We then consider the energy-level pattern of a simple molecule and the absorption (or emission) spectrum that results therefrom. We next examine the cohesive forces of the dominant crystal types; focusing attention on the electrical properties, we show how the differences between metallic, semiconducting, and insulating crystals can be explained in terms of the modern theory of solids. The chapter concludes with a brief qualitative description of a few semiconductor devices.

31.2 Molecular Bonds

The simple fact that two atoms combine to form a molecule tells us that there must be a net attractive force between them when their separation is greater than the equilibrium separation in the molecule. It also follows that since the molecule is stable, the energy of the molecule must be less than the energy of the separated atoms.

The fundamental mechanisms that give rise to this attractive interaction are largely, though not exclusively, of electrostatic origin. A complete description presents a highly complex problem in quantum mechanics, involving the mutual interactions of many particles–several electrons and two or more positively charged nuclei. An adequate physical picture can be formed, much computational drudgery avoided, and much better physical insight gained by constructing simplified models that highlight the most important interactions. It is wise to keep in mind, however, that conclusions and numerical results are based on a simplified model; more often than not, real systems do not conform strictly to one model but exhibit some features attributable to a different model or perhaps not contained in any of the standard models.

The models of molecular bonding are, in order of decreasing bond strength, the ionic bond, the covalent bond, the hydrogen bond, and the van der Waals bond.

31.2 (a) Ionic Bonds

The prototype of the ionic molecule is NaCl, ordinary table salt. In this molecule, sodium appears in the singly ionized state, as Na^+, and chlorine in the form of the singly ionized negative ion, Cl^-. The system as a whole is electrically neutral; in effect, the chlorine atom has accepted the extra electron donated by the sodium atom. The relatively weakly bound 3s valence electron can be detached from the sodium atom with an expenditure of 5.1 eV, leaving a positively charged Na^+ ion. The neutral chlorine atom is just one electron short of the closed-subshell, inert-gas configuration of argon (see Table 30.2). Because closed-shell structures are energetically favored, the Cl^- ion is more stable than the neutral Cl atom by about 3.7 eV. That is, the energy of the Cl^- ion is 3.7 eV *less* than the energy of the neutral Cl atom; chlorine is said to have an *electron affinity* of 3.7 eV.

Still, the overall energy balance does not favor formation of the isolated Na^+ and Cl^- ions by the transfer of one electron. Net energy reduction, a requisite for molecular stability, comes from the electrostatic interaction between the positive and negative ions. The potential energy of two charges, $+e$ and $-e$, separated a distance r, is

$$\mathrm{PE} = -\frac{ke^2}{r} \tag{31.1}$$

When the separation r is less than about 11 Å (1.1 nm), the total energy

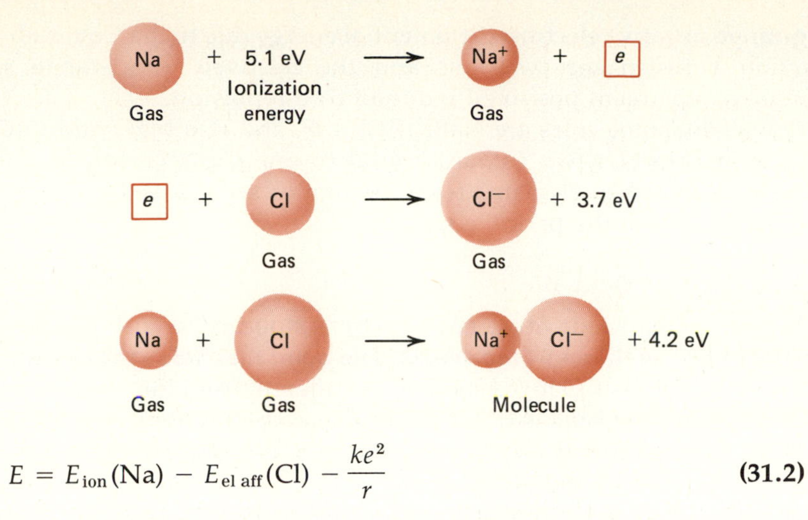

Figure 31.1 The energetics of formation of the NaCl molecule.

$$E = E_{\text{ion}}(\text{Na}) - E_{\text{el aff}}(\text{Cl}) - \frac{ke^2}{r} \qquad \textbf{(31.2)}$$

becomes negative. In Equation (31.2), E_{ion}(Na) is the ionization energy of sodium, 5.1 eV; $E_{\text{el aff}}$(Cl) is the electron affinity of chlorine, 3.7 eV; and the last term is the electrostatic potential energy of the two ions viewed as point charges.

The equilibrium distance between the Na and Cl nuclei in NaCl is 2.4 Å, and the total energy is then −4.2 eV. That is, 4.2 eV of energy are required to separate an NaCl molecule into a neutral sodium and a neutral chlorine atom.

According to Equation (31.2), it seems energetically advantageous to bring the two ions much closer together than 2.4 Å, indeed as close as possible. The two ions do not approach more closely because the electrons in the closed-shell configurations would then overlap significantly. The *core repulsion* between closed shells has its origin in the exclusion principle, which prohibits two electrons from occupying the same quantum state.

The potential energy (PE) of NaCl as a function of nuclear separation is shown in Figure 31.2. The dependence on r is primarily the sum of the

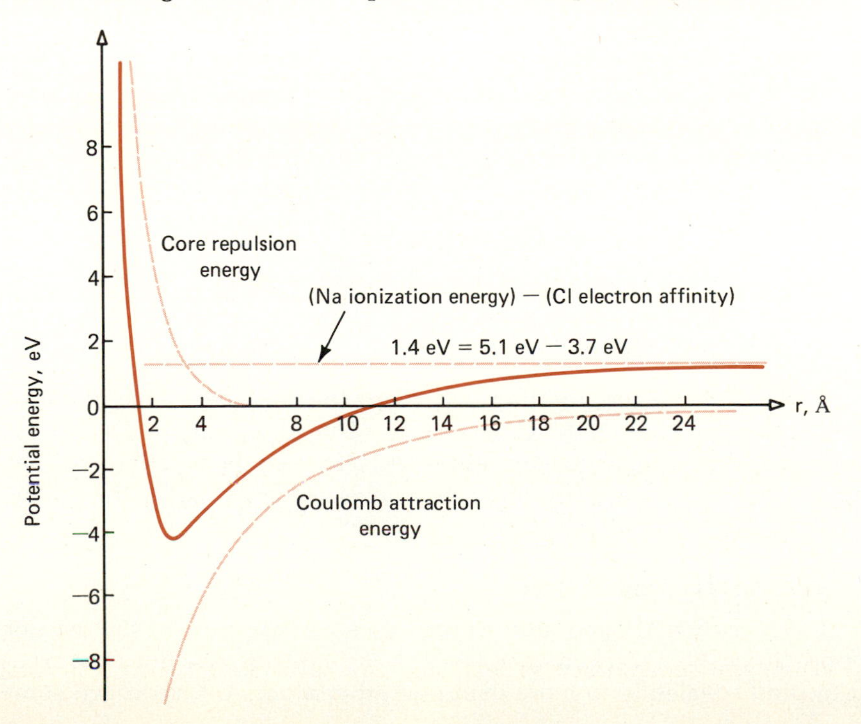

Figure 31.2 The potential energy of Na and Cl as a function of the internuclear distance, relative to the energy of the two separated neutral atoms. The three principal contributions are indicated.

long-range negative electrostatic potential energy due to the Coulomb interaction between the two ions and the relatively short-range and strongly r-dependent positive PE due to core repulsion.

Other ionic molecules are NaBr, KCl, LiF, and also MgO and $CaCl_2$. In calcium chloride, two Cl^- are bound to one Ca^{2+}. Generally, ionic bonds form between elements from the first or second and the sixth or seventh column of the periodic table.

31.2 (b) Covalent Bond

Many elements, for instance hydrogen, fluorine, chlorine, and oxygen, form stable diatomic molecules. The bond that holds the two atoms of H_2, F_2, as well as CO together is quite different from the ionic bond of NaCl. This *covalent bond* is the result of a *sharing* of electrons by the participating atoms. For example, in F_2 the two atoms share their valence electrons, thereby effectively forming a stable, closed-shell configuration for the combined two-atom system. In the molecule H_2, the wave functions of the two electrons are correlated so that their spins are antiparallel and the oribital motion is the same for both electrons, with their charge density concentrated predominantly in the midregion between the two protons. The effect of this charge distribution is a net attraction between the two positively charged protons mediated by the negatively charged electron concentration between them.

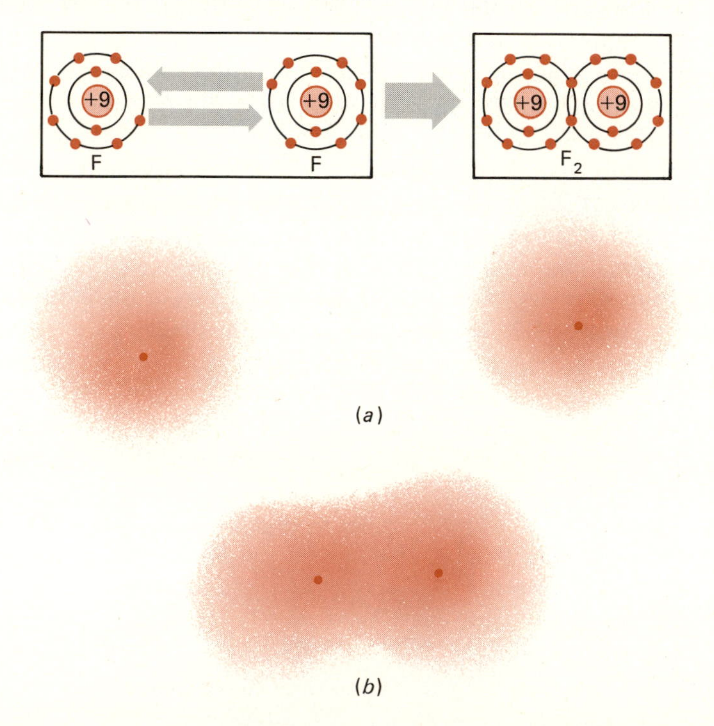

Figure 31.3 Schematic drawing, showing electron sharing in the formation of the F_2 molecule.

Figure 31.4 The electronic distribution about two isolated hydrogen atoms and about a hydrogen molecule (H_2).

Covalent molecular binding is perhaps the most pervasive in nature. The molecules H_2O, NH_3, CO_2, CH_4, and most organic molecules owe their stability largely to covalent bonds formed between constituent atoms. Even in nominally ionic molecules, such as MgO, the valence electron distribution indicates some covalent binding.

31.2 (c) Hydrogen Bond

In H_2 each hydrogen atom contributes one electron to the covalent bond. Since the hydrogen atom has but one electron, one does not expect it to bind covalently to more than one other atom. Yet hydrogen some-

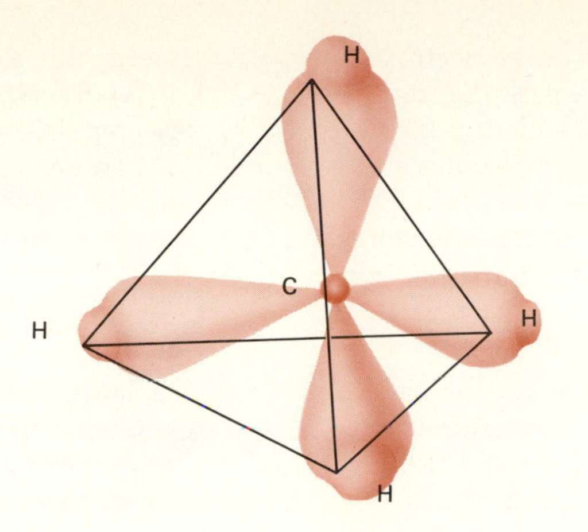

Figure 31.5 *Schematic drawing showing the electronic distribution of the four covalent bonds that bind the four H atoms to the tetravalent carbon atom in CH_4.*

times appears to be the "glue" that bonds *two* atoms or ions, as in the hydrogen difluoride ion, $(HF_2)^-$, shown schematically in Figure 31.6. The hydrogen bond, with a bond energy of only about 0.1 eV, can be viewed as the inverse of the covalent bond. Two negative charges, in this case the two F^- ions, are bound by the positively charged proton situated between them. Since negative ions are generally large, simple geometric considerations show that the H bond cannot be effective in binding more than two atoms. Nonetheless, hydrogen bonding is the critical binding force in many important molecular conformations, particularly biochemical ones. For instance, this is the binding force between the organic bases cytosine and guanine, and thiamine and adenine, of the famous DNA double helix.

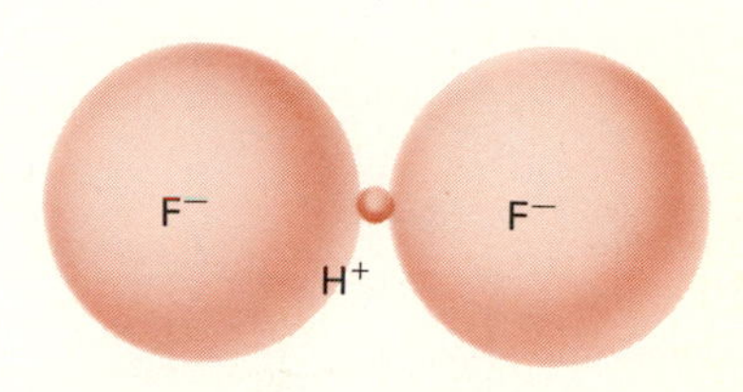

Figure 31.6 *Schematic drawing showing the proton H^+ in relation to the two F^- ions in the $(HF_2)^-$ molecular ion. The two negative ions are bound through the hydrogen bond, that is, through the positively charged proton between them.*

31.2 (d) Van der Waals Bond

In Chapter 17 we described how and why an electrically neutral object is attracted to a charged object: the electric field due to the charge induces or orients permanent dipoles in the neutral object. These dipoles then experience a net attractive force in the nonuniform field of the point charge. A net attractive force also acts between two dipoles oriented to minimize their interaction energy. (This was shown in Chapter 22 in connection with magnetic dipoles.) This dipole-dipole attraction is, however, much weaker than the Coulomb interaction between two charges.

The van der Waals interaction is weaker still, although it is also derived from the interaction of two dipoles. In this case, however, neither neutral entity has a permanent dipole. Random fluctuations of one entity result in randomly oriented dipoles of variable magnitude. The associated randomly varying electric field then induces a correlated dipole in the other neutral entity, and the two dipoles exert an attractive force on one another.

Though this van der Waals interaction is very weak, it is responsible for the low-temperature condensation of homopolar molecules such as H_2 and N_2 and of inert gases into the liquid state.

31.3 Molecular Spectra

As one might expect, the complicated structure of the polyatomic molecule gives rise to a complicated energy-level system and emission or absorption spectrum. We have seen how Bohr's angular momentum quantization condition led to discrete energy levels for an electron in the field

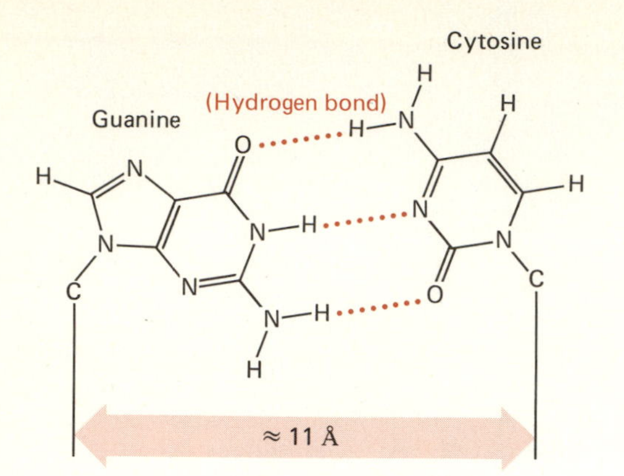

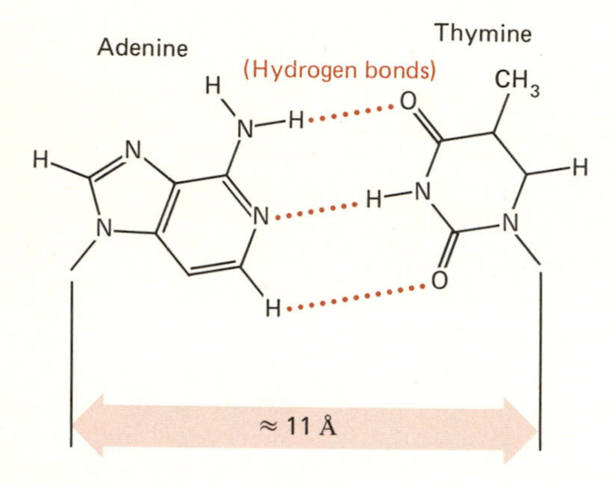

(a)

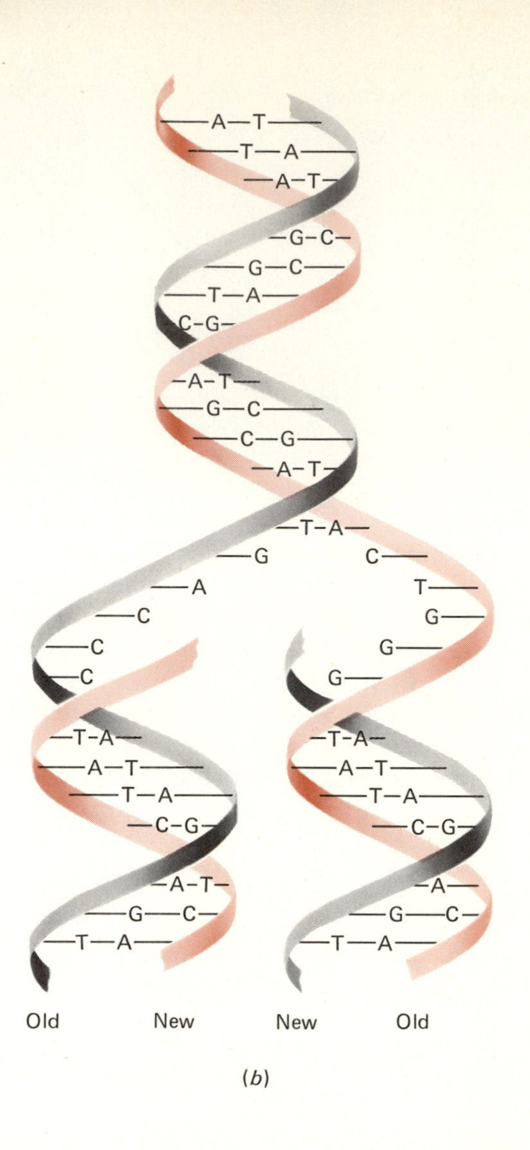

(b)

Figure 31.7 (a) Binding of the organic bases of the DNA molecule, guanine and cytosine, and of adenine and thymine, is achieved by hydrogen bonding. (b) Simplified representation of DNA replication, in which the specificity of the bonding involving the hydrogen bonds plays a crucial role.

of a nucleus. Quantization and discrete energy levels, however, are not limited to electronic motion. Whereas in classical dynamics, the energies of vibration and rotation of a diatomic molecule can assume any value, quantum mechanics imposes certain conditions that restrict also these energies. Indeed, Planck's initial quantization hypothesis, later formalized in the Schroedinger equation, limits the energies of a harmonic oscillator to

$$E_v(n) = (n + \tfrac{1}{2})h\nu \qquad n = 0, 1, 2, 3, \ldots \tag{31.3}$$

where $E_v(n)$ is the energy of the vibrating oscillator in the quantum state characterized by the integer n, and ν is the frequency of oscillation.

Typical vibrational frequencies of diatomic molecules are in the neighborhood of 2×10^{13} Hz, corresponding to an energy separation of $h\nu \simeq 10^{-1}$ eV.

The quantization rule for rotational energy can be made plausible, without solving the Schroedinger equation, by rewriting the classical expression for the kinetic energy (KE) of a rotating system,

$$\text{KE} = \tfrac{1}{2}I\omega^2 \tag{8.22}$$

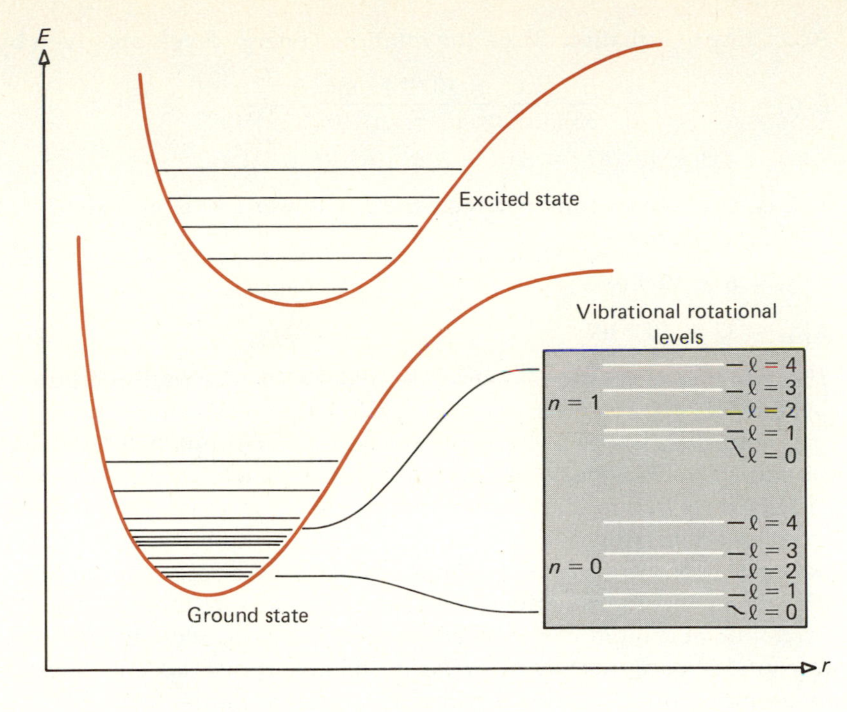

Figure 31.8 The potential energy and energy-level structure of a diatomic molecule. The largest separation is between the electronic ground state and the electronic excited state. The dependence of the potential energy on nuclear separation differs somewhat for these two states. Within each of these states there is a substructure of energy states corresponding to vibrational energy levels and the much more closely spaced rotational energy levels, shown in the inset.

in terms of the angular momentum, $L = I\omega$,

$$\text{KE} = \frac{L^2}{2I} \tag{31.4}$$

where I is the molecule's moment of inertia about the axis of rotation.

If we now replace the classical angular momentum by its quantum expression

$$L = \sqrt{l(l+1)}\hbar \tag{31.5}$$

we obtain the energy levels of the quantum mechanical rotator:

$$E_r(l) = \frac{l(l+1)\hbar^2}{2I} \qquad l = 0, 1, 2, 3, \ldots \tag{31.6}$$

Note that the separation between adjacent rotational levels is not constant but increases with increasing l. Hence, quantization of the rotator and of the harmonic oscillator leads to qualitatively different patterns. The energy separation between low-lying rotational levels is readily estimated; for typical diatomic molecules such as Cl_2, N_2, or O_2 it is in the neighborhood of 10^{-4} eV, or about one one-thousandth of the separation between adjacent vibrational levels.

● ***Example 31.1*** Estimate the separation between the $l = 0$ and the $l = 1$, and between the $l = 1$ and the $l = 2$, rotational levels of Cl_2. The separation between the two chlorine nuclei is 2 Å.

We regard this diatomic molecule as a symmetric dumbbell. The moment of inertia of the system for rotation about its center of mass is then

$$I = 2(M_{Cl}r^2)$$

where $M_{Cl} = 35 \times 1.67 \times 10^{-27}$ kg and $r = 1$ Å $= 10^{-10}$ m is half the internuclear separation.

According to Equation (31.6), the rotational energy levels are given by

$$E_r(l) = l(l+1)\frac{(1.06 \times 10^{-34}\ \text{J}\cdot\text{s})^2}{2[2(35)(1.67 \times 10^{-27}\ \text{kg})(10^{-10}\ \text{m})^2]}$$
$$= 4.8 \times 10^{-24}[l(l+1)]\text{J} = 3 \times 10^{-5}[l(l+1)]\ \text{eV}$$

The energy separations between the $l = 0$ and $l = 1$, and the $l = 1$ and $l = 2$, levels are

$$\Delta E_{01} = 6 \times 10^{-5}\ \text{eV}$$

$$\Delta E_{12} = 12 \times 10^{-5}\ \text{eV}$$

The vibrational-rotational energy-level substructure outlined here is associated with every electronic energy level of a diatomic molecule. Hence, optical emission or absorption due to transitions between electronic energy states appears not as sharp lines as in atoms, but as a series of fairly narrow bands. Each of these bands corresponds to a transition between two vibrational levels, and each band, when examined under higher resolution, is seen to be composed of many closely spaced lines, which correspond to transitions between different rotational levels.

Triatomic and more complicated polyatomic molecules display a far more complex pattern in their emission or absorption spectra because there are numerous vibrational and also rotational modes with a diversity of energy-level spacings. Nevertheless, it is possible to glean much detailed information about molecular structure and interatomic forces

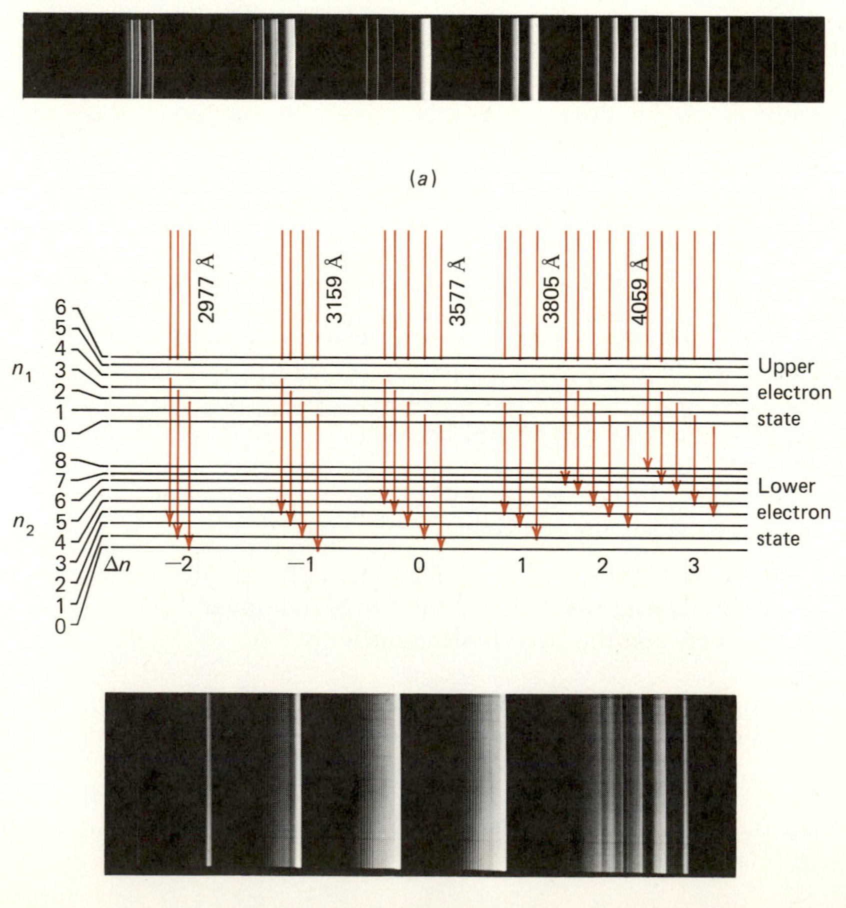

Figure 31.9 A portion of the emission spectrum of the N_2 molecule, corresponding to the transition between the same two electronic states. (a) The various broadened lines of the upper photograph can be attributed to transitions between vibrational levels in which the vibrational quantum number n changes by different integral values, as indicated. If the spacing between vibrational states in the two electronic states were the same, the groups of broad lines corresponding to a given value of Δn would coalesce. The vibrational level separation changes because the two potential functions are not identical, nor are they exactly the potentials associated with the perfect harmonic oscillator (see Chapter 14). (b) A part of the spectrum shown in (a) has been enlarged and now reveals that the broad lines are really a series of bands, each band consisting of numerous closely spaced lines. These lines arise from transitions between different rotational levels. The spacing between the lines varies because the spacing between rotational levels increases with increasing rotational quantum number l.

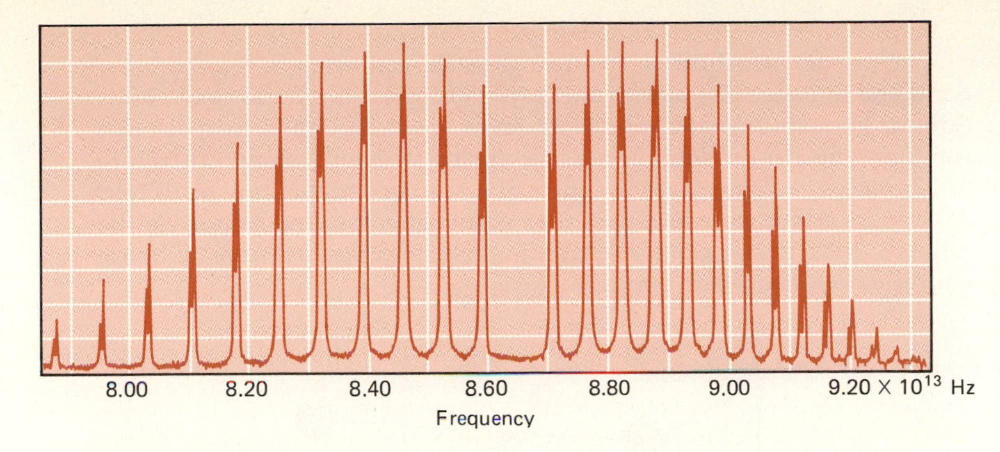

Figure 31.10 The infrared absorption spectrum (portion) of the molecule HCl. The two groups of lines to the right and left of the central space correspond to transitions in which the rotational quantum number l changes by $+1$ and -1, respectively. The splitting of each peak into a doublet arises because chlorine has two isotopes, ^{35}Cl and ^{37}Cl (see Chapter 32). The difference in nuclear mass accounts for the difference in molecular moment of inertia indicated by the splitting.

from a careful analysis of molecular spectra. Moreover, since a molecular spectrum is as characteristic a signature of a specific molecule as the line spectrum of an atom, the appearance of certain spectral patterns is generally good evidence that a particular molecular unit is present in a complex polymer. Molecular spectroscopy is an analytic tool that is widely used in organic chemistry and biochemistry.

31.4 Cohesion of Solids

The cohesive energy of atoms within a crystal depends on the dominant binding forces. The classification scheme that was appropriate for molecular binding is also useful in the theory of solids. One further category is needed, however, namely metallic bonding. This mechanism, effective in binding atoms of sodium, copper, silver, and other metals in the liquid and solid phase, does not produce molecular complexes in the gaseous phase.

31.4 (a) Ionic Solids

In crystals that form by ionic bonding, the dominant Coulomb interaction dictates the main features of the lattice structure. The structure fulfills the requirement that each positive ion is surrounded by nearest-neighbor negative ions, and vice versa. For example, as shown in Figure 31.11, every Na^+ ion in the NaCl crystal has six nearest-neighbor Cl^- ions, and each Cl^- ion is surrounded by six nearest-neighbor Na^+ ions.

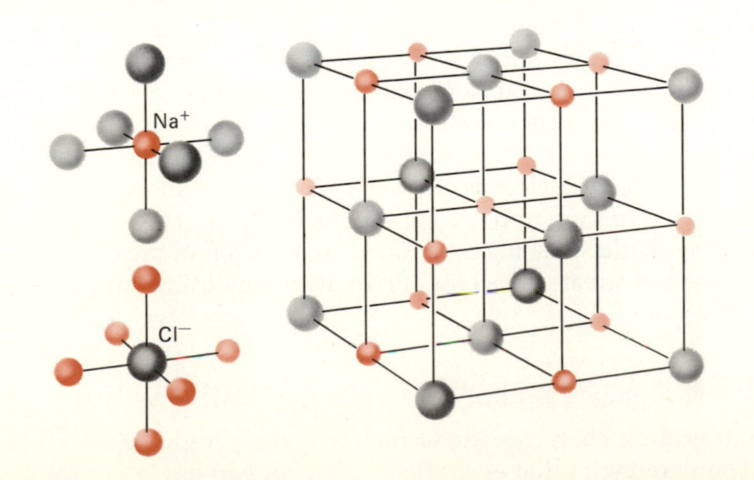

Figure 31.11 The crystal structure of NaCl. Each positive ion is surrounded by six nearest-neighbor negative ions, and each negative ion is similarly surrounded by positive ions.

31.4 (b) Covalent Crystals

The crystal structure of covalent crystals conforms to the covalent bond and enhances its effectiveness. For example, the tetravalent atoms carbon, germanium, and silicon all favor covalent bonds in molecular combinations. In the solid phase, the crystal structure, shown in Figure 31.12, places each atom at the center of a regular tetrahedron, with four like atoms at the corners. Thus, four equivalent covalent bonds can be formed, with every atom contributing one electron to each of these equivalent electron-pair bonds.

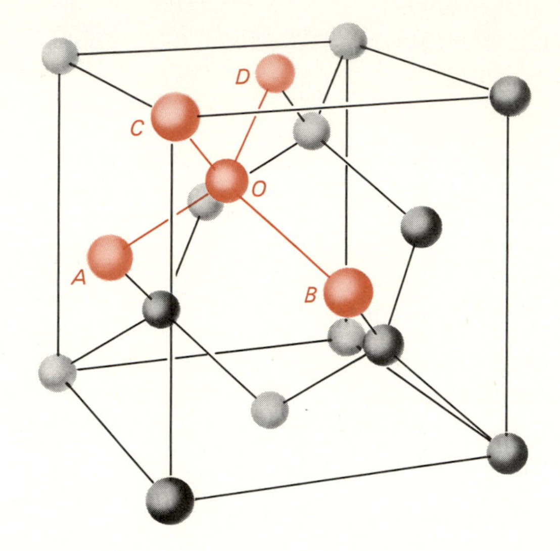

Figure 31.12 The diamond lattice. Each carbon atom, such as the one at O, is at the center of a tetrahedron, with four other carbon atoms at the corners (A, B, C, D). Germanium and silicon also crystallize in the diamond structure.

31.4 (c) Metallic Bonding

The metallic bond, which is somewhat weaker on the average than the ionic or covalent, but stronger than the hydrogen or van der Waals bond, does not have a molecular counterpart. In metals, the constituent atoms are ionized. However, in contrast to ionic crystals, wherein the valence electron liberated from one atom attaches itself to and is localized at another, in metals, valence electrons are relatively free to move throughout the material. The model of a metallic crystal is a lattice of positive ions, embedded in a sea of mobile, negatively charged electrons. It is this mobility of the valence electrons that lends metals their excellent electrical conductivity.

The cohesion of a metal arises from a delicate balance between large repulsive and attractive forces, with the attractive interactions proving a bit stronger. In sodium, for example, the valence electron density in the free atom extends much farther from the nucleus than it does in the solid; in the solid, the charge of one electron must be confined, on the average, to the volume occupied by one atom in the crystal. Consequently, in solid sodium this valence electron is, on the average, much closer to the strong, attractive Coulomb field of the nucleus, and this reduces the PE. On the other hand, this more restricted localization of the electron necessarily means a greater uncertainty in its momentum and therefore a greater average KE.

*31.5 Band Theory of Solids

The outstanding characteristic of metals is their high electrical conductivity compared with that of covalent, ionic, or van der Waals (molecular) crystals. Good conductivity is evidently consistent with the high mobility of valence electrons in metals, whereas the well-localized nature of valence electrons in ionic and covalent crystals mitigates against charge transfer due to electronic motion. These qualitative arguments can be made more precise by means of the so-called *band theory of solids,* which also provides an explanation for the optical properties of semiconductors and metals.

When N atoms condense into a crystalline solid, there is some overlap of electronic wave functions between neighboring atoms. Electrons in the most confined, tightly bound orbits (K and L shells in all but the lightest metals) are almost totally unaffected by the condensation of atoms, and their energy levels are nearly unchanged. Valence electron wave functions, on the other hand, may overlap significantly. As a result, the N identical sharp valence-electron energy levels of N isolated atoms broaden, on condensation, into a spectrum of N closely spaced energy levels. With about 10^{23} atoms/cm^3, these levels are so closely spaced in energy that for all practical purposes, they form a continuum, called an *energy band.*

The widths of these energy bands vary from one substance to another, depending on details of the atomic wave functions and on crystal structure. Energies between the bands are *forbidden* regions; they do not correspond to solutions of the wave equation for an electron in this crystal, and electrons of that energy simply cannot exist as stationary states in that solid. Moreover, just as atomic energy levels can accommodate only a finite number of electrons (for example, two in 3s states, six in 3p states), so the number of available energy levels in a band formed from these states is also limited.

What does this band picture have to do with the conductivities of metals, semiconductors, and insulators? Well, in a metal such as sodium or copper, the energy band formed from overlap of valence-electron wave functions is only partly filled. In that case, electrons in this valence band are free to move in response to an applied electric field, and the solid is a good conductor.

If, however, the valence band is completely filled, as is, for example, the 3p band in NaCl, electrons in that band have no other available states nearby and consequently cannot change their velocities and kinetic energies in response to an electric field. These crystals are therefore good insulators.

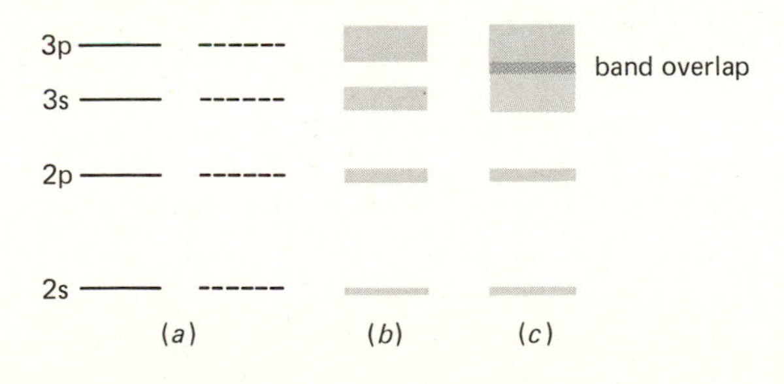

Figure 31.13 Broadening of the energy levels of isolated atoms (a) into bands (b) on condensation to a solid. Energy states between bands are "forbidden" energy regions. Broadening may be so great that neighboring bands overlap in energy (c), leaving no forbidden region.

31.5 (a) Semiconducting Crystals

Pure covalent crystals are also insulators at low temperatures because their valence bands are full, reflecting the saturation of the electron-pair bonds. In many of these materials, however, the forbidden region between the highest energy state of the valence band and the lowest energy state of the next-higher band, the *conduction band,* is relatively

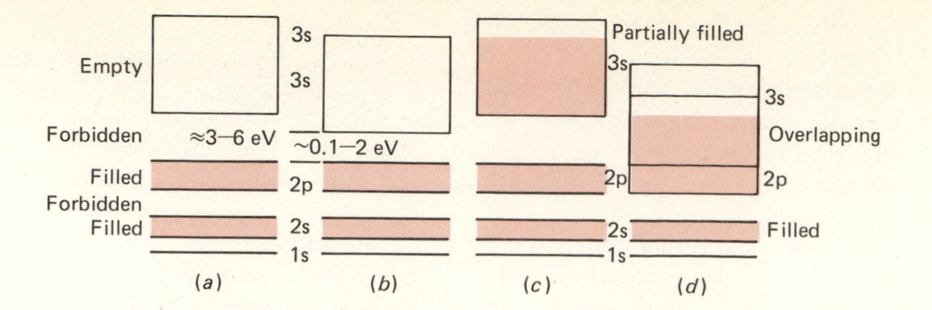

Figure 31.14 Classification of solids according to their band structure. (*a*) In insulators the valence band (and all lower energy bands) is completely filled and the energy gap between the top of the valence band and the next-higher band is several electron volts. (*b*) In a semiconductor, the valence band is also completely filled; however, the gap between the valence band and the next-higher band, the conduction band, is less than 2 eV. The energy gaps in germanium and silicon are about 0.65 eV and 1.14 eV, respectively. (*c*) In a metal, the valence band either is not full or overlaps another band so that both bands are only partially filled.

small. As the temperature of the solid is raised, electrons from the top of the valence band can gain enough energy to make the transition into a state of the conduction band. These substances, of which germanium and silicon are examples, are then able to transport charge, although their conductivity is much less than that of metals. The conductivity of such *semiconductors* increases rapidly with increasing temperature as more and more electrons are promoted from the valence to the conduction band.

Charge transport under these conditions is a result of electronic motion in the valence band as well as in the conduction band. Once a few states have been vacated in the valence band, electrons in that band can move about like cars in a parking lot with just a few open slots. In calculating the current carried by the valence-band electrons, it is far more

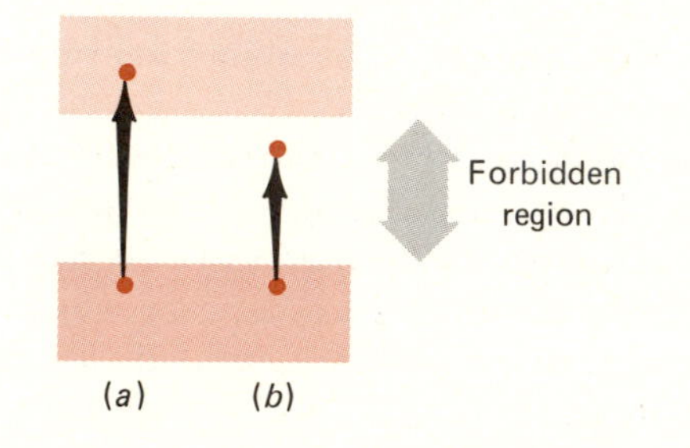

Figure 31.15 An electron can be promoted from the valence to the conduction band only if it can acquire enough energy to surmount the energy gap. (*a*) shows a possible transition; the transition (*b*) is forbidden.

convenient and completely equivalent to describe conduction in terms of the motion of the few *holes* (empty states) instead of the many remaining electrons. These holes must be assigned a *positive* charge, because in the presence of an applied electric field, as the electrons drift in the direction $-\mathbf{E}$, the holes drift in the opposite direction, $+\mathbf{E}$, as positively charged particles would.

Semiconductors owe their pivotal role in today's technology not to the intrinsic, thermally activated conductivity but to the dependence of their electrical properties on small and controllable amounts of impurities that can be introduced into the crystals. If, for example, a tetravalent germanium atom is replaced by a pentavalent arsenic atom, four of the five

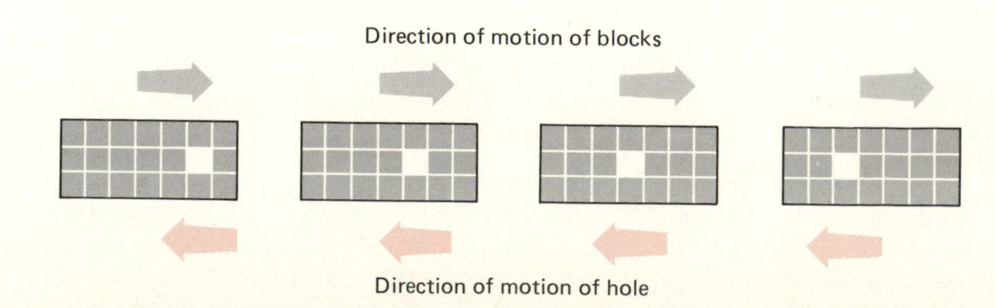

Figure 31.16 Schematic diagram illustrating that the motion of holes is opposite to the motion of the numerous electrons in the valence band. The solid blocks represent the abundant electrons. A net motion of blocks from left to right is equivalent to motion of the vacant site from right to left.

valence electrons of the arsenic atom participate in forming the four covalent bonds with the neighboring germanium atoms; the fifth valence electron is then loosely bound to the arsenic impurity and readily promoted into the conduction band.

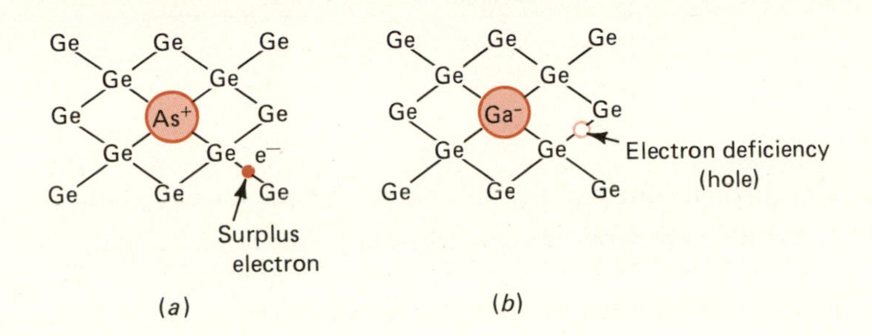

Figure 31.17 *Doping of germanium. (a) A pentavalent arsenic atom can complete the four covalent bonds with neighboring germanium atoms and has one more electron that is not needed in forming covalent bonds. This last electron is weakly bound to the arsenic impurity and can easily acquire enough energy to break free of this impurity and move about through the crystal; that is, this last electron is easily promoted into the conduction band. (b) A trivalent gallium atom can complete only three of the four covalent bonds with the neighboring germanium atoms. A fourth electron is easily pulled free from a nearby bond to complete the fourth covalent bond about the gallium impurity. This incomplete bond (hole) is free to wander about the crystal; that is, an electron is easily promoted from the valence band into a bound state at the impurity, leaving a hole in the valence band.*

When the electron leaves the arsenic impurity, that impurity is ionized positively; when the electron attaches itself to the gallium impurity, that impurity is ionized negatively.

Correspondingly, if a germanium atom is replaced by a trivalent gallium atom, there will be a deficiency of one electron in one of the covalent bonds about the gallium impurity. That bond can be saturated by removing an electron from a nearby site, leaving an unsaturated bond at that location. That is, an electron can be promoted with relative ease from the valence band into a localized state at the gallium impurity, ionizing that impurity negatively and leaving a positive hole in the valence band.

Impurities such as arsenic that contribute an electron to the conduction band are called donor impurities, or simply *donors*. Impurities such as gallium that can take electrons from the valence band are known as acceptor impurities, or simply *acceptors*. Other common donor elements are phosphorus and antimony, whereas aluminum, boron, and indium are alternative acceptor impurities. Forming dilute alloys of a pure semiconductor and a donor or an acceptor element is known as *doping* the host crystal. Since donor impurities produce conduction by *n*egatively charged electrons, one speaks of such semiconductors as being *n-type*. Correspondingly, doping with acceptors results in a *p-type* sample, the current being carried by *p*ositively charged holes.

*31.6 Semiconductor Devices

31.6 (a) Junction Diodes

Suppose a semiconductor crystal, such as germanium or silicon, is prepared so that one half is p-type, the other n-type, as shown schematically in Figure 31.20. In each half, the primary charge carriers are either holes (p-type region) or electrons (n-type region). In the immediate neighborhood of the *junction*, the plane separating the p- and n-type

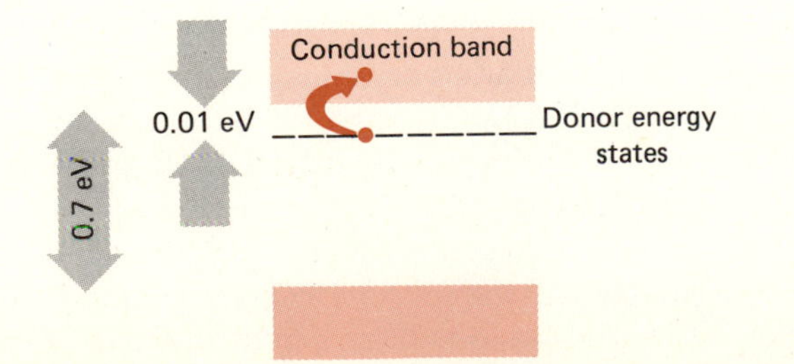

Figure 31.18 *The energy levels in an n-type sample of germanium. A donor energy level appears just below the bottom of the conduction band. Because the energy difference between the donor level and the conduction band is so small, most donor impurities are ionized (donor levels are empty) at all but very low temperatures. At room temperature, the number of electrons is nearly identical to the number of donor impurities (provided that there are no acceptor impurities).*

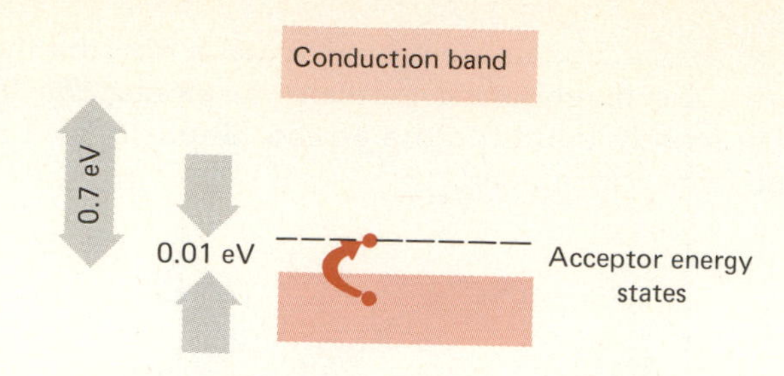

Figure 31.19 The energy levels in a p-type sample of germanium. An acceptor level appears just above the top of the valence band. In this case nearly all acceptors are ionized; that is, acceptor levels are filled at normal temperatures.

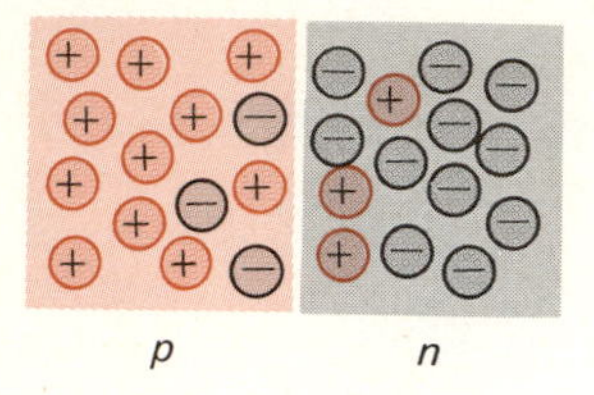

Figure 31.20 The p-n junction diode. In the immediate vicinity of the junction, there are both electrons and holes present as a result of diffusion from n and p regions.

regions, the situation is more complicated. Electrons from the n-type region will diffuse into the p-type region and vice versa for holes, until the electric field established by this layer of positive and negative charges near the junction prevents further diffusive charge flow.

Suppose we then connect a battery and load resistor to this *diode,* as in Figure 31.21(a). With this polarity of the battery, the negatively charged electrons will be pulled to the left across the junction and their number replenished from the large reservoir of electrons in the n-type region. Similarly, holes will flow in the opposite direction across the junction. The result is a net transport of positive charge from left to right across the junction and through the device. The diode is said to be *biased in the forward direction,* and a large forward current, usually limited by the magnitude of the load resistor, will flow in the circuit.

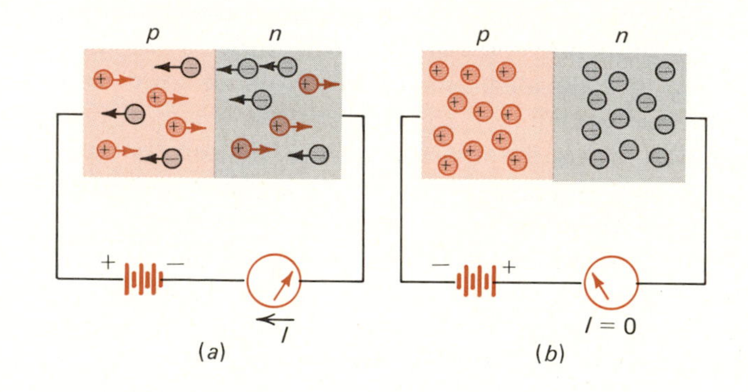

Figure 31.21 (a) A p-n junction diode biased in the forward direction. The field within the diode pulls electrons out of the n-type region into the p-type region, and pulls holes from the p-type to the n-type region. Under these conditions, current flows in the circuit. (b) A p-n junction diode biased in the reverse direction. The electric field inside the diode pulls electrons back into the n-type region, holes back into the p-type region. Thus the junction region is left devoid of charge carriers and acts as a good insulating layer, preventing flow of current in the circuit.

If we reverse the polarity of the battery (see Figure 31.21(b)), the electric field will pull both electrons and holes away from the junction, back toward the n- and p-regions, respectively. The junction region, now largely depleted of mobile charge carriers, acts as a very good insulating layer between the relatively low resistance portions of the doped semiconductor halves. The total current flow under this *reverse-bias* condition is therefore extremely small. The overall current-voltage characteristic of the diode is as shown in Figure 31.22.

31.6 (b) Junction Transistors

An n-p-n (or a p-n-p) *junction transistor* contains three regions, shown schematically in Figure 31.23(a). The central p-type region, called the *base* of the n-p-n transistor, is made very narrow so that electrons from one side, the *emitter,* have a reasonably high probability of drifting across the base to the *collector* on the opposite side. Under these conditions, the number of electrons that do get across the base depends sensitively on the potential difference between base and emitter. Hence, by changing this voltage, one can control the *collector* current. For reasonable values of load resistance, the variation in voltage across the load resistor R_L may be

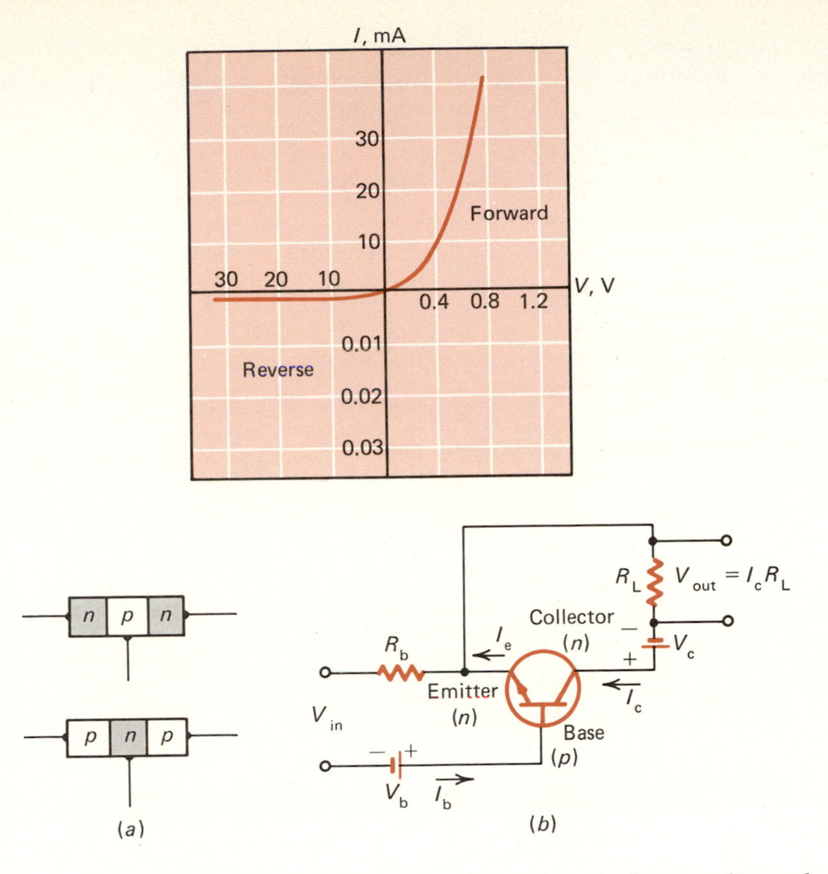

Figure 31.22 The I-V characteristic of a semiconductor p-n junction diode. Note the change in abscissa scales by a factor of 25 between the forward and reverse directions and the change in ordinate scale by a factor of 10^3.

Figure 31.23 (a) Schematic of an n-p-n and a p-n-p transistor. (b) Circuit diagram for a simple amplifier using an n-p-n transistor.

many thousands of times the variation in emitter-to-base voltage that caused the change in collector current. That is, the transistor acts as a high-gain voltage amplifier.

Transistors are not only cheaper, much smaller, and considerably more efficient than vacuum tubes, but also far more rugged and long-lived. There is no reason why a transistor should deteriorate measurably even after decades of constant use. Except in some high-power applications such as radio transmitters, transistors have completely displaced the older vacuum tube.

During the past two decades, development of the *integrated circuit* (IC) has further revolutionized the electronics industry. These minute *IC chips* are complete assemblies of complex circuits; they consist of numerous transistors, diodes, and passive circuit elements such as resistors

Figure 31.24 An integrated circuit (IC) chip. A chip this size can contain hundreds of transistors.

on a single semiconductor crystal, which can perform a great variety of intricate functions. IC chips are at the heart of digital wristwatches, hand-held calculators, and large computers.

31.6 (c) Photosensitive Devices and Light-Emitting Diodes (LED's)

In the preceding chapter we described how an electron in an atom can make a transition from one stationary state to another by absorbing or emitting a photon,

$$h\nu = |E_f - E_i|$$

where E_f and E_i are the energies of the final and initial states. Similarly, in an insulating solid, an electron can make a transition from the valence band to an empty state in the conduction band by absorbing a photon whose energy $h\nu$ matches the difference between the electronic energy levels in the two bands. In such a process, an electron is promoted to the conduction band, and simultaneously, a hole is created in the valence band. The inverse process is also possible; an electron in the conduction band may drop into an empty state in the valence band, releasing a photon of energy $h\nu$. One refers to such an event as *radiative electron-hole recombination.*

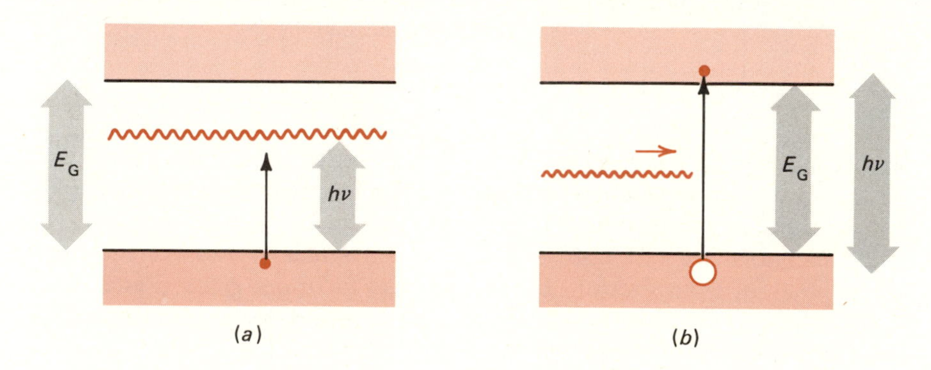

Figure 31.25 (*a*) *A photon of energy less than E_G is not absorbed in a semiconductor, since there are no available energy states to which an electron from the valence band could be excited.* (*b*) *If the photon energy exceeds E_G, optical absorption sets in as electrons are excited from the valence to the conduction band.*

These simple events embrace much interesting physics as well as a great diversity of technological applications. Here we can only touch on a few of them.

Suppose we illuminate a thin slab of pure silicon with monochromatic light, starting at the long-wavelength, low-frequency end of the spectrum. We shall find that this material is almost transparent in the far-infrared region. However, when $h\nu > E_G$, where E_G is the energy gap between the top of the valence and the bottom of the conduction bands, the material becomes opaque; that is, photons whose energy is greater than this energy gap E_G are strongly absorbed. Moreover, if one also measures the conductivity, one observes a very dramatic increase directly associated with the increase in the optical absorptivity.

One can understand these results using the band theory and the Planck-Einstein hypothesis. As long as $h\nu < E_G$, an electron in the full valence band cannot absorb an incident photon, because there is no allowed state at the energy $h\nu$ above the initial state to which this electron can make the transition. However, when $h\nu > E_G$, the transition from valence to conduction band is possible, and incident photons will be absorbed as they cause these transitions. The absorption edge, the thresh-

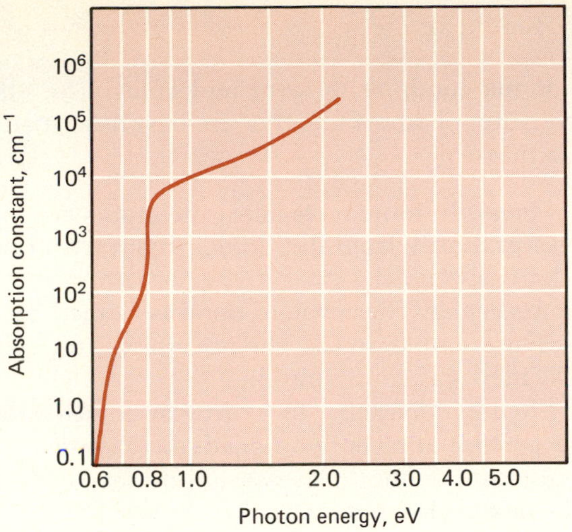

Figure 31.26 The absorption coefficient of germanium as a function of photon energy. Note the logarithmic ordinate scale. An increase in photon energy from 0.65 eV to 0.8 eV increases absorption by a factor of about one thousand.

old for photon absorption, is therefore a direct measure of the energy gap E_G.

The associated pronounced increase in conductivity is now readily understood. When $h\nu > E_G$, electrons and holes are created optically with a concentration that depends on the incident light intensity and is generally orders of magnitude greater than the concentration due to thermal activation. The light-enhanced carrier concentration brings about commensurate increases in conductivity. Photoconductive devices are widely used in photographic light meters, in alarm systems, computer card readers, etc.

The energy gaps of germanium, silicon, and most other semiconductors fall into the near- to far-infrared regions of the spectrum. These materials are often used as infrared detectors.

Lastly, if one could, without optical irradiation, produce an electron-and-hole concentration vastly in excess of that at thermal equilibrium, one should observe the emission of radiation as electrons and holes recombine. This can be achieved by impressing a large forward bias on a junction diode. The concentration of electrons on the p side and of holes on the n side can then be orders of magnitude greater than under equilibrium conditions. Under these circumstances, the diode emits light due to electron-hole recombination. Light-emitting diodes (LED's) find extensive application in hand-held calculators, digital clocks, and other digital displays.

● ***Example 31.2*** The energy gap in silicon is 1.14 eV. What is the threshold wavelength for photoconductivity of silicon?

Silicon will show photoconductive effects when $h\nu > E_G = 1.14$ eV. Since $h = 6.63 \times 10^{-34}$ J·s $= 4.14 \times 10^{-15}$ eV·s, we have as the condition for photoconductivity

$$\nu > \frac{E_G}{h} = \frac{1.14 \text{ eV}}{4.14 \times 10^{-15} \text{ eV}\cdot\text{s}} = 2.75 \times 10^{14} \text{ Hz}$$

The corresponding wavelength condition is

$$\lambda = \frac{c}{\nu} < \frac{3 \times 10^{8} \text{ m/s}}{2.75 \times 10^{14} \text{ Hz}} = 1.09 \times 10^{-6} \text{ m} = 1.09\ \mu\text{m}$$

●

Summary

Individual atoms combine to form molecules. The attractive forces that hold the constituent atoms together are best described using the following classifications.

(a) *Ionic Molecules.* In ionic molecules, the atoms are ionized and the cohesion derives primarily from the Coulomb attraction between opposite electrical charges.

(b) *Covalent Molecules.* The covalent bond is formed by sharing of valence electrons. For example, in Cl_2, the net effect is to give both atoms a nearly full-shell, inert-gas configuration.

(c) *Hydrogen-Bonded Molecules.* In hydrogen bonding, an ionized hydrogen atom (a proton) attracts two negatively charged ions and bonds them. Hydrogen bonding often plays an important role in the formation of organic molecules and polymers.

(d) *Van der Waals Molecules.* The van der Waals force is a very weak attractive force. It acts between inert gas atoms and nonpolar molecules and causes their condensation into the liquid phase.

Molecules have far more complex optical spectra than isolated atoms. Molecular spectra reflect not only the electronic energy levels but also the excitation energies associated with the vibrational and rotational motion of the molecule. Generally, the separation between quantized vibrational levels is one-tenth to one one-hundredth the separation of electronic energy levels; and the separation between adjacent rotational levels is about one one-thousandth of the separation between vibrational levels. For the vibrational levels, the energies are given by

$$E_v(n) = (n + \tfrac{1}{2})h\nu$$

For the rotational levels, the energies are given by

$$E_r(l) = \frac{l(l+1)\hbar^2}{2I}$$

The forces that cause condensation to the solid phase can also be classified as already noted, with the addition of one more force, that associated with the *metallic bond.*

The division of solids into metallic conductors, semiconductors, and insulators can best be understood in terms of the *band theory of solids.* In insulators, the atomic valence electrons form filled bands in the solid, and the energy difference between the top of the valence band and the next-higher unfilled energy band is 3 eV or more. In semiconductors, the valence band is also filled as one approaches $T = 0$ K. However, the energy gap between the valence and the conduction bands is 1.5 eV or less, with the result that with increasing temperature a significant number of electrons are thermally excited from the valence band into the conduction band, leaving *holes* in the valence band. In these materials, electrons (or holes) can also be introduced by adding *donor* (*acceptor*) impurities such as arsenic (gallium). The purposeful addition of donor or acceptor impurities is called *doping.* A semiconductor doped predominantly with donors is called *n-type;* one doped predominantly with acceptors is called *p-type.*

In a metal, the valence band is not full. Consequently, many electrons are relatively free to move throughout the interior of the crystal and can carry electric current.

Multiple Choice Questions

31.1 Which of the following molecules is most probably bound by an ionic bond?

(a) H_2
(b) CCl_4
(c) LiF
(d) Cl_2

31.2 Which of the following materials is most likely to have the lowest boiling point?

(a) argon
(b) water (H_2O)
(c) tin
(d) carbon tetrachloride (CCl_4)

31.3 The optical spectra of polyatomic molecules provide information on

(a) the electronic energy levels only.
(b) the electronic energy levels and also the vibrational energies of the molecule.
(c) the electronic energy levels and also the rotational energies of the molecule.
(d) the rotational and vibrational energies but not the electronic energy levels.
(e) the rotational and vibrational energies as well as the electronic energy levels.

31.4 The quantized energy levels of a rotating object with moment of inertia I are given by

(a) $l\hbar I$
(b) $\sqrt{l(l+1)}\hbar$
(c) $l(l+1)\hbar^2/2I$
(d) $(l+\frac{1}{2})\hbar\nu$
(e) None of the above

31.5 Which of the following are most likely to serve as donor impurities in silicon?

(a) sodium
(b) indium
(c) germanium
(d) antimony
(e) argon

31.6 The equilibrium separation between nuclei of some diatomic molecules is given in the accompanying table.

Molecule	*At. wt.*	*Equil. sep.*
H_2	1	0.075 nm
N_2	14	0.11 nm
O_2	16	0.15 nm
Cl_2	35	0.2 nm
Br_2	80	0.23 nm
I_2	127	0.27 nm

If near the equilibrium separation the forces between two Cl and two I atoms in Cl_2 and I_2 are the same, the spacing between corresponding vibrational levels in I_2 will be about

(a) four times
(b) twice
(c) equal to
(d) half
(e) one-fourth

that in Cl_2.

31.7 Using the same information as given for Problem 31.6, one can conclude that the separation between corresponding rotational energy levels is

(a) greatest for H_2.
(b) greatest for O_2.
(c) greatest for I_2.
(d) about the same for all these molecules.
(e) There is not enough information to make a reasonable estimate.

Problems

(Sections 31.1–31.3)

31.1 At equilibrium, the separation between Li^+ and Cl^- in LiCl is 0.257 nm. What is the electrostatic potential energy of this system?

31.2 Using the data given in Question 31.6, calculate the moments of inertia of the molecules H_2 and N_2.

• **31.3** The equilibrium separation between Rb^+ and Cl^- in RbCl is 0.329 nm. Calculate the electrostatic potential energy of this system and the electrostatic attractive force acting on the two ions. What is the strength of the core repulsive force at this equilibrium separation?

• **31.4** If the vibrational frequency of a KCl molecule is 4×10^{12} Hz, what is the value of the "spring constant" associated with the binding force? (Imagine the diatomic molecule as made up of two nearly equal masses held together by a spring.)

• **31.5** The equilibrium separation between K^+ and Cl^- in KCl is 0.315 nm. Calculate the electrostatic potential energy of the system and the electrostatic attractive Coulomb force at equilibrium separation.

• **31.6** The ionization energies of Li, K, and Rb are 5.39 eV, 4.34 eV, and 4.18 eV. The electron affinities of F and Cl are 3.5 eV and 3.7 eV. For the three chloride salts, the equilibrium separations have been given above; those of LiF, KF, and RbF are 0.201 nm, 0.267 nm, and 0.282 nm, respectively. Calculate approximate values for the binding energies of (a) LiF, (b) LiCl, (c) KF, (d) KCl, (e) RbF, and (f) RbCl.

• **31.7** Estimate the energy difference between the $l = 4$ and $l = 3$ rotational energy levels in O_2 and in I_2. (See the Table of Question 31.6.)

• **31.8** Sketch the rotational energy-level structure for N_2 and calculate the approximate energies of the first three rotational levels of this molecule.

• **31.9** Use the results of Problem 31.2 to calculate the energies of the three lowest-lying rotational energy levels of H_2 and of N_2. Also calculate the wavelengths for the $\Delta l = 1$ transitions involving these levels.

• **31.10** The vibrational frequency of the molecule H_2 is 1.3×10^{14} Hz. Determine the energy difference between adjacent vibrational levels and compare this energy difference with that calculated for the lower rotational levels of H_2 in Problem 31.9.

• **31.11** From the information provided in Problem 31.10, calculate the "spring constant" of the restoring force that acts between the two atoms of the H_2 molecule. Sketch a curve of the potential energy versus interatomic spacing for the molecule, for spacings close to the equilibrium separation of 0.075 nm.

• • **31.12** Prove that the energy spacing between adjacent rotational levels with quantum numbers l and $l - 1$ is proportional to l.

(Sections 31.4–31.6)

31.13 The energy gap between the valence and conduction bands in germanium is 0.67 eV. What is the threshold wavelength for optical absorption in germanium?

31.14 The threshold for optical absorption for a certain semiconductor is found to be 1.24 μm. What is the energy gap in this semiconductor?

• **31.15** From Figure 31.26, estimate the energy gap in germanium. What is the wavelength of the incident radiation for which the absorption coefficient is about 10^4 cm^{-1}?

• **31.16** What is the longest wavelength of the recombination radiation emitted by a forward-biased silicon *p-n* junction?

•32•

Nuclear Physics and Elementary Particles

Neutrinos, they are very small.
They have no charge and have no mass
And do not interact at all.
The earth is just a silly ball
To them, through which they simply pass,
Like dustmaids down a drafty hall
Or photons through a sheet of glass.
They snub the most exquisite gas,
Ignore the most substantial wall,
Cold-shoulder steel and sounding brass,
Insult the stallion in his stall,
And, scorning barriers of class,
Infiltrate you and me! Like tall
And painless guillotines they fall
Down through our heads into the grass.
At night they enter at Nepal
And pierce the lover and his lass
From underneath the bed—you call
It wonderful; I call it crass.

JOHN UPDIKE, *"Cosmic Gall"*

32.1 Introduction

The early years of the twentieth century not only saw the vindication of the atomic hypothesis through the work of Einstein, Boltzmann, and others, the first suggestion for a quantum theory of radiation by Planck, and Einstein's special theory of relativity; these years also marked the beginning of nuclear physics as a new and identifiable field of scientific research.

Radioactivity was first observed by Henri Becquerel in 1896.* The following year, J. J. Thomson announced the discovery of the electron, and

* The biography of Becquerel provides another example of science as a family tradition. Henri's father and grandfather were both respected scientists, the latter the recipient of the coveted Royal Society's Copley medal in 1837 for work on electrochemistry. Henri's son, Jean, made valuable contributions to physics until his death in 1953.

Becquerel soon showed that the radiations from his uranium sulfate crystals were fast-moving electrons. Thomson demonstrated that whatever the gas in his discharge tube, the particles that constituted the cathode rays were always electrons. Thus, within a few years, the notion of the atom as the ultimate indivisible unit of matter had to be discarded. Evidently, atoms included electrons among their constituents, and these electrons could be set free by electromagnetic excitation as in a gas discharge, or spontaneously as in radioactivity.

In Chapter 30 we surveyed the early development of atomic physics, culminating with the work of Bohr, Schroedinger, Heisenberg, and Pauli. Their theoretical insight was stimulated by the work of Rutherford and his collaborators, who had employed α particles emitted in radioactive decay as projectiles in scattering experiments to probe the structure of the atom. Rutherford, however, did not confine himself to using α particles as subatomic probes; he was also deeply concerned with understanding the phenomenon of radioactivity itself.

In 1902, while Rutherford was at McGill University, he and an ingenious young chemist, Frederick Soddy, showed that in radioactive decay an atom changed to a different chemical element. The following year, Rutherford, measuring the charge-to-mass ratio of α particles, inferred from the results that these particles were helium nuclei. This idea was convincingly proved by Rutherford and Royds in 1909, when they detected helium gas in a sealed glass vessel, previously free of that inert gas, following the insertion of some uranium salt. Ten years later, Rutherford reported the first artificially induced nuclear transmutation; this was the reaction in which nitrogen, upon bombardment by α particles emits a proton and is transmuted to an isotope of oxygen. Rutherford correctly concluded that "the hydrogen atom which is liberated forms a constituent part of the nitrogen nucleus." Previously, it had already been established that the nucleus contained α particles. Thus, the 1920 nuclear model was that of α particles and protons, sufficient to account for the total mass, and as many electrons inside the nucleus as might be required to neutralize the excess positive charge and give the correct charge-to-mass ratio. Yet this model soon ran into serious difficulties and had to be abandoned within a decade.

Rutherford's 1911 paper already included a correct estimate of nuclear size, a sphere having a radius about 10^{-14} m. According to the Heisenberg uncertainty relation, Equation (29.9), an electron confined to a region of 10^{-14} m would have a momentum uncertainty $\Delta p \geq \hbar/\Delta x = 10^{-20}$ kg·m/s and consequently, an average kinetic energy of about 10 MeV. Though possible, this seemed an unreasonably large energy, especially in light of the much smaller energies of electrons emitted in β decay, as the process first observed by Becquerel came to be known.

The other, more devastating argument against the model came from high-resolution atomic spectroscopy, which revealed that nuclei have an intrinsic spin and associated magnetic moment whose value depends on the nuclear species. The intrinsic spin of helium is zero; the proton, like the electron, has intrinsic spin $\frac{1}{2}$. It follows that the spin of a nucleus must be an integer, 0, 1, 2, . . . if the number of protons plus electrons is even, and must be $\frac{1}{2}, \frac{3}{2}, \frac{5}{2}$, . . . if that number is odd. Yet the nitrogen nucleus of atomic weight 14 and charge $7e$, for which the number of protons plus electrons must be odd, had a nuclear spin of 1.

These difficulties were resolved in 1932, when James Chadwick discovered the neutron. The nuclear model that emerged, namely a very small, roughly spherical, and enormously dense object, consisting of protons and neutrons, has survived to this day.

32.2 Nuclear Structure and Nuclear Forces

32.2 (a) Composition

Nuclei are composed of protons and neutrons. The *proton* is the nucleus of the hydrogen atom. Its mass is

$$m_p = 1.6726 \times 10^{-27} \text{ kg};$$

it carries a positive charge

$$e = 1.6 \times 10^{-19} \text{ C}$$

and its intrinsic spin quantum number is $\frac{1}{2}$.

The *neutron* has a mass slightly greater than that of the proton,

$$m_n = 1.6750 \times 10^{-27} \text{ kg}$$

It carries no charge; and its spin quantum number is also $\frac{1}{2}$. The term *nucleon* is used as a generic name for either proton or neutron.

The *atomic number* Z of an atom is the number of atomic electrons in the neutral atom; it is also *the number of protons in the nucleus*. The *atomic mass number A equals the number of nucleons in the nucleus*. The number of neutrons, called the *neutron number* N, is therefore

$$N = A - Z \tag{32.1}$$

A nucleus is completely specified by any two of those three numbers. It is conventional practice to identify a nucleus by the symbol

$${}^{A}_{Z}\text{X}$$

where X is the standard chemical abbreviation for the element. For example, a rare but natural isotope of carbon is ${}^{13}_{6}\text{C}$. This nucleus has 6 protons and $13 - 6 = 7$ neutrons. Evidently, the notation is redundant, since the identification of the chemical element fixes the number of protons. For that reason the atomic number, the subscript Z, is always omitted in oral (one speaks of "carbon-13" and not "carbon 6, 13") and sometimes also in written communication.

As Soddy first discovered, a given element (given Z) frequently has two or more stable nuclei, each containing different numbers of neutrons and therefore having different atomic masses. Nuclei with the same number of protons but different atomic mass numbers are called *isotopes* of the chemical element.*

32.2 (b) Mass

Since the masses of a proton and a neutron are nearly identical whereas the mass of an electron is much less ($m_e = m_p/1836$), the atomic mass of an isotopically pure element should be nearly an integral mul-

* *Isotope* is derived from the Greek: *iso*, "same," and *topos*, "place." It refers to nuclei occupying the same place in the periodic table.

tiple of the mass of hydrogen. If one set out to establish a new mass unit that might be convenient for atomic and nuclear calculations (the kilogram is a large and cumbersome unit in this field), it would seem reasonable to assign unit mass to the proton. However, a slightly different standard has been adopted internationally. The *atomic mass unit,* abbreviated "u" is defined as one-twelfth the mass of a carbon atom of the isotope $^{12}_{6}C$. Note that the atomic mass unit so defined includes the masses of the six atomic electrons.

Conversion factors and mass-energy equivalents are listed in Table 32.1. The last column gives the energy equivalent of the mass. This is the amount of energy that would be released if the entire rest mass were converted into energy according to the Einstein relation, $E = mc^2$.

Table 32.1 *Nucleon and electron masses and energy equivalents*

Particle	*Mass, u*	*Mass, kg*	*Energy equivalent, MeV*
	1	1.6606×10^{-27}	931.50
Proton	1.007276	1.6726×10^{-27}	938.28
Neutron	1.008665	1.6750×10^{-27}	939.57
Electron	5.486×10^{-4}	9.1095×10^{-31}	0.511

32.2 (c) Size

There are several ways of defining nuclear size. For example, experiments with very energetic electrons have shown that the charge density in the interior of a nucleus is constant but falls off gradually at the periphery (Figure 32.1). One definition of the average nuclear radius is the value at which the nuclear charge density has dropped to half the density in the interior. Another, earlier definition of nuclear size is based on scattering experiments of the sort Rutherford first performed. The nuclear radius is then defined as that distance at which the interaction between the target nucleus and an approaching proton departs significantly from the classical repulsive Coulomb force. Whatever definition is used, the results are substantially the same, namely,

$$R = R_0 A^{1/3} \tag{32.2}$$

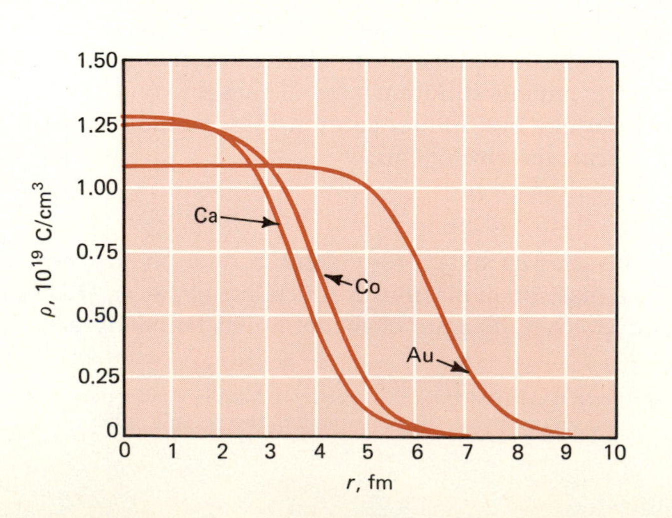

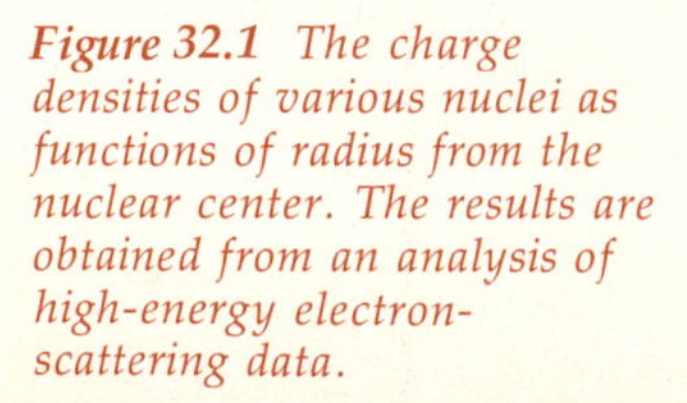

Figure 32.1 *The charge densities of various nuclei as functions of radius from the nuclear center. The results are obtained from an analysis of high-energy electron-scattering data.*

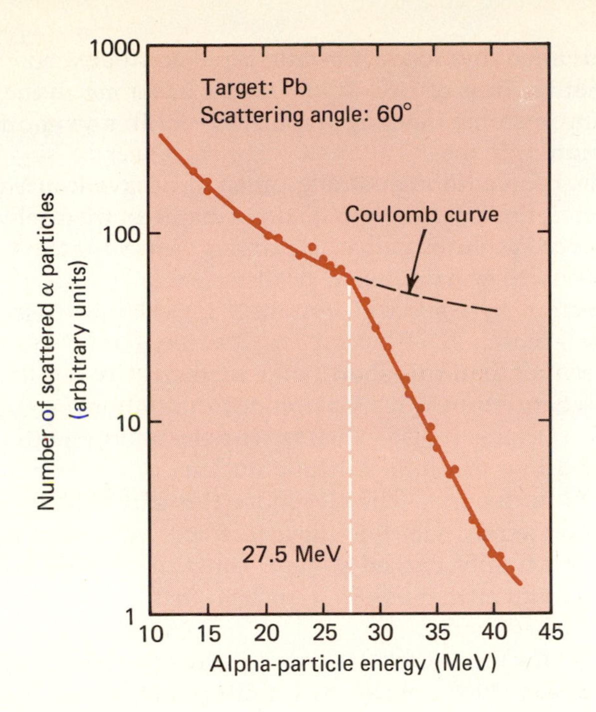

Figure 32.2 Elastic scattering of α particles by lead nuclei as a function of particle energy. For energies less than 27.5 MeV, the number scattered is in excellent agreement with the theoretical result based on Coulomb interaction between two charged particles. When the energy of the incident α particles exceeds 27.5 MeV they come within the range of the nuclear force of the lead nuclei and the experimental results then depart dramatically from the Coulomb curve.

where

$$R_0 \simeq 1.2 \times 10^{-15} \text{ m} = 1.2 \text{ fm} \tag{32.2a}$$

Since the volume of a sphere is proportional to the cube of its radius, it follows from Equation (32.2) that the nuclear volume is proportional to the number of nucleons. Equation (32.2) is therefore consistent with the picture of a nucleus as a collection of incompressible nucleons packed into a spherical volume, like marbles stuffed into a rubber balloon, except that the nucleon "marbles" are not at rest but move about within this volume with much agitation.

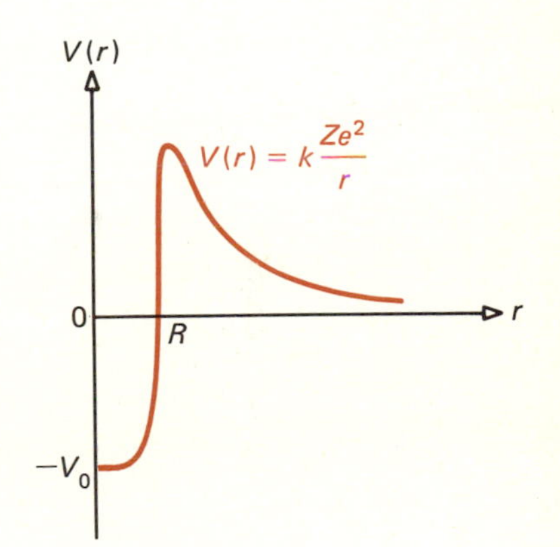

Figure 32.3 The PE of a proton as a function of distance from the center of another nucleus. For $r > R$, the potential energy is due to the Coulomb repulsion between the proton and the nucleus of charge Ze. Near the nuclear radius, the sharp drop in PE drops indicates that the strong nuclear force comes into play.

32.2 (d) The Nuclear Force

The central problem, which even today has not yet been satisfactorily solved, concerns the forces that hold the nucleons together. We do know many details and can state the general features for which the evidence is presented:

1. The nuclear force is attractive and is enormous compared with electrostatic or gravitational forces.

2. The nuclear force has a very short range and drops to zero very rapidly if the critical range is exceeded.

3. The nuclear force between two nucleons is charge-independent; that is, the same nuclear force acts between two protons, two neutrons, and a proton and neutron.

One can appreciate the fantastic strength of the nuclear force by calculating the electrostatic repulsion between two protons in a nucleus. The separation between the two positively charged particles is about $2R_0 = 2.4 \times 10^{-15}$ m. The mutual Coulomb repulsion is therefore

$$F = k\frac{q^2}{r^2} = \frac{(9 \times 10^9 \text{ N}\cdot\text{m}^2/\text{C}^2)(1.6 \times 10^{-19} \text{ C})^2}{(2.4 \times 10^{-15} \text{ m})^2} = 40 \text{ N}$$

Thus, the force due to electrostatic repulsion between the two protons, each with a mass of 1.67×10^{-27} kg, is as strong as the weight of a 4-kg mass! Overcoming that huge repulsion requires an enormous attractive interaction.

The nuclear force, though strong, must also have limited range. If it were otherwise, these forces would surely manifest themselves at macroscopic distances. Yet unless a particle comes very close to a nucleus, that nucleus appears to behave like a positive point charge, and the long-range interaction is correctly described by ordinary electrodynamic forces.

One consequence of this short range is referred to as *saturation of the nuclear force*. Saturation is responsible for the uniform charge distribution within a nucleus; it also explains why the energy with which a nucleon is bound in a medium-to-heavy nucleus is nearly independent of the number of nucleons A. The reason for this saturation is that a given nucleon can interact, via the nuclear force, only with its nearest neighbors. In a tightly packed sphere, containing more than about 30 particles, the number that adjoin a nucleon in the interior does not depend on the total number.

By contrast, the electrostatic energy required to add one more positive charge to a spherical volume already containing Z protons is proportional to Z. The reason for this difference in behavior is that the long range of the Coulomb force allows the newly added charge to interact with all the charges already present, not merely its nearest neighbors. This difference between nuclear and electrostatic interactions bears directly on nuclear stability.

The third property of the strong nuclear force, its charge independence, is indicated by certain systematics of nuclear binding energies. For instance, the difference in binding energy between ${}^{17}_{8}O$ ($Z = 8$, $N = 9$) and ${}^{17}_{9}F$ ($Z = 9$, $N = 8$) is just the calculated difference in electrostatic energy; the same is true for other pairs of so-called mirror nuclei. The most direct evidence, however, comes from analysis of neutron-proton (n-p) and proton-proton (p-p) scattering experiments. Once the effect of the additional Coulomb force operating in p-p scattering is subtracted, one finds that the nuclear interaction between two protons and between a proton and neutron are the same.

32.3 Stability of Nuclei

32.3 (a) Binding Energy

Since the strong nuclear force binds nucleons together, work must be done to separate them. Conversely, energy is released when nucleons join to form a stable nucleus. This energy difference between a nucleus and its separate constituent nucleons is called the *binding energy* of the nucleus.

It follows from the special theory of relativity that if the nucleons, when bound, have less energy than when they are free, the total mass of the nucleus must also be less than the sum of the masses of the isolated protons and neutrons that combine to form the nucleus. That is,

$${}^{A}_{Z}Mc^2 < Zm_pc^2 + Nm_nc^2$$

and therefore,

$${}^A_ZM < Zm_p + Nm_n$$

where ${}^A_ZM$ is the mass of the nucleus of atomic mass number $A = Z + N$.

By definition,

$$\text{BE} = (Zm_p + Nm_n - {}^A_ZM)c^2 \quad \textbf{(32.3)}$$

where BE is the binding energy of the nucleus of mass ${}^A_ZM$. The quantity within the parenthesis in Equation (32.3) is called the *mass defect*, and it is a measure of the binding energy. As implied by Equation (32.3), nuclear binding energies are determined by measuring nuclear masses with a mass spectrometer (see Problem 22.8).

● ***Example 32.1*** The atomic mass of ${}^{23}_{11}\text{Na}$ is 22.98977 u. What is the total BE of this nucleus and what is the average BE per nucleon?

The number of protons and neutrons in this nucleus are $Z = 11$ and $N = 23 - 11 = 12$. Since the atomic mass includes the mass of 11 electrons, we must use the mass of a hydrogen atom rather than the bare proton mass in writing the energy balance equation. The mass defect for this nucleus is therefore

$$\begin{aligned} 11m_H + 12m_n - {}^{23}_{11}M &= 11(1.007825 \text{ u}) + 12(1.008665 \text{ u}) - 22.98977 \text{ u} \\ &= 0.200285 \text{ u} \end{aligned}$$

and the corresponding total BE and the BE per nucleon are

$$\text{BE} = (0.200285 \text{ u})(931.5 \text{ MeV/u}) = 186.6 \text{ MeV, and}$$

$$\frac{\text{BE}}{\text{A}} = \frac{186.6 \text{ MeV}}{23} = 8.11 \text{ MeV} \quad ●$$

The *average binding energy per nucleon* is just the total binding energy divided by the number of nucleons, that is, by the atomic mass number A. For ${}^{23}_{11}\text{Na}$, the BE per nucleon is 8.11 MeV. The average binding energy per nucleon as a function of A is shown in Figure 32.4. The average BE increases rapidly though irregularly at first until $A \simeq 16$; then it rises more slowly to about 8.7 MeV per nucleon near $A = 60$, and thereafter decreases gradually to about 7.5 MeV per nucleon for the heaviest natu-

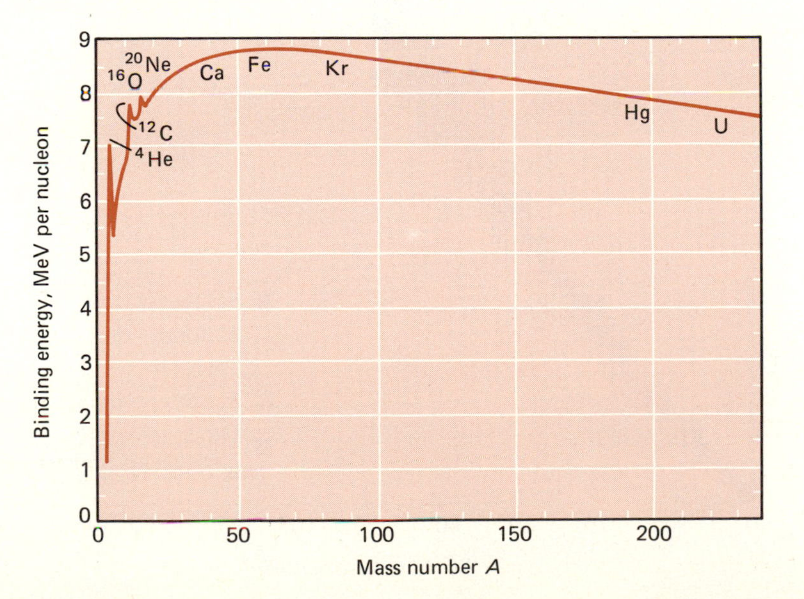

Figure 32.4 The average binding energy per nucleon as a function of atomic mass number.

rally occurring elements. The near constancy of the BE per nucleon over a wide range of A reflects the saturation of the nuclear force referred to earlier.

32.3 (b) Systematics of Stability

The lighter stable nuclei, for example, ${}^{4}_{2}\mathrm{H}$, ${}^{6}_{3}\mathrm{Li}$, ${}^{7}_{3}\mathrm{Li}$, ${}^{9}_{4}\mathrm{Be}$, ${}^{20}_{10}\mathrm{Ne}$, and ${}^{32}_{16}\mathrm{S}$, contain almost equal numbers of protons and neutrons; that is, $N \simeq Z \simeq A/2$. However, as A increases, the neutron number increases more rapidly than the proton number, as is shown in Figure 32.5. This pattern is understandable in light of the balance between the short-range attractive nuclear force and the long-range repulsive Coulomb force

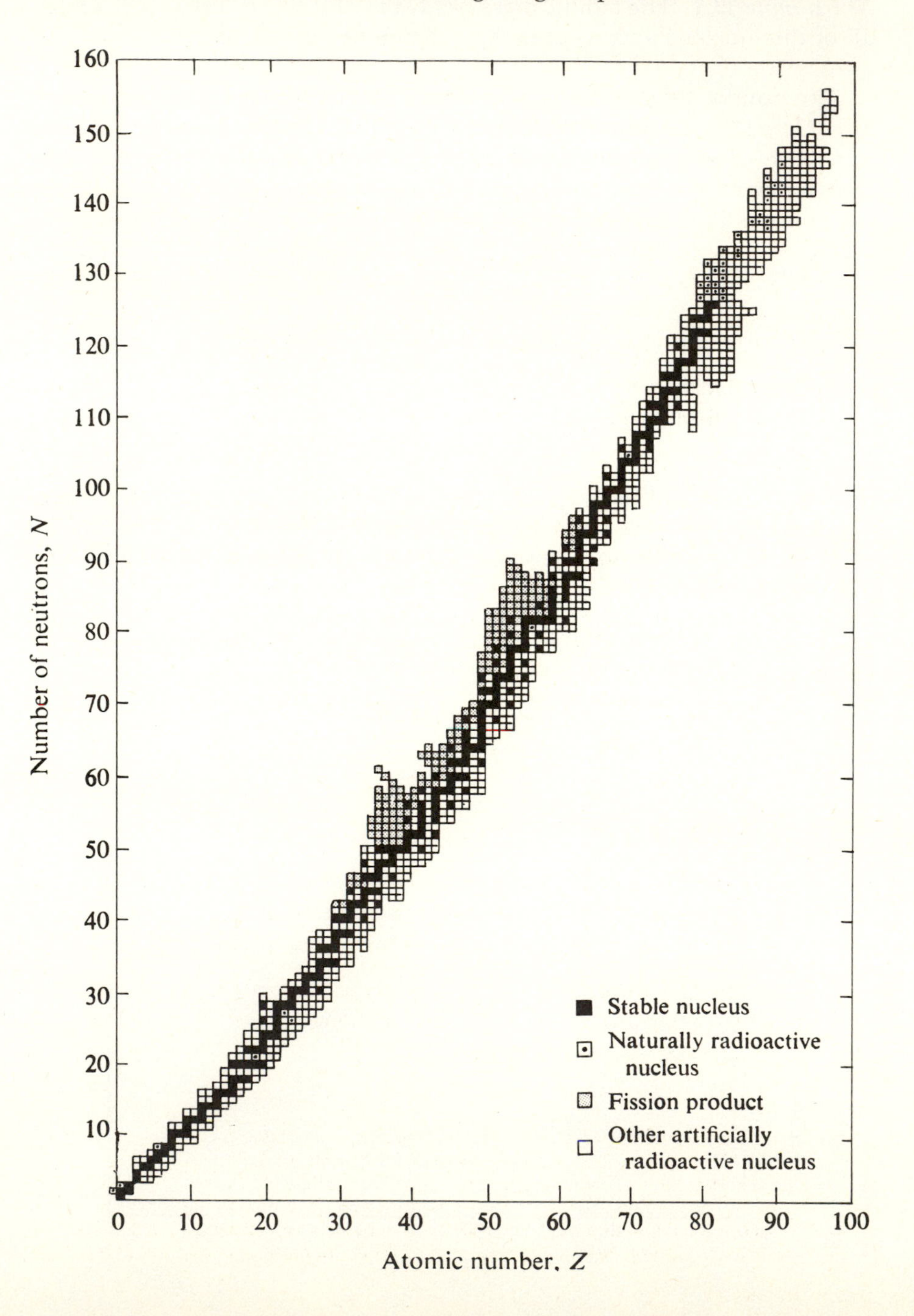

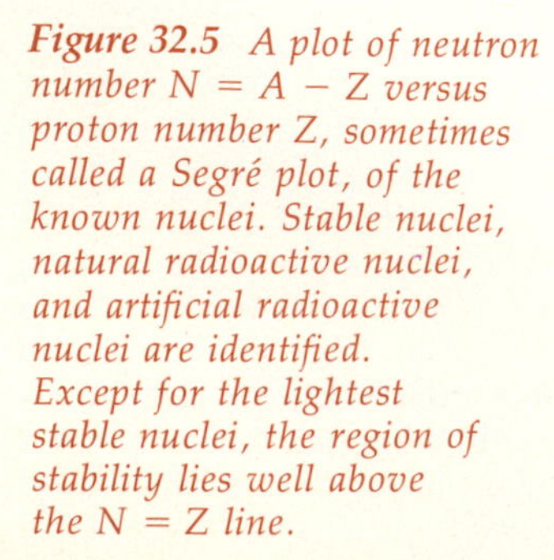
Figure 32.5 *A plot of neutron number $N = A - Z$ versus proton number Z, sometimes called a Segré plot, of the known nuclei. Stable nuclei, natural radioactive nuclei, and artificial radioactive nuclei are identified. Except for the lightest stable nuclei, the region of stability lies well above the $N = Z$ line.*

between protons. With increasing Z, the Coulomb force increasingly favors the accretion of neutrons instead of protons.

Ultimately, the Coulomb repulsion becomes so strong that a net reduction in energy can be achieved by the emission of the very stable α particle—with the simultaneous transmutation of the *parent nucleus,* with atomic number Z_P and atomic mass number A_P, to the *daughter nucleus,* with atomic number $Z_D = Z_P - 2$ and atomic mass number $A_D = A_P - 4$. The heaviest stable element is bismuth, $^{209}_{83}Bi$. Of the heavy elements with atomic number greater than 83, only thorium ($Z = 90$) and uranium ($Z = 92$) are found in nature. Those are the only elements with isotopes whose radioactive decay is so slow that since the formation of the solar system only a modest fraction has been transmuted by α decay.

32.3 (c) Radioactivity

The first observation of radioactivity by Becquerel was β decay, the emission of an electron in a chain of events that begins with the α decay of $^{238}_{92}U$. Some years later, a third type of nuclear radiation was detected, so-called γ rays, which are high-energy photons.

The basic rules for radioactive decay, and indeed for all nuclear reactions, are simple. All are based on conservation laws. Except for the first, they have already been considered earlier in the text.

1. The nucleon number A must be conserved.
2. Electric charge must be conserved.
3. Energy must be conserved.
4. Momentum must be conserved.
5. Angular momentum (including spin angular momentum) must be conserved.

The first rule is a generalization of a great many observations; the number of nucleons has not been known to increase or decrease.*

α Decay

To illustrate these simple rules, we consider the isotope of thorium, $^{227}_{90}Th$, which decays to the radium isotope $^{223}_{88}Ra$ with the emission of an α particle. In standard notation, this decay is written

$$^{227}_{90}\text{Th} \rightarrow {}^{223}_{88}\text{Ra} + {}^{4}_{2}\text{He}$$

or

$$^{227}_{90}\text{Th} \rightarrow {}^{223}_{88}\text{Ra} + \alpha$$

The mass of $^{227}_{90}Th$ is 227.027706 u, the mass of $^{223}_{88}Ra$ is 223.018501 u, and that of $^{4}_{2}He$ is 4.002603 u. The measured kinetic energy of the emitted α particle is 6.04 MeV. Let us see how these facts fit the rules.

1. Adding the number of nucleons (superscripts) on both sides of the reaction equation, we see that no nucleons have been created or destroyed; that is, $227 = 223 + 4$.

* As in electron-positron creation, if enough energy is available a nucleon-antinucleon pair can be created. The nucleon conservation condition is still satisfied; the antiparticle is always counted as a negative particle, so that the total number of nucleons remains constant. An antinucleon does not last long in this world. It soon interacts with a nucleon, and the two objects annihilate each other with the release of $2m_Nc^2$ of energy, where m_N is the mass of the nucleon.

2. The total nuclear charge is $90e$ before and after the decay; that is, $90 = 88 + 2$.

3. The difference in mass before and after the reaction is

$$227.027706\ \text{u} - 223.018501\ \text{u} - 4.002603\ \text{u} = 0.006602\ \text{u}$$

which corresponds to a difference in energy of

$$(0.006602\ \text{u})(931.5\ \text{MeV/u}) = 6.15\ \text{MeV}$$

This excess energy of 6.15 MeV should now appear as kinetic energy (KE) of the decay products. The α particle is observed to come away with an energy of 6.04 MeV, which is close to the calculated value but not close enough. The difference, though small, is well outside experimental uncertainty. The apparent discrepancy is resolved by applying the fourth condition, momentum conservation.

4. Since the initial momentum of the system was zero, the final momentum must also be zero. Therefore, the α particle's momentum must be compensated for by an equal but opposite momentum of the Ra nucleus; that is, the radium nucleus recoils just as a gun recoils on firing a high-speed bullet. Consequently, in the decay the daughter nucleus also acquires some KE. This recoil energy is readily determined.

Momentum conservation requires that

$$M_D v_D = M_\alpha v_\alpha$$

where D stands for daughter.

Squaring both sides and dividing by $2M_D$, we obtain

$$\text{KE(D)} = \frac{1}{2} M_D v_D^2 = \frac{1}{2}\left(\frac{M_\alpha}{M_D}\right) M_\alpha v_\alpha^2 = \left(\frac{M_\alpha}{M_D}\right) \text{KE}(\alpha)$$

In the present instance, the ratio $M_\alpha/M_D = 4/223$ and $\text{KE}(\alpha) = 6.04$ MeV. Hence, the KE of recoil of the daughter nucleus is

$$\text{KE(Ra)} = 6.04(\tfrac{4}{223}) = 0.11\ \text{MeV}$$

which just equals the difference between ΔMc^2 of the reaction and the KE of the α particles.

β Decay

The term *β decay* is used to denote any of three kinds of nuclear transmutations, involving an electron or a positron, in which the atomic numbers Z of parent and daughter nucleus differ by ± 1 and the atomic mass numbers remain unchanged. The most common is β^- decay, in which a neutron within the nucleus decays to a proton, emitting an electron. The process is represented by the scheme

$${}^{A}_{Z}\text{P} \rightarrow {}_{Z+1}^{\ \ A}\text{D} + {}_{-1}^{\ 0}\text{e} \quad \text{or} \quad {}^{A}_{Z}\text{P} \rightarrow {}_{Z+1}^{\ \ A}\text{D} + \beta^-$$

A particularly simple example of β^--decay is that of the free neutron,

$${}^{1}_{0}\text{n} \rightarrow {}^{1}_{1}\text{p} + {}_{-1}^{\ 0}\text{e}$$

The process is energetically possible, since the rest energy of the neutron, 939.57 MeV, is greater than the combined rest energies of proton and electron, 938.28 MeV + 0.511 MeV = 938.79 MeV, by 0.78 MeV.

In positron emission, also called β^+ decay, a proton in the nucleus is transformed to a neutron. The process is represented by

$${}^{A}_{Z}\text{P} \rightarrow {}_{Z-1}^{\ \ A}\text{D} + {}_{+1}^{\ 0}\text{e}(\beta^+) \text{ or } {}^{A}_{Z}\text{P} \rightarrow {}_{Z-1}^{\ \ A}\text{D} + \beta^+$$

Lastly, a change in Z without a change in A can also be accomplished by *electron capture*. In this process, an atomic electron—almost invariably the $n = 1$ (K-shell) electron—is captured by the nucleus. This process is therefore referred to as K-capture. Electron capture has the same net effect as positron emission; a proton in the nucleus is converted to a neutron.

One of the natural β^- emitters is an isotope of carbon, ${}^{14}_{6}\mathrm{C}$. This nucleus decays to ${}^{14}_{7}\mathrm{N}$ with the emission of an electron; that is,

$${}^{14}_{6}\mathrm{C} \rightarrow {}^{14}_{7}\mathrm{N} + {}^{0}_{-1}\mathrm{e}$$

Note that as required, charge and nucleon number are conserved. The atomic masses of ${}^{14}_{6}\mathrm{C}$ and ${}^{14}_{7}\mathrm{N}$ are 14.003242 u and 14.003074 u, respectively. It follows that the energy of the emitted electron should be*

$$\mathrm{KE} = (14.003242\ \mathrm{u} - 14.003074\ \mathrm{u})(931.5\ \mathrm{MeV/u})$$

$$= 0.156\ \mathrm{MeV} = 156\ \mathrm{keV}.\dagger$$

Now a curious thing was discovered when the energy of the emitted electrons was measured. Instead of a single value of 156 keV, the β^- energy covered a broad range, all the way from nearly zero energy to a maximum of 156 keV. Similar broad energy spectra were reported whenever energies of β particles were measured. Yet the rest energies of parent and daughter nuclei are well defined. It seemed as though in β decay, energy conservation was no longer valid!

There was another, equally perplexing problem related to conservation of angular momentum. Since a proton, a neutron, and an electron have spin quantum number $\frac{1}{2}$, the total spin angular momentum in the β^- decay of the neutron must change by $\frac{1}{2}\hbar$. Thus angular momentum conservation also appeared to be violated in this decay.

To rescue these fundamental conservation laws, Pauli proposed that in β decay another, as yet undetected particle was created. That particle had to be neutral to avoid violating charge conservation, had to have spin quantum number $\frac{1}{2}$, and had to have negligible rest mass, since a few electrons did emerge with just about the full anticipated KE. Most of the decays, however, involved a sharing of the available KE between the electron and this new particle, named the *neutrino* (little neutral one) by Enrico Fermi.

The symbol for the neutrino is the lowercase Greek letter nu, ν. The correct representation of the decay of ${}^{14}_{6}\mathrm{C}$ is therefore

$${}^{14}_{6}\mathrm{C} \rightarrow {}^{14}_{7}\mathrm{N} + {}^{0}_{-1}\mathrm{e} + \bar{\nu}$$

The bar over the ν is used to indicate that the particle is an antineutrino, a detail that need not concern us. (Just as there are antielectrons, namely positrons, there are antiparticles of other particles, for example, antiprotons, antineutrons, etc.)

Because the neutrino is massless, carries no charge, and does not interact with nucleons through the strong nuclear force, its detection is

* In calculating the energy balance in β decay, it is important to keep track of the electrons and remember that masses listed in the tables are *atomic masses*, that is, the masses of the respective neutral atoms, *including the Z electrons*. Thus the atomic mass of ${}^{14}_{6}\mathrm{C}$ includes the mass of six electrons, the atomic mass of ${}^{14}_{7}\mathrm{N}$ includes the mass of seven electrons. Hence we need not, indeed we must not, add another electron mass when writing the mass balance equation for this reaction.

† In β decay one can generally neglect the recoil energy of the daughter nucleus (see Problem 32.11).

extremely difficult. A beam of neutrinos would pass through the entire earth with only very small attenuation. It is therefore not surprising that decades passed before direct experimental evidence of neutrinos was obtained.

γ Decay

When $^{227}_{90}$Th decays by α emission, it is found that well over a dozen groups of α particles with different, sharply defined energies are released. At the same time, groups of energetic photons, called γ rays, are also emitted. Similarly, β decay is frequently accompanied by the emission of one or more γ rays.

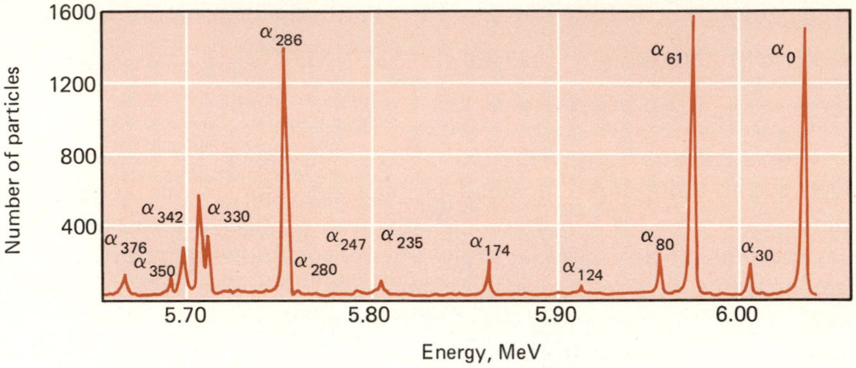

Figure 32.6 *The energy spectrum of α particles emitted from the ground state of $^{227}_{90}$Th. The different groups of α particles correspond to different energy levels of the daughter nucleus $^{223}_{88}$Ra as shown schematically in Figure 32.7.*

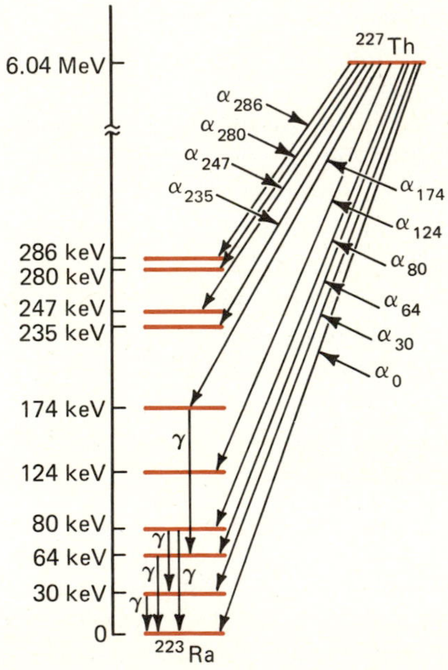

Figure 32.7 *The energy level diagram of $^{223}_{88}$Ra, showing the various α particles emitted by $^{227}_{90}$Th and the γ rays emitted by the daughter nucleus.*

These observations suggest that a nucleus, like an atom, can exist in excited states as well as in its ground state. Following α or β decay, the daughter nucleus may be left in one of these excited states. If that happens, the daughter nucleus usually decays quickly, in about 10^{-21} s, to the ground state, releasing electromagnetic radiation. In a few instances, the daughter nucleus may be left in a metastable state with a relatively long half-life (see Section 32.4 on half-life).

32.3 (d) Radioactive Decay Series

Earlier, we considered the α decay of $^{227}_{90}$Th. This radioactive nucleus is itself the product of a series of events that commenced with the decay

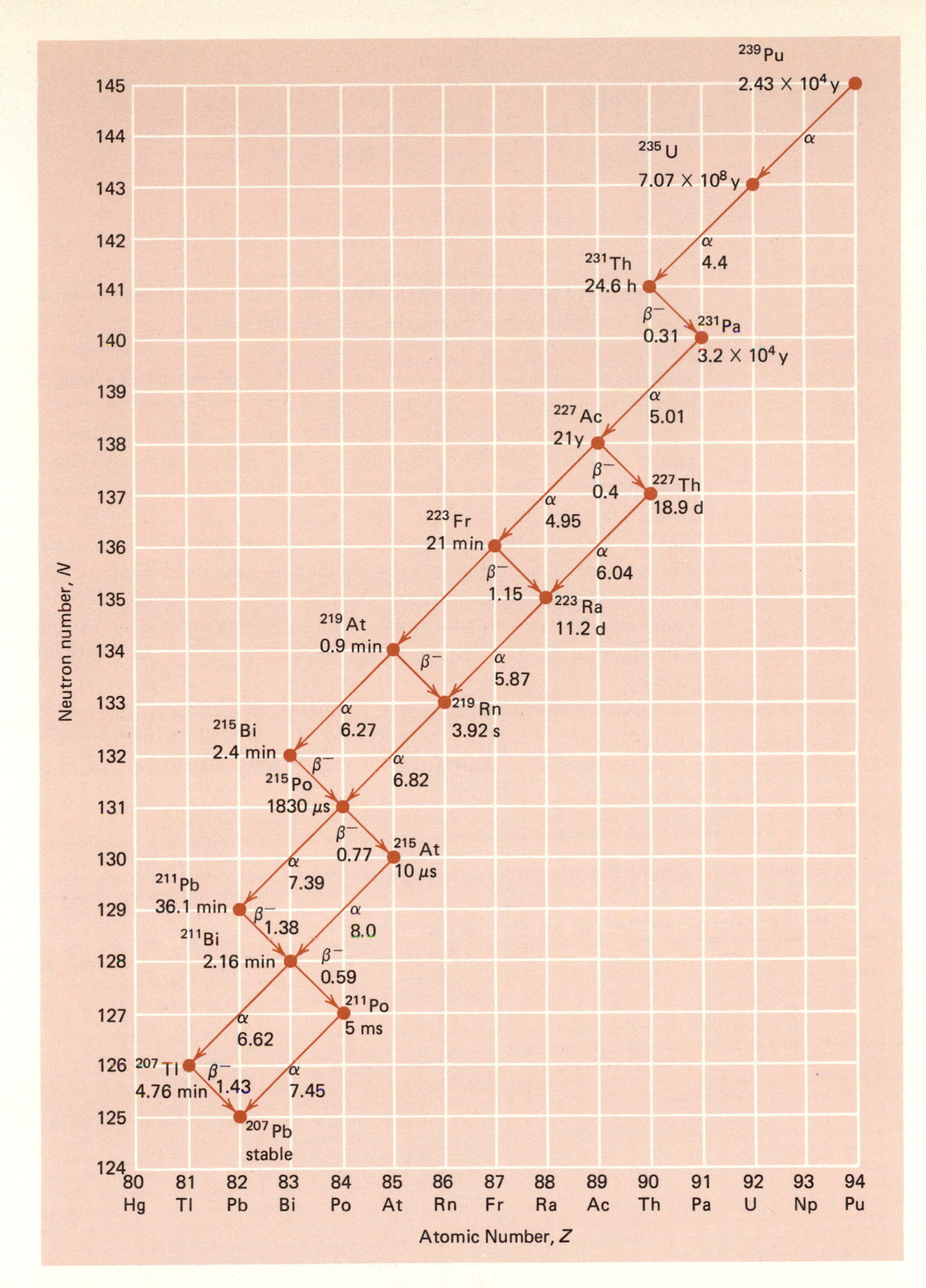

Figure 32.8 The decay series of $^{235}_{92}U$ on the Segré chart. The half-lives of the radioactive isotopes and the maximum energies in MeV of the decay products are given. The series terminates with the stable isotope $^{207}_{82}Pb$.

of $^{235}_{92}$U. The decay scheme is shown schematically on the N-vs-Z plot of Figure 32.8. The end result of this radioactive chain, which includes a number of α and β decays, and various γ emissions also, is the stable isotope of lead, $^{207}_{82}$Pb.

It is not difficult to see why the α decay of a heavy nucleus is generally followed by a β^- decay. On the N-vs-Z plot, the α decay is represented by an arrow pointing two units down and two units to the left; the daughter nucleus has two neutrons and two protons fewer than the parent nucleus. One can see from Figure 32.5 that such a move is likely to place the daughter into the neutron-rich region, off the line of stability. A more stable nucleus will follow β^- decay, which, since it leaves A constant, corresponds to a shift of one unit to the right and down on the N-vs-Z plot. (See Figure 32.6.) At some points in a decay scheme there may be branching. For instance, in the decay scheme of $^{235}_{92}U$, the nucleus $^{211}_{83}Bi$ can decay by α emission to $^{207}_{81}Tl$ or by β^- emission to $^{211}_{84}Po$.

32.4 Decay Constants, Half-Lives, and Activities

The moment of radioactive decay of an unstable nucleus is not a precisely predictable time but can only be assigned a statistical probability. For a particular nucleus of $^{227}_{90}Th$, it is impossible to predict when this nucleus will decay, although one can, on the basis of empirical data, state with confidence the probability that this nucleus will have decayed to $^{223}_{88}Ra$ at the end of a given time. This statistical information is quite sufficient because a radioactive sample of only a few micrograms usually contains a vast number of radioactive nuclei. Random fluctuations due to the statistical nature of the decay are then of little consequence. The situation is not unlike that encountered in thermodynamics (see Section 13.2), where the enormous number of gas molecules, even in a small container at normal temperature and pressure, makes observable departures from thermal equilibrium so improbable that they would never occur within a time comparable to the age of the earth, much less the lifetime of an observer.

Given a sample containing N radioactive nuclei, one can expect that the number ΔN that decay in a brief time Δt will be proportional to N, the number present. Consequently, the change in the number of radioactive nuclei can be written

$$\Delta N = -\lambda N \, \Delta t \tag{32.4}$$

where the negative sign appears because decay decreases the number of radioactive nuclei.

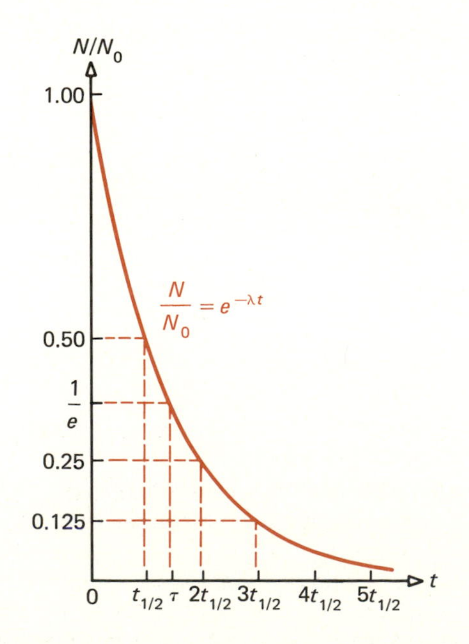

Figure 32.9 *The time course of radioactive decay. The plot shows the fraction of radioactive nuclei remaining at the end of a time interval t. The decay time $\tau = 1/\lambda$, where λ is the decay constant. The fraction remaining diminishes by a factor $\frac{1}{2}$ after each half life $t_{1/2} = 0.693\tau$.*

The constant λ in Equation (32.4) is called the *decay constant*. It is a measure of the rapidity with which the decay proceeds.

The solution of Equation (32.4) is

$$N(t) = N_0 e^{-\lambda t} \qquad \textbf{(32.5)}$$

where $N(t)$ is the number of radioactive nuclei remaining at time t, and N_0 is the number that were present at $t = 0$. Accordingly, the number of radioactive nuclei decreases exponentially.

Comparison of Equation (32.5) with Equation (24.3) shows that the decay constant λ is just the reciprocal of the time constant τ of the exponential curve. That is, after a time interval $\tau = 1/\lambda$, the number of remaining radioactive nuclei is $1/e$, or approximately 37 percent of those that were at hand at time $t = 0$.

The *half-life* $t_{1/2}$ is defined as the time interval in which the number of radioactive nuclei drops by a factor of 2. Referring to Section 24.2, we have

$$t_{1/2} = 0.693\tau \qquad \textbf{(32.6)}$$

The *activity* of a radioactive source is defined as the number of decays per second; that is,

$$\text{Activity} = -\frac{\Delta N}{\Delta t} = \lambda N \qquad \textbf{(32.7)}$$

The activity is therefore proportional to the amount of radioactive material and to the decay constant. For a given number of active nuclei, the shorter the half-life of the species, the greater the activity of the sample, and the faster the activity diminishes. It is general practice to state the activity of a source in curies (abbreviated Ci) or microcuries (abbreviated μCi),* where

$$1 \text{ Ci} = 3.70 \times 10^{10} \text{ disintegrations/second} \qquad \textbf{(32.8)}$$

● ***Example 32.2*** The nucleus $^{198}_{79}$Au is a β^- emitter with a half-life of 2.7 days. What is the activity of a sample containing 1 μg of pure $^{198}_{79}$Au? Estimate the activity of the sample at the end of 5 days and at the end of 4 weeks.

The number of $^{198}_{79}$Au atoms in 1 μg of the material is

$$\frac{(10^{-6}\text{ g})(6 \times 10^{23}\text{ atoms/mol})}{198\text{ g/mol}} = 3 \times 10^{15}\text{ atoms}$$

The decay constant of $^{198}_{79}$Au is

$$\lambda = \frac{1}{\tau} = \frac{0.693}{t_{1/2}} = \frac{0.693}{(2.7\text{ d})(8.64 \times 10^4\text{ s/d})} = 3 \times 10^{-6}\text{ s}^{-1}$$

The number of disintegrations per second is therefore

$$-\frac{\Delta N}{\Delta t} = \lambda N = 9 \times 10^9 \text{ disintegrations/second}$$

The activity, expressed in curies, is

* One curie is approximately equal to the number of disintegrations per second in 1 g of radium.

$$\frac{9 \times 10^9}{3.7 \times 10^{10}} = 0.243 \text{ Ci}$$

Five days corresponds to approximately two half-lives. At the end of two half-lives, the number of nuclei of $^{198}_{79}Au$ has diminished by the factor $(\frac{1}{2})^2 = \frac{1}{4}$. Consequently, the activity will then be only about

$$\frac{0.243}{4} = 0.061 \text{ Ci}$$

Four weeks later, at the end of about ten half-lives, the activity will be

$$\frac{0.243}{2^{10}} = \frac{0.243}{1024} = 240\ \mu\text{Ci}$$

32.5 Nuclear Reactions

The first artificially induced nuclear reaction was observed by Rutherford in 1919; he noted that when an α particle passed through a gas of nitrogen, very energetic protons were occasionally emitted. This reaction

$$^{14}_{7}N + ^{4}_{2}He \rightarrow ^{17}_{8}O + ^{1}_{1}H$$

is commonly written in the abbreviated form $^{14}_{7}N(\alpha,p)^{17}_{8}O$.

Since then, and with the aid of cyclotrons and other particle accelerators, literally thousands of nuclear reactions have been studied, among them other (α,p) reactions as well as (α,n), (n,p), (n,α), (n,γ), (p,γ), and so on. The end products of these reactions are often radioactive isotopes of practical value in medicine, biochemistry, or other fields. For example, the product of the reaction

$$^{27}_{13}Al(n,\alpha)^{24}_{11}Na$$

is a radioactive isotope of sodium that decays, with a half-life of 15 hours, to $^{24}_{12}Mg$ with the emission of a β^-. This isotope and a more long-lived isotope of sodium, $^{22}_{11}Na$, which decays by positron emission, are used in studying how sodium ions are transported across biological membranes. Another important atom in biology is phosphorus, which plays a critical role in all metabolic processes in which energy is either stored or used. The principal energy-storing molecule of a cell is adenosine triphosphate (ATP), and the involvement of ATP in metabolic pathways can be followed by using the radioactive tracer $^{32}_{15}P$, a β^- emitter with a half-life of 14.5 days. This isotope is produced in the reaction

$$^{32}_{16}S(n,p)^{32}_{15}P$$

We shall consider some other applications of induced radioactivity in a later section of this chapter.

32.6 Fission

In the summer of 1939, shortly before World War II started, the journal *Die Naturwissenschaften* (*Natural Philosophy*) contained a brief paper by O. Hahn and F. Strassmann. In it, they reported that when uranium was bombarded by neutrons, nuclei with atomic masses about half the mass of uranium were produced. Lise Meitner and Otto Frisch, two refugee scientists from Nazi Germany who had fled to Sweden, realized immediately the great significance of these results—the uranium nucleus had

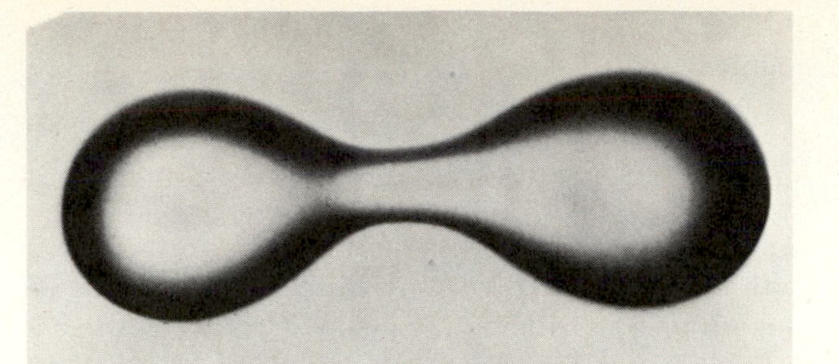

Figure 32.10 The liquid-drop model of fission showing a drop of water suspended in oil just before fission into two smaller drops. Fission was induced by application of a potential difference across the oil, thereby producing an initial deformation of the drop, giving the drop some excitation energy.

split into two fragments! Subsequent work showed that fission does not always yield the same reaction products. Instead, fission appears to be more in the nature of a statistical event, with atomic mass numbers of the products ranging between about $A = 80$ and $A = 160$. In fact, a highly successful model of fission proved to be the liquid-drop model, the splitting of a liquid drop into two parts as a result of a random disturbance.

A glance at Figure 32.4 shows that fission is energetically favorable because the binding energy of the product nuclei is greater than that of the uranium nucleus. Hence the system can reduce its potential energy by fission, releasing the energy difference as kinetic energy and electromagnetic radiation. The fission of one $^{235}_{92}U$ nucleus releases approximately 200 MeV.

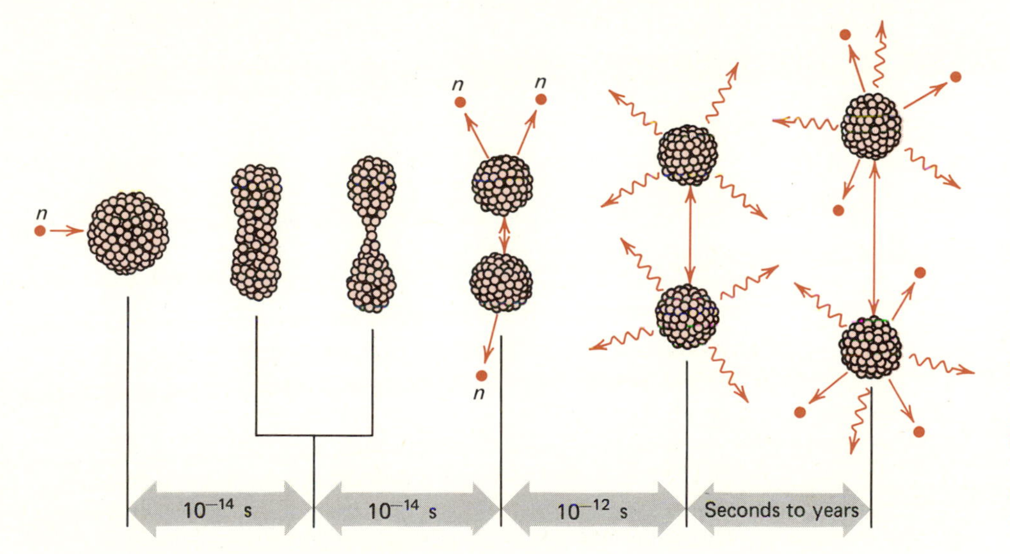

Figure 32.11 Schematic representation of neutron-induced fission of $^{235}_{92}U$. Following absorption of a neutron, the $^{236}_{92}U$ nucleus is left in an excited state and deforms somewhat like a drop of liquid. The nucleus fissions into two fragments in less than a picosecond, with the simultaneous emission of several neutrons. The fission fragments are highly excited states of unstable nuclei, whose lifetimes against β or γ decay range from picoseconds to years.

Since heavy nuclei are neutron-rich compared with nuclei near $A = 120$, fission is accompanied by the release of several neutrons, between 2 and 3 per fission. If these neutrons could be made to initiate further fission of other nuclei, a self-sustaining chain reaction would ensue.* If that chain reaction proceeded rapidly, an enormous amount of energy would be suddenly released—a titanic explosion would result.

Scientists both inside and outside Germany reached this conclusion. In the summer of 1939, even before the war had formally commenced, the German government strictly prohibited exportation of uranium ore from the Czech mines that, decades earlier, had supplied this raw material to

* The idea of a chain reaction as a source of nuclear energy had occurred several years earlier to Leo Szilard, who patented the concept in 1934, assigning the patent to the British Admiralty.

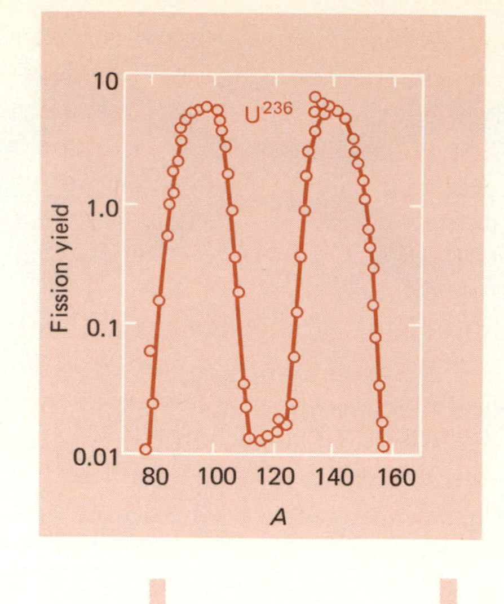

Figure 32.12 *Distribution of mass numbers of fission fragments from the fission of $^{236}_{92}U$ following absorption of a neutron by $^{235}_{92}U$. The most probable event is fission into two nuclei with mass numbers near 95 and 140.*

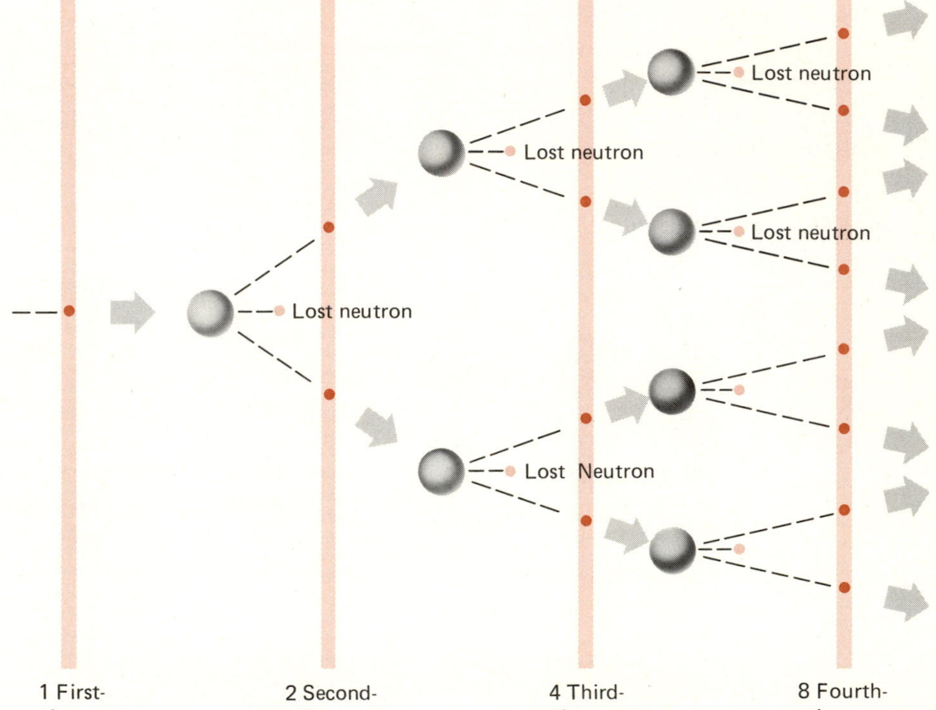

Figure 32.13 *Schematic representation of a fission chain reaction. Since each event requires less than 10^{-13} s, and releases approximately 200-MeV energy, an enormous amount of total energy can be liberated in an extremely brief time.*

Pierre and Marie Curie. The implication of that order was not lost on physicists in the United States.* On August 2, 1939, at the urging of his friend Leo Szilard, Einstein sent the now famous letter to President Roosevelt that initiated the Manhattan project and led to Hiroshima.

The first step in constructing a bomb was to show the feasibility of a chain reaction with uranium as the fuel. The problem is that neutron capture and subsequent fission of $^{235}_{92}U$ is a likely event only for neutrons of

* The final paragraph of Einstein's letter to President Roosevelt reads as follows:

> I understand that Germany has actually stopped the sale of uranium from the Czechoslovakian mines which she has taken over. That she should have taken such early action might perhaps be understood on the ground that the son of the German Under-secretary of State, von Weizsacker, is attached to the Kaiser Wilhelm Institut in Berlin, where some of the American work on uranium is now being repeated.

low energy. However, the two or three neutrons released in fission have very high energies. It was therefore necessary to devise a technique for slowing, or "moderating," these neutrons. That is accomplished by allowing them to collide with other particles. Protons, that is hydrogen, would be effective, but cannot be used because protons tend to capture neutrons and form deuterons, the nuclei of heavy hydrogen. Deuterons, however, would be effective, and the Germans did attempt to build a heavy-water reactor during the war.

The reactor Fermi constructed under the stands of the football stadium at the University of Chicago used graphite as the *moderator.* To prevent a disastrous runaway reaction, numerous cadmium rods were initially inserted into the reactor assembly. Cadmium is a very effective absorber of neutrons and would keep the reaction subcritical, that is, well below the self-sustaining level. On December 2, 1942, this first nuclear reactor was tested. The cadmium control rods were slowly withdrawn while the temperature of the reactor was carefully monitored, and at a certain point a self-sustaining chain reaction was achieved.

In addition to its potential for almost unimaginable destruction, nuclear fission also spawns a gamut of radioactive species that constitute a widespread hazard. For example, one typical fission reaction is

$$^{235}_{92}\mathrm{U} + ^{1}_{0}\mathrm{n} \rightarrow ^{144}_{56}\mathrm{Ba} + ^{89}_{36}\mathrm{Kr} + 3^{1}_{0}\mathrm{n}$$

$$^{89}_{36}\mathrm{Kr} \rightarrow ^{89}_{37}\mathrm{Rb} \rightarrow ^{89}_{38}\mathrm{Sr} \rightarrow ^{89}_{39}\mathrm{Y} \quad \text{(stable)}$$

$$^{144}_{56}\mathrm{Ba} \rightarrow ^{144}_{57}\mathrm{La} \rightarrow ^{144}_{58}\mathrm{Ce} \rightarrow ^{144}_{59}\mathrm{Pr} \rightarrow ^{144}_{60}\mathrm{Nd} \quad (\alpha \text{ decay}, \; t_{1/2} = 2 \times 10^{15} \text{ yr})$$

Other fission fragments with long half-lives are $^{99}_{43}\mathrm{Tc}$ ($t_{1/2} = 2.2 \times 10^5$ yr) and $^{98}_{43}\mathrm{Tc}$ ($t_{1/2} = 4.2 \times 10^6$ yr). One of the most insidious fission products is $^{90}_{38}\mathrm{Sr}$, whose half life is 29 years. This β emitter is chemically similar to calcium (see Table 30.3). Washed to earth by rain and consumed by cattle, it finds its way into milk and milk products. When these are ingested by humans, the strontium concentrates in normally calcium-rich regions of the body, such as bone. There the emitted electrons can do serious damage to cells, causing leukemia and bone cancer.

Figure 32.14 *Leo Szilard (1889–1964), who held the first patent to a nuclear chain reaction several years before uranium fission was first discovered.*

Figure 32.15 *Enrico Fermi (1901–1954), who supervised the construction of the first nuclear reactor and led the scientists of the Manhattan project.*

Figure 32.16 *A modern nuclear reactor.*

In the decades following Hiroshima, much effort has been expended toward developing nuclear reactors to exploit nuclear energy under controlled conditions. Today's power reactors are many times bigger and more elaborate than Fermi's "pile" under Stagg Field, but the basic principle has not changed. Generally the fuel is enriched uranium, that is, uranium containing between 2% and 4% $^{235}_{92}U$ (the natural abundance of $^{235}_{92}U$ is only 0.72%). The far more abundant isotope, $^{238}_{92}U$, although it too undergoes fission, is much more likely to enter into the reaction

$$^{238}_{92}U + ^{1}_{0}n \rightarrow ^{239}_{92}U \rightarrow ^{239}_{93}Np + ^{0}_{-1}e + \bar{\nu}$$
$$\hookrightarrow ^{239}_{94}Pu + ^{0}_{-1}e + \bar{\nu}$$

Plutonium, one of about a dozen known transuranic elements, is an α emitter with a half-life of 2.4×10^4 yr. It is also susceptible to fission by slow neutrons. In a *breeder reactor,* some of the neutrons from the fission of $^{235}_{92}U$ are used to transmute $^{238}_{92}U$ to $^{239}_{94}Pu$ according to the above reaction. Of course, the breeder reactor does not create fissionable material out of a vacuum or produce an inexhaustible supply of fission fuel. It converts the otherwise useless but abundant isotope of uranium into fissionable plutonium.

There are, however, still numerous problems with operating a breeder reactor. Some are strictly technical and will surely be solved in due time. One worrisome aspect is that plutonium is an extremely toxic chemical and once produced will be with us for millennia. It is also the active material of the most common fission bomb; a breeder reactor producing raw materials for nuclear bombs is a grave concern to many people.

Nuclear reactors are not only sources of useful power but also sources of useful neutrons, and therefore important research tools in physics, and also medicine, biology, biochemistry, and related disciplines. Since neutrons carry no charge, their interaction with matter (neglecting nuclear reactions) arises because neutrons have magnetic moments. This has made neutrons especially valuable as probes in studying the magnetic structures of ordered magnetic systems, such as ferromagnetic crystals.

The principal use of reactors in biological work is in producing various radioactive isotopes for research and treatment. The isotope $^{32}_{15}P$, a tracer in biological research, $^{99}_{43}Tc$ used in radiation therapy, and the isotope $^{131}_{53}I$, used in therapy of thyroid cancer, are produced by neutron-induced reactions; for example,

$$^{32}_{16}S(n,p)^{32}_{15}P \qquad \text{and} \qquad ^{130}_{52}Te(n,\beta)^{131}_{53}I$$

32.7 Fusion

Well before the turn of the century, scientists knew that the ancients' picture of the sun as a ball of fire, fueled by wood or some other combustible material, was incorrect. The principal difficulty lay in the apparently inexhaustible supply of fuel, but there were other perplexing problems. For example, all fuels known to man, such as wood, coal, and animal or vegetable fats, generate CO or CO_2 in oxidation. Yet from spectroscopic data, the atmosphere surrounding the sun appeared to contain none of these residues of combustion.

Years before the discovery of fission, scientists had recognized that the energy released in a nuclear reaction is a million times greater than that of a chemical reaction. For example, when two hydrogen atoms combine to form the H_2 molecule, about 4.5 eV are released. By contrast, in the reaction

$${}^{2}_{1}H + {}^{3}_{1}H \rightarrow {}^{4}_{2}He + {}^{1}_{0}n$$

in which a deuteron and a tritium nucleus fuse to form an α particle and a neutron, 17.6 million eV are liberated.

The presence of helium gas in the photosphere of the sun lent support to the hypothesis that the energy of the sun and of other stars is derived from the nuclear synthesis of helium. In 1938, H. A. Bethe, in a brief but justly famous paper, described the most probable fusion cycle, in which the ${}^{12}_{6}C$ nucleus acts as catalyst. The energy released in the completion of one cycle, during which four protons combine to form a helium nucleus and two positrons, is more than 26.7 MeV, that is, more than 6×10^8 kcal/mol of He.

Figure 32.17 *Hans Albrecht Bethe*

Since about 1948 there has been a concerted effort to produce a controlled fusion reaction on earth. The problem in arranging this can be seen by considering the deuteron-tritium reaction shown above. These two particles normally repel each other through the Coulomb interaction, and much energy must be given to the two nuclei to force them within the range of the strong nuclear force. In the interior of the sun, this great energy is provided by thermal agitation; the temperature at the core of the sun is probably more than 2×10^7 K (more than 20 million degrees Kelvin). It is clearly impossible to confine hydrogen gas within an ordinary container at temperatures even remotely approaching this. Various sophisticated schemes have been tried for the confinement of a hot, ionized gas (called a plasma), including the use of strong inhomogeneous magnetic fields. Progress toward controlled fusion has been made, though only with herculean effort and then at perhaps disappointing speed.

Uncontrolled fusion has been accomplished. It was soon realized that the temperature within a fission bomb during the first few nanoseconds approaches the enormous values required for spontaneous fusion. Fusion bombs triggered by fission, euphemistically called thermonuclear devices, have been constructed and tested. Unfortunately, the Promethean promise of virtually unlimited useful energy through fusion is still at best a dream, at worst a nightmare.

Figure 32.18 *An uncontrolled fusion reaction.*

*32.8 Radiation Detectors

The study of radiation can be carried forward only if there are instruments that can detect and measure that radiation. Such detectors are based on the fact that passage of an energetic particle through matter produces ionization or excitation of the atoms or molecules in the detector. With charged particles, such as electrons, protons, deuterons, and α particles, energy is transferred directly to atoms along their path by Coulomb interaction.

With γ rays, ionization or excitation is less direct. These photons lose energy by Compton scattering or pair production. It is the energetic charged particles produced thereby that cause the ionization within the detector.

One way of observing neutrons is by making the detector from a substance whose nuclei capture neutrons and subsequently decay by emitting a charged particle that ionizes the medium. Another way is by the ionization resulting from nuclei that recoil in a direct collision with an incident neutron.

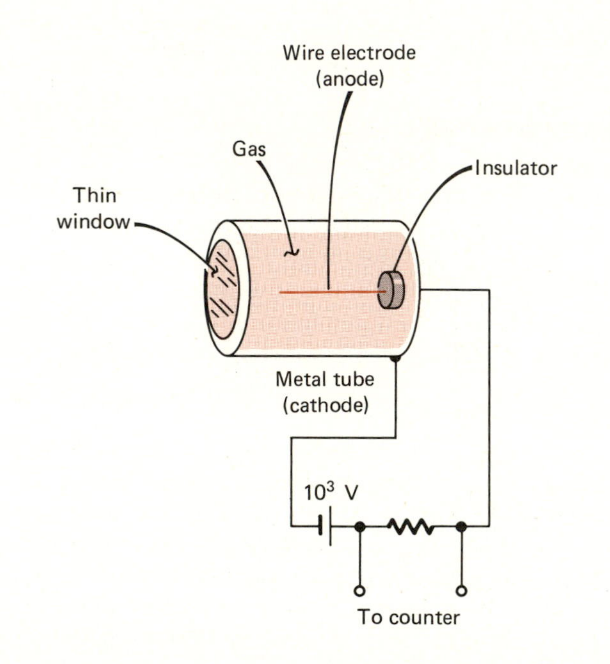

Figure 32.19 Schematic diagram of a Geiger counter.

One of the best-known ionization detectors is the *Geiger counter.* It consists of a cylindrical metal chamber, containing a gas at low pressure, and a thin wire, insulated from the cylinder and along its axis. One of the end windows of the counter is made thin to reduce absorption of incident particles. A potential difference of about 1000 V is maintained between the cylinder and central wire, with the wire positive. If an ionizing particle enters the chamber, the electrons liberated from gas molecules along its path acquire enough energy from the electric field to ionize other gas molecules. Thus a single energetic particle traversing the chamber produces an avalanche of charge, and a brief pulse of current results. That pulse is amplified and recorded or made audible as a click. The number of clicks is a direct measure of the number of energetic particles that enter the counter.

The *scintillation counter,* a direct descendant of the early phosphorescent screen used by Rutherford and collaborators, consists of a trans-

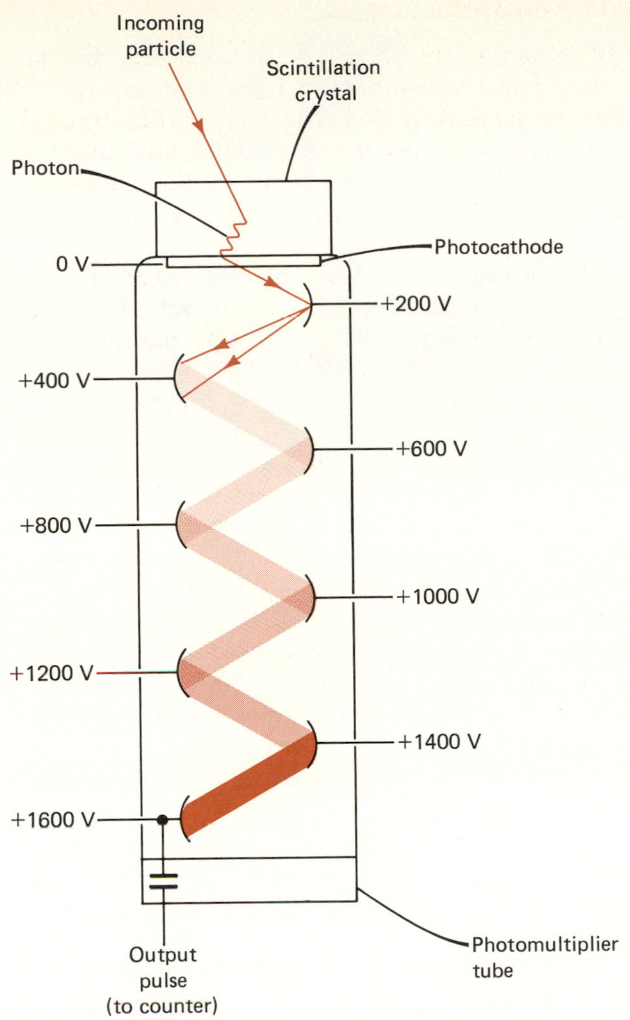

Figure 32.20 *Schematic diagram of a scintillation counter. The cathode of the photomultiplier tube, which is in direct contact with the scintillating crystal, releases an electron by photoemission when a photon of visible light strikes its surface. That electron is accelerated toward the first anode, which it strikes with an energy of 200 eV, releasing several secondary electrons from its surface. This process of secondary electron emission by collision is repeated several times, producing a significant pulse of charge from each electron liberated at the photocathode.*

parent crystal, such as NaI, and a photomultiplier tube, a sensitive electronic device that registers a current pulse even when only a single photon in the visible region strikes its sensitized surface. Except at the face in contact with the photomultiplier, the surface of the sodium iodide crystal is coated with a thin reflecting layer. When an energetic particle or a photon (γ ray) traverses the NaI crystal, some atoms along its path are excited and then decay to the ground state with the emission of visible light. The magnitude of the photomultiplier current pulse is proportional to the number of visible photons generated by the passage of the particle through the scintillator, and this, in turn, depends on the type of particle and its energy.

Semiconductor detectors have gained popularity in the past two decades. They are basically reverse-biased *p-n* junctions, and normally carry almost no current. If, however, ionizing radiation passes through the depleted junction region, this will create electron-hole pairs and allow a brief pulse of reverse current. The detectors are rugged and easy to operate but their active volume is quite small.

In addition to particle detectors that merely note the passage of an energetic particle, there are also devices that render a visual record of the particle's trajectory. The simplest of these is the *photographic emulsion.* An energetic particle traversing an emulsion leaves a latent image, just as a

Figure 32.21 Photograph showing several commercially available scintillation counters.

photon would. When the emulsion is later developed, silver precipitates along the particle's track. The disadvantage of the emulsion is that it records every event indiscriminately and cannot be reused.

The *cloud chamber*, invented by C. T. R. Wilson, a meteorologist and atmospheric physicist, consists of a gas under high pressure and at room temperature, maintained in a container whose volume can be suddenly increased by moving a piston. The chamber is usually shaped like a pillbox, with front and back surfaces of glass. If the piston is suddenly withdrawn, expanding the gas adiabatically, the temperature of the gas drops sufficiently to cause condensation. If an ionizing particle traverses the chamber at this time, droplets form along its trajectory. The path is made visible by illuminating the chamber; the droplets scatter the incident light and reveal the path.

The *bubble chamber* is based on the same principle as the cloud chamber, except that in this device the pressure above a liquid is suddenly reduced so that the liquid tends to boil. Ions produced along the path of an energetic particle become nucleation sites for bubbles, which then delineate the particle's trajectory. Since a liquid is much more dense than a supercooled gas, the bubble chamber is more efficient than the cloud chamber and has completely displaced it.

Figure 32.22 A cloud chamber photograph showing several electron-positron pairs produced by γ rays incident on a thin lead plate near the top of the photograph. The e^+ and e^- trajectories are bent by a **B** field pointing out of the page.

Bubble chambers are invariably placed between the pole faces of a magnet or in a large, usually superconducting solenoid, with the magnetic field oriented perpendicular to the direction of the incident beam. While the bubble density along the path indicates the particle's energy, the curvature of the path measures the particle's momentum. The direction of deflection by the magnetic field gives the sign of the charge on the particle.

Neutral particles and γ rays do not leave visible tracks in a bubble chamber. However, their presence can often be inferred indirectly, for example, from the appearance of Compton-scattered electrons or electron-positron pairs created by γ rays.

Spark chambers and *streamer chambers* are widely used in elementary-particle, high-energy research. A spark chamber consists of an array of closely spaced parallel metal plates inside a gas-filled enclosure. Even-numbered plates are grounded, odd-numbered plates are maintained at a potential of a few hundred volts. If an energetic particle passes through the chamber, an electric discharge appears between adjacent plates along the particle's path. A streamer chamber has only two plates, maintained at a substantial potential difference. An energetic charged particle traversing the chamber can then be seen by the glow, or streamer, along its path.

Ionization, which makes it possible to detect and observe energetic charged particles, is also responsible for damage to materials, especially biological organisms. As a charged particle traverses a cell, it disrupts the relatively weak chemical bonds of proteins, nucleic acids, and other complex organic molecules, and usually destroys the cell's viability. It may also happen that if only a few bonds are broken, the molecule re-forms with a slight rearrangement that can drastically alter its biological function. If such an alteration appears in a DNA strand, it can profoundly influence the entire organism, because this DNA molecule is replicated, propagating the mutant cell.

The destructive power of radiation can also be used to advantage, as in cancer therapy. The malignant region is subjected to intense nuclear bombardment with a view to killing the cancer cells.

*32.9 Radioactive Dating

32.9 (a) Carbon-14 Dating

The isotope ${}^{14}_{6}C$ (carbon-14) is unique among naturally occurring radioactive nuclei because its half-life of 5730 years is brief compared with the estimated age of the earth (about 4 billion years). Even though the concentration of carbon-14 is extremely small—for every carbon-14 atom in the atmosphere there are about 7.6×10^{11} atoms of ordinary carbon-12—it is nevertheless impossible that the carbon-14 remaining today, after the lapse of more than 100,000 half-lives, is due to the presence of that isotope in great abundance when the earth was first formed. Clearly, the concentration of carbon-14 must be due to some production mechanism that compensates for the loss due to radioactive decay.

● ***Example 32.3*** Suppose that at time $t = 0$ the carbon in the atmosphere were entirely the isotope ${}^{14}_{6}C$. How many half-lives would have to elapse before the carbon-14 concentration was diminished by a factor of approximately 10^{12}?

Each half-life reduces the number of carbon-14 nuclei by a factor of 2. In 10 half-lives, this leads to a reduction by a factor of $2^{10} = 1024$, or about 1000.

At the end of 20 half-lives, the number of carbon-14 nuclei would have been diminished by a factor of about $(1000)^2 = 10^6$; at the end of 30 half-lives, by 10^9; and at the end of 40 half-lives by about 10^{12}. ●

There is a production mechanism for $^{14}_{6}C$ that takes place in the upper atmosphere. There, neutrons produced by cosmic rays interact with nitrogen nuclei according to the reaction

$$^{14}_{7}N + ^{1}_{0}n \rightarrow ^{14}_{6}C + ^{1}_{1}H$$

The isotope carbon-14 then decays by β emission,

$$^{14}_{6}C \rightarrow ^{14}_{7}N + ^{0}_{-1}e + \bar{\nu}$$

with a half-life of 5730 ± 30 years.

Although the cosmic ray flux is not isotropic over the surface of the earth, the concentration of carbon-14 in the atmosphere is uniform due to the rapid (compared with the carbon-14 lifetime) mixing in the atmosphere and biosphere. The concentration has remained nearly constant since it represents an equilibrium between global production rates due to cosmic rays and the fixed decay rate.

The concentration of carbon-14 in living organisms—animal and plant life—which get carbon either directly from atmospheric CO_2 by photosynthesis (plants) or indirectly by ingesting photosynthetically produced carbohydrates and other organic molecules, is the same as the concentration in the atmosphere, 1 part in 7.6×10^{11}. However, once a living organism dies, the carbon-14 within it decays and is not replenished. Consequently, careful measurement of the relative concentration of carbon-14 in dead matter of organic origin can establish the time at which the organism ceased to take up atmospheric CO_2.

● ***Example 32.4*** (a) Determine the activity, in decays per minute per gram of natural carbon from a living tree. (b) Carbon from wood samples

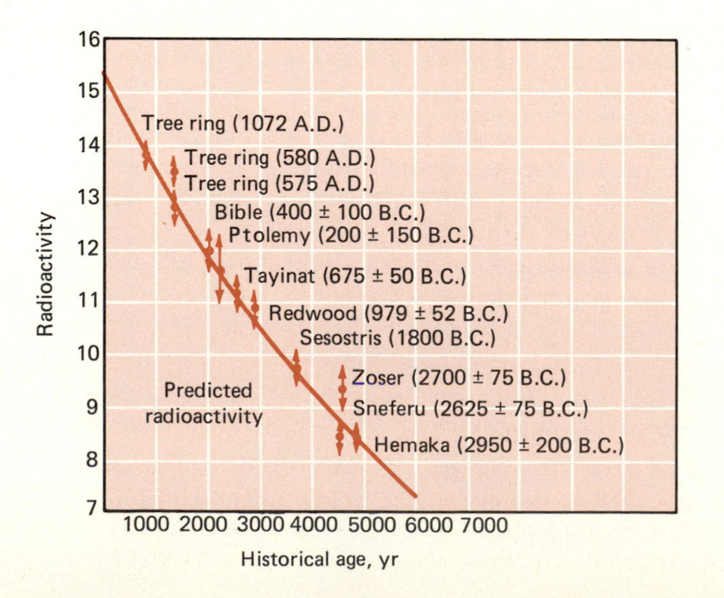

Figure 32.23 *Predicted and observed radioactivity from carbon samples of known age. (Note that the ordinate scale begins at a counting rate of seven counts per minute; that is, only a portion of the exponential decay curve of $^{14}_{6}C$ is shown.)*

taken from an Egyptian tomb give a reading of 7.6 counts per gram per minute. What is the approximate date of the tomb?

(a) Carbon has an atomic mass of 12.0. Hence 1 g of carbon contains

$$\frac{6 \times 10^{23} \text{ nuclei/mol}}{12 \text{ g/mol}} = 0.5 \times 10^{23} \text{ nuclei/g}$$

and consequently,

$$\frac{0.5 \times 10^{23} \text{ nuclei/g}}{7.6 \times 10^{11}\ {}^{12}\text{C}/{}^{14}\text{C}} = 6.6 \times 10^{10}\ {}^{14}\text{C nuclei/g}$$

The half-life of ^{14}C is

$$t_{1/2} = 5730 \text{ yr} = (5730 \text{ yr})(5.26 \times 10^5 \text{ min/yr}) = 3.01 \times 10^9 \text{ min}$$

The decay constant is therefore

$$\lambda = \frac{0.693}{t_{1/2}} = \frac{0.693}{3.01 \times 10^9 \text{ min}} = 2.3 \times 10^{-10} \text{ min}^{-1}$$

It follows that the number of disintegrations per minute per gram of natural carbon is

$$-\frac{\Delta N}{\Delta t} = \lambda N = (2.3 \times 10^{-10} \text{ min}^{-1})(6.6 \times 10^{10}\ {}^{14}\text{C nuclei/g})$$
$$= 15.2 \text{ counts/g·m.}$$

(b) Since the sample gave only 7.6 counts/g·min, that sample must have been dead for one half-life of carbon-14, or about 5700 yr. Therefore, the tomb can be placed chronologically near 3700 B.C. ●

In view of the low counting rate from atmospheric carbon, 15 counts per minute per gram, it is clear that extremely sensitive methods must be used to measure the decay of $^{14}_{6}$C from ancient samples of wood or bone, and extensive precautions must be taken to reduce or compensate for counts due to background sources. Currently, the upper limit for this dating procedure, even with large samples, is about 50,000 years, or about 9 half-lives. A sample of that age would yield only about two counts per gram per *hour*. Fortunately, radiocarbon dating is not the only archeological technique that relies on radioactivity.

32.9 (b) Thermoluminescence

If a small sample of ancient pottery is ground to a fine powder and then heated to about 500 °C, a small but readily measurable amount of light is emitted in addition to the normal blackbody radiation. If this same sample is subsequently heated once more, only the blackbody radiation is observed.

The additional light seen during the first heating is called *thermoluminescence*. The light originates from very long-lived excited states in some of the minerals of the clay; when the clay is heated, these excited states decay to the ground state with emission of light. The excited states are formed by absorption of energy from decay of natural radioactive impurities such as uranium, thorium, and potassium-40 in the pottery and possibly the surrounding soil. Even on archeological time scales, the activities of these radioactive impurities are practically constant since their half-lives are a billion years or longer. Consequently, the concentration of excited thermoluminescent centers in a piece of pottery depends

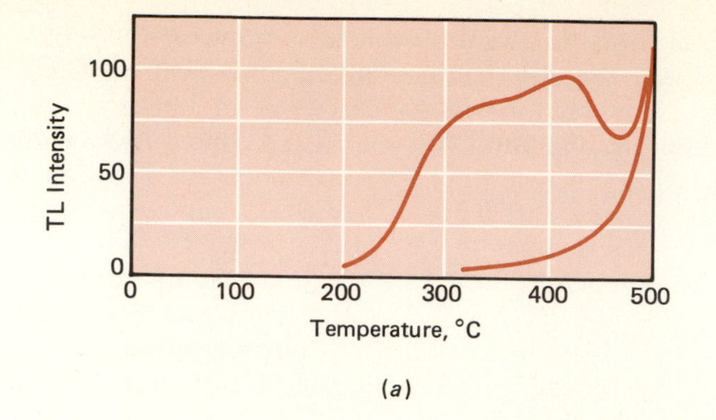

Figure 32.24 *Thermoluminescent glow curve from an ancient pottery sample. (a) Light emitted by the sample on initial heating. (b) Light emitted on subsequent heating. The difference is due to light released in the decay to the ground state of excited luminescent centers.*

on the concentration of radioactive elements and on the time that elapsed since the clay was last fired and the thermoluminescent centers "emptied." Thus, the age of the pottery can be established with reasonably good precision.

Different types of pottery clay have different susceptibilities to formation of thermoluminescent centers. Dating therefore involves three basic steps. First, the concentration of activated thermoluminescent centers stored in the clay is determined by sensitive photometric measurement and carefully controlled heating. Next, the thermoluminescent sensitivity of the clay sample is obtained. This involves preparing a concentrated source of radiation from artificial radioactive isotopes with α, β, and γ-ray emissions comparable to those of the natural elements present in the clay. The sample is exposed then to this radiation and the thermoluminescence induced thereby is subsequently measured. Last, the concentration of the natural radioactive sources present in the clay is determined by sensitive chemical, spectroscopic, or nuclear techniques.

Thermoluminescence, besides having archeological application, has also been used to detect modern forgeries of ancient Chinese pottery. Obviously, one would not want to grind up a substantial part of a valuable Ming vase to ascertain its authenticity; fortunately, in these situations a small sample, between 20 and 50 milligrams, obtained by drilling a tiny hole in an unobtrusive spot at the base of the vase, is normally sufficient.

32.9 (c) $^{206}_{82}Pb/^{238}_{92}U$ Dating

Natural lead contains four stable isotopes, $^{204}_{82}Pb$, $^{206}_{82}Pb$, $^{207}_{82}Pb$, and $^{208}_{82}Pb$. (Lead-204, once thought to be stable, is very weakly radioactive with a half-life of 1.4×10^{17} yr.) Of these four isotopes, all but lead-204 are the ultimate decay products of a uranium or a thorium isotope. If a rock sample that contains both uranium and lead is free of the isotope $^{204}_{82}Pb$, it is a safe presumption that the lead found in that rock is the result of the decay of uranium and thorium that were there at the time the rock was first formed, and that at that distant geological date, this rock contained no lead at all. Therefore, the ratio $^{206}_{82}Pb/^{238}_{92}U$ and $^{208}_{82}Pb/^{232}_{90}Th$ are measures of the geologic age of the rock. This method of dating is extensively used in geochronology.

32.10 Elementary Particles

For a few years it seemed that protons, neutrons, electrons, and neutrinos were the much-sought-after building blocks of matter. Yet their very existence and their interactions raised new and troublesome questions to which we still cannot give unequivocal answers. Foremost among these was what the forces are that come into play when two nucleons interact. In searching for an answer to this question, the Japanese physicist Hideki Yukawa proposed the existence of a particle of rest mass equal to about $300m_e$, where m_e is the mass of an electron. This particle was given the name *meson*. Two years later, in 1937, a meson was found in a cosmic ray reaction but was soon proved to be not the particle Yukawa had predicted. The Yukawa meson, known as the π meson, or *pion*, was observed in 1946; it can now be produced copiously in high-energy accelerating machines. The earlier meson, the μ meson, or *muon*, is the decay product of the pion.

With the development of ever more powerful particle accelerators, a truly bewildering array of so-called elementary particles has been discovered. Some live for a long time by nuclear standards, 10^{-10}–10^{-8} s, and others are more transitory, existing for only about 10^{-20} s or less, the time it takes light to travel from one side of a nucleus to the other. Today, scores of different elementary particles have been seen in bubble-chamber photographs of high-energy collisions.

The problem facing physics today is not unlike that which confounded chemists of the last century. First, common substances like ordinary table salt, saltpeter, quartz, and water were shown to be combi-

Figure 32.25 (*a*) Bubble-chamber photograph showing the production of an elementary particle and its antiparticle, the Ξ^- and the Ξ^+, as a result of annihilation of an energetic antiproton, $\bar{p}$, and a proton, p. Near the upper right portion of the photograph another $\bar{p}$-p annihilation gives rise to a group of charged particles, mostly π^+ and π^-, and also some unseen neutral particles. (*b*) Production and decay scheme of the Ξ particles.

nations of elements. Soon chemists learned how to synthesize new compounds, and became aware of certain regularities and rules that served as useful recipes to newcomers. These regularities in the properties of the elements and their manner of combining to form compounds led to the organizational chart we now call the periodic table, which the Russian chemist Mendeleev first proposed. But even then, the pattern remained mysterious until the principles of atomic structure were clarified between 1915 and 1930.

In the subatomic realm, there seem to be also certain rules that dictate how these numerous elementary particles interact. The rules, as in atomic physics, suggest a classification scheme that permits grouping the beasts of the elementary-particle zoo into families and classes. The broadest classification is based on the dominant mode of interaction. *Leptons* are particles that interact via the so-called *weak force,* the interaction responsible for β decay. *Hadrons* are particles that interact through the strong nuclear force. Hadrons are divided into mesons and *baryons.* The baryons are all particles with masses equal to or greater than the proton mass, and are, in turn, further separated into nucleons—protons and neutrons—and *hyperons,* particles heavier than nucleons.

The reason for this scheme is that in addition to those well-established conservation laws—conservation of energy, linear momentum, angular momentum, and charge—there seem to exist further conservation rules, for example, conservation of leptons and conservation of baryons. Moreover, fitting particles into proper categories often suggests a pattern that can be understood in terms of a suitable model. Over the years, many such models have been constructed—and demolished—with disconcerting regularity. However, the one most favored today, the quark model, has survived, with some modifications, for about two decades.

The quark model in its original form consisted of three basic building blocks, the u, d, and s quarks, with charges of $\frac{2}{3}e$, $-\frac{1}{3}e$, and $-\frac{1}{3}e$, respectively. Mesons are particles made up of two quarks; baryons consist of three quarks (or antiquarks). Between its conception and now, the model has become more elaborate. There are more quarks, and each quark has certain attributes that are whimsically referred to as color and charm. Quarks have not been observed, and theory suggests that they cannot be produced in isolation or stripped from a nucleon. That does not invalidate the quark model, any more than Schroedinger's difficulty in visualizing quantum jumps invalidated the quantum theory. There are, in fact, indications of a substructure in protons and neutrons that can be discerned from the scattering of very energetic electrons by nucleons. Moreover, the quark model has been quite successful in predicting the existence and properties of some elementary particles as well as in explaining their interactions.

Still, with new elementary particles appearing each year, one may well ask where it will end and what it all means. There is no ready answer. Almost certainly, as particles are pounded against one another with ever increasing energy, more "elementary" particles will be discovered. It seems as though nature produces all kinds of particles in great profusion if conditions are right. And yet, out of this bewildering abundance and diversity of fundamental particles a pattern has emerged based on the very few fundamental interactions that nature has so far revealed: gravitational interaction, the weak nuclear (leptonic) interaction, electromagnetic interaction, and the strong nuclear (hadronic) interaction. Perhaps these are, indeed, the only interactions in the universe. Perhaps they are but different manifestations of one universal force. Perhaps the vision of a unified-field theory that drove Einstein will someday become reality in the work of another intellectual giant. The search continues, and what the future holds none can predict.

Summary

A *nucleus* is composed of N *neutrons* and Z *protons*, with $N + Z = A$, where *A is the atomic mass number* of the nucleus, *Z the atomic number*, and *N the neutron number*. A nucleus is designated by the symbols

$${}^{A}_{Z}\mathrm{X}$$

where X is the standard chemical abbreviation for the element of atomic number Z.

Isotopes are nuclei with the same number of protons but different neutron numbers.

The mass of a nucleus is less than the sum of the masses of the constituent nucleons. The difference between these masses is known as the *mass defect* ΔM and is related to the binding energy (BE) of the nucleus by the Einstein relation

$$\mathrm{BE} = \Delta M c^2$$

The *nuclear radius* is given by

$$R = R_0 A^{1/3}$$

where $R_0 = 1.2 \times 10^{-15}$ m. It follows that, since the nuclear volume is proportional to R^3, the nuclear density is a constant.

Stable nuclei are found only for a narrow range of N and Z values. For the very light nuclei, stability occurs when $N \simeq Z$; with increasing A, the range of stability shifts to values of N that are somewhat greater than Z.

Nuclei that on an N-versus-Z plot are located above the line of stability are naturally radioactive and decay by emitting an electron. This process is called *beta decay*. Nuclei located below the line of stability on an N-versus-Z plot decay by *positron emission* or *electron capture;* the former is known as *β^+ decay*, the latter as K capture. Nuclei with Z greater than 83 (bismuth) are all radioactive.

In radioactive decay, energy, momentum, angular momentum, nucleon number (A), and electric charge must be conserved.

In β decay (β^+ and β^-), a neutrino (or an antineutrino) is also emitted.

Following α or β decay, the daughter nucleus is often in an excited state. It then decays to the ground state by emitting an energetic photon, called a *γ ray*.

The number of radioactive nuclei present at any time t is related to the number that were present at time $t = 0$ by

$$N(t) = N_0 e^{-\lambda t}$$

where λ is the *decay constant*.

The *half-life* is the time interval during which the number of radioactive nuclei is reduced by a factor of $\frac{1}{2}$ as a result of radioactive decay. The half-life $t_{1/2}$ is related to the decay constant by

$$t_{1/2} = \frac{0.693}{\lambda} = 0.693\tau$$

where

$$\tau = \frac{1}{\lambda}$$

is called the *decay time*.

The *activity* of a radioactive sample is defined as the number of radioactive decays per second. The unit of activity is the *curie* (Ci);

$$1 \text{ Ci} = 3.70 \times 10^{10} \text{ disintegrations per second.}$$

Fission is the disintegration of a heavy nucleus, such as a nucleus of uranium or plutonium, into two fragments of roughly equal masses. In the fission of uranium, about 200 MeV of energy is released.

Fusion is the process in which protons and neutrons, or very light nuclei, combine to form an α particle. In stars, energy is released as a result of fusion, in which, in effect, four protons combine to yield one helium nucleus and two positrons. Each such event liberates nearly 27 MeV.

Various kinds of *nuclear detectors* are used to observe and measure nuclear events. Among these devices are photographic emulsions, Geiger counters, scintillation counters, semiconducting *n-p* junctions, cloud and bubble chambers, and spark chambers.

There are numerous other *elementary particles* in addition to electrons, neutrinos, neutrons, and protons; among them are *mesons* and *hyperons*. Mesons have rest masses between that of an electron and that of a proton; hyperons have rest masses greater than that of a neutron. Elementary particles are classified according to the dominant interactions in which they can participate. *Leptons* are particles that interact via the weak force; *hadrons* are particles that interact via the strong nuclear force. According to current theory, all hadrons are composed of a combination of particles known as *quarks*, which carry charges of $\pm\frac{1}{3}$ and $\pm\frac{2}{3}$ of an electronic charge.

Multiple Choice Questions

32.1 If two atoms are each other's isotopes, they have

(a) the same atomic mass number.

(b) the same number of electrons in the neutral atom.

(c) the same number of neutrons.

(d) the same number of nucleons.

(e) the same lifetime against radioactive decay.

32.2 The nucleus $^{131}_{53}\text{I}$ has

(a) 53 protons and 131 neutrons.

(b) 131 protons and 53 neutrons.

(c) 78 protons and 53 neutrons.

(d) 53 protons and 78 neutrons.

(e) 78 protons and 131 neutrons.

32.3 A sample contains atoms of a radioactive isotope whose half-life is 30 s. The number of disintegrations from this sample

(a) depends on the total mass of the sample.

(b) depends on the number of disintegrations that have already taken place.

(c) depends on the number of radioactive nuclei present.

(d) depends on the energy of the emitted particles.

(e) depends on both (c) and (d).

32.4 Of the several nuclear reactions listed below, which one is possible?

(a) $^{10}_{5}\text{B} + ^{4}_{2}\text{He} \rightarrow ^{13}_{7}\text{N} + ^{1}_{1}\text{H}$

(b) $^{10}_{5}\text{B} + ^{1}_{0}\text{n} \rightarrow ^{11}_{5}\text{B} + \beta^- + \bar{\nu}$

(c) $^{23}_{11}\text{Na} + ^{1}_{1}\text{H} \rightarrow ^{20}_{10}\text{Ne} + ^{4}_{2}\text{He}$

(d) $^{14}_{7}\text{N} + ^{1}_{1}\text{H} \rightarrow ^{14}_{6}\text{C} + \beta^+ + \nu$

(e) None of the above is possible.

32.5 An α particle is

(a) an energetic neutron.

(b) an energetic photon emitted by a nucleus.

(c) an energetic electron emitted by a nucleus.

(d) an energetic proton emitted by a nucleus.

(e) none of the above.

32.6 A particle that has a mass 400 times that of an electron and a charge twice that of an electron is accelerated through a potential difference of 5 V. If the particle was initially at rest, its final kinetic energy will be

(a) 5 eV.

(b) 10 eV.

(c) 100 eV.

(d) 2000 eV.

(e) none of the above.

32.7 At a given instant, the counting rate from a radioactive source is 128,000 counts/s. After 24 min the counting rate has dropped to about 8000 counts/s. It follows that

(a) the half-life of the source is 1.5 min.

(b) the half-life of the source is 2 min.

(c) the half-life of the source is 6 min.

(d) the decay constant of the source is $\frac{1}{16}\ s^{-1}$

(e) none of these conclusions is correct.

32.8 A naturally radioactive nucleus decays by α emission. Mark the correct statement regarding this decay.

(a) Such an event does not occur; there are no natural α emitters.

(b) The parent nucleus has an atomic number greater than that of the daughter nucleus by 4.

(c) The mass of the parent nucleus is greater than the sum of the masses of the daughter nucleus and an α particle.

(d) The mass of the parent nucleus is less than the sum of the masses of the daughter nucleus and an α particle.

(e) Both (b) and (d) are correct.

32.9 A β ray is

(a) an energetic neutron emitted by a nucleus.

(b) an energetic photon emitted by a nucleus.

(c) an energetic electron emitted by a nucleus.

(d) an energetic proton emitted by a nucleus.

(e) none of the above.

32.10 A γ ray is

(a) an energetic neutron emitted by a nucleus.

(b) an energetic photon emitted by a nucleus.

(c) an energetic electron emitted by a nucleus.

(d) an energetic proton emitted by a nucleus.

(e) none of the above.

32.11 Which of the following is *not* used in detecting nuclear radiation:

(a) Photographic emulsion.

(b) Semiconducting *p-n* junction diode.

(c) Bubble chamber.

(d) Spark chamber.

(e) Copper-constantan thermocouple.

32.12 When $^{215}_{83}Bi$ decays to $^{215}_{84}Po$, the following is released:

(a) a proton.

(b) a neutron.

(c) an electron.

(d) positron.

(e) an α particle.

32.13 A neutrino

(a) always accompanies β decay.

(b) is a neutral particle of zero spin angular momentum.

(c) has never yet been detected experimentally.

(d) accompanies all radioactive decay.

(e) is among the most dangerous emissions following fission.

32.14 When two deuterons (^2_1H) combine to form a helium nucleus, much energy is released (approximately 24 MeV) because

(a) the helium nucleus immediately breaks apart, releasing much energy.

(b) the two deuterons have a smaller total mass than that of the helium nucleus.

(c) the two deuterons have a greater total mass than that of the helium nucleus.

(d) the electrons are more tightly bound to the helium nucleus.

(e) one electron carries away kinetic energy, leaving the helium singly ionized.

32.15 In the nuclear reaction $^{235}_{92}U + X \rightarrow {}^{236}_{93}Np$, the particle X is

(a) a proton.

(b) a neutron.

(c) an α particle.

(d) an electron.

(e) a deuteron.

32.16 One of the following nuclear reactions is *not* possible. That reaction is

(a) $^2_1H + {}^1_1H \rightarrow {}^3_2He + \gamma$

(b) $^{14}_7N + {}^4_2\alpha \rightarrow {}^{17}_8O + {}^1_1H$

(c) $^{226}_{88}Ra \rightarrow {}_{-1}^{0}e + {}^{223}_{86}Rn + \bar{\nu}$

(d) $^{214}_{82}Pb \rightarrow {}_{-1}^{0}e + {}^{214}_{83}Bi + \bar{\nu}$

(e) $^{32}_{16}S + {}^1_0n \rightarrow {}^{32}_{15}P + {}^1_1H$

32.17 Nucleus $^{228}_{90}X$ decays by α-particle emission. The daughter nucleus is

(a) $^{232}_{92}Y$

(b) $^{228}_{92}Y$

(c) $^{232}_{94}Y$

(d) $^{224}_{88}Y$

(e) $^{224}_{86}Y$

Problems

(Sections 32.2, 32.3)

32.1 Argon has several stable isotopes. The atomic mass of $^{40}_{18}Ar$ is 39.962383 u. What is the binding energy per nucleon for this isotope?

32.2 Use the data in Appendix E to compute the binding energy per nucleon of (a) $^{3}_{1}H$, (b) $^{4}_{2}He$, (c) $^{12}_{6}C$, (d) $^{56}_{26}Fe$.

32.3 What is the mass of a pion ($m_\pi = 237\ m_e$) expressed in MeV/c^2?

• **32.4** Aluminum has only one stable isotope, $^{27}_{13}Al$, whose atomic mass is 26.981541 u. The isotope $^{26}_{13}Al$ has an atomic mass of 25.986982 u, and the isotope $^{28}_{13}Al$ has an atomic mass of 27.981905 u.

Find the average binding energy per nucleon for each of the three isotopes. Both the lighter and the heavier isotopes are radioactive. What do you expect the decay processes might be for each of these isotopes?

• **32.5** Calculate the approximate nuclear radii of (a) $^{40}_{19}K$ (b) $^{136}_{58}Ce$ (c) $^{232}_{90}Th$.

• **32.6** Show that $^{8}_{4}Be$ is unstable against decay into two α particles. If a $^{8}_{4}Be$ nucleus at rest decays, what are the kinetic energies of the emitted α particles?

• **32.7** Nickel-64 ($^{64}_{28}Ni$) has an excited state 1.34 MeV above the ground state. The atomic mass of $^{64}_{28}Ni$ is 63.927968 u. What is the mass of the excited state of this nucleus? What is the wavelength of the γ ray emitted when the nucleus decays from the excited to the ground state?

• **32.8** Bismuth-214 ($^{214}_{83}Bi$) can decay by either α or β^- emission. What are the daughter nuclei for the two decay schemes? Using tabulated values of atomic masses, calculate the energies of the α particles and the maximum energy of the β particles.

• **32.9** Radium-226 ($^{226}_{88}Ra$) decays to $^{222}_{86}Rn$ with the emission of an α particle. Use the data in Appendix E to calculate the KE of the emitted α particle. What percent error results from neglecting the recoil energy of the daughter nucleus?

• **32.10** The isotope $^{64}_{29}Cu$ is unstable against β^- and β^+ decay. What are the daughter nuclei for these two decay modes? From the data provided in Appendix E, calculate the maximum energy of the β particles emitted in the two decay modes.

• • **32.11** In solving Problem 32.10 and similar β-decay problems, one usually neglects the recoil energy of the daughter nucleus. Estimate the percent error in the calculated maximum energy of the β particles due to this omission.

• • **32.12** A neutron star is an object of tightly packed neutrons. The density of such a neutron star is roughly the same as the density within a nucleus.

(a) Calculate the nuclear density.

(b) If our sun were to collapse to a neutron star, what would its radius be?

• • **32.13** What must the energy of bombarding protons be if these protons are to come within range of the nuclear forces of a target made of (a) zinc? (b) silver?

(Sections 32.4–32.5)

32.14 What is the range of half-life for radionuclides used in medical diagnosis? What determines that range?

• **32.15** Suppose nucleus A has a half-life that is twice that of nucleus B. If at some instant there are twice as many B nuclei in a sample as A nuclei, and the half-life of A is 1 hr, when will the number of A and B nuclei be the same? Or will this never happen?

• **32.16** If a certain radioactive material contains 2.56×10^{18} radioactive nuclei on a certain day, and 21 days later the number of radioactive nuclei has diminished to 2×10^{16}, what is the half-life of this species?

• **32.17** The unit of activity, the curie, is approximately equal to the number of disintegrations per second from a 1-g sample of radium, $^{226}_{88}Ra$. Calculate the half-life of $^{226}_{88}Ra$.

• **32.18** Many of the radioactive nuclides used in biological work either emit positrons or decay by electron capture (for example, $^{11}_{6}C$, $^{13}_{7}N$, $^{14}_{8}O$, $^{18}_{9}F$, and $^{68}_{31}Ga$). In every case, the decay is detected not by observing the emitted positron, but by identifying a characteristic γ ray. What is the γ-ray energy that one should look for when the radionuclide is a positron emitter?

• • **32.19** A radioactive sample has an activity A_0 at $t = 0$. At a later time $t = t_1$, the activity has diminished to A_1 ($A_1 < A_0$). Show that the decay constant for the radioactive material contained in the sample is given by $\lambda = [\ln(A_0/A_1)]/t_1$.

• **32.20** The activity of a sample is measured every 30 s with the following results (in counts per second): 2000, 1650, 1340, 1080, 896, 738, 600, 495, 404, 328, 274, 222, 180, 150, 120, 100, 82. (a) Plot these data versus time and use this to determine the half-life. (b) Replot on semilog paper. (c) Use the result of Problem 32.19 to determine the decay constant and half-life of this radioactive sample.

• **32.21** Is the reaction

$$^{7}_{3}Li + ^{1}_{1}H \rightarrow ^{4}_{2}He + ^{4}_{2}He$$

exothermic or endothermic? What is the amount of energy released or absorbed in this $^{7}Li(p,\alpha)^{4}He$ reaction?

• • **32.22** What is the minimum kinetic energy of the incident proton for the reaction of Problem 32.21? (Assume that the target Li nucleus does not recoil.)

• • **32.23** Thorium-232 ($^{232}_{90}Th$) has a half life of 1.4×10^{10} yr. This nucleus decays by α emission to $^{228}_{88}Ra$.

(a) What is the activity of 1 kg of pure thorium-232?

(b) What is the total mass of radium produced in 1 yr by the decay of thorium from this 1-kg block?

• • **32.24** Uranium-235 ($^{235}_{92}U$) has a half-life of 7.1×10^8 yr. The natural abundance of this isotope is only 0.72%. Uranium-238 ($^{238}_{92}U$), which has an abundance of 99.28%, is also radioactive with a half-life of 4.5×10^9 yr. Both isotopes emit α particles in their decay. If a sample of natural uranium, in its pure metallic form, shows an activity of 1.2 μCi, what is the total mass of the sample? What fraction of the total activity is due to each of the two isotopes?

(Sections 32.6–32.9)

32.25 A nuclear reactor which uses $^{235}_{92}U$ as a fuel has an output of 16 MW. If the energy released in the fission of one $^{235}_{92}U$ nucleus is 200 MeV, how many grams of $^{235}_{92}U$ are consumed in one day?

• **32.26** One of the several fusion reactions is

$$^{2}_{1}H + ^{2}_{1}H \rightarrow ^{3}_{1}H + ^{1}_{1}H$$

Compute the amount of energy released in this reaction. If the deuterium that is naturally in water were used in this fusion process, how much energy would be released from 1 L of water? Contrast this energy with that obtained when burning one gallon of gasoline, about 10^8 J.

• **32.27** Calculate the total energy released in the fission, described in Section 33.6, that yields $^{89}_{39}Y$ and $^{144}_{60}Nd$ as end products. What fraction of that energy is liberated during the initial step of the fission reaction, which yields the unstable isotopes of Ba and Kr and three neutrons?

• **32.28** The activity from a sample of carbon taken from the core of a giant redwood, that is, from the innermost rings, is about 12.8 counts/g-min. Assuming that as the tree grows the central part of the trunk becomes biologically inert, determine the approximate number of tree rings on this newly felled tree?

• **32.29** The activity from a 50-g sample of carbon taken from an ancient wooden casket is about 95 counts/min. Approximately when did the person buried in that casket live?

• **32.30** An archeologist finds a skeleton and claims that it is between 9000 and 10,000 years old. The bone contains 5 g of carbon. The counting rate from this 5 g of carbon is found to be 50 counts/min. Does that counting rate agree with the archeologist's estimate?

• **32.31** A sample of wood from an ancient site contains 50 g of carbon. The ^{14}C counting rate from this sample is 240 counts/min. How old is this sample?

• • **32.32** Suppose that 2 billion years ago 1 percent of the mass of the earth was carbon-14. Approximately how many nuclei of $^{14}_{6}C$ would be left today? (Assume that after the earth was formed, no more $^{14}_{6}C$ was produced.)

• APPENDIX A • Review of Mathematics

In Mathematics he was greater
Than Tycho Brahe or Erra Pater:
For he by Geometrick scale
Could take the size of Pots of Ale;
Resolve by Sines and Tangents straight,
If Bread or Butter wanted weight;
And wisely tell what hour o'th' day
The clock does strike, by Algebra.

SAMUEL BUTLER, from *Hudibras*

A.1 Equations

Relations between physical quantities are expressed by equations. For example, the equation

$$s = v_0 t + \tfrac{1}{2}at^2 \qquad \textbf{(A.1)}$$

relates the displacement s to the constant acceleration a, the initial speed v_0, and the elapsed time t.

"Solving an equation" means finding an expression for one variable, the unknown, in terms of the other variables whose numerical values may be known or can be determined. Equation (A.1) is the solution for s in terms of v_0, a, and t. However, the problem statement may give values of s, v_0, and t, and ask for the corresponding value of a. In that case, we rewrite Equation (A.1) so that the unknown, a, appears by itself on the left-hand side of the equation. From (A.1) we have

$$\tfrac{1}{2}at^2 = s - v_0 t \qquad \textbf{(A.2)}$$

and on multiplying both sides of (A.2) by $2/t^2$ we obtain the solution

$$a = \frac{2s}{t^2} - \frac{2v_0}{t} \qquad \textbf{(A.3)}$$

If, for instance, the values of s, v_0, and t are 40 m, 5 m/s, and 4 s, respectively, we find that the acceleration a is

$$a = \frac{2(40\ \text{m})}{(4\ \text{s})^2} - \frac{2(5\ \text{m/s})}{4\ \text{s}} = 5\ \text{m/s}^2 - 2.5\ \text{m/s}^2 = 2.5\ \text{m/s}^2$$

Equation (A.1) is a *linear* equation with respect to the variables s, v_0, and a. That is, these variables appear only to the first power. It is, however, a *quadratic* equation in the variable t. The general form of a quadratic equation is

$$ax^2 + bx + c = 0 \qquad \textbf{(A.4)}$$

where a, b, and c are constants. The solution of Equation (A.4) for x is

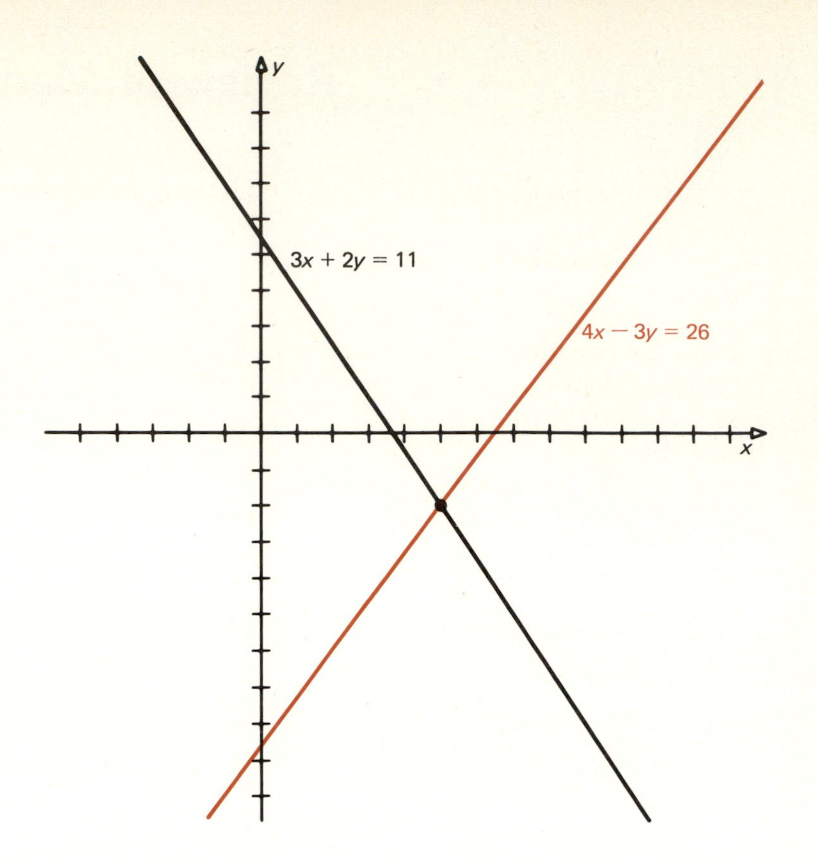

Figure A.1

$$x = \frac{-b \pm \sqrt{b^2 - 4ac}}{2a} \tag{A.5}$$

There are, in fact, *two* solutions, corresponding to the positive and negative values of the square root in Equation (A.5). Sometimes both, but frequently only one solution corresponds to a physically realizable situation. Both solutions of a quadratic equation must be examined to confirm that they conform with the problem statement and one solution may have to be discarded if it is inappropriate.

Another group of problems encountered in the text calls for the solution of simultaneous linear equations with two unknowns. The following is a typical example.

Determine the values of x and y that satisfy the equations

$$3x + 2y = 11 \tag{A.6}$$

$$4x - 3y = 26 \tag{A.7}$$

Equation (A.6) defines a straight line on an x,y plot. Consequently, an infinite number of x,y pairs will satisfy this equation, for instance, the pairs $x = 1$ and $y = 4$, or $x = 3$ and $y = 1$, or $x = -1$, $y = 7$. Similarly, Equation (A.7) also defines a straight line. The solution of the two simultaneous equations corresponds to the intersection of these two straight lines, as shown in Figure A.1.

There are several analytic procedures for solving simultaneous linear equations, two of which we now demonstrate.

Procedure I.

Step 1. Multiply each equation by a constant such that the coefficient of the unknown y has the same magnitude in both equations:

If we multiply (A.6) by 3 and (A.7) by 2 we obtain

$$9x + 6y = 33 \tag{A.6a}$$

$$8x - 6y = 52 \tag{A.7a}$$

Step 2. Add or subtract one equation to or from the other to eliminate the unknown y:

Adding (A.6a) to (A.7a) gives

$$17x = 85 \tag{A.8}$$

Step 3. Solve the equation for the unknown x:

Solving (A.8) yields

$$x = \frac{85}{17} = 5 \tag{A.9}$$

Step 4. Substitute the solution for x into one of the two original equations and solve for y:

We use Equation (A.6) and replace x by its value, 5. Thus

$$3(5) + 2y = 11$$

$$2y = 11 - 15 = -4$$

$$y = -2 \tag{A.10}$$

Step 5. Check the solutions. Substitute the results for x and y into the linear equation not used in step 4:

If we substitute the values $x = 5$ and $y = -2$ into Equation (A.7) we have

$$4(5) - 3(-2) = 20 + 6 = 26$$

Procedure II.

Step 1. Use one of the two equations to solve for x in terms of y:

From Equation (A.7) we obtain

$$4x = 26 + 3y$$

$$x = \frac{26 + 3y}{4} \tag{A.11}$$

Step 2. Substitute that expression for x into the other of the two equations to obtain a linear equation for the unknown y:

If we substitute (A.11) into (A.6) we obtain

$$3\left(\frac{26 + 3y}{4}\right) + 2y = 11 \tag{A.12}$$

Step 3. Solve the equation for y:

Multiplying both sides of (A.12) by 4, one obtains

$$\begin{aligned} 3(26 + 3y) + 8y &= 44 \\ 17y &= -34 \\ y &= -2 \end{aligned} \tag{A.13}$$

Step 4. Substitute the solution obtained in step 3 into the equation obtained in step 1 to determine the unknown x:

$$x = \frac{26 + 3(-2)}{4} = \frac{20}{4} = 5 \tag{A.14}$$

Step 5. Check the solutions. Substitute the results for x and y into the equation not used in step 1:

Since we used Equation (A.7) in step 1, we now substitute $x = 5$ and $y = -2$ into Eq. (A.6) with the result

$$3(5) + 2(-2) = 15 - 4 = 11$$

A.2 Exponents and Logarithms

The decimal system is based on multiples of 10. That is, $10 = 1 \times 10$, $100 = 1 \times 10 \times 10$, $1000 = 1 \times 10 \times 10 \times 10$, etc.

Multiplication of any number by itself n times is called raising that number to the n'th power. Thus

$$N \times N \times N = N^3$$

One refers to n as the exponent of N. In the above, the exponent of N is 3.

A negative exponent denotes the reciprocal of the number with positive exponent. That is,

$$N^{-1} = \frac{1}{N}, \; N^{-2} = \frac{1}{N^2}, \text{ etc.}$$

In particular,

$$0.1 = \frac{1}{10} = 10^{-1}, \; 0.01 = \frac{1}{100} = \frac{1}{10^2} = 10^{-2}, \text{ etc.}$$

Any quantity can be written as the product of a number between 1 and 10, and 10 raised to an integral power. For example,

$$21{,}600{,}000 = 2.16 \times 10^7,\text{*} \; 0.000825 = 8.25 \times 10^{-4}$$

This way of writing numbers is referred to as *scientific notation*.

Since

$$(10^n)(10^m) = 10^{n+m} \qquad \textbf{(A.15)}$$

the product of the two preceding numbers can be written (to three significant figures)

$$\begin{aligned}(21{,}600{,}000)(0.000825) &= (2.16 \times 10^7)(8.25 \times 10^{-4}) \\ &= (2.16)(8.25)(10^7)(10^{-4}) \\ &= 17.8 \times 10^3 = 1.78 \times 10^4\end{aligned}$$

Also, any number N^n that is raised to the mth power is

$$(N^n)^m = N^{nm} \qquad \textbf{(A.16)}$$

and, specifically,

$$(10^n)^m = 10^{nm}$$

Arithmetic operations involving powers are readily performed using scientific notation. For instance,

$$\begin{aligned}\frac{(21{,}600{,}000)^3}{(0.000825)^2} &= \frac{(2.16 \times 10^7)^3}{(8.25 \times 10^{-4})^2} = \frac{(2.16)^3(10^7)^3}{(8.25)^2(10^{-4})^2} \\ &= 0.148 \times \frac{10^{21}}{10^{-8}} = 0.148(10^{21})(10^8) \\ &= 0.148 \times 10^{29} = 1.48 \times 10^{28}\end{aligned}$$

* We have assumed here that none of the zeros in 21,600,000 is a significant figure. If, for instance, the first of the zeros were significant, we should then write this number as 2.160×10^7.

Clearly, keeping track of powers of ten—the location of the decimal point—is greatly facilitated by scientific notation.

In general, we can always write any number x as another number A raised to a power y, where y need not and usually is not an integral. That is,

$$x = A^y \tag{A.17}$$

The exponent y is called the logarithm of x to the base A and is written as follows

$$\log_A x = y \tag{A.18}$$

Logarithms to the base 10 are called *common logarithms,* and the symbol "log" is then used without indicating the base. Another base for logarithms that is widely used is the Napier or *natural* base, $e = 2.71828...$ Natural logarithms are indicated by the symbol "ln."

Since logarithms are exponents, the rules for multiplication using logarithms are

$$\begin{aligned} \log(xy) &= \log x + \log y \\ \log(x/y) &= \log x - \log y \\ \log(x^n) &= n\log x \end{aligned} \tag{A.19}$$

Logarithms were at one time extensively used in accurate calculations involving multiplication and division of numbers, and scientists and engineers of an earlier generation always carried devices based on logarithms, called slide rules. Today's electronic hand calculators have relegated the bulky seven-place log tables and the slide rules to museums of science and technology.

A.3 Trigonometry

The trigonometric functions are defined in terms of ratios of the lengths of the sides and hypotenuse of a right triangle. In Figure A.2, the right triangle has sides of lengths a and b and its hypotenuse is of length c. The trigonometric functions of the angle θ are defined by

$$\begin{aligned} \sin\theta &= a/c \qquad \csc\theta = (\sin\theta)^{-1} = c/a \\ \cos\theta &= b/c \qquad \sec\theta = (\cos\theta)^{-1} = c/b \\ \tan\theta &= a/b \qquad \cot\theta = (\tan\theta)^{-1} = b/a \end{aligned} \tag{A.20}$$

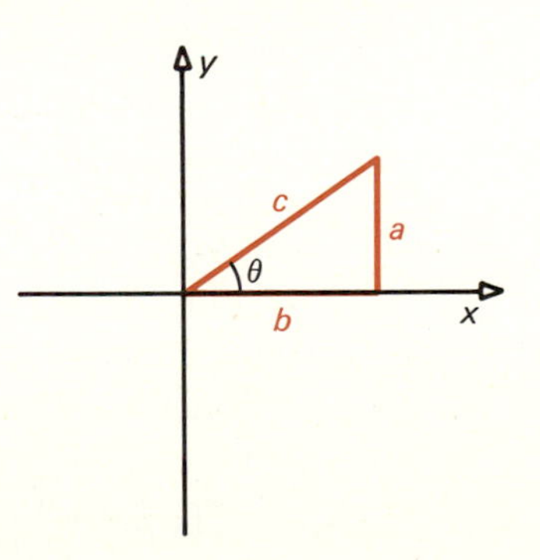

Figure A.2

All trigonometric functions are positive if θ lies in the first quadrant. For angles greater than $\pi/2$ (90°) or negative angles, the following rules apply:

Table A.1

Trigonometric function	Quadrant			
	I	II	III	IV
	$0 < \theta < \frac{\pi}{2}$	$\frac{\pi}{2} < \theta < \pi$	$\pi < \theta \lesssim \frac{3\pi}{2}$	$\frac{3\pi}{2} < \theta < 2\pi$
$\sin\theta$	+	+	−	−
$\cos\theta$	+	−	−	+
$\tan\theta$	+	−	+	−

1. The definition of the function is unchanged.

2. The hypotenuse c is always taken as positive.

3. The sides a and b are positive if they point in the positive direction of the x and y axes, negative if they point in the negative direction of these axes.

It follows from these rules that the signs of the sine, cosine and tangent are as indicated in Table A.1.

The trigonometric identities listed below are used in the text.

$$\sin^2\theta + \cos^2\theta = 1 \qquad \textbf{(A.21)}$$

$$\sin(\alpha + \beta) = \sin\alpha\cos\beta + \cos\alpha\sin\beta \qquad \textbf{(A.22)}$$

$$\cos(\alpha + \beta) = \cos\alpha\cos\beta - \sin\alpha\sin\beta \qquad \textbf{(A.23)}$$

$$\sin(2\theta) = 2\sin\theta\cos\theta \qquad \textbf{(A.24)}$$

$$\cos(2\theta) = \cos^2\theta - \sin^2\theta \qquad \textbf{(A.25)}$$

A.4 Series Expansions

In some instances, series expansions are valuable aids in obtaining analytic or numerical results. We mention here just a few series, of which the most useful is the *binomial expansion*

$$(1 + a)^n = 1 + na + \frac{n(n-1)}{2}a^2 + \frac{n(n-1)(n-2)}{6}a^3 + \cdots \qquad \textbf{(A.26)}$$

This series is especially useful when $a \ll 1$, for then all but the first two terms of the expansion can usually be neglected. To illustrate an application of this expansion we consider the following problem.

A metal cube measures exactly 18.6407cm on each side at 0°C. When the cube is warmed to 20°C, each side of the cube expands by 0.0084%. What is the fractional change in the volume of the cube?

One way to do this problem is to calculate the length of a side at 20°C by multiplying $L_0 = 18.6407$ cm by $(1 + 0.000084)$ giving

$$L_{20} = (18.6407 \text{ cm})(1.000084 \text{ cm}) = 18.64226582 \text{ cm}$$

One would then find the fractional change in volume, defined by

$$\frac{\Delta V}{V_0} = \frac{V_{20} - V_0}{V_0} = \frac{V_{20}}{V_0} - 1$$

by taking the cube of L_{20}, subtracting from this the cube of L_0, and dividing the difference by the cube of L_0. Since L_0 and L_{20} are not much different, one must carry the calculation to a large number of significant figures to be certain of a result accurate to only three significant figures.

The proper way to do this problem is with the help of Equation (A.26). We write

$$L_{20} = L_0(1 + a)$$

where $a = 8.4 \times 10^{-5} \ll 1$. The ratio of the two volumes is then

$$\frac{V_{20}}{V_0} = \frac{L_{20}^3}{L_0^3} = (1 + a)^3 \simeq 1 + 3a \qquad \textbf{(A.27)}$$

and the fractional change in volume is, therefore,

$$\frac{V_{20}}{V_0} - 1 \simeq 3a = 3(8.4 \times 10^{-5}) = 2.52 \times 10^{-4}. \qquad \textbf{(A.28)}$$

Note that the result, correct to three significant figures, was obtained without resorting to lengthy calculation and without the use of a hand calculator.

Other series expansions used in the text are

$$e^x = 1 + x + \frac{x^2}{2!} + \frac{x^3}{3!} + \cdots \qquad \textbf{(A.29)}$$

$$\sin\theta = \theta - \frac{\theta^3}{6} + \cdots \qquad \textbf{(A.30)}$$

$$\cos\theta = 1 - \frac{\theta^2}{2} + \cdots \qquad \textbf{(A.31)}$$

where the angle is expressed in radians. When $\theta \ll 1$,

$$\sin\theta \simeq \theta \qquad \textbf{(A.32)}$$

$$\cos\theta \simeq 1 \qquad \textbf{(A.33)}$$

$$\tan\theta = \frac{\sin\theta}{\cos\theta} \simeq \sin\theta \simeq \theta \qquad \textbf{(A.34)}$$

• APPENDIX B • Solar, Terrestrial, and Planetary Data

Object	*Mass (kg)*	*Radius (m)*	*Surface gravity (m/s^2)*	*Escape speed (m/s)*	*Orbit*	
					Radius (m)	*Period (yr)*
Sun	1.99×10^{30}	6.96×10^{8}	—	—	—	—
Earth	5.98×10^{24}	6.37×10^{6}	9.8	1.12×10^{4}	1.5×10^{11}	1.00
Moon	7.35×10^{22}	1.74×10^{6}	1.57	2.4×10^{3}	3.84×10^{8}	27.3 days
Mercury	3.3×10^{23}	2.43×10^{6}	3.72	4.2×10^{3}	5.8×10^{10}	0.241
Venus	4.87×10^{24}	6.05×10^{6}	8.92	1.03×10^{4}	1.08×10^{11}	0.615
Mars	6.4×10^{23}	3.39×10^{6}	3.82	5.1×10^{3}	2.28×10^{11}	1.88
Jupiter	1.90×10^{27}	6.87×10^{7}	26.9	6.1×10^{4}	7.78×10^{11}	11.86
Saturn	5.69×10^{26}	5.76×10^{7}	11.5	3.6×10^{4}	1.43×10^{12}	29.46
Uranus	8.7×10^{25}	2.51×10^{7}	9.2	2.2×10^{4}	2.87×10^{12}	84.01
Neptune	1.03×10^{26}	2.47×10^{7}	11.3	2.3×10^{4}	4.50×10^{12}	164.8
Pluto	6.6×10^{23}	$\leq 2.9 \times 10^{6}$	~ 5	—	5.90×10^{12}	248.4

• APPENDIX C • Currently Accepted Values of Fundamental Physical Constants

Quantity	*Symbol*	*Value†*	*Units*	*Uncertainty, parts in 10^6*
speed of light in vacuum	c	2.997924580(12)	10^8 m·s^{-1}	0.004
permeability of vacuum	μ_0	4π exactly	10^{-7} H·m^{-1}	—
permittivity of vacuum $1/\mu_0 c^2$	ϵ_0	8.854187818(71)	10^{-12} F·m^{-1}	0.008
Planck constant	h	6.626176(36)	10^{-34} J·Hz^{-1}	5.4
		4.135701(11)	10^{-15} eV·Hz^{-1}	2.6
$h/2\pi$	$\hbar$	1.0545887(57)	10^{-34} J·s	5.4
		6.582173(17)	10^{-16} eV·s	2.6
elementary charge (charge of the proton)	e	1.6021892(46)	10^{-19} C	2.9
Avogadro constant	N_A	6.022045(31)	10^{23} mol^{-1}	5.1
atomic mass unit, 10^{-3} kg·mol^{-1}·$N_A{}^{-1}$	u	1.6605655(86)	10^{-27} kg	5.1
		931.5016(26)	10^6 eV	2.8
Faraday constant of electrolysis $N_A e$	F	9.648456(27)	10^4 C·mol^{-1}	2.8
gravitational constant	G	6.6720(41)	10^{-11} N·m^2·kg^{-2}	615
Bohr radius $[\mu_0 c^2/4\pi]^{-1}(\hbar^2/m_e e^2)$	a_0	5.2917706(44)	10^{-11} m	0.82
charge to mass ratio of the electron	e/m_e	1.7588047(49)	10^{11} C·kg^{-1}	2.8
Compton wavelength of the electron $h/m_e c$	λ_C	2.4263089(40)	10^{-12} m	1.6
rest mass of the electron	m_e	9.109534(47)	10^{-31} kg	5.1
		5.4858026(21)	10^{-4} u	0.38
		0.5110034(14)	MeV	2.8
1 electron volt	eV	1.6021892(46)	10^{-19} J	2.9
in frequency units		2.4179696(63)	10^{14} Hz	2.6
in wavenumber units		8.065479(21)	10^5 m^{-1}	2.6
in temperature units		1.160450(36)	10^4 K	31
rest mass of proton	m_p	1.6726485(86)	10^{-27} kg	5.1
		1.007276470(11)	u	0.011
		938.2796(27)	MeV	2.8
ratio of proton mass to electron mass	m_p/m_e	1836.15152(70)	—	0.38
rest mass of neutron	m_n	1.6749543(86)	10^{-27} kg	5.1
		1.008665012(37)	u	0.037
		939.5731(27)	MeV	2.8
Rydberg constant (fixed nucleus) $[\mu_0 c^2/4\pi]^2(m_e e^4/4\pi\hbar^3 c)$	R_∞	1.097373177(83)	10^7 m^{-1}	0.075
		2.179907(12)	10^{-18} J	5.4
		13.605804(36)	eV	2.6
		3.28984200(25)	10^{15} Hz	0.075
		1.578885(49)	10^5 K	31

(*Continued*)

† The figures in parentheses represent the best estimates of the standard deviation uncertainties in the last two digits.

Quantity	*Symbol*	*Value*†	*Units*	*Uncertainty, parts in 10^6*
Boltzmann constant R/N_A	k	1.380662(44)	10^{-23} J·K^{-1}	32
		0.861735(28)	10^{-4} eV·K^{-1}	32
molar gas constant P_0V_m/T_0	R	8.31441(26)	J·mol^{-1}·K^{-1}	31
molar volume of ideal gas at STP (T_0 = 273.16 K; P_0 = 101325 Pa = 1 atm)	V_m	22.41383(70)	10^{-3} m^3·mol^{-1}	31
Stefan-Boltzmann constant $2\pi^5k^4/15h^3c^2$	σ	5.67032(71)	10^{-8} W·m^{-2}·K^{-4}	125

• APPENDIX D • The SI Units

Table D.1 SI base units

Base quantity	*SI base unit*	
	Name	*Symbol*
Length	meter	m
Mass	kilogram	kg
Time	second	s
Electric current	ampere	A
Temperature	kelvin	K
Amount of substance	mole	mol

The above base units are defined as follows.

1. The *meter* is the length equal to 1,650,763.73 wavelengths in vacuum of the radiation corresponding to the transition between the levels $2p_{10}$ and $5d_5$ of the krypton-86 atom.

2. The *kilogram* is the unit of mass; it is equal to the mass of the international prototype of the kilogram.

3. The *second* is the duration of 9,192,631,770 periods of the radiation corresponding to the transition between the two hyperfine levels of the ground state of the cesium-133 atom.

4. The *ampere* is that constant current which, if maintained in two straight parallel conductors of infinite length, of negligible circular cross

Table D.2 Derived SI units

Quantity	*Name*	*Symbol*	*Expression in terms of base units*	*Expression in terms of other SI units*
Plane angle	radian	rad	m/m	
Frequency	hertz	Hz	s^{-1}	
Force	newton	N	$kg{\cdot}m/s^2$	J/m
Pressure	pascal	Pa	$kg/m{\cdot}s^2$	N/m^2
Energy: work Quantity of heat	joule	J	$kg{\cdot}m^2/s^2$	$N{\cdot}m$
Power, radiant flux	watt	W	$kg{\cdot}m^2/s^3$	J/s
Electric charge	coulomb	C	$A{\cdot}s$	
Electric potential, emf	volt	V	$kg{\cdot}m^2/A{\cdot}s^3$	W/A
Capacitance	farad	F	$A^2{\cdot}s^4/kg{\cdot}m^2$	C/V
Electric resistance	ohm	Ω	$kg{\cdot}m^2/A^2{\cdot}s^3$	V/A
Magnetic flux	weber	Wb	$kg{\cdot}m^2/A{\cdot}s^2$	$V{\cdot}s$
Magnetic flux density	tesla	T	$kg/A{\cdot}s^2$	Wb/m^2
Inductance	henry	H	$kg{\cdot}m^2/A^2{\cdot}s^2$	Wb/A

section, and placed 1 meter apart in vacuum, would produce between these conductors a force equal to 2×10^{-7} newton per meter of length.

5. The *kelvin*, unit of thermodynamic temperature, is the fraction 1/273.16 of the thermodynamic temperature of the triple point of water.

6. The *mole* is the amount of substance of a system which contains as many elementary entities as there are atoms in 0.012 kilogram of carbon-12. When the mole is used, the elementary entities must be specified and may be atoms, molecules, ions, electrons, other particles, or specified groups of such particles.

• APPENDIX E • An Abbreviated Table of Isotopes

(1) *Atomic number Z*	(2) *Element*	(3) *Symbol*	(4) *Mass number, A*	(5) *Atomic mass‡*	(6) *Percent abundance, or decay mode if radioactive*	(7) *Half-life (if radioactive)*
0	(Neutron)	*n*	1	1.008665	β^-	10.6 min
1	Hydrogen	H	1	1.007825	99.985	
	Deuterium	D	2	2.014102	0.015	
	Tritium	T	3	3.016049	β^-	12.33 yr
2	Helium	He	3	3.016029	0.00014	
			4	4.002603	≈100	
3	Lithium	Li	6	6.015123	7.5	
			7	7.016005	92.5	
4	Beryllium	Be	7	7.016930	EC, γ	53.3 days
			8	8.005305	2α	6.7×10^{-17} s
			9	9.012183	100	
5	Boron	B	10	10.012938	19.8	
			11	11.009305	80.2	
6	Carbon	C	11	11.011433	β^+, EC	20.4 min
			12	12.000000	98.89	
			13	13.003355	1.11	
			14	14.003242	β^-	5730 yr
7	Nitrogen	N	13	13.005739	β^+	9.96 min
			14	14.003074	99.63	
			15	15.000109	0.37	
8	Oxygen	O	15	15.003065	β^+, EC	122 s
			16	15.994915	99.76	
			18	17.999159	0.204	
9	Fluorine	F	19	18.998403	100	
10	Neon	Ne	20	19.992439	90.51	
			22	21.991384	9.22	
11	Sodium	Na	22	21.994435	β^+, EC, γ	2.602 yr
			23	22.989770	100	
			24	23.990964	β^-, γ	15.0 h
12	Magnesium	Mg	24	23.985045	78.99	
13	Aluminum	Al	27	26.981541	100	

(*Continued*)

† Data are taken from *Chart of the Nuclides*, 12th ed., 1977, and from C. M. Lederer and V. S. Shirley, eds., *Table of Isotopes*, 7th ed., John Wiley & Sons, Inc., New York, 1978.

‡ The masses given in column (5) are those for the neutral atom, including the Z electrons.

(1) Atomic number Z	(2) Element	(3) Symbol	(4) Mass number, A	(5) Atomic mass‡	(6) Percent abundance, or decay mode if radioactive	(7) Half-life (if radioactive)
14	Silicon	Si	28	27.976928	92.23	
			31	30.975364	β^-, γ	2.62 h
15	Phosphorus	P	31	30.973763	100	
			32	31.973908	β^-	14.28 days
16	Sulfur	S	32	31.972072	95.0	
			35	34.969033	β^-	87.4 days
17	Chlorine	Cl	35	34.968853	75.77	
			37	36.965903	24.23	
18	Argon	Ar	40	39.962383	99.60	
19	Potassium	K	39	38.963708	93.26	
			40	39.964000	β^-, EC, γ, β^+	1.28×10^9 yr
20	Calcium	Ca	40	39.962591	96.94	
21	Scandium	Sc	45	44.955914	100	
22	Titanium	Ti	48	47.947947	73.7	
23	Vanadium	V	51	50.943963	99.75	
24	Chromium	Cr	52	51.940510	83.79	
25	Manganese	Mn	55	54.938046	100	
26	Iron	Fe	56	55.934939	91.8	
27	Cobalt	Co	59	58.933198	100	
			60	59.933820	β^-, γ	5.271 yr
28	Nickel	Ni	58	57.935347	68.3	
			60	59.930789	26.1	
			64	63.927968	0.91	
29	Copper	Cu	63	62.929599	69.2	
			64	63.929766	β^-, β^+	12.7 hr
			65	64.927792	30.8	
30	Zinc	Zn	64	63.929145	48.6	
			66	65.926035	27.9	
31	Gallium	Ga	69	68.925581	60.1	
32	Germanium	Ge	72	71.922080	27.4	
			74	73.921179	36.5	
33	Arsenic	As	75	74.921596	100	
34	Selenium	Se	80	79.916521	49.8	
35	Bromine	Br	79	78.918336	50.69	
36	Krypton	Kr	84	83.911506	57.0	
			89	88.917563	β^-	3.2 min
37	Rubidium	Rb	85	84.911800	72.17	
38	Strontium	Sr	86	85.909273	9.8	
			88	87.905625	82.6	
			90	89.907746	β^-	28.8 yr
39	Yttrium	Y	89	88.905856	100	
40	Zirconium	Zr	90	89.904708	51.5	

(Continued)

(1) *Atomic number Z*	(2) *Element*	(3) *Symbol*	(4) *Mass number, A*	(5) *Atomic mass‡*	(6) *Percent abundance, or decay mode if radioactive*	(7) *Half-life (if radioactive)*
41	Niobium	Nb	93	92.906378	100	
42	Molybdenum	Mo	98	97.905405	24.1	
43	Technetium	Tc	98	97.907210	β^-, γ	4.2×10^6 yr
44	Ruthenium	Ru	102	101.904348	31.6	
45	Rhodium	Rh	103	102.90550	100	
46	Palladium	Pd	106	105.90348	27.3	
47	Silver	Ag	107	106.905095	51.83	
			109	108.904754	48.17	
48	Cadmium	Cd	114	113.903361	28.7	
49	Indium	In	115	114.90388	95.7; β^-	5.1×10^{14} yr
50	Tin	Sn	120	119.902199	32.4	
51	Antimony	Sb	121	120.903824	57.3	
52	Tellurium	Te	130	129.90623	34.5; β^-	2×10^{21} yr
53	Iodine	I	127	126.904477	100	
			131	130.906118	β^-, γ	8.04 days
54	Xenon	Xe	132	131.90415	26.9	
			136	135.90722	8.9	
55	Cesium	Cs	133	132.90543	100	
56	Barium	Ba	137	136.90582	11.2	
			138	137.90524	71.7	
			144	143.922673	β^-	11.9 s
57	Lanthanum	La	139	138.90636	99.911	
58	Cerium	Ce	140	139.90544	88.5	
59	Praesodymium	Pr	141	140.90766	100	
60	Neodymium	Nd	142	141.90773	27.2	
			144	143.910096	α, 23.8	2.1×10^{15} yr
61	Promethium	Pm	145	144.91275	EC, α, γ	17.7 yr
62	Samarium	Sm	152	151.91974	26.6	
63	Europium	Eu	153	152.92124	52.1	
64	Gadolinium	Gd	158	157.92411	24.8	
65	Terbium	Tb	159	158.92535	100	
66	Dysprosium	Dy	164	163.92918	28.1	
67	Holmium	Ho	165	164.93033	100	
68	Erbium	Er	166	165.93031	33.4	
69	Thulium	Tm	169	168.93423	100	
70	Ytterbium	Yb	174	173.93887	31.6	
71	Lutecium	Lu	175	174.94079	97.39	
72	Hafnium	Hf	180	179.94656	35.2	
73	Tantalum	Ta	181	180.94801	99.988	
74	Tungsten	W	184	183.95095	30.7	
75	Rhenium	Re	187	186.95577	62.60, β^-	4×10^{10} yr

(Continued)

(1) *Atomic number Z*	(2) *Element*	(3) *Symbol*	(4) *Mass number, A*	(5) *Atomic mass‡*	(6) *Percent abundance, or decay mode if radioactive*	(7) *Half-life (if radioactive)*
76	Osmium	Os	191	190.96094	β^-, γ	15.4 days
			192	191.96149	41.0	
77	Iridium	Ir	191	190.96060	37.3	
			193	192.96294	62.7	
78	Platinum	Pt	195	194.96479	33.8	
79	Gold	Au	197	196.96656	100	
80	Mercury	Hg	202	201.97063	29.8	
81	Thallium	Tl	205	204.97441	70.5	
			210	209.990069	β^-	1.3 min
82	Lead	Pb	204	203.973044	β^-, 1.48	1.4×10^{17} yr
			206	205.97446	24.1	
			207	206.97589	22.1	
			208	207.97664	52.3	
			210	209.98418	α, β^-, γ	22.3 yr
			211	210.98874	β^-, γ	36.1 min
			212	211.99188	β^-, γ	10.64 h
			214	213.99980	β^-, γ	26.8 min
83	Bismuth	Bi	209	208.98039	100	
			211	210.98726	α, β^-, γ	2.15 min
			214	213.998702	β^-, α	19.7 min
84	Polonium	Po	210	209.98286	α, γ	138.38 days
			214	213.99519	α, γ	164 μs
85	Astatine	At	218	218.00870	α, β^-	$\approx$2 s
86	Radon	Rn	222	222.017574	α, γ	3.8235 days
87	Francium	Fr	223	223.019734	α, β^-, γ	21.8 min
88	Radium	Ra	226	226.025406	α, γ	1.60×10^3 yr
			228	228.031069	β^-	5.76 yr
89	Actinium	Ac	227	227.027751	α, β^-, γ	21.773 yr
90	Thorium	Th	228	228.02873	α, γ	1.9131 yr
			232	232.038054	100, α, γ	1.41×10^{10} yr
91	Protactinium	Pa	231	231.035881	α, γ	3.28×10^4 yr
92	Uranium	U	232	232.03714	α, γ	72 yr
			233	233.039629	α, γ	1.592×10^5 yr
			235	235.043925	0.72; α, γ	7.038×10^8 yr
			236	236.045563	α, γ	2.342×10^7 yr
			238	238.050786	99.275; α, γ	4.468×10^9 yr
			239	239.054291	β^-, γ	23.5 min
93	Neptunium	Np	239	239.052932	β^-, γ	2.35 days
94	Plutonium	Pu	239	239.052158	α, γ	2.41×10^4 yr
95	Americium	Am	243	243.061374	α, γ	7.37×10^3 yr
96	Curium	Cm	245	245.065487	α, γ	8.5×10^3 yr
97	Berkelium	Bk	247	247.07003	α, γ	1.4×10^3 yr

(Continued)

(1) *Atomic number Z*	(2) *Element*	(3) *Symbol*	(4) *Mass number, A*	(5) *Atomic mass‡*	(6) *Percent abundance, or decay mode if radioactive*	(7) *Half-life (if radioactive)*
98	Californium	Cf	249	249.074849	α, γ	351 yr
99	Einsteinium	Es	254	254.08802	α, γ, β^-	276 days
100	Fermium	Fm	253	253.08518	EC, α, γ	3.0 days
101	Mendelevium	Md	255	255.0911	EC, α	27 min
102	Nobelium	No	255	255.0933	EC, α	3.1 min
103	Lawrencium	Lr	257	257.0998	α	≈35 s
104	Rutherfordium(?)	Rf	261	261.1087	α	1.1 min
105	Hahnium(?)	Ha	262	262.1138	α	0.7 min
106			263	263.1184	α	0.9 s
107			261		α	1–2 ms

• APPENDIX F • Trigonometric Table

Angle				
Degrees	*Radians*	*Sine*	*Cosine*	*Tangent*
0°	0.000	0.000	1.000	0.000
1°	0.017	0.017	1.000	0.017
2°	0.035	0.035	0.999	0.035
3°	0.052	0.052	0.999	0.052
4°	0.070	0.070	0.998	0.070
5°	0.087	0.087	0.996	0.087
6°	0.105	0.105	0.995	0.105
7°	0.122	0.122	0.993	0.123
8°	0.140	0.139	0.990	0.141
9°	0.157	0.156	0.988	0.158
10°	0.175	0.174	0.985	0.176
11°	0.192	0.191	0.982	0.194
12°	0.209	0.208	0.978	0.213
13°	0.227	0.225	0.974	0.231
14°	0.244	0.242	0.970	0.249
15°	0.262	0.259	0.966	0.268
16°	0.279	0.276	0.961	0.287
17°	0.297	0.292	0.956	0.306
18°	0.314	0.309	0.951	0.325
19°	0.332	0.326	0.946	0.344
20°	0.349	0.342	0.940	0.364
21°	0.367	0.358	0.934	0.384
22°	0.384	0.375	0.927	0.404
23°	0.401	0.391	0.921	0.424
24°	0.419	0.407	0.914	0.445
25°	0.436	0.423	0.906	0.466
26°	0.454	0.438	0.899	0.488
27°	0.471	0.454	0.891	0.510
28°	0.489	0.469	0.883	0.532
29°	0.506	0.485	0.875	0.554
30°	0.524	0.500	0.866	0.577
31°	0.541	0.515	0.857	0.601
32°	0.559	0.530	0.848	0.625
33°	0.576	0.545	0.839	0.649
34°	0.593	0.559	0.829	0.675
35°	0.611	0.574	0.819	0.700
36°	0.628	0.588	0.809	0.727
37°	0.646	0.602	0.799	0.754
38°	0.663	0.616	0.788	0.781
39°	0.681	0.629	0.777	0.810
40°	0.698	0.643	0.766	0.839
41°	0.716	0.656	0.755	0.869
42°	0.733	0.669	0.743	0.900
43°	0.750	0.682	0.731	0.933
44°	0.768	0.695	0.719	0.966
45°	0.785	0.707	0.707	1.000

Angle				
Degrees	*Radians*	*Sine*	*Cosine*	*Tangent*
46°	0.803	0.719	0.695	1.036
47°	0.820	0.731	0.682	1.072
48°	0.838	0.743	0.669	1.111
49°	0.855	0.755	0.656	1.150
50°	0.873	0.766	0.643	1.192
51°	0.890	0.777	0.629	1.235
52°	0.908	0.788	0.616	1.280
53°	0.925	0.799	0.602	1.327
54°	0.942	0.809	0.588	1.376
55°	0.960	0.819	0.574	1.428
56°	0.977	0.829	0.559	1.483
57°	0.995	0.839	0.545	1.540
58°	1.012	0.848	0.530	1.600
59°	1.030	0.857	0.515	1.664
60°	1.047	0.866	0.500	1.732
61°	1.065	0.875	0.485	1.804
62°	1.082	0.883	0.469	1.881
63°	1.100	0.891	0.454	1.963
64°	1.117	0.899	0.438	2.050
65°	1.134	0.906	0.423	2.145
66°	1.152	0.914	0.407	2.246
67°	1.169	0.921	0.391	2.356
68°	1.187	0.927	0.375	2.475
69°	1.204	0.934	0.358	2.605
70°	1.222	0.940	0.342	2.748
71°	1.239	0.946	0.326	2.904
72°	1.257	0.951	0.309	3.078
73°	1.274	0.956	0.292	3.271
74°	1.292	0.961	0.276	3.487
75°	1.309	0.966	0.259	3.732
76°	1.326	0.970	0.242	4.011
77°	1.344	0.974	0.225	4.332
78°	1.361	0.978	0.208	4.705
79°	1.379	0.982	0.191	5.145
80°	1.396	0.985	0.174	5.671
81°	1.414	0.988	0.156	6.314
82°	1.431	0.990	0.139	7.115
83°	1.449	0.993	0.122	8.144
84°	1.466	0.995	0.105	9.514
85°	1.484	0.996	0.087	11.43
86°	1.501	0.998	0.070	14.30
87°	1.518	0.999	0.052	19.08
88°	1.536	0.999	0.035	28.64
89°	1.553	1.000	0.017	57.29
90°	1.571	1.000	0.000	

Answers to Multiple Choice Questions and Odd Numbered Problems

• CHAPTER 1 •

Problems*

1.1 (a) 56.4 km/hr
(b) 15.7 m/s

1.3 (a) 8.00 ft^3
(b) 0.227 m^3
(c) 2.27×10^5 cm^3
(d) 227 liters
(e) 59.8 gallons

1.5 (a) 2.50×10^{-2} amp
(b) 2.50×10^4 μamp
(c) 2.50×10^{-5} kamp

1.7 No, the student would not qualify.

1.9 6.328×10^{-7} m

1.11 5.60×10^3 km

1.13 17.9 m/s; 6.10 m/s; 13.9 m/s

1.15 4.807×10^3 m

1.17 $[E] = [M][v]^2$; $[P] = [M][v][a]$

1.19 $\frac{F}{A} L^3 = \frac{MLT^{-2}}{L^2} L^3 = ML^2T^{-2}$, which "is" energy

1.21 $v \, \alpha \, \frac{mg}{r\eta}$; yes.

1.23 $MLT^{-2} = LT^{-2}X$ and $X = M$

1.25 (varies from college to college)

1.27 7,000 – 10,000

1.29 40–50 million persons in age group 6 to 17; 2–3 million teachers.

1.31 0.3 m (rounded off from 0.25 m); 5×10^{-3} m^2

1.33 0.196 m^2; 8.18×10^{-3} m^3

1.35 2.31 m

1.37 5.49×10^3 kg/m^3

1.39 13.0 km, 67.3° north of east

1.41 $A = 5$
$B_x = -8$, $B_y = 0$
$C_x = 5.20$, $C_y = 3.00$

1.43 $\mathbf{F} = (1.20, 6.00)$ or $F = 6.12$, $\theta_F = 78.7°$

1.45 $\mathbf{P} = (2.8, -3.00)$ or $P = 4.1$, $\theta_P = -47°$

1.47 $\mathbf{Q} = (13.20, 3.00)$ or $Q = 13.5$, $\theta_Q = 12.8°$

1.49 Angle between x'' and x-axis is 36.9°
$B_{x''} = -6.40$, $B_{y''} = 4.80$

1.51 22.5 km, 58° south of west

1.53 60.3 m, 28° north of east

1.55 57.0 N, 128°

1.57 75° north of west; 5.12 km

1.59 29.5 km

* Problems demanding sketches, curves, etc. are not included among the answers.

• CHAPTER 2 •

Multiple Choice

2.1 a **2.2** b **2.3** d **2.4** c **2.5** d
2.6 d **2.7** c **2.8** d **2.9** a

Problems

2.1 5.33 m/s; 0

2.3 15 mph; 6.71 m/s

2.5 486 km/hr, 21.8° south of east; 205 km east and 120 km south of Springfield

2.7 2 m/s; −2 m/s; −8 m/s

2.9 160 m

2.11 4.05° north of east, 427 km/hr

2.13 27.4 s; 170 s (2 min., 50 s)

2.15 16.1° north of west; 0.346 hr = 10.4 min

2.17 1.23 km, 54.9° south of east

2.19 8.84° south of west; 10:23 a.m.

2.21 (a) 2.78 km/h; 19.9° east of south
(b) 53.5° north of west; 3.71 h = 3 h, 43 min.

2.23

$v = 0$	$v > 0$	$v < 0$	$a = 0$	$a > 0$	$a < 0$
CD, I	DI	AC	AB	BC	FG
		IK	CD	DE	HJ
			EF		
			GH		
			JK		

2.25 40 m/s; 15 m/s^2

2.27 7.31 m; 2.44 s

2.29 6.40 m

2.31 9.60 m/s, upward

2.33 (a) at $t = 0$: 0 m/s, 5.00 m/s^2
at $t = 2$s: 10.0 m/s, 5.00 m/s^2
(b) 15.0 m/s

2.35 at $t = 2$ s: $x = 13.0$ m, $y = 24$ m
at $t = 3$ s: $x = 10.5$ m, $y = 34$ m

2.37 −6.94 m/s^2; 31.2 m

2.39 $a = 0.250$ m/s^2, $\langle v \rangle = 0.5$ m/s, $v = 1.00$ m/s

2.41 $a = 1.06$ m/s^2, west, 15.1 s; 295 m

2.43 2.50 s, 30.6 m

2.45 32.1 m, 17.8 m/s, downward; 30.8 m/s

2.47 2.74 s, 18.9 m/s, 23.3 m

2.49 Yes, second stone will overtake first; $t = 2.69$ s, 35.5 m below cliff.
If second stone is thrown at $v_{02} = 5$ m/s, it will not catch up with the first.
Yes, there is a minimum initial downward speed of 14.7 m/s.

2.51 For $v_0 = 5.40$ m/s, $h = 1.49$ m; for $v_0 = 53.4$ m/s, $h = 145$ m

2.53 415 m

2.55 (a) no collision
(b) yes, they collide; 504 m; 4.7 m/s

2.57 1.21 s if thrown downward, 4.12 s if thrown upward

2.59 713 m from plane; 3.57 s, 272 m/s

2.61 15.7 m/s

2.63 978 m; 1.06×10^3 m

2.65 109 m, 35.9 m/s

2.67 20.6 m/s, 56.3° above horizontal; 11.4 m/s; 1.75 s

• CHAPTER 3 •

Multiple Choice

3.1 a **3.2** c **3.3** c **3.4** c **3.5** b
3.6 d **3.7** a **3.8** c **3.9** d **3.10** b
3.11 d **3.12** d **3.13** a

Problems

3.1 (a) The table pushes up on the ball, and the ball pushes down on the table.
(b) The earth pulls down on the ball, and the ball pulls up on the earth.
(c) The floor pushes up on the ball, and the ball pushes down on the floor.

3.3 2.50 kg

3.5 (a) Force of mass on hand: 7.84 N down; force of hand on mass: 7.84 N up
(b) 2.70 m/s² upward; the force of the earth on the mass is still 7.84 N downward, and is equal and opposite to the force exerted by the mass on the earth.

3.7 18.0 N

3.9 (a) 19.6 N
(b) 39.2 N

3.11 At point *A*: 490 N; at point *B*: 245 N; at point *C*: 245 N; at point *D*: 245 N

3.13 4.00 kg

3.15 1186 N; tension in string: 1200 N

3.17 tension in rope: 39.2 N; traction force: 58.8 N

3.19 494 N

3.21 4.90 kg; 48.5 N

3.23 1.07×10^5 N

3.25 23.1 N, 34.6 N, 73.2 N

3.27 At $t = 2$ s: 0.006 m below original position
At $t = 6$ s: 30.6 m above original position
$M_1 = 7.04$ kg

3.29 588 N, 408 N

3.31 (a) 1.80 m/s²
(b) 58.0 N, 96.0 N
(c) 2.33 s

3.33 0.714

3.35 Yes, it is 0.577. Yes, a 10-kg block would move at constant velocity once it was moving.

3.37 16.3 m, 4.08 s

3.39 16.74 N

3.41 (a) Relative to the block, the mass will slide down it. Relative to an outside observer, the block will move to the left and the mass to the right.
(b) $F = mg \cos 37° \sin 37°$
(c) $a = g \tan 37°$
(d) $F = (m + M)g \tan 37°$

3.43 6.75 m/s²; 69.4 N

3.45 0.94; 5.9 N

3.47 (a) 3.92 m/s²
(b) In 1 s, the box has traveled 1.275 m toward the rear of the truck. At that time, $v_{BT} = -2.55$ m/s. The box ends up 2.602 m "behind" its starting position, or 3.40 m from the truck's tailgate, at rest relative to the truck.
(c) Yes the box will slip off, after 2.17 s

3.49 (a) 0.510
(b) 0.510, 15.0 N
(c) 0.292, 23.6 N

3.51 $\mu_k = 0.37$; $F_f = 18.5$ N

• CHAPTER 4 •

Multiple Choice

4.1 b **4.2** b **4.3** c **4.4** c **4.5** c
4.6 a **4.7** b **4.8** c **4.9** c **4.10** b
4.11 b

Problems

4.1 44.1 J

4.3 1.55×10^7 m/s

4.5 KE = 5.79×10^5 J. Space heater: 6.00×10^5 J. Thus heater is slightly more.

4.7 Work = area under curve, or $\frac{1}{2}$ (height × base)
$W = \frac{1}{2}Fx$
$F = kx$, $W = \frac{1}{2}kx^2$

4.9 0.459 m

4.11 22.2 m/s

4.13 (a) 7.00 m/s
(b) 6.02 m/s

4.15 33.3 N, compared to $W = 39.2$ N

4.17 0.816 m

4.19 41.4°

4.21 0.349

4.23 3.67 m/s; 6.71 kg

4.25 833 N; 666 J;
525 N; 840 J

4.27 5.42 m/s; 5.79 m

4.29 4.17 kg; 5.00 s

4.31 (a) 4.312×10^5 J
(b) $v_C = 23.0$ m/s, $v_D = 18.0$ m/s
(c) 2.16×10^4 N

4.33 806 N

4.35 24.8 W

4.37 83.0 W

4.39 KE = 2.083×10^4 J, $P = 914$ W

4.41 4.61×10^3 hp. This is unrealistic because neglecting all friction losses is unrealistic.

4.43 After 3 s: 15.5 m/s
After 6 s: 21.9 m/s
$\langle a \rangle_{0-3} = 5.17$ m/s^2, $\langle a \rangle_{3-6} = 2.13$ m/s^2

4.45 19.6 s; 735 N

• CHAPTER 5 •

Multiple Choice

5.1 b 5.2 a 5.3 c 5.4 d 5.5 d
5.6 b 5.7 a 5.8 a 5.9 a 5.10 a
5.11 c 5.12 d 5.13 d 5.14 a

Problems

5.1 2.50×10^4 kg·m/s

5.3 90.0 N

5.5 132 N

5.7 19.14 kg·m/s; 1.28×10^4 N, 15° above horizontal.

5.9 (a) $a = 9.03 \times 10^5$ m/s^2, $\langle F \rangle = -3.61 \times 10^4$ N, $t = 4.21 \times 10^{-4}$ s
$\langle F \rangle t = 15.2$ N·s, $\Delta p = 15.2$ kg·m/s
(b) (i) 18.1 m/s
(ii) 66.9 m
(iii) 2.75×10^3 J, 138 J

5.11 570 m/s, 1.30×10^3 J

5.13 144 N, 10.95 m/s

5.15 (a) 4.74 m/s; 2.13×10^3 J
(b) 5.79 m/s; 3.18×10^3 J
(c) 5.26 m/s; 2.63×10^3 J

5.17 1.41 m

5.19 $(\frac{1}{2}, \frac{1}{2})$

5.21 8 kg·m/s, 5.33 J

5.23 12 km/h (3.33 m/s), 41.6° east of north; 1.13 m

5.25 1.06 m/s, $\theta = 45.0°$

5.27 $v_1 = 10$ m/s, $v_2' = 12$ m/s

5.29 (a) 18.03 m/s toward the west, 42.0° above the horizontal
(b) ΔKE = $+125M$ J
(c) The fragment of mass $M/3$ will strike the ground first

5.31 $v_2' = v_1 \left(\frac{2M}{M - m}\right)$; if $M \gg m$ $v_2' \sim 2v_1$

5.33 1.56×10^4 m from gun.

• CHAPTER 6 •

Multiple Choice

6.1 b 6.2 c 6.3 c 6.4 b 6.5 b
6.6 b 6.7 c 6.8 b 6.9 d 6.10 c

Problems

6.1 23.9 revolutions

6.3 4.71 rad/s

6.5 0.848 m

6.7 27.8 rad/s

6.9 0.532 m/s

6.11 16.3 rad/s

6.13 -1.80×10^{-2} rad/s^2

6.15 8.66 rad/s

6.17 3.86×10^3 m

6.19 12.1 cm (4.76 in.)

6.21 9.39 m/s, 4.50 m above top of loop or 16.5 m above bottom of loop

6.23 4.43 rad/s

6.25 2.88 g (his apparent weight is 3.88 mg)

6.27 2.42 m/s; 1.62 rad/s; 11 N

6.29 $\frac{5}{4}d$

6.31 At position where mass loses contact with globe, centripetal force is provided by gravity alone. $F_c = mv^2/R = mg \cos \theta$, $\frac{5}{3} R$

6.33 0.0885

6.35 $T = 2\pi \sqrt{\frac{L \cos \theta}{g}}$

6.37 92.2 km/h

6.39 5.40°

6.41 3.83×10^3 N toward center.
$\mu_s \geq 0.33$

6.43 1.97×10^3 N; 11.6×10^3 N

• CHAPTER 7 •

Multiple Choice

7.1 a 7.2 b 7.3 b 7.4 a 7.5 d
7.6 a 7.7 d

Problems

7.1 1.63 m/s^2

7.3 2.21 N, 40.7 h

7.5 3.38×10^{-8} N

7.7 Setting $M = M_s$ and $r = R$ yields
$M_S = 4\pi^2 R^3/GT^2$
1.97×10^{30} kg

7.9 $2.21 \times 10^{-2} F_{ME}$

7.11 1.89×10^{27} kg, 1.24×10^3 kg/m^3, 24.7 m/s^2

7.13 3.37×10^{-2} m/s^2. This is 3.44×10^{-3} g, or about $\frac{1}{3}$ of 1% of g

7.15 3.46×10^{-8} m from Earth.

• CHAPTER 8 •

Multiple Choice

8.1 a	**8.2** d	**8.3** b	**8.4** c	**8.5** d
8.6 b	**8.7** b	**8.8** b	**8.9** b	**8.10** b
8.11 b	**8.12** d	**8.13** b	**8.14** b	**8.15** c

Problems

8.1 The string should be attached 0.900 m from the left end of the rod; $T = 11.7$ N

8.3 1.33×10^3 N

8.5 566 N; 1.88×10^3 N upward at foot of mast

8.7 212 N

8.9 206 N

8.11 $T = 40.8$ N; $R_H = 35.4$ N, $R_V = 8.98$ N or $R = 36.5$ N, 14.2° above horizontal

8.13 He should stop at 4.21 m

8.15 $X_{50}^{max} = 8.02$ m; $X_{80}^{max} = 9.48$ m

8.17 (a) 0.048 kg·m²
(b) 0.048 kg·m²
(c) 0.096 kg·m²

8.19 $I = M_1L^2$

8.21 0.480 kg·m²

8.23 0.168 N·m, 75.8 W

8.25 200 kg·m²

8.27 (a) 16.4 N
(b) 8.04 rad/s²
(c) 0.408 kg·m²
(d) 20.4 kg

8.29 0.216 kg·m²; 0.108 kg·m²

8.31 11.6 N; 5.8 kg

8.33 1.28 kg·m²

8.35 2.13 rad/s; $KE_i = 1.07 \times 10^3$ J, $KE_f = 602$ J

8.37 After girl has swung onto the edge: 0.5 rev/s ($=\pi$ rad/s), 1.58×10^3 J
After girl has moved toward the center: 5.32 rad/s, 2.67×10^3 J

8.39 $I = 187$ kg·m², $KE_i = 163$ J, $KE_f = 286$ J

8.41 14 s; 1.42×10^5 J

8.43 2.86 rad/s; for an ordinary pendulum, 2.33 rad/s

8.45 $T = 13.4$ N, $v = 4.62$ m/s, KE = 58.8 J, $t = 0.865$ s

• CHAPTER 9 •

Multiple Choice

9.1 d	**9.2** a	**9.3** a	**9.4** d	**9.5** b

Problems

9.1 4.80 cm × 9.60 cm × 19.2 cm

9.3 253 kg

9.5 5.10 mm

9.7 0.377 N

9.9 7.00×10^6 Pa (about 70 atm)

9.11 7.5×10^5 Pa or ~7.5 atm

9.13 5.00 mm

9.15 11.5 cm

9.17 5.2 mm

9.19 1.44 mm

9.21 0.826 mm

9.23 $\frac{7}{19}$ supported by Al. $\frac{12}{19}$ supported by Cu.

• CHAPTER 10 •

Multiple Choice

10.1 a	**10.2** b	**10.3** c	**10.4** c	**10.5** b
10.6 b	**10.7** c	**10.8** d	**10.9** c	**10.10** d
10.11 b	**10.12** c	**10.13** b	**10.14** d	

Problems

10.1 0.9770×10^5 Pa; approx. 0.9767×10^5 Pa
0.8302×10^5 Pa; approx. 0.814×10^5 Pa
0.7565×10^5 Pa; approx. 0.721×10^5 Pa

10.3 0.950 atm; approx. 0.948 atm

10.5 1.79×10^5 Pa = 1.77 atm

10.7 1.14×10^5 Pa

10.9 328 cm²

10.11 $P(4) = 1.41 \times 10^5$ Pa, $P(3) = 1.31 \times 10^5$ Pa

10.13 2.79×10^6 Pa

10.15 1.21 cm

10.17 1.08×10^4 Pa

10.19 80 kg/m³

10.21 0.121 m

10.23 680 kg/m³; 0.731

10.25 0.107

10.27 0.80

10.29 1.0 kg

10.33 194 cm³/s = 0.194 L/s; 1.34 Pa

10.35 (a) 0.604 m/s
(b) 0.955 m/s

10.37 1.73×10^{-3} m³/s = 1.73 L/s; 29 s

10.39 $d = 0.372$ mm

10.41 71.4 cm

10.43 $d = 8.04$ cm, 61 atm; 29.4 atm

10.45 3.15 mm/s

10.47 0.68 mm/s

10.49 0.212 mm/s; 75.9 mm/s

10.51 1.19×10^3 rpm

• CHAPTER 11 •

Multiple Choice

11.1 b	**11.2** d	**11.3** b	**11.4** d	**11.5** c
11.6 b	**11.7** c	**11.8** b	**11.9** c	**11.10** d
11.11 a				

Problems

11.1 37.0 °C, 310.2 K
11.3 29.8 °C, 85.6 °F
11.5 0.253 °C; water cooled by evaporation.
11.7 48.8 min
11.9 107.6 °F; the patient is seriously ill.
11.11 −459.8 °F
11.13 8.87×10^3 kg/m^3
11.15 0.0324 liter of mercury will overflow the container
11.17 0.024% too high
11.19 62.9 °C
11.21 3600 N
11.23 2.40×10^5 cal
11.25 4.38×10^5 cal
11.27 0.196 cal/gm·C°
11.29 5.32×10^3 cal
11.31 45.45 °C
11.33 21.4 g
11.35 53.7 s
11.37 9.38%
11.39 1.97 μm/s or 7.1 mm/hr
11.41 100 − 200 W
11.43 127 K or −146 °C

• CHAPTER 12 •

Multiple Choice

12.1 d **12.2** b **12.3** c **12.4** c **12.5** d
12.6 d **12.7** a **12.8** d **12.9** c

Problems

12.1 1.42×10^{23} atoms
12.3 27.8 moles of O_2, 55.6 moles of H_2
12.5 1.65 moles of C_2H_5OH
12.7 9.79 L
12.9 288 K or 15 °C
12.11 3.29×10^{16} molecules/cm^3
12.13 8.97×10^5 Pa
0.394 m below top of bell
12.15 7.03×10^3 gm, or 7.03 kg
251 moles
12.17 61.4 N
12.19 0.43%
12.21 5.52×10^{-17} Pa
316 m/s
12.23 0.108 s
The time to diffuse will be longer because of the many collisions that H_2S molecules will encounter.
12.25 426 m/s
12.29 5.5 min
12.31 789 cal = 3.30×10^3 J
12.33 11.8 L; $P_{Ar} = \frac{1}{21}$ atm, $P_{H_2} = \frac{20}{21}$ atm
12.35 70 L
$P_{O_2} = 0.11$ atm, $P_{H_2} = 1.72$ atm, $P_{Ne} = 0.17$ atm
12.37 176 cal = 736 J

• CHAPTER 13 •

Multiple Choice

13.1 a **13.2** b **13.3** b **13.4** b **13.5** b
13.6 a **13.7** a **13.8** c **13.9** c **13.10** c

Problems

13.1 20.0
13.3 The most likely total is 7, which has 6 microstates.
13.5 1.13
13.7 754 J
13.9 $\Delta Q = 0$
$\Delta U = +80$ J
13.11 4.72×10^6 Pa
1096 °C
13.13 From A to C, no work is done; $W_{CB} = 1.50 \times 10^3$ J
$\Delta U = 1500$ J
$\Delta Q = 3000$ J
13.15 $\Delta U = 1500$ J
$\Delta Q = 3.89 \times 10^3$ J
13.17 0.626
13.19 123 °C
13.21 255 °C
13.23 10.3 kg
13.27 0.12; for a Carnot engine: 0.80
13.29 21.5 °C,
$\Delta S = 8.7 \times 10^{-3}$ cal/K

• CHAPTER 14 •

Multiple Choice

14.1 a **14.2** c **14.3** c **14.4** c **14.5** d
14.6 c **14.7** c

Problems

14.1 (a) $y(t) = 0.02 \sin (377t)$ m
(b) $y(t) = 0.02 \sin \left(377t + \frac{\pi}{2}\right)$ m
(c) $y(t) = 0.02 \sin \left(377t - \frac{\pi}{2}\right)$ m
14.3 (a) $x(t) = 0.03 \sin \left(10.47t + \frac{\pi}{2}\right)$ m
(b) $x(t) = 0.03 \sin (10.47t)$ m
(c) $x(t) = 0.03 \sin \left(10.47t - \frac{\pi}{6}\right)$ m
14.5 99.3 g
14.7 18 N/m

14.9 6.12 cm

14.11 0.291 Hz

14.13 $y(t) = 0.03\cos(14.0t)$ meters, where down is the assumed positive direction

14.15 55.8 g; 0.0425 J

14.17 (a) $\omega = \sqrt{k/M}$
(b) $\omega = \sqrt{k/M}$
(c) $\omega = \sqrt{k/2M}$
(d) $\omega = \sqrt{2k/M}$

14.19 18.18 m/s

14.21 $A = 0.6$ m, $f = 1.25$ Hz, $\omega = 7.85$ rad/s
At $t = 0$, $v = 0$
At $t = 0$, $a = -37.0\ \text{m/s}^2$

14.23 $\omega = 0.337\pi \sin(4.04t)$

14.25 24.8 cm

14.29 If it is climbing at constant speed up a 6° grade, the period will be the same as it is at rest. If it is accelerating, the tension in the pendulum support will be

$$T = \frac{mg}{\cos 6°}$$

$$g_{\text{eff}} = \frac{g}{\cos 6°} = 1.0055\ g$$

Thus $T' = 2\pi\sqrt{l/g_{\text{eff}}}$; $T = 2\pi\sqrt{l/g}$; $T' = T\sqrt{g/g_{\text{eff}}} = 1.2\sqrt{\frac{1}{1.0055}} = 1.1967$ s

14.31 0.01% smaller

14.33 1.75 kg·m/s

14.37 $\text{PE} = \frac{1}{2}mgL\theta^2$
$\text{KE} = \frac{1}{2}mL^2\omega^2$

14.39 $\frac{E_A}{E_B} = \frac{8}{5}$

14.41 0.042

14.43 2.35 W
Energy of vibration is 3.55 J

$$\frac{\text{energy given}}{\text{energy of vibration}} = 0.22$$

14.45 $T = 1.4$ s

• CHAPTER 15 •

Multiple Choice

15.1 e **15.2** b **15.3** b **15.4** a **15.5** e
15.6 b **15.7** c **15.8** e **15.9** b

Problems

15.1 0.02 s; 50 m/s

15.3 1.20 m/s

15.5 2.5 m/s

15.7 $x = 0.05\sin[2\pi(62.5t - 2.5z)]$ m

15.9 $\lambda = 5$ cm, $f = 0.25$ Hz, $T = 4$ s, $A = 2.5$ cm, $v = 1.25$ cm/s

15.11 1.73 A; 1.41 A; 0.52 A.

15.15 $\lambda = 0.5$ m $f = 1.35$ Hz

15.17 96 m/s

15.19 80 cm
25 Hz
First overtone will be the third harmonic, f = 75 Hz.

15.21 Increase by 0.68%

15.23 1.09×10^{-4} kg/m

15.25 330 Hz

15.27 230 N

15.29 2.96 g, 87.5 Hz

15.31 0.228 g

15.33 $\Delta f = -0.0165$ Hz

15.35 $E = 0.197$ J
Power transmitted = 0

• CHAPTER 16 •

Multiple Choice

16.1 c **16.2** b **16.3** b **16.4** d **16.5** a
16.6 b **16.7** c **16.8** e

Problems

16.1 7.49×10^{-3} m

16.3 −49 °C; −59 °C

16.5 1782 m/s

16.7 −51 °C

16.9 268 Hz

16.11 283 Hz; 567 Hz

16.13 The pipe is closed at one end, since it resonates only at odd harmonics.
L = 0.500 m
Yes, the temperature drop will affect the frequencies. Presuming the pipe length does not change, the frequencies drop by 1.7%.

16.15 350 m/s; 52.5 cm

16.17 The first is 40 dB "higher" in intensity than the second.

16.19 At 60 dB, 3×10^{-2} Pa; at 120 dB, 30 Pa

16.21 $2.65 \times 10^{-2}\ \text{W/m}^2 = 104$ dB

16.23 88.8 dB

16.25 0.503

16.27 403 Hz

16.29 386 Hz

16.31 No; 200 Hz

16.33 6660 Hz

16.35 61 dB

• CHAPTER 17 •

Multiple Choice

17.1 d **17.2** c **17.3** b **17.4** b **17.5** b
17.6 c **17.7** c **17.8** b **17.9** b **17.10** b
17.11 c **17.12** c **17.13** b

Problems

17.1 1.2×10^6 N (attractive)
17.3 9.63×10^4 C
17.5 1.28 N
17.7 9.216×10^{-5} N
17.9 57.6 N; weight of 1 mol of helium = 0.0392 N
17.11 6.83 μC; 1.17 μC
17.13 1.79×10^{-8} C
17.15 $F_{+4} = 29.3 \times 10^{-3}$ N, $\theta = 37.8°$; $F_{+2} = 11.8 \times 10^{-3}$ N, $\theta = 130°$, $F_{-3} = 31.2 \times 10^{-3}$ N, $\theta = 239°$
17.17 Yes, 3 m to the right of the -2 μC charge
17.19 $F_1 = 5.41 \times 10^{-3}$ N; $\theta = -54.1°$
$F_2 = 2.42 \times 10^{-3}$ N; $\theta = 135°$
$F_3 = 5.41 \times 10^{-3}$ N; $\theta = -35.9°$
$F_4 = 8.27 \times 10^{-3}$ N; $\theta = 135°$

(2) .3 μC .3 μC (3)
(1) (4)
.3 μC $-$.4 μC

• CHAPTER 18 •

Multiple Choice

18.1 b	**18.2** b	**18.3** c	**18.4** a	**18.5** a
18.6 d	**18.7** a	**18.8** a	**18.9** d	**18.10** b
18.11 c	**18.12** c	**18.13** a	**18.14** d	
18.15 c	**18.16** d	**18.17** b	**18.18** d	
18.19 a	**18.20** a	**18.21** b	**18.22** b	

Problems

18.1 20.4 g; **E** is directed vertically upward
18.3 $\mathbf{F} = 9.60 \times 10^{-17}$ N; $\mathbf{a} = 1.05 \times 10^{14}$ m/s² in negative x-direction
18.5 $x = 30$ cm, $y = 35.53$ cm
18.7 $E = 1.01 \times 10^5$ N/m; $\theta = -26.6°$
18.9 1.09×10^{-3} m/s; 3.27×10^{-3} m/s up
18.11 1.44×10^6 N/m
18.13 $E_{r<4} = 2.81 \times 10^8\, r$ V/m; $E_{r>4} = \dfrac{18 \times 10^9}{r^2}$ V/m
18.15 9.02 N; the weight in an actual experiment would be greater because the earth's density is not uniform, but increases toward the earth's center.
18.17 4.38×10^7 m/s; speed of proton = 8.76×10^7 m/s
18.19 $F = 8.20 \times 10^{-8}$ N directed toward the proton. PE of the electron = -27.2 eV = -4.35×10^{-18} J
18.21 2.0×10^5 V/m, $\theta = 201.6°$
18.23 4.77×10^7 N/C; $\theta = -45°$
$V_A = 1.80 \times 10^6$ V
18.25 9.58×10^{10} m/s²; 5.22 m
18.29 -10 μC
$E = 5.03 \times 10^6$ N/C; $\theta = 153°$ with positive x axis

• CHAPTER 19 •

Multiple Choice

19.1 b	**19.2** d	**19.3** c	**19.4** b	**19.5** b
19.6 b	**19.7** c	**19.8** c	**19.9** a	

Problems

19.1 2.21 mm
19.3 48 μC
19.5 1.5 μF
19.7 1.69
19.11 142 pF
19.13 $C = 4\pi\epsilon_0 \dfrac{R_b R_a}{R_b - R_a}$
19.15 1.05 mJ
19.17 $A = 1.13 \times 10^{-2}$ m², $C = 2.50 \times 10^{-10}$ F, PE = 2.00×10^{-5} J
19.19 Energy provided by the battery (to the capacitor) = 1.4×10^{-4} J
19.21 7.08×10^{-6} C/m²
19.23 $F \propto Q^2$
19.25 4.8×10^{-5} J
19.27 $PE_6 = 6.91 \times 10^{-5}$ J
$PE_3 = 7.78 \times 10^{-5}$ J
$PE_1 = 2.59 \times 10^{-5}$ J
19.29 (a) $C_{eq} = 400$ μF; PE = 2.88 J
(b) $C_{eq} = 1$ μF; $Q = 2.4$ mC
(c) No, there has been no change in energy.
(d) 9.60×10^5 W
19.31 12 μF; 0.192 mJ
19.33 16 μF

• CHAPTER 20 •

Multiple Choice

20.1 b	**20.2** c	**20.3** b	**20.4** b	**20.5** e
20.6 b	**20.7** a	**20.8** d	**20.9** c	**20.10** b

Problems

20.1 1.56×10^{19}
20.3 3.75×10^{13}
20.5 4.42×10^{-5} m/s
20.7 $\frac{2}{3}$A
20.9 67.5 V
20.11 0.879 mm
20.13 128 Ω
20.15 $[\rho] = \dfrac{[M][L]^3}{[Q]^2[T]}$
20.17 919 Ω

20.19 90.1 Ω; 0.072 Ω

20.21 $\frac{d_{Cu}}{d_{Al}} = 0.773; \frac{M_{Cu}}{M_{Al}} = 1.98$

20.23 $I^2R = IV \rightarrow (C/s)(J/C) = J/s =$ Watt

20.25 0.03 A

20.27 293 W, 0.491 Ω

20.29 10.91 A, 10.08 Ω, ΔP = 300 W

20.31 10.3 Ω

20.33 8.22 L/min

20.35 2.93 mm (Note: Conduit has 2 wires, each carrying a current of 20 A)

20.37 R increases by 19.2%; P decreases by 16.1%

20.39 $\frac{1}{16}R_0$

20.41 2 Ω

20.43 In the 3 Ω resistor: 6 A; in the 2 Ω resistor: 9 A

20.45 4 Ω

20.47 6 Ω

20.49 $R_1 = 20\ \Omega\ \mathcal{E} = 90$ V

20.51 $R_L = 2.5\ \Omega$; 202.5 W

20.53 9.84 V

• CHAPTER 21 •

Multiple Choice

21.1 c **21.2** b **21.3** a **21.4** c **21.5** d
21.6 c **21.7** a **21.8** a

Problems

21.1 $V_L = 15.8$ V if $R_L = 2000\ \Omega$; $V_L = 33$ V if $R_L = 200{,}000\ \Omega$
R of voltage divider should be much less than R_L.

21.3 0.776 A; 0.355 W; 2.45 V

21.5 I at 5 °C = 56.6 A, I at −15 °C = 35.3 A
P at 5 °C = 38.4 W, P at −15 °C = 174 W

21.7 (a) $I_2 = 2.56$ A; $I_{16} = 1.18$ A; $I_{18} = 1.38$ A
(b) same as (a); $I_{24} = 0.25$ A; $I_{36} = 0.167$ A

21.9 $I = 4$ A; current between A and B when S is closed = 1.5 A
When S is open, resistance between D and $C = 3\ \Omega$; when S is *closed*, resistance = $\frac{8}{3}\ \Omega$

21.11 $I_{10} = I_{14} = 0.5$ A; $I_5 = I_7 = 0.5$ A; $I_8 = 0.75$ A

21.13 $I_{24} = 0.156$ A; $I_6 = 0.125$ A; $I_8 = 0.281$ A

21.15 (a) 1 mΩ; (b) 6.67 Ω; (c) 20,000 Ω; (d) 180 Ω

21.17 44 Ω; 1.66 mW

21.19 $4 \times 10^{-3}\ \Omega$; $4 \times 10^{-4}\ \Omega$; $8 \times 10^{-5}\ \Omega$

21.21 $R_V = 1500\ \Omega$

21.23 2.79 V; 8.37 V

21.25 $r_{int} = 0.08\ \Omega$; $\mathcal{E} = 9$ V

• CHAPTER 22 •

Multiple Choice

22.1 c **22.2** c **22.3** d **22.4** c **22.5** c
22.6 c **22.7** c **22.8** b **22.9** c **22.10** a
22.11 d **22.12** c **22.13** a **22.14** b
22.15 b

Problems

22.1 8×10^{-15} N directed northward

22.3 6 N/m directed southward

22.5 4.75×10^{-18} N directed eastward

22.7 The particles lose energy and thus slow down; hence r in the relation $r = \frac{mv}{qB}$ also becomes smaller.

22.9 1.73 cm

22.11 1.024×10^{-25} kg = 61.7 u

22.13 2.39×10^{-3} T; directed to east

22.15 $r_P = 8.08 \times 10^{-3}$ m, $r_D = 11.4 \times 10^{-3}$ m

22.17 11.7 A

22.19 0.08 amp·m^2

22.21 0.048 N·m

22.23 7.11×10^{-3} T

22.25 5.03×10^{-4} T

22.29 120 A

• CHAPTER 23 •

Multiple Choice

23.1 c **23.2** a **23.3** a **23.4** c **23.5** c
23.6 c **23.7** b **23.8** a **23.9** b

Problems

23.1 0.0785 V, counterclockwise

23.3 (a) 3.82×10^{-3} T/s
(b) 7.64×10^{-3} T/s

23.5 0.5 V in both cases

23.7 0.3 V, but reversed in polarity

23.9 0.0563 W

23.11 The current will flow upward through the nearest vertical part of the loop.

23.13 $L = \frac{\mu_0 \pi R^2 N^2}{l}$; R = radius, l = length of solenoid

23.15 9.57 cm

23.17 2122 turns

23.19 8.94 A

23.21 $L = 5.05 \times 10^{-3}$ H; energy = 3.64 mJ

23.23 $M_{21} = 1.26$ mH

23.27 10.9 V

23.29 $\mathcal{E} = NAB\,\omega \sin \omega t$
$\mathcal{E}_{max} = 151$ V

• CHAPTER 24 •

Multiple Choice

24.1 b **24.2** b **24.3** c **24.4** a **24.5** c
24.6 a **24.7** b **24.8** a **24.9** c **24.10** b

Problems

24.1 $2.0 \times 10^6\ \Omega$

24.3 1.25×10^{-4} s

24.5 $C = 0.231\ \mu F$
$Q = 1.39\ \mu C$

24.7 0.75, $t = 0.520$ s

24.9 $t = 1.39$ s

24.11 192 Ω, 0.863 s

24.13 (a) 1.04×10^{-7} s
(b) 10,000 A
(c) 0.05625 J; $\overline{P} = 5.41 \times 10^5$ W

24.15 0.748 H

24.17 At 0 Hz: $X_L = 0$, $Z = 25\ \Omega$
At 60 Hz: $X_L = 56.55\ \Omega$, $Z = 61.8\ \Omega$
At 500 Hz: $X_L = 471\ \Omega$, $Z = 472\ \Omega$

24.19 923 Hz

24.21 53.1 mH

24.23 1460 Hz

24.25 L can be any value between 0.23 mH and 0.33 mH.

24.27 $R = 96.1\ \Omega$
$C = 1.04\ \mu F$

24.29 $f = 239$ Hz, $\overline{P} = 4.61$ W

24.31 4.42 μF

24.33 In 24.31: $P = 0$
In 24.32: $P = 8$ W

24.35 $V_R = 70.7$ V
$V_L = 2.12 \times 10^3$ V
$V_C = 2.12 \times 10^3$ V

24.37 398 Hz

24.39 $I_0 = 8.51$ A
$V_0 = 354$ V
$V_{max} = 250$ V
$\overline{P} = 1.45$ kW
Max energy stored = 0.869 J, Avg stored energy = 0 J.

24.41 $N_1 = 48$; $N_2 = 1700$

24.45 (a) $f = 1.59$ kHz
(b) $f = 497$ Hz
Filter will accentuate low frequencies. If $R_F = 200\ \Omega$, frequencies greater than 1.6 kHz will be much reduced. If $R_F = 2400\ \Omega$, frequencies greater than 500 Hz will be much reduced in intensity.

• CHAPTER 25 •

Multiple Choice

25.1 d **25.2** c **25.3** c **25.4** c **25.5** b
25.6 b **25.7** b **25.8** e **25.9** b
25.10 b **25.11** a **25.12** c

Problems

25.1 125 m

25.3 5×10^{17} Hz

25.5 3×10^{10} Hz

25.7 0.379 A

25.9 8.33 min

25.11 Lowest speed: 530 rps
Next lowest speed: 1060 rps

25.13 1.84×10^8 m/s

25.15 387 nm

25.17 1.51

25.19 1.22

25.23 Toward the west at an angle of 41.2° above the horizontal

25.25 55.8°

25.27 24.4°

25.29 1.66

25.31 49.9°

25.33 59.8°

25.37 1.29°

• CHAPTER 26 •

Multiple Choice

26.1 e **26.2** a **26.3** c **26.4** b **26.5** c
26.6 b **26.7** a **26.8** e **26.9** a
26.10 a **26.11** a **26.12** d **26.13** c
26.14 c

Problems

26.1 1.6 m

26.7 (a) i = 75 cm
(b) i = −60 cm

26.9 80/3 cm

26.11 $\frac{1}{p} + \frac{1}{i} = \frac{1}{f}$ For a plane mirror, R = ∞. Since R = 2f, f = ∞. Thus, $\frac{1}{p} = -\frac{1}{i}$ or $i = -p$. For $+p$, i = negative (virtual), and $m = -\frac{i}{p}$ is +1. Hence the image is erect and the same size as the object.

26.13 60 cm

26.15 5.44 cm

26.17 $i = -12$ cm, $m = +\frac{12}{30}$

26.19 $p = +20$ cm, $f = +40$ cm

26.21 The image is formed 120 cm behind the lens; it is real, inverted, and has a height of 8 cm, twice that of the object.

26.23 $i_1 = 66.7$ cm; $i_2 = -33.3$ cm; $i_3 = 30.8$ cm.

26.25 A converging lens with a power of +2D

26.27 15.4 cm

26.29 uncorrected near point: 50 cm
uncorrected far point: ∞

26.31 2.0 m

26.33 −4.5 D is the distance part, and −3.5 D is the reading part. The uncorrected far point is at

22.2 cm, and the uncorrected near point is at 13.3 cm.

26.35 13.5, or, using approximate formula: 12.5

26.37 1.7 mm

26.39 2.30 m wide

26.41 from ∞ to 2.7 m (range of object distances)

26.43 0.960 cm; or, using approximate formula: 1.04 cm

26.45 5.6

26.47 $m = -586$

• CHAPTER 27 •

Multiple choice

27.1 a **27.2** b **27.3** c **27.4** e **27.5** c
27.6 e

Problems

27.1 6.83 m

27.3 1000 nm

27.5 240 nm

27.7 114 nm

27.9 The only visible wavelength is 5.2×10^{-7} m

27.11 0.192 mm

27.13 1.0003

27.15 6.88 nm

27.17 Since the wavelength of the light in the oil will be smaller, conditions for interference will be met for smaller thicknesses, and the ring pattern will be shrunk slightly. The central region will continue to be dark.
$r_1 = 3.50 \times 10^{-4}$ m; $r_2 = 6.05 \times 10^{-4}$ m

27.19 400.6 m

27.21 3.09°

27.23 5940

27.25 8.44×10^{-6} m

27.27 $\lambda = 320$ nm
If the slit is increased to 0.16 mm, the pattern will "shrink" or become compressed by a factor of 2 (at least for small angles), with the central maximum having a width, for example, of 2 cm instead of 4 cm.

27.29 In *A* and *C* the pattern is repetitive, with the separation of the centers of each series of units being 3 cm.
A: Two-slit pattern; d = 0.08 mm
B: sharp lines, diffraction grating; 125 lines/cm
C: single-slit diffraction pattern; d = 0.08 mm

27.31 $\lambda = 288$ nm

27.33 $m_R = 3$, $m_G = 4$

27.35 In first order, angular separation = 2.6°
In second order, angular separation = 5.5°
Through order 4 can be observed

27.37 8.33×10^3 lines/cm

27.39 4 cm, 600 nm line; 400 lines/cm

27.41 No

27.43 23.6 km for $\lambda = 530$ nm

27.45 48.4°

27.47 53° above horizon in east

27.49 91.73° with vertical

27.51 0.0938 I_0

• CHAPTER 28 •

Multiple Choice

28.1 b **28.2** c **28.3** c **28.4** b **28.5** c
28.6 c **28.7** c **28.8** d

Problems

28.1 6.25 s

28.3 148 m

28.5 0.95 c

28.7 ~0.01 s

28.9 1.25×10^{-8} s; 3 m

28.11 0.88 c

28.15 0.88 c

28.17 9×10^7 s or 1042 days or 2.85 years

28.19 1.15 m_0

28.21 v = 0.98 c

28.23 v = 0.943 c

• CHAPTER 29 •

Multiple Choice

29.1 d **29.2** d **29.3** c **29.4** e **29.5** b
29.6 a **29.7** b **29.8** c **29.9** e **29.10** c
29.11 a

Problems

29.1 1.19×10^{-29} J

29.3 2.97×10^{19}

29.5 7.96 nm

29.7 flux = 2.51×10^{15} photons/m²·s
$P = 1.26 \times 10^4$ W

29.9 (a) For 630 nm: $h\nu <$ threshold, so no electrons are emitted
(b) For 420 nm: 0.572 eV

29.11 477 nm

29.13 1.73 eV
438 nm

29.15 $\lambda_{min} = 6.2 \times 10^{-2}$ nm

29.17 $V = 2000$ V

29.19 For 5 eV, fractional loss = 2×10^{-5}
For 500 eV, fractional loss = 1.96×10^{-3}
For 500,000 eV, fractional loss = 0.164

29.21 6.86 keV
$\lambda' \cong 0.186$ nm

29.23 0.020 eV

29.25 KE = 0.976 eV

29.27 0.0546 eV

29.29 electron: 2.4×10^9 eV
neutron: 1.3 MeV

29.31 $\Delta E = 3.59 \times 10^{-19}$ J
$\Delta p_x = 1.67 \times 10^{-24}$ kg·m/s
$\Delta v_y = 0.063$ m/s

• CHAPTER 30 •

Multiple Choice

30.1 c **30.2** b **30.3** d **30.4** a **30.5** c
30.6 e **30.7** c **30.8** d **30.9** d **30.10** c
30.11 e

Problems

30.1 2.11 eV

30.3 433 nm

30.5 122.4 eV

30.7 Transition from $n_i = 6$ to $n_f = 3$

30.9 from $n_i = 5$ to $n_f = 2$

30.11 1.43×10^{-8}

30.13 3.40 eV

30.15 $4 \rightarrow 3$: $\lambda = 1.88 \times 10^{-6}$ m
$4 \rightarrow 2$: $\lambda = 4.89 \times 10^{-7}$ m
$4 \rightarrow 1$: $\lambda = 9.75 \times 10^{-8}$ m
$3 \rightarrow 2$: $\lambda = 6.61 \times 10^{-7}$ m
$3 \rightarrow 1$: $\lambda = 1.03 \times 10^{-7}$ m
$2 \rightarrow 1$: $\lambda = 1.22 \times 10^{-7}$ m

30.17 Relativistic effects are of order $(\frac{1}{137})^2$, i.e., quite small. So it is reasonable to treat the hydrogen atom nonrelativistically. However, the very high precision of wavelength measurements in spectroscopy result in relativisitc effects being observed, and relativity must be considered in a more complete theory.

30.19 0.388 Tesla

30.21 $1s^2\ 2s^2\ 2p^3$

30.23 The most energetic characteristic X ray, $\lambda = 0.038$ nm, will not occur. For 20,000 V, $\lambda_{min} = 0.062$ nm

30.25 0.5 J; 1.62×10^{18}

30.27 $\lambda_{Fe} = 0.23$ nm; $\lambda_{Te} = 0.055$ nm

• CHAPTER 31 •

Multiple Choice

31.1 c **31.2** a **31.3** e **31.4** c **31.5** d
31.6 d **31.7** a

Problems

31.1 -5.60 eV

31.3 2.13×10^{-9} N

31.5 PE = 4.57 eV
$F_{attr} = 2.32 \times 10^{-9}$ N

31.7 O_2: 0.924×10^{-3} eV
I_2: 3.6×10^{-5} eV

31.9 H_2: $l = 2$: 4.48×10^{-2} eV
$l = 1$: 1.49×10^{-2} eV
$l = 0$: 0 eV
$\lambda_{1-0} = 8.34 \times 10^{-5}$ m
$\lambda_{2-1} = 4.17 \times 10^{-5}$ m
N_2: $l = 2$: 1.49×10^{-3} eV
$l = 1$: 4.98×10^{-4} eV
$l = 0$: 0 eV
$\lambda_{1-0} = 2.50 \times 10^{-3}$ m
$\lambda_{2-1} = 1.25 \times 10^{-3}$ m

31.11 $k = 5.57 \times 10^2$ N/m; near equilibrium, V is parabolic.

31.13 $\lambda = 1.77\ \mu$m

31.15 $E \sim 0.7$ eV
$\lambda = 1240$ nm

• CHAPTER 32 •

Multiple Choice

32.1 b **32.2** d **32.3** c **32.4** c **32.5** e
32.6 b **32.7** c **32.8** c **32.9** c **32.10** b
32.11 e **32.12** c **32.13** a **32.14** c
32.15 a **32.16** c **32.17** d

Problems

32.1 8.6 MeV/nucleon

32.3 135 MeV/c^2

32.5 (a) 4.10×10^{-15} m
(b) 6.17×10^{-15} m
(c) 7.37×10^{-15} m

32.7 63.929407 u; 9.25×10^{-4} nm

32.9 4.785 MeV; 1.77%

32.11 8.5×10^{-6} (about 1 part in 100,000)

32.13 (a) 7.16 MeV
(b) 9.78 MeV

32.15 1 hr

32.17 1580 yr

32.21 17.3 MeV released.

32.23 (a) 110 μCi; (b) 0.05 μg

32.25 16.9 g

32.27 196.2 MeV; initial release ~88.5% of total.

32.29 15,000 BC

32.31 9530 yr.

• INDEX •

CHAPTER 10 FIG. 10.1, p. 189—Culver Pictures. QUOTES, p. 189—W. F. Magie, *A Source Book in Physics*, Harvard University Press, Cambridge, Mass., 1963, p. 75. FIG. 10.6, p. 193—Copyright by Deutsches Museum, Munich. FIG. 10.12, p. 197—Courtesy Bettmann Archive. FIG. 10.19, p. 202—Courtesy of the late Professor Frank N. M. Brown Photographic Collection, University of Notre Dame. FIG. 10.29(d), p. 211—Courtesy of Bradford Washburn, Museum of Science, Boston. FIG. 10.32, p. 215—Fundamental Photos.

CHAPTER 11 FIG. 11.1, p. 227—The British Museum. QUOTES, p. 227—W. F. Magie, *A Source Book in Physics*, Harvard University Press, Cambridge, Mass., 1963, pp. 151, 160. FIG. 11.2, p. 228—Culver Pictures. FIG. 11.3(a), p. 228—Courtesy of the Science Museum, London; lent to the Science Museum by the late Dr. Joule. FIG. 11.4, p. 229—Courtesy Bettmann Archive. QUOTE, p. 229—Magie, p. 211. FIG. 11.5(a), p. 230—The Granger Collection. FIG. 11.5(b)—The Granger Collection. FIG. 11.10, p. 234—World Wide Photos. FIG. 11.14(c), p. 241—Soaring Society of America. FIG. 11.17, p. 243—St. Louis Post Dispatch. FIG. 11.23, p. 249—Courtesy of Dr. Norman Sadowsky, Faulkner Hospital, Boston.

CHAPTER 12 FIG. 12.1, p. 256—Courtesy of H. E. Harris. FIG. 12.2(a), p. 257—Culver Pictures. FIG. 12.2(b), p. 257—Courtesy of Cambridge University Library. QUOTE, p. 258—W. F. Magie, *A Source Book in Physics*, Harvard University Press, Cambridge, Mass., 1963, p. 169.

CHAPTER 13 FIG. 13.3, p. 281—Courtesy of Harvard University Cruft Photo Lab (A. J. Dionne). FIG. 13.6, p. 285—The Granger Collection. QUOTE, p. 285—W. F. Magie, *A Source Book in Physics*, Harvard University Press, Cambridge, Mass., 1963, p. 323 (*The Second Law of Thermodynamics*, W. F. Magie, ed., New York, 1899). FIG. 13.10, p. 289—Culver Pictures. FIG. 13.13, p. 291—BBC, London.

CHAPTER 14 FIG. 14.9, p. 304—Courtesy of Adriane Bishko. FIG. 14.13, p. 308—Courtesy of Dr. Harold E. Edgerton, M.I.T., Cambridge, Mass. FIG. 14.15, p. 310—from A. P. French, *Vibrations and Waves*, W. W. Norton & Co., New York, 1971. Used with permission.

CHAPTER 15 FIG. 15.14, p. 329—From D.C. Miller, *The Science of Musical Sounds*, Macmillan, New York, 1922.

CHAPTER 16 FIG. 16.6, p. 346—Wide World Photos. FIG. 16.9, p. 350—Educational Development Center. FIG. 16.11, p. 353—Angermayer, Tierbildarchiv.

CHAPTER 17 QUOTE, p. 364—W. F. Magie, *A Source Book in Physics*, Harvard University Press, Cambridge, Mass., 1963, p. 399. FIG. 17.3(b), p. 365—Courtesy of Central Scientific Company. FIG. 17.6, p. 369—United Press International.

CHAPTER 18 FIG. 18.4, p. 385—Educational Services, Inc. FIG. 18.22(a)(b), p. 401—U.S. Geological Survey, F. J. Blatt. FIG. 18.23(c), p. 401—Educational Services, Inc. FIG. 18.24(b), p. 402—Courtesy of Professor Harold M. Waage, Palmer Physical Laboratory, Princeton University.

CHAPTER 19 FIG. 19.1, p. 411—Smithsonian Institution. QUOTES, p. 411—W. F. Magie, *A Sourcebook in Physics*, Harvard University Press, Cambridge, Mass., 1963, pp. 403, 405. FIG. 19.4(a), p. 414—Courtesy of Larry Langrill, Oakland University. FIG. 19.44(c), p. 414—Courtesy of Sprague Electric Co.

CHAPTER 22 QUOTE, p. 465—W. F. Magie, *A Source Book in Physics*, Harvard University Press, Cambridge, Mass., 1963, p. 437. FIG. 22.1, p. 466—Culver Pictures. QUOTE, p. 466—Magie, p. 438. FIG. 22.3, p. 468—Varian Associates. FIG. 22.4, p. 469—Courtesy of Harvard University, Cruft Photo Lab. FIG. 22.7, p. 473—U.S. Geological Survey. FIG. 22.9, p. 475—Courtesy of Fermi National Accelerator Laboratory. FIG. 22.10(b), p. 477—Courtesy of Fermilab. FIG. 22.12(b), p. 478—Gould Medical Products Division. FIG. 22.13, p. 479—Courtesy of D. C. Heath. QUOTE, p. 482—Magie, p. 457. FIG. 22.19(b), p. 485—Courtesy of Kodansha, Tokyo. FIG. 22.23(b), p. 486—Courtesy of Pergamon Press. FIG. 22.24(c), p. 486—Courtesy of Kodansha, Tokyo.

CHAPTER 23 QUOTE, p. 495—S. Pearce Williams, *Michael Faraday*, Simon and Schuster, New York, 1971, p. 11 (Personal Remembrances of Sir Frederick Pollock, London, 1887, p. 247). FIG. 23.1, p. 496—Courtesy of Fisher Scientific Co. FIG. 23.2, p. 496—Royal Institute of Great Britain. QUOTE, p. 496—S. Pearce Williams, p. 109 (Burndy Library). QUOTE, p. 497—W. F. Magie, *A Source Book in Physics*, Harvard University Press, Cambridge, Mass., 1963, p. 475. FIG. 23.10(b), p. 503—Manufactured by J. W. Miller Division of Bell Industries. FIG. 23.16, p. 509—Argonne National Laboratory. FIG. 23.18, p. 510—Courtesy of Professor J. F. Allen, University of St. Andrews. FIG. 23.19, p. 510—Courtesy of Consulate General of Japan, New York.

CHAPTER 24 FIG. 24.18, p. 527—Courtesy of Boston Edison Co.

CHAPTER 25 FIG. 25.1, P. 458—Culver Pictures. FIG. 25.13, p. 555—J. F. Blatt. FIG. 25.20(d), p. 563—Courtesy Educational Development Center. FIG. 25.23, p. 565—Courtesy Educational Development Center. QUOTES, p. 569, 570—Isaac Newton, "Isaac Newton's Philosophiae Naturalis Principia Mathematica", London, 1687, edited by Alexandre Koyre and I. Bernard Cohen, 2 vols., Cambridge University Press, 3rd ed., 1972, quoted in Bronowski, *The Ascent of Man*.

CHAPTER 26 FIG. 26.26(a), p. 595—Courtesy of American Optical Co. FIG. 26.28, p. 598—O. K. Harter. FIG. 26.31, p. 599—Photo Courtesy of Bausch and Lomb. FIG. 26-34, p. 600—Photo courtesy of American Hospital Supply Corporation; Stamford, CT.

CHAPTER 27 FIG. 27.3(c), p. 610—Courtesy of Educational Development Center. FIG. 27.4, p. 612—Sears/Zemansky, *College Physics*, Addison-Wesley, Reading, Mass., 1980. FIG. 27.8(b), p. 616—E. Hecht and A. Zajac, *Optics*, Addison-Wesley, Reading, Mass., 1974. FIG. 27.9(b), p. 617—Courtesy of Central Scientific Co. FIG. 27.13, p. 621—Hecht and Zajac. FIG. 27.14, p. 622—Hecht and Zajac. FIG. 27.17, p. 625—Hecht and Zajac. FIG. 27.18, p. 626—Courtesy of E. O. Wollan, Oak Ridge National Laboratory. FIG. 27.21, p. 629—Allyn and Bacon. FIG. 27.23, p. 630—Crown copyright, Central Photographic Section, National Physical Laboratory, Teddington, Middlesex, England. FIG. 27.24(a), p. 630—Hecht and Zajac, *Optics*. FIG. 27.26, 632—Allyn and Bacon. FIG. 27.27, p. 633—Photo by David Mills. FIG. 27.3, p. 636—Richard T. Weidner and Robert L. Sells, *Elementary Classical Physics*, Allyn and Bacon, Boston, 1973. FIG. 27.33, p. 638—Greenberg, *Physics with Modern Applications*, W. B. Saunders, Philadelphia.

CHAPTER 28 QUOTE, p. 648—G. Holton, *Thematic Origins of Scientific Thought*, Harvard University Press, Cambridge, Mass., 1973. FIG. 28.1(a), p. 649—Courtesy of the American Friends of the Hebrew University and the American Institute of Physics. FIG. 28.1(b), p. 649—Wide World Photos. FIG. 28.1(c), p. 649—Wide World Photos. QUOTES, p. 650—Leon N. Cooper, *An Introduction to the Meaning and Structure of Physics*, Harper & Row, New York, 1968, p. 358. QUOTE, p. 651—Cooper, p. 362. QUOTE, p. 654—A. A. Michelson and E. W. Morley, *American Journal of Science*, Vol. 34 (1887), p. 341. FIG. 28.6, p. 655—Wide World Photos. QUOTE, p. 655—A. A. Michelson, *Light Waves and Their Uses*, University of Chicago Press, Chicago, 1902, p. 341. QUOTE, p. 657—W. F. Magie, *A Source Book in Physics*, Harvard University Press, Cambridge, Mass., 1963, p. 33. QUOTE, p. 658—A. Einstein, "On the Electrodynamics of Moving Bodies," *Ann. d. Physik*, 17, 891 (1905). QUOTE, p. 667—J. C. Hafee and R. E. Keating, "Around-the-World Atomic Clocks," *Science*, Vol. 177, 166, 168 (July 14, 1972).

CHAPTER 29 FIG. 29.2, p. 676—Wide World Photos. QUOTE, p. 676—A. Einstein, "Uber einen die Erzeugung und Verwandlung des Lichtes betreffenden heuristischen Gesichtspunkt," *Ann. d. Physik*, 17, 132 (1905). QUOTE, p. 678—Jeremy Bernstein, *Einstein*, Penguin, New York, 1976, p. 211 (Phillipp Frank, *Einstein: His Life and Times*, p. 232). FIG. 29.8, p. 682—Culver Pictures. FIG. 29.9(a), p. 682—The Smithsonian Institution. FIG. 29.9(b), p. 682—Deutsches Museum, Munich. FIG. 29.10, p. 683—New York City Dept.

of Health, Bureau of Radiation Control. FIG. 29.11, p. 683—Courtesy of E. O. Wollan, Oak Ridge Laboratory. FIG. 29.12(a), p. 684—Courtesy of General Electric. QUOTE, p. 688—quoted in George Trigg, *Crucial Experiments in Modern Physics* (Van Nostrand Reinhold, New York, 1971) p. 103. FIG. 29.17, p. 688—Wide World Photos. FIG. 29.18, p. 689—Educational Development Center; Allyn and Bacon; E. O. Wollan, Oak Ridge Laboratory. QUOTE, p. 690—*The Physical Review*, Vol. 29, p. 908 (1927). FIG. 29.19, p. 691—Courtesy of the Nobel Foundation.

CHAPTER 30 FIG. 30.4, p. 702—Radio Times, Hulton Picture Library. FIG. 30.5, p. 703—Courtesy of Neils Bohr Library, AIP. QUOTE, p. 703 (top)—*Philosophical Magazine*, Vol. 26, July 1913. QUOTE, p. 703 (footnote)—Werner Heisenberg, *Physics and Beyond*, Harper & Row, New York, 1971, p. 74. QUOTE, p. 705, *Philosophical Magazine*, Vol. 26, July 1913. QUOTE, p. 707—*Philosophical Magazine*, Vol. 26, July 1913. FIG. 30.10, p. 719—Courtesy of Los Alamos Scientific Laboratory.

CHAPTER 31 FIG. 31.9, p. 730—Spectra courtesy of J. A. Marquisee. FIG. 31.24, p. 737—Courtesy of Texas Instruments.

CHAPTER 32 FIG. 32.10, p. 759—Courtesy of the Berkeley Laboratory, University of California. QUOTE, p. 760, Franklin D. Roosevelt Library. FIG. 32.14, p. 761—Wide World Photos. FIG. 32.15, p. 761—Courtesy of the University of Chicago. FIG. 32.16, p. 761—Courtesy of General Electric Co. FIG. 32.17, p. 763—Wide World Photos. FIG. 32.18, p. 763—U.S. Navy Photo. FIG. 32.21, p. 766—Bicron Corp. FIG. 32.22, p. 766—Weidner and Sells, *Elementary Classical Physics*. FIG. 32.25(a), p. 771—Courtesy of Brookhaven National Library.

Periodic Table

Group I	*Group II*	*Transition elements*						
H 1 1.0080 $1s^{1}$								
Li 3 6.94 $2s^{1}$	**Be** 4 9.012 $2s^{2}$							
Na 11 22.99 $3s^{1}$	**Mg** 12 24.31 $3s^{2}$							
K 19 39.102 $4s^{1}$	**Ca** 20 40.08 $4s^{2}$	**Sc** 21 44.96 $3d^{1}4s^{2}$	**Ti** 22 47.90 $3d^{2}4s^{2}$	**V** 23 50.94 $3d^{3}4s^{2}$	**Cr** 24 51.996 $3d^{5}4s^{1}$	**Mn** 25 54.94 $3d^{5}4s^{2}$	**Fe** 26 55.85 $3d^{6}4s^{2}$	**Co** 27 58.93 $3d^{7}4s^{2}$
Rb 37 85.47 $5s^{1}$	**Sr** 38 87.62 $5s^{2}$	**Y** 39 88.906 $4d^{1}5s^{2}$	**Zr** 40 91.22 $4d^{2}5s^{2}$	**Nb** 41 92.91 $4d^{4}5s^{1}$	**Mo** 42 95.94 $4d^{5}5s^{1}$	**Tc** 43 (99) $4d^{5}5s^{2}$	**Ru** 44 101.1 $4d^{7}5s^{1}$	**Rh** 45 102.91 $4d^{8}5s^{1}$
Cs 55 132.91 $6s^{1}$	**Ba** 56 137.34 $6s^{2}$	57– 71§	**Hf** 72 178.49 $5d^{2}6s^{2}$	**Ta** 73 180.95 $5d^{3}6s^{2}$	**W** 74 183.85 $5d^{4}6s^{2}$	**Re** 75 186.2 $5d^{5}6s^{2}$	**Os** 76 190.2 $5d^{6}6s^{2}$	**Ir** 77 192.2 $5d^{7}6s^{2}$
Fr 87 (223) $7s^{1}$	**Ra** 88 (226) $7s^{2}$	89–103‖	**Rf** 104 (261) $6d^{2}7s^{2}$	**Ha** 105 (262) $6d^{3}7s^{2}$	106 (263)	107 (261)		

Symbol—**Cl** 17—Atomic number
Atomic mass†—35.453
$3p^{5}$ —Electron configuration

§ Lanthanide series	**La** 57 138.91 $5d^{1}6s^{2}$	**Ce** 58 140.12 $5d^{1}4f^{1}6s^{2}$	**Pr** 59 140.91 $4f^{3}6s^{2}$	**Nd** 60 144.24 $4f^{4}6s^{2}$	**Pm** 61 (147) $4f^{6}6s^{2}$	**Sm** 62 150.4 $4f^{6}6s^{2}$
‖ Actinide series	**Ac** 89 (227) $6d^{1}7s^{2}$	**Th** 90 (232) $6d^{2}7s^{2}$	**Pa** 91 (231) $5f^{2}6d^{1}7s^{2}$	**U** 92 (238) $5f^{3}6d^{1}7s^{2}$	**Np** 93 (239) $5f^{3}6d^{0}7s^{2}$	**Pu** 94 (239) $5f^{6}6d^{0}7s^{2}$

Notes * Atomic mass values given are averaged over isotopes in the percentages in which they exist in nature.

† For an unstable element, mass number of the most stable known isotope is given in parentheses.